Fig (*Ficus carica*): Production, Processing, and Properties

Mohamed Fawzy Ramadan
Editor

Fig (*Ficus carica*): Production, Processing, and Properties

Springer

Editor
Mohamed Fawzy Ramadan
Department of Clinical Nutrition
Faculty of Applied Medical Sciences
Umm Al-Qura University
Makkah, Saudi Arabia

ISBN 978-3-031-16492-7   ISBN 978-3-031-16493-4 (eBook)
https://doi.org/10.1007/978-3-031-16493-4

This Springer imprint is published by the registered company Springer Nature Switzerland AG
The registered company address is: Gewerbestrasse 11, 6330 Cham, Switzerland

Dedicated to the soul of my father, Professor Fawzy Ramadan Hassanien, and my beloved family.

Preface

Fig has been mentioned in the Holy Quran, in the form of the divine oath, in the words of God Almighty: "*By the Fig and the Olive, and by Mount Sinai*," "Surat Al-Tin: 1-2." God Almighty swore to fig in this noble verse because of its great importance.

Ficus carica L. illustrates a promising item of functional food and healthy products. There is a lack of books presenting complete information on the production, processing, chemistry, products, and medical traits of figs and the plant parts (i.e., fruit, skin, leaves, roots, latex, and by-products). This book is considered necessary since it covers all the information about fig.

This book project aims to create a multidisciplinary discussion forum on *Ficus carica* with particular emphasis on its horticulture, post-harvest, marketability, phytochemistry, extraction protocols, biochemistry, nutritional value, functionality, health-promoting traits, ethnomedicinal applications, technology, and processing. The impact of traditional and innovative processing on recovering high-added valuc compounds from *Ficus carica* wastes is reported. Besides, the book discusses the potential applications of *Ficus carica* in food, cosmetics, and pharmaceutical products.

Intending to provide a comprehensive contribution to the scientific community involved in clinical nutrition, food science, horticulture, phytochemistry, health, and pharmacology, this book comprehensively reviews the aspects that led to the recent advances in *Ficus carica* biochemistry, production, and functionality. The editor hopes the handbook will be a rich source for researchers and developers in related disciplines.

Book chapters have a diversity of developments in food science and horticultural research. The book contains comprehensive chapters under main sections, namely

- ***Ficus carica*: Cultivation, Species, and Cultivars**
- ***Ficus carica*: Chemistry, Functionality, and Health-Promoting Properties**
- ***Ficus carica*: Technology, Processing, and Applications**

The editor sincerely thanks all contributors for their valuable contributions and their cooperation. The Springer-Nature staff's help and support, especially Daniel Falatko and Sofia Valsendur, was essential for completing my task and is highly appreciated.

Makkah, Saudi Arabia Mohamed Fawzy Ramadan

Description

Fig (*Ficus carica* L.) is one of the oldest cultivated and consumed fruits worldwide. More than 800 varieties of the *Ficus carica* genus are cultivated in a warm climate. Fig is a seasonal fruit that could be harvested twice annually and could be consumed fresh, dried, and as a jam or juice. Fresh and dry figs are appreciated as food and for their health-promoting impacts. The syrup is used as a remedy for mild constipation. Leaves are used as fodder for animals. Fig latex is used as a curdling agent in dairy products.

The fruit is a rich source of health-enhancing phytochemicals (i.e., phenolics, organic acids, vitamin E, and carotenoids). Phytochemicals are influenced by the harvest time, variety, maturity, color, fruit part, and fruit processing. Amino acids, organic acids, fatty acids, sterols, hydrocarbons, anthocyanins, aliphatic alcohols, volatiles, and other secondary metabolites are found in fig fruit, latex, leaves, and root. In addition, figs are an essential source of minerals (i.e., potassium, iron, and calcium) and vitamins (i.e., riboflavin and thiamin).

Ficus carica fresh fruit, extracts, and isolated bioactive compounds exhibited a broad spectrum of health-promoting traits. Different nutrients and bioactive compounds, including free sugars, organic acids, tocopherols, phenolic components, and fatty acids, were detected in the *Ficus carica* peel extracts. *Ficus carica* leaves, roots, fruit, and latex are known for their biological and health traits, including antifungal, anti-helminthic, acetyl cholinesterase inhibition, and anticarcinogenic activities. Fig is used in traditional medicine for various reproductive, digestive, endocrine, and respiratory ailments. It is used in the gastrointestinal tract and urinary tract infections. Furthermore, the fig treats ailments such as diabetes, anemia, cancer, leprosy, liver diseases, skin diseases, and ulcers. Studies reported on the applications of fig extracts as functional edible ingredients, clinical trials to confirm the health effects of extracts, and the valorization of plant by-products. *Ficus carica* has been included in occidental pharmacopeias (i.e., British Pharmacopoeia and Spanish Pharmacopoeia) and therapeutic guides of herbal medicines.

The influences of processing on the *F. carica* product's quality and the amounts of individual phytochemicals in fruit were studied. Drying of *Ficus carica* has proven to be a reliable preservation method. The main factors that affect the drying process are temperature and the length of the process. Freeze-drying and microwave-drying for fig preserving are also essential technologies. Besides, the effect of post-harvest and packing methods on fruits' phytochemicals profile and biological traits were of interest.

Ficus carica is of importance due to its widespread food, industrial, and medicinal applications. Although *Ficus carica* products are already commercially available in the international market, it is hard to find in the bookstore works on the production, processing, chemistry, and properties of *Ficus carica*.

Key Features

- Explores the chemistry of *Ficus carica* phytochemicals and extracts
- Discusses *Ficus carica* active constituents and their health-enhancing traits
- Presents the applications of *Ficus carica* phytochemicals and extracts
- Authored by international scientists and industry experts
- Addresses the growing application areas, including horticulture, functional food, clinical nutrition, pharmaceuticals, and cosmetics

Readership

- Clinical nutrition, food chemistry, biochemistry, pharmacology, and horticulture researchers and students.
- Developers of nutraceuticals, novel food, and pharmaceuticals, as well as R&D researchers in different sectors, apply fruits and medicinal plants.

Contents

About the Editor

Mohamed Fawzy Ramadan is a Food Chemistry Professor at the Department of Clinical Nutrition, Faculty of Applied Medical Sciences, Umm Al-Qura University, Makkah, Saudi Arabia. Since 2014, he is Professor at Biochemistry Department, Faculty of Agriculture, Zagazig University, Egypt.Prof. Ramadan obtained his Ph.D. (*Dr. rer. nat.*) in Food Chemistry from the Berlin University of Technology (Germany, 2004). He continued his postdoctoral research at ranked universities such the University of Helsinki (Finland), Max-Rubner Institute (Germany), Berlin University of Technology (Germany), and the University of Maryland (USA). In 2012, he was appointed Visiting Professor (100% teaching) in the School of Biomedicine, Far Eastern Federal University in Vladivostok, Russian Federation.

Prof. Ramadan published more than 300 research papers, and reviews in peer-reviewed journals. He also edited and published several books (Scopus *h*-index is 44 and more than 6500 citations). In addition, he was an invited speaker at several international conferences. Since 2003, Prof. Ramadan is a reviewer and an editor in several highly-cited international journals such as *Journal of Umm Al-Qura University for Medical Sciences*, *eFood, Journal of Medicinal Food* and *Journal of Advanced Research.*

Prof. Ramadan received several prizes, including Abdul Hamid Shoman Prize for Arab Researcher in Agricultural Sciences (2006), Egyptian State Prize for Encouragement in Agricultural Sciences (2009), European Young Lipid Scientist Award (2009), AU-TWAS Young Scientist National Awards (Egypt) in Basic Sciences, Technology and Innovation (2012), TWAS-ARO Young Arab Scientist (YAS) Prize in Scientific and Technological Achievement (2013), and Atta-ur-Rahman Prize in Chemistry (2014).

Chapter 1
Introduction to Fig (*Ficus carica*): Production, Processing, and Properties

Mohamed Fawzy Ramadan

1 Fig (*Ficus carica*): Production, Processing, and Properties

The United Nations Sustainable Development Goals (SDGs) are recognized to offer a sustainable world vision (https://sustainabledevelopment.un.org). "*Good Health and Well-Being*" goal is associated with applying health-improving plants and environmental-friendly processes in food chain sectors (Ramadan, 2021). Due to their safety, fruits and vegetables have been utilized as pharmaceuticals and nutraceuticals. Therefore, there is a great interest in fruits and vegetables as phytoconstituents-rich sources for nutraceuticals and pharmaceuticals (McClements, 2019; Ramadan, 2020).

Fig (*Ficus carica* Linn) is considered one of the oldest cultivated and consumed fruit tree worldwide. Fig has been mentioned in the Holy Quran, in the form of the divine oath, in the words of God Almighty: "**By the Fig and the Olive, and by Mount Sinai**", "**Surat Al-Tin: 1-2**". God Almighty swore to fig in this noble verse because of their great importance.

Ficus is one of 37 genera (family Moraceae). Fig species of the highest commercial value is *Ficus carica* (syn. Ficus kopetdagensis Pachom.), which consists of varieties with high genetic diversity. About 800 varieties of the *Ficus carica* genus are cultivated in a warm climate. Fresh or dry figs have been appreciated as food and for their health-promoting impacts. Besides, *Ficus carica* leaves are utilized as fodder. The syrup is utilized as a remedy for mild constipation. In addition, fig latex is used as a curdling agent in dairy products. *Ficus carica* is a seasonal fruit that could be harvested twice annually and could be consumed fresh, dried, as a jam or juice

M. F. Ramadan (✉)
Department of Clinical Nutrition, Faculty of Applied Medical Sciences,
Umm Al-Qura University, Makkah, Saudi Arabia
e-mail: mfhassanien@uqu.edu.sa

M. F. Ramadan (ed.), *Fig* (Ficus carica)*: Production, Processing, and Properties*, https://doi.org/10.1007/978-3-031-16493-4_1

(Mawa et al., 2013; Barolo et al., 2014; Badgujar et al., 2014; Desa et al., 2019; Teixeira et al., 2019; Arvaniti et al., 2019; Yao et al., 2021; Salehi et al., 2021).

Ficus carica L. is a symbol of longevity and included in the human diet since ancient time. The fruit is a rich source of health-promoting phytochemicals (i.e., phenolics, organic acids, vitamin E, and carotenoids). Flavonoids and phenolic acids are the main phytochemicals in fresh and dried fruits. Chlorogenic acid, gallic acid, quercetin-3-*O*-rutinoside, rutin, and epicatechin are the main phenolic acids and flavonoids in *Ficus carica* fruit (fresh or dried). Phytochemical levels are influenced by the harvest time, variety, maturity, color, fruit part, and fruit processing. The analysis of flavonoids and phenolic acids in fresh and dried fig varieties and the distribution of these compounds between fruit pulp and skin were investigated. The antioxidant traits of fig are correlated with its content of bioactive phenolics. Amino acids, organic acids, fatty acids, sterols, hydrocarbons, anthocyanins, aliphatic alcohols, volatiles, and other secondary metabolites were found in fig fruit, latex, leave, and root. *Ficus carica* is an essential source of minerals (i.e., potassium, iron, and calcium) and vitamins (i.e., riboflavin and thiamin). In addition, fig fruits are sodium-, fat- and cholesterol-free and rich in fibers (Mawa et al., 2013; Barolo et al., 2014; Badgujar et al., 2014; Desa et al., 2019; Teixeira et al., 2019; Arvaniti et al., 2019; Yao et al., 2021; Salehi et al., 2021).

Ficus carica fresh fruit, crude extracts, and isolated bioactive compounds exhibited a broad spectrum of health-promoting traits. *Ficus carica* fruits, roots, leaves, and latex are known for their biological and health traits, including antimicrobial, anti-helminthic, acetylcholinesterase inhibition, and anticarcinogenic effects. Fig has been used for several types of disorders worldwide. Fig is used in traditional medicine for various reproductive, digestive, endocrine, and respiratory ailments. In addition, it is used for urinary tract and gastrointestinal tract infections. Furthermore, fig treats ailments such as cancer, diabetes, liver diseases, skin diseases, anemia, leprosy, and ulcers. Studies reported on the applications of fig extracts as functional edible ingredients, clinical trials to confirm the health effects of extracts, and the valorization of plant byproducts. *Ficus carica* is included in occidental pharmacopeias (i.e., British Pharmacopoeia and Spanish Pharmacopoeia) and therapeutic guides of herbal medicine. Therefore, the fig is a promising item in pharmaceutical biology for the formulations of novel drugs and clinical applications (Mawa et al., 2013; Barolo et al., 2014; Badgujar et al., 2014; Desa et al., 2019; Teixeira et al., 2019; Arvaniti et al., 2019; Yao et al., 2021; Salehi et al., 2021).

The influences of processing techniques on the *F. carica* product's quality, the phytochemicals profile, and the amounts of individual phenolics in fruit were studied. Drying of *Ficus carica* has proven to be a reliable preservation method. The main factors that affect the drying process are temperature and the length of the process. Freeze-drying and microwave-drying for fig preserving are also essential technologies. Besides, the effect of post-harvest and packing methods on fruits' phytochemicals profile and biological traits were of interest. On the other side, the impact of environmental factors on the phytochemicals content suggests the best production area and the optimum conditions for processing. On the other hand, different nutrients and bioactive compounds, including organic acids, free sugars, tocols, phenolic components, and fatty acids, were detected in the *Ficus carica* peel hydroethanolic extract. In addition, *Ficus carica* peel extract displayed promising antioxidant and antibacterial capacities (Mawa et al., 2013; Barolo et al., 2014; Badgujar et al., 2014; Desa et al., 2019; Teixeira et al., 2019; Arvaniti et al., 2019; Yao et al., 2021; Salehi et al., 2021).

2 Fig (*Ficus carica*) Market

Recent FAOSTAT (2022) statistics reported that the global harvested area of *Ficus carica* is 281,522 ha, global yield is 44,932 hg/ha, and production is 1,264,943 tonnes. Figure 1.1 presents the world's top producing countries of *Ficus carica* and

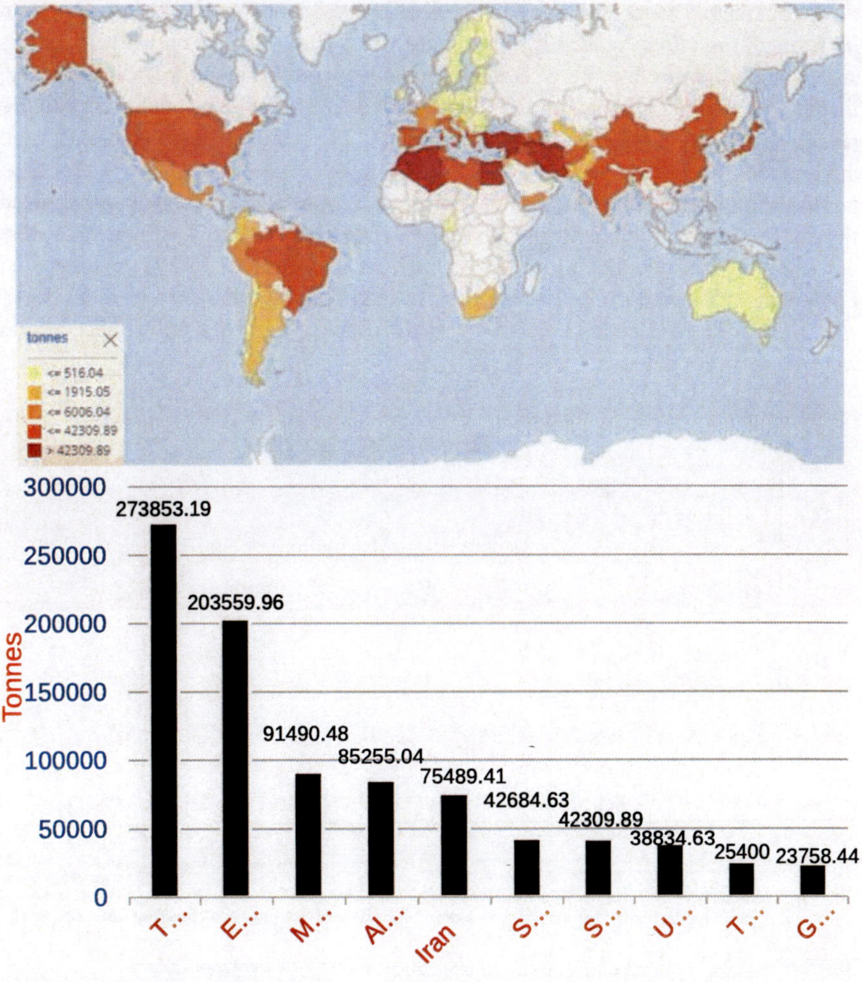

Fig. 1.1 Top producer countries and production quantities (tonnes) of fig. (FAOSTAT, 2022)

the production quantities (tonnes). *Ficus carica* tree is one of the oldest fruits cultivated in the Mediterranean area and presents an essential nutritional and economic value because of its high consumption worldwide (Melgarejo et al., 2003; Crisosto et al., 2011; Badgujar et al., 2014; Núñez-Gómez et al., 2021). Turkey, Egypt, Greece, Algeria, Italy, and Spain are among the Mediterranean producers, from which *ca.* 90% of fig yield is produced (Sadder & Ateyyeh, 2006; Melgarejo, 2017; Núñez-Gómez et al., 2021). Europe, Spain is the top fig-producing country in Europe, with about 56,600 tonnes in 2019, representing *ca.* 4% of the world's production and *ca.* 44% of European production (Núñez-Gómez et al., 2021).

3 Fig (*Ficus carica*) in the Scientific Literature

As a part of this work, a survey in the literature (Scopus and PubMed) has been performed. Fig (*Ficus carica* L.) highly attracts international scientific research. A survey and scientific literature search with the keyword "(Fig and *Ficus carica* L.)" in the PubMed database (June 2022) revealed 288 documents belonging to fig (*Ficus carica* L.) bioactivity, phyto-extracts, oils, bioactive constituents, and applications.

A careful search on fig (*Ficus carica* L.) in Scopus (www.scopus.com) showed that the number of documents is high (*approx.* 1500 till June 2022). Figure 1.2 presents the document counts on fig (*Ficus carica* L.) from 2001 to 2021. The annual documents published in fig (Ficus carica L.) significantly increased from 13 documents in 2001 to 187 documents s in 2021. These numbers reflect the current

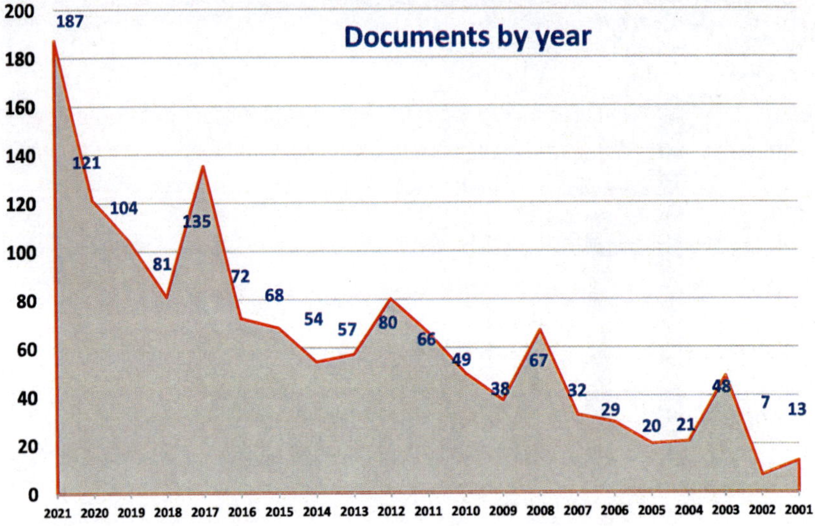

Fig. 1.2 Scholarly output on *Ficus carica* from 2001 to 2021. (www.scopus.com)

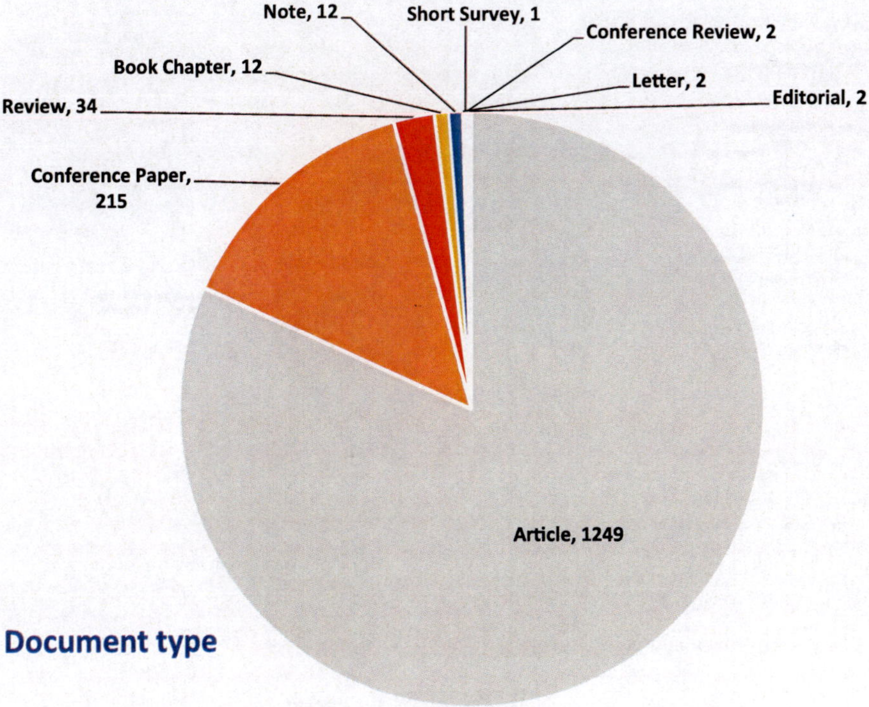

Fig. 1.3 Distribution by types of the document on *Ficus carica*. (www.scopus.com)

interest in fig (*Ficus carica* L.) as a topic in the scientific community. Figure 1.3 presents the distribution of the types of documents in fig (*Ficus carica* L.), which includes research articles (1249), conference papers (215), reviews (34), and book chapters (12). The contributions related to the subject fields (Fig. 1.4) of Agricultural and Biological Science (47%), Biochemistry, Genetics, and Molecular Biology (12%), Medicine (6%), Pharmacology, Toxicology, and Pharmaceutics (6%), Environmental Science (5%), and Chemistry (4%). Scientists from Turkey, Italy, Brazil, USA, Iran, Spain, Tunisia, China, India, Japan, France, Egypt, and Saudi Arabia emerged as principal authors (Fig. 1.5). Acta Horticulturae, Scientia Horticulturae, Revista Brasileira De Fruticultura, Journal Of The Japanese Society For Horticultural Science, Plant Disease, Journal Of Plant Pathology, International Journal Of Fruit Science, Hortscience, Food Chemistry, Frontiers In Plant Science, Fruits and South African Journal of Botany are the leading journals that published scientific research on the fig (*Ficus carica* L.).

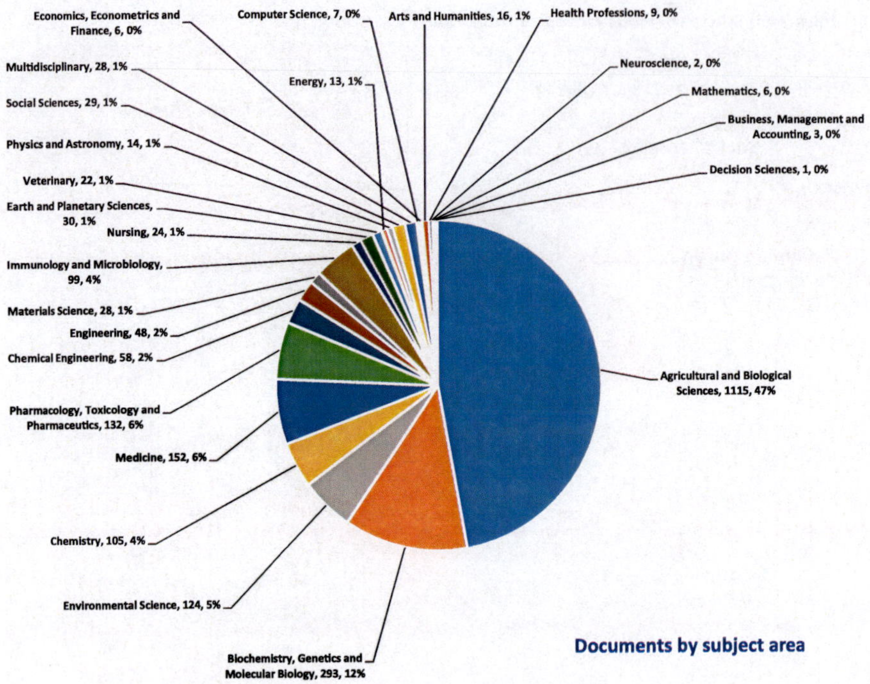

Fig. 1.4 Distribution by subject area of documents on *Ficus carica.* (www.scopus.com)

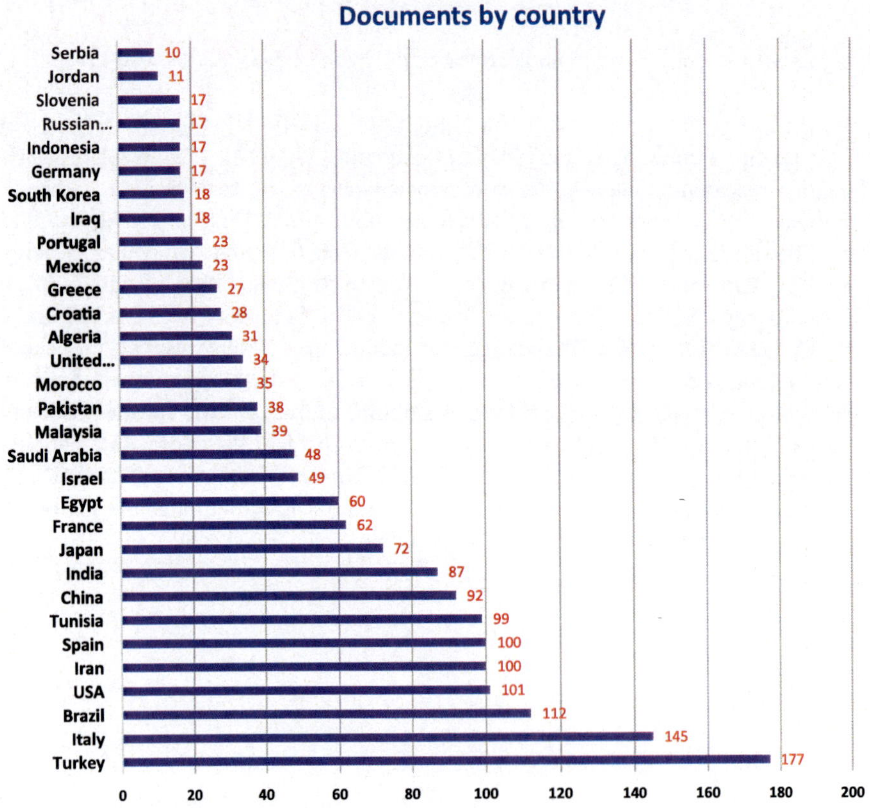

Fig. 1.5 Distribution by country of documents on *Ficus carica.* (www.scopus.com)

4 Aims and Features of This Book

Fig (*Ficus carica* L.) illustrates an excellent functional food and healthy product item. To the best of knowledge, there is a lack of books discussing complete information on the production, processing, chemistry, products, and medical properties of figs and the plant parts (i.e., fruit, skin, leaves, roots, latex, and byproducts). Therefore, this book is considered necessary since it will cover all the information about fig (*Ficus carica* L.). In addition, to the best of knowledge, this book presents and collects available scientific information mentioned above in one work.

Fig (Ficus carica): Production, Processing, and Properties create a multidisciplinary discussion forum on *Ficus carica* with particular emphasis on its horticulture, post-harvest, marketability, phytochemistry, extraction protocols, biochemistry, functionality, nutritional value, health-promoting traits, ethnomedicinal applications, technology, and processing. The impact of processing (traditional and innovative) on recovering value-added compounds from *Ficus carica* byproducts is reported. Also, the book discusses the novel applications of *Ficus carica* in foodstuffs, cosmetics, and pharmaceutical products.

The tentative manuscripts have a diversity of developments in nutrition, food science, and horticulture research. The book contains comprehensive chapters under main sections, namely

Part I: *Ficus carica*: Cultivation, Species, and Cultivars
Part II: *Ficus carica*: Chemistry, Functionality, and Health-Promoting Properties
Part III: *Ficus carica*: Technology, Processing, and Applications

References

Arvaniti, O. S., Samaras, Y., Gatidou, G., Thomaidis, N. S., & Stasinakis, A. S. (2019). Review on fresh and dried figs: Chemical analysis and occurrence of phytochemical compounds, antioxidant capacity and health effects. *Food Research International, 119*, 244–267. https://doi.org/10.1016/j.foodres.2019.01.055

Badgujar, S. B., Patel, V. V., Bandivdekar, A. H., & Mahajan, R. T. (2014). Traditional uses, phytochemistry and pharmacology of *Ficus carica*: A review. *Pharmaceutical Biology, 52*(11), 1487–1503. https://doi.org/10.3109/13880209.2014.892515

Barolo, M. I., Ruiz Mostacero, N., & López, S. N. (2014). *Ficus carica* L. (Moraceae): An ancient source of food and health. *Food Chemistry, 164*, 119–127. https://doi.org/10.1016/j.foodchem.2014.04.112

Crisosto, H., Ferguson, L., Bremer, V., Stover, E., & Colelli, G. (2011). Fig (*Ficus carica* L.). In *Postharvest biology and technology of tropical and subtropical fruits* (pp. 134e–160e). Elsevier.

Desa, W. N. M., Mohammad, M., & Fudholi, A. (2019). Review of drying technology of fig. *Trends in Food Science & Technology, 88*, 93–103. https://doi.org/10.1016/j.tifs.2019.03.018

FAOSTAT. (2022). Food and Agriculture Organization of the United Nations. Available online https://www.fao.org/faostat/en/#home. Accessed 10 June 2022.

Mawa, S., Husain, K., & Jantan, I. (2013). *Ficus carica* L. (Moraceae): Phytochemistry, traditional uses and biological activities. *Evidence-based Complementary and Alternative Medicine*, 974256. https://doi.org/10.1155/2013/974256

McClements, D. J. (2019). The science of foods: Designing our edible future. In D. J. McClements (Ed.), *Future foods: How modern science is transforming the way we eat*. Springer.

Melgarejo, P. (2017). El cultivo de la higuera (*Ficus carica* L.). In *Frutales de Zonas Áridas* (p. 118). A. Madrid Vicente, Ediciones. ISBN 84-89922-37-3.

Melgarejo, P., Sánchez, M., Hernández, F., & Martínez, J. (2003). Chemical and morphological characterization of four fig tree cultivars (*Ficus carica* L.) grown under similar culture conditions. *International Society for Horticultural Science, 605*, 33–36.

Núñez-Gómez, D., Legua, P., Martínez-Nicolás, J. J., & Melgarejo, P. (2021). Breba fruits characterization from four varieties (*Ficus carica* L.) with important commercial interest in Spain. *Food, 10*(12), 3138. https://doi.org/10.3390/foods10123138

Ramadan, M. F. (2020). Chapter 1 – Introduction to cold pressed oils: Green technology, bioactive compounds, functionality, and applications. In M. F. Ramadan (Ed.), *Cold pressed oils* (pp. 1–5). Academic. https://doi.org/10.1016/B978-0-12-818188-1.00001-3

Ramadan, M. F. (2021). Introduction to black cumin (*Nigella sativa*): Chemistry, technology, functionality and applications. In M. F. Ramadan (Ed.), *Black cumin (Nigella sativa) seeds: Chemistry, technology, functionality, and applications* (Food bioactive ingredients). Springer. https://doi.org/10.1007/978-3-030-48798-0_1

Sadder, M. T., & Ateyyeh, A. F. (2006). Molecular assessment of polymorphism among local Jordanian genotypes of the common fig (*Ficus carica* L.). *Scientia Horticulturae, 107*, 347–351.

Salehi, B., Prakash Mishra, A., Nigam, M., et al. (2021). Ficus plants: State of the art from a phytochemical, pharmacological, and toxicological perspective. *Phytotherapy Research, 35*, 1187–1217. https://doi.org/10.1002/ptr.6884

Teixeira, N., Melo, J. C. S., Batista, L. F., Paula-Souza, J., Fronza, P., & Brandão, M. G. L. (2019). Edible fruits from Brazilian biodiversity: A review on their sensorial characteristics versus bioactivity as tool to select research. *Food Research International, 119*, 325–348. https://doi.org/10.1016/j.foodres.2019.01.058

Yao, L., Mo, Y., Chen, D., Feng, T., Song, S., Wang, H., & Sun, M. (2021). Characterization of key aroma compounds in Xinjiang dried figs (*Ficus carica* L.) by GC-MS, GC-olfactometry, odor activity values, and sensory analyses. *LWT, 150*, 111982. https://doi.org/10.1016/j.lwt.2021.111982

Part I
Fig (*Ficus carica*): Cultivation, Species, and Cultivars

Chapter 2
Figs in Morocco: Diversity Patterns, Valorization Pathways and Value Chain Resilience

Lahcen Hssaini, Rachid Razouk, Aziz Fadlaoui, and Karim Houmanat

Abbreviations

ABTS	Ethylbenzothiazoline-6-sulfonic acid
CE	Catechin equivalent
cy-3 rutinoside	Cyanindin-3- rutinoside
DPPH	2,2-diphenyl1-picrylhydrazyl
dw	dry weight
FRSA	Free radical scavenging activity
GAE	Gallic acid equivalent
IC_{50}	Half maximum inhibitory concentration
ISSR	Inter-simple sequence repeat
mM	Millimole
PCA	Principal components (PC) analysis
RAPD	Random amplified polymorphic DNA
RFLP	Restriction fragment length polymorphism
SSC	Soluble sugars content
SSR	Simple sequence repeat
TA	Titratable acidity
TAC	Total anthocyanins
TFC	Total flavonoids content
TPAC	Total pro-anthocyanidins
TPC	Total phenolic content
TSS	otal soluble solids
β-Car	β-carotene

L. Hssaini (✉) · R. Razouk · A. Fadlaoui · K. Houmanat
National Institute of Agricultural Research, Avenue Ennasr, BP 415 Rabat Principale, Rabat, Morocco
e-mail: lahcen.hssaini@inra.ma

M. F. Ramadan (ed.), *Fig* (Ficus carica)*: Production, Processing, and Properties*, https://doi.org/10.1007/978-3-031-16493-4_2

1 Introduction

Morocco hosts large biodiversity and considerable within-species genetic diversity (Achtak et al., 2010). Located in the northwest of Africa, between the Strait of Gibraltar and the Mediterranean Sea to the north, the Sahara Desert to the south, with a long coastline in the west overlooking the Atlantic Ocean and crossed diagonally by the Atlas Mountains moving eastwards from the Atlantic, Morocco shelters an exciting diversity of both climates and agroecosystems hosting the second most diverse flora in the Mediterranean basin, after Turkey. The country domesticates around 4200 species, of which over 20% are endemic (Rankou et al., 2013). The fig (*Ficus carica* L., 2n = 26) tree is undoubtedly one of the most emblematic species of the Mediterranean landscape (Khadari et al., 2008; Hssaini et al., 2020b). Being the earliest domesticated fruit species of the Neolithic Revolution (7000–8000 B.P), the fig tree belongs to the *Moraceae* family, which is one of the most populous of all plant genera with over 800 species (Schmitzer et al., 2011; Pourghayoumi et al., 2017; Boudchicha et al., 2018). *Ficus carica* L. is thought to be native to the Middle East and has been cultivated for millennia in close association with grapevine and olive (Barolo et al., 2014; Pérez-Sánchez et al., 2016; Vitelli et al., 2017). In Morocco, the fig cultivation is ancestral and is mainly planted in mountains and family gardens (Hssaini et al., 2019b). The fig landscape is characterized by four cultivation systems based on applied agricultural practices and varietal associations. The first system, described as extensive, concerns plantations where no cultivation operations other than picking are practiced. These are generally small-sized farms located on steep slopes. The second system is semi-extensive, characterized by maintaining several agricultural practices, including tillage, pruning, impluviums, and caprification. These practices are generally applied for both fig trees and the intercropping system. The third system, semi-intensive, is represented by plantations that benefit, in addition to the cultivation practices adopted in the semi-extensive system, from manual weeding, maintenance pruning, and/or contribution of mineral or organic manure. These last two systems are the most common and are also distinguished from the extensive system by the very high number of varieties used.

The country ranks the third-largest producer of figs globally after Turkey and Egypt, with an annual production of more than 153,472 tons, representing about 11% of the worldwide production (FAOSTAT, 2019). This production is concentrated mainly in the Rif region (Northern Morocco), the main fig growing area (Hssaini et al., 2019b). Traditionally, local fig resources are defined based on qualitative criteria or the proper names associated with pruning, pollination, and protection practices that make them similar regardless of being genetically far from each other.

In the Moroccan diet, figs are consumed either fresh or dried during the whole year but abundantly during religious occasions (Hssaini et al., 2019b). This growing interest in fig consumption is related to the fruit's nutritional values, as they constitute an essential source of fiber, minerals (potassium, magnesium, calcium, iron, etc.), vitamins (vitamin E, vitamin C, thiamine, and riboflavin), carotenoids (lycopene as the most abundant, zeaxanthin, α-carotene, lutein, cryptoxanthin, and β-carotene) and antioxidant polyphenols which are more abundant in a fig than red wine and tea and concentrated mainly in the peel (chlorogenic acid, gallic acid, and syringic acid). In addition, figs are also rich in flavonoids (anthocyanins, catechin, and epicatechin), sugars (glucose and fructose, with minor traces of sucrose), organic acids, amino acids (aspartic acid and glutamine), and volatile molecules (Solomon & Golubowicz, 2006; Oliveira et al., 2009; Veberic & Mikulic-Petkovsek, 2015; Hernández, 2016; Hssaini et al., 2019a). Those compounds are responsible for the fig's pleasant taste, color, astringent flavors, and aroma (Veberic et al., 2008; Muji et al., 2012). They are also associated with preventing coronary diseases due to their numerous biological effects, such as scavenging unstable free radicals from anticarcinogenic and anti-inflammatory activities (Makris et al., 2007; Çalişkan & Polat, 2012; Palmeira et al., 2019). The aforementioned bioactive compounds' amounts strongly depend on the genetic, geographic, and growing factors (Solana et al., 2018; Hssaini et al., 2020b; Mirheidari et al., 2020). Figs are classified as highly perishable commodities because their high moisture content reaches up to 80% (Saki et al., 2019; Hssaini et al., 2020c). Therefore, maintaining their freshness has always been the adequate approach to keeping their health-promoting attributes. However, this decision implies sophisticated storage techniques and chain distribution with considerable energy consumption.

Nevertheless, figs cannot be stored for an extended period without a significant loss of nutritional value (Mat Desa et al., 2019). Therefore, processing remains the best way to extend a fig's shelf-life and preserve the nutritional quality of the end products. Traditionally, salting, dehydration, and fermentation, however, new processing techniques are acquiring popularity worldwide since they offer key advantages such as shelf-life extension, nutritional and sensory quality preservation, and food safety (Martins et al., 2019). In Morocco, the fig processing is an increasing challenge because of the low interest given to this sector and the low investment in existing infrastructure, which significantly impacts the fruits and their end products' marketability. So far, the fig processing is limited to drying, juice, and jam, even though the fruit has many other potentials to be exploited with a high value-added such as seeds (Hssaini et al., 2020a). The latter has been the less valorized part of the fruit, although some recent studies reported a satisfactory oil yield with high unsaturation level and antioxidant potency (Icyer et al., 2016; Hssaini et al., 2020a, b). On the other hand, improving the fig value chain resilience to changing environments and different chocks is a challenging endeavor nowadays since the sector is poorly organized and requires a holistic approach and coordinated policies to evade undesired chocks that may occur during specific supply chain steps or in other related sections.

This chapter focuses on the fig diversity in Morocco, highlights its valorization pathways, and examines the fig value chain vulnerability to shocks and the exploration of solutions aiming at improving the system's ability to adapt and to be resilient when single or multiple shocks by mapping the fig value chain weaknesses alongside potential consequences of various chocks that may occur taking into account the opportunities that can be captured during the fig value chain examination.

2 Genetic Diversity Patterns

2.1 Overview of Available Fig Germplasm

The fig tree is one of Morocco's most diversified fruit species, with over 130 morphotypes, mostly seedlings clones propagated by cuttings and exotic varieties introduced from neighboring Mediterranean countries (Achtak et al., 2010). The Moroccan fig genetic resources also include various unknown types and spontaneous forms in coexistence with cultivated types, which are few subjected to intensive breeding programs and therefore continue to exhibit a large genetic diversity (Hmimsa et al., 2012). This outstanding genetic diversity results from a series of domestication events, gene flow between wild and cultivated compartments, effects of natural adaptive selection, human selection, and large diffusion dynamics over long periods (Khadari et al., 2005a). The impact of these processes on diversity depends not only on the species' biology but also on social context, human practices, and local dietary preferences (Ater et al., 2008; Perez-Jimenez et al., 2010). Since 2008, the local fig diversity has been enriched by some exotic varieties, which were confirmed to be suitable for drying, particularly 'Sarilope' from Turkey, 'Kadota' from Italy, and 'Col de Dame Blanche' from France. These varieties were efficient under local conditions through behavioral studies undertaken by the National Agricultural Research Institute (INRA) in 1995 (Oukabli et al., 2003a).

Two types of edible figs are growing in Morocco: biferous (breba and main crop) and uniferous (main crop). The local biferous types are mainly represented by eleven cultivars, 'Ghouddane', 'Ournaksi', 'El Khal', 'Ember El Khal', 'Fassi', 'Messari', 'Filalia', 'Jeblia', 'Hamra', 'Ounq El Hmam' and 'Beida'. The uniferous types mainly concern 'Nabout', 'El Quoti Labied', 'Embar Labied', 'Hafer El Brhel', 'Chaari', 'Ferquouch Jmel', and 'Ferzaoui' (El Hajjam et al., 2018). Denominations of the aforementioned cultivars and others refer mainly to leaf shape and fruit morphometric traits (color, shape, or taste) and geographic origin or the production locality (Table 2.1). In addition, the exchange of plant material, which was accompanied by the flows of human populations, caused varietal confusion. This problem has led to the attribution of the same denomination to various genotypes, although they have different pomological characteristics (homonymy), or rather naming differently some genetically identical individuals (synonymy). The most remarkable case is that of 'Ghouddane' cultivar, known as polyclones (Achtak et al., 2009). The cases of polyclonality have also been revealed in the cultivars 'Nabout', 'Chaari', 'Ournaksi', 'Hamra', and 'Bioudi' (Khadari & Oukabli, 2005). This list of polyclonal cultivars is not exhaustive due to the lack of intra-varietal molecular authentication studies within other local fig ecotypes. Due to the

Table 2.1 Denomination meaning of some fig cultivars in Morocco

Cultivar	Denominations meaning
Nabout[a]	Generic Arabic word meaning a seedling plant
Gouddane[b]	Dark skin color
Zerqui[b]	Blue skin color
Sebti[b]	Originating from Sebta in northern Morocco
Messari[b]	Originating from Beni Messara in northern Morocco
Jaadi[b]	fruit with ribs over the skin
Ounq El Hmam[b]	Refers to the pigeon neck shape
Ferquouch Jmel[b]	Refers to the camel foot in Moroccan dialect, which describes a typical fruit shape

[a]Hmimsa et al. (2017)
[b]Hssaini et al. (2019b)

environmental effect, this confusion in the cultivar denomination is linked to the fruit phenotypic variation. In particular, the human factor's modest technicality during the exchange and reproduction of the plant material may also explain the observed varietal confusions (El Khaloui, 2010; Achtak et al., 2010). The morphological similarities and the inter-connections characterizing these species have influenced the human selection pressure exerted on the individuals, mainly resulting from seedlings, distinguished by the local term 'Nabout' (Hmimsa et al., 2017). The problems of homonymy and synonymy constitute a severe challenge in the management, use, study, and conservation of genetic resources of fig. In addition to Morocco, this problem has been reported in several other Mediterranean countries such as Tunisia (Chatti et al., 2003; Saddoud et al., 2011), France (Khadari, 2005b), Spain (Sanchez et al., 2016), Turkey (Caliskan & Polat, 2012) and Italy (Ciarmiello et al., 2015).

The Moroccan agroecosystems also include an important diversity within the caprifigs (male fig trees), which remains less documented. This diversity is dominated by spontaneous forms resulting from seedlings growing alongside water sources and frequently encountered near the riversides, streams, irrigation canals, and wells (Oukabli et al., 2003b; Hmimsa et al., 2012). The named types, multiplied by cuttings, remain less diversified. Hmimsa et al. (2017) counted less than 10 well-known denominations of caprifig in Rif areas in northern Morocco, of which the most popular are 'Ḥlu', 'Marr', 'Lwizi', 'Ahurri', 'Aharchiw', 'Ahfriw' and 'Azundri'. The later remains thus far less domesticated and still growing naturally in the wild, even though few can be found near houses and family gardens. Farmers give little importance to caprifig cultivation, although they know its role as a pollen source.

Morocco set up the first national ex-situ collection in 1995 at the National Institute for Agricultural Research (INRA) experimental station in north-central Morocco to conserve this genetic diversity while avoiding varietal confusion. This collection comprises 160 local clones prospected with 60 exotic varieties over the kingdom. Lately, it has been enriched by around sixty hybrids within a program of intraparietal diversification and around thirty caprifig trees. Thus far, a large part of this important fig germplasm has been screened for its morphometric, physio-biochemical, and molecular traits, of which the superior results are briefly reported throughout the following sections of this chapter.

2.2 *Fruit Morphometric Diversity*

The international guide for fig tree description includes 192 morphometric traits for the female type, among which 40 are specific for fresh fruit, while 14 are considered highly discriminating descriptors. The latter concerns fruit shape (width/length index and fruit shape according to the location of the maximum width), apex shape, fruit weight, skin ground color, internal pulp color, ease of peeling, skin cracks, ostiole width, resistance to ostiole-end cracks, fruit flesh thickness, pulp juiciness, fruit cavity and total soluble solids (IPGRI and CIHEAM, 2003). However, the use of morphometric descriptors to characterize the fig diversity faces two significant problems, firstly, the subjectivity in qualitative traits description (Khadari et al., 1995), and on the other hand, the influence of environmental conditions that induce fluctuations in the expressions of traits as well as uncertainties in identifying synonymies and homonymies, which are widespread within the local fig populations (Giraldo et al., 2010). Despite these inaccuracies, morphometric characterization continues to be widely used as a practical approach in fig germplasm screening in several Mediterranean countries (Essid et al., 2017; Khadivi et al., 2018). According to these studies, the fruit geometry, skin color, and total soluble solids were the most discriminant morphometric traits for fig screening. Such studies were also carried out in Morocco, *in-situ*, and *ex-situ*, which concerned the local fig germplasm. Regarding studies carried out in-situ, the available literature illustrates that several surveys were performed within local fig populations in some main producing areas to highlight the extent of existing diversity. However, it is essential to emphasize that these studies remain questionable as the environment and agricultural management practices are highly variable and significantly affect the phenotypic expression.

In the most recent screening studies carried out by Hssaini et al. (2020a, b, c, d) on the fig *ex-situ* collection previously mentioned, the large variability revealed was attributed to the genetic factor as the edaphoclimatic conditions alongside the orchard management practices were the same for all genotypes. This made it possible to fix the 'genotype x environment' interaction impact on the expression of the morphometric and physico-biochemical trait and thus compare the genetic potentialities among 135 fig accessions (Table 2.2). It is noteworthy to mention that the abovementioned collection was planted in 2003 following a completely random design.

The descriptive analysis highlighted a wide variability among genotypes for the observed traits, except for the drop at the ostiole, lenticels color, seed size, and ostiole width (Tables 2.3 and 2.4). Visually, 9 skin colors were distinguished within the collection, where yellow-green and light-green were the dominant ones (Fig. 2.1).

Table 2.2 List of local clones and exotic cultivars of fig in *ex-situ* collection screened for fruit morphometric and biochemical traits

Local clones		Exotic cultivars
1. Ahra 2870	48. Arguil_PS8	95. Abgaiti 2111
2. Aicha Moussa 2208	49. Chaari_PS15	96. Abiarous 3015
3. Amellal	50. ELQuoti Lbied_PS11	97. Bellone
4. Assel 2890	51. ELQuoti Lbied_PS20	98. Bougie
5. Ben_T1	52. ElQuoti Lbied_PS3	99. Breval Blanca 2736
6. Ben_T2	53. ElQuoti Lbied_PS6	100. Brown Turkey
7. Bioudi 2218	54. Ghani_PS2	101. Burjasot Blanca 3037
8. Bioudi 2878	55. Ghoudan_PS1	102. Cuello Dama Blanco 2233
9. Bioudie 2255	56. Ghoudan_PS17	103. Cuello Dama Blanco
10. Bousbati 2880	57. Ghoudan_PS4	104. Colle de dame blanc
11. Btitarbi	58. Jaadi_PS16	105. Conadria (Porquert)
12. Chaari 2881	59. Lamtal_PS9	106. Conidria
13. Chaari 2881	60. Mssari_PS13	107. Diamna
14. Chabaa Ourgoud	61. Nabout_PS12	108. Dottato Perguerolles
15. El Ghani	62. Nabout_PS7	109. Figue de Marseille
16. EL Hmiri 2224	63. Ounq Hmam_PS14	110. Grise St. Jean
17. EL Khal 2283	64. Sebti_PS10	111. Grosse dama Blanca
18. EL Qoti Lezreq 2883	65. Tabli _PS19	112. Grosse Dame Blanche 2953
19. EMBAR EL KHAL 2247	66. Tabli_PS18	113. Gulgium
20. EMBAR LEBIED 2240	67. Zerqui_PS5	114. Herida
21. Fassi	68. INRA 1301	115. III 31 Roger
22. Fassi 2267	69. INRA 1302	116. Khelema 3148
23. Filalia 2211	70. INRA 1303	117. Kodata
24. Hafer El Brhel	71. INRA 1304	118. MELISSOSYKI 3074
25. Hafer Jmel 2253	72. INRA 1305	119. Nardine
26. Hamra	73. INRA 1306	120. Palmeras
27. Hamra 2252	74. INRA 1308	121. Pingo de Mel
28. Hamra 2588	75. INRA 1314	122. Princesse
29. Hmidi 2250	76. INRA 1502	123. Rey Blanche
30. Kahoulta 2251	77. INRA 1503	124. Royal Blanck
31. Lamandar Noir	78. INRA 1506	125. Snowden
32. Lmandar Bied	79. INRA 1606	126. Sucre vert
33. Mendar 2891	80. INRA 2101	127. Sucre vert
34. Nabout 2893	81. INRA 2103	128. Tena
35. Noukali 2254	82. INRA 2105	129. Trojana
36. Noukalli	83. INRA 2201	130. VCR 153/17
37. Rhoudane 2227	84. INRA 2201	131. VII 1 Roger 3
38. Taranimt 2399	85. INRA 2204	132. Violette d'Agenteruil
39. V12	86. INRA 2205	133. White Adriatic_102
40. V2 (b)	87. INRA 2206	134. White Adriatic_13
41. V33(b)	88. INRA 2304	135. Palmares

(continued)

Table 2.2 (continued)

Local clones		Exotic cultivars
42. Zerouali	89. INRA 2305	
43. Ztaniti	90. INRA 2307	
44. INRA 2506	91. INRA 2307	
45. INRA 2508	92. INRA 2401	
46. INRA 2603	93. INRA 2404	
47. INRA 2802	94. INRA 2501	

Table 2.3 Descriptive analysis of qualitative morphometric traits within the *ex-situ* collection of INRA (135 fig cultivars)

Descriptor	Dominant character	Frequency	ANOVA signification
Fruit shape	Globose	76%	**
Fruit symmetry	Ovoid	70%	**
Shape around stalk	Rounded	55%	**
Stalk abscission	Medium	60%	**
Ostiole color	White	51%	**
Drop at the ostiole	Absent	65%	ns
Color of ostiole liquid	Pinkish	61%	**
Skin color	Green-yellow	46%	**
Skin ribs	Intermediate	81%	**
Easy of peeling	Medium	51%	**
Skin cracks	Cracked skin	72%	**
Juiciness	Little juicy	74%	**
Skin over color	Yellow	54%	**
Shape of over color	Absent	69%	**
Lenticels quantity	Scarce	43%	**
Lenticels color	White	66%	ns
Lenticels size	Medium	80%	*
Pulp color	Pink	45%	**
Pulp texture	Medium	52%	**
Fruit cavity	Small	52%	*
Seed quantity	Intermediate	50%	**
Seeds size	Medium	78%	ns

* Significant at 0.05; ** Significant at 0.01; *ns* non-significant

However, according to the International Commission on Illumination (CIE), the color coordinates of fruit skin showed much wider variability than the visual assessment. They made it possible to detect color gradients varying from bright yellow to black by revealing various intermediate colors (yellow-green, brown, purple, blue-purple, etc.). Most studied genotypes have a globose and symmetric shape of fruit and did not have a drop at the ostiole, with a considerable variation in fruit weight, from 12.40 to 87.03 g. The highest fruit weights were recorded in the local genotypes ‘Lmandar Noir’ (87.03 g), ‘Bougie’ (76.9 g), and ‘Chaaba Ourgoud’ (73.7 g) as well as in the imported varieties ‘Kadota’ (61.5 g) and ‘Brown Turkey’ (60.4 g).

Table 2.4 Descriptive analysis of quantitative morphometric and biochemical traits within the ex-situ fig collection of INRA (n = 135 cultivars)

	Range	Mean	Std. deviation	ANOVA signification
Fruit weight (g)	12.40–87.03	36.31	4.784	***
Fruit length (mm)	21.34–48.83	34.41	5.89	***
Fruit width (mm)	25.93–61.58	39.85	7.23	***
Stalk length (mm)	2.95–16.30	8.10	4.13	***
Stalk width	2.92–7.96	4.86	1.23	***
Neck length (mm)	1.64–9.52	7.01	2.16	***
Neck width (mm)	2.22–15.69	8.64	2.62	***
Ostiole width (mm)	2.34–8.79	5.13	1.14	ns
Fruit peel thickness (mm)	1.32–8.35	3.74	1.64	***
Volume (mm)	10.00–85.00	35.17	14.70	***
L*	16.71–83.64	52.36	19.32	***
c*	1.98–63.70	27.24	15.3	***
h*	1.24–360	120.23	25.51	***
Soluble Sugars (g/100 g_{dw})	10.08–15.10	12.08	1.26	***
Phenols (mg GAE/100 g_{dw})	25.33–322.00	142.74	9.78	***
Flavonoids (mg CE/100 g_{dw})	14.59–103.71	42.04	9.06	***
Anthocyanins (mg cy-3 rutinoside/100 g_{dw})	0.41–47.95	13.57	10.3	***
Proanthiocyanidins (mg/100 g_{dw})	0.18–8.51	2.24	1.82	**
DPPH (mM Trolox eq/100 g_{dw})	20.92–488.2	180.74	10.3	***
ABTS (mM Trolox eq/100 g_{dw})	31.32–658.21	425.5	14.97	***
β-Carotene (mM Trolox eq/100 g_{dw})	73.5–824.31	403.42	17.38	***
Titrable acidity (g citric acid/100 g juice)	0.07–1.66	0.77	0.4	***
TSS (%)	8.00–48.00	19.68	5.74	***
Maturity index	8.79–236.72	36.08	12.52	***
Titrable acidity (g citric acid / 100 g of juice)	0.07–1.72	0.76	0.31	***
pH	2.51–6.6	4.41	0.91	***

ns non-significant; **; *** denote significant of difference at level 0.01 and 0.001respectively
GAE Gallic acid equivalent, *CE* Catechin equivalent, *TSS* Total soluble solids

However, the lowest was recorded in the genotypes "INRA 1303", "Ounq Hmam_PS14", "Zerqui_PS5", "Ghoudan_PS1", and "Gulgium", for which fruit weight ranged between 12.4 and 16.4 g. The chemical screening highlighted a wide diversity in the sweet taste degree of fruit, determined by the balance between total soluble solids (TSS) and titratable acidity (TA). According to TSS/TA ratio (maturity index), the sweetest figs were of the local genotypes 'Sebti_PS10', 'Noukali 2254' and 'Chaari_PS15' with respective ratio values of 219.30, 138.35, and 104.50, while the most acidic were of 'Dottato Perguerolle' (7.85), 'V2 (b)' (8.81) and 'Hafer Bghal' (8.81). Overall, the variations of the whole observed morphometric traits were in agreement with those reported in other similar studies carried out in

Fig. 2.1 Fruit skin colors identified visually within the *ex-situ* fig collection of INRA (n = 135 cultivars). The percentage of each color within the collection is given in parentheses

other Mediterranean countries, with some differences related to the analyzed genotypes and the environmental conditions of the experimental sites (Ateyyeh & Sadder, 2006; Caliskan et Polat, 2008; Aljane et al., 2008; Podgornik et al., 2010; Jimenez, 2016).

Principal components analysis (PCA) determined the most discriminating morphometric descriptors among the observed traits. Thus, ten PCs explained a total variance of 60.2%, where 27.8% was attributed to the first three PCs. The PC1 explained 12.3% of total variance and was strongly determined by 7 morphometric traits, namely: fruit weight (r = 0.86), fruit dimensions (r = 0.69, r = 0.85 for length and width, respectively), stalk width (r = 0.72), neck dimensions (r = 0.58, r = 0.61 for length and width, respectively) and fruit volume (r = 0.74). The PC2 contributed 9.98% of the total variance and was influenced by 4 colorimetric traits of fruit skin, which are visual color (r = –0.66) and chromaticity coordinates, c* (r = –0.67), L* (r = –0.65), and a* (r = 0.58). The PC3 explained 5.51% of the total variance and was mainly linked to fruit symmetry (r = 0.60). These 12 characters were, therefore, the most effective descriptors to discriminate the studied fig accessions. Their correlations to the first three PCs are estimated to be very strong and significant, owing to the studied collection size and the large number of the variables involved. These findings agree with the previously reported results in edible figs (Ciarmiello et al., 2015; Khadivi et al., 2018). A cluster analysis was performed using Ward's method to identify the similarities and dissimilarities among the accessions, based on all qualitative and quantitative variables measured. The obtained clustering showed a high morphometric polymorphism within the fig collection, distinguishing five clusters, mostly differentiated by fruit weight, shape, and skin color (Fig. 2.2). Given the high number of local clones included in this study, which seems to cover a large part of the local fig genetic heritage, this investigation provides information on morphometric diversity amplitude in the Moroccan figs, allowing a wide diversity in terms of opportunities for their exploitation and valuation.

2.3 *Fruit Chemodiversity*

The chemo-diversity within the Moroccan fig germplasm is thus far poorly documented and still somewhat ambiguous. However, to the best of knowledge, there is a single exhaustive ex-situ study regarding this topic performed on the same collection (Hssaini, 2020). The ultimate objective was to produce a large database on the quality attributes of figs in Morocco, usable across various levels: varietal selection, agroindustry, fresh consumption, quality label development, etc., and identify discriminant biochemical markers for future research purposes. According to the internationally recognized methods, a total of 11 biochemical descriptors were involved; namely, the total contents of polyphenols (TPC), flavonoids (TFC), anthocyanins (TAC), soluble sugars (SSC) and proanthocyanidins (TPAC) as well as the antioxidant activity, assessed using three assays: (i) the DPPH test, (ii) the ABTS method and (iii) the β-carotene bleaching test with their respective half-maximal inhibitory concentration (IC_{50}) (Hssaini et al., 2019a).

The study displayed high significant ($p < 0.001$) differences among genotypes, with large variation amplitudes (Table 2.4). SSC, the main factor of taste and sweetness, ranged from 9.77% to 17.7%, with an average of 12.0%. Among the local genotypes, 'Zerqui_PS5', 'El Quoti Lbied_PS3', and 'Rhoudane 2227' were the sweetest, with respective SSC values of 15.1%, 14.53%, and 14%. TPC varied between 22 and 417.5 mg GAE/100 g_{dw}, with an average of 142.7. It was higher in 'White Adriatic', 'INRA 2304', and 'Ghoudan_PS1' genotypes, where the averages were 415, 322, and 288 mg GAE/100 g_{dw}, respectively. However, 'INRA 1606', 'INRA 1503', and 'Rey Blanche' were the least rich in polyphenols, with average contents of 25.33, 28.67, and 40.89 g GAE/100 g_{dw}. TFC was higher in 'White Adriatic_102' (103.7 mg CE/100 g_{dw}), 'INRA 1301' (101 mg CE/100 g_{dw}), and 'INRA 1306' (96.05 mg CE/100 g_{dw}), while lower in 'Kadota', 'Rey blanche' and 'Diamna', with the average TFC values were 14.6, 14.5 and 15.35 mg CE/100 g_{dw},

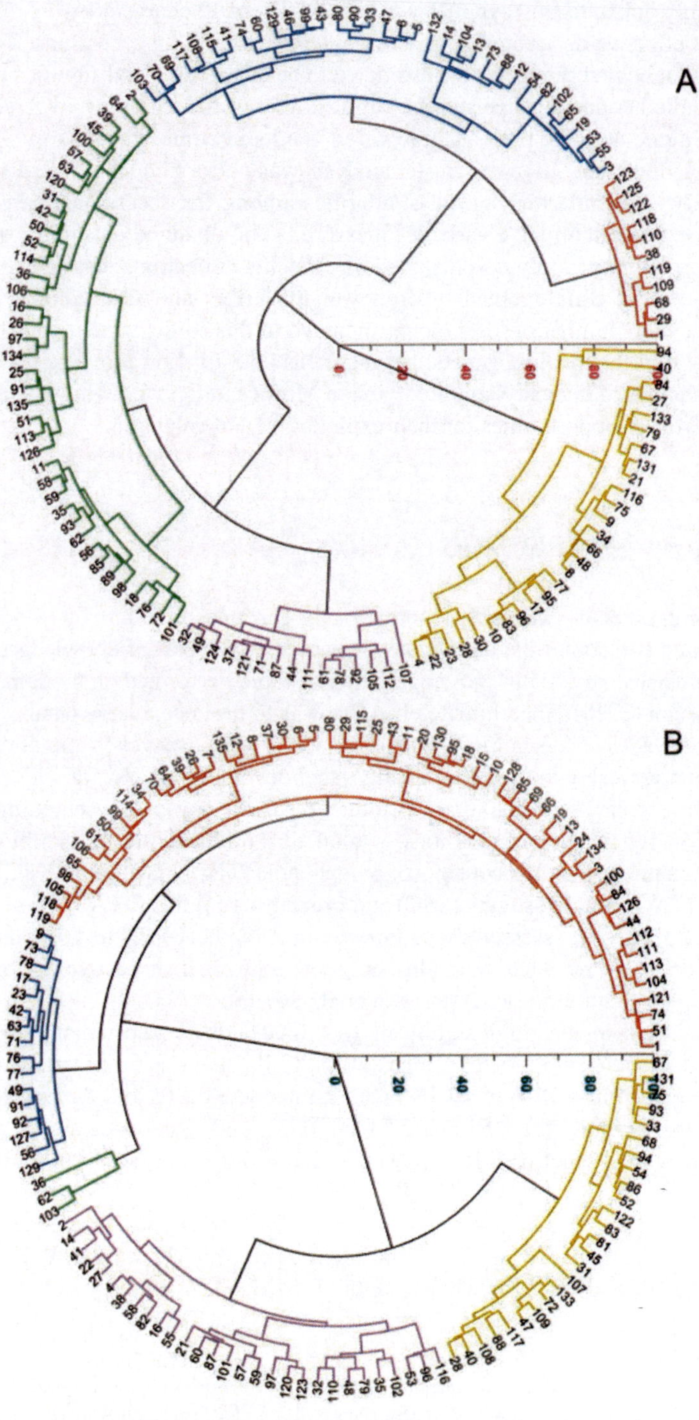

Fig. 2.2 Cluster analysis of 135 fig genotypes based on fruit morphometric (**a**) and biochemical (**b**) traits using Ward's method. Discriminated clusters were marked in different colors

respectively. TAC ranged between 0.45 and 57.47 mg cy-3-rutinoside/100 g_{dw}, with an average of 13.57 mg. The highest values were observed in 'INRA 2105' (52.95 mg), 'Burjasot Blanca' (42.57 mg), and 'INRA 1302' (40.92 mg), while the lowest were found in 'Sucre vert' (0.45 mg) and 'INRA 2307' (2.69 mg). Nevertheless, TAC was extremely low in some cultivars such as 'Felalia' and 'Trojana' to the point that it was not detected. This was also the same case for TPAC, which was not detected in some accessions, including 'INRA 1314' and 'Princesse'. TPAC was low within the studied genotypes (0.18–10.0 mg/100 g_{dw}). Thus, 'Mendar Noir' (8.51 mg/100 g_{dw}) and 'Cello Dama Blanca' (6.64 mg/100 g_{dw}) were the highest genotypes.

The antioxidant activity varied greatly across genotypes according to the three used assays. Indeed, the difference between the extreme values was 23 times for the DPPḤ assay, with an IC_{50} of 0.38–27.28 μg mL^{-1}, 21 times using the ABTS test allowing an IC_{50} of 0.20–5.78 μg mL^{-1}, and 11 times according to the β-carotene bleaching methods, for which IC_{50} ranged between 3.58 and 22.7 μg mL^{-1}. Overall, the observed variations agreed with similar previous studies (Oliveira et al., 2009; Wojdyło et al., 2016; Sedaghat & Rehemi, 2018). Thus, some genotypes were distinguished by their high antioxidant potential, such as 'Kahlouta' (488.2 mM Trolox eq/100 g_{dw}), 'INRA 2501' (451.36 mM Trolox), and 'INRA 2603' (433.5 mM Trolox) for the DPPḤ assay, 'V12' (658.2 mM Trolox) and 'El khal' (632 mM Trolox) for the ABTS method and 'Melissosyki', 'Hmidi' and 'El Qoti Lezreq' according to the β-carotene bleaching test. The highest values of the radical scavenging capacity were related to the genotype richness in phenolic compounds, particularly flavonoids and anthocyanins.

According to PCA based on correlation coefficients, six traits seemed to have the highest impact on discrimination between genotypes in terms of biochemical proprieties, namely, TFC, TAC, TPAC, radical DPPḤ scavenging activity (DPPḤ-RSA), IC_{50} of DPPḤ radicals (IC_{50}-DPPH) and IC_{50} using the β-Carotene bleaching test (IC_{50}-β-Car). These traits presented correlation coefficients of more than 0.5 to the first three PCs, with a cumulative variance of 40.91% (65.0% for six PCs). Indeed, TAC (r = 0.73) and DPPḤ-RSA (r = 0.58) were the most correlated to PC1, explaining 21.6% of the total variance. TFC (r = 0.81) and IC_{50}-DPPḤ (r = 0.70) were more linked to PC2, which contributed 10.0% of the total variance, while TPAC (r = 0.72) and IC_{50}-β-Car (r = 0.69) determined mostly PC3 (9.19%). Moreover, Ward's method's multivariate analysis highlighted a high polymorphism in fruit biochemical traits within the studied collection. Indeed, five clusters of genotypes were obtained, mainly distinguished by soluble sugar content, free radical scavenging activity (DPPḤ and ABTS), and content in polyphenols, particularly anthocyanins (Fig. 2.3). Further details on phenotypic pooling and involved variables discriminating strength could be found in the studies carried out by Hssaini et al. (2020a, b). Therefore, it emerges from this study that the analyzed fig accessions, including local genotypes, exhibit a wide diversity in fruit quality, thus offering multiple choices for clonal selection and commercial and industrial purposes.

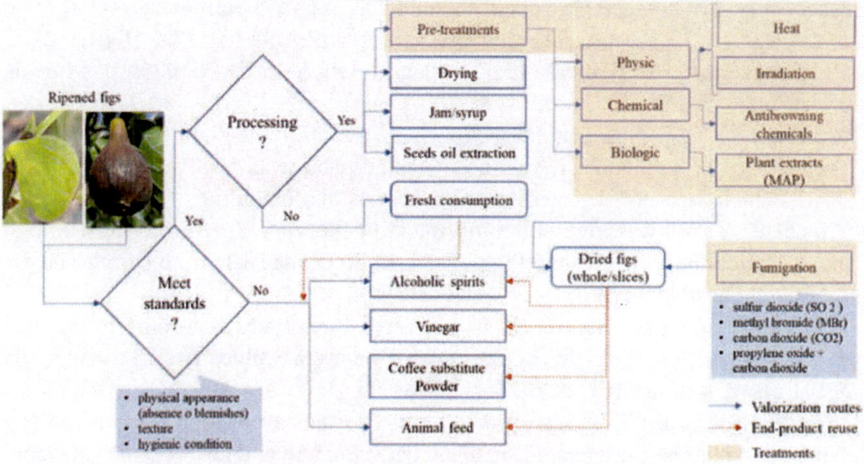

Fig. 2.3 Ripened figs pathways beyond the farm gate. (Source: Authors' construction)

2.4 Molecular Fingerprinting

On fig species, molecular fingerprinting investigations were carried out for a dual objective: (i) to categorize genetic similarities between genotypes and (ii) to identify the widely encountered cases of homonymy and synonymy. In Moroccan fig germplasm, molecular markers were fairly applied, including RFLP on mitochondrial DNA, RAPD, ISSR, and SSR (Khadari et al., 1995, 2005b). In the fig genetic diversity investigation of Khadari et al. (2004), 75 accessions collected from northern Morocco were analyzed using 8 ISSRs markers and 6 SSR loci. As a result, 52 genotypes, including 6 caprifig trees, were identified as genetically heterogeneous cultivars. The homonymy cases were frequently observed in 6 denominations of female figs, namely, 'Bioudi', 'Chaari', 'El khal', 'Hamra', 'Ournakssi', and 'Rhoudane'. In addition, the SSR markers, using 17 microsatellite loci, were applied to assess the genetic diversity within an ex-situ collection of INRA Morocco, which includes 75 accessions representing eight caprifigs, 51 local accessions, 11 exotic varieties, and 5 accessions of unknown origins (Achtak et al., 2009). Among the analyzed samples, 62 distinct genotypes were identified, classified into 54 female figs and 8 caprifigs. This study confirmed the high frequency of homonymy for the abovementioned female figs and highlighted synonyms for 7 genotypes, namely, 'Jeld Elhmar', 'Bousbati', 'El Har', 'Hamra', 'Fassi', 'Noukali' and 'Jeblia'.

These molecular analyzes, covering a large part of the local morphotypes, highlighted the high genetic diversity magnitude within the Moroccan fig germplasm and thus confirmed the phenotypic variation revealed by multivariate analysis of fruit morphological and biochemical traits. Some suspected cases of varietal confusion were also confirmed. However, the correlations between molecular and phenotypic traits remain unstudied to the best of knowledge. Such studies will significantly interest the early selection of efficient fig cultivars among local clones and created genotypes through hybridization or mutagenesis.

3 Valorization Pathways

There has been a notable increase in fig production in Morocco during the last decade. Thus, the production increased by about 29%, as it rose from about 109,200 tons in 2009 to 153,472 tons in 2019 (FAOSTAT, 2019). This increased rate is mainly related to the extension of area harvested from 47,600 to 62,969 ha during the same period, equivalent to 24.4%, as targeted by the national green Moroccan plan launched in 2008. A large amount of this production is usually oriented to processing, as fresh figs storage requires specific low-temperature facilities, which has been thus far difficult to maintain throughout the distribution chain. Moreover, fresh fig is very perishable under the ambient condition as it contains a large amount of moisture, up to 80% (Mat Desa et al., 2019). Therefore, fresh figs can be stored for up to 8 days under low temperature and high hygrometry (4.44–6.11 °C and 75%, respectively). However, they can only last for 2 days under ambient storage conditions (Mat Desa et al., 2019). Therefore, an extensive storage period of fresh figs could only be achieved through continuous freezing, which requires a significant investment and a considerable amount of energy, which is not feasible on a large scale (Hssaini et al., 2020c).

Figure 2.3 describes the Moroccan fig valorization pathways beyond the farm gates. Indeed, ripened figs that fail to meet the standards to direct consumption or manufacture, mainly if they present visual blemishes, injuries, bruises, or defects, are mainly exploited for alcoholic spirits or vinegar production or even used as animal feed. Figs meeting the aforementioned standards, including good physical appearance and texture with hygienic conditions, are selected for direct consumption or processing. For fresh consumption, several criteria are considered related mainly to morphological characteristics such as shape, color, and caliber, alongside taste and texture.

In Morocco, drying is the most practiced technique for fig shelf-life extending (Chimi et al., 2008). It is also one of the apparent techniques to meet the market demand by reducing wastage, mainly owing to the poorly established logistics for low-temperature distribution and handling facilities leading to improper sell realization (Mat Desa et al., 2019). Direct sun-drying along with solar-assisted drying is particularly the main employed methods to extend fig shelf-life since Morocco receives an intensive daily average solar radiation intensity of 5.3 kW h/m^2 and 5–6 h of sunshine duration in the winter and 9–10 h during the summer (Kousksou et al., 2015). Drying depicts a complex process of heat and mass transfer of removing moisture from biological products. In food processing, the leading role of drying technologies is to preserve agricultural commodities' quality, extend their shelf life and produce new products that would not otherwise be feasible (Nurlaila et al., 2019).

Furthermore, drying allows to obtain the desired moisture content, physical form, flavor, color, or texture and reduces the volume or the weight for low-cost transportation (da Silva et al., 2012). In the Moroccan context, using drying is technically convenient, and the cost associated with processing, packaging, transportation, and storage is sensitively less for dried figs than for canned and frozen products

(Rahman et al., 2016). Unfortunately, as far as we know, there are very few reports on drying kinetics and thermodynamic studies and their impact on the physical and biochemical attributes of figs growing in Morocco. Therefore, how different drying techniques impact the fig drying curve and its functional properties still need to be investigated. Also, energy efficiency and the simulation of fig drying curves with the help of empirical and semi-empirical models are valuable for designing new or improving existing drying systems. They are also helpful in drying process configuration and control (Adak et al., 2017).

In northern Morocco, the main fig cultivation area, open sun drying remains dominant compared to indirect solar drying techniques, as it is the most employed by local small-holder farmers and traditional cooperatives, which dominates the agricultural landscape. However, despite being an eco-friendly method, open sun drying might be challenging due to the occasional occurrence of wet periods during the fig drying period from mid-August to the last decade of September (Ekechukwu, 1999). In addition, this drying method is the slowest among other drying modes, leading to extending the fig's exposure to sunlight and wind. Furthermore, this lengthy process under ambient conditions has enormous drawbacks, some of which are related to the high probability of peel browning occurrence, microbial growth, pest disturbance, insect infestation, and dust contamination, which are disadvantageous to the final product quality. Furthermore, in previous studies, direct sun-drying of several fruits and beverages is associated with postharvest product loss, which is more problematic in developing countries (Mesta et al., 2013; Bradford et al., 2020; Jayaraman & Gupta, 2020). Today, there is growing awareness of the need to adopt an eco-friendly drying mode that fulfills market demands for economic value-added and consumer satisfaction by producing hygienic dried figs with high nutritional properties (Durance, 2002). Solar dryers can fulfill the abovementioned requirements and substitute direct sun-drying techniques (Chua & Chou, 2003). Unlike direct solar dryers that use solar radiation, solar dryers use heated air in active and passive circulation modes. For figs, these dryers can reach a temperature of 50–60 °C and thus produce a good product quality under constant temperature and moisture removal rate (Singh et al., 2018).

In Morocco, the overall quality of dried figs quality is most often determined through several key attributes related to the physical appearance (bright color and optimal hygienic aspect), chemical (high sugars, optimal water activity, and texture), and microbial (absence of microbial load, pests, and contaminants). However, some issues can occur during or after drying, diminishing the product quality and marketability (Tan, 2017; Aksoy, 2017). Peel browning related to enzymatic or Maillard reactions is among these issues which concern white peel-colored figs. Thus, the color change is the primary limitation on dried fig's shelf-life that influences consumer acceptance (Hssaini et al., 2020c). Enzymatic browning is caused by the conversion of polyphenols to quinones, which is catalyzed by polyphenol oxidases, while Maillard browning is a result of the crystallization of carbohydrates from inside to the external surface of the fruit (Tareen et al., 2012; Gamboa-Santos et al., 2014; Derardja et al., 2019; Mat Desa et al., 2019).

Moreover, both browning types speed up as the temperature increases (Bet, 2015). Several approaches are employed to deal with these issues, including the application of pretreatments which are divided into three groups: chemical, using organic acids, such as exogenous ascorbic and salicylic acids (Hssaini et al., 2021a), physical like using heat such as blanching in boiling water (Piga et al., 2004; Piazzolla et al., 2018) or by irradiation (i.e., gamma rays and ultraviolet) (Hamanaka et al., 2011; Afsah-Hejri et al., 2021), or finally biological through the application of plant extracts particularly aromatic and medicinal plants (Allegra et al., 2017, 2021). In Morocco, sulfites are well-known anti-browning agents and thus are extensively used to prevent browning reactions, despite their negative effect on consumer health (Oliphant et al., 2012; Stohs & Miller, 2014). Efforts are deployed to substitute sulfites with more healthy and safe anti-browning chemicals such as ascorbic and salicylic acids (Hssaini et al., 2021a). Considering those above, some consumers prefer dark-colored fruits, as they do not need a few additional treatments.

In the scope of circular processing, unsold dried figs are used either for wine or spirits production, with plurivarietal or monovarietal dried figs due to their reduced quality due to storage conditions or time or even both the raw inputs (Solana et al., 2018). They could also be roasted and used as a coffee substitute or animal feed (Ishurd et al., 2004; Liang et al., 2016; Mat Desa et al., 2019). Processing to jam and syrup comes second after drying but in few amounts and, similar to some other products, are in some cases meld with extracts of medicinal and aromatic plants, especially thyme (Allegra et al., 2017, 2021). Recently, some research outlined the great opportunity behind fig seeds processing as a novel source of value-added oil with a high unsaturated oil yield reaching up to 30% and presenting a promising antioxidant potency (Hssaini et al., 2020a, 2021b). Finally, as the third worldwide fig producer, Morocco still has millstones to achieve in fig processing as it hosts a wide range of diversity with excellent attributes that need to be further valorized so that the Moroccan fig and its derivates could be competitive in the worldwide market.

4 Value Chain and Its Resilience

In Morocco, no single study is interested in studying and assessing the fig value chain, probably because the sector remains poorly structured and lacks reliable data. Nevertheless, the fig supply chain value in Morocco is depicted in Fig. 2.4, based on observations made in the field. The figure draws a traditional food supply chain model, where vertical connections are spotted within the interactions between stakeholders and system operations. At the model upstream end, the classical supply chain is mainly driven by smallholder farmers and agroindustry local-oriented, providing fresh figs and its fig-based products. The main way fresh figs and processed ones reach the traditional markets and supermarkets is through a web of stakeholders, including small and medium-scale traders and trailers who purchase figs at the farmgate or traditional markets and transport these commodities to either rural or modern fresh fruits markets or local cooperatives for drying or producing traditional

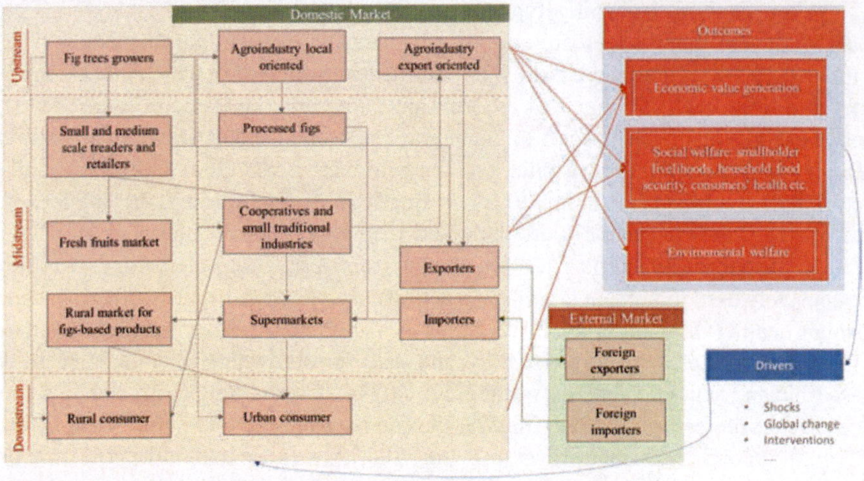

Fig. 2.4 Schematic representation of figs supply chains in Morocco, with potential outcomes and disturbances drivers. (Source: Authors' construction)

fig-based preparations. At the model downstream end, figs and fig-based products reach consumers through rural assembly markets and supermarkets.

Going back to the model upstream end, small and medium-scale agro-industrial firms provide consumers through modern markets with fig-based products. It is noteworthy that the market share of figs and fig-based products in the supermarkets remains, thus far, much smaller than the traditional markets. Export-oriented supply is mainly based on medium retailers who provide export-oriented agro-industry for fresh and dried figs. Between 2005 and 2020, these exports were 46 tons (t) for fresh figs, 455 t for dry figs for human consumption, and 250 t for damaged dry figs (completely discarded for not meeting consumption standards) for industrial use. Morocco imports mainly dried figs for human consumption through a tiny network of local and foreign importers, mainly from Turkey, the first worldwide dried fig exporter (FAOSATA, 2019). Imported figs are subjected to import duty set at 40% upon entry into the country (Moroccan Office of Change, 2021). These commodities are also subjected to value-added tax at a rate of 20% and the parafiscal import tax of 0.25%. However, thanks to free trade agreements, some countries are exonerated from these duties.

Based on the model depicted in Fig. 2.4, many weaknesses persist upstream and downstream, hindering its development. At the upstream level, the sector is subject to structural constraints, as it is characterized by the dominance of small farms, excessive fragmentation, and the multitude of legal systems in the fig landscape. Furthermore, most orchards are managed traditionally with old-aged trees. Except for some experiences that are beginning to emerge, the professional organization of producers is still very weak, limiting the possibilities to access production key factors at reduced prices and technical assistance services for modernization of management methods. At the midstream level, the markets for fig products have an oligopolistic nature, characterized by numerous small and medium-sized producers, who are poorly organized alongside a few large buyers who set prices. Thus, more than a third of the sector's total value added is captured by intermediaries. Besides, the insurance of fig producers regarding the impact of climate change risk appears to be limited only to six hazards, namely hail, frost, strong wind, dry-hot wind, overheating, and excess water. However, it does not cover the drought, pests, and diseases. At the downstream level, fig consumption, mainly dried figs, is still poor and follows a stagnant rate of about 0.15 Kg per capita per year (High-Commission for Planning of Morocco, 2014). Notably, the dried figs consumption rate follows the demographic growth besides being mainly concentrated during religious events, including the fasting month of Ramadan (Achtak, 2009). The combination of all the above mentioned factors has limited the possibilities of intensification and economies of scale, leading to the low profitability of this crop.

Some of the major limitations that face the modernization of the Moroccan fig supply chain are the lack of organization among fig-trees, and farmers, within a coordinated aggregative, alongside the absence of coordination between itinerant components involved in the fig supply chain, mainly traders and retailers. Therefore, we hypothesize that fig supply chain actors, including smallholder farmers, small and medium-scale traders, and rural assembly markets are more vulnerable when the fig supply chain is disrupted.

Today, there is a rising concern about bringing resilience thinking into the food system and the fig supply chain. The global food system is increasingly complex through transversal interdependencies across various stakeholders, locally or globally. Despite the very positive results, these diagonal interdependencies also engender system vulnerabilities to many factors and conditions, together with system shocks. According to Meuwissen et al. (2019), a food system resilience is defined as the system's ability to withstand disturbances and adapt to external stress over time. According to Tendall et al. (2015), disturbances in the food system can be sudden or gradual undesirable changes, internal or external, with structural or cyclical impact. These disturbances can separately or jointly impact food availability, access, and utilization. These three dimensions describe the three first components of food security (FAO, 2008). The social and economic impact can also be considered cumulative impacts of the aforementioned disturbances, where smallholder farmers and their livelihood may be the weakest link.

Bringing resilience to the food value chain toward unexpected shocks such as the recent crises caused by the COVID-19 pandemic, of which the impacts on human lives and livelihoods were and continue to be devastating, is an urgent challenge. Moroccan figs sector cannot withstand any disturbance because of the lack of strong structure and coordination among the system components and stakeholders. Therefore, many weaknesses can be spotted from the actual model, such as (1) lacking efficient marketing strategy and insufficient development of valorization routes, (2) small-sized farms, (3) weak coordination among stakeholders, and low connections strength in the value chain, (4) quality and quantity losses due to inadequate post-harvest facilities and infrastructures such as cold and storage, and (5) low

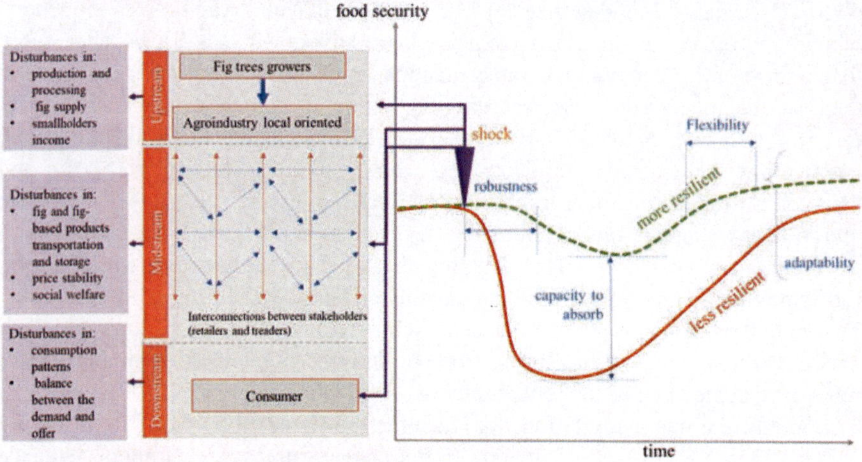

Fig. 2.5 Scheme showing the food system resilience to various shocks with their impacts over different fig value chain levels. (Based on the model proposed by Tendall et al., 2015)

investment in new agricultural technologies. These weaknesses can broadly weaken the fig value chain resilience and thus make it easily exposed to various shocks.

According to Simonovic and Peck (2013), the food system resilience pattern can be depicted in four main regions, including (1) robustness, which is the capability to withstand the disturbance occurrence for a consistent time before showing any food security loss, and (2) redundancy, which is the capacity of absorbing disturbance effect, adaptiveness describing the ability to recover the lost occurred in food security (3) resourcefulness and (4) adaptability, which defines how much and how fast the lost in food security is recovered. For the Moroccan fig value chain that lacks tangible indicators for resilience behavior, unforeseen disturbances impacts can be drawn following three levels: upstream, midstream, and downstream (Fig. 2.5), according to the scheme in Fig. 2.4. Thus, for the upstream end, consequences could be seen in production and processing disturbances besides a loss in smallholders' livelihoods. Downstream, interconnexions between retailers and treaders can interrupt figs and fig-based products transportation and storage. Disruptions can also induce great instability in product prices and, therefore, impact social welfare. At the consumer level, disturbances can occur in the consumption pattern and the balance between the demand and offer.

Finally, no attempt was previously made to scale the fig value chain and assess its resilience toward unforeseen shocks, like the ongoing one engendered by the COVID-19 pandemic. Therefore, this chapter tried to bring more attention to the poor connections within the Moroccan fig value chain, which makes its resilience very questionable. Bringing resilience to the fig value chain will build its ability to bounce back when confronted with a specific disruption and allow its webs of stakeholders and their interconnexions to return to their initial configuration. This will safeguard the continuous flow of goods through the system and drive the supply of fig and fig-based products. However, this is a challenging task that requires pertinent organization and collaborative coordination from all stakeholders.

5 Conclusion

Morocco hosts a fascinating fig diversity that has gone largely unheeded and is naturally preserved in traditional agroecosystems, particularly in the northern regions. Available research has mapped a large phenotypic and chemotypic diversity using high discriminating methods. However, this diversity pattern, particularly locally cultivated figs and wild-growing figs, know substantial homonymy and synonymy cases. This outstanding diversity remains confusing and undocumented and deserves to be further exploited. Even though this fruit is globally recognized as a traditional pantry for healthy eating, Moroccan fig and its end products are less competitive on the international market. As a highly perishable commodity, fig processing remains less developed, contributing to low marketability. The current chapter also aims to deepen our understanding of the resilience of fig chain value to changing environments and shocks, including the disruptions that emerged because of the ongoing covid pandemic. This study brings significant nuances to the debate that should be undertaken on Moroccan fig chain value resilience to various types of shocks and explores potential solutions to the benefit of policymakers, industry, and end-users.

Acknowledgments The authors are thankful to Zahra Oussi Ali and Ali Hssaini.

References

Achtak, H. (2009). *Domestication et diversification variétale chez le figuier au Maroc: Bases pour la conservation et la valorisation des ressources génétiques locales*. PhD thesis, University of Abdelmalek Essaadi, Morocco.

Achtak, H., Oukabli, A., Ater, M., Santoni, S., Kjellberg, F., & Khadari, B. (2009). Microsatellite markers as reliable tools for fig cultivar identification. *Journal of the American Society for Horticultural Science, 134*, 624–631. https://doi.org/10.21273/JASHS.134.6.624

Achtak, H., Ater, M., Oukabli, A., Santoni, S., Kjellberg, F., & Khadari, B. (2010). Traditional agroecosystems as conservatories and incubators of cultivated plant varietal diversity: The case of fig (*Ficus carica* L.) in Morocco. *BMC Plant Biology, 10*, 1–12. https://doi.org/10.1186/1471-2229-10-28

Adak, N., Heybeli, N., & Ertekin, C. (2017). Infrared drying of strawberry. *Food Chemistry, 219*, 109–116. https://doi.org/10.1016/j.foodchem.2016.09.103

Afsah-Hejri, L., Toudeshki, A., Homayouni, T., Mehrazi, S., Gholami Pareh, A., Gordon, P., & Ehsani, R. (2021). Potential of ozonated-air (OA) application to reduce the weight and volume loss in fresh figs (*Ficus carica* L.). *Postharvest Biology and Technology, 180*, 111631. https://doi.org/10.1016/J.POSTHARVBIO.2021.111631

Aksoy, U. (2017). The dried fig management and the potential for new products. *Acta Horticulturae, 1173*, 377–382. https://doi.org/10.17660/ACTAHORTIC.2017.1173.65

Aljane, F., Ferchichi, A., & Boukhris, M. (2008). Pomological characteristics of local fig (*Ficus carica*) cultivars in southern Tunisia. *Acta Horticulturae, 798*, 123–128. https://doi.org/10.17660/ActaHortic.2008.798.15

Allegra, A., Sortino, G., Inglese, P., Settanni, L., Todaro, A., & Gallotta, A. (2017). The effectiveness of Opuntia ficus-indica mucilage edible coating on post-harvest maintenance of 'Dottato' fig (*Ficus carica* L.) fruit. *Food Packaging and Shelf Life, 12*, 135–141. https://doi.org/10.1016/J.FPSL.2017.04.010

Allegra, A., Farina, V., Inglese, P., Gallotta, A., & Sortino, G. (2021). Qualitative traits and shelf life of fig fruit ('Melanzana') treated with Aloe vera gel coating. *Acta Horticulturae, 1310*, 87–92. https://doi.org/10.17660/ACTAHORTIC.2021.1310.14

Ater, M., El Oualkadi, A., Achtak, H., Oukabli, M., & Khadari, B. (2008). Diversity of the local varieties of the fig tree in the North-Western Morocco. *Acta Horticulturae, 789*, 69–76. https://doi.org/10.17660/ActaHortic.2008.798.7

Ateyyeh, A. F., & Sadder, M. T. (2006). Growth pattern and fruit characteristics of six common fig (*Ficus carica* L.) cultivars in Jordan. *Jordan Journal of Agricultural Sciences, 2*, 105–112.

Barolo, M. I., Ruiz Mostacero, N., & López, S. N. (2014). *Ficus carica* L. (Moraceae): An ancient source of food and health. *Food Chemistry, 164*, 119–127. https://doi.org/10.1016/j.foodchem.2014.04.112

Bet, K. (2015). *Changes in quality of dried fig (Ficus carica L.) delight in different packages under cold and ambient storage*. https://doi.org/10.20289/euzfd.72940

Boudchicha, R. H., Hormaza, J. I., & Benbouza, H. (2018). Diversity analysis and genetic relationships among local Algerian fig cultivars (*Ficus carica* L.) using SSR markers. *The South African Journal of Botany, 116*, 207–215. https://doi.org/10.1016/j.sajb.2018.03.015

Bradford, K. J., Dahal, P., Van Asbrouck, J., Kunusoth, K., Bello, P., Thompson, J., & Wu, F. (2020). The dry chain: Reducing postharvest losses and improving food safety in humid climates. In *Food industry wastes* (pp. 375–389). https://doi.org/10.1016/B978-0-12-817121-9.00017-6

Caliskan, O., & Polat, A. A. (2008). Fruit characteristics of fig cultivars and genotypes grown in Turkey. *Scientia Horticulturae, 115*, 360–367. https://doi.org/10.1016/j.scienta.2007.10.017

Çalişkan, O., & Polat, A. A. (2012). Effects of genotype and harvest year on phytochemical and fruit quality properties of Turkish fig genotypes. *The Spanish Journal of Agricultural Research, 10*, 1048. https://doi.org/10.5424/sjar/2012104-2652

Caliskan, O., & Polat, A. A. (2012). Morphological diversity among fig (*Ficus carica* L.) accessions sampled from the Eastern Mediterranean Region of Turkey. *Turkish Journal of Agriculture and Forestry, 36*, 179–193. https://doi.org/10.3906/tar-1102-33

Chatti, K., Salhi-Hannachi, A., Mars, M., Marrakchi, M., & Trifi, M. (2003). Analysis of genetic diversity of Tunisian fig tree cultivars (*Ficus carica* L.) using morphological characteristics. *Fruits, 59*, 49–61. https://doi.org/10.1080/15538360802365921

Chimi, H., Ouaouich, A., Semmar, M., & Tayebi, S. (2008). Industrial processing of figs by solar drying in Morocco. *Acta Horticulturae, 798*, 331–334. https://doi.org/10.17660/ACTAHORTIC.2008.798.48

Chua, K. J., & Chou, S. K. (2003). Low-cost drying methods for developing countries. *Trends in Food Science and Technology, 14*, 519–528. https://doi.org/10.1016/J.TIFS.2003.07.003

Ciarmiello, L. F., Piccirillo, P., Carillo, P., De Luca, A., & Woodrow, P. (2015). Determination of the genetic relatedness of fig (*Ficus carica* L.) accessions using RAPD fingerprint and their agro-morphological characterization. *South African Journal of Botany, 97*, 40–47. https://doi.org/10.1016/j.sajb.2014.11.012

da Silva, W. P., e Silva, C. M. D. P. S., Farias, V. S. O., & Gomes, J. P. (2012). Diffusion models to describe the drying process of peeled bananas: Optimization and simulation. *Drying Technology, 30*, 164–174. https://doi.org/10.1080/07373937.2011.628554

Derardja A. eddine, Pretzler, M., Kampatsikas, I., Barkat, M., & Rompel, A. (2019). Inhibition of apricot polyphenol oxidase by combinations of plant proteases and ascorbic acid. *Food Chemistry X, 4*, 100053. https://doi.org/10.1016/j.fochx.2019.100053

Durance, T. (2002). Handbook of food preservation (M. Shafiur Rahman, ed., pp. 809). Marcel Dekker, Inc., 1999. ISBN: 0-8247-0209-3. *Food Research International, 35*, 409. https://doi.org/10.1016/S0963-9969(00)00143-5

Ekechukwu, O. V. (1999). Review of solar-energy drying systems I: An overview of drying principles and theory. *Energy Conversion and Management, 40*, 593–613. https://doi.org/10.1016/S0196-8904(98)00092-2

El Hajjam, A., Ezzahouani, A., & Sehhar, E. A. (2018). Conduite technique et inventaire des variétés marocaines locales de figuier (*Ficus carica* L.) dans quatre principaux sites de production, provinces de Chefchaouen, El Jadida, Ouezzane, et Taounate. *Revue Marocaine des Sciences Agronomiques et Vétérinaires, 6*, 494–504.

El Khaloui, M. (2010). Valorisation de la figue au Maroc. *Transfert de Technologie en Agriculture, 186*, 1–4.

Essid, A., Aljane, F., & Ferchichi, A. (2017). Morphological characterization and pollen evaluation of some Tunisian *ex situ* planted caprifig (*Ficus carica* L.) ecotypes. *South African Journal of Botany, 111*, 134–143. https://doi.org/10.1016/j.sajb.2017.03.001

FAO. (2008). *An introduction to the basic concepts of food security. Practical guides* (p. 3). Food and Agriculture Organization of the United Nations.

FAOSTAT (2019) http://apps.fao.org

Gamboa-Santos, J., Megías-Pérez, R., Soria, A. C., Olano, A., Montilla, A., & Villamiel, M. (2014). Impact of processing conditions on the kinetic of vitamin C degradation and 2-furoylmethyl amino acid formation in dried strawberries. *Food Chemistry, 153*, 164–170. https://doi.org/10.1016/j.foodchem.2013.12.004

Gauthier, C. (2014). *L'Indice des Prix à la consommation (IPC) du mois de mars 2014*. Site institutionnel du Haut-Commissariat au Plan du Royaume du Maroc. Retrieved October 23, 2022, from https://www.hcp.ma/L-Indice-des-prix-a-la-consommation-IPC-du-mois-de-Mars-2014_a1386.html

Giraldo, E., Lopez-Corrales, M., & Hormaza, J. I. (2010). Selection of the most discriminating morphological qualitative variables for characterisation of fig germplasm. *Journal of the American Society for Horticultural Science, 135*, 240–249. https://doi.org/10.21273/JASHS.135.3.240

Hamanaka, D., Norimura, N., Baba, N., Mano, K., Kakiuchi, M., Tanaka, F., & Uchino, T. (2011). Surface decontamination of fig fruit by combination of infrared radiation heating with ultraviolet irradiation. *Food Control, 22*, 375–380. https://doi.org/10.1016/J.FOODCONT.2010.09.005

Hernández, F. (2016). Phenolic compounds, antioxidant and antidiabetic activity of different cultivars of *Ficus carica* L. fruits. *Journal of Functional Foods, 25*, 421–432. https://doi.org/10.1016/j.jff.2016.06.015

Hmimsa, Y., Aumeeruddy-Thomas, Y., & Ater, M. (2012). Vernacular taxonomy, classification and varietal diversity of fig (*Ficus carica* L.) among Jbala cultivators in northern Morocco. *Human Ecology, 40*, 301–313. https://doi.org/10.1007/s10745-012-9471-x

Hmimsa, Y., Aumeeruddy-Thomas, Y., & Ater, M. (2017). Une forme spontanée de figuier (*Ficus carica* L.), le nabut: Diversité de nomenclature, d'usage et de pratiques locales au Nord du Maroc. *Revue d'ethnoécologie*. https://doi.org/10.4000/ethnoecologie.3186

Hssaini, L. (2020). *Etude agro-morphologique, biochimique et chimiométrique d'une collection du figuier (Ficus carica L.): screening, modélisation des isothermes et cinétique de séchage*. PhD thesis, University of Sultan Moulay Slimane, Morocco.

Hssaini, L., Charafi, J., Hanine, H., Ennahli, S., Mekaoui, A., Mamouni, A., & Razouk, R. (2019a). Comparative analysis and physio-biochemical screening of an ex-situ fig (*Ficus carica* L.) collection. *Horticulture, Environment and Biotechnology, 60*, 671–683. https://doi.org/10.1007/s13580-019-00170-4

Hssaini, L., Hanine, H., Razouk, R., Ennahli, S., Mekaoui, A., & Charafi, J. (2019b). Characterization of local fig clones (*Ficus carica* L.) collected in Northern Morocco. *Fruits, 74*, 55–64. https://doi.org/10.17660/th2019/74.2.1

Hssaini, L., Hanine, H., Charafi, J., Razouk, R., Elantari, A., Ennahli, S., Hernández, F., & Ouaabou, R. (2020a). First report on fatty acids composition, total phenolics and antioxidant activity in seeds oil of four fig cultivars (*Ficus carica* L.) grown in Morocco. *OCL, 27*, 8. https://doi.org/10.1051/ocl/2020003

Hssaini, L., Hanine, H., Razouk, R., Ennahli, S., Mekaoui, A., Ejjilani, A., & Charafi, J. (2020b). Assessment of genetic diversity in Moroccan fig (*Ficus carica* L.) collection by combining morphological and physicochemical descriptors. *Genetic Resources and Crop Evolution, 67*, 457–474. https://doi.org/10.1007/s10722-019-00838-x

Hssaini, L., Hanine, H., Razouk, R., Ennahli, S., Mekaoui, A., Guirrou, I., & Charafi, J. (2020c). Diversity screening of fig (*Ficus carica* L.) germplasm through integration of morpho-agronomic and biochemical traits. *International Journal of Fruit Science, 20*, 939–958. https://doi.org/10.1080/15538362.2019.1700871

Hssaini, L., Ouaabou, R., Charafi, J., Idlimam, A., Lamharrar, A., Razouk, R., & Hanine, H. (2020d). Hygroscopic proprieties of fig (*Ficus carica* L.): Mathematical modelling of moisture sorption isotherms and isosteric heat kinetics. *South African Journal of Botany, 145*, 265–274. https://doi.org/10.1016/j.sajb.2020.11.026

Hssaini, L., Ouaabou, R., Charafi, J., Razouk, R., Houmanat, K., & Irchad, A. (2021a). Effects of pre-storage ascorbic and salicylic acids treatments on the enzymatic browning and nutritional quality of dried fig: Combined use of biochemical and ATR-FTIR analyses. *Vibrational Spectroscopy, 115*, 103269. https://doi.org/10.1016/J.VIBSPEC.2021.103269

Hssaini, L., Razouk, R., Charafi, J., Houmanat, K., & Hanine, H. (2021b). Fig seeds: Combined approach of lipochemical assessment using gas chromatography and FTIR-ATR spectroscopy using chemometrics. *Vibrational Spectroscopy, 114*(3), 103251. https://doi.org/10.1016/j.vibspec.2021.103251

Icyer, N. C., Said, O., Salih, T., & Fatih, K. (2016). Microencapsulation of fig seed oil rich in polyunsaturated fatty acids by spray drying. *Journal of Food Measurement and Characterization*. https://doi.org/10.1007/s11694-016-9370-8

IPGRI, CIHEAM. (2003). *Descriptors for fig*. International Plant Genetic Resources Institute/International Centre for Advanced Mediterranean Agronomic Studies.

Ishurd, O., Zgheel, F., Kermagi, A., Flefla, M., Elmabruk, M., Wu, Y., Kennedy, J. F., & Pan, Y. (2004). Microbial (1→3)-β-d-glucans from Libyan figs (*Ficus carica*). *Carbohydrate Polymers, 58*, 181–184. https://doi.org/10.1016/J.CARBPOL.2004.06.040

Jayaraman, K. S., & Das Gupta, D. K. (2020). Drying of fruits and vegetables. In *Handbook of industrial drying* (pp. 643–690). https://doi.org/10.1201/9780429289774-21

Jimenez, C. P. (2016). *Comportamiento agronómico y estudio del punto óptimo de maduración nutricional y funcional de variedades de higuera interesantes para consumo en fresco*. PhD thesis, University of Extremadura, Spain.

Khadari, B., Lashermes, P., & Kjellberg, F. (1995). RAPD fingerprints for identification and genetic characterization of fig (*Ficus carica* L.) genotypes. *Journal of Genetics and Breeding, 44*, 77–86.

Khadari, B., Oukabli, A., Ater, M., Mamouni, A., Roger, J. P., & Kjellberg, F. (2004). Molecular characterization of Moroccan fig germplasm using intersimple sequence repeat and simple sequence repeat markers to establish a reference collection. *HortScience, 40*, 29–32. https://doi.org/10.21273/HORTSCI.40.1.29

Khadari, B., Oukabli, A., Ater, M., Mamouni, A., Roger, J. P., & Kjellberg, F. (2005). Molecular characterization of Moroccan fig germplasm using intersimple sequence repeat and simple sequence repeat markers to establish a reference collection. *HortScience, 40*(1), 29–32.

Khadari, B., Roger, J. P., Ater, M., Achtak, H., Oukabli, A., & Kjellberg, F. (2005a). Moroccan fig presents specific genetic resources: A high potential of local selection. *III International Symposium on Fig, 798*, 33–37. https://doi.org/10.17660/ActaHortic.2008.798.3

Khadari, B., Grout, C., Santoni, S., & Kjellberg, F. (2005b). Contrasted genetic diversity and differentiation among Mediterranean populations of *Ficus carica* L.: A study using mtDNA RFLP. *Genetic resources and crop evolution, 52*, 97–109. https://doi.org/10.1007/s10722-005-0290-4

Khadari, B., Roger, J. P., Ater, M., Achtak, H., Oukabli, A., & Kjellberg, F. (2008). Moroccan fig presents specific genetic resources: A high potential of local selection. *Acta Horticulturae, 798*, 33–37. https://doi.org/10.17660/ActaHortic.2008.798.3

Khadivi, A., Anjam, R., & Anjam, K. (2018). Morphological and pomological characterization of edible fig (*Ficus carica* L.) to select the superior trees. *Scientia Horticulturae, 238*, 66–74. https://doi.org/10.1016/j.scienta.2018.04.031

Kousksou, T., Allouhi, A., Belattar, M., Jamil, A., El Rhafiki, T., Arid, A., & Zeraouli, Y. (2015). Renewable energy potential and national policy directions for sustainable development in Morocco. *Renewable and Sustainable Energy Reviews, 47*, 46–57. https://doi.org/10.1016/j.rser.2015.02.056

Liang, N., Lu, X., Hu, Y., & Kitts, D. D. (2016). Application of attenuated total reflectance-fourier transformed infrared (ATR-FTIR) spectroscopy to determine the chlorogenic acid isomer profile and antioxidant capacity of coffee beans. *Journal of Agricultural and Food Chemistry, 64*, 681–689. https://doi.org/10.1021/acs.jafc.5b05682

Makris, D. P., Boskou, G., & Andrikopoulos, N. K. (2007). Polyphenolic content and in vitro antioxidant characteristics of wine industry and other agri-food solid waste extracts. *Journal of Food Composition and Analysis, 20*, 125–132. https://doi.org/10.1016/j.jfca.2006.04.010

Martins, I. B. A., Oliveira, D., Rosenthal, A., Ares, G., & Deliza, R. (2019). Brazilian consumer's perception of food processing technologies: A case study with fruit juice. *Food Research International, 125*, 108555. https://doi.org/10.1016/j.foodres.2019.108555

Mat Desa, W. N., Mohammad, M., & Fudholi, A. (2019). Review of drying technology of fig. *Trends in Food Science and Technology, 88*, 93–103. https://doi.org/10.1016/j.tifs.2019.03.018

Mesta, R. K., Shivaprasad, M., Hegde, G. M., Nidoni, U., & Kurubar, A. R. (2013). Postharvest management of fruit rot of chilli using solar tunnel dryer. *Acta Horticulturae, 1012*, 755–758. https://doi.org/10.17660/ACTAHORTIC.2013.1012.101

Meuwissen, M. P. M., Feindt, P. H., Spiegel, A., Termeer, C. J. A. M., Mathijs, E., de Mey, Y., Finger, R., Balmann, A., Wauters, E., Urquhart, J., Vigani, M., Zawalińska, K., Herrera, H., Nicholas-Davies, P., Hansson, H., Paas, W., Slijper, T., Coopmans, I., Vroege, W., Ciechomska, A., Accatino, F., Kopainsky, B., Poortvliet, P. M., Candel, J. J. L., Maye, D., Severini, S., Senni, S., Soriano, B., Lagerkvist, C. J., Peneva, M., Gavrilescu, C., & Reidsma, P. (2019). A framework to assess the resilience of farming systems. *Agricultural Systems, 176*, 102656. https://doi.org/10.1016/J.AGSY.2019.102656

Mirheidari, F., Khadivi, A., Moradi, Y., & Paryan, S. (2020). Phenotypic variability of naturally grown edible fig (*Ficus carica* L.) and caprifig (*Ficus carica* var. caprificus Risso) accessions. *Scientia Horticulturae (Amsterdam), 267*, 109320. https://doi.org/10.1016/j.scienta.2020.109320

Muji, I., Kralj, M. B., & Joki, S. (2012). Characterisation of volatiles in dried white varieties figs (*Ficus carica* L.). *The Journal of Food Science and Technology, 51*(9), 1837–1846. https://doi.org/10.1007/s13197-012-0740-x

Nurlaila, W., Desa, M., Mohammad, M., & Fudholi, A. (2019). SC. *Trends in Food Science and Technology*. https://doi.org/10.1016/j.tifs.2019.03.018

Office des changes | maroc. (2021, December). Retrieved October 23, 2022, from https://www.oc.gov.ma/sites/default/files/reglementation/pdf/2022-01/IGOC%202022.pdf

Oliphant, T., Mitra, A., & Wilkinson, M. (2012). Contact allergy to sodium sulfite and its relationship to sodium metabisulfite. *Contact Dermatitis, 66*, 128–130. https://doi.org/10.1111/J.1600-0536.2011.02029.X

Oliveira, A. P., Valentão, P., Pereira, J. A., Silva, B. M., Tavares, F., Andrade, P. B., Faculdade, C., Ciências, D., Pessoa, U. F., & Maia, R. C. (2009). *Ficus carica* L.: Metabolic and biological screening. *Food and Chemical Toxicology, 47*, 2841–2846. https://doi.org/10.1016/j.fct.2009.09.004

Oukabli, A., Mamouni, A., Laghezali, M., Roger, J., Kjellberg, F., & Ater, M. (2003a). Genetic variability in Morrocan fig cultivars (*Ficus carica* L.) based on morphological and pomological data. *Acta Horticulturae, 605*, 311–318. https://doi.org/10.17660/ActaHortic.2003.605.47

Oukabli, A., Mamouni, A., Laghezali, M., Ater, M., Roger, J. P., & Khadari, B. (2003b). Local caprifig tree characterization and analysis of interest for pollination. *Acta Horticulturae, 605*, 61–64. https://doi.org/10.17660/ActaHortic.2003.605.7

Palmeira, L., Pereira, C., Dias, M. I., Abreu, R. M. V., Corrêa, R. C. G., Pires, T. C. S. P., Alves, M. J., Barros, L., & Ferreira, I. C. F. R. (2019). Nutritional, chemical and bioactive profiles of different parts of a Portuguese common fig (*Ficus carica* L.) variety. *Food Research International, 126*, 108572. https://doi.org/10.1016/j.foodres.2019.108572

Pérez-Sánchez, R., Morales-Corts, M. R., & Gómez-Sánchez, M. Á. (2016). Agro-morphological diversity of traditional fig cultivars grown in Central-Western Spain. *Genetika, 48*, 533–546. https://doi.org/10.2298/GENSR1602533P

Piazzolla, F., Amodio, M. L., & Colelli, G. (2018). Effects of thermal treatments on quality of 'Petrelli' figs during storage. *Acta Horticulturae, 1194*, 879–886. https://doi.org/10.17660/ACTAHORTIC.2018.1194.124

Piga, A., Pinna, I., Özer, K. B., Agabbio, M., & Aksoy, U. (2004). Hot air dehydration of figs (*Ficus carica* L.): Drying kinetics and quality loss. *International Journal of Food Science and Technology, 39*, 793–799. https://doi.org/10.1111/j.1365-2621.2004.00845.x

Podgornik, M., Vuk, I., Vrhovnik, I., & Mavsar, D. (2010). A survey and morphological evaluation of fig (*Ficus carica* L.) genetic resources from Slovenia. *Scientia Horticulturae, 125*, 380–389. https://doi.org/10.1016/j.scienta.2010.04.030

Pourghayoumi, M., Bakhshi, D., Rahemi, M., Noroozisharaf, A., Jafari, M., Salehi, M., Chamane, R., & Hernandez, F. (2017). Phytochemical attributes of some dried fig (*Ficus carica* L.) fruit cultivars grown in Iran. *Agriculturae Conspectus Scientificus, 81*, 161–166.

Rahman, N. F. A., Shamsudin, R., Ismail, A., & Karim Shah, N. N. A. (2016). Effects of post-drying methods on pomelo fruit peels. *Food Science and Biotechnology, 25*, 85–90. https://doi.org/10.1007/s10068-016-0102-y

Rankou, H., Culham, A., Jury, S. L., & Christenhusz, M. J. M. (2013). The endemic flora of Morocco. *Phytotaxa, 78*, 1–69. https://doi.org/10.11646/phytotaxa.78.1.1

Rothwell, J. A., Perez-Jimenez, J., Neveu, V., Medina-Remon, A., M'hiri, N., García-Lobato, P., ... & Scalbert, A. (2013). *Phenol-Explorer 3.0: a major update of the Phenol-Explorer database to incorporate data on the effects of food processing on polyphenol content*. Database, 2013.

Saddoud, O., Baraket, G., & Chatti, K. (2011). Using morphological characters and simple sequence repeat (SSR) markers to characterize Tunisian fig (*Ficus carica* L.) cultivars. *Acta Biologica Cracoviensia, 53*, 7–14. https://doi.org/10.2478/v10182-011-0019-y

Saki, M., ValizadehKaji, B., Abbasifar, A., & Shahrjerdi, I. (2019). Effect of chitosan coating combined with thymol essential oil on physicochemical and qualitative properties of fresh fig (*Ficus carica* L.) fruit during cold storage. *Journal of Food Measurement and Characterization, 13*, 1147–1158. https://doi.org/10.1007/s11694-019-00030-w

Sanchez, R., Morales-Corts, M. R., & Sanchez, M. A. (2016). Agro-morphological diversity of traditional fig cultivars grown in Central-Western Spain. *Genetika, 48*, 533–546. https://doi.org/10.2298/GENSR1602533P

Schmitzer, V., Slatnar, A., Mikulic-Petkovsek, M., Veberic, R., Krska, B., & Stampar, F. (2011). Comparative study of primary and secondary metabolites in apricot (*Prunus armeniaca* L.) cultivars. *Journal of the Science of Food and Agriculture, 91*, 860–866. https://doi.org/10.1002/jsfa.4257

Sedaghat, S., & Rahemi, M. (2018). Effects of physio-chemical changes during fruit development on nutritional quality of fig (*Ficus carica* L. var.'Sabz') under rain-fed condition. *Scientia Horticulturae, 237*, 44–50. https://doi.org/10.1016/j.scienta.2018.04.003

Simonovic, S. P., & Peck, A. (2013). Dynamic resilience to climate change caused natural disasters in coastal megacities quantification framework. *International Journal of Environment and Climate Change, 3*, 378–401. https://doi.org/10.9734/BJECC/2013/2504

Singh, P., Shrivastava, V., & Kumar, A. (2018). Recent developments in greenhouse solar drying: A review. *Renewable and Sustainable Energy Reviews, 82*, 3250–3262. https://doi.org/10.1016/J.RSER.2017.10.020

Solana, R. R., Galego, L. R., Pérez, E., & Anabela, S. (2018). Production method and varietal source influence the volatile profiles of spirits prepared from fig fruits (*Ficus carica* L.). *European Food Research and Technology*. https://doi.org/10.1007/s00217-018-3131-3

Solomon, A., & Golubowicz, S. (2006). Antioxidant activities and anthocyanin content of fresh fruits of common fig (*Ficus carica* L.). *The Journal of Agricultural and Food Chemistry, 54*(20), 7717–7723.

Stohs, S. J., & Miller, M. J. S. (2014). A case study involving allergic reactions to sulfur-containing compounds including, sulfite, taurine, acesulfame potassium and sulfonamides. *Food and Chemical Toxicology, 63*, 240–243. https://doi.org/10.1016/J.FCT.2013.11.008

Tan, N. (2017). Effect of cabinet drying method on dried fruit quality and functional properties of 'Sarilop' (*Ficus carica* L.) fig cultivar. *Acta Horticulturae, 1173*, 359–364. https://doi.org/10.17660/ActaHortic.2017.1173.62

Tareen, M. J., Abbasi, N. A., & Hafiz, I. A. (2012). Postharvest application of salicylic acid enhanced antioxidant enzyme activity and maintained quality of peach cv. "Flordaking" fruit during storage. *Scientia Horticulturae (Amsterdam), 142*, 221–228. https://doi.org/10.1016/j.scienta.2012.04.027

Tendall, D. M., Joerin, J., Kopainsky, B., Edwards, P., Shreck, A., Le, Q. B., Kruetli, P., Grant, M., & Six, J. (2015). Food system resilience: Defining the concept resilience sustainability. *Global Food Security, 6*, 17–23. https://doi.org/10.1016/j.gfs.2015.08.001

Veberic, R., & Mikulic-Petkovsek, M. (2015). Phytochemical composition of common fig (*Ficus carica* L.) cultivars. In *Nutritional composition of fruit cultivars* (pp. 235–255). Elsevier Inc.

Veberic, R., Colaric, M., & Stampar, F. (2008). Phenolic acids and flavonoids of fig fruit (*Ficus carica* L.) in the northern Mediterranean region. *Food Chemistry, 106*, 153–157. https://doi.org/10.1016/j.foodchem.2007.05.061

Vitelli, M., Vessella, F., Cardoni, S., Pollegioni, P., Denk, T., Grimm, G. W., & Simeone, M. C. (2017). Phylogeographic structuring of plastome diversity in Mediterranean oaks (Quercus Group Ilex, Fagaceae). *Tree Genetics & Genomes, 13*, 1–17. https://doi.org/10.1007/s11295-016-1086-8

Wojdyło, A., Nowicka, P., Carbonell-Barrachina, A. A., & Hernndez, F. (2016). Phenolic compounds, antioxidant and antidiabetic activity of different cultivars of *Ficus carica* L. fruits. *Journal of Functional Foods, 25*, 421–432. https://doi.org/10.1016/j.jff.2016.06.015

Chapter 3
Fig Tree Genome and Diversity

Dunja Bandelj, Alenka Baruca Arbeiter, and Matjaž Hladnik

Abbreviations

1-MCP	1-Methylcyclopropene
ABA	Abscisic acid
AFLP	Amplified Fragment Length Polymorphism
AGB	Active Germplasm Bank
BERC	Biodiversity and Environmental Research Center
CAPS	Cleaved Amplified Polymorphic Sequence
CIHEAM	International Centre for Advanced Mediterranean Agronomic Studies
CRDA	Commissariat Régional au Développement Agricole
CRPh	Centre de Recherches Phoénicicoles
DNA	Deoxyribonucleic acid
EST	Expressed Sequence Tag
EST-SSRs	Short Sequence Repeats derived from Expressed Sequence Tag
FEIS	Faculty of Engineering of Ilha Solteira, University of São Paulo
GA	Gibberellin
GWAS	Genome-wide association study
H_e	Expected heterozygosity
H_o	Observed heterozygosity
HTS	High-throughput sequencing
INRA	Institut National de la Recherche Agronomique
IPGRI	International Plant Genetic Resources Institute
IRA	Institut des Régions Arides
ISA	Institut Supérieur Agronomique
ISSR	Inter-Simple Sequence Repeats
ITS	Internal Transcribed Spacer
KASP	Kompetitive Allele Specific PCR
LTR-RE	Long terminal repeat retrotransposon
mtDNA	Mitochondrial DNA

D. Bandelj (✉) · A. Baruca Arbeiter · M. Hladnik
Faculty of Mathematics, Natural Sciences and Information Technologies, University of Primorska, Koper, Slovenia
e-mail: dunja.bandelj@famnit.upr.si

M. F. Ramadan (ed.), *Fig* (Ficus carica)*: Production, Processing, and Properties*, https://doi.org/10.1007/978-3-031-16493-4_3

n	Number of amplified alleles
NCGR	National Clonal Germplasm Repository
n_e	Number of effective alleles
NGS	Next-generation sequencing
ONT	Oxford Nanopore Technologies
PacBio	Pacific Biosciences
PC	Principal Component
PCA	Principal Component Analysis
PCR	Polymerase Chain Reaction
PDO	Protected Designation of Origin
PI	Probability of identity
PIC	Polymorphic information content
RAD-seq	Restriction site-associated DNA sequencing
RAMPO	Randomly Amplified Microsatellites Polymorphisms
RAPD	Random Amplified Polymorphic DNA
RFLP	Restriction Fragment Length Polymorphism
RNA	Ribonucleic acid
SMRT	Single-molecule real-time sequencing
SNP	Single nucleotide polymorphism
SSR	Simple sequence repeat
STMS	Sequence-tagged microsatellites
TE	Transposable element
TGS	Third-generation sequencing

1 Origin, Taxonomy and Biological Properties of the Common Fig

Undoubtedly, the common fig is one of the oldest fruits accompanying man and is associated with the beginning of horticulture in the Mediterranean basin and southwest Asia (Zohary et al., 2012). Archaeobotanic remains, dated back to 11,200–11,400 years ago, have confirmed its consumption among settlers in an early Neolithic village in the Lower Jordan Valley (Kislev et al., 2006). Next, the fig tree was spread from its center of origin to southern Arabia and western Asia, and introgression of local wild figs and landraces and human selection contributed to the development of diverse forms and varieties (Aradhya et al., 2010). Further spreading of the common fig from the east to the west Mediterranean basin into Greece, Italy, Spain, Portugal, and towards the south, into Egypt increased the number of varieties (Aradhya et al., 2010). Genetic diversity study of mtDNA of Mediterranean fig populations has revealed the existence of three gene pools -Balearic, West, and East Mediterranean- and significant genetic differentiation between the West and East Mediterranean (Khadari et al., 2005). In the mid-sixteenth century, Spanish missionaries brought the common fig into the New World, and Franciscans were the first to plant figs in California around the mid-nineteenth century (Aradhya et al., 2010). Finally, the fig tree was introduced from its native area to China about a thousand years ago, where significant progress in its research has been obtained in selecting the most suitable varieties, cultivation techniques, processing, and medical use (Lianju et al., 2003).

The fig tree (*Ficus carica* L.) belongs to the mulberry family (Moraceae), which includes about 39 genera (World Flora Online, 2021). The genus *Ficus* L. contains more than 841 species, mainly grown in the tropics (World Flora Online, 2021), and is considered one of the largest genera of flowering plants (Lansky & Paavilainen, 2011). The genus *Ficus* L. is currently divided into 17 sections, and the species *Ficus carica* L. is placed within the section *Ficus*. The species *F. carica* is represented by two subspecies, (1) *Ficus carica* L. subsp. *carica*, known as the common fig and native to Afghanistan, Cyprus, Greece, Iran, Iraq, Crete, Lebanon-Syria, the North Caucasus, Palestine, Tadzhikistan, Transcaucasia, Turkey, and Turkmenistan, and (2) *Ficus carica* subsp. *rupestris* (Hausskn. ex Boiss.) Browicz, native to Afghanistan, Iran, Iraq, Lebanon-Syria, Pakistan, Turkey, and the Western Himalayas. In addition to the native species, the common fig and numerous cultivars are spread and widely cultivated in the tropics and subtropics (GRIN, 2021; KEW, 2021).

Ficus species are gynodioecious and functionally dioecious with unique pollination biology. They have developed a mutual symbiosis with specific pollination wasps of the Agaonidae family. The advantage of highly specific symbiosis (a single specific agaonid wasp pollinates the majority of *Ficus* species) is that it allows the development of the wasp. At the same time, the wasp's presence allows pollination events inside the fig syconia. *F. carica* has developed a symbiotic relationship with its wasp pollinator *Blastophaga psenes* L., which enters the syconium through the ostiole to oviposit short-styled female flowers. The *F. carica -Blastophaga* symbiosis system has stabilized the fruiting phenophase of the common fig, and dioecy prevents inbreeding and is a result of adaptation to seasonal climate conditions (Kjellberg et al., 1987, 2005; Mars et al., 2017).

The common fig possesses two tree morphs with three functional floral forms (Fig. 3.1): (1) The edible fig produces syconia with long-styled female flowers (female tree); (2) The caprifig produces syconia with short-styled female flowers (serves as the oviposition site for the wasp) and male flowers (functionally male tree) (Beck & Lord, 1988). Based on the floral biology and cropping/pollination characteristics, figs are further classified into one of four types (Stover et al., 2007; Flaishman et al., 2008; Çalişkan et al., 2017): (a) the common type, also called 'persistent', which develops fruits parthenocarpically (one or two crops), (b) the Smyrna type, which requires pollination with pollen from (c) the caprifig (goat fig), and (d) the San Pedro type, which produces the first crop through parthenocarpy (breba), and a second crop (main crop) after caprification (Ikegami et al., 2013a). The caprifig, which is also considered as wild and an ancestor form of the female fig (Mori et al., 2017), is thus an important source of pollen in the caprification process in which the pollen from caprifig trees is transferred to Smyrna and San Pedro types by the pollination vector *B. psenes* (Çalişkan et al., 2017). Caprifigs produce three

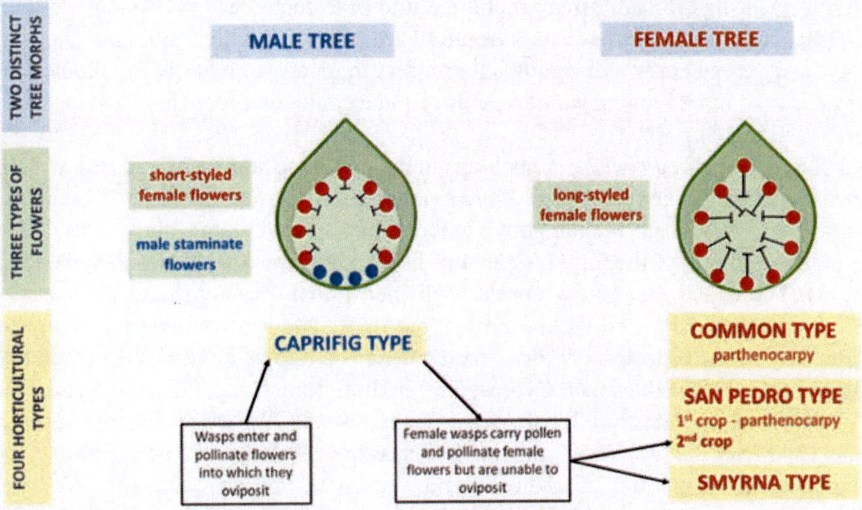

Fig. 3.1 Schematic representation of gynodioecious fig tree (*F. carica*) flower types and female fig types differentiation based on the need for caprification to set a crop

crops during each growing season: 'profichi', 'mammoni' and 'mamme'. Only 'profichi', which ripen in the early summer, are the source of pollen and the pollination occurs when female trees and caprifigs are present in the same orchard (Essid et al., 2015).

Domestication of the fig tree has been focused on the selection and vegetative propagation of female clones with desirable features of fruits and expressed parthenocarpy. Vegetative propagation with cuttings or other vegetative parts of the plant has preserved the original features of the maternal trees. In addition, domestication has increased fruit size, amount of flesh, and sugar content (Zohary et al., 2012).

Fig genetic resources can be divided into cultivated, wild, and feral forms. Wild populations of figs are privileged to a few isolated parts of native forests, such as rock crevices, springs, wet gorges, and stream banks, and their reproduction is based on seed dispersal. They are often complemented by feral forms, which are not exactly wild. Feral forms derived from seeds produced by local domesticated clones pollinated by the adjacent wild-growing caprifigs and distributed sporadically in secondary habitats, such as the edges of plantations, ruins, collapsed cisterns, or moist cave entrances (Zohary et al., 2012). Both feral and wild forms usually have relatively small, hardly-edible syconia and show phenotypic plasticity with considerable variation in their morphology and vegetative traits (Zohary, 1983). Reproduction is exclusively sexual and depends on elaborate symbiosis with the *B. psenes* wasp (Zohary, 1983; Galil & Neeman, 1977). After random cross-pollination, plants developed from seeds are highly diverse with distinct characteristics (Zohary & Spiegel-Roy, 1975).

2 Fig Genetic Resources: The Importance of Research

At the end of the twentieth century, the common fig was classified as an underutilized and neglected fruit species. However, in the last twenty years, its status has changed since many published research articles are now available, and there are many ongoing activities orientated towards the characterization and conservation of fig genetic resources at the national and international levels.

The economic value and functional properties of common fig fruits (Vinson, 1999) and other phytochemicals with potential health-promoting effects found in figs have contributed to its broader interest in the investigation (Arvaniti et al., 2019; Hssaini et al., 2020a). A review of worldwide ethnomedical uses of *F. carica* reported its traditional use in healing more than 40 disorders (Badgujar et al., 2014). Phytochemicals with biological activities are found in fig leaves, fruits, latex, and roots and places the common fig among promising candidates for the pharmaceutical industry to develop new drugs and for clinical uses (Badgujar et al., 2014). Moreover, recent studies have shown that figs can also be used for new purposes. The fig seed oil contains unsaturated fatty acids, gamma-tocopherol and is a rich source of *omega*-3 and *omega*-6 fatty acids (Kucukerdonmez et al., 2021). Soltana et al. (2015) confirmed the favorable nutritional properties and dietary importance of oil from fig achenes. Their results have also opened up new possibilities for valorizing waste material generated from fig fruit processing, which can be used in the cosmetic industry. These findings support the integrated research of fig genetic resources patrimony involving genomic, transcriptomic, and proteomic levels.

The genetic patrimony of the common fig has been developed through millennia of cultivation. The unique and special pollination biology of the common fig and early domestication and propagation practices focused on selecting genotypes based on morphological or other prospective traits have contributed to and influenced the development of fig genetic resources and diversity. In addition, outcrossing the cultivated common fig with wild individuals and other artificial selections of interesting phenotypes has contributed to common fig diversity. Conversely, the different ecological factors, the selection criteria, and the vegetative propagation of a limited number of fig genotypes could have led to reduced diversity and have been reported in some fig-producing countries (Saddoud et al., 2007; Boudchicha et al., 2018).

Characterization of common fig varieties at the national and international level contributes to understanding geographic patterns of genetic diversity distribution. The relationships between fig varieties may be unclear due to the exchange of planting material among fig-producing countries and their export from the Mediterranean basin to other continents. The result of these transfers is a generation of many synonyms (several denominations of the same genotype) and homonyms (several genotypes under the same denomination). Dispersal of varieties and the lax use of names by users or just because of incorrect translation of original names (Aradhya et al., 2010) have contributed to the variety denomination confusion. For example, the study of fig varietal diversity and denomination in Northern Morocco has demonstrated that out of 133 fig varieties, 39% had one or more synonyms (Hmimsa et al.,

2012). Clear distinction of varieties is also essential for developing fig products with quality labels such as the protected designation of origin (PDO), which provides added value and is highly recognized by consumers. Accurate identification of varieties is of essential importance during the propagation process in nurseries to ensure the identity of the planting material and thus the production economy.

In the light of changing climate conditions, the emergence of new pests and diseases, and demanding marketing standards, genetic diversity represents an essential source for plant breeders. The common fig has not been subjected to intensive breeding programs, and from this point of view, it is essential to fully characterize its genetic patrimony (Perez-Jiménez et al., 2012). The main breeding programs are oriented to achieve farm and market requests in terms of biotic and abiotic resistance, the improvement of nutritional value, higher productivity, elimination of caprification, the persistence of mature syconia, and bridging the gap between first and second crop fruitfulness (Bohanec, 2008). From this perspective, phenotyping of fig genetic resources, together with molecular characterization, is essential for the effectiveness of breeding programs and germplasm conservation efforts. Furthermore, millennia of cultivation, propagation practices, and environmental-ecological factors have contributed to the development of local ecotypes in a given geographical area. These genotypes could be the carriers or reservoirs of some important genes related to the quality traits and resistance to abiotic and biotic stress factors and are important to breeders looking for a unique combination of traits or genes.

The awareness of the importance of preserving the diversity of crops contributed to the establishment of national and international germplasm banks where all diversity of the target crop is collected and maintained. The inventory of local genetic diversity of figs in a defined geographical region is usually the first step in cataloging all different ecotypes/varieties. Identified, distinct phenotypes are then clonally propagated and planted in regional or international collections (germplasm banks), where accessions are further characterized with morphological descriptors, agronomical traits, and quality parameters under uniform ecological conditions and with molecular markers. Gene banks or *ex-situ* collections, therefore, represent a unique repository of plant diversity, which significantly contribute to effective conservation, management, and utilization of germplasm. They offer insight into the amount and patterns of genetic diversity distribution and permit germplasm classification based on genetic similarities and differences (Aradhya et al., 2010). The accurate identification and valorization of collected accessions is an important task, leading to better management of collections by identifying gaps, including deficiencies and redundancies of the accessions. In addition, the collections contribute effectively to the conservation of rare, local genotypes and prevent the loss of individual genes, sets of genes, or in the extreme scenario, genetic erosion of the population of the species.

3 Tools for Fig Genetic Resources Diversity Studies

3.1 Morphological Evaluation

The traditional approach of fig germplasm characterization is based on the description of morphological characters and traits. One of the first major works published in the twentieth century presenting a review and descriptions of fig varieties was the monograph written by Condit (1955). His work was based on his observations and studies extending over more than thirty years. In 2003, the Descriptor List for *Ficus carica* L. was released by the International Plant Genetic Resources Institute (IPGRI) and International Centre for Advanced Mediterranean Agronomic Studies (CIHEAM) as a result of the collaborative work of scientists from different countries (IPGRI and CIHEAM, 2003). The main goal was to provide a universal, standard tool for fig genetic resources investigation. The list of descriptors is divided into several categories. Characterization descriptors, which are usually highly heritable and less dependent on environmental conditions, allow easy and quick discrimination between phenotypes. The list of characterization descriptors for figs involves plant descriptors, i.e., biological characters (9 characters), growth descriptors (16 characters with some subcharacters), and leaf (21 characters) and fruit descriptors (43 characters). For evaluation of caprifig, 26 characters related to the crop setting fruits ('mamme', 'profichi' and 'mammoni') were selected: fruit yield, amount of gall flowers, presence of male flowers, pollen maturation, *Blastophaga*'s exit, presence of parasitic insects, and germination rate of pollen. The list of evaluation descriptors includes plant descriptors and agronomic performance, including yield (8 characters), abiotic (7 characters), and biotic (2 characters) stress susceptibility. Evaluation descriptors are more related to the information about traits needed for crop improvement, and many of the descriptors in this category are more dependent on environmental conditions. This category of descriptors also includes biochemical (isoenzymes and polyphenol profiling, for example) and molecular markers (RFLP (Restriction Fragment Length Polymorphism), AFLP (Amplified Fragment Length Polymorphism), RAPD (Random Amplified Polymorphic DNA), STMS (sequence-tagged microsatellites), PCR-sequencing, and others). The authors also proposed a list of highly discriminating descriptors of figs, including fruit and leaf characters, maturity, crop setting, need for pollination, rooting ability, yield, etc.

Although the use of DNA markers represents one of the basic approaches and good practice in evaluating and managing plant genetic resources today, the morphological description and phenotyping remain the first step in evaluating germplasm, allowing the first insight into the diversity of the plant. Studies have reported evaluation of fig genetic resources with morphological descriptors or combined (morphological markers together with physiological, phenological and biochemical traits) in the common fig. This information is valued in pre-breeding programs and helps select the most appropriate genotypes with desired traits.

IPGRI and CIHEAM fig descriptors (2003), as well as some others additionally implemented by different research groups (Giraldo et al., 2010; Podgornik et al.,

2010), have been widely used for fig germplasm characterization in many fig producing countries such as Turkey (Çalişkan & Polat, 2012a, b), Slovenia (Podgornik et al., 2010), Spain (Giraldo et al., 2010), Morocco (Hssaini et al., 2020a, b), Palestine (Ali-Shtayeh et al., 2014), Iran (Fatahi et al., 2017), and Brazil (Rodrigues et al., 2019). Studies reported their usefulness for discriminating fig varieties and accessions in collections and their capability to determine the degree of diversity discovered in a defined geographical region or country. In some studies, several morphological and chemical traits were identified as the most informative for discriminating fig varieties (Çalişkan & Polat, 2012b; Hssaini et al., 2020a, b).

Çalişkan and Polat (2012a), among 64 IPGRI and CIHEAM fig descriptors, including plant and fruit characters, identified 37 characters as more beneficial for separating 76 Turkish fig accessions. Traits that showed higher discrimination capacity among fig accessions were: (a) for the description of plant and leaf: apical dominancy, lateral shoot formation, leaf shape, number of lobes, length of the central lobe, leaf area, and leaf width; and (b) for a description of fruit: fruit length and width, pH, fruit flesh color, abscission of the stalk from the twig, fruit neck length, fruit weight, and antioxidant capacity. The authors highlighted that the phytochemical properties could also be success criteria for describing fig genetic resources. Antioxidant capacity, anthocyanins, phenols, glucose, and fructose have been reported helpful descriptors. These characters are, in fact, economically essential and contribute to the nutritional value of the fig accessions and thus represent critical selective markers for the selection of genotypes during the breeding program. In another study, Çalişkan and Polat (2012b) confirmed the significant influence of genotype on phytochemical and fruit quality properties. The Turkish variety 'Bursa Siyahı' with dark black fruit skin, for example, has the highest levels of total anthocyanins, total phenolics, and antioxidant capacity. It has also been observed that total phenolics correlate moderately with antioxidant capacity and less with total anthocyanins. These observations confirm the importance of a comprehensive study from genotype to phenotype, including pomological and biochemical characters of individual accession. Biochemical attributes provide additional information about the nutritional value of a defined accession, and only complete information about the individual genotype allows the selection of the most appropriate accessions for cultivation.

Hssaini et al. (2020b), in the analysis of 96 Moroccan and 44 introduced fig varieties, demonstrated that fruit dimensions and skin color coordinates were the most informative for discrimination of fig accessions. In another study of five Moroccan and six introduced fig accessions, Hssaini et al. (2020a) also confirmed the effectiveness of combined morphological and biochemical characters in pre-breeding programs. The most relevant pomological characters were reported fruit as weight, size, stalk, and neck dimensions, whereas among biochemical descriptors, organic acids, fructose, lightness (L), Chroma (c*), and free radical scavenging activity based on ABTS assay were the most informative.

A Brazilian study of 42 fig accessions from the Active Germplasm Bank (AGB) of Fig belonging to the Faculty of Engineering of Ilha Solteira (FEIS) of the University of Sao Paulo State, among fig leaf measures studied, identified the number of leaf lobes as the most efficient marker for accession discrimination with high heritability (0.85) (Rodrigues et al., 2019) which is in agreement with Çalişkan and Polat (2012a). The Iranian study reported that most discriminator parameters depended on a leaf, fruit size, and growth form (Fatahi et al., 2017). In an extensive study of Spanish figs, Giraldo et al. (2010) analyzed 134 quality parameters of the tree, leaf, fruit, and two biological variables of 35 accessions grown in the fig collection in the Finca La Orden-Valdesequera in Badajoz, Spain. Most of the descriptors have been selected from the list of descriptors developed for the common fig by IPGRI and CIHEAM (2003), and, in addition, 43 new descriptors have been employed as informative for germplasm evaluation. The aim was to select a subset of highly informative descriptors that were able to distinguish accessions according to the importance of: (1) production type; (2) morphology of the main and breba crop; (3) morphology of five-lobed and three-lobed leaves; (4) morphology of entire leaves, and (5) tree morphology. They performed PCA analysis and a total of 34 variables were selected and broken down into 97 characters based on the highest discrimination ability in the first four PCs. The first 11 PCs explained 93.3% of the total variability, while the first four PCs explained 79%. The authors have shown that for the generation of representative data with morphological evaluation, the optimization of the approach is needed, including careful selection and reduction of the number of morphological parameters. Such optimization prevents redundant information and contributes to saving time and resources.

Morphological evaluation has also provided information on the diversity of the common fig. Hssaini et al. (2020a, b) reported that the Moroccan local fig clones are a rich source of variability. Similar conclusions have been reported for wild Iranian figs (Fatahi et al., 2017). In contrast, in Brazil, fig accessions showed medium to high diversity expressed with morphological descriptors and contributed to identifying accessions with attractive agronomic traits with the potential for their use in further genetic improvement programs (Rodrigues et al., 2019).

3.2 Molecular Evaluation

The development of DNA markers has further enriched the characterization approaches of fig genetic resources by an additional description of each sample with the tree's accurate fingerprint (DNA profile). These markers have complemented classical, morphological markers traditionally used to evaluate individuals and biochemical markers (allozymes, isozymes), considered the first molecular markers able to detect amino acid sequence variances (Adhikari et al., 2017). The main features of DNA markers are that they are not influenced by environmental conditions or the developmental stage of individuals and are detectable in all tissues of an organism. In addition, the variation of alleles makes them a powerful genotyping tool for the distinction of varieties and assessment of genetic diversity within and between populations. The potential that DNA markers can be linked to a particular trait or locus is also used in plant breeding programs for marker-assisted selection.

They, therefore, support many research areas, from taxonomy, phylogeny, population, and ecology genetics to biotechnology, plant breeding, and forensics. However, DNA markers have their strengths and weaknesses, and their selection depends on the research's purpose and informativeness (Riaz et al., 2021).

Most of the studies of the common fig involving DNA markers have aimed to evaluate the genetic diversity of local germplasm, to establish the identification key for varieties which allow fast distinguishing of varieties present in a collection/country/region with a minimum number of DNA markers, and to resolve variety denomination. The diversity of the common fig has been investigated with several molecular markers such as RAPD (Random Amplified Polymorphic DNA), AFLP (Amplified Fragment Length Polymorphism), RAMPO (Randomly Amplified Microsatellites Polymorphisms), ISSR (Inter-Simple Sequence Repeats), and SSR (Simple Sequence Repeats or microsatellites). In addition, some genetic studies have employed different DNA marker systems and presented their comparison and capability to discriminate among fig varieties (Cabrita et al., 2001; Ali-Shtayeh et al., 2014).

Dominant, multilocus PCR-based DNA markers such as AFLP (Vos et al., 1995) and RAPD (Williams et al., 1990; Welsh & McClelland, 1990) amplify more loci in a single analysis simultaneously, and it is expected that they cover a higher proportion of the genome. The AFLP method detects polymorphisms generated with selective PCR amplification of previously digested genomic DNA with two restriction enzymes. This method generates 50–100 fragments per assay, a unique genome fingerprint under investigation. In contrast, RAPD detects polymorphisms of DNA fragments generated with short, arbitrary primers in the PCR, resulting from changes in nucleotide sequences at or between the primer binding sites. The RAPD method amplifies several loci in a single assay, ranging from 0.5 to 5 kb (Adhikari et al., 2017). The main feature of these two markers is that they can be used in genetic studies without any prior knowledge of the DNA sequence and can detect polymorphisms between closely related genotypes generating unique DNA fingerprints. However, their limitation is based on the fact that we cannot distinguish heterozygous from homozygous individuals, and the methods demand strict control of protocol conditions to ensure the repeatability of the results.

For example, in the common fig, RAPDs combined with pomological markers have been employed to assess Palestinian ecotypes collected in the *ex-situ* field gene bank BERC-Til Botanic Gardens, Biodiversity, and Environmental Research Center (BERC), Til, Nablus (Ali-Shtayeh et al., 2014). Analysis has demonstrated considerable genetic diversity with the potential to improve the crop. Researchers have also reported a higher discrimination power of combined pomological and molecular markers, in contrast to using each marker alone. RAPDs have also been employed to study the genetic structure of 14 individuals representing cultivated Iranian fig populations (Zolfaghari et al., 2019). Dominant AFLPs, RAPDs, and biochemical markers were also used in Turkey to evaluate a clone of a variety 'Sarizeybek' and eleven clones of 'Sarilop', recognized as the main commercial variety used for dried figs in Turkey (Cabrita et al., 2001). Five isoenzyme systems discriminated against two varieties, while cluster analysis based on RAPDs divided clones into two groups, although discrimination among all the clones was not possible. Higher discrimination power of clones was demonstrated with AFLPs, where 8 combinations of *Eco*RI/*Mse*I primers clearly distinguished all eleven 'Sarilop' clones. The authors concluded that both RAPD and isozymes are useful for distinguishing varieties. Although RAPDs demonstrated lower polymorphism between clones, they were able to discriminate, whereas AFLPs were appropriate markers for discrimination of closely related clones within the fig variety.

The RAMPO method combines RAPD primer and hybridization of amplified RAPD fragments with microsatellite probes. Fingerprints that are microsatellite-based may display polymorphisms (Sharma & Singh, 2021). However, the complexity of banding patterns is low, and the method is susceptible (Adhikari et al., 2017). Chatti et al. (2008) designed the RAMPO method based on RAPD and ISSR assays to test the method's capability for differentiation of Tunisian fig varieties. The method showed a good capability for identity determination and resolved the mislabel of varieties.

The ISSR method includes amplifying the DNA region positioned between the two microsatellites with microsatellite-complementary PCR primers that can be used in an anchored or unanchored version (Sharma & Singh, 2021). The amplicons' length is usually between 200 and 2000 bp, and their polymorphisms depend on the abundance and hypervariability of microsatellites in the genome (Adhikari et al., 2017). The method is usually used for cultivar distinction (Sharma & Singh, 2021), although the dominant inheritance and homoplasy are the main limitations (Adhikari et al., 2017). ISSR markers have been used to investigate genetic relationships among cultivated and wild figs from Jordan, showed intermediate to high discriminating power, and thus provided the basic information necessary for conserving accessions in gene banks (Almajali et al., 2012).

DNA barcoding is another approach preferentially used to identify plants at the species level. The method involves the production of PCR amplicons from the particular regions and their sequencing. For example, in plants DNA regions from the chloroplast genome (for example, *matK*, *rbcL*) and internal transcribed spacer (ITS) of nuclear ribosomal DNA are usually analyzed (Cheng et al., 2016).

DNA barcoding, involving amplification of the ITS regions, *matK* and *trnH-psbA*, has been used in an attempt to differentiate seven fig accessions (varieties) from the Centro de Investigaciónen Biotecnología of Instituto Tecnológico de Costa Rica (Castro et al., 2015). In general, the authors reported low genetic variation observed between the genotypes, but the matK barcode was a reliable and cost-efficient alternative for identifying samples. The authors also presented the strengths and weaknesses of barcodes for *F. carica* genotype identification at the intraspecific level, which should be considered before their use.

The potential use of ITS sequences for diversity study was tested in a set of twelve economically important fig varieties from Tunisia (Baraket et al., 2009). The variation of ITS sequences allowed the establishment of relationships between closely related genotypes, indicating ITS as a valuable tool for clarifying relationships at a lower taxonomic level.

Oliveira et al. (2012), in an attempt to link the genetic profile with the metabolome of the fig variety, analyzed five Portuguese *F. carica* cultivars with two plastid regions (*trnH-psbA*, *rbcL*), six leaf morphological descriptors, and chemical compounds (phenols, sterols, and triterpenes). As a result, the correlation between genetics and metabolome has been confirmed for the variety 'Preta Tradicional' only. In addition, the chloroplastic *rbcL* region was reported as potentially useful for fig authenticity purposes.

3.2.1 Development of Fig Microsatellites

Microsatellites, or simple sequence repeats (SSRs), have an important place in plant genome research (Baruca Arbeiter et al., 2021) and are believed to be the most powerful technique for identification and variety protection (Sharma & Singh, 2021). Microsatellite sequence consists of repetitive DNA, made of short motifs usually long from one to six bp, which is repeated repeatedly. The variable number of repeats enables accurate genetic profiling of each individual. Microsatellites combine several properties of an ideal marker; they are codominant, highly polymorphic, abundant in genome (genomic SSRs) and transcriptome (genic SSRs), their amplification is fast and easy, alleles are easily scored and interpreted, and genotyping results are highly reproducible, allowing an easy exchange of data among laboratories (Kumar et al., 2009). Although most of the studies have been focused on the development of genomic SSRs, genic SSRs usually derived from Expressed Sequence Tags (ESTs) (EST-SSRs) are also attractive for gene mapping and functional studies (Rudd, 2003; Ozgenturk et al., 2010; Baruca Arbeiter et al., 2017). They are positioned in expressed sequences, generated with next-generation sequencing technology, and represent a snapshot of gene expression during a particular developmental stage and under defined conditions when sampling in cells, tissues, or organs of an organism (Rudd, 2003; Kalia et al., 2011). Therefore, their value relies on the potential that they are linked to a defined function or locus of agronomic interest. The largest set of EST-SSRs has been obtained from the fruit syconium, leaf, and stem transcriptome (Mori et al., 2017). The authors designed 3467 locus specific primer pairs for EST-SSR markers of which 58 were integrated into a high-density consensus linkage map.

Genomic microsatellites are routinely and most frequently used for genetic studies and management of plant genetic resources. However, the successes of genetic analysis rely on selecting the most informative microsatellite markers that allow the detection of polymorphisms. The selection of markers should be based on the essential parameters of diversity, such as the number of amplified alleles (n), effective alleles (n_e), observed (H_o) and expected heterozygosity (H_e), polymorphic information content (PIC), and probability of identity (PI). Therefore, the screening process and quality testing of candidate loci are essential steps in genetic analysis (Selkoe & Toonen, 2006).

Several microsatellite markers have been published for closely related species such as *Morus* sp. and *Ficus* sp., but their informativeness for diversity studies and transferability to the common fig should be first tested. The homologous DNA sequences in microsatellite flanking regions and evolution distance of species may influence the transferability (Knap et al., 2016). Only three sets of genomic microsatellite markers have been developed and fully characterized for common fig diversity studies.

The first set of microsatellites, named MFC, was released by French researchers in 2001 (Khadari et al., 2001). From 20 developed locus-specific primer pairs, 8 produced polymorphic alleles in 14 varieties and two wild populations, sampled in the fig germplasm collection of the Conservatoire Botanique National Méditerranéen de Porquerolles, France. The MFC set amplified from 2 to 6 alleles, with 4.25 alleles per locus. The highest number of alleles was obtained with MFC1 (6 alleles) and MFC2 (5 alleles), and others amplified 4 or 3 alleles. The authors have demonstrated the transferability of MFC microsatellites to other 17 *Ficus* species.

Giraldo et al. (2005), a research team from Spain, developed the second, LMFC, set of fig microsatellites. They obtained 26 polymorphic microsatellites from a genomic library enriched for CT/AG repeats. The set of microsatellite markers was characterized in a group of 15 fig varieties from Spain and France. The highest power for distinguishing fig varieties has been demonstrated with locus LMFC30, with which 8 alleles have been amplified, and the lowest PI was obtained (0.15). Comparison of DNA profiles allowed discrimination of nine varieties out of 15. Six samples (two groups of three varieties) remained undistinguished and showed identical allelic profiles. Molecular data corresponded with morphological evaluation and confirmed the inconsistency in the naming of fig varieties. The authors assigned the low values of variability parameters and the lack of clear geographical groupings to a narrow genetic base in the cultivated common fig. Giraldo et al. (2005) confirmed the transferability of some LMFC microsatellite markers to another 13 *Ficus* species and *Morus alba* L. and *Morus nigra* L. species.

The limited number of highly polymorphic microsatellite markers in the common fig encouraged a Slovenian research group to develop another set of microsatellites. Fifteen FCUP microsatellites were released in 2007 (Bandelj et al., 2007) and characterized 22 wild and 19 cultivated fig samples. The highest number of amplified alleles in cultivated fig varieties was obtained at loci FCUP8 (7 alleles), FCUP68 (6 alleles), FCUP70 (5 alleles), and FCUP44 (5 alleles). Nine out of 15 loci had PIC higher than 0.5, classifying loci as informative markers. The lowest PI was detected at locus FCUP8, FCUP44, and FCUP68, indicating that loci are appropriate for genotype identification. An additional sixteen primer pairs of the FCUP series were published in 2016 (Knap et al., 2016). These markers genotyped 18 fig varieties and 65 wild samples from Slovenia and Croatia. Twelve loci were informative with PIC higher than 0.5. The highest number of polymorphic alleles was detected with FCUP40 and FCUP95 (n = 8), 7 alleles were amplified with FCUP1 and FCUP3, locus FCUP24 amplified 6 alleles, and five polymorphic alleles were found on loci FCUP17, FCUP18, FCUP26, FCUP28, FCUP59, FCUP75, FCUP81, and FCUP91. These markers demonstrated high discrimination power and represent an efficient tool for verifying the origin of seedling material in nurseries, distinguishing closely related varieties, and managing fig genetic resources collections. All markers were also successfully cross-amplified in five closely related species belonging to the Moraceae family, namely *Ficus* (*F. barteri* Sprague, *F. rubiginosa* Desf. ex Vent., *F. benjamina* L.) and *Morus* (*M. alba* L. and *M. nigra* L.).

3.3 Application of Microsatellites in Fig Diversity Studies

Microsatellites of the common fig have been used in many fig producing countries, mainly for evaluation of genetic diversity of fig varieties and other accessions present in *ex-situ* collections or maintained *in-situ* at growing sites, for the establishment of the identification key of varieties, and to assess the diversity of caprifigs (Table 3.1). Knowledge of the genetic diversity range represents a valuable contribution to the optimization of fig tree breeding programs (Ganopoulos et al., 2015; Essid et al., 2015) and the conservation and improvement of local fig germplasm (Ben Abdelkrim et al., 2015; Baraket et al., 2011; Saddoud et al., 2011; Essid et al., 2015). Interlaboratory comparison of fig varieties' genotyping profiles with microsatellites has also been shown (Hladnik et al., 2018). Some studies compared different marker systems and presented their potential for discrimination of fig accessions (Ikegami et al., 2009; Chatti et al., 2010; Ikten et al., 2010; Baraket et al., 2011; Abou-Ellail et al., 2014).

Microsatellites have been used to characterize the Moroccan fig germplasm collection managed by the Institut National de la Recherche Agronomique (INRA) Meknes. Achtak et al. (2009) genotyped 75 accessions, including eight caprifigs, 51 local accessions, 11 foreign accessions, and five accessions of unknown origin. Among 17 microsatellite loci used in the study, two optimal combinations of primer pairs have been proposed to discriminate all accessions. The combination LMFC3 0 + MFC2 + MFC3 + FSYC01 + MFC9 allowed almost all analyzed samples to be identified except for two genotypes, which were further differentiated by adding the locus MFC11 to the combination. The lowest PI values have been reported for loci LMFC30, MFC2, MFC3, and MFC1 ($PI < 0.2$), which confirms the good ability of loci to distinguish varieties included in the analysis. In addition, with the help of microsatellites, cases of synonyms and homonyms, and mislabels in the collection have been resolved. The study of the Moroccan fig collection was further upgraded with microsatellite profiling of 277 individual fig trees from 40 sites distributed in 6 geographical regions (Achtak et al., 2010), representing traditional Moroccan agroecosystems. With the same set of 17 microsatellite primer pairs, the highest discrimination power has been noted for loci LMFC30, MFC1, MFC2, and MFC3 ($D > 0.80$). The extensive study has demonstrated that traditional Moroccan agroecosystems represent active varietal and genetic diversity incubators and are an important repository of local varieties. Although the varieties are clonally maintained, small differences from 1 to 3 alleles among the trees were assigned to somatic mutations accumulated through the long period of variety cultivation. The higher number of allele differences among genotypes (4–34 alleles) has been explained by the sexual reproduction of cultivated figs and wild growing figs in the

Table 3.1 The list of published scientific articles involving the common fig SSRs in diversity studies

Country	Common fig SSRs (set)	Fig genetic material analysed (accessions)	Research aims	Reference
Algeria	24 SSRs (LMFC, FCUP)	34 varieties	Fig diversity and denomination	Boudchicha et al. (2018)
Brazil	5 SSRs (MFC)	11 varieties (collection)	Fig evaluation and identification	do Val et al. (2013)
France	SSRs (MFC)	14 varieties and 2 wild populations	SSRs development and characterization: 20 primers	Khadari et al. (2001)
Greece	7 SSRs (LMFC, MFC)	90 varieties from different countries (collection)	Fig diversity, genetic differentiation	Ganopoulos et al. (2015)
Italy	7 SSRs (LMFC, MFC)	79 parthenocarpic accessions (collection)	Fig diversity and denomination, creation of fig database	Rodolfi et al. (2018)
Italy	18 SSRs (LMFC, MFC, FCUP)	181 traditional varieties	Fig diversity and denomination	Costa et al. (2017)
Morocco	17 SSRs (LMFC, MFC)	277 fig trees from 6 regions	Fig diversity in traditional agroecosystems	Achtak et al. (2010)
Morocco	17 SSRs (LMFC, MFC)	8 caprifigs, 51 local,11 introduced, 5 unknown accessions (collection)	Fig diversity and denomination	Achtak et al. (2009)
Slovenia	12 SSRs (FCUP, LMFC)	23 Adriatic varieties, 218 accessions from NCGR collection	Fig diversity and genetic relationships	Knap et al. (2018)
Slovenia	SSRs (FCUP)	18 varieties and 65 wild forms	SSRs development and characterization: 16 primers	Knap et al. (2016)
Slovenia	SSRs (FCUP)	19 varieties and 22 wild forms	SSRs development and characterization: 15 primers	Bandelj et al. (2007)
Slovenia & France	6 SSRs (LMFC, MFC, FSYC01)	90 fig accessions (84 from the East Adriatic, 6 from the French collection)	Comparison of the genotyping results of two laboratories	Hladnik et al. (2018)
Spain	9 SSRs (LMFC, MFC)	229 accessions from Spain and abroad	Establishment of core collection	Balas et al. (2014)
Spain	21 SSRs (LMFC, MFC)	42 local accessions belonging to 12 local denominations and two caprifigs	Genetic relationships among fig accessions	Perez-Jiménez et al. (2012)

(continued)

Table 3.1 (continued)

Country	Common fig SSRs (set)	Fig genetic material analysed (accessions)	Research aims	Reference
Spain	11 SSRs (LMFC, MFC)	196 samples from different growing areas in Spain and 13 samples from abroad	Establishment of a genotyping database of reference profiles	Giraldo et al. (2008)
Spain	SSRs (LMFC)	15 varieties from Spain and France	SSRs development and characterization: 26 primers	Giraldo et al. (2005)
Tunisia	13 SSRs (LMFC, MFC)	30 cultivated accessions	Differentiation of accessions	Essid et al. (2021)
Tunisia	17 SSRs (LMFC, MFC)	19 varieties, 52 wild figs from 4 locations	Identification, diversity study	Ben Abdelkrim et al. (2015)
Tunisia	13 SSRs (LMFC, MFC)	20 caprifigs	Differentiation of caprifigs, fig diversity	Essid et al. (2015)
Tunisia	6 SSRs (MFC)	72 local ecotypes	Identification	Saddoud et al. (2007)
Turkey	15 SSRs (LMFC, MFC, FCUP)	96 native caprifigs	Genetic population structure	Çalişkan et al. (2018)
USA	15 SSRs (LMFC, MFC)	194 accessions from different countries (NCGR collection)	Fig diversity and genetic relationships	Aradhya et al. (2010)

agroecosystem. This spontaneous hybridization represents an important driving force for developing varieties with new qualitative properties.

Severe genetic erosion of local fig germplasm and the confusion in the denomination of varieties in Tunisia have encouraged the evaluation programs of fig genetic patrimony (Saddoud et al., 2007). The first attempt at the preservation of fig ecotypes included the inventory and propagation of cuttings of around 100 ecotypes that are *ex-situ* maintained in different collections (Institut Supérieur Agronomique (ISA), Chott Meriem; Institut des Régions Arides (IRA), Medenine; Centre de Recherches Phoénicicoles (CRPh), Degache; Commissariat Régional au Développement Agricole (CRDA), Gafsa; Groupement Interprofessionnel des Fruits (GIF), Sbikha) or are *in-situ* preserved (Saddoud et al., 2007). Six MFC markers have been used for genotyping 72 local ecotypes originating from different Tunisian areas. Fingerprinting allowed the establishment of the first identification key for Tunisian varieties.

Ben Abdelkrim et al. (2015) performed genotyping of Tunisian fig germplasm, including 19 fig varieties and 52 wild fig trees from four different locations with the identical primer pairs set of 17 microsatellites as previously reported in an analysis of Moroccan fig germplasm (Achtak et al., 2009, 2010). On average, the number of amplified alleles per locus was similar to that reported in other fig diversity studies

(4.35 alleles/locus). In Tunisian fig samples, the highest PIC was found in loci MFC3, LMFC30, and MFC11. These three loci, together with MFC19, constitute the identification key for Tunisian fig germplasm. Although some of the samples were not discriminated with this combination, the authors found other loci that could distinguish these genotypes. This result suggests the careful selection of microsatellite loci for identification purposes and the narrow genetic base of some fig genotypes. The newest study, conducted by Essid et al. (2021), employed 13 microsatellites LMFC and MFC loci and presented their characteristics in 30 cultivated Tunisian fig accessions. PI values of loci were relatively high, indicating a lower capability to differentiate accessions under evaluation. Similar results were reported in Tunisian caprifigs from different geographic regions showing distinct phenotypic traits (Essid et al., 2015). The same set of 13 microsatellite LMFC and MFC loci generated a low number of alleles (2 or 3 alleles per locus), and diversity parameters were below the range observed in other studies. Although the authors attributed this result to a low genetic diversity of the Tunisian male trees, the limited number of analyzed caprifig trees (20) and the selection of loci with lower diversity information content could have influenced the final observations. The authors highlighted the need for additional studies of caprifigs to obtain a better insight into their genetic diversity.

The study of the genetic population structure of caprifigs from Turkey showed the opposite result analyzing 96 native caprifigs with 15 loci of series MFC, LMFC, and FCUP, which were chosen based on their high PIC values (Çalişkan et al., 2018). On average, 8.3 alleles were amplified, ranging from 3 (LMFC18, LMFC23) to 14 alleles (FCUP38, FCUP8). The study showed a great genetic diversity of caprifigs assigned to generative propagation with seeds in nature. The population genetic structure analysis of caprifigs demonstrated grouping according to the geographical region of sampling.

The presence of figs from ancient times in Italian territory has been supported with archaeobotanical and historical evidence, and the long tradition of cultivation has contributed to a large number of varieties in the country. Rodolfi et al. (2018) genotyped 79 parthenocarpic fig accessions from a private, *ex-situ* collection in Pescia (Tuscany, IT) with MFC and LMFC microsatellites. The analysis demonstrated the existence of 56 different genotypes and has contributed to resolving cases of homonyms, synonyms, and mistaken denominations. As a result, a databank of Tuscan and other Italian fig varieties has been created. In another study of Italian fig germplasm, Costa et al. (2017) employed 18 FCUP, LMFC, and MFC microsatellites and presented their characterization in 181 traditional fig varieties and genotypes cultivated in Apulia, Basilicata, Calabria, Campania, and Sicily (south Italy). Nine out of 18 loci displayed the PIC value higher than 0.5; the most informative markers were FCUP27, FCUP38, MFC1, MFC2, MFC3, MFC11 (PIC > 0.6), LMFC28, MFC7 (PIC > 0.5) and LMFC30 (PIC > 0.7). Their study demonstrated the richness of the fig genetic resources in Southern Italy and helped resolve fig varieties denomination.

In Spain, Giraldo et al. (2008) analyzed 196 samples from different growing areas and 13 samples from the USA, France, Israel, Portugal, and Turkey. Among

20 polymorphic SSR markers (series LMFC, MFC) tested, eleven amplified one locus, and these markers were further characterized for their polymorphisms. Microsatellite data helped establish the database with DNA profiles of 104 reference genotypes from 209 starting accessions. The redundancy of samples in a germplasm collection is not desirable, and before incorporating new accessions into a collection, it is worth testing their identity. The study demonstrated that microsatellites are an excellent tool for collection management and can identify synonyms and homonyms. The applicability of microsatellites for the evaluation of local fig germplasm has been demonstrated by Perez-Jiménez et al. (2012). In their study, 42 trees belonging to 12 local denominations and two caprifigs were genotyped with 21 LMFC and MFC microsatellite primers. Their allelic profiles were compared with 15 reference varieties from collections of the Conservatoire Botanique National Méditerranéen de Porquerolles (France) and Estación Experimental Agraria de Elche (Spain), and a local nursery from Córdoba (Spain). The highest PIC (higher than 0.5) was obtained with loci MFC2, MFC3, MFC6, LMFC12, LMFC13, LMFC30, and LMFC38. The clustering method separated the studied Andalusian fig trees into 11 distinct groups corresponding to their local denominations. These data will serve for fig germplasm conservation strategy in Spain.

The National Clonal Germplasm Repository (NCGR) at Davis, California (USA) is one of the largest collections of fruit and nut crops that collect, preserve, evaluate, and distribute the genetic resources of crops for current and future needs. The collection is home to numerous fig accessions from all over the world. Aradhya et al. (2010) have analyzed 194 germplasm accessions representing all four types of figs with fifteen MFC and LMFC loci. Cluster analysis based on microsatellite profiles revealed ten groups, which have been confirmed with PCA. With PCA, more differentiation has been noted in some groups. The gene diversity was higher within the groups than among the groups. The study's main conclusion is that fig germplasm possesses substantial genetic polymorphism but exhibits narrow differentiation. In addition, Turkmenistan figs showed different genetic backgrounds compared to Mediterranean and Caucasian fig germplasm.

Knap et al. (2018) have compared genotyping profiles of 23 Adriatic fig varieties with 218 accessions from the Californian NCGR collection obtained with 12 FCUP and LMFC SSR loci. In this study, the number of amplified alleles per locus ranged from 4 to 14. The average number of amplified alleles was relatively high (8.33) compared with other fig diversity studies. In addition, seven loci out of twelve had a PIC value higher than 0.5, indicating the high polymorphic information content of the loci used. With eight loci, it was possible to discriminate 166 genotypes. The analysis confirmed significant genetic diversity in the common fig, attributed to the careful selection of highly informative SSR markers and large sample size with redundant worldwide accessions. The important achievement of the study was the identification of several international synonyms of varieties expanded worldwide, originating from the United States of America ('Ventura', 'Calvert', 'Figue D'Or', 'Capitola Long'), Georgia ('Asashendu green figs', 'Shilda Purple-Bronze Fig', 'GIHVO Italian Green Fig'), Azerbaijan ('Belaya Fraga', 'Blanche Fig') and Turkey ('Patlican fig'). In addition, 17 out of 23 North Adriatic varieties had distinct genotypes compared to the NCGR collection, indicating that fig varietal structure and diversity in Slovenia is unique and represents an important reservoir of genetic variation for cultivated figs. Local varieties with unique genotypes can significantly promote the region and produce valuable local products.

Algerian cultivation of figs provides basic income in the rural and marginal areas, and the country is ranked third among the most important producing countries of figs in the world, after Turkey and Egypt (Boudchicha et al., 2018). Despite the importance of its cultivation, genetic erosion has been reported, and the Algerian Ministry of Agriculture launched a developmental program to improve fig production in the country. Boudchicha et al. (2018) presented the first molecular report of fig diversity. The authors evaluated 34 fig varieties of San Pedro, Smyrna, and common types from *ex-situ* and *in situ* with 24 microsatellite primers. With the highest PIC, five markers (LMFC30, FCUP038, FS4-11, FCUP044, FCUP070, and FM4-11) showed an excellent discriminatory capacity for Algerian fig varieties. Genotyping revealed cases of synonyms, homonyms, and mislabelling errors in the collections and demonstrated the significant genetic variability with a narrow genetic structure.

In Brazil, the cultivation of the common fig is increasing due to the growing demand for its fruits in the food industry (do Val et al., 2013). To overcome some pest and disease problems related to the cultivation of a single variety, 'Roxo de Valinhos', and to diversify cultivation with other fig varieties, an evaluation of eleven varieties maintained in the germplasm bank located in Lavras, Minas Gerais has been investigated by the use of five MFC loci. The allelic variation allowed the identification of 10 genotypes and two synonymous individuals. In this study, the lowest PI values were obtained for markers MFC1, MFC3, and MFC7, indicating good discriminatory capability for the Brazilian collection.

In Greece, seven LMFC and MFC microsatellites have been used to characterize ninety varieties from Greece, Italy, Cyprus, Spain, Turkey, and France in an *ex-situ* collection (Ganopoulos et al., 2015). The diversity parameters of all selected markers were highly informative, with PIC higher than 0.5. Although substantial genetic polymorphism has been reported, limited genetic differentiation among the different geographical groups has been noted, probably due to the common origin of analyzed varieties. For example, fig genotypes from Cyprus were genetically different from the rest of the accessions.

Microsatellites are confirmed as a very useful tool for collection management. The high costs of managing a large number of accessions in the collection, primarily due to redundancies and duplications, led to the search for new solutions to reduce accessions without the decline in the overall diversity of the plant. One of the approaches includes creating a core collection composed of a reference subset of genotypes selected from a larger set of varieties. Balas et al. (2014) found that the maximum diversity could be achieved with a maximization strategy (M strategy) based on microsatellite data. Their study was conducted using nine LMFC and MFC microsatellites on 229 fig accessions from Spain and abroad, maintained at Centro de Investigación La Orden fig germplasm bank (Spain). M strategy maximizes the number of alleles sampled at each locus, and an iterative process eliminates the redundant accessions. It was found that with the subset of 30 fig accessions (only 13% of the entire collection), it is possible to cover all the SSR alleles. In addition, in comparing allele frequency distribution, H_e and H_O showed no significant differences between the whole and core collection. M strategy seems to be an efficient method for reducing fig accessions in the collection. The main consideration is that some locally spread fig ecotypes with a rare combination of alleles could be lost with this strategy; thus, they require special attention for conservation. According to Balas et al. (2014), economically important varieties can be additionally included in the core collection.

Some genetic studies of figs have employed different marker systems for diversity or identification proposes. Chatti et al. (2010) compared the ability of RAPD, ISSR, RAMPO, and SSR to detect polymorphism and to establish the genetic relationships among fig varieties. A comparison of the number of polymorphic markers obtained with RAPD, ISSR, and RAMPO demonstrated a similar and efficient capability of polymorphism detection. SSR analysis employed six MFC loci, of which five were able to discriminate fig varieties except for Kahli1 and Soltani2 samples. These two samples were distinguished with combined RAPD, ISSR, and RAMPO markers. This result is expected and can be assigned to the nature of the aforementioned markers, which cover a few random regions in a single PCR reaction, whereas one microsatellite locus covers a very small part of the genome. Difficulties with discriminating samples with SSRs were already reported by Knap et al. (2017), when the set of seven FCUP loci was not able to discriminate Slovenian varieties 'Zeleni (green) Matalon' and 'Črni (black) Matalon' despite their apparent differences in the color of the fruit peel. However, these two varieties were previously successfully distinguished by selecting other combinations of loci (Bandelj et al., 2008). This result additionally confirmed that the selection of markers for identification procedures is crucial, and besides the informativeness of markers, it depends on the genetic background and closes the genetic relationship of the plant material under investigation.

Tunisian fig germplasm was evaluated through the comparison of AFLPs and SSRs. Baraket et al. (2011) genotyped 38 accessions, including 30 varieties and eight male trees from different regions. This study used six MFC microsatellites and six EcoRI/MseI primer pairs. Both marker systems showed a large genetic diversity of Tunisian figs and gave good and comparable results, although their nature differed. According to the authors' opinion, the advantage of AFLPs is a high effective multiplex ratio value (the number of bands analyzed simultaneously per experiment) and estimated marker index (the amount of information obtained per experiment for a given marker system). Furthermore, varieties were clustered independently from their geographical origin, horticultural classifications, and tree sex phenotype with both markers.

In Egypt, some local fig names have been generated by farmers based on fruit color and flavor, and the presence of homonyms and synonyms characterizes the local germplasm. Abou-Ellail et al. (2014) used two biochemical markers, isozymes

Polyphenol Oxidase, Peroxidase, and 6 SSR (MFC) loci, to establish the molecular key for fig identification to provide necessary information for breeding programs and conservation of local germplasm. The analysis was performed on 7 fig varieties sampled in different growing areas. The study has demonstrated that isozymes are suitable, but SSRs were more efficient for genetic variability and fig cultivars fingerprinting.

Ikegami et al. (2009) analyzed the genetic diversity of 8 varieties from Japan, 4 from the USA, 5 from France, and 2 from Spain with ISSR (13 primer combinations), RAPD (19 primers), and SSR (13 loci) markers. Examination of each marker system demonstrated differences in the separation of samples according to genetic relationships, and a weak correlation among marker systems was detected. The study's main conclusions were that the genetic diversity of analyzed fig samples was low and that the multiple marker utilization is critical to estimating the genetic relationships at a varietal level.

4 Gene, Genomic and Transcriptomic Studies of Figs

Technologies for high-throughput sequencing (HTS) have enabled the generation of large amounts of data compared to technologies based on Sanger's method. Sequencing DNA or RNA offers deeper insight into the genome structure and identification of new genes and mechanisms involved in gene expression. The first high-throughput sequencing technologies are classified as next-generation sequencing technologies (NGS), e.g., Illumina and Ion Torrent, whereas the most recent are designated as third-generation sequencing technologies (TGS) (van Dijk et al., 2018).

The common fig has received significant attention for analysis with HTS technologies, which several published scientific articles have confirmed in recent years. 'Short read' sequencing was applied for genomic sequences and transcriptome (reviewed in the next chapters), whereas 'long read' sequencing technology was recently used for genome sequencing (Usai et al., 2020).

The 'short read' NGS technologies involve massively parallel sequencing of short reads, meaning that millions of individual sequencing reactions occur in parallel (Hu et al., 2021). However, reconstructing regions with repetitive elements and identifying large structural variants is often difficult with short reads. The TGS technologies, on the other hand, including 'single-molecule real-time' (SMRT) sequencing technology developed by Pacific Biosciences (PacBio) and Oxford Nanopore Technologies (ONT), can sequence longer fragments which leads to superior performance in the analysis of repeated regions, structural variants, haplotype phasing, transcriptome analysis, and other applications such as closing the gaps of previous short-read assemblies (van Dijk et al., 2018).

4.1 Genomic Studies

4.1.1 Genome Characterization

The common fig is a diploid species with 2n = 26 chromosomes (Flaishman et al., 2008). The estimated genome size with flow cytometry is 356 Mbp (Loureiro et al., 2007). The first drafts of the common fig genome sequence were generated by Mori et al. (2017) and Barghini et al. (2017) using NGS sequencing technologies. Mori et al. (2017) sequenced DNA of the traditional Japanese variety 'Horaishi', whereas Barghini et al. (2017) sequenced DNA of an old Italian variety 'Dottato'. Both varieties are classified as common-type figs.

Recently, the genome of the variety 'Dottato' was sequenced with PacBio SMRT technology (Usai et al., 2020), which enabled the better reconstruction of the genome, especially parts represented by repetitive elements. The final length of the assembled genome sequence was ~333 Mbp, corresponding to ~95% of the estimated size. A total of 37,840 protein-coding genes were identified, and repeated represented 37.39% of the genome assembly. The most abundant repeats were transposable elements (TE), covering 33.5% of the assembly. Tandem repeats represent 3.82% of the assembly. Long terminal repeat retrotransposons (LTR-REs) were the most abundant among the transposable elements, representing 28.03% of the genome assembly. Altogether, 1867 LTR-REs were revealed (Vangelisti et al., 2021). *Gypsy* elements (62.5%) and *Copia* elements (33.3%) were annotated, and the rest of the full-length LTR-REs remained unknown.

4.1.2 New Research Approaches and Findings Based on the Available Genomes

The available genomes have opened the possibilities for new research approaches, like genome-wide gene identification and characterization (Song et al., 2021), identification of genome-wide single nucleotide polymorphism (SNP) markers, and generation of high-density genetic maps (Mori et al., 2017). In addition, available genomes are an essential platform for transcriptome studies.

Mori et al. (2017) performed a genome-wide association study (GWAS) and identified two missense mutations in the *RESPONSIVE-TO-ANTAGONIST1* (*RAN1*) gene, which were highly associated with sex phenotypes of the investigated figs. Several analyses were performed to obtain this result. Genome-wide SNPs and indels polymorphisms were identified based on the 'Horaishi' variety draft genome and genome sequences of five different varieties. Restriction site-associated DNA sequencing (RAD-seq) analyses were performed on plants of the 'Horaishi' and Caprifig6085 which were cross-pollinated, on its F1 population, and additional plants with known sex phenotype (22 males, 97 females) and on 3 individuals with unknown sex phenotype for genotyping and the construction of the high-density genetic map. *RAN1* is involved in the transduction of ethylene, and a suggested mechanism is that male plants with normal *RAN1* have active ethylene receptors resulting in the formation of stamens, whereas, in the females, these receptors are nonfunctional. The role of phytohormones in the differentiation of floral organs was observed in other plant species (Feng et al., 2020). Kompetitive allele specific PCR (KASP) and cleaved amplified polymorphic sequence (CAPS) markers were developed to detect A or G alleles in the *RAN1* gene (G/A in males, G/G in females) or all three codons that could result in a combination of neighboring SNP, respectively (Ikegami et al., 2021; Mori et al., 2017). Surprisingly, segregation distortion was observed in a specific parental cross on this locus, resulting in a highly male-biased segregation ratio. Several candidate genes possibly responsible for the segregation distortion were identified (Ikegami et al., 2021). Recently, three *AGAMOUS* genes were recruited from the *F. carica* genome. FcAG2-Gall_Stamen and FcAG3-Gall_Stamen were male-specific (they had 15 bp in the 7th exon absent in *FcAG-Pistil*) and were only expressed in the gall and stamen of the male tree fig fruit. A method for detecting male plants with PCR was developed (AG-Marker) (Wang et al., 2021b). The same authors also designed the marker based on 6 SNPs of the *RAN1* gene, and both markers were confirmed as an accurate approach for sex determination.

Publicly available genomes are an essential base for genome-wide gene identification and characterization. Song et al. (2021) studied genes from the basic helix-loop-helix (bHLH) transcription factor family involved in the anthocyanin biosynthesis. In total, 118 *FcbHLH* genes were identified in reference genomes. Fifteen *FcbHLH* genes showed increased expression in the peel and female flower tissues at different main crop developmental stages. The common fig varieties 'Purple-Peel' ('Zibao') and 'Green-Peel' ('Qingpi') were used. The 'Purple-Peel' variety is a bud mutation of 'Green-Peel', a major common fig variety in China (Wang et al., 2017). The gene *FcbHLH42* synchronicity between expression and the corresponding anthocyanin content in the female flower and peel was observed. The impact of the *FcbHLH42* gene on anthocyanin accumulation was also demonstrated with transient expression in tobacco leaves.

Due to the relatively small genome size, the common fig has become a new model species for fruit trees, and recently methylation-based studies have been performed (Usai et al., 2021). The study aimed to analyze the pattern of unconventional epigenetic changes, including N4-methylcytosine (4mC) and N6-methyladenine (6mA) modifications within the transposable elements (TE) and the impact of TE methylation on the expression of the neighboring genes. However, an additional analysis should be implemented to fully elucidate the role of TE methylation patterns on the neighboring gene expression (Usai et al., 2021).

4.2 Transcriptomic and Gene-Based Studies

4.2.1 Mechanisms Related to Fruit Development and Ripening

To elucidate biological mechanisms related to plant and fruit development, several transcriptomes were sequenced (Chai et al., 2017, 2018, 2019; Cui et al., 2019, 2020; Freiman et al., 2014; Ikegami et al., 2013a; Li et al., 2020; Marcotuli et al., 2020; Qiao et al., 2021; Rosianski et al., 2016a; Wang et al., 2017, 2019).

The first transcriptome dataset for multiple developmental stages of common fig fruit was provided by Freiman et al. (2014). They sequenced RNA obtained from the red flash variety fruits, classified as common type biferous fig, 'Brown Turkey'. As a result, the differential activity of several metabolic processes was observed, and new ethylene-synthesis genes and the involvement of several transcription factors during the ripening process were determined.

In order to identify genes involved in fruit-quality formation control transcriptome analysis between receptacle tissues of commercial ripe (first half of stage III) and tree-ripe (late-stage III) fruits were performed by Cui et al. (2019). Differentially regulated genes related to ficins, sugar accumulation, cell-wall modification, ethylene pathway, and other biological processes involved in fruit quality and ripening were observed. In addition, genes involved in cell wall modification and their expression during ripening were studied by Owino et al. (2004), and it was suggested that they are coordinated in time and their intensity of expression to achieve fruit softening.

To discover the key candidate genes in ripening regulation to achieve more simultaneous ripening of all fruits on a tree, consequently shortening the harvest season and making it less labor-intensive, an experiment was performed with ethephon-treated fruits (Cui et al., 2020). Ethephon stimulates fig fruits' growth and ripening. It was observed that ethephon mainly regulates genes related to the expression of plant hormones, cell-wall modification, and sugar accumulation in female flowers and receptacles.

Since fruits harvested before ripening onset will not ripen postharvest, Freiman et al. (2015) studied the influence of 1-methylcyclopropene (1-MCP) (an inhibitor of ethylene), which improves fruit-storage abilities after treatment of fruits on trees, on ethylene production, and other mechanisms related to fruit ripening. The behaviour of genes involved in fruit ripening and genes involved in the prolonged storage abilities after treatment with 1-MCP was characterized. In addition, genes involved in the ethylene synthesis were isolated, and their expression pattern during ripening was also studied by Owino et al. (2006). In addition, it has been shown that on-tree abscisic acid (ABA) application synchronizes fruit ripening and maintains its quality (Lama et al., 2019), and a model for the role of the ABA-ethylene interaction during fruit ripening was proposed by Qiao et al. (2021).

Flaishman et al. (2020) proposed ABA and ethylene crucial biosynthetic genes, *FcNCED2* and *FcACS4*, as candidate genes for genome editing to alter the ripening process of fruits. An attempt to edit the *FcNCED2* gene was already performed, using *Agrobacterium*-mediated transformation of a construct carrying the CRISPR/Cas9 complex components. However, the desired mutation has to be confirmed.

For different common type varieties, it was observed that pollinated fruits had larger diameters, weight, improved firmness, and more commercially desirable characteristics compared to the parthenocarpic fruit. This was observed for the 'Brown Turkey' variety (Rosianski et al., 2016b) and recently for fruits of the 'Dottato' variety (Marcotuli et al., 2020). To study gene expression during fruit ripening in pollinated and parthenocarpic fruits, a transcriptome comparison of 'Brown Turkey' fruits was performed by Rosianski et al. (2016a). Results indicate that the reproductive part of the syconium coordinates fruit ripening, and the differences in ABA production between pollinated and parthenocarpic fruit might be the key to their different ripening characteristics.

4.2.2 Secondary Metabolites

Transcriptome studies related to the flavonoid biosynthetic pathways were conducted using fruit peels (Li et al., 2020; Wang et al., 2017). Wang et al. (2017) analyzed 'Green Peel' and its color mutant 'Purple Peel'. Significant and simultaneous upregulation was revealed in almost all of the flavonoid and anthocyanin pathway components and relevant transcription factors in the mature stage of 'Purple Peel' fruits. Li et al. (2020) used plant material of the variety 'Bojihong' and compared transcriptomes at the yellow and red stages. Differentially expressed anthocyanin-related genes were observed. Three of them (*FcCHS1*, *FcCHI1*, and *FcDFR1*) and two genes from the R2R3 MYB family of transcription factors (*FcMYB21* and *FcMYB123*) were additionally characterized after cloning. The regulatory role in the anthocyanin accumulation of the last two genes was confirmed with transient expression in apple fruits and calli. In addition, a full-length MADS-box gene, *FcMADS9*, significantly up-regulated during the pigmentation process, was identified (Li et al., 2021).

A study performed by Wang et al. (2019) focused on the anthocyanins content and gene expression in fruit peel and female flower tissue of a variety 'Zibao' at different time points. In addition, treatment with light deprivation was included, and anthocyanin content in the peel was repressed (suppression of anthocyanin synthesis was also observed on a gene expression level) compared to the control, confirming the statement that fruit coloration can be firmly and differentially affected by light. Overall, transcriptome analysis revealed differentially expressed pathways, specific structural anthocyanin-biosynthesis genes, and several potential transcription factors, particularly MYBs. Promoter sequences of *FcCHS1* and *FcDFR1* were cloned, and elements related to light response were identified in their promoter regions. G-box, L-box, Box 4, and GTGA-box were found in the promoter region of *FcCHS1*, indicating that *FcCHS1* could be regulated by light-induced signal-transduction elements such as HY5. On the other hand, H-box, I-box, and G-box elements, required for binding MYB transcription factors, were identified in the *FcDFR1* promoter. The importance of *FcHY5* and *FcMYB114* in the expression of

FcCHS1 and *FcDFR1* was confirmed with a yeast one-hybrid screen. Recently transcriptome analysis was performed to compare gene expression before and after fruit bagging of the same variety (Wang et al., 2021a).

Cao et al. (2016) studied anthocyanidin synthase (ANS), a key enzyme in anthocyanin biosynthesis cloned from syconia (*FcANS1*). They performed a structural analysis of the gene and monitored its expression in different fruit tissues. Their work provided an essential basis for further studies on anthocyanin accumulation and regulation in fruit. Additionally, it was observed that ABA could initiate anthocyanin biosynthesis, which can improve the color and nutritional value of fig fruit (Lama et al., 2020).

The first gene of the carotenoid pathway, lycopene beta-cyclase, which is responsible for converting lycopene to beta-carotene, was identified in the common fig, and a full-length cDNA was cloned by Araya-Garay et al. (2011). In addition, Zeta carotene desaturase, involved in carotene desaturation, was also cloned from the common fig (Araya-Garay et al., 2012).

4.2.3 Physiological Differences Between Common Fig Types

With the aim to reveal changes in fig fruit physiology of different fig types, Ikegami et al. (2013a) sequenced the transcriptome of caprifig (Caprifig 6085) and a common type variety 'Houraishi'. Among the differentially expressed genes, chalcone synthase (CHS) was identified, which was previously involved in the development of parthenocarpic fruit (Schijlen et al., 2007).

Fig varieties classified as San Pedro type were also studied to infer the mechanism underlying the development of the parthenocarpic breba and non-parthenocarpic main crop (Chai et al., 2017). Fruits of the 'Asteran' variety were used for gene expression studies. Gibberellin (GA)- and auxin-biosynthesis genes were repressed, and GA and ABA response in main crop female flowers, whereas ABA- and ethylene-biosynthesis genes were enhanced. The impact of gibberellin and cytokinin treatments on parthenocarpy induction of the main crop was studied on fruits from the same variety (Chai et al., 2018, 2019).

Fruit buds developed in the current season can differentiate between developing the main crop or developing the breba crop in the following year; therefore, the differences in gene expression between fruit buds differentiation in common type variety and San Pedro type variety were studied by Marcotuli et al. (2020). The two fig varieties used in the study were the common type variety 'Dottato', mainly producing main crops (rarely brebas), and 'Petrelli', belonging to the San Pedro type, which mainly produces brebas and sometimes a few main crop fruits. Variability in gene expression patterns was observed between the varieties and between the fruit buds of the breba and the main crop of the same variety. Particularly in the 'Petrelli' main crop, *flowering locus T* showed a high expression level, whereas in 'Dottato', especially in brebas, the *AGAMOUS* gene, which is required for normal development of stamens and the carpel in flower, was highly expressed compared to the 'Petrelli'. In addition, several genes related to auxin and gibberellin synthesis were also differentially expressed. A study of *FcFT1*, a *FLOWERING LOCUS T*, was previously performed by Ikegami et al. (2013b). They observed higher mRNA levels in the 'Houarishi' variety leaves at the 1st to 6th nodes than higher nodes. However, fruit-bearing and *FcFT1*-expressing nodes did not correspond completely, as *FcFT1* expression was detected even in basal nodes that usually bear no fruits in figs. An additional flowering-related gene *FLORICAULA/LEAFY* (*FcLFY*), was identified and characterized in the 'Brunswick' variety by Ma et al. (2020). The highest expression was observed in apical buds, followed by flower buds and leaf buds, which confirmed their involvement in flower bud differentiation and apical meristem formation.

4.2.4 Response to Abiotic Factors

In addition to previously mentioned studies that tackle the plant and fruit developmental biology, studies related to the physiological response to abiotic and biotic factors represent an essential part of analyses (Kim et al., 2003; Kitajima et al., 2018; Mascellani et al., 2021; Nawade et al. 2020a, b; Sadder et al., 2021; Vangelisti et al., 2019; Zhai et al., 2021).

In order to study the mechanism involved in response to salinity, Vangelisti et al. (2019) analyzed transcriptomes from leaves of the variety 'Dottato' after irrigation of the trees with 100 mM salinity concentration. As a result, already known salt-regulated genes and some genes that have not been previously reported to be involved in plant salinity stress were identified. The results can be used to develop potential markers in plant breeding programs.

The effect of salinity stress on genes involved in the synthesis, accumulation, and transport of soluble carbohydrates in ripe fruits of the variety 'Dottato' was investigated by Mascellani et al. (2021), where a general increase in the transcript levels of genes involved in the transport of soluble carbohydrates was observed.

Sadder et al. (2021) studied the differential expression of five putative salinity responsive genes in three common fig landraces ('Mwazi', 'Zraki', and 'Khdari'). Based on the combined physiological and molecular data, *FcSORD* and *FcDHN* genes were identified as potential molecular markers associated with stress tolerance response as they were able to distinguish tolerance levels.

4.2.5 Response to Biotic Factors

Plants pose different mechanisms to protect themselves against pests and diseases. Latex, which is exuded from the wounded sites of the fig plant, contains several toxic compounds. A cDNA for chitinase (*CHI*), a pathogenesis-related protein, was isolated from the fig tree latex, and it was suggested that it could be involved in the stress response caused by insect or pathogen attack (Kim et al., 2003). To infer latex-associated defense mechanisms, a comparative study of transcriptomes obtained from latex from immature fruit, young petioles, and lignified trunks was performed by Kitajima et al. (2018). Differentially expressed unigenes coding proteins with known toxicity to pests and unigenes with no previously reported anti-pest functions were identified, representing possible candidates for novel defense-related proteins or chemicals.

It was observed previously that ficins are, in addition to 20% gum, 70% water, and 1% other dry materials, a major component of fig latex (Zare et al., 2013). Ficins are a class of papain-like cysteine proteases (PLCPs) which are the most abundant family of cysteine proteases in plants, involved in abiotic and biotic stress responses, growth, and senescence. Increased abundance in receptacles of commercial-ripe fruit was observed, and genes encoding the main ficin isoforms were characterized by Zhai et al. (2021). Ficin is used in the food and pharmaceutical industry, and to optimize the production of ficin, a gene was cloned from the variety 'Sabz,' and recombinant protein was expressed in *E. coli* (Sattari et al., 2020).

Insights into fig-insect interactions (i.e., the black fig fly, *Silba adipata*) and molecular mechanisms of terpene biosynthesis related to insect-plant interaction were provided by Nawade et al. (2020a). In addition, Nawade et al. (2020b) studied the emission of apocarotenoid volatiles and identified and characterized carotenoid cleavage dioxygenase (*FcCCD*) genes involved in apocarotenoid biosynthesis. They observed significant changes in apocarotenoids emitted over the development process, which may have consequences for the interaction of fig-insect (including fig wasp).

5 Conclusions

The genetic resources of figs consist of many varieties and ecotypes that arose spontaneously under distinct ecological pressures and through the human selection of trees with attractive traits. The spontaneous hybridization of cultivated and wild types and their further propagation also contributed to today's varieties' diversity. Despite the research efforts in individual countries, a global catalog of figs has not yet been prepared. Therefore, it is necessary to emphasize that comparison of fig diversity from distinct countries is a difficult task, and the differences may be reflected in: (1) different forms of fig under investigation (cultivated vs. wild), (2) the number of analyzed samples, (3) selection of DNA markers and protocols, and (4) history, practices, and traditions of fig cultivation. Therefore, establishing a consortium of the fig research community with more target-oriented and coordinated activities regarding the management of global common fig genetic resources is necessary. Such collaborative organization would contribute to faster progress and novel achievements in global fig genetic resources.

In general, fig genetic resources are rich, although different levels of genetic diversity of cultivated figs and caprifigs have been reported in fig-producing countries. The genetic structure of figs has evolved differently over millennia of cultivation and reflects human activities, including selection and propagation processes, the pressure of ecological factors, and, ultimately, the introduction of varieties from other countries in a defined region. As a result, genetic diversity studies do not necessarily show a clear distinction or similarity of varieties according to geographical origin. The lack of standardized protocol and a recommendation list of highly informative DNA markers appropriate for fig genetic resources evaluation also prevents direct comparison of fig genetic diversity and the discovery of international synonyms and homonyms.

SSRs were the most frequently used in fig diversity studies among all available DNA markers. A review of the fig studies involving microsatellite markers has shown that individual countries have used different combinations of microsatellites, and in some studies, less informative loci were employed. In these cases, the choice of loci certainly influenced the assessment of the degree of diversity of figs. The latter suggests that a standardized list of the most informative microsatellite markers for genetic studies of figs should be prepared. Uniform protocol and a standardized list of microsatellites would better compare fig genetic resources among fig producing regions around the world and help address international synonyms of varieties. The choice of the most appropriate markers is an important step in identifying varieties. It affects the potential for discrimination of fig varieties, the rapidity and extent of molecular analysis, and, consequently, the cost of the procedure. Microsatellites are still the choice marker for variety identification, mainly due to their high reproducibility and unambiguously readable allelic patterns. Among all available microsatellites, choosing loci with the highest PIC and the lowest PI values from the published scientific reports is recommended.

Some studies have reported that the discrimination power of a defined DNA marker system could be different concerning the genetic background of genotypes. When closely related individuals such as clones are under investigation, finding differences among them is higher with DNA markers that, in a single assay, cover a larger part of the genome (for example, AFLPs). However, the complexity of interpretation of banding patterns makes these dominant markers a less-preferred tool for variety identification. With the increasing availability of fig genome and transcriptome sequences, SNPs will undoubtedly contribute to identifying varieties and genetically close-related genotypes by identifying novel alleles. A few studies also reported that with the combination of different DNA marker systems, higher discrimination power was obtained compared to a single DNA marker system used. Despite the popularity of DNA markers in genetic studies, phenotyping with detailed morphological description remains the essential and 'must' step in identifying and characterization fig genetic resources.

The availability of high-throughput sequencing technologies has contributed to the generation of new knowledge related to the biological process involved in plant growth, stress response, development of different fig fruit types, buds differentiation, and fruit ripening. A relatively small genome has also made the common fig attractive as a model organism for studies focused on basic research. The high number of genomic and transcriptomic studies performed in recent years and the results will contribute to the more efficient development of new varieties either with conventional breeding techniques or with techniques based on recombinant DNA technology (e.g., CRISPR/Cas9) as first attempts have already been performed.

References

Abou-Ellail, M., Sherin, A., Mahfouze, S. A., El-Enany, M. A. M. M., & Mustafa, N. S. A. (2014). Using biochemical and simple sequence repeats (SSR) markers to characterize (*Ficus carica* L.) cultivars. *World Applied Sciences Journal, 29*(3), 313–321. https://doi.org/10.5829/idosi.wasj.2014.29.03.13835

Achtak, H., Oukabli, A., Ater, M., Santoni, S., Kjellberg, F., & Khadari, B. (2009). Microsatellite markers as reliable tools for fig cultivar identification. *Journal of the American Society for Horticultural Science, 134*(6), 624–631. https://doi.org/10.21273/JASHS.134.6.624

Achtak, H., Ater, M., Oukabli, A., Santoni, S., Kjellberg, F., & Khadari, B. (2010). Traditional agroecosystems as conservatories and incubators of cultivated plant varietal diversity: The case of fig (*Ficus carica* L.) in Morocco. *BMC Plant Biology, 10*(1), 28. https://doi.org/10.1186/1471-2229-10-28

Adhikari, S., Saha, S., Biswas, A., Rana, T. S., Bandyopadhyay, T. K., & Ghosh, P. (2017). Application of molecular markers in plant genome analysis: A review. *Nucleus, 60*, 283–297. https://doi.org/10.1007/s13237-017-0214-7

Ali-Shtayeh, M., Jamous, R., Abu Zaitoun, S., Mallah, O., & Mubaslat, A. (2014). Genetic diversity of the palestinian fig (*Ficus carica* L.) collection by pomological traits and RAPD markers. *American Journal of Plant Sciences, 5*, 1139–1155. https://doi.org/10.4236/ajps.2014.59127

Almajali, D., Abdel-Ghani, A. H., & Migdadi, H. (2012). Evaluation of genetic diversity among Jordanian fig germplasm accessions by morphological traits and ISSR markers. *Scientia Horticulturae, 147*, 8–19. https://doi.org/10.1016/j.scienta.2012.08.029

Aradhya, M. K., Stover, E., Velasco, D., & Koehmstedt, A. (2010). Genetic structure and differentiation in cultivated fig (*Ficus carica* L.). *Genetica, 138*, 681–694. https://doi.org/10.1007/s10709-010-9442-3

Araya-Garay, J. M., Feijoo-Siota, L., Veiga-Crespo, P., & Villa, T. G. (2011). CDNA cloning of a novel gene codifying for the enzyme lycopene β-cyclase from *Ficus carica* and its expression in Escherichia coli. *Applied Microbiology and Biotechnology, 92*(4), 769–777. https://doi.org/10.1007/s00253-011-3488-8

Araya-Garay, J. M., Feijoo-Siota, L., Veiga-Crespo, P., Sanchez-Perez, A., & Gonzalez Villa, T. (2012). Cloning and functional expression of Z-carotene desaturase, a novel carotenoid biosynthesis gene from *Ficus carica*. *Journal of Bioprocessing & Biotechniques, 02*(04). https://doi.org/10.4172/2155-9821.1000125

Arvaniti, O. S., Samaras, Y., Gatidou, G., Thomaidis, N. S., & Stasinakis, A. S. (2019). Review on fresh and dried figs: Chemical analysis and occurrence of phytochemical compounds, antioxidant capacity and health effects. *Food Research International, 119*, 244–267. https://doi.org/10.1016/j.foodres.2019.01.055

Badgujar, S. B., Patel, V. V., Bandivdekar, A. H., & Mahajan, R. T. (2014). Traditional uses, phytochemistry and pharmacology of *Ficus carica*: A review. *Pharmaceutical Biology, 52*(11), 1487–1503. https://doi.org/10.3109/13880209.2014.892515

Balas, F. C., Osuna, M. D., Domínguez, G., Pérez-Gragera, F., & Corrales, M. L. (2014). Ex situ conservation of underutilised fruit tree species: Establishment of a core collection for *Ficus carica* L. using microsatellite markers (SSRs). *Tree Genetics & Genomes, 10*, 703–710. https://doi.org/10.1007/s11295-014-0715-3

Bandelj, D., Javornik, B., & Jakše, J. (2007). Development of microsatellite markers in the common fig, *Ficus carica* L. *Molecular Ecology Notes, 7*, 1311–1314. https://doi.org/10.1111/j.1471-8286.2007.01866.x

Bandelj, D., Jakše, J., & Javornik, B. (2008). Development of molecular markers for identification of fig varieties in Istria. In D. Bandelj, M. Bučar-Miklavčič, & I. Vrhovnik (Eds.), *The common fig (Ficus carica L.) in Istria: Morphological, molecular and some chemical characteristics* (pp. 84–89). Annales.

Baraket, G., Saddoud, O., Chatti, K., Mars, M., Marrakchi, M., Trifi, M., & Salhi-Hannachi, A. (2009). Sequence analysis of the internal transcribed spacers (ITSs) region of the nuclear ribosomal DNA (nrDNA) in fig cultivars (*Ficus carica* L.). *Scientia Horticulturae, 120*(1), 34–40. https://doi.org/10.1016/j.scienta.2008.09.013

Baraket, G., Chatti, K., Saddoud, O., Abdelkarim, A. B., Mars, M., Trifi, M., & Hannachi, A. S. (2011). Comparative assessment of SSR and AFLP markers for evaluation of genetic diversity and conservation of fig, *Ficus carica* L., genetic resources in Tunisia. *Plant Molecular Biology Reporter, 29*, 171–184. https://doi.org/10.1007/s11105-010-0217-x

Barghini, E., Mascagni, F., Giordani, T., Solorzano Zambrano, L. J., Natali, L., & Cavallini, A. (2017). An insight into structure and composition of the fig genome. *Acta Horticulturae, 1173*, 69–74. https://doi.org/10.17660/ActaHortic.2017.1173.12

Baruca Arbeiter, A., Hladnik, M., Jakše, J., & Bandelj, D. (2017). Identification and validation of novel EST-SSR markers in olives. *Scientia Agricola, 74*, 215–225. https://doi.org/10.1590/1678-992x-2016-0111

Baruca Arbeiter, A., Hladnik, M., Jakše, J., & Bandelj, D. (2021). First set of microsatellite markers for immortelle (*Helichrysum italicum* (Roth) G. Don): A step towards the selection of the most promising genotypes for cultivation. *Industrial Crops and Products, 162*, 113298. https://doi.org/10.1016/j.indcrop.2021.113298

Beck, N. G., & Lord, E. M. (1988). Breeding system in *Ficus carica*, the common fig. I. Floral diversity. *American Journal of Botany, 75*, 1904–1912. https://doi.org/10.1002/j.1537-2197.1988.tb11272.x

Ben Abdelkrim, A., Baraket, G., Essalouh, L., Achtak, H., Khadari, B., & Salhi-Hannachi, A. (2015). Use of morphological traits and microsatellite markers to characterize the Tunisian cultivated and wild figs (*Ficus carica* L.). *Biochemical Systematics and Ecology, 59*. https://doi.org/10.1016/j.bse.2015.01.026

Bohanec, B. (2008). Characteristic of fig breeding and biotechnological techniques of propagation. In D. Bandelj, M. Bučar-Miklavčič, & I. Vrhovnik (Eds.), *The common fig (Ficus carica L.) in Istria: Morphological, molecular and some chemical characteristics* (pp. 76–83). Annales.

Boudchicha, R. H., Hormaza, J. I., & Benbouza, H. (2018). Diversity analysis and genetic relationships among local Algerian fig cultivars (*Ficus carica* L.) using SSR markers. *South African Journal of Botany, 116*, 207–215. https://doi.org/10.1016/j.sajb.2018.03.015

Cabrita, L. F., Aksoy, U., Hepaksoy, S., & Leitão, J. M. (2001). Suitability of isozyme, RAPD and AFLP markers to assess genetic differences and relatedness among fig (*Ficus carica* L.) clones. *Scientia Horticulturae, 87*(4), 261–273. https://doi.org/10.1016/S0304-4238(00)00181-3

Çalişkan, O., & Polat, A. A. (2012a). Morphological diversity among fig (*Ficus carica* L.) accessions sampled from the Eastern Mediterranean Region of Turkey. *Turkish Journal of Agriculture and Forestry, 36*, 179–193. https://doi.org/10.3906/tar-1102-33

Çalişkan, O., & Polat, A. A. (2012b). Effects of genotype and harvest year on phytochemical and fruit quality properties of Turkish fig genotypes. *Spanish Journal of Agricultural Research, 10*(4), 1048–1058. https://doi.org/10.5424/sjar/2012104-2652

Çalişkan, O., Bayazit, S., Ilgin, M., & Karatas, N. (2017). Morphological diversity of caprifig (*Ficus carica* var. *caprificus*) accessions in the eastern Mediterranean region of Turkey: Potential utility for caprification. *Scientia Horticulturae, 222*, 46–56. https://doi.org/10.1016/j.scienta.2017.05.008

Çalişkan, O., Bayazit, S., Ilgin, M., Karatas, N., & Ergül, A. (2018). Genetic diversity and population structure in caprifigs (*Ficus carica* var. *caprificus*) using SSR markers. *Spanish Journal of Agricultural Research, 16*(3), e0703. https://doi.org/10.5424/sjar/2018163-11662

Cao, L., Xu, X., Chen, S., & Ma, H. (2016). Cloning and expression analysis of Ficus carica anthocyanidin synthase 1 gene. *Scientia Horticulturae, 211*, 369–375. https://doi.org/10.1016/j.scienta.2016.09.015

Castro, C., Hernández, A., Alvarado, L., & Flores, D. (2015). DNA barcodes in fig cultivars (*Ficus carica* L.) using ITS regions of ribosomal DNA, the psbA-trnH spacer and the matK coding sequence. *American Journal of Plant Sciences, 6*, 95–102.

Chai, L., Wang, Z., Chai, P., Chen, S., & Ma, H. (2017). Transcriptome analysis of San Pedro-type fig (*Ficus carica* L.) parthenocarpic breba and non-parthenocarpic main crop reveals divergent phytohormone-related gene expression. *Tree Genetics & Genomes, 13*(4), 1–14. https://doi.org/10.1007/S11295-017-1166-4

Chai, L., Chai, P., Chen, S., Flaishman, M. A., & Ma, H. (2018). Transcriptome analysis unravels spatiotemporal modulation of phytohormone-pathway expression underlying gibberellin-induced parthenocarpic fruit set in San Pedro-type fig (*Ficus carica* L.). *BMC Plant Biology, 18*(1), 1–15. https://doi.org/10.1186/s12870-018-1318-1

Chai, P., Dong, S., Chai, L., Chen, S., Flaishman, M., & Ma, H. (2019). Cytokinin-induced parthenocarpy of San Pedro type fig (*Ficus carica* L.) main crop: Explained by phytohormone assay and transcriptomic network comparison. *Plant Molecular Biology, 99*(4-5), 329–346. https://doi.org/10.1007/s11103-019-00820-2

Chatti, K., Saddoud, O., Salhi-Hannachi, A., Marrakchi, M., Trifi, M., & Mars, M. (2008). RAMPO: A powerful molecular marker in fig (*Ficus carica*) diversity analysis. *Acta Horticulturae, 798*, 183–189. https://doi.org/10.17660/ActaHortic.2008.798.25

Chatti, K., Baraket, G., Abdelkrim, A. B., Saddoud, O., Mars, M., Trifi, M., & Hannachi, A. S. (2010). Development of molecular tools for characterization and genetic diversity analysis in Tunisian fig (*Ficus carica*) cultivars. *Biochemical Genetics, 48*, 789–806. https://doi.org/10.1007/s10528-010-9360-1

Cheng, T., Xu, C., Lei, L., Li, C., Zhang, Y., & Zhou, S. (2016). Barcoding the kingdom Plantae: New PCR primers for ITS regions of plants with improved universality and specificity. *Molecular Ecology Resources, 16*, 138–149. https://doi.org/10.1111/1755-0998.12438

Condit, I. (1955). Fig varieties: A monograph. *Hilgardia, 23*(11), 323–538. https://doi.org/10.3733/hilg.v23n11p323

Costa, F., Marchese, A., Mafrica, R., Di Vaio, C., Ferrara, G., Fretto, S., Quartararo, A., Marra, F. P., Mennone, C., Vitale, F., Reale, S., & Caruso, T. (2017). Genetic diversity of fig (*Ficus carica* L.) genotypes grown in Southern Italy revealed by the use of SSR markers. *Acta Horticulturae., 1173*, 75–80. https://doi.org/10.17660/ActaHortic.2017.1173.13

Cui, Y., Wang, Z., Chen, S., Vainstein, A., & Ma, H. (2019). Proteome and transcriptome analyses reveal key molecular differences between quality parameters of commercial-ripe and tree-ripe fig (*Ficus carica* L.). *BMC Plant Biology, 19*(1), 1–16. https://doi.org/10.1186/s12870-019-1742-x

Cui, Y., Zhai, Y., Flaishman, M., Li, J., Chen, S., Zheng, C., & Ma, H. (2020). Ethephon induces coordinated ripening acceleration and divergent coloration responses in fig (*Ficus carica* L.) flowers and receptacles. *Plant Molecular Biology, 105*(4), 347–364. https://doi.org/10.1007/S11103-020-01092-X

do Val, A. D. B., Souza, C. S., Ferreira, E. A., Salgado, S. M., Pasqual, M., & Cançado, G. M. (2013). Evaluation of genetic diversity in fig accessions by using microsatellite markers. *Genetic and Molecular Researches, 12*(2), 1383–1391. https://doi.org/10.4238/2013.April.25.9

Essid, A., Aljane, F., Ferchichi, A., & Hormaza, J. I. (2015). Analysis of genetic diversity of Tunisian caprifig (*Ficus carica* L.) accessions using simple sequence repeat (SSR) markers. *Hereditas, 152*, 1. https://doi.org/10.1186/s41065-015-0002-9

Essid, A., Aljane, F., Neily, M. H., Ferchichi, A., & Hormaza, J. I. (2021). Assessment of genetic diversity of thirty Tunisian fig (*Ficus carica* L.) accessions using pomological traits and SSR markers. *Molecular Biology Reports, 48*, 335–346. https://doi.org/10.1007/s11033-020-06051-9

Fatahi, S., Cheghamirza, K., Arji, I., & Zarei, L. (2017). Evaluation of genetic variation of common fig (*Ficus carica* L.) in West of Iran. *Journal of medicinal plants and by-products, 6*, 229–240.

Feng, G., Sanderson, B. J., Keefover-Ring, K., Liu, J., Ma, T., Yin, T., Smart, L. B., DiFazio, S. P., & Olson, M. S. (2020). Pathways to sex determination in plants: How many roads lead to Rome? *Current Opinion in Plant Biology, 54*, 61–68. https://doi.org/10.1016/j.pbi.2020.01.004

Flaishman, M. A., Rodov, V., & Stover, E. (2008). The fig: Botany, horticulture and breeding. In J. Janick (Ed.), *Horticultural Reviews* (pp. 113–197). Wiley. https://doi.org/10.1002/9780470380147.ch2

Flaishman, M. A., Peer, R., Raz, A., Cohen, O., Izhaki, K., Bocobza, S., Lama, K., Pliner, M., & Levy, A. (2020). Advanced molecular tools for breeding in Mediterranean fruit trees: Genome editing approach of *Ficus carica* L. *Acta Horticulturae, 1280*, 1–10. https://doi.org/10.17660/ACTAHORTIC.2020.1280.1

Freiman, Z. E., Doron-Faigenboim, A., Dasmohapatra, R., Yablovitz, Z., & Flaishman, M. A. (2014). High-throughput sequencing analysis of common fig (*Ficus carica* L.) transcriptome during fruit ripening. *Tree Genetics & Genomes, 10*(4), 923–935. https://doi.org/10.1007/S11295-014-0732-2

Freiman, Z. E., Rosianskey, Y., Dasmohapatra, R., Kamara, I., & Flaishman, M. A. (2015). The ambiguous ripening nature of the fig (*Ficus carica* L.) fruit: A gene-expression study of potential ripening regulators and ethylene-related genes. *Journal of Experimental Botany, 66*(11), 3309–3324. https://doi.org/10.1093/jxb/erv140

Galil, J., & Neeman, G. (1977). Pollen transfer and pollination in the common fig (*Ficus carica* L.). *New Phytologist, 79*(1), 163–171. https://www.jstor.org/stable/2433728

Ganopoulos, I., Xanthopoulou, A., Molassiotis, A., Karagiannis, E., Moysiadis, T., Katsaris, P., Aravanopoulos, F., Tsaftaris, A., Kalivas, A., & Madesis, P. (2015). Mediterranean basin *Ficus carica* L.: From genetic diversity and structure to authentication of a Protected Designation of Origin cultivar using microsatellite markers. *Trees, 29*, 1959–1971. https://doi.org/10.1007/s00468-015-1276-2

Giraldo, E., Viruel, M., López-Corrales, M., & Hormaza, J. I. (2005). Characterisation and cross-species transferability of microsatellites in the common fig (*Ficus carica* L.). *Journal of Horticultural Science and Biotechnology, 80*, 217–224. https://doi.org/10.1080/14620316.2005.11511920

Giraldo, E., Lopez-Corrales, M., & Hormaza, J. I. (2008). Optimization of the management of an ex-situ germplasm bank in common fig with SSRs. *Journal of American Society for Horticultural Science, 133*, 69–77. https://doi.org/10.21273/JASHS.133.1.69

Giraldo, E., López-Corrales, M., & Hormaza, J. I. (2010). Selection of the most discriminating morphological qualitative variables for characterization of fig germplasm. *Journal of the American Society for Horticultural Science, 135*(3), 240–249. https://doi.org/10.21273/JASHS.135.3.240

GRIN. (2021). Available at: https://npgsweb.ars-grin.gov/gringlobal/taxon/taxonomydetail?id=16801

Hladnik, M., Jakše, J., Khadari, B., Santoni, S., & Bandelj, D. (2018). Interlaboratory comparison of fig (*Ficus carica* L.) microsatellite genotyping data and determination of reference alleles. *Acta Agriculturae Slovenica, 111*(1), 143–159. https://doi.org/10.14720/aas.2018.111.1.14

Hmimsa, Y., Aumeeruddy-Thomas, Y., & Ater, M. (2012). Vernacular taxonomy, classification and varietal diversity of fig (*Ficus carica* L.) among Jbala cultivators in Northern Morocco. *Human Ecology, 40*(2), 301–313. https://doi.org/10.1007/s10745-012-9471-x

Hssaini, L., Charafi, J., Razouk, R., Hernández, F., Fauconnier, M.-L., Ennahli, S., & Hanine, H. (2020a). Assessment of morphological traits and fruit metabolites in eleven fig varieties (*Ficus carica* L.). *International Journal of Fruit Science, 20*(2), 8–28. https://doi.org/10.1080/15538362.2019.1701615

Hssaini, L., Hanine, H., Razouk, R., Ennahli, S., Mekaoui, A., Ejjilani, A., & Charafi, J. (2020b). Assessment of genetic diversity in Moroccan fig (*Ficus carica* L.) collection by combining morphological and physicochemical descriptors. *Genetic Resources and Crop Evolution, 67*, 457–474. https://doi.org/10.1007/s10722-019-00838-x

Hu, T., Chitnis, N., Monos, D., & Dinh, A. (2021). Next-generation sequencing technologies: An overview. *Human Immunology, 82*(11), 801–811. https://doi.org/10.1016/j.humimm.2021.02.012

Ikegami, H., Nogata, H., Hirashima, K., Awamura, M., & Nakahara, T. (2009). Analysis of genetic diversity among European and Asian fig varieties (*Ficus carica* L.) using ISSR, RAPD, and SSR markers. *Genetic Resources and Crop Evolution, 56*, 201–209. https://doi.org/10.1007/s10722-008-9355-5

Ikegami, H., Habu, T., Mori, K., Nogata, H., Hirata, C., Hirashima, K., Tashiro, K., & Kuhara, S. (2013a). De novo sequencing and comparative analysis of expressed sequence tags from gynodioecious fig (*Ficus carica* L.) fruits: Caprifig and common fig. *Tree Genetics & Genomes, 9*(4), 1075–1088. https://doi.org/10.1007/S11295-013-0622-Z

Ikegami, H., Nogata, H., Inoue, Y., Himeno, S., Yakushiji, H., Hirata, C., Hirashima, K., Mori, M., Awamura, M., & Nakahara, T. (2013b). Expression of FcFT1, a flowering locus T-like gene, is regulated by light and associated with inflorescence differentiation in fig (*Ficus carica* L.). *BMC Plant Biology, 13*(1), 1–12. https://doi.org/10.1186/1471-2229-13-216

Ikegami, H., Shirasawa, K., Yakushiji, H., Yabe, S., Sato, M., Hayashi, T., Tashiro, K., & Nogata, H. (2021). Analysis of the segregation distortion of FcRAN1 genotypes based on whole-genome resequencing of fig (*Ficus carica* L.) breeding parents. *Frontiers in Plant Science, 12*, 1–8. https://doi.org/10.3389/fpls.2021.647599

Ikten, H., Mutlu, N., Gulsen, O., Kocatas, H., & Aksoy, U. (2010). Elucidating genetic relationships, diversity and population structure among the Turkish female figs. *Genetica, 138*, 169–177. https://doi.org/10.1007/s10709-009-9400-0

IPGRI & CIHEAM. (2003). *Descriptors for fig*. International Plant Genetic Resources Institute/International Centre for Advanced Mediterranean Agronomic Studies. Available at: https://www.bioversityinternational.org/fileadmin/_migrated/uploads/tx_news/Descriptors_for_fig__Ficus_carica__907.pdf

Kalia, R. K., Rai, M. K., Kali, S., Singh, R., & Dhawan, A. K. (2011). Microsatellite markers: An overview of the recent progress in plants. *Euphytica, 177*, 309–334. https://doi.org/10.1007/s10681-010-0286-9

KEW. (2021). Available at: http://plantsoftheworldonline.org/taxon/urn:lsid:ipni.org:names:852556-1

Khadari, B., Hochu, I., Santoni, S., & Kjellberg, F. (2001). Identification and characterization of microsatellite loci in the common fig (*Ficus carica* L.) and representative species of the genus Ficus. *Molecular Ecology Notes, 1*, 191–193. https://doi.org/10.1046/j.1471-8278.2001.00072.x

Khadari, B., Grout, C., Santoni, S., & Kjellberg, F. (2005). Contrasted genetic diversity and differentiation among Mediterranean populations of *Ficus carica* L.: A study using mtDNA RFLP. *Genetic Resources and Crop Evolution, 52*, 97–109. https://doi.org/10.1007/s10722-005-0290-4

Kim, J. S., Kim, Y. O., Ryu, H. J., Kwak, Y. S., Lee, J. Y., & Kang, H. (2003). Isolation of stress-related genes of rubber particles and latex in fig tree (*Ficus carica*) and their expressions by abiotic stress or plant hormone treatments. *Plant and Cell Physiology, 44*(4), 412–414. https://doi.org/10.1093/PCP/PCG058

Kislev, M. E., Hartmann, A., & Bar-Yosef, O. (2006). Early domesticated fig in the Jordan Valley. *Science, 312*, 1372–1374. https://doi.org/10.1126/science.1125910

Kitajima, S., Aoki, W., Shibata, D., Nakajima, D., Sakurai, N., Yazaki, K., Munakata, R., Taira, T., Kobayashi, M., Aburaya, S., Savadogo, E. H., Hibino, S., & Yano, H. (2018). Comparative multi-omics analysis reveals diverse latex-based defense strategies against pests among latex-producing organs of the fig tree (*Ficus carica*). *Planta, 247*(6), 1423–1438. https://doi.org/10.1007/S00425-018-2880-3

Kjellberg, F., Gouyon, P. H., Ibrahim, M., Raymond, M., & Valdeyron, G. (1987). The stability of the symbiosis between dioecious figs and their pollinators: A study of *Ficus carica* L. and

Blastophaga psenes L. *Evolution, 41*(4), 693–704. https://doi.org/10.1111/j.1558-5646.1987.tb05846.x

Kjellberg, F., Jousselin, E., Hossaert-McKey, M., & Rasplus, J.-Y. (2005). Biology, ecology and evolution of fig-pollinating wasps (Chalcidoidea, Agaonidae). In A. Raman, C. W. Schaefer, & T. M. Withers (Eds.), *Biology, ecology, and evolution of gall-inducing arthropods* (pp. 539–572). Science Publishers.

Knap, T., Jakše, J., Cregeen, S., Javornik, B., Hladnik, M., & Bandelj, D. (2016). Characterization and defining of a core set of novel microsatellite markers for use in genotyping and diversity study of Adriatic fig (*Ficus carica* L.) germplasm. *Brazilian Journal of Botany, 39*, 1095–1102. https://doi.org/10.1007/s40415-016-0299-2

Knap, T., Baruca Arbeiter, A., Jakše, J., Čizmović, M., Adakalić, M., Popović, R., Lazović, B., Strikić, F., Podgornik, M., & Bandelj, D. (2017). Diversity of figs (*Ficus carica* L.) from the east Adriatic coast. *Acta Horticulturae, 1173*, 11–16. https://doi.org/10.17660/ActaHortic.2017.1173.2

Knap, T., Aradhya, M., Baruca Arbeiter, A., Hladnik, M., & Bandelj, D. (2018). DNA profiling of figs (*Ficus carica* L.) from Slovenia and Californian USDA collection revealed the uniqueness of some North Adriatic varieties. *Genetic Resources and Crop Evolution, 65*(5), 1503–1516. https://doi.org/10.1007/s10722-018-0634-5

Kucukerdonmez, O., Baygeldi, N., Akder, R. N., & Özlem, Ç. (2021). Medicinal and nutritional analysis of fig (*Ficus carica*) seed oil: A New gamma tocopherol and omega-3 source. *Progress in Nutrition, 23*(2), e2021052. https://doi.org/10.23751/pn.v23i2.9980

Kumar, P., Gupta, V., Misra, A., Modi, D., & Pandey, B. (2009). Potential of molecular markers in plant biotechnology. *Plant Omics, 2*(4), 141–162. https://search.informit.org/doi/10.3316/informit.090706285698938

Lama, K., Modi, A., Peer, R., Izhaki, Y., & Flaishman, M. A. (2019). On-tree ABA application synchronizes fruit ripening and maintains keeping quality of figs (*Ficus carica* L.). *Scientia Horticulturae, 253*, 405–411. https://doi.org/10.1016/j.scienta.2019.04.063

Lama, K., Harlev, G., Shafran, H., Peer, R., & Flaishman, M. A. (2020). Anthocyanin accumulation is initiated by abscisic acid to enhance fruit color during fig (*Ficus carica* L.) ripening. *Journal of Plant Physiology, 251*(May), 153192. https://doi.org/10.1016/j.jplph.2020.153192

Lansky, E. P., & Paavilainen, H. M. (2011). *Figs: The genus Ficus*. CRC Press. https://doi.org/10.1201/9781420089677

Li, J., An, Y., & Wang, L. (2020). Transcriptomic analysis of *Ficus carica* peels with a focus on the key genes for anthocyanin biosynthesis. *International Journal of Molecular Sciences, 21*(4). https://doi.org/10.3390/ijms21041245

Li, J., Ma, N., An, Y., & Wang, L. (2021). FcMADS9 of fig regulates anthocyanin biosynthesis. *Scientia Horticulturae, 278*. https://doi.org/10.1016/j.scienta.2020.109820

Lianju, W., Weibin, J., Kai, M., Zhifeng, L., & Yelin, W. (2003). The production and research of fig (*Ficus carica* L.) in China. *Acta Horticulturae, 605*, 191–196. https://doi.org/10.17660/ActaHortic.2003.605.28

Loureiro, J., Rodriguez, E., Doležel, J., & Santos, C. (2007). Two new nuclear isolation buffers for plant DNA flow cytometry: A test with 37 species. *Annals of Botany, 100*(4), 875. https://doi.org/10.1093/AOB/MCM152

Ma, N., An, Y., Li, J., & Wang, L. (2020). Cloning and characterization of a homologue of the FLORICAULA/LEAFY gene in *Ficus carica* L., FcLFY, and its role in flower bud differentiation. *Scientia Horticulturae, 261*, 1–10. https://doi.org/10.1016/j.scienta.2019.109014

Marcotuli, I., Mazzeo, A., Colasuonno, P., Terzano, R., Nigro, D., Porfido, C., Tarantino, A., Aiese Cigliano, R., Sanseverino, W., Gadaleta, A., & Ferrara, G. (2020). Fruit development in *Ficus carica* L.: Morphological and genetic approaches to fig buds for an evolution from monoecy toward dioecy. *Frontiers in Plant Science, 0*, 1208. https://doi.org/10.3389/FPLS.2020.01208

Mars, M., Trad, M., & Gaaliche, B. (2017). The unique fig caprification system and its effects on productivity and fruit characteristics. *Acta Horticulturae, 1173*, 127–136. https://doi.org/10.17660/ActaHortic.2017.1173.22

Mascellani, A., Natali, L., Cavallini, A., Mascagni, F., Caruso, G., Gucci, R., Havlik, J., & Bernardi, R. (2021). Moderate salinity stress affects expression of main sugar metabolism and transport genes and soluble carbohydrate content in ripe fig fruits (*Ficus carica* L. cv. dottato). *Plants, 10*(9). https://doi.org/10.3390/plants10091861

Mori, K., Shirasawa, K., Nogata, H., Hirata, C., Tashiro, K., Habu, T., Kim, S., Himeno, S., Kuhara, S., & Ikegami, H. (2017). Identification of RAN1 orthologue associated with sex determination through whole genome sequencing analysis in fig (*Ficus carica* L.). *Scientific Reports, 7*(1), 1–15. https://doi.org/10.1038/srep41124

Nawade, B., Shaltiel-Harpaz, L., Yahyaa, M., Kabaha, A., Kedoshim, R., Bosamia, T. C., & Ibdah, M. (2020a). Characterization of terpene synthase genes potentially involved in black fig fly (*Silba adipata*) interactions with *Ficus carica*. *Plant Science, 298*, 110549. https://doi.org/10.1016/j.plantsci.2020.110549

Nawade, B., Shaltiel-Harpaz, L., Yahyaa, M., Bosamia, T. C., Kabaha, A., Kedoshim, R., Zohar, M., Isaacson, T., & Ibdah, M. (2020b). Analysis of apocarotenoid volatiles during the development of Ficus carica fruits and characterization of carotenoid cleavage dioxygenase genes. *Plant Science, 290*, 110292. https://doi.org/10.1016/j.plantsci.2019.110292

Oliveira, A. P., Baptista, P., Andrade, P. B., Martins, F., Pereira, J. A., Silva, B. M., & Valentão, P. (2012). Characterization of Ficus carica L. cultivars by DNA and secondary metabolite analysis: Is genetic diversity reflected in the chemical composition? *Food Research International, 49*(2), 710–719. https://doi.org/10.1016/j.foodres.2012.09.019

Owino, W. O., Nakano, R., Kubo, Y., & Inaba, A. (2004). Coordinated expression patterns of genes encoding cell wall modifying enzymes during ripening in distinct anatomical tissue regions of the fig (*Ficus carica* L.) fruit. *Postharvest Biology and Technology, 32*(3), 253–261. https://doi.org/10.1016/j.postharvbio.2004.01.003

Owino, W. O., Manabe, Y., Mathooko, F. M., Kubo, Y., & Inaba, A. (2006). Regulatory mechanisms of ethylene biosynthesis in response to various stimuli during maturation and ripening in fig fruit (*Ficus carica* L.). *Plant Physiology and Biochemistry, 44*(5-6), 335–342. https://doi.org/10.1016/j.plaphy.2006.03.009

Ozgenturk, N. O., Oruc, F., Sezerman, U., Kucukural, A., Korkut, S. V., Toksoz, F., & Cemal, U. (2010). Generation and analysis of expressed sequence tags from Olea europaea L. *Comparative and Functional Genomics, 2010*, 757512. https://doi.org/10.1155/2010/757512

Perez-Jiménez, M., López, B., Dorado, G., Pujadas-Salvá, A., Guzmán, G., & Hernandez, P. (2012). Analysis of genetic diversity of southern Spain fig tree (*Ficus carica* L.) and reference materials as a tool for breeding and conservation. *Hereditas, 149*, 108–113. https://doi.org/10.1111/j.1601-5223.2012.02154.x

Podgornik, M., Vuk, I., Vrhovnik, I., & Bandelj, D. (2010). A survey and morphological evaluation of fig (*Ficus carica* L.) genetic resources from Slovenia. *Scientia Horticulturaea, 125*, 380–389. https://doi.org/10.1016/j.scienta.2010.04.030

Qiao, H., Zhang, H., Wang, Z., & Shen, Y. (2021). Fig fruit ripening is regulated by the interaction between ethylene and abscisic acid. *Journal of Integrative Plant Biology, 63*(3), 553–569. https://doi.org/10.1111/jipb.13065

Riaz, A., Muhammad, A. A., Naz, S., & Balal, R. M. (2021). Applications of molecular markers in fruit crops for breeding Programs – A review. *Phyton, 90*(1), 17–34. https://doi.org/10.32604/phyton.2020.011680

Rodolfi, M., Ganino, T., Chiancone, B., & Petrucceli, R. (2018). Identification and characterization of Italian common figs (*Ficus carica*) using nuclear microsatellite markers. *Genetic Resources and Crop Evolution, 65*, 1337–1348. https://doi.org/10.1007/s10722-018-0617-6

Rodrigues, M. G. F., Monteiro, L. N. H., Ferreira, A. F. A., Santos, T. P., Pavan, B. E., Neves, V. A. B., & Boliani, A. C. (2019). Biometric characteristics among fig tree genotypes in Brazil. *Genetics and Molecular Research, 18*(2), GMR18191. https://doi.org/10.4238/gmr18191

Rosianski, Y., Doron-Faigenboim, A., Freiman, Z. E., Lama, K., Milo-Cochavi, S., Dahan, Y., Kerem, Z., & Flaishman, M. A. (2016a). Tissue-specific transcriptome and hormonal regulation of pollinated and parthenocarpic fig (*Ficus carica* L.) fruit suggest that fruit ripening is coordinated by the reproductive part of the syconium. *Frontiers in Plant Science, 7*, 1–14. https://doi.org/10.3389/fpls.2016.01696

Rosianski, Y., Freiman, Z. E., Cochavi, S. M., Yablovitz, Z., Kerem, Z., & Flaishman, M. A. (2016b). Advanced analysis of developmental and ripening characteristics of pollinated common-type fig (*Ficus carica* L.). *Scientia Horticulturae, 198*, 98–106. https://doi.org/10.1016/j.scienta.2015.11.027

Rudd, S. (2003). Expressed sequence tags: Alternative or complement to whole genome sequences? *Trends in Plant Science, 8*, 321–329. 10.1016/S1360-1385(03)00131-6.

Sadder, M. T., Alshomali, I., Ateyyeh, A., & Musallam, A. (2021). Physiological and molecular responses for long term salinity stress in common fig (*Ficus carica* L.). *Physiology and Molecular Biology of Plants, 27*(1), 107–117. https://doi.org/10.1007/s12298-020-00921-z

Saddoud, O., Chatti, K., Salhi-Hannachi, A., Mars, M., Rhouma, A., Marrakchi, M., & Trifi, M. (2007). Genetic diversity of Tunisian figs (*Ficus carica* L.) as revealed by nuclear microsatellites. *Hereditas, 144*, 149–157. https://doi.org/10.1111/j.2007.0018-0661.01967.x

Saddoud, O., Baraket, G., Chatti, K., Trifi, M., Marrakchi, M., Mars, M., & Salhi-Hannachi, A. (2011). Using morphological characters and simple sequence repeat (SSR) markers to characterize Tunisian fig (*Ficus carica* L.) cultivars. *Acta Biologica Cracoviensia Series Botanica, 53*(2), 7–14. https://doi.org/10.2478/v10182-011-0019-y

Sattari, F., Rigi, G., & Ghaedmohammadi, S. (2020). The first report on molecular cloning, functional expression, purification, and statistical optimization of Escherichia coli-derived recombinant Ficin from Iranian fig tree (*Ficus carica* cv. Sabz). *International Journal of Biological Macromolecules, 165*. https://doi.org/10.1016/j.ijbiomac.2020.10.050. Elsevier B.V.

Schijlen, E. G. W. M., De Vos, C. H. R., Martens, S., Jonker, H. H., Rosin, F. M., Molthoff, J. W., Tikunov, Y. M., Angenent, G. C., Van Tunen, A. J., & Bovy, A. G. (2007). RNA interference silencing of chalcone synthase, the first step in the flavonoid biosynthesis pathway, leads to parthenocarpic tomato fruits. *Plant Physiology, 144*(3), 1520–1530. https://doi.org/10.1104/pp.107.100305

Selkoe, K. A., & Toonen, R. J. (2006). Microsatellites for ecologists: A practical guide to using and evaluating microsatellite markers. *Ecology letters, 9*(5), 615–629. https://doi.org/10.1111/j.1461-0248.2006.00889.x

Sharma, M., & Singh, A. (2021). DNA based molecular markers in plant varietal identification: A review. *Plant Archives, 21*(1), 1973–1980.

Soltana, H., Tekaya, M., Amri, Z., El-Gharbi, S., Nakbi, A., Harzallah, A., Mechri, B., & Hammami, M. (2015). Characterization of fig achenes' oil of *Ficus carica* grown in Tunisia. *Food Chemistry, 196*, 1125–1130. https://doi.org/10.1016/j.foodchem.2015.10.053

Song, M., Wang, H., Wang, Z., Huang, H., Chen, S., & Ma, H. (2021). Genome-wide characterization and analysis of bHLH transcription factors related to anthocyanin biosynthesis in fig (*Ficus carica* L.). *Frontiers in Plant Science, 12*, 1–14. https://doi.org/10.3389/fpls.2021.730692

Stover, E., Aradhya, M., Ferguson, L., & Crisosto, C. (2007). The fig: Overview of an ancient fruit. *Journal of the American Society for Horticultural Science, 42*(5), 1083–1087. https://doi.org/10.21273/HORTSCI.42.5.1083

Usai, G., Mascagni, F., Giordani, T., Vangelisti, A., Bosi, E., Zuccolo, A., Ceccarelli, M., King, R., Hassani-Pak, K., Zambrano, L. S., Cavallini, A., & Natali, L. (2020). Epigenetic patterns within the haplotype phased fig (*Ficus carica* L.) genome. *The Plant Journal, 102*(3), 600–614. https://doi.org/10.1111/TPJ.14635

Usai, G., Vangelisti, A., Simoni, S., Giordani, T., Natali, L., Cavallini, A., & Mascagni, F. (2021). DNA modification patterns within the transposable elements of the fig (*Ficus carica* L.) genome. *Plants (Basel, Switzerland), 10*(3), 1–13. https://doi.org/10.3390/PLANTS10030451

van Dijk, E. L., Jaszczyszyn, Y., Naquin, D., & Thermes, C. (2018). The third revolution in sequencing technology. *Trends in Genetics, 34*(9), 666–681. https://doi.org/10.1016/j.tig.2018.05.008

Vangelisti, A., Zambrano, L. S., Caruso, G., Macheda, D., Bernardi, R., Usai, G., Mascagni, F., Giordani, T., Gucci, R., Cavallini, A., & Natali, L. (2019). How an ancient, salt-tolerant fruit crop, *Ficus carica* L., copes with salinity: A transcriptome analysis. *Scientific Reports, 9*(1), 1–13. https://doi.org/10.1038/s41598-019-39114-4

Vangelisti, A., Simoni, S., Usai, G., Ventimiglia, M., Natali, L., Cavallini, A., Mascagni, F., & Giordani, T. (2021). LTR-retrotransposon dynamics in common fig (*Ficus carica* L.) genome. *BMC Plant Biology, 21*(1), 1–18. https://doi.org/10.1186/S12870-021-02991-X

Vinson, J. (1999). The functional food properties of figs. *Cereal Foods World, 44*(2), 82–87.

Vos, P., Hogers, R., Bleeker, M., Reijans, M., van de Lee, T., Hornes, M., Frijters, A., Pot, J., Peleman, J., Kuiper, M., & Zabeau, M. (1995). AFLP: A new technique for DNA fingerprinting. *Nucleic Acids Research, 23*(21), 4407–4414. https://doi.org/10.1093/nar/23.21.4407

Wang, Z., Cui, Y., Vainstein, A., Chen, S., & Ma, H. (2017). Regulation of fig (*Ficus carica* L.) fruit color: Metabolomic and transcriptomic analyses of the flavonoid biosynthetic pathway. *Frontiers in Plant Science, 0*, 1990. https://doi.org/10.3389/FPLS.2017.01990

Wang, Z., Song, M., Li, Y., Chen, S., & Ma, H. (2019). Differential color development and response to light deprivation of fig (*Ficus carica* L.) syconia peel and female flower tissues: Transcriptome elucidation. *BMC Plant Biology, 19*(1), 1–15. https://doi.org/10.1186/s12870-019-1816-9

Wang, Z., Song, M., Wang, Z., Chen, S., & Ma, H. (2021a). Metabolome and transcriptome analysis of flavor components and flavonoid biosynthesis in fig female flower tissues (*Ficus carica* L.) after bagging. *BMC Plant Biology, 21*(1), 1–14. https://doi.org/10.1186/s12870-021-03169-1

Wang, X., Song, M., Flaishman, M. A., Chen, S., & Ma, H. (2021b). AGAMOUS gene as a new sex-identification marker in fig (*Ficus carica* L.) is more efficient than RAN1. *Frontiers in Plant Science, 12*(October), 1–12. https://doi.org/10.3389/fpls.2021.755358

Welsh, J., & McClelland, M. (1990). Fingerprinting genomes using PCR with arbitrary primers. *Nucleic Acids Research, 18*, 7213–7218.

Williams, J. G. K., Kubelik, A. R., Livak, K. J., Rafalski, J. A., & Tingey, R. S. (1990). DNA polymorphisms amplified by arbitrary primers are useful as genetic markers. *Nucleic Acids Research, 18*(22), 6531–6535. https://doi.org/10.1093/nar/18.22.6531

World Flora Online. (2021). Available at: http://www.worldfloraonline.org/taxon/wfo-0000687690

Zare, H., Moosavi-Movahedi, A. A., Salami, M., Mirzaei, M., Saboury, A. A., & Sheibani, N. (2013). Purification and autolysis of the ficin isoforms from fig (*Ficus carica* cv. Sabz) latex. *Phytochemistry, 87*, 16. https://doi.org/10.1016/J.PHYTOCHEM.2012.12.006

Zhai, Y., Cui, Y., Song, M., Vainstein, A., Chen, S., & Ma, H. (2021). Papain-like cysteine protease gene family in fig (*Ficus carica* L.): Genome-wide analysis and expression patterns. *Frontiers in Plant Science, 12*, 1–14. https://doi.org/10.3389/fpls.2021.681801

Zohary, D. (1983). Wild genetic resources of crops in Israel. *Israel Journal of Plant Sciences, 32*, 97–127.

Zohary, D., & Spiegel-Roy, P. (1975). Beginnings of fruit growing in the old world. *Science, 187*, 319–327. https://doi.org/10.1126/science.187.4174.319

Zohary, D., Hopf, M., & Weiss, E. (2012). *Domestication of plants in the old world: The origin and spread of domesticated plants in Southwest Asia, Europe, and the Mediterranean Basin* (4th ed.). Oxford University Press. https://doi.org/10.1093/acprof:osobl/9780199549061.001.0001

Zolfaghari, M., Salimpour, F., Sharifnia, F., & Marandi, S. (2019). Analysis of genetic diversity among Iranian populations of *Ficus carica* L. (Moraceae). *Acta Biologica Sibirica, 5*(3), 131–138. https://doi.org/10.14258/abs.v5.i3.6534

Chapter 4
Genetic Diversity of Fig Varieties

Rim Ben Abdallah, Imed Othmani, Amel Lagha, and Sami Fattouch

Abbreviations

AMOVA	Molecular Variance Analysis
ANOVA	Analysis of variance
GO	Genetic Ontology
ISSR	Inter-Simple Sequence Repeats
Itraq	Isobaric Tags for Relative and Absolute Quantitation
LTR	Long Terminal Repeat
NCGR	National Clonal Germplasm Repository
PCA	Principal Component Analysis
PCR	Polymerase Chain Reaction
RAPD	Random Amplified Polymorphism DNA
RFLP	Restriction Fragment Length Polymorphism
SAS	Statistical Analysis System
SSR	Simple-Sequence Repeats
UPGMA	Unweighted Pair-Group Approach of Arithmetic Averages

R. Ben Abdallah (✉) · I. Othmani · A. Lagha · S. Fattouch
EcoChem Laboratory, Department of Biological and Chemical Engineering, National Institute of Applied Sciences and Technology (INSAT), University of Carthage (UCAR), Tunis, Tunisia
e-mail: rim.benabdallah@fst.utm.tn; sami.fattouch@insat.ucar.tn

M. F. Ramadan (ed.), *Fig* (Ficus carica)*: Production, Processing, and Properties*, https://doi.org/10.1007/978-3-031-16493-4_4

1 Introduction

The fig (*Ficus carica* L.) is one of the world's oldest cultivated fruit species. The English word fig comes from the Latin word ficus, which means "fig tree". Caria, a historic location of Asia Minor known for figs, is referenced in the species name carica (Ferguson et al., 1990). The cultivation of figs was developed by the Phoenicians, Carthaginians, and Greeks. The Romans spread it throughout the Mediterranean basin (Roger, 2003). The fig tree progressively spread worldwide, reaching far-flung places like China, India, Ethiopia, Arabia, Australia, and the Americas, where it was introduced and widely propagated. The fig tree *Ficus carica L.* developed from a monoecious species to a functionally dioecious species. Mgyz (Persians), erineos (Greeks), teb (Egyptians), caprificus (Romans), and ettine or dhokkar (Egyptians) are some of the names for the fig tree (Arabs). In Mediterranean regions, the Arabs developed fig production. The Portuguese named them figo de toco, while the Italians called profico or caprificus (Lahbib, 1984). *Ficus carica* L., a subtropical fruit tree, belongs to the genus Ficus, mostly found in the tropics (Ferguson et al., 1990). Figs are grown in most warm and temperate regions and have been for many years as a Mediterranean fruit tree (Sahin, 1997). Figs are commonly grown with other fruit trees, including olives, almonds, pomegranates, and grapevines (Aljane, 2011). The common fig (*Ficus carica* L.) is a member of the *Moraceae* family, which includes approximately 1400 species divided into 60 genera. 2n 2x 26 is the number of somatic chromosomes (Ferguson et al., 1990). There are about 700 species of *Ficus* in the genus *Ficus* (Berg, 2003). The fruits of this species are drupelets that emerge from the ovaries in a closed syconium. The tiny fig wasp pollinator (*Blastophaga psenes* L.) enters the fruit through a small opening (ostiole). The only member of the genus *Ficus* farmed for its fruit is *Ficus carica* (Ferguson et al., 1990). According to Smartt (1976), four types of figs are grown worldwide: (a) The most familiar variety, which produces infertile seeds and produces figs without pollination. Some cultivars produce an excellent first crop (breba crop) and then a good second crop (main crop); (b) the San Pedro type produces a first crop (breba crop) that persists to ripening and is parthenocarpic. Pollination is required to maturation the second crop (main crop). This variety's specialty is the breba crop; (c) Smyrna type provides full fruit with viable seeds and requires pollination with pollen from caprifigs (caprification). Though, the fruits will fall from the tree before they mature if pollination is not done; (d) The Caprifig type produces syconia flowers, which contain pollen that will pollinate the Smyrna in addition to San Pedro varieties.

Several prospections and alternative methodologies for genetic resource management were studied to protect fig genetic resources. Pomological and morphological features, molecular markers, Biochemical traits, phytochemical contents diversity, and proteomic and transcriptomic analysis were utilized to describe local cultivars and assess genetic diversity. Various investigations on fig genetic resources have recently been conducted around the regions. This chapter aims to present an outline of fig agriculture, germplasm diversity, fig conservation, and usage improvement using both traditional and modern approaches.

2 Genetic Variability in Fig Germplasm

2.1 Morphological Variability

Morphological markers have typically been used to determine the amount and distribution of genetic diversity. Even though environmental factors and agronomic methods substantially influence the expression of these markers, morphological characterization is a highly suggested initial step to take before attempting more in-depth biochemical or molecular studies. As a result, the morphological description remains the initial stage in agricultural plant genetic variety preservation. Furthermore, plant inventories based on morphological descriptions provide the first glimpse into the degree of phenotypic variation, which is critical for planning genetic resource preservation strategies and establishing national collections (Podgornik et al., 2010).

Botanical Description *F. carica* L. is a tree that grows to be 15–20 feet tall with multiple spreading branches and a trunk that is more than 7 feet in diameter. The plant's latex is milky white and mostly made up of ficin, a protein hydrolytic enzyme. Carica refers to the papaya-like leaves of the plant.

Leaf the leaves are bright green, single, alternating, and enormous, reaching a length of up to one foot. They have a rough upper surface and a soft bottom and are deeply lobed with 1–5 sinuses. Flowers are observed in containers; they emerge from the axils of old leaves.

Flower male flowers occupy the lower part of the receptacle, while female flowers occupy the upper section. The ripe receptacle, Saikonium, has a huge quantity of little whitish seeds.

Seeds seeds range from 30 to 1600 per fruit and can be very small, small, medium, or large. Unless fertilized, the seeds are edible and empty. The fertilized seeds give the dried figs their distinctive nutty flavor. A white inner ring contains a seed mass bonded with jelly-like flesh on the inside.

Fruit fruit figs are paired or solitary, axillary on leafy branchlets, and generally pear-shaped. The fruit (fig) and reproductive systems of the *Ficus* species are mutually exclusive. Only their related agaonid wasps (*Hymenoptera*: *Chalcoidea*: *Agaonide*) may pollinate it, and the wasps can only lay eggs in their linked fruit. Pollinator wasps are required for successful pollination and reproduction of *F. carica* species. On the other hand, Agaonid wasps require the presence of their linked species of *F. carica* to reproduce successfully. In comparison to apple pulp, fruit skin provided the majority of phytochemicals and antioxidant activity.

Root The plant's root system is usually shallow and spread out. Roots can sometimes cover 50 feet of ground, although some roots can reach up to 20 feet in porous soil.

Bark The bark is silky and smooth. The exterior bark is ash to silvery grey, with irregular spherical flakes exfoliated. The center region of the bark is brownish or light reddish-brown. Layers of pale yellowish or orange-brown colored granular tissue make up the interior section (Gafoor et al., 2019).

The majority of descriptor lists for the characterization of genetic resources in plants comprise many features whose evaluation is time-consuming and costly, making the characterization of large germplasm collections challenging. As a result, to make germplasm research and conservation easier, the most valuable factors for each species must be carefully chosen. In addition, morphological characterization is time-consuming; hence, descriptor lists for a wide range of species have been created to help.

In order to make the characterization procedure in this species easier, a group of international experts created and published a list of descriptors (International Plant Genetic Resources Institute and Centre International de Hautes Etudes Agronomiques Méditerranéennes, 2003). However, because of some unique characteristics of this species, such as the frequent production of two different crops (breba and main crop), the presence of four different types of leaves, and its complex floral biology, morphological trait analyses in fig are even more complicated than in most fruit tree species.

As a result, the current descriptors list has 192 properties, 126 of which are qualitative and 66 of which are quantitative. Because of the large number of features and the requirement to replicate observations for at least two years, phenotypic characterization in this species is time- and resource-intensive, particularly in developing nations with limited resources.

Several studies have been carried out to determine the morphological and pomological properties of cultivars of *F. carica* L. Darjazi (2011) reported physical characteristics and quality parameters of some local cultivars of fig (*Ficus carica* L.) grown in Varamin, Iran. His results are presented in Table 4.1. At least ten mature fruits and ten mature leaves from each of the fig cultivars were taken from several areas of the same trees in Varamin Research Station during the last week of July 2001. Darjazi (2011) conducted another study on the morphological and pomological diversity of figs, in which 76 fig accessions were gathered in Hatay, Turkey's Eastern Mediterranean Region, in 2008 and 2009. Results are shown in Table 4.2.

Table 4.1 Variation in morphological characteristics of fig cultivars in Varamin, Iran

	Cultivar								
Traits	Bidaneh	Paizeh	Zard	Siah bolbol riz	Siah zoodras	Siah diras	Morabaii	Hallavi riz	Hallavi dourosht
Leaf trait									
Leaf length (cm)	18.75	20.503.	17.00	12.00	25.20	26.50	25.00	24.00	21.50
Leaf width (cm)	12.60	16.18	13.75	8.25	14.80	17.00	18.60	16.20	14.60
Leaf area (LxW) (cm2)	135	187	85	40	185	220	280	225	165
Number of lobes	3	3	5	3	3	5	3	3	3
Length of central lobe (cm)	8.5	9.8	9.5	5.5	11.3	12.50	13	11.9	10.51
Leaf margin	Regular toothed	Regular toothed	Irregular toothed	Regular toothed	Irregular toothed	Irregular toothed	Regular toothed	Irregular toothed	Irregular toothed
Leaf color	Light green	Dark green	Dark green	Dark green	Dark green	Dark green	Light green	Dark green	Dark green
Petiole length (cm)	4.75	5.50	3.75	4.10	7.00	12.00	6.00	4.75	4.75
Petiole thickness (mm)	2.60	3.60	2.75	3.25	3.75	3.50	3.40	3.60	2.80
Depth of upper sinuses (cm)	2.40	3.20	3.75	2.25	3.50	4.70	4.40	2.80	2.80
Leaf trace (mm)	4.25	4.70	3.90	3.45	4.40	4.25	5.30	4.10	3.10

(continued)

Table 4.1 (continued)

	Cultivar								
Traits	Bidaneh	Paizeh	Zard	Siah bolbol riz	Siah zoodras	Siah diras	Morabaii	Hallavi riz	Hallavi dourosht
Midvein length (cm)	14	15	13	7.50	18.50	17.00	19.00	19.00	17.00
Mid lobe base width (cm)	4.00	5.00	2.50	1.40	4.20	3.20	4.75	4.90	4.50
Angle of central lobe tip	110	76	90	110	0	0	85	0	0
Angle of lateral lobe tip	102	64	70	100	0	0	76	0	0
Length of lateral lobe (cm)	10.50	12.30	10.50	6.00	12.70	12.70	14.60	13.00	11.80
Leaf blade surface	Soft	Soft	Medium	Soft	Rough	Rough	Rough	Medium	Medium
Leaf blade texture	Fragile	Flexible	Flexible	Fragile	Fragile	Fragile	Fragile	Fragile	Fragile
Upper sinuses shape	U	U	V	U	V	U	U	V	V
Density of hairs on leaf upper surface	None	Intermediate	None	Intermediate	None	None	None	None	None
Bud trait									

Length of current season growth (cm)	19.50	12.00	7.40	6.25	9.50	9.50	11.00	18.75	5.50
Diameter of current season growth (mm)	8.25	7.75	8.00	7.50	9.50	9.50	8.50	8.75	6.25
Terminal bud length (mm)	9.50	11.90	10.30	6.40	10.80	9.50	12.50	8.30	8.50
Terminal bud width (mm)	3.60	4.30	3.90	3.80	5.40	5.00	5.00	4.30	3.70
Tbl / Tbw	2.63	2.76	2.64	1.68	2.00	1.90	2.50	1.93	2.29
Shoot internode length (mm)	48	27	15	12	23	22	23	37	15
Internode number	8	6	7	8	8	8	7	9	5
Tree traits									
Tree growth habit	Weeping	Erect	Spreading	Spreading	Weeping	Erect	Weeping	Weeping	Weeping
Tree vigor	Medium	Medium	Low	Low	Medium	Low	High	Medium	Medium

Darjazi (2011)

Table 4.2. Variation in morphological characteristics of fig cultivars from the Eastern Mediterranean Region of Turkey

Accession	Tree growth habit	Tree vigor	Shoot length (cm)	Number of leaves on a shoot	Number of fruits on a shoot	Leaf length (cm)	Leaf width (cm)	Length of center lobe (cm)	Leaf area (cm²)	Petiole length (cm)	Petiole thickness (mm)	Leaf shape	Number of leaf lobes	Apical dominancy	Full maturity	Harvesting period
Mor 1	Open	High	40.0	11.2	7.2	22.8	21.8	13.3	320.1	8.5	5.7	D	5	Present	1–30 sept.	21–40
Mor 2	Open	Intermediate	23.1	7.3	4.4	22.4	18.2	13.8	282.2	7.8	5.3	A	5	Present	1–30 sept.	41–60
Mor 3	Open	Intermediate	11.8	6.3	5.4	20.5	17.1	12.6	256.4	6.2	4.3	C	5	Absent	11–31 Aug.	21–40
Mor 4	Weeping	Intermediate	29.4	7.4	6.2	20.1	15.8	10.1	246.6	8.2	6.0	G	3	Present	1–30 sept.	>60
Mor 5	Semierect	Intermediate	15.9	6.1	3.5	19.8	17.2	11.9	259.8	5.4	6.1	D	5	Absent	1–10 Aug.	21–40
Mor 6	Open	High	8.8	5.2	2.7	18.3	16.6	10.9	223.6	5.8	4.8	D	5	Absent	1–10 Aug.	21–40
Sultani 1	Open	Intermediate	7.8	4.4	2.4	18.9	14.8	12.4	204.5	7.1	4.7	B	5	Absent	11–31 Aug.	21–40
Sultani 2	Open	Intermediate	9.4	4.7	2.5	16.4	15.0	11.2	163.4	6.0	3.6	C	5	Absent	11–31 Aug.	21–40
Sultani 3	Open	Intermediate	10.2	5.7	4.1	17.9	15.7	11.5	213.0	7.1	4.3	D	5	Present	11–31 Aug.	21–40
Kabak 1	Open	High	11.5	6.3	4.7	26.6	22.1	15.3	348.6	8.5	5.9	C	5	Absent	1–10 Aug.	41–60
Kabak 2	Spreading	Intermediate	10.6	6.5	4.8	23.4	21.2	12.8	325.8	7.3	5.6	C	5	Absent	1–10 Aug.	41–60
Şeble 1	Spreading	High	26.1	8.8	5.7	20.6	18.8	12.8	278.5	8.1	4.6	C	5	Present	11–31 Aug.	21–40
Şeble 2	Open	Intermediate	26.4	9.1	5.3	21.5	17.5	12.2	269.4	6.9	5.4	C	5	Present	11–31 Aug.	21–40
Kıreni 1	Open	Intermediate	15.2	10.5	2.6	21.0	19.1	16.1	263.4	8.0	5.9	F	5	Absent	11–31 Aug.	21–40
Kıreni 2	Spreading	Intermediate	13.6	6.5	4.5	21.4	18.0	12.6	280.4	9.1	5.8	B	5	Absent	1–10 Aug.	41–60
Sehli 1	Open	Intermediate	9.3	5.2	3.4	27.6	19.5	13.0	276.3	7.4	5.1	C	5	Absent	1–10 Aug.	41–60
Sehli 2	Open	Intermediate	9.0	7.5	2.6	17.1	14.6	11.0	176.6	5.9	5.0	D	5	Absent	1–10 Aug.	21–40
Meryemi 1	Open	Low	18.0	7.1	4.9	19.9	17.9	12.2	221.7	7.5	4.4	B	5	Absent	1–10 Aug.	21–40
Meryemi 2	Spreading	Intermediate	11.9	6.1	3.7	23.4	17.5	14.0	269.6	6.6	5.2	G	3	Absent	11–31 Aug.	21–40
Kuruye 1	Semierect	Intermediate	11.9	4.9	4.2	20.2	17.5	12.4	235.8	6.9	5.1	D	5	Present	11–31 Aug.	41–60
Kuruye 2	Semierect	Intermediate	12.7	7.3	5.2	18.8	16.5	12.1	219.7	5.9	4.8	D	5	Present	11–31 Aug.	41–60
Kırmızı 1	Semierect	Intermediate	31.8	9.7	7.2	21.1	18.8	13.5	264.0	7.5	4.7	C	5	Present	11–31 Aug.	21–40
Kırmızı 2	Erect	Low	8.1	5.1	2.7	16.5	13.7	10.3	172.5	5.2	4.6	G	3	Absent	1–10 Aug.	15–20
Lopkara 1	Open	High	10.9	6.3	5.4	22.8	20.1	13.6	302.6	7.5	5.3	D	5	Absent	11–31 Aug.	21–40

Lopkara 2	Semierect	Intermediate	33.3	7.5	5.2	24.1	21.1	13.2	346.2	9.6	5.5	G	3	Present	11–31 Aug.	21–40
Ramlı 1	Open	High	10.7	5.4	2.7	16.6	14.6	10.0	166.0	4.7	4.1	C	5	Absent	1–10 Aug.	15–20
Ramlı 2	Semierect	Low	9.7	5.4	3.3	17.2	14.1	10.5	166.8	4.8	4.2	G	3	Absent	1–10 Aug.	15–20
Bığrasi 1	Open	High	20.9	7.2	5.3	24.3	22.6	13.2	358.8	7.3	5.3	G	3	Absent	11–31 Aug.	41–60
Bakrasi 2	Open	Intermediate	15.2	4.2	4.2	17.5	17.4	9.8	205.3	5.0	4.3	G	3	Absent	11–31 Aug.	21–40
Bakras 3	Spreading	Low	8.5	7.2	5.9	22.5	20.1	13.1	283.8	7.5	5.2	G	3	Present	11–31 Aug.	41–60
Bakrasi 4	Weeping	High	18.6	7.4	6.1	21.0	21.1	11.0	306.5	7.1	4.5	G	3	Present	11–31 Aug.	41–60
Bakrasi 5	Semierect	Intermediate	25.7	9.4	8.1	23.6	22.5	11.8	371.6	8.9	5.0	G	3	Present	11–31 Aug.	40–60
Şami	Open	Intermediate	24.5	10.4	7.5	23.5	20.7	12.8	280.5	8.2	5.4	C	5	Present	11–31 Aug.	41–60
Fransavi	Open	Low	20.2	6.0	3.7	20.3	17.1	11.4	250.3	8.4	4.8	G	3	Present	11–31 Aug.	41–60
Hılvıni	Weeping	High	15.9	5.8	3.4	19.3	16.5	13.4	219.4	5.1	4.6	A	5	Absent	11–31 Aug.	21–40
Büyük Siyahlop	Semierect	Intermediate	18.5	7.2	3.7	21.3	17.9	13.1	262.7	7.0	5.4	C	5	Present	11–31 Aug.	41–6
Sıhle	Open	High	9.1	6.3	3.9	20.4	17.5	13.3	243.7	5.3	4.8	C	5	Absent	11–31 Aug.	41–60
Kilis	Erect	Low	16.8	9.5	4.7	22.8	21.6	12.5	334.1	7.0	5.0	C	5	Present	1–10 Aug.	21–40
Ahmediye	Weeping	High	11.3	6.9	5.6	23.5	21.2	14.5	332.4	6.1	5.0	G	3	Absent	11–31 Aug.	41–60
Burnu Kızıl	Semierect	Low	9.0	5.4	4.3	19.9	17.4	11.0	283.2	6.0	4.5	G	3	Absent	11–31 Aug.	41–60
Allene Karası	Semierect	Intermediate	11.1	5.8	3.4	23.1	18.9	15.1	280.3	6.7	6.6	D	5	Present	1–10 Aug.	21–40
Beyaz Fahli	Open	Intermediate	22.2	7.5	3.9	21.6	16.7	13.6	253.5	6.0	5.4	A	5	Present	11–31 Aug.	21–40
Kandamık	Spreading	Low	9.4	6.5	3.3	21.5	18.0	12.4	316.1	5.3	5.6	C	5	Present	1–10 Aug.	21–40
Fahli	Semierect	Intermediate	10.9	5.8	2.6	19.0	16.0	12.5	221.2	6.1	5.6	C	5	Present	11–31 Aug.	21–40
Fetike	Open	Intermediate	9.8	6.2	3.0	19.4	16.2	12.3	246.1	7.3	5.2	B	5	Present	11–31 Aug.	40–60
Armut Sapı	Open	Low	6.5	5.8	2.8	17.2	16.1	10.9	195.7	6.5	4.5	D	5	Absent	1–10 Aug.	21–40
Payas	Spreading	Intermediate	8.4	4.8	2.6	19.6	15.9	12.0	223.8	6.8	3.7	B	5	Present	11–31 Aug.	41–60
Gud Yeniği	Semierect	Low	9.2	6.4	4.3	21.1	17.0	14.1	237.3	6.4	4.8	D	5	Present	1–10 Aug.	21–40
Baldır	Open	Intermediate	9.4	5.7	4.1	20.1	15.1	11.5	207.0	9.8	4.1	G	3	Absent	11–31 Aug.	41–60
Halep	Semierect	Intermediate	25.8	6.6	4.5	25.1	23.5	13.5	347.0	11.4	6.4	G	3	Present	11–31 Aug.	41–60
Erkenci	Semierect	Intermediate	8.9	6.6	4.3	20.7	16.2	14.6	223.6	5.8	4.1	C	5	Absent	End of July	21–40

(continued)

Table 4.2 (continued)

Accession	Tree growth habit	Tree vigor	Shoot length (cm)	Number of leaves on a shoot	Number of fruits on a shoot	Leaf length (cm)	Leaf width (cm)	Length of center lobe (cm)	Leaf area (cm^2)	Petiole length (cm)	Petiole thickness (mm)	Leaf shape	Number of leaf lobes	Apical dominancy	Full maturity	Harvesting period
Yeşil İncir	Open	Intermediate	32.1	9.2	5.3	21.6	18.5	12.6	303.2	5.3	5.4	A	7	Present	11–31 Aug.	21–40
Şebli	Spreading	Intermediate	21.8	5.6	3.3	19.9	17.9	11.8	271.2	6.1	4.7	G	3	Absent	11–31 Aug.	21–40
Tınesvit	Weeping	Intermediate	15.0	5.8	4.3	21.7	18.4	13.0	254.9	6.6	4.7	C	5	Absent	11–31 Aug.	21–40
Sütlü Sarı	Open	Intermediate	18.0	7.5	4.2	22.8	19.5	12.4	281.5	6.7	5.1	G	3	Present	11–31 Aug.	21–40
Mersinli	Open	Intermediate	9.6	5.1	3.0	23.0	16.6	14.3	231.6	6.9	5.4	C	5	Present	1–10 Aug	21–40
Zırhıni	Weeping	High	19.1	7.2	5.1	22.4	19.0	13.3	259.5	8.6	5.1	A	5	Absent	11–31 Aug.	21–40
Şibili	Open	Intermediate	15.7	8.5	6.6	21.6	19.9	12.2	311.8	7.1	5.7	F	5	Absent	11–31 Aug.	>60
Karagöz	Erect	Intermediate	25.8	8.1	5.4	20.8	18.4	10.7	273.9	10.2	4.8	G	3	Present	11–31 Aug.	41–60
Beyaz İncir	Open	Intermediate	14.2	6.3	4.2	21.8	17.3	13.5	280.1	7.4	4.8	D	5	Absent	11–31 Aug.	21–40
Sarı 1	Spreading	Intermediate	20.9	8.2	6.0	23.9	21.0	13.5	336.0	8.1	6.0	C	5	Present	11–31 Aug.	21–40
Sarı 2	Open	Intermediate	16.6	6.9	2.8	19.3	16.5	10.9	287.3	7.4	5.4	D	5	Present	1–10 Aug.	21–40
Sarı 3	Open	Low	7.9	6.8	4.3	20.9	19.0	13.5	259.3	10.1	4.1	B	5	Absent	1–10 Aug.	21–40
Sarı 4	Semierect	High	19.4	10.4	7.2	22.8	21.6	11.9	356.9	9.7	5.0	G	3	Present	11–31 Aug.	41–60
Sarı 5	Open	Intermediate	33.3	10.3	6.4	21.9	18.3	12.2	249.1	5.0	5.0	A	7	Present	11–31 Aug.	21–40
Sarı 6	Open	High	15.4	7.1	4.7	21.2	19.2	12.4	288.1	8.2	5.3	G	3	Absent	1–30 sept.	21–40
Siyah 1	Open	Low	15.4	6.4	3.8	23.5	20.8	13.5	296.7	8.4	5.4	G	3	Absent	11–31 Aug.	21–40
Siyah 2	Semierect	Low	7.5	4.9	3.1	21.1	17.9	14.0	277.6	6.0	5.6	D	5	Present	11–31 Aug.	21–40
Siyah 3	Weeping	Intermediate	17.8	6.6	5.0	21.8	18.0	14.0	280.4	8.0	4.6	C	5	Present	11–31 Aug.	21–40
Siyah 4	Spreading	Intermediate	19.8	7.0	5.4	22.2	18.6	12.7	301.0	8.5	5.7	C	5	Absent	1–30 sept.	41–60
Siyah 5	Open	Low	14.9	5.7	3.6	21.6	19.8	12.3	275.5	7.1	4.4	G	3	Absent	11–31 Aug.	21–40
Siyah 6	Weeping	High	22.1	5.0	3.7	19.0	17.2	11.6	221.6	6.7	4.6	C	5	Absent	11–31 Aug.	21–40
Siyah 7	Weeping	High	19.3	7.4	5.2	23.1	18.4	13.8	273.3	11.0	5.2	A	5	Present	11–31 Aug.	21–40
Siyah 8	Spreading	High	32.1	6.4	3.6	20.7	18.1	11.1	262.3	10.4	4.8	G	3	Present	1–30 sept.	>60

Çalişkan and Atila (2012)

2.2 *Pomological Traits*

Darjazi (2011) has reported pomological properties and quality parameters of some local cultivars of fig (*Ficus carica* L.) grown in Varamin, Iran. Fruits were evaluated for pomological purposes using a set of criteria with physical attributes and qualitative factors. Three replications are considered; in each replication, there are 10 fruits. Results are reported in Table 4.3 in a study of the morphological and pomological variety of figs in Hatay, Turkey's Eastern Mediterranean Region, in which 76 fig accessions were gathered in 2008 and 2009. Results are reported in Table 4.4.

2.3 *Agronomical Parameters*

Different agronomical parameters such as annual yield, cumulative yield, yield efficiency, ripening period, harvesting of brebas and figs, and trunk cross-sectional area were studied (Pereira et al., 2015, 2017) to investigate the agronomic behavior and quality of different fig cultivars produced in Spain for fresh consumption. The researched fig cultivars were 'Cuello Dama Blanco', 'Brown Turkey', 'Colar Elche', 'SanAntonio', 'Banane', and 'Blanca Bétera'.

Results showed that the ripening period for brebas was 9–21 days, while the main crop was 67–75 days. The earliest maturing cultivar was 'San Antonio,' which reached maturity in the first week of June. 'Brown Turkey' and 'Blanca Bétera' breba fruit ripened in mid-June, followed by 'Banane', 'Cuello Dama Blanco', and 'Colar Elche' breba fruit in the third week of June. San Antonio, Brown Turkey, and Blanca Bétera's second crops matured in mid-July, while 'Banane,' 'Cuello Dama Blanco,' and 'Colar Elche' matured at the end of July. The maturation period for these six cultivars lasted until early October, when temperatures decreased and the first autumn rains began. Each cultivar and block's annual breba crop production (kg/tree) was much lower than the main crop's yearly yield, with unpredictable values between years and blocks. The cultivars 'San Antonio' and 'Brown Turkey' had the largest annual breba crop yield, with maximum values of 5.5 and 5.4 kg/tree on the fifth green, respectively (2012). Annual primary crop production improved year after year, reaching the sixth green (2013) with values of 83.2, 74.6, 60.2, 56.7, 53.7, 27.2 kg/tree for 'Banane', 'BrownTurkey', 'Colar Elche', 'Cuello Dama Blanco', 'San Antonio', and 'Blanca Bétera'. Furthermore, a high percentage of damaged fruit was discovered during the harvest of brebas and main crop figs, ranging from 30% to 11% depending on the brebas and main crop figs; the dark cultivars had the highest percentage of damaged fruit.

The annual yield also differed between blocks, which delayed the trees' entry into production due to harsh pruning throughout the winter. During the study period of four years (2010–2013), the cumulative yield of the breba crop (kg/tree) was considerably lower than the cumulative yield of main crop figs during the same period. The cultivars 'San Antonio,' 'Banane,' and 'Brown Turkey' yielded the most brebas, with 17.4, 17.2, and 10.8 kg/tree, respectively, while the cultivar 'Cuello

Table 4.3 Variation in pomological characteristics of fig cultivars in Varamin, Iran

	Cultivar								
Traits	Bidaneh	Paizeh	Zard	Siah bolbol riz	Siah zoodras	Siah diras	Morabaii	Hallavi riz	Hallavi dourosht
Fruit trait									
Fresh fruit weight (g)	27.9	43.5	24.4	8	33.5	30.7	8.6	19.7	20.7
Dried fruit weight (g)	5	7.4	4.1	1.6	4.9	4.9	1.6	3.4	4.4
Ffw /Dfw	0.17	0.16	0.16	0.19	0.14	0.15	0.18	0.16	0.21
Fruit volume (cm3)	31.2	46.6	24.3	7.4	33.7	31.4	9.8	19.7	23.7
Fruit diameter (mm)	38	45	35	21	40	38	24	32	37
Fruit length (mm)	27	35	23	20	31	25	25	36	22
Fruit shape index(Fd/Fl)	1.40	1.52	1.28	1.05	1.29	1.52	0.98	0.88	1.68
Fruit shape	Oblate	Oblate	Oblate	Globose	Oblate	Oblate	Globose	Oblong	Oblate
Ostiole width (mm)	2.9	2.7	4.2	1.6	4.6	4.2	0	3.1	4
Stalk length (mm)	7.00	6.00	11.5	15	4.00	5.00	5.80	11.0	12.0
Stalk diameter (mm)	5.20	5.30	4.00	1.70	4.80	4.50	3.80	4.30	4.60
Sugar (%)	15.3	9.8	16.2	16.5	15	16.8	16.5	18.9	17.9
Total soluble solid (%)	18	13.3	18.4	28.5	20	22.5	28.2	24	17.1

Fruit skin over color	Light-green	Canary yellow	Yellow-green	Purplish brown	Purple	Black	Canary yellow	Light-brown	Light-brown
Fruit ribs	None	None	None	None	None	None	None	Prominent	Prominent
Bloom	Medium	None	None	None	None	None	None	None	None
Fruit skin cracks	None	None	Minute	None	None	None	None	None	None
Ease of peeling	Easy	Easy	Easy	Easy	Difficult	Easy	Difficult	Easy	Easy
Fruit flesh color	Light yellow	Light yellow	Light yellow	White	White	White	Light yellow	Light yellow	Light yellow
Pulp internal color	Amber	Light strawberry red	Dark strawberry red	Amber	Amber	Amber	Amber	Amber	Amber
Fruit cavity	Medium	Medium	Small	Small	Small	Small	None	Small	Medium
Pulp juiciness	Juicy	Little juicy	Little juicy	Little juicy	Little juicy	Little juicy	Little juicy	Little juicy	Little juicy
Shape of the fruit stalk	Short thick	Short thick	Long slender	Long slender	Short thick	Short thick	Short thick	Long slender	Short thick
Pulp flavor	Aromatic	None	None	None	None	None	None	None	None
Seed number	Low	Medium	High	Low	Low	Medium	Low	Low	Medium
Neck length	None	None	None	None	None	None	Short	Short	None
Scale color of eye	Light-orange	Purplish pink	Red with white margins	Light yellow- white	Dark- purple with white margins	Dark- purple	Canary yellow	Light-brown	Light-brown
Stalk color	Green	Yellow-green	Yellow-green	Light-yellow	Dark-green	Dark-green	Light-yellow	Dark-green	Dark-green
Number of fruit per node	1	1	1	1	1	1	2	1	1
Breba crop	High	Low	Meduim	None	None	None	High	None	None
Skin thickness	Medium	Medium	Medium	Thin	Medium	Thin	Thin	Thin	Thin

Darjazi (2011)

Table 4.4 Variation in pomological characteristics of fig cultivars from the Eastern Mediterranean Region of Turkey

Accession	Fruit weight (g)	Fruit diameter (mm)	Fruit length (mm)	Index	Neck length (mm)	Ostiole width (mm)	Fruit skin thickness (mm)	Total soluble solids (%)	TSS/ Acidity	Fruit shape	Skin color	Pulp color
Şami	53.1	49.1	56.3	0.9	9.0	4.1	1.8	20.6	176.3	Globose	Green	Amber
Fransavi	48.9	46.1	46.4	1.0	5.3	4.0	1.4	19.8	77.1	Globose	Green	Red
Hılvıni	21.6	33.3	32.4	1.0	3.4	1.3	1.2	22.7	163.9	Globose	Green	Amber
Büyük Siyahlop	40.4	41.3	42.8	1.0	4.6	5.1	1.3	20.6	141.4	Globose	Purple	Amber
Sıhle	38.2	40.6	45.7	0.9	7.2	1.9	1.4	22.5	117.9	Globose	Brown	Amber
Kilis	71.6	55.5	47.1	1.2	4.1	3.5	1.1	20.2	81.1	Oblate	Green	Amber
Ahmediye	78.2	52.6	52.3	1.0	4.7	1.3	1.6	20.8	115.3	Globose	Green	Pink
Burnu Kızıl	52.5	46.2	44.5	1.0	3.0	8.1	1.2	19.0	68.5	Globose	Green	Red
Allene Karası	45.6	47.0	38.7	1.2	4.0	6.2	1.3	19.2	129.6	Oblate	Black	Amber
Beyaz Fahli	40.7	47.2	36.2	1.3	3.5	21.0	1.1	23.7	127.5	Oblate	Green	Pink
Kandamık	58.1	49.2	39.4	1.2	1.9	1.9	1.1	20.0	94.2	Oblate	Green	Pink
Fahli	37.0	42.1	34.9	1.2	3.9	6.0	1.4	23.4	127.8	Oblate	Green	Pink
Fetike	59.4	49.0	42.9	1.1	5.7	8.3	1.6	20.5	156.3	Globose	Green	Pink
Armut Sapı	34.0	40.0	41.1	1.0	6.0	4.6	1.5	20.0	87.6	Globose	Yellow	Pink
Payas	18.9	32.5	31.0	1.0	0.5	1.9	1.2	20.4	108.4	Globose	Green	Dark red
Gud Yeniği	32.7	40.2	29.8	1.4	0.0	5.4	1.0	22.5	131.5	Oblate	Green	Amber
Baldır	40.5	42.3	39.6	1.1	5.6	3.2	1.6	20.5	81.1	Globose	Green	Red
Halep	56.2	50.3	44.6	1.1	8.1	4.5	1.6	18.4	110.9	Globose	Green	Amber
Erkenci	39.6	41.8	44.8	0.9	5.7	4.9	1.6	20.9	150.0	Globose	Brown	Amber
Yeşil İncir	66.5	51.8	43.0	1.2	2.3	3.3	1.9	22.5	70.9	Oblate	Green	Dark red
Şebli	38.6	39.1	42.3	0.9	2.3	3.7	1.0	23.5	155.1	Globose	Brown	Amber
Tınesvit	37.1	38.4	51.4	0.7	8.5	1.5	1.5	22.0	119.5	Oblong	Brown	Amber
Sütlü Sarı	51.1	47.0	50.2	0.9	7.5	2.4	1.4	20.4	168.6	Globose	Yellow	Pink
Mersinli	56.6	48.5	44.6	1.1	2.9	4.9	1.4	20.7	123.9	Globose	Yellow	Red

Zırhını	22.9	34.7	46.6	0.7	7.1	1.3	1.1	22.4	118.2	Oblong	Brown	Red
Şibili	42.9	41.7	49.5	0.8	13.0	4.7	1.1	20.2	172.8	Oblong	Purple	Pink
Karagöz	46.6	42.9	47.4	0.9	5.8	1.1	1.2	20.2	69.5	Globose	Purple	Dark red
Beyaz İncir	46.8	45.7	41.7	1.1	5.1	5.8	1.4	20.4	79.0	Globose	Green	Dark red
Sarı 1	81.6	56.7	50.3	1.1	9.5	4.5	2.3	18.4	75.9	Globose	Yellow	Dark red
Sarı 2	62.3	50.1	52.8	0.9	7.8	6.7	1.8	19.7	199.8	Globose	Yellow	Pink
Sarı 3	54.3	48.6	43.1	1.1	8.3	6.8	1.7	20.8	104.1	Globose	Yellow	Pink
Sarı 4	37.5	38.0	40.3	0.9	4.0	5.9	1.9	17.5	82.0	Globose	Green	Pink
Sarı 5	74.0	53.8	44.1	1.2	3.1	3.0	2.0	21.4	80.7	Oblate	Yellow	Red
Sarı 6	87.4	58.7	48.4	1.2	4.3	5.1	1.4	23.6	132.0	Oblate	Yellow	Amber
Siyah 1	58.4	50.2	47.6	1.1	5.5	2.5	1.7	19.9	107.9	Globose	Brown	Red
Siyah 2	36.0	42.5	36.5	1.2	2.7	3.8	1.0	21.9	102.4	Oblate	Black	Pink
Siyah 3	24.8	36.1	31.3	1.2	0.5	3.2	1.0	19.7	59.5	Oblate	Black	Dark red
Siyah 4	80.4	55.4	47.5	1.2	3.7	6.0	1.5	18.3	55.1	Oblate	Black	Red
Siyah 5	21.3	32.6	36.1	0.9	2.8	0.6	0.6	22.3	117.7	Globose	Black	Pink
Siyah 6	29.0	38.0	44.6	0.9	6.4	2.2	1.4	22.4	156.2	Globose	Brown	White
Siyah 7	35.6	38.8	47.8	0.8	7.8	2.4	1.8	21.0	91.5	Oblong	Brown	Pink
Siyah 8	40.1	40.6	44.3	0.9	5.0	1.1	1.2	21.5	67.2	Globose	Purple	Dark red
Mor 1	66.8	52.2	50.7	1.0	8.2	3.6	2.1	20.7	100.0	Globose	Brown	Red
Mor 2	65.3	49.9	48.4	1.0	3.3	4.6	1.8	19.4	104.8	Globose	Purple	Pink
Mor 3	36.2	39.5	47.4	0.8	7.4	2.0	1.7	20.8	147.1	Oblong	Brown	Amber
Mor 4	46.6	44.1	43.7	1.0	4.6	1.9	1.1	21.8	64.3	Globose	Purple	Red
Mor 5	30.9	37.0	38.3	1.0	2.7	3.6	1.3	22.6	204.3	Globose	Purple	Amber
Mor 6	45.3	45.7	36.1	1.3	2.5	2.8	1.4	18.5	108.1	Oblate	Brown	Red
Sultani 1	35.3	40.1	40.4	1.0	4.2	3.2	1.9	25.5	170.5	Globose	Yellow	Pink
Sultani 2	50.4	46.6	46.7	1.0	6.0	5.0	2.2	23.0	114.7	Globose	Yellow	Red
Sultani 3	44.1	45.8	43.7	1.0	6.3	3.5	1.3	19.2	81.9	Globose	Green	Red

(continued)

Table 4.4 (continued)

Accession	Fruit weight (g)	Fruit diameter (mm)	Fruit length (mm)	Index	Neck length (mm)	Ostiole width (mm)	Fruit skin thickness (mm)	Total soluble solids (%)	TSS/Acidity	Fruit shape	Skin color	Pulp color
Kabak 1	81.7	55.8	54.5	1.0	8.0	2.2	1.7	17.9	51.4	Globose	Green	Dark red
Kabak 2	99.4	61.1	46.4	1.3	7.5	4.6	2.2	19.0	81.2	Oblate	Green	Dark red
Şeble 1	51.9	43.7	50.8	0.9	3.9	5.0	1.7	19.1	105.9	Globose	Brown	Pink
Şeble 2	50.7	44.5	46.8	1.0	4.7	6.7	1.2	19.4	99.5	Globose	Brown	Pink
Kıreni 1	47.7	46.9	43.4	1.1	3.9	10.3	1.8	21.2	112.1	Globose	Yellow	Pink
Kıreni 2	69.7	51.8	52.4	1.0	6.4	4.4	1.3	22.0	195.4	Globose	Yellow	Pink
Sehli 1	33.7	38.1	43.0	0.9	5.5	2.1	1.4	23.4	107.3	Globose	Brown	Amber
Sehli 2	23.1	32.2	38.9	0.8	4.8	2.1	1.2	24.3	144.0	Oblong	Brown	Amber
Meryemi 1	55.3	49.2	50.1	1.0	8.7	6.8	1.8	19.2	94.4	Globose	Yellow	Dark red
Meryemi 2	47.6	45.5	39.4	1.2	3.1	5.1	1.2	20.6	77.7	Oblate	Green	Pink
Kuruye 1	25.2	34.5	42.2	0.8	7.0	2.6	1.5	23.0	145.8	Oblong	Brown	Amber
Kuruye 2	21.2	30.9	43.4	0.7	8.7	2.0	1.4	24.5	230.6	Oblong	Brown	Amber
Kırmızı 1	28.6	36.9	46.9	0.8	7.9	1.3	1.6	22.4	149.6	Oblong	Brown	Red
Kırmızı 2	12.3	27.7	28.4	1.0	1.7	2.8	1.5	27.1	136.8	Globose	Brown	Red
Lopkara 1	30.5	35.4	42.7	0.8	6.8	1.4	1.9	22.4	161.1	Oblong	Brown	Pink
Lopkara 2	54.4	46.6	47.1	1.0	5.8	6.3	1.7	19.5	69.5	Globose	Black	Red
Ramlı 1	24.6	33.3	37.9	0.9	5.4	4.9	1.1	20.0	145.4	Globose	Green	Pink
Ramlı 2	18.0	29.7	34.6	0.9	2.3	4.4	0.9	22.3	127.8	Globose	Green	Pink
Bığrasi 1	29.3	37.0	37.3	1.0	2.0	3.4	1.0	21.2	74.2	Globose	Green	Pink
Bakrasi 2	22.4	37.5	36.6	1.0	1.3	4.6	1.2	19.7	90.2	Globose	Green	Red
Bakras 3	34.0	38.8	39.7	1.0	2.8	2.8	1.2	20.4	82.0	Globose	Green	Pink
Bakrasi 4	45.0	43.0	47.0	0.9	4.5	4.4	1.8	16.0	63.6	Globose	Green	Pink
Bakrasi 5	39.3	41.8	49.8	0.8	3.4	5.0	1.7	19.5	100.9	Oblong	Green	Pink

Çalişkan and Atila (2012)

Dama Blanco' yielded the least, with 4.2 kg/tree. The cultivars 'Banane' and 'Brown Turkey' had the largest cumulative yields of main crop figs, with

229 and 186 kg/tree, respectively. The cultivar 'Blanca Bétera', on the other hand, had the lowest cumulative yield of 65.8 kg/tree.

During the sixth green, the trunk cross-sectional area connected to tree vigour differed amongst cultivars. With a trunk section of 93.3 cm^2, the cultivar 'Cuello Dama Blanco' was the most vigorous, followed by 'Banane' and 'Colar Elche' with 90.5 and 79 cm^2, respectively. According to the International Plant Genetic Resources Institute (IPGRI), these three cultivars have only moderate vigour under these growing conditions. Regardless, these varieties are regarded as prolific cultivars. The 'Colar Elche' cultivar, on the other hand, has a spreading growth tendency. Hence commercial plantations for this cultivar would need to have larger tree spacing, such as 6 m x 6 m, to minimize tree overlap.

Yield efficiency (kg/cm^2), calculated as the ratio between the cumulative yield and the trunk cross-sectional area, differed significantly among cultivars. Brebas presented values lower than main crop figs due to the low annual yields. The 'Brown Turkey' and 'Banane' cultivars exhibited good yield efficiency with 2.7 and 2.6 kg/cm^2, respectively, followed by the 'San Antonio' and 'Colar Elche' cultivars with 1.9 and 1.6 kg/cm^2, respectively.

2.4 Molecular and Biochemical Variability Assessment

2.4.1 Structure and Composition of Fig Genome

Sequencing of the genome of *Ficus carica* showed a genome size estimated at 333 Mbp, of which 80% is anchored on 13 chromosomes, very rich in repeated sequences, including LTR-RE (retro-transposons with long terminal repeats) and which are the major constituent of the whole genome (Vangelisti et al. 2021).

LTR-REs are mobile DNA sequences, and this characteristic gives them a vital role in producing an immense diversity of expression and gene function in plant species and consequently leads to a significant contribution to the evolution of flora (Lisch, 2013). Indeed, they can promote chromosomal rearrangements by allowing illicit recombination.

(Devos et al., 2002) through distant locations on the same or other chromosomes, its coding sequence, or splicing model (Dubin MJ et al., 2018).

In the same context, a study conducted by Usai and his research team in 2019 aims to highlight the abundance of truncated trees in the fig tree genome. In this study, the entire genome was scanned using predictive tools based on structure and homology, and they were able to identify a total of 123.8 Mbp of repeated sequences, representing 37.3% of the genome assembly (Fig. 4.1). TEs were the most abundant, covering 33.57% of the assembly (11.06 Mbp), while tandem repetitions represented 3.82% (12.74 Mbp). The most abundant TEs are retrotransposons or PCR-RFLP fingerprinting.

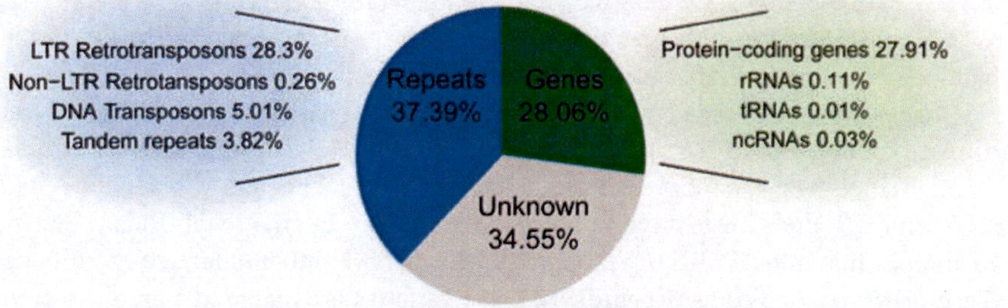

Fig. 4.1 Annotation of the fig genome. Pie chart showing the percentage of genes (i.e. protein-coding genes, rRNAs, tRNA and ncRNAs) (left), repeats (i.e. LTR Retrotransposons, Non-LTR Retrotransposons, DNA transposons and tandem repeats) (right) and unknown sequences. (Usai et al., 2020)

The rapid and precise identification of various plants was made possible by molecular fingerprinting employing genes and non-coding areas. Specific primers can be created by aligning the separated nucleotide sequences, allowing for the identification of high variation in genes and areas. A PCR product can also be fragmented by restriction enzymes, resulting in fragment length polymorphisms. RFLP-PCR analysis has been regarded as a good tool for distinguishing closely related genotypes and a quick and accurate method for plant identification.

RAPD (Random Amplified Polymorphism DNA) is a PCR reaction in which the experimenter does not choose the amplified DNA segments but is amplified "at random". This tool has been a prime tool for considering fig genetic variability in various studies. Khadari et al. (1994), in their research work entitled Varietal Identification and Genetic Resources in Fig (*Ficus carica* L.) using RAPD markers, succeeded in identifying 21 figs (*Ficus carica* L.) samples representing different cultivated varieties. In this study, 19 markers were obtained using 12 primers, making it possible to characterize 17 genotypes. The primers used revealed a significant polymorphism. This makes these markers an effective tool for varietal identification in fig trees.

They could be used to analyze the genetic diversity of more cultivated varieties and natural populations of *Ficus carica.* According to another study conducted by Ikegami and his research team in 2009, which used more than one molecular marker among other ISSR, RAPD, and SSR. The results obtained suggest that the genetic diversity of the fig population is not too great; this requires the use of several molecular markers to estimate the fig kinship at the level of the variety. In addition, it was assumed that 'Houraihi', the oldest variety in Japan, was distributed independently of other foreign varieties in the seventeenth century or earlier.

Indeed, each marker system produced incompletely separated clusters, although a weak linkage group based on race type appeared in the combined data set. In addition, Molecular Variance Analysis (AMOVA) showed that most of the total polymorphism (Fig. 4.2) was attributable to intra-group variance (ISSR + RAPD, 97.4%; SSR, 90.1%).

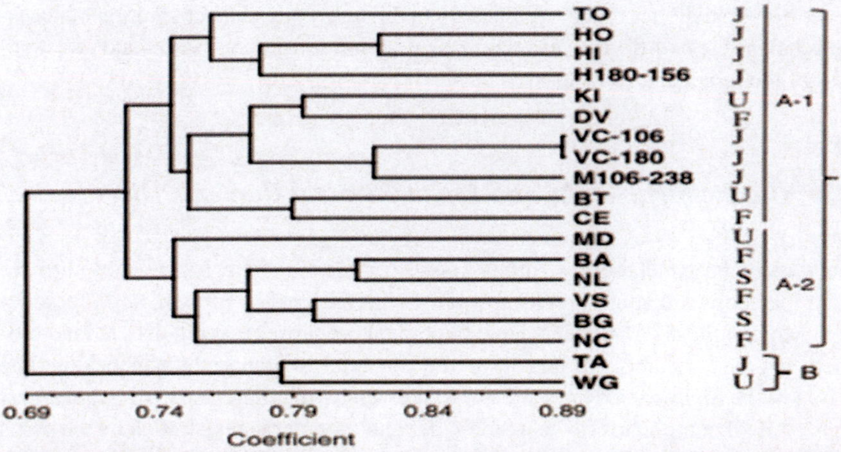

Fig. 4.2 Dendrogram generated using UPGMA; the method used a simple matching similarity coefficient estimated on 19 fig varieties based on combined ISSR, RAPD, and SSR data (227markers). Geographic origins are given with the following abbreviations: Japan (J), USA (U), France (F), and Spain (S). The scale indicates the genetic distance. (Ikegami et al., 2009)

2.4.2 SSR Markers

The microsatellites consist of two-, three- or four-nucleotide sequences repeated in tandem (always in the same direction). They are uniformly distributed in several copies over the whole genome of a species and have a high polymorphism rate. This polymorphism is based on the number of repetition units constituting the microsatellite variation. It is a molecular analysis tool of choice for studying genetic diversity, especially in plant species, since they are an essential component of their genomes and are distributed throughout the genome (Deng et al., 2016). Several studies have been based on SSR as a powerful molecular tool to elucidate genetic variability fig. Among other things, the work of Boudchicha, in 2019, in the context of her thesis, which was based on molecular characterization by the use of 22 SSR-type molecular markers on 34 cultivars represented by 77 fig accessions made it possible for the first time to remove the veil on the state of the diversity of this species in Algeria.

Unlike another study carried out by Essid and her research team in 2015, which aimed to Analyze the genetic diversity of accessions of Tunisian caprifig (*Ficus caric* L.) using simple sequence repeat markers (SSR). However, the results of this study showed a low genetic diversity. Indeed, even though the sampling sites were very diverse and distributed over different territories, no clear grouping based on geographical origin is observed, suggesting a generalized exchange of caprifig plant material by vegetative multiplication between different areas of the Tunisia territory.

In the same context, another study conducted by Saddoud and his collaborators in 2005 used microsatellites to characterize 16 cultivars (*Ficus carica* L.). This study suggests a high polymorphism rate: 4 to 12 alleles per locus, and mean heterozygosity of 0.656 were noted. The resolution power (Rp) of the six microsatellites tested ranged from 2.12 to 3.87 for the 16 cultivars studied, showing significant genetic diversity (Ht = 0.762). According to this study, the characterization of accessions belonging to different varieties was possible, showing the power and effectiveness of the molecular tools used.

2.5 *Biochemical Traits and Phytochemical Contents Divers*

Plant secondary metabolites, often known as phytochemicals, are non-nutritive plant metabolites that are crucial for plant survival, healthy growth, and reproduction. Most of these components have biological properties that regulate animal biochemistry and metabolism and have the potential to impact human health. The leaves of the *Ficus carica* have the greatest diversity of compounds, with the largest density of all compounds (except aldehydes and monoterpenes), trailed by the pulps and peels of the fruits (Barolo et al., 2014; Salem et al., 2013). Numerous bioactive substances, for example, phenolic compounds, anthocyanin structures, triterpenoids, coumarins, phytosterols, organic acids, and volatile composites such as aliphatic alcohols, hydrocarbons, and several more chemicals, were identified in phytochemical research on *Ficus carica* L. secondary metabolites from several portions of the *Ficus carica* L. (Fig. 4.3). The majority of fig species include phenolic composites, organic acids, and volatile complexes. The volatile compounds found in the fruits (Trad et al., 2012; Mawa et al., 2013; Ware et al., 1993; Gibernau et al., 1997) and leaves (Mawa et al., 2013; Wang et al., 2017) of *F. carica* are classified as terpenes (monoterpenes and sesquiterpenes), alcohols, aldehydes, ketones, esters, and some other compounds. Terpenes, including monoterpenes (C10) and sesquiterpenes (C15), are the most common plant secondary metabolites. In typical atmospheric circumstances, these chemicals' high vapor pressures allow considerable escape into the air (Trad et al., 2012). Monoterpenes like linalool and, more prominently, epoxylinalool are linked to the fig/wasp linkage (Trad et al., 2012), where they play a crucial role in attracting certain pollinators. Although sesquiterpenes made up only 3% of total volatiles in Tunisian cultivars, they were the most common chemicals found in leaves, with germacrene D, β-caryophyllene, and τ-elemene being the most common (Barolo et al., 2014). Different portions of the *F. carica* were found to contain prevalent terpenes.

α-Pinene, one of the principal monoterpenes reported in many studies, has only been detected in fruits, while sesquiterpenes have only been found in monovarietal fig spirits from Portugal (Rodríguez-Solana et al., 2018). Fruit and leaves include common chemicals such as menthol, τ-muurolene, and τ-cadinene [2, (Barolo et al., 2014). The more advanced chemical groups in matured fruits include alcohols, esters, and ketones, which account for 41% of the total aroma.

Different alcohols were identified in fruits [(Z)-3-hexen-1-ol], leaves [2-methyl-1-butanol and 1-heptanol], spirits [methanol, ethanol, 1-propanol, 1-butanol, isobutyl alcohol, 1-hexanol, octanol, and decanol, among others], in both fruits and leaves [1-penten-3-ol, benzyl alcohol, and (E)-2-nonen-1-ol] (Mawa et al., 2013; Barolo et al., 2014), leaves and spirits [1-heptanol] (Barolo et al., 2014; Rodríguez-Solana

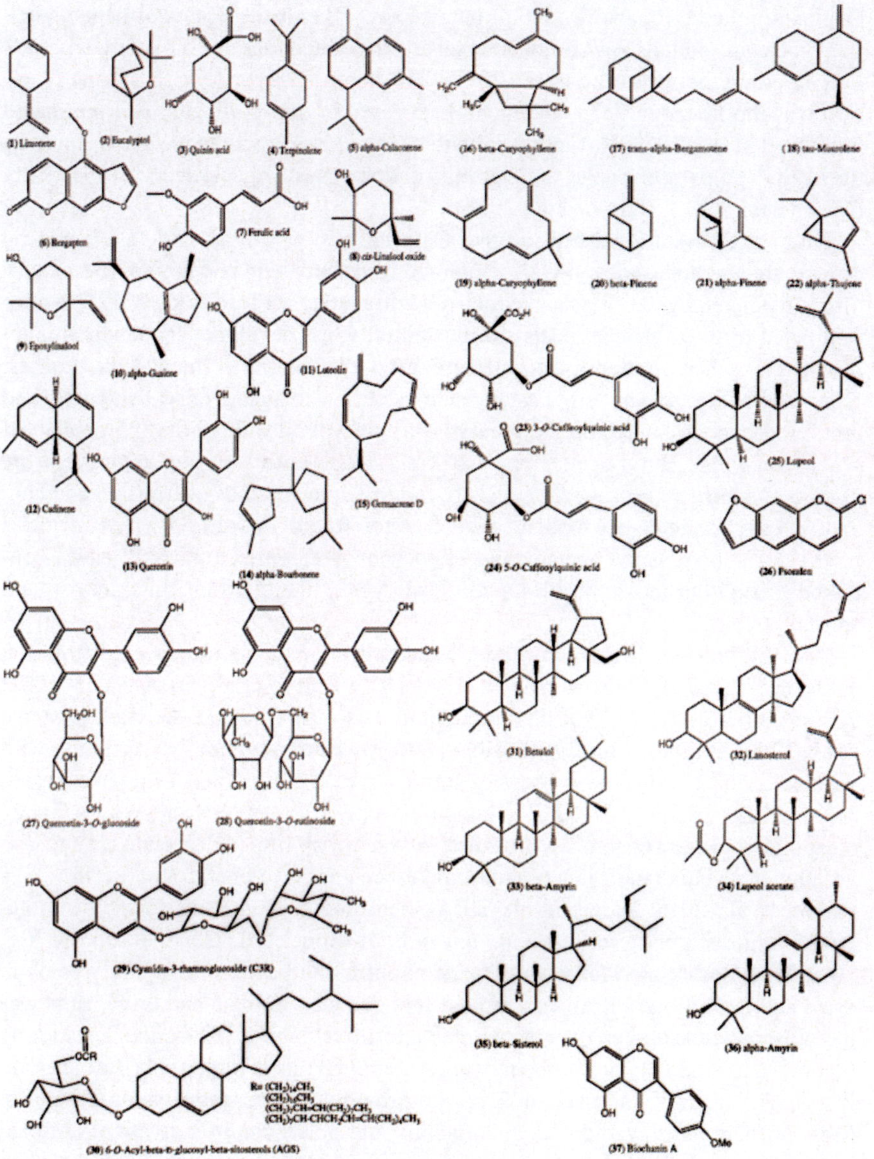

Fig. 4.3 Chemical structures of compounds from *Ficus carica*. (Mawa et al., 2013)

et al., 2018), and lastly in raw materials then spirits [3-methylbutanol, phenylethyl alcohol] (Mawa et al., 2013; Barolo et al., 2014; Miličević et al., 2017). It is essential to remind that the production process's processing conditions influence the concentration. Since this component is naturally present in fruits, higher concentrations were detected in spirits made from fresh figs, while lower amounts were identified in spirits made utilizing immobilized yeast cell technology (Miličević et al., 2017).

Heptanal, octanal, nonanal, 2-methyl-butanal, (Z)-2-heptenal, (E,E)-2,4-heptadienal, (E)-2-octenal, and (E,Z)-2,6-nonadienal are the aldehydes found solely in fruits (Mawa et al., 2013; Barolo et al., 2014). The ketone 3-hydroxy-2-butanone (acetoin) was the first significant chemical discovered in non-pollinated and pollinated figs (Trad et al., 2012). 6-Methyl-5-hepten-2-one and 3-pentanone were found in the fruits (pulps and peels) and leaves of Portuguese fig cultivars, respectively (Mawa et al., 2013; Barolo et al., 2014).

Esters are essential to fruit aroma, especially in ripe figs. They are made by esterifying alcohols and acyl-CoAs obtained from fatty acid and amino acid metabolism, mediated by the enzyme alcohol-o-acyltransferase (Trad et al., 2012). In non-pollinated fruits, these chemicals are substantially less developed, and their amount diminishes when immobilized cells are used during the fermentation process. Esters, notably fatty acid ethyl esters such as ethyl decanoate, ethyl octanoate, and ethyl dodecanoate, make up the greatest class of volatiles in spirits (95 percent of whole volatiles). Butyl acetate and isoamyl acetate with a banana odor were the second and third significant chemicals found in Tunisian kinds of fruits (Trad et al., 2012). Ethyl salicylate was also another ester found in fruits. Leaves included methyl butanoate, hexyl acetate, and ethyl benzoate, while fruits and leaves contained methyl hexanoate besides methyl salicylate, identified in fruits, leaves, and spirits.

The shikimic, malic, oxalic, fumaric, and citric acids were recovered from the fruits and leaves of *F. carica* (Mawa et al., 2013; Barolo et al., 2014; Salem et al., 2013; Soni et al., 2014), but the quinic acid was only found in the leaves (Mawa et al., 2013; Barolo et al., 2014). Most plant foodstuffs include phytosterols, with vegetable oils getting the highest quantities. Sterols (modified triterpenes) such as -sitosterol (Salem et al., 2013), as well as the triterpenoids methyl maslinate, oleanolic acid, taraxasterol, w-taraxasterol ester, calotropenyl acetate, bauerenol, 24-ethylenecycloartanol, lupeol, and lupeol acetate, have been found in fig leaves (Mawa et al., 2013; Barolo et al., 2014), betulinic acid in fruits (Wojdyło et al., 2016), while stigmasterol was reported in both (Soni et al., 2014; Joseph & Raj, 2011). Polyunsaturated fatty acids made about 84 and 69% of total fatty acids in dried and fresh *F. carica* fruits. Linoleic acid detected in fresh and dried fruits was the only polyunsaturated fatty acid detected in fresh and dried fruits. Considering monounsaturated fatty acids, the most prevalent in fruits is oleic acid (Barolo et al., 2014). The fig fruit and bark of *F. carica* provided fifteen anthocyanin pigments. Most of them have cyanidin as an aglycone and pelargonidin derivatives (Ahmad et al., 2013). Overall, individual phenolic compounds, including phenolic acid, chlorogenic acid, flavones, and flavonols, were extracted from fresh and dried fig skins of *F. carica.*

Dried figs had total greater levels of phenolics than fresh fruit pulp, owing to the input of the dry skin. The main significant phenolic was quercetin rutinoside (Soni et al., 2014), while microbial β-D-glucans were obtained from Libyan figs of *F. carica* (Joseph & Raj, 2011). The pulps and peels of figs were used to extract phenolic acids such as 3-*O*- and 5-*O*-caffeoylquinic acids, ferulic acid, quercetin-3-*O*-glucoside, quercetin-3-*O*-rutinoside, psoralen, and bergapten, as well as organic acids such as oxalic, citric, malic, shikimic, and fumaric acids (Salem et al., 2013). Many biological activities have been considered and validated on *F. carica* extracts, and bioassay-guided fractionation has, in most cases, permitted the chemical structures involved for such biological effects to be assigned, finalizing much of its traditional applications (Barolo et al., 2014). Phenolic compounds are among the most significant phytochemicals investigated in *F. carica,* having an antioxidant activity (AC). Many of these chemicals can operate as antioxidants in various ways, including reducing agents, hydrogen donors, free radical scavengers, and singlet oxygen quenchers (Mawa et al., 2013). In ethanol (40%) leaf extracts of *F. carica* (solid to liquid ratio 1:60 g/mL, temperature extraction of 60 °C, and 50 minutes of ultrasonic treatment), the highest total flavonoid content (25.0 mg/g) with noticeable scavenging activities against hydroxyl and superoxide anion free radicals in a dependent way were reported (Ahmad et al., 2013).

Different studies have looked at the AC of fruit extracts. Total polyphenolic content (TPC) and total flavonoid content (TFC), as well as the quantity and pattern of anthocyanins, were measured in extracts from six commercial fig types using the ferric reducing antioxidant technique (FRAP). TPC, TFC, and anthocyanins (cyanidin-3-*O*-rutinoside as the major constituent) were found in abundance in the samples with the greatest AC (Mawa et al., 2013; Ahmad et al., 2013). Using *in vitro* scavenging properties on DPPH·, superoxide, and hydroxyl radicals as well as reducing power tests, two fruit extracts [water (WE) and crude hot water-soluble polysaccharide (PS)] were assessed for AC in another study. Both extracts demonstrate significant DPPH· scavenging activity [WE (EC_{50}, 0.72 mg/ml) and PS (EC_{50}, 0.61 mg/ml)], while PS has the greatest superoxide radical scavenging activity (EC_{50}, 0.95 mg/ml) and the hydroxyl anion radical (43.4% at 4 mg/ml concentration) (Ahmad et al., 2013). Holm oak acorns, in addition to carob pods, were compared to ethanolic extracts from white Beni Maouche Algerian figs (Amessis-Ouchemoukh et al., 2017).

Fig extracts had lower DPPH· scavenging efficacy (20.5%) than ABTS radicals (68.9%) but better phosphomolybdenum lowering ability (638 mg Gallic Acid Equivalent, GAE, per 100 g). This extract (73.1%) also prevented the generation of the Fe2 + −ferrozine complex and effectively scavenged H_2O_2. The nitric oxide (NO) radical scavenging capacities of the extracts from the three studied (carob, acorns, and figs) fruits were not significant (Amessis-Ouchemoukh et al., 2017). Chemiluminescence utilizing lucigenin was used to assess the generation of reactive oxygen species (ROS) in the presence of ethanolic fig extract. After 15 minutes of treatment, the extract at the highest concentration (250 μg/mL) appeared to achieve its higher level of lucigenin inhibition, with a value 44% lower than that procured with diphenylene iodonium (0.2 mM), the standard preferential inhibitor of nicotinamide adenine dinucleotide phosphate (NADPH) oxidase tested (Amessis-Ouchemoukh et al., 2017).

The enzyme xanthine oxidase (XO), which produces reactive oxygen species, was inhibited by ethanolic fig extracts. As a result of its prooxidant impact, the extracts evaluated at 500 μg/mL exhibited a reduction in the inhibition of XO activity. The high correlation coefficients between XO inhibitory activity and phenolic substances and flavonoids show XO's inhibition activity. The equilibrium between the synthesis of reactive oxygen species (free radicals) and antioxidant defenses are disrupted in oxidative stress, resulting in the formation of reactive oxygen species (ROS). Four groups: Streptozotocin-induced diabetic rats, diabetic rats given a single dose of a basic fraction of *F. carica* leaf extract, diabetic rats given a single dose of a chloroform fraction of the extract, and normal rats were used to study the role of these free radicals in the creation of tissue destruction in diabetes mellitus. Diabetic animals had increased erythrocyte catalase activity, vitamin E plasma levels, monounsaturated and polyunsaturated fatty acids, saturated fatty acids, and linoleic acid levels than control animals. All *F. carica fractions* were shown to correct the fatty acid and plasma vitamin E levels of diabetic animals. The isolation of a few groups of plant metabolites has resulted from phytochemical studies on *F. carica.* Most phytochemical studies on *F. carica* have focused on the leaves and fruits, with little knowledge of the stem and root phenolic profiles. However, given the wide range of traditional applications and well-established pharmacological effects of *F. carica,* there is still unlimited attention for phytochemical research employing bioassay-guided isolation. Future studies in the areas above will provide convincing support for the future therapeutic usage of *F. carica* in modern medicine.

2.6 *Proteomic and Transcriptomic Analysis*

Multi-omics provides data integration and processing to acquire insights into the interrelationships, functioning, and biological processes at several levels of biological systems (Fabres et al., 2017; Padden et al., 2014). In recent years, multiple research projects on the transcriptome elements of fig fruit growth and ripening have been conducted. Transcriptomic comparison of young San Pedro type fig and common fig fruit revealed that zeatin biosynthesis and plant hormone signaling pathways are differentially regulated (Chai et al., 2017), and ethylene synthesis and phytohormone signaling were found to be differentially expressed between caprifig and common fig fruit (Ikegami et al., 2013). Fruit cell wall-modification enzymes, genes encoding ethylene-response factors (ERFs), and ascorbate oxidase were significantly increased during fig ripening at the transcriptome level (Freiman et al., 2014). The final biofunction executor is the protein, the end outcome of gene expression. Proteomic research of essentially biological processes, including fruit growth and ripening, may give more detailed insights (Palma et al., 2011). Isobaric tags for relative and absolute quantitation (iTRAQ) are a commonly utilized quantitative method in proteomics investigations that enhances protein quantification accuracy and reliability over traditional 2D-polyacrylamide gel electrophoresis (PAGE) (Miyagi & Rao, 2007). As a result, it might help reveal and speculate on the extensive and widespread alterations that underpin critical fruit ripening, quality determination, and horticultural trait creation, making commercial and market-driven fig quality management easier for innovation.

In recent work, researchers used BLASTN (Basic Local Alignment Search Tool for nucleotide sequence analysis) algorithm to compare the projected transcriptomes of two fig cultivars: the Italian cv Dottato and the Japanese cv. Horaishi (Raskovic et al., 2016). While most genes were discovered in both cultivars, a small number of genes were detected only in the Japanese or Italian cultivars. Phosphoglycerolipid metabolism, which is involved in membrane composition and signal transmission, and cyano amino acid metabolism, which is engaged in chemical defense against herbivores and pathogens, were considerably over-represented in the cv. Dottato projected transcriptome Among Kyoto Encyclopedia of Genes and Genomes (KEGG) Pathways. For RNA-seq gene expression investigations, the availability of a gDNA-based reference transcriptome is the ideal choice. For example, in tree species under abiotic stress, such a transcriptome was applied (Fabres et al., 2017; Sun & Hu, 2016). Such reference transcriptome is available for fig (Raskovic et al., 2016). The availability of a genotype's expected transcriptome allows for a more accurate and thorough examination of gene expression in that genotype. Furthermore, increasing the number of reference transcriptomes for a species allows for the characterization of that species' pan-genome. The fig reference transcriptome had 41,857 predicted genes from *F. carica* cv. Dottato.

Gene ontology and metabolic pathways were used to classify predicted genes. There were disparities in the cultivars'anticipated gene repertoires, with 4803 and 2383 genes discovered in the Horaishi and Dottato predicted transcriptomes. Several genes are unique to the cv. Dottato projected transcriptomes are involved in chemical resistance against herbivores and pathogens, which is essential. Those findings provide a resource for fig functional genomics and may be used to answer issues about growth, protection, and environmental adaptability in this plant species.

3 Multivariate Statistical Analysis of Qualitative and Quantitative Traits

The analysis of the joint behavior of more than one random variable is done using multivariate statistical approaches. Multivariate approaches come in a variety of forms. Multivariate statistical methods have been employed in several research on the genetic diversity of figs. A study was conducted to assess the genetic variability of a fig tree (*Ficus carica* L.) gene bank in the Tunisian Sahel region (Chatti et al., 2004). It was based on morphological characteristics related to the tree's vegetative development. The data obtained were subjected to various statistical analyses using SAS (Statistical Analysis System) software: principal component analysis (PCA) and canonical discriminant analysis, providing the Mahalanobis phenotypic distances for the parameters taken together. A classification dendrogram was established from these distances to translate the phylogenic relationships between the cultivars studied. Means were compared using Duncan's test.

Pereira et al. (2015) investigated the agronomic behavior and quality of six fig cultivars for fresh consumption grown in Extremadura, Spain (the breba and major crops). SPSS for Windows, version 19.0, was used to do a statistical analysis of the data. Multivariate analysis of variance was used to investigate the mean values of the interaction factors 'cultivar-year,' 'cultivar-block,' and 'year-block' for pomological and quality aspects (ANOVA). In addition, Tukey's honestly significant difference (HSD) test ($p \leq 0.05$) was used to examine mean values.

The principal phenolic components, phenolic profiles and antioxidant activity in nine sun-dried fig cultivars with varying skin colors originating from South-Eastern and Middle-Eastern Tunisia are reported (Khadhraoui et al., 2019). Statistical analyses were carried out using an ANOVA with SPSS version 22.0 software. The Duncan test was used to assess significant differences, with a significance level of $p < 0.05$, and Pearson's correlation coefficient (r) was employed to determine correlations between analyzed parameters. Principal component analysis was used to examine the structure of genetic variability across fig cultivars based on mean values of total polyphenol content, phenolic compounds such as phenolic acids and flavonoids, fatty acids, and antioxidant activity (PCA). To examine the role of phenolic chemicals in the differentiation of Tunisian dried fig samples, these multivariate analyses were carried out using the Xlstat 9.0 software.

The morphological and pomological variability of edible fig genotypes belonging to the Smyrna-type was investigated (Khadivi et al., 2018) from the Estahban region in Iran's Fars province. For statistical analysis, the average values of phenotypic data were employed. The mean, minimum, maximum, standard deviation, and coefficient of variation for the measured features were calculated. SAS software was used to assess the variance for all of the characters. SPSS software was used to calculate Pearson correlation coefficients between the attributes. Using the SPSS software, the correlations between the genotypes were determined using principal component analysis (PCA). Using PAST statistics software, a distance matrix created from phenotypic data was used for Ward method cluster analysis to understand the patterns of variability among genotypes better.

Ahmed et al. (2015) characterized 71 farmed, and wild Tunisian fig trees using eight morphological features and 17 simple sequence repeats loci. Statistical evaluation of physical characteristics, the principal component analysis (PCA) was used. A phenetic analysis was conducted using the coordinates of the three first axes of this PCA. The function "hclust" was used to do hierarchical clustering using the unweighted pair-group approach of arithmetic averages (UPGMA). The Kelly Gardnere Sutcliffe penalty function for a hierarchical cluster tree was calculated using the "kgs" package maptree, Version 1.4–7, and a dataset with several clusters. The distance (or similarity) level at which two observations become part of the same cluster is defined as the cophenetic distance between them.

4 Variation in Graft Compatibility of Wild Fig Species as Rootstocks

Grafting is a technique that has been mastered for a long time to cultivate fruit trees and conifers. It was known at least six thousand years BC in the Far East. When we study its long history, we discover that grafting was applied even to herbaceous plants both in China and ancient Greece, far from being limited to fruit trees. Nevertheless, grafting is an essentially conservative operation. This opinion was established little by little because most of our fruit varieties are hybrids, which result from a crossing of one or more other forms and whose qualities cannot be reproduced by sowing their seeds. Thus, many improved varieties have also become incapable of reproducing other than by vegetative means because they no longer even give fertile seeds (Pierre, 1959).

In other contexts, grafting has become an indisputable solution to solve problems for certain fruit varieties which are a severe threat to their crops because of several factors such as soil disturbances, for example, the presence of nematodes (Condit, 1933), soil disease (Hosomi, 2015) and ceratocyst cancer (Figueiredo & Pinheiro, 1969); therefore rootstocks resistant to such soil disorders have long been desired. Some *F. carica* rootstock cultivars are adapted to grow under these unfavorable conditions (Valarini & Tokeshi, 1980; Hosomi, 2015). Nevertheless, these rootstocks are not immune and cannot completely prevent the adverse effects of such soil disorders. In this context, the work of Hosomi and his collaborators in 2015 takes on its value in trying to assess the compatibility of grafts. Their study concluded that A possible compatibility of the common fig tree scion was anticipated on the rootstocks of F. *palmata*, *F. septica*, *F. pumila,* and *F. stipulata*. These rootstocks have previously been classified in the same subgenus as *Ficus* (Corner, 1965), and the observed compatibility probably reflects a taxonomic relationship. Interspecific hybridization of common fig trees and wild Ficus species has been attempted to create a truly resistant rootstock (Yakushiji et al., 2012). According to their results, evaluating the compatibility of species grafts from the various taxonomic groups would help identify the best possible cross parents for hybridization. In the same context, another study was carried out by Yakushiji and his collaborators in 2012 to seek a solution to Ceratocystis canker disease. It is a severe disease of the fig tree (*Ficus carica* L.) which can even kill the tree, caused mainly by the soil fungus Ceratocystis ficicola. During this study, crosses were made, and they concluded that the backcrossing of *F. carica* with the interspecific hybrid between *F. carica* and *F. erecta* contributes to establishing practical fig tree rootstocks with resistance Ceratocystis canker.

5 Trends in Fig Germplasm Conservation

For centuries, fig germplasm has been damaged by severe genetic loss. As a result, it is essential to build up programs worldwide to restore and typify fig genetic resources, and such initiatives have evolved in many countries. As a result, fig germplasm enlistments have been documented in Asia, Europe, North Africa, and North America. With 442, 314, 190, 180, and 154 accessions, respectively, Italy, Turkey (Elbreyli Fig Research Institute), California (National Clonal Germplasm Repository (NCGR) Davis), Turkmenistan (TES-PGR), and France (CBNM Porquerolles) have the greatest collections. In addition, distinct fig cultivars (154) have been gathered from Italy, North Africa, Greece, Turkey, France, Tunisia, and Spain for the CBNM Porquerolles collection.

There are 190 accessions in the NCGR fig gathering. The Institute of Arid Regions of Medenine, Tunisia's Arid Land and Oasis Cropping Laboratory germplasm collection contains 119 native fig accessions and 20 caprifigs gathered from Tunisia, Algeria, and Egypt. In situ or on-farm conservation of fig cultivars in Tunisia is an issue that research institutes, local nongovernmental organizations (NGOs), development agencies, and farmers are tackling. Specific geographic sites were chosen on the way to protect and transplant native fig varieties. In southern Tunisia, the Zammour and Douiret districts were designated traditional places to conserve fig varieties (Mars et al., 2008). The evolving climatic circumstances linked with climate change have become deeply critical in vitro conservation of fig germplasm. Tissue culture is used to maintain genomic resource collections *in vitro* (IPGRI & CIHEAM, 2003).

Cryopreservation and the slow-growth approach are two in vitro conservation procedures that have been established (Engelmann, 2000). Cryopreservation is the preservation of germplasm in liquid nitrogen at 196 °C for long-term storage; cellular divisions and metabolic functions are halted at such temperatures. Genetic instability, on the other hand, poses a challenge to preserved plant materials (Panis et al., 2000). For cryopreservation of fig materials, two procedures are used: the traditional technique and the vitrification technique (Engelmann, 2000). The traditional approach entails two steps: gradual controlled chilling followed by rapid immersion of plant components in liquid nitrogen. A penetrating cryoprotectant, such as dimethyl sulfoxide, ethylene glycol, or polyethylene glycol, is used to treat samples in this approach. This method keeps water from crystallizing inside cells and preventing cell damaging (Panis et al., 2000).

Furthermore, traditional cryopreservation is inefficient for fig tissue such as callus and cell suspension cultures. The vitrification approach was used to preserve plant tissue cultures and is the most commonly used in the plant cryopreservation procedures (Panis et al., 2000). Plant components such as shoot tips, buds, and meristem are cultivated under freezing temperatures in the slow growth technique, which keeps the plant material alive but drastically slows the growth (Engelmann, 2000). The slow-growth approach is the most effective for preserving shoot tips in vitro. Treatment of shoot tips with cryoprotectant for 20 minutes and subsequently chilling to 40 °C resulted in a 50% survival rate for plant regeneration. The most frequent method for cryopreservation of shoot tips, meristems, cell cultures, and somatic embryos is slow growth. Plant regeneration from cryopreserved root tips was produced by adventitious bud development; however, no difference was identified between the cryopreservation and slow-growth procedures for root tip explants (Towill & Bajaj, 2002).

6 Conclusion

Understanding the genetic basis of the complete metabolic system involved in caprification will be critical for developing cultivars with greater yield and longer shelf-life products. Findings show that when studying genetic diversity in fig trees, two methodologies should be considered: morphological and molecular. Molecular analyses should be used in conjunction with morphological and horticultural evaluations because they give additional information, improve the resolving power of genetic diversity investigations, and help understand the domestication process. This has crucial consequences for fig management, necessitating strategies to ensure the species' durability and genetic variety. In addition, knowing about the genetic diversity in this germplasm will make it easier to employ in breeding operations and enhance its management in the face of climate change.

References

Ahmad, S., Bhatti, F. R., Khaliq, F. H., Irshad, S., & Madni, A. (2013). A review on the prosperous phytochemical and pharmacological effects of *Ficus carica*. *International Journal of Bioassays, 2*(5), 843–849.

Ahmed, B. A., Ghada, B., Laila, E., Hafid, A., Bouchaib, K., & Amel, S. H. (2015). Use of morphological traits and microsatellite markers to characterize the Tunisian cultivated and wild figs (*Ficus carica* L.). *Biochemical Systematics and Ecology, 59*, 209–219.

Aljane, F. (2011). Caractérisation et évaluation des accessions locales de figuier *(Ficus carica L,) en Tunisie et sélection des plus performantes* (Doctoral dissertation, Thèse de doctorat, Faculté des Sciences).

Amessis-Ouchemoukh, N., Ouchemoukh, S., Meziant, N., Idiri, Y., Hernanz, D., Stinco, C. M., Rodríguez-Pulido, J. R., Heredia, J. F., Madani, K., & Luis, J. (2017). Bioactive metabolites involved in the antioxidant, anticancer and anticalpain activities of *Ficus carica* L., *Ceratonia siliqua* L. and *Quercus ilex* L. extracts. *Industrial Crops and Products, 95*, 6–17.

Barolo, M. I., Mostacero, N. R., & López, S. N. (2014). *Ficus carica* L. (Moraceae): An ancient source of food and health. *Food Chemistry, 164*, 119–127.

Berg, C. C. (2003). Flora Malesiana precursor for the treatment of Moraceae 1: The main subdivision of *Ficus*: The subgenera. *Blumea-Biodiversity, Evolution and Biogeography of Plants, 48*(1), 166–177.

Boudchicha, R. H. (2019). *Etude de la diversité génétique de quelques variétés locales de figuier (Ficus carica* L.) *en Algérie*. Doctoral dissertation, Université de Batna 2

Çalişkan, O., & Polat, A. A. (2012). Morphological diversity among fig (*Ficus carica* L.) accessions sampled from the eastern Mediterranean region of Turkey. *Turkish Journal of Agriculture and Forestry, 36*(2), 179–193.

Chai, L., Wang, Z., Chai, P., Chen, S., & Ma, H. (2017). Transcriptome analysis of San Pedro-type fig (*Ficus carica* L.) parthenocarpic breba and non-parthenocarpic main crop reveals divergent phytohormone-related gene expression. *Tree Genetics & Genomes, 13*(4), 1–14.

Chatti, K., Salhi-Hannachi, A., Mars, M., Marrakchi, M., & Trifi, M. (2004). Analyse de la diversité génétique de cultivars tunisiens de figuier (*Ficus carica* L.) à l'aide de caractères morphologiques. *Fruits, 59*(1), 49–61.

Condit, I. J. (1933). *Fig culture in California* (Vol. 77). Рипол Классик.

Corner, E. J. H. (1965). Check-list of *Ficus* in Asia and Australasia with keys to identification. *Garden's Bulletin Singapore, 21*, 1–186.

Darjazi, B. B. (2011). Morphological and pomological characteristics of fig (*Ficus carica* L.) cultivars from Varamin, Iran. *African Journal of Biotechnology, 10*(82), 19096–19105.

Deng, P., Wang, M., Feng, K., Cui, L., Tong, W., Song, W., & Nie, X. (2016). Genome-wide characterization of microsatellites in Triticeae species: Abundance, distribution and evolution. *Scientific Reports, 6*(1), 1–13.

Devos, K. M., Brown, J. K., & Bennetzen, J. L. (2002). Genome size reduction through illegitimate recombination counteracts genome expansion in Arabidopsis. *Genome Research, 12*(7), 1075–1079.

Dubin, M. J., Scheid, O. M., & Becker C. (2018). Transposons: A blessing curse. *Current Opinion in Plant Biology, 42*, 23–29.

Engelmann, F. (2000). *Importance of cryopreservation for the conservation of plant genetic resources*. In Cryopreservation of tropical plant germplasm—Current research progress and applications, IPGRI, Rome & JIRCAS (pp. 8–20)

Essid, A., Aljane, F., Ferchichi, A., & Hormaza, J. I. (2015). Analysis of genetic diversity of Tunisian caprifig (*Ficus carica* L.) accessions using simple sequence repeat (SSR) markers. *Hereditas, 152*(1), 1–7.

Fabres, P. J., Collins, C., Cavagnaro, T. R., & Rodríguez López, C. M. (2017). A concise review on multi-omics data integration for terroir analysis in *Vitis vinifera*. *Frontiers in Plant Science, 8*, 1065.

Ferguson, L., Michailides, T. J., & Shorey, H. H. (1990). *The California fig industry*. Horticultural reviews (USA).

Figueiredo, P., & Pinheiro, E. D. (1969). Uma nova doença da figueira (*Ficus carica* L.) na região de Valinhos, SP. *Biológico, 35*, 227–203.

Freiman, Z. E., Doron-Faigenboim, A., Dasmohapatra, R., Yablovitz, Z., & Flaishman, M. A. (2014). High-throughput sequencing analysis of common fig (*Ficus carica* L.) transcriptome during fruit ripening. *Tree Genetics & Genomes, 10*(4), 923–935.

Gafoor, H., Bhattacharjee, A., Hegde, K., & Shabaraya, A. R. (2019). *Ficus carica*: A brief review. *International Journal of Pharmaceutical Sciences Review and Research, 59*(2), 40–43.

Gibernau, M., Buser, H. R., Frey, J. E., & Hossaert-McKey, M. (1997). Volatile compounds from extracts of figs of *Ficus carica*. *Phytochemistry, 46*(2), 241–244.

Hosomi, A. (2015). Variation in graft compatibility of wild *Ficus* species as rootstock for common fig trees (*Ficus carica*). *V International Symposium on Fig, 1173*, 199–206.

Ikegami, H., Nogata, H., Hirashima, K., Awamura, M., & Nakahara, T. (2009). Analysis of genetic diversity among European and Asian fig varieties (*Ficus carica* L.) using ISSR, RAPD, and SSR markers. *Genetic Resources and Crop Evolution, 56*(2), 201–209.

Ikegami, H., Habu, T., Mori, K., Nogata, H., Hirata, C., Hirashima, K., Tashiro, K., & Kuhara, S. (2013). De novo sequencing and comparative analysis of expressed sequence tags from gynodioecious fig (*Ficus carica* L.) fruits: Caprifig and common fig. *Tree Genetics & Genomes, 9*(4), 1075–1088.

IPGRI & CIHEAM. (2003). Descriptors for figs. International Plant Genetic Resources Institute (IPGRI) and the International Centre for Advanced Mediterranean Agronomic Studies (CIHEAM).

Joseph, B., & Raj, S. J. (2011). Pharmacognostic and phytochemical properties of *Ficus carica* Linn-an overview. *International journal of pharmtech research, 3*(1), 8–12.

Khadari, B., Lashermes, P., & Kjellberg, F. (1994). *Identification variétale et ressources génétiques chez le figuier (Ficus carica L.): utilisation des marqueurs RAPD*. AUPELF-UREF.

Khadhraoui, M., Bagues, M., Artés, F., & Ferchichi, A. (2019). Phytochemical content, antioxidant potential, and fatty acid composition of dried Tunisian fig (*Ficus carica* L.) cultivars. *Journal of Applied Botany and Food Quality, 92*, 143–150.

Khadivi, A., Anjam, R., & Anjam, K. (2018). Morphological and pomological characterization of edible fig (*Ficus carica* L.) to select the superior trees. *Scientia Horticulturae, 238*, 66–74.

Lahbib, T. (1984). Etude pomologique des variétés de figuier (*Ficus carica* L.) répertoriées au Sahel tunisien. *Mémoire de Fin d'Études du Cycle de Spécialisation.*

Lisch, D. (2013). How important are transposons for plant evolution? *Nature Reviews Genetics, 14*(1), 49–61.

Mars, M., Chatti, K., Saddoud, O., Salhi-Hannachi, A., Trifi, M., & Marrakchi, M. (2008). Fig cultivation and genetic resources in Tunisia, an overview. *Acta Horticulturae, 798*, 27–32.

Mawa, S., Husain, K., & Jantan, I. (2013). *Ficus carica L.(Moraceae): Phytochemistry, traditional uses and biological activities*. Evidence-Based Complementary and Alternative Medicine.

Miličević, B., Ačkar, Đ., Babić, J., Jozinović, A., Oroz, M., & Šubarić, D. (2017). Impact of the fermentation process with immobilized yeast cells on the aroma profile and sensory quality of distillates produced from two fig (*Ficus carica* L.) cultivars. *Poljoprivreda, 23*(1), 49–55.

Miyagi, M., & Rao, K. S. (2007). Proteolytic 18O-labeling strategies for quantitative proteomics. *Mass Spectrometry Reviews, 26*(1), 121–136.

Padden, J., Megger, D. A., Bracht, T., Reis, H., Ahrens, M., Kohl, M., Eisenacher, M., Schlaak, J. F., Canbay, A. E., Weber, F., Hoffmann, A. C., Kuhlmann, K., Meyer, H. E., Baba, H. A., & Sitek, B. (2014). Identification of novel biomarker candidates for the immunohistochemical diagnosis of cholangiocellular carcinoma. *Molecular & Cellular Proteomics, 13*(10), 2661–2672.

Palma, J. M., Corpas, F. J., & Luís, A. (2011). Proteomics as an approach to the understanding of the molecular physiology of fruit development and ripening. *Journal of Proteomics, 74*(8), 1230–1243.

Panis, B., Swennen, R., & Engelmann, F. (2000, July). Cryopreservation of plant germplasm. In *IV International Symposium on In Vitro Culture and Horticultural Breeding, 560* (pp. 79–86).

Pereira, C., Serradilla, M. J., Martín, A., del Carmen Villalobos, M., Pérez-Gragera, F., & López-Corrales, M. (2015). Agronomic behaviour and quality of six fig cultivars for fresh consumption. *Scientia Horticulturae, 185*, 121–128.

Pereira, C., Serradilla, M. J., Pérez-Gragera, F., Martín, A., Villalobos, M. C., & López-Corrales, M. (2017). Evaluation of agronomic and fruit quality traits of fig tree varieties (*Ficus carica* L.) grown in Mediterranean conditions. *Spanish Journal of Agricultural Research, 15*(3), e0903–e0903.

Pierre, B. (1959). La greffe, important facteur de l'augmentation du rendement des plantes et des animaux. *IMPACT science et société*, IX, n 3.

Podgornik, M., Vuk, I., Vrhovnik, I., & Mavsar, D. B. (2010). A survey and morphological evaluation of fig (*Ficus carica* L.) genetic resources from Slovenia. *Scientia Horticulturae, 125*(3), 380–389.

Raskovic, B., Lazic, J., & Polovic, N. (2016). Characterisation of general proteolytic, milk clotting and antifungal activity of *Ficus carica* latex during fruit ripening. *Journal of the Science of Food and Agriculture, 96*(2), 576–582.

Rodríguez-Solana, R., Galego, L. R., Pérez-Santín, E., & Romano, A. (2018). Production method and varietal source influence the volatile profiles of spirits prepared from fig fruits (*Ficus carica* L.). *European Food Research and Technology, 244*(12), 2213–2229.

Roger, J. P. (2003). L'origine des arbres fruitiers. *Conservatoire botanique national méditerranéen de Porquerolles. Antenne Provence-Alpes, Côte d'Azur.*

Saddoud, O., Salhi-Hannachi, A., Chatti, K., Mars, M., Rhouma, A., Marrakchi, M., & Trifi, M. (2005). Tunisian fig (*Ficus carica* L.) genetic diversity and cultivar characterization using microsatellite markers. *Fruits, 60*(2), 143–153.

Sahin, N. (1997, June). Fig adaptation studies in Western Turkey. In *I International Symposium on Fig, 480* (pp. 61–70).

Salem, M. Z., Salem, A. Z. M., Camacho, L. M., & Ali, H. M. (2013). Antimicrobial activities and phytochemical composition of extracts of *Ficus* species: An overview. *African Journal of Microbiology Research, 7*(33), 4207–4219.

Smartt, J. (1976). *Evolution of crop plants* (pp. 205–208). Longman.

Soni, N., Mehta, S., Satpathy, G., & Gupta, R. K. (2014). Estimation of nutritional, phytochemical, antioxidant and antibacterial activity of dried fig (*Ficus carica*). *Journal of Pharmacognosy and Phytochemistry, 3*(2), 158–165.

Sun, Y. V., & Hu, Y. J. (2016). Integrative analysis of multi-omics data for discovery and functional studies of complex human diseases. *Advances in Genetics, 93*, 147–190.

Towill, L. E., & Bajaj, Y. P. S. (2002). *Biotechnology in agriculture and forestry*. Cryopreservation of plant germplasm II.

Trad, M., Ginies, C., Gaaliche, B., Renard, C. M., & Mars, M. (2012). Does pollination affect aroma development in ripened fig (*Ficus carica* L.) fruit? *Scientia Horticulturae, 134*, 93–99.

Usai, G., Mascagni, F., Giordani, T., Vangelisti, A., Bosi, E., Zuccolo, A., Ceccarelli, M., King, R., Hassani-Pak, K., Zambrano, L. S., Cavallini, A., & Natali, L. (2020). Epigenetic patterns within the haplotype phased fig (*Ficus carica* L.) genome. *The Plant Journal, 102*(3), 600–614.

Valarini, P. J., & Tokeshi, H. (1980). *Ceratocystis fimbriata*, causal agent of fig dieback, and its control. *Summa Phytopathologica, 6*(3/4), 102–106.

Vangelisti, A., Simoni, S., Usai, G., Ventimiglia, M., Natali, L., Cavallini, A., Mascagni, F., & Giordani, T. (2021). LTR-retrotransposon dynamics in common fig (*Ficus carica* L.) genome. *BMC Plant Biology, 21*(1), 1–18.

Wang, T., Jiao, J., Gai, Q. Y., Wang, P., Guo, N., Niu, L. L., & Fu, Y. J. (2017). Enhanced and green extraction polyphenols and furanocoumarins from fig (*Ficus carica* L.) leaves using deep eutectic solvents. *Journal of Pharmaceutical and Biomedical Analysis, 145*, 339–345.

Ware, A. B., Kaye, P. T., Compton, S. G., & Van Noort, S. (1993). Fig volatiles: Their role in attracting pollinators and maintaining pollinator specificity. *Plant Systematics and Evolution, 186*(3), 147–156.

Wojdyło, A., Nowicka, P., Carbonell-Barrachina, Á. A., & Hernández, F. (2016). Phenolic compounds, antioxidant and antidiabetic activity of different cultivars of *Ficus carica* L. fruits. *Journal of Functional Foods, 25*, 421–432.

Yakushiji, H., Morita, T., Jikumaru, S., Ikegami, H., Azuma, A., & Koshita, Y. (2012). Interspecific hybridization of fig (*Ficus carica* L.) and Ficus erecta Thunb., a source of Ceratocystis canker resistance. *Euphytica, 183*(1), 39–47.

Chapter 5
Bud Structure and Evolution

Giuseppe Ferrara and Andrea Mazzeo

1 Fig Buds Description

In temperate climates, the buds of deciduous fruit trees are dormant during the autumn and winter when the coldest temperatures occur. It is commonly assumed that this rest period consists of an endo-dormancy phase, followed by an eco-dormancy phase, and finishes with para-dormancy during the growing season (Lang et al., 1987; Crabbe, 1994; Faust et al., 1995, 1997). Para-dormancy is synonymous with correlative inhibition as manifested in apical dominance (AD) (Crabbe & Barnola, 1996) and occurs during the vegetative/reproductive phases in the growing season. In contrast, endo-dormancy requires a cold period to release buds from an 'inactive' stage. Apart from overcoming AD, lateral buds still have to meet their specific chilling requirement (Faust et al., 1995). Terminal buds are generally easier to release from dormancy than auxiliary buds (Crabbe, 1994). The fig tree (*Ficus carica* L.) is one of the oldest perennial fruit species cultivated in the Mediterranean. According to Eisen (1901), *Ficus carica* (common fig) is a deciduous species in most of the world (Fig. 5.1) but is evergreen in tropical areas (Condit, 1947). The common fig produces both vegetative and flower buds at the axil of the leaves (Fig. 5.2) in summer-fall, and some of them evolve into fruits during the current season (main crop), and others could overwinter (breba crop in the successive season or dormant and latent mixed/vegetative buds). Vegetative buds are generally smaller, thinner, and conical (Fig. 5.3), whereas the flower buds are larger and rounded (Fig. 5.4), although often the difference is relatively small (Condit, 1947; Sarkhosh et al., 2022). The terminal bud is a mixed bud, which is larger and longer than the other two, conical, and has an attenuate tip (Fig. 5.5). This bud develops in

G. Ferrara (✉) · A. Mazzeo
Department of Soil, Plant and Food Sciences, University of Bari "Aldo Moro", Bari, Italy
e-mail: giuseppe.ferrara@uniba.it; andrea.mazzeo@uniba.it

M. F. Ramadan (ed.), *Fig (Ficus carica): Production, Processing, and Properties*, https://doi.org/10.1007/978-3-031-16493-4_5

Fig. 5.1 Fig orchard in July (**a**) and during winter rest (**b**)

spring, producing leaves and flower buds, either producing fruit (main crop), or overwintering flower and vegetative buds. Sometimes sub-apical or lateral buds are also mixed buds.

Unlike many temperate species that have a single blooming season and differentiate the buds, for the successive spring flowering, during the previous summer or autumn, the common fig continually produces flower buds during the growing season (Fig. 5.6). Fig buds require little or no winter chilling to break endo-dormancy (Ferguson et al., 1990), and growth resumes in early spring (Flaishman et al., 2008). Under temperate growth conditions, with no limiting water and nutrient conditions, the common fig produces several flower buds during the growing season until the autumn developing into fruits (main crop), or these buds then overwinter and produce a first spring fruit crop (called brebas or breba figs). The brebas crop may be destroyed in cold climates by late spring frosts (Fig. 5.7). The brebas are borne on the previous year or "old" wood, the 1-year shoot, whereas the second fruit crop or main crop is borne on the shoot grown during the current season (Fig. 5.8). The leaves fall at the end of the growth period, and the tree enters the endo-dormancy

Fig. 5.2 Developing fruits (from flower buds) and vegetative buds at the axil of the leaves of the shoot

Fig. 5.3 Small and thin vegetative buds between two developing fruits

period. The length of the dormant season and the budburst time in spring depend upon local climatic conditions. Due to strong apical dominance in many cultivars, the seasonal growth is either from the terminal bud or the adjacent one or two buds (Fig. 5.9), even though apical dominancy varies depending on the cultivar (Ersoy et al., 2008).

The number and sizes of flower buds are closely related either to the vegetative growth of the tree or to the size of the crop maturing in summer and fall. Caprifig trees, which have a light summer and fall crop, commonly produce an enormous number of flower buds that grow in the spring and mature a crop in June, the profichi

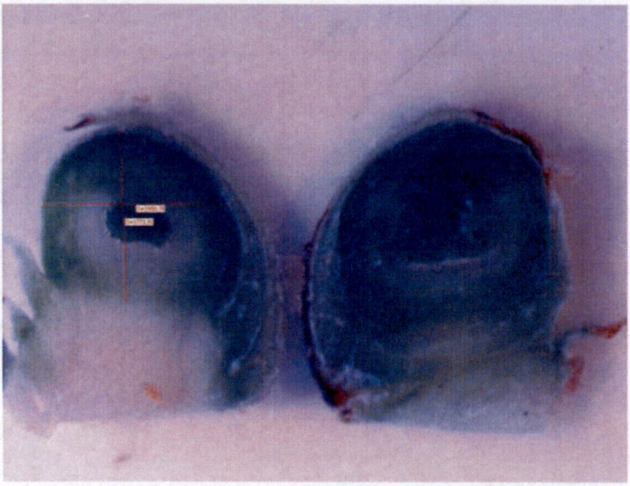

Fig. 5.4 Longitudinal section of a rounded fruit bud. Measures are indicated in μm

Fig. 5.5 Longitudinal section of a conical terminal mixed bud

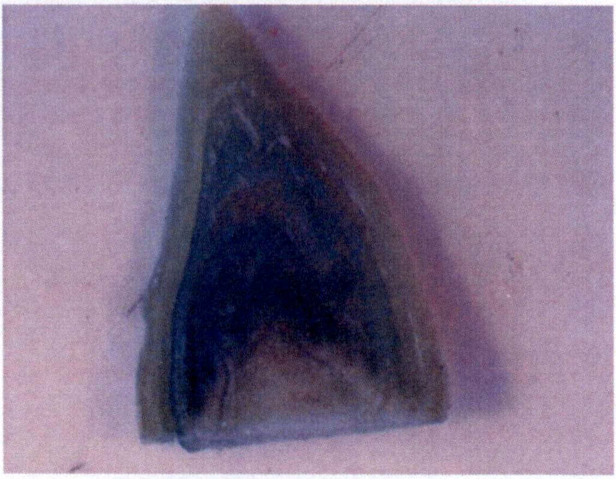

crop (Condit, 1947). The color of the fruit bud is approximately similar to that of the terminal mixed bud, from greenish to reddish. The color of the terminal bud is largely correlated with the color of the fruit, most green-fruited cultivars have green buds, and most dark-fruited cultivars have buds with the shade of red-violet color (Fig. 5.10). Some cultivars having dark figs, such as Col de Dame Noir and Gouraud Rouge, show green terminal buds; vice versa, some with green figs, notably Genoa, have colored buds. All the fig buds are developed on the same shoot, i.e., fruit (flower) bud, vegetative bud, and mixed bud. In the common fig, the basal portion of the shoot holds flower buds evolving into the main crop, whereas the distal part of the shoots holds the overwintering flower buds developing (or not) the brebas in the successive spring. The buds can be isolated but often are aggregated with two

Fig. 5.6 Main crop fruits developing along the growing shoot

Fig. 5.7 Buds injured by late spring frost. Necrosis is visible on flower, vegetative and mixed buds

flower buds and a vegetative bud in the middle as in the cultivar Sultani (Fig. 5.11) or only a flower bud and a vegetative bud or even one/two flower bud (Fig. 5.12). When two flower buds and one vegetative bud are found in the axil of each leaf, rarely more than one of the flower buds ever develops into a fig in the Mission

Fig. 5.8 Presence of both brebas (bigger and on the 1-year shoot) and main crop fruits (smaller and on the current year shoot)

cultivar, and the remaining bud generally remains latent, while in Kodota and Calimyrna or some other cultivars it is quite common for two figs in the axil of an individual leaf to develop to maturity (Petrucci & Crane, 1950). Some fig cultivars also have flower buds in the distal portion, which may develop in late summer-autumn into the fruit as a third crop (or even the only main crop), in some areas defined as the late main crop (Christmas or winter fruits).

The fig bears numerous latent buds, generally protected by tick scales, abundant on 2–3-year-old branches, and which sometimes can burst and produce fruits but of poor quality (Grassi, 1991); these latent buds are often permanently arrested in their development and never burst.

The breba or first crop of figs is borne laterally on the wood of the previous season's growth from buds produced in the axils of leaves along the distal portion of the shoot. These buds enlarge the following spring and mature into fruit during the month of May–June. Morphological and developmental studies conducted by Duric et al. (1992) during dormancy and growth on cv. Tenica bearing a single crop and cv. Petrovaca bearing two crops per year reveal that three types of buds occur on 1-year-old shoots. (1) Flower buds which are spherical, have 3–5 scales, with the

Fig. 5.9 Growth of apical and lateral mixed and vegetative buds. The growing brebas are also visible

inflorescence axis in the internal cavity enclosing female primordia differentiated to varying degrees depending on the position of the bud on the shoot. These buds produce parthenocarpic fruits only in cv. Petrovaca. The number and the size of flower buds are closely correlated with the vegetative growth and determine the size of the summer and the fall crops (Flaishman et al., 2008). (2) Mixed buds are conical, made up of 5–8 scales, and inflorescences on shoots arising from these buds produce fruit in both cultivars. (3) Vegetative buds, made up of 3–5 scales, usually remain dormant and may produce shoots in the successive season. The scales covering the vegetative bud absciss as the bud elongates and develops into a shoot. The distal buds that do not produce fruit during the current season are in para-dormancy and overwinter, giving rise to the brebas the following summer.

2 Development and Evolution of Inflorescence

The initiation and development of flower buds was macroscopically examined by Tanrisever (1998). Primordia, which will develop into inflorescence, were generally covered with three but sometimes with four scales. Petrucci & Crane (1950) reports that the first visible evidence of inflorescence differentiation is the broadening and elongation of the axis and the initiation of scales that eventually surround the ostiole of the syconium. The initial 3–4 scales that cover the primordium are persistent and, upon maturity of the fig, are located at the apex of the stalk. As observed at the microscope, at the beginning of the growing season, in the cultivar Petrelli, the

a b

Fig. 5.10 Greenish scales of flower and mixed buds of a cultivar producing fruits with a green skin (a) and reddish scales of flower and mixed buds of a cultivar producing fruits with a red-violet skin (b)

Fig. 5.11 Presence of three buds, two growing flower buds (main crop) and a vegetative bud in the middle (cultivar Sultani)

Fig. 5.12 Presence of two buds at each node, a single flower bud and a small vegetative bud

Fig. 5.13 Microscopic observation of a flower bud in the distal portion of the current season shoot with small, short and curved ostiolar scales and very small flower primordia (cultivar Petrelli)

flower buds in the distal portion of the current season shoot have small, elongated, and curved ostiolar scales with very small female flower primordia (Fig. 5.13).

In contrast, in the basal flower buds, the ostiolar scales were much larger, with female flowers more developed in the cavity of the syconium (Fig. 5.14) (Marcotuli et al., 2020). In the Dottato cultivar, differences in the development of the flower buds were also evident; in the smaller flower buds at the apical position, the ostiolar scales were still underdeveloped, and the female flower primordia were small and almost rounded (Fig. 5.15); whereas at the basal nodes the flower buds had ostiolar scales completely developed and the structures of the female flower primordia were more evident (Fig. 5.16) (Marcotuli et al., 2020). Further development of the inflorescence primordium consists of a continued broadening of the apex, the cells around the periphery divide and give rise to many ostiolar scales, and the single flowers continue to grow, assuming a more cylindrical shape (Fig. 5.17). At this stage of development, the apex of the axis changes from convex to a concave shape. Following

Fig. 5.14 Microscopic observation of a flower bud in the basal portion of the current season shoot with large, elongated and curved ostiolar scales and developed female flower primordia (cultivar Petrelli)

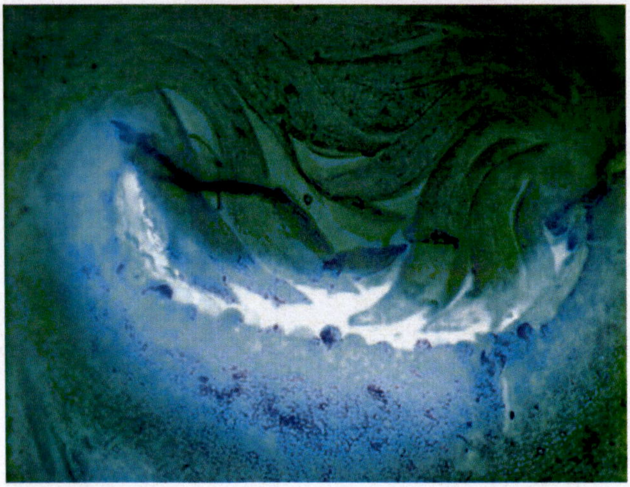

Fig. 5.15 Microscopic observation of a flower bud in the distal portion of the current season shoot with small, underdeveloped ostiolar scales and very small female flower primordia (cultivar Dottato)

this stage, inflorescence primordia broaden and elongate, forming the neck and stalk of the syconium. The formation of the syconium was completed when the apical portion of the cup-shaped structure grows towards the center and forms an ostiole closed with numerous scales. Buds destined to become flower or vegetative buds are identical in their initial stages of development. There are generally no visible inflorescence primordia in the mixed buds before bud burst (Fig. 5.18), as observed in summer for the flower buds, but only an undifferentiated structure until the end of winter with few exceptions (Fig. 5.19). The mixed bud will first differentiate inflorescence primordia after bursting, and the flower primordia will completely differentiate into the developing syconium for the main crop in spring-summer. In the

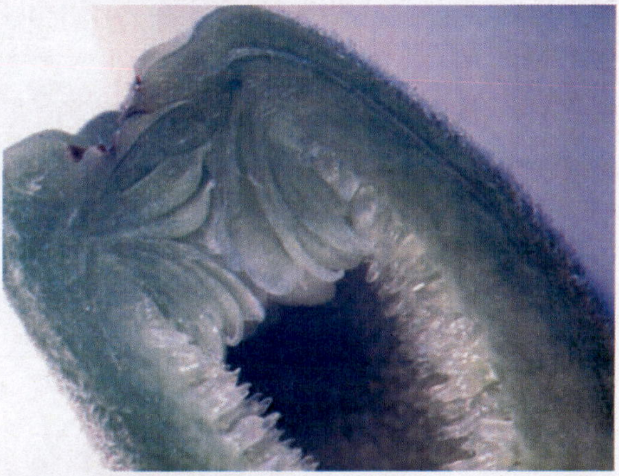

Fig. 5.16 Stereomicroscopic observation of a flower bud in the basal portion of the current season shoot with large, elongated and curved ostiolar scales and developed female flower primordia with evident ovary and style (cultivar Dottato)

Fig. 5.17 Microscopic observation of developing female flowers in the receptacle of main crop

distal flower buds, flower primordia will develop into the buds but will rest their differentiation during the winter and complete the development in the successive spring with the production of brebas (Marcotuli et al., 2020).

Flower buds (of brebas) had already formed florets (not completely differentiated) at the beginning of summer, with a receptacle characterized by masses of cylindrical pistillate primordia (Fig. 5.20). They remained at this stage from summer until the end of winter before bud burst (Marcotuli et al., 2020). The X-ray analyses of flower buds either developing into brebas or the main crop showed these buds very similar at the first stages of development with the difference only in the number and thickness of bud scales; the flower primordia in the receptacle are almost identical. This light difference may suggest that the flower buds of the brebas are prone to overwinter (number and thickness of scales), whereas the flower buds of the main crop have lighter scales because they have to develop in the current season and do not need to overwinter (Marcotuli et al., 2020). In Dottato, the flower buds of the brebas (Fig. 5.21) and main crop (Fig. 5.22) are almost identical, taken at the same stage.

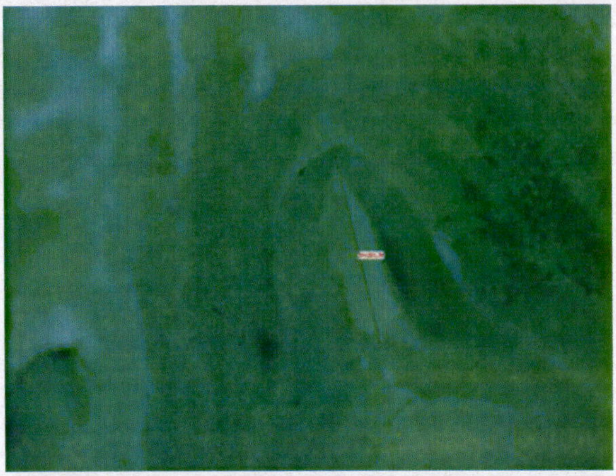

Fig. 5.18 Microscopic longitudinal section of a mixed bud covered with scales and the apex is visible at the bottom during winter rest

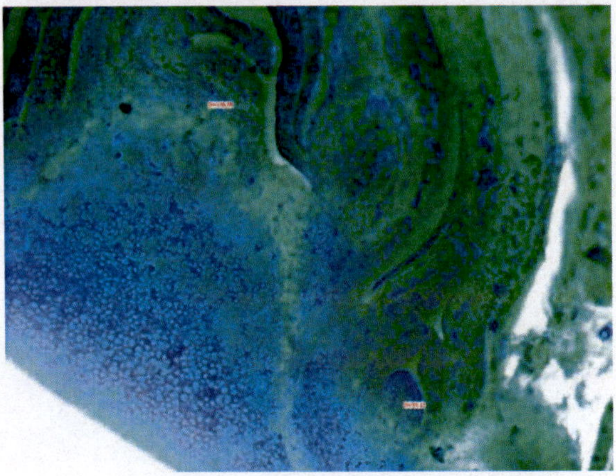

Fig. 5.19 Microscopic longitudinal section of a mixed bud covered with scales and the apex is visible at the bottom before bursting. The meristematic apex and a possible main crop are also indicated

The flower bud of breba is covered with several bud scales since it is an overwintering bud. In the case of Petrelli, the two types of flower buds also differentiate for the more elongated shape and less bud scales of the main crop (Fig. 5.23) with respect to the brebas (Fig. 5.24) (Marcotuli et al., 2020).

Bud's position on a shoot about fruitfulness has been investigated, and Petrucci & Crane (1950) reported that although inflorescence primordia are located at each node on the shoot, the buds at the distal position of the shoot are less fruitful. The fruitfulness increases from the third node, where together with nodes 4 and 5, the main crop is mainly produced. However, fruitfulness strongly depends on other factors such as crop load, vegetative activity, cultural practices, cultivar, etc. (Ferrara et al., 2016, 2017; Marcotuli et al., 2019).

The flower bud differentiation begins at the basal nodes at the beginning of the growing season and proceeds in the successive nodes depending on the cultivar, climatic conditions, cultural practices, and resource availability (water, nutrients). The first nodes will differentiate into flower buds and develop the syconium

Fig. 5.20 Microscopic longitudinal section of a flower bud (brebas) in summer with evident female flowers primordia

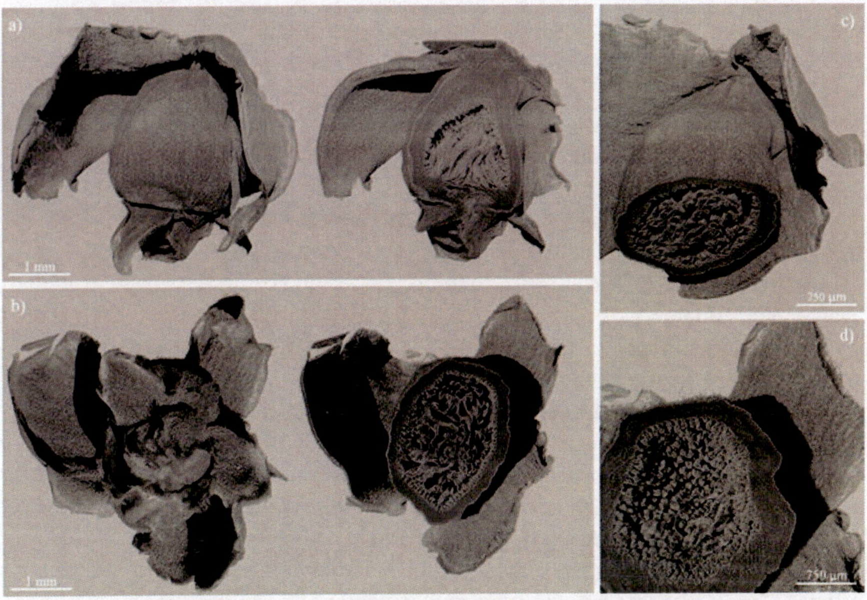

Fig. 5.21 X-ray picture of a small Dottato first crop (brebas) at the beginning of development in spring. Bud scales, florets and ostiolar scales are clearly evident. (From Marcotuli et al., 2020)

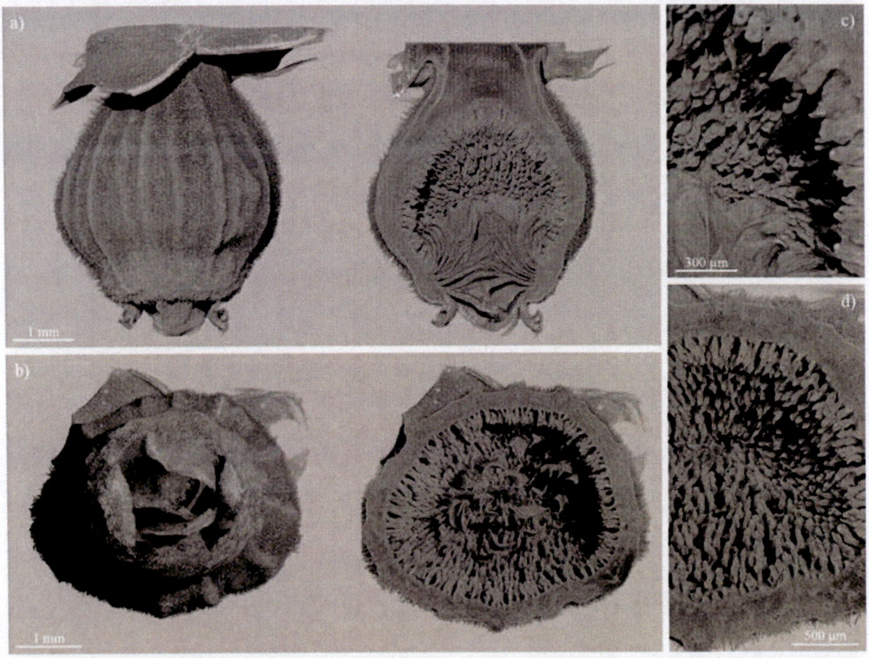

Fig. 5.22 X-ray picture of a small Dottato main crop at the beginning of development. Florets and ostiolar scales are clearly evident but the bud scales are much less developed than in the first crop (brebas). (From Marcotuli et al., 2020)

containing the inflorescences until the ripening of the fruit (main crop). In the distal nodes, the flower bud differentiation will start during the season but stop at the summer's end (Marcotuli et al., 2020). The flower buds of brebas are mainly located in the distal nodes, a portion of the shoot which grows at the end of the growing season, as indicated by the shorter length of the internodes, and these buds may be more susceptible to erratic climatic conditions at the end of summer (rain, cold, heat waves, etc.). However, as previously reported, flower buds are already differentiated even at the distal nodes in summer (July–August). In a 3-year study carried out in Tunisia, the main crop amount exerted a strong influence on the fruiting intensity of the successive year; in particular, a heavy main crop load decreased the number of brebas in the following year, while a heavy breba crop load reduced the main crop load of the current season (Gaaliche et al., 2011) in a sort of productive balance by the tree for the use of its resources. In the case of a heavy main crop, the number of flower buds (brebas) overwintering is clearly lower since almost all buds developed into the main crop during the season. Conversely, the higher number of flower buds (brebas) retained on one-year shoots and developing in the successive season will compete with both the current year's shoot growth and the main crop fruits, thus reducing both the yield of the main crop and the shoot length (Marcotuli et al., 2020).

The most relevant fruit tree species, such as sweet cherry, apple, peach but also grape, have flower (mixed) buds which start to differentiate during the summer of

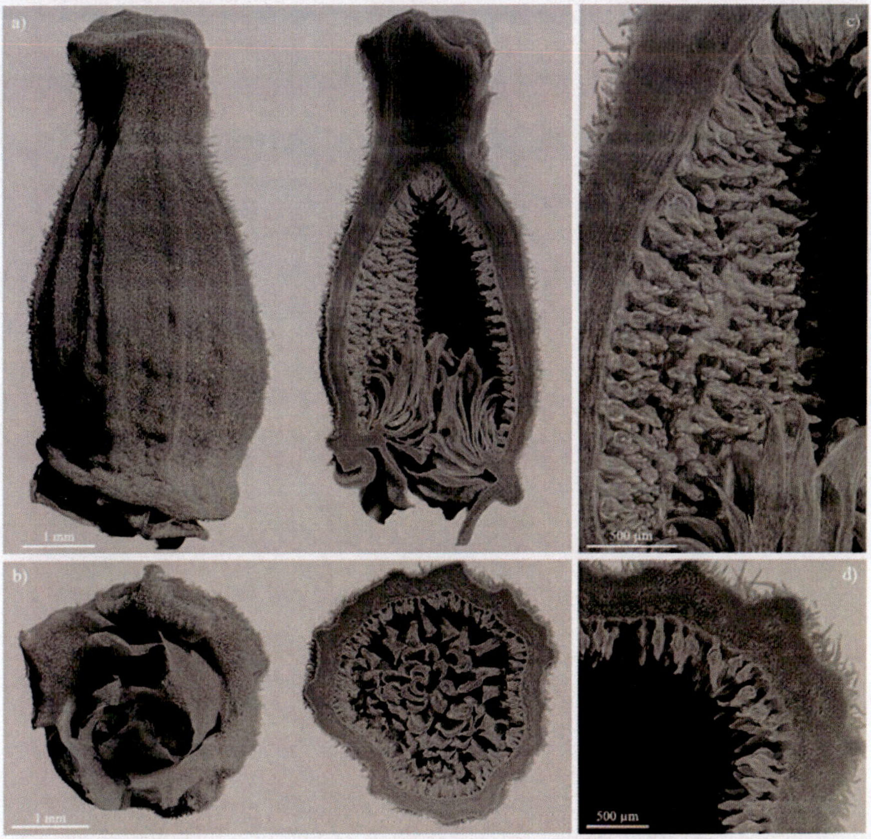

Fig. 5.23 X-ray picture of a small and elongated Petrelli main crop at the beginning of development. Florets and ostiolar scales are clearly visible. (From Marcotuli et al., 2020)

the year 1 (such as the flower buds of brebas), and the process is completed at bud burst in the successive season for a two-year process (Koutinas et al., 2010). The distal flower buds (brebas) differentiate in two seasons, whereas the inflorescence of the main crop differentiates in the same year of formation since, at the end of winter, only rudimentary primordia are visible in the basal nodes of the apical/lateral mixed buds (Marcotuli et al., 2020).

Molecular analyses also indicated a higher expression level of APETALA1 and APETALA2, AGAMOUS genes involved in flower organ formation, in brebas of both Dottato and Petrelli concerning the main crop. The higher expression of AGAMOUS (required for stamen development) in breba may suggest a possible presence of staminate flowers at the ostiole of brebas during its evolution, as the profichi fruits in the caprifig.

Phylogenetic studies have shown that the common fig has a monoecious ancestor (Machado et al., 2001) and successively evolved in a gynodioecious species. Therefore, the presence of these dormant flower buds evolving in brebas at the

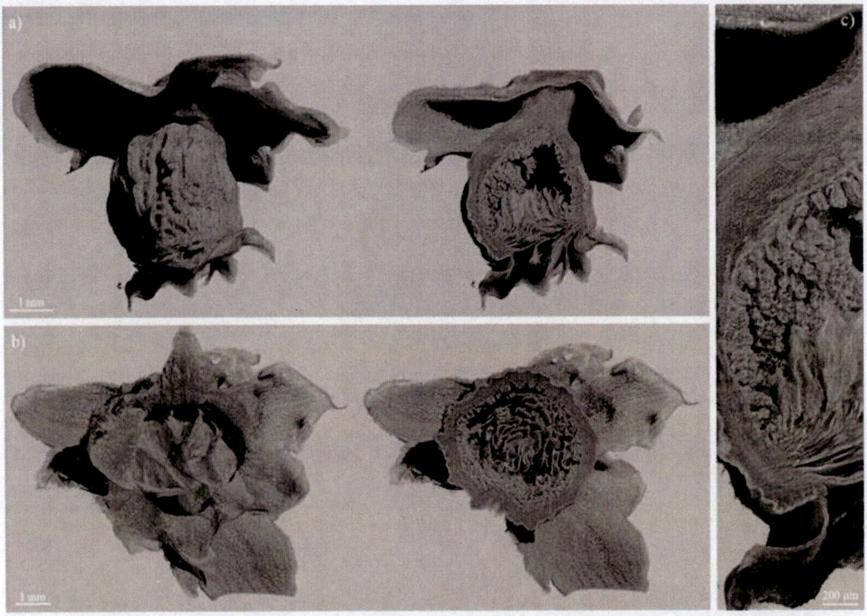

Fig. 5.24 X-ray picture of a small Petrelli first crop (brebas) at the beginning of development in spring. Florets and ostiolar scales are clearly evident and also the bud scales are well developed. (From Marcotuli et al., 2020)

beginning of the season may be a relic of the ancient monoecy, with the wasps entering the different inflorescences (main summer crop, main late crop, breba) during the year to lay the eggs, as they nowadays do in the caprifigs (mammoni, mamme, profichi).

Dioecy was considered as an adaptation to seasonal climates in order to support the *Ficus-Blastophaga* symbiosis (Kjellberg et al., 1987) or also to stimulate the inter-crosses to increase the variability of the offsprings.

A study of the genes involved in the natural process of flowering and flower differentiation in flower buds (breba or main crop) showed twelve genes expressed in the two crops (breba and main crop) and in the two cultivars analyzed (Dottato and Petrelli) involved in the formation of organs of the fig inflorescence (sepals, petals, stamens, ovary) (Marcotuli et al., 2020).

3 Influence of Cultural Practices and Plant Growth Regulators (PGRs) on the Buds

The common fig is a species with a strong apical dominance; thus, pruning and other cultural practices are required to control the growth of leader shoots and promote the emergence of lateral shoots that bear large amounts of fruit. Between flower bud initiation and fruit maturity, it takes around 90–120 days. Pruning and other techniques can maximize the production of flower buds and help timing the production to market needs, but this practice should be done with care. Severe pruning in 'Col de Dame Noire' and 'Bourjasotte Noire' caused a reduction in the average number of flower buds on shoots (Gerber et al., 2010). The flower buds of brebas may wait until the season is somewhat advanced (i.e., warmer) before they start active growth, depending on the world region. If the growth and ripening of brebas are delayed, the tree may put much of its initial resources into vegetative growth upon breaking dormancy. The use of plant growth regulators (PGRs) and other means to control bud burst can be an important factor. Cyanamide salts, such as calcium cyanamide, are PGRs that are effective regulators of flower buds (brebas) break (Shulman et al., 1986; Yablowitz et al., 1998). Applying 1–5% cyanamide advances bud break by 14–20 days compared with non-treated trees. Applications of 1–5% cyanamide with ethylene (400 ppm) further advance bud break by 20–30 days compared to non-treated trees. In addition to, or as a replacement for cyanamide treatment, GA (gibberellic acid) can be applied to stimulate the breba fruit and temporarily slow vegetative growth of the terminal mixed bud. Once a terminal bud begins active vegetative growth, it suppresses (or slows) those breba fruits located on the 1-year-old wood. However, if the tree is sprayed with gibberellins (for example, GA_3 at a concentration of about 40–50 ppm), usually two times at 1-week interval, the growth of brebas will be significantly stimulated. The optimal time for application is as soon as swelling of the terminal mixed bud is observed.

TDZ (N-phenyl-N-1,2,3,-thidiazol-5-ylurea), a cytokinin analogue, and mineral oil have been used as an exogenous application, especially in apple, to overcome dormancy (Steffens & Stutte, 1989; Wang et al., 1991; Honeyborne, 1996).

Girdling consists of removing a ring of bark (phloem tissue) from the shoot or branch to restrict the movement of assimilates from the aerial portion of the tree to the basal portion (trunk and roots), so the sap flow (mainly carbohydrates but also hormones) is directed to buds, flowers, and fruits above the girdled zone. The interruption of phloem vessels is temporary (several days), and the tree then restores the vessels through the production of a callus, a mass of parenchymatic cells (Davie et al., 1995; Theron & Steyn, 2008; Ferrara et al., 2014). Interruption of the auxin transport causes a sharp rise in cytokinin content leading to a stimulation of the meristematic activity, which could cause an increase in flower bud induction and initiation (Theron & Steyn, 2008).

Notching involves cutting through or removing a strip of bark directly above a bud, effectively disrupting transport in the phloem. Notching involves an interruption of the influx of auxins which do not exert their inhibiting action and allow the bud to develop (Greene & Autio, 1994). Notching was successfully used in the fig 'Brown Turkey' to increase bud break (Valia et al., 1994). When main management factors (i.e., irrigation, fertilization) are in a proper balance, the new shoots will continually produce new flower buds, leaves, and fruits. Overabundance of nitrogen fertilizers can accelerate vegetative growth during the season but delaying the development of the flower buds.

References

Condit, I. J. (1947). *The fig* (pp. 1–222). Chronica Botanica.

Crabbe, J. (1994). Dormancy. In *Encyclopedia of agricultural science* (Vol. 1). Academic.

Crabbe, J., & Barnola, P. (1996). A new conceptual approach to bud dormancy in woody plants. In G. A. Lang (Ed.), *Plant dormancy: Physiology, biochemistry and molecular biology* (pp. 83–113). CAB International.

Datwyler, S. L., & Weiblen, G. D. (2004). On the origin of the fig: phylogenetic relationships of Moraceae from ndhF sequences. *American Journal of Botany, 91*, 767–777. https://doi.org/10.3732/ajb.91.5.767

Davie, S. J., Stassen, P. J. C., & Van Der Walt, M. (1995). Girdling for increased fruit size and its effect on carbohydrate production and storage. *Proceedings of the World Avocado Congress, III*, 25–28.

Duric, G., Micic, N., & Dabie, G. (1992). *Jugoslovensko Vocarstvo, 26*, 15–21.

Eisen, G. (1901). *The fig: Its history, culture, and curing*. Government Printing Office.

Ersoy, N., Gozlekci, S., Kaynak, L., Dal, B. and Yazıcı, K. (2008). Relationship among the shoot, leaf and fruit growth of some FIG (Ficus carica) cultivars grown in antalya conditions. *Acta Horticulturae, 798*, 143–148 https://doi.org/10.17660/ActaHortic.2008.798.19

Faust, M., Liu, D., Wang, S. Y., & Stutte, G. W. (1995). Involvement of apical dominance in winter dormancy of apple buds. *Acta Horticulturae, 395*, 47–56.

Faust, M., Erez, A., Rowland, L. J., Wang, S. Y., & Norman, H. A. (1997). Bud dormancy in perennial fruit trees: Physiological basis for dormancy induction, maintenance, and release. *HortScience, 32*(4), 623–629.

Ferguson, L., Michailides, T. J., & Shorey, H. H. (1990). The California fig industry. *Horticultural Reviews, 12*, 409–490.

Ferrara, G., Mazzeo, A., Netti, G., Pacucci, C., Matarrese, A. M. S., Cafagna, I., Mastrorilli, P., Vezzoso, M., & Gallo, V. (2014). Girdling, gibberellic acid, and forchlorfenuron: Effects on yield, quality, and metabolic profile of table grape cv. Italia. *American Journal of Enology and Viticulture, 65*, 381–387.

Ferrara, G., Mazzeo, A., Pacucci, C., Matarrese, A. M. S., Tarantino, A., Crisosto, C., Incerti, O., Marcotuli, I., Nigro, D., Blanco, A., Gadaleta. A. (2016). Characterization of edible fig germplasm from Puglia, southeastern Italy: is the distinction of three fig types (Smyrna, San Pedro and Common) still valid? *Scientia Horticulturae, 250*, 52–58. https://doi.org/10.1016/j.scienta.2016.04.0016.

Ferrara, G., Mazzeo, A., Gallotta, A., Pacucci, C., Matarrese, A. M. S., Tarantino, A., Incerti, O., Marcotuli, I., Nigro, D., Blanco, A., Gadaleta, A. (2017). Fruit-set and SSR markers of fig cultivars from Puglia region, Southeastern Italy. *Acta Horticulturae, 1173*, 39-43. https://doi.org/10.17660/ActaHortic.2017.1173.7.

Flaishman, M. A., Rover, V., & Stover, E. (2008). The fig: Botany, horticulture, and breeding. *Horticultural Reviews, 34*, 113–197.

Gaaliche, B., Lauri, P. E., Trad, M., Costes, E., and Mars, M. (2011). Interactions between vegetative and generative growth and between crop generations in fig tree (Ficus carica L.). *Scientia Horticulturae, 131*, 22–28. https://doi.org/10.1016/j.scienta.2011.09.022

Gerber, H. J., Steyn, W. J., & Theron, K. I. (2010). The optimum one-year-old shoot length to optimize yield and fruit size of three fig (*Ficus carica* L.) cultivars. *HortScience, 45*, 1321–1326.

Grassi G., (1991). *Il Fico. Collana Manuali pratici.* Ed. Reda, Roma 1991. (pg.128).

Greene, D. W., & Autio, W. R. (1994). Notching increases branching of young apple trees. *Journal of the American Society for Horticultural Science, 119*(4), 678–682.

Honeyborne, G. E. (1996). Chemical restbreaking in pink lady, sundowner and fiesta: A preliminary report. *Deciduous Fruit Grower, 46*, 135–141.

Kjellberg, F., Gouyon, P. H., Ibrahim, M., Raymond, M., and Valdeyron, G. (1987). The stability of the symbiosis between Dioecious figs and their pollinators: a study of Ficus carica L. and

Blastophaga psenes L. *Evolution 41*, 693–704. https://doi.org/10.1111/j.1558-5646.1987.tb05846.x

Koutinas, N., Pepelyankov, G., & Lichev, V. (2010). Flower induction and flower bud development in apple and sweet cherry. *Biotechnology & Biotechnological Equipment, 24*, 1549–1558. https://doi.org/10.2478/V10133-010-0003-9

Lang, G. A., Early, J. D., Martin, G. C., & Darnell, R. L. (1987). Endo-, para-, and ecodormancy: Physiological terminology and classification for dormancy research. *HortScience, 22*, 371–377.

Machado, C. A., Jousselin, E., Kjellberg, F., Compton, S. G., and Herre, E. A. (2001). Phylogenetic relationships, historical biogeography and character evolution of fig-pollinating wasps. *Proceedings Biological Sciences, 268*, 685–694. https://doi.org/10.1098/rspb.2000.1418

Marcotuli, I., Mazzeo, A., Nigro, D., Giove, S.L., Giancaspro, A., Colasuonno, P., Prgomet, Ž., Prgomet, I., Tarantino, A., Ferrara, G., Gadaleta. A. (2019). Analysis of genetic diversity of Ficus carica L. (Moraceae) collection using Simple Sequence Repeat (SSR) markers. *Acta Scientiarum Polonorum Hortorum Cultus, 18*(4), 93–109. https://doi.org/10.24326/asphc.2019.4.9.

Marcotuli, I., Mazzeo, A., Colasuonno, P., Terzano, R., Nigro, D., Porfido, C., Tarantino, A., Aiese Cigliano, R., Sanseverino, W., Gadaleta, A., & Ferrara, G. (2020). Fruit development in *Ficus carica* L.: Morphological and genetic approaches to fig buds for an evolution from monoecy toward dioecy. *Frontiers in Plant Science, 11*, 1208. https://doi.org/10.3389/fpls.2020.01208

Petrucci, V. E., & Crane, J. C. (1950). Fruit bud initiation and differentiation in the fig. *Proceedings of the American Society for Horticultural Science, 56*, 86–92.

Sarkhosh, A., Yavari, A. M., & Ferguson, L. (2022). *The fig: Botany, production and uses*. CABI.

Shulman, Y., Nir, G., & Lavee, S. (1986). Oxidative processes in bud dormancy and the use of hydrogen cyanamide in breaking dormancy. *Acta Horticulturae, 179*, 141–148.

Steffens, G. L., & Stutte, G. W. (1989). Thidiazuron substitution for chilling requirement in apple cultivars. *Journal of Plant Growth Regulation, 8*, 801–808.

Tanrisever, A., Özeker, E. & Isfendiyaroglu, M. (1998). Anatomical and characterization of Sarilop (Calimyrna) fig cultivars in pictures. *Acta Horticulturae, 480*, 109–114. https://doi.org/10.17660/ActaHortic.1998.480.16

Theron, K. I., & Steyn, W. J. (2008). Girdling: Science behind the age-old technique. *Acta Horticulturae, 800*, 51–60.

Valia, R. Z., Bhatt, J. P., Jadav, R. G., & Kapadia, P. K. (1994). Effect of horticultural practices and fertilization on the production of fig (*Ficus carica* L.). *South Indian Horticulture, 42*(5), 321–323.

Wang, S. Y., Jiao, H. J., & Faust, M. (1991). Changes in metabolic enzyme activities during thidiazuron-induced lateral budbreak of apple. *HortScience, 26*, 171–173.

Yablowitz, Z., Nir, G., & Erez, A. (1998). Breba fig production in Israel: Regular and pesticide-free systems. *Acta Horticulturae, 480*, 137–141.

Zohary, D., & Spiegel-Roy, P. (1975). Beginnings of fruit growing in the old world. *Science, 187*(4174), 319–327.

Chapter 6
Phenotypic Variability of Fig (*Ficus carica* L.)

Ali Khadivi and Farhad Mirheidari

1 Introduction

Fig (*Ficus carica* L.) is one of the oldest domesticated fruit species in the world. The English word fig is of ancient origin, derived from the Latin *ficus*. The species name *carica* refers to Caria, an ancient region of Asia Minor noted for figs (Ferguson et al., 1990). The genus *Ficus* is one of 40 genera belonging to the Moraceae family with more than 800 species and most found in the tropics or subtropics, the most important of which are *F. carica*, *F. religiosa*, *F. elastica*, *F. benghalensis*, and *F. rumphi* (Condit, 1969; Woodland, 1997; Berg & Corner, 2005).

The genus *Ficus* is found mainly in the tropics, but *Ficus carica* L. is a subtropical fruit tree (Ferguson et al., 1990). Figs are currently cultivated in warm and temperate climates and have grown as a Mediterranean fruit tree for many years (Sahin, 1998). The ancestral fig (*Ficus carica* L.) tree was monoecious and later evolved into gynodioecious species with bisexual trees (functional male figs or caprifigs) and unisexual female trees (Nabli, 1989; Machado et al., 2001). Both female and male flowers are organized inside the syconium. A female fruit contains hundreds of female flowers. A male fig plant, called caprifig, has hundreds of separated male (staminate) flowers and female (pistillate) flowers. The differentiation between the hermaphroditic and female strains forms establishes the basis for maintaining the close symbiotic relationship between *Ficus* plants and the Blastophaga wasps (Galil & Neeman, 1977).

The common fig tree is indigenous to Persia and Syria (Ferguson et al., 1990). The fig tree has been distributed from Persia, Asia Minor, and Syria by people

A. Khadivi (✉) · F. Mirheidari
Department of Horticultural Sciences, Faculty of Agriculture and Natural Resources, Arak University, Arak, Iran
e-mail: a-khadivi@araku.ac.ir

M. F. Ramadan (ed.), *Fig* (Ficus carica)*: Production, Processing, and Properties*, https://doi.org/10.1007/978-3-031-16493-4_6

throughout the Mediterranean area. Fig is considered the oldest cultivated crop (Storey, 1975; Kislev et al., 2006). Fig cultivation has been related to its pleasant taste and medicinal properties (Khadivi et al., 2018). Thousands of cultivars, primarily unnamed, have been developed or came into existence as human migration brought the fig to many places outside its natural range. They are now grown in most Mediterranean countries, Europe, North and South Africa, the Americas, and many other Asian countries (Flaishman et al., 2017).

Variability within fig accessions is attributed to the wide genetic diversity of fig germplasm, as revealed by studies based on morphometric parameters (Mars et al., 1998; Caliskan & Polat, 2008; Gaaliche et al., 2012; Khadivi et al., 2018; Mirheidari et al., 2020; Hssaini et al., 2020b) and molecular markers (Khadari et al., 1995a; Aksoy et al., 2003; Stover & Aradhya, 2008; Baraket et al., 2009; Simsek & Yildirim, 2010; Podgornik et al., 2010; Giraldo et al., 2010; Almajali et al., 2012; Simsek et al., 2017). In addition, researching various genotypes of wild edible figs will allow the fruit growers and breeders to select the most exciting and appropriate range of wild edible fig accessions based on market needs taking into account the ripening season and fruit color (Mirheidari et al., 2020; Khadivi & Mirheidari, 2022). Furthermore, the cultivated fig cultivars' performance varies under different climatic conditions (Ge & Zhang, 2006; Podgornik et al., 2010; Jacob et al., 2014; Gruda & Tanny, 2015; Saadi et al., 2015; Caliskan et al., 2017). Cultivars also vary in susceptibility to diseases (Yakushiji et al., 2012), with yield quantity and fruit quality depending enormously on cultivation practices and climatic conditions (Gaaliche et al., 2011).

A crucial genetic erosion occurs due to biotic and abiotic processes (urbanization, extension of intensive crops, absence of caprification, water deficit in marginal areas, etc.) (Mars et al., 1998; Salhi-Hannachi et al., 2004). To better conserve and use genetic resources, characterization designs of morphological variability within the collections and selecting the most significant variables to establish strategies for better conservation and use of genetic resources should be performed carefully (Giraldo et al., 2010). Despite the advances in molecular characterization in fig (Achtak et al., 2009; Giraldo et al., 2005, 2008a; Ikegami et al., 2009; Khadari et al., 2005), phenotypic characterization is always needed, and it should be included in any program of conservation and use of genetic resources (Giraldo et al., 2008b). Ideally, this characterization should be made under the same edaphoclimatic conditions.

2 Botany and Biology in Different Types of Figs

Most species of the genus *Ficus* are monoecious and have male and female flowers within single syconia. *Ficus carica* is a gynodioecious woody perennial species with two tree types: male fig trees (the inedible caprifig) produce male flowers and short-style female flowers, while female fig trees (the edible fig trees, i.e., Smyrna, San Pedro, and common type) only produce long-style female flowers (Lisci &

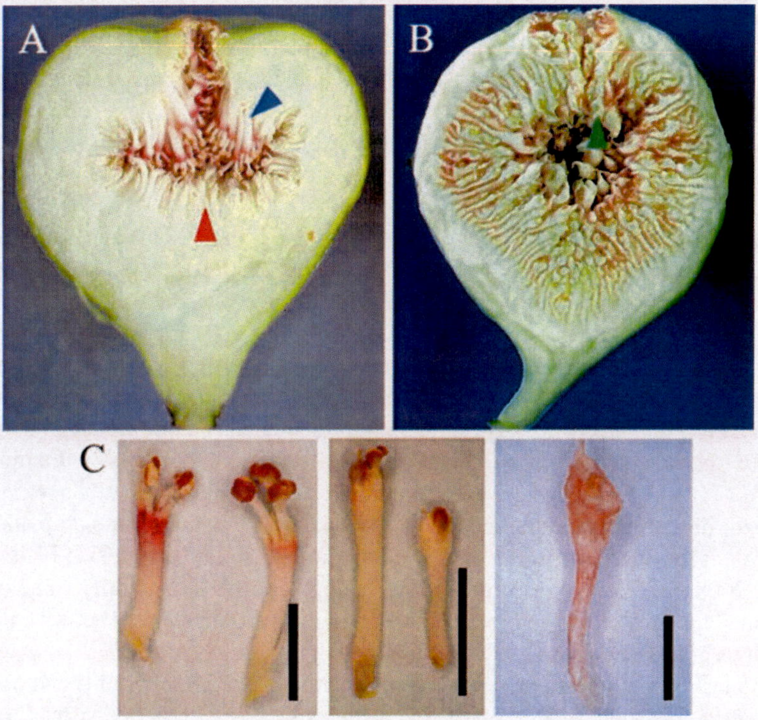

Fig. 6.1 (**a**) Monoecious caprifig fruit and (**b**) Female fig fruit. Blue, red, and green arrowheads indicate staminate flower, short-style pistillate flower, and long-style pistillate flower. (**c**) Caprifig staminate flowers (left), caprifig short-style pistillate flowers (center), and fig long-style pistillate flower (right) (Mori et al., 2017)

Pacini, 1994; Valdeyron & Lloyd, 1979; Stover et al., 2007a) (Fig. 6.1). Although, some so-called caprifigs are reported to be edible and have a more succulent fruitlet than typical caprifigs (Storey, 1975).

Caprifigs (male figs) of *Ficus carica* var. *caprificus* Risso produce three crops of syconia per year: the *profichi* that ripen in early summer, the pollen source for the edible fig, are produced in large numbers on wood from the previous season; the *mammoni* that ripen in fall; and the *mamme* that over-winter on the tree and ripen in spring. Fig pollen is transferred from male flowers to female flowers by an insect known as the fig wasp (*Blastophaga psenes*, Hymenoptera: Chalcidoidea: Agaonidae); they represent perhaps the most extreme and ancient (~75 million years) obligate pollination mutualisms known (Machado et al., 2001; Cruaud et al., 2012). These wasps live, lay their eggs, and feed within the fig syconia. Only the winged female wasps emerge and leave the fig through the ostiole (wingless males cannot leave the syconia). The female wasps fly to new syconia to oviposit eggs in some female flowers, one egg per ovary. However, the *profichi* caprifig has many male flowers near the ostiole, and the pollen from these flowers is carried by the female wasps to the next syconium, pollinating the ovaries. Although caprifig syconia incubate and perpetuate these tiny fig wasps, seeds may also develop in the female flowers. This is especially true of *mammoni* syconia in which pollination results in some ovaries (without wasps) developing seeds with viable embryos (Condit, 1955; Galil, 1977).

Three different types of female figs could be distinguished depending on the cropping/pollination characteristics, common type, Smyrna, and San Pedro (Stover et al., 2007a). Most fig genotypes are common (e.g., 'Brown Turkey', 'Mission', 'Adriatic', Dottato, etc.) that produce parthenocarpic fruit without caprification. The drupelets inside develop parthenocarpically, and the endocarps are generally hollow without fertile seeds. Common type figs can produce one (unifera types) or two crops (bifera types). The Smyrna types (e.g., 'Calimyrna' ('Sarilop'), 'Black Bursa', 'Zidi', 'Sabz', etc.) require pollination with pollen from caprifigs (caprification) for fruit development to occur. These trees produce only two crops of syconia annually, a *breba* crop that ripens in early summer and the main crop that ripens in autumn. Finally, the San Pedro types (e.g., 'Dauphine', 'King', 'San Pedro', 'Lampiera', 'Petrelli', etc.) can produce two crops; the first crop (*breba*) is parthenocarpic (it proceeds directly without pollination and fecundation) and the second crop, like the Smyrna, is produced only after caprification (Condit, 1955; Galil, 1977; Flaishman et al., 2008a). This complex reproductive biology allows us to identify some of the botanical features of figs and use them to distinguish more precisely between different types of figs (Table 6.1).

The last two types of edible figs are not parthenocarpic (persistent) and require pollination to set the main crop of figs. Botanists use the term "persistent" rather than "parthenocarpic" because the fig is not a true fruit. Botanically, these nonpersistent types are classified as "caducous". Of the cultivars described by Condit (1955), 78% are common types, less than 4% are San Pedro types, and the remaining 18% are Smyrna types.

In some varieties, the main crop could be divided into two sub-groups, the (summer) main crop (ripening in July-September) and the late main crop ripening in autumn and borne on the trees up to December. Some varieties produce only the main crop ripening late (cv. Natalino) and are eaten almost at Christmas (Marcotuli et al., 2020). In the two-crop variety, breba fruits are considerably heavier and have a shorter ripening time than the main crop, but in some years, there are not present (Podgornik et al., 2010). However, this difference in crops leads to the distinction of the varieties in uniferous (only one crop, main crop), biferous (two crops, breba and main), and even triferous (breba, summer, and late main crop) (Ferrara et al., 2017; Marcotuli et al., 2019).

3 Germplasm Diversity and Conservation

The increasing concerns about reduced levels of genetic diversity in crop species (Esquinas-Alcazar, 2005; Tanksley & McCouch, 1997) have led to the need to preserve as much genetic diversity as possible both in situ and ex situ not only for the

Table 6.1 Fig types and sexual characteristics (Bearing time can slightly vary in different ecological areas)

	Male figs	Female figs		
		Caducous		
Fig types	Caprifig	Smyrna	San Pedro	Common type
Flower morphology	Male flower with five stamens, female flower with short style	Male flower not presented, female flower with long style	Male flower not presented, female flower with long style	Male flower not presented, female flower with long style
Presence of wasp	Male and female wasp (wasps' eggs can develop only in short style flowers)	Only female wasp, from a caprifig syconia (wasp cannot lay its eggs in long style flowers)	Only female wasp, from a caprifig syconia (in the *mammoni* crop)	No wasps involved
Mamme (in male figs) (autumn-winter)	Not fecundation (pollen is not viable), small fig not edible	Not developed	Not developed	Not developed
Profichi (in male figs) *Breba* (in female figs) (spring-beginning summer)	Production of pollen in the male flower but no fecundation, fig not edible	Fecundation by pollen from a carprifig, edible fig, but it cannot be dried	Not fecundation (persistent), setting on wood from the previous year, edible fig, but it cannot be dried.	Not fecundation (persistent), setting on wood from the previous year, edible fig, but it cannot be dried (developed only in bifera types)
Mammoni (in male figs) *Main crop* (in female figs) (summer-autumn)	Fecundation by pollen from *profichi* (*mammoni* does not produce pollen), small and hardly edible fig	Fecundation by pollen from a caprifig, edible fig and it can be dried	Fecundation by pollen from a carprifig, setting on current season's wood, edible fig, can be dried.	Not fecundation (persistent), setting on current season's wood, edible fig and it can be dried (developed in both unifera and bifera types)
Seed characteristic	Fertile seed (embryo and endosperm developed)	Fertile seed (embryo and endosperm developed)	Infertile seed *profichi* (hollow seed), fertile seed *mammoni* (embryo and endosperm developed)	Infertile seed (hollow seed)

Adapted from Mateos and Gil (2014)

long term survival of the species but also to have enough variability available for breeding programs. Although several efforts have been carried out to conserve fig genetic resources in germplasm banks (Flaishman et al., 2008b; Giraldo et al., 2008a; Stover & Aradhya, 2008), appropriate management of the germplasm collections requires defining patterns of phenotypic variability within the collections and selecting the most significant variables to establish strategies for better conservation and use of genetic resources. For this purpose, the analysis of genetic diversity in germplasm collections facilitates reliable classification of the accessions conserved and the identification of subsets or core accessions with possible use for specific breeding purposes (Mohammadi & Prasanna, 2003; Escribano et al., 2008). Therefore, it is imperative to establish programs to preserve and characterize fig genetic resources globally, and programs have been developed in several countries. Fig germplasm accessions are reported in Asia, Europe, North Africa, and North America. The largest collections are in Italy, Turkey (Elbreyli Fig Research Institute), California (National Clonal Germplasm Repository (NCGR) Davis), Turkmenistan (TES-PGR), and France (CBNM Porquerolles), containing 442, 314, 190, 180 and 154 accessions, respectively. The CBNM Porquerolles collection has 154 different fig cultivars from Italy, North Africa, Greece, Turkey, France, Tunisia, and Spain. The NCGR fig collection contains 190 accessions. The Arid Land and Oasis Cropping Laboratory germplasm collection of the Institute of Arid Regions of Medenine, Tunisia, holds 119 local fig accessions and 20 caprifigs collected from Tunisia, Algeria, and Egypt (Aljane et al., 2012).

To better conserve and utilize genetic resources, characterization designs of morphological variability within the collections and selection of the most significant variables should be carefully performed (Giraldo et al., 2010). This characterization usually involves a wide range of data which include both qualitative and quantitative traits (Khadivi-Khub et al., 2012). It is worth noting that some of these parameters are generally influenced by environmental conditions and agronomic practices. Nevertheless, morphological characterization is a highly recommended first step before starting biochemical or molecular studies (Hoogendijk & Williams, 2001).

Several studies have reported the morphological characterization of female fig cultivars and proved the sustainability of the use of morphological parameters to evaluate and establish a description of genotypes (Condit, 1941; Mars et al., 1998; Oukabli et al., 2002; Hedfi et al., 2003; Salhi-Hannachi et al., 2003; Aljane et al., 2012; Gaaliche et al., 2012; Caliskan & Polat, 2008, 2012; Almajali et al., 2012; Khadivi et al., 2018; Rodrigues et al., 2019a; Mirheidari et al., 2020; Hssaini et al., 2020b, c; Moniruzzaman et al., 2020; Khadivi & Mirheidari, 2022).

Authors stress the relevance of morphoagronomic variability in cultivars' identification and breeding programs and claim such analysis should be performed before molecular studies are carried out (Podgornik et al., 2010; Khadivi-Khub et al., 2012; Djordjević et al., 2014; Khadivi et al., 2018).

4 Wild Figs, Cultivars, Local Clones, and Hybrids

4.1 Wild Fig

Fig (*Ficus carica* L.) is a traditional fruit from West Asia and the eastern portion of the Mediterranean, including Turkey and Iran, in which wild variants of this species are found (Ikegami et al., 2009). These wild species consist of ecotypes called common figs (unisexual female trees) and caprifigs (bisexual with functional male trees) occurring in similar frequencies in wild populations (Valdeyron & Lloyd, 1979). In Tunisia, the wild fig is mainly found in the north, and the center, such as in the islands of Galite and Kerkennah, Djebba, El Haouaria, and Kesra and is widely spread over all the climatic stages (Mars et al., 1998; Aljane et al., 2005). Wild fig is seed-propagated, and seeds are disseminated by birds (Lisci & Pacini, 1994). Therefore, it is primarily present in rocks, usually along the riverbanks, in steep-sloped valleys (Khadari et al., 2005).

4.2 Cultivars and Local Clones

Many of the existing fig tree cultivars may result from selection focusing on agronomic characteristics and/or selection and transportation to distant regions from breeders and growers (Condit, 1955). Stem-cuttings usually carry out fig propagation, and this procedure has contributed to synonymy and homonymy because misidentification and somaclonal variations are common events in sexually propagated species such as fig (do Val et al., 2013). The long period of domestication, the usefulness of the perpetuation of certain traits through asexual propagation, and the exchange of materials also contributed to synonymy (Aradhya et al., 2010).

Fig cultivars are numerous, and genotypes are either misidentified or called by different denominations in different regions of the world (Aljane, 2016; Aljane et al., 2012; Essid et al., 2017). Fig names were mainly given based on phenotypic and morphological characters such as fruit color, fruit shape (width and neck length), fruit taste and flavor, local geography, and maturity period (IPGRI and CIHEAM, 2003; Caliskan & Polat, 2008; Aljane & Ferchichi, 2009; Hssaini et al., 2019). For example, the variety 'Ghoudan' indicates the darkness of skin color, and 'Zerqui' refers to the blueness of fruit skin. 'Ounq Hmam' refers to the length of the neck (long fruit neck) and means in the local dialect "pigeon neck." Denomination of 'Jaadi' describes genotypes having ribs over the skin. 'Nabout' is a generic term that means "tree that grows spontaneously" (Hmimsa et al., 2012). Moreover, the same variety may have different names depending on the cultivated region. For example, the cultivar 'Sari-Lop' ('Smyrna') from Asia referred to California as 'Calimyrna'.

Regarding the phenotypic diversity between fig cultivars, Condit (1955) has already provided a comprehensive list of fig varieties, and recently the names and characteristics of some cultivated accessions/cultivars in different countries have been reviewed (Singh et al., 2015; Hiwale, 2015; Núñez-Gómez et al., 2021).

In Turkey, fig culture is seen in almost every part of the country except the very continental climatic areas of the Middle and East Anatolia. Almost all the coastal lines of the country, such as the Aegean, Mediterranean, Marmara, and Black Sea regions, are the main areas for fig production (Arpaci, 2017).

Most Turkish dried and fresh fig cultivars, 'Sarilop' and 'Bursa Siyahi' are caducous and classified as Smyrna types, respectively (Ozen et al., 2007; Arpaci, 2017). 'Bursa Siyahi' is Turkey's most commonly grown fig cultivar (Aksoy et al., 2003). Fruit characteristics of the fig cultivars grown in Antalya conditions (Bursa Siyahi, Yesilguz, Bardakci, Sultan Selim, Karabakunva, Beyaz Orak, and Sarilop) were studied by Kaynak et al. (1998). Bursa Siyahi had the heaviest fruits and Karabakunya the lightest. Fruit width was greatest in Beyaz Orak, Sarilop, and Bursa Siyahi, and Bursa Siyahi had the longest fruits. Bursa Siyahi, Sarilop, and Yesilguz were the easiest to peel. Bardakci, Bevaz Orak, Bursa Siyahi and Yesilguz had the least ostiole and skin cracking. Also, five local cultivars (Degirmen, Ipek, Istanbul, Pamuk, and Pathcan) in Northern Turkey are predominantly consumed as fresh fruit. In addition, some of these cultivars are used for making jams. The harvest period is long, from early September to late October. Therefore, single cropping systems of large fig plantations are essential in optimizing economic production (Bostan et al., 1998).

El-Kassas et al. (1992a) evaluated the growth of four local and introduced cultivars in Egypt. The cultivars DiRedo, Conadria, Sultani, and Black Mission, were best in terms of tree vigour. However, the yield and quality of syconium also showed marked variation in different cultivars (El-Kassas et al., 1992b).

The most common cultivars grown in Morocco are 'Nabout', 'Ghoudan', 'El-quoti lbied', 'Ghani', 'Chaari', 'Fassi', and 'Ounq Hmam' (Hssaini et al., 2020b). The Riffian Mountains in Northern Morocco are the focal area of fig planting and production, which is considered a source of income for the local farmers (Hmimsa et al., 2012). Also, Hssaini et al. (2019) indicated that the most dominant varieties in Northern Morocco (Taounate, Ouzzane, and Meknes regions) were: 'Nabout', 'Ghoudan', 'El Quouti lbied', 'Zerqui,' 'Ghani', 'Sbeti', 'Tabli', 'Lamtal', 'Mssari', 'Ounq Hmam', 'Arguil', 'Chaari', and 'Jaadi'. According to their survey results, these varieties were all propagated using hardwood stem cuttings. In addition, Hmimsa et al. (2012) counted about 133 denominations in the Rif Mountains in Northern Morocco. Furthermore, they reported 191 lexemes corresponding to 133 morphotypes or varieties reproduced by vegetative propagation in the Rif area. Furthermore, they concluded that names might vary from one locality to another for the same morphotype, which explains the problem of mislabeling that characterized the cultivated varieties in the Rif mountains.

Iran is a favorable country for fig growing. In Iran, fig culture is seen in almost every country. It is also found in the wild type (edible fig and caprifig) in most parts of the country. Dry farming of figs in the Fars province of Iran started about 650–700 years ago, and most of the land cultivated with dry-farmed figs is located in Fars province (Rahemi & Jafari, 2008). The Smyrna-type cultivars represent the main type commercially grown in Iran. Estahaban is the main region of fig fruit production in Iran (Khadivi et al., 2018). The fig landrace 'Sabz', the main dried fig in Iran, belongs to the Smyrna type figs that bear only pistillate flowers and require the wasp pollinator to pollinate its fruit. The fruit crop of 'Sabz' ripens and partially dries on the tree. The'Sabz' landrace fruit crops are dried directly on the tree and do not fall until perfectly ripened. Instead, they are gathered after fallen and left in the sun for a few hours before being conditioned for sale (Khadivi et al., 2018). Also, other commercial cultivars include 'Payves', 'Sigoto', 'Khafrak', 'Rowno', 'Shahanjir', 'Atabaki', 'Kashki', 'Matti', 'Manbili', and 'Seyah' in Fars province in the south-west (Pourghayoumi et al., 2017; Khadivi et al., 2018) and 'Lashei', 'Majifi', 'Bar-anjir', 'Sham', 'Bavameli', 'Shamamleh', 'Zardleh', 'Siavleh', 'Estahban-e-Sabz' in Kermanshah and Ilam province in the west (Fatahi et al., 2017), 'Siah Esfehan', 'Emadi Sabze Esfahan', 'Poost sabze eghda', 'Peyvandiye Varamin', 'Siyahe Mahali', 'Zard Bongah Varamin' in the center and 'Peykami kashmar', 'Zarde Boshghabi Kashmar' in east of Iran (Mahdavian et al., 2008).

In 1850, more than 70 fig varieties were introduced to Algeria. However, these were not adopted by the local farmers, who continued to grow the fig trees that were familiar to them (Mahmoudi et al., 2018). The Technical Institute of Fruit-bearing Arboriculture in Algeria described 40 varieties, including comestible and caprifig types (Chouaki, 2006; Meziant et al., 2015). The cultivar 'Kalamon' is the main variety cultivated to produce dried figs in southern Peloponnese and is represented in FTC-IK by three accessions originating from Messinia. On the other hand, 'Kalamatiani Evias' that represents another accession cultivated in different parts of Greece for the same commercial purpose, shares identical morphological characteristics with 'Kalamon' and is, therefore, believed to be the same cultivar (i.e., 'Kalamon') (Papadopoulou et al., 2002).

Prospection undertaken in different regions in Tunisia contributed to identifying and describing numerous cultivars (Ben Salah & Lejri, 1995; Mars et al., 1998; Chatti et al., 2004; Mars et al., 2009). Some, like 'Bither' and 'Bither besbessi', are of the common type (parthenocarpic) that produce figs without caprification (pollination). Many others, like 'Bidhi', 'Zidi', 'Kahli' and 'Soltani', are Smyrna types that need caprification. Smyrna types are predominant in the south while, in the north, cultivars of common type are equally represented (Mars, 2003). The fig germplasm consists of many ecotypes selected for their fruit traits ('Bidhi' and 'Zidi') or their high adaptative potentialities ('Bouhouli' and 'Wahchi'). The cultivar Zidi (Smyrna type) has a large geographical distribution in the country and is known for its high commercial value and best fruit taste for fresh consumption (Mars et al., 2008). Mars et al. (1998) analyzed the diversity among 22 cultivars of the southern arid regions in Tunisia using morphometric studies. Cultivar characterization and multivariate analysis based on 18 physical and chemical fruit characters of 22 main crop varieties suggested that fig germplasm is diverse. It was also possible to differentiate three cultivar groups and two distinct cultivars.

In Spain, almond and fig tree cultivation coexist with grape cultivation, introduced during the twentieth century by immigrant farmers from the 'Levante' region, eastern Spain (Perez-Jiménez et al., 2012). The main fig fruit-producing areas in

Spain are Extremadura-Andalusia and neighbouring regions, the Levantine area, and Galicia (Pérez-Sánchez et al., 2016). Most of the varieties cultivated in Spain (e.g., 'Colar', 'Super-Fig', 'Cuello de Dama Negro' and 'San Antonio') belong to the parthenocarpic and biferous group, but there are also those of the San Pedro type (i.e., 'Nazaret' and 'Tiberio' varieties) mainly cultivated in Spanish western (López-Corrales et al., 2011; Núñez-Gómez et al., 2021). In addition, 'Carballar Negra' and 'Moscatel' are the only cultivars with one crop per year (unifera type) (Pérez-Sánchez et al., 2016).

Despite its socioeconomic and historical importance, in Italy, the fig is considered a minor fruit species. The regions in which the fig tree assumes economic importance are Campania (Cilento-SA) and Calabria (province of Cosenza), followed by Puglia (Salento area-LE) and Sicily (Ciarmiello et al., 2015). In Calabria and Campania, the main production of dried figs is based on the cultivar 'Dottato', admitted in two different protected designation of origin (PDO): "Fichi di Cosenza" (Calabria) and "Fico Bianco del Cilento" (Campania). The cultivar 'Dottato', syns. = 'Ottato', 'Uttato', 'Ortata', 'Fico Bianco', 'Kadota', (Gallesio, 1820; Vallese, 1909; Condit, 1955; Grassi, 1991) is of ancient origin. 'Dottato' develops parthenocarpically both the first crop (breba) and second-crop (figs), which represent the main production.

The most famous Greek fig tree cultivar, 'Kymis', which is widely cultivated on the island of Evia, has a central market position and high potential since it is in the process to be appointed a Protected Designation of Origin (PDO) EU mark. Currently, it is highly-priced, offering a significant income to local farmers (Vlachos, 2013). 'Dauphine' is another fig cultivar similar to 'Masui Dauphine', an important Japanese cultivar popular in Japan. Masui Dauphine was evaluated for 5 years in the Zhenjiang area in China. It produced large (80–100 g), elongate-oval, purple-red fruits with 15% soluble solids content and high aspartic acid content (0.22 mg/100 g). The fruit was suitable for fresh consumption and processing (Yang et al., 1994).

The main varieties cultivated in Cyprus are 'Vazanata', 'Vasilika', 'Smyrneika', 'Vardika', 'Kadota', 'Napolitana negra', 'Progontto Blanco' and 'Gentile Bianco' (Gregoriou, 1995). The main cultivars planted in California are "Calimyrna," "Mission," "Brown Turkey," and "Kadota" (Stover et al., 2007b). Most fig production in California is used for the dried market; only a small percentage is harvested for fresh fig consumption (Stover et al., 2007b). Also, Marseilles (White Marseilles, or Lemon) is commonly found in California (Singh et al., 2015). 'Black Mission' is cultivated in California on a large scale. This variety is old but highly productive. As per the name, the fruits are black. The fruits are sweet and tasty and suitable for drying and preparing jam. The first season's fruits are bigger than the second season (Hiwale, 2015). The Tree is a vast, elongated fruit, with Flesh watermelon to pink and a reasonably good taste. Breba crop is prolific (Singh et al., 2015).

'Adriatic fig' trees could sometimes be found in the state of Washington. This is a common variety in Chile and Argentina (Singh et al., 2015). 'Celeste figs' are most commonly grown in Texas. 'Brown Turkey' is grown on all the islands of Hawaii.

In India, fig farming is limited to a few states such as Maharashtra, Gujarat, Uttar Pradesh, Karnataka, and Tamil Nadu. In northern India, its cultivation is spread to some parts of Uttar Pradesh, Uttarakhand, Punjab, and Himachal Pradesh (Sharma & Badiyala, 2006). 'Poona' and 'Dinkar' are widely acceptable varieties of fig. Apart from these, the important varieties of figs cultivated commercially in India are; Conadria, Deanna, Excel, Celeste, Brown Turkey, and Brunswick (Magnolia) (Gani et al., 2018). Also, At Saharanpur, India, 'Brown Turkey', 'Bangalore', 'Black Ischia' and 'Lucknow' are successfully grown. Around Mumbai, there is only one variety, 'Poona' (Singh et al., 2015).

The fig tree has been cultivated in Brazil since the beginning of the last century, and owing to the growing demand for these fruits by the food industry, the cultivated area continues to increase each year, especially in the States of São Paulo, the Rio Grande do Sul, and Minas Gerais (Paula et al., 2009). Fig cultivation in Brazil is mainly based on a single cultivar, 'Roxo de Valinhos', introduced by Italian immigrants who settled in the State of São Paulo in the late nineteenth century (Paula et al., 2009; Francisco et al., 2011). According to Ferreira et al. (2009), the predominance of this cultivar in Brazilian orchards is due to its desired traits of vigor, productivity, rusticity, and wide acceptance by consumers and the industry.

In Costa Rica, the main fig cultivar produced is 'Brown Turkey', and it has been being found in Tierra Blanca and Pacayas (in the north region of the Cartago province) since the colonial period as a complementary agricultural activity to the potato and onion crops (Flores et al., 2011).

4.3 Hybrids and Transgenic Varieties

The classical breeding approaches in fig were summarized by Condit (1947), Storey (1975), Obenauf et al. (1978), Ferguson et al. (1990), and Mars (2003). However, fig breeding is complex because of involving two tree morphs (Caprifig and edible fig), three floral forms (long-styled female, short-styled female, and male flowers), and the insect pollinator (Beck & Lord, 1988). These cause problems in breeding procedures like; the occurrence of parthenocarpy and seedlessness, complex pollination mechanism, i.e., caprification, lack of reference collections, lack of universal descriptor list to differentiate different varieties/species, presence of large intra-variety diversity, the closeness of wild types to cultivated plants in some regions (Beck & Lord 1988).

Fig crosses and their evaluation has been described in Japan (Awamiira et al., 1996). In addition, two breeding programs were conducted at the beginning of the twentieth century at the University of California. The most significant achievement of the California fig breeding was the development of 5 hybrid cvs.; Conadria, Di Redo, Flanders, Tena and Excel. Furthermore, a common cv. Sierra was developed with late summer maturity and green color skin (Doyle & Ferguson, 2005).

Transgenic varieties are limited in fruit trees, and their efficiency is very diverse according to the fruit species, some being recalcitrant (Chevreau, 2009).

Nevertheless, transgenic varieties have been developed to improve agronomic traits, fruit storability, and disease resistance. The transgenic approach for fig-tree breeding was described by Yancheva et al. (2005), wherein an *Agrobacterium*-mediated transformation of the cv. Brown Turley and Smyrna were realized. Technological progress enabled gene transfer and permitted the production of desirable proteins in fig fruits to enhance the nutritional and pharmaceutical constitution (Flaishman et al. 2008b). Also, the objective of Ozen et al. (2017) research was to investigate the segregation of leaf characteristics among 'Sarilop' (female parent) × 'Kaba ilek' (male parent) hybrids. The importance of their study was to acquire knowledge about the inheritance of some leaf characteristics in *Ficus carica*. In conclusion, except for central lobe length, leaf length, width, area, petiole length, and thickness were lower than parents. This might indicate negative transgressive segregation in the leaf characteristics mentioned in the 'Sarilop' × 'Kaba ilek' fig cross.

5 Morphological Characteristics

The characterization and evaluation of plant species consist of establishing an identity for each accession through the knowledge of a series of data aiming to describe the different accessions of a collection of germplasm, using characteristics of interest. Cruz et al. (2014) mentioned that the characters studied may be morphological, physiological, cytological, biochemical, or molecular and should include attributes, characteristics, or traits of a culture that are informative and useful for the characterization and evaluation of specific types (i.e., accessions) of the studied species. Currently, morphological markers are still important and widely used in different fruit species, cultivars and genotypes and are appropriate for classification (Poledica et al., 2012; Yilmaz et al., 2012; Koc & Bilgener, 2013).

Selecting highly discriminant descriptors is important to optimize the resources necessary for a viable and reliable morphological characterization (Giraldo et al., 2005). In addition, an adequate descriptor should allow the distinction between different accessions from the same culture to be feasible and valuable, avoid redundancy, be environmentally stable, and be easily manipulated by the breeder (Grum & Atieno, 2007).

Growers and breeders appreciate morphological traits such as fruit dimensions and stalk shape and, nevertheless, good adaptation to the environment and good commercial value (Trad et al., 2013b). Besides the fruit size and shape, ostiole width, acidity, and TSS (Mars et al., 1998), plant and leaf characteristics such as plant growth habit, plant vigor, number of leaf and lobe per shoot, leaf shape are essential for genotype selection by grower and breeders (Papadopoulou et al., 2002).

The researchers who have suggested that fruit and leaf traits are essential factors in differentiating and analyzing breeding materials, and addressing the morphological characterization of fig cultivars, are Mars et al. (1998), Hedfi et al. (2003), Saddoud et al. (2008b, 2011), Gaaliche et al. (2012), Aljane et al. (2012) and Khadivi-Khub and Anjam (2014). Also, many studies (Condit, 1955; Mars et al.,

1998; Papadopoulou et al., 2002; Mars, 2003; Aljane & Ferchichi, 2007; Saddoud et al., 2008b, 2011; Gaaliche et al., 2012; Almajali et al., 2012) revealed that morphological traits are beneficial in identification and evaluation of fig germplasm by estimating genetic relationships among fig genotypes.

Furthermore, high phenotypic variability was reported in female figs from Iran (Mahdavian et al., 2006; Baziar et al., 2018; Khadivi et al., 2018; Mirheidari et al., 2020), Tunisia (Mars et al., 1998; Chatti et al., 2003; Aljane et al., 2012; Gaaliche et al., 2012; Essid et al., 2017), Turkey (Caliskan & Polat, 2008, 2012; Caliskan et al., 2017), Morroco (Oukabli et al., 2002; Hssaini et al., 2019, 2020b), Spain (Sanches et al., 2002; Pérez-Sánchez et al., 2016), Lebanon (Chalack et al., 2005), Palestine (Ali-Shtayeh et al., 2014), Jordan (Almajali et al., 2012), and Slovenia (Podgornik et al., 2010). These studies indicated that high diversity in pomological and leaf-related traits could be used as an efficient marker system to discriminate between fig genotypes (Khadivi et al., 2018).

5.1 Biological Characteristics

Biological characteristics include traits like crop setting fruit, pollination requirement for fruit set, date of terminal bud-burst (leafing) in Breba, main crop and late crop, beginning of fruit maturation, Full maturity, harvest period, onset of caprification, length of the caprification period, onset of leaf fall. Among them, crop setting fruit (bifera, unifera, san pedro, smyrna, caprifig), pollination requirement for fruit set in breba (caducous, persistent), and main crop (deciduous, persistent) are more discriminated traits as discussed in the text. Other traits are unstable in different climatic conditions and can change according to environmental factors, and they can be used more for accession comparison in germplasm collections or a specific climatic region.

One of the most important aims is to expand the sales time thanks to the cultivation of genotypes/cultivars with different ripening times (Khadivi et al., 2018). The ripening time of the fruits in the 87 Iranian edible fig genotypes was variable and ranged between very early and late. Ripening time was early in 66 genotypes, late in 15 genotypes, very early in three genotypes, and intermediate in three genotypes (Khadivi et al., 2018). Also, Mirheidari et al. (2020) reported that the ripening time of Iranian wild edible figs ranged from early July to early September.

The time of fruits ripening for breba was evaluated by Ferrara et al. (2016) in some Italian accessions; they divided the biotypes into three classes: early (Columbro, Mariana and Fiorone nero; 10–15 June), medium-late (Fico del Vescovo and Tauro; 20–30 June) and late (Arodio; 5–10 July).

In Lebanese fig accessions studied by (Chalak et al., 2008), maturity dates were determined in comparison to Biadi accessions, which were considered as mid-season (11–31 August) in all fig growth areas in Lebanon, despite their different ecogeographic conditions and only one accession (American) was classified very late (after 1 October). Also, in Slovenian fig accession studied by (Podgornik et al.,

2010), the ripening period of all accessions lasted 5–15 days for the first crop and 15–60 days for the second crop. Breba fruits of most accessions mature in mid-late June, except accessions 'Zelena Bonka', 'Madona Rjava' and 'Madnici', which ripen in early July. These three accessions are mainly cultivated in the hinterland zone. The second crop ripens in early August and continues to ripen into September. Accessions with early second crop (ripening between 1st and 10th August) were: 'Greka crna, 'Belica-2', 'Belica-1", 'Belica-3', 'Kanora', 'Zuccherina', and 'Bjelica', while late second crop (ripening between 1st and 30th September) is characteristics to accessions 'Kamberij-1', 'Madona Rjava', 'Zelena Bonka', 'Crnica' and 'Zimica' (Podgornik et al., 2010).

5.2 *Vegetative Characteristics*

5.2.1 Growth Traits

The fig tree's growth is seasonal and related to diverse climatic factors. This includes a period of more growth where the vegetative and productive cycles are completed, followed by a dormancy stage, influenced by the region's low temperatures and the length of the winter season (Flaishman et al., 2008a). The adaptation to the region can be monitored through various indicators; for example, if there are adverse environmental conditions, the growth of the trees will slow down (Rostami & Rahemi, 2013; Gholami et al., 2012). In addition, other factors can affect the growth and development of the fig tree (such as the type of agronomical management, nutrient availability, the soil conditions, the hydric balance, among others) (Melgarejo, 2000; Botti et al., 2001; Hallac Turk & Aksoy, 2011; Crisosto et al., 2015) which are characteristic of a given location (Botti et al., 2001).

Tree growth habit, size of the tree, degree of branching, number of lobes per leaf (Giraldo et al., 2010), the number and shape of lobes (Saddoud et al., 2008a), leaf length, leaf width, leaf area, the density of hairs/spicules on the leaf's upper surface, and petiole thickness (Podgornik et al., 2010) were the vegetative traits used for the discrimination of fig accessions.

Fig trees vary in their growth habits, ranging from open and drooping to upright and compact (Ferguson et al., 1990). 'Brown Turkey' has a weeping willow type of growth with compact, down-spreading branches, while the cultivars 'Sierra' and 'Autumn Honey' have a vigorous growth habit with upright-rising branches. Individual trees in a favorable environment often reach a large size, while fig trees in orchards are usually compact (Flaishman et al., 2008a). The most frequent tree vigor observed by Khadivi et al. (2018) in Iranian fig cultivars was intermediate, and the majority of the studied trees had an open and spreading habit. Also, the Shoot color examined by Mirheidari et al. (2020) in Iranian wild edible figs was highly variable and included white, light brown, brown, dark brown, gray-white, and gray. In addition, in another study on Iranian figs by Darjazi (2011), Plant growth habits were obtained spreading (for Zard, Siah bolbol riz), erect (for Paizeh, Siah diras), and weeping habit (for Bidaneh, Siah zoodras, Morabaii, Hallavi riz and Hallavi dourosht) and tree vigor was found high (for Morabaii), medium (for Bidaneh, Paizeh, Siah zoodras, Hallavi riz and Hallavi dourosht) and low (for Zard, Siah bolbol riz and Siah diras).

Podgornik et al. (2010) reported intermediate tree vigor and spreading habit for Slovenian figs. Very diverse vegetative habits were observed by Pérez-Sánchez et al. (2016) in traditional fig cultivars grown in central-western Spain, ranging from semi-erect to weeping, and cultivars vigor ranged between weak ('Carballar Blanca') and strong ('Antigua', 'Blanca Común' and 'Prieto'). López-Corrales et al. (2011) also observed considerable variability in the growth habit character of Spanish fig cultivars.

5.2.2 Leaf Traits

Leaf biometric markers exhibit drastic shape variation within the same species, reflecting both the heteroblastic development of the apical meristem from which they are derived and the ontogeny of individual leaves as they expand allometrically (Chitwood et al., 2016). Each aspect of leaf morphology provides different points of view about its pattern. The number of lobes is a highly relevant factor for the differentiation of "toothed" plants, as is also the case with the fig tree, which proved to be a promising feature of selection of leaf morphology among the features evaluated by Rodrigues et al. (2017), because it presents high heritability and high genetic correlation, being little influenced by the environment.

Fig has typical bright green, single, alternate, and large leaves. Leaf characters are quite stable and serve as an essential parameter in cultivar identification (Ferguson et al., 1990). In addition, the traits related to leaves are important for the assessment of landrace and taxonomic investigations (Papadopoulou et al., 2002).

Leaf characters include the number of leaves per shoot, leaf shape, lobes number and shape, location of little lateral lobes, degree of leaf lobation/incision, shape of leaf base, leaf length and width, leaf area, length of leaf stalk/length of leaf, leaf margin dentation, leaf margin, density of hairs/spicules on leaf upper surface, density of hairs or spicules on lower surface, leaf color, petiole length, petiole thickness, petiole cross-section, petiole color (Condit, 1947; IPGRI and CIHEAM, 2003), vein number, central and lateral lobe depth (Khadivi et al., 2018), leaf venation clarity, central lobe length, upper lateral vein length, and upper lateral lobe base width (Mirheidari et al., 2020).

Different leaf shapes in fig trees, according to Condit (1947), are (1) Base calcarate, lobes linear, (2) Base cordate, five-lobed, lobes spatulate, (3) Base calcarate, lobes lyrate, (4) Base calcarate, lobes latate, (5) Base cordate, three-lobed, (6) Base truncate, (7) Base decurrent and (8) Leaf not lobed. In addition, according to IPGRI and CIHEAM, 2003), various leaf base shapes (petiole sinus) include truncate, cordate, calcarate, decurrent, and open calcarate. Also, the shape of lobes in fig leaves in different genotypes can be spatulate (narrower at the base and wider at the top), linear (more slender and regular in shape), latate (wider lobes), or Lyrate (as in

Ficus lyrata) and different leaf margin shapes in fig leaves are crenate, dentate, serrate, double serrate and undulate.

In 92 Iranian wild edible figs, leaf shape showed substantial diversity among the accessions, including base calcarate/lobes linear (13 accessions), base cordate/five-lobed/lobes spatulate (11), base calcarate/lobes lyrate (5), base calcarate/lobes latate (10), base cordate/three-lobed (12), base truncate (6), base decurrent (31), and leaf not lobed (4). In addition, leaf base shape (petiole sinus) was highly variable and included truncate (8 accessions), cordate (42), calcarate (30), and decurrent (12). Leaf margin showed six forms, although the crenate form (54 accessions) was predominant. Petiole cross-section was round and flattened (57 and 35 accessions, respectively) (Mirheidari et al., 2020). Also, Khadivi et al. (2018) found high variabilities among 87 Iranian edible fig genotypes detected in the leaf's traits such as leaf dimensions, leaf lobe depth, and the leaf lobe number. Most of the genotypes had five lobes per leaf. Central lobe depth ranged from 16.8 to 62.0 mm, while lateral lobe depth ranged from 16.8 to 92.0 mm. Petiole length ranged from 16.8 to 58.0 mm, and petiole thickness ranged between 1.70 and 6.30 mm.

In 21 local cultivated fig accessions from Egypt and Libya studied by Mohamed et al. (2017), leaf length ranged from 5.4–23.5 cm, while leaf width varied from 6–23 cm. Four different leaf edge shape in accessions was strait (e.g., Kahramany, Abodey-Giza), waved (e.g., Sultani black, Sultani-Giza), zigzag (e.g., Hamouri, San-Badr), and serrated (e.g., Barry, Mejahal). Also, leaf top shape showed three forms, although round form (e.g., Bioudi, Aswany) was predominant. In their studied accessions, the number of leaf lobes ranged from one lobe (e.g., Green-yellow, Sultani Red Siwa), three lobes (e.g., Mejahal, San-Badr), five lobes (e.g., Sultani black, Sultani yellow), six lobes (e.g., Hamouri), seven lobes (e.g., Adsey-Giza, Fayoumi), eight lobes (e.g., Black-Mission) and ten lobes (e.g., Aswany).

According to Medeiros (2002), the 'Roxo-de- Valinhos' cultivar presents large leaves with five large and two smaller lobes; dark green color; compact texture, somewhat rigid; creased margin; petiole sinus in the form of lyre and long petiole. Rodrigues et al. (2019b) mentioned that the number of lobes was the most promising leaf morphological character among those fig tree genotypes examined in their study due to its high heritability and high genetic correlation, corroborant to Caliskan and Polat (2012), which used the number of lobes per leaf as one of the essential parameters for the differentiation of fig genotypes in Turkey. The differences between Slovenian figs have been reported in the degrees of leaf lobation (the central lobe length/the leaf length) (Podgornik et al., 2010).

Khadivi et al. (2018) studied edible Smyrna-type fig clones from the Estahban region in Iran and observed the range of 62.20–138 mm and 41–153 mm for leaf length and leaf width. Also, the highest values of fruit stalk length and fruit stalk diameter were 20.10 mm and 2.95 mm, respectively. Also, in examined fig cultivars from Varamin in Iran, the leaf blade length of the genotypes was changed between 12.00 and 26.50 cm, and leaf blade width differed between 8.25 and 18.60 cm (Darjazi, 2011).

In Spanish figs cultivars studied by Pérez-Sánchez et al. (2016), leaf length was varied from 24.7 to 36.3 cm, 'Tardía Portuguesa' being the cultivar with the longest leaf. Its blade size was also significant, with a length of 29.4 cm and a width of 24.2 cm. Stalk length ranged from 4.89 ('Moscatel') to 13.3 cm ('Carballar Negra'). It was observed that the leaf blades were longer than the stalks in all the fig cultivars (stalk length/blade length ratio: 0.22–0.65). Concerning the number of main lobes, most cultivars showed three/five-lobed leaves. These results are consistent with those reported by López and Guzmán (2007) and González and Grajal (2012) for Spanish fig cultivars. 'Carballar Blanca', 'Cuarterón', and 'Moscatel' were the only cultivars that showed entire leaves. The shape of central lobe varied between triangular ('Carballar Blanca' and 'Cuarterón') and lyrate ('Cuello de Dama Negro'). López-Corrales et al. (2011) also observed a lyrate central lobe shape in leaves of this latter cultivar. The shape of the leaf base was highly variable among cultivars (truncate-strongly calcarate). Most cultivars did not have additional lobes regarding the basal lateral lobes on the petiole sinus. They were only observed in the leaves of the 'Blanca Común', 'Carballar Negra' and 'Cuello de Dama Negro' cultivars. The most frequent petiole color was green. Other researchers who have addressed the leaf morphological characterization of edible fig cultivars are Giaccone et al. (2017), Baziar et al. (2018), Rodrigues et al. (2019a), Abdelsalam et al. (2019), Shahinuzzaman et al. (2020).

5.3 Pomological Characteristics

Common fig cultivars are facultatively persistent and may produce persistent pollinated main-crop figs. Morphologically, the pollinated fig syconium creates true fruits, while the non-pollinated fig syconium presents an enlarged inflorescence with multiple long-styled pistillate flowers. 'Autumn Honey' and the 'Brown Turkey' are two cultivars that produce caprified and non-pollinated figs on the current year's wood. The caprified 'Autumn Honey' fruits have darker purple skin color and red pulp than the white, pink pulp of the non-caprified fruit. Usually, caprified fruits are bigger and have longer storage ability (Rodov et al., 2005). Where *B. psenes* wasps are present, caprified and non-caprified fruit may develop on the same branch in common-type fig fruits (Flaishman et al., 2008a). In the case of the 'Smyrna' cultivar, its shape tends to change due to the caprification process indispensable for the normal development of fruits of this pomological class (Aradhya et al., 2010).

Traditionally, the amplitude and distribution of fig tree genetic diversity have been widely assessed based on conventional pomological markers, even though the expression of these markers is strongly affected by environmental and agronomic conditions of the species (Caliskan et al., 2017; Rodolfi et al., 2018). Therefore, fig assessment using morphological, pomological, and physicochemical traits is essential for the inventory, assessment, preservation, and breeding programs of genetic resources (Podgornik et al., 2010). Also, Caliskan et al. (2017) and Khadivi et al. (2018) have reported that the pomological characteristics are essential to evaluate the variation in traits of edible fig accessions.

The most important discriminators of fig fruits were the fruit weight, fruit length, fruit diameter, fruit skin and flesh color (Saddoud et al., 2008a; Aljane & Ferchichi, 2009; Podgornik et al., 2010), fruit shape (Giraldo et al., 2010; Podgornik et al., 2010), firmness of the fruit skin (Saddoud et al., 2008a; Podgornik et al., 2010), fruit skin cracks (Saddoud et al., 2008a), production type, skin firmness (Giraldo et al., 2010), fruit neck length (Aljane & Ferchichi, 2009; Podgornik et al., 2010), abscission of the stalk from the twig (Podgornik et al., 2010), stalk diameter, neck diameter, ostiole diameter, ostiole opening, and flesh thickness (Aljane and Ferchichi 2009).

Fruit Shape Different fruit shapes in fig trees, base on fruit index [(width/length) = I], are mainly Oblong ($I < 0.9$), Globose ($I = 0.9–1.1$) and Oblate ($I > 1.1$). Also, fruit shapes, according to the location of the maximum width, include; Ovoid (in the middle), Bell-shaped (nearer to the neck), and Pyriform (nearer to the ostiole-end) (IPGRI and CIHEAM, 2003).

Fig fruits can be vertically and horizontally symmetric or asymmetric shape. However, symmetric fruits are more acceptable for consumers and more suitable for packaging. Fruit index (width/length) is important in packing and transportation (Condit, 1941). Also, the fig shape and its index are of great importance when it comes to trade and is prized by consumers. Aljane et al. (2008) indicated that fruit shape is important for packing and transportation. However, the globose shape is the most suitable fruit shape (Condit, 1941; Caliskan & Polat, 2008) and a more dominant character in fig accessions (Gozlekci, 2011; Caliskan & Polat, 2012, Mirheidari et al., 2020; Hssaini et al., 2020b) and also, it is preferred for its suitability for packaging and transportation (Condit, 1941; Benettayeb et al., 2017). Although, other authors reported that oblate fruit shape is the more dominant character (Aljane et al., 2008; Darjazi, 2011; Caliskan & Polat, 2012). The harvesting dates and growing areas can explain these contradictory reports.

Most of the analyzed cultivars had an oblate fruit shape among some Iranian cultivars. Fruits of Siah bolbol riz and Morabaii were globose while those of Hallavi riz were oblong. (Darjazi, 2011). Pérez-Sánchez et al. (2016) reported that the most frequent fruit shapes in Spanish figs were turbinate or spherical-turbinate. 'Cuello de Dama Negro' was the only cultivar that showed pyriform fruits in their study. López-Corrales et al. (2011) observed that this cultivar has pyriform breba. The fruit skin ground color was generally green-yellow, 'Prieto' (purple), 'Cuarterón' (purple-black), and 'Cuello de Dama Negro' (black) being the only exceptions. Fruit flesh color ranged from amber to red. Finally, some relevant cultivars with regard to breba quality were 'Antigua', 'Cuarterón', 'Cuello de Dama Blanco', 'Gota de Miel', 'Pringo de Mel' and 'Tardía Portuguesa'. Their fruits were very juicy and easily peeled. González and Grajal (2012) also observed different fruit shapes with Spanish fig cultivars.

The morphological characterization of 45 accessions in Active Germplasm Bank Fig tree accessions of the Faculty of Agricultural and Technological Sciences (FCAT/UNESP) was studied by Rodrigues et al. (2019a). They reported that the shape of accession fruits varied from elongated to oval. The fruit shapes for

'Pingo-de-Mel', 'Irradiated Plant 440', 'Roxo-de-Valinhos', and 'White Genova' accessions were oval, elongated, oblong pyriform, and flattened, respectively. 'Roxo-de-Valinhos' cultivar has a predominance of fruits with oblong-pyriform shape, with a short and thick neck, almost without limit of separation with the body from the receptacle, as described by Rigitano (1955).

Fruit apex shape was flat in 87 out of 92 accessions studied in Iranian wild edible figs. Conversely, the fruit was symmetric in 83 out of 92 accessions investigated (Mirheidari et al., 2020). Hssaini et al. (2019) studied local fig clones from Northern Morocco and reported that most studied genotypes had the symmetric shape of the fruit.

Fruit Weight, Width and Length Fig yield is influenced by several factors, including cultivar, growth conditions, and harvest. Typical mature fig yield varies between 5 and 12 t/ha (Flaishman et al., 2008a). Sahin et al. (2001) explained that fig plants produced optimum yields at 7 years. Besides other criteria, weight is the most important fruit dimension, and fruit prices depend mainly on this (Pérez-Sánchez et al., 2016). Previous studies of pomological assessment of figs demonstrated that the most important characteristics affecting the commercial value of fruits for fresh consumption are fruit weight (Aksoy et al., 1992), width, and length (Can et al., 2001; Ateyyeh & Sadder, 2006; Podgornik et al., 2010). Also, Fruit dimensions are of great importance in the packing and transportation of fig fruits (Condit, 1947). Hence, small fruits are generally used for canning, whereas big ones are consumed as fresh. Moreover, the fruit size reflects the tree's proper maintenance (Tamboli et al., 2015). However, these character is usually negatively influenced by the fruit load on the tree. It is noteworthy that besides the genetic effect, fruit weight also depends on the growing location, growing season conditions, agricultural practices, as well as the interaction between the genotype and the maturity stage, which is properly explained by the index of maturity (Gozlekci, 2011; Benettayeb et al., 2017).

In general, fruits developing along the shoot are larger at the lower nodes and become smaller toward the top of the shoot. Selected cultural practices may exert a positive impact on fruit size. For example, in 'Black Mission', cultivar girdling increased fruit size, although late girdling produced less marked effects (Ferguson et al., 2003).

Uniformity of the final fruit size in fig trees is very important in the sales market. Particularly for fresh products, due to the short post-harvest life of the fig and its perishability, the more uniform the product, the less time it takes to sort. On the other hand, it does not reduce the price of a part of the product due to its smaller size.

In Moroccan figs studied by Hssaini et al. (2020b), fig height values ranged from 21.3 to 48.8 mm with an average of 34 mm. At the same time, the fruit width was 25.93–61.58 mm, with an average of 39.8 mm. These results agreed with previous works (Gozlekci, 2011; Papadopoulou et al., 2002; Ferrara & Papa, 2003; Rodrigues et al., 2018).

In wild Iranian fig accessions, fruit length varied from 11.2 to 34.1 mm, and fruit width ranged from 14.1 to 39.3 mm. Also, Fruit fresh weight ranged between 0.56 and 28.69 g with an average of 8.33 (Mirheidari et al., 2020). Aljane and Ferchichi

(2010) studied fig cultivars from Tunisia and reported that Mean fruit lengths varied from 51 mm in cv. Wahchi to 80 mm in cv. Faouari and mean fruit diameter was between 47 mm in cv. Soltani Abiadh and 76 mm in cv. Faouari

In Turkish fig accessions studied by Gozlekci (2011), fruit width was between 28.6 and 51.1 mm, and fruit length was between 28.3 and 54.7 mm. Khadivi et al. (2018) studied edible Smyrna-type fig clones from the Estahban region in Iran and reported a range of 1.86–7.15 g for dried fruit weight. Also, in their study, Dried fruit length ranged from 19.8 to 36.2 mm, and dried fruit width ranged from 15.7 to 25.2 mm.

Darjazi (2011) reported that Fruit diameter ranged from 21 to 45 mm, and fruit length was varied from 20 to 36 mm in fig cultivars from Varamin in Iran. Also, in their study, the fresh fruit weight ranged from 8.0 to 43.5 g, and the 'Paizeh' cultivar showed the highest fruit diameter and fresh fruit weight.

Among eleven local and introduced fig varieties in Morocco that were studied by Hssaini et al. (2020a), the variety 'Kadota' had the heaviest fruit (61.5 g), followed by 'Breval Blanca' and 'Snowden', which the average weights were, respectively, 46.4 and 44.0 g. Furthermore, the latter recorded the highest

Caliskan and Polat (2008) reported fruit weight ranged from 22.2 to 52.5 g on many fig accessions sampled in the East Mediterranean region. Gozlekci (2011) carried out a selection study on figs in the Kemer and Alanya districts, which belong to Antalya provinces located west Mediterranean Region in Turkey, and found that fruit weight was between 14.7 and 60.5 g in the Kemer district, while the fruit weight of fig accessions from Alanya district varied from 13.8 to 48.5 g. Similar work carried out on a large part of fig diversity in different countries showed wide variability in fruit weights that varied from 30 to 90 g (Ozeker & Isfandiyaroglu, 1998), 24 to 92 g (Sahin et al., 2001), 28 to 107 g (Polat & Ozkaya, 2005), 12.3 to 99.4 g (Caliskan & Polat, 2008) and 12.3 to 99.4 g (Caliskan & Polat, 2012) in Turkey, 24 to 58 g (Aljane et al., 2005), 24 to 58 g (Aljane et al., 2008) and 63 to 172 g (Aljane & Ferchichi, 2010) in Tunisia, 30 to 85 g (Podgornik et al., 2010) in Slovenia, 16.4 to 77.7 g (Moniruzzaman et al., 2020) in Malaysia and 12.4 to 87.0 g (Hssaini et al., 2020c) in Morocco.

Fruit Stalk (Shape, Cross-Section, Length and Width) and Fruit Neck (Length, Width) Different shapes of the fruit stalk in fig trees are short and thick, enlarged, and long and slender. Also, the fruit stalk cross-section includes Circular, flattened, rhomboid, and triangular (IPGRI and CIHEAM, 2003).

In Iranian wild edible fig accessions studied by Mirheidari et al. (2020), fruit stalk shapes ranged from variously enlarged to long and slender. Hssaini et al. (2020b) studied local Moroccan figs and reported that most genotypes had a rounded shape at the stalk.

Also, Darjazi (2011) reported that the fruit stalk was long and slender for three cultivars (Siah bolbol riz, Zard, and Hallavi riz), and it was short and thick for the remaining fig cultivars from Varamin province in Iran (Darjazi, 2011). In Iranian fig cultivars studied by Khadivi et al. (2018), the highest values of fruit stalk length and fruit stalk diameter were 20.1 mm and 2.95 mm, respectively.

A neck in figs facilitates picking the fruit from the tree and is thus associated with easier harvesting (Trad et al., 2012). In addition, short neck length is reported to cause fig fruit damage during harvest (Ozeker & Isfandiyaroglu, 1998; Gozlekci, 2011; Darjazi, 2011). In Iranian wild edible fig accessions studied by Mirheidari et al. (2020), the highest value of fruit stalk length, fruit stalk width, and fruit neck length was 34.97, 4.75, and 8.31 mm, respectively.

Among the evaluated Moroccan fig germplasm, the stalk length and width ranged between 2.95 and 16.3 mm and 2.92 and 7.96 mm, respectively. The largest size was recorded with 'Mnedar 289' and 'Figue de Marseille' followed by 'Trojana' and then 'El Ghani'. Neck Length ranged from 1.64 to 9.52 mm, and Neck Width ranged from 2.22 to 15.6 mm. The longest fruit neck was recorded in Bioudi 2878 (9.52 mm), followed by Nardine (9.4 mm) and Kodota (9.2 mm) (Hssaini et al., 2020c).

In the same study by Hssaini et al. (2020a), 'Fassi' had the highest fig stalk (10.58 ± 3.5 mm) while 'kadota' had the highest stalk width (6.59 mm). Furthermore, the fruit neck was long in the cultivars 'Kadota' (9.17 mm) and 'Palmeras' (7.28 mm) and short in 'Fassi' (2.29 mm), whereas it was absent in others such as 'Ghoudan', 'Nabout' and 'Palmeras'. Also, in another study by Hssaini et al. (2020b), for the neck length, the values ranged between 1.64 and 9.52 mm with an average of 7.01. At the same time, neck width varied from 2.22 to 15.7 mm, with an average of 8.64 mm. The longest fruit neck was recorded respectively by "Bioudi 2878" (9.52 mm), "Nardine" (9.4 mm), and "Kadota" (9.2 mm).

In Turkish fig accessions studied by Polat and Ozkaya (2005), fruit neck length in fig accessions ranged between 1.0 and 8.9 mm. neck length was longest on 31-IN-15 (8.86 mm) and Sarilop (6.39 mm). Gozlekei (2011) also reported fruit neck length between 1.97 and 17.4 mm among fig accessions from Turkey. In other studies, the longest neck was 14.5 mm (Ilgın, 1995), 8.70 mm (Ozeker & Isfandiyaroglu, 1998) and 8.01 mm (Polat & Ozkaya, 2005). The cultivar 'Cuello Dama Blanca' was closed to 'White Adriatic', 'Nabout', and 'Noukali', since they all had no fruit neck. They may be used for canning, whereas big ones like 'Kadota' may be recommended for fresh consumption.

Ostiole Width The ostiole is characterized as the opening for the entrance of the *Blastophaga psene* wasp so that inflorescence pollination within the siconium can occur to develop fig fruits (Lamas et al., 2009). The ostiole provides entry for fungal decay and moisture loss during the maturity period. It also provides an entryway for insects that vector diseases such as endosepsis that spread to healthy fruits (Michailides & Morgan, 1998; Crisosto et al., 2011). Thus, a large ostiole in the fig is an undesirable characteristic (Caliskan & Polat, 2008; Can, 1993). The smaller ostiole width, the better fruit can be stored and protected from infectious agents (Trad et al., 2012). In the specific case of the fig tree crop, an essential aspect of determining the quality of fruits and their size is the small ostiole opening (Aljane et al., 2013).

Cultivars with a large ostiole (Doster et al., 2002) are susceptible to ostiole splitting (Michailides, 2003) and are more susceptible to decay. The high susceptibility of figs to decay and their fast ripening accelerates fruit softening (Çelikel & Karaçalı,

1998), making fresh fig shelf life extremely short, lasting only 1–2 days (Morton, 2000).

In Iranian fig cultivars studied by Khadivi et al. (2018), fruit ostiole width ranged from 0.00 to 10.8 mm. Also, in Iranian wild edible fig accessions studied by Mirheidari et al. (2020), the range of ostiole width was 1.06–6.9 mm. In addition, Darjazi (2011) reported that Fruits ostiol width was changed between 0.0 (Morabaii) and 4.6 mm (Siah zoodras) in fig cultivars from Varamin province in Iran. Fruit ostiole width has been reported as 2.34–8.79 mm with an average of 5.13 mm (Aksoy et al., 1992), 1.50–4.00 mm (Koyuncu et al., 1998), 1.00–9.4 mm (Polat & Ozkaya, 2005), 1.1–4.9 mm (Caliskan & Polat, 2008), and 2.25–8.93 mm (Gozlekci, 2011) in different fig growing areas in Turkey. Also, fruit ostiole width has been reported as 0.00–8.20 mm (Hssaini et al., 2019), 0.60–9.10 mm (Hssaini et al., 2020c), in different fig accessions from Morocco. In addition, Hssaini et al. (2020a), who studied eleven fig varieties from Northern Morocco, reported that the ostiole was generally large fruits such as ‘Kadota’ and ‘Palmeras’ (7.69 and 5.81 mm, respectively).

Fruit Skin Cracks Cracking around the ostiole is undesirable because the pathogens and pests enter fruit via these cracks (Can, 1993). Skin cracks influence consumer acceptance because consumers think that this character indicates the degree of fruit ripening (Mahmoudi et al., 2018). Many factors (i.e., genetic, morphological, environmental, and physiological) influence fruit cracking. Under the same environmental conditions, fruits from different cultivars show differences in cracking susceptibility. Some correlations have been observed between susceptibility of fruit cracking and some fruit traits (fruit shape, fruit size, fruit firmness; anatomy and strength of the fruit skin, stomata in fruit skin, cuticular properties, osmotic concentration, water capacity of the fruit pulp and growth stage of the fruit). Also, orchard management (such as irrigation and nutrition) and environmental condition (such as temperature, wind, and light) can influence fruit cracking. Besides, fruit cracking is a quantitative trait controlled by several genes. Therefore, the best way to reduce fruit cracking would be suitable for orchard management that takes into account and minimizes the stress of the water, nutrition, and physiological factors that contribute to fruit cracking. Also, the most resistant cultivars to fruit cracking that have desirable fruit quality can be selected for cultivation (Khadivi-Khub, 2015). Fruit splitting in figs is due to sudden changes in the internal pressure of the fruit that can be the result of cool temperatures and high humidity while the fruit matures, precipitation at fruit maturity, or over-caprification (excessive pollination) (Crisosto et al., 2011).

In Iranian fig cultivars studied by Khadivi et al. (2018), fruit cracking percentage varied from 0.00% to 70.00%. The seven genotypes, including ‘Siah-1’, ‘Siah-2’,’Sabz-22’, ‘Sabz-40’, ‘Sabz-60’, ‘Choopani’, and ‘Mati’ had no fruit cracking. The highest width of the cracked fruit section was 9.10 mm and was observed in the ‘Sabz-71’ genotype, while the lowest value of this trait was 0.00 mm. Also, in 92 Iranian wild edible fig accessions studied by Mirheidari et al. (2020), fruit skin cracks were not observed in the majority of accessions (72), and among the accessions having fruit skin cracks, the cracking area was predominantly on the base of fruits (16). Furthermore, Hssaini et al. (2019) reported that the skin cracks were primarily absent in local fig clones from Northern Morocco.

Fruit Flesh and Skin Thickness and, Ease of Peeling The thickness of the flesh and its hydration rate can effectively crack the skin. In Moroccan fig accessions studied by Hssaini et al. (2020c), fruit peel thickness (mm) was 1.32–8.35. Hssaini et al. (2020a), by evaluating 11 local and introduced fig cultivars cultivated in the Moroccan climate, reported that 'Breval Blanca' and 'Nabout' cultivars had the thick-skinned fruit, whichthe average values were, respectively, 8.35 and 5.91 mm.

Also, the Ease of peeling is critical to satisfying local and global customer preferences (Caliskan & Polat, 2008; Can, 1993). Easy skin peeling is an essential characteristic of consumer acceptance (especially for the cultivars that have thick skin, such as 'Bakkor Biadh' and 'Onk Elhamam') because consumers tend to peel figs before eating (Mahmoudi et al., 2018). Interesting cultivars in fig cultivars grown in Central-Western Spain, with respect to fruit quality were 'Antigua', 'Cuarterón', 'Cuello de Dama Blanco', 'Gota de Miel' and 'Pringo de Mel'. These developed very juicy and easily peeled fruits (Pérez-Sánchez et al., 2016).

Fruit Skin and Flesh Color Fruit skin and flesh colors determine ripening time. These characteristics are also used with other features to determine the promising fig genotypes in breeding studies and the selection of accessions used in breeding studies (Tsantili, 1990; Sacks & Shaw, 1994). In addition, the fruit skin color of fresh figs is essential for consumer preferences (Mirheidari et al., 2020), and fresh figs with pink and red flesh colors are preferred by consumers (Caliskan & Polat, 2008). The peels and pulps of different cultivars had variable levels of polyphenols, flavonoids, anthocyanins, tannins, and antioxidant activity. Generally, fig fruit peels, especially those with a dark color, contained the highest concentrations of phytochemicals and exhibited the highest antioxidant activity compared to fig fruit pulps. Based on mineral and phytochemicals, it is recommended consumption of the whole fig fruit (Mahmoudi et al., 2018).

Fig skin color varies from dark purple to green. According to their color, figs are categorized into 3 groups: White, Red, and Black. The skin of the white variety is white, yellowish, or even green, e.g., the Kadota variety. Another variety known as Dotato has yellowish-green skin and purple-colored flesh. The white variety is either seeded or colored seedless. The reddish variety is distinguished through its bluish brown color. The black variety includes those figs with dark red to black skin color. This group includes varieties like Brown Turkey, Celeste, and Sari Cob (Stover et al., 2007b). Gozlekci (2011) indicated that fruit skin color in fig accessions ranged from green-yellow to black among fig accessions from Turkey. In 92 Iranian wild edible fig accessions studied by Mirheidari et al. (2020), fruit skin ground color was strongly variable. It included light violet (3 accessions), violet (20), dark violet (35), red (5), yellow (7), yellow-violet (7), yellow-red (1), violet-cream (6), cream (6), and light brown (2) and internal pulp color included cream (1), gold (1), red-gold (6), light red (3), red (7), dark red (13), light violet (8), violet (29), light brown (16), brown (6), and dark brown (2). Also, Darjazi (2011) reported that the fruit skin color varied from yellow to black in fig cultivars from Varamin province in Iran. 'Paizeh' and 'Morabaii' cultivars had canary yellow colors. 'Zard' cultivar was yellow-green, and 'Bidaneh' cultivar was light green, while 'Hallavi riz' and 'Hallavi dourosht' cultivars were light-brown, 'Siah bolbolriz' cultivar was purplish-brown, 'Siah zoodras' cultivar was purple and 'Siah diras' cultivar was black. Also, in their study, the skin ribs were Dark-brown (for 'Hallavi riz' and 'Hallavi dourosht') and absent for the remaining cultivars.

In local fig clones collected in Northern Morocco by Hssaini et al. (2019), the skin colors of sampled fruits varied from the blue-purple to the green-yellow, and the predominant internal pulp color was dark-red and purple. Also, fruit flesh color in fig cultivars grown in Central-Western Spain ranged from white to red (Pérez-Sánchez et al., 2016).

In 82 Lebanese fig accessions studied by (Chalak et al., 2008), skin ground color ranged from a purplish black to yellow with all the intermediate scales. The skin over color was absent or present as irregular patches with yellow, purple, or green sectors. Internal color varied between white, amber, pink, dark pink, red and dark red. In Iranian fig cultivars studied by Khadivi et al. (2018), Fruit skin color was most often yellow (40 genotypes, 46%) and followed by dark yellow (31 genotypes, 35.6%), and also the fruit pulp internal color was yellow in most of the genotypes (63 genotypes, 72.4%). In 38 Slovenian fig accessions described by Podgornik et al. (2010), the fruit skin ground color of the main crop varied from black for 'Kamberij-1' to purple for 'Flazana-2', 'Crni Matalon', 'Crnica', 'Miljska figa' and 'Kamberij-2' through purple-brown for 'Madona Rjava' and 'Rjavi Matalon', brown for 'Kanor', dark green for 'Crna Petrovka' and 'Petrovaca crna' to green and yellow-green for all others varieties and the internal color varied from amber to dark red. Eleven accessions were dark red, nine accessions were pink, and only three accessions were amber-colored.

Fruitlet Amount, Size and Weight In Iranian fig cultivars studied by Khadivi et al. (2018), the 'Mati' genotype was seedless, while the range of seed number in other genotypes ranged from 24 (in 'Siah-1' and 'Siah-2' genotypes) to 575 (in 'Sabz-11' genotype).

Also, Darjazi (2011) reported that the number of seeds per fruit was high (for 'Zard'), medium (for 'Paizeh', 'Siah diras', 'Hallavi dourosht'), and low for the remaining fig cultivars from Varamin province in Iran. The fruit of 'Bidaneh' had the strongest aromatic flavor. In 92 Iranian wild edible fig accessions studied by Mirheidari et al. (2020), the seed amount was high in most of the accessions (61). The seed length, seed width, and seed thickness range were 1.21–2.13 mm, 1.00–1.90 mm, and 0.69–1.52 mm, respectively. Hssaini et al. (2019) reported that most of the genotypes from Northern Morocco had a medium seed size.

5.3.1 Environmental Condition's Effects on Fruit Characteristics

Climate markedly affects figs' size, shape, skin, and pulp color (Condit, 1947). Cooler climates produce greener than yellow skins, more vivid pulp colors, and larger, more elongated fruits. Crane (1986) has suggested that the larger individual size of the first breba crop, which competes with shoot growth and the second crop for available carbohydrates, is due to its development during a cooler period. In addition, climate may also affect pollination requirements. Other environmental conditions such as rain, hail, and wind can reduce fruit quality and production. Rain may cause fruit splits. Splitting results from sudden changes in the internal fruit pressure brought on by cool temperatures and/or high humidity as the fruit matures (Ferguson et al., 1990). Splitting in 'Calimyrna' and other varieties may also result from excessive pollination and the growth of too many developing seeds during fruit ripening. Strong winds during the ripening season whip the foliage and cause scarring of fruits such as 'Kadoti' and 'BrownTurkey' (Flaishman et al., 2008a).

A comparison of the role of climatic conditions in dry and fresh fig production revealed that dry fig production is strongly dependent on climatic conditions and is successful mostly under dry and warm-temperate climates. Fresh figs, however, can be cultivated under a wider range of ecological conditions (Sahin, 1998).

Spontaneous fig trees can potentially exchange genetic material each year (Kjellberg et al., 1987). A study of enzymatic polymorphism revealed that the spontaneous fig trees of southern France show that alleles are not randomly distributed; their frequencies vary regularly with the climate and reflect the role of adaptation (Kjellberg & Valdeyron, 1984).

The fig weight (20–60 g) in a study quoted by Crisosto et al. (2011) on ten fig varieties (152-4s, 'Brown Turkey', 'Kadota', 'Mission', 'Orphan', 'Panachee', UCR 276-14, UCR 291, 'White Texas Everbearing' and 'Zidi') grown under California conditions, was comparable with that of Turkish varieties [22–52 g (Caliskan & Polat, 2008), 40–65 g (Bostan et al., 1998), and 30–90 g (Ozeker & Isfandiyaroglu, 1998)]. The varieties grown in California had similar fruit length (41-63.5 mm) to varieties from Turkey (38.5–62.0 mm) but were smaller in fruit diameter (35.6–48.4 mm versus 45.0–55.0 mm) (Bostan et al., 1998). Soluble solids concentrations of these varieties grown in California (13–31%), though with more extreme values, were in general comparable to the Turkish varieties [15.1–21% (Bostan et al., 1998), 16–27% (Ozeker & Isfandiyaroglu, 1998), and 20.1–27.4% (Caliskan & Polat, 2008)]. Finally, the TA of the varieties grown in California (0.14–1% citric acid) was, in general, higher than the same varieties from Turkey [0.09–0.26% (Caliskan & Polat, 2008), 0.14–0.22% (Bostan et al., 1998), and 0.06–0.15% (Ozeker & Isfandiyaroglu, 1998)].

5.3.2 Caprification Effects on Fruit Morphological Characteristics

Caprification is the pollination of long-styled female flowers of the edible fig by wasps carrying the pollen from the profichi borne on the caprifig (hermaphroditic, with male and female flowers in the profichi fruits). The number of male figs dispensed and length of pollination period depends on weather conditions (Condit, 1947), age, size, the yield of female cultivar, and type of caprifig.

The common practice of caprification is to distribute the male figs at intervals of a few days over about 3 weeks (Khadari et al., 1995b; Rahemi & Jafari, 2008; Zare, 2008). In Tunisia, two to six male figs are connected with a wire, or a stick passed through their neck and hung onto branches of female fig trees of the Smyrna-type (Mars et al., 2008). In California, ripe caprifigs are cut and placed in bags or baskets that are stapled on female trees of the Smyrna and San Pedro-type (Stover et al., 2007a). In Iran, caprifigs are put inside containers hung on the female fig trees of the Smyrna-type (Zare, 2008).

The selection of pollen sources with good quality and quantity of pollen is an important and essential practice in fig orchards (Küden & Tanriver, 1998; Khadivi-Khub & Anjam, 2016). In addition, it is well known that the type of caprifig can have a significant effect on the skin color, ostiole width, total soluble solids, time of ripening, seed germination, size and shape of syconia (Janick & Moore, 1975; Rahemi & Jafari, 2008; Gaaliche et al., 2011).

1. Caprification and Fruit-Set and Yield Fig caprification is a complex system for fruit production and is a model system for studying the mutualism and coevolution between figs and pollinating wasps (Bandelj et al., 2007; Giraldo et al., 2008a). Despite its importance, this mechanism remains unclear, and the classification of fig types based on pollination and fruit production is still ambiguous. Ferrara et al. (2016) reported that varieties such as Rosso Comune, Nero Terlizzi, Nero Sava, etc., clearly needed profichi (wasps) to set figs and obtain good yields. Other varieties in their study, such as Dottato Marchese, Dottato Sava, Nero Crotone, Regina Gioia, Rosso Triggiano, etc., were not mainly influenced by showing satisfactory fruit-set and yield with or without caprification. Their results indicated that fruit-set was significantly influenced by the presence/absence of profichi in the orchard (i.e., caprifigs), with varieties producing well yield only when caprificated and varieties setting figs parthenocarpically, thus not strictly needing profichi for the pollination. The fruit set of brebas was highly variable, with some varieties setting in significant numbers of fruits (Dottato Marchese, Dottato Sava, Monaca, etc.) and others showing potential for brebas production Petrelli, Sant'Antonio, Rosso Triggiano, etc. (Ferrara et al., 2016).

Pollination by caprification greatly influences fruit size, which can double, while the fruit set is also noticeably increased (Gaaliche et al., 2011). Ferrara et al. (2016) indicated that yield was dramatically reduced when figs were non-caprificated (either manually or naturally). Indeed, non-caprificated figs show an internal cavity, dry matter content reduction, a collapse of pedicel development, and lack of seeds, and the fruit is significantly firmer at harvest (Trad et al., 2013a). Physiologically, caprification acts indirectly as a precursor of several hormone activities. Pourghayoumi et al. (2012) showed a significant difference in the time of fruit ripening in Iranian cultivars, but the effects of the pollen sources on the time of fruit ripening were not significant. Their results were in contrast to the results found by Rahemi and Jafari (2008). It is well known that ethylene plays a significant role in ripening (Galil, 1968). Fertilized seeds produce a large amount of auxin, which may stimulate ethylene production within the tissue (Ryugo, 1988). Ethylene, responsible for physiological and biochemical changes in fruit development and maturation, may be enhanced by pollination. Ferrara et al. (2016) data on caprificated and non-caprificated fruits are quite similar to the percentages reported by others (Rahemi & Jafari, 2008; Zare, 2008). This demonstrated that caprification is necessary for San Pedro and Smyrna type and Common type varieties. Whether required or not, caprification can increase fruit-set and improve several organoleptic quality aspects (Trad et al., 2013a). Even though caprification enhances fig production and yield, little information is available on the genetic basis of this complex and exciting system. Large-scale comparative gene expression studies have provided comprehensive data on its effect on fruit physiology and elucidated genetic factors underlying the traits involved in type differentiation (Ikegami et al., 2013).

2. Caprification and Fruit Morphometric Characteristics Pollinated figs are usually considered larger greener, and darker interior pulp color than unpollinated figs (Condit, 1947; Oukabli et al., 2003; Michailides et al., 2008). Flaishman et al. (2008a) reported that the caprified fruits of cultivar 'Autumn Honey' were bigger, with darker purple skin color and red pulp, and have longer storability than non-caprified fruits. In Tunisian cultivars, fruit weight, length and diameter, ostiole diameter, flesh thickness, and the number of fertile seeds were significantly increased by pollination rate (Gaaliche et al., 2011).

Pourghayoumi et al. (2012) confirmed that fruit length was significantly affected by pollen sources and female cultivars, and also, fruit diameter was significantly affected by female cultivars. Their results are in parallel with those reported by Rahemi and Jafari (2008) and Gaaliche et al. (2011). In addition, these findings agree with those of Bose and Mitra (1990), who indicated that the length of the fruit neck of Smyrna-type figs was affected by pollen sources. The specific pollens with increased IAA (indole-3-acetic acid) may affect the cell number in early fruit development and stimulate cell elongation (El-Hamady et al., 2010; Khan & Chaudhry, 2010).

Pourghayoumi et al. (2012) found that pollen sources did not significantly affect fruit weight and ostiole width. Due to fruit weight, their results follow other reports as in Iran; pollen sources had no significant effect on fruit weight (Rahemi & Jafari, 2008). However, the results contrast the results found by Gaaliche et al. (2011). Pollen source affects the growth of ovarian tissues by stimulating hormone production and exerting a specific effect on fruit growth (Shafique et al., 2011). Due to ostiol width, the results of other reports in Iran and Tunisia confirmed that pollen source had a significant effect on ostiol width (Rahemi & Jafari, 2008; Gaaliche et al., 2011).

In other experimental works, caprification was evident in fruit size, weight, and flesh thickness (Trad et al., 2013a). Caprified figs have almost twice the weight of non-caprified fruits. Caprification increased fig weight with a proportional increase in the fruit's flesh. Increased flesh thickness undoubtedly improves the consistency, taste, and flavor of figs when ripe Non-caprified figs showed a larger internal cavity and lack of seeds and the fruit was significantly firmer at harvest time than caprified fruits (Trad et al., 2013a).

3. Caprification and Dry Matter Content Pollination increases the dry matter content of the fruit (Trad et al., 2013a, 2014). This is an important fruit qualitative parameter when figs are destined for drying (Piga et al., 2003). Pollen source affects ovarian tissue growth and fruits probably by stimulating hormone synthesis. The increase in the synthesis of indole-3-acetic acid may stimulate cell elongation (Pourghayoumi et al., 2012). Thus, caprification may indirectly act as a precursor of several hormonal activities (Trad et al., 2013a).

4. Caprification and Skin/Pulp Color It is well known that the type of caprifig can significantly affect the color of both the fruit skin and its interior edible flesh (Janick & Moore, 1975). Skin color and external visual appearance exhibited no change in an experiment with deficient caprification. However, good pollination generally leads to pronounced pulp color, and caprified figs are richer in anthocyanins (Trad et al., 2013a). This suggests that biosynthesis of phenolic compounds occurs later after pollination and can be affected by caprification (Trad et al., 2013a). Janick and Moore (1975) reported that the type of caprifig might significantly affect the color of both the fruit skin and edible flesh and the size and shape of syconia. In addition, Bose and Mitra (1990), Rahemi and Jafari (2008), and Pourghayoumi et al. (2012) indicated that pollen sources had a significant effect on skin color of fig fruit.

5. Caprification and Fruit Firmness Both caprified and non-caprified figs lose their firmness, but caprification enhances fruit softening (Trad et al., 2013a). Although some authors reported that total fibers were somewhat fewer in caprified figs than in non-caprified fruits, Trad et al. (2013a) found that caprification did not affect the dietary fiber content of figs. Caprified fruits, having better pollination, contain less aborted flowers and more seeds. Thus, the development of the flesh appeared to be proportional to the number of seeds (Trad et al., 2013a). Pectin, estimated by the uronic acid concentration, was less abundant and had a higher methylation degree after pollination, while the fruit had softer textures (Trad et al., 2014). Pollination may allow the proper ripening of fruits. Enzymes like polygalacturonase or pectin-methyl esterase efficiently proceed to fruit softening (Gross & Sams, 1984). It is essential to highlight that caprification improves the fig receptacle's textural properties by increasing the pectin's methylation degree. Fertile seeds resulting from pollination may be involved in hormonal activity stimulating many related enzymes of the wall matrix depolymerization, in particular polygalacturonase (PG) and pectin methylesterase (PME) (Trad et al., 2014).

6. Caprification and Fruit Cracking and Splitting Excessive pollination increases fruit splitting and incidence of internal defects incited by microorganisms (Stover et al., 2007a). The high number of fertilized seeds may be a reason for ostiole cracking (Crisosto et al., 2010). Zare (2008) showed that using 'Daneh-Sefid' as caprifig to pollinate Iranian cultivar 'Sabz' increased the number and diameter of seeds, resulting in a rapid ostiole-end cracking before ripening of the fruits. As a result, the female trees closest to the caprifigs can be overfertilized, and their figs may split (Flaishman et al., 2008a). In addition, Increase in fruit size and tissue softening occurs during the ripening of fig fruit. Increased risk of splitting could be linked to the cell wall compositions and/or to the relative ratio of peel, the resistant tissue, to a pulp (Trad et al., 2013a). Softening is due to cell wall modification in both the pectic and hemicellulosic polymers in the receptacle and drupelets tissues. Quantitative changes in polysaccharides occurred in both tissues. Qualitative variations between the two tissues concerned only the pectic but not the hemicellulosic polymers (Owino et al., 2004a, b). Fruit ripening is a complex metabolic event and includes increased cell wall hydration, disassembly of cell wall polysaccharides, reduction in cell-to-cell adhesion, and decline in turgor pressure. Eleven genes encoding cell wall modifying enzymes during the ripening of the fig fruit (*endo*-polygalacturonases (PGs), *endo*-glucanases (EGases), ß-galactosidases (ß-Gals), xyloglucan endotransglycosylases (XTHs), expansins (Exps) and α-arabinofuranosidases (Arabfs)) were reported (Owino et al., 2004b). These genes may act together in an interdependent way to achieve softening. The average nitrate content of persistent figs was much higher than that of non-persistent ones. The regulation of indole-acetic acid oxidase is involved in the process, and persistent cultivars are expected to have higher auxin levels (Flaishman et al., 2008b).

6 Descriptor's Optimization Tools

Some optimization tools for descriptors that are empirically useful for selecting more viable and reliable morphological variables and detecting the distinction among the genotypes are analysis of correlation coefficients and principal component analysis (PCA). In addition, phenotypic and genetic studies from a heritability point of view can also help select more stable morphological characteristics in the description and discrimination of genotypes. The higher the percentage of heritability in a variable, the more stable and reliable the traits. However, traits with a low percentage of genetic inheritance and those most affected by environmental conditions can also be used to identify different ecotypes of a genotype/cultivar.

Morphological characterization is a complex process, and, to optimize it, descriptor lists for a high number of species have been developed. For example, in fig, a group of international experts developed and published a list of descriptors (International Plant Genetic Resources Institute and Centre International de Hautes Etudes Agronomiques Mediterraneennes, 2003) to facilitate the characterization process of this species. However, the analyses of morphological traits in fig are even more complex than in most fruit tree species as a result of some particular characteristics of this species, such as the frequent production of two different crops (breba and main crop), the presence of four different types of leaves, or its complex floral biology. Consequently, the current descriptors list available has 192 traits, of which 126 are qualitative and 66 are quantitative. This high number of traits and the need to replicate observations for at least 2 years make phenotypic characterization in this species a slow time-consuming, and resource-consuming process, especially in developing countries with scarce economic resources (Giraldo et al., 2010).

6.1 Principal Component Analysis (PCA)

Principal component analysis (PCA) is a technique for reducing the dimensionality of such datasets, increasing interpretability while minimizing information loss. It does so by creating new uncorrelated variables that successively maximize variance. In other words, it is the process of computing the principal components and using them to perform a change of basis on the data, sometimes using only the first few principal components and ignoring the rest.

Principal component analysis (PCA) is used to identify patterns of variability among the genotypes and the traits with the highest variation between cultivars and the greatest impact on their separation in the data set (Iezzoni & Pritts, 1991).

The approach used in Giraldo et al. (2010) work has suggested selecting a subset of highly discriminant qualitative traits by finding a correlation between variables and cultivars based on qualitative trait frequencies and multivariate analysis followed by PCA in fig that, in order of importance, are the following: (1) production type: common type (bifera and unifera type), Smyrna, San Pedro, and caprifig type; (2) morphology of the main crop and the breba crop [fruit shape, lenticels color and abundance, skin ground color, overcolor (regular bands and irregular patches), pulp and pulp cavity color, stalk color, skin firmness, and presence of male flowers]; (3) morphology of five-lobed and three-lobed leaves: central lobe shape, petiole sinus shape, petiole color; (4) morphology of entire leaves: leave shape, petiole sinus shape; and (5) tree morphology: tree growth habit, size of the tree, degree of branching, and bark tubers. Most of the variables are related to fruit characteristics (breba and main crops), highlighting the importance of correct fruit characterization to identify fig accessions correctly.

Also, in another study by Mirheidari et al. (2020) on 92 Iranian accessions of wild edible figs with a total of 55 quantitative and qualitative morphological and pomological traits were used for phenotypic evaluations. The PCA showed that the traits measured were subdivided into 15 independent components, explaining 78.9% of the total variance. The PC1 was positively correlated with 13 traits, including leaf length, leaf width, central lobe length, upper lateral vein length, hair density on leaf upper surface, petiole length, petiole thickness, ripening time, fruit length, fruit width, fresh fruit weight, fruit stalk width, and fruit flesh thickness that explained 21.3% of the total variance. The vegetative-related traits, including branching, branch density, canopy density, and leaf density, loaded significantly on PC2 and accounted for 7.29% of the total variance. The PC3 was correlated with the number of main leaf venation, lobe number, upper lateral lobe depth, and upper lateral lobe base width, which accounted for 6.40% of the total variance. These findings, in most cases, corresponded with the previous edible fig morphological studies (Mars et al., 1998; Saddoud et al., 2011; Darjazi, 2011; Gaaliche et al., 2011; Ciarmiello et al., 2015; Khadivi et al., 2018; Hssaini et al., 2020b) that have reported the importance of morphological characterization as the main factor in discriminating and assessing breeding materials of fig trees. Khadivi et al. (2018) have reported that the pomological characteristics are essential to evaluate the variation in traits of edible fig accessions.

Similarly, many studies revealed that principal component analysis in qualitative and quantitative morphological traits (vegetative and pomological) is very helpful in evaluating the phenotypic diversity and correctly identifying fig accessions.

6.2 Correlation Coefficients

Overall, the inventory of plant material based on morphological traits is important for managing genetic resources, maintaining the existing genetic variability, and establishing a germplasm collection (Podgornik et al., 2010). Authors stress the relevance of morphological assessment in the selection and breeding programs of cultivars and claim such analysis should be performed before biochemical or molecular studies are carried out (Djordjević et al., 2014; Khadivi et al., 2018; Khadivi-Khub et al., 2012; Podgornik et al., 2010).

The correlation coefficient is a statistical measure of the strength of the relationship between the relative movements of two variables. In other words, it reflects how similar the measurements of two or more variables are across a dataset. The values range between −1.0 and 1.0. A calculated number greater than 1.0 or less than −1.0 means an error in the correlation measurement.

The correlation coefficients can provide information on the most important traits in assessing genotypes (Norman et al., 2011; Khadivi-Khub & Anjam, 2016). In addition, the correlation coefficients may provide information on the descriptors that are potentially important in assessing fig genotypes (Norman et al., 2011). Correlated variables can also be used to predict other ones and could be considered of importance for the characterization and the discrimination and fig genotypes (Podgornik et al., 2010; Khadivi-Khub & Anjam, 2016; Hssaini et al., 2020a). In addition, a close relationship between traits could facilitate or hinder gene introgression since strong selection for a desirable trait could favor the presence of another desirable trait from germplasm (Dicenta & Garcia, 1992).

The results reported in previous female fig studies have suggested that fruit and leaf traits are important factors in differentiating and analyzing breeding materials dealing with the morphological characterization of fig (Mars et al., 1998; Hedfi et al., 2003; Saddoud et al., 2008b, 2011; Gaaliche et al., 2012; Aljane et al., 2012;

Caliskan & Polat, 2012; Khadivi-Khub & Anjam, 2014; Khadivi et al., 2018; Hssaini et al., 2020b; Mirheidari et al., 2020). Also, Papadopoulou et al. (2002) and Almajali et al. (2012) reported that the fruit size and shape, ostiole width, leaf characteristics, and leaf shape are very important for genotype selection. Therefore, factor analysis had great potential to differentiate the highlighted distinctions between studied genotypes and agreed with the findings of others in the study of fig (Mars et al., 1998; Saddoud et al., 2008b, 2011; Gaaliche et al., 2012). Correlations between characteristics revealed by the PCA method may correspond to a genetic linkage between loci of controlling traits or a pleiotropic effect (Iezzoni & Pritts, 1991).

Among the examined morphological and pomological characters of 87 edible fig genotypes in Iran by Khadivi et al. (2018), leaf length showed positive correlations with leaf width, lobe number, central lobe depth, lateral lobe depth, petiole length, and petiole thickness, and agreed with the finding of Khadivi-Khub and Anjam (2014) in caprifig. The existence of close positive correlations among leaf traits indicates that more leaf expansion leads to stronger aerial growth (Khadivi-Khub & Anjam, 2014, 2016; Anjam et al., 2017). Also, in their study, fruit yield showed positive correlations with leaf density, dried fruit width, and dried fruit weight. In addition, dried fruit weight was positively correlated with leaf density, leaf length, leaf width, leaf color, vein number, central lobe depth, and lateral lobe depth, and also highly correlated with dried fruit length, and dried fruit width, following the results of Khadivi-Khub and Anjam (2014, 2016) in caprifig. Khadivi et al. (2018) found that fruit cracking percentage showed negative correlations with leaf color, fruit fall type, and fruit skin color, while it showed positive collations with fruit yield, ostiole width, and fruit cracking width, seed number, and seed weight. Also, fruit cracking width was negatively correlated with fruit fall type, while this trait indicated positive correlations with ostiole width, seed number, and seed weight. Also, fruit quality was positively correlated with leaf density, dried fruit width, dried fruit weight, ostiole width, fruit cracking width, and fruit cracking percentage (Khadivi et al., 2018).

Among the examined characters for fruits of 22 cultivars in South Tunisia by Mars et al. (1998), fruit weight was highly correlated with fruit diameter and flesh thickness. Fruit diameter was also correlated with stalk diameter and flesh thickness. Neck length was correlated to fruit length. Ostiole type and ostiole diameter were also well correlated. Also, in Iranian fig cultivars studied by Khadivi et al. (2018), dried fruit weight was positively correlated with leaf density, leaf dimensions, dried fruit length, and dried fruit width.

According to five-year data of Turkish cultivars, fruit weight was found to have positive correlation by fruit diameter ($r = 0.92$; $P < 0.01$), fruit length ($r = 0.81$; $P < 0.01$), neck length ($r = 0.35$; $P < 0.01$), ostiolium width ($r = 0.23$; $P < 0.01$), trunk diameter ($r = 0.26$; $P < 0.01$), fruit skin cracks ($r = 0.26$; $P < 0.01$) and negative correlation by TSS ($r = -0.26$; $P < 0.01$) and fruit ribs ($r = -0.21$; $P < 0.01$). It was also found out in that period that as the trunk diameter increased, so did the fruit size while the ostiolium width was smaller. This could be related to the increase in the fruit size with age and the bigger trunk diameter (Polat & Caliskan, 2017).

Indeed, Botti et al. (2003) reported that fruit size could continue to grow until the tree gains its final shape.

Hssaini et al. (2020b) confirmed that Fruit weight was positively correlated to stalk width, neck length, width, ostiole width, and peel thickness. These results were similar to those reported in several works (Mahdavian et al., 2006; Gozlekci, 2011; Darjazi, 2011; Gaaliche et al. 2011; Aljane et al., 2012; Polat & Caliskan, 2017. Khadivi et al., 2018). In addition, they concluded that all fruit dimensions were positively and strongly correlated to fig volume except for stalk length and ostiole width.

Hssaini et al. (2019) concluded that total soluble solids (TSS) are negatively and significantly correlated to fruit weight, width, and the volume, which means that cultivars with high content of total soluble solids would tend to present a reduction in the fruit weight, width, and volume. Similar results were demonstrated in figs (Gozlekci, 2011) and other fruits such as yellow passion (Viana et al., 2003).

In the past decades, strong attention has been given to combining several descriptors of interest as an important quality parameter in fig selection. Morphological and biochemical traits seem to have attracted particular substantial interest since some of them can be used to predict each other based on correlations. In addition, it can provide information about the relationships between variables that are potentially important in assessing fig genotypes (Hssaini et al., 2020a).

6.3 Heritability

The main role of heritability is related to the fact that it expresses the reliability of the phenotypic value as an estimator of the genotypic value, such that the higher the heritability in selection (Acquaah, 2012). Fruit weight and size (diameter and length), SSC, TA, and SSC:TA ratio, skin color (luminosity, chroma, and hue), ethylene production and respiration rates (CO2 production), firmness, ostiole diameter, internal cavity diameter, shriveling, growth cracks, shelf life and susceptibility to blemishes are affected by genotype (Crisosto et al., 2011).

Although the findings of Cowart and Graham (1999) indicated that the length and width of the leaves are not good biometric characteristics, it is worth noting that higher values of these traits in combination with other characteristics can help determine possible selection characters. For example, Castellen et al. (2007), characterizing 68 accessions of the Active Germplasm Bank of Papaya of Embrapa Mandioca and Tropical Fruit through multivariate analysis, detected that the components 1 (petiole color) and 2 (leaf length) explained most of the total observed genetic variation (67.80%). Also, in fig accession studied by Rodrigues et al. (2019b), despite the low heritability observed for length and width of the leaves, the number of leaf lobes showed high heritability (0.85), with a CVg/Cve ratio higher than 1. The values observed for the number of leaf lobes characters ranged from three non-pronounced to seven well-defined lobes; these extremes were observed in 'Troyano' and 'Brunswich' accessions, respectively.

7 Conclusions

Traditionally, the amount and distribution of genetic diversity have been determined by morphological markers. Although environmental conditions and agronomic technologies strongly influence the expression of these markers, the morphological characterization is a highly recommended first step that should be made before more in-depth biochemical or molecular studies are attempted (Hoogendijk & Williams, 2001).

The fig tree has not been subjected to intensive plant breeding programs, and thus many fig tree populations exhibit rich genetic biodiversity that can only be fully exploited once it is properly identified and classified. Traditionally, the plant germplasm characterization with the aim of its conservation has been carried out using morphological or agronomical traits. Nevertheless, despite the progress in elaborating descriptors, fluctuations among years, environments, or repetitions have made application difficult until recently (Giraldo et al., 2010). However, selecting highly discriminant variables is important to optimize resources for a feasible morphological characterization. This is especially important in a crop such as a fig fruit with hundreds of genotypes described worldwide in which many synonymies and homonymies may be observed (Khadivi et al., 2018).

Presumably, the most reliable method for accurate investigation and optimizing resources of reliable and highly discriminant variables for a feasible morphological characterization and to detect separation of genotypes (varieties, cultivars, and accessions) is vegetative propagation and thus cloning of a homogeneous and uniform genotype and planting them in different regions with different climatic conditions and of course with a specific nutritional program. Then, under different environmental and geographical conditions, it is possible to diagnose stability and instability in the measured qualitative and quantitative morphological traits in that cultivar. In this way, the altered and environmentally affected traits can be ignored, and as a result, valid and stable morphological and distinctive traits in that cultivar can be obtained. Also, in this method, we can define a pattern for unstable traits in different climatic conditions as additional side information for different cultivars.

References

Abdelsalam, N. R., Awad, R. M., Ali, H. M., Salem, M. Z., Abdellatif, K. F., & Elshikh, M. S. (2019). Morphological, pomological, and specific molecular marker resources for genetic diversity analyses in fig (*Ficus carica* L.). *HortScience, 54*(8), 1299–1309.

Achtak, H., Oukabli, A., Ater, M., Santoni, S., Kjellberg, F., & Khadari, B. (2009). Microsatellite markers as reliable tools for fig cultivar identification. *Journal of the American Society for Horticultural Science, 134*(6), 624–631.

Acquaah, G. (2012). *Principles of plant genetics and breeding* (2nd ed.). Wiley-Blackwell, 664p.

Aksoy, U., Seferoglu, G., Misirli, A., Kara, S., Sahin, N., Bulbul, S., & Duzbastilar, M. (1992). Selection of the table fig genotypes suitable for Egean region. In *1st Turkish National Horticultural Congress Proceedings* (Vol. 1, pp. 545–548).

Aksoy, U., Balci, B., Can, H. Z., & Hepaksoy, S. (2003). Some significant results of the research-work in Turkey on fig. *Acta Horticulturae, 605*, 173–181.

Ali-Shtayeh, M. S., Jamous, R. M., Zaitoun, S. Y. A., Mallah, O. B., & Mubaslat, A. K. (2014). Genetic diversity of the Palestinian fig (*Ficus carica* L.) collection by pomological traits and RAPD markers. *American Journal of Plant Sciences, 5*, 1139–1155.

Aljane, F. (2016). Analysis of genetic diversity in Tunisian fig (*Ficus carica* L.) germplasm bank revealed by RAPD markers and morphological characters. *European Journal of Scientific Research, 142*(2), 172–192.

Aljane, F., & Ferchichi, A. (2007). Characterization and evaluation of six cultivars of caprifig (*Ficus carica* L.) in Tunisia. *Plant Genetic Resources Newsletter, 151*, 22–26.

Aljane, F., & Ferchichi, A. (2009). Assessment of genetic diversity among some southern Tunisian fig (*Ficus carica* L.) cultivars based on morphological descriptors. *Jordan Journal of Agricultural Sciences, 5*(1), 1–16.

Aljane, F., & Ferchichi, A. (2010). Assessment of genetic diversity of Tunisian fig (*Ficus carica* L.) cultivars using morphological and chemical characters. *Acta Bot Gallica, 157*(1), 171–182.

Aljane, F., Ferchichi, A., & Boukhris, M. (2005). Pomological characteristics of local fig (*Ficus carica*) cultivars in Southern Tunisia. *Acta Horticulturae (ISHS), 798*, 123–128.

Aljane, F., Ferchichi, A., & Boukhris, M. (2008). Pomological characteristics of local fig (*Ficus carica*) cultivars in southern Tunisia. *Acta Horticulturae, 798*, 123–128.

Aljane, F., Nahdi, S., & Essid, A. (2012). Genetic diversity of some accessions of Tunisian fig tree (*Ficus carica* L.) based in morphological and chemical traits. *Journal of Natural Product and Plant Resources, 2*(3), 350–359.

Aljane, F., Essid, A., & Nahdi, S. (2013). Improvement of Fig (*Ficus carica* L.) by conventional breeding and biotechnology. In J. M. Al-Khayri, S. M. Jain, & D. V. Johnson (Eds.), *Advances in plant breeding strategies: Fruits* (Vol. 3, pp. 343–375). Springer.

Almajali, D. A., Abdel-Ghani, A. H., & Migdadi, H. (2012). Evaluation of genetic diversity among Jordanian fig germplasm accessions by morphological traits and ISSR markers. *Scientia Horticulturae, 147*, 8–19.

Anjam, K., Khadivi-Khub, A., & Şarkhosh, A. (2017). The potential of caprifig genotypes for sheltering *Blastophaga psenes* L. for caprification of edible figs. *Erwerbs-Obstbau, 59*(1), 45–49.

Aradhya, M. K., Stover, E., Velasco, D., & Koehmstedt, A. (2010). Genetic structure and differentiation in cultivated Fig (*Ficus carica* L.). *Genetica, 138*, 681–694. https://doi.org/10.1007/s10709-010-9442-3

Arpaci, S. (2017). An overview on fig production and research and development in Turkey. *Acta Horticulturae, 1173*, 57–61.

Ateyyeh, A. F., & Sadder, M. T. (2006). Growth pattern and fruit characteristics of six common fig (*Ficus carica* L.) cultivars in Jordan. *Jordan Journal of Agricultural Sciences, 2*(2), 105–112.

Awamiira, M., Shoda, K., & Yahata, D. (1996). Effect of various seed parents on frequency distribution of parthenocarpy among seedling progenies of *fig (F. carica* L.*). Journal of the Japanese Society for Horticultural Science, **65***, 21–26.

Bandelj, D., Javornik, B., & Jakse, J. (2007). Development of microsatellite markers in the common fig, Ficus carica L. *Molecular Ecology Resources, 7*, 1311–1314.

Baraket, G., Chatti, K., Saddoud, O., Mars, M., Marrakchi, M., Trifi, M., & Hannachi, A. S. (2009). Genetic analysis of Tunisian fig (*Ficus carica* L.) cultivars using amplified fragment length polymorphism (AFLP) markers. *Scientia Horticulturae, 120*, 487–492.

Baziar, G., Jafari, M., Noori, M. S. S., & Samarfard, S. (2018). Evaluation of Genetic Diversity among Persian Fig cultivars by morphological traits and RAPD markers. *HortScience, 53*(5), 613–619.

Beck, N. G., & Lord, E. M. (1988). Breeding system in *Ficus carica*, the common fig. I., II. Floral diversity. *American Journal of Botany, 75*(12), 1904–1922.

Ben Salah, M., & Lejri, M. H. (1995). Description phénopomologique de quatre variétés de figuier (*Ficus carica* L.) dans l'oasis de Gafsa. Rev. *Régions Arides, 8*(1/95), 3–15.

Benettayeb, Z. E., Bencheikh, M., Setti, B., & Chaillou, S. (2017). Genetic diversity of Algerian fig (*Ficus carica* L.) cultivars based on morphological and quality traits. *The Horticultural Society Of India (Regd.), 74*(3), 311–316. https://doi.org/10.5958/0974-0112.2017.00064.0

Berg, C. C., & Corner, E. J. H. (2005). *In Flora Malesiana Series I -Seed Plants* (Vol. 17, pp. 1–702) (H. P. Nooteboom, Ed.). Nationaal Herbarium.

Bose, T. K., & Mitra, S. M. (1990). *Tropical and subtropical fruits* (p. 823). Naya Proksh Publication.

Bostan, S. Z., Islam, A., & Aygün, A. (1998). A study pomological characteristics of local fig cultivars in Northern Turkey. *Acta Horticulturae, 480*, 71–73.

Botti, C., Carrasco, O., Escobar, B., Estévez, A. M., Franck, N., Osses, D., & Prat, L. (Eds.). (2001). *La Higuera (Ficus carica L.)* (p. 39). Serie Ciencias Agronómicas Universidad de Chile.

Botti, C., Franck, N., Prat, L., Ioannidis, D., & Morales, B. (2003). The effect of climatic conditions on fresh fig fruit yield, quality and type of crop. *Acta Horticulturae, 605*, 37–43.

Caliskan, O., & Polat, A. (2008). Fruit characteristics of fig cultivars and genotypes grown in Turkey. *Scientia Horticulturae, 115*, 360–367. https://doi.org/10.1016/j.scienta.2007.10.017

Caliskan, O., & Polat, A. A. (2012). Morphological diversity among fig (*Ficus carica* L.) accessions sampled from the Eastern Mediterranean Region of Turkey. *Turkish Journal of Agriculture and Forestry, 36*, 179–193.

Caliskan, O., Bayazit, S., Ilgin, M., & Karatas, N. (2017). Morphological diversity of caprifig (*Ficus carica* var. *caprificus*) accessions in the eastern Mediterranean region of Turkey: Potential utility for caprification. *Scientia Horticulturae, 222*, 46–56.

Can, H. Z., 1993. *The investigation of some horticultural characteristics of some selected fig genotypes in Aegean Region*. Master thesis, Ege University, Izmir, Turkey (in Turkish, with English abstract).

Can, H. Z., Misirli, A., Kara, S., Seferoglu, G., Sahin, N., & Aksoy, U. (2001, May). Fig (*Ficus carica* L.) selection study for fresh market in western Turkey. In *II International symposium on Fig 605* (pp. 197–203).

Castellen, M. D. A. S., Ledo, C. A. D. A. S., Oliveira, E. J. D. E., Monteiro, F. L. S., et al. (2007). Caracterização de acessos do Banco Ativo de Mamão por meio de análise multivariada. *Magistra, 19*, 299–303.

Çelikel, F. G., & Karaçalı, I. (1998). Effects of harvest maturity and precooling on fruit quality and longevity of 'Bursa Siyahi' figs (*Ficus carica* L). *Acta Horticulturae, 480*, 283–288. https://doi.org/10.17660/ActaHortic.1998.480.55

Chalack, L., Chehade, A., Mattar, E., & Khadari, B. (2005). Morphological characterization of fig accessions cultivated in Lebanon. *Acta Horticulturae, 798*, 54–61.

Chalak, L., Chehade, A., Mattar, E., & Khadari, B. (2008). Morphological characterization of fig accessions cultivated in Lebanon. *Acta Horticulturae, 798*, 49.

Chatti, K., Salhi-Hannachi, A., Mars, M., Marrakchi, M., & Trifi, M. (2003). Analysis of genetic diversity of Tunisian fig tree cultivars (*Ficus carica* L.) using morphological characteristics. *Fruits, 59*, 49–61.

Chatti, K., Hannachi-Salhi, A., Mars, M., Marrakchi, M., & Trifi, M. (2004). Analyse de la diversité génétique de cultivars tunisiens de figuier (*Ficus carica* L.) à l'aide de caractères morphologiques. *Fruits, 59*, 49–61.

Chevreau, E. (2009). La transgénèse pour l'innovation variétale fruitière: état des lieux et perspectives. *Innovations Agronomiques, 7*, 153–163.

Chitwood, D. H., Klein, L. L., O'hanlon, R., Chacko, S., Greg, M., Kitchen, C., Miller, A. J., & Londo, J. P. (2016). Latent developmental and evolutionary shapes embedded within the grapevine leaf. *New Phytologist, Cambridge, 210*, 343–355.

Chouaki, S. (2006). *Deuxième Rapport National sur l'État des Ressources Phytogénétiques (INRAA)*, 91 pp.

Ciarmiello, L. F., Piccirillo, P., Carillo, P., De Luca, A., & Woodrow, P. (2015). Determination of the genetic relatedness of fig (*Ficus carica* L.) accessions using RAPD fingerprint and their agro-morphological characterization. *South African Journal of Botany, 97*, 40–47.

Condit, I. J. (1941). Fig characteristics useful in the identification of varieties. *Hilgardia, 14*, 1–69.

Condit, I. J. (1947). The Fig Chronica Botanica. *Whaltan. Massachusetts*, 25–80.

Condit, I. (1955). Fig varieties: A monograph. *Hilgardia, 23*(11), 323–539.

Condit, I. J. (1969). *Ficus: The exotic species* (363 pp). Division of Agricultural Sciences. University of California, Berkeley.

Cowart, N. M., & Graham, J. H. (1999). Within- and among-individual variation in fluctuating asymmetry of leaves in the fig (*Ficus carica* L.). *International Journal of Plant Sciences, 160*, 116–121.

Crane, J. C. (1986). Fig. In S. P. Monselise (Ed.), *Handbook of fruit set and development*. CRC Press.

Crisosto, C. H., Bremer, V., Ferguson, L., & Crisosto, G. M. (2010). Evaluating quality attributes of four fresh fig (*Ficus carica* L.) cultivars harvested at two maturity stages. *HortScience, 45*(4), 707–710.

Crisosto, C. H., Bremer, V., & Stover, E. (2011). Fig (*Ficus carica* L.). In E. E. Yahia (Ed.), *Postharvest biology and technology of tropical and subtropical fruits. Vol. 3, Cocona to Mango* (pp. 134–158). Woodhead Publishing.

Crisosto, C. H., Ferguson, L., Norton, M., & Crisosto, G. (2015). *Studies on fertirrigation with calcium and potassium to reduce culls in fig*. Paper presented at: V International Symposium on Fig. ISHS.

Cruaud, A., et al. (2012). An extreme case of plant-insect co-diversification: Figs and fig-pollinating wasps. *Systematic Biology, 61*, 1029–1047.

Cruz, S. M., Nery, M. C., Von Pinho, E. V. R., & Laia, M. L. (2014). Molecular characterisation of radish cultivars. *Revista Ciência Agronômica, 45*, 815–822.

Darjazi, B. B. (2011). Morphological and pomological characteristics of fig (*Ficus carica* L.) cultivars from Varamin, Iran. *African Journal of Biotechnology, 10*(82), 19096–19105.

Dicenta, F., & Garcia, J. E. (1992). Phenotypical correlations among some traits in almond. *Journal of Genetics and Breeding, 46*, 241–246.

Djordjević, B., Rakonjac, V., Akšić, M. F., Šavikin, K., & Vulić, T. (2014). Pomological and biochemical characterization of European currant berry (*Ribes* sp.) cultivars. *Scientia Horticulturae, 165*, 156–162. https://doi.org/10.1016/j.scienta.2013.11.014

Do Val, A. D. B., Souza, C. S., Ferreira, E. A., Salgado, S. M. L., Pasqual, M., & Cançado, G. M. A. (2013). Evaluation of genetic diversity in fig accessions by using microsatellitemarkers. *Genetics and Molecular Research, 12*, 1383–1391.

Doster, M., Michailides, T., Doyle, J., Cotty, P., Morgan, D., & Boeckler, L. (2002). Aflatoxin control in figs: Development of resistant cultivars, identification of contaminated fruit and biocontrol. Aflatoxin/Fumonisin elimination and fungal genomics workshops (Phoenix, AZ). *Mycopathologia, 155*, 45.

Doyle, J. F., & Ferguson, L. (2005, May 16–20). Sierra: A new non-caprifying Calimyrna. In *Abstracts of the third international symposium on Fig*. University of Algarve.

El-Hamady, M., Hamdia, M., Ayaad, M., Salama, M. E., & Omar, A. K. H. (2010). Metaxenic effects as related to hormonal changes during date palm (*Phoenix dactylifera* L.) fruit growth and development. *Acta Horticulturae, 882*, 155–164.

El-Kassas, S. E., Mahmoud, H. M., Amen, K. I. A., & Badawy, A. A. (1992a). Evaluation of some introduced and local fig cultivars under Assiut climatic conditions. 4. Yield and syconia (fruit) quality. *Assiut Journal of Agricultural Sciences (Egypt), 23*, 270–280.

El-Kassas, S. E., Mahmoud, H. M., Amen, K. I. A., & Badawy, A. A. (1992b). Evaluation of some introduced and local fig cultivars under Assiut climatic conditions. 2. Time of leafing out and leaf characters. *Assiut Journal of Agricultural Sciences (Egypt), 23*, 281–294.

Escribano, P., Viruel, M. A., & Hormaza, J. I. (2008). Comparison of different methods to construct a core germplasm collection in woody perennial species with simple sequence repeat markers. A case study in cherimoya (*Annona cherimola*, Annonaceae), an underutilised subtropical fruit tree species. *Annals of Applied Biology, 153*(1), 25–32.

Esquinas-Alcázar, J. (2005). Protecting crop genetic diversity for food security: Political, ethical and technical challenges. *Nature Reviews Genetics, 6*(12), 946–953.

Essid, A., Aljane, F., & Ferchichi, A. (2017). Morphological characterization and pollen evaluation of some Tunisian *ex situ* planted caprifig (*Ficus carica* L.) ecotypes. *South African Journal of Botany, 111*, 134–143.

FAOSTAT agricultural data (2019). http://faostat.fao.org/site/570/DesktopDefault.aspx. Accessed 5 Aug 2021.

Fatahi, S., Cheghamirza, K., Arji, I., & Zarei, L. (2017). Evaluation of genetic variation of common Fig (*Ficus carica* L.) in West of Iran. *Journal of Medicinal Plants and By-Product, 6*(2), 229–240.

Ferguson, L., Michailides, T. J., & Shorey, H. H. (1990). The California fig industry. *Horticultural Reviews, 12*, 409–490.

Ferguson, L., Mariscal, M., Reyes, H., & Metheney, P. (2003). Using girdling to improve Black Mission Fig Size. *Acta Horticulturae, 605*, 167–169.

Ferrara, E., & Papa, G. (2003). Evaluation of fig cultivars for breba crop. *Acta Horticulturae, 605*, 91–93.

Ferrara, G., Mazzeo, A., Pacucci, C., Matarrese, A. M. S., Tarantino, A., Crisosto, C., et al. (2016). Characterization of edible fig germplasm from Puglia, southeastern Italy: Is the distinction of three fig types (Smyrna, San Pedro and Common) still valid? *Scientia Horticulturae, 205*, 52–58.

Ferrara, G., Mazzeo, A., Gallotta, A., Pacucci, C., Matarrese, A. M. S., Tarantino, A., et al. (2017). Fruit-set and SSR markers of fig cultivars from Puglia region, Southeastern Italy. *Acta Horticulturae, 1173*, 39–44. https://doi.org/10.17660/ActaHortic.2017.1173.7

Ferreira, E. A., Pasqual, M., & Tulmann, N. (2009). A. *In vitro* sensitivy of fig plantlets to gamma ray. *Science in Agriculture, 4*, 540–542.

Flaishman, M. A., Rodov, V., & Stover, E. (2008a). The fig: Botany, horticulture, and breeding. *Horticultural Reviews (American Society for Horticultural Science), 34*, 113–196.

Flaishman, M. A., Yablovich, Z., Golobovich, S., Salamon, A., Cohen, Y., Perl, A., Yancheva, S. D., Kerem, Z., & Haklay, E. (2008b). Molecular breeding in fig (*Ficus carica*) by the use of genetic transformation. *Acta Horticulturae, 798*, 151–158.

Flaishman, M. A., Peer, R., Freiman, Z. E., Izhaki, Y., & Yablovitz, Z. (2017). Conventional and molecular breeding systems in fig (*Ficus carica* L.). *Acta Horticulturae, 1173*, 1.

Flores, D., Arguello, J. F., Orozco, R., Jiménez, V., & Rosales, J. (2011). Manejo agronómico de la higuera. In *El Cultivo del Higo (Ficus carica) en Costa Rica. Aspectos Generales para el Manejo de una Plantación de Higo (Ficus carica)* (pp. 91–100). Editorial EUNED.

Francisco, V. L. F. S., Baptistella, C. S. L., Amaro, A. A., & Fagundes, P. R. S. (2011). A evolução e os aspectos socioeconômicos da cultura do figo no Estado de São Paulo. *Information Economics, 41*, 13–22.

Gaaliche, B., Trad, M., & Mars, M. (2011). effect of pollination intensity, frequency and pollen source on fig (*Ficus carica* L.) productivity and fruit quality. *Scientia Horticulturae, 130*(4), 737–742.

Gaaliche, B., Saddoud, O., & Mars, M. (2012). Morphological and pomological diversity of fig (*Ficus carica* L.) cultivars in northwest of Tunisia. *ISRN Agronomy*. https://doi.org/10.5402/2012/326461

Galil, J. (1968). An ancient technique for ripening sycomore fruit in east-mediterranean countries. *Economic Botany, 22*, 178–190.

Galil, J. (1977). Fig biology. *Endeavour, 1*(2), 52–56.

Galil, J., & Neeman, G. (1977). Pollen transfer and pollination in the common fig (*Ficus carica* L.). *The New Phytologist, 79*(*1*), 163–171. https://doi.org/10.1111/j.1469-8137.1977.tb02192.x

Gallesio, G. (1820). *Pomona Italiana* (Vol. I).

Gani, G., Fatima, T., Qadri, T., Gulzar, B., Jan, N., & Bashir, O. (2018). Phytochemistry and pharmacological activities of fig (*Ficus carica*): A review. *International Journal of Research in Pharmacy and Pharmaceutical Sciences, 3*, 80–82.

Ge, H., & Zhang, Y. (2006). Characteristics of climate variations affected winter solar greenhouse production in east of Guanzhong areas of Shaanxi province. *Chinese Journal of Agrometeorology, 3*, 187–190.

Gholami, M., Rahemi, M., & Rastegar, S. (2012). Use of rapid screening methods for detecting drought tolerant cultivars of fig (*Ficus carica* L.). *Scientia Horticulturae (Amsterdam), 143*, 7–14. https://doi.org/10.1016/j.scienta.2012.05.012

Giaccone, M., Pannico, A., Scognamiglio, P., Rivera, C. M., Cirillo, C., Rouphael, Y., et al. (2017). Regression model for leaf area estimation in *Ficus carica* L. *Acta Horticulturae, 2017*, 163.

Giraldo, E., Viruel, M. A., López-Corrales, M., & Hormaza, J. I. (2005). Characterisation and cross-species transferability of microsatellites in the common fig (*Ficus carica* L.). *The Journal of Horticultural Science and Biotechnology, 80*(2), 217–224.

Giraldo, E., Lopez-Corrales, M., & Hormaza, J. I. (2008a). Optimization of the management of an ex-situ germplasm bank in common fig with SSRs. *Journal of the American Society for Horticultural Science, 133*(1), 69–77.

Giraldo, E., Lopez-Corrales, M., & Hormaza, J. I. (2008b). Selection of morphological quantitative variables in fig characterization. *Acta Horticulturae, 798*, 103.

Giraldo, E., López-Corrales, M., & Hormaza, J. I. (2010). Selection of the most discriminating morphological qualitative variables for characterization of fig germplasm. *Journal of the American Society for Horticultural Science, 135*(3), 240–249.

González, A. M., & Grajal, M. J. (2012). *Higueras de Canarias. Caracterización morfológica de variedades*. Instituto Canario de Investigaciones Agrarias, Gobierno de Canarias.

Gozlekei, S. (2011). Pomological traits of fig (*Ficus carica* L) genotypes collected in the west Mediterranean region in Turkey. *Journal of Animal and Plant Sciences, 21*(4), 646–652.

Grassi, G. (1991). *Il Fico* (p. 128). Manuale Pratico (REDA).

Gregoriou, C. (1995). Cultivation of fig (*Ficus caríca*), Loquat (*Eriobotrya japonica*), Japanese persimmon (*Díospyros kaki*), Pomegranate (*Puníca granafum*) and Barbary fig (*Opunfía fícus-indica*) in Cyprus. *Cahiers Options Méditerranéennes, 13*, 9–12.

Gross, K. C., & Sams, C. E. (1984). Changes in cell wall neutral sugar composition during fruit ripening: A species survey. *Phytochemistry, 23(11)*, 2457–2461. https://doi.org/10.1016/S0031-9422(00)84075-3

Gruda, N., & Tanny, J. (2015). Protected crops a recent advances, innovative technologies and future challenges. In *Ed 1107* (pp. 271–278). International Society for Horticultural Science (ISHS).

Grum, M., & Atieno, F. (2007). *Statistical analysis for plant genetic resources: Clustering and indices in R made simple* (Handbooks for Genebanks, n.9). Bioversity International.

Hallac Turk, F., & Aksoy, U. (2011). Comparison of organic, biodynamic and conventional fig farms under rainfed conditions in Turkey. *Cell Plant Science, 2(3)*, 22–33.

Hedfi, J., Trifi, M., Salhi-Hannachi, A., Ould Mohamed Salem, A., & Marrakchi, M. (2003). Morphological and isoenzymatic polymorphism in Tunisian fig (*Ficus carica* L.) collection. *Acta Horticulturae, 605*, 319–325.

Hiwale, S. (2015). *Sustainable horticulture in semiarid dry lands* (pp. 135–152). Springer.

Hmimsa, Y., Aumeeruddy-Thomas, Y., & Ater, M. (2012). Vernacular taxonomy, classification and varietal diversity of fig (*Ficus carica* L.) among Jbala cultivators in Northern Morocco. *Human Ecology, 40*(2), 301–313.

Hoogendijk, M., & Williams, D. E. (2001). Characterizing the genetic diversity of home garden crops: Some examples from the Americas. In J. W. Watson & P. B. Eyzaguirre (Eds.), *Proceedings of the second international home gardens workshop: Contribution of home gardens and in situ conservation of plant genetic resources in farming systems* (pp. 34–40). IPGRI.

Hssaini, L., Hanine, H., Razouk, R., Ennahli, S., Mekaoui, A., & Charafi, J. (2019). Characterization of local fig clones (*Ficus carica* L.) collected in Northern Morocco. *Fruits, The International Journal of Tropical and Subtropical Horticulture, 74*(2), 55–64.

Hssaini, L., Charafi, J., Razouk, R., Hernández, F., Fauconnier, M. L., Ennahli, S., & Hanine, H. (2020a). Assessment of morphological traits and fruit metabolites in eleven Fig varieties (*Ficus carica* L.). *International Journal of Fruit Science, 20*(sup2), 8–28.

Hssaini, L., Hanine, H., Razouk, R., Ennahli, S., Mekaoui, A., Ejjilani, A., & Charafi, J. (2020b). Assessment of genetic diversity in Moroccan fig (*Ficus carica* L.) collection by combining morphological and physicochemical descriptors. *Genetic Resources and Crop Evolution, 67*(2), 457–474.

Hssaini, L., Hanine, H., Razouk, R., Ennahli, S., Mekaoui, A., Guirrou, I., & Charafi, J. (2020c). Diversity Screening of Fig (*Ficus carica* L.) Germplasm through Integration of Morpho-agronomic and Biochemical Traits. *International Journal of Fruit Science, 20*(4), 939–958.

Iezzoni, A. F., & Pritts, M. P. (1991). Applications of principal components analysis to horticultural research. *HortScience, 26*, 334–338.

Ikegami, H., Nogata, H., Hirashima, K., Awamura, M., & Nakahara, T. (2009). Analysis of genetic diversity among European and Asian fig varieties (*Ficus carica* L.) using ISSR, RAPD, and SSR markers. *Genetic Resources and Crop Evolution, 56*(2), 201–209.

Ikegami, H., Habu, T., Mori, K., Nogata, H., Hirata, C., Hirashima, K., Tashiro, K., & Kuhara, S. (2013). De novo sequencing and comparative analysis of expressed sequence tags from gynodioecious fig (*Ficus carica* L.) fruits: Caprifig and common fig. *Tree Genetics & Genomes, 9*, 1075–1088.

Ilgın, M. (1995). *The investigation of fertilization biology of some fig genotypes selected from Kahramanmaras¸ region*. PhD thesis, Cukurova University, Adana, Turkey (in Turkish, with English abstract).

International Plant Genetic Resources Institute and Centre International de Hautes Etudes Agronomiques Méditerranéennes. (2003). *Descriptor for fig*. IPGRI/CIHEAM.

IPGRI and CIHEAM. (2003). *Descriptors for fig (Ficus carica L.)*. International Plant Genetic Resources Institute (IPGRI)/International Center for Advanced Mediterranean Agronomic Studies (CIHEAM).

Jacob, D., Petersen, J., Eggert, B., Alias, A., Christensen, O. B., Bouwer, L. M., Braun, A., Colette, A., Déqué, M., et al. (2014). EURO-CORDEX: New high-resolution climate change projections for European impact research. *Regional Environmental Change, 14*, 563–578.

Janick, J., & Moore, J. N. (1975). *Advances in fruit breeding* (p. 464, 623). Purdue University Press.

Kaynak, L., Gozlekci, S., Ersoy, N., & Aksoy, N. (1998). A research on storage and pomological properties of some fig (*Ficus carica* L.) cultivars grown in Antalya conditions. *Acta Horticulturae, 480*, 277–282.

Khadari, B., Lashermes, P., & Kjellberg, F. (1995a). RAPD fingerprints for identification and genetic characterization of fig (*Ficus carica* L.) genotypes. *Journal of Genetics and Breeding, 49*, 77–77.

Khadari, B., Gibernau, M., Anstett, M. C., Kjellberg, F., & Hossaert-McKey, M. (1995b). When figs wait for pollinators: The length of fig receptivity. *American Journal of Botany, 82*(8), 992–999.

Khadari, B., Oukabli, A., Ater, M., Mamouni, A., Roger, J. P., & Kjellberg, F. (2005). Molecular characterization of Moroccan fig germplasm using intersimple sequence repeat and simple sequence repeat markers to establish a reference collection. *HortScience, 40*(1), 29–32.

Khadivi, A., & Mirheidari, F. (2022). Selection of the promising fig (*Ficus carica* L.) accessions using fruit-related characters. *Food Science & Nutrition, 10*, 2911–2921.

Khadivi, A., Anjam, R., & Anjam, K. (2018). Morphological and pomological characterization of edible fig (*Ficus carica* L.) to select the superior trees. *Scientia Horticulturae, 238*, 66–74. https://doi.org/10.1016/j.scienta.2018.04.031

Khadivi-Khub, A. (2015). Physiological and genetic factors influencing fruit cracking. *Acta Physiologiae Plantarum, 37*(1), 1718.

Khadivi-Khub, A., & Anjam, K. (2014). Characterization and evaluation of male fig (caprifig) accessions in Iran. *Plant Systematics and Evolution, 300*(10), 2177–2189.

Khadivi-Khub, A., & Anjam, K. (2016). The relationship of fruit size and light condition with number, activity and price of *Blastophaga psenes* wasp in caprifigs. *Trees, 30*(5), 1855–1862.

Khadivi-Khub, A., Zamani, Z., & Fatahi, M. R. (2012). Multivariate analysis of *Prunus* subgen. *cerasus* germplasm in Iran using morphological variables. *Genetic Resources and Crop Evolution, 59*(5), 909–926. https://doi.org/10.1007/s10722-011-9733-2

Khan, A. S., & Chaudhry, N. Y. (2010). Florlgenic effects of IAA for improving pistillate and staminate flowering in some cucurbits under Pb stress. *Pakistan Journal of Botany, 42*, 1835–1840.

Kislev, M. E., Hartmann, A., & Bar-Yosef, O. (2006). Early domesticated fig in the Jordan Valley. *Science, 312*, 1372–1374.

Kjellberg, F., & Valdeyron, G. (1984). The pollination of fig tree (*Ficus carica* L.) and its control in horticulture. *Acta Oecologica, 5*(4), 407–412.

Kjellberg, F., Gouyon, P. H., Ibrahim, M., et al. (1987). The stability of the symbiosis between dioecious figs and their pollinators: A study of *Ficus carica* L. and *Blastophaga psenes* L. *Evolution, 41*, 653–660.

Koc, A., & Bilgener, S. (2013). Morphological characterization of cherry rootstock candidates selected from Samsun Province in Turkey. *Turkish Journal of Agriculture and Forestry, 37*, 575–584.

Koyuncu, M. A., Bostan, S. Z., Islam, A., & Koyuncu, F. (1998). Investigation on physical and chemical characteristics in fig cultivars grown in Ordu. *Acta Horticulturae, 480*, 87–89.

Küden, A., & Tanriver, E. (1998). Plant genetic resources and selection studies on figs in the East Mediterranean and South East Anatolia Regions. *Acta Horticulturae, 480*, 49–54.

Lamas, L., Elias, G., & Pereira, R. A. S. (2009). Comportamento agressivo em femeas de vespas polinizadoras (Hymenoptera: Agaonidae) de *Ficus* (Moraceae). In *CONGRESSO DE ECOLOGIA DO BRASIL*, 9., 2009. São Lourenço. Anais [...]

Lisci, M., & Pacini, E. (1994). Germination ecology of drupelets of the fig (*Ficus carica* L.). *Botanical Journal of the Linnean Society, 114*(2), 133–146.

López, B., & Guzmán, G. (2007). *Catálogo de variedades locales de higuera (Ficus carica L.) de la Sierra de la Contraviesa (Granada)*. Consorcio Centro de Investigación y Formación en Agricultura Ecológica y Desarrollo Rural.

López-Corrales, M., Gil, M., Pérez, F., Cortés, J., Serradilla, M. J., & Chome, P. M. (2011). *Variedades de higuera: Descripción y registro de variedades*. Ministerio de Medio Ambiente y Medio rural y Marino. http://www.magrama.gob.es/

Machado, C. A., Jousselin, E., Kjellberg, F., Compton, S. G., & Herre, E. A. (2001). Phylogenetic relationships, historical biogeography and character evolution of fig-pollinating wasps. *Proceedings of the Royal Society B: Biological Sciences, 268(1468)*, 685–694. https://doi.org/10.1098/rspb.2000.1418

Mahdavian, M., Lessani, H., Kuhi, M., Zare, H., Krami, M., & Tabatabei, Z. (2006). Assessment of genetic diversity among different fig (*Ficus carica* L.) collection from Iran. *Acta Horticulturae, 760*, 29–37.

Mahdavian, M., Lessani, H., Ebadi, A., Fatah, R., & Habibi Kuhi, M. (2008). Morphological study of genetic variation among Iranian figs (*Ficus carica* L.) cultivars. In *Agronomy and horticulture*. Taylor & Francis.

Mahmoudi, S., Khali, M., Benkhaled, A., Boucetta, I., Dahmani, Y., Attallah, Z., & Belbraouet, S. (2018). Fresh figs (*Ficus carica* L.): Pomological characteristics, nutritional value, and phytochemical properties. *European Journal of Horticultural Science, 83*(2), 104–113.

Marcotuli, I., Mazzeo, A., Nigro, D., Giove, S., Giancaspro, A., Colasuonno, P., et al. (2019). Analysis of genetic diversity of *Ficus carica* L. (Moraceae) collection using simple sequence repeat (SSR) markers. *Acta Scientiarum Polonorum, Hortorum Cultus, 18*, 93–109. https://doi.org/10.24326/asphc.2019.4.9

Marcotuli, I., Mazzeo, A., Colasuonno, P., Terzano, R., Nigro, D., Porfido, C., et al. (2020). Fruit development in *Ficus carica* L.: Morphological and genetic approaches to Fig buds for an evolution from Monoecy toward Dioecy. *Frontiers in Plant Science, 11*, 1208.

Mars, M. (2003). Fig (*Ficus carica* L.) genetic resources and breeding. *Acta Horticulturae, 605*, 19–27. https://doi.org/10.17660/ActaHortic.2003.605.1

Mars, M., Chebli, T., & Marrakchi, M. (1998). Multivariate analysis of fig (*Ficus carica* L.) germplasm in southern Tunisia. *Acta Horticulturae, 480*, 75–82. https://doi.org/10.17660/ActaHortic.1998.480.10

Mars, M., Chatti, K., Saddoud, O., Salhi-Hannachi, A., Trifi, M., & Marrakchi, M. (2008). Fig Cultivation and Genetic Resources in Tunisia an Overview. *Acta Horticulturae (ISHS), 798*, 27–32. https://doi.org/10.17660/ActaHortic.2008.798.2

Mars, M., Gaaliche, B., Ouerfelli, I., & Chouat, S. (2009). Systèmes de Production et Ressources Génétiques du figuier (*Ficus carica* L.) à Djebba et Kesra, deux villages de montagne au nord ouest de la Tunisie. *Revue des Régions Arides, 22*, 33–45.

Mateos, J. M., & Gil, J. (2014). Fruit as staple food: The role of fig (*Ficus carica* L.) during the Pre-Hispanic period of the Canary Islands, Spain (from the 3rd–2nd centuries BCE to the 15th century CE). In *Early agricultural remnants and technical heritage (EARTH): 8,000 years of resilience and innovation* (pp. 182–190). Oxbow Books Oxford.

Medeiros, A. R. M. (2002. 16p). *Figueira (Ficus carica l.) do plantio ao processamento caseiro* (Circular técnica, 35). Embrapa.

Melgarejo, P. (2000). *Tratado de Fruticultura para Zonas Áridas y Semiáridas. El Medio Ecológico, la Higuera, el Alcaparro y el Nopal* (p. 375). S.A Mundi-Prensa.

Meziant, L., Saci, F., Bachir Bey, M., & Louaileche, H. (2015). Varietal influence on biological properties of Algerian light figs (*Ficus carica* L.). *International Journal of Bioinformatics and Biomedical Engineering, 1*, 237–243.

Michailides, T. J. (2003). Diseases in fig. In R. C. Ploetz (Ed.), *Diseases of tropical fruit crops* (pp. 253–274). Wallingford, UK.

Michailides, T. J., & Morgan, D. P. P. (1998). Spread of endosepsis in Calimyrna fig orchards. *Phytopathology, 88*(7), 637–647. https://doi.org/10.1094/PHYTO.1998.88.7.637

Michailides, T. J., Morgan, D. P., Felts, D., & Doster, M. A. (2008). Control of decay in caprifigs and Calimyrna figs with fungicides. *Acta Horticulturae, 798*, 269–275. https://doi.org/10.17660/ActaHortic.2008.798.39

Mirheidari, F., Khadivi, A., Moradi, Y., & Paryan, S. (2020). Phenotypic variability of naturally grown edible fig (*Ficus carica* L.) and caprifig (*Ficus carica* var. *caprificus* Risso) accessions. *Scientia Horticulturae, 267*, 109320.

Mohammadi, S. A., & Prasanna, B. M. (2003). Analysis of genetic diversity in crop plants-salient statistical tools and considerations. *Crop Science, 43*(4), 1235–1248.

Mohamed, Z. R., Abdelsalam N. R., Abdel Latif K. F., & Abdelhady, R. M., (2017). Genetic diversity of fig (*Ficus carica L.*) based on morphological characters and two-way hierarchical cluster analysis. *Alexsandria Science Exchange Journal, 38*, 168–174.

Moniruzzaman, M., Anuar, N., Yaakob, Z., Islam, A. A., & Al-Khayri, J. M. (2020). Performance evaluation of seventeen common fig (*Ficus carica* L.) cultivars introduced to a tropical climate. *Horticulture, Environment, and Biotechnology, 61*(5), 795–806.

Mori, K., Shirasawa, K., Nogata, H., Hirata, C., Tashiro, K., Habu, T., et al. (2017). Identification of RAN1 orthologue associated with sex determination through whole genome sequencing analysis in fig (*Ficus carica* L.). *Scientific Reports, 7*(1), 1–15.

Morton, J. F. (2000). *Fruits of warm climates. Fig. Ficus carica*. Purdue University, NewCROP. http://www.hort.purdue.edu/newcrop/default.html. Accessed 25 Nov 2007.

Nabli, M. A. (1989). *Essai de synthèse sur la végétation et la phyto-écologie tunisienne. Elément de Botanique et de phyto-écologie*. Faculté des Sciences de Tunis et UNESCO.

Norman, P. E., Tongoona, P., & Shanahan, P. E. (2011). Determination of interrelationships among agr-morphological traits of yams (*Discorea* spp.) using correrlation and factor analyses. *Journal of Applied Biosciences, 45*, 3059–3070.

Núñez-Gómez, D., Legua, P., Martínez-Nicolás, J. J., & Melgarejo, P. (2021). Breba fruits characterization from four varieties (*Ficus carica* L.) with important commercial interest in Spain. *Food, 10*(12), –3138.

Obenauf, C., Gerdts, M., Leavitt, G., & Crane, J. (1978). *Commercial dried fig production in California*. Univ. Calif. Agr. Ext. l.eafl. 21051: 30pp.

Oukabli, A., Mamouni, A., Laghezali, R., Khadari, B., Roger, J. P., Kjellberg, F., & Ater, M. (2002). Genetic variability in Morrocan fig cultivars (*Ficus carica*) based on morpholigical and pomological data. *Acta Horticulturae, 605*, 54–60.

Oukabli, A., Mamouni, A., Laghezali, M., Ater, M., Roger, J. P., & Khadari, B. (2003). Local caprifig tree characterization and analysis of interest for pollination. *Acta Horticulturae, 605*, 61–64.

Owino, W. O., Nakano, R., Kubo, Y., & Inaba, A. (2004a). Alterations in cell wall polysaccharides during ripening in distinct anatomical tissue regions of the fig (*Ficus carica* L.) fruit. *Postharvest Biology and Technology, 32(1)*, 67–77. https://doi.org/10.1016/j.postharvbio.2003.10.003

Owino, W. O., Nakano, R., Kubo, Y., & Inaba, A. (2004b). Coordinated expression patterns of genes encoding cell wall modifying enzymes during ripening in distinct anatomical tissue regions of the fig (*Ficus carica* L.) fruit. *Postharvest Biology and Technology, 32(3)*, 253–261. https://doi.org/10.1016/j.postharvbio.2004.01.003

Ozeker, E., & Isfandiyaroglu, M. (1998). Evaluation of table fig cultivars in Cesme Peninsula. *Acta Horticulturae, 480*, 55–60.

Ozen, M., Çobanoğlu, F., Kocataş, H., Tan, N., Ertan, B., Şahin, B., Konak, R., Doğan, Ö., Tutmuş, E., Kösoğlu, İ., & Özkan, R. (2007). *Fig Growing (in Turkish)*. Erbeyli İncir Araştırma Enstitüsü, Tuna Matbaacılık.

Ozen, M., Dakilic, Z., Kocatas, H., Belge, A., & Ertan, B. (2017). Investigation of leaf characteristics in 'Sarılop' and 'Kaba ilek' fig hybrids Acta Horticulturae, 1173, 35.

Papadopoulou, K., Ehaliotis, C., Tourna, M., Kastani, P., Karydis, I., & Zervakis, G. (2002). Genetic relatedness among diocious *Ficus carica* L. cultivars by random amplified polymorphic DNA analysis, and evaluation of agronomic and morphological characters. *Genetica, 114*(2), 183–194.

Paula, L. A., Corrêa, L. S., Boliani, A. C., & Santos, P. C. (2009). Efeito do ácido indolbutírico e épocas de estaqueamento sobre o enraizamento de estacas herbáceas de figueira (*Ficus carica* L.). *Acta Scientiarum Agronomy, 31*, 87–92.

Perez-Jiménez, M., López, B., Dorado, G., Pujadas-Salvá, A., Guzmán, G., & Hernandez, P. (2012). Analysis of genetic diversity of southern Spain fig tree (*Ficus carica* L.) and reference materials as a tool for breeding and conservation. *Hereditas, 149*(3), 108–113.

Pérez-Sánchez, R., Morales-Corts, R. M., & Gómez-Sánchez, Á. M. (2016). Agro-morphological diversity of traditional fig cultivars grown in Central-Western Spain. *Genetika, 48*(2), 533–546.

Piga, A., Agabbio, M., & Farris, G. A. (2003). Dehydration performance of local fig cultivars. *Acta Horticulturae, 605*, 241–245. https://doi.org/10.17660/ActaHortic.2003.605.37

Podgornik, M., Vuk, I., Vrhovnik, I., & Mavsar, D. B. (2010). A survey and morphological evaluation of fig (*Ficus carica* L.) genetic resources from Slovenia. *Scientia Horticulturae, 125*(3), 380–389.

Polat, A. A., & Caliskan, O. (2017). Correlations among important fruit quality and plant characteristics of some fig genotypes. *Journal of Life Sciences, 11*, 141–144.

Polat, A. A., & Ozkaya, M. (2005). Selection studies on fig in the Mediterranean region of Turkey. *Pakistan Journal of Botany, 37*(3), 567.

Poledica, M. M., Milivojevic, J. M., Radivojevic, D. D., & Dragisic-Maksimovic, J. J. (2012). Prohexadione-Ca and young cane removal treatments control growth, productivity, and fruit quality of the Willamette raspberry. *Turkish Journal of Agriculture and Forestry, 36*, 680–687.

Pourghayoumi, M., Bakhshi, D., Rahemi, M., & Jafari, M. (2012). Effect of pollen source on quantitative and qualitative characteristics of dried figs (*Ficus carica* L.) cvs 'Payves' and 'Sabz' in Kazerun – Iran. *Scientia Horticulturae (Amsterdam), 147*, 98–104. https://doi.org/10.1016/j.scienta.2012.08.026

Pourghayoumi, M., Bakhshi, D., Rahemi, M., Noroozisharaf, A., Jafari, M., Salehi, M., Chamane, R., & Hernandez, F. (2017). Phytochemical attributes of some dried fig (*Ficus carica* L.) fruit cultivars grown in Iran. *Agriculturae Conspectus Scientificus, 81*(3), 161–166.

Rahemi, M., & Jafari, M. (2008). Effect of caprifig type on quantity of Estahaban dried fig *Ficus carica* cv. Sabz. *Acta Horticulturae, 798*, 249–252.

Rigitano, O. (1955). *A figueira cultivada no Estado de São Paulo*. Tese (Doutorado em Fitotecnia) – Escola Superior de Agricultura "Luiz de Queiroz", Universidade de São Paulo, Piracicaba.

Rodolfi, M., Ganino, T., Chiancone, B., & Petruccelli, R. (2018). Identification and characterization of Italian common figs (*Ficus carica*) using nuclear microsatellite markers. *Genetic Resources and Crop Evolution, 65*(5), 1337–1348.

Rodov, V., Horev, B., Goldman, G., Vinokur, Y., Yablowich, Z., Golubowich, S., & Flaishman, M. A. (2005). Purple fig: Pollination effects on fruit quality and storage potential (in Hebrew). *Alon Hanotea, 60*, 110–112.

Rodrigues, M. G. F., Monteiro, L. N. H., Ferreira, A. F. A., Pavan, B. E., & Boliani, A. C. (2017). Caracterização morfológica de acessos de banco de germoplasma de figueira. In *CONGRESSO BRASILEIRO DE FRUTICULTURA, 25., 2017. Porto Seguro. Anais [...]*. SBF.

Rodrigues, M. G. F., Monteiro, L. N. H., Flã, A., Dos Santos, T. P., Pavan, B. E., & Conceiã, A. (2018). Genetic variability in morphological characters among Fig tree accessions Fig genetic conservation. *Genetics and Molecular Research, 17*(4), 1.

Rodrigues, M. G. F., Santos, T. P. D., Ferreira, A. F. A., Monteiro, L. N. H., Nakanishi, E. S., & Boliani, A. C. (2019a). Morphological characterization of active germoplasm bank fig tree accessions. *Revista Brasileira de Fruticultura, 41*, e-074.

Rodrigues, M. G. F., Monteiro, L. N. H., Ferreira, A. F. A., Dos Santos, T. P., Pavan, B. E., Neves, V. A. B., & Boliani, A. C. (2019b). Biometric characteristics among fig tree genotypes in Brazil. *Genetics and Molecular Research, 18*(2), gmr18191.

Rostami, A. L., & Rahemi, M. (2013). Responses of caprifig genotypes to water stress and recovery. *Journal Biology Environmental Science, 7(21)*, 131–139.

Ryugo, K. (1988). *Fruit Culture: Its science and art* (p. 321). Wiley.

Saadi, S., Todorovic, M., Tanasijevic, L., Pereira, L. S., Pizzigalli, C., & Lionello, P. (2015). Climate change and Mediterranean agriculture: Impacts on winter wheat and tomato crop evapotranspiration, irrigation requirements and yield. *Agricultural Water Management, 147*, 103–115.

Sacks, E. J., & Shaw, D. V. (1994). Optimum allocation of objective color measurements for evaluating fresh strawberries. *Journal of the American Society for Horticultural Science, 119*(2), 330–334.

Saddoud, O., Chatti, K., Salhi-Hannachi, A., Mars, M., Rhouma, A., Marrakchi, M., & Trifi, M. (2008a). Genetic diversity of Tunisian figs (*Ficus carica* L.) as revealed by nuclear microsatellites. *Hereditas, 144*, 149–157.

Saddoud, O., Baraket, G., & Chatti, K. (2008b). Morphological variability of fig (*Ficus carica* L.) cultivars. *International Journal of Fruit Science, 8*(1–2), 35–51.

Saddoud, O., Baraket, G., & Chatti, K. (2011). Using morphological characters and simple sequence repeat (SSR) Markers to characterize Tunisian fig (*Ficus carica* L.) cultivars. *Acta Biologica Cracoviensia, 53*(2), 7–14.

Sahin, N. (1998). Fig adaptation studies in Western Turkey. *Acta Horticulturae, 480*, 61–70.

Sahin, N., Cabanoglu, F., & Sahin, B., 2001. *Fig report. T.R. Prime ministry state planning organization. Planning of development with five years. Plant production (fruits)* (Report of special committee, pp. 548). Ankara (in Turkish).

Salhi-Hannachi, A., Mars, M., Chatti, K., Marrakchi, M., & Trifi, M. (2003). Specific genetic markers for Tunisian fig germplasm: Evidence of morphological traits, random amplified polymorphic DNA and inter simple sequence repeats markers. *Journal of Genetics and Breeding, 57*(2), 125–136.

Salhi-Hannachi, A., Trifi, M., Zehdi, S., Hedfi, J., Mars, M., Rhouma, A., & Marrakchi, M. (2004). Inter-Simple Sequence Repeat fingerprints to assess genetic diversity in Tunisian fig (*Ficus carica* L.) germplasm. *Genetic Resources and Crop Evolution, 51*(3), 269–275.

Sanches, J., Melgarejo, P., Hemandz, F., & Martienz, J. J. (2002). Chemical and morphological characterization of four fig tree cultivars (*Ficus carica* L.) grown under similar culture conditions. *Acta Horticulturae, 605*, 33–36.

Shafique, M., Khan, A. S., Malik, A. U., Shahid, M., Rajwana, I. A., Saleem, B. A., Amin, M., & Ahmad, I. (2011). Influence of pollen sorce and pollination frequency on fruit drop, yield and quality of date palm (*Phoenix dactylifera* L.) CV. Dhakki. *Pakistan Journal of Botany, 43*, 831–839.

Shahinuzzaman, M., Yaakob, Z., Anuar, F. H., Akhtar, P., Kadir, N. H. A., Hasan, A. K. M., Sobayel, K., et al. (2020). In vitro antioxidant activity of *Ficus carica* L. latex from 18 different cultivars. *Scientific Reports, 10*(1), 1–14.

Sharma, S. K., & Badiyala, S. D. (2006). Variability studies in common fig in Hamirpur district of Himachal Pradesh. *Indian Journal of Horticulture, 63*(2), 159–161.

Simsek, M., & Yildirim, H. (2010). Fruit characteristics of the selected fig genotypes. *African Journal of Biotechnology, 9*(37), 6056–6060.

Simsek, M., Gulsoy, E., Kirar, M. Z., Turgut, Y., & Yucel, B. (2017). Identification and selection of some female fig (*Ficus carica* L.) genotypes from Mardin province of Turkey. *Pakistan Journal of Botany, 49*(2), 541–546.

Singh, A., Prakash, J., Meghawal, P. R., & Ranpise, S. A. (2015). The fig (*Ficus carica*). In S. N. Ghosh (Ed.), *Breeding of underutilized fruit crops Part I* (pp. 149–179). Jaya Publishing House.

Storey, W. B. (1975). Figs. In J. Janick & J. N. Moore (Eds.), *Advances and fruit breeding Indiana* (pp. 568–589). Purdue University Press.

Stover, E., & Aradhya, M. (2008). Fig genetic resources and research at the US National Clonal Germplasm Repository in Davis, California. *Acta Horticulturae, 798*, 57–68.

Stover, E., Aradhya, M., Ferguson, L., & Crisosto, C. H. (2007a). The fig: Overview of an ancient fruit. *HortScience, 42*(5), 1083–1087.

Stover, E., Aradhya, M., Crisosto, C., & Ferguson, F. (2007b). Overview of the California fig industry and new interest in varieties for fresh fruit. In *Proceedings California Plant Soil Conference: Opportunities for California Agriculture* (pp. 169–175). Sacramento, CA, USA.

Tamboli, B. D., Sawale, D. D., Jagtap, P. B., Nimbalkar, R. U., & Teke, S. R. (2015). Effect of micronutrients on yield and fruit quality of fig on Inceptisol. *Indian Journal of Horticulture, 72*(3), 419–422.

Tanksley, S. D., & McCouch, S. R. (1997). Seed banks and molecular maps: Unlocking genetic potential from the wild. *Science, 277*(5329), 1063–1066.

Trad, M., Gaaliche, B., Renard, C., & Mars, M. (2012). Quality performance of 'Smyrna' type figs grown under Mediterranean conditions of Tunisia. *Journal of Ornemental and Horticultural Plants, 2*(3), 139–146.

Trad, M., Le Bourvellec, C., Gaaliche, B., Ginies, C., Renard, C. M. G. C., & Mars, M. (2013a). Caprification modifies polyphenols but not cell wall concentrations in ripe figs. *Scientia Horticulturae (Amsterdam), 160*, 115–122. https://doi.org/10.1016/j.scienta.2013.05.037

Trad, M., Gaaliche, B., Renard, C. M. G. C., & Mars, M. (2013b). Plant natural resources and fruit characteristics of fig (*Ficus carica* L.) change from coastal to continental areas of Tunisia. *Journal of Agricultural Research and Development, 3*(2), 022–025.

Trad, M., Ginies, C., Gaaliche, B., Renard, C. M. G. C., & Mars, M. (2014). Relationship between pollination and cell wall properties in common fig fruit. *Phytochemistry, 98*, 78–84. PubMed. https://doi.org/10.1016/j.phytochem.2013.12.011

Tsantili, E. (1990). Changes during development of 'Tsapela'fig fruits. *Scientia Horticulturae, 44*(3–4), 227–234.

Valdeyron, G., & Lloyd, D. G. (1979). Sex differences and flowering phenology in the common fig, *Ficus carica* L. *Evolution, 33*, 673–685.

Vallese, F. (1909). *Il Fico* (p. 381). Francesco Battiato.

Viana, A. P., Pereira, T. N. S., Pereira, M. G., de Souza, M. M., Maldonado, J. F. M., & do Amaral Júnior, A. T. (2003). Simple and canonic correlation between agronomical and fruit quality traits in yellow passion fruit (*Passiflora edulis* f. *flavicarpa*) populations. *Crop Breeding and Applied Biotechnology, 3*(2), 133–140. https://doi.org/10.12702/1984-7033.v03n02a06

Vlachos, I. P. (2013). *Consumers' perceptions of food quality products: Greece's experiences. Consumer attitudes to food quality products* (pp. 247–259). Springer.

Woodland, D. W. (1997). *Contemporary plant systematics*. Andrews University Press.

Yakushiji, H., Morita, T., Jikumaru, S., Ikegami, H., Azuma, A., & Koshita, Y. (2012). Interspecific hybridization of fig (*Ficus carica* L.) and *Ficus erecta* Thunb., a source of Ceratocystis canker resistance. *Euphytica, 183*, 39–47.

Yancheva, D., Golubowicz, S., Yablowicz, Z., et al. (2005). Efficient *Agrobacterium*-mediated transformation and recovery of transgenic fig (*Ficus carica* L.) plants. *Plant Science, 168*, 1433–1441. https://doi.org/10.1016/j.plantsci.2004.12.007

Yang, J. S., Zhao, Y. F., Zhou, Y. H., & Mi, L. (1994). Observation on the performance of introduced promising fig cultivar Masui Dauphine. In *CAB Abstracts An* (Vol. 21191).

Yilmaz, K. U., Paydas Kargi, S., & Kafkas, S. (2012). Morphological diversity of the Turkish apricot (*Prunus armeniaca* L.) germplasm in the Irano-Caucasian ecogeographical group. *Turkish Journal of Agriculture and Forestry, 36*, 688–694.

Zare, H. (2008). Comparison of fig caprification vessels, period and caprifig cultivar usable in Iran. *Acta Horticulturae, 798*, 233–239. https://doi.org/10.17660/ActaHortic.2008.798.33

Chapter 7
Morpho-Chemical Characteristics Useful in the Identification of Fig (*Ficus carica* L.) Germplasm

Oguzhan Caliskan, Safder Bayazit, and Derya Kilic

Abbreviations

CIHEAM	International center for advanced Mediterranean agronomic studies
IPGRI	International plant genetic resources institute
PCA	Principal component analyzes
TSS	Total soluble solids
UPOV	The international union for the protection of new varieties of plants

1 Introduction

Fig is one fruit species with a history as much as human history. The spread of the figs from Anatolia, the Caucasus, where it originated, to the whole world is accepted as a sacred fruit in the monotheistic religions, a fruit that is fondly consumed, and its nutritional properties. However, after this spread, important morphological differences emerged both in the origin areas and in the areas to which it was moved (Çalışkan & Dalkılıç, 2022; Condit, 1947). Figs can be grown in tropical and subtropical regions, although it is cultivated mainly in the mild temperate climate of the Mediterranean basin. The fig is probably the only edible fruit in a family of many giant rubber tree species growing in tropical and subtropical climates and has, therefore, been traded for thousands of years. It is considered an exotic fruit, especially in countries where cultivation is not possible (Aksoy et al., 2007).

For ages, the fig tree's fruit has been valued in its fresh and dry forms. Seventy percent of the world's fig production is produced in the countries along the Mediterranean coast. In these countries, figs are an essential component of the

O. Caliskan (✉) · S. Bayazit · D. Kilic
Department of Horticulture, Faculty of Agriculture, Hatay Mustafa Kemal University, Hatay, Turkey
e-mail: ocaliskan@mku.edu.tr; sbayazit@mku.edu.tr; deryakilic@mku.edu.tr

M. F. Ramadan (ed.), *Fig* (Ficus carica)*: Production, Processing, and Properties*, https://doi.org/10.1007/978-3-031-16493-4_7

Mediterranean diet, which is considered one of the healthiest in the world and associated with long life (Caliskan, 2015; Solomon et al., 2006). Especially the polyphenol and antioxidant contents of black-colored figs are richer than the light-colored ones, increasing the interest in colored individuals because they are healthier (Caliskan & Polat, 2011).

Ficus species are gynodioecious, bearing either hermaphroditic or edible figs on separate plants. Figs have different pollination biology from other fruit species. Male fig plants, caprifigs, produce male and short-styled (gall) female flowers in syconium (Fig. 7.1). Edible figs produce only female flowers (Beck & Lord, 1988), but their pollination requirement differs. The female flowers in edible figs are long-styled and occur as succulent fruitlet (Stover et al., 2007).

Three types of female figs are defined based on their cropping. The common fig develops fruit parthenocarpically without pollination and could produce one (unifera cultivars) or two (bifera cultivars) crops. The Smyrna-type fig requires pollination with pollen from caprifigs. Furthermore, the San Pedro-type fig produces a first crop parthenocarpically (breba) and a second main crop of fruit only after pollination with pollen from caprifigs (Flaishman et al., 2008). Condit (1955) indicated that 78% are common types, less than 4% are San Pedro types, and the remaining 18% are Smyrna types of edible figs. Besides, cultivars are diverse in such properties as leaf morphology, plant vigor, fruit skin and internal color, fruit flavor, total soluble

Fig. 7.1 The appearance of ripe fruits and fruit flesh differences in caprifig (left) and edible fig (right)

solids content, titratable acidity, fruit shape, skin thickness, ostiole width, and duration of fruit production (Flaishman et al., 2008).

A caprifig tree gives three crop periods of fruit each year. Profichi fruit in summer, mammoni fruit in fall, and mamme fruit in winter. The profichi, the main caprifig crop, comes across with the main summer crop of edible fig trees. The caprifig trees supply a pollen source for caprification, which is the transfer of pollen from caprifig trees to edible fig trees via the wasp vector *Blastophaga psenes*. Caprification is a traditional practice in all fig-growing regions and is essential in edible figs' fruit set and quality (Condit, 1947; Ferrara et al., 2016; Caliskan et al., 2017a).

The fig fruits, primarily used for fresh consumption, have various colors, sizes, shapes, and flavors in origin areas. The names of these figs are mainly given based on local geographic origin, fruit size, fruit shape, fruit skin color, ostiole color, maturity dates, or the name of the orchard owner. Researchers may encounter similar synonym and homonym genotypes because of the interchange of fig plant material from nearby locations (Caliskan & Polat, 2012a).

The fig plants remain mostly as traditional crops as in other Mediterranean countries, and important genetic variations have been taking place due to biotic and abiotic processes such as urbanization, the extension of intensive crops, and fig mosaic disease (Salhi-Hannachi et al., 2004). Losses in the genetic diversity of crop species due to commercialization have led to the need to preserve the present genetic resources as much as possible, not only for the long-term survival of the species but also to ensure enough variability for breeding programs (Esquinas Alcazar, 2005). There has been some research dealing with the genetic diversity in edible fig germplasm (Aksoy et al., 2003; Giraldo et al., 2010; Podgornik et al., 2010; Şimşek & Yildirim, 2010; Stover & Aradhya, 2008); however, to date, there has been little research into the genetic resources available for caprifig germplasm (Dalkılıç et al., 2011; Khadivi-Khub & Anjam 2014; Caliskan et al., 2017b).

Plant genetic resources should be identified and protected due to their advantages, such as being compatible with different ecological environments and containing genes resistant to different diseases and pests. For this reason, in addition to the determination of morphological, phenological, and pomological data of genotypes, the development of technology, the popularity of gene science, and the use of genetic studies, especially DNA markers, to identify individuals have led many developed countries to carry out genetic studies to identify their plant gene resources. Characterization designs of morphological variability within the collections and selection of the most significant variables should be carefully preserved to conserve better and utilize genetic resources (Giraldo et al., 2010). This characterization is a highly recommended first step before starting biochemical or molecular studies (Hoogendijk & Williams, 2001). Recent developments show that the most important natural resource of the current century plants genetic resources. Protecting these resources necessitates preserving the genetic materials as they are today and transforming these resources into benefits. Improving the use of plant genetic resources for food and agriculture can be achieved by determining the essential properties of the material during preservation (Inal, 2002).

In recent years, principal component analysis (PCA) has been used to identify fruit genetic resources and in different stages of breeding programs. This method determines the most effective ones among hundreds of plant characteristics and saves time and labor by reducing the number of examined attributes (Giraldo et al., 2010; Saddoud et al., 2008; Simsek et al., 2020). This chapter presents the results of the studies examining the morpho-chemical properties of figs using the principal component analysis.

2 Plant and leaf Characteristics

The first detailed study on the plant and leaf description of fig genetic resources was carried out by Condit (1941). He identified more than 600 fig genotypes, especially leaf and fruit characteristics (Condit, 1955). In the following years, the characteristics were evaluated by different researchers, and a fig descriptor was formed by IPGRI and CIHEAM (2003). In addition, another fig descriptor was published by UPOV in 2007. In both fig descriptors, nearly 200 features, including all plant and fruit quality characteristics used to identify fig genetic resources, have been defined, and expensive processes make the characterization of large germplasm collections difficult. Therefore, useful variables in distinguishing fig genotypes have been determined (Table 7.1).

The majority of the edible figs have an open and spreading canopy for tree growth (Condit, 1941; Giraldo et al., 2010). Furthermore, the leaf shape is mainly *G* with three lobes or *C* with five lobes (Caliskan & Polat, 2012a). The plant and leaf characteristics that effectively distinguish the genotypes of edible figs are presented (Table 7.2). The number and shape of lobes per leaf (Saddoud et al., 2008), tree growth habit, size of the tree, degree of branching (Giraldo et al., 2010; Pérez-Sánchez et al., 2016), terminal bud length and diameter (Essid et al., 2017), leaf length, leaf width, leaf area, the density of hairs/spicules on the leaf's upper surface, and petiole thickness (Caliskan & Polat, 2012a; Ciarmiello et al., 2015; Khan et al., 2022; Kokaj, 2021; Mirheidari et al., 2020; Podgornik et al., 2010) are the traits used for the discrimination of edible figs. Although there are different leaf forms on the same shoot, leaf shape is an essential morphological criterion in identifying fig genotypes. In addition, apical dominancy, the number of leaves per shoot, the number of fruit per shoot (Caliskan et al. 2012a; Kokaj, 2021), and the beginning of fruit maturation (Mirheidari et al., 2020; Simsek et al., 2020) also are very useful to use for the identification of edible fig germplasm.

There are significant differences among the plant characteristics of caprifigs, similar to edible figs (Table 7.2). Khadivi-Khub and Anjam (2014) reported that caprifigs from Iran have numerous variables for tree shape and vigor, tree habit, the number of lobes per leaf, leaf shape (Fig. 7.2), leaf length, leaf width, length of the central lobe, little lateral lobe number, petiole length, and petiole thickness. Caliskan et al. (2017b) showed that high diversity is determined in mamme and mammoni number per shoot, ostiole width, terminal bud color, pollen number per fruit, fruit

Table 7.1 The most diverse plant and leaf characteristics in the edible figs

Variable	References
Beginning of fruit maturation	Mirheidari et al. (2020) and Simsek et al. (2020)
Cropping efficiency	Simsek et al. (2020)
Tree growth habit	Giraldo et al. (2010) and Pérez-Sánchez et al. (2016)
Size of the tree	Giraldo et al. (2010) and Pérez-Sánchez et al. (2016)
Tree vigor	Simsek et al. (2020)
Branching	Mirheidari et al. (2020)
Apical dominancy	Caliskan and Polat (2012a)
Degree of the branching	Giraldo et al. (2010)
Terminal bud length	Gaaliche et al. (2012)
Shoot length	Simsek et al. (2020)
Shoot color	Mirheidari et al. (2020)
Leaf shape	Saddoud et al. (2008), Caliskan and Polat (2012a), Baziar et al. (2018) and Kokaj (2021)
Number of lobes	Caliskan and Polat (2012a) and Rodrigues et al. (2019)
Shape of lobes	Saddoud et al. (2008)
Upper lateral lobe depth	Mirheidari et al. (2020)
Upper lateral lobe base width	Mirheidari et al. (2020)
Leaf length	Podgornik et al. (2010), Caliskan and Polat (2012a), Gaaliche et al. (2012), Ciarmiello et al. (2015) and Kokaj (2021)
Leaf width	Podgornik et al. (2010), Caliskan and Polat (2012a), Gaaliche et al. (2012), Ciarmiello et al. (2015) and Kokaj (2021)
Leaf area	Podgornik et al. (2010), Caliskan and Polat (2012a), Ciarmiello et al. (2015), Kokaj (2021) and Khan et al. (2022)
Length of the central lobe	Caliskan and Polat (2012a) and Pérez-Sánchez et al. (2016)
Density of hairs on the leaf	Podgornik et al. (2010) and Mirheidari et al. (2020)
Petiole length	Khan et al. (2022)
Petiole thickness	Podgornik et al. (2010), Mirheidari et al. (2020)
The number of leaves per shoot	Caliskan and Polat (2012a) and Simsek et al. (2020)
The number of fruits per shoot	Caliskan and Polat (2012a) and Simsek et al. (2020)

skin color, internal pulp color, leaf shape, shoot length, and tree growth habit among the caprifigs from Turkey. In addition, Mirheidari et al. (2020) indicated that ripening time, upper lateral lobe depth, and upper lateral lobe base width are important variables for identifying caprifigs.

Generally, plant growth habit mainly has open and spreading habits; tree vigor is most frequently classified as intermediate; shoot color is most often grey or brown;

Table 7.2 The most diverse plant and leaf characteristics in the caprifigs

Variable	References
Ripening time	Mirheidari et al. (2020) and Khan et al. (2022)
Tree shape	Khadivi-Khub and Anjam (2014) and Caliskan et al. (2017a)
Tree vigor	Khadivi-Khub and Anjam (2014) and Caliskan et al. (2017a)
Tree habit	Essid et al. (2017) and Mirheidari et al. (2020)
Shoot length	Caliskan et al. (2017a)
Shoot color	Caliskan et al. (2017b)
Terminal bud length	Essid et al. (2017)
Terminal bud diameter	Essid et al. (2017)
The number of lobes per leaf	Khadivi-Khub and Anjam (2014) and Caliskan et al. (2017b)
Leaf shape	Khadivi-Khub and Anjam (2014), Caliskan et al. (2017b), Essid et al. (2017) and Fatahi et al. (2017)
Leaf length	Khadivi-Khub and Anjam (2014), Caliskan et al. (2017b), Essid et al. (2017) and Mirheidari et al. (2020)
Leaf width	Khadivi-Khub and Anjam (2014), Caliskan et al. (2017b), Essid et al. (2017) and Mirheidari et al. (2020)
Leaf area	Caliskan et al. (2017b)
Length of the central lobe	Khadivi-Khub and Anjam (2014), Essid et al. (2017) and Mirheidari et al. (2020)
Little lateral lobe number	Khadivi-Khub and Anjam (2014) and Mirheidari et al. (2020)
Upper lateral lobe depth	Mirheidari et al. (2020)
Upper lateral lobe base width	Mirheidari et al. (2020)
Petiole length	Khadivi-Khub and Anjam (2014)
Petiolethicknesss	Khadivi-Khub and Anjam (2014)
Leaf number per shoot	Caliskan et al. (2017a)

and the leaf area values of caprifigs are generally smaller than those of edible figs (Caliskan et al., 2017b; Mirheidari et al., 2020). In addition, most caprifigs have five or three lobes per leaf in Turkey (Caliskan et al., 2018), whereas it has mostly three lobes per leaf in Iran (Khadivi-Khub & Anjam, 2014). This situation clearly shows that the origin areas affect the difference among caprifig genetic resources. Besides, Özen et al. (2017) indicated that the leaf length, width, area, petiole length, and thickness of hybrid figs are lower than parents due to negative transgressive segregation.

3 Fruit Characteristics

Fig occurs in many genotypes, which have evolved mainly as natural seedlings during the many centuries in fig-growing origin areas and in other continental countries where it spread. Several cultivated and wild forms of fig are found in the origin areas with a great diversity of color, shape, and flavor primarily for fresh consumption

Fig. 7.2 Leaf shapes of figs: (**a**) Entire, (**b**) Base decurrent, (**c**) Base truncate, (**d**) base cordate, (**e**) Base calcarate, lobes latete, (**f**) Base calcarate, lobes lyrate, (**h**) Base calcarate, lobes lineate

(Baziar et al., 2018; Caliskan & Polat, 2012a; Fatahi et al., 2017; Gozlekci, 2011; Khadivi-Khub & Anjam, 2014; Sezen et al., 2014; Simsek et al., 2017). Fruit shape, fruit size, small ostiole opening, ease of peeling (Aksoy et al., 2003; Condit, 1941), total soluble solids content (TSS), TSS/acidity (Caliskan & Polat, 2012b; Crisosto et al., 2010), flavor and fruit skin and flesh color (Ercisli et al., 2012; Flaishman et al., 2008; Gozlekci et al., 2011) are evaluated as the main quality criteria of fresh figs. Although these attributes are affected by ecology, they are mainly controlled by the genetic capacity of the genotype (Caliskan & Polat, 2012b). Therefore, these fruit quality characteristics are important in identifying fig genotypes. Distinct fruit characteristics from these studies to date are summarized in Table 7.3.

The most important discriminators of edible figs are the fruit weight, fruit length, fruit diameter, fruit shape (Giraldo et al., 2010; Podgornik et al., 2010), firmness of the fruit skin (Podgornik et al., 2010; Saddoud et al., 2008), production type (Aljane & Ferchichi, 2009; Podgornik et al., 2010), abscission of the stalk from the twig (Podgornik et al., 2010), stalk diameter, neck diameter, ostiole diameter, ostiole opening, and flesh thickness (Aljane & Ferchichi, 2009). In addition, crossing and soma-clonal differences contribute to such diversity among fig genotypes (Ergül et al., 2021).

Longitudinal cracks in the fruit peel and the ostiole opening (Fig. 7.3) are valuable features in identifying genotypes (Caliskan & Polat, 2012a; Saddoud et al., 2008). Although accepted as a maturity criterion, the first is an undesirable feature in the fresh fig trade. The second one, besides the genotype, may occur due to water stress and air humidity, especially close to harvest.

Table 7.3 The most diverse physical and chemical characteristics in the edible figs

Variable	References
Production type	Aljane and Ferchichi (2009) and Podgornik et al. (2010)
Abscission of the stalk from the twig	Podgornik et al. (2010)
Fruit shape	Giraldo et al. (2010), Ciarmiello et al. (2015) and Baziar et al. (2018)
Fruit weight	Podgornik et al. (2010), Caliskan and Polat (2012a), Fatahi et al., 2017; Abdelsalam et al. (2019), Hssaini et al. (2020a), Simsek et al. (2020)
Fruit diameter	Saddoud et al. (2008), Pérez-Sánchez et al. (2016), Fatahi et al., 2017; Hssaini et al. (2020a) and Simsek et al. (2020)
Fruit length	Podgornik et al. (2010), Pérez-Sánchez et al. (2016), Fatahi et al., 2017; Hssaini et al. (2020a), Simsek et al. (2020) and Khan et al. (2022)
Fruit flesh thickness	Mirheidari et al. (2020)
Fruit neck length	Podgornik et al. (2010), Ciarmiello et al. (2015), Pérez-Sánchez et al. (2016), Hssaini et al. (2020a) and Simsek et al. (2020)
Ostiole opening	Aljane and Ferchichi (2009)
Ostiole width	Aljane and Ferchichi (2009) and Gaaliche et al. (2012)
Fruit skin color	Saddoud et al. (2008), Caliskan and Polat (2011), Gozlekci (2011) and Hssaini et al. (2020a)
Fruit flesh color	Saddoud et al. (2008)
Firmness of the fruit skin	Saddoud et al. (2008) and Podgornik et al. (2010)
Fruit skin thickness	Saddoud et al. (2008)
Fruit skin cracks	Saddoud et al. (2008) and Caliskan and Polat (2012a)
Easy of peeling	Ciarmiello et al. (2015) and Pérez-Sánchez et al. (2016)
Drop at the eye	Simsek et al. (2020)
Color of liquid drop at the ostiole	Simsek et al. (2020)
Total soluble solids (TSS)	Giraldo et al. (2010), Caliskan and Polat (2012a), Ciarmiello et al. (2015), Pérez-Sánchez et al. (2016) and Hssaini et al. (2020a)
pH	Simsek et al. (2020)
Acidity	Messaouidi and Haddadi (2008), Gaaliche et al. (2012) and Simsek et al. (2020)
TSS/acidity	Caliskan and Polat (2012a) and Simsek et al. (2020)
Total antioxidant capacity	Caliskan and Polat (2012a)
Total phenolic	Caliskan and Polat (2012a) and Hssaini et al. (2020b)
Total anthocyanin	Caliskan and Polat (2012a) and Hssaini et al. (2020a)
Glucose	Gozlekci (2011), Caliskan and Polat (2012a) and Hssaini et al. (2020b)
Fructose	Gozlekci (2011), Caliskan and Polat (2012a) and Hssaini et al. (2020b)

Fruit skin and flesh color of fresh figs are very important for consumer preferences. Consumers prefer fresh figs with dark skin color and red or pink flesh color. However, the main crops' most frequent fruit skin color is green, and the pulp is red and pink (Caliskan & Polat, 2012a; Condit, 1941; Giraldo et al., 2010). In addition,

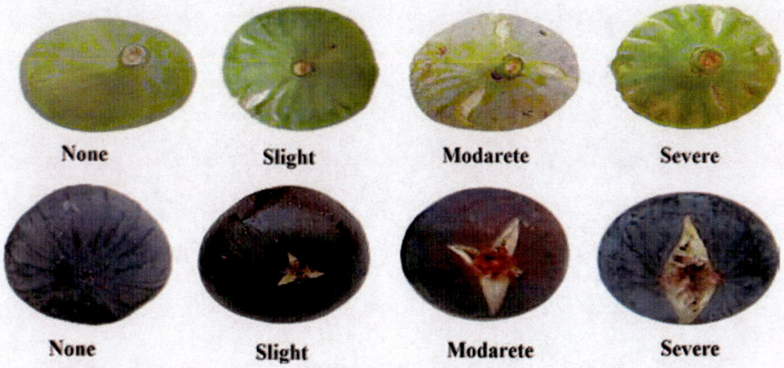

Fig. 7.3 Classification of cracking in fig fruit peel (upper) and ostiole width (lower)

Flaishman et al. (2017) reported that the black color of the skin is dominant over the green and red flesh color is dominant over pale color. Moreover, figs are one of the richest fruits in terms of skin color and flesh color diversity. According to the fig descriptor, there are seven different fig skin colors and five fruit flesh color (Fig. 7.4) groupings. This rich color difference in figs constitutes one of the essential criteria for identifying fig genotypes (Caliskan & Polat, 2011; Gozlekci, 2011; Hssaini et al., 2020a; Saddoud et al., 2008). Also, the skin coloration of figs is not homogeneous, especially in colored genotypes, which can cause differences in color evaluation such as green-brown and purple-black. However, the more objective conversion of these color observations into unbiased values results in more successful discrimination of various genotypes. For this purpose, using a colorimeter instead of a color chart can be preferred in descriptive and selective research to obtain more reproducible results. Indeed, Gozlekci (2011) indicated that fig genotypes' lightness (L), chroma, and hue° values of skin color vary greatly. Also, Caliskan and Polat (2011) reported that for figs with a vast diversity of fruit color among genotypes, objective investigations were very appropriate due to the researcher's subjective observation changes.

In describing the genetic sources of figs, the physical measurements and phytochemical properties can be used as suitable criteria. Fig cultivars with dark skin contain higher levels of polyphenols, anthocyanins, and flavonoids accompanied by higher antioxidant activity than fig genotypes with lighter skin (Aljane et al., 2020; Caliskan & Polat, 2011; Solomon et al., 2006). Caliskan and Polat (2011) reported that the dark genotypes contained the highest amounts of total antioxidant capacity, total phenolics, and total anthocyanins, whereas the yellow and green genotypes contained the lowest amounts. Previous studies on figs have focused on selecting and describing plant characteristics, fruit quality characteristics, and genetic markers. In recent years; however, because of the increasingly apparent importance of a healthy diet, researchers may also need to include phytochemical analysis in germplasm evaluation. The total antioxidant capacity, total phenolics, anthocyanins, fructose, and glucose contents of the fig fruits are; besides, Caliskan and Polat (2012b) showed that the sugar content of figs negatively correlated with total

Fig. 7.4 Fruit skin and flesh color grouping in edible figs. Fruit skin color groups: 1 Black, 2 Purple, 3 Brown, 4 Green, 5 Light green, 6 Yellow-green, 7 yellow. Fruit flesh groups: 1 White, 2 Amber, 3 Pink, 4 Red, 5 Dark red

anthocyanins due to the sugar content of the cultivars with yellow and green fruit skin color being higher than those with dark colors.

Caprification is a common application in fig-growing areas and significantly affects the fruit set in edible figs (Condit, 1947). Caprifigs have a positive effect not only on fruit set and yield but also on quality characteristics such as fruit size, fruit flesh color, TSS content (Gaaliche et al., 2011; Ferrara et al., 2016), aroma and phytochemical content (Pourghayoumi et al., 2012; Trad et al., 2012) of edible figs. Important variables of fruit characteristics in the caprifigs are shown in Table 7.4.

Acarsoy Bilgin et al. (2020) indicated that identifying caprifig genetic material is important for caprification and breeding studies. Unfortunately, there have been a few studies into the genetic resources possible for caprifig germplasm. Further, caprifig accessions' morphological and pollinizer parameters have not been detailed. On the other hand, all caprifigs are not helpful in caprification because of important parameters in their unique fruit characteristics.

The critical characteristics of the caprifig for profichi crops (Table 7.4) include fruit size, fruit number per shoot, the time of *Blastophaga* wasps exit from caprifig fruits, ripening period, coinciding with the female flowers in edible figs (Caliskan & Bayazit, 2012), amount of gall, and male flowers, amount of pollen production (Ahi Koşar et al., 2022; Yaman & Caliskan, 2016a), pollen viability and germination ratios (Ilgin et al., 2007; Yaman & Caliskan, 2016b), the inclusion of the mammoni

Table 7.4 Important variables of fruit characteristics in the caprifigs

Variable	References
Fruit shape	Caliskan et al. (2017b)
Fruit weight	Khadivi-Khub and Anjam (2014), Caliskan et al. (2017b), Essid et al. (2017) and Acarsoy Bilgin et al. (2020)
Fruit width	Khadivi-Khub and Anjam (2014), Essid et al. (2017) and Acarsoy Bilgin et al. (2020)
Fruit length	Caliskan et al. (2017b) and Acarsoy Bilgin et al. (2020)
Fruit neck length	Caliskan et al. (2017b)
Fruit stalk length	Khadivi-Khub and Anjam (2014) and Essid et al. (2017)
Ostiole width	Caliskan et al. (2017a) and Essid et al. (2017)
Fruit internal pulp width	Khadivi-Khub and Anjam (2014)
Fruit skin color	Khadivi-Khub and Anjam (2014) and Caliskan et al. (2017b)
Fruit internal color	Caliskan et al. (2017b)
Profichi: Date of *Blastophaga* emergence	Caliskan et al. (2017b)
Profichi: Duration of *Blastophaga* emergence	Caliskan et al. (2017b)
Profichi: The number of gall flowers	Caliskan et al. (2017b)
Mammoni number per shoot	Caliskan et al. (2017a)
Mamme number per shoot	Caliskan et al. (2017a)
The numbers of pollen grains per fruit	Caliskan et al. (2017b)
The numbers of pollen grains per flower	Caliskan et al. (2017b)
The numbers of pollen grains per anther	Caliskan et al. (2017b)
Pollen viability	Caliskan et al. (2017b)
Pollen germination	Caliskan et al. (2017b)
Pollen shape	Caliskan et al. (2021)
Pollen diameter	Acarsoy Bilgin et al. (2020) and Caliskan et al. (2021)
Pollen length	Acarsoy Bilgin et al. (2020) and Caliskan et al. (2021)
Colpus width	Caliskan et al. (2021)
Exine thickness	Caliskan et al. (2021)
Number of single porates	Caliskan et al. (2021)
Number of double porates	Caliskan et al. (2021)
Number of triple porates	Caliskan et al. (2021)
Porate width	Acarsoy Bilgin et al. (2020) and Caliskan et al. (2021)
Abnormal pollen ratio	Caliskan et al. (2021)

and mamme crops, and free from disease and pests (Caliskan et al., 2016; Yaman & Caliskan, 2016a).

The fruit shapes of the caprifigs are mostly oblong, globose, or oblate. After pollination, the flowers within fruits of edible figs may fill the syconium cavity as they mature and form a solid pulp that constitutes about 83% of the weight of a mature fig. However, this does not occur in caprifigs (Condit, 1947), which have low specific gravity due to the large intercellular air spaces in the fruit's peel compared to edible figs (Neeman & Galil, 1978). Thus, the fruit weights, particularly caprifigs, can be lower than edible figs. The caprifigs have green skin color and white internal color. In addition, Caliskan et al. (2017b) reported that mamme and mammoni crops could be varied by genotypes. All caprifigs have a profichi crop set; however, the numbers of fruit differed depending on genetic and ecological conditions. Also, the profichi fruit yield is higher than that of edible figs.

Khadivi-Khub and Anjam (2014) showed that fruit and leaf size, fruit shape, leaf shape, and fruit skin color are important parameters for the morphological characterization of caprifigs. Caliskan et al. (2017b) stated that the fruit skin color of male figs is not as rich as that of female figs and that 90% of the genotypes are green in color, and there are a small number of black, purple, and brown skin colored genotypes (Fig. 7.5). Besides, the internal pulp color of caprifigs is most often white, whereas a color distribution ranging from dark purple to light purple can be seen in the fruit pulp color (Fig. 7.6).

Caliskan et al. (2017b) reported that the number of gall flowers per fruit, date of *Blastophaga* emergence, duration of *Blastophaga* emergence, percentages of pollen viability and germination, the number of pollen grains per anther, the number of pollen grains per flower, and the number of pollen grains per fruit that could be successfully used to characterize caprifig accessions. Besides, caprifig classes according to the number of gall flowers per fruit are mostly medium (250–499) and high (500–749), and the number of male flowers per profichi fruit ranges between 50 and 150 (Caliskan et al., 2018). They revealed great genetic diversity and intensive

Fig. 7.5 Fruit skin color differences in caprifigs

Fig. 7.6 Fruit internal color differences in caprifigs

differentiation of caprifigs. This rich genetic variation could have been due to the establishment of caprifig populations propagated by seed.

Caliskan et al. (2017b) reported a few morphological characteristics such as the fruit size, the number of gall flowers per fruit, pollen viability and germination, and pollen number per anther that could be used for the differentiation of caprifigs. Thus, more distinctive variables are essential to define germplasm and establish a breeding program for caprifigs.

Pollen grains have specific morphological characteristics and demonstrate powerful genetic preservation; therefore, they can be used in taxonomy, phylogeny, morphological characterization, and breeding studies of fruit species (Shivanna, 2003; Arzani et al., 2005; Evrenosoglu & Misirli, 2009). Caliskan et al. (2021) showed that polar length, equatorial diameter, colpus width, pollen shape, number of porates (Fig. 7.7), porate width, exine thickness, and abnormal pollen ratio are the most important properties in distinguishing caprifigs from each other. Therefore, these pollen characteristics can also be used to identify caprifigs.

4 Conclusions

Selecting highly distinctive variables for appropriate morphological characterization of fruit genetic resources is invaluable for sustainably optimizing resources. This should be necessary for fig germplasm studies where many similar,

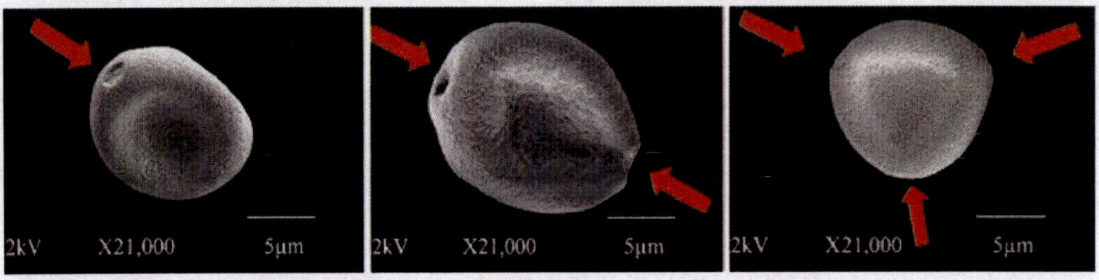

Fig. 7.7 Porate number of caprifig pollens: Single-porate (left), double-porate (middle), and triple-porate (right)

homonymous, and synonymous genotypes can be observed. Some morpho-chemical characters show significant variability in edible figs and caprifigs. Plant characteristics such as ripening time and leaf characteristics, fruit characteristics such as size, shape, skin and ostiole cracking, skin and flesh color, and chemical properties such as TSS, acidity, TSS/acidity, sugar profile, and phytochemical contents were found to be very successful in distinguishing genotypes from each other in edible figs. These characteristics can be used together to identify genotypes and the selection of individuals before breeding. In addition, fruit characteristics such as fruit size, skin and pulp color, the number of gall flowers, pollen number per fruit, and pollen characteristics such as pollen size and shape and pore number are successful in distinguishing individuals in a profichi crop of caprifigs. This review will be an important contribution to selecting highly discriminative variables to optimize resources to evaluate fig germplasm.

References

Abdelsalam, N. R., Awad, R. M., Ali, H. M., Salem, M. Z. M., Abdellatif, K. F., & Elshikh, M. S. (2019). Morphological, pomological, and specific molecular marker resources for genetic diversity analyses in fig (*Ficus carica* L.). *HortScience, 54*, 1299–1309.

Acarsoy Bilgin, N., Mısırlı, A., Belge, A., & Özen, M. (2020). The pollen and fruit properties of *Ficus carica Caprificus*. *International Journal of Fruit Science, 20*, S1696–S1705.

Ahi Koşar, D., Aktepe Tangu, N., Gencer, N. S., Durgut, E., & Ertürk, Ü. (2022). Some characteristics and caprification potentials of caprifigs (*Ficus carica* var. *caprificus* L.) grown in Bursa, Turkey. In Z. Dalkılıç (Ed.), *Ficus carica: Production, cultivation and uses*. Nova Science Publishers, Inc.

Aksoy, U., Balci, B., Can, H. Z., & Hepaksoy, S. (2003). Some significant results of the research work Turkey on fig. *Acta Horticulturae, 605*, 173–181.

Aksoy, U., Zafer, H. C., Meyvacı, B., & Şen, F. (2007). Kuru incir: Türk sultanları çekirdeksiz kuru üzüm, kuru incir ve kuru kayısı. *Ege Kuru Meyve ve Mamulleri İhracatçıları Birliği*, 139s.

Aljane, F., & Ferchichi, A. (2009). Postharvest chemical properties and mineral contents of some fig (*Ficus carica* L.) cultivars in Tunisia. *Journal of Food Agriculture and Environment, 7*(2), 209–212.

Aljane, F., Neily, M. H., & Msaddak, A. (2020). Phytochemical characteristics and antioxidant activity of several fig (*Ficus carica* L.) ecotypes. *Italian Journal of Food Science, 32*, 755–768.

Arzani, K., Nejatian, M. A., & Karimzadeh, G. (2005). Apricot (*Prunus armeniaca*) pollen morphological characterisation through scanning electron microscopy, using multivariate analysis. *New Zealand Journal of Crop and Horticultural Science, 33*, 381–388.

Baziar, G., Jafari, M., Sharifi Noori, M. S., & Samarfard, S. (2018). Evaluation of genetic diversity among Persian fig cultivars by morphological traits and RAPD markers. *HortScience, 53*, 613–619.

Beck, N. G., & Lord, E. M. (1988). Breeding system in *Ficus carica*, the common fig. I. floral diversity. *American Journal of Botany, 75*, 1904–1912.

Caliskan, O., & Polat, A. A. (2011). *Investigation by subjective and objective methods of fruit coloring in fig (Ficus carica L.), Türkiye VI* (pp. 758–764). International Horticultural Congress.

Caliskan, O., & Bayazit, S. (2012). Caprification and its role in fig cultivation. *Journal of Agricultural Faculty, MKU, 17*(1), 47–61.

Caliskan, O., & Polat, A. A. (2012a). Morphological diversity among fig (*Ficus carica* L.) accessions sampled from the Eastern Mediterranean Region of Turkey. *Turkish Journal of Agriculture and Forestry, 36*, 179–193.

Caliskan, O., & Polat, A. A. (2012b). Effects of genotype and harvest year on phytochemical and fruit quality properties of Turkish fig genotypes. *Spanish Journal of Agricultural Research, 10*, 1048–1058.

Caliskan, O. (2015). Mediterranean figs (*Ficus carica* L.) functional food properties. In V. R. Preedy & R. R. Watson (Eds.), *The Mediterranean diet an evidence-based approach* (pp. 629–637).

Caliskan, O., Bayazit, S., Ilgin, M., Kocatas, H., & Karatas, N. (2016). *Caprification and it's important for edible fig cultivation* (p. 22). Hatay.

Caliskan, O., Bayazit, S., Ilgin, M., Karatas, N., & Kocatas, H. (2017a). Preliminary results on morpho-pomological traits and pollinizer characterization of some caprifig genotypes. *Acta Horticulturae, 1173*, 45–50.

Caliskan, O., Bayazit, S., Ilgin, M., & Karatas, N. (2017b). Morphological diversity of caprifig (*Ficus carica* var. *caprificus*) accessions in the eastern Mediterranean region of Turkey: Potential utility for caprification. *Scientia Horticulturae, 222*, 46–56.

Caliskan, O., Bayazit, S., Ilgin, M., Karatas, N., & Ergul, A. (2018). Genetic diversity and population structure in caprifigs (*Ficus carica* var. *caprificus*) using SSR markers. *Spanish Journal of Agricultural Research, 16*(3), e0703.

Caliskan, O., Bayazit, S., Kilic, D., Ilgin, M., & Karatas, N. (2021). Pollen morphology and variability of caprifig (*Ficus carica* var. *caprificus*) genetic resources in Turkey using multivariate analysis. *Scientia Horticulturae, 287*, 110283.

Çalışkan, O., & Dalkılıç, Z. (2022). Ancient history and cultural heritage of Ficus carica in Turkey. In Z. Dalkılıç (Ed.), *Ficus carica: Production, Cultivation and Uses* (pp. 1–20). Nova Science Publishers, Inc..

Ciarmiello, L. F., Piccirillo, P., Carillo, P., De Luca, A., & Woodrow, P. (2015). Determination of the genetic relatedness of fig (*Ficus carica* L.) accessions using RAPD fingerprint and their agro-morphological characterization. *South African Journal Botany, 97*, 40–47.

Condit, I. J. (1941). Fig characteristics useful in the identification of varieties. *Hilgardia, 14*, 1–69.

Condit, I. J. (1947). *The Fig* (p. 220). The Chronica Botanica Company.

Condit, I. J. (1955). Fig varieties: A monograph. *Hilgardia, 23*, 323–538.

Crisosto, C. H., Bremer, V., Ferguson, L., & Crisosto, G. M. (2010). Evaluating quality attributes of four fresh figs (Ficus carica L.) cultivars harvested at two maturity stages. *HortScience, 45*(4), 707–710.

Dalkılıç, Z., Mesav, H. O., Günver-Dalkılıç, G., & Kocataş, H. (2011). Genetic diversity of male fig (*Ficus carica caprificus* L.) genotypes with random amplified polymorphic DNA (RAPD) markers. *African Journal of Biotechnology, 10*, 519–526.

Ercisli, S., Tosun, M., Karlidag, H., Dzubur, A., Hadziabulic, S., & Aliman, Y. (2012). Color and antioxidant characteristics of some fresh fig (*Ficus carica* L.) genotypes from Northeastern Turkey. *Plant Foods for Human Nutrition, 67*, 271–276.

Ergül, A., Büyük, B. P., Hazrati, N., Yılmaz, F., Kazan, K., Arslan, N., Özmen, C. Y., Aydın, S. S., Bakır, M., Tan, N., Kösoğlu, İ., & Çobanoğlu, F. (2021). Genetic characterisation and population structure analysis of Anatolian figs (*Ficus carica* L.) by SSR markers. *Folia Horticulturae, 33*, 49–78.

Esquinas Alcazar, J. (2005). Protecting crop genetic diversity for food security: Political, ethical and technical challenges. *Nature Reviews Genetics, 6*, 946–953.

Essid, A., Aljane, F., & Ferchichi, A. (2017). Morphological characterization and pollen evaluation of some Tunisian ex situ planted caprifig (*Ficus carica* L.) ecotypes. *South African Journal of Botany, 111*, 134–143.

Evrenesoglu, Y., & Mısırlı, A. (2009). Investigations on the pollen morphology of some fruit species. *Turkish Journal of Agricultural and Forestry, 33*, 181–190.

Fatahi, S., Cheghamirza, K., Arji, I., & Zarei, L. (2017). Evaluation of genetic variation of common fig (*Ficus carica* L.) in west of Iran. *Journal of Medicinal Plants and By-products, 2*, 229–240.

Ferrara, G., Mazzeo, A., Pacucci, C., Matarrese, A.M.S., Tarantino, A., Crisosto, C., Incerti, O., Marcotuli, I., Nigro, D., Blanco, A. & Gadaleta, A. (2016). Characterization of edible fig germplasm from Puglia, southeastern Italy: Is the distinction of three fig types (Smyrna, San Pedro and Common) still valid? *Scientia Horticulturae*, 205, 52–58.

Flaishman, M. A., Rodov, V., & Stover, E. (2008). The fig: Botany, horticulture, and breeding. *Horticultural Review, 34*, 113–197.

Flaishman, M. A., Peer, R., Freiman, Z. E., Izhaki, Y., & Yablovitz, Z. (2017). Conventional and molecular breeding systems in fig (*Ficus carica* L.). *Acta Horticulturae, 1173*, 1–9.

Gaaliche, B., Trad, M., & Mars, M. (2011). Effect of pollination intensity, frequency and pollen source on fig (*Ficus carica* L.) productivity and fruit quality. *Scientia Horticulturae, 130*, 737–742.

Gaaliche, B., Saddoud, O., & Mars, M. (2012). Morphological and Pomological diversity of fig (*Ficus carica* L.) cultivars in northwest of Tunisia. *ISRN Agronomy, 326461*, 9. https://doi.org/10.5402/2012/326461

Giraldo, E., López-Corrales, M., & Hormoza, J. I. (2010). Selection of the most discriminating morphological qualitative variables for characterization of fig germplasm. *Journal of the American Society for Horticultural Science, 135*, 240–249.

Gozlekci, S. (2011). Pomological traits of fig (*Ficus carica* L.) genotypes collected in the west Mediterranean region in Turkey. *The Journal of Animal & Plant Sciences, 21*(4), 646–652.

Gozlekci, S., Kafkas, E., & Ercisli, S. (2011). Volatile compounds determined by HS/GC-MS technique in peel and pulp of fig (*Ficus carica* L.) cultivars grown in Mediterranean region of Turkey. *Notulae Botanicae Horti Agrobotanici Cluj-Napoca, 9*, 105–108.

Hoogendijk, M., & Williams, D. E. (2001). Characterizing the genetic diversity of home garden crops: Some examples from the Americas. In J. W. Watson & P. B. Eyzaguirre (Eds.), *Proceedings of the second international home gardens workshop: Contribution of home gardens and in situ conservation of plant genetic resources in farming systems* (pp. 34–40). IPGRI.

Hssaini, L., Hanine, H., Razouk, R., Abderrahman, S. E., Ibtissam Guirrou, M., & Charafi, J. (2020a). Diversity screening of fig (*Ficus carica* L.) germplasm through integration of morpho-agronomic and biochemical traits. *International Journal of Fruit Science, 20*, 939–958.

Hssaini, L., Charafi, J., Razouk, R., Hernanadez, F., Fauconnier, M. L., Said, E., & Hanine, H. (2020b). Assessment of morphological traits and fruit metabolites in 11 fig varieties (*Ficus carica* L.). *International Journal of Fruit Science, 20*, 8–28.

Ilgin, M., Ergenoglu, F., & Caglar, S. (2007). Viability, germination and amount of pollen in selected caprifig types. *Pakistan Journal of Botany, 39*, 9–14.

Inal, A. (2002). Yerel çeşitlerin önemi ve korunması. T. C. Tarım ve Köyişleri Bakanlığı Ege Tarımsal Araştırma Enstitüsü Müdürlüğü, Teknik Broşür, No: 3, Menemen, İzmir.

IPGRI & CIHEAM. (2003). *Descriptors for figs*. International Plant Genetic Resources Institute (IPGRI) and the International Centre for Advanced Mediterranean Agronomic Studies (CIHEAM).

Khadivi-Khub, A., & Anjam, K. (2014). Characterization and evaluation of male fig (caprifig) accessions in Iran. *Plant Systematics and Evolution, 10*, 2177–2189.

Khan, M. R., Khan, M. A., Habib, U., Maqbool, M., Rana, R. M., Awan, S. I., & Duralija, B. (2022). Evaluation of the characteristics of native wild himalayan fig (*Ficus palmata* Forsk.) from Pakistan as a potential species for sustainable fruit production. *Sustainability, 14*, 468. https://doi.org/10.3390/su14010468

Kokaj, T. (2021). Investigation and value of qualitative and quantitative characteristics 10 of genotypes of figs. *International Journal of Food Science and Agriculture, 5*(3), 448–454.

Messaouidi, Z., & Haddadi, L. (2008). Morphological and chemical characterization of fourteen fig trees cultivated in Qulmes area, Morocco. *Acta Horticulturae, 798*, 83–86.

Mirheidari, F., Khadavi, A., Moradi, Y., & Paryan, S. (2020). Phenotypic variability of naturally grown edible fig (*Ficus carica* L.) and caprifig (*Ficus carica* var. *caprificus* Risso) accessions. *Scientia Horticulturae, 267*, 109320. https://doi.org/10.1016/j.scienta.2020.109320

Neeman, G., & Galil, J. (1978). Seed set in the 'male syconia' of the common fig *Ficus carica* L. (caprificus). *New Phytologist, 81*, 375–380.

Özen, M., Dalkılıç, Z., Kocataş, H., Belge, A., & Ertan, B. (2017). Investigation of leaf characteristics in 'Sarılop' and 'Kaba ilek' fig hybrids. *Acta Horticulturae, 1173*, 35–38.

Pourghayoumi, M., Bakhshi, D., Rahemi, M., & Jafari, M. (2012). Effect of pollen source on quantitative and qualitative characteristics of dried figs (*Ficus carica* L.) Cvs 'Payves' and 'Sabz' ın Kazerun – Iran. *Scientia Horticulturae, 147*, 98–104.

Pérez-Sánchez, R., Morales-Corts, M. R., & Gómez-Sánchez, M. A. (2016). Agro-morphological diversity of traditional fig cultivars grown in central-western Spain. *Genetika, 48*, 533–546.

Podgornik, M., Vuk, I., Vrhovnik, I., & Mavsar, D. B. (2010). A survey and morphological evaluation of fig (*Ficus carica* L.) genetic resources from Slovenia. *Scientia Horticulturae, 125*, 380–389.

Rodrigues, M. G. F., Monterio, L. N. H., Ferreira, A. F. A., Dos Santos, T. P., Pavan, B. E., Neves, V. A. B., & Boliani, A. C. (2019). Biometric characteristics among fig tree genotypes in Brazil. *Genetics and Molecular Research, 18*, gmr18191. https://doi.org/10.4238/gmr18191

Saddoud, O., Baraket, G., Chatti, K., Trifi, M., Marrakchi, M., Salhi-Hannachi, A., & Mars, M. (2008). Morphological variability of fig (*Ficus carica* L.) cultivars. *International Journal of Fruit Science, 8*(1–2), 35–51.

Salhi-Hannachi, A., Trifi, M., Zehdi, S., Hedfi, J., Mars, M., Rhouma, A., & Marrakchi, M. (2004). Inter-simple sequence repeat fingerprints to assess genetic diversity in Tunisian fig (*Ficus carica* L.) germplasm. *Genetic Resources and Crop Evolution, 51*, 269–275.

Sezen, I., Ercisli, S., & Gozlekci, S. (2014). Biodiversity of figs (*Ficus carica* L.) in Coruh valley of Turkey. *Erwerbs-obstbau, 56*, 139–146.

Shivanna, K. R. (2003). *Pollen biology and biotechnology* (p. 316). Science Publisher. https://doi.org/10.1201/9780429187704

Şimşek, M., & Yildirim, H. (2010). Fruit characteristics of the selected fig genotypes. *African Journal of Biotechnology, 9*, 6056–6060.

Simsek, M., Gulsoy, E., Kirar, M. Z., Turgut, Y., & Yucel, B. (2017). Identification and selection of some female fig (*Ficus carica* L.) genotypes from Mardin province of Turkey. *Pakistan Journal of Botany, 49*, 541–546.

Simsek, E., Kilic, D., & Caliskan, O. (2020). Phenotypic variation of fig genotypes (*Ficus carica* L.) in the Eastern Mediterranean of Turkey. *Genetika, 52*, 957–972.

Solomon, A., Golubowicz, S., Yablowicz, Z., Grossman, S., Bergman, M., Gottlieb, H., Altman, A., Kerem, Z., & Flaishman, M. A. (2006). Antioxidant activities and anthocyanin content of fresh fruits of common fig (*Ficus carica* L.). *Journal of Agricultural and Food Chemistry, 54*, 7717–7723.

Stover, E., Aradhya, M., Ferguson, L., & Crisosto, C. H. (2007). The fig: Overview of an ancient fruit. *HortScience, 42*, 1083–1087.

Stover, E., & Aradhya, M. (2008). Fig genetic resources and research at the US National Clonal Germplasm Repository in Davis, California. *Acta Horticulturae, 798*, 57–68.

Trad, M., Ginies, C., Gaaliche, B., Renard, C. M. G. C., & Mars, M. (2012). Does pollination affect aroma development in ripened fig (*Ficus carica* L.) fruit? *Scientia Horticulturae, 134*, 93–99.

Yaman, S., & Caliskan, O. (2016a). Pollinizer characteristics of some caprifig genotypes (*Ficus carica* var. *caprificus*) selected from Hatay. *Anadolu Journal of Agricultural Sciences, 31*, 315–320.

Yaman, S., & Caliskan, O. (2016b). Investigations on pollen viability and germination of some caprifig genotypes (*Ficus carica* var. *caprificus*). *Journal of Agriculture Faculty of Ege University, 53*(4), 461–467.

Chapter 8
Agronomic Strategies for Fig Cultivation in a Temperate-Humid Climate Zone

Norma Micheloud, Paola Gabriel, Juan Carlos Favaro, and Norberto Gariglio

1 Introduction

The fig (*Ficus carica* L.), apparently the earliest known cultivated fruit crop (Kislev et al., 2006), is a small deciduous tree native to Western Asia, which has been distributed and cultivated throughout the Mediterranean region, being Turkey, Egypt, Iran, Greece, Algeria, Morocco, Syria, Italy, and Spain the main producing countries (Flaishman et al., 2008). In the northern hemisphere, it is cultivated between 35 and 40° S. In the American continent, its cultivation is widespread mainly in the United States and Brazil (Nieto et al., 2007), Peru, Bolivia, and Argentina (Morton, 2013).

The species of the genus *Ficus* have fruits of little commercial interest, except for the fig (Sánchez de Lorenzo, 2009), which is considered one of the healthiest fruits associated with longevity (Trichopoulou et al., 2007) and food security (Barolo et al., 2014). In addition, this fruit plays an essential role in human nutrition due to its high vitamin, mineral, and fiber content (Venu et al., 2005). Although figs can be used for direct consumption as fresh fruits, since they are highly perishable even in refrigerated conditions (Piga et al., 1995), they are preferentially eaten dried as the main or auxiliary ingredient in many desserts (Uzundumlu et al., 2018) or as fruit powder (Khapre et al., 2015). Figs are also considered excellent for processing fruit juice and fruit tea (Lim, 2016). Dried figs can also be mixed with nuts for snacks or

N. Micheloud · N. Gariglio (✉)
ICiAgro Litoral, UNL, CONICET, FCA, Esperanza, Santa Fe, Argentina
e-mail: nmicheloud@fca.unl.edu.ar; ngarigli@fca.unl.edu.ar

P. Gabriel · J. C. Favaro
Facultad de Ciencias Agrarias, Universidad Nacional del Litoral, Esperanza, Santa Fe, Argentina
e-mail: pgabriel@fca.unl.edu.ar; jcfavaro@fca.unl.edu.ar

M. F. Ramadan (ed.), *Fig* (Ficus carica)*: Production, Processing, and Properties*, https://doi.org/10.1007/978-3-031-16493-4_8

cereals for muesli, while dried fig puree can be used as a raw material for ice cream and marmalades (Aksoy, 2017).

Although the fig may be the oldest cultivated fruit species, little is known about its production practices, and the low number of available cultivars limits its global expansion (Botti et al., 2003). However, it is worth noting that figs are relatively easy to produce in an agro-ecological way (López-Corrales & Balas, 2014) and that the fig has excellent commercial potential due to its high market prices, the increases in demand, and the health benefits of its fresh and dried fruits (Abbas et al., 2019; López-Corrales & Balas, 2014).

2 Eco-physiological Requirements of Fig Tree

The fig is a functionally dioecious species with numerous small flowers and fruits located inside a fleshy receptacle called syconium (López-Corrales & Balas, 2014). Fig growth and production are strongly dependent on climatic conditions. Generally, the fig grows best and produces high-quality fruit in the Mediterranean and dry warm-temperate climates, with hot summers and mild winters, low rainfall, and relative humidity (Gaaliche et al., 2011). To express maximum fruit yield, the fig tree also requires high radiation. However, fig trees can also grow adequately under less favorable conditions (Leonel & Tecchio, 2010), although crop production is affected by the decrease in temperature during autumn, the cold winter conditions, and rainfall mainly during fruit maturation (Flaishman et al., 2008). Under colder weather, the tree stops growth, becomes defoliated, develops a typical terminal bud, and enters a dormancy period (Kawamata et al., 2002).

The fig tree has low chilling requirements to break dormancy and tolerates light frost (Limeira Da Silva et al., 2016). However, low winter temperatures constitute a limiting factor for its commercial cultivation, being −12.2 °C is the threshold of tree death by frost, although young trees may be damaged by higher frost temperatures between −5 and −10 °C (Ferguson et al., 1990). Tolerance to cold dramatically depends on the cultivar‘s origin (López & Salazar, 2010).

In contrast, despite its deciduous characteristics, under hot climatic conditions such as those in Brazil, the tree can grow continuously and behave as evergreen because with night temperatures above 12 °C, the plant continues to grow, and fruits ripen.

The climatic conditions also affect the crop type (Breba or Main), the harvesting time, and the fruit quality. For example, in cultivars sensitive to frost and/or low winter temperature conditions, early or late frosts kill the terminal part of the current year shoots, affecting the Breba crop (Ferguson et al., 1990); consequently, Breba production is successful only under relatively moderate winters.

The annual fig growth and phenology also respond to thermal accumulation, calculated by considering a minimum basal temperature of 8 °C and a maximum temperature of 36 °C (Souza et al., 2009). For example, in the municipality of Botucatu, in the state of São Paulo, Brazil, Ferraz et al. (2020) found that the cv.

'Roxo de Valinhos' requires 167 days or 2458 degree-days from pruning to the beginning of the harvest period. Temperature conditions also affect the color and shape of the fruits. In colder climates, the pulp takes on more intense color, and the fruits are more elongated; and since the fruits do not suffer damage by cooling, they can be stored at up to −1 °C and 90–95% of relative humidity (Caliskan & Polat, 2011). On the other hand, a comparison between dried and fresh figs has shown that dried fig production is highly dependent on climatic conditions, successful in dry and warm temperate climates. In contrast, fresh figs can grow in broader ecological conditions (López-Corrales et al., 2011; Sahin, 1998).

Fig trees adapt well to various soil types, except those with poor drainage (Flaishman et al., 2008). They can be grown in heavy clays, barns, and light sands, but ideally, the soil should be well-drained and with good fertility. Fig trees are tolerant to high soil calcium content (Aksoy, 1998) and are among the few fruit trees of temperate zones with moderate resistance to salinity (Ayers & Westcost, 1985). Upon root exposure to 100 mM sodium chloride (NaCl), fig leaves remain healthy and green, and gas exchange parameters also remain relatively high (Caruso et al., 2017). Their resistance to salinity is comparable to that exhibited by olive plants (Vangelisti et al., 2019). Fig genes may thus be useful as targets for gene editing to increase salt tolerance in figs or other tree species (Vangelisti et al., 2019).

Fig trees are also resistant to drought and perform well in semi-arid climate regions. Although rainfed cultivation is widely spread, the growth and productivity of fig trees respond positively to irrigation (Tapia et al., 2003). Furthermore, the extensive root system of the fig tree allows it to be used for both soil restoration and erosion control, as well as for rainforest regeneration because of the seed rain, seed bank, and vegetation established beneath the canopies of fig trees (Guevara et al., 2004).

In fig plantations, rootstocks are not usually used because the commercial variety is usually propagated by cuttings (López-Corrales & Balas, 2014). Wood cuttings are taken from one-year-old shoots, and rooting is not a problem for growers; the main adversity is the loss of plants by soil pathogens and nematodes (Kilinç et al., 2007). Therefore, soilless cultivation in nursery production as one of the safest alternatives to reduce plant loss has expanded enormously since the 1980s (Chong & Lumis, 2000; Evans et al., 1996; Şirin et al., 2010), and studies have been conducted to adapt the use of widely available regional substrates for fig cultivation (Wright et al., 2006).

3 Climatic Conditions of the Central Area of Santa Fe and Their Suitability for Fig Cultivation

The fig is mainly cultivated in semiarid regions (Gaaliche et al., 2011; Limeira Da Silva et al., 2016). Thus, in Argentina, the main fig cultivation areas are located in the country's northwest, in the Cuyo region, and in Buenos Aires province

(Prataviera, 2003), with the predominance of three varieties: Brown Turkey, Guarinta, and Kadota. Since the production is not enough to supply the domestic market, dried figs are imported from Turkey and Chile annually. In addition, new plantations in Argentina have been adapted to high tree densities (>1000 plants per hectare), and the canopy is annually renewed by pruning (Prataviera, 2003), which allows trees to remain small, facilitating the horticultural practices.

Fig crops have also been spread as a complementary commercial activity of traditional horticultural crops in the horticultural area around Santa Fe city (31°26′ S; 60°56′ W), in the central-east region of Argentina (Travadelo et al., 2017), although the climate of this area is temperate humid mesothermal (Fig. 8.1) classified as Cfa (Köppen, 1948).

In this region, the mean annual global radiation is 16 MJ m^{-2} day^{-1}, with maximum and minimum values of 23 and 9 MJ m^{-2} day^{-1} for January and July, respectively. In the summer and winter solstice, the astronomical heliophany ranges between 14 and 10 h, respectively (García et al., 2014). The average temperatures of the warmest (January) and coolest months (July) are 26.7 °C and 11.7 °C, respectively (García et al., 2014). The absolute minimum temperature reaches an average value of −3.9 °C, and frosts occur between May and September (García et al., 2013). The availability of cold has an average value of 301 chilling hours, with a high coefficient of variation (31%) (Gariglio et al., 2012).

Precipitations have an irregular distribution, with a tendency to the Monsoon pattern (Papadakis, 1952). The annual precipitation is near 1200 mm on average. The minimum rainfall values occur during the winter (June, July, and August) with a monthly average value of 23 mm. Precipitations increase in spring and reach their

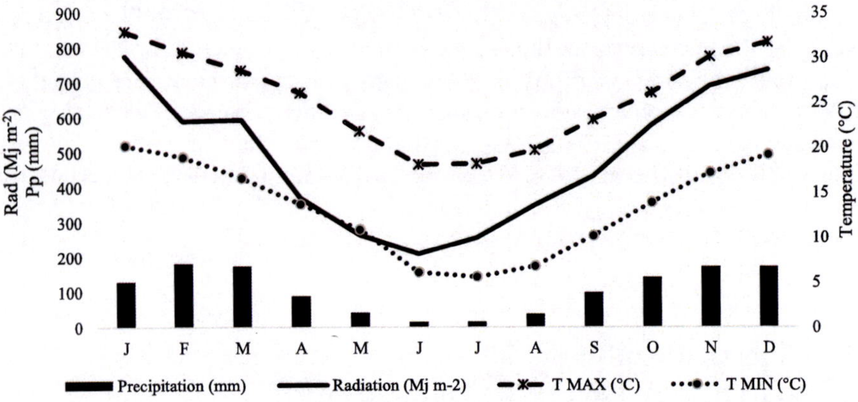

Fig. 8.1 Monthly mean values of accumulated solar radiation (Rad) and precipitation (Pp), maximum temperature (TMAX), and minimum temperature (TMIN) recorded in the central area of Santa Fe province (Argentina) during 2007–2016. (Adapted from Micheloud et al., 2018)

maximum during the summer, with a monthly average of 164 mm. The annual number of rainy days is 92 ± 10 days, being March and November the months with the highest value, and floods usually occur during February and March. The relative humidity is higher during summer, ranging between 72 and 80%, with an annual value of 76% (García et al., 2014).

The soils have subnormal to normal relief, and it is common to find vast plains with slopes less than 1%. The predominant soils are typical Argiudolls, aquic Argiudolls, and Argialbolls; the best soils in this area show a slight limitation due to the clayey texture of the subsoil (B horizon). On the other hand, the soils near the Paraná River are sandy, with slight to moderate limitations due to their low water retention capacity; however, they are very suitable for horticultural crops with drip irrigation (García et al., 2014).

4 Climatic Conditions with Adverse Effects of Humidity on the Fig Tree and Its Fruits

Figs are climacteric fruits slightly sensitive to the action of ethylene on promoting softening and decay, especially when fruits are kept over 5 °C (Gözlekçi et al., 2008). Since fruit maturation induces important structural and chemical changes the harvesting time is a determining factor of fruit quality attributes, mainly flavor, with the particularity that overripe fruits can become undesirable due to fermentative products (Martínez-González et al., 2017; Owino et al., 2006). The manifestation of these changes depends on the cultivar and the agro-ecological conditions (López-Corrales et al., 2011; Morton, 2013).

Fruit quality is also affected by high humidity and frequent rains; mainly if ripening occurs under these unfavorable conditions, in which case, fruits develop vinegary odors that affect their commercialization (Nienow et al., 2006; Sala, 2000). Furthermore, fig fruits are prone to cracking or splitting at maturity due to rains or abundant irrigation (Ferguson et al., 1990; Nienow et al., 2006) (Fig. 8.2). Fruit quality and production can also be reduced by other environmental conditions such as hail and wind (Martínez-González et al., 2017).

Very little research has been done to extend the postharvest life of fresh figs; however, low temperatures (0–2 °C) and high relative humidity (90–95%) are recommended. Fruit deterioration occurs mainly by the incidence of microbial molds and rots (Cantín et al., 2011). It has been found that figs kept at 2 °C with modified atmosphere packaging can be stored for 21 days and have better internal and external appearance than control fruits, which show significant loss of quality by dehydration (18%) (Bouzo et al., 2012).

To avoid water logging in the base of the trunk and principal roots during abundant rainfall periods and possible plant flooding damage, fig trees are cultivated on 20-cm-high ridges (Fig. 8.3). Low oxygen supply to the roots caused by soil

Fig. 8.2 Fig fruits affected by 'splitting' caused by excess water during maturation in Santa Fe (Argentina)

flooding greatly reduces the energy status of root cells, limiting the growth, development, and survival of fig trees (An et al., 2016).

5 Alternatives to Reduce the Adverse Effects of Rains During the Fruit Ripening Period

Green figs are hard, have a rubbery feel, and cannot ripen off the tree; however, fruit picked just before full ripeness will continue to soften and become soft to the touch and sweet at room temperature, and the riper the fig, the sweeter it is. Consequently, the adverse effects of rains can be diminished by harvesting the fruits for the fresh market just before full ripeness. Accordingly, at the time of full harvest and high-temperature conditions, as it occurs during summer in the central area of Santa Fe, harvesting must be carried out daily (January and February in the southern hemisphere), while when the temperature decreases toward the end of summer and autumn (March to May), the frequency between harvests may decrease up to a once a week (Gariglio et al., 2014). Therefore, the organization of this cultural practice and the estimation of its labor demand are crucial aspects of the fig crop because of its perishable fruit and long harvesting period.

The application of ethylene to advance fig fruit ripening proves suitable (Owino et al., 2006). Ethylene is considered the plant hormone responsible for fruit

Fig. 8.3 Fig trees, cv. 'Brown Turkey' cultivated on raised ridges to avoid excess water accumulation in the base of the trunk due to the abundant rainfall that commonly occurs during autumn in Santa Fe (Argentina)

ripening, and some of its physiological effects include the modification of fruit color due to the degradation of chlorophylls and the synthesis of carotenoids and flavonoids; the modification of the fruit texture by changes in the cell turgor and the cell wall structure; the modification of the concentration of sugars, organic acids and volatile compounds that affect the fruit nutritional quality, flavor and aroma; and the increase in the susceptibility to be attacked by opportunistic pathogens that are associated with the loss of integrity of the cell wall (Martínez-González et al., 2017).

Applying ethylene to fig fruits during the late stage of their development accelerates the ripening process by stimulating endogenous ethylene synthesis. As a result, treated fruits reach a greater size, have higher soluble solid content, and have better coloration, texture, and flavor, being able to be harvested 7–14 days after the application according to the temperature (Melgarejo-Moreno, 1999). Because of the fast effect of ethylene on fig fruit ripening, it is possible to advance fruit harvesting when a prolonged rainy period is predicted and consequently reduce the adverse effect on fruit quality. The treatment must be directed only to the fruits that are going to be harvested in that period (unpublished data).

6 The Extension of the Harvest Period as an Alternative to Reduce the Risk of Fruit Damage by Rains During fruit Maturation

In areas where rainfall does not have a seasonal distribution or shows high variability between years, as in the central area of Santa Fe (Argentina), an alternative to reduce the risk of fruit damage by rainfall during ripening is the extension of the harvesting period. A period of excessive rains in coincidence with a short harvest period can damage a greater proportion of the annual fruit production. In contrast, only the proportion of fruits that ripen during the rainy period will be damaged in a long harvesting strategy. Furthermore, a long harvesting period is favorable for fresh market supply. Since figs ripen sequentially along with the shoot, picking should be done repeatedly during harvest (Flaishman et al., 2008).

6.1 *Promotion of Stem Growth as a Strategy to Extend the Harvest Period*

Fig cultivars may produce one or two crops a year: the Breba crop and the Main crop. The fruit of the Breba crop is born laterally along with the shoot of the previous season, whereas the fruit of the Main crop is produced laterally along with the shoot of the current season (Flaishman et al., 2008). As a consequence of the axillary fruit location, fig yield strongly depends on the vegetative growth, so the number of fruits of each crop is closely related to shoot length (Gaaliche et al., 2011). The Breba crop depends on the length of one-year-old shoots, whereas the Main crop depends on the length of the shoot of the current season; furthermore, the Main crop indirectly affects the Breba crop for the next year (Gaaliche et al., 2011). In the warm climate region of São Paulo, Brazil, for example, since figs of the Main crop ripen sequentially along with the current shoot, longer shoots cause the extension of the harvesting period (Leonel & Tecchio, 2010). Therefore, regular and high fig tree productivity requires the promotion of shoot growth, which can be achieved by applying effective cultural practices, such as irrigation, fertilization, and pruning.

6.1.1 Irrigation

As mentioned above, fig trees adapt to different soils, being the most appropriate those of clayey-sandy texture and rich in organic matter (Pereira, 1981). Since the fig is considered a rustic plant (Rodrigues et al., 2012), the largest traditional cultivation areas in the world are located in Mediterranean arid regions under rainfed conditions (Çalişkan & Polat, 2011).

Despite its rusticity, excessively dry soil causes plants to remain resting, developing few leaves without producing fruits (Pereira, 1981). However, the fig is currently one of the fruit species with great global expansion and a favorable response to technology incorporation (Mendoza-Castillo et al., 2017). Under this context, irrigation is an important technology for achieving high yield and fruit quality (Melgarejo et al., 2007). For example, in the state of São Paulo, Brazil, irrigation of fig trees with a water supply of 75% of the Class A pan evaporation showed positive effects on fruit yield, branch length, and fruit length and diameter (Hernández et al., 1994). However, the irrigation frequency can also influence the water requirement; for example, for 2- and 4-day irrigation intervals, the adequate water supply can be reduced from a mean crop coefficient of 0.71 to 0.51, respectively (Andrade et al., 2014).

In combination with severe pruning (see below), irrigation increases the diameter of the current shoots, the number and length of internodes, and consequently, the length of shoots (Norberto et al., 1998). A good positive correlation between the irrigation rate and shoot length has also been observed in fig crops in the semi-arid north-central zone of Chile (Tapia et al., 2003).

6.1.2 Nitrogen Fertilization

The literature concerning fig tree fertilization is scarce. The N:P:K ratio commonly used in fig tree nutrition is approximately 20:5:20 (Flaishman et al., 2008). Under the traditional cultivation system for dry figs in Greece and very calcareous, clayey, and alkaline soils, Sotiropoulos et al. (2020) found that the treatment that resulted in the highest yield was the single annual application of 1 kg of N, 1.2 kg of P and 0.6 kg of K per tree, during February. These authors found that, despite the adequate fertilization with N, the trees showed a leaf N content lower than the critical value of 1.7% d.w. In contrast, leaf K concentrations were found to be sufficient (K > 1% d.w.), and leaf P content generally showed a marginal sufficiency (P > 0.1% d.w.) (Sotiropoulos et al., 2020). In another experiment under hydroponics and greenhouse conditions in Mexico, Mendoza-Castillo et al. (2019) found that the total dry matter accumulation 135 days after pruning was 12.7 t ha^{-1}, with a mineral extraction of 250 kg ha^{-1} of N, 80 kg ha^{-1} of P and 110 kg ha^{-1} of K. In this experiment, the authors generated the curves of dry matter accumulation and mineral extraction, which are essential to establish fertilization programs based on the extraction curves and crop phenology.

6.1.3 Pruning Intensity

Traditionally, the fig tree was trained in an open-vase shape with three main branches to obtain a tree less than 3 m in height (Melgarejo-Moreno, 1999). However, tree height could be reduced, and plant density could be increased by minimal pruning, which consists of the removal and/or head back of some branches (30%) to prevent the basal and internal canopy senescence and allow tree rejuvenation and shoot renovation (Lodolini et al., 2021).

In different fruit tree crops, severe pruning increases the vigor of vegetative shoots and reduces the number of flower buds (Greene, 1999). This effect is due to the production of large amounts of gibberellins in young leaves and apical meristems (Jones & Phillips, 1966), which inhibit flower induction (Cleland, 1969) and reduce the flowering intensity the following year. In the fig winter pruning stimulates the growth of the shoot, which is monopodial with the apical meristem developing leaves and inflorescences along the axis (Alvarenga-Gonçalves et al., 2006; Gaaliche et al., 2011; Lodolini et al., 2021). However, shoot growth stimulated by pruning does not affect fruit production of fig, as observed in other fruit trees; in the fig, greater shoot growth should favor greater fruit production per shoot and a longer harvesting period (Alvarenga-Gonçalves et al., 2006; Micheloud et al., 2018). This behavior tends to spread the increases in pruning intensity of the fig tree, even with an annual renewal of the canopy. In situations where this pruning and training system is adopted, it has been associated with small plant size and high plant density. Furthermore, this system has been adapted to greenhouse cultivation, allowing its large expansion in different countries such as Mexico (Mendoza-Castillo et al., 2017), Japan (Hosomi et al., 2015), and Malaysia (Shamin-Shazwan et al., 2019). In fig orchards under rainfed conditions in Iran, severe pruning intensity has also been long-term beneficial effects in reducing drought damages (Zare, 2021).

6.1.4 Other Strategies to Improve Stem Growth

Regarding the cold requirements of the fig tree, there are certain contradictions. Its buds require little or no winter chilling to break endodormancy (Ferguson et al., 1990). However, the evolution of dormancy in different cultivars under Moroccan conditions studied using the biological test (single node cuttings) showed that the removal of dormancy required the accumulation of 450 and 400 chill hours during the first and second year of study, respectively (Oukabli & Mekaoui, 2012). This cold requirement justifies the application of hydrogen cyanamide in warm areas to increase fig bud break and stimulate earlier bud break and harvest of the Breba crop (Gerber et al., 2010; Yablowitz et al., 1998). In Tunisia, the main fig cultivar of the Smyrna type grown commercially is 'Zidi', a cultivar that produces only one Main crop (Gaaliche et al., 2017). By studying this cultivar, Gaaliche et al. (2017) found that hydrogen cyanamide (1.5%) advanced bud break by 10 days and increased shoot length, fruit number per shoot, and yield. They also found that fruit ripening occurred 7 days earlier than in controls. These authors thus concluded that, in this warm area, hydrogen cyanamide could be used to improve vegetative growth and fig productivity and promotes early fruit maturity, allowing the grower to take advantage of the premium prices of the early season (Gaaliche et al., 2017).

According to the previous evidence, adopting the pruning system that causes the annual canopy renewal of the fig tree requires ensuring adequate shoot growth since the number of nodes and reproductive buds is related to shoot length (Alvarenga-Gonçalves et al., 2006; Gaaliche et al., 2011). The longer the current shoots, the greater the potential number of fruits per shoot and the longer the harvesting period.

Therefore, although the fig tree is a rustic plant that grows in arid areas, changing the training system to annual canopy renewal requires good agronomic practices and technological application.

7 Effects of Annual Renewal of the Canopy

As previously explained, the annual renewal of the canopy by pruning is a simple strategy to promote shoot growth and extend the fig's harvesting period. Under the traditional pruning and training system of the Mediterranean area and rainfed conditions, the current shoots of the fig tree have an annual growth of about 10 cm (Ateyyeh & Sadder, 2006), and trees can produce two crops a year (Breba and Main) depending on the cultivar (Gaaliche et al., 2011). In the temperate climate conditions of the central area of Argentina, in both the 'Brown Turkey' and 'Guarinta' cultivars, the annual renewal of the canopy causes the loss of the Breba crop and dramatically increases the current shoot length over 139 cm, without affecting the fruit set (Micheloud et al., 2018). Under the tropical conditions of Minas Gerais, Brazil, and a severe pruning system, the cv. 'Roxo de Valinhos' develops one fruit per node, reaching an annual shoot growth of 189 cm (Alvarenga-Gonçalves et al., 2006). In Japan, in the cv. 'Masui Dauphine' under greenhouse cultivation, one fruit is borne on every node of the shoots, and the harvesting period is extended from the base to approximately the 20th node (August to November in the northern hemisphere) (Hosomi et al., 2013). Under Mediterranean climate conditions and light pruning, the percentage of nodes with fruit in the Main crop of five fig varieties ranged between 50.2 and 88.5% (Gaaliche et al., 2011), whereas in the temperate climate conditions of central Argentina and intense pruning, the percentage of nodes with fruit ranged between 45.0 and 49.9% (Micheloud et al., 2018). Concerning harvest distribution, in this region of Argentina and under severe pruning, it is abundant from January to the end of March (Micheloud et al., 2018) and declines during April and May (Fig. 8.4).

In the fig, the annual duration of the harvest period ranges from 8 to 21 weeks, with modification of its ending but not of its beginning date when different years are compared. Furthermore, the duration of harvesting shows a great linear relationship with the annual fruit yield, which, in the central area of Santa Fe, Argentina, ranges between 4.43 and 12.1 t ha^{-1} (Micheloud et al., 2018). This fruit yield is in accordance with that observed in different areas of Brazil using the cv. 'Roxo de Valinhos' and severe winter pruning (Leonel & Tecchio, 2010; Nienow et al., 2006).

The duration of the harvesting period and fruit yield are also affected by tree age. In Santa Fe, the longest harvesting period was observed in the fourth growing season after planting, with a later decline (Micheloud et al., 2018). This effect was not observed under tropical conditions of Minas Gerais, Brazil, where fruit production and the duration of the harvesting period are stabilized after the fifth year of plantation (Alvarenga-Gonçalves et al., 2006). The duration of the harvesting period is also affected by meteorological conditions, decreasing one week for every 5–6 rainy

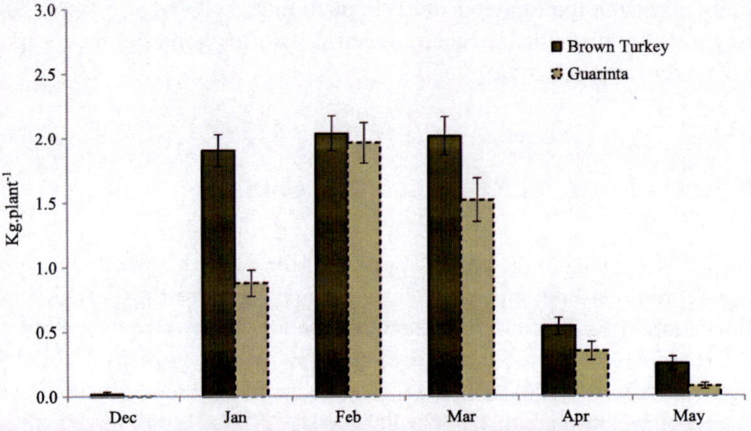

Fig. 8.4 Monthly distribution of the annual fruit harvest (Kg plant^{-1}) of two fig (*Ficus carica* L.) cultivars, 'Brown Turkey' and 'Guarinta', under a severe pruning system with annual renewal of the canopy in the central-east area of Santa Fe, Argentina. The vertical lines indicate the standard error. Data are the mean values of ten years of experimentation. (Adapted from Micheloud et al., 2018)

days (>75 mm of accumulated precipitations) from January to May (Micheloud et al., 2018).

It is possible that, under severe pruning, the development of the root system makes the fig tree too vigorous, with a later fruit set that is unable to reach maturity under temperate climate conditions. Thus, reducing the harvest period occurs not because of the absence of fruits but because of the lack of ripeness (Micheloud et al., 2018). In contrast, the ripening of later fruits normally occurs under tropical climate conditions (Alvarenga-Gonçalves et al., 2006), explaining the different evolution of fruit yield and the duration of the harvesting period with tree age under severe pruning intensity under temperate and tropical climates. Furthermore, the fig fruit has the particularity that, to ripen successfully, it needs high temperatures, preferably above 30 °C, which explains why it is common to observe a large number of immature fruits during the autumn period in temperate zones (Lavín & Reyes, 2004). It also explains why the harvest period ends despite fruits on the plant.

8 Fig Production Strategies

Fig trees can be adapted to marginal conditions and tolerant of high soil calcium content, salinity, and drought (Golombek & Lüdders, 1990). As a result, figs have recently attracted a great deal of attention and are becoming widespread worldwide.

In Asia, several countries such as China, Japan, Thailand, Indonesia, and Malaysia have started to plant figs (Isa et al., 2020). In Mexico, fig cultivation has also increased both under open field and greenhouse conditions because it responds favorably to agronomic management (Mendoza-Castillo et al., 2017) and provides economic viability to the rural sector (Macías et al., 2014). However, since open field production in Mexico has a higher incidence of pests and diseases, fruit rotting during the rainy season, and frost damage, in comparison with production under greenhouse conditions, the latter is becoming more expanded than the former (Mendoza-Castillo et al., 2017).

Currently, contrasting strategies can achieve fig production (Table 8.1). One of these strategies is the traditional system of semi-arid regions, mainly used in Mediterranean countries, characterized by open field cultivation, rainfed conditions, and low tree density (156 pl ha^{-1}). Other features of this system are a slight pruning, one (Main) or two (Breba and Main) crops per year according to the variety, low shoot growth around the year (6.8–to 14.2 cm), and several shoots per plant (non-controlled), three (Breba crop) to six (Main crop) fruits per shoot, and a fruit yield of 7–8 t ha^{-1} for the Breba crop (Gaaliche et al., 2011) and 15 t ha^{-1} for the Main crop (Melgarejo-Moreno, 1999). In addition, the use of different organic and inorganic mulches under rainfed conditions improves soil moisture content, the number, and width of leaves, shoot length and diameter, and fruit quality, and results in milder and more stabilized soil temperature throughout the year (Jafari et al., 2012).

Figs can also be cultivated under an intensive system in greenhouse conditions, where plants are spaced at 1.6 × 0.5 m (12,500 plants ha^{-1}) using 40-L containers per plant, substrate (soilless cultivation), and conducted with 3–8 stems per plant (6 shoots on average with 20 fruits per shoot), with severe pruning and annual renewal of the canopy. In this strategy, the nutritive solution is applied by drip irrigation during the whole growing season. In Chapingo, Mexico, by using the native cultivar 'Netzahualcoyotl', the potential yield of this production system is over 60 t ha^{-1}. This fresh fruit yield results up to 20 times higher than that achieved by open field plantations in the same area of Mexico. Total dry mass production reaches near 30 t ha^{-1}, and dry matter partition to fruit ranges from 38 to 57% between extreme treatments (3 and 7 shoots per plant, respectively). In the northern hemisphere, the harvesting period lasts 3 months, from July to September (Mendoza-Castillo et al., 2017).

In Japan, fig cultivation under greenhouse is widespread using the "straight line training" system, adopted mainly for the cv. 'Masui Dauphine'. The system involves bilateral horizontal limbs of 40 cm height with shoots growing upward from them. Shoots are cut off at the end of the growing season, similar to that performed in grapevines. This system greatly improves cultural practices such as pruning, disbudding, and harvesting (Hosomi et al., 2013).

Finally, in the central area of Santa Fe and other regions of Argentina, growers adopt an intermediate fig production system (Table 8.1). This system is characterized by open field conditions with complementary drip irrigation, a tree density near

Table 8.1 Description of different production systems for the fig tree crop

	Traditional	Intermediate	Intensive
Location	M'hamdia (Tunisia)	Santa Fe (Argentina)	Chapingo (Mexico)
Coordinates	35°55′N, 10°34′E, 18 masl	31°26′ S, 60°56′ W, 20 masl	19°20′N, 98°53′W, 2240 masl
Climate	Semi-arid. Mild winter and hot, dry summer.	Temperate humid with no dry season	–
Annual precipitation (mm)	439	1100	–
Soil	Silty-clay, saline	Sand-loamy	Soilless (40 l plant^{-1} container)
Cultivation methods	Open field	Open field	Greenhouse
Irrigation	Rainfed	Complementary drip irrigation	Drip irrigation with nutrient solution
Cultivar	'Bither Abiadh', 'Besbessi', 'Khedhri', 'Chetoui', 'Zidi'	'Brown Turkey', 'Guarinta'	'Netzahualcoyotl'
Crops per year	2 (Breba and Main crops) (San Pedro cvs.) 1 (Main crop) (Smyrna cvs.).	1 (Main crop)	1 (Main crop)
Plant density	8.0 × 8.0 m (156 pl ha^{-1})	4.0 × 2.5 m (1000 pl ha^{-1})	1.6 × 0.5 m (12,500 pl ha^{-1})
Plant age	15	7	–
Shoots per plant	Non-controlled	25–30	3–8
Shoot length (cm)	6.8–14.2	160	–
Fruit per stem	3 (Breba crop) 6 (Main crop)	25	20
Pruning system	Light	Annual renewal of the canopy	Annual renewal of the canopy
Dry matter production (t ha^{-1})	–	–	15.25–29.6
Dry matter partition to fruits (%)	–	–	38–57%
Fruit production (t ha^{-1})	7–8 (Breba crop) 15 (Main crop)	7–15	40.5–109.5 (60 on average)
Fruit harvest period (months)	2 (June-July; Breba; north hemisphere) 3 (July-September; Main crop; north hemisphere)	4–5 (January-May; south hemisphere). Induction of last fruit maturation during autumn (May).	3 (July-September; north hemisphere)
Fresh fruit weight (g fruit^{-1})	–	–	52.37–58.45
Fruit length (cm)	–	–	6.57–7.56

Traditional: Adapted from Gaaliche et al. (2011); Intermediate: adapted from Micheloud et al. (2018); Intensive: Adapted from Mendoza-Castillo et al. (2017)

Fig. 8.5 Fig crop for fresh market and industry after winter pruning with an annual canopy renewal in central Argentina. Tree density 6 × 3 m. (Photo: Obrador, C.)

1000 pl ha^{-1} (4.0 × 2.0 m; 4.0 × 2.5 m), severe pruning with annual renewal of the canopy, 25–30 shoots per plant, 25 fruits per shoot, and a fruit yield of 7–15 t ha^{-1}. In addition, the harvesting period is extended from December to May (south hemisphere) (Micheloud et al., 2018). However, the fig tree responds favorably to an increase in nitrogen fertilization, with farms that achieve commercial yields up to 35 t ha^{-1}, even with a lower tree density (6 × 3 m, 555 pl ha^{-1}) (Fig. 8.5).

Unlike many other fruit trees (Berman & Dejong, 2003; Bussi et al., 2005; Mahmood et al., 2000), the fig tree has the particularity that vegetative growth does not affect the fruit set of the Main crop (Hosomi et al., 2015). Consequently, the fig tree has great reproductive stability under different growth conditions, and, according to our observations, even an increase in the tree vigor does not delay the beginning of harvesting.

8.1 *Stimulation of Fruit Growth and Ripening*

Since the third Century BC, it has been known that fig growth and ripening could be stimulated by applying olive oil and other vegetable oils to the ostiole of the fruit (oleification) (Chessa, 1997; Hirai et al., 1966) because this induces ethylene production. Furthermore, fruit growth and earlier ripening can also be stimulated by auxin (Miller et al., 1987) and ethephon. However, it has been recently demonstrated that ethylene production in response to these compounds is induced by 1-methyl cyclopropane and inhibited by propylene, indicating a negative feedback regulation mechanism (Owino et al., 2006).

These techniques used for the anticipation of fig fruit ripening have been used for the Breba crop because the advance of its maturity is justified by the higher prices of the earliest fruit in the fresh market; furthermore, the uniformity of maturity reduces the labor costs of harvesting (Melgarejo-Moreno, 1999).

It is essential to consider that, in the temperate climate conditions of Santa Fe, Argentina, near eight fruits per shoot are not able to reach maturity because they set later in the growing season (Micheloud et al., 2018), and the fig requires lots of heat units to reach maturity and achieve good fruit quality (Stover et al., 2007), limiting fig cultivation in low-temperature regions. Furthermore, these late fruits that remain on the plant are later affected by frost and then abort, resulting in a considerable loss of fruit yield because they represent up to one-third of the total fruit sets annually by the fig tree (Main crop).

The stimulation of fruit growth and ripening used traditionally for the Breba crop has been evaluated with satisfactory results in a temperate climate to ripen the fruits that set later and remain green at the end of the growing season, increasing fruit yields under a severe pruning system. Under unfavorable autumn temperature conditions, olive oil and ethephon (200–400 ppm) can ripen the fruits, allowing them to reach similar size and weight but lower total soluble solid content (°Brix) than fruits that ripen early during normal temperature conditions (unpublished data). Ethephon is applied to the plant but directed to the most basal fruits of each current shoot. A weekly application is carried out for 3 weeks, and one-third of the remaining fruits are treated at each application. Treated fruits ripen during the week of treatment, previous to the next ethephon application (unpublished data).

9 Conclusions

In temperate humid areas, such as Santa Fe, in central Argentina, fig tree cultivation for the fresh market can be carried out successfully. The main strategies that have been proved to be more effective in obtaining high-quality fruit and yield are; (i) high-density plantations (>1000 pl ha^{-1}); (ii) severe pruning with annual canopy renewal; (iii) stem growth stimulation through irrigation and fertilization, which allows a long harvest period of the Main crop (3–5 months); (iv) fruit ripening stimulation by fruit-localized ethephon application either to advance fruit harvest prior to a rainy period or for the ripening of the fruits that remain green on the plant at the end of the growing season, and (v) high-technology application such as greenhouse cultivation with plantations over 10,000 pl ha^{-1} and fruit yield that could exceed 100 t ha^{-1}. However, further research on fig cultivation in temperate-humid areas should be carried out to adjust these incipient production strategies.

References

Abbas, M., Hussain, S., Ahmad, M., Junaid, M., Ali, Y., Alamgir, M., & Jalal, A. (2019). Testing the performance of fig (*Ficus carica*) cultivars planted in high density system under agro-climatic condition of Islamabad. *Scholars Bulletin, 5*(2), 37–41. Available at https://www.saudijournals.com/media/articles/SB_52_37-41_c.pdf

Aksoy, U. (1998). Why figs? An old taste and a new perspective. *Acta Horticulturae, 480*, 25–27.

Aksoy, U. (2017). The dried fig management and the potential for new products. *Acta Horticulturae, 1173*, 377–382. https://doi.org/10.17660/ActaHortic.2017.1173.65

Alvarenga-Gonçalves, C. A., De Oliveira-Lima, L. C., Nascimento-Lopes, P. S., & Teixeira De Souza, M. (2006). Poda e sistemas de condução na produção de figos verdes. *Pesquisa Agropecuária Brasileira, 41*, 955–961. https://doi.org/10.1590/S0100-204X2006000600009

An, Y., Qi, L., & Wang, L. (2016). ALA pre-treatment improves waterlogging tolerance of fig plants. *PLoS One, 11*(1), e0147202. https://doi.org/10.1371/journal.pone.0147202

Andrade, I. P. S., Carvalho, D. F., Almeida, W. S., Silva, J. B. G., & Silva, L. D. B. (2014). Water requirement and yield of fig trees under different drip irrigation management. *Eng Agríc Jaboticabal, 34*(1), 17–27. https://doi.org/10.1590/S0100-69162014000100003

Ateyyeh, A. F., & Sadder, M. T. (2006). Growth pattern and fruit characteristics of six common fig (*Ficus carica* L.) cultivars in Jordan. *Jordan Journal of Agricultural Sciences, 2*(2), 105–112. Available at https://journals.ju.edu.jo/JJAS/article/view/1261/5957

Ayers, S. R., & Westcost, W. D. (1985). *Water quality for agriculture* (FAO irrigation and drainage paper 29). Organization of the United Nations. Available at https://www.fao.org/3/t0234e/t0234e00.htm

Barolo, M. I., Ruiz-Mostacero, N., & López, S. N. (2014). *Ficus carica* L. (Moraceae): An ancient source of food and health. *Food Chemistry, 164*, 119–127. Available at https://www.academia.edu/8882646/Ficus_carica_L_Moraceae_An_ancient_source_of_food_and_health

Berman, M. E., & Dejong, T. M. (2003). Seasonal patterns of vegetative growth and competition with reproductive sinks in peach (*Prunus persica*). *Journal of Horticultural Science and Biotechnology, 78*(3), 303–309. https://doi.org/10.1080/14620316.2003.11511622

Botti, C., Franck, N., Prat, L., & Ioannidis, D. (2003). The effect of climatic conditions on fresh fig fruit yield, quality and type of crop. *Acta Horticulturae, 605*, 37–42. https://doi.org/10.17660/ActaHortic.2003.605.4

Bouzo, C. A., Travadelo, M., & Gariglio, N. F. (2012). The effect of different packaging materials on postharvest quality of fresh fig fruit. *International Journal of Agriculture and Biology, 14*, 821–825.

Bussi, C., Lescourret, F., Genard, M., & Habib, R. (2005). Pruning intensity and fruit load influence vegetative and fruit growth in an early-maturing peach tree (cv. Alexandra). *Fruits, 60*(2), 133–142. https://doi.org/10.1051/fruits:2005017

Çalişkan, O., & Polat, A. (2011). Phytochemical and antioxidant properties of selected fig (*Ficus carica* L.) accessions from the eastern Mediterranean region of Turkey. *Scientia Horticulturae, 128*(4), 473–478. https://doi.org/10.1016/j.scienta.2011.02.023

Cantín, C. M., Palou, L., Bremer, V., Michailides, T. J., & Crisosto, C. H. (2011). Evaluation of the use of sulfur dioxide to reduce postharvest losses on dark and green figs. *Postharvest Biology and Technology, 59*, 150–158. https://doi.org/10.1016/j.postharvbio.2010.09.016

Caruso, G., Gennai, C., Ugolini, F., Marchini, F., Quartacci, M. F., & Gucci, R. (2017). Tolerance and physiological response of young *Ficus carica* L. plants irrigated with saline water. *Acta Horticulturae, 1173*, 137–141. https://doi.org/10.17660/ActaHortic.2017.1173.23

Chessa, I. (1997). Fig. In S. Mitra (Ed.), *Postharvest physiology and storage of tropical and subtropical fruits* (pp. 245–268). CAB International.

Chong, C., & Lumis, G. P. (2000). Mixtures of paper mill sludge, wood chips, bark, and peat in substrates for pot-in-pot shade tree production. *Plant Science, 80*, 669–675. https://doi.org/10.4141/P99-083

Cleland, R. E. (1969). The gibberellins. In M. B. Williams (Ed.), *Physiology of plant growth and development* (pp. 49–81). McGraw Hill.

Evans, M. R., Konduru, S., & Stamps, R. H. (1996). Source variation in physical and chemical properties of coconut coir dust. *HortScience, 31*, 965–967. Available at http://www.planetcococoir.com/docs/Source-Variation-in-Physical.pdf

Ferguson, L., Michailides, T. J., & Shorey, H. H. (1990). The California fig industry. *Horticultural Reviews, 12*, 409–490.

Ferraz, R. A., Leonel, S., Azevedo-Souza, J. M., Ferreira, R. B., Modesto, J. H., & Lencioni-Arruda, L. (2020). Phenology, vegetative growth, and yield performance of fig in Southeastern Brazil. *Pesquisa Agropecuária Brasileira, 55*, e01192. https://doi.org/10.1590/S1678-3921.pab2020.v55.01192

Flaishman, M. A., Rodov, V., & Stover, E. (2008). The fig: Botany, horticulture, and breeding. In J. Janick (Ed.), *Horticultural reviews 34* (pp. 113–197). Willey.

Gaaliche, B., Lauri, P., Trad, M., Costes, E., & Messaoud, M. (2011). Interactions between vegetative and generative growth and between crop generations in fig tree (*Ficus carica* L.). *Scientia Horticulturae, 131*, 22–28. https://doi.org/10.1016/j.scienta.2011.09.022

Gaaliche, B., Ghrab, M., & Mimoun, M. B. (2017). Effect of Hydrogen Cyanamide on vegetative growth, yield, and fruit quality of fig cv. Zid in a warm production area. *International Journal of Fruit Science, 17*(1), 63–71. https://doi.org/10.1080/15538362.2016.1202169

García, M. S., Micheloud, N. G., Leva, P. E., Tóffoli, G., Gariglio, N. F., & Pilatti, R. A. (2013). Efecto de las heladas tardías en la producción de durazneros precoces cultivados en la región central de la provincia de Santa Fe (Argentina). *Revista FAVE Sección Ciencias Agrarias, 12*(1-2), 135–143. https://doi.org/10.14409/fa.v12i1/2.5123

García, M. S., Leva, P. E., Toffoli, G., Pilatti, M. A., & Alesso, C. A. (2014). Caracterización agroclimática y edafológica del centro de la provincia de Santa Fe. In N. F. Gariglio, C. A. Bouzo, & M. R. Travadelo (Eds.), *Cultivos frutales y ornamentales para zonas templado-cálidas. Experiencias en la zona central de Santa Fe* (pp. 9–23). Ediciones UNL. Available at https://www.researchgate.net/publication/273960524_Cultivos_frutales_y_ornamentales_para_zonas_templado-calidas_Experiencias_en_la_zona_central_de_Santa_Fe

Gariglio, N. F., Weber, M., Castro, D., & Micheloud, N. (2012). Influence of the environmental conditions, the variety, and different cultural practices on the phenology of peach in the central area of Santa Fe (Argentina). In X. Zhang (Ed.), *Phenology and climatic change* (pp. 217–240). InTech. Available at http://cdn.intechopen.com/pdfs/32931/InTech-Influence_of_the_environmental_conditions_the_variety_and_different_cultural_practices_on_the_phenology_of_peach_in_the_central_area_of_santa_fe_argentina_.pdf

Gariglio, N. F., Favaro, J. C., & Forte, R. (2014). Higuera. In N. F. Gariglio, C. A. Bouzo, & M. R. Travadelo (Eds.), *Cultivos frutales y ornamentales para zonas templado-cálidas. Experiencias en la zona central de Santa Fe.* (pp. 109–127). Ediciones UNL. Available at https://www.researchgate.net/publication/273960524_Cultivos_frutales_y_ornamentales_para_zonas_templado-calidas_Experiencias_en_la_zona_central_de_Santa_Fe

Gerber, H. J., Steyn, W. J., & Theron, K. I. (2010). The optimum one-year-old shoot length to optimize yield and fruit size of three fig (*Ficus carica* L.) cultivars. *HortScience, 45*, 1321–1326. https://doi.org/10.21273/HORTSCI.45.9.1321

Golombek, S. D., & Lüdders, P. (1990). Gas exchange of *Ficus carica* in response to salinity. In M. L. Van Beusichem (Ed.), *Plant nutrition, physiology and applications* (pp. 487–493). Springer.

Gözlekçi, S., Erkan, M., Karaşahin, I., & Şahin, G. (2008). Effect of 1-MethylCycloPropene (1-MCP) on fig (*Ficus carica* cv. Bardakci) storage. *Acta Horticulturae, 798*, 325–330. https://doi.org/10.17660/ActaHortic.2008.798.47

Greene, D. W. (1999). Tree growth management and fruit quality of apple trees treated with prohexadione-calcium (BAS 125). *HortScience, 34*, 1209–1212. Available at file:///D:/Docs/Downloads/_journals_hortsci_34_7_article-p1209-preview.pdf.

Guevara, S., Laborde, J., & Sánchez-Ríos, G. (2004). Rain Forest regeneration beneath the canopy of fig trees isolated in pastures of Los Tuxtlas, Mexico. *Biotropica, 36*(1), 99–108. Available at http://www.jstor.org/stable/30043093

Hernández, F. B. T., Suzuki, M. A., Buzetti, S., & Corrêa, L. S. (1994). Resposta da figueira (*Ficus carica* L.) ao uso da irrigação e nitrogênio na região de Ilha Solteira. *Scientia Agricola, 51*(1), 99–104. https://doi.org/10.1590/S0103-90161994000100016

Hirai, J., Hirata, N., & Tada, H. (1966). Effect of oleification on hastening the maturity of the fig fruits. I. On time of application and kinds of oils. *Journal of the Japanese Society for Horticultural Science, 35*, 354–360.

Hosomi, A., Isobe, T., & Miwa, Y. (2015). Shoot growth and fruit production of the 'Masui Dauphine' variety of fig (*Ficus carica* L.) undergoing renewal long pruning. *JARQ, 49*(4), 391–397. https://doi.org/10.6090/jarq.49.391

Hosomi, A., Miwa, Y., & Mano, T. (2013). Shoot growth and fruit production of 'Masui Dauphine' fig trees having high limb position with downward shoots. *Journal of the Japanese Society for Horticultural Science, 82*(3), 215–221. https://doi.org/10.2503/jjshs1.82.215

Isa, M. M., Jaafar, M. N., Kasim, K. F., & Mutalib, M. F. A. (2020). Cultivation of Fig (*Ficus carica* L.) as an alternative high value crop in Malaysia: A brief review. *Materials Science and Engineering, 864*, 012134. https://doi.org/10.1088/1757-899X/864/1/012134

Jafari, M., Haghighi, J. A. P., & Zare, H. (2012). Mulching impact on plant growth and production of rainfed fig orchards under drought conditions. *Journal of Food, Agriculture & Environment, 10*(1), 428–433.

Jones, R. L., & Phillips, D. J. (1966). Organs of gibberellin synthesis in light-grown sunflower plants. *Plant Physiology, 41*, 1381–1386.

Kawamata, M., Nishida, E., Ohara, H., Ohkawa, K., & Matsui, H. (2002). Changes in the intensity of bud dormancy and internal compositions of current shoot in fig. *Journal of the Japanese Society for Horticultural Science, 71*, 177–182. https://doi.org/10.2503/jjshs.71.177

Khapre, A. P., Satwadhar, P. N., & Syed, H. M. (2015). Studies on processing technology and cost estimation of fig (*Ficus carica* L.) fruit powder enriched Burfi (Indian cookie). *Journal of Applied and Natural Science, 7*(2), 621–624.

Kilinç, S. S., Ertan, E., & Seferoglu, S. (2007). Effects of different nutrient solution formulations on morphological and biochemical characteristics of nursery fig trees grown in substrate culture. *Scientia Horticulturae, 113*, 20–27. https://doi.org/10.1016/j.scienta.2007.01.032

Kislev, M. V., Hartmann, A., & Bar-Yosef, O. (2006). Early domesticated fig in Jordan Valley. *Science, 312*, 1372–1374. https://doi.org/10.1126/science.1125910

Köppen, W. (1948). *Climatología: un estudio de los climas de la tierra.* Trad. P. H. R. Pérez. Fondo de Cultura Económica.

Lavín, A., & Reyes, M. (2004). Higuera (*Ficus carica* L.). In A. Lavín, & K. Matsuya (Eds.), *Frutales: especies con potencial en el Secano Interior. Boletín INIA n° 120.* (pp. 82–94). Ed. INIA. Available at https://frutales.files.wordpress.com/2011/01/g-26-frutales-especies-con-potencial.pdf

Leonel, S., & Tecchio, M. A. (2010). Épocas de poda e uso da irrigação em figueira 'Roxo de Valinhos' na região de Botucatu, SP. *Bragantia, 69*(3), 571–580. https://doi.org/10.1590/S0006-87052010000300008

Lim, T. K. (2016). *Edible medicinal and non-medicinal plants*. Springer.

Limeira Da Silva, F., De Araújo Viana, T. V., Gomes De Sousa, G., Caminha Costa, S., & Moreira De Azevedo, B. (2016). Yield of common fig fertigated with bovine biofertilizer in the semi-arid region of Ceará. *Revista Caatinga, 29*, 425–434. https://doi.org/10.1590/1983-21252016v29n219rc

Lodolini, E. M., Ferlito, F., & Neri, D. (2021). Pruning fig (*Ficus carica* L.) during the early stages after planting. *Acta Horticulturae, 1310*, 193–198. https://doi.org/10.17660/ActaHortic.2021.1310.29

López-Corrales, M., & Balas, F. (2014). La higuera. In J. J. Hueso-Martín, & J. Cuevas-González (Eds.), *La fruticultura del siglo XXI en España* (pp. 241–260). Cajamar Caja rural. Available at https://www.publicacionescajamar.es/publicacionescajamar/public/pdf/series-tematicas/agricultura/la-fruticultura-del-siglo-xxi-en-espana-2.pdf

López-Corrales, M., Gil, M., Pérez, F., Cortés, J., Serradilla, M. J., & Chomé, P. M. (2011). *Variedades de higuera. Descripción y registro de Variedades*. Ministerio de Medio Ambiente y Medio Rural y Marino.

López, D. M., & Salazar, A. (2010). *Estados fenológicos de la higuera*. Available at http://www.afrasa.es/utilidades/estados-fenologicos/id_13/estados-fenolgicos-de-la-higuera

Macías, R. H., Velásquez, M. A., Villa, M. M., Rivera, M., & Muñoz, A. (2014). Evaluación experimental en higuera para implementar poda inicial en altas densidades de plantación con macrotúnel en la región lagunera. *AGROFAZ, 14*(2), 33–38.

Mahmood, K., Carew, J., Hadley, P., & Battey, N. (2000). The effect of chilling and post-chilling temperatures on growth and flowering of sweet cherry (*Prunus avium* L.). *Journal of Horticultural Science & Biotechnology, 75*(5), 598–601. https://doi.org/10.1080/14620316.2000.11511292

Martínez-González, M. E., Balois-Morales, R., Alia-Tejacal, I., Cortes-Cruz, M. A., Palomino-Hermosillo, Y. A., & López-Gúzman, G. G. (2017). Poscosecha de frutos: Maduración y cambios bioquímicos. *Revista Mexicana de Ciencias Agrícolas, 19*, 4075–4087. Available at https://www.redalyc.org/pdf/2631/263153823018.pdf

Melgarejo-Moreno, P. (1999). Frutales en zonas áridas. El cultivo de la Higuera (*Ficus carica* L.). A. Madrid Vicente, Ediciones. Available at http://dspace.umh.es/bitstream/11000/4945/1/HIGUERA.pdf

Melgarejo, P., Martínez, J. J., Hernández, F., Salazar, D. M., & Martínez, R. (2007). Preliminary results on fig soil-less culture. *Scientia Horticulturae, 111*, 255–259. https://doi.org/10.1016/j.scienta.2006.10.032

Mendoza-Castillo, V. M., Pineda-Pineda, J., Vargas-Canales, J. M., & Hernández-Arguello, E. (2019). Nutrition of fig (*Ficus carica* L.) under hydroponics and greenhouse conditions. *Journal of Plant Nutrition, 42*(11-12), 1350–1365. https://doi.org/10.1080/01904167.2019.1609510

Mendoza-Castillo, V. M., Vargas-Canales, J. M., Calderón-Zavala, G., Mendoza-Castillo, M. D. C., & Santacruz-Varela, A. (2017). Intensive production systems of fig (*Ficus carica* L.) under greenhouse conditions. *Experimental Agriculture, 53*, 339–350. https://doi.org/10.1017/S0014479716000405

Micheloud, N., Favaro, J. C., Castro, D., Buyatti, M., Favaro, M. A., García, M. S., & Gariglio, N. (2018). Fig production under an intensive pruning system in the moist central area of Argentina. *Scientia Horticulturae, 234*, 261–266. https://doi.org/10.1016/j.scienta.2018.02.035

Miller, A. N., Walsh, C. S., & Cohen, J. D. (1987). Measurement of indole-3-acetic acid in peach fruits (*Prunus persica* L. Batsch cv. Redhaven) during ripening. *Plant Physiology, 83*, 491–494.

Morton, J. F. (2013). *Fruits of warm climates*. Echo Point Books & Media, LLC.

Nienow, A. A., Chaves, A., Lajús, C. R., & Calvete, E. O. (2006). Produção da figueira em ambiente protegido submetida a diferentes épocas de poda e número de ramos. *Revista Brasileira de Fruticultura, 28*, 421–424. https://doi.org/10.1590/S0100-29452006000300018

Nieto, C., Jarrin, P., & Pinto, N. (2007). *El Higo. Manual de producción uso y aprovechamiento*. Available at http://repositorio.educacionsuperior.gob.ec/bitstream/28000/946/1/L--SENESCYT-0070.pdf

Norberto, P. M., Chalfun, N. N. J., Pasqual, M., Delly-Veiga, R., & Mota, J. H. (1998). Efeito de época de poda, cianamida hidrogenada e irrigação na produção antecipada de figos verdes. *Revista Brasileira de Fruticultura, 36*, 1363–1369. https://doi.org/10.1590/S0100-204X2001001100006

Oukabli, A., & Mekaoui, A. (2012). Dormancy of fig cultivated under Moroccan conditions. *American Journal of Plant Sciences, 2*(3), 473–479. https://doi.org/10.4236/ajps.2012.34056

Owino, W. O., Manabe, Y., Mathooko, F. M., Kubo, Y., & Inaba, A. (2006). Regulatory mechanisms of ethylene biosynthesis in response to various stimuli during maturation and ripening in fig fruit (*Ficus carica* L.). *Plant Physiology and Biochemistry, 44*(5-6), 335–342. https://doi.org/10.1016/j.plaphy.2006.03.009

Papadakis, J. (1952). Mapa ecológico de la República Argentina. In A. Amols (Ed.), *Geografía económica Argentina* (pp. 177–206). Editorial Kapeluz.

Pereira, F. M. (1981). *Cultura da figueira*. Livroceres.

Piga, A., D'Aquino, S., Agabbio, M., & Pappoff, C. (1995). Influenza del confezionamento con film plastici sulla conservazione del fico. *Italus Hortus, 2*, 3–7.

Prataviera, A. (2003). Una producción alternativa en marcha. El cultivo de la Higuera. *Idia, XXI, 5*, 142–126. Available at http://www.biblioteca.org.ar/libros/210394.pdf

Rodrigues, D. N. B., Viana, T. V. A., Marinho, A. B., Ferreira, T. T. S., Azevedo, B. M., & Gomes-Filho, R. R. (2012). Fertirrigação potássica na cultura da figueira no semiárido Cearence. *Revista Brasileira de Agricultura Irrigada, 6*(3), 176–183. https://doi.org/10.7127/rbai.v6n300082

Sahin, N. (1998). Fig adaptation studies in Western Turkey. *Acta Horticulturae, 480*, 61–73. https://doi.org/10.17660/ActaHortic.1998.480.8

Sala, F. (2000). *El cultivo de la higuera breval*. Ministerio de Agricultura.

Sánchez de Lorenzo, J. C. (2009). *Catálogo descriptivo e ilustrado. Obtenido de Aportación al conocimiento del género Ficus (Moraceae) en Murcia*. Available at http://www.ayto-murcia.es/medio-ambiente/parquesyjardines/material/EL%20GENERO%20FICUS%20EN%20MURCIA.pdf

Shamin-Shazwan, K., Shahari, R., Amri, C. N. A. C., & Tajuddin, N. S. M. (2019). Figs (*Ficus carica* L.): Cultivation method and production based in Malaysia. *Engineering Heritage Journal (GWK), 3*(2), 6–8. https://doi.org/10.26480/gwk.02.2019.06.08

Şirin, U., Ertan, E., & Ertan, B. (2010). Growth substrates and fig nursery tree production. *Science in Agriculture, 67*(6), 633–638. https://doi.org/10.1590/S0103-90162010000600003

Sotiropoulos, S., Paschalidis, C., Kavvadias, V., Chatzissavvidis, C., Koriki, A., & Xirogiannis, G. (2020). Effect of N-P-K fertilization on the yield and nutrient status of fig (*FicusCarica* L. cv. Kalamon) trees grown under Mediterranean conditions. *International Journal of Fruit Science, 20*(3), 1920–1928. https://doi.org/10.1080/15538362.2020.1836706

Souza, A. P. D., Da Silva, A. C., Leonel, S., & Escobedo, J. F. (2009). Temperaturas basais e soma térmica para a figueira podada em diferentes épocas. *Revista Brasileira de Fruticultura, 31*, 314–322. https://doi.org/10.1590/S0100-29452009000200005

Stover, E., Aradhya, M., Ferguson, L., & Crisosto, C. H. (2007). The fig: Overview of an ancient fruit. *HortScience, 42*, 1083–1087. https://doi.org/10.21273/HORTSCI.42.5.1083

Tapia, R., Botti, C., Carrasco, O., Prat, L., & Franck, N. (2003). Effect of four irrigation rates on growth of six fig tree varieties. *Acta Horticulturae, 605*, 113–118. https://doi.org/10.17660/ActaHortic.2003.605.17

Travadelo, M., Bortoluzzi, A., Maina, M., Justo, A., & Micheloud, N. (2017). Localización de planta de empaque asociativa de productores frutícolas del centro de Santa Fe, Argentina. *Agroalimentaria, 23*(44), 123–131. Available at http://erevistas.saber.ula.ve/index.php/agroalimentaria

Trichopoulou, A., Soukara, S., & Vasilopoulou, E. (2007). Traditional foods: A science and society perspective. *Trends in Food Science & Technology, 18*(8), 420–427. Available at https://www.academia.edu/20746039/Traditional_foods_a_science_and_society_perspective

Uzundumlu, A. S., Oksuz, M. E., & Kurtoglu, S. (2018). Future of fig production in Turkey. *Journal of Tekirdag Agricultural Faculty, 15*(2), 138–146. Available at https://dergipark.org.tr/tr/download/article-file/484884#:~:text=According%20to%20the%20findings%2C%20it,1%20percent%20in%20fig%20production

Vangelisti, A., Zambrano, L. S., Caruso, G., Macheda, D., Bernardi, R., Usai, G., Mascagni, F., Giordani, T., Gucci, R., Cavallini, A., & Natali, L. (2019). How an ancient, salt-tolerant fruit crop, *Ficus carica* L., copes with salinity: A transcriptome analysis. *Scientific Reports, 9*, 2561. https://doi.org/10.1038/s41598-019-39114-4

Venu, D. K., Munjal, S. V., Waskar, D. P., Patil, S. R., & Kale, A. A. (2005). Biochemical changes during growth and development of fig (*Ficus carica* L.) fruits. *The Journal of Food Science and Technology, 42*(3), 279–282.

Wright, R. D., Browder, J. F., & Jackson, B. E. (2006). Ground pine chips as a substrate for container-grown woody nursery crops. *Journal of Environmental Horticulture, 24*, 181–184.

Yablowitz, Z., Nir, G., & Erez, A. (1998). Breba fig production in Israel: Regular and pesticide free systems. *Acta Hort, 480*, 137–141. https://doi.org/10.24266/0738-2898-24.4.181

Zare, H. (2021). Effects of different methods of pruning intensity on old fig (Sabz cultivar) trees under rainfed conditions. *International Journal of Fruit Science, 21*(1), 379–391. https://doi.org/10.1080/15538362.2021.1892011

Chapter 9
Cultivars and Agriculture Practice of Fig (*Ficus carica*)

Walid Nosir

1 Introduction

The fig tree has a lengthy history; archaeologists have discovered evidence of its cultivation dating back to 5000 BC. Fig trees are tiny, warm-climate trees that could grow practically everywhere, with some types surviving in temperatures (−12 to −6 °C.). For around 15 years, fig-trees will yield well. The fig is mentioned in the holy books, such as Bible and the Quran. It is also shown in Egyptian hieroglyphics. It is adaptable to moderate climates and has been highly valued and widely cultivated in most Mediterranean countries for millennia. The first figs were reportedly delivered to Parris Island, South Carolina, in 1577, where they were found thriving. They were discovered 2 years later near Saint Augustine, Florida. Figs were introduced to Virginia from the Bermudas in 1621 and were noted to be nutritious there in 1629. In 1560.

There are four types of figs in the horticultural world. Capri figs yield a small, non-edible fruit, but their blossoms produce pollen. Pollen from Smyrna and San Pedro figs is necessary to fertilize their flowers (Ferguson et al., 1990; McEachern, 1996). Smyrna figs (Calimyrna figs in the United States) produce huge, edible fruit with viable seeds (Condit, 1955). Normal fruit development necessitates pollination. San Pedro figs have two crops: a parthenocarpic breba crop (first crop) and a pollinated main crop (Tous & Ferguson, 1996). Although several common figs generate a breba crop, one trait that distinguishes San Pedro-type figs is the setting of a breba crop (Flaishman et al., 2008). The common fig, which includes the varieties 'Bourjasotte Noire,' 'Col de Dame Noire,' and 'Noire de Caromb,' develops parthenocarpically and rarely requires pollination (Ozeker & Isfendiyaroglu, 1997). As a

W. Nosir (✉)
Vertical Farm Academy, Newburgh, IN, USA
e-mail: w.nosir@vfarm.academy

M. F. Ramadan (ed.), *Fig* (Ficus carica)*: Production, Processing, and Properties*, https://doi.org/10.1007/978-3-031-16493-4_9

result, the fruit usually has underdeveloped, sterile pips (Caliskan & Polat, 2008). Some common fig cultivars, such as Bourjasotte Noire, shed all or virtually all of the breba crop's fruit buds before producing a main productive crop. 'Noire de Caromb,' show almost complete parthenocarpic growth in both crops (Ferrara & Papa, 2001). Fig trees can bear two crops/ a year (Bostan et al., 1997). The first crop, the breba, is borne laterally on the previous season's growth from buds produced in the leaf axils. The buds start developing with the onset of the following spring, and fruit usually matures within the first month of summer. The second crop, consider the main crop, also arises from buds in the axils of leaves but on shoots of the current season. Fruit maturation may start in the middle of summer and can last several weeks (Flaishman et al., 2008). Production practices must be created to produce figs profitably for the local growth circumstances. Understanding the phenology of 'Bourjasotte Noire,' 'Col de Damme Noire,' and 'Noire de Caromb' as expressed under local growing conditions is required for tree training using pruning and rest breaking agents (RBAs). When left to grow organically, fig tree canopies can reach a height of 15 meters. Allowing trees to reach this size is ideal for producing dry figs harvested off the ground. However, in most modern orchards where figs are grown for fresh consumption, trees are pruned to keep their height under three meters to provide fruit pickers easy access (Flaishman et al., 2008). Fig tree leaves have been shown in several experiments to lower blood sugar and total cholesterol while increasing antioxidant levels.

According to FAO, the area planted with *F. carica* in 2021 exceeded 457,737 hectares, with an output of over one million metric tons per year (FAOSTAT). Table 9.1 presents the top fig-producing countries. The countries of the European Union produce the majority of the world's *F. carica*. Fig fruits are essential components of the Mediterranean diet, which is often regarded as one of the healthiest on the planet. It is a dietary habit that's also linked to longevity. The cultivation of fig trees is widespread. They thrive in the Mediterranean and are well-adapted to drought and high temperatures. The nations with the most increased production of *F. carica* are Turkey and Egypt, Algeria, Morocco, and Iran, which account for 43%

Table 9.1 Top fig growing countries

Rank	Country	Annual production of fresh figs in metric tons
1	Turkey	262,644
2	Egypt	203,238
3	Morocco	76,625
4	Iran	75,834
5	Algeria	72,898
6	Greece	50,368
7	Syria	44,032
8	United States	43,001
9	Spain	42,856
10	Tunisia	25,786

of global production. Turkey is the most important country. Austria, Spain, Italy, and the Netherlands are the biggest exporters of *F. carica*, whereas the Netherlands is the primary importer. France, Germany, Austria, and the United Kingdom are home to *F. carica*. Commercial fig production is carried out in Gozlekci, Gok, and Yilmaz. California, Australia, and South America, among other places, have a Mediterranean climate (Pereira et al., 2020). Figs have recently gained much popularity and are grown worldwide. Several Asian countries, including China, Japan, and Thailand, have begun cultivating *F. carica*.

The commercial was assessed by Lianju, Weibin, Kai, Zhifeng, and Yelin. The output of figs grew in China due to scientific studies into this fruit, including cultivar selection, cultivation methodologies, training, pruning techniques, and soil management. In addition, the production of figs was aided by fertilizer, irrigation, processing techniques, and medical applications. As a result, the fig is widely favored by growers. Furthermore, before 1985, the area dedicated to fig cultivation was approximately 350 hm^2, and currently, there are 6667 hm^2 available, which is 20 times more than 20 years ago.

Meanwhile, the figs industry is growing. However, the industry is only concentrated in a few areas in Japan, with "Masui Dauphine" being the dominant cultivar. Thailand's fig industry concentrates on developing new traditional agricultural goods to sell in local markets.

The *Ficus carica* L. tree is an underutilized and minor fruit species in Malaysia, with cultivation primarily for hobby purposes. Currently, there are only a few *F. carica* orchards scattered over Malaysia. However, although *F. carica* farming is still in its infancy, it is progressively gaining traction among those captivated by this ancient plant (Vemmos et al., 2013). As a result, *F. carica* has been proposed as an alternative high-value crop that provides farmers with a new source of income because it is easier to produce, could be harvested in a short period, and provides a faster return on investment than fruit crops. Furthermore, the cultivation of *F. carica* has grown worldwide, owing to its uses as a food, medicine, and cultural symbol. This issue has sparked much interest in looking into its production processes.

2 Fig Cultivars

The mandatory outcrossing in this species increases genetic variety in figs, resulting in developing new varieties with potentially advantageous features from seeds. Because fig is easily propagated through cuttings and is routinely re-propagated to preserve suitable cultivars, natural mutations within a cultivar could produce significant phenotypic heterogeneity. The practice of naming excellent fig cultivars dates back to the fourth century BCE. Flaishman et al. (2008) listed 607 varieties that produce fruit. However, only a few cultivars are used in commercial production. The California fig industry, for example, is based mainly on five cultivars:

'Calimyrna' ('Sarilop'), 'Adriatic,' 'Mission,' 'Brown Turkey,' and 'Kadota' (California Fig Advisory Board, 2006; California Fresh Fig Growers Association, 2006).

Leaf morphology, plant vigor, external and internal color, flavor, percentage soluble solids, titratable acidity, seed properties, fruit shape, skin thickness, ostiole diameter, and fruit production time are all qualities cultivars differ. Traditionally, distinct cultivars were classified based on fruit and tree features. This method is practical and sensible, especially when marketing fruit or selecting planting material. However, other cultivars have similar descriptions, and a single genotype is thought to be associated with dozens of names in certain circumstances (Flaishman et al., 2008). More precise data on brebas, main-crop fig leaves, and fruit has been proven to characterize practically all cultivars in the Extremadura collection in Spain (Giraldo et al., 2010). This description has the advantage of requiring less specialist equipment, but it is more prone than molecular markers to be altered by environmental variability. Isozyme analysis has been used to sort genotypes for many years among molecular approaches. Even though isozymes have been used to differentiate fig cultivars in several studies, only a few cultivars have been studied in practice. Pereira et al. (2020) reported six enzyme systems to identify a more significant number of cultivars, but just the three most helpful enzyme systems were needed to generate distinct patterns for each of the 31 genotypes studied. Using 25 primers to synthesize random amplified polymorphic DNA, these researchers discovered comparable discriminating (RAPDs). DNA microsatellites, also known as simple sequence repeats (SSRs), have proven to be particularly valuable in fingerprinting genotypes in various animals. Harrison et al. (2003) studied RAPDs, inter simple sequence repeats (ISSR), and microsatellite markers for molecular characterization of 30 cultivars, finding that RAPDs were the least effective technique for distinguishing fig genotypes. In a study of 70 fig accessions at the Conservatoire Botanique National Méditerranéen de Porquerolles in France (Harrison et al., 2003), 52 genotypes were discovered using five SSR loci. While the authors speculated that these were most likely 52 different genotypes with several duplicates labeled differently, they also proposed that more markers be used. Six SSR loci were used to analyze 75 fig accessions, resulting in the identification of 72 genotypes (Khadari et al. 2005). Giraldo et al. (2010) used 25 polymorphic SSR loci to describe 15 accessions. The SSRs revealed an average of 3 alleles per locus and 11 distinct genotypes among the 15 accessions, indicating a small genetic base in cultivated figs. 16 SSR loci were recently utilized to analyze 181 fig accessions in the Germplasm Repository in Davis, California, yielding 128 distinct genotypes (Aradhya and Stover, personal comm.). In an ideal scenario, fig researchers worldwide would agree on a set of markers that could be used to verify the authenticity of figs across different collections.

Characteristics that should be considered in the selection of fig cultivars include the following: cold hardiness, the ability to set fruit without pollination (parthenocarpy), the fruit having a closed eye, or ostiole, a long peduncle that allows the fruit to droop and shed moisture, and a green skin on the fruit to minimize bird herbivory.

2.1 Fig Cultivars

Alma This cultivar is reported to produce high yields. Alma is moderately cold tolerant and produces a medium-sized fruit with brown skin and a light tan pulp (Fig. 9.1). The eye is medium open, and the fruit has few seeds and is very sweet. Fruit ripens from late July through August and is good fresh or processed. Considered a high-quality fig tree, the Alma was bred at Texas A&M University and has adapted well to the southeast. Alma trees have dark green foliage and require full to partial sun. Trees thrive in well-drained, sandy, moist, loam soils but are highly tolerant of poor soil conditions. The Alma is an extremely productive tree and produces figs that ripen in late June and again in August. The fruit is small to medium and has a pear shape, golden-brown skin, and amber pulp. Alma figs are sweet, delicate caramel-flavored figs best served fresh, dried, and preserved.

A small to medium, golden-brown, pear-shaped (pyriform) fig with amber/light yellow pulp. Sweet and delicate flavor. Well-adapted in the Southeast. Highly resistant to fruit rots. Wood is very hardy. Excellent quality, succulent and sweet, with barely noticeable seeds and a well-sealed eye. Very productive, compact tree is cold, hardy, and has a prolonged spring dormancy.

Black Spanish (California Brown Turkey, San Pedro): This cultivar yields medium-sized, purplish-brown fruit with a red center (Fig. 9.2). Black Spanish has an eye open. Fruit ripens in June and in August.

Brown Turkey (Brunswick, Eastern Brown Turkey, Harrison, Lees Perpetual, Ramsey, Texas Everbearing). This cultivar is probably the most popular fig in the southeastern United States. In the spring, a small fruit crop (called breba fruit) ripens in July, followed by the main crop one month later (Fig. 9.3). The fruit is

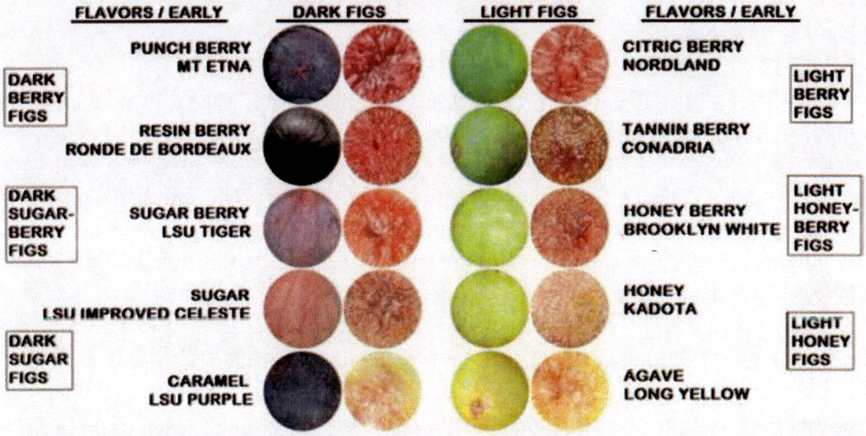

Fig. 9.1 Fig flavors, pulp and skin color examples. (Credit: https://mountainfigs.net/)

Fig. 9.2 Fig Alma

Fig. 9.3 Black Spanish Fig

medium-large in size and has bronze skin and amber pulp. The fruit has a small to the medium closed eye and is good fresh, or processed.

Celeste (Blue Celeste, Celestial, and Little Brown Sugar). This cultivar is probably the second most common fig in the southeastern United States. Celeste is fairly cold and hardy; the fruit is small to medium in size and purplish bronze to light brown (Fig. 9.4). Celeste has a closed eye and begins ripening in early July.

Champagne (Golden Celeste). This newly released cultivar from Louisiana State University produces a medium-sized fruit with yellow skin, tan-colored pulp, and a closed ostiole. Fruit ripening is in early July (Fig. 9.5).

Conadria (Genoa). This cultivar is a vigorous tree that produces green to yellow fruit (Fig. 9.6). The flesh of the fruit is pink to red with a good flavor. The eye is small and tight. Fruit ripening is in June and again in August. Conadria fruit is good fresh and excellent dried.

Green Ischia (Ischia Green, Ischia Verte, and White Ischia). This cultivar produces a green small to medium-sized fruit with a strawberry center and a closed eye (Fig. 9.7). Fruit of Green Ischia ripens in late July to early August.

Hunt This cultivar is very cold and hardy and produces a small pear-shaped, violet-brown fruit with a long neck (Fig. 9.8). Hunt has a closed eye, amber-colored flesh, and few seeds; the fruit ripens in July.

Jelly (Mary Lane Seedless). This cultivar produces a longnecked, yellow fig medium in size with clear amber flesh and very few seeds (Fig. 9.9). Fruit of Jelly is good for eating fresh and preserving, although the skin is soft. Jelly ripens from late July to August.

Kadota (Florentine). This cultivar produces a medium-large, yellow fruit with an open ostiole partially sealed with a honey-like substance (Fig. 9.10). Fruit quality declines with extremely wet weather. Although Kadota figs can be eaten fresh, they are better suited for canning and preserves. Fruit ripens in July.

LSU Gold This cultivar is a Louisiana State University release producing a large, yellow fig with pink-to-red pulp. The fruit of LSU Gold should be picked as soon as it is mature because this fruit has an open eye, and fruit spoilage may occur (Fig. 9.11). Fruit ripens in July through August. Fruit is of good quality for eating fresh and for preserving.

LSU Purple This cultivar is a Louisiana State University release that produces a medium-small, glossy purple fig with amber-to-pink flesh and a closed eye

Fig. 9.4 Celeste Fig

Fig. 9.5 Champagne (Golden Celeste) Fig

Fig. 9.6 Conadria (Genoa) Fig

(Fig. 9.12). The main crop ripens in August, but some fruit can ripen well into the fall. Fruit is of good quality for eating fresh and for preserving.

Magnolia (Brunswick, Madonna). This cultivar is common in some southeastern United States but not in Florida. Magnolia is cold, hardy down to 50 °F, and often produces the largest fig. The fruit is asymmetric, bronze in color, and has an open ostiole (Fig. 9.13). The flesh is amber to strawberry in color. Ripening is from mid-July through August. Fruit should be picked early because it may split and turn sour under wet conditions.

Mission (Black Mission, Franciscana). A large, black fig with reddish-pink pulp (Fig. 9.14). The mission is an ever-bearing fig that produces fruit from summer to winter. Not sufficiently cold hardy for the southeastern United States.

Fig. 9.7 Green Ischia Fig

Fig. 9.8 Hunt Fig

O'Rourke (Improved Celeste). This cultivar is a Louisiana State University fig that produces a small to medium-sized fruit brown with tan pulp (Fig. 9.15). The eye is partially closed with the aid of a honey-like substance. Fruit ripens in early July. Fruit is of good quality for eating fresh and for preserving.

Osborn Prolific (Arachipel, Hardy Prolific, Neveralla, Osborne, Rust). This cultivar produces a medium-large fruit with reddish-brown skin and light-colored flesh (Fig. 9.16). The fruit is sweet with few seeds. The eye is partially closed. Osborn Prolific is reported to perform better in cooler climates. Fruit ripens in August, and the fruit is best eaten fresh.

Fig. 9.9 Jilly Fig

Fig. 9.10 Kadota Fig

Pasquale (Natalino, Vernino). This cultivar produces a sweet, small, purple fig with amber-to-pink pulp (Fig. 9.17). Pasquale ripens from late November to December and is often damaged by frost. However, Pasquale is not cold-hardy. A medium-large fig that is greenish-yellow in color with light strawberry pulp. The fruit has a closed ostiole. Tena thrives in hot, dry weather. The fruit is good for eating fresh or preserving.

Tiger (Giant Celeste). This new cultivar released by Louisiana State University has large brown fruit, yellow pulp, and a partially closed eye (Fig. 9.18). Fruit of Tiger ripens in early July.

Fig. 9.11 LSU Gold Fig

Fig. 9.12 LSU Purple Fig

3 Production Process

F. carica cultivation is rapidly increasing in the open field and under greenhouse conditions because it is a species that responds well to agronomic management, making it an excellent production alternative. However, *F. carica* planted in the open field often encounters high pest and disease incidence, rotting of the fruit during the rainy season, and poor plantation management. These harmful conditions

Fig. 9.13 Magnolia fig

Fig. 9.14 Mission Fig

will have a significant impact on crop growth and fruiting. Thus, the shielded culture of *F. carica* will be the solution amid severe growing conditions. Greenhouse horticulture is the most intensive mode of crop production, yielding up to 10 times more per cultivated unit area than a field crop. Growing crops in a protected environment protect them from natural hazards. It allows for artificial manipulation of the crop micro-environment to facilitate optimal plant performance, extend production duration, induce early flowering, and improve production and product quality (Figs. 9.19 and 9.20). Protected cultivation allows for year-round production, which is impossible in open field farming due to excessive rainfall and wind, especially in tropical locations (Pereira et al., 2015).

3.1 Nutrient Control

Fertilization management is critical in agriculture because of economic and environmental concerns. Over the last few decades, agricultural crop nutritional requirements mismanagement has resulted in significant health and environmental concerns. The fertilizer should be applied based on the location's variety, soil type, and climate. The nutritional needs of the plants are usually met by proper fertilization. However, it is vital to apply fertilizers sparingly to lessen fertilization's environmental impact. In a previous study by Lianju et al., nitrogen fertilizers such as urea and NH_4HCO_3 were utilized in orchards where the soils were infertile. The weather was less rainy during the early growing season. Because fig fruits at the lower regions of shoots began to mature while those at the top were just at the stage of differentiation, potassium and phosphate fertilizers were frequently sprayed more in fig orchards and sooner than in other fruits. One month before harvesting summer and fall fruits, potassium and phosphate fertilizers were often applied. According to a study on the effects of potassium fertigation on yield, marketable output, and fruit size in two different fig cultivars, 'Black Mission' and 'Sierra,' considerable fertigation of potassium may enhance the marketable yield; of 'Black Mission.' And 'Sierra' fresh fig cultivars are critical for new fig growers because the commercial output is revenue (Islam et al., 1997). The fertigation treatment 'Sierra' figs also boosted fruit size, and this rise is linked to high leaf K concentrations. The administration of correct fertilizers could result in increased growth with enhanced fruit output and quality because any nutrient, whether lacking or excessive, could result in a drop in crop yield and inferior fruit quality.

Little research has been done on the use of fertilizers for figs. However, there is certain basic information that is useful. Fig trees characteristically produce good crops under relatively low fertility levels. However, increased vigor and fruiting can often be achieved by fertilizers. This is particularly true where figs are grown on infertile, sandy soils and grown in competition with lawn grasses and ornamental

Fig. 9.15 O'Rourke (improved Celeste)

Fig. 9.16 Osborn Prolific Fig

Fig. 9.17 Pasquale Fig

plants. Fig trees habitually grow vigorously, and the leaves are large. Below-normal leaf size and terminal growth indicate the need for fertilizing. Fig trees should be fertilized, as should other fruit trees such as peaches and plums, except that fig trees rarely experience iron and zinc deficiencies. Nitrogen is usually the only mineral in short supply. For large fig bushes, 1 to 1.5 pounds of nitrogen per tree per year was applied in late 18 AGR. Winter or early spring is generally adequate, but more frequent applications may be required in heavy rainfall areas. The fertilizer amount can be increased or decreased as indicated by the bush size and the response obtained. In those areas where complete fertilizers are needed, a mixed fertilizer applied in an amount that will contain 1–2 pounds of nitrogen should be used (Fig. 9.21). The phosphorus and potassium ratios to nitrogen should be about the same as commercial acreages.

Fig. 9.18 Tiger Fig

Fig. 9.19 Fig propagation steps using one-year cuttings

3.2 *Water Requirement*

Because water is a scarce resource in many parts of the world, producing more with less water in agriculture is critical. Irrigation water is constrained by two factors: water scarcity and inefficient water utilization. The irrigation system is one of the most critical factors determining the yield and quality of agricultural produce in a greenhouse farming system. Water should be given in the right amount and at the right time. As a result, water management is critical for avoiding plant moisture

Fig. 9.20 Fig air layering

stress during the growth stages of the crop (Saddoud et al., 2008). As a result, the development of water-efficient agriculture and crop yield and water usage efficiency improvements have much potential as practical approaches for developing sustainable agriculture in irrigated areas.

F. carica plants prefer drier growth conditions than most fruit trees. However, this plant still needs adequate water for optimal performance and production. The frequency is determined by the trees' size, vigor, soil type, and rainfall. Because *F. carica* trees are susceptible to root rot, it is best to prevent severe moisture swings, such as excessive irrigation water, as *F. carica* cannot thrive in these conditions. Caruso et al. (2022) collaborated on a study to assess the gas exchange and growth response of *F. carica* plants to water deficit and relief. According to the study, mild water deficiency substantially impacts shoot growth, leaf biomass, and gas exchange parameters. El-Shazly et al. (2014) got similar results on many *F. carica* cultivars grown under various irrigation regimens. Compared to 100% ETc, watering at 75% ETc and 50% ETc resulted in lower growth rate values.

Treating the crop with water without first understanding the parameters involved will result in drainage and salinity issues and a failure to reap the benefits of irrigation. As a result, knowledge of proper irrigation management is critical for the growth of *F. carica*, as it will result in increased crop production and more efficient use of water resources.

3.3 Soil Requirement

According to Aljane and Ferchichi (2006), the fig could be cultivated on various soils, including light sand, rich loams, clay, or limestone, as long as there is sufficient depth and proper drainage. Excessive plant growth is encouraged at the expense of fruit yield in heavy, damp soil. Soils with a high acidity level are inappropriate. Between 6.0 and 6.5 is the optimal pH range. Figs thrive in soils that are well-drained, fertile, and high in organic matter (Bostan et al., 1997). They indicate that fig trees may adapt to various soils, but the best are those with a clayey-sandy texture, high organic matter, and a pH of 6.0–6.8. In poorly drained soils, root rots can occur, and in overly dry soils, plants go into a dormant state, producing few leaves and no fruits. In recent years, the harvested area and volume of fig production have remained stable globally, with Turkey, Egypt, Morocco, Algeria, and Iran serving as the primary producers. Fig cultivation has increased tendency in some nations, such as Mexico, with an average yearly growth rate of 6% in harvested area and 13% in output volume.

In 2017, 1440 ha of land were harvested in Mexico, with an average yield of 5.6 t Darjazi, 2011. Under greenhouse conditions, however, products of above 100 t hal have been reported. Crop development can be influenced by various physical, chemical, and biological factors, including plant nutrition. In plants, nutrients play a crucial and unique role. When one of these elements is not present in sufficient amounts in the plant's tissues, it causes alterations in the plant's metabolism, affecting vegetative growth.

Fig. 9.21 Fig growing in containers using soilless culture technique

4 Fig Production in Soilless Culture

4.1 *Propagation*

Starting a fig tree from fig cuttings is easy and can be done in three ways. Each of these ways of rooting figs is basic and straightforward, and the weather conditions in your area will likely determine your choice during the dormant season. Fig propagation layering, the first method of propagating fig trees outside, relies on temperatures that never drop below freezing during the dormant season. Ground layering is a method of rooting figs that involves burying a piece of a low-growing branch 15–20 cm of the tip visible above ground, allowing the buried portion to root, and then removing the buried portion from the parent tree. This is the basic way of fig growth; keeping the ground clean while the branches root can be challenging. Finally, planting fig cuttings in the garden fig cuttings is a more popular method of rooting figs outside. Take fig cuttings from little branches that are 2–3 years old, late in the dormant season, when the threat of frost has passed. They should be 12–34 inches thick (1.3–1.9 cm) in diameter, about the breadth of your pinky, and 8–12 inches (20–30 cm). The bottom end should be flat, while the branch tip should be one-sided. Use a sealant on the slanted end and rooting hormone on the flat end to prevent disease. Fig cuttings can grow 36–48 inches in a year (91–122 cm.). The new trees will be ready to be transplanted during the following dormant season.

4.2 *Soilless Media for Growing Fig Indoor*

The variety of figs, the pH of irrigation water, the cost, the product's shelf life, the type of system utilized, and the grower's preference go into selecting a growing medium. When selecting a medium, a grower should search for specific traits. For example, soilless media, like soil, must offer oxygen, water, nutrients, and support for plant roots (Stone, 2014).

All or some of the following properties should be present in an optimal growing medium:

aeration and drainage are also important. While the medium must be able to retain water, it must also be able to drain well. Excessively fine materials should be avoided to minimize water retention and a lack of aeration within the medium. The soilless media should contain the following characteristics;

- *Durability*. The medium must be long-lasting. Avoid using soft aggregates that readily dissolve.
- *Porosity*. Between cycles, the medium must remain wet from the nutrient flow long enough for the plants to receive their essential nutrients.
- *Sterile*. Diseases and pests will be less likely to spread if the growing medium is clean and sterile. The roots do not receive any more nutrients from a clean

medium. Some media can be reused by pasteurizing for 30 min or soaking in a 10% bleach solution for 20 min and then rinsing with tap water many times.

- *Chemical characteristics*. Good cation-exchange capacity and a neutral pH (the ability to hold nutrients).
- *Functionality*. Lightweight, simple to handle, reusable, and long-lasting.

4.2.1 Rockwool

The most common fig-growing medium is Rockwool, a sterile, porous, non-degradable material made mostly of granite and limestone that is melted and spun into small threads before being molded into blocks, sheets, cubes, slabs, or flocks. It absorbs water quickly and has good drainage, so it is commonly used as a starting medium, a rooting media for cuttings, and a large biomass crop.

Advantages It has a high water retention capacity of 18% to 25% air, providing enough oxygen to the root system if the medium is not fully submerged.

It co.mes in various sizes and forms to suit various hydroponic applications. From 1-inch cubes to massive slabs, there is something for everyone.

Mineral wool slabs can be steam-sterilized between crops and reused. However, it takes three to four years for it to fall structurally.

Disadvantages It has a high pH, which necessitates adjusting nutritional solutions to account for it. In addition, because the commercial material's initial pH is high (7.0–8.0), continual pH correction to a more suitable range (5.5–6.0) is required, or the medium must be conditioned before soaking in a low-pH solution.

Because mineral wool does not biodegrade, it is an environmental hazard when discarded. As a result, mineral wool has become less popular in recent years.

It has a confined root habitat and low water and nutrient buffering capacity. Even though the water content appears high, the water passage to plant roots may be hampered.

4.2.2 Coconut Coir

Trade names for coconut coir include Ultrapeat®, Cocopeat®, and Coco-tek®. It is an all-natural medium comprised of shredded coconut husks. End products vary significantly due to different suppliers and production techniques. The compressed briquette is the most common, and it must be soaked in water before use. The coir rehydrates during the soaking process and expands up to six times its original size.

Advantages Coconut coir is slightly acidic, holds much moisture, and provides efficient root aeration. In addition, there are suggestions that root-promoting chemicals in coir dust aid rooting.

Coir could be used as a stand-alone medium or as part of a mixed medium for growing vegetables and cut flowers. Under mist and high humidity conditions, it can also be used as a rooting medium for cuttings. It is biodegradable, organic, and non-toxic, making it easy to dispose of and good for the environment.

Because it is tiny, it may be purchased compressed and inflated at home, saving money on delivery.

Disadvantages High salinity could be a concern if the husks are soaked in saltwater during manufacture and not washed with fresh water.

Coconut coir is high in sodium and chlorine, harming plants, so it must be washed. Calcium and magnesium are commonly added to help remove salt and provide minerals.

4.2.3 The Aggregate of Expanded Clay

Heat dried heavy clay and stretched it to form spherical porous balls to make expanded clay pellets. Expanded clay aggregate **(ECA)**, grow rocks, or Hydroton® are some of the frequent names. They are heavy enough to provide strong plant support while remaining lightweight. In addition, ECA spherical form and porosity maintain an excellent oxygen/water balance, preventing the roots from drying out or drowning.

Advantages Expanded clay pellets have a pH of about 7.0 and release essentially no nutrients into the water stream. They feature a large pore space, which allows for improved solution flow. They rarely clog or become blocked, so water drains quickly, making them ideal for ebb and flow systems. The pellets can be rinsed and sterilized after use and reused.

Disadvantages Compared to many other substrates, clay pellets do not have a high water-holding capacity. As a result, they drain and dry quickly, potentially drying out the roots.

They are somewhat pricey. In Dutch bucket systems, they frequently bond tightly around roots and are difficult to remove. In addition, clay pellets can get sucked into filters or drain lines and cause blockages since they float for the first few months until they are saturated.

4.2.4 Perlite

Perlite is a naturally occurring volcanic material that expands when heated to extremely high temperatures, becoming extremely light, porous, and absorbent. It comes in various diameters, the most common being 0–2 mm and 1.5–3 mm. Perlite can be used alone or in combination with other growing media.

Advantages It has one of the highest percentages of oxygen retention of any growing media.

It has a high capillary activity and is highly porous. It can hold three to four times its weight in water. Its sterility makes it ideal for germinating seeds. Root rot and damping-off are unlikely to occur. It is reasonably priced and can be reused. It can be steam pasteurized after use.

Disadvantages It is readily washed away due to its small weight. Because of this flaw, perlite is an unsuitable medium for flood-and-flush hydroponic systems.

It does not retain water well when used alone in hydroponic systems such as drip systems.

Perlite dust can cause respiratory issues and eye discomfort. Therefore it is important to take precautions like wearing goggles and a mask when working with it. Fans may blow it about the greenhouse once it is dry. Perlite is prone to algae growth, which can cause problems with watering and fungus gnats.

4.2.5 Vermiculite

Is a micaceous mineral that swells into pebbles when heated to 2000 degrees Fahrenheit. It has thought to be a tremendous rooting medium. As a result, seedlings are frequently combined with other media types, such as coconut coir or peat moss. It comes in various diameters, 0–2 mm, 2–4 mm, and 4–8 mm.

Advantages It has a high cation exchange capacity and can store nutrients for later use. In addition, it is incredibly porous, has a strong capillary action, and can hold much water.

Disadvantages When used alone, it can hold too much moisture, resulting in soggy conditions that invite bacterial and fungal growth. It cannot be steam-sterilized because it breaks down when heated. It is a bit pricey and may contain a small quantity of asbestos.

4.2.6 Cubes of Oasis

Cubes of Oasis are water-absorbent phenolic foam, often known as floral foam, used to make media. It is a seed and cutting growing media mostly used for plant propagation. Oasis cubes are most commonly used for seeding lettuce and cole crops (cabbage, collards, and kale), onions and alliums, herbs, and occasionally tomato and eggplant seedlings.

Advantages It has a pH of 7.0 and a high water-holding capacity.

It is a very adaptable plant that may be transferred into various hydroponic systems and grow mediums. It is cheap, and there is no need to soak it first. It is available in a variety of sizes.

Disadvantages There is no buffering capacity, no cation exchange capacity, and no starting nutritional charge. It has limited utility beyond seed germination and propagation. In addition, the foam may break off and clog the pump filters.

4.2.7 Sand

Sand is unquestionably the oldest and most ubiquitous hydroponic media. Other substrates, such as vermiculite, perlite, and coconut coir, are frequently blended. Growers frequently prefer coarse sand as a growing medium because it helps enhance the roots' aeration by increasing the air spaces between the grains of sand.

Advantages It is relatively affordable and widely available in most areas.

Because the finer sand particles allow lateral water circulation through capillary action, the solution supplied to each plant is evenly distributed throughout the root zone.

It helps aerate the mix for roots when combined with vermiculite, perlite, and coconut coir.

Sand is highly durable because it is not chemically or biologically impacted.

It may be steam-sterilized and reused with ease.

Disadvantages It has low water and nutrient retention capacity, quickly exacerbating deficiencies. During the growth season, salt can build up in the sand. This can be remedied by flushing the medium with purified water regularly. It is quite hefty.

4.2.8 Peat Moss

Sedges, grasses, and mosses are among the partially degraded marsh vegetation that makes up peat. In horticultural taxonomy, sphagnum peat moss, hypnum peat moss, and reed and sedge peat moss are the three forms of peat. The most desirable and popular form is sphagnum peat moss, which has a higher moisture-holding capacity and does not break down as quickly as other varieties of peat.

Advantages Peat moss has a high moisture capacity, holding up to 10 times its dry weight in water. With a pH of 3.8–4.5, most peat mosses are acidic, which can benefit some acid-loving plants. Peat moss can drain freely despite its incredible ability to absorb water. Excess water quickly evaporates as it passes through the material.

The disposal of used peat moss does not harm the environment.

Disadvantages It is widely considered a suitable environment for various soil-borne diseases. Although peat can be sterilized, this does not solve the problem since sterilization creates a biological vacuum that pathogenic fungi can quickly fill. Because its acidic nature can be detrimental to some crops, lime or dolomite is commonly added to raise the pH. Peat moss extraction from bogs is a disruptive operation that eliminates layers that have taken centuries to form.

4.2.9 Growstones

Recycled glass is used to make Growstones. They are thin, irregularly shaped, porous, and reusable. They can wick water up to 4 inches above the waterline and have good wicking ability. To keep the stems from decaying, proper drainage is essential.

Advantages Because grow stone is inert, it provides no additional inputs or elements to the plants that could interfere with the nutrient solution in the system.

It has a high porosity and allows for many aerations of the roots.

It is non-toxic and guaranteed to be free of pollutants such as germs because it is constructed of glass. In addition, grow stones can be reused or repurposed again.

Disadvantages They tend to hold the plant roots too tightly, so growing stones can sometimes cause root injury. This also makes moving the plants from one medium or grow space to another challenging.

Growstones are coated in fine silica dust that must be carefully wiped away. Because the dust can clog drains and is dangerous to breathe, it is best to do this outside or in a well-ventilated area.

4.2.10 Husks of Rice

Rice hulls are a rice industry waste. Even though it is organic plant material, it degrades slowly, similar to coconut coir, making it an ideal hydroponic growing substrate. It is frequently used as part of a growth media mix of 30–40% rice hulls and pine bark. The different types are fresh, aged, composted, parboiled, or carbonized rice hulls. As a medium modification, parboiled hulls have outperformed other hulls.

Advantages Parboiled and composted rice hulls have an overall pH of 5.7–6.5, which is right in the sweet spot for most hydroponically grown plants.

They have a water-holding capacity per weight comparable to perlite but a higher air-porosity ratio and can hold more oxygen in the root zone.

They drain well and, in general, hold very little water.

Disadvantages Manganese levels in fresh and composted rice hulls are frequently high. Manganese poisoning can occur if the pH is not correctly regulated.

Rice hulls perform well when blended with peat or coir, but they do not work as well as a stand-alone medium. This is because it has a poor capability for cation exchange.

4.2.11 Pine needles

Pine bark that had been composted and aged was one of the first growing media used in hydroponics. It was once thought to be a waste product, but it has since been helpful as a ground mulch and a substrate for hydroponically grown vegetables.

Advantages Pine bark resists decomposition better than other tree bark and contains fewer organic acids that can seep into the nutrient solution.

Used bark is a naturally biodegradable substance that can be reused in various ways, including as mulch. It keeps nutritional solution and oxygen well due to its fibrous structure with pockets of different sizes.

Disadvantages It readily absorbs water, which can lead to flooded circumstances.

Pine bark floats, which can cause issues with ebb and flow systems. It is more suited to a drip or wick setup. Pine bark has an acidic pH, which could be a drawback.

4.2.12 Pumice

Pumice is a volcanically formed siliceous substance. It has been graded and kiln-dried to an internal temperature of 80 degrees Fahrenheit, making it sterile and ready to use. Pumice media Promote aeration and drainage; it can be blended with other forms of growing media like vermiculite or coir.

Advantages It degrades slowly and is relatively light.

Its light tint is an excellent medium for summer growing because it does not absorb heat. In addition, it has a high level of oxygen retention.

Disadvantages It has almost identical qualities to perlite, although it does not absorb water.

It may be too light for some hydroponics systems if purchased in little bits.

4.2.13 Sawdust

Many factors influence how sawdust performs; the most important is the type of wood utilized and its cleanliness. The best results have been observed using sawdust from Douglas fir and western hemlock, but western red cedar is harmful and should never be utilized. In addition, water distributes better laterally through moderately fine sawdust or sawdust with a higher proportion of planer shavings than coarse sawdust.

Advantages The best part about sawdust is that it is frequently inexpensive or free.

Because it holds much moisture, it must be watered with caution.

Disadvantages Sawdust may accumulate salt levels that are harmful to plants. As a result, the sodium chloride content should be determined before using the samples. Sawdust should be completely leached with fresh water if a substantial amount of sodium chloride is discovered (more than ten ppm). In addition, growers must guarantee that their sawdust is free of soil, diseases, and chemicals from wood-processing industries and undesired tree species.

4.2.14 Polyurethane Slabs

Polyurethane grows slabs and cubes are an uncommon hydroponics medium that can be utilized as a beginning cubes alternative to oasis cubes or Rockwool. It is available in the form of poly foam in hobby and fabric stores. It comes in a variety of thicknesses and widths in rolls or sheets. Self-made starter cubes can be created from 1- to 2-inch-thick polyfoam sheets/rolls.

Advantages It is a less expensive alternative to Rockwool or oasis cubes for seed beginning.

Disadvantages It could include potentially dangerous substances. On the other hand, it is unlikely to contain predefined seed germination holes.

4.2.15 Gravel

Gravel has proven very effective, especially in ebb and flow systems. It is a fragmented media made of rocks such as sandstone, limestone, or basalt, with huge gaps between each particle. This provides a good supply of oxygen to the roots, but the medium does not keep water effectively, causing the roots to dry out soon.

Advantages Gravel usually is inexpensive, works well as a beginning media, and is relatively easy to get by. In addition, it is durable and reusable as long as it is washed and sterilized between crops.

Its structure does not degrade, and it can be reused.

Disadvantages It is tough to handle due to its hefty weight. Gravel is not suited for plant roots with much weight. Shale that has been expanded. Expanded shale is formed when quarried shale is heated to temperatures above 2000 degrees Fahrenheit. The shale is rendered chemically and physiologically inert due to the treatment. The water in the heated shale evaporates, causing the shale to expand. It is regarded as one of the most effective aquaponics grow media. It is light and easy to use in aquaponic grow beds. Each stone has a wide surface area for the microorganisms that convert ammonia to nitrates to thrive.

Fig trees are unlike other deciduous fruit trees in that the shoots continue to elongate as long as favorable environmental conditions for growth. Moreover, the fruit continues to set in the axils of the new leaves throughout the growing season. For this reason, conditions good for shoot elongation should, if possible, be maintained throughout the summer.

4.3 Fig Cultivation

Fig should be cultivated lightly because many feeder roots are near the surface and may be damaged if deep tillage is used. In addition, light cultivation is necessary to break up surface crusts caused by heavy rains followed by intense sunlight. Cultivation close to the tree's base is difficult if low branches are allowed to develop parallel to the ground.

4.4 Mulching

Mulches of hay, corn cobs, pine needles, sawdust, and other coarse organic matter have been used with figs, and the results have been excellent. Mulching generally results in excellent soil moisture, temperature, and aeration. In addition, under a hay mulch, available mineral elements are leaching from the slowly decomposing organic material. Mulching also helps overcome or prevent the ill effects of nematodes. Since some roots grow on the soil's surface under a mulch, the mulch should be at least 4–6 inches deep and extend from the trunk to several feet beyond the extremities of the branches. However, good results are obtained even where the mulch is placed only under the branches.

4.5 Irrigation

In fig cultivation, pay careful attention to soil moisture management. The roots of most fig trees are close to the soil's surface and can quickly dry out. In addition, figs are particularly vulnerable to soil-borne nematodes, which eat tiny roots and limit water flow into the tree. Apply water to the trees when the drought worsens for these reasons. Water stress is indicated by slight leaf wilting in the afternoon.

Mulching with straw or grass clippings helps keep soil moisture consistent and decreases weed competition for water. Due to water stress, premature fruit drop is common in fig cultivars with no genuine seeds. When the fig tree is cultivated in shallow soil, and the roots are nematode-infested, this problem is widespread in hot, dry climates. In locations with poor drainage, do not overwater.

This depletes the soil's oxygen supply, causing the tree to be harmed or die. Regular irrigation and mulching, as well as good water management, help keep trees healthy and vigorous and decrease fruit drops. The factors influencing a fig tree's vulnerability to cold harm have to do with the tree's dormancy. A mature tree that has lost all of its leaves and is completely dormant at the first frost can survive significantly lower temperatures than a quickly growing tree.

Inhibit irrigation in the fall to support the onset of dormancy and reduce growth. A fully dormant fig tree can tolerate temperatures as low as 10 degrees F. Plant fig trees along the south side of a structure in north Texas to help prevent ice damage. Place straw mulches around the tree's base to keep the warm soil temperature from harming the tree's crown during freezes. Allow time for trees or limbs to grow before removing the frozen limbs. After that, fresh wood can be made.

Figs need a large, continuous water supply for the highest yields throughout the summer. This is more important for Magnolia and Brown Turkey varieties, which tend to bear continuously during the growing season, than Celeste, which produces little fruit in late summer and early fall. Rainfall is usually sufficient in the South to produce a good crop, even when lack of water in late summer and fall results in discontinuing shoot elongation and fruiting. The commercial Magnolia orchards are located where the average rainfall is high and little irrigation is used, although supplemental irrigation would frequently be beneficial. The amount and frequency of irrigation vary depending on the soil. Irrigation is required more frequently in shallow or moisture-retentive soils. If shoot elongation continues and leaf size is normal, moisture is sufficient. The grass beneath the grass in lawns, bushes, or trees may wilt in the heat, while the rest of the lawn does not. This indicates the need for water. Some fig trees grown in lawn grasses need one or more watering. From the standpoint of survival, figs are surprisingly drought resistant. During severe droughts, trees shed their leaves before twig injury occurs and into a drought-induced dormancy. Subsequent rains often result in a late fall growth, which is likely to be injured by cold weather.

4.6 Pruning

The main reasons for pruning are to produce mechanically strong bushes, open the frontiers to sunlight and air movement, facilitate spraying and harvesting, lengthen the harvest season, and make cultivation easier. To avoid decay, care should be taken in pruning to cut back to a bud or lateral growing branch. Pruning differs somewhat with the variety; however, all suckers and low-growing lateral branches should be removed. The center should be thinned out to permit the entrance of sunlight and air. This helps to control leaf and fruit diseases and makes spraying easier. From the standpoint of pruning, figs grew in the South fall loosely FIG GROWING IN THE SOUTH 19 into two categories: (1) Those that will not tolerate heavy annual pruning and (2) those that can be pruned heavily and still produce large crops. The first group includes Celeste, Green Ischia, Mission, and Hunt. Except for Hunt, these varieties branch freely and produce an abundance of laterals. Since breba crops are insignificant and vigorous, the previous season's wood is unnecessary. However, if the branches are cut back heavily, the resulting vigorous sprouts are only lightly productive, severely reducing yield.

For this group, only a general maintenance type pruning is needed. This includes heading the tree back occasionally to keep it in bounds, an annual thinning out of weak growth in the center, and removing deadwood. The Hunt variety produces long, unbranched growth that should be headed back to make it a branch. The Celeste variety produces a dense bush, and the inside growth should be thinned out. The second group of varieties includes Magnolia, Brown Turkey, and Kadota. These varieties will produce good crops even if killed to the ground. Thus, they can be pruned severely and still produce well. The main crop is produced on new wood, but an appreciable breba crop will also be produced on the old wood. Thus, some vigorous previous season's wood is desirable. However, except for Kadota, dooryard trees of these varieties are best handled similarly to those of the first group. Kadota does not branch freely and often fruits excessively in relation to the available leaf surface. They were heading back all vigorous growth from one-third to one-half of its length, resulting in branching and a more favorable leaf-fruit ratio. Commercially, Magnolia trees are pruned more heavily to facilitate harvesting and spraying than are dooryard trees. Also, the harvest season is lengthened because the vigorous shoots from heavy pruning continue to grow longer. A longer harvest period is essential because nearly all of the crop is canned or made into preserves, and it cannot be processed economically in a short time.

On the other hand, heavy pruning reduces total yield considerably, and a smaller portion of the crop is produced early in the season when prices are highest in years past. Fig trees were pruned back annually to short stubs. Now, most growers head back to the previous season's growth of about one-third to one-half. This lengthens the season appreciably but does not reduce total yield as drastically as more severe pruning. The bushes are kept low by frequently cutting back to upright growing laterals from older wood closer to the ground. Vigorous suckers arising from below the ground level are used to replace old leaders that have become damaged (Karami et al., 2018).

4.7 Fruit Development and Growth

Crane, 1986 studied the formation and development of fruit. The fig tree's terminal bud opens and grows in the spring. Both vegetative and reproductive buds are present at the same lateral position. In cultivars like 'Mission' and 'Brown Turkey,' only one inflorescence develops into a syconium, whereas in 'Kadota' and 'Calimyrna,' both inflorescences grow into syconium. Both inflorescences at a node may emerge in some cultivars. Comparable to other fruits, the fig syconium has three distinct growth periods. Crane et al. (1964) reported that the initial growth stage, Stage I, is marked by a quick increase in diameter and a slower rate of fresh and dry growth with almost no change in sugar accumulation; you can lose weight.

Fruit diameter, dry and fresh weights, and sugar content have changed.

Over 70% of the entire dry weight and 90% of the total sugar are produced during the growing process. During this time, the chlorophyll level in many dark cultivars increased. As a result, the thickness of the fruit skin reduces rapidly, and the color of the fruit skin changes from green to bluish.

In addition, the size of the fruit grows more extensive, and the tissue softens occurs at the end of the fig fruit's growth (Chessa, 1997). In common type fig fruits, caprified and noncaprified fruit can emerge on the same branch. In most cultivars, the first ripening period occurs within a single fruit. The second period lasts 5–6 weeks, while the third lasts 3–5 weeks. However, there are significant variances in the life of fig cultivars. The second stage of development is called quiescence. When it comes to cultivars like 'Mission'. In the autumn-producing cultivars'Siera' or autumn Honey,' the second phase lasts 3–4 weeks, while in the spring-producing cultivars'Siera' or 'Autumn Honey,' the second period lasts 3–4 weeks. The fruit harvest is shorter and lasts only 2–3 weeks since all breba fig syconia on the same branch are at comparable developmental stages. With or without pollination, different fig varieties can produce fruit. There is a consistent variation in nitrate levels between persistent and nonpersistent fig varieties (El-Shazly et al., 2014). During Stages I and II of the summer main crop figs, the average nitrate content of persistent figs is three times that of nonpersistent figs. Nitrate is not discovered in nonpersistent fruit by Stage III. El-Shazly et al. (2014) discovered that as nitrate levels rise, indoleacetic acid (IAA) is inhibited, implying that the difference in nitrate levels between persistent and nonpersistent figs is due to the regulation of indoleacetic acid oxidase. As a result, larger auxin levels are expected in persistent cultivars. In nonpersistent Smyrna-type figs, auxin treatment enhanced fruit sets (Aljane & Ferchini, 2009). Auxins, gibberellins, and cytokinins, among other growth regulators, can cause persistence in the Smyrna-type 'Calimyrna' cultivar (Crane, 1986).

Auxin-induced persistent fruits take a little longer to mature than caprifled ones, but their shape is very similar. Because they are facultatively persistent, common fig cultivars can yield both persistent and pollinated main-crop figs. The pollinated fig syconium produces genuine fruits, whereas the nonpollinated fig syconium has an expanded inflorescence with several long-styled pistillate blooms. Two varieties, 'Autumn Honey' and 'Brown Turkey,' yield caprified and nonpollinated figs on the current year's wood. The caprified 'Autumn Honey fruits have a darker purple skin color and a red pulp, unlike the noncaprified fruit's white, pink pulp. Caprifed fruits are often larger and have a longer shelf life (Rodov et al., 2005; Yablowicz et al., 2005). The quantity of crops produced by fig trees directly impacts the carbohydrate balance of the tree.

Early spring, midsummer, and late fall are the peak times for starch concentration in Smyrna-type figs, which produce a single main crop (Aljane & Ferchichi, 2006). When the shoot and breba fruit begin to form in March, there is a drop in sugar and a parallel increase in starch concentration. Shortages of accessible carbohydrates and competition between new foliage and the breba syconium can result in syconium drop and the breba crop's extinction.

Applying gibberellins to the San Pedro-type 'Nazareth' cultivar, or nipping of the terminal bud, temporarily stops new foliage development, allowing breba growth and reducing competition and syconium loss (Bostan et al., 1997).

4.8 Harvesting

The fig is a climacteric fruit that is particularly perishable and prone to physiological disintegration. Even when stored at low temperatures, the fruit's postharvest life is estimated to be between 7 and 10 days (Aljane & Ferchichi, 2006). During the development of tissues, extensive cell wall alteration occurs (Aljane & Ferchichi, 2006). Therefore, fundamental research into the biological and physiological processes that occur during ripening is necessary for investigating systems in which maturation-related biological and physiological processes are implicated in postharvest degradation.

To accelerate growth and ripening, ethanol is applied to fig fruits during their late Stage 11 development. However, the receptacle tissue and the pulpy tissue of the drupelets within mature and ripe fig fruit are visibly separate. As a result, analyzing both tissues as a single combination may hide some cell wall modifications critical to understanding cell wall alteration processes, Particularly during ripening.

During sequential ripening in fig fruit, Islam et al. (1997) studied the alterations in cell wall polysaccharides that occur inside the distinct and independent tissues of the receptacle and the pulpy drupelets. The pectic extracts included a lot of uronic acids and a lot of neutral sugars. The uronic acid and total sugars in the drupelets were higher than in the receptacle at the start of fig ripening. Fig fruit XETs and EGases are part of a gene family with divergent members regulated differently during fig fruit ripening.

The expression of these genes is influenced by ethylene and other developmental variables, implying that numerous actions are necessary for the cooperative remodeling of the hemicellulose network during fig fruit softening. The scientists (Owino et al., 2004) concluded that, like most other fruit species investigated, the gene products of the recovered 11 cDNAs, putatively encoding cell wall-related enzymes, are coordinated in time and amount during fig fruit production. Allow figs to ripen completely on the tree for the best results. However, they must be plucked as soon as they ripen to avoid spoiling by the dried fruit beetle. Microorganisms in fully ripe fruit induce on-the-tree deterioration or souring.

Insects, particularly the dried fruit beetle, carry these organisms into the open eye of the Fig. Daily harvests, and the removal of overripe, damaged figs can help decrease spoiling significantly. This is especially true for kids with wide eyes. When collecting figs, wear gloves and long sleeves to avoid skin irritation from the fig latex.

4.9 Disease Prevention and Control

Fig rust is a severe fungus that damages the leaves of fig trees. Physopella fici is to blame.

Small yellowish-orange dots on the leaves are the first signs of fig rust. As the season passes, these expand somewhat and may become quite numerous. Every year, rust causes the full defoliation of many trees in the state, leaving them with a ragged appearance. Furthermore, trees that have been defoliated early in the season may produce new foliage sensitive to cold harm. However, defoliation does not usually occur early enough to cause fruit loss except in late-ripening types.

Neutral copper sprays are used to keep rust at bay.

A few treatments in May or early June usually keep the trees in good shape until the fruit ripens. One or two more applications may be required during particularly wet seasons. When the tree's initial leaves have achieved full size, it is a good indicator to start spraying. In 3–4 weeks, the second spray should be used. It is critical to ensure that the spray substance covers all the leaves. Growing suggested types with a closed eye, a drooping fruit feature, and fruit-splitting resistance is the first step in preventing losses due to souring. Most fruit sourness concerns are prevented for most of the season by pest control and the use of resistant varietals. This organism is a fungus that is mainly found in alkaline soils. The plant's roots are killed by this organism, causing the plant to wither and perish quickly. Unfortunately, there is no resistant rootstock or cultivar. The only control is reconditioning the soil before planting, which is impractical at best. This entails using a soil acidifier to completely change the soil's pH in the area.

References

Aljane, F., & Ferchichi, A. (2006). Morphological, chemical, and sensory characterization of Tunisian Fig (Ficus carica L.) cultivars based on dried fruits. In *I International Symposium on Fresh Food Quality Standards: Better Food by Quality and Assurance 741* (pp. 81–85).

Aljane, F., & Ferchichi, A. (2009). Postharvest chemical properties and mineral contents of some fig (*Ficus carica* L.) cultivars in Tunisia. *Journal of Food Agricultural Environmental, 7*(209), e212.

Bostan, S. Z., Islam, A., Koyuncu, F., & Koyuncu, M. A. (1997). Investigations on some physical and chemical characteristics in fig cultivars grown in Ordu. In *I International Symposium on Fig 480* (pp. 87–90).

Caruso, G., Palai, G., Macheda, D., Tozzini, L., & Gucci, R. (2022). Growth, gas exchange, water relations, fresh and dry matter partitioning in young fig (*Ficus carica* L.) plants irrigated with saline water. *European Journal of Horticultural Science, 87*(2), 1–9. https://doi.org/10.17660/eJHS.2022/023

Çalişkan, O., & Polat, A. A. (2008). Fruit characteristics of fig cultivars and genotypes were grown in Turkey. *Scientia Horticulture, 115*(4), 360–367.

Chessa, I., Nieddu, G. & Serra, P. (1997). *June. Fig germplasm characterization using isozyme analysis*. In I International Symposium on Fig 480 (pp. 143–148).

Condit, I. (1955). Fig varieties: A monograph. *Hilgardia, 23*(11), 323–538. https://doi.org/10.3733/hilg.v23n11p323

Crane, J. C. (1964). Growth substances in fruit setting and development. *Annual Review of Plant Physiology, 15*(1), 303–326.

Darjazi, B. B. (2011). Morphological and pomological characteristics of Fig (*Ficus carica* L.) cultivars from Varamin, Iran. *African Journal of Biotechnology, 10*(82), 19096–19105.

El-Shazly, S. M., Mustafa, N. S., & El-Berry, I. M. (2014). Evaluation of some fig cultivars grown under water stress conditions in newly reclaimed soils. *Middle-East Journal of Scientific Research, 21*(8), 1167–1179.

Ferrara, E., & Papa, G. (2001). Evaluation of fig cultivars for breba crop. In *II International Symposium on Fig 605* (pp. 91–93).

Flaishman, M. A., Rodov, V., & Stover, E. (2008). Fig: botany, horticulture, and breeding.

Giraldo, E., López-Corrales, M., & Hormaza, J. I. (2010). Selection of the most discriminating morphological qualitative variables for characterization of fig germplasm. *Journal of the American Society for Horticultural Science, 135*(3), 240–249.

Harrison, R. D., & Yamamura, N. (2003). A few more hypotheses for the evolution of dioecy in figs (Ficus, Moraceae). *Oikos, 100*(3), 628–635.

Islam, A., Aygün, A., & Bostan, S. Z. (1997). A study on pomological characteristics of local fig cultivars in Northern Turkey. In *I International Symposium on Fig 480* (pp. 71–74).

Karami, H., Rezaei, M., Sarkhosh, A., Rahemi, M., & Jafari, M. (2018). Cold hardiness assessment in seven commercial fig cultivars (*Ficus carica* L.). *Gesunde Pflanzen, 70*(4), 195–203.

Khadari, B., Roger, J. P., Ater, M., Achtak, H., Oukabli, A., & Kjellberg, F. (2005). Moroccan Fig presents specific genetic resources: a high potential of local selection. In *III International Symposium on Fig 798* (pp. 33–37).

Mendoza-Castillo, V. M., Pineda-Pineda, J., Vargas-Canales, J. M., & Hernández-Arguello, E. (2019). Nutrition of fig (*Ficus carica* L.) under hydroponics and greenhouse conditions. *Journal of Plant Nutrition, 42*(11–12), 1350–1365.

Mendoza-Castillo, V. M., Vargas-Canales, J. M., Calderon-Zavala, G., Mendoza-Castillo, M. D. C., & Santacruz-Varela, A. (2017). Intensive production systems of fig (*Ficus carica* L.) under greenhouse conditions. *Experimental Agriculture, 53*(3), 339–350.

Mustafa, N. S., & Taha, R. A. (2012). Influence of plant growth regulators and subculturing on in vitro multiplication of some fig (*Ficus carica*) cultivars. *Journal of Applied Sciences Research, 8*(8), 4038–4044.

Owino, W. O., Nakano, R., Kubo, Y., & Inaba, A. (2004). Alterations in cell wall polysaccharides during ripening in distinct anatomical tissue regions of the fig (*Ficus carica* L.) fruit. *Postharvest Biology and Technology, 32*(1), 67–77.

Özeker, E., & Isfendiyaroglu, M. (1997). Evaluation of table fig cultivars in Cesme Peninsula. In *I International Symposium on Fig 480* (pp. 55–60).

Pereira, C., Martín, A., López-Corrales, M., Córdoba, M. D. G., Galván, A. I., & Serradilla, M. J. (2020). Evaluation of the physicochemical and sensory characteristics of different fig cultivars for the fresh fruit market. *Food, 9*(5), 619.

Pereira, C., Serradilla, M. J., Martín, A., del Carmen Villalobos, M., Pérez-Gragera, F., & López-Corrales, M. (2015). Agronomic behaviour and quality of six fig cultivars for fresh consumption. *Scientia Horticulturae, 185*, 121–128.

Polat, A. A., & Caliskan, O. (2008). Fruit characteristics of table fig (*Ficus carica*) cultivars in subtropical climate conditions of the Mediterranean region. *New Zealand Journal of Crop and Horticultural Science, 36*(2), 107–115.

Puebla, M., Toribio, F., & Montes, P. (2001). Determination of fruit-bearing pruning date and cutting intensity in "San Pedro" (*Ficus carica* L) type fig cultivars. In *II International Symposium on Fig 605* (pp. 147–157).

Saddoud, O., Baraket, G., Chatti, K., Trifi, M., Marrakchi, M., Salhi-Hannachi, A., & Mars, M. (2008). Morphological variability of Fig (*Ficus carica* L.) cultivars. *International Journal of Fruit Science, 8*(1–2), 35–51.

Stone, M. (2014). *How to hydroponics: A beginner's and intermediate's in depth guide to hydroponics*. Martha Stone.

Tous, J., & Ferguson, L. (1996). Mediterranean fruits. *Progress in New Crops, 416*, 430.

Vemmos, S. N., Petri, E., & Stournaras, V. (2013). Seasonal changes in photosynthetic activity and carbohydrate content in leaves and fruit of three fig cultivars (*Ficus carica* L.). *Scientia Horticulturae, 160*, 198–207.

Yancheva, S. D., Golubowicz, S., Yablowicz, Z., Perl, A. & Flaishman, M.A. (2005). Efficient Agrobacterium-mediated transformation and recovery of transgenic fig (Ficus carica L.) plants. *Plant Science, 168*(6), 1433–1441.

Chapter 10
Physiological Behaviour of Fig Tree (*Ficus carica* L.) Under Different Climatic Conditions

Aroua Ammar, Imed Ben Aissa, Faten Zaouay, Mohamed Gouiaa, and Messaoud Mars

Abbreviations

E	transpiration rate
F_V/F_M	maximum quantum yield of PSII
g_s	leaf stomatal conductance
P_N	photosynthesis rate
RH	relative humidity

1 Environmental Constraints on Fig Growth and Fruiting

Fig (*Ficus carica* L.) is one of the oldest fruit trees grown in the world (Mars, 2003; Oukabli et al., 2008), especially in the Mediterranean basin and the Middle East regions, which cover more than 90% of fresh fig production (FAOstat, 2020). Fig growth and production are mainly dependent on climatic conditions. As limited water resources have constantly threatened the stability of fig production, especially in arid and semi-arid regions (Can et al., 2000), interest is increasing in evaluating fig adaptation capacity to climatic variability. It aims to manage better water

A. Ammar (✉) · F. Zaouay · M. Mars
Research Laboratory on Agrobiodiversity & Ecotoxicology (LR21AGR02), Higher Agronomic Institute, Chott-Mariem, IRESA – University of Sousse, Sousse, Tunisia

I. Ben Aissa
Research Laboratory on Production & Protection for Sustainable Horticulture (LR21AGR03), Regional Research Centre on Horticulture and Organic Agriculture, BP 57, 4042, Chott-Mariem, IRESA – University of Sousse, Sousse, Tunisia

M. Gouiaa
Research Unit on Conservation and Valorization of Plant Resources (UR13AGR07), Higher Agronomic Institute, Chott-Mariem, IRESA- University of Sousse, Sousse, Tunisia

M. F. Ramadan (ed.), *Fig* (Ficus carica)*: Production, Processing, and Properties*, https://doi.org/10.1007/978-3-031-16493-4_10

resources, higher quality maintenance, and improved productivity. Traditionally, fig has been mainly cultivated under rain-fed conditions (Can et al., 2008); however, irrigation practices have improved fruit yield and quality (Melgarejo et al., 2007; Al-Desouki et al., 2009). It has been reported that the productive orientation of the fig orchard, timing of production, and fresh fig yield and quality are conditioned by climatic conditions, and fig grows well in regions with intense solar radiance, hot, dry summer, mild winters, and low relative humidity (Botti et al., 2003). However, high temperatures could accelerate vegetative growth and fig fruit maturity and decrease the active adult lifespan of the fig pollinator wasps (Jevanandam et al., 2013). Under hot weather in some regions of South America and when temperatures during winter reach more than 10 °C, dormancy and defoliation could be eliminated, and figs could continuously grow and produce fruits (Flaishman et al., 2008). The rainfall occurring in spring is useful for female fig tree growth but could negatively affect fig yield due to the interference with the life cycle of the fig wasp (*Blastophaga psenes*).

Furthermore, variable annual rainfall in rain-fed fig production areas leads to yield fluctuation (Bagheri & Sepaskhah, 2014). Fruit split may result from rain, and strong winds may cause fruit scarring (Flaishman et al., 2008). Moreover, young fig plants are susceptible to winter frost, and the increase in relative humidity associated with low temperatures could lead to fruit quality deterioration. Indeed, fig production is unsuitable for regions where the minimum temperatures could reach less than −8 °C (Roger, 2002). On the other hand, the fig tree is generally grown under marginal conditions such as drought and saline soils in some countries (Can et al., 2008); that is why organic fertilization is recommended to improve tree vigor and fruit quality (Mordoğan et al., 2013).

2 Drought Stress Mitigation by Different Approaches in Rain-Fed Fig Cultivation

The irrigation scheduling in fig orchards has gained significant importance, particularly in arid and semi-arid regions characterized by rainfall fluctuations and heatwaves. Despite the importance of irrigation for fig yield and quality, few studies were done worldwide on the quantification of fig water requirement. Previous research confirmed the positive effect of supplemental irrigation in improving fig tree morphological and physiological characteristics under drought conditions. According to Roger (2002), theoretical water requirements have been estimated at 600–700 mm per year. De Sousa Andrade et al. (2014) studied the effect of drip irrigation management on the growth and yield of the 'Roxo de Valinhos' fig cultivar grown in Brazil. It has been demonstrated that fig water requirement was affected by irrigation frequency, and mean crop coefficients Kc were about 0.71 and 0.51 for 2- and 4-days irrigation intervals, respectively.

In comparison with non-irrigated plants, irrigation increased about 27% in fruit total yield of irrigated fig trees at a 2-days interval. Abdolahipour et al. (2019) showed that the application of supplemental irrigation with 2000 l of water per tree in early spring near tree trunk during drought periods is recommended to provide high soil water status, promote vegetative growth, and increase the number and the size of fruits for fig trees grown in rain-fed orchards. However, in the case of limited water resources, the application of 1000 l of water per tree could also be efficient. Al-Desouki et al. (2009) also reported that applying supplemental irrigation by 80–90 mm/ tree in the first of May, June, July, or August and using anti-transpirant agents, namely folicote and vapor-guard, increased vegetative growth, fruit quality, and yield of the Egyptian 'Sultani' fig cultivar grown under rain-fed conditions. Honar et al. (2021) had recommended applying supplemental irrigation in March for rain-fed fig orchards in severe drought conditions to increase soil water content, new shoot length and diameter, number of leaves per new shoot, leaf width, and number of fruits on new shoots, fruit diameter, and fruit yield. Also, irrigation in August gave the best fruit yield because irrigation timing was close to the fruit ripening. Besides, it has been approved that irrigation in summertime improved shoot growth to prepare for the upcoming season's first crop proactively. Another study on 'Sultani' fig cultivar grown in Egypt indicated that foliar application of ascorbic acid and citric acid up to 300 ppm had improved vegetative growth as well as fruit number and characteristics such as total soluble solids, fruit dimension, and fresh and dry weights of the leaves, suggesting that foliar spray of ascorbic and citric acids could be used to enhance water stress tolerance of fig trees (Mohamed & El-Berry, 2021). Although the total yield and some quality characteristics were improved through irrigation (Melgarejo et al., 2007), the excess irrigation water must be avoided due to fig sensitivity to root rot (Tapia et al., 2003). Several additional methods to minimize drought risks in rain-fed fig orchards, including severe pruning and mulching to reduce leaf chlorosis, mechanical weed control and trunk whitening to enhance fruit weight (Zare et al., 2021). These horticultural practices were effective in conserving soil moisture and reducing evapotranspiration.

3 Physiological Characteristics of Fig Grown in Different Regions

Various fig cultivars have been described worldwide, showing an impressive physiological and behavioral plasticity under different climatic conditions. In semi-arid and arid ecosystems, the fig tree has characteristic adaptations spanning from physiological (photosynthesis, stomatal conductance, transpiration rate) to morphological (vegetative growth, leaf development) and phenological (dormancy, period of physiological activity) features (Table 10.1). It has been reported that the seasonal variation of ecophysiological traits may depend on various factors such as cultivar type, age, growing area, edapho-climatic conditions, and period of measurements (Table 10.1).

Table 10.1 Ecophysiological characteristics of different fig cultivars from different origins under different climatic conditions

Fig type and/or cultivar	Context of the study	Brief description of ecophysiological characteristics
Clones of 'Sarilop' (*Smyrna type*)	Evaluation of gas exchange capacity of fig grown under rain-fed conditions in Turkey (Kutlu et al., 2000)	Leaf characteristics determining photosynthesis capacity (stomatal density, leaf area, relative water content, and index of leaf succulence) were significantly different among all clones. The mean photosynthesis values were higher in August than in September, reaching a maximum of 32.15 μmol m^{-2} s^{-1}. The maximum value of E recorded in August was 22.62 mmol m^{-2} s^{-1}.
'Bursa Black', 'Gokalp', 'Morozer', 'Patlican', 'Sultan Selim', Morguz, 'Halebi', 'Sultani', 'Sarilop'	The gas exchange capacity of 15-years-old fig cultivars from different origins and grown under the same rain-fed conditions in Turkey (Can et al., 2000)	Significant differences between fig cultivars were revealed. The transpiration rate reached its maximum values at 11 am in 'Sarilop' and 'Sultani'. The highest transpiration rates were found in cultivars adapted to a milder and humid climate ('Patlican', 'Black Bursa' and 'Sultan Selim'). E and P_N rates were significantly higher in August compared to September (Maximum E was 20.5 mmol m^{-2} s^{-1} for 'Bursa Black' and maximum P_N was 29.4 20 μmol m^{-2} s^{-1} for 'Gokalp').
Caprifig cultivars	Measurement of mean leaf photosynthesis of 30-year-old trees located in France (Pigé et al., 2001)	CO_2 assimilation rate at 1500 mmol photons m^{-2} s^{-1} varied among caprifig cultivars from 9.4 to 13.3 μmol m^{-2} s^{-1}.
Clones of 'Sarilop' (*Smyrna type*)	Evaluation of seasonal and diurnal photosynthetic and transpiration rates of fig under semi-arid and rain-fed conditions in Turkey (Can and Aksoy, 2007)	Significant differences in P_N throughout the season, with maximum values for southern and northern parts being in August (36.3 and 22.8 μmol m^{-2} s^{-1}) and September (37 and 24.8 μmol m^{-2} s^{-1}) under relatively lower air temperatures (30–33°C). Leaf temperature was the primary factor limiting the gas exchange capacity of fig trees. g_s was the major control mechanism of P_N and E, particularly in the northern parts of the trees. Maximum g_s reached 600 mmol m^{-2} s^{-1} around mid-July.

(continued)

Table 10.1 (continued)

Fig type and/or cultivar	Context of the study	Brief description of ecophysiological characteristics
'Yediveren', 'Banana Fig', 'Nazareth'(*San Pedro*) 'Siyah Çiçek' and 'Beyaz Çiçek','Noire de Coromb'(*Common type*)	Determination of gas exchange capacity of different fig cultivars in Turkey during June (time of harvest) and August (period of heaviest drought stress) (Can et al., 2008)	The measurements showed higher photosynthetic activity in June and higher transpiration rates in August P_N varied between 18 to 25 μmol m^{-2} s^{-1}, E varied from 10 to 20 mmol m^{-2} s^{-1} and g_s ranged from 300 to 400 mmol m^{-2} s^{-1} in June. In August, P_N varied between 15 to 20 μmol m^{-2} s^{-1}, E ranged from 11 to 25 mmol m^{-2} s^{-1} and g_s reached 270 to 370 mmol m^{-2} s^{-1}. Significant differences were detected in gas exchange parameters among fig cultivars. 'Nazareth', 'Siyah Çiçek' and 'Beyaz Çiçek' were group together; whereas 'Banana Fig', 'Noire de Coromb' and 'Yediveren' comprised another second group. Stomatal conductance depended on tree water status, leaf and air temperatures, and photosynthetically active radiation.
'De a Libra' (*Common type*)	The gas exchange capacity of 9-years-old fig in Spain (González-Rodríguez and Peters, 2009)	P_N varied during leaf expansion. P_N ranged from 4μmol m^{-2} s^{-1} (13 days after bud break), when leaf area was 23% of maximum, to 20 μmol m^{-2} s^{-1} on day 51. Maximum g_s and E, at full leaf expansion were 0.4 ± 0.01 mol m^{-2} s^{-1} and 7.6 ± 0.09 mmol m^{-2} s^{-1}, respectively. A marked decrease in P_N and g_s was registered on days 30 and 85, correlated with low relative humidity and high temperature.
'Kymis', 'Fracasana', 'San Pedro' and 'Politiko' (white cultivars) and 'Vasilika Black', 'Vasilika Melia' and 'Mission' (black cultivars)	Evaluation of the photosynthetic activity in 5-year-old fig cultivars in Greece (Pisimisi et al., 2012)	P_N varied among cultivars and with the season. 'Vasilika Black' had the highest value of P_N (17 μmol m^{-2} s^{-1}). Most of the cultivars had the highest P_N values in June, coinciding with early fruit development.

(continued)

Table 10.1 (continued)

Fig type and/or cultivar	Context of the study	Brief description of ecophysiological characteristics
European fig cultivars: 'Mission' and 'Kalamon' (*Smyrna type*) 'Fracasana' (*San Pedro type*)	Investigation of the seasonal changes in photosynthetic activity and carbohydrate content in leaves and fruit of three fig cultivars grown under rain-fed conditions in Greece (Vemmos et al., 2013)	P_N and g_s reached maximum values in May when leaves have reached maximum development and increased again during fruit maturation. 'Kalamon' had significantly higher P_N (18.2 μmol m^{-2} s^{-1}) than the other two cultivars, attributed to the high g_s of this cultivar (350 mmol m^{-2} s^{-1}) as to the better leaf development. Breba crop in 'Fracasana' early in the season (April) did not affect leaf $P_{N,}$ so leaves are not assimilated exporters, and the main carbohydrate source for this early fruit growth was that stored in shoots and other parts of the fig tree.
'Bither Abiadh' (*San Pedro type*) and 'Zidi' (*Smyrna type*)	Evaluation of seasonal variation of ecophysiological characteristics of young potted- fig plants fully irrigated in Tunisia (Ammar et al., 2020a)	The vegetative cycle of a fig tree under normal conditions 'without water stress' comprises 4 phases of growth: a phase of intense growth (May-June: the beginning of the summer season), a slower growth phase (July-August), a medium growth phase (September-October) and "rest" phase during winter. The ecophysiological behaviour of fig cultivars was influenced by climate change. Changes in P_N, $g_{s,}$ and chlorophyll content were not significantly affected by cultivar, and their variation was attributed to year. Maximum g_s reached 435.4 and 370.6 mmol m^{-2} s^{-1} for 'Zidi' and 'Bither Abiadh' in the spring season (81–88 days after bud break). E rates fluctuated during all the growing seasons and did not exceed 2.5 mmol m^{-2} s^{-1} for both cultivars. The lowest values of P_N and g_s were recorded in summer (156 days after bud break) and at the end of the growing season in autumn.

E (mmol m^{-2} s^{-1}): Transpiration rate, P_N (μmol m^{-2} s^{-1}): Photosynthesis rate, g_s (mmol m^{-2} s^{-1}): Leaf stomatal conductance

4 Fig Responses and Resilience Towards Drought Under Different Climatic Conditions

To help predict future impacts of climate change on fig tree growth and yield, more and more profound knowledge of the ecophysiological mechanisms of the fig tree is needed. Simple stress indicators based on easily measurable traits are needed to overcome this. Different physiological behaviors under water stress have been observed in fig; according to the cultivar, the growing conditions, and the degree of stress imposed on fig. Genotype related differences have been described in fig responses to drought in terms of vegetative growth (El-Shazly et al., 2014; Rostami & Rahemi, 2013a; Shahidi-Rad et al., 2015), fruit production (Jafari et al., 2012), biochemical changes including increased activity of radical scavenging enzymes such as peroxidase (POD), superoxide dismutase (SOD), ascorbic peroxidase (APX) and catalase (CAT), osmolytes accumulation in particular proline (Shirbani et al. 2013; Rostami & Rahemi, 2013b; Shahidi-Rad et al., 2015) and producing proteolytic enzymes such as chlorophyllase (Gholami et al., 2012). Some different cultivars may behave similarly. For example, Ammar et al. (2020b) found that 'Zidi' as *Smyrna type* and 'Bither Abiadh' as *San Pedro type* exhibited the same response to water stress and recovery.

Studies aiming to understand fig tree gas exchange capacity in response to water deficit are not frequent. However, Ammar et al. (2020b) demonstrated that stomatal control of gas exchange was crucial for fig drought response. Indeed, the reduction of stomatal conductance has been considered an immediate response to water stress before any change was detectable in leaf water content (Pirasteh-Anosheh et al., 2016), showing the important role played by abscisic acid (ABA) in promoting stomatal closure (Bondada & Shutthanandan, 2012).

To cope with water stress, fig trees developed two different strategies to overcome the dry and hot summer period and complete their life cycle. In this sense, an adaptive strategy relies on maintaining low rates of active photosynthesis and reducing water loss. This took place when moderate water stress was applied by reducing fig water requirements by 40 and 60% (Caruso et al., 2017) or by 50% (Ammar, 2020), which induced a decrease in stomatal conductance (less than 100 mmol m^{-2} s^{-1}) associated with a reduction of photosynthesis rates. Besides, moderate water stress also affected the plant water status by decreasing leaf relative water content and transpiration rate and an increase of leaf temperature owing to stomatal closure, with not very high rates of leaf senescence. Furthermore, regarding fig fruit quality, regulated deficit irrigation with 55% ETc significantly reduced fruit skin side cracking and ostiole-end splitting at harvest, improving marketable yield (Kong et al., 2013).

Prolonged drought by suspending irrigation for 2 weeks (Ammar et al., 2020b) or applying prolonged moderate water stress by reducing to 50% the water supply combined with high temperatures exceeding 45 °C (Ammar, 2020) induced a sharp decrease in stomatal conductance, photosynthesis activity, transpiration rate and maximum quantum yield of the photosystem II (PSII), chlorophyll degradation and necrotic spots in leaves, resulting in growth inhibition and leaf abscission. This corresponded to an avoidance strategy to overcome drought periods associated with extreme heat stress during summer. This pattern could be considered 'summer rest' when the fig tree exhibits a temporary growth suspension and massive leaf abscission to evade the unfavorable hot and dry season (Shahidi-Rad et al., 2015; Ammar et al., 2020b).

Fig has revealed 'drought stress memorizing' by showing a rapid re-growth after subsequent re-irrigation, extending the annual vegetative cycle and biomass accumulation. Indeed, plants could survive after leaf fall by keeping the stem alive during the drought period and accumulating carbohydrates in the stem and roots to stimulate the emergence of new leaves once the stress is relieved (Xu et al., 2008, 2012; Naidoo & Naidoo, 2018). Furthermore, plant recovery after re-irrigation induced an increase in fructose in leaves related to a decrease of sucrose in leaves and stems, suggesting sucrose cleavage and mobilization for the growing buds after leaf abscission (Xu et al., 2008).

Fig plants, which were more affected by high air temperatures combined with water stress, required less time to resume vegetative development (Ammar, 2020). Furthermore, new leaves of re-watered fig plants showed high stomatal conductance reaching more than 300 mmol m^{-2} s^{-1}, associated with high photosynthesis activity and chlorophyll content compared to leaves of non-stressed plants. Also, water stress recovery resulted in an extended vegetative growth until December (Ammar et al., 2020b). The cycle length of the fig could vary according to the genotype, solar radiation, accumulated degree-days, and air temperature as well (Ferraz et al., 2020).

Thus, the strategy to cope with water scarcity is mainly oriented toward enduring the summer period under different climatic conditions, as summarized in Fig. 10.1.

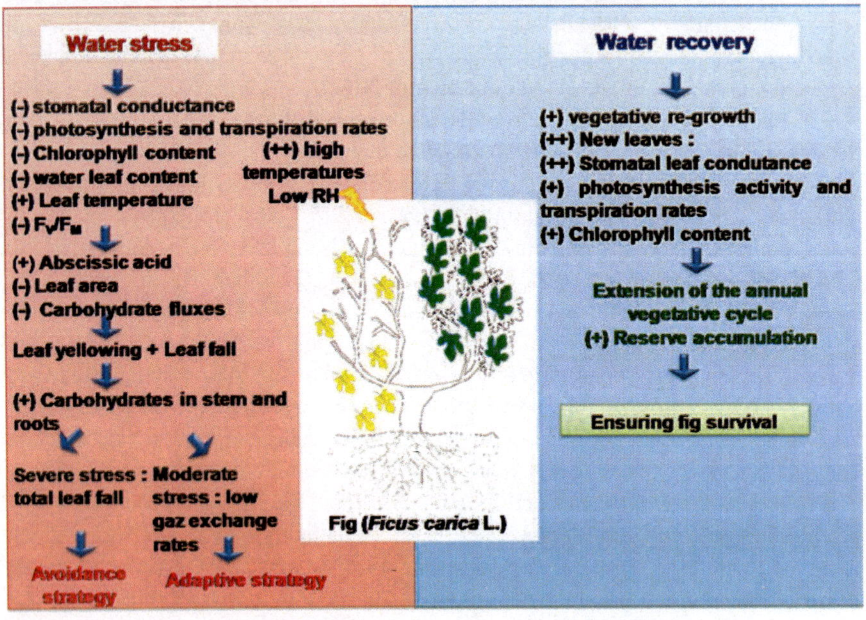

Fig. 10.1 Fig ecophysiological responses and strategies to water stress and recovery (–): negative effect, (+): positive effect

Regardless of water stress intensity and duration, fig could be classed as a drought-stress resilient species considering its rapid growth recovery. In this sense, the interest in understanding fig physiological control of abiotic stresses has increased. Nevertheless, most published research was conducted on potted-fig plants and covered just a short period of exposure to water stress in experimental conditions, which is not comparable to the real field conditions. Furthermore, as fig represents a vast genetic diversity (Moniruzzaman et al., 2020), more effort is needed better to understand different fig cultivars'behavior under field conditions and identify the most suitable cultivars for cultivation in a given region.

References

Abdolahipour, M., Kamgar-Haghighi, A. A., Sepaskhah, A. R., Zand-Parsa, S., Honar, T., & Razzaghi, F. (2019). Time and amount of supplemental irrigation at different distances from tree trunks influence morphological characteristics and physiological responses of rainfed fig trees under drought conditions. *Scientia Horticulturae, 253*, 241–254.

Al-Desouki, M., Abd El-Rahman, I., & Sahar, A. (2009). Effect of some antitranspirants and supplementary irrigation on growth, yield and fruit quality of Sultani fig (*Ficus carica*) grown in the Egyptian western coastal zone under rainfed conditions. *Research Journal of Agriculture and Biological Sciences, 5*, 899–908.

Ammar, A. (2020). Caractérisation écophysiologique et adaptation au stress hydrique du figuier (*Ficus carica* L.). Ph.D dissertation, Higher Institute of Agronomy (ISA) of Chott Mariem, IRESA-University of Sousse, Tunisia. 163pp.

Ammar, A., Ben Aissa, I., Mars, M., & Gouiaa, M. (2020a). Seasonal variation of fig tree (*Ficus carica* L.) physiological characteristics reveals its adaptation performance. *South African Journal of Botany, 132*, 30–37. https://doi.org/10.1016/j.sajb.2020.04.020

Ammar, A., Ben Aissa, I., Mars, M., & Gouiaa, M. (2020b). Comparative physiological behavior of fig (*Ficus carica* L.) cultivars in response to water stress and recovery. *Scientia Horticulturae, 260*, 108881. https://doi.org/10.1016/j.scienta.2019.108881

Bagheri, E., & Sepaskhah, A. R. (2014). Rain-fed fig yield as affected by rainfall distribution. *Theoretical and Applied Climatology, 117*, 433–439.

Bondada, B., & Shutthanandan, J. (2012). Understanding differential responses of grapevine (*Vitis vinifera* L.) leaf and fruit to water stress and recovery following re-watering. *American Journal of Plant Science, 3*(9), 1232–1240.

Botti, C., Franck, N., Prat, L., Ioannidis, D., & Morales, B. (2003). The effect of climatic conditions on fresh fig fruit yield, quality, and type of crop. *Acta Horticulturae, 605*, 37–42.

Can, H. Z., & Aksoy, U. (2007). Seasonal and diurnal photosynthetic behavior of fig (*Ficus carica* L.) under semi-arid climatic conditions. *Acta Agriculturae Scandinavica, Section B – Soil & Plant Science, 57*(4), 297–306.

Can, H. Z., Hepaksoy, S., Akoy, U., & Kutlu, E. (2000). Leaf characteristics and net gas exchange of fig cultigens adapted to different climatic zones. *Acta Horticulturae, 516*, 131–137.

Can, H. Z., Meyyaci, K. B., & Balci, B. (2008). Determination of gas exchange capacity of some Breba fig cultivars. *Acta Horticulturae, 798*, 117–112.

Caruso, G., Gennai, C., Ugolini, F., & Gucci, R. (2017). Gas exchange and growth response of young *Ficus carica* L. plants to water deficit and relief. *Acta Horticulturae, 1173*, 157–161.

De Sousa Andrade, I. P., De Carvalho, D., De Almeida, W. S., Gonçalves Silva, J. B., & Da Silva, L. D. B. (2014). Water requirement and yield of fig trees under different drip irrigation management. *Engenharia Agrícola, 34*(1), 17–27.

El-Shazly, S. M., Mustafa, N. S., & El-berry, I. M. (2014). Evaluation of some fig cultivars grown under water stress conditions in newly reclaimed soils. *Middle East Journal of Scientific Research, 21*, 1167–1179.

FAOstat. (2020). URL:http://www.fao.org/faostat/fr

Ferraz, R. A., Leonel, S., Souza, J. M. A., Ferreira, R. B., Modesto, J. H., & Arruda, L. L. (2020). Phenology, vegetative growth, and yield performance of fig in Southeastern Brazil. *Pesquisa Agropecuária Brasileira, 55*, e01192.

Flaishman, M. A., Rodov, V., & Stover, E. (2008). The fig: Botany, horticulture, and breeding. *Horticultural Reviews, 34*, 113–197.

Gholami, M., Rahemi, M., & Rastegar, S. (2012). Use of rapid screening methods for detecting drought tolerant cultivars of fig (*Ficus carica* L.). *Scientia Horticulturae, 143*, 7–14.

González-Rodríguez, A. M., & Peters, J. (2009). Strategies of leaf expansion in *Ficus carica* under semiarid conditions. *Plant Biology, 12*, 469–474.

Honar, T., Shabani, A., Abdolahipour, M., Dalir, N., Sepaskhah, A. R., Haghighi, A. A., & Jafari, M. (2021). Rain-fed fig trees response to supplemental irrigation timing and potassium fertiliser in micro-catchment. *The Journal of Horticultural Science and Biotechnology.* https://doi.org/10.1080/14620316.2021.1923409

Jafari, M., Abdolahi Pour Haghighi, J., & Zare, H. (2012). Mulching impact on plant growth and production of rainfed fig orchards under drought conditions. *Journal of Food, Agriculture and Environment, 10*, 428–433.

Jevanandam, N., Goh, A. G., & Corlett, R. T. (2013). Climate warming and the potential extinction of fig wasps, the obligate pollinators of figs. *Biology Letters, 9*(3), 20130041. https://doi.org/10.1098/rsbl.2013.0041

Kong, M., Lampinen, B., Shackel, K., & Crisosto, C. H. (2013). Fruit skin side cracking and ostiole-end splitting shorten postharvest life in fresh figs (*Ficus carica* L.), but are reducedby deficit irrigation. *Postharvest Biology and Technology, 85*, 154–161.

Kutlu, E., Can, H. Z., Aksoy, U., & Hepaksoy, S. (2000). Evaluation of gas exchange capacity and physiological responses of selected Sarilop (Caliyrma) fig clones. *Acta Horticulturae, 517*, 59–63.

Mars, M. (2003). Fig (*Ficus carica* L.) genetic resources and breeding. *Acta Horticulturae, 605*, 19–27.

Melgarejo, P., Martínez, J. J., Hernández, F., Salazar, D. M., & Martínez, R. (2007). Preliminary results on fig soil-less culture. *Scientia Horticulturae, 111*, 255–259.

Mohamed, Y. I., & El-Berry, I. M. (2021). Alleviation of drought stress on rainfed fig (*Ficus carica* L.) trees in Egypt using the foliar application of Ascorbic and Citric Acids. *Acta Horticulturae, 1310*, 235–242. https://doi.org/10.17660/ActaHortic.2021.1310.36

Moniruzzaman, M., Anuar, N., Yaakob, Z., Aminul Islam, A. K. M., & Al-Khayri, J. M. (2020). Performance evaluation of seventeen common fig (*Ficus carica* L.) cultivars introduced to a tropical climate. *Horticulture, Environment, and Technology, 61*, 795–806.

Mordoğan, N., Hakerlerler, H., Ceylan, Ş., Aydın, Ş., Yağmur, B., & Aksoy, U. (2013). Effect of organic fertilization on fig leaf nutrients and fruit quality. *Journal of Plant Nutrition, 36*, 1128–1137.

Naidoo, G., & Naidoo, K. K. (2018). Drought stress effects on gas exchange and water relations of the invasive weed *Chromolaena odorata*. *Flora, 248*, 1–9.

Oukabli, A., Mekaoui, A., Ibnouali-El-Aloui, M., & Bari, A. (2008). Contribution to identification of fig (*Ficus carica* L.) genotypes tolerant to drought. *Acta Horticulturae, 798*, 87–93.

Pigé, L. C., Salager, J.-L., Hossaert-Mckey, M., & Roy, J. (2001). Carbon allocation to volatiles and other reproductive components in male *Ficus carica* (Moraceae). *American Journal of Botany, 88*, 2214–2220.

Pirasteh-Anosheh, H., Saed-Moucheshi, A., Pakniyat, H., & Pessarakli, M. (2016). *Stomatal responses to drought stress. Water stress and crop plants: A sustainable approach* (pp. 24–40). Wiley.

Psimisi, E., Vemmos, S. N., & Mili, E. (2012). The photosynthetic activity and evaluation of fruit quality in seven fig cultivars (*Ficus carica* L.). *Acta Horticulturae, 940*, 341–348. https://doi.org/10.17660/ActaHortic.2012.940.48

Roger, J.P. (2002). La conduite du figuier *Ficus carica* L. famille des moracées genre *Ficus*. Synthèse. In *Acte de la journée figuier. Potentialités et perspectives de développement de la figue sèche au Maroc* (pp. 32–41).

Rostami, A. A., & Rahemi, M. (2013a). Responses of caprifig genotypes to water stress and recovery. *Journal of Biological & Environmental, 7*(21), 131–139.

Rostami, A. A., & Rahemi, M. (2013b). Screening drought tolerance in caprifig varieties in accordance to responses of antioxidant enzymes. *World Applied Sciences Journal, 21*(8), 1213–1219.

Shahidi-Rad, K., Shekafandeh, A., & Jamali, B. (2015). Physiological and antioxidant enzymes responses of two fig cultivars under drought stress. *Jordan Journal of Agricultural Sciences, 11*(2), 381–391. https://doi.org/10.12816/0030433

Shirbani, S., Pour Hagigi, J. A., Jafari, M., & Davarynejad, G. H. (2013). Physiological and biochemical responses of four edible fig cultivars to water stress condition. *Scholarly Journal of Agricultural Science, 3*(11), 473–479.

Tapia, R., Botti, C., Carrasco, O., Prat, L., & Franck, N. (2003). Effect of four irrigation rates on growth of six fig tree varieties. *Acta Horticulturae, 605*, 113–118.

Vemmos, S. N., Petri, E., & Stournaras, V. (2013). Seasonal changes in photosynthetic activity and carbohydrate content in leaves and fruit of three fig cultivars (*Ficus carica* L.). *Scientia Horticulturae, 160*, 198–207.

Zare, H., Zare, E., Sedaghat, S., & Jafari, M. (2021). Investigation of different horticultural practices to minimize drought impacts in rainfed fig (*Ficus carica* L. 'Sabz'). *Acta Horticulturae, 1310*, 211–216. https://doi.org/10.17660/ActaHortic.2021.1310.32

Chapter 11
Fig (*Ficus carica* L.) Production and Yield

Malek Ben Temessek, Wafa Talbi, Hana Chrifa, and Sami Fattouch

Abbreviations

CEC	Cation Exchange Capacity
GA	Gibberellic acid
Kg ha	Kilogram per hectare
SSC	Soluble solid concentration
TA	Titratable acidity
TSS	Total soluble solid

1 Conventional and Molecular Breeding Systems in Fig (*Ficus carica* L.)

The fig tree (*Ficus carica* L.) is a diploid heterozygous (2n = 26) plant. The DNA 2C levels are quite low, ranging from 1.37 pg to 1.37 pg, corresponding to 270 million base pairs and 50,000 genes. People from Persia, Asia Minor, and Syria spread the fig tree throughout the Mediterranean region. It has been an essential food culture for thousands of years, and it is considered highly good in the Mediterranean diet (Solomon et al., 2006). Thousands of cultivars, most unnamed, have been developed or introduced, and the fig has spread far beyond its natural area due to human migration. Most Mediterranean countries, Europe, North, South Africa, America, and many other Asian countries now grow them. The monoic ancestor of

W. Talbi

M. Ben Temessek (✉) · H. Chrifa · S. Fattouch
EcoChem Laboratory, Department of Biological and Chemical Engineering, National Institute of Applied Sciences and Technology (INSAT), University of Carthage (UCAR), Tunis, Tunisia
e-mail: sami.fattouch@insat.ucar.tn

M. F. Ramadan (ed.), *Fig* (Ficus carica)*: Production, Processing, and Properties*, https://doi.org/10.1007/978-3-031-16493-4_11

the common fig tree (Machado et al., 2001) evolved into gynodio species with bisexual (functioning) male figs or caprifigs and unisexual female trees. Female and male flowers are arranged inside the syconium, a constructed fig fruit structure. Hundreds of female flowers are included in a female fruit. A male fig plant, called caprifig, has hundreds of separated male (staminate) flowers and female (pistillate) flowers. The differentiation between the hermaphroditic and female strains forms establishes the basis for maintaining the close symbiotic relationship between *Ficus* plants and the Blastophaga wasps (Galil & Neeman, 1977).

Plant breeding is a dynamic area of applied horticultural science. It relies on genetic variation and uses breeding to enhance traits and characteristics of interest to growers and consumers. The genus *Ficus* has two reproductive systems, the monocecity (approx. 400 species) and gynodioecy (350 species, including *Ficus carica* L.) (Berg, 1989). In *Ficus*, monoecy is hypothesized to be ancestral, and gynodioecy derived. In monoecious figs, each fruit (syconium) has both male and female florets functioning to produce pollen and viable seed. Some florets are parasitized to produce wasps that carry the pollen to other trees. In gynodioecious figs, separate trees produce syconia creating either viable seed or pollen and wasps. The studies of fig floral biology and the understanding of the coevolution with the fig allowed the establishment of fig breeding techniques (Weiblen, 2002). The juvenile stage can be reduced by grafting seedling buds into older trees or pumping and anchoring them in a vertical position at the beginning of development (Stover et al., 2007) and then growing them intensively, using the last method under the warm climate conditions of Israel. Indeed, Flaishman et al. (2008) obtained reproductive development and performed fruit evaluation as early as one year after seed extraction. The majority of today's commercial fig varieties are thought to be the product of unintentional selections made by unknown growers. Local fig collections exist in several places, used to select superior clones (Mars et al., 1997; Mars et al., 2009). Due to its very narrow industry, fig breeding projects have a restricted budget and a small number of breeding programs.

In the United States, two breeding programs were conducted at the beginning of the twentieth century; the largest was at the University of California, led by I.J. Condit and Storey (begun by R.E. Smith in 1922). At the same time, Slykov led another breeding program in the Soviet Union. Doyle and Ferguson (1997) continued the fig breeding program in California. In addition, fig crosses and the heritage of various fig features have been described in many countries like Japan (Awamura et al., 1996, 1997), Turkey, and other nations.

2 Fig Production, Research, and Development

Fig fruits are transformed into several processing products, such as dried figs, preserved fruits, jam, juice, wine, powder, and others, but the most popular ones are preserved fruits and jam. There are nearly 20 figs-processing factories on the mainland. Fig fruits and leaves contain many amino acids and inorganic elements, showing their high nutritional value. Furthermore, fig contains several medical components such as flavones, rutin, and quercetin, which could be used in cardiovascular disease medicine production. Additionally, the cooperative research showed that the extracts from fig had significant inhibition in the growth of several cancer cells in mouse bodies, the inhibition for spleen cancer high up to 64.0%, EAC 53.8%, Lewis lung cancer 48.8%, HAC liver cancer 44.4% and S180 cancer 41.8%. Furthermore, the nontoxic extracts of figs could increase the body's immunity and postpone the life span (Arpaci, 2015).

3 Fig Fruit Latex Yield and Protease Activity as Influenced by Cultivar, Maturity Stage and Harvesting Time

Laticiferous cells generate fig latex, commonly coupled with other botanical components like a blue flag, barley, fenugreek, ginger, and hot pepper to soften solid tumors and treat skin problems (Lansky et al., 2008). Most species' latex contains a wide range of physiologically active chemicals. Many of these chemicals provide herbivore resistance through toxicity or antinutritive effects, while others contribute to the stickiness that can ensnare insect herbivores (Lazreg-Aref et al., 2011). Several defense-related components (e.g., rubber, cysteine protease, alkaloids) appear in the latex of distant phylogenetic groups, suggesting common functions and convergent evolution (Agrawal & Konno, 2009). The volatile compounds released by fig latex comprise mainly mono- and sesquiterpenes (Hossaert-McKey et al., 2010). Oliveira et al. (2010) identified 34 compounds distributed by distinct chemical classes: five aldehydes, seven alcohols, one ketone, nine monoterpenes, nine sesquiterpenes, and three other compounds. Moreover, thirty-six coumarins were identified from Tunisian caprifig latex, which caused a strong bactericidal effect (Lazreg-Aref et al., 2012). The volatile compounds are critical factors in the fig-fig wasp mutualism. Indeed, each wasp species seems to be attracted to specific chemical compounds of its associated fig tree species (Chen & Song, 2008; Chen et al., 2009). Maffei (2010) demonstrated the involvement of the volatile organic compounds in the attraction of pollinating and predatory insects.

In some cases, pollinators of some species are stimulated by the emitted volatile compounds of their associated fig species and generally not by the volatile compounds of another species (Grison-Pigé et al., 2002). Fig latex is involved in defense mechanisms by protecting ripening fruits against plant pathogens like fungi and insects, which they can attack directly (Konno et al., 2004). There was a substantial difference in latex yield for the several fig cultivars between harvesting time and the fruit maturity process. The highest values were achieved in the mid maturity stage for all cultivars, with no significant changes between stages (Table 11.1).

Table 11.1 Protease activity and latex yield of Smyrna-type fig cultivars during three times a day

Cultivar	Maturity stage	Latex harvesting time	Protease activity (U/ml)	Latex yield (ml)	Cultivar	Maturity stage	Latex harvesting time	Protease activity (U/ml)	Latex yield (ml)
Khedri	Unripe	1	324685^{a}	0.21^{b}	Bidhi	Unripe	1	523363^{b}	0.20^{NS}
		2	204671^{b}	0.22^{b}			2	556677^{ab}	0.00^{NS}
		3	419712^{a}	0.44^{a}			3	598765^{c}	0.40^{NS}
	Mid-maturity	1	229760^{a}	0.50^{b}		Mid-maturity	1	191212^{a}	0.50^{NS}
		2	303214^{b}	0.49^{b}			2	102312^{b}	0.50^{NS}
		3	330329^{a}	0.80^{a}			3	113321^{b}	0.80^{NS}
	Ripe	1	11563^{c}	0.10		Ripe	1	89754^{a}	0.12^{b}
		2	102866^{b}	0.10			2	34254^{c}	0.10^{b}
		3	228427^{a}	0.10			3	66654^{b}	0.20^{a}
Chetoui	Unripe	1	1,008,181	0.10^{NS}	Zidi	Unripe	1	6940901^{b}	0.10^{NS}
		2	201,315	0.10^{NS}			2	443312^{c}	0.10^{NS}
		3	1,141,515	0.10^{NS}			3	798865^{a}	0,30 ^{NS}
	Mid-maturity	1	131,567	0.20^{NS}		Mid-maturity	1	988076^{a}	0,30 ^{NS}
		2	123,899	0.10^{NS}			2	808765^{b}	0,30 ^{NS}
		3	1,107,181	0.40^{NS}			3	1012345^{a}	0,60 ^{NS}
	Ripe	1	1205181b	0.10^{NS}		Ripe	1	990865^{a}	0,20 ^{NS}
		2	123,498	0.05^{NS}			2	809743^{b}	0,17 ^{NS}
		3	191,418	0.20^{NS}			3	988065^{a}	0,30 ^{NS}
Soltani Abiadh	Unripe	1	634090^{a}	0.70^{NS}	Delgane	Unripe	1	580909^{b}	0.20^{NS}
		2	609854^{ab}	0.30^{NS}			2	446785^{c}	0.10^{NS}
		3	568743^{b}	0.30^{NS}			3	665432^{a}	0.40^{NS}
	Mid-maturity	1	556698^{b}	0.50^{NS}		Mid-maturity	1	764367^{b}	0.50^{NS}
		2	412211^{c}	0.40^{NS}			2	708743^{b}	0.30^{NS}
		3	709854^{a}	0.90^{NS}			3	806534a	0.70^{NS}
	Ripe	1	209812b	0.35^{NS}		Ripe	1	664365^{a}	0.10^{NS}
		2	123421a	0.20^{NS}			2	609865^{a}	0.10^{NS}
		3	223254a	0.40^{NS}			3	667898^{a}	0.20^{NS}

Results are expressed as means ± deviation (n = 3). Different small letters within one cultivar are significantly ($P < 0.05$) different according to Duncan test. Capital letters within cultivars are significantly different ($P < 0.05$) according to Duncan test. *Significant at $P < 0.05$. *NS* non-significant. (1: morning, 2: mid-day and 3: late afternoon) afternoon). (Lazreg-Aref et al., 2018)

4 Fig Production in Subtropical South-Western Saudi Arabia

Fig has been considered an essential crop for the people of southwest Asia for quite a long time. Persians, Arabs, Canaanites, and Egyptians cultivated figs for more than 3500 years (Obenauf et al., 1978). Despite being cultivated in this part of the world for millennia, fig Production remains limited in Saudi Arabia. Two main cultivars of figs could be found in the southwestern part of the Kingdom, 'Baladi' and 'Brown Turkey'. Both cultivars are of the common type. The first is a local cultivar with small sweet fruit a yellowish-green skin. The latter, introduced in the early eighties, has a medium-size fruit with brownish skin. The fruit matures in late summer at high elevation (Assir) and winter (December–February) (off-season) in the plains and foot hills. Fruit quality in Assir is good, but farm size is limited by the lack of water for irrigation. Therefore, most acreage is concentrated in the coastal region, where water for irrigation is available. The Saudi Ministry of Agriculture does not publish separate statistics for fig production or acreage.

However, the terminal report of the FAO/Saudi Arabian Government Cooperation Program (UTFN/SAU.013/SAU) gives an estimate for the number of fig trees in the Jazan province alone to be about 311,300 (FAO, 2007). However, heavy tree losses to termites and diseases were reported by growers during this last decade. Nevertheless, a sizable crop is produced with small quantities (150–200 t) of fresh fruit exported to Europe and Japan. Presently, the crop is mainly sold locally for reasonable prices; no drying or canning is practiced. However, as new plantings come to bearing, such infrastructure will be needed to maintain orchard profitability. Furthermore, the lack of an efficient marketing strategy thwarting fig production progress despite the region's enormous potential.

5 Fig Pollination and Caprification System

The fig tree (*Ficus carica* L.) is a functionally dioecious species. The male fig tree or caprifig ensures the pollen production and survival of the symbiotic pollinator, blastophagus (*Blastophaga psenes* L.). The blastophage reproduces exclusively in caprifig. Female or domestic fig trees produce figs pollinated in July and ripe in autumn: these are the autumn figs. A few fig buds do not develop during the summer but only begin to grow the following spring.

On most domestic figs, these figs fall before the stage of pollen receptivity; it is the fall early. Nevertheless, some particular genotypes retain figs to receptivity. At this stage, the figs cannot be pollinated because there is no pollen available at this time of year, but they usually continue to develop by parthenocarpic way to give ripe figs in July; fig flowers. These fig trees are called bifères as opposed to fig trees that lose their flower figs before receptivity, said unifying. Much has been written about the extraordinary complexity of the biological reproduction of Ficus species, particularly regarding pollination (Dickson & Dickson, 1996). Ficus species are gynodious and functionally dioecious. Some of them are functionally female and produce only one fruit seed, while others are functionally male and produce only pollen and pollen wasp offspring (pollen carrier); it is then Blastophaga psenes that brings the pollen of the male flower to the female flower (Janzen, 1979; Weibes, 1979). On figs, receptivity is essential in developing the fruit distinguished by volatile emissions to attract pollinators.

Pollination is essential for proper fruit development (Wagner et al., 1999). Blastophaga wasp and caprifig are necessary for caprification and normal fruit development. If this fertilization process does not occur, the fruits do not develop properly and will fall from the tree. Many fig cultivars are totally or partially parthenocarps; the synconium swells without pollination and does not have only underdeveloped infertile seeds (Dickson & Dickson, 1996). Pollinated figs are generally larger and greener than unpolluted figs, with a darker pulp (Oukabli et al., 2003). In fig orchards, for high economic return, the pollination process must be repeated two or three times as the figs' synconium gradually becomes receptive. Thus, it is essential to have two or three-period cultivars (Zare et al., 2013). The number of spent caprifigs and the pollination period depend on the female cultivar's weather conditions, age, size, and yield. The common practice of caprification is to distribute the male figs at intervals of a few days over approximately 3 weeks (Rahemi & Jafari, 2005; Zare et al., 2013).

6 Micropropagation of *Ficus carica* L. Studies on the Propagation of Fig Cultivars by Budding

In India, figs are considered small commercial fruit because their output cannot compete with other commercial fruit crops. The major causes for insufficient yield include a lack of knowledge on varietal introduction and characterization and a scarcity of high-quality planting material. Tip cuttings are used to grow fig trees (Dolgun & Tekintas, 2008; Lionakis, 1995; Mansour, 1995; Kai et al., 1997). Fig cuttings usually are simple to root; however rapid variations in air temperature and soil moisture conditions impair rooting and shoot growth. Furthermore, soil-borne bacterial and fungal infections impact production cost and quantity during growth in the field (Kabasakal, 1990).

Consequently, nursery plant production efficiency in the field is around 50–60% (Cobanoglu et al., 2004). Cuttings are frequently used to propagate fig plants (Dolgun & Tekintas, 2008). However, fewer nursery plants are created because of the restricted branches generated by traditional layering and bottom shoots procedures. Thus, these approaches are not commonly used in commercial nursery plant production (Dolgun & Tekintas, 2008).

Because of the constraints above in fig nursery plants, intended success and sustainability cannot be guaranteed. Furthermore, 5–10 cm long shoots are abundant; however, they are unsuitable. However, these shoots can be used for budding with success. New fig cultivars have been introduced at the Punjab Agricultural University in Ludhiana. Some of them show great promise in terms of yield and fruit quality. However, the introduction of fig as a new crop in Punjab necessitates many plants, which may be difficult to meet using traditional techniques (Rattanpal et al., 2014). Furthermore, several potential genotypes grow poorly on their roots under Punjab circumstances. As a result, the ability to swiftly and cheaply produce fig plants would have significant commercial value.

Thus, the research conducted at the College Orchard of Punjab Agricultural University was based on analyzing thirteen fig tree cultivars obtained from various

Indian and alien sources for their capacity to be propagated by the budding method. The thirteen cultivars of fig (*Ficus carica L.*): 'Black Fig', 'Brown Turkey', 'California Brown Turkey', 'Conadria', 'Deanna', 'Dinkar', 'Genoa', 'Golden Celeste', 'King', 'Panache', 'Poona', 'San Piero' and 'Texas', were budded on 'Brown Turkey' seedlings, during the last week of August in 2012 and 2013. The cultivar 'Brown Turkey' had the highest mean budding success (93.4%), while 'Black fig' had the lowest (70.3%) (Rattanpal et al., 2015). There was no significant difference between the cultivars regarding the number of days it took from budding to sprouting. Furthermore, no incompatibility was found in any cultivar even 2 years after budding. The union of most cultivars got smoother with time.

Moreover, none of the plants split from the bud union; instead, the union smoothed out over time. This method not only saved the propagules, but preliminary investigations suggested that budded plants were more vigorous than cuttings-propagated plants. The cultivar 'Genoa' plants were far more robust on 'Brown Turkey' rootstock than on their roots. This is verified by the findings of Krezdorn and Glasgow (1970), who budded or grafted various *F. carica* cultivars onto *F. glomerata, F. cocculifolia, F. graphalocarpa,* and *F. palmate* and found no incompatibility with any of the rootstock species. Similarly, in Japan, Hosomi et al. (2002) used rootstock to gain an advantage in fig varieties by grafting sensitive cultivars on resistant rootstocks. Although fig culture in Indian Punjab is new, this approach will facilitate the rapid dissemination of cultivars discovered to be appropriate for Punjab conditions (Rattanpal et al., 2015).

7 Fertigation with Potassium Increases Size and Yield in Fresh Figs

California is the major producer of figs (*Ficus carica* L.) in the United States, and the state's fresh fig industry has lately expanded. In addition, several countries have reported the benefits of nitrogen (N), phosphorus (P), and potassium (K) fertilization with additional calcium (Ca) or K treatments during the growing season, increasing overall quality by reducing the number of fruit with ostiole-end cracking and sunscald on 'Calimyrna' or 'Sarlop' figs (İrget et al., 2008), and others (Antunes et al., 2008; Irfan et al., 2013).

Indeed, studies carried out by Holstein et al. (2015) conducted two seasons of research to assess the effects of K and Ca, both foliar and fertigation treatments, on fresh fig fruit size, yield, and soundness. Using a foliar spray containing K and Ca to 'Black Mission,' 'Brown Turkey,' and 'Sierra' decreased side cracking and enhanced fruit size. The response to foliar treatments, on the other hand, was very varied amongst plants, suggesting that absorption issues exist. Furthermore, in 'Sierra' figs, an increase in the percentage of fruit with ostiole-end splitting was found. Fertigation with K and Ca enhanced yield in both cultivars. In 'Black Mission,' the percentage of culls was 49% for control trees and 43% for treated trees. Similar findings were achieved in 'Sierra,' with 42% for control trees and 36% for treated trees. As a result, marketable production was greater for treated trees than for control trees for both varieties. In 'Sierra,' only fruit size was enhanced by fertigation (40.4% compared with control 38.7%), but soluble solids concentration (SSC) and titratable acidity (TA) were unaffected in either cultivar.

In terms of leaf macronutrient concentrations, nutritional development was comparable in both cultivars, with leaf N, P, and K levels dropping from late spring (May) through fall (October) (Tables 11.2). These low N and K leaf levels in control plants late in the season point to the possible advantages of spraying N and K during the season. On the other hand, Ca levels increased a little late in the season. After the June sampling, all leaf nutrients hit a plateau, suggesting that leaf sampling was done in midsummer (July–September) will be more stable and representative than earlier leaf sampling, as previously advocated for California dry fig orchards. Except for late in the season for 'Black Mission', leaf sample nutrients revealed no differences between trees treated and controlled with K and Ca (Holstein et al., 2015). This study found that dramatically increasing K and Ca fertigation may greatly boost marketable production of the fresh fig cultivars 'Black Mission' and 'Sierra,' which is critical for fresh fig growers since marketable yield determines income. In 'Sierra' figs, irrigation treatment boosted fruit size, linked to high leaf K concentrations.

As well as, Soliman et al. (2018) investigated the effects of potassium fertilization on the fruit quality of figs grown in Riyadh. The findings of this study show that potassium fertilization treatment had a substantial impact on the enhancement of the physical qualities of fruits in both seasons (during the 2015 and 2016 seasons). The maximum fruit quality was achieved by applying K_2O at 400 g per tree, followed by 200 g per tree. The findings are consistent with Kassem (2012), who found that potassium fertilization enhanced the fruit's weight, volume, and dimensions.

Table 11.2 Mean macronutrient concentrations (% dry weight) of 'Black Mission'(a) fig and 'Sierra' (b) fig leaves collected on four sampling dates from trees fertigated with 12.6 units of K ha^{-1} and 2.5 units of Ca ha^{-1} grown at the Kearney Agricultural Research and Extension Center, Parlier, CA. Each mean is a composite sample of 10 trees

Date	Control N	Treated N	Control P	Treated P	Control K	Treated K	Control Ca	Treated Ca
(a)	'Black mission' fig treatment							
May 12	2.89	2.99	0.156	0.164	1.27	1.36	2.40	2.24
Jun 29	2.24	2.30	0.119	0.123	1.23	1.35	3.12	2.90
Aug 10	2.14	2.21	0.122	0.115	1.21	1.20	2.46	2.94
Sep 11	1.98	2.13	0.109	0.109	1.08	1.42	3.33	3.32
(b)	'Sierra' fig treatment							
May 12	3.21	3.15	0.166	0.172	1.64	1.44	2.79	2.49
Jun 29	2.34	2.41	0.136	0.133	1.79	1.88	3.35	3.41
Aug 10	2.20	2.00	0.132	0.114	1.65	1.54	3.21	3.51
Sep 11	2.13	2.09	0.118	0.107	1.66	1.52	3.50	3.80

Holstein et al. (2015)

According to Ganeshamurthy et al. (2011), high K concentrations are required to obtain the largest overall fruit yield of grapes and bananas with good maintaining quality and marketability. Thus, potassium (K) has a tight interaction with water and plays an active part in cell swelling and expansion. More than one study found that K supplementation increased the volume of fruits and the thickness of the crust in citrus (Mattos et al., 2005).

Furthermore, potassium plays a vital function in soil pH adjustment, osmotic regulation, protein synthesis, stomatal movement, photosynthesis, and cell expansion (Läuchli & Pflüger, 1980). As a result, they demonstrated that potassium fertilization was the most successful fertilization method for fig. Furthermore, it improved fruits'physical and chemical quality by increasing their weight, size, dimensions, total sugars, reducing and non-reducing sugars, total soluble solids, and leaf mineral content (Soliman et al., 2018).

8 Effect of Gibberellic Acid on Growth, Yield and Fruit Quality of Fig (*Ficus carica* L.)

Because of its drought tolerance and simplicity of control, fig cultivation is becoming more popular in arid areas, where it may be grown as a monoculture or in a mixed crop with other fruit trees. Its incorporation into the agricultural land use system aids crop diversity (Kurubar et al., 2017). Commercial products such as fig syrups, jam, jelly, and spiced or pickled figs are made. Fresh or dried fig fruit is prized for its laxative properties. Its latex is used to coagulate milk in various countries. However, the fig fruit is fragile and perishable by nature; therefore, there is a risk of rotting after harvest.

Aside from nutrients, growth regulators have been shown to have favorable impacts on various fruit crops. Scientists have reported that different plant growth substances such as Parachlorophenoxy acetic acid (PCPA), indole-3-n-butyric acid (IBA), naphthalene acetic acid (NAA), 2,4,5-trichlorophenoxy acetic acid (2,4,5-T), gibberellic acid (GA), ethylene (C_2H_4), 2-(chloroethyl) phosphonic acid (ethephon) increased yield and quality while decreasing maturity period (Kurubar et al., 2017). The studies carried out by Kurubar et al. (2017) were based on the evaluation of the effect of different concentrations of GA with an increased number of spray applications at vegetative bud initiation after annual pruning, 15 days after bud initiation, and 30 days after bud initiation. It is inferred that increasing the concentration of GA with increasing the number of sprays led to increased shoot length (2.95 cm in Treatment 9) (Table 11.3). Similarly, for the number of functional leaves (616.2) and the length of the fruit (5.72 cm), the behavior is reversed for the diameter, fruit weight, and number of fruits per plant. Increased GA concentration and spray frequency reduce the fruit diameter and weight, whereas the fruit diameter and weight increase where GA concentration is lower.

Table 11.3 Effect of treatments with gibberellic acid (GA) on vegetative and fruit characters, quality, and number of days taken for maturity and fruit yield of fig

	Average of 2 years													
Treatment	Shoot length (cm)	Number of functional leaves	Number of fruits	Fruit length (cm)	Fruit diameter (cm)	Fruit weight (g)	Number of days to maturity	Fruit yield (kg plant^{-1})	Fruit yield (t ha^{-1})	Total soluble solids (%)	Titratable acidity (%)	Reducing sugars (%)	Non reducing sugars (%)	Total sugars (%)
T_1	2.28	477.87	322.08	5.22	4.48	39.23	108.87	12.63	8.40	16.21	0.198	14.59	1.96	16.54
T_2	2.40	521.43	329.36	5.48	4.13	38.10	105.80	12.54	8.35	16.50	0.203	14.79	1.86	16.64
T_3	2.49	555.88	357.56	5.59	4.06	36.68	102.21	13.11	8.73	16.90	0.212	15.10	1.80	16.90
T_4	2.39	516.52	335.77	5.27	4.18	39.23	108.37	13.17	8.76	16.38	0.202	15.00	1.84	16.84
T_5	2.48	554.84	353.52	5.47	4.04	37.72	99.99	13.38	8.91	16.97	0.215	15.43	1.78	17.20
T_6	2.50	577.43	352.86	5.62	3.94	36.98	97.79	13.09	8.71	17.27	0.230	15.57	1.72	17.29
T_7	2.70	601.53	361.15	5.54	4.06	36.76	93.46	13.27	8.83	17.08	0.212	15.23	1.83	17.06
T_8	2.80	600.22	360.71	5.69	3.90	36.71	93.63	13.24	8.81	17.18	0.231	15.69	1.72	17.40
T_9	2.95	616.23	351.11	5.72	3.93	36.81	91.23	12.92	8.60	17.28	0.230	15.78	1.62	17.40
T_{10}	2.22	462.62	314.73	5.08	4.67	39.26	125.83	12.35	8.22	17.39	0.196	16.02	1.98	18.00

Kurubar et al. (2017)

T1: GA 20 ppm at bud initiation (BI) stage; T2: GA 40 ppm at BI stage; T3: GA 60 ppm at BI stage; T4: GA 20 ppm at BI stage and 15 days after bud initiation (DABI); T5: GA 40 ppm at BI stage and at 15 DABI; T6: GA 60 ppm at BI stage and at 15 DABI; T7: GA 20 ppm at BI +15 DABI +30 DABI; T8: GA 40 ppm at BI +15 DABI +30 DABI; T9: GA 60 ppm at BI +15 DABI +30 DABI; T10: Control

This is due to GA, which plays a key role in shoot elongation, breaks dormancy, and utilizes carbohydrates for shoot growth rather than fruit development, and a larger concentration of GA accelerates the initiation of leaf formation (Gaikwad, 1975). T9 demonstrated that sprays' enhanced concentration and quantity resulted in early fruit maturity. The fruit production per plant and hectare showed a non-significant relationship between GA content and spray frequency. Among the different concentrations of GA spray-on fig, the highest rate for yield per plant and yield per hacter were noted in T5 (13.3 kg and 8.91 t ha^{-1}, respectively). Thus, the total soluble solid (TSS), acidity, reducing sugars, and total sugars, were improved with the addition of GA concentration and frequency of sprays, but, in the case of non-reducing sugars, the effect was reversed. However, the control treatment outperformed the chemical spray in statistical significance.

The rise in concentration causes an increase in fruit length, which has no economic benefit because it negatively influences fruit weight, which is a significant contributing yield characteristic. Therefore, GA spray may not be helpful unless paired with other growth regulators that might cause fruit thickness after elongation.

9 Increasing the Tolerance of Fig Trees to Drought Stress by Trunk Thinning

The fig tree is indigenous to Iran, East Asia, and Syria. One of the most serious issues in rain-fed fig production areas is a lack of precipitation and drought. Most fig trees in Iran are grown in Estahban, a semi-arid area in Fars province's southeast. The production of Iran dried figs is dominated by this area, which produces 90% of figs (Javanmard & Mahmoudi, 2008; Javanmard, 2010). Extreme temperatures in this region range between −7 and + 41 °C in winter and summer, respectively. The annual rainfall averages around 350 mm $years^{-1}$. The majority of the precipitation falls in the late fall and winter (Rahemi & Jafari, 2005).

Brady and Weil (2002) found that more than half of the precipitation received in semi-humid, semi-arid locations evaporates and returns to the atmosphere. In the tropics, particularly in semiarid locations, water is frequently a key limiting factor for agricultural production (Scopel et al., 2004). Furthermore, water shortage is the most critical stress factor impacting plant development in most terrestrial ecosystems (Kozlowski & Pallardy, 1996). Zare et al. (2009) performed socioeconomic research in this region to investigate the effects of drought on rainfed fig farming. The findings revealed that the damaged rain-fed orchards of Estahban had a 50% loss in fig production and price, implying that the fig tree is directly or indirectly responsible for 85% of the economy in the city. The quantity of precipitation in Estahban is usually about 350 mm, and the amount of fig output is around 17,000 t. In 2007, 64 mm annual rainfall in certain locations resulted in more than 80% crop decrease. Due to a lack of water, around 10% of trees died this year, and the remainder had major physiological difficulties (Jafari et al., 2015).

However, recent severe droughts have had a considerable impact on fig productivity. Furthermore, world water supplies are currently constrained because of global climate change. Therefore, many researches have been applied to examine the response of fig trees to drought stress and additional watering (Tapia et al., 2001; Mi et al., 2009).

According to the authors, drought significantly reduces growth and productivity. Furthermore, there had been a positive and significant association between irrigation rates and vegetative growth or yield in all supplemental irrigation treatments (Jafari et al., 2012). Researchers studied the edible fig 'Sabz' at the Estahban Fig Research Station in Fars province from February 2008 to September 2011. The effect of trunk thinning on drought diseases was studied using a random complete block design with three replications.

Because 'Sabz' fig trees are trained with numerous trunks, 16 intensities of trunk thinning were compared during winter pruning. The treatments were achieved by trimming five separate trees to leave 0, 1, 2, 3, or 4 trunks $tree^{-1}$, respectively. The pruning intensity of each of the 16 treatments was calculated by dividing the total circumference of the residual trunks by the total trunk circumference before the thinning treatment was applied (Jafari et al., 2015). Different levels of trunk thinning improved tree performance in terms of leaf width, leaf number, leaf water potential, leaf temperature, leaf chlorosis, leaf necrosis, fruit color, and size.

The statistical analysis of the 2 years findings revealed that trunk thinning substantially impacted the quantity of leaves per tree. The number of leaves per plant in 0 T trees (trees with all trunks thinned) was much higher than in control trees and other treatments. Trunk thinning had a considerable impact on leaf width (which varied from 9.4 to 13.6 cm); the 0 T (one treatment contained trees with no trunks remaining and all trunks removed from the soil surface using pruning) treatment raised this parameter much more than the other treatments. Improvements in tree water status caused by trunk trimming may have created a more favorable environment for these trees to produce bigger leaves. For growth to occur, turgor must be maintained (Prajapati et al., 2003). Partial tree wilting can be caused by a reduction in turgor. The loss of turgor, which affects the rate of cell growth and final cell size, is the initial result of water stress. Loss of turgor pressure is probably the most sensitive process to water stress. As a result, growth rate, stem elongation, leaf dilation, and stomatal opening are reduced. It also significantly impacts yield quality (Prajapati et al., 2003). They increased the root-to-shoot ratio by severe pruning (trunk thinning); thus, leaf water potential increased.

Because of increased leaf transpiration, which can lower leaf temperature, greater leaf water potential leads to reduced leaf sunburn (Mi et al., 2009). The more leaves create a better cover and lower the temperature of the leaves under the canopy, so the sunburn on the leaves is also reduced. Although reducing the number of trunks per tree under drought stress increased fig tree resistance to water stress, thinning all the trunks enhanced the vegetative look more than the other treatments. This treatment guarantees the survival of rain-fed fig trees in dry years. In general, trunk thinning of rainfed fig trees improves vegetative and reproductive qualities by reducing overall tree transpiration requirement and sink organs in drought situations. This demonstrates that this kind of pruning (thinning the whole trunk) had no positive influence on fruit growth in fruit yield and growth. Fruit size decreases due to reduced cell size and division due to lower turgor pressure and internal auxin content (Jones et al., 2008; Prajapati et al., 2003). Since the significant increase in weight and size of edible figs occurs in the third phase of fruit development, a favorable tree water status plays a major role during this phenological stage (Mitra, 1997).

Tree water status plays a vital role in cultivating fig trees under rain-fed conditions, improved by trunk thinning in rain-fed figs in dry years. Trunk thinning also led to increased vegetative growth and fruit quality. In this study, thinning all of the trunks produced the greatest advantages in terms of tree appearance. Furthermore, it had a good impact on both tree and fruit qualities. As a result, producers working in dry years are advised to use the later pruning method (Jafari et al., 2015).

10 Effect of *In Vitro* Thermotherapy Application on Shoot Development in Some Fig Cultivars

The fig tree (*Ficus carica* L.) has piqued interest due to its economic and therapeutic benefits. The fig tree has a great genetic variety in Tunisia, and all cultivars are well adapted to local circumstances (Mars et al., 2008, 2009). However, several abiotic and biotic restrictions hinder the growth of this fruit crop, contributing to a decline in revenue and the progressive removal of plantations (Saddoud et al., 2011). The scarcity of pollinators that have been specifically chosen to meet the demands of female cultivars and the challenges in multiplying them in the field are severe restrictions (Gaaliche et al., 2013). In addition, virus infections particularly fig mosaic disease, pose a problem and have spread throughout the industry (Saddoud et al., 2007), inhibiting the growth of healthy orchards and limiting production (Ashihara et al., 2004). According to Gella et al. (1997), fig mosaic disease is frequent in Mediterranean regions with temperate and subtropical climates. By defoliating leaves and dropping fruit, the disease causes economic losses. Thermotherapy and meristem culture are effective strategies for growing virus-free plants. Thermotherapy can be used in one of three ways: (1) Thermotherapy is administered to a plant first, and then meristem culture is formed; (2) Thermotherapy is applied to *in vitro* shoots first, and then meristem/shoot tip culture is established; (3) Thermotherapy is applied to cultured meristem/shoot tip (George, 1993). *In vitro* thermotherapy on fig shoot cultures infected with fig, mosaic has demonstrated that fig shoots produced in vitro are heat tolerant (Gella et al., 1997).

As well as that, Günver et al. (2015) used a thermal approach to treat the tips of fig shoots, and they assessed the shoot viability ratio and the condition of shoot development. So, 40 shoot tips (0.7–1.0 cm) of fig cultivars 'Sarlop', 'Bursa Siyah' female and 'Ak ilek', 'Kaba ilek' male was by hardwood cuttings with mosaic symptoms. The explants were grown in MS (Murashige & Skoog, 1962) medium, which included 2.0 mg L^{-1} benzyladenine (BA), and 0.2 mg L^{-1} gibberellic acid

Table 11.4 Shoot ratio and new shoot emergence ratio (%) after thermotherapy applications

Treatment	Survival ratio (%) Sarılop	Bursa Siyahı	New shoot emergence ratio (%) Sarılop	Bursa Siyahı
Control	67.00	69.69	50.00	46.00
Thermotherapy	35.00	55.00	10.00	23.00

Günver et al. (2015)

(GA_3), 0.1 mg L^{-1} indole butyric acid (IBA), and 89 mg L^{-1} phloroglucinol (PG). As a result, first, due to severe contamination, shoot explants of the male fig cultivars 'Ak ilek,' and 'Kaba ilek' were eliminated from the experiment.

Table 11.4 shows the shoot survival ratios of the female fig cultivars 'Sarlop' and 'Bursa Siyah' 3 weeks after thermotherapy treatments. After 5 eeks of thermotherapy administration, the control had the highest survival ratio and the fresh shoot emergence ratio excised from living shoot tips. Thus, the control group had a higher shoot emergence ratio than the thermotherapy group. Bursa Siyah outperformed Sarlop in thermotherapy applications among cultivars. Finally, when thermotherapy was applied, the shoot survival ratio and new shoot emergence ratio were reduced compared to the control. These ratios varied depending on the cultivar. On the other hand, Gella et al. (1997) acquired a very high survival rate in fig shoot cultures the following thermotherapy. The benefit of in vitro thermotherapy is that it produces larger explants (>1 mm) than meristem culture.

11 GIS-Based Soil Suitability Evaluation for Fig Orchards in Southern Peloponnese, Greece

Most fig orchards in the Southern Peloponnese are located on various soil types, and fig production is generally regarded as a low input in a dry-farmed system. However, as production costs continue to rise and harvesting accounts for more than half of total production costs, farmers must adopt more efficient and cost-effective cultivation and management practices (Melgarejo et al., 2007). According to the study of Melgarejo et al. (2007), fig cultivation is most often associated with a low profile and marginal lands (Because fresh figs have a concise shelf life and are fragile, the majority of the world's fig production is dried or otherwise processed (Stover et al., 2007). However, researchers and growers should pay more attention to this fruit's excellent nutritional worth.

In recent decades, tree-planting systems have seen a trend toward reducing tillage or maintaining vegetation cover to reduce soil erosion and degradation and improve water retention (Landa et al., 2014). However, the soil's physicochemical properties and fertility parameters must be considered to apply this soil management practice. Therefore, soil samples were collected from 45 fig orchards in the Messinia and Laconia regions of the Southern Peloponnese. Soil samples were dried to constant weight using forced air at ambient temperature < 36 °C, then passed through a 2 mm sieve (fine soil). Nine soil physicochemical parameters were measured for individual GIS-based map references: pH, saturation percent (SP), particle size analysis (PSa), organic matter (OM), cation exchange capacity (CEC), calcium (Ca), Magnesium (Mg), Potassium (K) and Phosphorus (P).

The results of the nine soil parameter studies provided information on the suitability of the studied areas for fig cultivation. Only for the soil parameters that show inhomogeneity among the studied fig orchards are tabulated data or maps provided (Kotsiras et al., 2017).

In terms of pH suitability, the findings show that the majority (73.3%) of the 45 orchards fall into category 4 (suitable), and 17.8% fall into category 5 (highly suitable). This indicates that the pH levels of the majority of the soils studied (91.1%) are suitable for this crop (Kotsiras et al., 2017). The cation exchange capacity of soil is an essential indicator of its productivity and can be used to make nutrition recommendations for phosphorus, potassium, and magnesium levels in soils of various textures (Adams et al., 1994). According to this study's CEC findings, 75.5% of fig orchards are moderate to very appropriate (Kotsiras et al., 2017).

Compared to other fruits, figs are considered extremely high in calcium (Rao et al., 2002). When examining the individual values of the fig orchards, it is clear that a large percentage of the soils (43%) have Ca concentration levels greater than 4000 ppm, which is consistent with the finding that a large percentage of the studied soils (approximately 60%) have a pH value between 7.6 and 8.0 (Kotsiras et al., 2017). On the other hand, most fig orchards (51.1%) have low amounts of exchangeable Mg, and only 17.8% of the soil had appropriate and highly suitable Mg levels (Fig. 11.1).

12 Effect of Organic and Inorganic Fertilizers on Fruit Characters, Quality, and Economics of Fig Production

The fig (*Ficus carica*) is a nutritional fruit high in fiber, potassium, calcium, and iron, with values higher than bananas, grapes, oranges, strawberries, and apples (Vinson, 1999). Therefore, a field experiment was carried out to investigate the effect of organic and inorganic fertilizers on the fruit characteristics, quality, and yield of fig. Kurubar et al. (2015) used soil with a pH of 8.29 and 0.44% organic carbon, 265 kg ha^{-1} of available N, 38 kg ha^{-1} of available P_2O_5, and 325 kg ha^{-1} of available K_2O in their experiment. In addition, the control inorganic fertilizer (RDF) containing NPK was compared to the other twelve treatments based on different combinations and doses of farmyard manure (FYM), poultry manure, vermicompost, and RDF (Kurubar et al., 2017).

The application of organic manure and chemical fertilizers through T9 and T11 resulted in significantly higher fruit length (5.82 cm), diameter (5.57 cm), number of fruits plant^{-1} (430.6), fruit yield plant^{-1} (19.4 kg), and ha^{-1} (12.97 t), as well as improved quality in the form of titratable acidity (0.24%) and TSS: acid ratio (78.5%) (Kurubar et al., 2017).

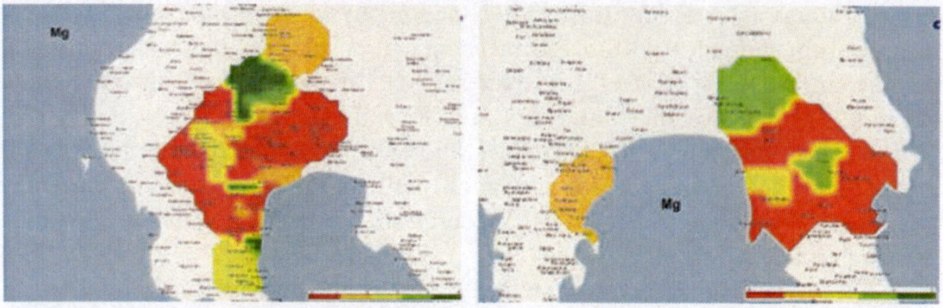

Fig. 11.1 Exchangeable Mg mapping in Messinia and Laconia, respectively. On the maps, suitability is indicated by a five-color coded key ranging from unsuitable (red) to highly acceptable (dark green). (Kotsiras et al., 2017)

According to the results of the Kurubar and Allolli (2015), the crop's economic concept revealed that both treatments (T_9 and T_{11}) were lucrative, with the maximum net profits of Rs.198207 (T_9) and Rs.185102 (T_{11}), respectively, reflecting a higher Benefit: Cost (B: C) ratio of Rs.6.58 and Rs.6.26 (Table 11.5). The use of RDF alone resulted in increased yield and quality of fruits; however, it is also clear that RDF used in combination and smaller quantities results in better size, quality, and quantity of fruits, as well as higher profit with increased yield by the use of T_9 (Kurubar et al., 2017).

Various workers have reported the beneficial effects of farm yard manure and poultry manure, either alone or in combination with inorganic fertilizers, in various crops such as guava, acid lime, coconut, banana, papaya, and guava (Corrales Garriga et al., 2000; Dubey & Yadava, 2001; Ray et al., 2008).

13 Shelf-Life Extension Using Modified Atmosphere Packaging and Moisture Absorption

Due to their rapid ripening and great susceptibility to rot, fig has a short shelf life after harvest. As a result, postharvest strategies that allow for the extension of post-harvest life in fresh figs, such as modified atmosphere packing, are required (MAP) (Colelli et al., 1991). In addition, moisture absorbers for relative humidity (RH) control could be a viable strategy for maintaining fruit quality and safety by delaying fruit rotting and softening.

Shirazi and Cameron (1992) found that 10 g of sorbitol, xylitol, NaCl, KCl, and CaCl2 allowed ripe green tomato fruit to be kept at 20 °C while maintaining a consistent relative humidity. The results obtained by Villalobos and Ansah (2015) show that no significant difference was identified between fruits stored under MAP (gas concentrations of 20 kPa CO_2 and 20 kPa O_2) and fruits stored in MAP with 3% (w/w) of $CaSO_4$, indicating that the use of this moisture absorber did not appear to affect the fruit respiration rate. Other authors (Villaescusa & Gil, 2003) reported similar results, stating that using moisture absorbers such as sorbitol or silica gel in

Table 11.5 Effect of organic manures and inorganic fertilizers on the economics of fig cultivation

Treatment	Cost of cultivation (Rs. ha^{-1})	Gross returns (Rs. ha^{-1})	Net returns (Rs. ha^{-1})	B:C ratio
T_1	37,096	195,030	157,934	5.26
T_2	35,259	152,100	116,841	4.31
T_3	36,487	171,090	134,603	4.69
T_4	34,650	128,970	94,320	3.72
T_5	33,950	176,850	142,900	5.21
T_6	32,113	153,810	121,697	4.79
T_7	36,792	210,240	173,448	5.71
T_8	34,955	145,080	110,125	4.15
T_9	35,523	233,730	198,207	6.58
T_{10}	33,686	179,820	146,134	5.34
T_{11}	35,218	220,320	185,102	6.26
T_{12}	33,380	133,020	99,640	3.98
T_{13}	33,248	167,130	133,882	5.03

T1: FYM at 16.5 t ha^{-1} + 75% RDF; T2: FYM at 16.5 t ha^{-1} + 50% RDF; T3: Vermicompost at 5 t ha^{-1} + 75% RDF; T4: Vermicompost at 5 t ha^{-1} + 50% RDF; T5: Poultry manure at 5 t ha^{-1} + 75% RDF; T6: Poultry manure at 5 t ha^{-1} + 50% RDF; T7: FYM at 8.25 t ha^{-1} + vermicompost at 2.5 t ha^{-1} + 75% RDF; T8: FYM at 8.25 t ha^{-1} + vermicompost at 2.5 t ha^{-1} + 50% RDF; T9: FYM at 8.25 t ha^{-1} + poultry manure at 2.5 t ha^{-1} + 75% RDF; T10: FYM at 8.25 t ha^{-1} + poultry manure at 2.5 t ha^{-1} + 50% RDF; T11: Vermicompost at 2.5 t ha^{-1} + poultry manure at 2.5 t ha^{-1} + 75% RDF; T12: Vermicompost at 2.5 t ha^{-1} + poultry manure at 2.5 t ha^{-1} + 50% RDF; T13: RDF alone (200:132:132 NPK kg ha^{-1})

conjunction with MAP did not affect the headspace gas concentration. Del carmen et al. (2014) reported minor weight losses in figs under MAP due to the film's effect on increasing vapor pressure. Furthermore, after 14 days of cold storage, the addition of $CaSO_4$ showed no significant interactions with the type of film used. The most significant weight loss was observed when fruits were stored with a humidity absorb combined with a micro-perforated film.

High levels of CO_2 and low levels of O_2 have been shown to suppress mould and yeast growth in fruit, particularly figs (Jacxsens et al., 2001; Crisosto & Kader 2007; Bouzo et al., 2012). Furthermore, moisture may have antibacterial properties since the amount of water available on the skin surface is likely to be lower than the essential water activity of microorganisms, which is an efficient approach to inhibiting their growth (Falconi, 2008). Shirazi and Cameron (1992) reported similar results, claiming that using NaCl as a moisture absorber in red ripe tomato fruit prevented surface mold growth.

The use of MAP has been proven to be beneficial in maintaining the quality of fig fruit and, as a result, extending its potential storability, reducing weight and acceptability losses, and regulating microbial growth. However, even though $CaSO_4$ application controlled fungal growth and resulted in the highest firmness, it negatively influenced fruit attractiveness. As a result, no synergistic effects were detected with moisture absorbers, which had minimal benefits when used without MAP (Villalobos & Ansah, 2015).

14 Developing the Fresh Fig Industry

Fig fruit, one of the world's first cultivated trees, is grown in most temperate climates. Figs are sodium-free, fat-free, and cholesterol-free, just like other fruits. Figs are also high in vitamins, amino acids, phenolics, and antioxidants (Solomon et al., 2006). Unfortunately, fresh figs are extremely vulnerable to physical damage and post-recollection infections, which cause significant losses during commercialization. Therefore, the conditions of verging before and after harvesting are critical for improving fruit quality and shelf life after harvesting (Preece et al., 2015). An in-store consumer review of four California fresh fig cultivars at two maturity stages found that cultivar and maturity at harvest were the most critical factors affecting like (acceptance).

Figs picked at tree-ripe maturity received a considerably higher acceptance rating than figs harvested at commercial maturity in all four cultivars. The average acceptance rate for the four cultivars was 86% for tree-ripened figs and 55% for commercially ripened figs (Crisosto et al., 2010). Therefore, it is time to find new cultivars that can be harvested at an advanced or commercial maturity stage and withstand postharvest processing. Unfortunately, there is a scarcity of high-quality consumer cultivars that can be picked at an advanced maturity stage and remain solid throughout postharvest handling. As a result, we looked for genotypes and cultivars with low decay susceptibility and desirable sensory traits such as sweetness, pleasant volatiles, and firm texture during postharvest handling (King et al., 2012; Haug et al., 2013).

To promote fresh fig consumption in the United States and abroad, it is necessary to choose high consumer-grade figs with good flavor that remain firm during ripening and survive postharvest handling (Crisosto et al., 2017). Therefore, the National Clonal Germplasm Repository (NCGR) advised a breeding program for generating firm-fleshed cultivars with high consumer approval based on postharvest analyses of consumer acceptance, firmness, and decay susceptibility of promising fresh figs (Preece et al., 2015).

15 Fresh, Semi-Dried and Sun-Dried Fig (*Ficus carica* 'Sarılop') Fruit Production

Due to the diversity of ecological conditions and geographical structures in Turkey, fig cultivation is practiced in various regions of the country. It is an economically important fruit species in Turkey's Western Aegean region, particularly in the Great and Lesser Meander Valleys (Arpaci, 2015).

The average fruit weight, the ratio of defective fruit (sunscald, cracked, or black neck), and total soluble solids and glucose levels in fresh, semi-dried, and dried fig fruit were significantly affected by climatic circumstances during fruit growth, maturation, and drying. Furthermore, skin color C* and h° values in fresh figs; skin h°

value and total sugar contents in semi-dried figs; and firmness and total sugar contents in dried figs were affected positively or negatively by climatic conditions.

The best results were obtained in 2011 when temperatures did not surpass 40 °C, and relative humidity was between 40 and 70%. Between the beginning of August and the end of September, the temperature/relative humidity regime significantly impacts overall dried fig quantity and quality in the Region, particularly fruit weight and defect ratio. Some of the unfavorable consequences, such as high relative humidity levels or rain during the sun-drying period, can be partially mitigated by using highly aerated tunnels for drying. Sun drying is a cost-effective and ecologically friendly preservation method, and modest management changes can improve dried fig fruit's ultimate quality and marketability (Şen et al., 2017).

16 Conclusion

The global production of the fig (*Ficus carica* L.) has grown in prominence, owing to its significant function as a dietary supplement. This circumstance has sparked much curiosity about its manufacturing, processing, and preservation methods. Growing fig trees in poor circumstances can result in crop loss and numerous fruit damage. Although this crop may grow in various climates, agronomic considerations such as production method, fertilizer management, water demand, and soil requirement can all impact fig tree yield and output. Increasing the value of agricultural products cultivated, inefficient production systems are required to provide a good return on investment. Thus, proper agronomic management must be employed to eliminate difficulties caused by internal and external causes, creating an environment suited for operating efficiency and improved *Ficus carica* L. output.

References

Adams, J. F., Bryant, H. H., & Mitchell, C. C. (1994). Soil test fertilizer recommendations for Alabama crops.

Agrawal, A. A., & Konno, K. (2009). Latex: A model for understanding mechanisms, ecology, and evolution of plant defense against herbivory. *Annual Review of Ecology, Evolution, and Systematics, 40*, 311–331.

Antunes, M. D. C., Mendonça, V., Miguel, M. G., Neves, M. A., Martins, M. A., Gomes, C., Candeias, M. F., & Pica, M. C. (2008). The effect of irrigation level and pre-harvest treatments with calcium oxide on storage ability of fig fruits (*ficus carica* 'lampa preta' and 'bêbera branca'). *Acta Horticulturae, 798*, 335–339.

Arpaci, S. (2015). An overview on fig production and research and development in Turkey. *V International Symposium on Fig, 1173*, 57–62.

Ashihara, W., Kondo, A., Shibao, M., Tanaka, H., Hiehata, K., & Izumi, K. (2004). Ecology and control of eriophyid mites injurious to fruit trees in Japan. *Japan Agricultural Research Quarterly, 38*(1), 31–41.

Awamura, M., Shoda, K., & Yahata, D. (1996). Effect of various seed parents on frequency distribution of parthenocarpy among seedling progenies of fig (*Ficus carica* L.). *Journal of the Japanese Society for Horticultural Science, 65*(1), 21–26.

Awamura, M., Yahata, D., & Shoda, K. (1997). Differences in fruit weight and soluble solids content in the juice from caprifig and fig types, and first or second crops among fig (*Ficus carica* L.) hybrid seedlings. *Journal of the Japanese Society for Horticultural Science, 66*(1), 77–84.

Berg, C. C. (1989). Classification and distribution of. *Ficus. Experientia, 45*, 605–611.

Bouzo, C., Travadelo, M., & Gariglio, N. F. (2012). The effect of different packaging materials on postharvest quality of fresh fig fruit. *International Journal of Agriculture & Biology, 14*, 821–825.

Brady, N. C., & Weil, R. R. (2002). *The nature and proprieties of soils*. Upper Saddle River.

Chen, C., & Song, Q. (2008). Responses of the pollinating wasp Ceratosolen solmsi marchali to odor variation between two floral stages of *Ficus hispida. Journal of Chemical Ecology, 34*(12), 1536–1544.

Chen, C., Song, Q., Proffit, M., Bessière, J. M., Li, Z., & Hossaert-McKey, M. (2009). Private channel: A single unusual compound assures specific pollinator attraction in *Ficus semicordata. Functional Ecology, 23*(5), 941–950.

Cobanoglu, F., Sahin, B., Kocatas, H., & Ozen, M. (2004). Productivity and quality parameters on fig young plant in tube. Gazi Osmanpasa University. *Journal of Agricultural Faculty, 21*(1), 1–8.

Colelli, G., Mitchell, F. G., & Kader, A. A. (1991). Extension of postharvest life of mission' figs by CO_2-enriched Atmospheres. *Horticultural Science, 26*(9), 1193–1195.

Corrales Garriga, I., Lopez Labarta, P., & Guerra Gomez, A. (2000). Use of poultry manure and mineral fertilizer on guava (*Psidium guajava* L.). *Centro Agricola, 27*(4), 47–57.

Crisosto, C., & Kader, A. A. (2007). *Figs. Postharvest quality maintenance guidelines. Postharvest information for fruits and nuts.*

Crisosto, C. H., Bremer, V., Ferguson, L., & Crisosto, G. M. (2010). Evaluating quality attributes of four fresh fig (*Ficus carica* L.) cultivars harvested at two maturity stages. *Horticultural Science, 45*(4), 707–710.

Crisosto, C. H., Ferguson, L., Preece, J. E., Michailides, T. J., Haug, M. T., López Corrales, M., & Crisosto, G. M. (2017). Developing the California fresh fig industry. *Journal of Acta Horticulturae*, 285–292.

Del Carmen Villalobos, M., Serradilla, M. J., Martín, A., Ruiz-Moyano, S., Pereira, C., & de Guía Córdoba, M. (2014). Use of equilibrium modified atmosphere packaging for preservation of 'San Antonio'and 'Banane'breba crops (*Ficus carica* L.). *Postharvest Biology and Technology, 98*, 14–22.

Dickson, J. H., & Dickson, C. (1996). Ancient and modern occurrences of common fig (*Ficus Carica* L.) in the British Isles. *Quaternary Science Reviews, 15*(5–6), 623–633.

Dolgun, O., & Tekintas, F. E. (2008). Production of fig (*Ficus carica* L.) nursery plants by stem layering method. *Agriculturae Conspectus Scientificus, 73*(3), 157–160.

Doyle, J., & Ferguson, L. (1997). Breeding persistent figs with Calimyrna quality. *I International Symposium on Fig, 480*, 259–264.

Dubey, A. K., & Yadav, D. S. (2001). Effects of pig manure, nitrogen and phosphorus on fruit yield and quality of acid lime (Citrus aurantifolia Swingle) fruits. *Progressive Horticulture, 33*(2), 147–150.

Falconí, M. D. (2008). Efectos del escaldado y de la adición de metabisulfito de sodio en la evaluación física y sensorial de aguacate Hass (Persea americana) deshidratado.

FAO. (2007). *Renforcement des systèmes nationaux de contrôle alimentaire: directives pour l'évaluation des besoins en renforcement des capacités*. Food & Agriculture Organisation.

Flaishman, M. A., Rodov, V., & Stover, E. (2008). The fig: Botany, horticulture, and breeding. *In Horticultural Reviews*, 113–196.

Gaaliche, B., Majdoub, A., Trad, M., & Mars, M. (2013). Assessment of pollen viability, germination, and tube growth in eight Tunisian caprifig (*Ficus carica* L.) cultivars. *International Scholarly Research Notices, 2013.*

Gaikwad, T. S. (1975). *Effect of various growth regulators on growth, maturity period and quality of fig cv. Poona Fig.* M. Sc. (Agri.) thesis. Mahatma Phule Agricultural University.

Galil, J., & Neeman, G. (1977). Pollen transfer and pollination in the common fig (*Ficus carica* L.). *New Phytologist, 79*(1), 163–171.

Ganeshamurthy, A. N., Satisha, G. C., & Patil, P. (2011). Potassium nutrition on yield and quality of fruit crops with special emphasis on banana and grapes. *Karnataka Journal of Agricultural Sciences, 24*(1).

Gella, R., Lopez Corrales, M., Toribio, F., & Marin, J. A. (1997). Elimination of fig mosaic from fig shoot-tip cultures by thermotherapy. *I International Symposium on Fig, 480*, 173–178.

George, E. F. (1993). Plant propagation by tissue culture. Part 1. *The Technology*.

Grison-Pigé, L., Bessière, J. M., & Hossaert-McKey, M. (2002). Specific attraction of fig-pollinating wasps: Role of volatile compounds released by tropical figs. *Journal of Chemical Ecology, 28*(2), 283–295.

Günver Dalkılıç, G., Ertan, B., Dağ, S., & Görücüoğlu, P. (2015). Effect of in vitro thermotherapy application on shoot development in some fig cultivars. *V International Symposium on Fig, 1173*, 195–198.

Haug, M. T., King, E. S., Heymann, H., & Crisosto, C. H. (2013). Sensory profiles for dried fig (*Ficus carica* L.) cultivars commercially grown and processed in California. *Journal of Food Science, 78*(8), 1273–1281.

Holstein, H., Crisosto, G. M., & Crisosto, C. H. (2015). Fertigation with potassium increases size and yield in fresh figs growing in California. *V International Symposium on Fig, 1173*, 177–182.

Hosomi, A., Dan, M., & Kato, A. (2002). Screening of fig varieties for rootstocks resistant to soil sickness. *Journal of the Japanese Society for Horticultural Science, 71*(2), 171–176.

Hossaert-McKey, M., Soler, C., Schatz, B., & Proffit, M. (2010). Floral scents: Their roles in nursery pollination mutualisms. *Chemoecology, 20*(2), 75–88.

Irfan, P. K., Vanjakshi, V., Prakash, M. K., Ravi, R., & Kudachikar, V. B. (2013). Calcium chloride extends the keeping quality of fig fruit (*Ficus carica* L.) during storage and shelf-life. *Postharvest Biology and Technology, 82*, 70–75.

İrget, M. E., Aksoy, U., Okur, B., Ongun, A. R., & Tepecik, M. (2008). Effect of calcium based fertilization on dried fig (*Ficus carica* L. cv. Sarılop) yield and quality. *Scientia Horticulturae, 118*(4), 308–313.

Jacxsens, L., Devlieghere, F., Van der Steen, C., & Debevere, J. (2001). Effect of high oxygen modified atmosphere packaging on microbial growth and sensorial qualities of fresh-cut produce. *International Journal of Food Microbiology, 71*(2–3), 197–210.

Jafari, M., Haghighi, J. A. P., & Zare, H. (2012). Mulching impact on plant growth and production of rainfed fig orchards under drought conditions. *Journal of Food, Agriculture & Environment, 10*(1), 428–433.

Jafari, M., Rahemi, M., & Zare, H. (2015). Increasing the tolerance of fig tree to drought stress by trunk thinning. *V International Symposium on Fig, 1173*, 189–194.

Janzen, D. H. (1979). How to be a fig. *Annual Review of Ecology and Systematics, 10*(1), 13–51.

Javanmard, M. (2010). Occurrence of mould counts and Aspergillus species in Iranian dried figs at different stages of production and processing. *Journal of Agricultural Science and Technology, 12*(3), 331–338.

Javanmard, M., & Mahmoudi, H. (2008). *A SWOT analysis of organic dried fig production in Iran.*

Jones, H. G., Flowers, T. J., & Jones, M. B. (2008). *Plants under stress: Biochemistry, physiology and ecology and their application to plant improvement.* Cambridge University Press.

Kabasakal, A. (1990). *Fig Cultivation, Foundation of Agricultural Research and Promotion, Publishing No: 20.* Yalova, Turkey.

Kai, M., Zhifen, L., Yan, T., & WeiBing, J. (1997). The green cutting propagation techniques for fig trees. *China Fruits, 3*, 32–38.

Kassem, H. A. (2012). The response of date palm to calcareous soil fertilisation. *Journal of Soil Science and Plant Nutrition, 12*(1), 45–58.

King, E. S., Hopfer, H., Haug, M. T., Orsi, J. D., Heymann, H., Crisosto, G. M., & Crisosto, C. H. (2012). Describing the appearance and flavor profiles of fresh fig (*Ficus carica* L.) cultivars. *Journal of Food Science, 77*(12), S419–S429.

Konno, K., Hirayama, C., Nakamura, M., Tateishi, K., Tamura, Y., Hattori, M., & Kohno, K. (2004). Papain protects papaya trees from herbivorous insects: Role of cysteine proteases in latex. *The Plant Journal, 37*(3), 370–378.

Kotsiras, A., Amorgianiotis, T., Tsiavtari, E., Antonopoulos, D. F., & Demopoulos, V. (2017). GIS based soil suitability evaluation for fig orchards in Southern Peloponnese, Greece. *Acta Horticulturae*.

Kozlowski, T. T., & Pallardy, S. G. (1996). *Physiology of woody plants*. Elsevier.

Krezdorn, A. H., & Glasgow, S. K. (1970). Propagation of *Ficus carica* on tropical species of Ficus. *Proceedings of the Tropical Region, American Society for Horticultural Science, 14*, 156–164.

Kurubar, A. R., Allolli, T. B., Naik, M. K., & Angadi, S. G. (2015). Effect of organic and inorganic fertilizers on fruit characters, quality and economics of fig production (Ficus carica L.). *In V International Symposium on Fig* 1173, 213–216.

Kurubar, A. R., Allolli, T. B., Naik, M. K., & Angadi, S. G. (2017). Effects of gibberellic acid on growth, yield and fruit quality of fig (*Ficus carica* L.). *Acta Horticulturae*.

Landa, B. B., Montes-Borrego, M., Aranda, S., Soriano, M. A., Gómez, J. A., & Navas-Cortés, J. A. (2014). Soil factors involved in the diversity and structure of soil bacterial communities in commercial organic olive orchards in Southern Spain. *Environmental Microbiology Reports, 6*(2), 196–207.

Lansky, E. P., Paavilainen, H. M., Pawlus, A. D., & Newman, R. A. (2008). Ficus spp. (fig): Ethnobotany and potential as anticancer and anti-inflammatory agents. *Journal of Ethnopharmacology, 119*(2), 195–213.

Lauchli, A., & Pfluger, R. (1980). Potassium transport through plant cell membranes and metabolic role of potassium in plants. *Potassium transport through plant cell membranes and metabolic role of potassium in plants, 6*, 5–57.

Lazreg-Aref, H., Gaaliche, B., Ladhari, A., Hammami, M., & Hammami, S. O. (2018). Co-evolution of enzyme activities and latex in fig (*Ficus carica* L.) during fruit maturity process. *South African Journal of Botany, 115*, 143–152.

Lazreg Aref, H., Gaaliche, B., Fekih, A., Mars, M., Aouni, M., Pierre Chaumon, J., & Said, K. (2011). In vitro cytotoxic and antiviral activities of *Ficus carica* latex extracts. *Natural Product Research, 25*(3), 310–319.

Lazreg-Aref, H., Mars, M., Fekih, A., Aouni, M., & Said, K. (2012). Chemical composition and antibacterial activity of a hexane extract of Tunisian caprifig latex from the unripe fruit of *Ficus carica*. *Pharmaceutical Biology, 50*(4), 407–412.

Lionakis, S. M. (1995). Present status and future prospects of the cultivation in Greece of the plants: fig, loquat, Japanese persimmon, pomegranate and Barbary fig. *Underutilized fruit crops in the Mediterranean region. Zaragoza: CIHEAM-IAMZ*, 21–30.

Machado, C. A., Jousselin, E., Kjellberg, F., Compton, S. G., & Herre, E. A. (2001). Phylogenetic relationships, historical biogeography and character evolution of fig-pollinating wasps. *Proceedings of the Royal Society of London. Series B: Biological Sciences, 268*(1468), 685–694.

Maffei, M. E. (2010). Sites of synthesis, biochemistry and functional role of plant volatiles. *South African Journal of Botany, 76*(4), 612–631.

Mansour, K. M., Llácer, G., Aksoy, U., & Mars, M. (1995). *Underutilized fruit crops in the Mediterranean region* (pp. 13–19). CIHEAM-IAMZ.

Mars, M., Chebli, T., & Marrakchi, M. (1997, June). Multivariate analysis of fig (*Ficus carica* L.) germplasm in southern Tunisia. *I International Symposium on Fig, 480*, 75–82.

Mars, M., Chatti, K., Saddoud, O., Salhi-Hannachi, A., Trifi, M., & Marrakchi, M. (2008). Fig cultivation and genetic resources in Tunisia, an overview. *Acta Horticulturae*.

Mars, M., Gaaliche, B., Ouerfelli, I., & Chouat, S. (2009). Systèmes de production et ressources génétiques du figuier (*Ficus carica* L.) à Djebba et Kesra, deux villages de montagne au nord-ouest de la Tunisie. *Revue des Régions Arides, 22*(1), 33–45.

Mattos, D., Quaggio, J., & Cantarella, H. (2005). Nitrogen and potassium fertilization impacts fruit yield and quality of citrus. *Better Crops, 89*(2), 17–19.

Melgarejo, P., Martínez, J. J., Hernandez, F., Salazar, D., & Martínez, R. (2007). Preliminary results on fig soil-less culture. *Scientia Horticulturae (Amsterdam), 111*(3), 255–259.

MI, A. D., Abd El-Rhman, I. E., & Sahar, A. F. (2009). Effect of some antitranspirants and supplementery irrigation on growth, yield and fruit quality of Sultani fig (*Ficus carica*) grown in the Egyptian western coastal zone under rainfed conditions. *Research Journal of Agriculture and Biological Sciences, 5*, 899–908.

Mitra, S. K. (1997). Postharvest physiology and storage of tropical and subtropical fruits. *Centre for Agricultural Bioscience International, 4*(359), 6.

Murashige, T., & Skoog, F. (1962). A revised medium for rapid growth and bio assays with tobacco tissue cultures. *Physiologia Plantarum, 15*(3), 473–497.

Obenauf, C. D., Shaw, H. E., Sydnor, C. F., & Klintworth, G. K. (1978). Sarcoidosis and its ophthalmic manifestations. *American Journal of Ophthalmology, 86*(5), 648–655.

Oliveira, A. P., Silva, L. R., De Pinho, P. G., Gil-Izquierdo, A., Valentão, P., Silva, B. M., Pereira, J. A., & Andrade, P. B. (2010). Volatile profiling of *Ficus carica* varieties by HS-SPME and GC–IT-MS. *Food Chemistry, 123*(2), 548–557.

Oukabli, A., Bartolini, S., & Viti, R. (2003). Anatomical and morphological study of apple (Malus× domestica Borkh.) flower buds growing under inadequate winter chilling. *The Journal of Horticultural Science and Biotechnology, 78*(4), 580–585.

Prajapati, N. D., Purohit, S. S., Sharma, A. K., & Kumar, T. (2003). *Medicinal plants* (Vol. 353, 3rd ed.). Agrobios Published Company.

Preece, J. E., Crisosto, C. H., Ferguson, L., Michailides, T. J., Haug, M. T., López Corrales, M., & Crisosto, G. M. (2015). Developing the California fresh fig industry. *V International Symposium on Fig, 1173*, 285–292.

Rahemi, M., & Jafari, M. (2005). Effect of caprifig type on quantity and quality of Estahban dried fig *Ficus carica* cv. Sabz. *III International Symposium on Fig, 798*, 249–252.

Rao, R. B., Anupama, K., Swaroop, K. A., Murugesan, T., Pal, M., & Mandal, S. C. (2002). Evaluation of anti-pyretic potential of Ficus racemosa bark. *Phytomedicine, 9*(8), 731–733.

Rattanpal, H. S., & Sidhu, G. S. (2015). Studies on propagation of fig (*Ficus carica* L.) cultivars by budding. *V International Symposium on Fig, 1173*, 173–176.

Rattanpal, H. S., Singh, G., & Singh, G. (2014). Standardization of rapid propagation technique for fig (*Ficus carica* l.) cultivars by budding. *HortFlora Research Spectrum, 3*(1), 94–96.

Ray, P. K., Singh, A. K., & Kumar, A. (2008). Performance of Pusa Delicious papaya under organic farming. *Indian Journal of Horticulture, 65*(1), 100–101.

Saddoud, O., Chatti, K., Salhi-Hannachi, A., Mars, M., Rhouma, A., Marrakchi, M., & Trifi, M. (2007). Genetic diversity of Tunisian figs (*Ficus carica* L.) as revealed by nuclear microsatellites. *Hereditas, 144*(4), 149–157.

Saddoud, O., Baraket, G., Chatti, K., Trifi, M., Marrakchi, M., Mars, M., & Salhi-Hannachi, A. (2011). Using morphological characters and simple sequence repeat (SSR) markers to characterize Tunisian fig (*Ficus carica* L.) cultivars. *Acta Biologica Cracoviensia s. Botanica, 53*(2).

Scopel, E., Da Silva, F. A., Corbeels, M., Affholder, F., & Maraux, F. (2004). Modelling crop residue mulching effects on water use and production of maize under semi-arid and humid tropical conditions. *Agronomie, 24*(6–7), 383–395.

Şen, F., Aksoy, U., Özer, K. B., Can, H. Z., Köseoğlu, İ., & Konak, R. (2017). Impact of yearly conditions on major physical and chemical properties of fresh, semi-dried and sun-dried fig (*Ficus carica* L. 'Sarılop') fruit. *Acta Horticulturae*, (1173), 309–314.

Shirazi, A., & Cameron, A. C. (1992). Controlling relative humidity in modified atmosphere packages of tomato fruit. *HortScience, 27*(4), 336–339.

Soliman, S. S., Alebidi, A. I., Al-Obeed, R. S., & Al-Saif, A. M. (2018). Effect of potassium fertilizer on fruit quality and mineral composition of fig (*Ficus carica* L. cv. Brown Turky). *Pakistan Journal of Botany, 50*(5), 1753–1758.

Solomon, A., Golubowicz, S., Yablowicz, Z., Grossman, S., Bergman, M., Gottlieb, H. E., Altman, A., Kerem, Z., & Flaishman, M. A. (2006). Antioxidant activities and anthocyanin content of fresh fruits of common fig (*Ficus carica* L.). *Journal of Agricultural and Food Chemistry, 54*(20), 7717–7723.

Stover, E., Aradhya, M., Ferguson, L., & Crisosto, C. H. (2007). The fig: Overview of an ancient fruit. *Journal of horticultural science, 42*(5), 1083–1087.

Tapia, R., Botti, C., Carrasco, O., Prat, L., & Franck, N. (2001). Effect of four irrigation rates on growth of six fig tree varieties. *II International Symposium on Fig, 605*, 113–118.

Villaescusa, R., & Gil, M. I. (2003). Quality improvement of Pleurotus mushrooms by modified atmosphere packaging and moisture absorbers. *Journal of Postharvest Biology and Technology, 28*(1), 169–179.

Villalobos, M. C., Ansah, F., Amodio, M. L., Serradilla, M. J., & Colelli, G. (2015). Application of modified atmosphere packaging with moisture absorber to extend the shelf life of 'Domenico Tauro' breba fruit. *V International Symposium on Fig, 1173*, 365–370.

Vinson, J. A. (1999). The functional food properties of figs. *Cereal Foods World, 44*(2), 82–87.

Wagner, R. G., Mohammed, G. H., & Noland, T. L. (1999). Critical period of interspecific competition for northern conifers associated with herbaceous vegetation. *Canadian Journal of Forest Research, 29*(7), 890–897.

Weiblen, G. D. (2002). How to be a fig wasp. *Annual Review of Entomology, 47*(1), 299–330.

Wiebes, J. T. (1979). Co-evolution of figs and their insect pollinators. *Annual Review of Ecology and Systematics, 10*(1), 1–12.

Zare, H., Zare, E., Pir Moradiyan, N., Joukar, L., Mobayen, K., Nazari, A., & Kamkar Haghighi, A. A. (2009). *Reducing damages caused by drought on the rainfed fig cultivation and its economic-social impacts*. Seed and Plant Improvement Institute.

Zare, H., Moosavi-Movahedi, A. A., Salami, M., Mirzaei, M., Saboury, A. A., & Sheibani, N. (2013). Purification and autolysis of the ficin isoforms from fig (*Ficus carica* cv. Sabz) latex. *Phytochemistry, 87*, 16–22.

Chapter 12
Defense Mechanism of Fig (*Ficus carica*) Against Biotic Stresses: An Advanced Role Model Under Moraceae

Sudeepta Pattanayak, Siddhartha Das, and Suryakant Manik

Abbreviations

ISR	Induced Systemic Resistance
PAG	Psoralic acid glucoside
ROS	Reactive oxygen species
SAR	Systemic Acquired Resistance

1 Introduction

The genus *Ficus* is one of the widely versatile members, which comprises approximately 800 species of herb, shrub, trees, and epiphytes and are distributed cosmopolitical (Singh et al., 2011). *Ficus carica* (*Moraceae*) is a large deciduous member belonging to *Moraceae* with a unique defense system. *Ficus* has wide adaptability through its various lifeforms like epiphytes, trees, and vines in various climatic and geographic regions (Pierantoni et al., 2018). Morphological uniqueness is visible with its unique inflorescence type (syconium). The most well-known *Ficus* species is *Ficus carica,* which is deciduous and maintains a height of up to 5 m. The origin *Ficus* thinks to be Middle East countries which latterly distributed in warmer regions across the globe. A maximum diverse population exists on the Asian and Australian continents. Different names of *Ficus* in different locations are reported (Table 12.1). In alternate *Ficus* population in Africa is less, approximately 100–150 (Badgujar et al., 2014). Fruits of *Ficus* are edible and could be consumed in fresh,

S. Pattanayak · S. Manik
Division of Plant Pathology, ICAR-Indian Agricultural Research Institute, New Delhi, India

S. Das (✉)
Department of Plant Pathology, M.S. Swaminathan School of Agriculture, Centurion University of Technology and Management, Paralakhemundi, Odisha, India

M. F. Ramadan (ed.), *Fig* (Ficus carica)*: Production, Processing, and Properties*, https://doi.org/10.1007/978-3-031-16493-4_12

Table 12.1 Various names of *Ficus carica* across the globe

Sl. No	Country/location	Name
1.	India	Fig, Dumur, Anjir, Angir, Shimayatti
2	China	Mo Fa Guo, Wu Hua Guo
3.	Czech	Smokvon
4.	German	Echter, Feigenbaum, Essfeige
5.	Italian	Fico, Fico Comune
6.	Iran	Anjeer
7.	Korea	Mu Hwa Gwa, Mu Hwa Gwa Na Mu
8.	Russia	Inzir
9.	Sweden	Fikon, Fikontrad
10.	Turkey	Incir, Yemis

dried, and processed forms by all classes of people. Although, different countries like Turkey, Egypt, and Algeria are the pioneer in promoting *Ficus* cultivation commercially.

Ficus carica, therefore, is considered an important species with both agronomical and economic impacts. Several *Ficus* species showed wide genetical diversity with excellent pharmacological potentiality. Like other plant species, *Ficus* is also infected by several destructive insect pests (Novtny et al., 2010), fungi (Hosomi et al., 2012), and viruses (Bayoudh et al., 2017). Though a few numbers of pathogenic microbes infect *Ficus* severity or aggressiveness of the pathogen turned to blur by its unique defense mechanism (Chamont, 2014). Phytochemical studies revealed that it is a rich source of phenolics, organic acids, and various volatile compounds (Mawa et al., 2013). In the red cultivar of figs higher level of anthocyanin was observed. Carotenoids are mainly stored on the fruit peel. *Ficus* species have developed various physical defense strategies like mineral-fiber and rigidity levels. Chemical defense strategies encompass target-based metabolites and PR proteins to combat foreign pathogens. Secretion of volatile organic compounds may help *Ficus racemosa* attract parasitoids like *Apocryptophagus agraensis* and *Apocrypta westwoodi* (Borges et al., 2013), which indirectly helps to control various destructive insect pests. In this present chapter, we focus on a unique defense system as a role model under *Moraceae.*

2 Physical Barriers or Structural Defenses

2.1 Hard, Tough and Mineralized Leaf Texture

Leaves are the herbivores' main target, but various associated factors are directly or indirectly responsible for the herbivores' edibility property or food quality. Leaf rigidity and abrasiveness are important defense indicators for a plant's defense strategy. Toughness creates a major barrier to herbivore crop consumption, while host resistance might be related to storing biochemical substances such as lignin, fiber, and phenolic content (Xiang & Chen, 2004). Toughness comes from the rebuilt prolonged storage of calcium oxalate, silica, and calcium carbonate deposition. In 2018, Pierantoni and his co-workers reported that aforesaid three components were impregnated in various orientations among different species of *Ficus* in various environmental conditions.

2.2 Calcium Oxalate Crystal-Based Defense

Calcium oxalate is deposited as a solid mineral on the plant surface (Korth et al., 2006; Bauer et al., 2011). Calcium oxalate presented intra-cellularly and extra-cellularly within the plant cells and gathered and impregnated in the vacuole, known as idioblasts (Franceschi and Nakata 2005; Bauer et al., 2011). Idioblasts cells are structurally and functionally different from the neighbor cells. These idioblasts cells can develop in any part of the host. They can participate in multifold activities like the calcium regulatory pathway and metal detoxifying pathway, which have a broader defense mechanism impact (Franceschi & Nakata, 2005; Pierantoni et al., 2018). Those crystals can develop a protective shield (Ruiz et al., 2002; Hudgins et al., 2003) against destructive pests, and herbivorous infestation (Ward et al., 1997) and are an effective corrading agent for the mouthparts of suckers and chewing insects (Korth et al., 2006). After all, these calcium oxalate crystals turn into a tough barrier in the defense mechanism (Finley, 1999). In *Ficus*, a fruitful eradication of pest infestation has been reported through the aforesaid defensive mechanisms (Sosnovsky, 2016).

Calcium oxalate crystals are impregnated in parallel lines nearer to the main veins of *Ficus* leaves. Hence the crystals are associated with the vein bundles, the phloem, or the basal tissue of the veins and veinlets (Pierantoni et al., 2018). As per the existence in different locations within the plant body, those calcium oxalate crystals are well-known as prismatic crystals, extended to mesophyll and bundle sheath cells (Wu & Kuo-Huang, 1997; Pierantoni et al., 2018). Occurrence of calcium oxalate crystal founds in bundle sheath cell and mesophyll cells in the form of druses in *Ficus variagata.* In the *case of Ficus elastica,* the entire mesophyll layer and bundle sheath surface are covered with the sharper needle-like calcium oxalate crystals, which are also dispersed throughout the epidermal region of leaves (Wu & Kuo-Huang, 1997). In *F. carica*, calcium oxalate was noticed in the phloem region in two different shapes – druses and prismatic crystals. Developmental and depositional orientation patterns revealed that calcium oxalate was stored on both dorsiventral sides of the leaf vein and covered the entire surface (Pierantoni et al., 2018). In *Ficus* species, calcium oxalate crystals exist as prismatic crystals for their localization towards bundle sheath. However, similar several other variations are also observed. In the case of *Ficus elastica,* two different types of crystals are found; prismatic crystals are larger numbers towards bundle sheath cells, whereas needle-shaped in fewer numbers (Wu & Kuo-Huang, 1997). Occurrence of calcium oxalate crystal found in *F. carica* in the phloem. Calcium oxalate crystals are deposited on the dorsi-ventral surface of the leaf.

2.3 Calcium Carbonate (Cystoliths) Based Structural Defense

Calcium carbonate is one of the major components which originates and provides structural defense in some specific members of *Moraceae*. These calcium-oriented depositions made up a unique layer of cellulose microfibrils with pectin. Cystoliths are oblong, reticulate, cigar, 'Y- shaped', I shaped, bean-shaped, grapes-shaped, etc. The size of the cystolith crystal varies from 20–80 μm in wide and 10–120 μm in length. The shape of cystolith depends on its localized occurrence in a specific part of adaxial tissue or abaxial tissue. In the case of *Ficus benghalensis,* cystolith is found in adaxial condition, and Ficus religiosa is in abaxial condition (Pierantoni et al., 2018). Based on that, it transformed into an elongated or spherical cystolith. Mainly, the occurrence of Cystolith is noticed in idioblastic cells and is well known as lithocysts. Lithocysts frequently showed an externally projected sharper tail made up of silica whose size depends on silica deposition on the surface.

2.4 Silica Compositions

Silicon is easily available under the soil and easily intake by plants (Strömberg et al., 2016). Silicon helps in multifold activities like in various physiological processes and makes a protection shield against various biotic stresses. Various polymerized silica in plant bodies was the toughest element that helped resist plants.

2.5 Crystal Enriched Idioblast Cell

In *Ficus carica* leaves, several idioblasts that comprise spiny crystals were found along with the primary layer of palisade cells. Those crystals are unique structures in the cystoliths, but characterization has not been diagnosed (Mamoucha et al., 2016). However, those sharper crystals would possibly create leaf acerbity and harshness.

3 Specialized Transformed Structures

3.1 Glandular Trichomes

Glandular trichomes are multi-celled structures that project externally from the main plant surface, secreting viscous substances of defense-related metabolites. Morphological variability is diverse among the *Ficus* species. However, characteristic features of glandular trichomes are uniseriate, capitates, and filamentous. Capitate types of trichomes bear a short stalk (nearly 5–6 celled). In the case of *Ficus erecta,* glandular trichomes are elliptic, and in *Ficus sagitata,* it is spherical. In *Ficus*, glandular trichomes are present in both dorsi-ventral surfaces, but the maximum occurrence is found on the ventral surface (Sosnovsky, 2016). In the case of *Ficus carica,* glandular trichomes are very small, approximately 4 celled structures. The origin of these small appendages occurred at the time of maturity of the leaves.

3.2 Non-glandular Trichomes

Trichomes are generally two types; glandular types and non-glandular types. Non-glandular trichomes are one of the essential structures and cosmopolitical present in almost all the *Moraceae* members, which helps in structural defense mechanisms. These celled structures are also well-known as idioblasts. Non-glandular trichomes make one structural shield against herbivores and other similar agents. Within the *Ficus* genus, initially growing lengthy trichomes seem to be major and extra ample on the abaxial facet than on the adaxial side of the leaf; however, their precise shape, length, and density were found to be stronger with inter- and intra-specific variations (Klimko & Truchan, 2006). Small unicellular hairs containing silica deposition covered the leaf's surfaces, although maximum growth is found on the adaxial surface. Some peltate non-glandular trichomes are found on the abaxial surface. In *F. carica* leaves, minute non-glandular trichomes appear at initial developmental stages and alter their external appearance by competitive look earlier than they progressively appear on the leaf surfaces.

3.3 Laticifers

Laticifers are a specialized secretory structure or a group of connected cells containing a liquid-like substance known as latex. These are a complex network of elongated living cells spread through the veins in entire plants. The physical structure and anatomy mainly vary from species to species. Similarly, the concentration of latex secretion varies from low to high among species (Kim et al., 2003). Latex is a major constituent of many potential proteins and enzymes that play a key role in the plant-pathogen war. These laticifers are reported mainly from tropical plants having a significant role in defense strategy (Hagel et al., 2008; Konno, 2011). For example, in the plant *Ficus*, a cysteine protease (ficin) plays a major role in defense against several pests (Kajii et al., 2014). Though laticifers can be articulated or non-articulated (Fig. 12.1), in the genus *Ficus*, these are non-articulated and contain more than one nucleus. These laticifer cells are mainly observed during the embryonic stage and confined to regions around the leaf veins only where the secretory system of capitate trichome is also present. Therefore, it clarifies the presence of these structures just above the xylem and near the nerves (Mamoucha et al., 2016).

Latex is a constituent of various chemicals that can be a potential weapon against pests, pathogens, and herbivores; it can act as a physical barrier and chemical defense simultaneously. Even very few amounts of latex are present in some Species if the *Ficus* plant can effectively control or defend the pests and herbivores. For instance, latex is present in laticifers cells under high pressure (Agarwal & Konno, 2009). When an injury occurs by any biotic agents, a sudden rupture of laticifer cells occurs, resulting in the secretion of latex in a high amount (Farrell et al., 1990). The sudden secretion of lates in a larger concentration can potentially control pests and mammalian herbivores (Konno et al., 2004).

Moreover, when the latex comes in contact with the external environment or air, it becomes dry and sticky, forming a physical defense barrier against a wide range of pests. The stickiness and color of the latex are due to the presence of a terpenoid or rubber, which is secreted from laticifer cells. This may be closing the injuries to stop the entry of harmful biotic agents or immobilizing harmful insects by trapping them or their other body parts (Dussourd & Denno, 1994; Kim et al., 2003).

Additionally, the presence of chemicals such as secondary metabolites, irritant properties, enzymes, etc., in higher concentrations than other plant parts has made

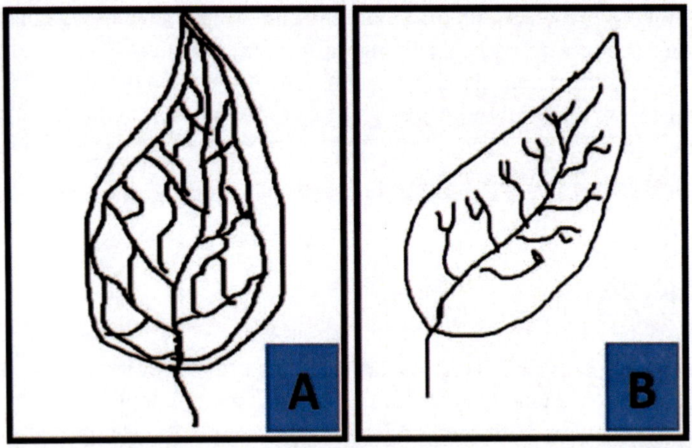

Fig. 12.1 Structure and anatomy of articulated leaves (**a**) and non-articulated leaves (**b**). *Ficus* plants have non-articulated laticifer

them more potential defense structure in the whole plant. The chemical defense of latex has also been reported from *Ficus carica* (Kitajima et al., 2018) and *Ficus virgata* (Konno et al., 2004). In the case of *Ficus carica*, more than 50 proteins have been observed, and many of them are reported to have defensive proteins (Kitajima et al., 2018). Similarly, in the *Ficus virgata* plant, the compounds present in latex can effectively control lepidopteran larvae (Konno et al., 2004).

4 Cellular Defense Responses

During host-pathogen/pest interaction, the physical and chemical barrier of the host serves as the first line of defense and protects itself from the harmful effect of such introducing or interacting pests. However, in some cases, or due to a lack of the first line of defense, one or both defensive barriers fail(s) to counteract such virulent or aggressive pests. Then, the host plant develops an active and dynamic mechanism to gain cellular resistance, which is termed cellular defense. Cellular defense responses are categorized into two groups (Fig. 12.2) based on their time and duration of occurrence: – (i) Rapid active cellular defense and (ii) Delayed active cellular defense.

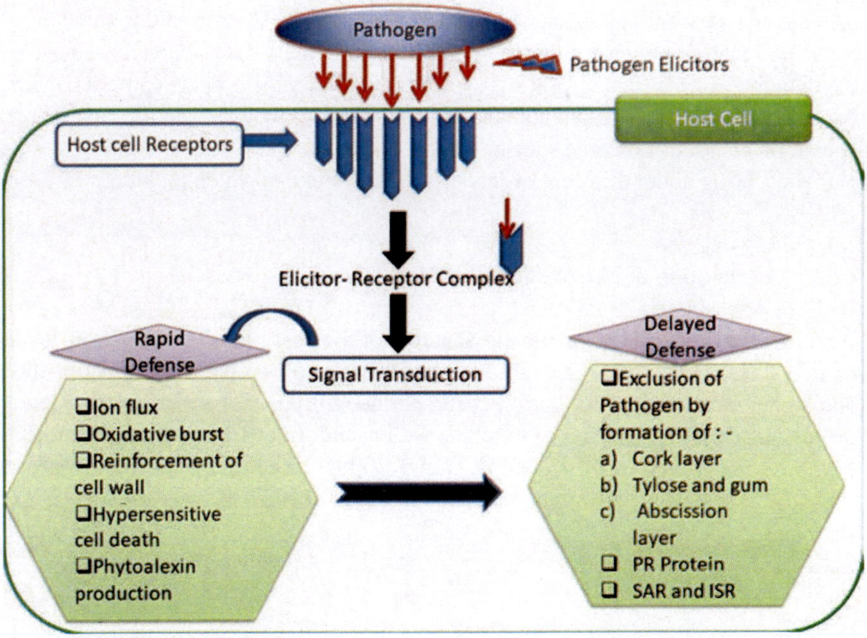

Fig. 12.2 Mechanism of host-pathogen interaction and development of rapid and delayed defence responses in plants against the pathogen. [SAR (Systemic Acquired Resistance) and ISR (Induced Systemic Resistance)]

4.1 Rapid Active Cellular Defense

Activation of rapid cellular defense resides in the consequences of host-pathogen interaction. Understanding their molecular level interaction, there is a need to recognize elicitors (Pathogen's avirulent gene product) and with the corresponding receptors (host's R gene product) and formation of receptor- elicitor complex, which will trigger rapid response cellular defense. This complex causes quick and delayed responses by triggering signal transduction reactions at the local and systemic levels, respectively.

4.2 Delayed Active Cellular Defense

This defense system is activated in the host later to the activation of rapid cellular defense, and it is systemic and persists for a more extended period. It occurs in plants within minutes to hours of establishing the elicitor-receptor complex.

Events of Rapid Active Cellular Defense

Plants have an integrated signalling system that regulates plant growth and development by mediating perceptions and responses to hormones, nutrients, the environment, and stressors. Plant-pathogen interactions trigger a range of plant defense mechanisms (Hammond-Kosack & Jones, 1996), and these interactions are complicated, and there is no general model or sequence of events that accurately represents the dynamics of resistance. When the elicitor ligand binds to its receptor, it triggers the cascade of signalling phenomena such as protein phosphorylation, ion fluxes, activation of reactive oxygen species (ROS), nitric oxide (NO), etc. These signalling pathways bring about different events of cellular defense, which include:

4.2.1 Alternation of Membrane Function

According to most studies on the initial stages of the host-parasite interaction, it was demonstrated that the host membrane plays a vital role in pathogen identification and signal transduction. Alteration in the permeability of plasma membrane led to proton (H^+) and calcium (Ca^{2+}) influx as well as chloride (Cl^-) and potassium (K^+) outflow in plant cells (McDowell & Dangl 2000). In addition, ion fluxes have a significant role in activating reactive oxygen species during host-pathogen interaction.

4.2.2 Oxidative Burst

Oxidative burst is significant defense machinery in host-pathogen conflict (Alvarez et al., 1998). Extracellular synthesis of reactive oxygen intermediates such as superoxide (O_2^-), hydroxyl free radical (OH^-), and hydrogen peroxide (H_2O_2), known as oxidative burst. It is induced by ion fluxes and mediated by NADPH oxidase and/or apoplastic-localized peroxidases (plasma membrane-located) (Somssich & Hahlbrock, 1998; Abdin et al., 2013). Following pathogen attack, cell wall peroxidases have also been implicated in producing reactive oxygen species (ROS) (Bolwell et al., 1995). The fate of oxidative burst in plant cell include:

- Expression of defense genes
- Cross-linking of host cell wall
- Direct killing of invading pathogen at their site of infection
- Triggering of the hypersensitive response in plant cell
- Moderate to severe host cell damage

4.2.3 Reinforcement of Cell Wall

To establish a parasitic relationship with the host plant, most pathogens must breach host cell walls at some stages, either as germ tubes, hyphae, or haustoria. To counteract the ingress of the pathogen, there is an increase in cytoplasmic streaming and accumulation of such cytoplasmic aggregates in infected cells of the host, which are promised to strengthen the cell wall. Cell wall strengthening or fortification of the cell wall is carried out by forming cell aggregates, including papilla, lignotubers (lignified callose deposits), and hydroxyproline-rich glycoproteins. Reinforcements of the cell wall, due to infection of the pathogen, in a resistant host plant are rapid, and the structure formed is larger, where an incompatible host-pathogen interaction takes place, and in the case of a susceptible host, this process is delayed, and structure formed is smaller in size providing a narrow range of protection (Guest & Brown, 1997).

4.2.4 Hypersensitive Response (HR)

Hypersensitive Response is defined as the rapid death of cells in the immediate and surrounding regions of an infection, which acts to limit the growth and development of the pathogen and transmission to other regions of the plant. E.C. Stakman (Stakman, 1915) is known to be the first person to coin the word "*hypersensitive*." He stated that the 'host plant in such situations is hypersensitive to the fungus' when discussing the extreme resistance of particular grass hosts to *Puccinia graminis*. This phenomenon of disease resistance development in plants is similar to apoptosis, or programmed cell death, which is observed during the development of disease defense in mammals. Hypersensitive cell death typically occurs before oxidative burst, increased cytoplasmic aggregation, streaming, granulation, membrane breakdown, and cellular decompartmentalization, all occurring within 12–24 h of attempted entry (Guest & Brown, 1997). HR is a common occurrence in most, if not all, higher plants, and a variety of pathogens can trigger it. HRs caused by a fungus, oomycetes, bacteria, and viruses are the most prevalent, but HRs can also be generated by other species that create persistent intimate connections with the host plant, such as insects (Rossi et al., 1998) and nematodes (Dropkin, 1969). For example, a link between necrotic mesophyll cells in *Bromus* sp. and the fungus *Puccinia recondite* was noted, attempting to infect resistance cultivars (Guest & Brown, 1997).

4.2.5 Phytoalexin Accumulation

One of the most significant advances in physiological plant pathology in the last 25 years has probably been the phytoalexin paradigm of disease resistance. Plants produce phytoalexins, which are low-molecular-weight secondary metabolites that serve as antibiotics in response to pathogen infection, and they are undetectable in uninfected plant tissues. The accumulation of defensive chemicals is generally followed by the hypersensitive response of infected host cells. This phytoalexin-mediated resistance response of the host is considered the last event of rapid cellular defense and subsequently activates a wide range of delayed resistance phenomena in the host plant. More than 300 phytoalexin-like compounds have been identified from plants belonging to over 30 different families. Phytoalexins produced by plants in the same family usually have comparable chemical structures; for example, phytoalexins produced by most legumes are isoflavonoids, while phytoalexins produced by the Solanaceae are terpenoids. The majority of phytoalexins are produced in plants in response to fungus infection. However, phytoalexins have also been induced by bacteria, viruses, and nematodes, and their mode of action includes: (i) disruption of membrane integrity and function and (ii) Inhibition of the electron transport system (Huang, 2001).

Ficus species are infected by fungal, viral, and bacterial pathogens, which cause minor to major crop losses. The most common diseases of fig include Bacterial canker, fig rust, anthracnose, and fig mosaic, and the occurrence and severity of diseases vary with different places. It is reported that fig canker (by bacteria *Xanthomonas campestris*) and fig rust (by fungus *Cerotelium fici*) are the most common diseases in Europe and come to insect pest attack severe damage to fig is caused by fig wax scale. In order to acclimatize themselves to the adverse biotic stresses, *Ficus* species have modified their physical structures (like tough and mineralized leaves, glandular and non-glandular trichome, and laticifer) and also chemical compositions (alkaloids, terpenoids, and phenolics) and accumulation of defensive proteins like chitinase, protease inhibitors and PR (Pathogenesis Related) proteins (Villard et al., 2019). Such a wide range of defensive activities of Ficus may be achieved by constitutively and or by induced defenses. Accumulation of some potent shielding chemicals and proteins in *Ficus* is mainly followed by a rapid induced defense strategy method and results in delayed systemic and persistent induced defenses. So far, there are no reports of the specific molecular mechanisms unraveling the host-pathogen and host insect relationship in *Ficus*, but the occurrence of this is the most needed phenomenon to bring about rapid active cellular defenses to ward off the pest attack.

5 Formation of Defensive Metabolites and Proteins as Biochemical Defense

The genus *Ficus* contains many specialized potential chemicals and secondary metabolites that act against various microbes and nematodes (Table 12.2) (Jeong et al., 2009; Liu et al., 2011). In addition, these metabolites also confer other biological roles, such as irritant and antipyretic properties (Saeed & Sabir, 2002; Patil et al., 2010). The number of proteins and unique compounds like terpenoids, alkaloids, phenolics, etc., varies depending on the plant parts, species, cultivars, and the existing environmental factors.

5.1 The potential Metabolites

Past research has proved that more than 100 metabolites have a role in defense-related activities (Sirisha et al., 2010; Marrelli et al., 2014), while others are associated with specific biological properties like antioxidant activities (Oliveira et al., 2009, 2012). Furthermore, the metabolites responsible for defense-related actions are specialized, such as alkaloids, phenolics, terpenoids, etc.

5.1.1 Alkaloids

Alkaloid metabolites present in the *Ficus* genus have a wide range of defense activities such as antifungal, antibacterial, antiviral, anti-amoebic, and against a wide range of insect pests (Bhutani et al., 1987; Baumgartner et al., 1990; Li et al., 2006). Among all the alkaloid metabolites, the most studied one is the phenanthroindolizidines, mainly found in *Moraceae* and *Asclepiadaceae* families (Al-Khdhairawi et al., 2017). However, some less explored also hold potential defensive roles against fungi and bacteria (Yap et al., 2016). These molecules are mainly confined to stems and roots, sometimes leaves of *Ficus septica* (Wu et al., 2002; Damu et al., 2009), *Ficus hispida* (Yap et al., 2015), *Ficus fistulosa* (Peraza-Sanchcz et al., 2002; Yap et al., 2016). It is also observed that the concentration of phenanthroindolizidines in the latex and other plant exudates of some of the *Ficus* spp. much higher than it can quickly kill herbivores (Konno, 2011). These compounds also display a strong cytotoxic effect on cancer cells (Peraza-Sanchez et al., 2002; Yap et al., 2016).

Table 12.2 Species of *Ficus* plants and their associated secondary metabolites with a defensive role

Plant	Secondary metabolites or compounds	Type	Sub type	Defensive Function	References
Ficus septica	Alkaloids	Phenanthroindolizidine alkaloids	Antofine	Antifungal and antibacterial	Damu et al. (2009)
			Ficuseptines A and B, 13aR-tylocrebrine N-oxide	Toxicity toward cells (Cytotoxicity)	Wu et al. (2002) and Damu et al. (2005)
		Indolizidine alkaloids	4,6-bis-(4-Methoxyphenyl)-1,2,3-trihydroIndolizidinium chloride	Anti-bacterial and anti-fungal	Baumgartner et al. (1990)
Ficus carica		Triterpenoids	Methyl maslinate, oleanolic acid	Irritant	Jaina et al., (2013)
		Steroids	6-O-acyl-β-d-Glucosyl-β-sitosterol	Toxicity toward cells (Cytotoxicity)	Rubnov et al. (2001)
		Monoterpenes, sesquiterpenes and esters,	–	–	Ahmad et al. (2011) and Ojo et al. (2014)
	Phenolics	Phenolic acids	Chlorogenic acid	Antioxidant	Lazreg Aref et al. (2011)
	Flavonoids	Flavones, Flavanones, anthocyanidins	Rutin	Antioxidant	Vaya and Mahmood (2006) and Takahashi et al. (2014)
		Furanocoumarins	Bergapten	Nematicidal	Alam et al. (2015) and Guo et al. (2016)
			Bergaptol, Xanthotoxin, Xnthotoxol Chalepin,	–	Mamoucha et al. (2016) and Wang et al. (2017)
Ficus fistulosa	Alkaloids	Chlorophenanthroindolizidine alkaloids	Tengechlorenine	Toxicity toward cells (Cytotoxicity)	Al-Khdhairawi et al. (2017)
		Septicine-type alkaloids	Fistulopsine A and B	Toxicity toward cells (Cytotoxicity)	Yap et al. (2016)
Ficus nervosa	Flavonoids	Pyranocoumarins	3-Hydroxyxanthyletin	Antimycobacterial	Chen et al. (2010)

5.1.2 Terpenoids

The potential terpenoids such as norisoprenoids, esters, steroids, monoterpenoids, triterpenoids, and sesquiterpenoids are present in most *Ficus* spp. These terpenoids have been studied and observed to carry potential bioactive roles. Among all the terpenoid compounds, the most studied one is sesquiterpenes and monoterpenes, which are abundantly present in the leaves and fruits of *Ficus carica,* respectively. The most significant example of defensive terpenoids in Ficus is cis-1,4-isoprene polymer, popularly known as rubber, responsible for latex stickiness (Oliveira et al., 2010; Konno, 2011).

Triterpenoids have also shown their promising roles in host defense. The three most studied triterpenoids in *Ficus* spp. are methyl maslinate, lupeol acetate, and calotropenyl acetate, which are present in the leaves and exhibit irritant activities (Saeed & Sabir, 2002).

Similarly, antiproliferative roles were found in *Ficus* spp. due to the presence of potential steroids. Many researches have been carried out to check the potentiality of terpenoids. It was found that cardiac steroids, *i.e.*, cardenolides are present in the leaves of *Ficus asperifolia* (Roberts et al., 2016) while 6-*O*-acyl-β-d-glucosyl-β-sitosterol is present in the latex of *Ficus carica* (Ojo et al., 2014; Rubnov et al., 2001). Cardenolides show high toxicity against animals (Roberts et al., 2016) and caterpillars (Konno, 2011). Cardenolides can cause several symptoms in caterpillars, such as unresponsiveness, regurgitation, and convulsions (Dussourd & Hoyle, 2000). Another exciting fact observed by scientists is that many potential volatile terpenoid compounds are also capable of showing significant defensive activity in other plants. For example, antibacterial and anti-insecticidal activity were exerted by monoterpenoids limonene which is present in the fruits of *Ficus carica* (Hebeish et al., 2008). Other monoterpenoids present in the fruit are β-caryophyllene and menthol, which can also act against bacteria (Kim et al., 2008; Skalicka-Wozniak et al., 2009).

5.1.3 Phenolics

Phenolic compounds are mainly uniformly distributed in fruits and carry specific properties like color, flavor, and defense against biotic stresses (Badgujar et al., 2014; Villard et al., 2019). *Ficus* genus's most studied phenolic compounds are tannins, flavonoids, phenolic acids, coumarins, and derivatives. These are also known for their antioxidant (Marrelli et al., 2014), antibacterial (Debib et al., 2014), and anti-mycobacterial activity (Chen et al., 2010). For instance, flavonoids class of compounds like genistein and pyranocoumarins such as 3-hydroxyxanthyletin present in the root region of *Ficus nervosa* are known to have anti-mycobacterial activity (Chen et al., 2010). Similarly, the antibacterial activity of some of the fig extracts may be due to the presence of flavonoids and tannins (Debib et al., 2014). Moreover, one of the coumarins derivative, furanocoumarins, plays a potential defensive compound against various plants (Ojala et al., 2000), herbivores, and pathogens (Berenbaum, 2002; Guo et al., 2016).

The two significant furanocoumarins, such as Bergapten and Psoralen, are reported from the leaves of *Ficus* spp. (Oliveira et al., 2012). This compound is mainly located surrounding the nerves having glandular trichomes and laticifer cells (Wang et al., 2017; Mamoucha et al., 2016). They could activate toxic reactive oxygen species and cause mutation or cell death through a series of events (Dardalhon et al., 1998; Berenbaum, 2002). Additionally, the compounds Bergapten and Psoralen in figs plant can inhibit digestive enzymes such as amylase and cellulase, resulting in a nematicidal effect (Guo et al., 2016). The photosensitizing action of fig plants can also cause allergy-like contact dermatitis (Pathak et al., 1962). The recent discovery of the presence of psoralic acid glucoside (PAG) in large concentrations many opened the path to find its physiological function (Takahashi et al., 2017) and defensive role (Wang et al., 2017).

5.2 *Defensive Proteins Present in* Ficus *species*

The proteins such as chitinase, oxidase, protease, and protease inhibitors are present in the Ficus plant lates and are very stable. Moreover, their properties are independent and not affected by herbivores. Therefore, these proteins play a significant role against pathogens and herbivores (Zhu-Salzman et al., 2008; Konno, 2011).

5.2.1 Chitinases

Chitinase is widely known to inhibit the growth of fungi by degrading the cell wall (Taira et al., 2005). These can also show insecticidal activities sometimes by slowing down their developmental process (Lawrence & Novak, 2006). Chitinase is mainly found in the latex of *Ficus microcarpa* and *Ficus carica* (Kitajima et al., 2018).

5.2.2 Oxidases

The main oxidase proteins, such as polyphenol oxidases and peroxidases, are present in the *Ficus* genus, and they are associated with rubber and the rate of latex coagulation, respectively (Wahler et al., 2009). The protein Peroxidases are mainly found in the latex region of *Ficus sycomorus, Ficus carica,* and *Ficus glabrata,* while polyphenol oxidases have been found only in *Ficus elastica* (Kim et al., 2008; Mohamed et al., 2011). These proteins are considered one of the potential defensive proteins due to their anti-insecticidal and antinutritive effects (Zhu-Salzman et al., 2008; Konno, 2011). However, insects consume the Ficus plant parts, and a cascade of signaling events occurs. This involves the production of ROS (reactive oxygen species) for disruption of major amino acids and nutrients for the insects (Sethi et al., 2009).

5.2.3 Proteases

Plant genomes encode some proteases whose key role is in host defense against various harmful biotic and abiotic stress agents through generating signaling molecules. For example, ficins, one cysteine class of protease, have been reported from the latex of the *Ficus* family (Sgarbieri et al., 1964; Villard et al., 2019). This can control many insects by disturbing their feeding habit, digestion, growth rate, and development (Konno et al., 2004). For example, ficins in *Ficus virgata* can effectively control lepidopteran insects in the larval stage (Konno et al., 2004).

5.2.4 Protease Inhibitors

Many plant species produce protease inhibitors, a major class of defensive protein that is activated upon sensing a wound response initiated by an insect, pathogen, or nematode. These are highly potential in defending the biotic agents, especially restricting the infection by nematodes (Koiwa et al., 1997; Hellinger & Gruber, 2019). The major protease inhibitors are cysteine and serine protease inhibitors acting as an important weapon against pests and pathogens (Kitajima et al., 2018; Villard et al., 2019). Several evidences suggest the potential defensive role of trypsin, a serine protease against insects and fungi (Ryan, 1990; Terras et al., 1993). The latex of the *Ficus carica* plant contains different protease inhibitors, but the major one is trypsin inhibitors, whose concentration is more and can target the proteases present in insect gut (Huynh et al., 1992). Along with ficins, more than 10 isoforms of trypsin inhibitors can be found in the latex of leaves.

5.3 Some Important Pathogenesis-Related Proteins (PRs)

Pathogenesis-related proteins, or PR proteins, have a significant role in the host defense mechanism and are produced in host plants during various events of a pathogen attack. The function and mechanism of these proteins are widely studied. Significant PR proteins have a defensive role against fungi, bacteria, virus, and insects (Ali et al., 2018). The major PR proteins found in *Ficus* plants belong to PR-1, PR-4, and PR-5, and these are present mostly on the leaf latex (de Freitas et al., 2011). In addition, these proteins can inhibit or restrict the growth of many pathogens, mostly fungi (Bai et al., 2013), hence considered fungicides.

Table 12.3 Chemicals and proteins present in the latex of the *Moraceae* family having a defensive role against insects

Plant family	Scientific name of the plant	Category	Compound	Sub-category	References
	Ficus carica	Protein	Proteases	Cysteine proteases	Ramos et al. (2010)
Moraceae	*Ficus elastica*			Serine proteases	Patel et al. (2007) and Tomar et al. (2008)
	Ficus carica		Oxidase	Peroxidase (POD)	Saby et al. (2003) and Sethi et al. (2009)
	Ficus carica		Protease inhibitors	Serine Protease inhibitors	Azarkan et al. (2004) and Kehr (2006)
	Antiaris toxicaria	Chemicals	Cardenolide	Toxicariosides	Carter et al. (1997)
	Morus australis		Alkaloids	Sugar-mimic alkaloides	Konno et al. (2006)

5.4 *Different Associative Proteins*

In addition to these PR proteins, many additional proteins and enzymes are also present in the latex of *Ficus carica* leaves which may have a potential role in plant defense mechanism. For example, lectins belonging to the carbohydrate-binding protein group have been detected in the latex and are known to have anti-viral, anti-bacterial, and anti-fungal roles (Macedo et al., 2015). Furthermore, though the study on phosphatases is very few, some studies have shown the disturbance in the growth and development of insect pests (Table 12.3) (Liu et al., 2011; Konno, 2011).

6 Role of Genotype and Environmental Factors in *Ficus* Defense Mechanisms

Phenotypic variability consequences from interactions between genotype and environment and is a first-rate driver of ecological and evolutionary interactions. Measuring the relative contributions of a genetic variant, the environment, and their interaction with phenotypic variation remains an essential intention of evolutionary ecology. The genetic version is widely related to trait expression (Agrawal and Hastings, 2019a; Geber & Griffen, 2003), supplying the variant on which choice may also affect the duration of evolution. Environmental variation contributes to the phenotypic version (Decker et al., 2019; Couture et al., 2015; Davis et al., 2005), which influences specific genotypes to prosper (Beemelmanns & Roth, 2017; Potts & Hunter, 2021).

6.1 Role of Genotype in Modifying the Defense Mechanism

The type of defense mechanisms in *Ficus* depends on several factors, such as an individual organism, its variety, or species. However, the genotype of Ficus is the most responsible for the differentiation of morphological and biochemical factors, which include the color and concentration of latex; morphology of trichomes; size, shape, location, and amount of chemical and mineral present.

The genetic elements are the significant player in determining the *Ficus* species specific mineral system, which could elaborate the unique features of leaf minerals (Pierantoni et al., 2018). As an example, the morphological features of phytoliths could be an identifying factor for phylogenetic classification (Prychid et al., 2003).

6.2 Role of Plant Age and Season in Modifying the Defense Mechanism

The organs in Ficus are not protected throughout the year equally as the aging and occurrence of seasons have a major contribution towards altering the defense strategy. It has been observed that the furanocoumarin content of *F. carica* leaves and latex rises in the spring and summer but not in the autumn season (Zaynoun et al., 1984). Marrelli and co evaluated the chemical composition, and concentration of fruits of *F. carica* harvested in different months intervals, corresponding to distinct phenological stages of the fig. The total amount of phenolics in figs increased as they ripened, but the content of pyranocoumarin and furanocoumarin declined (Marrelli et al., 2012). Another research discovered that the chemical content of different parts of *F. carica* (such as bark, leaves, and woody components) varies over time and that these modifications are unique based on individual compounds and organs (Marrelli et al., 2014). Moreover, the range or the amount of latex produced varies according to the age of the plant part, i.e., young Ficus organs produce more latex than older parts. Both seasonality and plant growth stages are expected to play a major role in modulating the *Ficus* defense mechanism, and their effects are likely to be coupled (Konno, 2011).

7 Inducible Defense Mechanism

Against insect herbivores, plants have developed a variety of defense systems. These defense systems can be divided into two types: (1) constitutive or permanent defenses and (2) inducible or transient (Karban & Baldwin, 1997). Both defenses are achieved through comparable methods, but constitutive protections are inactive form prior to an herbivore attack, whereas induced defenses or transient defenses are activated only after an injury occurs (Gatehouse, 2002; Dicke et al., 2003; Chen,

2008). In addition to intrinsic defenses, initiating specialized herbivory-specific defense responses is a key strategy for plant survival and persistence (Karban & Baldwin, 1997).

Host plants can detect the type of injury or wound caused by mechanical means or insects or herbivores to activate the type of induced defense strategy. Elicitors like biotic (Insect, microbe), abiotic (Drought, UV, temperature, mechanical damage, etc.), and chemical (Jasmonic acid, Salicylic acid, ethylene, metallic ion, salt, etc.) are prerequisites for the induction of resistance response in plants (Fig. 12.3). As a result, plants respond to many stimuli, including contact and the attacking herbivore's salivary enzymes. Physical and molecular mechanisms are involved in this induction. Insect harm or mechanical damage to plants activates several signal transduction pathways (Korth & Dixon, 1997). Sometimes, a minor alteration in the host gene transcription can cause induced defenses. Although the genes involved in this process differ between species, systemically triggered defenses arise due to modification of the expression of genes (Walling, 2000). For example, photosynthetic genes in *Nicotiana tobacum* are down-regulated in response to insect attack, whereas defense genes are up-regulated (Hermsmeier et al., 2001). Induced resistance mediated defense responses in plants are carried out in two broad aspects or pathways, which include: – (i) direct defense against insects (secretion of inhibitory

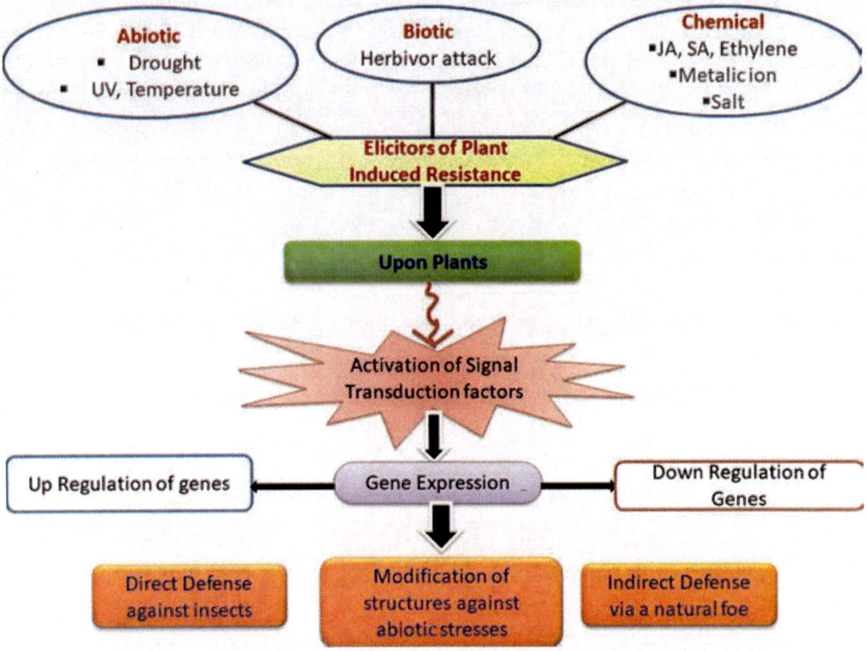

Fig. 12.3 Induction of resistance responses concerning abiotic, biotic, and chemical elicitors in *Ficus*

proteins, enzymes, and toxins) (ii) indirect defense (secretion of plant volatiles attracting natural predator and parasites of harmful insects).

Physical and chemical factors are involved in the induction of defense systems. Long-term consumption of plant parts by herbivores appears to strengthen the toughness of Ficus leaf and results in a high rise in the density of calcium oxalate crystal in a few species of *Ficus* (Xiang & Chen 2004). This herbivore injury can lead to silicification or an increase in trichome density and number (Baur et al., 1991; Massey et al., 2007; Strömberg et al., 2016; Liu et al., 2018) in some plants. The expression of chemicals such as chitinase, trypsin inhibitor, and peroxidase can be elevated in the latex of *F. carica* after several hours of wounding treatment (Kim et al., 2003). However, several investigations on other plants have led us to believe that herbivore or pathogen infection may affect the activation of the secretion of specific *Ficus* defense chemicals. It was also reported that mechanical injury and pathogen infection were shown to induce the synthesis of certain furanocoumarins in parsnip (Krieger et al., 2018). Other factors may influence *Ficus* defenses, but limited research on this area.

8 Evolutionary Trends Towards Complex Adaptation

The mechanisms through which complex adaptations develop are a fundamental challenge in evolutionary biology. Complex adaptations are thought to take a long time to develop because they require unique population genetic or environmental variables to evolve. Complex adaptation blueprints in molecular systems, on the other hand, are standard, implying that they can quickly evolve (Pal & Papp, 2017). If complex adaptation of a plant is observed, then it would be found that each plant organ has its unique histology and defensive architecture, and the amount of defensive chemicals varies from one organ to the next (Villard et al., 2019). Insects and pests coevolved with *Ficus* defense apparatus do so primarily through various tactics to attack the host and intricate adaptability. Ficus are thus plants that shed light on the exciting arms race those plants engage in to adapt to their surroundings. Evolutionary trends toward complex adaptability are ongoing and will continue to evolve in tandem with nature.

9 Evolution and Coevolution

Biotic stresses are one of the most powerful selective pressures on the plant, and their interaction with the host assists in improving both organisms' genetic diversity. Host and pathogen selection toward each other is highly significant as if a pathogen chooses its host and can adapt to the host system. Then the host also imposes the same levels of behaviour or selection pressure on its host, resulting in continuous mutual adaptation between them. Co-evolutionary interactions are highly related to organism evolution.

Various generalist and specialist insects prey on *Ficus* species (Novotny et al., 2010; Volf et al., 2018). It has been claimed that the generalist insects' defense mechanisms are often harmful compounds, resulting in higher mortality. Plants use a variety of tactics to defend themselves, including chemical diversity pyramiding and mechanical defenses, which reduce insect feeding efficiency and increase feeding duration resulting in a long time of contact with toxic compounds and harmful biotic agents, which lead to a high increase in insect mortality (Koricheva et al., 2004; Volf et al., 2015, 2018). In 2018, the link between *Ficus* pests and several *Ficus* defense features was explored, which showed the critical role of chemicals like protease, alkaloid, polyphenol oxidative activities, and structural features such as trichome length, trichome density.

On the other hand, in this insect-plant war, certain insects are smart enough to overcome the host's defense activities by secreting specific enzymes that can neutralize these host chemical weapons (Volf et al., 2018). To avoid latex toxicity, certain insects discovered the piercing technique in canal branches resulting emission of plant latex, freeing the distal tissues from toxic chemicals (Konno, 2011), after which it is safer to consume the distal tissues. Though this activity is not examined in *Ficus*, other latex-producing plants have been extensively discussed (Dussourd, 2017). The host plant and its corresponding parasites reciprocally affect each other's evolution during their interaction period and ultimately lead to the convolution of both the organisms.

10 Pattern of Defense Mechanism in Various Parts of *Ficus*

A wide range of biotic agents chooses their infection site according to their preference and suitability. The pests attacking the leaf may be or not be the same as the pests attacking the other parts. Similarly, the pests attacking fruits or floral parts may be or may not be the same as other pests infecting the same host. The histology, chemical constituent, and concentration are not the same for different parts of the plant, so the infecting pests and their infection strategy also vary. Moreover, there is no doubt that the mechanism of defense by the host in various parts also differs. For example, the herbivores mainly target the foliar part of the plant, among other parts.

Similarly, ficin is the main PR protein in trunk and fruits, while trypsin inhibitors are mostly present on the latex of leaves. Additionally, physical barriers such as trichomes and a wide variety of chemicals and minerals help the host plant *Ficus* get rid of herbivores. For example, Lignification, or the presence of lignin, is another important physical barrier in figs plant that increases mechanical protection and keeps the herbivores away from its trunk, bark, and other woody parts. The other chemical factors that help in defense mechanisms are ficin, PR proteins, chitinase acid phosphatases, furanocoumarins, wide antifungal, antibacterial, antiviral, insecticidal, and insecticidal nematicidal activities (Kitajima et al., 2018).

A different set of defense mechanisms has been observed in different plant parts dependent on the function of those plant parts. For example, the role of the insect aganoid wasp and vertebrate species are essential for the pollination of figs and dissemination of seeds, respectively. Therefore, the toxic chemicals should be present in an average concentration to not be more toxic to these host partners (Cruaud et al., 2012; Kitajima et al., 2018).

11 Conclusion and Future Perspectives

For years tremendous research has been carried out to understand the defense mechanism strategy in various plants. The mechanism of defense highly varies from plant to plant. The chemicals involved in the whole defense process are diverse and may or may not be similar in all host plants. Similarly, the cascade of signaling to defense mechanism varies among the members of the family *Moraceae* and different species of *Ficus*. Moreover, the events of defense mechanisms are also highly varied and different based on the plant's diverse parts (leaves, stem, floral, etc.). These strategies may be inducible or constitutive. Due to the new strain development in the pest population, environmental function, and synergistic action of all mechanisms, the host defense strategy has also evolved, resulting in broad resistance to various pests and pathogens. One of the most studied defense strategies is the latex of the fig plant, which can act as both mechanical and chemical means of defense strategy. However, the concentration of latex and the size of the aggressor are highly related as a small concentration of latex can be harmful to a tiny insect but may not be harmful against a large size animal herbivore. Therefore, the *Ficus* group of plants can provide some novel in-depth conclusions about the fascinating concept or strategies for host adaptation to the external environment. Exploiting omics data like genomes or transcriptomes of *Ficus religiosa* and *Ficus carica* which can be accessed from public databases such as Genbank, can open up the path for new advancements shortly.

References

Abdin, M. Z., Khan, M. A., Ali, A., Alam, P., Ahmad, A., & Sarwat, M. (2013). Signal transduction and regulatory networks in plant-pathogen interaction: A proteomics perspective. In *Stress signaling in plants: Genomics and proteomics perspective* (Vol. 1, pp. 69–90). Springer.

Agrawal, A. A., & Konno, K. (2009). Latex: A model for understanding mechanisms, ecology, and evolution of plant defense against herbivory. *Annual Review of Ecology, Evolution, and Systematics, 40*, 311–331.

Alam, P., Siddiqui, N., Basudan, O., et al. (2015). Comparative profiling of biomarker psoralen in antioxidant active extracts of different species of genus *Ficus* by validated HPTLC method. *African Journal of Traditional, Complementary, and Alternative Medicines, 12*, 57.

Ali, S., Ganai, B. A., Kamili, A. N., Bhat, A. A., Mir, Z. A., Bhat, J. A., Tyagi, A., Islam, S. T., Mushtaq, M., Yadav, P., Rawat, S., & Grover, A. (2018). Pathogenesis-related proteins and peptides as promising tools for engineering plants with multiple stress tolerance. *Microbiological Research, 212-213*, 29–37.

Al-Khdhairawi, A. A. Q., Krishnan, P., Mai, C. W., et al. (2017). A bis-benzopyrroloisoquinoline alkaloid incorporating a cyclobutane core and a chlorophenanthroindolizidine alkaloid with cytotoxic activity from *Ficus fistulosa var. tengerensis*. *Journal of Natural Products, 80*, 2734–2740.

Alvarez, M. E., Pennell, R. I., Meijer, P. J., Ishikawa, A., Dixon, R. A., & Lamb, C. (1998). Reactive oxygen intermediates mediate a systemic signal network in the establishment of plant immunity. *Cell, 92*(6), 773–784.

Ahmad, S., Rao, H., Akhtar, M., et al. (2011). Phytochemical composition and pharmacological prospectus of *Ficus benghalensis* Linn. (*Moraceae*)-a review. *Journal Medicinal Plants Research, 5*, 6393–6400.

Azarkan, M., Wintjens, R., Looze, Y., & Baeyens-Volant, D. (2004). Detection of three wound-induced proteins in papaya latex. *Phytochemistry, 65*, 525–534.

Badgujar, S. B., Patel, V. V., Bandivdekar, A. H., & Mahajan, R. T. (2014). Traditional uses, phytochemistry and pharmacology of *Ficus carica*: A review. *Pharmaceutical Biology, 52*(11), 1487–1503.

Bai, S., Dong, C., Li, B., Dai, H. (2013) A PR-4 gene identified from Malus domestica is involved in the defense responses against Botryosphaeria dothidea. *Plant Physiology and Biochemistry, 62*, 23–32. https://doi.org/10.1016/j.plaphy.2012.10.016

Bauer, P., Elbaum, R., & Weiss, I. M. (2011). Calcium and silicon mineralization in land plants: Transport, structure and function. *Plant Science, 180*, 746–756. https://doi.org/10.1016/j.plantsci.2011.01.019

Baur, R., Binder, S., Benz, G. (1991) Nonglandular leaf trichomes as short-term inducible defense of the grey alder, Alnus incana (L.), against the chrysomelid beetle. *Agelastica alni L. Oecologia 87*, 219–226. https://doi.org/10.1007/BF00325259

Baumgartner, B., Erdelmeier, C. A. J., Wright, A. D., Rali, T., & Sticher, O. (1990). An antimicrobial alkaloid from *Ficus septica*. *Phytochemistry, 29*, 3327–3330.

Bayoudh, C., Elair, M., Labidi, R., et al. (2017). Efficacy of tissue culture in virus elimination from caprifig and female fig varieties (*Ficus carica* L.). *Plant Pathology Journal, 33*, 288–295.

Beemelmanns, A., & Roth, O. (2017). Grandparental immune priming in the pipefish Syngnathus typhle. *BMC Evolutionary Biology, 17*, 1–15.

Berenbaum, M. R. (2002). Postgenomic chemical ecology: From genetic code to ecological interactions. *Journal of Chemical Ecology, 28*, 873–896.

Bhutani, K., Sharma, G., & Ali, M. (1987). Plant based antiamoebic drugs; part I. Antiamoebic activity of phenanthroindolizidine alkaloids; common structural determinants of activity with emetine. *Planta Medica, 53*, 532–536.

Bolwell, G. P., Butt, V. S., Davies, D. R., & Zimmerlin, A. (1995). The origin of the oxidative burst in plants. *Free Radical Research, 23*, 517–532.

Borges, R. M., Bessière, J.-M., & Ranganathan, Y. (2013). Diel variation in fig volatiles across syconium development: Making sense of scents. *Journal of Chemical Ecology, 39*, 630–642.

Carter, C. A., Forney, R. W., Gray, E. A., Gehring, A. M., Schneider, T. L., Young, D. B., Lovett, C. M., Jr., Scott, L., Messer, A. C., & Richardson, D. P. (1997). Toxicarioside A. A new cardenolide isolated from *Antiaris toxicaria* latex-derived dart poison. Assignment of the 1H- and 13C-NMR shifts for an antiarigenin aglycone. *Tetrahedron, 53*, 13557–13566.

Chamont, S. (2014). *Ficus carica*. Les maladies et ravageurs. In: Ephytia. http://ephytia.inra.fr/fr/C/18200/Hypp-encyclopedie-en-protection-des-plantes-Les-maladies-et-ravageurs. Accessed 17 Sep 2018

Chen, M. S. (2008). Inducible direct plant defense against insect herbivores: A review. *Insect Sci., 15*, 101–114.

Chen, L. W., Cheng, M. J., Peng, C. F., & Chen, I. S. (2010). Secondary metabolites and antimycobacterial activities from the roots of *Ficus nervosa*. *Chemistry & Biodiversity, 7*, 1814–1821.

Couture, J. J., Serbin, S. P., & Townsend, P. A. (2015). Elevated temperature and periodic water stress alter growth and quality of common milkweed (*Asclepias syriaca*) and monarch (*Danaus plexippus*) larval performance. *Arthropod-Plant Interactions, 9*, 149–161.

Cruaud A, Rønsted N, Chantarasuwan B et al (2012). An extreme case of plant–insect codiversification: figs and fig-pollinating wasps. *Systematic Biology, 61*, 1029–1047. https://doi.org/10.1093/sysbio/sys068

Dardalhon, M., de Massy, B., Nicolas, A., & Averbeck, D. (1998). Mitotic recombination and localized DNA double-strand breaks are induced after 8-methoxypsoralen and UVA irradiation in *Saccharomyces cerevisiae*. *Current Genetics, 34*, 30–42.

Damu, A. G., Kuo, P. C., Shi, L. S., et al. (2005). Phenanthroindolizidine alkaloids from the stems of *Ficus septica*. *Journal of Natural Products, 68*, 1071–1075.

Damu, A., Kuo, P. C., Shi, L. S., Li, C. Y., Su, C. R., & Wu, T. S. (2009). Cytotoxic phenanthroindolizidine alkaloids from the roots of *Ficus septica*. *Planta Medica, 75*, 1152–1156.

Davis, A. K., Farrey, B. D., & Altizer, S. M. (2005). Variation in thermally induced melanism in monarch butterflies (Lepidoptera: *Nymphalidae*) from three North American populations. *Journal of Thermal Biology, 30*, 410–421.

Debib, A., Tir-Touil, A., Mothana, R. A., Meddah, B., & Sonnet, P. (2014). Phenolic content, antioxidant and antimicrobial activities of two fruit varieties of Algerian *Ficus carica* L.: Antioxidant and antimicrobial activity of Algerian figs. *Journal of Food Biochemistry, 38*, 207–215.

Decker, L. E., Soule, A. J., de Roode, J. C., & Hunter, M. D. (2019). Phytochemical changes in milkweed induced by elevated CO_2 alter wing morphology but not toxin sequestration in monarch butterflies. *Functional Ecology, 33*, 411–421.

De Freitas, C. D. T., Sousa Nogueira, F. C., Vasconcelos, I. M., Oliveira, J. T. A., Domont, G. B., Ramos M. V. (2011). Osmotin purified from the latex of Calotropis procera: biochemical characterization, biological activity and role in plant defense. *Plant Physiology and Biochemistry, 49*, 738–743. https://doi.org/10.1016/j.plaphy.2011.01.027

Dicke, M., van Poecke, R. M. P., & de Boer, J. G. (2003). Inducible indirect defense of plants: From mechanisms to ecological functions. *Basic and Applied Ecology, 4*, 27–42.

Dropkin, V. H. (1969). The necrotic reaction of tomatoes and other hosts resistant to Meloidogyne: Reversal by temperature. *Phytopathology, 59*, 1632–1637.

Dussourd, D. E. (2017). Behavioral sabotage of plant defenses by insect folivores. *Annual Review of Entomology, 62*, 15–34. https://doi.org/10.1146/annur-ento-031616-035030

Dussourd, D. E., & Hoyle, A. M. (2000). Poisoned plusiines: Toxicity of milkweed latex and cardenolides to some generalist caterpillars. *Chemoecology, 10*, 11–16.

Dussourd, D. E., & Denno, R. F. (1994). Host range of generalist caterpillars: Trenching permits feeding on plants with secretory canals. *Ecology, 75*, 69–78.

Farrell, B. D., Dussourd, D. E., & Mitter, C. (1990). Escalation of plant defense: Do latex and resin canals spur plant diversification? *The American Naturalist, 138*, 881–900.

Finley, D. S. (1999). Patterns of calcium oxalate crystals in young tropical leaves: A possible role as an anti-herbivory defense. *Revista de Biología Tropical, 47*, 27–31.

Franceschi, V. R., & Nakata, P. A. (2005). Calcium oxalate in plants: formation and function. *Annual Review of Plant Biology, 56*, 41–71. https://doi.org/10.1146/annurev.arplant.56.032604.144106

Gatehouse, J. A. (2002). Plant resistance towards insect herbivores: A dynamic interaction. *New Phytologist, 156*, 145–169.

Geber, M. A., & Griffen, L. R. (2003). Inheritance and natural selection on functional traits. *International Journal of Plant Sciences, 164*, S21–S42.

Guest, D., & Brown, J. (1997). Plant defenses against pathogens. *Plant pathogens and plant diseases, 263*, 286.

Guo, Q., Du, G., He, H., Xu, H., Guo, D., & Li, R. (2016). Two nematicidal furocoumarins from *Ficus carica* L. leaves and their physiological effects on pine wood nematode (*Bursaphelenchus xylophilus*). *Natural Product Research, 30*, 1969–1973.

Hagel, J., Yeung, E., & Facchini, P. (2008). Got milk? The secret life of laticifers. *Trends in Plant Science, 13*, 631–639.

Hammond-Kosack, K. E., & Jones, J. D. G. (1996). Resistance gene dependent plant defense responses. *Plant Cell, 8*, 1773–1791.

Hebeish, A., Fouda, M. M. G., Hamdy, I. A., EL-Sawy, S. M., & Abdel-Mohdy, F. A. (2008). Preparation of durable insect repellent cotton fabric: Limonene as insecticide. *Carbohydrate Polymers, 74*, 268–273.

Hellinger, R., & Gruber, C. W. (2019). Peptide-based protease inhibitors from plants. *Drug Discovery Today, 24*(9), 1877–1889.

Hermsmeier, D., Schittko, U., & Baldwin, I. T. (2001). Molecular interactions between the specialist herbivore *Manduca sexta* (Lepidoptera, *Sphingidae*) and its natural host *Nicotiana attenuata*. 1. Large-scale changes in the accumulation of growth- and defense-related plant mRNAs. *Plant Physiology, 125*, 1875–1887.

Hosomi, A., Miwa, Y., Furukawa, M., & Kawaradani, M. (2012). Growth of fig varieties resistant to ceratocystis canker following infection with *Ceratocystis fimbriata*. *Journal of the Japanese Society for Horticultural Science, 81*, 159–165.

Huang, J. S. (2001). Accumulation of phytoalexins as a resistance mechanism. In *Plant pathogenesis and resistance* (pp. 525–621). Springer.

Hudgins, J. W., Krekling, T., & Franceschi, V. R. (2003). Distribution of calcium oxalate crystals in the secondary phloem of conifers: A constitutive defense mechanism? *The New Phytologist, 159*, 677–690.

Huynh, Q. K., Borgmeyer, J. R., & Zobel, J. F. (1992). Isolation and characterization of a 22 kDa protein with antifungal properties from maize seeds. *Biochemical and Biophysical Research Communications, 182*, 1–5.

Jaina, R., Jain, S. C., Bhagchandani, T., & Yadav, N. (2013). New furanocoumarins and other chemical constituents from *Ficus carica* root heartwood. *Z Für Naturforschung, 68*, 3–7.

Jeong, M. R., Kim, H. Y., & Cha, J. D. (2009). Antimicrobial activity of methanol extract from *Ficus carica* leaves against oral bacteria. *Journal of Bacteriology and Virology, 39*, 97–102.

Kajii, C., Morita, T., & Kuroda, K. (2014). Laticifers of *Ficus carica* and their potential role in plant defense. *IAWA Journal, 35*(2), 109–115.

Karban, R., & Baldwin, I. T. (1997). *Induced responses to herbivory*. Chicago University Press.

Kehr, J. (2006). Phloem sap proteins: Their identities and potential roles in the interaction between plants and phloem-feeding insects. *Journal of Experimental Botany, 57*, 767–774.

Kim, J. S., Kim, Y. O., Ryu, H. J., Kwak, Y. S., Lee, J. Y., & Kang, H. (2003). Isolation of stress-related genes of rubber particles and latex in fig tree (*Ficus carica*) and their expressions by abiotic stress or plant hormone treatments. *Plant & Cell Physiology, 44*, 412–414.

Kim, Y. S., Park, S. J., Lee, E. J., Cerbo, R. M., Lee, S. M., Ryu, C. H., Kim, G. S., Kim, J. O., & Ha, Y. L. (2008). Antibacterial compounds from rose bengal-sensitized photooxidation of β-caryophyllene. *Journal of Food Science, 73*, C540–C545.

Kitajima, S., Aoki, W., Shibata, D., et al. (2018). Comparative multi-omics analysis reveals diverse latex-based defense strategies against pests among latex-producing organs of the fig tree (*Ficus carica*). *Planta, 247*, 1423–1438.

Klimko, M., & Truchan, M. (2006). Morphological variability of the leaf epidermis in selected taxa of the genus *Ficus* L. (Moraceae) and its taxonomic implication. *Acta Societatis Botanicorum Poloniae, 75*, 309–324. https://doi.org/10.5586/asbp.2006.038

Koiwa, H., Ray, A., Bressan, P., & Hasegawa, M. (1997). Regulation of protease inhibitors and plant defense. *Trends in Plant Science, 2*(10), 379–384.

Konno, K. (2011). Plant latex and other exudates as plant defense systems: Roles of various defense chemicals and proteins contained therein. *Phytochemistry, 72*, 1510–1530.

Konno, K., Hirayama, C., Nakamura, M., Tateishi, K., Tamura, Y., Hattori, M., & Kohno, K. (2004). Papain protects papaya trees from herbivorous insects: Role of cysteine proteases in latex. *The Plant Journal, 37*, 370–378.

Konno, K., Ono, H., Nakamura, M., Tateishi, K., Hirayama, C., Tamura, Y., Hattori, M., Koyama, A., & Kohno, K. (2006). Mulberry latex rich in antidiabetic sugar-mimic alkaloids forces dieting on caterpillars. *Proceedings of the National Academy of Sciences of the United States of America, 103*, 1337–1341.

Koricheva, J., Nykänen, H., & Gianoli, E. (2004). Meta-analysis of trade-offs among plant antiherbivore defenses: Are plants jacks-of all-trades, masters of all? *The American Naturalist, 163*, E64–E75.

Korth, K. L., & Dixon, R. A. (1997). Evidence for chewing insect-specific molecular events distinct from a general wound response in leaves. *Plant Physiology, 115*, 1299–1305.

Korth, K. L., Doege, S. J., Park, S.-H., et al. (2006). *Medicago truncatula* mutants demonstrate the role of plant calcium oxalate crystals as an effective defense against chewing insects. *Plant Physiology, 141*, 188–195. https://doi.org/10.1104/pp.106.076737

Krieger, C., Roselli, S., Kellner-Thielmann, S., et al. (2018). The CYP71AZ P450 subfamily: A driving factor for the diversification of coumarin biosynthesis in apiaceous plants. *Frontiers in Plant Science, 9*, 820.

Lazreg-Aref, H., Gaaliche, B., Fekih, A., et al. (2011). *In vitro* cytotoxic and antiviral activities of *Ficus carica* latex extracts. *Natural Product Research, 25*, 310–319.

Lawrence, S. D., & Novak, N. G. (2006). Expression of poplar chitinase in tomato leads to inhibition of development in Colorado Potato Beetle. *Biotechnology Letters, 28*, 593–599.

Liu, F., Yang, Z., Zheng, X., et al. (2011). Nematicidal coumarin from *Ficus carica* L. *Journal of Asia-Pacific Entomology, 14*, 79–81.

Liu, M., Zhou, F., Pan, X., Zhang, Z., Thraw, M.B., LI, Bo. (2018) Specificity of herbivore-induced responses in an invasive species, *Alternanthera philoxeroides* (alligator weed). *Ecology and Evolution, 8*, 59–70. https://doi.org/10.1002/ece3.3615

Li, H., Hu, T., Wang, K., et al. (2006). Total synthesis and antiviral activity of enantioenriched (+)-deoxytylophorinine. *Letters in Organic Chemistry, 3*, 806–810.

Macedo, M. L. R., Oliveira, C. F. R., & Oliveira, C. T. (2015). Insecticidal activity of plant lectins and potential application in crop protection. *Molecules, 20*, 2014–2033.

McDowell, J. M., & Dangl, J. L. (2000). Signal transduction in the plant immune response. *Trends in Biochemical Sciences, 25*, 79–82.

Mamoucha, S., Fokialakis, N., & Christodoulakis, N. S. (2016). Leaf structure and histochemistry of *Ficus carica* (*Moraceae*), the fig tree. *Flora – Morphology, Distribution, Functional Ecology of Plants, 218*, 24–34.

Marrelli, M., Menichini, F., Statti, G. A., Bonesi, M., Duez, P., Menichini, F., & Conforti, F. (2012). Changes in the phenolic and lipophilic composition, in the enzyme inhibition and antiproliferative activity of *Ficus carica* L. cultivar Dottato fruits during maturation. *Food and Chemical Toxicology, 50*, 726–733.

Marrelli, M., Statti, G. A., Tundis, R., Menichini, F., & Conforti, F. (2014). Fatty acids, coumarins and polyphenolic compounds of *Ficus carica* L. cv. Dottato: Variation of bioactive compounds and biological activity of aerial parts. *Natural Product Research, 28*, 271–274.

Massey, F. P., Roland Ennos, A., Hartley, S. E. (2007). Herbivore specific induction of silica-based plant defences. *Oecologia, 152*, 677–683. https://doi.org/10.1007/s00442-007-0703-5

Mohamed, S. A., Abdel-Aty, A. M., Hamed, M. B., EL-Badry, M. O., & Fahmy, A. S. (2011). Ficus sycomorus latex: A thermostable peroxidase. *African Journal of Biotechnology, 10*, 17532–17543.

Mawa, S., Husain, K., & Jantan, I. (2013). *Ficus carica* L. (Moraceae): Phytochemistry, traditional uses and biological activities. *Evidence-based Complement Alternative Medicine*, 1–8.

Novotny, V., Miller, S. E., Baje, L., et al. (2010). Guild-specific patterns of species richness and host specialization in plant-herbivore food webs from a tropical forest: Plant-herbivore food webs in tropical forest. *The Journal of Animal Ecology, 79*, 1193–1203.

Ojo, O. A., Basiru, A., Ojo, A. B., et al. (2014). Phytochemical, proximate analysis and mineral composition of aqueous crude extract of *Ficus asperifolia* Miq. *Journal of Advances in Medical and Life Sciences, 1*, 1–4.

Ojala, T., Remes, S., Haansuu, P., Vuorela, H., Hiltunen, R., Haahtela, K., & Vuorela. (2000). Antimicrobial activity of some coumarin containing herbal plants growing in Finland. *Journal of Ethnopharmacology, 73*, 299–305.

Oliveira, A. P., Valentão, P., Pereira, J. A., Silva, B. M., Tavares, F., & Andrade, P. B. (2009). *Ficus carica* L.: Metabolic and biological screening. *Food and Chemical Toxicology, 47*, 2841–2846.

Oliveira, A. P., Silva, L. R., Pinho, P. G., et al. (2010). Volatile profiling of *Ficus carica* varieties by HS-SPME and GC-IT-MS. *Food Chemistry, 123*, 548–557.

Oliveira, A. P., Baptista, P., Andrade, P. B., et al. (2012). Characterization of *Ficus carica* L. cultivars by DNA and secondary metabolite analysis: Is genetic diversity reflected in the chemical composition? *Food Research International, 49*, 710–719.

Pal, C., & Papp, B. (2017). Evolution of complex adaptations in molecular systems. *Nature Ecology & Evolution, 1*(8), 1084–1092.

Patel, A. K., Singh, V. K., & Jagannadham, M. V. (2007). Carnein, a serine protease from noxious plant weed *Ipomoea carnea* (morning glory). *Journal of Agricultural and Food Chemistry, 55*, 5809–5818.

Pathak, M. A., Daniels, F., & Fitzpatrick, T. B. (1962). The presently known distribution of furocoumarins (psoralens) in plants. *Journal of Investigative Dermatology, 39*, 225–239.

Patil, V. V., Bhangale, S. C., & Patil, V. R. (2010). Evaluation of anti-pyretic potential of *Ficus carica* leaves. *Evaluation, 2*, 48–50.

Peraza-Sánchez, S. R., Chai, H. B., Shin, Y. G., et al. (2002). Constituents of the leaves and twigs of *Ficus hispida*. *Planta Medica, 68*, 186–188.

Pierantoni, M., Tenne, R., Rephael, B., et al. (2018). Mineral deposits in Ficus leaves: Morphologies and locations in relation to function. *Plant Physiology, 176*, 1751–1763.

Potts, A. S., & Hunter, M. D. (2021). Unraveling the roles of genotype and environment in the expression of plant defense phenotypes. *Ecology and Evolution, 11*(13), 8542–8561.

Prychid, C. J., Rudall, P. J., & Gregory, M. (2003). Systematics and biology of silica bodies in monocotyledons. *The Botanical Review, 69*, 377–440.

Ramos, M. V., Grangeiro, T. B., Freire, E. A., Sales, M. P., Souza, D. P., Araújo, E. S., & Freitas, C. D. T. (2010). The defensive role of latex in plants: Detrimental effects on insects. *Arthropod-Plant Interact, 4*, 57–67.

Roberts, D. M., Gallapatthy, G., Dunuwille, A., & Chan, B. S. (2016). Pharmacological treatment of cardiac glycoside poisoning: Cardenolide poisoning. *British Journal of Clinical Pharmacology, 81*, 488–495.

Rossi, M., Goggin, F. L., Milligan, S. B., Kaloshian, I., Ullman, D. E., & Williamson, V. M. (1998). The nematode resistance gene Mi of tomato confers resistance against the potato aphid. *The Proceedings of the National Academy of Sciences USA, 95*, 9750–9754.

Rubnov, S., Kashman, Y., Rabinowitz, R., Schlesinger, M., & Mechoulam, R. (2001). Suppressors of cancer cell proliferation from fig (*Ficus carica*) resin: isolation and structure elucidation. *Journal of Natural Products, 64*, 993–996.

Ruiz, N., Ward, D., & Saltz, D. (2002). Calcium oxalate crystals in leaves of *Pancratium sickenbergeri*: Constitutive or induced defence? *Functional Ecology, 16*, 99–105.

Ryan, C. A. (1990). Protease inhibitors in plants: Genes for improving defenses against insects and pathogens. *Annual Review of Phytopathology, 28*, 425–449.

Saby, J. K., Bhat, S. G., & Prasada Rao, U. J. S. (2003). Biochemical characterization of sap (latex) of a few Indian mango varieties. *Phytochemistry, 62*, 13–19.

Saeed, M. A., & Sabir, A. W. (2002). Irritant potential of triterpenoids from *Ficus carica* leaves. *Fitoterapia, 73*, 417–420.

Sethi, A., McAuslane, H. J., Rathinasabapathi, B., Nuessly, G. S., & Nagata, R. T. (2009). Enzyme induction as a possible mechanism for latex-mediated insect resistance in romaine lettuce. *Journal of Chemical Ecology, 35*, 190–200.

Sgarbieri, V. C., Gupte, S. M., Kramer, D. E., & Whitaker, J. R. (1964). Ficus enzymes I. Separation of the proteolytic enzymes of *Ficus carica* and *Ficus glabrata* latices. *The Journal of Biological Chemistry, 239*, 2170–2177.

Singh, D., Singh, B., & Goel, R. K. (2011). Traditional uses, phytochemistry and pharmacology of Ficus religiosa: A review. *Journal of Ethnopharmacology, 134*, 565–583.

Sirisha, N., Sreenivasulu, M., Sangeeta, K., & Chetty, C. M. (2010). Antioxidant properties of Ficus species-a review. *International Journal of Pharmtech Research, 2*(4), 2174–2182.

Skalicka-Wozniak, K., Los, R., Glowniak, K., & Malm, A. (2009). Volatile compounds in fruits of *Peucedanum cervaria* (Lap.) L. *Chemistry & Biodiversity, 6*, 1087–1092.

Somssich, I., & Hahlbrock, K. (1998). Pathogen defense in plants – a paradigm of biological complexity. *Trends in Plant Science, 3*, 86–90.

Sosnovsky, Y. (2016). Sucking herbivore assemblage composition on greenhouse *Ficus* correlates with host plant leaf architecture. *Arthropod-Plant Interactions, 10*, 55–69.

Stakman, E. C. (1915). Plants highly resistant to its attack. *Journal of Agricultural Research, 4*, 193.

Strömberg, C. A. E., Di Stilio, V. S., & Song, Z. (2016). Functions of phytoliths in vascular plants: An evolutionary perspective. *Functional Ecology, 30*, 1286–1297. https://doi.org/10.1111/1365-2435.12692

Taira, T., Ohdomari, A., Nakama, N., Shimoji, M., & Ishihara, M. (2005). Characterization and antifungal activity of Gazyumaru (*Ficus microcarpa*) latex chitinases: Both the chitin-binding and the antifungal activities of class I chitinase are reinforced with increasing ionic strength. *Bioscience, Biotechnology, and Biochemistry, 69*, 811–818.

Takahashi, T., Okiura, A., Saito, K., & Kohno, M. (2014). Identification of phenylpropanoids in fig (*Ficus carica* L.) leaves. *Journal of Agricultural and Food Chemistry, 62*, 10076–10083.

Takahashi, T., Okiura, A., & Kohno, M. (2017). Phenylpropanoid composition in fig (*Ficus carica* L.) leaves. *Journal of Natural Medicines, 71*, 770–775.

Terras, F., Schoofs, H., Thevissen, K., et al. (1993). Synergistic enhancement of the antifungal activity of wheat and barley thionins by radish and oilseed rape 2S albumins and by barley trypsin inhibitors. *Plant Physiology, 103*, 1311–1319.

Tomar, R., Kumar, R., & Jagannadham, M. V. (2008). A stable serine protease, wrightin, from the latex of the plant Wrightia tinctoria (Roxb.) R. Br.: Purification and biochemical properties. *Journal of Agricultural and Food Chemistry, 56*, 1479–1487.

Vaya, J., & Mahmood, S. (2006). Flavonoid content in leaf extracts of the fig (*Ficus carica L.*), carob (*Ceratonia siliqua L.*) and pistachio (*Pistacia lentiscus L.*). *Bio Factors, 28*, 169–175.

Villard, C., Larbat, R., Munakata, R., & Hehn, A. (2019). Defence mechanisms of Ficus: Pyramiding strategies to cope with pests and pathogens. *Planta, 249*(3), 617–633.

Volf, M., Hrcek, J., Julkunen-Tiitto, R., Novotny, V. (2015). To each its own: differential response of specialist and generalist herbivores to plant defence in willows. *The Journal of Animal Ecology, 84*, 1123–1132. https://doi.org/10.1111/1365-2656.12349

Volf, M., Segar, S.T., Miller, S.E., Isua, B., Sisol, M., Aubona, G., Simek, P., Moos, M., Laitila, J., Kim, J., Jr Zima, J., Rota, J., Weiblen, G.D., Wossa, S., Salminen, J.P., Basset, Y, Novotny, V. (2018). Community structure of insect herbivores is driven by conservatism, escalation and divergence of defensive traits in Ficus. *Ecology Letters, 21*(1), 83–92.

Wahler, D., Gronover, C. S., Richter, C., et al. (2009). Polyphenol oxidase silencing affects latex coagulation in *Taraxacum* species. *Plant Physiology, 151*, 334–346.

Walling, L. L. (2000). The myriad plant responses to herbivores. *Journal of Plant Growth Regulation, 19*, 195–216.

Wang, T., Jiao, J., Gai, Q. Y., et al. (2017). Enhanced and green extraction polyphenols and furanocoumarins from fig (*Ficus carica* L.) leaves using deep eutectic solvents. *Journal of Pharmaceutical and Biomedical Analysis, 145*, 339–345.

Ward, D., Spiegel, M., & Saltz, D. (1997). Gazelle herbivory and interpopulation differences in calcium oxalate content of leaves of a desert lily. *Journal of Chemical Ecology, 23*, 333–346.

Wu, C. C., & Kuo-Huang, L. L. (1997). Calcium crystals in the leaves of some species of Moraceae. *Botanical Bulletin of Academia Sinica, 38*, 97–104.

Wu, P. L., Rao, K. V., Su, C. H., et al. (2002). Phenanthroindolizidine alkaloids and their cytotoxicity from the leaves of *Ficus septica. Heterocycles, 57*, 2401–2408.

Xiang, H., & Chen, J. (2004). Interspecific variation of plant traits associated with resistance to herbivory among four species of Ficus (*Moraceae*). *Annals of Botany, 94*, 377–384.

Yap, V. A., Loong, B. J., Ting, K. N., Loh, S. H. S., Yong, K. T., Low, Y. Y., Kam, T. S., & Lim, K. H. (2015). Hispidacine, an unusual 8,4′-oxyneolignan-alkaloid with vasorelaxant activity, and hispiloscine, an antiproliferative phenanthroindolizidine alkaloid, from *Ficus hispida* Linn. *Phytochemistry, 109*, 96–102.

Yap, V. A., Qazzaz, M. E., Raja, V. J., et al. (2016). Fistulopsines A and B antiproliferative septicine-type alkaloids from *Ficus fistulosa. Phytochemistry Letters, 15*, 136–141.

Zaynoun, S. T., Aftimos, B. G., Ali, L. A., et al. (1984). *Ficus carica*; isolation and quantification of the photoactive components. *Contact Dermatitis, 11*, 21–25.

Zhu-Salzman, K., Luthe, D. S., & Felton, G. W. (2008). Arthropod-inducible proteins: Broad spectrum defenses against multiple herbivores. *Plant Physiology, 146*, 852–858.

Part II
Fig (*Ficus carica*): Chemistry, Functionality and Health-Promoting Properties

Chapter 13
Chemistry and Nutritional Value of Fresh and Dried Fig (*Ficus carica*)

Mohamed Fawzy Ramadan

1 Introduction

The fruit of the *Ficus carica* tree, which is part of the plant family Moraceae, is sweet and nutritious (Solomon et al., 2006; Arvaniti et al., 2019; Khadivi & Mirheidari, 2022). The *Ficus carica* genus contains more than 800 different kinds, most of which are grown in regions with hot, dry climates like the Middle East and the Mediterranean (Ouchemoukh et al., 2012; Harzallah et al., 2016; Meziant et al., 2015; Solomon et al., 2006). Figs are a seasonal fruit that could be picked twice a year, depending on the variety, either in the spring and summer or at the start and end of the summer (Ouchemoukh et al., 2012; Vallejo et al., 2012; Arvaniti et al., 2019). It is edible fresh, dried, jam, juice, and even in other forms (Solomon et al., 2006; Hoxha & Kongoli, 2016; Harzallah et al., 2016; Arvaniti et al., 2019).

Figs include beneficial natural antioxidant substances like carotenoids, organic acids, vitamin E, and phenolic compounds. Due to their antioxidant traits, phenolic compounds are the most well-liked; they are also significant contributors to color, flavor, and scent. Therefore, phenolic acids and flavonoids are two primary subcategories of phenolic substances (Chang et al., 2016; Gharras, 2009; Shahidi & Ambigaipalan, 2015; Arvaniti et al., 2019).

Different species of dried and fresh figs have undergone qualitative and quantitative examination of their phenolic content (Solomon et al., 2006; Bachir et al., 2013; Kamiloglu & Capanoglu, 2013; Bachir & Louaileche, 2015; Ajmal et al., 2016; Harzallah et al., 2016; Maghsoudlou et al., 2017; Arvaniti et al., 2019). In addition, researchers have investigated how phenolic component concentrations are affected

M. F. Ramadan (✉)
Department of Clinical Nutrition, Faculty of Applied Medical Sciences, Umm Al-Qura University, Makkah, Saudi Arabia
e-mail: mfhassanien@uqu.edu.sa

M. F. Ramadan (ed.), *Fig* (Ficus carica)*: Production, Processing, and Properties*, https://doi.org/10.1007/978-3-031-16493-4_13

by fruit color, fruit variety, harvest season, and drying techniques (Caliskan & Polat, 2011; Harzallah et al., 2016; Konak et al., 2017; Pereira et al., 2017; Ersoy et al., 2017; Slatnar et al., 2011; Arvaniti et al., 2019).

The primary goals of this work were to summarize the levels of phenolics and carotenoids in fresh and dried figs and to present the main components and phytochemicals in figs.

2 Nutrients and Phytochemicals in Fresh and Dried Figs

Figs, both fresh and dried, are a good source of amino acids, vitamins (thiamin and riboflavin), and trace minerals (iron, calcium, and potassium) (Solomon et al., 2006; Ouchemoukh et al., 2012; Viuda-Martos et al., 2015). In addition, figs are high in fiber and antioxidant components but free of sodium, fat, and cholesterol (Solomon et al., 2006; Viuda-Martos et al., 2015; Veberic et al., 2008; Arvaniti et al., 2019). Figure 13.1 shows fresh and dried figs' chemical makeup and nutritional value. It is

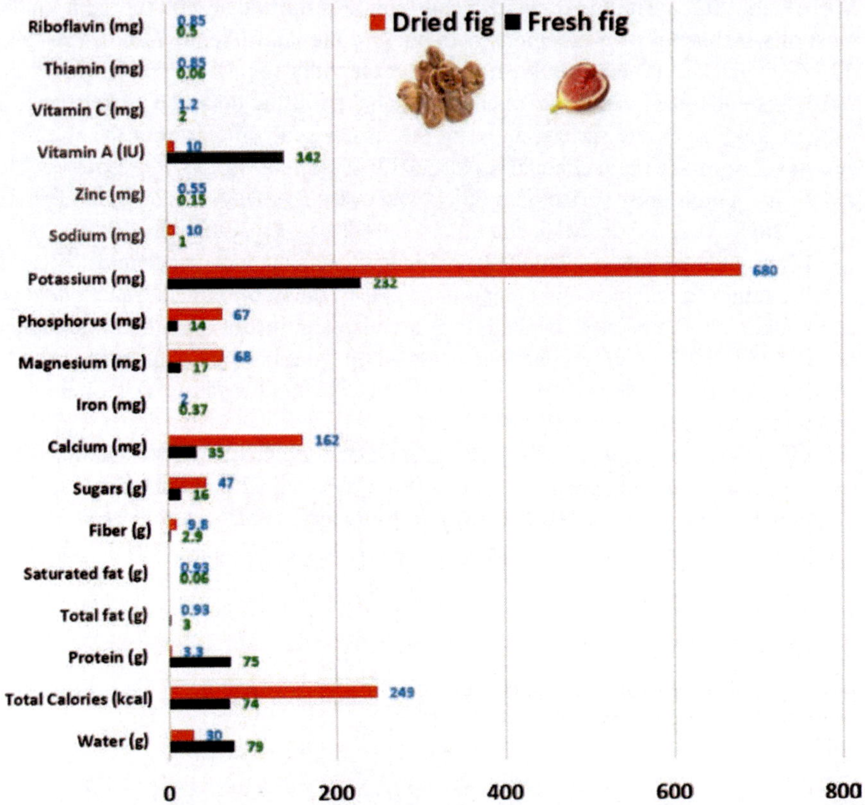

Fig. 13.1 Composition and nutrient content (value/100 g) of fresh and dried figs. (Source: Data from the USDA National Nutrient Database for Standard References 2018 and Arvaniti et al., 2019)

evident that the chemical composition of fresh and dried figs differs, particularly in terms of the amounts of total calories (kcal), sugars (g), calcium (mg), phosphorus (mg), magnesium (mg), and potassium (mg).

Due to a large number of phytochemicals present in the fruit tissues, peel (skin), and pulp, *Ficus carica* phytochemistry is complex. The fruits of the *Ficus carica* tree contain a variety of phytochemicals, such as phenolics, anthocyanins, and carotenoids (Jagtap & Bapat, 2019). Therefore, consuming whole, ripe fruits are advised because fruit skins are frequently the primary source of phenolic chemicals (Vinson et al., 2005).

Studies have reported the phytochemicals content of dried and fresh *Ficus carica* entire fruits. Individual phenolic compounds and carotenoids, as well as total phenolics (TPC), total flavonoids (TF), and total anthocyanins (TA) have been identified. The majority of the research showed that the phytochemical profile is substantially influenced by the fig variety as well as by other elements like color, fruit portion, level of maturity, and method of drying (Solomon et al., 2006; Ouchemoukh et al., 2012; Yemis et al., 2012; Nakilcioglu & Hısıl, 2013; Bachir et al., 2013; Bachir et al., 2014; Debib et al., 2014; Kamiloglu & Capanoglu, 2015; Bachir & Louaileche, 2015; Bachir et al., 2016; Harzallah et al., 2016; Pereira et al., 2017; Amessis-Ouchemoukh et al., 2017; Konak et al., 2017; Ersoy et al., 2017; Sedaghat & Rahemi, 2018; Arvaniti et al., 2019).

TPC levels in fresh fig fruit ranged from 19 mg GAE/100 g fresh weight (FW) to 6147 mg GAE/100 g FW (Caliskan & Polat, 2011; Harzallah et al., 2016; Arvaniti et al., 2019). TPC concentrations in dark-colored fig types ranged from 53 to 6147 mg GAE/100 g FW (Solomon et al., 2006; Caliskan & Polat, 2011; Harzallah et al., 2016; Arvaniti et al., 2019). It has been demonstrated that fig peels contain significantly more phytochemical components than fig pulp. In addition, it is commonly acknowledged that fruit and vegetable skin contains most anthocyanins, contributing to their high phytochemical content and nutritional worth (Solomon et al., 2006; Arvaniti et al., 2019).

Phenolic acids, flavonoids, and carotenoids are abundant in dried *Ficus carica*. TPC of dried fig extracts has been estimated to be between 45 and 644 mg GAE/100 g dry weight (DW) for light-coloured dried *Ficus carica* types and between 399 and 756 mg GAE/100 g DW for dark-colored dried *Ficus carica* varieties (Bachir et al., 2013; Debib et al., 2014; Bachir & Louaileche, 2015; Arvaniti et al., 2019). Additionally, studies have shown that anthocyanin pigments, such as peonidin, cyanidin, petudinin, delphinidin, and malvidin, are primarily found in purple, red, and other dark-coloured fruit (Khoo et al., 2017; Wang et al., 2017; Arvaniti et al., 2019).

3 Specific Phytochemicals in Fresh and Dried Figs

Russo et al. (2014) assessed the phenolic composition of dried and fresh figs (cultivar Dottato) from various local origins and contrasted the phenolic content of fresh figs with dried figs from various origins (Turkey, Italy, and Greece). The Cilento figs' phenolic content was decreased by sun-drying. Therefore, to prevent an excessive loss of phenolics, it is necessary to carefully control all the parameters involved in fig drying, especially the amount of time the fruit takes is exposed to the sun and other variables like environmental humidity.

The most common substances in dried and fresh *Ficus carica* are phenolic acids, including gallic acid, ferulic acid, syringic acid, chlorogenic acid, caffeic acid, and cinnamic acid. Gallic acid (5 mg/100 g DM) and chlorogenic acid (32 mg/100 g DM) are the two phenolic acids that are most prevalent in dried *Ficus carica*, although the same compounds were detected in fresh *Ficus carica* with quantities of 6.9 mg (gallic acid)/100 g DM and 28 mg (chlorogenic acid)/100 g DM, respectively (Arvaniti et al., 2019).

The highest concentrations of rutin, quercetin-3-*O*-rutinoside, and epicatechin were 14.6, 10.8, and 36.6 mg/100 g DM in dried *Ficus carica*. These substances accounted for 68 mg/100 g DM for rutin, 38 mg/100 g DM for quercetin-3-*O*-rutinoside, and 76.9 mg/100 g DM for epicatechin in fresh fig (Arvaniti et al., 2019).

The two primary anthocyanin pigments in all *Ficus carica* cultivars were cyanidin-3-*O*-rutinoside and cyanidin-3-*O*-glucoside. In dried and fresh *Ficus carica*, cyanidin-3-*O*-rutinoside concentrations varied between 73.0 and 91.0 mg/100 g DM (Arvaniti et al., 2019). Proanthocyanidins, in particular, are found in high concentrations in figs (Vinson, 1999; Russo et al., 2014), wherein the cultivar and time of harvest are just two examples of the many variables that affect the concentration of these compounds (Burda et al., 1990; Kennedy et al., 2001; Crisosto et al., 2010; Russo et al., 2014).

Both dried and fresh figs have been found to contain lutein, beta-carotene, zeaxanthin, and beta-cryptoxanthin. However, fresh figs were shown to have the most concentrations of beta-carotene, zeaxanthin, and lutein, whereas dried figs had the most concentrations of beta-cryptoxanthin and lutein (Arvaniti et al., 2019).

4 Future Perspectives

This work reported the nutrients, and phytochemical composition of *Ficus carica* dried and fresh figs. Figs, both fresh and dried, are a great source of bioactive phytochemicals. Topics of particular concern are how figs' harvesting practices and packing techniques affect the fruits' phytochemical makeup and antioxidant capability. In addition, it is crucial to look into alternative drying methods (such as freeze-drying and microwave-drying) for dried figs to preserve their phytochemical content and lengthen their shelf life.

To recommend the ideal circumstances and processes for *Ficus carica*, it is essential to investigate further the impact of various drying techniques on the phytochemical levels and nutritional value of *Ficus carica and* environmental factors. Furthermore, the potential uses of *Ficus carica* bioactive compounds and extracts as functional constituents of food items should be investigated in light of the existence of phytochemical substances in *Ficus carica* phytochemicals and their association with the bioactive potential. Finally, future clinical and dietary studies might help us better comprehend figs' positive impacts on human health.

References

Ajmal, M., Arshad, M. U., Saeed, F., Ahmed, T., Khan, A. U., Bader-ul-Ain, H., & Suleria, H. A. R. (2016). Exploring the nutritional characteristics of different parts of fig in relation to hypoglycemic potential. *Pakistan Journal of Life and Social Sciences, 14*, 115–122.

Amessis-Ouchemoukh, N., Ouchemoukh, S., Meziant, N., Idiri, Y., Hernanz, D., Stinco, C. M., et al. (2017). Bioactive metabolites involved in the antioxidant, anticancer and anticalpain activities of *Ficus carica* L., *Ceratonia siliqua* L. and *Quercus ilex* L. extracts. *Industrial Crops and Products, 95*, 6–17. https://doi.org/10.1016/j.indcrop.2016.10.007

Arvaniti, O. S., Samaras, Y., Gatidou, G., Thomaidis, N. S., & Stasinakis, A. S. (2019). Review on fresh and dried figs: Chemical analysis and occurrence of phytochemical compounds, antioxidant capacity and health effects. *Food Research International, 119*, 244–267.

Bachir, B. M., & Louaileche, H. (2015). A comparative study of phytochemical profile and in vitro antioxidant activities of dark and light dried fig (*Ficus carica* L.) varieties. *The Journal of Phytopharmacology, 4*, 41–48.

Bachir, B. M., Louaileche, H., & Zemouri, S. (2013). Optimization of phenolic compound recovery and antioxidant of light and dark fig (*Ficus carica* L.) varieties. *Food Science and Biotechnology, 22*, 1613–1619. https://doi.org/10.1007/s10068-013-0258-7

Bachir, B. M., Meziant, L., Benchikh, Y., & Louaileche, H. (2014). Deployment of response surface methodology to optimize recovery of dried dark fig (*Ficus carica* L. var. Azenjar) total phenolic compounds and antioxidant activity. *International Food Research Journal, 21*, 1477–1482. https://doi.org/10.1016/j.foodchem.2014.04.054

Bachir, B. M., Richard, G., Meziant, L., Fauconnier, M. L., & Louaileche, H. (2016). Effects of sun-drying on physicochemical characteristics, phenolic composition and in vitro antioxidant activity of dark fig varieties. *Journal of Food Processing and Preservation, 41*, 1–8. https://doi.org/10.1111/jfpp.13164

Burda, S., Oleszek, W., & Lee, C. (1990). Phenolic compounds and their changes in apples during maturation and cold storage. *Journal of Agricultural and Food Chemistry, 38*, 945–948.

Caliskan, O., & Polat, A. (2011). Phytochemical and antioxidant properties of selected fig (*Ficus carica* L.) accessions from ten eastern Mediterranean region of Turkey. *Scientia Horticulturae, 128*, 473–478. https://doi.org/10.1016/j.scienta.2011.02.023

Chang, S. K., Alasalvar, C., & Shahidi, F. (2016). Review of dried fruits: Phytochemicals, antioxidant efficacies, and health benefits. *Journal of Functional Foods, 21*, 113–132. https://doi.org/10.1016/j.jff.2015.11.034

Crisosto, C. H., Bremer, V., Ferguson, L., & Crisosto, G. M. (2010). Evaluating quality attributes of four fresh fig (*Ficus carica* L.) cultivars harvested at two maturity stages. *Hortscience, 45*, 707–710.

Debib, A., Tir-Touili, A., Mothana, R. A., Meddah, B., & Sonnet, P. (2014). Phenolic content antioxidant and antimicrobial activities of two fruit varieties of Algerian *Ficus carica*. *Journal of Food Biochemistry, 38*, 207–215. https://doi.org/10.1111/jfbc.12039

Ersoy, N., Gozlekci, S., Gok, V., & Yilmaz, S. (2017). Fig (*Ficus carica* L.) fruit some physical and chemical properties. *Acta Horticulturae, 1173*, 329–334. https://doi.org/10.17660/ActaHortic.2017.1173.57

Gharras, H. E. (2009). Polyphenols: Food sources, properties and applications-A review. *International Journal of Food Science and Technology, 44*, 2512–2518. https://doi.org/10.1111/j.1365-2621.2009.02077.x

Harzallah, A., Bhouri, A. M., Amri, Z., Soltana, H., & Hammami, M. (2016). Phytochemical content and antioxidant activity of different fruit parts juices of three figs varieties grown in Tunisia. *Industrial Crops and Products, 83*, 255–267. https://doi.org/10.1016/j.indcrop.2015.12.043

Hoxha, L., & Kongoli, R. (2016). Evaluation of antioxidant and potential of Albanian fig varieties "Kraps Zi" and "Kraps Bardhe" cultivated in the region of Tirana. *Journal of Hygienic Engineering and Design, 16*, 70–74.

Jagtap, U. B., & Bapat, V. A. (2019). Exploring phytochemicals of *Ficus carica* L. (Fig). In H. Murthy & V. Bapat (Eds.), *Bioactive compounds in underutilized fruits and nuts* (Reference series in phytochemistry). Springer. https://doi.org/10.1007/978-3-030-06120-3_19-1

Kamiloglu, S., & Capanoglu, E. (2013). Investigating the in vitro bioaccessibility of polyphenols in fresh and sun-dried figs (*Ficus carica* L.). *International Journal of Food Science and Technology, 48*, 2621–2629. https://doi.org/10.1111/ijfs.12258

Kamiloglu, S., & Capanoglu, E. (2015). Polyphenol content in figs (*Ficus carica* L.): Effect of sun drying. *International Journal of Food Properties, 18*, 521–535. https://doi.org/10.1080/10942912.2013.833522

Kennedy, J., Hayasaka, Y., Vidal, S., Waters, E., & Jones, G. (2001). Composition of grape peel proanthocyanidins at different stages of berry development. *Journal of Agricultural and Food Chemistry, 49*, 5348–5355.

Khadivi, A., & Mirheidari, F. (2022). Selection of the promising fig (*Ficus carica* L.) accessions using fruit-related characters. *Food Science & Nutrition*, 1–11. https://doi.org/10.1002/fsn3.2886

Khoo, H. E., Azlan, A., Tang, S. T., & Lim, S. M. (2017). Anthocyanidins and anthocyanins: Colored pigments as food, pharmaceutical ingredients, and the potential health benefits. *Food & Nutrition Research, 61*, 1361779. https://doi.org/10.1080/16546628.2017.1361779

Konak, R., Kosoglu, I., & Yemenıcıoglu, A. (2017). Effects of different drying methods on phenolic content antioxidant capacity and general characteristics of selected Turkish fig cultivars. *Acta Horticulturae, 1173*, 335–340. https://doi.org/10.17660/ActaHortic.2017.1173.58

Maghsoudlou, E., Kenari, R. E., & Amiri, Z. R. (2017). Evaluation of antioxidant activity of fig (*Ficus carica*) pulp and skin extract and its application enhancing stability of canola oil. *Journal of Food Processing and Preservation, 41*, 1–11. https://doi.org/10.1111/jfpp.13077

Meziant, L., Saci, F., Bachir Bey, M., & Louaileche, M. (2015). Varietal influence on biological properties of Algerian light figs (*Ficus carica* L.). *International Journal of Bioinformatics & Biomedical Engineering, 1*, 237–243.

Nakilcioglu, E., & Hısıl, Y. (2013). Research on the phenolic compounds in Sarilop (*Ficus carica* L.) fig variety. *The Journal of Food, 38*, 267–274.

Ouchemoukh, S., Hachoud, S., Boudraham, H., Mokrani, A., & Louaileche, H. (2012). Antioxidant activities of some dried fruits consumed in Algeria. *Food Science and Technology, 49*, 329–332. https://doi.org/10.1016/j.lwt.2012.07.022

Pereira, C., Lopez-Corrales, M., Serradilla, M. J., Del Carmen Villalobos, M., RuizMoyano, S., & Martín, A. (2017). Influence of ripening stage on bioactive compounds and antioxidant activity in nine fig (*Ficus carica* L.) varieties grown in Extremadura, Spain. *Journal of Food Composition and Analysis, 64*, 203–212. https://doi.org/10.1016/j.jfca.2017.09.006

Russo, F., Caporaso, N., Paduano, A., & Sacchi, R. (2014). Phenolic compounds in fresh and dried figs from Cilento (Italy), by considering breba crop and full crop, in comparison to Turkish and Greek dried figs. *Journal of Food Science, 79*(7), C1278–C1284.

Sedaghat, S., & Rahemi, M. (2018). Effects of physio-chemical changes during fruit development on nutritional quality of fig (*Ficus carica* L. var. 'Sabz') under rain-fed condition. *Scientia Horticulturae, 237*, 44–50. https://doi.org/10.1016/j.scienta.2018.04.003

Shahidi, F., & Ambigaipalan, P. (2015). Phenolics and polyphenolics in foods, beverages and spices: Antioxidant activity and health effects- A review. *Journal of Functional Foods, 18*, 820–897. https://doi.org/10.1016/j.jff.2015.06.018

Slatnar, A., Klancar, U., Stampar, F., & Veberic, R. (2011). Effect of drying of figs (*Ficus carica* L.) on the contents of sugars, organic acids and phenolic compounds. *Journal of Agricultural and Food Chemistry, 59*, 11696–11702. https://doi.org/10.1021/jf202707y

Solomon, A., Golubowicz, S., Yablowicz, Z., Grossman, S., Bergman, M., Gottlieb, H. E., et al. (2006). Antioxidant activities and anthocyanin content of fresh fruits of common fig (*Ficus carica*). *Journal of Agricultural and Food Chemistry, 54*, 7717–7723. https://doi.org/10.1021/jf060497h

Vallejo, F., Marin, J. G., & Tomas-Barberan, F. A. (2012). Phenolic compound content of fresh and dried figs (*Ficus carica* L.). *Food Chemistry, 130*, 485–492. https://doi.org/10.1016/j.foodchem.2011.07.032

Veberic, R., Colaric, M., & Stampar, F. (2008). Phenolic acids and flavonoids of fig fruit (*Ficus carica* L.) in northern Mediterranean region. *Food Chemistry, 106*, 153–157. https://doi.org/10.1016/j.foodchem.2007.05.061

Vinson, J. A. (1999). The functional foods: Properties of figs. *Cereal Foods World, 44*, 82–87.

Vinson, J. A., Zubik, L., Bose, P., Samman, N., & Proch, J. (2005). Dried fruits: Excellent in vitro and in vivo antioxidants. *Journal of the American College of Nutrition, 24*, 44–50.

Viuda-Martos, M., Barber, X., Perez-Alvarez, J. A., & Fernandez-Lopez, J. (2015). Assessment of chemical, physico-chemical, techno-functional and antioxidant properties of fig (*Ficus carica* L.) powder co-products. *Industrial Crops and Products, 69*, 472–479. https://doi.org/10.1016/j.indcrop.2015.03.005

Wang, Z., Cui, Y., Vainstein, A., Chen, S., & Ma, H. (2017). Regulation of fig (*Ficus carica* L.) fruit color: Metabolomic and transcriptomic analyses of the flavonoid biosynthetic pathway. *Frontiers in Plant Science, 8*, 1990. https://doi.org/10.3389/fpls.2017.01990

Yemis, O., Bakkalbas, E., & Artık, N. (2012). Changes in pigment profile and surface colour of fig (*Ficus carica* L.) during drying. *International Journal of Food Science and Technology, 47*, 1710–1719. https://doi.org/10.1111/j.1365-2621.2012.03025.x

Chapter 14
Fig Seeds: Source of Value-Added Oil Within the Scope of Circular Economy

Lahcen Hssaini

Abbreviations

ABTS	3-ethylbenzothiazoline-6-sulfonic acid
ALA	alpha-linolenic acids
CE	circular economy
DPPH	2,2-diphenyl1-picrylhydrazyl
DR	desaturation ratio
EFAs	essential fatty acids
FA(s)	fatty acid (s)
FSO	fig seeds oil
GAE	gallic acid equivalent
IC_{50}	half maximum inhibitory concentration
IV	iodine value
LA	linoleic acid
LCPUFA	long chain polyunsaturated fatty acids
LDR	linoleic desaturation ratio
MUFA	monounsaturated FA
ODR	oleic desaturation ratio
PUFA	polyunsaturated FA
SCPUFA	short chain polyunsaturated fatty acids
SV	saponification value
TSFA	total saturated FA
TUFA	total unsaturated FA

L. Hssaini (✉)
National Institute of Agricultural Research, Avenue Ennasr, BP 415 Rabat Principale, Rabat, Morocco
e-mail: lahcen.hssaini@inra.ma

M. F. Ramadan (ed.), *Fig* (Ficus carica)*: Production, Processing, and Properties*, https://doi.org/10.1007/978-3-031-16493-4_14

1 Fig Seeds: A Novel Source of Oil

Figs (*Ficus carica* L.) are whispered to be the oldest of all cultivated fruit species and the first domesticated tree of the Neolithic Revolution (Kislev et al., 2006). Since then, figs have been a major food source and have been utilized alongside other tree parts such as leaves, latex, bark, and roots for medicinal purposes (Badgujar et al., 2014). Due to its multiple health-promoting properties and biological activities, figs invite the attention of the scientists worldwide and have been widely studied, as a whole, for their biological compounds and activity along with their antidiabetic properties (Debib et al., 2014; Del Caro & Piga, 2008; Hssaini et al., 2019; Jeong & Lachance, 2001; Kamiloglu & Capanoglu, 2013; Schmitzer et al., 2011; Veberic et al., 2010). However, seeds remain the less studied part of the species (Hssaini et al., 2020b, 2021b). Thus, very few reports have been undertaken on their lipochemical and nutraceutical properties (Ben-Othman et al., 2020). Figs produce a considerable amount of yellowish and round-shaped seeds, with numbers fluctuating from 30 to 1600 and an average thousand seed weight of about 1.14 g equivalent (Hssaini et al., 2020a). The formation of seeds is dependent on the mutualism between the tree and the wasp pollinator. Therefore, the seeds are generally hollow without receiving the caprifig (male tree) pollen on the female of *blastophaga psenes* body (Kato & Kawakita, 2017). These seeds are known to contribute to the fruit's taste and flavor along with its health-promoting attributes (Hssaini et al., 2020a; Nakilcioğlu-Taş & Ötleş, 2021). Only very few studies have been interested in exploring fig seeds' lipochemical and bioactive compounds (Badgujar et al., 2014; Hssaini et al., 2020a, 2021b; Jeong & Lachance, 2001). In the studies of Jeong and Lachance (2001), Hssaini et al. (2020a, 2021a), results showed that fig seeds account for approximately 20 to 30% of oil (Hssaini et al. 2020a, 2021b), which is were comparable to oil content levels found in other fruits seeds such as kiwi (28.3%) (Taylor et al., 2011), papaya (27.87%) (Senrayan & Venkatachalam, 2019), pumpkin (27.83-%) (Wang et al., 2018), Mangosteen (21.68%) (Ajayi et al., 2007) and honeydew (25%) (Yanty et al. 2008). These yield levels are much higher than those found in prickly pear seeds, which exhibited an oil content of 5.4–9.9% (Chougui et al., 2013; Mannoubi et al. 2020), guava (12.6–16%) (Piombo et al., 2006; Prasad & Azeemoddin, 1994), grape (8.66–13.6%) (Lucarini et al., 2020; Tangolar et al., 2009) (Fig. 14.1).

As the first macro quality component of interest for all fruit seeds, the oil yielding is mainly dependent on the varietal factor, geographical variations, or pollen source (Raihana et al., 2015). In the first report of Hssaini et al. (2020a) on seeds extracted from four fig cultivars in Morocco, the impact of the phenotypic factor was highly significant on oil yield and chemical attributes. Authors attributed the large variation in those variables to the cultivar factor, as the sampled figs belonged to an *ex-situ* collection conducted under the same condition. In the same way, several studies have indicated that the genetic factor has a greater influence on seed oil yield and quality (Górnaś & Rudzińska, 2016). With a highly promising oil yield, fig seeds have all the potential to be economically attractive for food and non-food

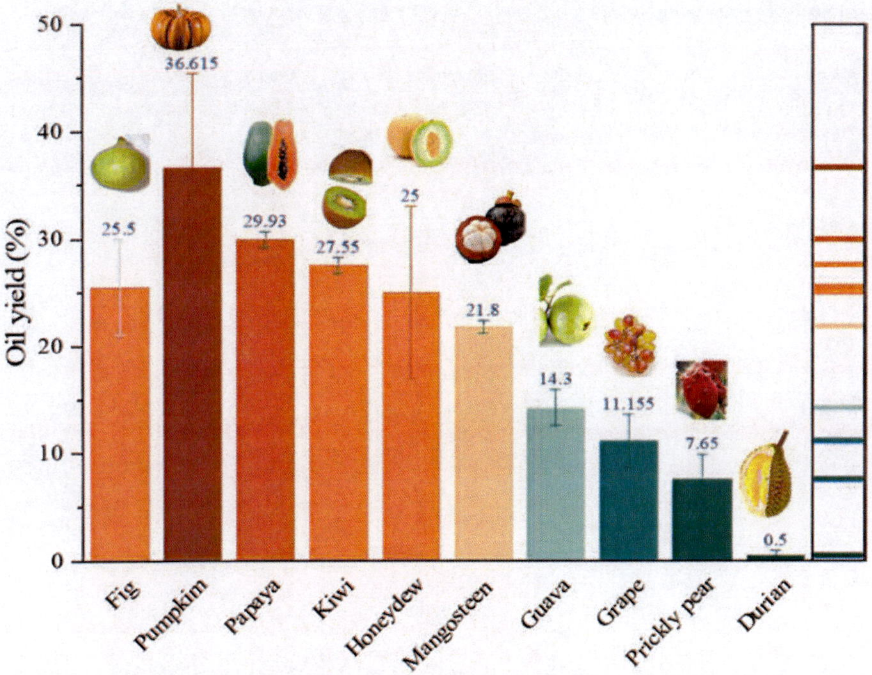

Fig. 14.1 Oil yield (%) of several fruit seeds compared to figs. Orange color indicates high oil yield, while turquoise refers to yield level. The low color intensity means the lower value and vice versa

applications, similar to other fruits' seeds oil such as grape, pomegranate, guava, honeydew, and kiwi (Górnaś & Rudzińska, 2016; Hssaini et al., 2020a; Piombo et al., 2006; Raihana et al., 2015). Besides, encouraging bringing this novel oil to the worldwide markets at favorable prices could lead to a tremendous economic impact for communities where this species is abundantly growing and significantly contribute to the household economy.

2 Fatty Acids Profile

As a minor vegetable oil from atypical sources, fig seeds oil **(FSO)** was less documented, and therefore very few data are available on this concern. Fig seeds produce a light yellowish colored oil with a high unsaturation rate reaching up to 88%, with a remarkable phenotypic effect (Hssaini et al., 2020a; Hssaini et al., 2021b). Linolenic (C18:3) and linoleic (C18:2) acids are the major fatty acids in FSO present in the following ranges 38.4–43.5% and 28.9–34.5%, respectively (Table 14.1). Oleic acid (C18:1) is the third predominant fatty acid (FA), with rates varying between 13.4 and 15.6% (Hssaini et al., 2020a). High levels of unsaturated fats are

Table 14.1 Fatty acid profiles of fig seed oil in comparsion with other fruit seed oils

Source of seeds	Fig	grape	honeydew	Kiwi	Guava	Prickly pear	Durian
Oil content (%)	21–30	–	13–37	26.8–28.3	12.6–16	5.4–9.9	0.5–0.5
Fatty acids (%):	–	–	–	–	–	–	–
C12:0	–	0.24–0.24	0.14–0.23	–	–	–	–
C13:0	–	–	0.06–0.24	–	–	–	–
C14:0	–	0.03–0.3	0.07–0.25	–	0.1–0.2	0.11–0.11	–
C15:0	0.02–0.02	–	0.03–0.13	–	–	–	–
C15:1	–	–	0.08–0.31	–	6.6–7.8	–	–
C16:0	8.5–9.11	8–11.55	8.51–23.88	5.72–8.02	–	12.76–26.8	3.5–9.7
C16:1	0.06–0.17	0.07–0.39	0.08–0.1	–	–	0.75–2.01	0.21–0.3
C17:0	0.03–0.07	0.03–0.14	0.08–0.19	–	–	–	–
C17:1	0.04–0.07	–	0.02–0.03	–	–	–	–
C18:0	2.51–3.3	4.24–4.96	4.89–6.09	3.2–4.55	4.5–5.8	3.2–4.32	3–14
C18:1	13.32–15.66	18.14–23.09	0.1–31.5	14–18.79	8.2–10.8	16.41–19.3	7–26.2
C18:2	28.85–34.56	59.02–67.34	51.6–61.1	14.08–18.43	76.4–77	49.7–60.69	44–80.07
C18:3	38.43–43.61	0.21–1.04	0.19–0.87	51.2–61.41	0.1–0.3	1.56–3.9	0.1–0.13
C20:0	0.02–0.17	0.05–0.4	0.22–0.29	–	0.3–0.5	0.3–0.3	–
C20:1	0.04–0.2	0.03–0.15	0.08–0.1	–	–	–	–
C20:2	–	–	–	–	–	–	–
C22:1	–	–	–	–	–	–	–
C24:0	–	–	–	–	–	–	–
DR	4.93–5.04	2.6–3.69	1.96–297.64	4.25–4.66	7.16–9.33	2.78–3.79	3.06–6.3
LDR	0.87–0.91	0.01	0.01	1.65–1.82	–	0.02–0.06	–
ODR	0.83–0.83	0.72–0.79	0.66–1	0.81–0.82	0.88–0.9	0.74–0.79	0.75–0.86
ω-6/ω-3	0.75–0.79	56.75–320.67	70.15–271.58	0.28–0.3	256.67–764	12.74–38.9	440–615.92
MUFA	13.45–16.1	18.36–23.51	0.28–31.73	14–18.79	8.2–10.8	17.16–21.31	7.21–26.5
PUFA	67.27–78.17	60.06–67.55	51.79–61.97	65.28–79.84	76.5–77.3	53.6–62.25	44.1–80.2

(continued)

Table 14.1 (continued)

Source of seeds	Fig	grape	honeydew	Kiwi	Guava	Prickly pear	Durian
TUFA	80.72–94.27	83.57–85.91	52.07–93.7	79.28–98.63	84.7–88.1	74.91–79.41	51.31–106.7
TSFA	11.07–12.67	13.07–16.87	13.95–31.06	8.92–12.57	4.9–6.5	16.37–31.12	12.7–17.5
MUFA/PUFA	0.18–0.22	0.27–0.39	0.01–0.51	0.21–0.24	0.11–0.14	0.28–0.4	0.16–0.33
TSFA/TUFA	0.13–0.14	4.95–6.57	0.27–0.33	0.11–0.13	0.06–0.07	0.21–0.42	0.16–0.25
Iodine value (IV) (g/100 g oil)	76.26–76.63	–	97.1–97.1	–	13.4–13.4	107.38–107.38	–
Saponification value (SV) (mg KOH/g)	201.49–201.67	–	116–116	–	196–196	173–173	–
References	Hssaini et al., (2020a, 2021b) and Jeong and Lachance, (2001)	Tangolar et al. (2009); Lucarini et al. (2020)	Yanty et al. (2008); Mallek-Ayadi et al. (2018)	Piombo et al. (2006); Wang et al. (2018)	Narváez-Cuenca et al. (2020); Kapoor et al. (2020)	Ennouri et al. (2005); Karabagias et al. (2020)	Ho and Bhat (2015); Mohd Ali et al. (2020)

MUFA Monounsaturated fatty acids, *PUFA* Polyunsaturated fatty acids, *TSFA* Total saturated fatty acids, *TUFA* Total unsaturated fatty acids

$$DR(desaturation\ ratio) = \frac{\%C20:1 + \%C22:1}{\%C20:1 + \%C22:1 + \%C18:1 + \%C18:2 + \%C18:3};$$

$$ODR(oleic\ desaturation\ ratio) = \frac{\%C18:1 + \%C18:2}{\%C18:1 + \%C18:2 + \%C18:3}$$

$$LDR(linoleic\ desaturation\ ratio) = \frac{\%C18:3}{\%C18:2 + \%C18:3};\ \omega - 6 / \omega - 3 = 18:2/18:3$$

important for the balance in the human immune system as it has a natural defensive function against cardiovascular diseases and in mitigation of some other health issues (Chougui et al., 2013; Ramadan 2019). Thus, C18:3 plays a significant role in reducing inflammatory disorder's severity, whereas C18:2 is involved in enhancing some physiological functions, including immune response and blood clotting, due to its ability to convert to eicosanoids hormone. Furthermore, C18:1 has a role in lowering plasma cholesterol (Gonçalves-de-Albuquerque et al., 2016; Park et al., 2018). Unlike other fatty acids, Linoleic acid (C18:2n-6) and α-linolenic acid (C18:3n-3) are essential FAs (EFAs) that must be regularly consumed through food, including oilseeds and oil-bearing crops, since they are not synthesized in the body and a deficiency may leads to transepidermal water loss increases, skin dryness, nails crack as well as hair-loss (Sassa & Kihara, 2014; Vermaak et al., 2011). Regarding saturated fats, palmitic acid (C16:0) was predominant in FSO within a narrow range of variation of 8.54 and 9.05%. Stearic acid (C18:0) was the second major saturated FA ranging from 2.50 to 3.30% (Table 14.1). According to Hssaini et al. (2020a), pentadecylic (C15:0), palmitic (C16:0), palmitoleic (C16:1), margaric acid (C17:0), arachidic acid (C20:0) and gondoic acid (C20:1) were only identified in minor levels, generally between 0.8 and 0.2% (Table 14.1). The same authors reported highly significant differences among sampled fig trees based on all FAs, with the exception of margaric, arachidic and pentadecylic acids besides the TSFA. Looking at the lipid profile, FSO seemed fairly similar to the oil extracted from kiwi seeds, which displays an unsaturation rate in the range of 79.2 and 98.6%, which is comparable to FSO unsaturation rate that falls in the range of 80.7 and 94.2%. Besides, C18:3 and C18:2 are the major FAs in both species seeds followed by oleic acid (Table 14.1). Unlike FSO, oil extracted from seeds of grape, honeydew, guava, prickly pear and durian have high levels of monounsaturated fatty acids (MUFA), a fraction dominated by C18:1 (59.0–67.3%, 51.6–61.1%, 76.4–77%, 49.7–60.6% and 44–80.0%, respectively) (Table 14.1), which is rather similar to olive oil. It is noteworthy that no data were reported on the following FA C20:2, C22:1, C24:0.

3 Fatty Acids Ratios

Assessment of seeds from different phenotypes and highlighting any potential diversity is not eventual for minor oil-bearing crops breeding. According to Velasco et al. (1998), emphasizing the lipchemical diversity through fatty acid profiling of considerable sample length of oil-bearing crops is difficult because they are intercorrelated, and any intensive selection or breeding modification may significantly affect the whole system. For this sake, a set of several FAs ratios, such as elongation ratio (ER), desaturation ratio (DR), oleic desaturation ratio (ODR), and linoleic desaturation ratio (LDR), are usually used to overcome this issue. The ER ratio evaluates the relative weight of the elongation pathway in the chain length of monounsaturated FA, from oleic acid (C18:1) to eicosenoic and erucic acids (C20:1and C22:1, respectively). Desaturation ratio (DR) describes the efficiency of the desaturation pathway in the chain length of polyunsaturated FAs. Thus it estimates the weight length in elongation from C18:1 to C18:2 (ODR) and from C18:1 to C18:3 (LDR) within the overall FAs biosynthetic system. It is noteworthy that the FAs desaturation cycle is mainly driven by genetic and environmental factors, including temperature (Clemente & Cahoon, 2009; Sakamoto & Murata, 2002). The overall n-3 and n-6 polyunsaturated FAs (PUFA) pathways are summarized in Fig. 14.2, where several enzymes are involved in chain elongation and desaturation. The desaturation efficiency in seed oils is essential since it has an important role in several metabolism functions, including immune system stability (Chougui et al., 2013;

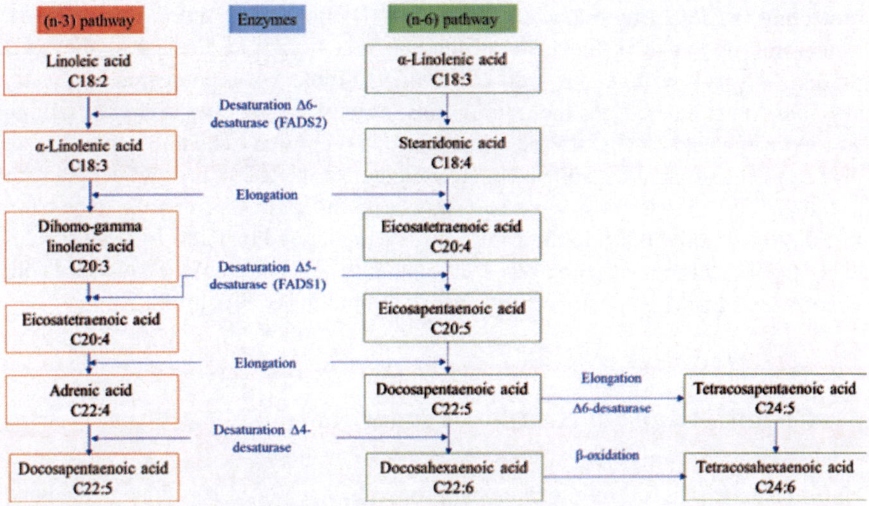

Fig. 14.2 n-3 and n-6 fatty acids pathway desaturation and elongation. The enzymes Δ6 and Δ5 desaturases are encoded by FADS2 and FADS1, respectively (Brayner et al. 2018)

Ramadan, 2019). Growing evidence suggests that increasing dietary intake of n-6 and n-3 long-chain polyunsaturated fatty acids (LCPUFA) has significant effects on neuronal metabolism by lowering neuroinflammatory processes and enhancing synaptic plasticity and neurogenesis. In FSO, higher desaturation efficiency in the LCPUFA pathway was reported, explaining the significant increase in the desaturation rate recorded in all investigated samples. Thus DR, LDR, and ODR values were in the range of 4.93–5.04, 0.87–0.91, and 0.83–0.83, respectively, which is quite similar to the kiwi seeds oil desaturation efficiency (4.25–4.66 and 0.81–0.82 for DR and ODR, respectively), with the exception of LDR ratio that lower in FSO compared to kiwi seeds oil (1.65–1.82). It is noteworthy that, based on the data summarized in Table 14.1, FOS seemed to exhibit high efficiency in LCPUFAs pathways compared to the fruits' seeds listed in Table 14.1. However, guava seeds oil appeared to display high levels for both DR (7.16–9.33) and LDR (0.8–0.9) ratios compared to FSO. Besides FAs desaturation ratios, the balance between omega-6 (ω-6) and omega-3 (ω-3), represented by linoleic acid (LA) and α-linolenic acids (ALA), respectively, constitutes an important determinant of seeds oils quality and is a critical component in human diet (Simopoulos & Gene, 2016). Generally, FSO displays a very low ω-6/ω-3 ratio which falls in the range of 0.75 and 0.79. Due to competing functions of both n-6 (LA) and n-3 (ALA) PUFAs in the formation of highly proinflammatory eicosanoids, balanced intake of LA and ALA FAs is required to avoid chronic diseases and to uphold good health (Saini & Keum, 2018). It is worth mentioning that nutrition societies acclaim a lower value of ω-6/ω-3 ratio for a balanced and healthy diet. Compelling evidence advises that low levels in the balance between ω-6 and ω-3 is significantly associated with reducing the risk of chronic cardiovascular disease and enhancing neuroinflammation alongside protecting and repairing brain cells protection (Dyall. 2017; Saini & Keum, 2018). Therefore, the intake of fig seeds oil may be very beneficial for human diet as it presents a very low ω-6/ω-3 ratio (0.75–0.79) (Table 14.1) compared to several vegetable oil sources, since the ratio in most of western societies is overall as high as 15:1. For blackberry seeds this ratio (3.63) was reported to be three times higher than that in elderberry seed oil (1.19) (Domínguez et al., 2020). In other berries such as bilberry, cranberry, gooseberry, and strawberry this ratio was reported to be close to 1.0 (Alves et al. 2020). In the fruits' seeds summarized in Table 14.1, all species exhibited very high levels of ω-6/ω-3 ratio, with the exception of the kiwi seeds oil, which seems once again very close to the FSO in terms of FAs profile.

4 Chemical and Biochemical Properties

Unfortunately, fig seed oil biochemical properties have been less documented and almost gone unheeded. As far as we know, the only available report on this aspect is the one carried out by Hssaini et al. (2020b), where the results are summarized in Table 14.1 and Fig. 14.3. Iodine value (IV) showed a significantly narrowed variation range of 76.2–76.6 g/100 g of fig seeds oil, classifying it as non-drying oil, owing to the low IV determined in fig seeds from four cultivars. These IV levels indicate the high unsaturation degree found in FSO fats. It is noteworthy that IV determines the number of mg of iodine required to saturate the fatty acids through the ability of an unsaturated carbon to carbon bond (Anang et al., 2019). Saponification value (SV) displayed high levels of 201, suggesting a high amount of FA with low molecular weight. Thus, glycerides made of short-chain polyunsaturated FAs (SCPUFA) exhibit higher SV than LCPUFA. SV indicates oil alkali-reactive groups (Cerchiara et al., 2010), which helps predict the type of glycerides in fats. According to Table 14.1, IV of honeydew, guava, and prickly pear oils are 97.1–97.1, 13.4–13.4, and 107.3–107.3 g/100 g of oil, respectively. The same seed types exhibited PV levels in 116–116, 196–196, and 173–173 mg KOH/g. Among the latter, guava seemed to show comparable physicochemical attributes to FSO. Furthermore, compared to FSO, rambutan seed oil showed IV in 41.8–49.6 g I_2/100 g and an SV ranging from 157 to190 mg KOH/g, respectively (Leong & Shui, 2002).

Primary and secondary oxidation of FSO evaluated through ultraviolet (UV) absorbency at 232 and 270 nm suggests the occurrence of some hydrolytic reactions during the oil chemical extraction (method 659:199) (Hssaini et al., 2021b). UV absorbance at the following wavelengths, 232 and 270 nm, were in the range of 1.83–5.36 and 0.21–0.7, respectively, showing a significant impact of the cultivar factor (Fig. 14.3). The only available data regarding FSO total phenolics, antioxidant activity, and IC_{50} are reported by Hssaini et al. (2020a), summarized in Fig. 14.3. Total phenolics were reported in 69.8 and 100 mg GAE/100 g of oil. These amounts are comparable with those found in prickly pear (48–89 mg GAE/100 g) (Chougui et al., 2013). However, they are higher than levels reported in other fruit seed oils

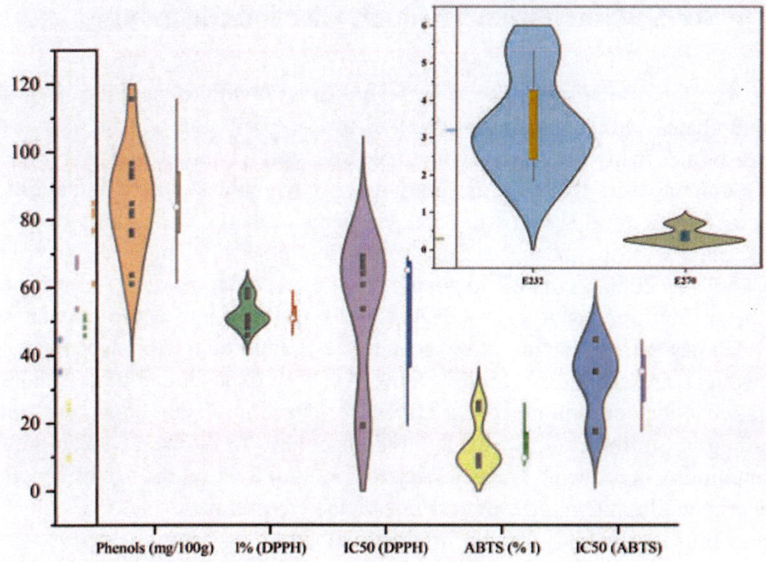

Fig. 14.3 Violin and box plots demonstrate the density of the scores within variables related to the biochemical properties of FSO. Total phenols (mg GAE/100 g), DPPH· and ABTS antioxidant activity (mg Trolox equivalent/g oil) and the half-maximum inhibitory concentration IC_{50} (mg/mL). The embedded graph at the top right shows the ultraviolet oil absorbance at 232 and 270 nm

like mango, avocado, and jackfruit (1.17, 0.88, and 27.7 mg GAE/100 g, respectively) (Wang et al., 2018). Furthermore, FSO exhibited high antioxidant potency through DPPH· (2,2-diphenyl1-picrylhydrazyl) and $ABTS^{o+}$ (3-ethylbenzothiazoline-6-sulfonic acid) free scavenging assays. Available results showed interesting antioxidant activity ranging from 226.4 to 294.3 and 136.6 to 22.65 mg Trolox equivalent/g oil for DPPH· and ABTS assays, respectively (Fig. 14.3). Both assays' half-maximum inhibitory concentration (IC_{50}) was 19.2–68.7 and 17.4–44.8 mg/mL, respectively. It is noteworthy that the aforementioned results depend on the phenolic profile more than phenols amounts, owing particularly to the fact that samples extracts contain a mixture of different phenolic compounds, making it difficult to attribute the radicals scavenging ability to one or a few active principles (Wu et al., 2009). Owing to data reported above, FSO is found to be rich in phenolics and present high free radical scavenging activity even more important than what have been reported over prickly pear seeds oil (Chougui et al., 2013).

All biochemical attributes aforementioned strongly suggest that FSO can be a potential natural source of antioxidants for the food industry and pharmaceutical and cosmetic applications. However, no single oil can accomplish all the requirements and standards of the growing oil sector. Since there has been a renewable nutritional interest for unsaturated oils with higher antioxidant capacity. Consequently, Further investigations on other aspects such as FSO vitamins, phenolics, and volatile compounds are required to evaluate their potential applications as an atypical source of highly nutritive oil, functional food ingredients, additives, or feed.

5 Fig Seeds Valorization Through Circular Economy

Today, the global food system is under severe stress owing to inequities in production and chain value about the misuse and over exploitation of natural and man-made resources, which are highly vulnerable to ongoing climate change. Nowadays, it is a common belief that there is an urgent need to increase the sustainability and resilience of our food systems to meet the demand of the exponentially growing population. Therefore, the global food production major challenge is to increase by at least 60% by 2050, as the world population is expected to reach 9.7 billion people (European Environment Agency (EEA), 2016). In this context, the circular economy (CE) has gained traction in the scientific community as a pioneer for innovative, resilient, and sustainable food systems (Ancut, 2020). CE embodies an ultimate alternative to the conventional linear take-make-consume-dispose economic model. According to Steffen et al. (2015) and Fidelis et al. (2019), the CE is an umbrella concept aiming at reducing waste generation and minimizing the use of new inputs and energy while mitigating the pressures on the environment.

Fruits processing, like other agriproducts valorization systems, generates considerable amounts of waste, which is a challenging endeavor that, if not accurately addressed, may negatively impact the environment and the economy alongside the ability to meet the population's needs in new food sources. Daily, the entire food system loses about 30% of all produced foods, where the mass of by-products generated by the fruit processing industry is becoming higher among all types of agrifoods, reaching up to more than 190 million tons of by-products worldwide (Campos et al., 2020; Gençdağ et al., 2020). In the fruit sector, more than 45% of the total production is wasted during the post-harvest valorization chain; For western Asia and northern Africa, this loss is about 11% (FAOSTAT, 2021). Seeds constitute a relevant part of this waste, as being usually discarded after fruit processing to jam, juice, syrup, or must (Ben-Othman et al., 2020). Therefore, there is an urgent need to transition from current linear food production to a circular economy model to recycle these seeds and recover resources for high-value-added products (Ancut, 2020; Campos et al., 2020; Fidelis et al., 2019; Lucarini et al., 2018). Large seeds generated from the conventional linear processing model are not yet appropriately exploited for figs. Thus, various potential products from this waste, particularly oil, press cake, or meal, are not yet systematically available on the market or perhaps not easily marketed as common oilseeds used by the food industry (i.e., sesame, sunflower, and soybean), that have been promoted for a long time (Alves et al., 2020; Ramadan, 2019).

Unfortunately, there is a severe lack of consistent statistics regarding waste amounts generated during fig processing, including peels, latex, and seeds. On the contrary, for other fruits, like mangos and oranges, available reports stressed a waste level in a range of 30–50%, while it can be in a higher range in the case of durian, jackfruit, and mangosteen, which can generate up to 70% of waste (Fidelis et al., 2019). Figure 14.4 describes the linear model of fig processing, explores potential products and byproducts produced, and proposes a circular model by reprocessing and reusing wastes for some potential industrial applications. Generally, figs are harvested for fresh consumption or transformation at the full ripening stage or left on the tree until they shrivel or become half-dried. In some cases, figs can also be harvested at their full maturity stage and then dried with industrial ovens or dryers with appropriate pretreatments to prevent browning. Dried figs, as a whole or sliced, can be roasted and used as a coffee substitute (Ishurd et al., 2004; Liang et al., 2016; Mat Desa et al., 2019) or used to produce wine and spirits, with either plurivarietal or monovarietal dried figs as the raw inputs (Solana et al., 2018).

Being a highly perishable commodity, a huge part of harvested fresh figs are processed into jam, syrup, juice, wine, or must. During which seeds received no intention and almost underwent unheeded. Seeds are usually crushed during processing, and thus their potential importance goes unnoticed or is often discarded as solid waste. As previously stressed, these seeds can be exploited as an essential source of highly unsaturated oil with high antioxidant potency. Owing to FSO lipochemical and biologic properties, it can be brought to the international markets at favorable prices (Hssaini et al., 2021b) as it can be successfully used for food and cosmetic applications. During oil extraction, press-cakes or meals are produced from which can be extracted high-value components such as fats, antioxidants, proteins, dietary fiber, and phytochemicals. They can also be used as fiber-based foods and dietary supplements to formulate new products or for animal feed (Fig. 14.4). So far, available literature has proposed several uses of fruit seeds byproducts in industrial formulations, such as muffins and cookies enriched with fruits seeds such as grape seeds and date seeds (Karnopp et al., 2015, 2017; Maqsood et al., 2020). These studies, among others, have shown that the incorporation of fruit seeds in newly formulated products resulted in enhancing their nutritional virtues (Bouaziz

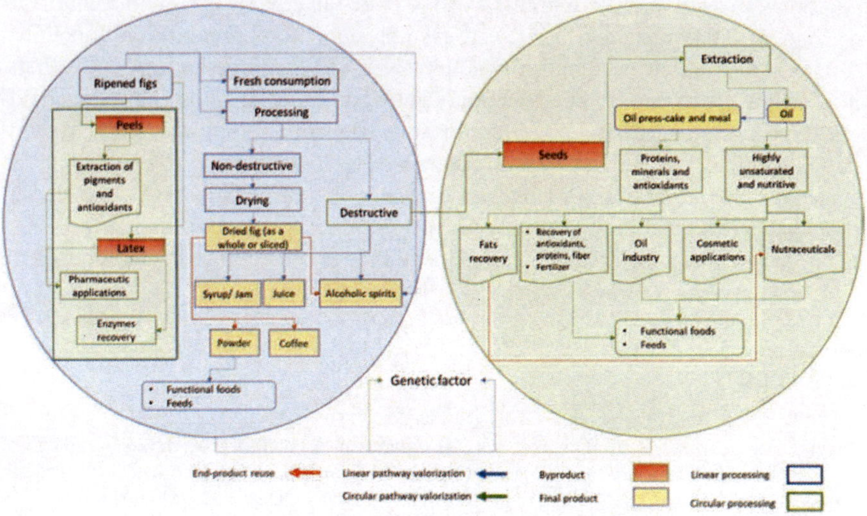

Fig. 14.4 Global diagram of fig processing and circular economy of waste materials focused on seeds

et al., 2010; Platat et al., 2015). In the proposed CE, we could also spot other wastes, solid like peels and liquid such as latex, which can be an important source of essential bioactives (pigments, proteins, enzymes, dietary fibers, etc.) for many industrial applications. This waste is reported in several previous studies as potential nutraceutical and functional food ingredients (Backes et al., 2018; Barolo et al., 2014a, b; Hssaini, et al., 2021a; Milošević et al., 2020; Sirisha et al., 2010). Figure 14.4 provides a holistic overview of the fig processing pathways, focusing on seeds valorization as they remain the less studied and exploited part of the species. It is noteworthy that all valorization pathways of fig products and byproducts fundamentally depend on the egentic factor.

6 Conclusions and Future Research Directions

Given the continuous progress of the oilseed industry, an important interest is currently being given to valorizing novel and unexploited sources for vegetable oils. Fig seed represents an interesting atypic source of oil with highly promising lipochemical and biochemical attributes, which can provide many virtues beyond balanced nutrition. Although fig seeds attributes are scarcely documented, oil and press-cake or meal can be value-added products to be brought to the markets and used as functional food or for cosmetic applications. Notwithstanding, more research is required to investigate the properties above further, explore other research fields such as the varietal effect on phytochemical properties, recovery of bioactive compounds with different extraction techniques, particularly green ones, and finally, the stability of FSO according to the above-cited factors. Furthermore, additional investigations are necessary to explore the effectiveness and technological opportunities of using fig oil and cake-press or meal produced as functional foods or by-products used in formulating new foods. Such research, among others, might motivate to meet circular economy goals and fulfill the food chain stakeholders' expectations regarding opportunities to create value for the waste and byproducts. Therefore, the purpose of this chapter was to draw attention to fig seeds in the scope of the circular economy by providing a review of the very few available relevant papers on the topic and simply proposing potential valorization pathways, and to encourage researchers to give more attention to seeds of this outstanding fruit.

References

Ajayi, I. A., Oderinde, R. A., Ogunkoya, B. O., Egunyomi, A., & Taiwo, V. O. (2007). Chemical analysis and preliminary toxicological evaluation of *Garcinia mangostana* seeds and seed oil. *Food Chemistry, 101*, 999–1004. https://doi.org/10.1016/J.FOODCHEM.2006.02.053

Alves, E., Simoes, A., & Domingues, M. R. (2020). Fruit seeds and their oils as promising sources of value-added lipids from agro-industrial byproducts: Oil content, lipid composition, lipid analysis, biological activity and potential biotechnological applications. *Critical Reviews in Food Science and Nutrition, 61*, 1305–1339.

Anang, M. A., Oteng-Peprah, M., & Opoku-Boadu, K. (2019). Extraction and characterisation of African star apple (*Chrysophyllum albidum*) seed oil and the adsorptive properties of the fruit shell in Ghana. *International Journal of Food Science*. https://doi.org/10.1155/2019/4959586

Ancut, P. (2020). Applied sciences oil press-cakes and meals valorization through circular economy approaches: A review. *Applied Sciences, 10*, 7432.

Backes, E., Pereira, C., Barros, L., Prieto, M. A., Genena, A. K., Barreiro, M. F., & Ferreira, I. C. F. R. (2018). Recovery of bioactive anthocyanin pigments from *Ficus carica* L. peel by heat, microwave, and ultrasound based extraction techniques. *Food Research International, 113*, 197–209. https://doi.org/10.1016/J.FOODRES.2018.07.016

Badgujar, S. B., Patel, V. V., Bandivdekar, A. H., & Mahajan, R. T. (2014). Traditional uses, phytochemistry and pharmacology of *Ficus carica*: A review. *Pharmaceutical Biology, 52*, 1487–1503. https://doi.org/10.3109/13880209.2014.892515

Barolo, M. I., Ruiz Mostacero, N., & López, S. N. (2014a). *Ficus carica* L. (Moraceae): An ancient source of food and health. *Food Chemistry, 164*, 119–127. https://doi.org/10.1016/j.foodchem.2014.04.112

Barolo, M. I., Ruiz Mostacero, N., & López, S. N. (2014b). *Ficus carica* L. (Moraceae): An ancient source of food and health. *Food Chemistry, 164*, 119–127.

Ben-Othman, S., Jõudu, I., & Bhat, R. (2020). Bioactives from agri-food wastes: Present insights and future challenges. *Molecules, 25*, 510.

Bouaziz, M. A., Amara, W. B., Attia, H., Blecker, C., & Besbes, S. (2010). Effect of the addition of defatted date seeds on wheat dough performance and bread quality. *Journal of Texture Studies, 41*, 511–531. https://doi.org/10.1111/J.1745-4603.2010.00239.X

Brayner, B., Kaur, G., Keske, M. A., & Livingstone, K. M. (2018). FADS polymorphism, Omega-3 fatty acids and diabetes risk: A systematic review. *Nutrients*. https://doi.org/10.3390/NU10060758

Campos, D. A., Ricardo, G., Vilas-boas, A. A., Madureira, A. R., & Pintado, M. M. (2020). Management of fruit industrial By-Products-A case. *Molecules, 25*(2), 320.

Cerchiara, T., Chidichimo, G., Ragusa, M. I., Belsito, E. L., Liguori, A., & Arioli, A. (2010). Characterization and utilization of Spanish Broom (*Spartium junceum* L.) seed oil. *Industrial Crops and Products, 31*, 423–426. https://doi.org/10.1016/j.indcrop.2009.11.003

Chougui, N., Tamendjari, A., Hamidj, W., Hallal, S., Barras, A., Richard, T., & Larbat, R. (2013). Oil composition and characterisation of phenolic compounds of *Opuntia ficus-indica* seeds. *Food Chemistry, 139*, 796–803. https://doi.org/10.1016/j.foodchem.2013.01.054

Clemente, T. E., & Cahoon, E. B. (2009). Soybean oil: Genetic approaches for modification of functionality and total content. *Plant Physiology, 151*, 1030–1040. https://doi.org/10.1104/PP.109.146282

Debib, A., Tir-Touil, A., Mothana, R. A., Meddah, B., & Sonnet, P. (2014). Phenolic content, antioxidant and antimicrobial activities of two fruit varieties of Algerian *Ficus carica* L. *Journal of Food Biochemistry, 38*, 207–215. https://doi.org/10.1111/jfbc.12039

Del Caro, A., & Piga, A. (2008). Polyphenol composition of peel and pulp of two Italian fresh fig fruits cultivars (*Ficus carica* L.). *European Food Research and Technology, 226*, 715–719. https://doi.org/10.1007/s00217-007-0581-4

Domínguez, R., Zhang, L., Rocchetti, G., Lucini, L., Pateiro, M., Munekata, P. E. S., & Lorenzo, J. M. (2020). Elderberry (*Sambucus nigra* L.) as potential source of antioxidants. Characterization, optimization of extraction parameters and bioactive properties. *Food Chemistry, 330*, 127266. https://doi.org/10.1016/j.foodchem.2020.127266

Dyall, S. C. (2017). Interplay between n-3 and n-6 long-chain polyunsaturated fatty acids and the endocannabinoid system in brain protection and repair. *Lipids, 52*, 885–900. https://doi.org/10.1007/S11745-017-4292-8

El Mannoubi, I, Barrek, S., Skanji, T., Casabianca, H., & Zarrouk, H. (2020) Seed oil from Tunisia 45:521–524

Ennouri, M., Evelyne, B., Laurence, M., & Hamadi, A. (2005). Fatty acid composition and rheological behaviour of prickly pear seed oils. *Food Chemistry, 93*, 431–437. https://doi.org/10.1016/J.FOODCHEM.2004.10.020

European Environment Agency (EEA). (2016). *Circular economy in Europe- developing the knowledge base*. European Environment Agency.

Fidelis, M., De Moura, C., Kabbas, T., Pap, N., Mattila, P., Mäkinen, S., Putnik, P., Kovačević, D. B., Tian, Y., Yang, B., & Granato, D. (2019). Fruit seeds as sources of bioactive compounds: Sustainable production of high value-added ingredients from by-products within circular economy. *Molecules, 24*, 1–54.

Gençdağ, E., Görgüç, A., & Yılmaz, F. M. (2020). Recent advances in the recovery techniques of plant-based proteins from agro-industrial by-products. *Food Review International, 1-22*.

Gonçalves-de-Albuquerque, C. F., Medeiros-de-Moraes, I. M., Oliveira, F. M. D. J., Burth, P., Bozza, P. T., Castro Faria, M. V., et al. (2016). Omega-9 oleic acid induces fatty acid oxidation and decreases organ dysfunction and mortality in experimental sepsis. *PLoS One, 11*, e0153607. https://doi.org/10.1371/journal.pone.0153607

Górnaś, P., & Rudzińska, M. (2016). Seeds recovered from industry by-products of nine fruit species with a high potential utility as a source of unconventional oil for biodiesel and cosmetic and pharmaceutical sectors. *Industrial Crops and Products, 83*, 329–338. https://doi.org/10.1016/J.INDCROP.2016.01.021

Ho, L. H., & Bhat, R. (2015). Exploring the potential nutraceutical values of durian (*Durio zibethinus* L.)-An exotic tropical fruit. *Food Chemistry, 168*, 80–89. https://doi.org/10.1016/J.FOODCHEM.2014.07.020

Hssaini, L., Charafi, J., Hanine, H., Ennahli, S., Mekaoui, A., Mamouni, A., & Razouk, R. (2019). Comparative analysis and physio-biochemical screening of an ex-situ fig (*Ficus carica* L.) collection. *Horticulture, Environment, and Biotechnology*, 60. https://doi.org/10.1007/s13580-019-00170-4

Hssaini, L., Hanine, H., Charafi, J., Razouk, R., Elantari, A., Ennahli, S., Hernández, F., & Ouaabou, R. (2020a). First report on fatty acids composition, total phenolics and antioxidant activity in seeds oil of four fig cultivars (*Ficus carica* L.) grown in Morocco. *OCL, 27*, 8. https://doi.org/10.1051/ocl/2020003

Hssaini, L., Razouk, R., Elantari, A., Ennahli, S., & Ouaabou, R. (2020b). First report on fatty acids composition, total phenolics and antioxidant activity in seeds oil of four fig cultivars (*Ficus carica* L.) grown in Morocco ☆. *OCL, 27*, 8.

Hssaini, L., Hernandez, F., Viuda-Martos, M., Charafi, J., Razouk, R., Houmanat, K., Ouaabou, R., Ennahli, S., Elothmani, D., Hmid, I., Fauconnier, M. L., & Hanine, H. (2021a). Survey of phenolic acids, flavonoids and in vitro antioxidant potency between fig peels and pulps: Chemical and chemometric approach. *Molecules, 26*, 2574. https://doi.org/10.3390/MOLECULES26092574

Hssaini, L., Razouk, R., Charafi, J., Houmanat, K., & Hanine, H. (2021b). Fig seeds: Combined approach of lipochemical assessment using gas chromatography and FTIR-ATR spectroscopy using chemometrics. *Vibrational Spectroscopy, 114*, 103251. https://doi.org/10.1016/j.vibspec.2021.103251

Ishurd, O., Zgheel, F., Kermagi, A., Flefla, M., Elmabruk, M., Wu, Y., Kennedy, J. F., & Pan, Y. (2004). Microbial (1→3)-β-d-glucans from Libyan figs (*Ficus carica*). *Carbohydrate Polymers, 58*, 181–184. https://doi.org/10.1016/J.CARBPOL.2004.06.040

Jeong, W. S., & Lachance, P. A. (2001). Phytosterols and fatty acids in fig (*Ficus carica*, var. Mission) fruit and tree components. *Journal of Food Science, 66*, 278–281. https://doi.org/10.1111/j.1365-2621.2001.tb11332.x

Kamiloglu, S., & Capanoglu, E. (2013). Investigating the in vitro bioaccessibility of polyphenols in fresh and sun-dried figs (*Ficus carica* L.). *International Journal of Food Science and Technology, 48*, 2621–2629. https://doi.org/10.1111/ijfs.12258

Kapoor, S., Gandhi, N., Tyagi, S. K., Kaur, A., & Mahajan, B. V. C. (2020). Extraction and characterization of guava seed oil: A novel industrial byproduct. *LWT, 132*, 109882. https://doi.org/10.1016/J.LWT.2020.109882

Karabagias, V. K., Karabagias, I. K., Gatzias, I., & Badeka, A. V. (2020). Prickly pear seed oil by shelf-grown cactus fruits: Waste or maste? *Process, 8*, 132. https://doi.org/10.3390/PR8020132

Karnopp, A. R., Figueroa, A. M., los PR, T. J. C., DRS, S., Barana, A. C., Kubiaki, F. T., Oliveira, J. G. B. d., & Granato, D. (2015). Effects of whole-wheat flour and bordeaux grape pomace (*Vitis labrusca* L.) on the sensory, physicochemical and functional properties of cookies. *Journal of Food Science and Technology, 35*, 750–756. https://doi.org/10.1590/1678-457X.0010

Karnopp, A. R., Oliveira, K. G., de Andrade, E. F., Postingher, B. M., & Granato, D. (2017). Optimization of an organic yogurt based on sensorial, nutritional, and functional perspectives. *Food Chemistry, 233*, 401–411. https://doi.org/10.1016/J.FOODCHEM.2017.04.112

Kato, M., & Kawakita, A. (Eds.). (2017). *Obligate pollination mutualism*. Springer.

Kislev, M. E., Hartmann, A., & Bar-Yosef, O. (2006). Early domesticated fig in the Jordan Valley. *Science, 80*(312), 1372–1374. https://doi.org/10.1126/SCIENCE.1125910

Leong, L. P., & Shui, G. (2002). An investigation of antioxidant capacity of fruits in Singapore markets. *Food Chemistry, 76*, 69–75. https://doi.org/10.1016/S0308-8146(01)00251-5

Liang, N., Lu, X., Hu, Y., & Kitts, D. D. (2016). Application of attenuated total reflectance-fourier transformed infrared (ATR-FTIR) spectroscopy to determine the chlorogenic acid isomer profile and antioxidant capacity of coffee beans. *Journal of Agricultural and Food Chemistry, 64*, 681–689. https://doi.org/10.1021/acs.jafc.5b05682

Lucarini, M., Durazzo, A., Romani, A., Campo, M., Lombardi-boccia, G., & Cecchini, F. (2018). *Bio-based compounds from grape seeds: A biorefinery approach* (pp. 1–12). https://doi.org/10.3390/molecules23081888

Lucarini, M., Durazzo, A., Kiefer, J., Santini, A., Lombardi-Boccia, G., Souto, E. B., Romani, A., Lampe, A., Nicoli, S. F., Gabrielli, P., Bevilacqua, N., Campo, M., Morassut, M., & Cecchini, F. (2020). Grape seeds: Chromatographic profile of fatty acids and phenolic compounds and qualitative analysis by FTIR-ATR spectroscopy. *Food, 9*. https://doi.org/10.3390/foods9010010

Mallek-Ayadi, S., Bahloul, N., & Kechaou, N. (2018). Chemical composition and bioactive compounds of *Cucumis melo* L. seeds: Potential source for new trends of plant oils. *Process Safety and Environment Protection, 113*, 68–77. https://doi.org/10.1016/J.PSEP.2017.09.016

Maqsood, S., Adiamo, O., Ahmad, M., & Mudgil, P. (2020). Bioactive compounds from date fruit and seed as potential nutraceutical and functional food ingredients. *Food Chemistry, 308*, 125522. https://doi.org/10.1016/J.FOODCHEM.2019.125522

Mat Desa, W. N., Mohammad, M., & Fudholi, A. (2019). Review of drying technology of fig. *Trends in Food Science and Technology, 88*, 93–103. https://doi.org/10.1016/j.tifs.2019.03.018

Milošević, J., Vrhovac, L., Đurković, F., Janković, B., Malkov, S., Lah, J., & Polović, N. D. (2020). Isolation, identification, and stability of Ficin 1c isoform from fig latex. *New Journal of Chemistry, 44*, 15716–15723. https://doi.org/10.1039/d0nj02938f

Mohd Ali, M., Hashim, N., Aziz, S. A., & Lasekan, O. (2020). Exploring the chemical composition, emerging applications, potential uses, and health benefits of durian: A review. *Food Control, 113*, 107189. https://doi.org/10.1016/J.FOODCONT.2020.107189

Nakilcioğlu-Taş, E., & Ötleş, S. (2021). Influence of extraction solvents on the polyphenol contents, compositions, and antioxidant capacities of fig (*Ficus carica* L.) seeds. *Anais da Academia Brasileira de Ciências, 93*. https://doi.org/10.1590/0001-3765202120190526

Narváez-Cuenca, C. E., Inampues-Charfuelan, M. L., Hurtado-Benavides, A. M., Parada-Alfonso, F., & Vincken, J. P. (2020). The phenolic compounds, tocopherols, and phytosterols in the edible oil of guava (*Psidium guava*) seeds obtained by supercritical CO_2 extraction. *Journal of Food Composition and Analysis, 89*, 103467. https://doi.org/10.1016/J.JFCA.2020.103467

Park, H. G., Engel, M. G., Vogt-Lowell, K., Lawrence, P., Kothapalli, K. S., & Brenna, J. T. (2018). The role of fatty acid desaturase (FADS) genes in oleic acid metabolism: FADS1 Δ7 desaturates 11-20:1 to 7,11-20:2. *Prostaglandins, Leukotrienes and Essential Fatty Acids, 128*, 21–25. https://doi.org/10.1016/j.plefa.2017.11.004

Piombo, G., Barouh, N., Barea, B., Boulanger, R., Brat, P., Pina, M., & Villeneuve, P. (2006). Characterization of the seed oils from kiwi (*Actinidia chinensis*), passion fruit (*Passiflora edulis*) and guava (*Psidium guajava*). *Oléagineux, Corps Gras, Lipides, 13*, 195–199. https://doi.org/10.1051/OCL.2006.0026

Platat, C., Habib, H. M., Hashim, I. B., Kamal, H., AlMaqbali, F., Souka, U., & Ibrahim, W. H. (2015). Production of functional pita bread using date seed powder. *Journal of Food Science and Technology, 52*, 6375–6384. https://doi.org/10.1007/S13197-015-1728-0

Prasad, N. B. L., & Azeemoddin, G. (1994). Characteristics and composition of guava (*Psidium guajava* L.) seed and oil. *Journal of the American Oil Chemists' Society, 71*, 457–458. https://doi.org/10.1007/BF02540531

Raihana, A. R. N., Marikkar, J. M. N., Amin, I., & Shuhaimi, M. (2015). A review on food values of selected tropical fruits' seeds. *International Journal of Food Properties, 18*, 2380–2392. https://doi.org/10.1080/10942912.2014.980946

Ramadan, M. F. (2019). Chemistry and functionality of fruit oils: An introduction. In *Fruit oils: Chemistry and functionality* (pp. 3–8). Springer International Publishing.

Saini, R. K., & Keum, Y. S. (2018). *Omega-3 and omega-6 polyunsaturated fatty acids: Dietary sources, metabolism, and significance-a review*. Elsevier Inc.

Sakamoto, T., & Murata, N. (2002). Regulation of the desaturation of fatty acids and its role in tolerance to cold and salt stress. *Current Opinion in Microbiology, 5*, 206–210. https://doi.org/10.1016/S1369-5274(02)00306-5

Sassa, T., & Kihara, A. (2014). Metabolism of very long-chain fatty acids: Genes and pathophysiology. *Biomolecules & Therapeutics, 22*, 83. https://doi.org/10.4062/BIOMOLTHER.2014.017

Schmitzer, V., Slatnar, A., Mikulic-Petkovsek, M., Veberic, R., Krska, B., & Stampar, F. (2011). Comparative study of primary and secondary metabolites in apricot (*Prunus armeniaca* L.) cultivars. *Journal of the Science of Food and Agriculture, 91*, 860–866. https://doi.org/10.1002/jsfa.4257

Senrayan, J., & Venkatachalam, S. (2019). A short extraction time of vegetable oil from *Carica papaya* L. seeds using continuous ultrasound acoustic cavitation: Analysis of fatty acid profile and thermal behavior. *Journal of Food Process Engineering, 42*, e12950. https://doi.org/10.1111/JFPE.12950

Simopoulos, A. P., & Gene, O. (2016). An increase in the Omega-6/Omega-3 fatty acid ratio increases the risk for. *Obesity, 1-17*. https://doi.org/10.3390/nu8030128

Sirisha, N., Sreenivasulu, M., Sangeeta, K., & Madhusudhana Chetty, C. (2010). Antioxidant properties of Ficus species – A review. *International Journal of Pharmaceutical Technology Research, 2*, 2174–2182.

Solana, R. R., Galego, L. R., Pérez, E., & Anabela, S. (2018). Production method and varietal source influence the volatile profiles of spirits prepared from fig fruits (*Ficus carica* L.). *European Food Research and Technology*. https://doi.org/10.1007/s00217-018-3131-3

Steffen, W., Richardson, K., Rockström, J., Cornell, S. E., Fetzer, I., Bennett, E. M., Biggs, R., Carpenter, S. R., de Vries, W., de Wit, C. A., Folke, C., Gerten, D., Heinke, J., Mace, G. M., Persson, L. M., Ramanathan, V., Reyers, B., & Sörlin, S. (2015). Planetary boundaries: Guiding human development on a changing planet. *Science, 347*. https://doi.org/10.1126/SCIENCE.1259855

Tangolar, S. G., özoğul, Y., Tangolar, S., & Torun, A. (2009). Evaluation of fatty acid profiles and mineral content of grape seed oil of some grape genotypes. *International Journal of Food Sciences and Nutrition, 60*, 32–39. https://doi.org/10.1080/09637480701581551

Taylor, P., Cravotto, G., Bicchi, C., Mantegna, S., Binello, A., Tomao, V., & Chemat, F.. (2011). *Natural product research: Formerly natural product letters extraction of kiwi seed oil: Soxhlet versus four different non-conventional techniques*. (pp. 37–41). https://doi.org/10.1080/14786419.2010.524162

Veberic, R., Jurhar, J., Mikulic-Petkovsek, M., Stampar, F., & Schmitzer, V. (2010). Comparative study of primary and secondary metabolites in 11 cultivars of persimmon fruit (*Diospyros kaki* L.). *Food Chemistry, 119*, 477–483. https://doi.org/10.1016/j.foodchem.2009.06.044

Velasco, L., Goffman, F. D., & Becker, H. C. (1998). Variability for the fatty acid composition of the seed oil in a germplasm collection of the genus brassica. *Genetic Resources and Crop Evolution, 45*, 371–382.

Vermaak, I., Kamatou, G. P. P., Komane-mofokeng, B., Viljoen, A. M., & Beckett, K. (2011). African seed oils of commercial importance-cosmetic applications. *South African Journal of Botany, 77*, 920–933. https://doi.org/10.1016/j.sajb.2011.07.003

Wang, Y., Li, L., Liu, H., Zhao, T., Meng, C., Liu, Z., & Liu, X. (2018). Bioactive compounds and in vitro antioxidant activities of peel, flesh and seed powder of kiwi fruit. *International Journal of Food Science and Technology, 53*, 2239–2245. https://doi.org/10.1111/IJFS.13812

Yanty, N. A. M., Lai, O. M., Osman, A., Long, K., & Ghazali, H. M. (2008). Physicochemical properties of Cucumis melo var. inodorus (honeydew melon) seed and seed oil. *Journal of Food Lipids, 15*, 42–55. https://doi.org/10.1111/J.1745-4522.2007.00101.X

Chapter 15
Fig (*Ficus carica*) Leaves: Composition and Functional Properties

Rashida Bashir, Samra Tabassum, Ayoub Rashid, Shafiqur Rehman, and Ahmad Adnan

Abbreviations

TPC	Total phenolic content
TFC	Total flavonoid content
TTC	Total tannin content
TCC	Total carotenoid content
DW	Dry weight
FW	Fresh weight

1 Extraction of Bioactive Compounds

Maceration is the simplest method of compound extraction. It has been applied to extract the latex of wild fig "caprifig" using 1 L of *n*-hexane for 10 days (Lazreg-Aref et al., 2012); however, simple maceration of latex can be modified with the help of mechanical shaking. A study conducted on fig latex has shown that maceration with shaking provided better total phenolic content and antioxidants than ultrasound-assisted extraction at the same reaction conditions especially for the white genoa variety (Shahinuzzaman et al., 2020).

R. Bashir (✉) · S. Tabassum
Division of Science and Technology, University of Education Lahore, Lahore, Pakistan
e-mail: rashida.bashir@ue.edu.pk

A. Rashid · A. Adnan
Department of Chemistry, GC University Lahore, Lahore, Pakistan
e-mail: ayoubrashid@gcu.edu.pk; ahmadadnan@gcu.edu.pk

S. Rehman
Department of Chemistry, Faculty of Sciences, University of Central Punjab, Lahore, Pakistan
e-mail: shafiq.rehman@ucp.edu.pk

M. F. Ramadan (ed.), *Fig* (Ficus carica)*: Production, Processing, and Properties*, https://doi.org/10.1007/978-3-031-16493-4_15

Magnetic stirring is also a standard method used to extract bioactive compounds from plants; optimum conditions of 115.14 min, 48.6 °C temperature, and 63.4% acetone have provided high recovery of phenolic compounds and antioxidants (Bachir Bey et al., 2014). However, Soxhlet extraction requires a higher temperature than a maceration. Soxhlet extraction has been carried out at 60–80 °C temperature using petroleum ether to recover hepatoprotective compound fraction from fig leaves (Gond & Khadabadi, 2008). In a study, Soxhlet extraction extracted polar compounds with antimicrobial activity from fig leaves by using ethanol in a solid/liquid ratio of 1:7, with 24 h extraction period (Al Askari et al., 2013).

Decoction also requires heat treatment but is confined to water-soluble fractions only (Hamedi et al., 2015). It has been carried out to recover water-soluble polar compounds with antibacterial activity; however, did not the antifungal activity was not reported in the decoction extract of fig leaves (Al Askari et al., 2013). The hydrodistillation technique also uses water as a solvent but is confined to essential oils extraction, and GC-MS analysis reported 108 and 121 volatile phytoconstituents in leaves and fruit, respectively (Li et al., 2012). Therefore, a solid-liquid phase has been commonly used for compound extraction. However, solid matrix disruption extraction (sea sand disruption method) of fig leaves provided better reproducibility (RSD < 5%) and recovery (above 85%) of phenolic acids, flavonols, and coumarins (Teixeira et al., 2006).

Phenolic acids, flavonols, and coumarins have been efficiently extracted using ultrasound-assisted extraction using ionic-liquid, which provided better recovery and reduced extraction time at the optimum conditions of 30 min extraction, 30 °C temperature, 1:50 solid-liquid ratio, and 1.0 mol/L of imidazolium ionic liquid concentration (Hongying, 2015). A tailored deep eutectic solvent system comprising glycerol, xylitol, and D-(-)-fructose in the molar ratio of 3:3:3 has also been used for fig extraction; results have indicated that DES assisted UAE and DES assisted MAE provided a better yield of polyphenolic and furanocoumarin compound as compared to methanolic and non-tailor DES assisted UAE and MAE. The yield reached 6.582 mg/g, 5.307 mg/g, 16.10 mg/g, 2.575 mg/g, 15.02 mg/g, and 2.575 mg/g for caffeoylamlic acid, rutin psoralic acid-glucoside, bergapten and, psoralen, respectively, under optimal conditions (Wang et al., 2017).

Supercritical fluid extraction has been used for defatting fig powder before polysaccharide extraction, using optimum conditions of 45 °C and 30 MPa (Yang et al., 2009). Microwave-assisted extraction has been used to extract flavonoids from fig leaves at 70 °C for three times using water and 70% ethanol. Water provided lesser yield (5.5% w/w) as compared to 70% ethanol (yield = 10% w/w) (Trifunschi & Ardelean, 2013). Surfactant based microwave-assisted extraction replaces organic solvent and provides better recovery of bioactive compounds in less time; polyphenolic compounds and furanocoumarins have been successfully extracted from fig leaves using this technique with optimal conditions of 40 °C temperature, 10.27 min, solid/liquid ratio of 1:19.95 (Yu et al., 2020). Enzyme assisted extraction is a sustainable green method for bioactive compounds; however, the combination of ultrasound-assisted extraction with enzymatic extraction gives rise to an innovative technology that reinforces the recovery of polyphenolic compounds; Ultrasonic-enzyme assisted two-phase aqueous extraction of fig leaves has provided higher flavonoid yield of 60.22 mg/g as compared to ultrasound-assisted ATPE, enzyme-ATPE, ATPE and Soxhlet extraction (Zhao et al., 2021). Solid-state fermentation using fungi is an emerging non-conventional technique reported to enhance the phenolic compound recovery and anti-oxidant activity up to 5.48 and 98.54 folds from fig processing by-products (Buenrostro-Figueroa et al., 2017).

The choice of extraction technique and conditions also vary depending upon target compounds; according to a study, 24 h maceration with ethanol in a 1:50 solid/liquid ratio was used to get extracts for flavonoids, flavonols, ascorbic acid, proanthocyanidin, and ortho-diphenols estimation however for the extraction of carotenoids, 5 min ultrasonication using a solvent system of hexane/acetone/methanol was used (Amessis-Ouchemoukh et al., 2017).

2 Phenolic Content and Common Phenolic Acids

Phenolic content is reported to vary among different cultivars of fig. For example, a study has shown that the total phenolic content ranged between 24–237 mg GAE/100 g fresh weight for 24 different genotypes and two standard cultivars (Ercisli et al., 2012). In another study, dark color varieties have shown high TPC, Fassi variety having the highest TPC of 524 mg GAE/100 g DW, followed by Noukali, which showed a TPC of 478.6 mg GAE/100 g DW; Snowden also showed a considerable amount of phenolic compounds showing TPC of 42.52 mg GAE/100 g DW) (Hssaini et al., 2020a).

The variation in phenolic content is not limited to different varieties only, as a study has reported that the different parts of the fig had a varied range of phenolic content, skin being the primary source of phenolic compounds with TPC range from 41.7 to 463 mg GAE/100 g FW followed by fruit 48.6–281 mg GAE/100 g FW while pulp was experienced low phenolic contents with TPC range of 36.5–100.6 mg GAE/100 g FW(Solomon et al., 2006).

Extraction temperature and the solvent system have a pronounced effect on the TPC of fig, a study has shown that 80% ethanol and 80 °C extraction temperature significantly enhanced the recovery of phenolic compounds (3.7 mg GAE/g dry basis) from *Ficus carica* L. cv. Šaraguja (Bucic-Kojic et al., 2011). Methanol extract of fig latex also has shown a high TPC value of 50.20 mg GAE/g of latex; this value was approximately five times higher than the TPC of methanol extract of *Euphorbia tirucalli* latex (Abdel-Aty et al., 2019). Aqueous methanol extract of whole fruit has also been reported to have a high TPC value of 31.8% w/w) (Alamgeer et al., 2017).

Extraction technique has pronounced effect on the recovery of phenolic compounds from fig, as a study conducted on eighteen fig varieties has shown that maceration extract of latex gave higher TPC value ranging from 233.1–311.8 μg GAE/mL as compared to UAE which provided the range TPC range of 224.37–291.98 μg GAE/mL, however in both types of extraction White Genoa variety yielded the highest amount of phenolic compounds. In contrast, the Download variety yielded the lowest TPC, again proving the dependence of TPC on a variety of figs as well (Shahinuzzaman et al., 2020). Polyphenolic content is reported to vary depending upon the maturation stages; samples of Dottato variety from June, July, and September has shown polyphenolic content of 3.9, 4.4, and 11.9 mg chlorogenic acid equivalent/g of dried material, respectively) (Marrelli et al., 2012).

Plants part, extraction technique, and the solvent system affect the phytochemical profile (Table 15.1). For example, HPLC analysis of fig fruit has shown the existence of chronotropic acid, caffeine acid, vanillic acid, syringic acid, gallic acid, and m-coumaric acid (Alamgeer et al., 2017).

HPLC quantification of phenolic acids of white and dark cultivars of fig fruit has shown chlorogenic acid to be the most abundant phenolic acid, with the content ranging from 0.46 to 1.71 mg/100 g FW, followed by gallic acid and synergic acid. Besides fruit variety and color parameters, phenolic acids were varied in crop cultivations from two different months, i.e., July and September (Veberic et al., 2008). However, another study has not detected gallic acid in fig varieties' pulp (Hssaini et al., 2021).

HPLC analysis of methanol extracts of fig latex also has not shown the presence of gallic acid; however, *p*-protocatechuic acid 8%, chlorogenic acid 59%, vanillic acid 1%, caffeic acid 3%, sinapic acid 1% were present along with other polyphenolic compounds (Abdel-Aty et al., 2019). When compared among peels and pulp of two different varieties of fig, protocatechuic acid was present in a higher amount in the pulp of both varieties (8.11 and 7.45 mg/100 g sample in Calor and Cuello dama variety, respectively) as compared to peels (6.22 and 3.27 mg/100 g in Calor and Cuello dama respectively). Chlorogenic acid ranged from 0.92–3.51 mg/100 g in all samples; however, the pulp of the Calor variety contained chlorogenic acid in the highest amount of 3.51 mg/100 g (Viuda-Martos et al., 2015). Leaves and twig of thirteen fig varieties have also been compared for their phenolic acids content, HPLC analysis identified caffeoylmalic as a predominant phenolic acid in leaves ranging from 0.79 to 5.97 mg/g DW, 3-*O*-caffeoylquinic acid (leaves = 1.31–3.54 mg/

Table 15.1 Major phenolic acids in fig

Major compound	Extraction technique	Plant part	Reference
Chlorogenic acid[a]	Shaking	Latex	Abdel-Aty et al. (2019)
Chlorogenic acid[b]	Ultrasonication	Fruit	Veberic et al. (2008)
Protocatechuic acid[a]		Peels	Viuda-Martos et al. (2015)
Protocatechuic acid[a]		Pulp	
Gallic acid[a]	Sonication followed by maceration	peels	Hssaini et al. (2021)
Chlorogenic acid[a]		pulp	
Chlorogenic acid[b]	Maceration	Leaves	Ladhari et al. (2020)
Dihydroxy benzoic acid[b]		Twig	

[a]80% methanol as solvent
[b]Methanol as solvent

DW, Twig = 0.97–1.62 mg/g DW), 5-*O*-caffeoylquinic acid (leaves = 0.34–0.59 mg/g DW, twig = 0.16–0.36 mg/g DW), dihydroxy benzoic acid(leaves = 1.15–2.15 mg/g DW, twig = 0.94–2.31 mg/g DW) and caffeoylmalic acid (twig = 0.75–3.66 mg/g DW) (Ladhari et al., 2020).

3 Flavonoid Content and Common Flavonoids

Flavonoid content estimation of skins, pulp, and fruit has shown a varied range of TFC depending upon the part of fig understudy, just as TPC. For example, dark-colored varieties of fig are reported to have higher flavonoid content as compared to light-colored varieties; however, skins of all varieties have shown higher flavonoid content (ranged from 2.2 to 45.6 mg catechin/100 g FW) as compared to the pulp (1.6–5.7 mg CE/100 g FW) and whole fruit (2.1–21.5 mg CE/100 g FW) (Solomon et al., 2006).

Another study has reported higher TFC in three dark fig varieties (Bouankik, Azandjar, and Aberkane) with a mean value of 126.55 mg quercetin equivalent/100 g of dry mass as compared to six light varieties, which showed a mean TFC value of 87.24 mg QE/100 g DM (Bey & Louaileche, 2015). However, another study has reported higher TFC values for lighter varieties with an average TFC of 37.71 mg CE/100 g DW, Kadota variety showing the highest TFC value of 83.71 mg CE/100 g DW while white Adriatic is showing least TFC value of 14.59 mg CE/100 g DW (Hssaini et al., 2020a). Finally, a study showed the significant influence of steam explosion pretreatment on flavonoid content of fig leaves extract. Mild and drastic steam explosion pretreatment resulted in the flavonoid yield of 51.03 mg/g and 44.77 mg/g, respectively. In comparison, untreated leaves showed a flavonoid yield of 32.73 mg/g after 6 h extraction period (Qin & Chen, 2015).

TFC has shown a positive correlation between extraction temperature and the percentage of ethanol used for extraction. TFC increased from 0.68 CE/g DB to 2.5 CE/g DB using temperatures from 25 °C to 80 °C; likewise, TFC was observed from 0.44 mg CE/g DB to 2.5 CE/g DB using 50% to 80% ethanol, respectively (Bucic-Kojic et al., 2011). The type of extraction solvent also affects the flavonoid content; ethyl acetate fraction of leaves extract has higher flavonoid content of 83.92 mg/g than water, and petroleum ether extract petroleum-ether gave the least TFC value of 18.71 mg/g (Li et al., 2021). Using an aqueous methanolic solvent system has resulted in a high recovery of flavonoids. A study has shown a high TFC value of 538.20% w/w using aqueous methanol in 70:30 (Alamgeer et al., 2017).

Rutin is a major phytoconstituent of fig (Table 15.2). HPLC analysis of white and dark cultivars from July and September has identified rutin as a major flavonoid with a content range from 4.89 to 28.7 mg/100 g FW, followed by catechin, which showed full content of 4.03 mg/100 g FW(Veberic et al., 2008). Methanol extract of fig latex also showed a similar trend, with rutin being in the highest percentage of 20%, followed by catechin, which showed a percentage composition of 8%; however, quercetin and apigenin were not detected in latex at all (Abdel-Aty et al., 2019).

Table 15.2 Major flavonoids compounds in fig

Major compound	Extraction technique	Plant part	Reference
Rutin[a]	Shaking	Latex	Abdel-Aty et al. (2019)
Rutin[b]	Ultrasonication	Fruit	Veberic et al. (2008)
Epichtechin[a]	Sonication followed by maceration	Pulp	Hssaini et al. (2021)
Rutin[a]		Peels	
Rutin[a]	Ultrasonication	Peels	Viuda-Martos et al. (2015)
Catechin[a]		Pulp	
Rutin[b]	Maceration	Leaves	Ladhari et al. (2020)
Rutin[b]		Twig	
Rutin[c]	Hot plate stirring	leaves	Belguith-Hadriche et al. (2017)

[a]80% methanol as solvent
[b]Methanol as solvent
[c]70% ethanol as solvent

Another study conducted on 25 different cultivars has shown epicatechin to be a significant flavonoid in the pulp of all cultivars, with a mean value of 5.23 μg/g DW. However, its content was most prominent in the darkest fig cultivar, where it showed a high content value, up to 54.66 μg/g DW; peels have also shown a considerable amount of epicatechin; other than this, catechin was present in both peels and pulp; however, quercetin and apigenin were not detected in the pulp (Hssaini et al., 2021). The absence of quercetin and apigenin accords with the study on methanolic extract of latex (Abdel-Aty et al., 2019). However, another study has shown the presence of apigenin along with other flavonoids in peels and pulp of powder co-product of Calor and Cruella dama fig varieties. Rutin was found to be the most abundant flavonoid in the peels of both Calor and Cruella dama varieties (15.3 and 14.2 mg/100 g sample, respectively), followed by catechin, which showed the content value of 6.10 and 14.20 mg/100 g in the peels of Calor and Cruella dama respectively. Interestingly catechin was the most abundant flavonoid in pulp (9.53 mg/100 g). Other than these two compounds, luteolin, apigenin, luteolin-7-*O*-glucoside, and epicatechin were also present in peels in considerable amounts; however, luteolin-7-*O*-glucoside was not detected in the pulp of both varieties (Viuda-Martos et al., 2015).

UPLC-QTOF-MS analysis has identified 11 flavonoid glycosides in leave extract obtained by ultrasound-enzyme assisted two-phase aqueous extraction. Only one compound has already been identified in leave extract; other newly identified compounds include quercetin glycosides, apigenin, kaempferol, luteolin, and napthohydroquinone (Zhao et al., 2021). Study of leaves and twig of 13 fig varieties has identified three flavonoid glycosides including quercetin-3-*O*-glucoside (leaves = 4.73–7.51 mg/g DW, twing = 2.49–6.45 mg/g DW), quercetin-3-*O*-rutinoside, (leaves = 12.43-22.70 mg/g DW, twig = 8.14–17.23 mg/g DW) and kaempferol-3-*O*-glucoside (leaves = 5.36–12.45 mg/g DW, twig = 637–10.23 mg/g DW) (Ladhari et al., 2020). RP-LC-DAD-QTOF-MS analysis of leaves extract of two Tunisian cultivars has identified thirteen flavonoids molecules of different subclasses, including flavonols (rutin and isoquercetin), flavones (luteolin C-hexoside C-pentoside I, luteolin C-hexoside C-pentoxide II, Apigenin C-hexoside C-pentoside I, Apigenin C-hexoside C-pentoxide II, vitexin, luteolin) and isoflavones (cajanin, prenylhydroxygenistein I, prenylhydroxygenistein II, prenylgenistein I, prenylgenistein II) (Belguith-Hadriche et al., 2017).

4 Anthocyanin

Anthocyanin content in fig also varied between different varieties as a study showed, anthocyanin content varied in the range of 0–42 μ cyanidine-3-rutinoside/g FW in different genotypes and cultivars of figs. It also found that the presence of anthocyanins influenced the color of figs, as black figs showed higher anthocyanin content than purple or green colored figs (Ercisli et al., 2012). Another study carried out on nine different fig varieties has also indicated higher anthocyanin content in dark varieties with the highest anthocyanin content in the Bouankik variety, which showed anthocyanin content of 20.78 mg quercetin-3-glucoside equivalent/100 g DM followed by Aberkane (18.73 mg Q3GE/100 g DM) and Azandjar (17.18 mg Q3GE/100 g DM), this content range is much higher than that was observed for light varieties which showed the mean content of 4.9 mg Q3GE/100 g DM (Bey & Louaileche, 2015). Hssaini reported similar results, where black-colored varieties Noukali, Fassi, and Ghoudan showed high anthocyanin content of 23.77, 14.88 11.16 mg Cy-3-rutinoside equivalent/100 g DW, respectively. Anthocyanin content comparison in skin and pulp of five varieties; Colar, Cuello de Dama (green and dark purple), Granilla, and Bursa, have shown anthocyanin content to be several times higher in the skin (range of 32–97 μg/g FW) with highest anthocyanin content in the skin of Bursa siyahi variety of fig as compared to a pulp (1.5–15 μg/g FW) where the pulp of Colar variety showed highest AC (Dueñas et al., 2008). This study is followed by the findings of Solomon et al. (2006), where the skin extract samples of fig showed the highest range of anthocyanin content (0.7–27.3 mg cyanidin-3-glucoside/100 g), followed by fruit extract(0.3–10.9 mg cyanidin-3-glucoside/100 g). Another study has shown the highest AC in peels of purple variety, 196 mg Cy-3-G/100 g DW; however, drying of fig can cause loss of anthocyanin content; according to a study, the anthocyanin content decreased up to 83 and 98% upon drying in purple and yellow fig variety respectively. AC ranged from 0.8 to 5.9 mg Cy-3-G/100 g DW in different fig parts of the yellow variety, which decreased to 0.1 mg Cy-3-G/100 g DW upon drying, while in the purple variety, the AC ranged from 38 mg to 196 Cy-3-G/100 g DW that declined to 14.5 mg Cy-3-G/100 g DW (Kamiloglu & Capanoglu, 2015).

HPLC-DAD-MS analysis of different fig varieties comprehensively profiled the anthocyanin composition of fig. Primary aglycon moiety was reported to be cyanidin, and glycon substituents were found to be rutinose and glucose. Cyanidin-3-rutinoside was the most abundant anthocyanin in both skin and pulp, constituting

48–81% of total anthocyanins in the skin and 68–79% of total anthocyanins in the pulp. Cyanidin-3-glucoside was the second prominent anthocyanin after cyanidin-3-rutinoside (Dueñas et al., 2008). These results are per a study conducted on twenty-five fig varieties by Hssaini et al. (2021), which also identified cyanidin-3-rutinoside as the most abundant anthocyanin in both plant parts, i.e., peel and pulp; however, higher content of cyanidin-3-rutinoside (77.97 μg/g DW) was present in peels of as compared to the pulp (9.01 μg/g DW) other than cyanidin-3-rutinoside, cyanidin-3,5-diglcoside was also present in high amount in peels (75.90 μg/g DW). It also found that the darkest colored cultivar provided the most promising anthocyanin profile with high content values of anthocyanin compounds, most prominently cyanidin-3-rutinoside and cyanidin-3, 5-diglcoside with the content value of 478.6 μg/g DW and 494.0 μg/g DW, respectively. Another study conducted on powder co-product of Calor and Cuello dama fig varieties has shown higher cynaidin-3-rutinoside and cyanidin-3-glucoside (2.01 and 1.33 mg/100 g respectively) in peels of Calor variety as compared to Cuello dama variety (0.43 and 0.12 mg/100 g respectively), however, both of these significant anthocyanidins were not detected in the pulp of both varieties (Viuda-Martos et al., 2015). Cyanidin-3-rutinoside and cyanidin-3-glucoside accumulation in the skin highly impact the fruit color; the accumulation of these compounds could be enhanced by the treatment of exogenous abscisic acid and ethephon during fruit ripening. Cyanidin-3-*O*-glucoside increased up to 31.66 and 12.35 μg/g FW after 72 h of exogenous ABA and ethephon treatment, respectively, compared to control (5.47 μg/g FW). Cyanidin-3-*O*-rutinoside reached the level of 10.39 μg/g FW after 72 h of exogenous-ABA treatment, while ethephon treatment showed cyaniding-3-*O*-rutinoside content of 4.84 μg/g FW as compared to control (2.45 μg/g FW) (Lama et al., 2020).

5 Proanthocyanidins

Proanthocyanidins are also a prominent class of bioactive compounds present in figs, and a research study has shown high proanthocyanidin content of 40 mg/g in a fig variety of Spain (Vallejo et al., 2012). Furthermore, the proanthocyanidin content of fig (*Ficus carica*) was reported to be even higher (496.6 mg CE/100 g DW) than *Ceratonia siliqua* (140 mg CE/100 g DW) and *Quercus ilex* (254.5 mg CE/100 g DW) (Amessis-Ouchemoukh et al., 2017).

Proanthocynidin content varies between different colored varieties, just as anthocyanins; when dark and light varieties were compared, Aberkane (a dark variety showed the highest proanthocyanidin content of (179.8 mg cyanidin equivalent/100 g dry mass); however, Abiarous (a light-colored variety) showed lowest proanthocyanidin content of 103.3 mg cyanidin equivalent/100 g DM (Bey & Louaileche, 2015). However, another study has shown that light color varieties contained higher proanthocyanidin content than dark varieties; Cuello Dama Blanca and Snowden varieties showed the highest proanthocyanin content of 6.64 cyanidin equivalent/100 g DW, respectively (Hssaini et al., 2020a).

Like anthocyanins, drying of figs is also reported to impact proanthocyanidin content. The purple fig variety was found to have higher proanthocyanidin content than the yellow variety, but drying decreased the content up to 74% in purple fig. Conversely, an increase of proanthocyanin content up to 71% was witnessed in yellow varieties upon drying. In yellow varieties, the content in different parts varied from 6.6 to 7.2 mg CyE/100 g DW, which increased to 12 CyE/100 g DW in dried fig. Similarly, in the purple fig, the content varied from 34 to 220 CyE/100 g DW, which upon drying declined to 16 CyE/100 g DW (Kamiloglu & Capanoglu, 2015).

6 Coumarins

Psoralen and bergapten are major coumarins that have been detected in the aqueous methanolic extract of fig leaves (Teixeira et al., 2006). An HPLC-UV analysis conducted on leaves, pulp, and peels of fig also has shown the highest psoralen and bergapten content in leaves (67.8 µg/0.1 g sample and 20.2 µg/0.1 g sample, respectively), followed by pulp (psoralen, 5.56 and bergapten, 2.49) however peels showed the lowest content (psoralen, 2.59 µg/0.1 g sample and bergapten 1.99 µg/0.1 g sample) (Hongying, 2015).

Bark and wood of fig plant have reported high levels of psoralen and bergapten, and psoralen ranged from 146.6–1110.3 µg/g DW in bark and 395.7–1671.8 µg/g DW in wood, while bergapten was lower in concentration as compared to psoralen, ranging from 144.2 to 718.6 µg/g DW in wood while 114.3 to 524.0 µg/g DW in the bark.(Rouaiguia-Bouakkaz et al., 2013) Other than psoralen and bergapten, 8-methoxypsoralen, angelicin, rutaretin, pimpinellin, and seselin have also been detected in the Dottato variety of fig. Psoralen was predominant coumarin, just as in other studies (23% peak area in GC-FID); however, it was followed by seselin in its content value (19.50% peak area) rather than bergapten (15.20% peak area). Extraction solvent affected the recovery of coumarins as *n*-hexane provided better recovery than dichloromethane (Marrelli et al., 2012).

RP-LC-DAD-QTOF-MS of leaves extract of Temri, and Tounsi varieties of fig have shown three hydroxycoumarins including Phellodenol A(hydrated 4′,5′-dihydropsoralen), Murrayacarpin B/dihydrated bergapten) and prenyl-7-hydroxycoumarin and two furanocoumarins hydroxylpsoralen hexoside and prenyl methoxypsoralen (Belguith-Hadriche et al., 2017). GC-MS profile of n-hexane extract of latex of caprifig (wild fig variety) has shown coumarins to be the major constituents of extracts; fourteen coumarins were identified, constituting 90.56% of total identified compounds (Lazreg-Aref et al., 2012).

7 Tannins

Tannins are complex bioactive compounds derived from phenolic acids; a study carried out on three fig varieties has shown higher total tannin content in green varieties (Bidhi) as compared to purple (Hamri) and black varieties (Kohli). Bidhi peels contained the highest full tannin content of 75.98 mg tannic acid equivalent/g FW, followed by pulp (65.87 mg TAE/g FW), while whole fruit represented TTC of 60.35 mg TAE/g FW. Peels of the Kohli variety also contained more tannins than pulp and whole fruit; however, the Hamri variety showed the highest concentration of tannins in the entire fruit (Harzallah et al., 2016).

Studies have shown that Taamriout and Azendjar varieties are rich in condensed tannins compared to hydrolyzable tannins. Methanol extracts have shown the highest TTC ranging from 160 to 194 mg GAE/100 g DW. In comparison, petroleum ether extract of Azendjar has shown the most negligible TTC value (below 100 mg GAE/100 g DW) (Debib et al., 2014). In another study carried out on the estimation of condensed tannin, peels of Bakkor Khal and pulp of Zarrouk showed the highest condensed tannin content of 18.4 and 4.4 μg catechol equivalent/g FW, respectively; however, the Bither variety showed the lowest CT content range of 0.388–0.607 μg CE/g FW) (Mahmoudi et al., 2018).

8 Carotenoids

Carotenoid's presence is also identified in fig fruit; a study has identified three carotenoids in fig fruit, including β-carotene, α-carotene derivative, and β-cryptoxanthin. β-carotene was present in the highest amount of 4.32 μg/100 g DE followed by β-cryptoxanthin (2.14 μg/100 g DE) (Amessis-Ouchemoukh et al., 2017).

Yellow fig varieties have reported four carotenoids: lutein, zaexanthin, β-cryptoxanthin, and β-carotene. Upon comparing fresh samples, the Sarilop variety showed higher total carotenoid content (29.6 μg lutein equivalent/g DW) as compared to the Sarizeybek variety (23.7 μg LE/g DW); however, lutein content was higher in Sarizeybek (7.15 μg/g DW) than Sarilop variety (6.14 μg/g DW). It also proved that fig drying negatively affects the total carotenoid content as TCC decreased up to 80% upon seven-day drying (Yemis et al., 2012).

A study reported the effect of storage duration on total carotenoid content; it was shown that Peels of the Dottato variety had higher total carotenoid content (17.8 μg/g) as compared to Melanzana at five day storage time. On the other hand, Melanzana showed an evident decline in leaves TCC over the storage period of 10 days. However, carotenoids in the pulp of both varieties were stable upon storage even for 10 days showing a total carotenoid content of approximately 3.0 μg/g (Allegra et al., 2017).

Carotenoid content tends to be different in hand-pollinated and parthenocarpic figs, which decreases during the maturation of fig. In pollinated figs, the carotenoid content decreased from 18.39 to 7.71 μg/g FW, while in parthenocarpic figs, the carotenoid accumulation reduced from 16.62 to 6.64 μg/g FW. Individual carotenoids were found to be lutein, b-carotene, zeaxanthin, and violaxanthin, lutein being the major carotenoid with the content value of 9.63 μg/g FW at 0 weeks of pollination, which increased to 12.58 μg/g FW after 1 week and decreased to 4.43 ug/g FW after 5-week maturation (Nawade et al., 2020).

9 Terpenes and Other Volatile Compounds

GC-MS analysis of fig leaves essential oil has shown the presence of 66 volatile compounds, among which the oxygenated compounds were most the abundant, constituting the 46.3% of volatile compounds composition, followed by aliphatic hydrocarbons (28.2%), fatty acids (9.9%) and sesquiterpenes (3.1%) however monoterpenes were least abundant exhibiting 0.6% of the total composition. The prominent compounds belonging to these classes included oxygenated compounds; (*Z*)-3-hexeny benzoate, (*E*)-2-hexenal, phytol and *n*-nonanal, aliphatic hydrocarbons; *n*-tetracosane and *n*-docosane, sesquiterpene; (*E, E*)-α-farnesene and *cis*-muurola-4(15),5-diene, monoterpene; α-pinene (only monoterpene in understudy oil), fatty acid; *n*-hexadecanoic acid (Ayoub et al., 2010).

Fig fruits samples from different localities have been studied for their flavoring agents, and primary flavoring agents included *p*-cymene, decanol, carvacrol, tricosane, tetracosane, pentacosane, heptacosane, octacosane, and nonacosane. It also found that geographical location influences the volatile composition of figs (Darjazi & Larijani, 2012).

Solid-phase micro extraction-GC-MS has been used for identification and quantification of volatile compounds from fructus (fruit) and folium (leaves) of fig; fig fructus has shown the presence of only one volatile compound, 2,-butanediol by using the SPME technique; however, SPME-GC-MS has demonstrated the presence of tetramethyl-decane, trimethyl-undecane, octadecane, carvacrol, caryophyllene oxide, and β-caryophyllene. SPME detection of folium has shown 3, 5-octadiene-2-on, trimethyl-methylene-vinibicyclononane, hexanal, undecane, dimethyl-undecane, and ethyl-cyclohexanol; however, folium steam distillate detected by SPME-GC-MS has identified methyl-undecane, dimethyl-undecane, nanodecane, β-caryophyllene, caryophyllene-oxide and a trace amount of β-ionone (Ficsor et al., 2012).

Hydrodistillate of fig fruit of six different cultivars (Bither Abiadh, Bidi, Bither Kholi, Himri, Kholi, and Tchich Asal) have shown higher cedrol (43.8%) and α-pinene (9.3%) as compared to leave distillate which showed the percentage of 38.9 and 6.1% for cedrol and α-pinene respectively, however many oxide (24.8%) and abietatriene (11.8%) was higher in leaves than fruits (12.9 and 8.1% of manoyl oxide and abietatriene respectively). Other than these compounds, leaves hydrodistillate also showed the presence of α-terpineol acetate (21.7%), γ-muurolene (7.4%), pentadecanal (5.2%), and nonadecanal (2.3%) however, fruits showed the presence of α-terpinyl acetate (22.5%), *tans*-calamenene (3.9%) and *n*-heneicosane (3.5%) (Soltana et al., 2017).

GC-MS-Olfactometric analysis of fig leaves pulp has identified almost 58 aroma compounds collectively in peels and pulp of Sarilop and Bursa Siyahi varieties of fig. That composition mainly comprised this composition of terpenes followed by ester and alcohols. The highest terpenes content was seen in peels of Sarilop variety that showed the content value of 7691 μg/kg FW; however, among 21 individually detected terpenes DL-limonene showed the highest content, ranging from 441 to 3320 μg/kg FW, followed by β-caryophyllene that delivered the content range of 556–2757 μg/kg FW. Other terpenes include D-3-carene, *p*-xylene, γ-terpinene, *o*-cymene, citronellal, α-cubebene, α-copaene, linalool, α-guaiene, humulene, germacrene, alloaromadendrene, β-bisbolene, α-farnesene, (*Z*)-α-bisabolene, caryophyllene oxide, and carvacrol (Sertkaya et al., 2021).

GC-MS analysis of *n*-hexane extract of caprifig also identified eight terpene compounds constituting 4.81% of the total identified compounds. Among these terpenes, sesquiterpenes were most abundant, with a percentage composition of 2.83% (germacrene-D, delta-cadiene, 2,6,10-Dodecatrien-1-ol), followed by two oxygenated triterpenes (0.74%) and two triterpenes (0.46%); however, bornanone-3 was the only identified monoterpene present in the extract (Lazreg-Aref et al., 2012). β-bourbonene, α-guaene, α-muurolene, τ-muurolene, τ-cadinene, α-cubenene, β-cubenene, menthol, 1-penten-3-ol, β-pinene, hydroxyl caryophyllene, and eucalyptol are terpenoids derived from leaves, fruit, and bark of fig which have been studied with the help of molecular docking and dynamic simulation for understanding the anticancer activity of the plant. Among all mentioned compounds, β-bourbonene, a sesquiterpenoid, has shown promising results (Gurung et al., 2021).

10 Fatty Acids and Phytosterols

Latex of fig has been reported as a source of fatty acids. Saturated fatty acids majorly contribute to latex's total fatty acid profile, constituting approx. 86% of it, arachidic being the most abundant saturated fatty acid (44.1%), followed by palmitic acid (21.4%) and behenic acid (13.1%) (Oliveira et al., 2010).

GC analysis of seeds oil of four fig varieties has shown a high amount of linolenic (38.4–43.5%) followed by linoleic acid (28.9–34.5%), both of which are polyunsaturated fatty acids. Among saturated fatty acids, palmitic acid was predominant, showing a content range from 8.54 to 9.05%, followed by stearic acid, which showed a content range from 2.59 to 3.3% (Hssaini et al., 2020b). Not only seeds and latex fig leaves are also found to be a source of fatty acids; however, the content percentage is low (18% of total ion current, identified by GC-MS), and leaves extracts analysis has shown the presence of myristic acid, palmitic acid, margaric acid, linoleic acid, α-linolenic acid, stearic acid, arachidic acid, behenic acid, lignoceric acid, and cerotic acid. Palmitic and stearic acids were predominant among saturated acids, showing content from 4% to 14% and 1% to 4% of TIC, respectively. In comparison, primary polyunsaturated acids were found to be α-linolenic acid (5–13.8% of TIC) and linoleic acid (1.8–3.9% of TIC) (Ivanov et al., 2018).

The fig fruit has also shown 20 fatty acids collected during June, July, and September, including methyl ester of hexanoic acid, the ethyl ester of (*E*)-2-heptenoic acid, caprylic acid ethyl ester, decanoic acid, pentadecanoic acid, ethyl ester, palmitic acid methyl ester, palmitic acid, myristic acid, palmitoleic acid ethyl ester, palmitic acid ethyl ester, 14-hexadecanoic acid methyl ester, pentadecanoic acid, margaric acid ethyl ester, linoleic acid methyl ester, stearic acid methyl ester, stearic acid, oleic acid ethyl ester, linoleic acid ethyl ester and stearic acid ethyl ester (Marrelli et al., 2012).

Phytosterols are also reported in fig latex extracts. GC-ITMS and HPLC analysis of fig latex has identified seven phytosterols, including b-amyrin, a-amyrin, b-sitosterol, lupeol, lupeol acetate, betulol, and lanosterol. Among these identified compounds, b-sitosterol was the most abundant phytosterol representing 54% of the total estimated phytosterols. In contrast, the least abundant phytosterol was a-amyrin, constituting only 0.4% of total phytosterols (Oliveira et al., 2010). GC-MS analysis of leaves extracts of fig has also identified the presence of phytosterols (18% of total ion current), including stigmasterol, β-sitosterol, lanosterol, and cycloartenol. Among all reported phytosterols, β-sitosterol was found to be the most abundant (11–14% of TIC), followed by cycloatenol acetate (5–9% of TIC) (Ivanov et al., 2018). Fig achenes oils have been investigating the presence of phytosterols; according to a study, fig achene oil of the kholi variety has shown a high phytosterol content of 1061 mg/100 g, where stigmastadienol was identified as the most abundant phytosterol with the content value of 73.7% (Soltana et al., 2016).

11 Vitamins and Other Compounds

α-Tocopherol is a kind of vitamin E reported in fig by many studies; according to a study, n-hexane leaves extract of fig contains α-tocopherol in a high amount of 57 mg/g dry leaves (Konyalıoğlu et al., 2005). Ascorbic has also been identified in fig leaves extract in 21.7 to 22.4 mg ascorbic acid/100 g DW. In the same study, vitamin E content showed variation according to the size of the leaves; large leaves showed slightly higher vitamin E content (1.9 mg vitamin E/100 g DW) as compared to small leaves (1.8 mg vitamin E/100 g DW) (Ghazi et al., 2012).

Turkish variety of figs has shown a diverse range of vitamins; however, the content varied in different sections of figs. γ-Tocopherol, δ-tocopherol, vitamin D2, vitamin D3, α-tocopherol, α-tocopherol acetate (acetate), and vitamin K1were identified in fruit, peel, red and white sections of fig. γ-tocopherol was predominant vitamin showing the highest content in the skin followed by a white section of fig (Guvenc et al., 2009).

Vitamin figs have also shown the presence of crude alkaloids (9.6 g/100 g DM) and saponins (0.59 g/100 g DM) in the edible portion; however, not much data is available in this instance (Soni et al., 2014).

References

Abdel-Aty, A. M., Hamed, M. B., Salama, W. H., Ali, M. M., Fahmy, A. S., & Mohamed, S. A. (2019). *Ficus carica, Ficus sycomorus*, and *Euphorbia tirucalli* latex extracts: Phytochemical screening, antioxidant and cytotoxic properties. *Biocatalysis and Agricultural Biotechnology, 20*, 101199. https://doi.org/10.1016/j.bcab.2019.101199

Al Askari, G., Kahouadji, A., Khedid, K., Ouaffak, L., Mousaddak, M., Charof, R., & Mennane, Z. (2013). In vitro antimicrobial activity of aqueous and ethanolic extracts of leaves of *Ficus carica* collected from five different regions of Morocco. *Journal of Materials and Environmental Science, 4*(1), 33–38.

Alamgeer, I. S., Asif, H., & Saleem, M. (2017). Evaluation of antihypertensive potential of *Ficus carica* fruit. *Pharmaceutical Biology, 55*(1), 1047–1053. (1744–5116, Electronic).

Allegra, A., Alfeo, V., Gallotta, A., & Todaro, A. (2017). Nutraceutical content in 'Melanzana' and 'Dottato' fig fruit (*Ficus carica* L.). *Acta Horticulturae*, (1173), 319–322. https://doi.org/10.17660/ActaHortic.2017.1173.55

Amessis-Ouchemoukh, N., Ouchemoukh, S., Meziant, N., Idiri, Y., Hernanz, D., Stinco, C. M., et al. (2017). Bioactive metabolites involved in the antioxidant, anticancer and anticalpain activities of *Ficus carica* L., *Ceratonia siliqua* L. and *Quercus ilex* L. extracts. *Industrial Crops and Products, 95*, 6–17. https://doi.org/10.1016/j.indcrop.2016.10.007

Ayoub, N., Singab, A. N., Mostafa, N., & Schultze, W. (2010). Volatile constituents of leaves of *Ficus carica* Linn. grown in Egypt. *Journal of Essential Oil Bearing Plants, 13*(3), 316–321. https://doi.org/10.1080/0972060X.2010.10643827

Bachir Bey, M., Meziant, L., Benchikh, Y., & Louaileche, H. (2014). Deployment of response surface methodology to optimize recovery of dark fresh fig (*Ficus carica* L., var. Azenjar) total phenolic compounds and antioxidant activity. *Food Chemistry, 162*, 277–282. https://doi.org/10.1016/j.foodchem.2014.04.054

Belguith-Hadriche, O., Ammar, S., Contreras, M. D. M., Fetoui, H., Segura-Carretero, A., El Feki, A., & Bouaziz, M. (2017). HPLC-DAD-QTOF-MS profiling of phenolics from leaf extracts of two Tunisian fig cultivars: Potential as a functional food. *Biomedicine & Pharmacotherapie, 89*, 185–193. https://doi.org/10.1016/j.biopha.2017.02.004

Bey, M. B., & Louaileche, H. (2015). A comparative study of phytochemical profile and in vitro antioxidant activities of dark and light dried fig (*Ficus carica* L.) varieties. *The Journal of Phytopharmacology, 4*(1), 41–48.

Bucic-Kojic, A., Planinic, M., Tomas, S., Jokic, S., Mujic, I., Bilic, M., & Velic, D. (2011). Effect of extraction conditions on the extractability of phenolic compounds from lyophilised fig fruits (*Ficus carica* L.). *Polish Journal of Food and Nutrition Sciences, 61*(3).

Buenrostro-Figueroa, J. J., Velázquez, M., Flores-Ortega, O., Ascacio-Valdés, J. A., Huerta-Ochoa, S., Aguilar, C. N., & Prado-Barragán, L. A. (2017). Solid state fermentation of fig (*Ficus carica* L.) by-products using fungi to obtain phenolic compounds with antioxidant activity and qualitative evaluation of phenolics obtained. *Process Biochemistry, 62*, 16–23. https://doi.org/10.1016/j.procbio.2017.07.016

Darjazi, B. B., & Larijani, K. (2012). The effects of climatic conditions and geographical locations on the volatile flavor compounds of fig (*Ficus carica* L.) fruit from Iran. *African Journal of Biotechnology, 11*, 9196–9204.

Debib, A., Tir-Touil, A., Mothana, R. A., Meddah, B., & Sonnet, P. (2014). Phenolic content, antioxidant and antimicrobial activities of two fruit varieties of Algerian *Ficus carica* L. *Journal of Food Biochemistry, 38*(2), 207–215. https://doi.org/10.1111/jfbc.12039

Dueñas, M., Pérez-Alonso, J. J., Santos-Buelga, C., & Escribano-Bailón, T. (2008). Anthocyanin composition in fig (*Ficus carica* L.). *Journal of Food Composition and Analysis, 21*(2), 107–115. https://doi.org/10.1016/j.jfca.2007.09.002

Ercisli, S., Tosun, M., Karlidag, H., Dzubur, A., Hadziabulic, S., & Aliman, Y. (2012). Color and antioxidant characteristics of some fresh fig (*Ficus carica* L.) genotypes from northeastern Turkey. *Plant Foods for Human Nutrition, 67*(3), 271–276. https://doi.org/10.1007/s11130-012-0292-2

Ficsor, E., Szentmihályi, K., Lemberkovics, É., Blázovics, A., & Balázs, A. (2012). Analysis of Ficus carica l. – Volatile components and mineral content. *European Chemical Bulletin, 2*(3), 126–129.

Ghazi, F., Rahmat, A., Yassin, Z., Ramli, N. S., & Buslima, N. A. (2012). Determination of total polyphenols and nutritional composition of two different types of *Ficus carica* leaves cultivated in Saudi Arabia. *Pakistan Journal of Nutrition, 11*(11), 1061–1065. https://doi.org/10.3923/pjn.2012.1061.1065

Gond, N. Y., & Khadabadi, S. S. (2008). Hepatoprotective activity of Ficus carica leaf extract on rifampicin-induced hepatic damage in rats. *Indian Journal of Pharmaceutical Sciences, 70*(3), 364–366. https://doi.org/10.4103/0250-474X.43003

Gurung, A. B., Ali, M. A., Lee, J., Farah, M. A., & Al-Anazi, K. M. (2021). Molecular docking and dynamics simulation study of bioactive compounds from *Ficus carica* L. with important anticancer drug targets. *PLoS One, 16*(7), e0254035. https://doi.org/10.1371/journal.pone.0254035

Guvenc, M., Tuzcu, M., & Yilmaz, O. (2009). Analysis of fatty acid and some lipophilic vitamins found in the fruits of the *Ficus carica* variety picked from the Adiyaman District. *Research Journal of Biological Sciences, 4*(3), 320–323.

Hamedi, A., Farjadian, S., & Karami, M. R. (2015). Immunomodulatory properties of Trehala manna decoction and its isolated carbohydrate macromolecules. *Journal of Ethnopharmacology, 162*, 121–126. https://doi.org/10.1016/j.jep.2014.12.064

Harzallah, A., Bhouri, A. M., Amri, Z., Soltana, H., & Hammami, M. (2016). Phytochemical content and antioxidant activity of different fruit parts juices of three figs (*Ficus carica* L.) varieties grown in Tunisia. *Industrial Crops and Products, 83*, 255–267. https://doi.org/10.1016/j.indcrop.2015.12.043

Hongying, Q. (2015). Application of ionic liquid-based ultrasound-assisted extraction of five phenolic compounds from fig (*Ficus carica* L.) for HPLC-UV. *Food Analytical Methods, 8*(7), 1673–1681. https://doi.org/10.1007/s12161-014-0047-9

Hssaini, L., Charafi, J., Razouk, R., Hernández, F., Fauconnier, M. L., Ennahli, S., & Hanine, H. (2020a). Assessment of morphological traits and fruit metabolites in eleven fig varieties (*Ficus carica* L.). *International Journal of Fruit Science, 20*, 8–28. https://doi.org/10.1080/15538362.2019.1701615

Hssaini, L., Hanine, H., Charafi, J., Razouk, R., Elantari, A., Ennahli, S., et al. (2020b). First report on fatty acids composition, total phenolics and antioxidant activity in seeds oil of four fig cultivars (*Ficus carica* L.) grown in Morocco. *OCL, 27*, 8. https://doi.org/10.1051/ocl/2020003

Hssaini, L., Hernandez, F., Viuda-Martos, M., Charafi, J., Razouk, R., Houmanat, K., et al. (2021). Survey of phenolic acids, flavonoids and in vitro antioxidant potency between fig peels and pulps: Chemical and chemometric approach. *Molecules, 26*(9), 2574.

Ivanov, I., Dincheva, I., Badjakov, I., Petkova, N., Denev, P., & Pavlov, A. (2018). GC-MS analysis of unpolar fraction from *Ficus carica* L. (fig) leaves. *International Food Research Journal, 25*.

Kamiloglu, S., & Capanoglu, E. (2015). Polyphenol content in figs (*Ficus carica* L.): Effect of sun-drying. *International Journal of Food Properties, 18*(3), 521–535. https://doi.org/10.1080/10942912.2013.833522

Konyalıoğlu, S., Sağlam, H., & Kıvçak, B. (2005). α-Tocopherol, flavonoid, and phenol contents and antioxidant activity of *Ficus carica*. leaves. *Pharmaceutical Biology, 43*(8), 683–686. https://doi.org/10.1080/13880200500383538

Ladhari, A., Gaaliche, B., Zarrelli, A., Ghannem, M., & Ben Mimoun, M. (2020). Allelopathic potential and phenolic allelochemicals discrepancies in *Ficus carica* L. cultivars. *South African Journal of Botany, 130*, 30–44. https://doi.org/10.1016/j.sajb.2019.11.026

Lama, K., Harlev, G., Shafran, H., Peer, R., & Flaishman, M. A. (2020). Anthocyanin accumulation is initiated by abscisic acid to enhance fruit color during fig (*Ficus carica* L.) ripening. *Journal of Plant Physiology, 251*, 153192. https://doi.org/10.1016/j.jplph.2020.153192

Lazreg-Aref, H., Mars, M., Fekih, A., Aouni, M., & Said, K. (2012). Chemical composition and antibacterial activity of a hexane extract of Tunisian caprifig latex from the unripe fruit of Ficus carica. *Pharmaceutical Biology, 50*(4), 407–412. https://doi.org/10.3109/13880209.2011.608192

Li, J., Tian, Y.-Z., Sun, B.-Y., Yang, D., Chen, J.-P., & Men, Q.-M. (2012). Analysis on volatile constituents in leaves and fruits of Ficus carica by GC-MS. *Chinese Herbal Medicines, 4*(1), 63–69. https://doi.org/10.3969/j.issn.1674-6384.2012.01.010

Li, C., Yu, M., Li, S., Yang, X., Qiao, B., Shi, S., et al. (2021). Valorization of fig (*Ficus carica* L.) waste leaves: HPLC-QTOF-MS/MS-DPPH system for online screening and identification of antioxidant compounds. *Plants, 10*(11). https://doi.org/10.3390/plants10112532

Mahmoudi, S., Khali, M., Benkhaled, A., Boucetta, I., Dahmani, Y., Attallah, Z., & Belbraouet, S. (2018). Fresh figs (*Ficus carica* L.): Pomological characteristics, nutritional value, and phytochemical properties. *European Journal of Horticultural Science, 83*(2), 104–113. https://doi.org/10.17660/eJHS.2018/83.2.6

Marrelli, M., Menichini, F., Statti, G. A., Bonesi, M., Duez, P., Menichini, F., & Conforti, F. (2012). Changes in the phenolic and lipophilic composition, in the enzyme inhibition and antiproliferative activity of *Ficus carica* L. cultivar Dottato fruits during maturation. *Food and Chemical Toxicology, 50*(3), 726–733. https://doi.org/10.1016/j.fct.2011.12.025

Nawade, B., Shaltiel-Harpaz, L., Yahyaa, M., Bosamia, T. C., Kabaha, A., Kedoshim, R., et al. (2020). Analysis of apocarotenoid volatiles during the development of *Ficus carica* fruits and characterization of carotenoid cleavage dioxygenase genes. *Plant Science, 290*, 110292. https://doi.org/10.1016/j.plantsci.2019.110292

Oliveira, A. P., Silva, L. R., Andrade, P. B., Valentao, P., Silva, B. M., Goncalves, R. F., et al. (2010). Further insight into the latex metabolite profile of Ficus carica. *Journal of Agricultural and Food Chemistry, 58*(20), 10855–10863. https://doi.org/10.1021/jf1031185

Qin, L., & Chen, H. (2015). Enhancement of flavonoids extraction from fig leaf using steam explosion. *Industrial Crops and Products, 69*, 1–6. https://doi.org/10.1016/j.indcrop.2015.02.007

Rouaiguia-Bouakkaz, S., Amira-Guebailia, H., Rivière, C., Delaunay, J.-C., Waffo-Téguo, P., & Mérillon, J.-M. (2013). Identification and quantification of Furanocoumarins in stem bark and wood of eight Algerian varieties of *Ficus carica* by RP-HPLC-DAD and RP-HPLC-DAD-MS. *Natural Product Communications, 8*(4), 1934578X1300800420. https://doi.org/10.1177/1934578X1300800420

Sertkaya, M., Guclu, G., Buyukkurt, O. K., Kelebek, H., & Selli, S. (2021). GC-MS-olfactometric screening of potent aroma compounds in pulps and peels of two popular Turkish fig (*Ficus carica* L.) cultivars by application of aroma extract dilution analysis. *Food Analytical Methods, 14*(11), 2357–2366. https://doi.org/10.1007/s12161-021-02057-6

Shahinuzzaman, M., Yaakob, Z., Anuar, F. H., Akhtar, P., Kadir, N. H. A., Hasan, A. K. M., et al. (2020). In vitro antioxidant activity of *Ficus carica* L. latex from 18 different cultivars. *Scientific Reports, 10*(1), 10852. https://doi.org/10.1038/s41598-020-67765-1

Solomon, A., Golubowicz, S., Yablowicz, Z., Grossman, S., Bergman, M., Gottlieb, H. E., et al. (2006). Antioxidant activities and anthocyanin content of fresh fruits of common fig (*Ficus carica* L.). *Journal of Agricultural and Food Chemistry, 54*(20), 7717–7723. https://doi.org/10.1021/jf060497h

Soltana, H., Tekaya, M., Amri, Z., El-Gharbi, S., Nakbi, A., Harzallah, A., et al. (2016). Characterization of fig achenes' oil of *Ficus carica* grown in Tunisia. *Food Chemistry, 196*, 1125–1130. https://doi.org/10.1016/j.foodchem.2015.10.053

Soltana, H., Flamini, G., & Hammami, M. (2017). Volatile compounds from six varieties of *Ficus carica* from Tunisia. *Records of Natural Products, 11*(2), 157–165.

Soni, N., Mehta, S., Satpathy, G., & Gupta, R. K. (2014). Estimation of nutritional, phytochemical, antioxidant and antibacterial activity of dried fig (*Ficus carica*). *Journal of Pharmacognosy and Phytochemistry, 3*, 158–165.

Teixeira, D. M., Patao, R. F., Coelho, A. V., & da Costa, C. T. (2006). Comparison between sample disruption methods and solid-liquid extraction (SLE) to extract phenolic compounds from Ficus carica leaves. *Journal of Chromatography A, 1103*(1), 22–28. https://doi.org/10.1016/j.chroma.2005.11.047

Trifunschi, S., & Ardelean, D. (2013). Flavonoid extraction from Ficus carica leaves using different techniques and solvents. *Zbornik Matice Srpske za Prirodne Nauke*, 81–86. https://doi.org/10.2298/ZMSPN1325081T

Vallejo, F., Marín, J. G., & Tomás-Barberán, F. A. (2012). Phenolic compound content of fresh and dried figs (*Ficus carica* L.). *Food Chemistry, 130*(3), 485–492. https://doi.org/10.1016/j.foodchem.2011.07.032

Veberic, R., Colaric, M., & Stampar, F. (2008). Phenolic acids and flavonoids of fig fruit (*Ficus carica* L.) in the northern Mediterranean region. *Food Chemistry, 106*(1), 153–157. https://doi.org/10.1016/j.foodchem.2007.05.061

Viuda-Martos, M., Barber, X., Pérez-Álvarez, J. A., & Fernández-López, J. (2015). Assessment of chemical, physico-chemical, techno-functional and antioxidant properties of fig (*Ficus carica* L.) powder co-products. *Industrial Crops and Products, 69*, 472–479. https://doi.org/10.1016/j.indcrop.2015.03.005

Wang, T., Jiao, J., Gai, Q.-Y., Wang, P., Guo, N., Niu, L.-L., & Fu, Y.-J. (2017). Enhanced and green extraction polyphenols and furanocoumarins from Fig (*Ficus carica* L.) leaves using deep eutectic solvents. *Journal of Pharmaceutical and Biomedical Analysis, 145*, 339–345. https://doi.org/10.1016/j.jpba.2017.07.002

Yang, X. M., Yu, W., Ou, Z. P., Ma, H. L., Liu, W. M., & Ji, X. L. (2009). Antioxidant and immunity activity of water extract and crude polysaccharide from *Ficus carica* L. fruit. *Plant Foods for Human Nutrition, 64*(2), 167–173. https://doi.org/10.1007/s11130-009-0120-5

Yemis, O., Bakkalbaşı, E., & Artik, N. (2012). Changes in pigment profile and surface colour of fig (*Ficus carica* L.) during drying. *International Journal of Food Science & Technology, 47*. https://doi.org/10.1111/j.1365-2621.2012.03025.x

Yu, L., Meng, Y., Wang, Z.-L., Cao, L., Liu, C., Gao, M.-Z., et al. (2020). Sustainable and efficient surfactant-based microwave-assisted extraction of target polyphenols and furanocoumarins from fig (*Ficus carica* L.) leaves. *Journal of Molecular Liquids, 318*, 114196. https://doi.org/10.1016/j.molliq.2020.114196

Zhao, C., Li, S., Li, C., Wang, T., Tian, Y., & Li, X. (2021). Flavonoids from fig (*Ficus carica* Linn.) leaves: The development of a new extraction method and identification by UPLC-QTOF-MS/MS. *Applied Sciences, 11*(16). https://doi.org/10.3390/app11167718

Chapter 16
Fig (*Ficus carica*) Seed Oil

Emi Grace Mary Gowshika Rajendran

Abbreviations

FAO	Food and Agricultural Organization
SFA	Saturated Fatty Acid
PUFA	Polyunsaturated Fatty Acid
MUFA	Monounsaturated Fatty Acid
LDL	Low Density Lipoprotein
GAE	Gallic Acid Equivalent
TH	Taamriwthe
AZ	Azegzaw
AV	Averkane
GC/MS	Gas Chromatography/Mass Spectrometry

1 Fig Seeds and Its Oil

More than 750 varieties of fig fruits, which come under the Moraceae family, were identified (Guvenc 2009). The edible fruit species among these varieties were *Ficus carica* L. The forms of consumption of figs can be dried, raw, canned or through other preserved forms (Mawa et al., 2013). The countries (Fig. 16.1) with hot and dry summers and very mild winters are the notable producers of this fruit, including Egypt, Spain, Brazil, Turkey, California, Morocco, Italy, Greece, and other countries. It is reported that 30% of the world's fig production comes from Turkey (Nakilcioglu & Hisil, 2013; Mehta et al., 2014).

The fig seeds are the most attractive part, providing effective health benefits and nutritional value. A jelly-like flesh is bound within a seed mass, which is contained

E. G. M. G. Rajendran (✉)
Women's Christian College, Chennai, India
e-mail: emigracerajendran@gmail.com

M. F. Ramadan (ed.), *Fig* (Ficus carica)*: Production, Processing, and Properties*, https://doi.org/10.1007/978-3-031-16493-4_16

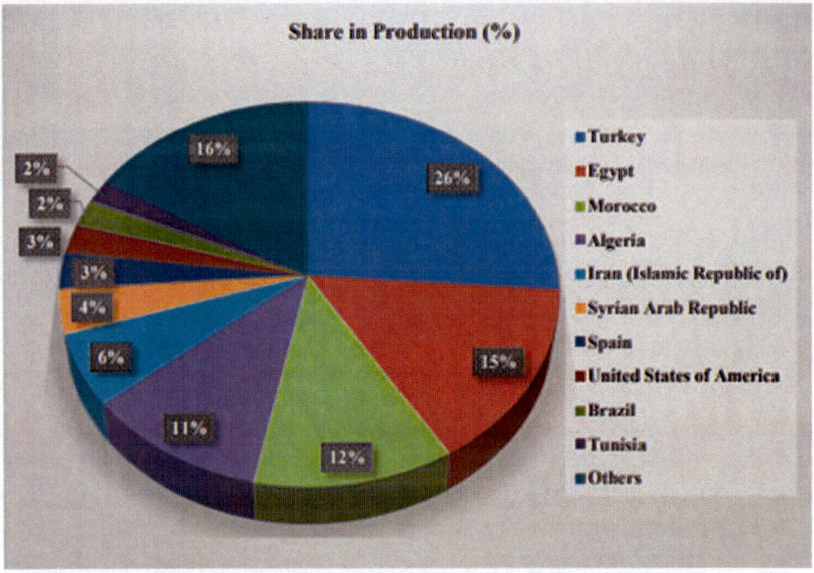

Fig. 16.1 Distribution of fig production by Country (2017). (Source: FAO, 2020)

in the white inner ring on the interior portion of the fig (Badgujar et al., 2014). The seed number varies between 30–1600 per fruit and is minute, small, medium, or large (Lansky et al., 2008; Badgujar et al., 2014). The fig seeds are hollow unless pollinated, and numerous edible seeds are found. In addition, the seeds provide the characteristic nutty flavor of the dried figs, which are now used as lubricants and edible oils (Badgujar et al., 2014).

The nutritive value of this fruit makes it an important constituent in the Mediterranean diet, which includes energy, vitamins, dietary fiber, minerals, antioxidant compounds (phenolics), and amino acids (Oliveira et al., 2009; Bucic-kojic et al., 2011; Amessis-Ouchemoukh et al., 2017). In addition, many usages in therapeutics from the roots, fruits, and leaves of the fig are widely known, especially for respiratory (cough, sore throat, bronchial issues), anti-inflammatory, cardiovascular, gastrointestinal disorders (indigestion, diarrhea, colic, loss of appetite) and antispasmodic remedies (Mawa et al., 2013; Amessis-Ouchemoukh et al., 2017).

The studies on the chemical composition (Table 16.1) of figs (*Ficus carica*) have been extensively done due to its traditional curative properties and its reputation in various fields (Penelope, 1997; Duke et al., 2002; Lansky et al., 2008; Joseph & Raj, 2011; Badgujar et al., 2014; Rahmani & Aldebasi, 2017). However, using fig seeds as a beneficial food source is not entirely possible due to their small size and indigestibility, leading to elimination via feces.

The fig seeds contained 18 fatty acids, and statistically significant differences were found between different locality seeds (Table 16.2). The fatty acid composition of fig seeds was comparatively less than fresh whole figs (Guvenc 2009).

Table 16.1 Chemical Composition of Stem, Leaves, Latex and Seeds of *Ficus carica*

Stem		Leaves		Latex		Seeds	
Campesterol	T	Moisture (%)	67.6	Caoutchouc	2.4%	Oleic acid	18.9%
Hentriacontanol	T	Protein (%)	4.3	Resin	T	Linoleic	33.7%
Stigmasterol	T	Fat (%)	1.7	Albumin	T	Linolenic	32.9%
Euphorbol	T	Crude fiber	4.7%	Cerin	T	Palmitic	5.23%
Hexacosanate	T	Ash (%)	5.3	Sugar	T	Stearic	2.18%
Ingenol	T	N-free extract	16.4%	Malic acid	T	Arachidic	1.05%
Taraxerone	T	Pentosans	3.6%	Rennin	T		
		Carotene	T	Proteolytic enzymes	T		
		Bergaptene	T	Diastase	T		
		Stigmasterol	T	Esterase	T		
		Sitosterol	T	Lipase	T		
		Rutin	T	Catalase	T		
		Sapogenin	T	Peroxidase	T		
		Calotropenyl	T				
		Lepeolacetate	T				
		Oleanolic	T				

T present in trace amounts
Adapted from Joseph and Raj (2011)

The γ-linolenic acid and linoleic acid were higher in fig seeds (41.8%) than in fresh figs. (1.46%). The Germencik location fig seeds contained the highest amount of PUFA and the lowest SFA. Fig seed oil is nutrient-rich, and a healthier source of oil as the essential fatty acids contribute to human health.

Higher contents of α-linolenic and linoleic acids, 32%–50% and 20%–35%, respectively, are found in fig seed oil. These PUFA acids are not synthesized in mammals but should be directly consumed through the diet, thereby classifying them as essential fatty acids (Field, 2003). *Omega*-3 fatty acids enhance ocular health and reduce the risk of age-related macular degeneration and cardiovascular diseases through their anti-inflammatory properties. In addition, fig seed oil reduces C-reactive protein levels through endothelial functions and protects the cell membranes from free radical damage through its vitamin E content (Richer et al, 2016).

New and under-utilized sources having adequate PUFA content are being discovered and formulated due to the rise in nutritional interest for healthy oils. Fig seed oil plays a preventive role in reducing LDL and total cholesterol levels, further preventing cardiovascular diseases (Ajayi & Ajayi, 2009). The increase in unsaturated fatty acid content terms fig seed oil as an important source of healthy essential oils. The oil could be obtained by processing figs and separating the seeds from the pulp. This could be used to formulate many dietary supplements and food products.

Table 16.2 Fatty acid compositions (%) of *Ficus carica* seeds

Fatty acid	Common name	Location			
		Turkey	Incirliova	Germencik	Nazilli
C16:0	Palmitic acid	5–9%	7.4%	3.58%	7.0%
C16:1	Palmitoleic acid	–	0.06%	0.06%	0.05%
C17:0	Heptadecanoic acid	–	0.06%	0.06%	0.06%
C18:0	Stearic acid	2–4%	3.38%	3.73%	2.97%
C18:1 (n-9)	Oleic acid	14–24%	16.8%	17.7%	16.9%
C18:2 (n-6)	Linoleic acid	20–35%	32.8%	37.9%	31.8%
C18:3 (n-3)	α-Linolenic acid	32–50%	0.30%	0.25%	0.30%
C18:3 (n-6)	γ-Linolenic acid	0.17%	37.8%	41.8%	39.1%
C20:0	Arachidic acid	Max. 0.7%	0.02%	0.03%	0.02%
C20:1 (n-9)	Eichosenoic acid	0.5%	0.09%	0.04%	0.09%
C22:0	Behenic acid	Max. 0.5%			
C21:0	Heneicosanoic acid	–	0.04%	0.08%	0.04%
C20:3n6	Eicosatrienoic acid	–	0.09%	0.04%	0.09%
C24:0	Lignoceric acid	–	0.03%	0.04%	0.03%
C20:5n3	Eicosapentaenoic acid	–	0.39%	0.24%	0.11%
SFAs		–	10.9%	7.10%	10.2%
MUFAs		–	17.2%	17.0%	17.4%
PUFAs		–	71.8%	75.8%	72.2%

SFAs saturated fatty acids; *MUFAs* monounsaturated fatty acids; *PUFAs* polyunsaturated fatty acids
Adapted from Guven et al. (2019) and Tas (2018)

2 Extraction and Processing of Fig Oil

The ripe figs were plucked from the branches, peeled manually, and dried using a grill. After drying the figs, they are cut into small pieces (5 mm), rinsed, precipitated, and dried. The dried fig pulp is separated from the seeds and dried naturally. These seeds are converted into pure seed oil through an extrusion process from the cold-extrusion oil processing machine with a temperature below 25 °C. The pure *Ficus carica* seed oil (Fig. 16.2) is preserved at −18 °C for further processing and formulation into different food products and supplements.

From the estimation of sterol composition (Table 16.3), the highest composition of sterol present in the fig seed oil was β-sitosterol (66.4%). On the contrary, the lowest was clerosterol (0.68%), and the total sterol content in the fig seed oil was 6516 mg/kg.

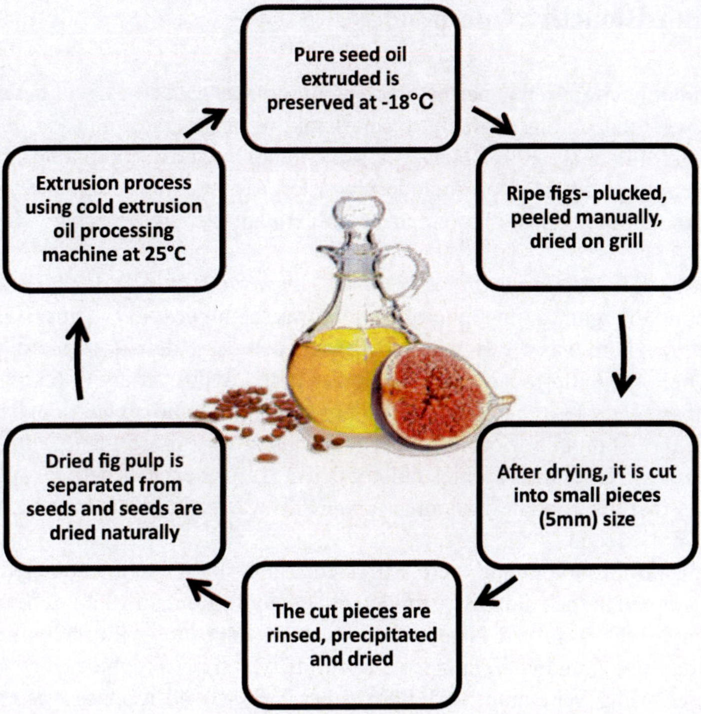

Fig. 16.2 Processing and Extraction of fig (*Ficus carica*) seeds into pure fig oil. (Adapted from Baygeldi et al. 2021)

Table 16.3 Sterol composition in Fig (*Ficus carica*) seed oil

Sterols	Sterol Composition %
Beta-sitosterol	66.4
Delta-5-avenasterol	20.1
Delta-7-avenasterol	5.07
Campesterol	2.99
Sigmasterol	2.18
Delta-7-sigmasterol	1.28
Sitosterol	1.18
Clerosterol	0.68
Total Sterol content	6516 mg/kg

Reproduced from Guvenc (2009)

3 Minor Bioactive Compounds

The fat-soluble vitamin that acts as a potent antioxidant reducing agent in vegetable oils, known to occur through the chain-breaking of saturated fats, is known as tocopherol (Shin et al., 2009). However, some minor bioactive compounds that exist predominantly in fig seed oil include phytosterols, phenolic compounds, organic acids, and anthocyanins that enhance antioxidant activity (Mawa, Husain & Jantan, 2013).

From the two location variation of *Ficus carica* seed oil in Turkey (Table 16.4), it was found that gamma-tocopherol content was the highest (391.8 mg/100 g) and (4267 ppm). There was a variation between the delta and alpha-tocopherol composition, where the delta-tocopherol was higher in the Aydin variety (7.65 mg/100 g) and comparatively less than Istanbul variety (147 ppm), and alpha-tocopherol was higher in the Istanbul variety (157 ppm) comparatively more than Aydin variety (4.6 mg/100 g). Both the varieties did not detect the presence of beta tocopherol. It was found that the content of gamma-tocopherol was present due to the extraction of oil from the seeds.

The phenolic compounds were extracted using slight modifications from the method adopted in the study by Tsimidou et al. (1992). First, in 10 ml of hexane, 2 g of oil sample was dissolved. Next, this mixture was added to 4 ml of 60% methanol. The mixture was kept in darkness for 2 h, during which it was subjected to agitation and filtered using Whatman No.1 filter paper. The second mixture was extracted using the same method and under the same conditions. First, the two filtrates obtained were mixed. Then, using a rotary evaporator, the mixture was concentrated using a vacuum and was reconstituted using 10 ml of pure methanol. This was stored at 20 °C.

The total phenolic content was determined based on the Folin-Ciocalteu colorimetric method (Favati et al., 1994). This reaction mixture comprises 250 μL of Folin-Ciocalteu reagent, 50 μL of extract, 750 μL of sodium carbonate (7%), and 3 ml of distilled water. A 950 μL of distilled water was added after 8 min of stirring under ambient temperature. The absorbance was measured against a 760 nm blank after 1 h incubation in the dark. The standard used for the calibration curve was gallic acid. The analysis was carried out in triplicates and was expressed as mg equivalent of Gallic acid (GAE) per 100 g of oil. Therefore, fig seed oil's total phenolic content was 79.5 mg (GAE)/ 100 g of the oil (Guvenc 2009).

Table 16.4 Tocopherol content in fig (*Ficus carica*) seed oil

	Aydin, Turkey	Istanbul, Turkey
Type of tocopherol	mg/100 g	ppm
Gamma-tocopherol	391.8	4267
Delta-tocopherol	7.65	147
Alpha-tocopherol	4.6	157
Beta-tocopherol	ND	ND

ND not detected, *ppm* parts per million
Adapted from Guvenc (2009) and Baygeldi et al. (2021)

4 Antioxidant Activity and Other Health-Promoting Effects

In a study, six different varieties of figs varying in color (yellow, black, red, and green) were studied for the antioxidant properties present in the oil when seeds are extruded from dried fig pulp. They were analyzed for their anthocyanin profile, total polyphenols, antioxidant capacity, and flavonoids. It was reported that 28% and 36% of the total antioxidant capacity was contributed by anthocyanin fraction in red Brown-Turkey and dark-colored mission varieties of fig oils. In addition, 92% of the antioxidant capacity in the anthocyanin fraction was contributed by cyaniding-3-*O*-rutinoside (C3R). The mission variety fig oils exhibited the highest antioxidant capacity with the highest flavonoids, polyphenols, and anthocyanin levels (Joseph & Raj, 2011). Studies evaluating the numerous health benefits of Fig (Ficus carica) seed oil have been tabulated (Table 16.5).

Fig is a small deciduous plant with immense therapeutic effects owing mostly to its antioxidant capacity. However, more animal and human clinical trials are required to confirm and confer other health benefits.

5 Volatile Compounds' Contribution to Organoleptic Properties

From a study conducted by Zidi et al. (2021), the volatile compounds identified from three cultivars of figs through which the oil was extruded from their seeds had identified almost 29 compounds (Table 16.6) during three different ripening stages

Table 16.5 Therapeutic roles of fig (*Ficus carica*) seed oil from clinical studies

Therapeutic Role	The outcome of the Study	References
Anti-oxidative	Effective scavenging activity was observed in fig oils	Yang et al. (2009)
Anti-cancer	Stomach cancer cell line growth was inhibited	Hashemi et al. (2011)
Anti-cancer	Inhibitory effects on proliferation of cancer cell lines by 6-O-acyl-beta-D-glucosyl-beta-sitosterols were noticed	Rubnov et al. (2001)
Anti-inflammatory	A reduction in granuloma weight of 71.66% in the chronic study and 75.90% in acute inflammation confirmed the anti-inflammatory properties of fig oil	Patil and Patil (2011)
Anti-diabetic	A hypoglycemic effect was observed in rats on treatment with fig oil	Perez et al. (1998)
Immunomodulatory	Amelioration of both humoral and cellular antibody responses was observed	Patil et al. (2010)
Sperm parameters	An improvement in gonadosomatic index, sperm count, and non-progressive spermatozoa motility was found with fig oil ingestion	Naghdi et al. (2016)

Adapted from (Rahmani and Aldebasi (2017)

Table 16.6 Volatile Organic Compounds identified through GC/MS in fig oils from three ripening stages and cultivars of figs

Volatile Compounds	Aroma characteristics	Odor threshold in water (μg/kg)	Taamriwthe			Azegzaw			Averkane		
			Unripe	Ripe	Fully ripe	Unripe	Ripe	Fully ripe	Unripe	Ripe	Fully ripe
Alcohol											
Pent-1-en-3-ol	Pungent link green vegetable and fruity nuances	400	nd	0.034	nd	0.035	0.026	0.020	nd	nd	0.035
Hexan-1-ol	Pungent, ethereal, fruity, and alcoholic	500	nd	nd	nd	0.026	0.024	0.024	nd	nd	nd
(E) hex-2-en-1-ol	Fruity and herbal nuances	100	nd	nd	nd	nd	0.020	nd	nd	nd	nd
Benzyl alcohol	Sweet, floral, fruity, and chemical nuances	10,000	nd	nd	nd	0.166	nd	nd	nd	nd	nd
Pentan-1-ol	Pungent, fermented bready, yeasty, moldy	nd	nd	nd	nd	nd	nd	nd	nd	0.020	nd
3-methylbutan-1-ol	Pungent, ethereal, fruity, cognac and alcoholic	nd	nd	0.050	0.085	0.028	nd	0.037	nd	nd	0.032
2-ethylhexan-1-ol	Citrus, fresh, floral, oily and sweet	nd	0.020	nd	nd	0.050	0.054	0.028	nd	nd	0.033
Aldehydes											
Acetaldehyde	Pungent, ethereal, fresh, fruity	15	0.124	0.148	0.273	0.075	0.064	0.126	nd	0.058	0.164
Heptanal	Fresh, citrus, green odor	3	nd	nd	0.032	nd	nd	nd	nd	nd	nd
Nonanal	Fresh, green lemon, citrus	1-98	nd	0.02	nd	0.026	nd	0.026	nd	nd	nd
Benzaldehyde	Fruity, almond, nutty, powdery	350	nd	0.093	0.075	0.937	0.424	0.35	nd	nd	0.09
Hexanal	Fatty, green, vegetative, leafy, fruity and woody nuance	5	0.311	0.553	0.359	0.788	0.33	0.258	0.226	0.152	0.141
(E)-2-methylbut-2-enal	Strong green fruit	nd	0.102	0.157	0.078	nd	nd	nd	nd	nd	nd
(E)-hex-2-enal	Aldehydic fatty cheese, green banana	17-316	0.115	0.214	0.311	nd	nd	nd	nd	0.05	0.175
(E)-oct-2-enal	Fatty green herbal banana, fresh cucumber, waxy green leaf	3	nd	nd	nd	0.016	nd	nd	nd	nd	nd

Esters											
Ethyl acetate	Ethereal, sweet, fruity, grape and rum-like	5000	0.216	0.159	0.135	0.137	nd	nd	nd	0.198	0.333
Methyl hexanoate	Fermented, fusel, banana, fruity, ethereal	84	nd	0.093	nd	nd	0.035	nd	nd	nd	nd
Methyl octanoate	Sweet, green, vegetative, orange nuances	nd	nd	nd	nd	nd	0.026	nd	nd	nd	0.067
Ethyl 2-methylbutanoate	Berry, fruity, fresh tropical nuances	nd	nd	nd	0.037	nd	nd	nd	nd	nd	nd
Ethyl hexanoate	Fruity, sweet, pineapple	14	nd	nd	0.034	nd	nd	nd	nd	nd	nd
Terpenes											
Limonene	Citrus orange fresh, sweet	10	nd	nd	nd	nd	0.215	0.109	nd	nd	nd
Linalool	Floral, citrus, orange	6	nd	0.021	nd	1.049	0.498	0.892	nd	nd	0.043
Epoxylinalol	Floral honey	nd	nd	0.025	nd	nd	nd	0.028	nd	nd	nd
α-Santalene	Woody	nd	nd	0.033	nd	nd	nd	nd	nd	nd	nd
1,8-cineol	Herbal eucalyptus, camphor medicinal	1,3-12	nd	nd	nd	nd	nd	nd	0.03	nd	nd
β-caryophyllene	Sweet, spice, woody, clove dry	150	0.292	0.254	nd	nd	nd	nd	nd	nd	nd
Ketones											
Acetone	Ethereal, solvent, apple, pear	nd	nd	0.059	0.1	0.034	0.05	0.045	0.116	0.079	0.043
Heptan-2-one	Fruity, cheese, ketonic, creamy nuances	140	nd	nd	0.073	nd	nd	nd	0.107	nd	0.152
6-methylhept-5-en-2--one	Apple, fruity, ketonic, musty, cheesy nuances	160	nd	nd	nd	0.022	0.022	nd	nd	nd	nd

nd not detected

Description for Aroma was obtained from (https://www.thegoodscentscompany.com/); thresholds were obtained from Tamura et al. (2001); Yang et al. (2019); Noguerol-Pato et al. (2012); Buttery et al. (1969); Wu et al. (2016); Takeoka et al. (1990); Czerny et al. (2008). Volatile compounds are classified according to chemical classes. The mean ± SD of three samples in triplicate were analyzed and obtained from, Zidi et al. (2021).

classified according to visual maturity characteristics based on fruit firmness and color change from Taamriwthe (TH), Azegzaw (AZ), and Averkane (AV) cultivars.

A significant increase in aldehyde content was observed in TH and AV fig oils. AZ had a decline in them. Totally six aldehydes were observed in these fig oils. The extract of terpenes was found only at the ripe stages in the TH cultivar. Linalool was observed to be the most abundant terpene in the AZ cultivar. AV fig oils contained only two terpenes. Esters were significantly increased obtaining from figs during the ripening process for oil extraction. TH and AV had increased total ester, whereas AZ had overall less proportion of esters. A significant increase was observed in alcohol content during the ripening process of AV cultivars. Whereas AZ had a significant decrease in alcohol content, and TH had only three alcohol compounds that showed a significant increase in the oils obtained from fully ripe figs. A total of three ketones were observed in this study, and only two were found in all three cultivars.

References

Ajayi, O. B., & Ajayi, D. D. (2009). Effect of oilseed diets on plasma lipid profile in albino rats. *Pakistan Journal of Nutrition, 8*(2), 116–118.

Amessis-Ouchemoukh, N., Ouchemoukh, S., Meziant, N., Idiri, Y., Hernanz, D., Stinco, C. M., Rodriguez-Pulido, F. J., Heredia, F. J., Madani, K., Luis, J., et al. (2017). Bioactive metabolites involved in the antioxidant, anticancer and anticalpain activities of *Ficus carica L., Ceratonia siliqua L.* and *Quercus ilex L.* extracts. *Industrial Crops and Products- Elsevier, 95*, 6–17.

Badgujar, S. B., Patel, V. V., Bandivdekar, A. H., & Mahajan, R. T. (2014). Traditional uses, phytochemistry and pharmacology of *Ficus carica*: A review. *Pharmaceutical Biology, 52*(11), 1487–1503.

Baygeldi, N., Kucukerdonmez, O., Akder, R. N., & Cagindi, O. (2021). Medicinal and nutritional analysis of fig (*Ficus carica)* seed oil; a new gamma tocopherol and omega-3 source. *Progress in Nutrition, 23*(2), 2–6.

Bucic-Kojic, A., Planinic, M., Tomas, S., Jokic, S., Mujic, I., et al. (2011). Effect of extraction conditions on the extractability of phenolic compounds from lyophilised fig fuits (*Ficus carica* L.). *Polish Journal of Food and Nutrition Sciences, 61*(3), 195.

Buttery, R. G., Seifert, R. M., Guadagni, D. G., & Ling, L. C. (1969). Characterization of some volatile constituents of bell peppers. *Journal of Agriculture and Food Chemistry, 17*, 1322–1327.

Czerny, M., Christlbauer, M., Christlbauer, M., Fischer, A., Granvogl, M., Hammer, M., et al. (2008). Re-investigation on odour thresholds of key food aroma compounds and development of an aroma language based on odour qualities of defined aqueous odorant solutions. *European Food Research and Technology, 228*, 265–273.

Duke, J. A., Bugenschutz-Godwin, M. J., duCellier, J., & Duke, P. K. (2002). *Handbook of medicinal herbs* (2nd ed.). CRC Press.

FAO. (2020). *Food and Agriculture Organization of The United Nations*. http:faostat3.fao.org/download/Q/QC/E.

Favati, F., Caporale, G., & Bertuccioli, M. (1994). Rapid determination of phenol content in extra virgin olive oil. *Grasas y Aceites., 45*(1–2), 68–70.

Field, C. J. (2003). *Fatty Acids: Dietary Importance*. Elsevier. https://doi.org/10.1016/B0-12-227055-X/00450-8

Guven, N., Gokyer, A., Koc, A., et al. (2019). Physicochemical composition of fig seed oil from Turkey. *Journal of Pharmacy and Pharmacology, 7*, 541–545.

Guvenc, M. (2009). Analysis of fatty acid and some lipophilic vitamins found in the fruits of the Ficus carica variety picked from the Adiyaman District. *Research Journal of Biological Sciences, 4*(3), 320–323.

Hashemi, S. A., Abediankenari, S., Ghasemi, M., Azadbakht, M., Yousefzaideh, Y., & Dehpour, A. A. (2011). The effect of fig tree latex (*Ficus carica)* on stomach cancer line. *Iranian Red Crescent Medical Journal, 13*(4), 272–275.

Joseph, B., & Raj, J. S. (2011). Pharmacognostic and phytochemical properties of *Ficus carica* Linn – An overview. *International Journal of PharmTech Research, 3*(1), 8–12.

Lansky, E. P., Paavilainen, H. M., Pawlus, A. D., & Newman, R. A. (2008). Ficus spp. (fig): Ethnobotany and potential as anticancer and anti-inflammatory agents. *Journal of Ethnopharmacology, 119*(2), 195–213.

Mawa, S., Husain, K., & Jantan, I. (2013). *Ficus carica* L. (Moraceae): Phytochemistry, traditional uses and biological activities. *Evidence-Based Complementary and Alternative Medicine, 974256*.

Mehta, S., Soni, N., Satpathy, G., & Gupta, R. K. (2014). Evaluation of nutritional, phytochemical, antioxidant and antibacterial activity of dried plum (*Prunes domestica). Journal of Pharmacognosy and Phytochemistry, 3*(2), 158–165.

Naghdi, M., Maghbool, M., Seifalah-Zade, M., Mahaldashtian, M., Makoolati, Z., Kouhpayeh, S. A., et al. (2016). Effects of common fig *(Ficus carica)* leaf extracts on sperm parameters and testis of mice intoxicated with formaldehyde. *Evidence-Based Complementary and Alternative Medicine, 9*.

Nakilcioglu, E., & Hisil, Y. (2013). Research on the phenolic compounds in Sarilop (*Ficus carica* L.) fig variety. *Gida, 38*(5), 267–274.

Noguerol-Pato, R., González-Barreiro, C., Cancho-Grande, B., Martínez, M. C., Santiago, J. L., & Simal-Gándara, J. (2012). Floral, spicy and herbaceous active odorants in Gran Negro grapes from shoulders and tips into the cluster, and comparison with Brancellao and Mouratón varieties. *Food Chemistry, 135*, 2771–2782.

Oliveira, A. P., Valentao, P., Pereira, J. A., Silva, B. M., Tavares, F., & Andrade, P. B. (2009). *Ficus carica* L.*:* Metabolic and biologic screening. *Food and Chemical Toxicology, 47*(11), 2841–2846.

Patil, V. V., & Patil, V. R. (2011). Evaluation of anti-inflammatory activity of *Ficus carica* leaves. *International Journal of Pharmaceutical Sciences, Reviews and Research, 2*, 48–50.

Patil, V. V., Bhangale, S. C., & Patil, V. R. (2010). Evaluation of anti-pyretic potential of *Ficus carica* leaves. *International Journal of Pharmaceutical Sciences, Review and Research, 2*, 48–50.

Penelope, O. (1997). *100 great natural remedies using healing*. Kyle Cathic Limited.

Perez, C., Dominguez, E., Ramiro, J. M., Romero, A., Campillo, J. E., & Torres, M. D. (1998). A study on the glycemic balance in streptozotocin-diabetic rats treated with and aqueous extract of *Ficus carica* (fig tree) leaves. *Phytotherapy Research, 10*, 82–83.

Rahmani, A. H., & Aldebasi, Y. H. (2017). Ficus carica and its constituents role in management of disease. *Asian Journal of Pharmaceutical and Clinical Research, 10*(6), 49–53.

Richer, S., Ulanski, L., & Popenko, N. A. (2016). Age-related macular degeneration beyond the age-related eye disease study II. *Advances in Ophthalmology and Optometry, 1*, 335–369.

Rubnov, S., Kashman, Y., Rabinowitz, R., Schlesinger, M., & Mechoulam, R. (2001). Suppressors of cancer cell proliferation from fig *(Ficus carica)* resin: Isolation and structure elucidation. *Journal of Natural Products, 64*(7), 993–996.

Shin, E. C., Huang, Y. Z., Pegg, R. B., Phillips, R. D., & Eitenmiller, R. R. (2009). Commercial runner Peanut cultivars in the United States: Tocopherol composition. *Journal of Agricultural and Food Chemistry, 57*(21), 10289–10295.

Takeoka, G. R., Flath, R. A., Mon, T. R., Teranishi, R., & Guentert, M. (1990). Volatile constituents of apricot (*Prunus Armeniaca*). *Journal of Agriculture and Food Chemistry, 38*, 471–477.

Tamura, H., Boonbumrung, S., Yoshizawa, T., & Varanyanond, W. (2001). The volatile constituents in the peel and pulp of a Green Thai Mango, Khleo Sawoei cultivar (*Mangifera indica L.). Food Science and Technology Research, 7*, 72–77.

Tas, E. N. (2018). Biochemical characterization of (*Ficus carica* L.) seeds. *Journal of Agricultural Sciences, 25*, 232–237.

Tsimidou, M., Papadopoulos, G., & Boskou, D. (1992). Determination of phenolic compounds in virgin olive oil by reversed-phase HPLC with emphasis on UV detection. *Food Chemistry, 44*(1), 53–60.

Wu, Y., Duan, S., Zhao, L., Gao, Z., Luo, M., Song, S., et al. (2016). Aroma characterization based on aromatic series analysis in table grapes. *Scientific Reports, 6*, 1–16.

Yang, X. M., Yu, W., Ou, Z. P., Ma, H. L., Liu, W. M., & Ji, Z. L. (2009). Antioxidant and immunity activity of water extract and crude polysaccharide from *Ficus carica* L. fruit. *Plant Foods and Human Nutrition, 64*(2), 167–173.

Yang, Y. N., Zheng, F. P., Yu, A. N., & Sun, B. G. (2019). Changes of the free and bound volatile compounds in *Rubus corchorifolius* L. f. Fruit during ripening. *Food Chemistry, 287*, 232–240.

Zidi, K., Kati, D. E., Bachir-bey, M., Genva, M., & Fauconnier, M. L. (2021). Comparative study of fig volatile compounds using headspace solid-phase microextraction-gas chromatography/mass spectrometry: Effects of cultivars and ripenings stages. *Frontiers in Plant Science, 12*, 1312.

Chapter 17
Composition and Functional Properties of Fig (*Ficus carica*) Phenolics

Mustafa Kiralan, Onur Ketenoglu, Sündüz Sezer Kiralan, and Fatih Mehmet Yilmaz

1 Introduction

Fig (*Ficus carica* L.) is a member of the Moraceae family and grown in warm, dry climates (Lo Turco et al., 2020). According to the official data belonging to 2019, the total world fig production is 1,315,588 tonnes. Turkey (310,000 tonnes), Egypt (225,295 tonnes), Algeria (114,092 tonnes), Iran (130,328 tonnes), and Morocco (153,472 tonnes) are the major producer countries, yielding almost 80% of the total production (FAO, 2021). *Ficus carica* Linn is the best-known member of the *Ficus* genus, known by over 135 names. It has been the subject of worldwide studies due to the positive effects of its various biological activities on the endocrine system, respiratory system, gastrointestinal system, and infectious diseases (Badgujar et al., 2014). Figs are processed into wine, liquor, fruit juice, and jelly products and are used fresh and dried (Backes et al., 2020).

Fig is an important constituent of a Mediterranean diet due to its nutritional value and phytochemicals. Fig is an essential source of dietary fiber, amino acids, carbohydrates, minerals, organic acids, and phenolic compounds (Slatnar et al., 2011; Slavin, 2006; Wojdyło et al., 2016). Additionally, fig fruits are important sources of vitamins (Vitamin C, thiamin, riboflavin) and trace minerals (potassium, calcium,

M. Kiralan (✉) · S. S. Kiralan
Department of Food Engineering, Faculty of Engineering, Balikesir University, Balikesir, Turkey

O. Ketenoglu
Department of Food Engineering, Faculty of Agriculture, Eskisehir Osmangazi University, Eskisehir, Turkey

F. M. Yilmaz
Department of Food Engineering, Faculty of Engineering, Aydin Adnan Menderes University, Aydin, Turkey

M. F. Ramadan (ed.), *Fig* (Ficus carica)*: Production, Processing, and Properties*, https://doi.org/10.1007/978-3-031-16493-4_17

magnesium). Furthermore, figs are cholesterol-free and contain many essential amino acids, fiber, and antioxidant compounds (Görgüç et al., 2021). Antioxidant compounds such as phenolics (phenolic acids, flavonoids, carotenoids) are the major constituents that provide the fig fruits'characteristic color, flavor, and aroma (Fateh Aljane & Ferchichi, 2009; Arvaniti et al., 2019). The overall composition of fresh fig is summarized in Table 17.1. Considering dried figs, total carbohydrate, fat, protein, dietary fiber, and moisture content were 73.5%, 0.56%, 4.67%, 3.68%, and 16.6%, respectively (Soni et al., 2014). Therefore, dried and fresh figs are good carbohydrates and dietary fiber sources.

Besides macronutrients, fig contains different phytochemicals such as anthocyanins, polyphenols, and carotenoids (Jagtap & Bapat, 2020). Phenolic compounds are secondary metabolites synthesized through the shikimic acid and phenylpropanoid pathways (de la Rosa et al., 2019). Fruits are rich in phenolic compounds that contribute to the prevention of several degenerative diseases (Renard, 2018). The health effects of phenolic compounds are largely attributed to the antioxidant activity and other mechanisms of action (Shahidi et al., 2019). These activities are closely related to the chemical diversity of phenolic compounds, especially the hydroxyl groups on the aromatic rings. In addition, the structure of hydroxyl groups attached to phenolic compounds influences their activities (Vuolo et al., 2019).

Figs exhibit numerous biological activities such as antioxidant, antimicrobial, and anti-diabetic (Arvaniti et al., 2019; Benmaghnia et al., 2019; Oliveira et al., 2009; Wojdyło et al., 2016), in particular, the different parts are reported to show antioxidant (Ahmad et al., 2013; Mostapha Bachir Bey & Louaileche, 2015; Solomon et al., 2006), antimicrobial (Aref et al., 2010; Jeong et al., 2009; Lazreg-Aref et al., 2012; Mahmoudi et al., 2016), anticancer (Jasmine et al., 2015; Purnamasari et al., 2019; Rubnov et al., 2001), and antidiabetic (Zaffer Ahmad et al., 2013) activities. The phenolic compounds, which are mainly responsible for these biological activities, have recently gained considerable attention, and the phenolic content of fig changes depending on various factors such as fruit quality, processing, and extraction techniques. This chapter discusses the composition of phenolic compounds and the biological activities of fig phenolics.

Table 17.1 The composition of some nutritients in fresh fig (g/100 g)

Nutrition	Quantity
Water	79.1
Protein	0.75
Total lipid (fat)	0.3
Carbohydrate	19.2
Fiber, total dietary	2.9
Sugars	16.3

Source: U.S. Department of Agriculture; FDC ID: 173021; NDB Number: 9089; FDC: Published:4/1/2019

2 Total Phenolics, Total Flavonoids and Total Anthocyanins in Figs

Fig fruits contain phenolic acids such as chlorogenic acid, flavanols such as (+)-catechin and (−)-epicatechin, flavonols such as quercetin-3-*O*-glucoside, kaempferol-3-O-glucoside, and rutin, flavons such as luteolin-8-C-glucoside, and anthocyanins such as cyanidin-3-*O* rutinoside (Slatnar et al., 2011). A general representation of fig phenolics is given in Fig. 17.1.

Total phenolics are determined using a colorimetric method called Folin-Ciocalteu (F-C) assay. This method is based on electron transfer reactions between the F-C reagent and polyphenols (Ainsworth & Gillespie, 2007; Sánchez-Rangel et al., 2013).

A summarizing list of total phenols in fresh and dried figs is given in Table 17.2, and total flavonoids and total anthocyanins in figs, depending on the part, and the origin of the fruit, are summarized in Table 17.3. The part of the fruit is one of these factors affecting the phenolic content. In general, leaves have higher total phenols values than other parts such as peels and pulps (Oliveira et al., 2009). These differences are attributed to especially flavonoids which have photoprotection properties in plants (Treutter, 2006). Besides, the skin part of figs contains a higher content of total phenols than the extracts from the pulp part. Solomon et al. (2006) explain these differences and emphasize that anthocyanins are the most abundant phenolics in the skin part of figs, so they contribute to the total phenolics and phytochemicals.

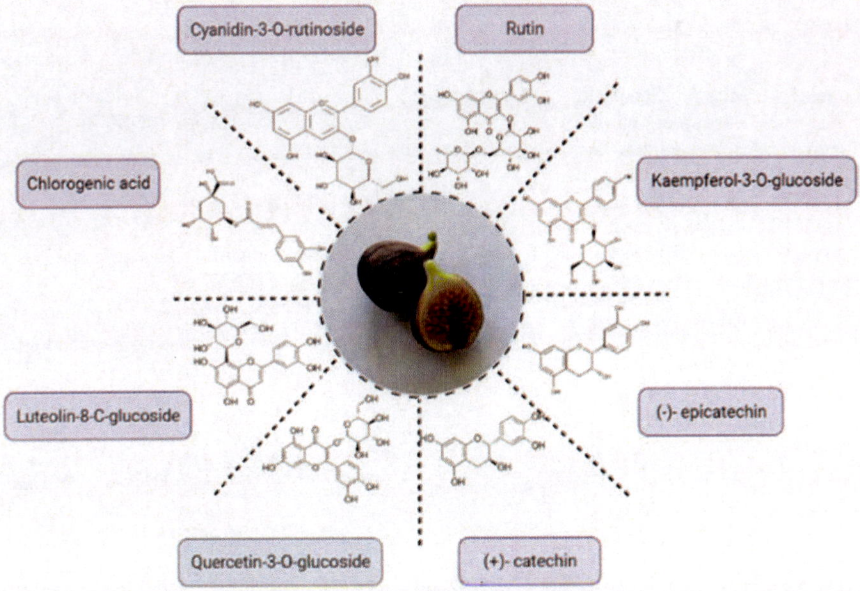

Fig. 17.1 The main phenolic compounds determined in fig fruits

Table 17.2 Total phenols in fresh and dried figs

Sample (Part/ Tissue)	Variety, origin, color	Assay method	Concentration	References
Fruit (whole)	76 local Mediterranean fig accessions from Turkey	Spectrometric-Folin-Ciocalteau	28.6–211.9 mg GAE/100 g FW	Çalişkan & Aytekin Polat (2011)
Leaves		Spectrometric-Folin-Ciocalteau	4.363–6.909 mg/ dry mass	Konyalıoğlu et al. (2005)
Peel Pulp Leaves	White varieties of *F. carica* (Pingo de Mel and Branca Tradicional) from Portugal	HPLC-DAD	600.6, 723 mg/kg lyophilized extract (Peel) 36.6, 130.3 mg/kg lyophilized extract (pulp) 32412.0, 34473.6 mg/kg lyophilized extract (leaves)	Oliveira et al. (2009)
Fruit (whole)	Škofjotka (with the synonym 'Zuccherina') a white type fruit, 'Črna petrovka' and 'Miljska figa', both dark type fruit from Slovenia	HPLC-DAD	Miljska cultivar had higher values in total phenol content, while Škofjotka cultivar is poor	Veberic et al. (2008)
Skin and pulp	Eighteen fig cultivars from Spain	HPLC-DAD	19.1–140.2 mg/ mg/100 g FW in fig skins 0.0–11.3 mg/100 g FW in fig pulps	Vallejo et al. (2012)
Fruit, skin and pulp	Mission, Chechick, Bursa (dark purple) Brown Turkey (purple), Brunswick (red yellow), and Kadota (green)	Spectrometric-Folin-Ciocalteau	48.6–281.1 mg of GAE/100 g of FW in fruits 41.7–463 mg of GAE/100 g of FW in skins 36.5–100.6 mg of GAE/100 g of FW in pulps	Solomon et al. (2006)

(continued)

Table 17.2 (continued)

Sample (Part/ Tissue)	Variety, origin, color	Assay method	Concentration	References
Leaves	Ten Algerian cultivars Onk Elhamam Hamra Zarrouk Boughandjo Safra Bidha Chatwi Bither Bakkor Dhokkar	Spectrometric-Folin-Ciocalteau	42.889–58.704 mg GAE/g DE	Mahmoudi et al. (2016)
Whole, skin, pulp	Two Turkish cultivars Sarilop-yellow Bursa siyahi- purple	Spectrometric-Folin-Ciocalteau	*Sarilop-yellow* Skin-351 mg GAE /100 g FW Pulp-167 mg GAE /100 g FW Whole-211 mg GAE /100 g FW *Bursa siyahi- purple* Skin-930 mg GAE /100 g FW Pulp-351 mg GAE /100 g FW Whole-493 mg GAE /100 g FW	Kamiloglu & Capanoglu (2015)
Whole	24 Turkish fig genotypes and two standard fig (*Ficus carica* L.) cultivars (Sarilop and Bursa Siyahi)	Spectrometric-Folin-Ciocalteau	24–237 mg GAE /100 g FW	Ercisli et al. (2012)
Whole	Ten different Spanish fig cultivars (*Ficus carica* L.)	UPLC-PDA-FL	715 to 186 mg/100 g dm	Wojdyło et al. (2016)
Whole	27 Tunisian ecotypes	Spectrometric-Folin-Ciocalteau	51.50–100.23 mg GAE /100 g FW	F Aljane et al. (2020)
Whole	11 local and introduced cultivars from Morocco	Spectrometric-Folin-Ciocalteau	90.75–524.74 mg GAE /100 g DW	Hssaini et al. (2020)

(continued)

Table 17.2 (continued)

Sample (Part/Tissue)	Variety, origin, color	Assay method	Concentration	References
Dried Leaves and fruits (whole)	Tounsi and Temri cultivars from Tunisia	Spectrometric-Folin-Ciocalteau	*Tounsi cultivar* Leaves-618.29 mg GAE /100 g DW Fruits-200.18 mg GAE /100 g DW *Temri cultivar* Leaves-686.88 mg GAE /100 g DW Fruits-124.48 mg GAE /100 g DW	Viuda-Martos et al. (2015)
Latex	18 diferent cultivars of F. carica from Malaysia	Spectrometric-Folin-Ciocalteau	The maceration extraction method 233.14–311.83 μg GAE/mL Ultrasound-assisted Extraction (224.37–291.98 μg GAE/mL	Shahinuzzaman et al. (2020)

FW Fresh weight, *GAE* gallic acid equivalents, *DE* dry plant extract, *dm* dry matter

Additionally, location is another influential factor in total phenols. According to Kamiloglu and Capanoglu (2015), the figs that grew in Turkey generally had higher total phenolic values than in other countries.

Latex is another important part of the fig, and it contains various phenolic compounds. Shahinuzzaman et al. (2020) extracted phenolic compounds from 18 different fig cultivars using two different extraction techniques: maceration and ultrasound-assisted extraction (UAE) with 75% ethanol. The latex was reported as a valuable natural source of phenolics. The latex from the *White Genoa* cultivar included higher amounts of total phenols.

Regarding color, the extracts from dark-colored fig cultivars contain higher levels of total phenols than the light-colored fig cultivars. For example, Solomon et al. (2006) stated that a dark-colored variety of *Mission* contained a high content of total phenols (281 mg/100 g of fresh weight (FW)), while a green variety, namely *Kadota*, was poor for total phenols (49 mg/100 g of FW).

Ripening is another factor in the phenolics of figs. Pereira et al. (2017) found that concentrations of phenolics increased during the ripening process. However, differences in the concentrations of total phenols were observed during ripening. Moreover, the effect of ripening on total phenolics of dark- and brown-colored cultivars was stronger than green- and yellow-green-colored cultivars. This effect could be related to the changes of anthocyanins accumulated in the skin during the ripening stage of dark-colored cultivars.

The drying process is also another essential factor in the total phenols of figs. As seen in Table 17.4, the content of total phenolics, total flavonoids, and total

Table 17.3 Total flavonoids and total anthocyanins in figs

Sample (Part/ Tissue)	Variety, origin, color	Total flavonoids	Total anthocyanins	References
Leaves	Ten Algerian cultivars Onk Elhamam Hamra Zarrouk Boughandjo Safra Bidha Chatwi Bither Bakkor Dhokkar	11.667–16.211 mg QE/g DE	–	Mahmoudi et al. (2016)
Whole, skin, pulp	Two Turkish cultivars Sarilop-yellow Bursa siyahi- purple	*Sarilop-yellow* 7–63 mg of CE/100 g DW *Bursa siyahi- purple* 66–234 mg of CE/100 g DW	*Sarilop-yellow* 0.8–5.9 mg of C3G/100 g DW *Bursa siyahi- purple* /100 g DW	Kamiloglu & Capanoglu (2015)
Whole	76 local Mediterranean fig accessions from Turkey	–	0.0–298.9 μg C3R /g FW	Çalişkan et al. (2011)
Whole	24 Turkish fig genotypes and two standard fig (*Ficus carica* L.) cultivars (Sarilop and Bursa Siyahi)	–	0–42 μg C3R /g FW	Ercisli et al. (2012)
Whole	Ten different Spanish fig cultivars (*Ficus carica* L.)	–	0.4–122 mg/100 g dm	Wojdyło et al. (2016)
Whole	27 Tunisian ecotypes	0.33–17.59 mg QE/100 g FW	1.61–11.67 mg C3G/ 100 g FW	Aljane et al. (2020)
Whole	11 local and introduced cultivars from Morocco	14.59–83.71 mg CE/100 g DW	3.1–24.8 mg C3R/100 g DW	Hssaini et al. (2020)

QE quercetin equivalents, *DE* dry plant extract, *FW* fresh weight, *C3G* cyanidin-3-glycoside, *C3R* cyanidin -3-rutinoside, *QE* quercetin, *CE* Catechin

anthocyanins differ in the literature. Slatnar et al. (2011) used two different drying processes: sun-drying and oven-drying, to obtain dried figs. The total phenolic compounds were 74.9 mg GAE/kg FW in fresh figs to 530 mg GAE/kg FW in dried figs. The higher values for total phenols were observed in the samples using the oven-drying process compared to sun-drying. Kamiloglu and Capanoglu (2015) found a decrease in total phenols of yellow and purple figs after the sun-drying process (8 and 15%, respectively). Their study evaluated total flavonoids, total

Table 17.4 Effect of drying on the content of total phenols, total flavonoids and total anthocyanins of figs

		Total phenols				Total flavonoid content				Total anthocyanin content			References
			Dried				Dried				Dried		
Sample (part/tissue)	Variety, origin, color	Fresh	Sun-drying	Oven-drying	Lyophilized	Fresh	Sun-drying	Oven-drying	Lyophilized	Fresh	Sun-drying	Oven-drying	
Whole	Sarilop-yellow	211 mg of GAE/100 g DW)	193 mg of GAE/100 g DW)	–	–	8 mg of CE/100 g DW	14 mg of CE/100 g DW	–	–	4.6 mg of C3G/100 g DW	0.1 mg of C3G/100 g DW	–	Kamiloglu and Capanoglu (2015)
Whole	Bursa siyahi-purple	493 mg of GAE/100 g DW)	417 mg of GAE/100 g DW)	–	–	66 mg of CE/100 g DW	52 mg of CE/100 g DW	–	–	83 mg of C3G/100 g DW	14.5 mg of C3G/100 g DW	–	Kamiloglu and Capanoglu (2015)
Whole		198.81–307.64 mg GAE/100 g DM	81.77–212.36 mg GAE/100 g DM	–	–	67.33–147.51 mg RE/100 g DM	32.78–52.52 mg RE/100 g DM	–	–	–	–	–	Nakilcioğlu and Hışıl (2013)
Whole	Aberkane (black), Azandjar, and Bouankik (dark-purple)	107.08–181.68 mg/100 g DM	–	–	30.81–41.91 mg/100 g DM	84.63–151.47 mg/100 g DM	–	–	12.09–20.82 mg/100 g DM	–	–	–	Mostapha Bachir Bey et al. (2017)

DW Dry weight, *GAE* gallic acid equivalents, *CE* (+)-catechin equivalent, *C3G* cyanidin-3-glycoside proanthocyanidin content, and total anthocyanin content depending on the sun-drying process. There were no significant changes between the total flavonoid content of fresh and dried fruits of purple figs. However, dried fruits from yellow figs contained more flavonoids than the fresh ones. The drying process showed a decrease in total proanthocyanidins in purple figs, while an increase was determined for yellow figs. The highest decrease was observed in the total anthocyanin content for purple and yellow figs depending on the sun-drying process (83% in purple figs, 98% in yellow figs). Lachtar et al. (2021) used two different systems: direct solar drying (open-air) and solar drying in a greenhouse to dry *Bither Abiadh* and *Bidhi* fig cultivars from Tunisia. The two cultivars dried in a greenhouse had higher total phenols than open air-dried samples. Mostapha Bachir Bey, Richard, Meziant, Fauconnier, and Louaileche (Bey et al., 2017) found a drastic reduction in total phenols and total flavonoids of figs dried by lyophilization.

Besides drying, the freezing process negatively influences the total phenol, total flavonoid, and total anthocyanins. A dramatic decrease was observed in the total phenol, total flavonoid, and total anthocyanins of the frozen figs in comparison to fresh figs, where the concentration of these phytochemicals in fresh figs was 31.0 mg GAE/100 g, 11.4 mg QE/100 g, and 0.8 mg C3G/100 g, while in frozen figs their concentrations were the lowest (14.0 mg GAE/100 g, 3.6 mg QE/100 g, and not detected, respectively) (Petkova et al., 2019).

3 Individual Phenolic Compounds in Figs

Individual phenolic compounds exhibit a wide variety in figs, and their amount changes with different fruit cultivars, color, ripening, and drying methods. Table 17.5 summarizes the most common individual phenolics in fresh and dried samples. Wojdyło et al. (2016) identified 11 polyphenolic compounds in such chemical classes as flavan-3-ols, phenolic acids, flavonols, flavons, and anthocyanins from different fig cultivars grown under the same conditions in Spain. The predominant flavanol was quercetin-3-*O*-rutinoside, with the highest concentration of 328 mg/100 g dry matter (DM) in the *Verdal brevas* cultivar. Chlorogenic acid was the only phenolic acid present in fig fruits, and the highest concentration of this compound was reported as 124 mg/100 g DM in *Verdal* fig. Catechin was a major compound among flavon-3-ols with a 20.6 to 95.7 mg/100 g DM concentration. Cyanidin-3-*O*-rutinoside was the main anthocyanin in figs, and *Cuello Dama Negra* cultivar contained the highest amount (116 mg/100 g DM).

Kamiloglu and Capanoglu (2015) studied individual phenolics of fruits with different colors. The predominant flavanoid was rutin for yellow figs. (12 mg/100 g DM), whereas cyanidin-3-rutinoside was found as the major anthocyanin for purple figs. (36 mg/100 g DM).

Faleh et al. (2012) evaluated the phenolic composition of three Tunisian fig phenotypes with different colors: black, green, and red. Quercetin-3-*O*-rutinoside was the major phenolic in all phenotypes, and its concentration was higher in the red

Table 17.5 The most common individual phenolics in fresh and dried samples

Subclass	Component	Dried	Fresh	References
Phenolic acids	Chlorogenic acid	1.4–2.0 mg/100 g	2–5.8 mg/100 g FW (skin) nd-1.3 mg/100 g FW (pulp)	Vallejo et al. (2012)
			0.46–1.71 mg/100 g FW (fruit)	Veberic et al. (2008)
		3.42–15.88 mg/100 g (sun-drying) 13.96–32.42 mg/100 g (oven-drying)	1.33–4.91 mg/100 g FW	Slatnar et al. (2011)
		8.8–124.5 mg/100 g DM	–	Wojdyło et al. (2016)
		0.40–2.21 mg/100 g DM	2.97–6.54 mg/100 g DM	Nakilcioğlu and Hışıl (2013)
		0.8–1.1 mg/100 g DM	1.9–8 mg/100 g DM (whole) 12–21 mg/100 g DM (skin) 0.5–6 mg/100 g DM (pulp)	Kamiloglu and Capanoglu (2015)
		0.06–0.55 mg/100 g DM	–	Pourghayoumi et al. (2016)
	Gallic acid		0.14–0.38 mg/100 g FW (fruit)	Veberic et al. (2008)
		0.47–2.18 mg/100 g DM	1.15–6.98 mg/100 g DM	Nakilcioğlu and Hışıl (2013)
		0.1–0.5 mg/100 g DM	0.3–0.3 mg/100 g DM (whole) 0.2–0.3 mg/100 g DM (skin) 0.3–0.5 mg/100 g DM (pulp)	Kamiloglu and Capanoglu (2015)
	Syringic acid		0.022–0.104 mg/100 g FW (fruit)	Veberic et al. (2008)
		0.41–2.83 mg/100 g DM	3.03–4.45 mg/100 g DM	Nakilcioğlu and Hışıl (2013)

(continued)

Table 17.5 (continued)

Subclass	Component	Dried	Fresh	References
	Ellagic acid	0.2–0.3 mg/100 g DM	0.2–0.2 mg/100 g DM (whole) 0.3–0.7 mg/100 g DM (skin) 0.1–0.2 mg/100 g DM (pulp)	Kamiloglu and Capanoglu (2015)
	p-Coumaric acid	1.1–3.1 mg/100 g DM	2.7–4.7 mg/100 g DM (whole) 5.1–18 mg/100 g DM (skin) 2.2–3.2 mg/100 g DM (pulp)	Kamiloglu and Capanoglu (2015)
Flavones	Luteolin 6C-hexose-8C-pentose	–	0.1–1.9 mg/100 g FW (skin)	Vallejo et al. (2012)
	Apigenin-rutinoside	–	0.8–2.4 mg/100 g FW (skin)	Vallejo et al. (2012)
	Apigenin-C-hexoside-pentoside	0.5–24.3 mg/100 g DM	–	Wojdyło et al. (2016)
Flavonols	Kaempferol-rutinoside	1.0–2.0 mg/100 g	0.2–0.9 mg/100 g FW (skin)	Vallejo et al. (2012)
		0.2–0.9 mg/100 g DM	0.3–0.3 mg/100 g DM (whole) 1.4–4 mg/100 g DM (skin) nd-0.1 mg/100 g DM (pulp)	Kamiloglu and Capanoglu (2015)
	Quercetine-acetilglucoside	2.1–2.8 mg/100 g	Nd-1.7 mg/100 g FW (skin)	Vallejo et al. (2012)
	Quercetin-rutinoside	10.2–13.0 mg/100 g	2.9–15.8 mg/100 g FW (skin) Nd-1.8 mg/100 g FW (pulp)	Vallejo et al. (2012)
	Quercetin-glucoside		0.2–3.2 mg/100 g FW (skin)	Vallejo et al. (2012)
	Rutin	–	4.89–28.7 mg/100 g FW (fruit)	Veberic et al. (2008)
		1.38–12.06 mg/100 g (sun-drying) 3.75–14.62 mg/100 g (sun-drying)	0.61–1.86 mg/100 g FW	Slatnar et al. (2011)
		5.75–12.35 mg/100 g DM	15.16–68.21 mg/100 g DM	Nakilcioğlu and Hışıl (2013)

(continued)

Table 17.5 (continued)

Subclass	Component	Dried	Fresh	References
		9–15 mg/100 g DM	12–13 mg/100 g DM (whole) 86–180 mg/100 g DM (skin) 0.9–2 mg/100 g DM (pulp)	Kamiloglu and Capanoglu (2015)
	Kaempferol-3-O-glucoside	0.31–0.59 mg/100 g (sun-drying) 0.56–1.43 mg/100 g (oven-drying)	0.04–0.13 mg/100 g FW	Slatnar et al. (2011)
	Luteolin-8-C-glucoside	0.13–0.16 mg/100 g (sun-drying) 0.21–0.45 mg/100 g (oven-drying)	–	Slatnar et al. (2011)
	Quercetin-3-O-glucoside	0.56–3.55 mg/100 g (sun-drying) 1.10–2.98 mg/100 g (oven-drying)	0.18–0.60 mg/100 g FW	Slatnar et al. (2011)
		0.2–0.7 mg/100 g DM	0.3–1.2 mg/100 g DM (whole) 1.9–9.4 mg/100 g DM (skin) 0.1–0.2 mg/100 g DM (pulp)	Kamiloglu and Capanoglu (2015)
		0.12–1.27 mg/100 g DM		Pourghayoumi et al. (2016)
	Keampferol-3-O-rutinoside	0.6–30.5 mg/100 g DM	–	Wojdyło et al. (2016)
	Quercetin-3-O-rutinoside	39.0–328.3 mg/100 g DM	–	Wojdyło et al. (2016)
		4.33–88.40 mg/kg DM		Faleh et al. (2012)
	Quercetin-3-Galactoside	4.6–14.2 mg/100 g DM	–	Wojdyło et al. (2016)
	Quercetin-3-O-malonyl-galactoside	5.3–58.0 mg/100 g DM	–	Wojdyło et al. (2016)
Anthocyanins	Cyanidin-3-glucoside		Nd-11.1 mg/100 g FW (skin)	Vallejo et al. (2012)

(continued)

Table 17.5 (continued)

Subclass	Component	Dried	Fresh	References
		Nd-0.1 mg/100 g DM	0.1–6 mg/100 g DM (whole) Nd-23 mg/100 g DM (skin) 0.1–2.8 mg/100 g DM (pulp)	Kamiloglu and Capanoglu (2015)
	Cyanidin-3-rutinoside		Nd-108.9 mg/100 g FW (skin) Nd-9.5 mg/100 g FW (pulp)	Vallejo et al. (2012)
		0.12–0.26 mg/100 g (sun-drying) 0.12–0.31 mg/100 g (oven-drying)	0.21–0.62 mg/100 g FW	Slatnar et al. (2011)
		Nd-1.5 mg/100 g DM	2.1–36 mg/100 g DM (whole) 0.1–180 mg/100 g DM (skin) 2.2–18 mg/100 g DM (pulp)	Kamiloglu and Capanoglu (2015)
	Cyanidin-3,5-O-diglucoside	0.0–1.9 mg/100 g DM	–	Wojdyło et al. (2016)
	Pelargonidin-3-O-rutinoside	0.0–3.8 mg/100 g DM	–	Wojdyło et al. (2016)
	Cyanidin-3-O-rutinoside	0.4–116.2 mg/100 g DM		Wojdyło et al. (2016)
Flavanols	(−)-Epicatechin		0.34–0.97 mg/100 g FW (fruit)	Veberic et al. (2008)
		10.44–23.30 mg/100 g (sun-drying) 26.66–36.65 mg/100 g (oven-drying)	7.11–8.67 mg/100 g FW	Slatnar et al. (2011)
		9.0–43.1 mg/100 g DM		Wojdyło et al. (2016)
		14.76–23.42 mg/100 g DM	42.29–76.90 mg/100 g DM	Nakilcioğlu and Hışıl (2013)

(continued)

Table 17.5 (continued)

Subclass	Component	Dried	Fresh	References
		0.6–2.2 mg/100 g DM	1.1–6.9 mg/100 g DM (whole) 2.7–29 mg/100 g DM (skin) 0.9–4.8 mg/100 g DM (pulp)	Kamiloglu and Capanoglu (2015)
	(+)-Catechin	–	1.07–4.03 mg/100 g FW (fruit)	Veberic et al. (2008)
		5.88–11.46 mg/100 g (sun-drying) 15.57–19.75 mg/100 g (oven-drying)	1.36–2.88 mg/100 g FW	Slatnar et al. (2011)
		20.6–95.7 mg/100 g DM	–	Wojdyło et al. (2016)
		1.5–2.5 mg/100 g DM	2.3–13 mg/100 g DM (whole) 6–8 mg/100 g DM (skin) 2.1–27 mg/100 g DM (pulp)	Kamiloglu and Capanoglu (2015)
		1.25–4.77 mg/100 g DM		Pourghayoumi et al. (2016)

phenotypes (12.3–88.4 mg/kg DM) than in the green (4.33–55.5 mg/kg DM) and black ones (17.5–39.6 mg/kg DM).

Vallejo et al. (2012) evaluated the individual phenolics from different parts of fig. According to the results, cyanidin-3-rutinoside was found as a major phenolic compound among anthocyanins in the skin part of fig, reaching up to 108.9 mg/100 g, while fig pulp was poor in cyanidin-3-rutinoside (9.5 mg/100 g).

Kamiloglu and Capanoglu (2015) evaluated the individual phenolics for skin and pulp parts of figs. The skin part was rich in chlorogenic acid, ellagic acid, and *p*-coumaric acid than pulp, whereas gallic acid was found at a higher pulp concentration than skin. Rutin was the predominant flavonol, identified mainly in the fruit skins. Among flavon-3-ols, catechin was found mainly in the skin part of yellow figs and the pulp of purple figs. The skin part contained higher anthocyanin levels than the purple figs pulp, with eight and ten folds of cyanidin-3-glucoside and cyanidin-3-rutinoside, respectively.

The ripening stage is another factor that affects the individual content of phenolics in figs. Slatnar et al. (2011) evaluated the effect of sampling date (ripening stage) on individual phenolics. The highest content of individual phenolics was observed at the second sampling of the first crop. Epicatechin was the predominant individual phenolic compound, with a concentration ranging between 7.11 and 8.67 mg 100 g FW in fresh figs.

The rain-fed condition also affects the phenolic composition of figs. Sedaghat and Rahemi (2018) found that (+) catechin, chlorogenic acid, (−) epicatechin, and quercetin-3-*O*- glucoside exhibited higher values in rain-fed figs during fruit ripening. In the last ripening stage, the concentration of (+) catechin, Chlorogenic acid, (−) epicatechin, and Quercetin-3-*O*-glucoside reached up to 17.06, 15.51, 33.3, and 4.19 mg/g, respectively.

Drying methods are also influential factors in the concentration of individual phenolic compounds. Slatnar et al. (2011) reported that the oven-drying process increased individual phenolics other than cyanidin-3-O-rutinoside compared with the sun-drying process. Kamiloglu and Capanoglu (2015) reported that the chlorogenic acid and *p*-coumaric acid contents decreased, while gallic acid increased after drying. After drying, the decreasing trend was observed in all flavonols of yellow figs. (24% for rutin, 33% both for kaempferol-rutinoside and Q3G, and 50% for quercetin derivatives). A similar loss was observed in Q3G and quercetin derivatives 1 and 2 for purple figs. On the other hand, for purple figs, rutin and kaempferol rutinoside increased with drying process (16% and 200%, respectively).

4 Extraction Methods

Several extraction methods, including conventional and novel techniques, have been used to extract the phenolic compounds from the fig fruit. Maghsoudlou et al. (2016) investigated the effects of different extraction techniques on the phenolic compounds of fig. The authors evaluated the subcritical water extraction, ultrasound-assisted extraction (UAE), and shaker solvent extraction techniques, and the antioxidant activities of the extracted compounds were determined using DPPH·, reducing power (RP), and rancimat tests. According to their results, subcritical water extraction was the most appropriate technique for higher extraction abilities for the total phenolics and flavonoids and obtaining the lowest values in the DPPH· and RP tests (in the skin part). The authors also stated that the ultrasound-assisted extraction gave better results when the Rancimat test was used to determine the antioxidant activity of the extracts.

Alcántara et al. (2020) studied the effects of UAE and solvent on the phenolic profile of fig leaves extracts. The researchers deployed the liquid solvent medium in two ways: a) distilled water and b) 50% (v/v) of water-alcohol solution. In comparing UAE and conventional extraction techniques, the researchers stated that UAE extracted more carotenoids while the extraction performance of total phenolics using conventional extraction was higher than the UAE. In conclusion, the authors stated that the highest anti-inflammatory effect was observed when the extractions were carried out using the hydro-ethanolic solvent.

Qin et al. (2015) used ionic liquid-based UAE to extract such phenolic components as rutin, psoralen, chlorogenic acid, gallic acid, and bergapten from different parts of fig. In addition, the researchers performed an optimization in their UAE procedure. In conclusion, they stated that the highest extraction yields of the phenolic compounds were obtained at the optimal conditions of 1.0 M of 1-butyl-3-methylimidazolium hexafluorophosphate [BMIM][PF6] concentration, the solid-liquid ratio of 1:50 and the extraction temperature of 30 °C. Furthermore, according to their findings, the ionic liquid usage achieved an increase in the extraction yield and a decrease in the extraction time.

Jokić et al. (2014) also compared the use of UAE and classic solid-liquid extraction and determined their effects on the total phenolics, total flavonoids, and total color change of fig extracts. Their results showed that higher total phenolic and flavonoid concentrations (up to 13.7 and 18.5%, respectively) could be obtained using the UAE. Besides, UAE achieved a decrease in the total color changes of the fig extracts compared to classical extraction. The researchers concluded that higher quality of extracts and higher extraction yields could be achieved by using UAE.

5 Biological Activities of Figs

5.1 Antioxidant Activity

The antioxidant effect of phenolic compounds is based on their capacity to reduce oxygen concentration, which has a degenerative effect on the body, to prevent the initiation of oxidation, and to scavenge free radicals (Karabulut & Yemiş, 2019; Minatel et al., 2017). Several studies indicated that phenolic compounds found in plant materials positively affect human health, attributed mainly to their antioxidant activities (Bucić-Kojić et al., 2011). Dried figs are good sources of non-enzymatic antioxidants that positively impact human health (Mostapha Bachir Bey & Louaileche, 2015), and dark-colored fig varieties contain more polyphenols flavonoids, flavonols, anthocyanins, and proanthocyanidins than lighter varieties (Mostapha Bachir Bey & Louaileche, 2015). The different antioxidant activities of various extracts from figs depending on the cultivar and applied antioxidant tests are presented in Table 17.6.

Slatnar et al. (2011) evaluated the antioxidant activity with the DPPH method in different drying processes. Antioxidant activities of dried figs using the oven and sun-drying methods exerted almost two-fold higher values than fresh figs.

F Aljane et al. (2020) examined the antioxidant activities of 27 Tunisian ecotypes using DPPH· and ABTS assays. The inhibition percentage in the DPPH assay changed between 11.3% for *Besbessi* and 64.7% for *Bouharrag*. The scavenging activity of ABTS radical ranged from 263 mg equivalent vitamin C/100 g FW (*Mahdoui*) to 676 mg equivalent vitamin C/100 g FW (*Nemri*).

Nakilcioğlu and Hışıl (2013) determined the antioxidant activities of ten fresh figs from the *Sarilop* cultivar and also evaluated the effect of the drying process on the antioxidant activity. The antioxidant activities of dried samples were higher than fresh samples. DPPH· scavenging activity of fresh samples was between 1657–2241 mg Trolox equivalent/100 g DM, while it was found between

Table 17.6 Studies on the antioxidant activities of figs

Antioxidant test	Fresh Fruit	Dried Fruit	References
ABTS	–	IC_{50} value: 19.8 mg/mL	Soni et al. (2014)
	263.7–676.13 mg equivalent vitamin C/100 g FW	–	F Aljane et al. (2020)
FRAP	–	60.48 µg BHT equivalent/mg sample	Soni et al. (2014)
	4.6–18.7 mmol Fe^{2+}/ kg FW	–	Ercisli et al. (2012)
DPPH		EC_{50}: 0.72 mg/mL (WE) EC_{50}:0.61 mg/mL (PS)	Yang et al. (2009)
	Peel EC_{50}: 4.52–26.7 mg/mL Pulp EC_{50}: 10.59–15.45 mg/mL Whole fruit EC_{50}: 7.04–14.6 mg/mL	–	Harzallah et al. (2016)
	–	IC_{50} value: 0.45–3.45 mg/mL	Elahe Maghsoudlou et al. (2017)
	Inhibition percentage: 11.37–64.73%	–	F Aljane et al. (2020)
	–	28.33% -45.25%	Bey and Louaileche (2015)
Superoxide radical scavenging activity		EC_{50}:0.95 mg/mL (PS) EC_{50}:2.81 mg/mL (WE)	Yang et al. (2009)
		137.42–315.70 mg GAE/100 g	Bey & Louaileche (2015)
Hydroxyl radical scavenging activity		At the concentration of 4 mg/mL −43.4% for PS −31.0% for WE	Yang et al. (2009)
Hydrogen peroxide scavenging activity		92.21–210.77 mg GAE/100 g	Bey and Louaileche (2015)
Phosphomolybdenum antioxidant activity		621.72–1024.37 mg GAE/100 g	Bey and Louaileche (2015)

(continued)

Table 17.6 (continued)

Antioxidant test	Fresh Fruit	Dried Fruit	References
Reducing power		At the concentration of 4 mg/mL – OD value: 0.58 for WE – OD value: 0.35 for PS	Yang et al., 2009)
		IC_{50} value: 0.90–4.62 mg/ml	Maghsoudlou et al. (2017)
ORAC		0.46–1.33 mM Trolox/100 g DM	Wojdyło et al. (2016)
TEAC	–	Whole 25.0–716.3 μmol of TE/100 g Skin 82.0–1987.0 μmol of TE/100 g Pulp 20.8–357.5 μmol of TE/100 g	Solomon et al. (2006)
	36–623 μg Trolox Equivalent/100 g	–	Ercisli et al. (2012)
	1.44–3.14 μmol TE/g FW	–	Crisosto et al. (2010)

WE water extract, *PS* crude hot-water soluble polysaccharide

663.2–1103 mg Trolox equivalent /100 g DM in dried samples. FRAP scavenging capacity of fresh samples was between 553.4–914.6 mg $FeSO_4$/100 g DM, while the scavenging capacity of dried samples changed between 176.49 and 290.94 mg $FeSO_4$/100 g DM.

Yang et al. (2009) used different antioxidant assays such as DPPH·, superoxide radical scavenging activity, hydroxyl radical scavenging activity, and RP to evaluate the activities of water extract (WE) and crude hot-water soluble polysaccharide (PS) from fig fruit. Both WE and PS showed good scavenging activities on DPPH· (EC_{50} values for WE and PS were 0.72 and 0.61 mg/ml, respectively). Besides, PS had higher antioxidant activity than WE on superoxide radical, hydroxyl anion radical, and reducing power assays.

Harzallah et al. (2016) examined the antioxidant properties of juices of peel, pulp, and total fruit of figs from three different varieties with three different colors (green, purple, and black) grown in Tunisia via DPPH· scavenging capacity and RP. Black peel juice showed the highest antioxidant activities (DPPH and RP), followed by black fig total fruit juice.

Pereira et al. (2017) reported that the hydrophilic fraction had higher antioxidant activity than the lipophilic fraction in the skin and flesh parts from nine different cultivars. Considering the parts of the fruit, the skin part had higher values (between two and ten times) than the flesh, depending on the cultivar. Besides, the dark-colored cultivars showed higher antioxidant activity (16.3–177 mmol Trolox equivalents per 100 g of FW) than the lighter ones. Solomon et al. (2006) emphasized that the effect of the skin part (82–2000 μmol/100 g of FW) on antioxidant activity was higher than that of the pulp (21–358 μmol/100 g of FW).

Besides fruit, latex is also rich in antioxidant compounds. Shahinuzzaman et al. (2020) extracted bioactive compounds from fig fruit using maceration (ME) and ultrasound-assisted extraction (UAE) methods. Amongst the 18 cultivars collected in Malaysia, the latex from the *White Genoa* cultivar exhibited the highest antioxidant activity on DPPH· (65.91% for ME, 61.07% for UAE), ABTS (98.9% for ME, 83.0% for UAE), FRAP (27.0 mg TE/g latex for ME, 24.9 mg TE/g latex for UAE).

Elahe Maghsoudlou et al. (2017) used pulp and skin extracts from two fig cultivars (*Sabz* and *Siyah*) to enhance the oxidative stability of canola oil. The skin extract from the *Siyah* cultivar enhanced the oxidative stability, and its effect was comparable with a synthetic antioxidant, TBHQ. Therefore, the authors declared that the *Siyah* skin extract *via* subcritical water extraction could be used as a natural antioxidant in the food system.

5.2 *Antimicrobial Activity*

There is an increasing interest in plants due to their resisting attributes to antimicrobial drugs and the adverse effects of synthetic agents. For this reason, the search for new antimicrobial compounds, such as phenolic compounds obtained from different plants, is inevitable (Mahmoudi et al., 2016). Some studies regarding the antimicrobial effect of fig, mainly originating from its phenolic compounds, are presented in Table 17.7. Jeong et al. (2009) reported that MeOH extract of *Ficus carica* leaves was inhibitory against some oral bacteria such as *S. gordonii, S. anginosus, P. intermedia, A. actinomycetemcomitans,* and *P. gingivalis*. It was suggested that the flavonoids in *Ficus carica* might be related to the antimicrobial effect observed in this study, as some flavonoid compounds had intense antimicrobial activity against oral bacteria. Mahmoudi et al. (2016) extracted phenolic compounds from ten *Ficus carica* L. leaves by the soxhlet method. They reported that the tested fig leaves were good sources of polyphenol and flavonoid compounds and had antimicrobial effects against Gram-positive and Gram-negative bacteria. Lazreg-Aref et al. (2012) determined that hexane extract contained high levels of coumarin and reported that pure coumarins had a strong antimicrobial effect against *S. saprophyticus, S. aureus, E. coli, S. epidermidis, E. cloacae, E. faecalis*, and *P. aeruginosa*. Soni et al. (2014) evaluated the antibacterial activity of fig extract against *Proteus mirabilis* and *Bacillus subtilis*. Fig extracts exhibited promising antimicrobial activity at 100 mg/ml concentration against *Proteus mirabilis* (18.5 mm) and *Bacillus subtilis* (16 mm).

Table 17.7 Studies on the antimicrobial activities of fig

Part used	Bioassay	Bioactivity	References
Leaves	Broth dilution	The MeOH extract showed strong antibacterial activity against oral bacteria.	Jeong et al. (2009)
Leaves	Disc diffusion, Macrodilution broth	The extract of fig leaves showed bactericidal activity against G(+) and G(−) bacteria and moderate antifungal activity.	Mahmoudi et al. (2016)
Leaves	Agar well diffusion	Aqueous extract was active against G(+) bacteria more than G(−) bacteria but not active against yeast. The ethanol extracts presented high antimicrobial properties against G(+) and G(−).	Al Askari et al. (2013)
Latex	Disc diffusion, Micro-well dilution	Ethyl acetate extract inhibited the growth of five bacterial species.	Aref et al. (2010)
Latex	Disc diffusion, Micro-dilution	The methanol extract inhibited the growth of *Staphylococcus aureus*.	Lee and Cha (2010)
Dried fig	Well dilution	The dried fig extract showed inhibition against *Bacillus subtilis* and *Proteus mirabilis*.	Soni et al. (2014)

5.3 *Anticancer Activity*

Plants have been used for medicinal purposes since ancient times and are still used in alternative medicine today. Studies have shown that phytochemicals in plants have many biological activities such as antioxidant, anticancer, antibacterial, and anti-inflammatory activities (Mawa et al., 2013). For example, anticancer activity is agents' ability with anticancer properties to reverse, stop or slow down the progression of carcinogens (Chanda & Nagani, 2013). For this reason, plants used in the treatment of many diseases also play an essential role in treating cancer, which causes the most deaths in the world in terms of their anticancer properties. Therefore, many anticancer drugs have been derived from plant-based sources.

In addition to herbs and plant-based drugs, fruits and vegetables also act as anticarcinogenic agents to prevent cancer. Fig is an excellent source of phytochemicals and anticancer medicine regarding seeds, leaves, and roots (Mawa et al., 2013). The fig fruit that does not contain fat has a rich content of amino acids, minerals, vitamins, phenolic compounds, and fiber. No substantial loss was reported in terms of phenolic compounds and phytochemicals in dried figs (Gençdağ et al., 2021), and it can maintain its anticancer properties for a long time (Hashemi et al., 2011). Many studies have been conducted on the anticancer properties of figs and their effects on health. For example, the effects of fig leaf extract on four types of human cancer cells: colon cancer, laryngeal carcinoma, hepatocellular carcinoma, and breast cancer were investigated and it was stated that this extract had a great potential inhibitory effect, especially on laryngeal carcinoma and hepatocellular carcinoma (Abdel-Rahman et al., 2021).

Similarly, the extract obtained from the fig fruit was reported to be effective on breast cancer. Moreover, the fruit extract was also reported to have inhibitory properties on breast cancer (Jasmine et al., 2015). It was also reported that silver nanoparticles synthesized using fig extract showed an anti-carcinogenic effect against breast cancer cells (Jacob et al., 2017).

Purnamasari et al. (2019) studied the growth inhibition in liver cancer cells using the fruit and leaf extracts obtained from fig. According to their results, it was determined that the extract obtained from the leaf inhibited the growth of liver cancer cells more than that of the fruit extract.

The anticancer properties of figs are based on the bioactive compounds and antioxidant properties that it contains (Soltana et al., 2019). Antioxidants have the protective mechanisms to alleviate cancers by improving the immune system. Therefore, the fig is a natural remedy that helps treat various ailments related to the immune system, such as gastrointestinal, respiratory, and cardiovascular disorders and cancer (Mawa et al., 2013). Furthermore, Jasmine et al. (2015) highlighted a strong correlation between the antioxidant activity and the anticancer activity of fig.

On the other hand, the factors that inhibit the growth of cancer cells could be due to other bioactive compounds that are naturally present in fig. It was observed that fig extracts, which do not show anti-proliferative effects in normal cell lines, cause cytotoxicity in breast cancer cells (Uz et al., 2015). Therefore, it can be stated that the bioactive compounds found in the fig fruits exhibit anti-proliferative properties acting directly on the cancer cells. Moreover, the inhibition of the growth of cancer cells might occur due to the individual bioactive compounds contained in the fig and their interaction with each other (Boyacıoğlu et al., 2021). Studies in the literature support using figs for therapeutic purposes, especially cancer, due to their phytochemical content.

6 Conclusion

This chapter focused on phenolic compounds'presence and biological activities in *Ficus carica* fruits. More than 800 species with 2000 varieties of the *Ficus* genus are defined by different researchers. In this chapter, a thorough review has been performed by considering the research findings up to this time. In this respect, phenolic constituents, i.e., phenolic acids, flavonoids, and anthocyanins isolated by different studies for various species, were compiled and presented. It can be concluded that the phenolic composition of fig fruits is highly dependent upon cultivar, harvesting practices, soil type, geography, and the extraction and analysis methods used. The novel extraction techniques such as supercritical fluid extraction, microwave-assisted extraction, and ultrasound-assisted extraction have simultaneous advantages in the extraction of fig phenolics by being more efficient and environmental-friendly. The figs' flesh and peel, leaf, and seed parts contain many phenolic compounds with potent antioxidant, anti-microbial and anticarcinogenic activities. Therefore, industrial by-products or wastes of fig fruits should also be evaluated and valorized. Although there are several studies regarding the biological activities of fig fruits, future investigations are needed on *in vitro* and *in vivo* studies to reveal the further beneficial properties of phenolic compositions of fig fruits.

References

Abdel-Rahman, R., Ghoneimy, E., Abdel-Wahab, A., Eldeeb, N., Salem, M., Salama, E., & Ahmed, T. (2021). The therapeutic effects of *Ficus carica* extract as antioxidant and anticancer agent. *South African Journal of Botany, 141*, 273–277. https://doi.org/10.1016/j.sajb.2021.04.019

Ahmad, J., Khan, I., Khan, S., & Iqbal, D. (2013). Evaluation of antioxidant and antimicrobial activity of *Ficus carica* leaves: An in vitro approach. *Journal of Plant Pathology and Microbiology, 4*(1), 1000157. https://doi.org/10.4172/2157-7471.1000157

Ainsworth, E. A., & Gillespie, K. M. (2007). Estimation of total phenolic content and other oxidation substrates in plant tissues using Folin-Ciocalteu reagent. *Nature Protocols, 2*(4), 875–877. https://doi.org/10.1038/nprot.2007.102

Al Askari, G., Kahouadji, A., Khedid, K., Ouaffak, L., Mousaddak, M., Charof, R., & Mennane, Z. (2013). In vitro antimicrobial activity of aqueous and ethanolic extracts of leaves of *Ficus carica* collected from five different regions of Morocco. *Journal of Materials and Environmental Science, 4*(1), 33–38.

Alcántara, C., Žugčić, T., Abdelkebir, R., García-Pérez, J. V., Jambrak, A. R., Lorenzo, J. M., et al. (2020). Effects of ultrasound-assisted extraction and solvent on the phenolic profile, bacterial growth, and anti-inflammatory/antioxidant activities of Mediterranean olive and fig leaves extracts. *Molecules, 25*(7), 1718. https://doi.org/10.3390/molecules25071718

Aljane, F., & Ferchichi, A. (2009). Postharvest chemical properties and mineral contents of some fig (*Ficus carica* L.) cultivars in Tunisia. *Journal of Food, Agriculture & Environment, 7*(2), 209–212.

Aljane, F., Neily, M., & Msaddak, A. (2020). Phytochemical characteristics and antioxidant activity of several fig (*Ficus carica* L.) ecotypes. *Italian Journal of Food Science, 32*(4), 755–768. https://doi.org/10.14674/IJFS.1884

Aref, H. L., Salah, K. B., Chaumont, J. P., Fekih, A., Aouni, M., & Said, K. (2010). In vitro antimicrobial activity of four *Ficus carica* latex fractions against resistant human pathogens (antimicrobial activity of *Ficus carica* latex). *Pakistan Journal of Pharmaceutical Sciences, 23*(1), 53–58.

Arvaniti, O. S., Samaras, Y., Gatidou, G., Thomaidis, N. S., & Stasinakis, A. S. (2019). Review on fresh and dried figs: Chemical analysis and occurrence of phytochemical compounds, antioxidant capacity and health effects. *Food Research International, 119*, 244–267. https://doi.org/10.1016/j.foodres.2019.01.055

Backes, E., Leichtweis, M. G., Pereira, C., Carocho, M., Barreira, J. C. M., Kamal Genena, A., et al. (2020). *Ficus carica* L. and *Prunus spinosa* L. extracts as new anthocyanin-based food colorants: A thorough study in confectionery products. *Food Chemistry, 333*, 127457. https://doi.org/10.1016/j.foodchem.2020.127457

Badgujar, S. B., Patel, V. V., Bandivdekar, A. H., & Mahajan, R. T. (2014). Traditional uses, phytochemistry and pharmacology of *Ficus carica*: A review. *Pharmaceutical Biology, 52*(11), 1487–1503. https://doi.org/10.3109/13880209.2014.892515

Benmaghnia, S., Meddah, B., Tir-Touil, A., & Gabaldón Hernández, J. A. (2019). Phytochemical analysis, antioxidant and antimicrobial activities of three samples of dried figs (*Ficus carica* L.) from the region of mascara. *Journal of Microbiology, Biotechnology and Food Sciences, 9*(2), 208–215. https://doi.org/10.15414/jmbfs.2019.9.2.208-215

Bey, M. B., & Louaileche, H. (2015). A comparative study of phytochemical profile and in vitro antioxidant activities of dark and light dried fig (*Ficus carica* L.) varieties. *The Journal of Phytopharmacology, 4*(1), 41–48.

Bey, M. B., Richard, G., Meziant, L., Fauconnier, M.-L., & Louaileche, H. (2017). Effects of sun-drying on physicochemical characteristics, phenolic composition and *in vitro* antioxidant activity of dark fig varieties. *Journal of Food Processing and Preservation, 41*(5), e13164. https://doi.org/10.1111/jfpp.13164

Boyacıoğlu, O., Kara, B., Tecimen, S., Kılıç, M., Delibaş, M., Erdoğan, R., et al. (2021). Antiproliferative effect of *Ficus carica* latex on cancer cell lines is not solely linked to peroxidase-like activity of ficin. *European Journal of Integrative Medicine, 45*, 101348. https://doi.org/10.1016/j.eujim.2021.101348

Bucić-Kojić, A., Planinić, M., Tomas, S., Jokić, S., Mujić, I., Bilić, M., & Velić, D. (2011). Effect of extraction conditions on the extractability of phenolic compounds from lyophilised fig fruits (*Ficus carica* L.). *Polish Journal of Food and Nutrition Sciences, 61*(3), 195–199. https://doi.org/10.2478/v10222-011-0021-9

Çalişkan, O., & Aytekin Polat, A. (2011). Phytochemical and antioxidant properties of selected fig (*Ficus carica* L.) accessions from the eastern Mediterranean region of Turkey. *Scientia Horticulturae, 128*(4), 473–478. https://doi.org/10.1016/j.scienta.2011.02.023

Chanda, S., & Nagani, K. (2013). In vitro and in vivo methods for anticancer activity evaluation and some Indian medicinal plants possessing anticancer properties: An overview. *Journal of Pharmacognosy and Phytochemistry, 2*(2), 140–152.

Crisosto, C. H., Bremer, V., Ferguson, L., & Crisosto, G. M. (2010). Evaluating quality attributes of four fresh fig (*Ficus carica* L.) cultivars harvested at two maturity stages. *HortScience, 45*(4), 707–710. https://doi.org/10.21273/hortsci.45.4.707

de la Rosa, L. A., Moreno-Escamilla, J. O., Rodrigo-García, J., & Alvarez-Parrilla, E. (2019). Phenolic compounds. In E. M. Yahia (Ed.), *Postharvest physiology and biochemistry of fruits and vegetables* (pp. 253–271). Woodhead Publishing.

Ercisli, S., Tosun, M., Karlidag, H., Dzubur, A., Hadziabulic, S., & Aliman, Y. (2012). Color and antioxidant characteristics of some fresh fig (*Ficus carica* L.) genotypes from northeastern Turkey. *Plant Foods for Human Nutrition, 67*(3), 271–276. https://doi.org/10.1007/s11130-012-0292-2

Faleh, E., Oliveira, A. P., Valentão, P., Ferchichi, A., Silva, B. M., & Andrade, P. B. (2012). Influence of Tunisian *Ficus carica* fruit variability in phenolic profiles and in vitro radical scavenging potential. *Revista Brasileira de Farmacognosia, 22*(6), 1282–1289. https://doi.org/10.1590/S0102-695X2012005000132

FAO. (2021). *Food and agriculture organization of the United Nations*. FAOSTAT Data: Crops and livestock products. https://www.fao.org/faostat/en/#data/. Accessed 23 Sept 2021.

Gençdağ, E., Görgüç, A., Okuroğlu, F., & Yılmaz, F. M. (2021). The effects of power – Ultrasound, peroxyacetic acid and sodium chloride washing treatments on the physical and chemical quality characteristics of dried figs. *Journal of Food Processing and Preservation, 45*(1), e15009. https://doi.org/10.1111/jfpp.15009

Görgüç, A., Gençdağ, E., Okuroğlu, F., Yılmaz, F. M., Bıyık, H. H., Öztürk Köse, S., & Ersus, S. (2021). Single and combined decontamination effects of power-ultrasound, peroxyacetic acid and sodium chloride sanitizing treatments on *Escherichia coli*, *Bacillus cereus* and *Penicillium expansum* inoculated dried figs. *LWT, 140*, 110844. https://doi.org/10.1016/j.lwt.2020.110844

Harzallah, A., Bhouri, A. M., Amri, Z., Soltana, H., & Hammami, M. (2016). Phytochemical content and antioxidant activity of different fruit parts juices of three figs (*Ficus carica* L.) varieties grown in Tunisia. *Industrial Crops and Products, 83*, 255–267. https://doi.org/10.1016/j.indcrop.2015.12.043

Hashemi, S. A., Abediankenari, S., Ghasemi, M., Azadbakht, M., Yousefzadeh, Y., & Dehpour, A. A. (2011). The effect of fig tree latex (*Ficus carica*) on stomach cancer line. *Iranian Red Crescent Medical Journal, 13*(4), 272–275.

Hssaini, L., Charafi, J., Razouk, R., Hernández, F., Fauconnier, M. L., Ennahli, S., & Hanine, H. (2020). Assessment of morphological traits and fruit metabolites in eleven fig varieties (*Ficus carica* L.). *International Journal of Fruit Science, 20*(sup2), 8–28. https://doi.org/10.1080/15538362.2019.1701615

Jacob, S. J. P., Prasad, V. L. S., Sivasankar, S., & Muralidharan, P. (2017). Biosynthesis of silver nanoparticles using dried fruit extract of *Ficus carica* – Screening for its anticancer activity and toxicity in animal models. *Food and Chemical Toxicology, 109*, 951–956. https://doi.org/10.1016/j.fct.2017.03.066

Jagtap, U. B., & Bapat, V. A. (2020). Exploring phytochemicals of *Ficus carica* L. (fig). In H. N. Murthy & V. A. Bapat (Eds.), *Bioactive compounds in underutilized fruits and nuts* (pp. 353–368). Springer International Publishing.

Jasmine, R., Manikandan, K., & Karthikeyan, K. (2015). Evaluating the antioxidant and anticancer property of *Ficus carica* fruits. *African Journal of Biotechnology, 14*(7), 634–641. https://doi.org/10.5897/AJB2014.13742

Jeong, M.-R., Kim, H.-Y., & Cha, J.-D. (2009). Antimicrobial activity of methanol extract from *Ficus carica* leaves against oral bacteria. *Journal of Bacteriology and Virology, 39*(2), 97–102. https://doi.org/10.4167/jbv.2009.39.2.97

Jokić, S., Mujić, I., Bucić-Kojić, A., Velić, D., Bilić, M., Planinić, M., & Lukinac, J. (2014). Influence of extraction type on the total phenolics, total flavonoids and total colour change of different varieties of fig extracts. *Hrana u zdravlju i bolesti, 3*(2), 90–95.

Kamiloglu, S., & Capanoglu, E. (2015). Polyphenol content in figs (*Ficus carica* L.): Effect of sun-drying. *International Journal of Food Properties, 18*(3), 521–535. https://doi.org/10.1080/10942912.2013.833522

Karabulut, G., & Yemiş, O. (2019). Fenolik Bileşiklerin Bağlı Formları ve Biyoyararlılığı. *Akademik Gıda, 17*(4), 526–537. https://doi.org/10.24323/akademik-gida.667270

Konyalıoğlu, S., Sağlam, H., & Kıvçak, B. (2005). α-Tocopherol, flavonoid, and phenol contents and antioxidant activity of *Ficus carica* leaves. *Pharmaceutical Biology, 43*(8), 683–686. https://doi.org/10.1080/13880200500383538

Lachtar, D., Zaouay, F., Pereira, C., Martin, A., Ben Abda, J., & Mars, M. (2021). Combined effects of drying method and variety on the quality of dried figs. *Journal of Materials and Environmental Science, 12*(2), 217–223.

Lazreg-Aref, H., Mars, M., Fekih, A., Aouni, M., & Said, K. (2012). Chemical composition and antibacterial activity of a hexane extract of Tunisian caprifig latex from the unripe fruit of Ficus carica. *Pharmaceutical Biology, 50*(4), 407–412. https://doi.org/10.3109/13880209.2011.608192

Lee, Y.-S., & Cha, J.-D. (2010). Synergistic antibacterial activity of fig (*Ficus carica*) leaves extract against clinical isolates of methicillin-resistant Staphylococcus aureus. *Microbiology and Biotechnology Letters, 38*(4), 405–413.

Lo Turco, V., Potortì, A. G., Tropea, A., Dugo, G., & Di Bella, G. (2020). Element analysis of dried figs (*Ficus carica* L.) from the Mediterranean areas. *Journal of Food Composition and Analysis, 90*, 103503. https://doi.org/10.1016/j.jfca.2020.103503

Maghsoudlou, E., Esmaeilzade Kenari, R., & Raftani Amiri, Z. (2016). The effects of extraction technique on phenolic compounds extracted from fig (*Ficus carica*) pulp and skin. *Iranian Food Science and Technology Research Journal, 11*(6), 758–769.

Maghsoudlou, E., Esmaeilzadeh Kenari, R., & Raftani Amiri, Z. (2017). Evaluation of antioxidant activity of fig (*Ficus carica*) pulp and skin extract and its application in enhancing oxidative stability of canola oil. *Journal of Food Processing and Preservation, 41*(4), e13077. https://doi.org/10.1111/jfpp.13077

Mahmoudi, S., Khali, M., Benkhaled, A., Benamirouche, K., & Baiti, I. (2016). Phenolic and flavonoid contents, antioxidant and antimicrobial activities of leaf extracts from ten Algerian *Ficus carica* L. varieties. *Asian Pacific Journal of Tropical Biomedicine, 6*(3), 239–245. https://doi.org/10.1016/j.apjtb.2015.12.010

Mawa, S., Husain, K., & Jantan, I. (2013). *Ficus carica* L. (Moraceae): Phytochemistry, traditional uses, and biological activities. *Evidence-Based Complementary and Alternative Medicine, 2013*, 974256. https://doi.org/10.1155/2013/974256

Minatel, I. O., Borges, C. V., Ferreira, H. A. G. G. M. I., Chen, C.-Y. O., & Lima, G. P. P. (2017). Phenolic compounds: Functional properties, impact of processing and bioavailability. In M. P.-T. Soto-Hernandez & M. D. R. Garcia-Mateos (Eds.), *Phenolic compounds-biological activity*. IntechOpen.

Nakilcioğlu, E., & Hışıl, Y. (2013). Research on the phenolic compounds in Sarilop (*Ficus carica* L.) fig variety. *Gida, 38*(5), 267–274. https://doi.org/10.5505/gida.2013.08208

Oliveira, A. P., Valentão, P., Pereira, J. A., Silva, B. M., Tavares, F., & Andrade, P. B. (2009). *Ficus carica* L.: Metabolic and biological screening. *Food and Chemical Toxicology, 47*(11), 2841–2846. https://doi.org/10.1016/j.fct.2009.09.004

Pereira, C., López-Corrales, M., Serradilla, M. J., Villalobos, M., Ruiz-Moyano, D. C., & Martín, A. (2017). Influence of ripening stage on bioactive compounds and antioxidant activity in nine fig (*Ficus carica* L.) varieties grown in Extremadura, Spain. *Journal of Food Composition and Analysis, 64*, 203–212. https://doi.org/10.1016/j.jfca.2017.09.006

Petkova, N., Ivanov, I., & Denev, P. (2019). Changes in phytochemical compounds and antioxidant potential of fresh, frozen, and processed figs (*Ficus carica* L.). *International Food Research Journal, 26*(6), 1881–1888.

Pourghayoumi, M., Bakhshi, D., Rahemi, M., Noroozisharaf, A., Jafari, M., Salehi, M., et al. (2016). Phytochemical attributes of some dried fig (*Ficus carica* L.) fruit cultivars grown in Iran. *Agriculturae Conspectus Scientificus, 81*(3), 161–166.

Purnamasari, R., Winarni, D., Permanasari, A. A., Agustina, E., Hayaza, S., & Darmanto, W. (2019). Anticancer activity of methanol extract of *Ficus carica* leaves and fruits against proliferation, apoptosis, and necrosis in Huh7it cells. *Cancer Informatics, 18*, 1–7. https://doi.org/10.1177/1176935119842576

Qin, H., Zhou, G., Peng, G., Li, J., & Chen, J. (2015). Application of ionic liquid-based ultrasound-assisted extraction of five phenolic compounds from fig (*Ficus carica* L.) for HPLC-UV. *Food Analytical Methods, 8*(7), 1673–1681. https://doi.org/10.1007/s12161-014-0047-9

Renard, C. M. G. C. (2018). Extraction of bioactives from fruit and vegetables: State of the art and perspectives. *LWT, 93*, 390–395. https://doi.org/10.1016/j.lwt.2018.03.063

Rubnov, S., Kashman, Y., Rabinowitz, R., Schlesinger, M., & Mechoulam, R. (2001). Suppressors of cancer cell proliferation from fig (*Ficus carica*) resin: Isolation and structure elucidation. *Journal of Natural Products, 64*(7), 993–996. https://doi.org/10.1021/np000592z

Sánchez-Rangel, J. C., Benavides, J., Heredia, J. B., Cisneros-Zevallos, L., & Jacobo-Velázquez, D. A. (2013). The Folin-Ciocalteu assay revisited: Improvement of its specificity for total phenolic content determination. *Analytical Methods, 5*(21), 5990–5999. https://doi.org/10.1039/C3AY41125G

Sedaghat, S., & Rahemi, M. (2018). Effects of physio-chemical changes during fruit development on nutritional quality of fig (*Ficus carica* L. var. 'Sabz') under rain-fed condition. *Scientia Horticulturae, 237*, 44–50. https://doi.org/10.1016/j.scienta.2018.04.003

Shahidi, F., Varatharajan, V., Oh, W. Y., & Peng, H. (2019). Phenolic compounds in agri-food by-products, their bioavailability and health effects. *Journal of Food Bioactives, 5*, 57–119. https://doi.org/10.31665/JFB.2019.5178

Shahinuzzaman, M., Yaakob, Z., Anuar, F. H., Akhtar, P., Kadir, N. H. A., Hasan, A. K. M., et al. (2020). In vitro antioxidant activity of *Ficus carica* L. latex from 18 different cultivars. *Scientific Reports, 10*(1), 10852. https://doi.org/10.1038/s41598-020-67765-1

Slatnar, A., Klancar, U., Stampar, F., & Veberic, R. (2011). Effect of drying of figs (*Ficus carica* L.) on the contents of sugars, organic acids, and phenolic compounds. *Journal of Agricultural and Food Chemistry, 59*(21), 11696–11702. https://doi.org/10.1021/jf202707y

Slavin, J. L. (2006). Figs: Past, present, and future. *Nutrition Today, 41*(4), 180–184.

Solomon, A., Golubowicz, S., Yablowicz, Z., Grossman, S., Bergman, M., Gottlieb, H. E., et al. (2006). Antioxidant activities and anthocyanin content of fresh fruits of common fig (*Ficus carica* L.). *Journal of Agricultural and Food Chemistry, 54*(20), 7717–7723. https://doi.org/10.1021/jf060497h

Soltana, H., Pinon, A., Limami, Y., Zaid, Y., Khalki, L., Zaid, N., & Hammami, M. (2019). Antitumoral activity of *Ficus carica* L. on colorectal cancer cell lines. *Cellular and Molecular Biology, 65*(6), 6–11.

Soni, N., Mehta, S., Satpathy, G., & Gupta, R. K. (2014). Estimation of nutritional, phytochemical, antioxidant and antibacterial activity of dried fig (*Ficus carica*). *Journal of Pharmacognosy and Phytochemistry, 3*(2), 158–165.

Treutter, D. (2006). Significance of flavonoids in plant resistance: A review. *Environmental Chemistry Letters, 4*(3), 147. https://doi.org/10.1007/s10311-006-0068-8

Uz, R., Abu Bakar, N. H., Swethadri, G. K., Baig, A., Idris, M. A., & Maryam, I. U. (2015). Non-toxic antiproliferative effect of *Ficus carica* fruit extracts on estrogen receptor positive breast cancer cell (MCF-7). *Journal of Chemical and Pharmaceutical Research, 7*(10), 815–821.

Vallejo, F., Marín, J. G., & Tomás-Barberán, F. A. (2012). Phenolic compound content of fresh and dried figs (*Ficus carica* L.). *Food Chemistry, 130*(3), 485–492. https://doi.org/10.1016/j.foodchem.2011.07.032

Veberic, R., Colaric, M., & Stampar, F. (2008). Phenolic acids and flavonoids of fig fruit (*Ficus carica* L.) in the northern Mediterranean region. *Food Chemistry, 106*(1), 153–157. https://doi.org/10.1016/j.foodchem.2007.05.061

Viuda-Martos, M., Barber, X., Pérez-Álvarez, J. A., & Fernández-López, J. (2015). Assessment of chemical, physico-chemical, techno-functional and antioxidant properties of fig (*Ficus carica* L.) powder co-products. *Industrial Crops and Products, 69*, 472–479. https://doi.org/10.1016/j.indcrop.2015.03.005

Vuolo, M. M., Lima, V. S., & Maróstica Junior, M. R. (2019). Phenolic compounds: Structure, classification, and antioxidant power. In M. R. S. Campos (Ed.), *Bioactive compounds* (pp. 33–50). Woodhead Publishing.

Wojdyło, A., Nowicka, P., Carbonell-Barrachina, Á. A., & Hernández, F. (2016). Phenolic compounds, antioxidant and antidiabetic activity of different cultivars of *Ficus carica* L. fruits. *Journal of Functional Foods, 25*, 421–432. https://doi.org/10.1016/j.jff.2016.06.015

Yang, X.-M., Yu, W., Ou, Z.-P., Ma, H.-L., Liu, W.-M., & Ji, X.-L. (2009). Antioxidant and immunity activity of water extract and crude polysaccharide from *Ficus carica* L. fruit. *Plant Foods for Human Nutrition, 64*(2), 167–173. https://doi.org/10.1007/s11130-009-0120-5

Zaffer Ahmad, M., Ali, M., & Mir, S. R. (2013). Anti-diabetic activity of *Ficus carica* L. stem barks and isolation of two new flavonol esters from the plant by using spectroscopical techniques. *Asian Journal of Biomedical and Pharmaceutical Sciences, 3*(18), 22–28.

Chapter 18
Phenolic Compounds of Fresh and Dried Figs: Characterization and Health Benefits

Aicha Debib and Soumaya Menadi

Abbreviations

ABTS	2,2-azinobis-(3-ethylbenzothiazoline-6-sulphonic acid)
BHT	1% tert-butyl-4-methyl-phenol
DM	dry matter
DPPH·	2,2-diphenyl-1-picrylhydrazyl
DW	Dry weight
FRAP	Ferric ion reducing antioxidant power
FW	Fresh weight
GC-MS	Gas chromatography-mass spectrometric
HPLC-DAD	High-performance liquid chromatography coupled to diode array detection
MDA	Malondialdehyd
MS	Mass spectroscopy
ORAC	Oxygen radical absorbance capacity assay.
ROS	Reactive oxygen species
TFC	Total flavonoids content
TPC	Total phenolics content
UHPLC-QTOF-MS and MS/MS	Ultra-high-performance liquid chromatography coupled to quadrupole-time-of-flight mass spectrometry
UV-Vis	Ultraviolet-visible

A. Debib (✉)
Management and Valuation of Agricultural and Aquatic Ecosystems Laboratory, Morsli Abdellah Tipaza University, Tipaza, Algeria

S. Menadi
Department of Molecular Biology and Genetics, Tokat Gaziosmanpasa University, Tokat, Turkey
e-mail: soumaya.menadi9321@gop.edu.tr

M. F. Ramadan (ed.), *Fig* (Ficus carica)*: Production, Processing, and Properties*, https://doi.org/10.1007/978-3-031-16493-4_18

1 Introduction

The idea that food could have medicinal properties is not new. Hippocrates, the father of medicine, proposed the concept "Let food be thy medicine and medicine be thy food" 2500 years ago. Fruits and vegetables are primary food sources providing essential nutrients for sustaining life. They also contain a variety of phytochemicals, such as phenolics and flavonoids, which provide important health benefits (Liu, 2013; Samtiya et al., 2021). Hence, regular consumption of fruits and vegetables is associated with reduced risks of chronic diseases, such as cancers and cardiovascular disease (Dragsted et al., 1993; Jideani et al., 2021; Kaur & Kapoor, 2001). Recently, attention has focused more on assessing the distribution of biologically active compounds among different cultivars (Elkordy et al., 2021). In addition, increasing recent interest in nutraceuticals and functional foods has led plant breeders to initiate the selection of crops with higher-than-normal phenolic antioxidant contents (Hernandez-Ledesma & Martinez-Villaluenga, 2021). Figs fruits are rich sources of bioactive compounds such as phenolics, anthocyanins, carotenoids, and minerals (Cano-Lamadrid & Artés-Hernández, 2021; Meziant et al., 2021). They are infructescences of the fig (*Ficus carica* L.), a deciduous tree belonging to the *Moraceae* family (**Fig.** 18.1). Based on archeological findings, figs were probably one of the first domestically used plants, *ca.* 12,000 years ago (Abbo et al., 2015; Kislev et al., 2006). It is an important crop worldwide for both fresh and dry consumption. In 2020, the total world production of figs was 1,260,000 metric tonnes (FAOSTAT, 2020).

The countries along the Mediterranean coast produce most of the world's figs. In these countries, figs are important constituents of the Mediterranean diet, which is considered one of the healthiest nutritional habits and associated with

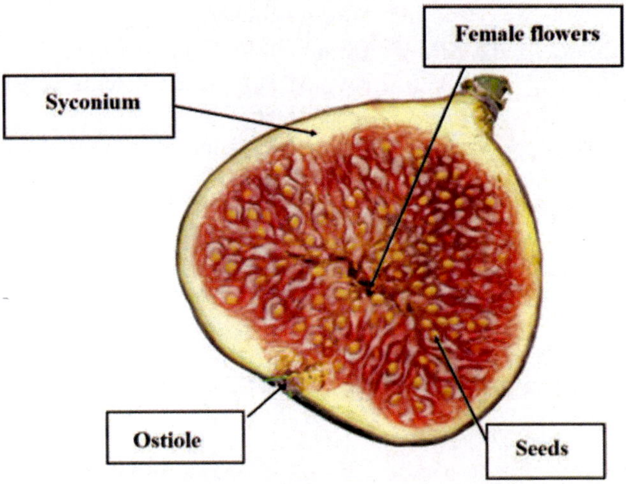

Fig. 18.1 Fig fruit anatomy

longevity(Veberic & Mikulic-Petkovsek, 2016). Bioactive compounds of figs fruits, their characterization and utilization in functional foods, and clinical assessment for human health are among the major targets of contemporary research. There is currently heightened interest focused on figs as major sources of antioxidants that may protect against a wide variety of human diseases (Caliskan, 2015). Previously, several studies have reported on the health-promoting potential of figs due to the presence of high concentrations of polyphenols, specifically in dried figs (Arvaniti et al., 2019; Chang et al., 2016; Debib et al., 2014; dos Anjos Cruz et al., 2021). According to our knowledge, the detailed information about the health-promoting components of figs is needed to better understand their use as functional foods and as ingredients in pharmaceuticals, nutraceuticals and in medicine. Thus, in this review we presents the main phenolic compounds found in fresh and dried figs from different countries, describes the analytical methods used for their identification and discuss theirs biological activities.

2 Research Methods

The search for articles (original or review) was performed using PubMed, Web of Science, Google Scholar, and Scopus electronic databases until March 2022. The main keywords used were: "phenolic components"; "Fig fruit"; "*Ficus carica*"; "Biological activities"; "antioxidant activity"; "antiviral activity"; "hepatoprotective effects,"; "anti-microbial activity" and "health benefits", among others. All the included references were manually selected and reviewed by the authors.

3 Fig Phytochemicals

3.1 Extraction and Quantification Processes

Extraction is a crucial step in isolating phenolic chemicals, as well as later identifying and quantifying them (Konak et al., 2015). Various extraction methods have been used, ranging from traditional solvent extraction using water or organic solvents to alternative techniques that use greener solvents or intensification tools to improve yields and rates (Chemat et al., 2019). From literature data, it is well known that the solvent and temperature are the process parameters that have the largest impact on the efficiency of extracting bioactive chemicals from plant material (Bucic-Kojic et al., 2011). Indeed, the selection of the solvent is one of the most important steps in extraction. Methanol, ethanol, and their mixtures in water, as well as the acetone, petroleum ether, and dimethylformamide, are the most commonly used solvents in the extraction of fig fruit phenolic compounds (Naczk & Shahidi, 2006; Nakilcioğlu-Taş & Ötleş, 2021). Moreover, ethanol and water are much safer and more suitable for the food and pharmaceutical industry than the other organic solvents from a toxicological point of view. According to Bucic-Kojic et al. (2011) and Debib et al. (2014), the fig variety and extraction process conditions significantly affected the content of phenolic compounds in the fig fruits. These two studies reported that the best extraction efficiency was obtained with 80% (v/v) aqueous ethanol solvent. The presence of the highest total phenolic content in aqueous solutions could be attributed to the increase in the polarity of the solvents by the addition of water. On the other hand, several studies reported that ultrasonic extraction generally improves extraction efficiency (Bimakr et al., 2017; Júnior et al., 2021; Vardanega et al., 2014). Thongson et al. reported that the ultrasonic extraction method only took 5 min to obtain the bioactive components of a medicinal plant (Thongson et al., 2004). Although their structural complexity and diversity hamper the quantitative determination of polyphenols, several methods have been used to determine polyphenols in fig fruit extracts. The colorimetric Folin-Ciocalteu essay is the most basic method used; however, modern analytical tools have helped to reveal the highly diverse and complicated chemical structure of phenolic fig compounds, usually using advanced and combined techniques such as GC-MS, HPLC-DAD-ESI/MS, RP-HPLC-PDA, and LC-ESI-MS/MS (Tenório et al., 2022). Many studies have discussed the use of the Folin-Ciocalteau reagent to determine polyphenols and the general or a specific value of the method (Blainski et al., 2013; Margraf et al., 2015; Míguez et al., 2022). In alkaline solutions, this reagent decomposes quickly, necessitating the use of a large amount of the reagent to complete the reaction. Precipitation and severe turbidity can ensue from this excess, making spectrophotometric analysis unfeasible. Folin and Ciocalteu solved this problem by include lithium salts in the reagent, which prevented turbidity. The reaction generally provides accurate and specific data for several groups of phenolic compounds, because many compounds change color differently due to differences in unit mass and reaction kinetics (Blainski et al., 2013).

3.2 *Phenolic Composition of Fig Fruit*

The term phenolic compound involves a wide range of various solvent-soluble compounds, having at least one aromatic ring with one or more hydroxyl groups attached. Based on their structural skeletons, these compounds are broadly divided into simple phenolics, phenolic acids, coumarins, flavonoids, stilbenes, tannins, lignans, and lignins (Miklavčič Višnjevec & Schwarzkopf, 2020; Vuolo et al., 2019). During the last decade, phytochemical studies on dried and fresh figs revealed the presence of numerous bioactive compounds such as phenolic compounds, phytosterols, organic acids, anthocyanin composition, triterpenoids, coumarins, and volatile compounds such as hydrocarbons, and a few other classes of secondary metabolites. However, the fruit peel presented the highest value of phenolic compounds. Indeed, the amount of fresh figs was higher than dried figs. In literature, the reason for this was explained as the demolition of some polyphenols of the fresh figs during the drying process or the transformation to non-antioxidant forms (Nakilcioğlu & Hışıl, 2013). The majority of studies on phenolic compounds in figs originates from Algeria and Tunisia (Ammar et al., 2015; Bachir Bey et al., 2017; Aicha Debib et al., 2016, 2014; Faleh et al., 2012; Kehal et al., 2021; Khadhraoui et al., 2019; Mahmoudi et al., 2016; Meziant et al., 2021; Soltana et al., 2018) and Turkey (Çalişkan & Polat, 2011; Ercisli et al., 2012; Kamiloglu & Capanoglu, 2013; Konak et al., 2015; Nakilcioğlu-Taş & Ötleş, 2021; Nakilcioğlu & Hışıl, 2013; Yemiş, Bakkalbaşı, & Artık, 2012). While there is less data available from Spain (Dueñas et al., 2008; Vallejo et al., 2012; Viuda-Martos et al., 2015; Wojdyło et al., 2016), Slovenia (Slatnar et al., 2011; Veberic et al., 2008), Italy (Del Caro & Piga, 2008; Piga et al., 2005), Croatia (Bucic-Kojic et al., 2011; Jokić et al., 2014), China (Feng et al., 2015; Qin et al., 2015), Iran (Maghsoudlou et al., 2017), United States (Crisosto et al., 2010; Kehal et al., 2021; Vinson et al., 2005) and Portugal (Oliveira et al., 2009; Palmeira et al., 2019,

3.2.1 Phenolic Acids

Phenolic acids consist of two subgroups, hydroxybenzoic and hydroxycinnamic acids. Hydroxybenzoic acids include gallic, *p*-hydroxybenzoic, protocatechuic, vanillic, and syringic acids having a C6-C1 structure, whereas hydroxycinnamic are aromatic compounds with a three-carbon side chain (C6–C3), with caffeic, ferulic, *p*-coumaric, ellagic and chlorogenic acids. The main phenolic acid compounds identified in the *Ficus carica* fruits are described in Table 18.1 According to Arvaniti et al. (2019), the most predominant phenolic acids found in dried figs are gallic acid up to 5.04 mg/100 g Dry Matter (DM) and chlorogenic acid up to 32.4 mg/100 g DM, while the same target compounds in whole fresh figs have been determined at concentrations up to 6.98 mg/100 g DM (gallic acid) and 28.8 mg/100 g DM (chlorogenic acid) (Arvaniti et al., 2019).

Apart from the aforementioned phenolic acids, figs have also been reported to have the presence of various coumarins such as psoralen, bergapten, marmesin, umbelliferone, and angelicin (Veberic & Mikulic-Petkovsek, 2016). In addition, methanolic extracts of fig were shown to possess good psoralen levels, methoxypsoralen, and bergapten (Oliveira et al., 2010; Ammar et al., 2015).

3.2.2 Flavonoids

The generic structure of flavonoids consists of two aromatic rings (A and B rings) linked by three carbons, usually in an oxygenated heterocycle ring called a C ring (Fig. 18.2). Based on differences in the heterocycle C ring, flavonoids are categorized as flavonols (quercetin, kaempferol, and myricetin), flavones (luteolin, and apigenin), flavanols (catechins, epicatechin, epigallocatechin, and epicatechin gallate), flavanones (naringenin), anthocyanidins, or isoflavonoids (genistein, daidzein, dihydrodaidzein, and equol) (Yang & Xiao, 2013). Figs flavonoids are

Table 18.1 Phenolics acids compounds from *Ficus carica* fruits

Structure	Compound	Sample preparation	Analytical method	Reference
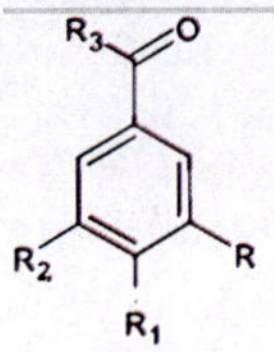	*p*-hydroxybenzoic acid R, R2 = H, R1, R3 = OH	Methanolic extract	HPLC-DAD- ESI/MS	Aicha Debib et al. (2016)
	Gallic acid R, R1, R2, R3 = OH	Aqueous extract Hydromethanolic extract Hydromethanolic extract Pressurized liquid extraction and solid-phase extraction	HPLC HPLC-DAD HPLC/UV-DAD LC-ESI-MS	Rtibi et al. (2018), Alamgeer et al. (2017), De Masi et al. (2015), and Khadhraoui et al. (2019)
		Hydromethanolic extract	HPLC	Bachir Bey et al. (2017)
		Ultrasonic extraction	HPLC-PDA	Veberic et al. (2008)
	Vanillic acid R2 = OCH3	Hydroethanolic extract	RP-UHPLC-DAD-QTOF-MS and MS/ MS	Belguith-Hadriche et al. (2016)
		Hydromethanolic extract	HPLC-DAD	Alamgeer et al. (2017)
		Methanolic extract	HPLC-DAD- ESI/MS	Aicha Debib et al. (2016)
		Hydromethanolic extract	HPLC	Bachir Bey et al. (2017)
	Syringic acid R, R2 = OCH3, R1 = OH	Aqueous extract	HPLC	Rtibi et al. (2018)
		Hydromethanolic extract	HPLC-DAD	Alamgeer et al. (2017)
		Aqueous extract	HPLC	Nakilcioğlu and Hışıl, (2013)
		Hydromethanolic extract	HPLC	Bachir Bey et al. (2017)
		Ultrasonic extraction	HPLC-PDA	Veberic et al. (2008)
	Protocatechuic acid R = H, R1, R2, R3 = OH	Hydromethanolic extract	HPLC/UV-DAD	De Masi et al. (2015)
		Pressurized liquid extraction and solid-phase extraction	LC-ESI-MS	Khadhraoui et al. (2019)
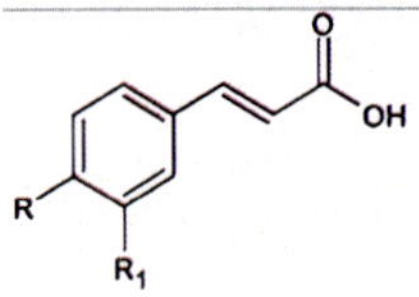	Cinnamic acid R, R1 = H	Pressurized liquid extraction and solid-phase extraction	LC-ESI-MS	Khadhraoui et al. (2019)
		Ethyl acetate extract	GC-MS	Ahmad et al. (2016)

(continued)

Table 18.1 (continued)

Structure	Compound	Sample preparation	Analytical method	Reference
		Hydromethanolic extract	HPLC	Bachir Bey et al. (2017)
		Methanolic extract	HPLC-DAD- ESI/MS	Aicha Debib et al. (2016)
	p-coumaric acid R = OH, R1 = H	Methanolic extract	HPLC-DAD- ESI/MS	Aicha Debib et al. (2016)
		Hydromethanolic ultrasonic extraction	HPLC	Kamiloglu and Capanoglu (2013)
	m-coumaric acid R = H, R1 = OH	Hydromethanolic extract	HPLC-DAD	Alamgeer et al. (2017)
	Caffeic acid R, R1 = OH	Ethyl acetate extract	GC-MS	Ahmad et al. (2016)
		Hydromethanolic extract	HPLC	Bachir Bey et al. (2017)
	Caffeic acid hexoside R = OH, R1 = *O*-hexoside	Hydroethanolic extracts	LC-DAD-ESI/MSn	Palmeira et al. (2019)
	Ferulic acid R = OH, R1 = OCH3	Ethyl acetate extract	GC-MS	Ahmad et al. (2016)
		Acetone extract	HPLC/DAD and HPLC/MS	Yemiş et al. (2012)
		Hydromethanolic extract	HPLC	Bachir Bey et al. (2017)
		Hydroethanolic extract	RP-UHPLC-DAD-QTOF-MS and MS/ MS	Belguith-Hadriche et al. (2016)
		Pressurized liquid extraction and solid-phase extraction	LC-ESI-MS	Khadhraoui et al. (2019)
		Hydromethanolic extract	HPLC/UV-DAD	De Masi et al. (2015)
		Hydromethanolic extract	HPLC	Bachir Bey et al. (2017)
		Acid water (pH 2 with HCl)	HPLC-DAD	Faleh et al. (2012)
	Chlorogenic acid R = OH	Hydroethanolic extracts	LC-DAD-ESI/MSn	Palmeira et al. (2019)
		Ultrasonic extraction	HPLC-PDA	Veberic et al. (2008)

(continued)

Table 18.1 (continued)

Structure	Compound	Sample preparation	Analytical method	Reference
		Solid-phase extraction	UHPLC-MS-CID	Yemiş et al. (2012)
		Aqueous extract	HPLC	Nakilcioğlu and Hışıl (2013)
		Aqueous extract Hydromethanolic extract Hydromethanolic extract Pressurized liquid extraction and solid-phase extraction	HPLC HPLC/UV-DAD LC-ESI-MS	Rtibi et al. (2018), Alamgeer et al. (2017), De Masi et al. (2015), and Khadhraoui et al. (2019)
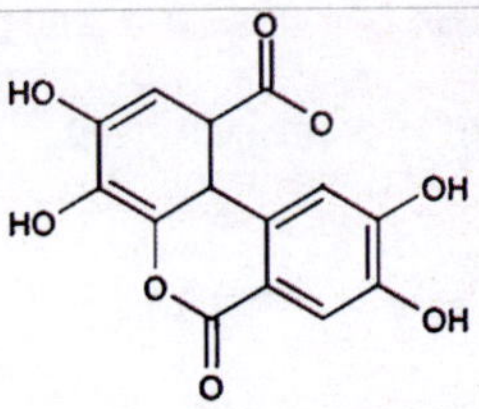	Ellagic acid	Aqueous extract	HPLC	Rtibi et al. (2018)
		Acetone/ formic acid extract	HPLC-DAD/ESI-MS	Pereira et al. (2020)
		Hydromethanolic ultrasonic extraction	HPLC	Kamiloglu and Capanoglu (2013)
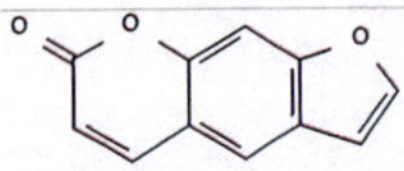 Furanocoumarins	Psoralen	Aqueous extract	HPLC/DAD and HPLC/UV	Oliveira et al. (2009)
		Hydroethanolic extract	RP-UHPLC-DAD-QTOF-MS and MS/ MS	Belguith-Hadriche et al. (2016)
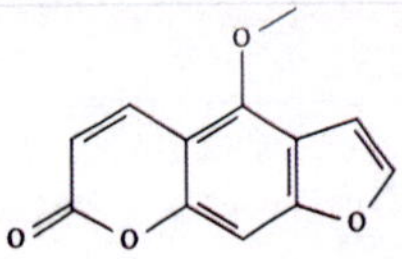	Bergapten	Aqueous extract	HPLC/DAD and HPLC/UV	Oliveira et al. (2009)
		Hydroethanolic extract	RP-UHPLC-DAD-QTOF-MS and MS/ MS	Belguith-Hadriche et al. (2016)
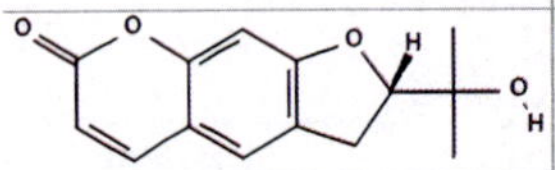	Marmesin	Ultrasonic extraction	HPLC-PDA	Veberic et al. (2008)
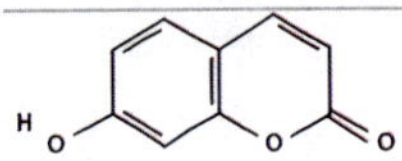	Umbelliferone	Ultrasonic extraction	HPLC-PDA	Veberic et al. (2008)
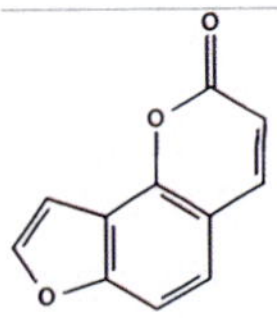	Angelicin	Ultrasonic extraction	HPLC-PDA	Veberic et al. (2008)

primarily categorized into flavanols, flavonols, flavones, and anthocyanidins (Table 18.2).

4 Health Benefits of Figs

The positive health effects of fresh and dried figs have been attributed to many factors, mainly phenolic compounds. In addition, numerous epidemiological data, animal work, and cell culture evidence support the health benefits of figs in preventing cardiovascular disease and certain cancers (Alamgeer et al., 2017).

4.1 Antioxidant Activity

From the chemical standpoint, the antioxidants act by neutralizing products of metabolism, such as free radicals, thus reducing degenerative processes related to aging phenomena. It is well known that free radicals can cause damage to cells and their constituents that can lead to the onset and development of chronic degenerative diseases. In fact, under normal physiological conditions, cellular metabolism produces reactive oxygen species (ROS) in equilibrium with the antioxidant defense system. When this homeostasis is interrupted by excess ROS, oxidative stress. In this situation, ROS can cause lipid peroxidation (e.g., increased malondialdehyde [MDA]) and oxidative damage to DNA molecules (e.g., cancer), membrane phospholipids, and proteins, which constitute the molecular basis of several diseases (Khutami, Sumiwi, Khairul Ikram, & Muchtaridi, 2022; Niemann et al., 2017). In

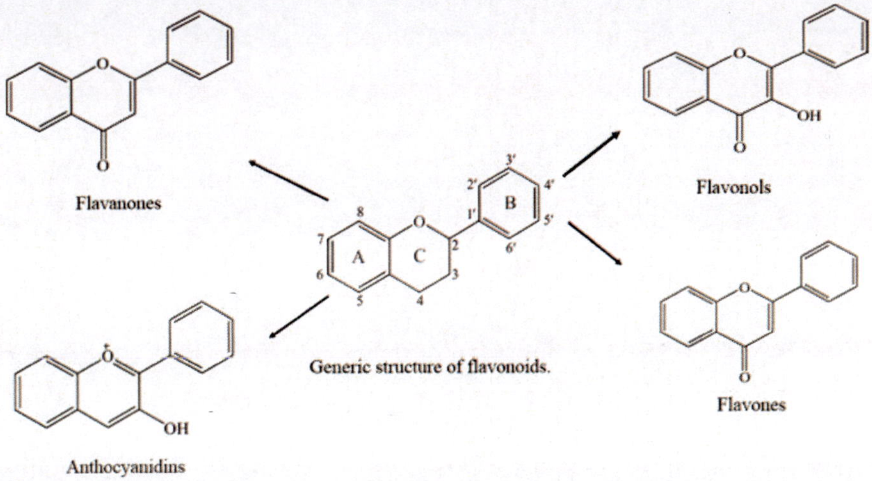

Fig. 18.2 Structure of main classes of figs flavonoids

Table 18.2 Flavonoids compounds from *Ficus carica* fruits

Flavonoids class	Compound	Sample preparation	Analytical method	Reference
Flavones	Luteonil-8-Cglucoside	Methanolic extract	HPLC-MS	Slatnar et al. (2011)
		Methanolic extract	HPLC-DAD-ESI/MS	Aicha Debib et al. (2016)
	Luteolin-6-C-Apentose Apigeniun	Pressurized liquid extraction and solid-phase extraction	LC-ESI-MS	Khadhraoui et al. (2019)
		Aqueous extract	HPLC	Rtibi et al. (2018)
	Apigenin-2″ -O-rhamnose-C-acetylhexoside	Hydroethanolic extracts	LC-DAD-ESI/MSn	Palmeira et al. (2019)
	Orientin	Hydroethanolic extract	RP-UHPLC-DAD-QTOF-MS and MS/ MS	Belguith-Hadriche et al. (2016)
	Isorientin	Hydroethanolic extract	RP-UHPLC-DAD-QTOF-MS and MS/ MS	Belguith-Hadriche et al. (2016)

(continued)

Table 18.2 (continued)

Flavonoids class	Compound	Sample preparation	Analytical method	Reference
Flavonols	Rutin	Acetone/ formic acid extract	HPLC-DAD/ ESI-MS	Pereira et al. (2020)
		Aqueous extract	HPLC	Rtibi et al. (2018)
		Hydroethanolic extracts	LC-DAD-ESI/MSn	Palmeira et al. (2019)
		Hydroethanolic extract	RP-UHPLC-DAD-QTOF-MS and MS/ MS	Belguith-Hadriche et al. (2016)
		Hydromethanolic extract	HPLC/ UV-DAD	De Masi et al. (2015)
	Quercetin	Hydromethanolic extract	HPLC/ UV-DAD	De Masi et al. (2015)
		Hydroethanolic extract	RP-UHPLC-DAD-QTOF-MS and MS/ MS	Belguith-Hadriche et al. (2016)
		Hydromethanolic extract	HPLC-DAD	Alamgeer et al. (2017)
		Methanolic extract	UHPLC-QTOF-MS/	Soltana et al. (2018)
		Methanolic extract	HPLC-DAD-ESI/MS	Aicha Debib et al. (2016)
	Quercetin-3-*O*-(malonyl)-glucoside	Hydroethanolic extract	RP-UHPLC-DAD-QTOF-MS and MS/ MS	Belguith-Hadriche et al. (2016)
		Methanolic extract	UHPLC-QTOF-MS/	Soltana et al. (2018)
	Quercetin-3-*O*-(acetyl)- glucoside	Acetone/ formic acid extract	HPLC-DAD/ ESI-MS	Pereira et al. (2020)
	Isoquercitrin	Methanolic extract	HPLC-DAD-ESI/MS	Aicha Debib et al. (2016)
		Methanolic extract	UHPLC-QTOF-MS	Soltana et al. (2018)
		Hydroethanolic extract	RP-UHPLC-DAD-QTOF-MS and MS/ MS	Belguith-Hadriche et al. (2016)
	Quercetin-3-Orutinoside	Hydromethanolic ultrasonic extraction	HPLC	Kamiloglu and Capanoglu (2013)
	Kaempeferol-3-O-glucoside	Methanolic extract	HPLC-MS	Slatnar et al. (2011)

(continued)

Table 18.2 (continued)

Flavonoids class	Compound	Sample preparation	Analytical method	Reference
Flavanols	Catechin	Hydromethanolic ultrasonic extraction	HPLC	Kamiloglu and Capanoglu (2013)
		Methanolic extract	HPLC-MS	Slatnar et al. (2011)
		Hydromethanolic extract	HPLC	Bachir Bey et al. (2017)
	Epicatechin	Acetone/ formic acid extract	HPLC-DAD/ ESI-MS	Pereira et al. (2020)
		Aqueous extract	HPLC	Rtibi et al. (2018)
	Pelargonidin-3-O-rutinoside	Acetone/ formic acid extract	HPLC-DAD/ ESI-MS	Pereira et al. (2020)
Anthocyanidins	Cyanidin-3-O-rutinoside	Acetone/ formic acid extract	HPLC-DAD/ ESI-MS	Pereira et al. (2020)
		Methanolic extract	HPLC-MS	Slatnar et al. (2011)
		Hydromethanolic ultrasonic extraction	HPLC	Kamiloglu and Capanoglu (2013)
		Hydromethanolic extract	HPLC	Bachir Bey et al. (2017)
	Cyanidin-3-*O*-glucoside	Hydromethanolic extract	HPLC	Bachir Bey et al. (2017)
		Solid-phase extraction	UHPLC-MS-CID	Yemiş et al. (2012)
		Hydromethanolic ultrasonic extraction	HPLC	Kamiloglu and Capanoglu (2013)
	Cyanidin-3,5-Diglucoside	Solid-phase extraction	UHPLC-MS-CID	Yemiş et al. (2012)
	Peonidin 3-O-glucoside	Methanolic extract	HPLC-DAD-ESI/MS	Dueñas et al. (2008)
	Pelargonidin 3-*O*-glucoside	Methanolic extract	HPLC-DAD-ESI/MS	Dueñas et al. (2008)
	Pelargonidin 3-O-rutinosid	Acetone/ formic acid extract	HPLC-DAD/ ESI-MS	Pereira et al. (2020)

addition, antioxidants inhibit the action of enzymes that catalyze the detoxification pathway in the body upon exposure to mutagens. Based on literature data, several studies reported that flavonoids have an antioxidant action that attenuates ROS by enhancing antioxidant enzymes or by inhibiting enzymes that employ a pro-oxidant effect on oxidative stress (e.g., NADPH oxidase, xanthine oxidase, and catalase) (Yan et al., 2013; Zhang et al., 2022).

The antioxidant capacity of different components of fresh figs (skin, pulp, entire fruit) and dried figs have been intensely studied using a variety of antioxidant assays. Some of the methods that have been widely used in selected fig varieties include the 1,1-diphenyl-2-picrylhydrazyl radical assay (DPPH·), 2,2-azino-di-(3-ethylbenzothialozine-sulphonic acid) assay (ABTS), ferric ion reducing antioxidant power assay (FRAP), cupric ion reducing capability assay (CUPRAC), and oxygen radical absorbance capacity assay (ORAC). According to the most published results, the antioxidant capacity of figs has been highly correlated with the amount of phenolic compounds (Ouchemoukh et al., 2012; Vinson et al., 2005; Wojdyło et al., 2016). Furthermore, according to khadhraoui et al. (2017), there is a strong correlation between skin color, total phenolic content, and antioxidant activity. Thus, figs from dark-colored cultivars exhibited a higher antioxidant capacity than those from light-colored cultivars, with 3–4 fold differences between cultivars. These findings were confirmed by several studies (Faleh et al., 2012; Konak et al., 2015; Pereira et al., 2020). Regarding the impact of drying on the antioxidant capacity, Chang et al. (2016) reported that the sun-drying process had a positive effect on antioxidant activity, which increases the antioxidant activity of dried figs compared to that of fresh figs (Chang et al., 2016). While in other studies, it has been reported that the sun-drying procedure reduced the antioxidant capacity of figs (Bachir Bey et al., 2017; Kamiloglu & Capanoglu, 2015) (Fig. 18.3).

4.2 Antimicrobial Activity

Antibacterial activity has been demonstrated in several varieties of *Ficus carica* fruit worldwide. Debib et al. (2014) reported that four different extracts of two Algerian figs varieties (Azendjar and Taamriout) showed antimicrobial effects against gram-positive bacteria such as *Staphylococcus aureus* (ATCC 25923), *Enterococcus faecalis* (ATCC 29212) and *Bacillus subtilis* subsp. *spizizenii* (ATCC 6633) and six gram-negative bacteria, *Pseudomonas aeruginosa* (ATCC 27853), *P. aeruginosa* (ATCC 10145), *Escherichia coli* (ATCC 25922), *Enterobacter cloacae* (ATCC 13047), *Salmonella enterica* subsp. *heindelberg* (ATCC 8326), *Klebsiella pneumoniae* (clinical isolated), and one yeast, *Candida albicans* (clinical isolated). According to these authors, this effect could be attributable to these varieties'high flavonoid and tannin levels. Their ability to combine with extracellular and soluble proteins and bacterial cell walls is most probably the cause of their activity. Indeed, microbial membranes may be disrupted by lipophilic flavonoids. In another study, the intense antibacterial activity of the figs methanolic extract was found against *Staphylococcus aureus, Bacillus subtilis, Bacillus cereus, Listeria monocytogenes, Salmonella typhimurium, and Escherichia coli,* mainly against gram-positive bacteria (Shahbazi, 2017). The peel and pulp that have been separated from the *F. carica* fruit (ethanol/water extract, 80:20 v/v) showed similar results in most bacterial strains tested. The best results were obtained against gram-positive

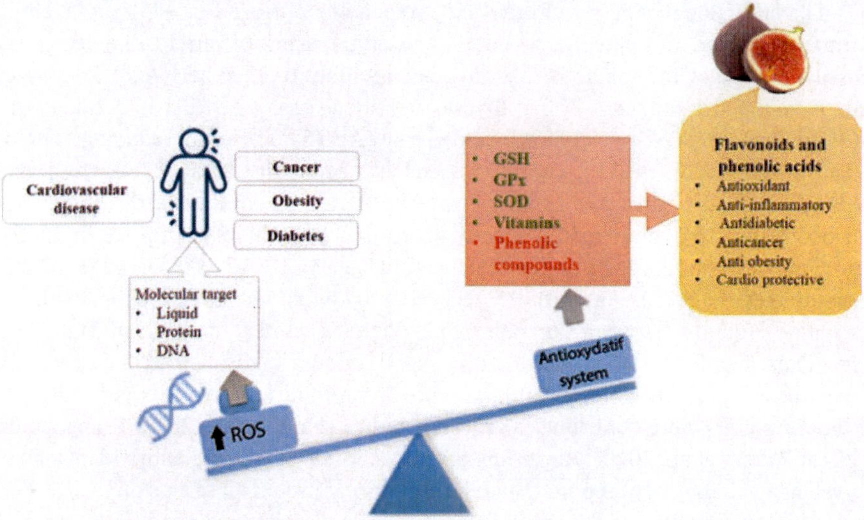

Fig. 18.3 Imbalance between reactive oxygen species (ROS), a possible antioxidant effect of phenolic compounds of figs

bacteria, especially methicillin-sensitive *S. aureus*, with a MIC of 2.5 mg/mL (Palmeira et al., 2019).

The oily dried figs macerates extracts possessed significantly higher antimicrobial activity than the oily extracts against *Staphylococcus aureus*, *Enterococcus faecalis, Bacillus subtilis Pseudomonas aeruginosa*, *Escherichia coli*, *Enterobacter cloacae*, *Salmonella enterica.* Antimicrobial activity results of this investigation validate the use of olive oily macerates of dried figs by Algerian people to fight infectious diseases (Debib et al., 2018).

4.3 Antidiabetic Effect

It has been revealed that long-term use of anti-diabetic medications is linked to several toxicities harmful to human health. Anti-diabetic fruits and vegetables could be a valuable tool in developing safer and more effective oral hypoglycemic treatments; similarly, the discovery of anti-diabetic therapies has turned to natural plant sources with few adverse effects (Gothai et al., 2016). Different extracts of fruits and leaves showed potent inhibitory activity against α-amylase and α-glucosidase enzymes (Wojdyło et al., 2016). According to El-Shobaki et al. (2010), fruits and leaves of *F. carica* showed intense antidiabetic activity in alloxon-induced diabetic rats by increasing antioxidant levels. In other studies, the methanol extract of fruits, aqueous leaves, and decoction obtained from the leaves also controlled the diabetic complications in streptozotocin-induced diabetic rats (Belguith-Hadriche et al., 2016; Mopuri & Islam, 2016). Several *in vivo* investigations have proposed mechanisms of action (Meziant et al., 2021; Perez et al., 2003). Pèrez et al. (2003) and Belguith-Hadriche et al. (2017) demonstrated that *Ficus carica* extracts have positive effects on insulin secretion and blood glucose levels (Perez et al., 2003). Song et al. (2002) reported flavonoids as hypoglycaemic agents because they improved glucose metabolism and reduced oxidative stress in diabetic rats (Song et al., 2002). When the diabetic rats were given glucose with quercetin, hyperglycemia significantly decreased compared with glucose administered alone, and flavonoids modulated glucose transport *via* their intestinal transporter. The same results were obtained in human subjects, *F. carica* fruit extract standardized in abscisic acid was added to a glucose beverage, reducing the postprandial glycemic and insulinemic values in healthy adult subjects when compared to the normal drink (Atkinson et al., 2019).

The hydromethanolic extract of *F. carica* has also been shown to have anti-obesity properties in recent *in vivo* investigations. The results were mainly evidenced by the reduction in the animals' body mass and the serum levels of low-density lipoprotein (LDL) (Arafa et al., 2020).

4.4 Anticancer Activity

The methanol extract of *F. carica* at a dose of 1000 μg/mL revealed low anti-cancer activity through apoptosis (0.44%) and necrosis (2.26%) of liver cancer cells (Huh7it). The positive control (doxorubicin, 10 μg/mL) reached 65.6 and 34.1%, respectively (Purnamasari et al., 2019). Sixteen prenylated isoflavone derivatives isolated from the ethanol extract presented an antiproliferative effect against human cancer cell lines, HL-60 (Liu et al., 2019). In another study, cytotoxicity of fruit and leaf extracts, as well as the latex on HeLa cell line, was examined, and results showed that latex and different extracts could reduce the viability of HeLa cells at concentrations as low as 2 μg/mL in a dose-dependent manner (Khodarahmi et al., 2011).

4.5 Anti-Inflammatory Activity

Inflammation is a natural response of the body against microorganisms and toxic materials. Various natural products have proven their role as anti-inflammatory agents without severe side effects. Debib et al. (2016) reported that the methanolic-dried figs fruit extract at a dose of 38.07 mg/kg/day enhanced the protection of CCl_4-induced oxidative stress and hepatotoxicity in rats. According to these studies, these hepatoprotective effects could be attributed to the inhibitory activity of flavonoids on free radical production since exogenous antioxidants may counteract the damaging effects of oxidative stress, cooperating with natural systems like glutathione tocopherol or protective enzymes.

4.6 *Effect on Alzheimer's Disease*

Subash et al. (2016) have studied the effect of dietary supplementation with 4% figs grown in Oman on the memory, anxiety, and learning skills in APPsw/Tg2576 (Tg mice) mice model for Alzheimer's disease. The authors have assessed spatial memory and learning ability, psychomotor coordination, and anxiety-related behavior in Tg and wild-type mice at the age of 4 months and after 15 months using the Morris water maze test, rota-rod test, elevated plus maze test, and open-field test. Results of these study revealed that Tg mice fed a control diet without figs showed significant memory deficits, increased anxiety-related behavior, and severe impairment in spatial, position discrimination learning ability, and motor coordination compared to the wild-type control mice on the same diet, and Tg mice fed on 4% fig diet supplementation for 15 months. These studies suggest that dietary supplementation of figs may help improve cognitive and behavioral deficits in Alzheimer's disease (Subash et al., 2016). Another study investigated the impact of dietary supplements, including pomegranate, figs, or dates, on inflammatory cytokines suppression. The researchers discovered that using the food supplements reduced inflammatory cytokines and eotaxin activity (Essa et al., 2015).

4.7 *Antispasmodic Activity*

The antispasmodic efficacy of an aqueous-ethanolic extract of dried figs was investigated. The research revealed the presence of spasmolytic activity in the ripe dried fruit, which could be mediated by the activation of K + ATP channels (Gilani et al., 2008).

4.8 *Anti-Osteoporotic Potential*

Another study found that in RANKL-stimulated RAW264.7 cells and bone marrow-derived macrophages, the hexane soluble extract of the fig is a potent inhibitor of osteoclastogenesis (Park et al., 2009). These results corroborate those of Mohammed et al. (2018), who revealed that 50 mg/kg raw extract solution showed potential against osteoporosis in ovariectomized rats, comparable to estrogen replacement therapy (Mohammad et al., 2018; Park et al., 2009).

5 Conclusion

The popularity of alternative medicine and natural products is increasing worldwide due to inexpensive and health-promoting activities. In this vista, *Ficus carica* Linn fruit, generally known as figs, is associated with a lowered risk of chronic diseases. In vitro and in vivo studies have demonstrated that figs phenolic compounds are involved in a broad spectrum of biological activities, including antioxidant, antimicrobial, antidiabetic, anticancer, anti-inflammatory, antispasmodic, and anti-osteoporotic activities. In addition, fresh and dried fig extract inhibits cancer cell proliferation and reduces Alzheimer's disease complications. Pharmacological studies on the fresh and dried fruit, crude extracts, and isolated components of *Ficus carica* provide experimental support for its numerous traditional uses. In this review, 89 papers described the chemical characterization and biological activities of *Ficus carica* L. fruit. However, there is a vast scope for establishing a relation between phytoconstituents and biological activities. The outcome of the future research in the areas mentioned above will provide convincing support for the future clinical uses of *Ficus carica* in modern medicine.

References

Abbo, S., Gopher, A., & Lev-Yadun, S. (2015). Fruit domestication in the near east. *Plant Breeding Reviews, 39*, 325–377.

Ahmad, N., Zuo, Y., Lu, X., Anwar, F., & Hameed, S. (2016). Characterization of free and conjugated phenolic compounds in fruits of selected wild plants. *Food Chemistry, 190*, 80–89.

Alamgeer, I. S., Asif, H., & Saleem, M. (2017). Evaluation of antihypertensive potential of *Ficus carica* fruit. *Pharmaceutical Biology, 55*(1), 1047–1053.

Ammar, S., del Mar Contreras, M., Belguith-Hadrich, O., Segura-Carretero, A., & Bouaziz, M. (2015). Assessment of the distribution of phenolic compounds and contribution to the antioxidant activity in Tunisian fig leaves, fruits, skins and pulps using mass spectrometry-based analysis. *Food and Function, 6*(12), 3663–3677.

Arafa, E.-S. A., Hassan, W., Murtaza, G., & Buabeid, M. A. (2020). *Ficus carica* and *Sizigium cumini* regulate glucose and lipid parameters in high-fat diet and streptozocin-induced rats. *Journal of Diabetes Research, 2020*, 6745873.

Arvaniti, O. S., Samaras, Y., Gatidou, G., Thomaidis, N. S., & Stasinakis, A. S. (2019). Review on fresh and dried figs: Chemical analysis and occurrence of phytochemical compounds, antioxidant capacity and health effects. *Food Research International, 119*, 244–267.

Atkinson, F. S., Villar, A., Mulà, A., Zangara, A., Risco, E., Smidt, C. R., . . . Bassaganya-Riera, J. (2019). Abscisic acid standardized fig (*Ficus carica*) extracts ameliorate postprandial glycemic and insulinemic responses in healthy adults. Nutrients, 11(8), 1757.

Bachir Bey, M., Richard, G., Meziant, L., Fauconnier, M. L., & Louaileche, H. (2017). Effects of sun-drying on physicochemical characteristics, phenolic composition and in vitro antioxidant activity of dark fig varieties. *Journal of Food Processing and Preservation, 41*(5), e13164.

Belguith-Hadriche, O., Ammar, S., Contreras, M. D. M., Turki, M., Segura-Carretero, A., El Feki, A., et al. (2016). Antihyperlipidemic and antioxidant activities of edible Tunisian *Ficus carica* L. fruits in high fat diet-induced hyperlipidemic rats. *Plant Foods for Human Nutrition, 71*(2), 183–189.

Belguith-Hadriche, O., Ammar, S., del Mar Contreras, M., Fetoui, H., Segura-Carretero, A., El Feki, A., & Bouaziz, M. (2017). HPLC-DAD-QTOF-MS profiling of phenolics from leaf extracts of two Tunisian fig cultivars: Potential as a functional food. *Biomedicine & Pharmacotherapy, 89*, 185–193.

Bimakr, M., Ganjloo, A., Zarringhalami, S., & Ansarian, E. (2017). Ultrasound-assisted extraction of bioactive compounds from Malva sylvestris leaves and its comparison with agitated bed extraction technique. *Food Science and Biotechnology, 26*(6), 1481–1490.

Blainski, A., Lopes, G. C., & De Mello, J. C. P. (2013). Application and analysis of the folin ciocalteu method for the determination of the total phenolic content from *Limonium brasiliense* L. *Molecules, 18*(6), 6852–6865.

Bucic-Kojic, A., Planinic, M., Tomas, S., Jokic, S., Mujic, I., Bilic, M., & Velic, D. (2011). Effect of extraction conditions on the extractability of phenolic compounds from lyophilised fig fruits (*Ficus carica* L.). *Polish Journal of Food and Nutrition Sciences, 61*(3).

Caliskan, O. (2015). *Mediterranean figs (Ficus carica L.) functional food properties the Mediterranean diet* (pp. 629–637). Elsevier.

Çalişkan, O., & Polat, A. A. (2011). Phytochemical and antioxidant properties of selected fig (*Ficus carica* L.) accessions from the eastern Mediterranean region of Turkey. *Scientia Horticulturae, 128*(4), 473–478.

Cano-Lamadrid, M., & Artés-Hernández, F. (2021). By-products revalorization with non-thermal treatments to enhance phytochemical compounds of fruit and vegetables derived products: A review. *Food, 11*(1), 59.

Chang, S. K., Alasalvar, C., & Shahidi, F. (2016). Review of dried fruits: Phytochemicals, antioxidant efficacies, and health benefits. *Journal of Functional Foods, 21*, 113–132.

Chemat, F., Abert Vian, M., Ravi, H. K., Khadhraoui, B., Hilali, S., Perino, S., & Fabiano Tixier, A.-S. (2019). Review of alternative solvents for green extraction of food and natural products: Panorama, principles, applications and prospects. *Molecules, 24*(16), 3007.

Crisosto, C. H., Bremer, V., Ferguson, L., & Crisosto, G. M. (2010). Evaluating quality attributes of four fresh fig (*Ficus carica* L.) cultivars harvested at two maturity stages. *HortScience, 45*(4), 707–710.

De Masi, L., Vella, F., Laratta, B., Volpe, M., Tiseo, M., & La Cara, F. (2015). Biochemical and genetic characterization of a red fig cultivar (Ficus carica) from Southern Italy. Paper presented at the V International Symposium on Fig 1173.

Debib, A., Tir-Touil, A., Mothana, R., Meddah, B., & Sonnet, P. (2014). Phenolic content, antioxidant and antimicrobial activities of two fruit varieties of Algerian *Ficus carica* L. *Journal of Food Biochemistry, 38*(2), 207–215.

Debib, A., Dueñas, M., Boumediene, M., Mothana, R. A., Latifa, A., & Tir-Touil, M. A. (2016). Synergetic Hepatoprotective effect of phenolic fractions obtained from *F icus carica* dried fruit and extra virgin olive oil on CCL4-induced oxidative stress and hepatotoxicity in rats. *Journal of Food Biochemistry, 40*(4), 507–516.

Debib, A., Tir-Touil, M., Meddah, B., Hamaidi-Chergui, F., Menadi, S., & Alsayadi, M. (2018). Evaluation of antimicrobial and antioxidant activities of oily macerates of Algerian dried figs (*Ficus carica* L.). *International Food Research Journal, 25*(1), 351–356.

Del Caro, A., & Piga, A. (2008). Polyphenol composition of peel and pulp of two Italian fresh fig fruits cultivars (*Ficus carica* L.). *European Food Research and Technology, 226*(4), 715–719.

dos Anjos Cruz, J. M., Corrêa, R. F., Lamarão, C. V., Kinupp, V. F., Sanches, E. A., Campelo, P. H., & de Araújo Bezerra, J. (2021). Ficus spp. fruits: Bioactive compounds and chemical, biological and pharmacological properties. *Food Research International, 152*, 110928.

Dragsted, L. O., Strube, M., & Larsen, J. (1993). Cancer-protective factors in fruits and vegetables: Biochemical and biological background. *Pharmacology & Toxicology, 72*, 116–135.

Dueñas, M., Pérez-Alonso, J. J., Santos-Buelga, C., & Escribano-Bailón, T. (2008). Anthocyanin composition in fig (*Ficus carica* L.). *Journal of Food Composition and Analysis, 21*(2), 107–115.

El-Shobaki, F. A., El-Bahay, A. M., Esmail, R. S. A., El-Megeid, A. A. A., & Esmail, N. S. (2010). Effect of figs fruit (Ficus carica L.) and its leaves on hyperglycemia in alloxan diabetic rats. *World Journal of Dairy & Food Sciences, 5*(1), 47–57.

Elkordy, A. A., Haj-Ahmad, R. R., Awaad, A. S., & Zaki, R. M. (2021). An overview on natural product drug formulations from conventional medicines to nanomedicines: Past, present and future. *Journal of Drug Delivery Science and Technology, 63*, 102459.

Ercisli, S., Tosun, M., Karlidag, H., Dzubur, A., Hadziabulic, S., & Aliman, Y. (2012). Color and antioxidant characteristics of some fresh fig (*Ficus carica* L.) genotypes from northeastern Turkey. *Plant Foods for Human Nutrition, 67*(3), 271–276.

Essa, M. M., Subash, S., Akbar, M., Al-Adawi, S., & Guillemin, G. J. (2015). Long-term dietary supplementation of pomegranates, figs and dates alleviate neuroinflammation in a transgenic mouse model of Alzheimer's disease. *PLoS One, 10*(3), e0120964.

Faleh, E., Oliveira, A. P., Valentão, P., Ferchichi, A., Silva, B. M., & Andrade, P. B. (2012). Influence of Tunisian *Ficus carica* fruit variability in phenolic profiles and in vitro radical scavenging potential. *Revista Brasileira de Farmacognosia, 22*(6), 1282–1289.

FAOSTAT, Food and Agriculture Organization: FAOSTAT data. (2020). Available at https://www.fao.org.

Feng, Y.-C., Li, W.-L., He, F.-M., Kong, T.-T., Huang, X.-W., Gao, Z.-H., et al. (2015). Aqueous two-phase system as an effective tool for purification of phenolic compounds from fig fruits (*Ficus carica* L.). *Separation Science and Technology, 50*(12), 1785–1793.

Gilani, A. H., Mehmood, M. H., Janbaz, K. H., Khan, A. U., & Saeed, S. A. (2008). Ethnopharmacological studies on antispasmodic and antiplatelet activities of *Ficus carica*. *Journal of Ethnopharmacology, 119*(1), 1–5.

Gothai, S., Ganesan, P., Park, S.-Y., Fakurazi, S., Choi, D.-K., & Arulselvan, P. (2016). Natural phyto-bioactive compounds for the treatment of type 2 diabetes: Inflammation as a target. *Nutrients, 8*(8), 461.

Hernandez-Ledesma, B., & Martinez-Villaluenga, C. (2021). *Current advances for development of functional foods modulating inflammation and oxidative stress*. Academic.

Jideani, A. I., Silungwe, H., Takalani, T., Omolola, A. O., Udeh, H. O., & Anyasi, T. A. (2021). Antioxidant-rich natural fruit and vegetable products and human health. *International Journal of Food Properties, 24*(1), 41–67.

Jokić, S., Mujić, I., Bucić-Kojić, A., Velić, D., Bilić, M., Planinić, M., & Lukinac, J. (2014). Influence of extraction type on the total phenolics, total flavonoids and total colour change of different varieties of fig extracts. *Hrana u zdravlju i bolesti: znanstveno-stručni časopis za nutricionizam i dijetetiku, 3*(2), 90–95.

Júnior, M. E. S., Araújo, M. V. R., Santana, A. A., Silva, F. L. H., & Maciel, M. I. S. (2021). Ultrasound-assisted extraction of bioactive compounds from ciriguela (*Spondias purpurea* L.) peel: Optimization and comparison with conventional extraction and microwave. *Arabian Journal of Chemistry, 14*(8), 103260.

Kamiloglu, S., & Capanoglu, E. (2013). Investigating the in vitro bioaccessibility of polyphenols in fresh and sun-dried figs (*Ficus carica* L.). *International Journal of Food Science and Technology, 48*(12), 2621–2629.

Kamiloglu, S., & Capanoglu, E. (2015). Polyphenol content in figs (*Ficus carica* L.): Effect of sun-drying. *International Journal of Food Properties, 18*(3), 521–535.

Kaur, C., & Kapoor, H. C. (2001). Antioxidants in fruits and vegetables-the millennium's health. *International Journal of Food Science and Technology, 36*(7), 703–725.

Kehal, F., Chemache, L., Chaalal, M., Benbraham, M., Capanoglu, E., & Barkat, M. (2021). A comparative analysis of different varietal of fresh and dried figs by bioaccessibility of phenolic compounds and antioxidant activities. *Acta Universitatis Cibiniensis. Series E. Food Technology, 25*(1), 15–30.

Khadhraoui, M., Bagues, M., Artés, F., & Ferchichi, A. (2019). Phytochemical content, antioxidant potential, and fatty acid composition of dried Tunisian fig (*Ficus carica* L.) cultivars. *Journal of Applied Botany and Food Quality, 92*, 143–150.

Khodarahmi, G. A., Ghasemi, N., Hassanzadeh, F., & Safaie, M. (2011). Cytotoxic effects of different extracts and latex of *Ficus carica* L. on HeLa cell line. *Iranian Journal of pharmaceutical research: IJPR, 10*(2), 273.

Khutami, C., Sumiwi, S. A., Khairul Ikram, N. K., & Muchtaridi, M. (2022). The effects of antioxidants from natural products on obesity, dyslipidemia, diabetes and their molecular signaling mechanism. *International Journal of Molecular Sciences, 23*(4), 2056.

Kislev, M. E., Hartmann, A., & Bar-Yosef, O. (2006). Early domesticated fig in the Jordan Valley. *Science, 312*(5778), 1372–1374.

Konak, R., Kösoğlu, İ., & Yemenıcıoğlu, A. (2015). *Effects of different drying methods on phenolic content, antioxidant capacity and general characteristics of selected dark colored Turkish fig cultivars.* Paper presented at the V International Symposium on Fig 1173.

Liu, R. H. (2013). Health-promoting components of fruits and vegetables in the diet. *Advances in Nutrition, 4*(3), 384S–392S.

Liu, Y.-P., Guo, J.-M., Yan, G., Zhang, M.-M., Zhang, W.-H., Qiang, L., & Fu, Y.-H. (2019). Anti-inflammatory and antiproliferative prenylated isoflavone derivatives from the fruits of *Ficus carica. Journal of Agricultural and Food Chemistry, 67*(17), 4817–4823.

Maghsoudlou, E., Esmaeilzadeh Kenari, R., & Raftani Amiri, Z. (2017). Evaluation of antioxidant activity of fig (*Ficus carica*) pulp and skin extract and its application in enhancing oxidative stability of canola oil. *Journal of Food Processing and Preservation, 41*(4), e13077.

Mahmoudi, S., Khali, M., Benkhaled, A., Benamirouche, K., & Baiti, I. (2016). Phenolic and flavonoid contents, antioxidant and antimicrobial activities of leaf extracts from ten Algerian *Ficus carica* L. varieties. *Asian Pacific Journal of Tropical Biomedicine, 6*(3), 239–245.

Margraf, T., Karnopp, A. R., Rosso, N. D., & Granato, D. (2015). Comparison between Folin-Ciocalteu and Prussian blue assays to estimate the total phenolic content of juices and teas using 96-well microplates. *Journal of Food Science, 80*(11), C2397–C2403.

Meziant, L., Bachir-bey, M., Bensouici, C., Saci, F., Boutiche, M., & Louaileche, H. (2021). Assessment of inhibitory properties of flavonoid-rich fig (*Ficus carica* L.) peel extracts against tyrosinase, α-glucosidase, urease and cholinesterases enzymes, and relationship with antioxidant activity. *European Journal of Integrative Medicine, 43*, 101272.

Míguez, C., Cancela, Á., Álvarez, X., & Sánchez, Á. (2022). The reuse of bio-waste from the invasive species *Tradescantia fluminensis* as a source of phenolic compounds. *Journal of Cleaner Production, 336*, 130293.

Miklavčič Višnjevec, A., & Schwarzkopf, M. (2020). Phenolic compounds in poorly represented Mediterranean plants in Istria: Health impacts and food authentication. *Molecules, 25*(16), 3645.

Mohammad, A., Razaly, N. I., Rani, M. D. M., Aris, M. S. M., Dom, S. M., & Effendy, N. M. (2018). A micro-computed tomography (micro-CT) analysis of postmenopausal osteoporotic rat models supplemented with *Ficus carica. Journal of Applied Pharmaceutical Science, 8*, 39.

Mopuri, R., & Islam, M. S. (2016). Antidiabetic and anti-obesity activity of *Ficus carica*: In vitro experimental studies. *Diabetes & Metabolism, 42*(4), 300.

Naczk, M., & Shahidi, F. (2006). Phenolics in cereals, fruits and vegetables: Occurrence, extraction and analysis. *Journal of Pharmaceutical and Biomedical Analysis, 41*(5), 1523–1542.

Nakilcioğlu, E., & Hışıl, Y. (2013). Research on the phenolic compounds in Sarilop (*Ficus carica* L.) fig variety. *Gida, 38*(5), 267–274.

Nakilcioğlu-Taş, E., & Ötleş, S. (2021). Influence of extraction solvents on the polyphenol contents, compositions, and antioxidant capacities of fig (*Ficus carica* L.) seeds. *Anais da Academia Brasileira de Ciências, 93.*

Niemann, B., Rohrbach, S., Miller, M. R., Newby, D. E., Fuster, V., & Kovacic, J. C. (2017). Oxidative stress and cardiovascular risk: Obesity, diabetes, smoking, and pollution: Part 3 of a 3-part series. *Journal of the American College of Cardiology, 70*(2), 230–251.

Oliveira, A. P., Valentão, P., Pereira, J. A., Silva, B. M., Tavares, F., & Andrade, P. B. (2009). Ficus carica L.: Metabolic and biological screening. *Food and Chemical Toxicology, 47*(11), 2841–2846.

Oliveira, A. P., Silva, L. R., Andrade, P. B., Valentao, P., Silva, B. M., Pereira, J. A., & de Pinho, P. G. (2010). Determination of low molecular weight volatiles in Ficus carica using HS-SPME and GC/FID. *Food Chemistry, 121*, 1289–1295.

Ouchemoukh, S., Hachoud, S., Boudraham, H., Mokrani, A., & Louaileche, H. (2012). Antioxidant activities of some dried fruits consumed in Algeria. *LWT-Food Science and Technology, 49*(2), 329–332.

Palmeira, L., Pereira, C., Dias, M. I., Abreu, R. M., Corrêa, R. C., Pires, T. C., . . . Ferreira, I. C. (2019). Nutritional, chemical and bioactive profiles of different parts of a Portuguese common fig (*Ficus carica* L.) variety. Food Research International, 126, 108572.

Park, Y. R., Eun, J. S., Choi, H. J., Nepal, M., Kim, D. K., Seo, S.-Y., . . . Cho, S.-D. (2009). Hexane-soluble fraction of the common fig, *Ficus carica,* inhibits osteoclast differentiation in murine bone marrow-derived macrophages and RAW 264.7 cells. *The Korean* Journal of Physiology & Pharmacology, 13(6), 417–424.

Pereira, C., Martín, A., López-Corrales, M., Córdoba, M. D. G., Galván, A. I., & Serradilla, M. J. (2020). Evaluation of the physicochemical and sensory characteristics of different fig cultivars for the fresh fruit market. *Food, 9*(5), 619.

Perez, C., Canal, J., & Torres, M. (2003). Experimental diabetes treated with Ficus carica extract: Effect on oxidative stress parameters. *Acta Diabetologica, 40*(1), 3–8.

Piga, A., Del Caro, A., Milella, G., Pinna, I., Vacca, V., & Schirru, S. (2005). *HPLC analysis of polyphenols in peel and pulp of fresh figs.* Paper presented at the III International Symposium on Fig 798.

Purnamasari, R., Winarni, D., Permanasari, A. A., Agustina, E., Hayaza, S., & Darmanto, W. (2019). Anticancer activity of methanol extract of Ficus carica leaves and fruits against proliferation, apoptosis, and necrosis in Huh7it cells. *Cancer Informatics, 18*, 1176935119842576.

Qin, H., Zhou, G., Peng, G., Li, J., & Chen, J. (2015). Application of ionic liquid-based ultrasound-assisted extraction of five phenolic compounds from fig (*Ficus carica* L.) for HPLC-UV. *Food Analytical Methods, 8*(7), 1673–1681.

Rtibi, K., Grami, D., Wannes, D., Selmi, S., Amri, M., Sebai, H., & Marzouki, L. (2018). Ficus carica aqueous extract alleviates delayed gastric emptying and recovers ulcerative colitis-enhanced acute functional gastrointestinal disorders in rats. *Journal of Ethnopharmacology, 224*, 242–249.

Samtiya, M., Aluko, R. E., Dhewa, T., & Moreno-Rojas, J. M. (2021). Potential health benefits of plant food-derived bioactive components: An overview. *Food, 10*(4), 839.

Shahbazi, Y. (2017). Antibacterial and antioxidant properties of methanolic extracts of apple (*Malus pumila*), grape (*Vitis vinifera*), pomegranate (*Punica granatum* L.) and common fig (*Ficus carica* L.) fruits. *Pharmaceutical Sciences, 23*(4), 308–315.

Slatnar, A., Klancar, U., Stampar, F., & Veberic, R. (2011). Effect of drying of figs (*Ficus carica* L.) on the contents of sugars, organic acids, and phenolic compounds. *Journal of Agricultural and Food Chemistry, 59*(21), 11696–11702.

Soltana, H., De Rosso, M., Lazreg, H., Vedova, A. D., Hammami, M., & Flamini, R. (2018). LC-QTOF characterization of non-anthocyanic flavonoids in four Tunisian fig varieties. *Journal of Mass Spectrometry, 53*(9), 817–823.

Song, J., Kwon, O., Chen, S., Daruwala, R., Eck, P., Park, J. B., & Levine, M. (2002). Flavonoid inhibition of sodium-dependent vitamin C transporter 1 (SVCT1) and glucose transporter isoform 2 (GLUT2), intestinal transporters for vitamin C and glucose. *Journal of Biological Chemistry, 277*(18), 15252–15260.

Subash, S., Essa, M. M., Braidy, N., Al-Jabri, A., Vaishnav, R., Al-Adawi, S., et al. (2016). Consumption of fig fruits grown in Oman can improve memory, anxiety, and learning skills in a transgenic mice model of Alzheimer's disease. *Nutritional Neuroscience, 19*(10), 475–483.

Tenório, C. J. L., Silva, S. L., Raimundo e Silva, J. P., Tavares, J. F., Ferreira, M. R. A., & Soares, L. A. L. (2022). HPLC-ESI-MSn analysis and validation of UV-vis and RP-HPLC-PDA methods for polyphenols quantification from Hymenaea eriogyne. *Revista Brasileira de Farmaco*, 1–10.

Thongson, C., Davidson, P., Mahakarnchanakul, W., & Weiss, J. (2004). Antimicrobial activity of ultrasound-assisted solvent-extracted spices. *Letters in Applied Microbiology, 39*(5), 401–406.

Vallejo, F., Marín, J., & Tomás-Barberán, F. A. (2012). Phenolic compound content of fresh and dried figs (*Ficus carica* L.). *Food Chemistry, 130*(3), 485–492.

Vardanega, R., Santos, D. T., & Meireles, M. A. A. (2014). Intensification of bioactive compounds extraction from medicinal plants using ultrasonic irradiation. *Pharmacognosy Reviews, 8*(16), 88.

Veberic, R., & Mikulic-Petkovsek, M. (2016). *Phytochemical composition of common fig (Ficus carica L.) cultivars nutritional composition of fruit cultivars* (pp. 235–255). Elsevier.

Veberic, R., Colaric, M., & Stampar, F. (2008). Phenolic acids and flavonoids of fig fruit (*Ficus carica* L.) in the northern Mediterranean region. *Food Chemistry, 106*(1), 153–157.

Vinson, J. A., Zubik, L., Bose, P., Samman, N., & Proch, J. (2005). Dried fruits: Excellent in vitro and in vivo antioxidants. *Journal of the American College of Nutrition, 24*(1), 44–50.

Viuda-Martos, M., Barber, X., Pérez-Álvarez, J. A., & Fernández-López, J. (2015). Assessment of chemical, physico-chemical, techno-functional and antioxidant properties of fig (*Ficus carica* L.) powder co-products. *Industrial Crops and Products, 69*, 472–479.

Vuolo, M., Lima, V., & Junior, M. M. (2019). *Chapter 2-phenolic compounds: Structure, classification, and antioxidant power.* Woodhead Publishing.

Wojdyło, A., Nowicka, P., Carbonell-Barrachina, Á. A., & Hernández, F. (2016). Phenolic compounds, antioxidant and antidiabetic activity of different cultivars of *Ficus carica* L. fruits. *Journal of Functional Foods, 25*, 421–432.

Yan, J., Zhang, G., Hu, Y., & Ma, Y. (2013). Effect of luteolin on xanthine oxidase: Inhibition kinetics and interaction mechanism merging with docking simulation. *Food Chemistry, 141*(4), 3766–3773.

Yang, J., & Xiao, Y.-Y. (2013). Grape phytochemicals and associated health benefits. *Critical Reviews in Food Science and Nutrition, 53*(11), 1202–1225.

Yemiş, O., Bakkalbaşı, E., & Artık, N. (2012). Changes in pigment profile and surface colour of fig (*Ficus carica* L.) during drying. *International Journal of Food Science & Technology, 47*(8), 1710–1719.

Zhang, J., Yang, Z., Liang, Z., Wang, M., Hu, C., Chang, C., et al. (2022). Secreted frizzled-related protein 4 exerts anti-atherosclerotic effects by reducing inflammation and oxidative stress. *European Journal of Pharmacology, 923*, 174901.

Chapter 19
Ficus carica L. as a Source of Natural Bioactive Flavonoids

Leila Meziant and Mostapha Bachir-bey

Abbreviations

C3G	Cyanidin-3-Glucoside
C3R	Cyanidin-3-Rutinoside
CE	Capillary Zone Electrophoresis
DEEs	Deep Eutectic Solvents
EAE	Enzyme-Assisted Extraction
ILs	Ionic-Liquids
IRAE	Infrared-Assisted Extraction
LLE	Liquid-Liquid Extraction
MAE	Microwave-Assisted Extraction
MALDI	Matrix Assisted Laser Desorption
MAP	Modified Atmosphere Packaging
MCAE	MechanoChemical-Assisted Extraction
MS	Mass Spectrometry
NIRS	Near-Infrared Reflectance Spectroscopy
NMR	Nuclear Magnetic Resonance
PAL	Phenylalanine Ammonia-Lyase
PEF	Pulsed-Electric Field Extraction
PLE	Pressurized-Liquid Extraction
RP-HPLC	Reverse-Phase High Performance
SFE	Supercritical-Fluid Extraction
SPE	Solid-Phase Extraction
UAE	Ultrasound-Assisted Extraction
UHPLC	Ultra-High Performance Liquid Chromatography

L. Meziant · M. Bachir-bey (✉)
Laboratory of Applied Biochemistry, Faculty of Natural and Life Sciences, University of Bejaia, Bejaia, Algeria
e-mail: lilameziant@rocketmail.com, mostapha.bachirbey@univ-bejaia.dz

M. F. Ramadan (ed.), *Fig* (Ficus carica)*: Production, Processing, and Properties*, https://doi.org/10.1007/978-3-031-16493-4_19

1 Introduction

Phenolic compounds are by far the most documented phytochemicals in plants. These heterocyclic molecules (more than 8000 structures found in nature) were at first considered of relevance just for the plants since phenolic compounds are key molecules in plant physiology, acting as insect and herbivore repellents, signaling mediators, and contributing to defense mechanisms against external biotic pathogens and non-biotic conditions (Treutter, 2006). However, new applications in food and health domains have been considered with the discovery of their biological activities. Such interest in phenolic compounds came from their ability to prevent and treat different health disorders related to oxidative stress. This high antioxidant activity is exerted through different mechanisms, including free radical scavenging, protecting and regenerating other dietary antioxidants, and reducing and chelating pro-oxidant metal ions (Khan et al., 2019). On the chemical structure basis, phenolic compounds could be divided into nine distinct groups, namely (1) phenolic acids (hydroxybenzoic acids, hydroxycinnamic acids), (2) flavonoids (anthocyanins, flavanols, flavonols, dihydroflavonols, flavones, isoflavones, flavanones, dihydrochalcones), (3) coumarins (coumestans, furanocoumarines), (4) stilbenoids, (5) quinones (benzoquinones, naphtoquinones, anthraquinones), (6) lignans, (7) xanthones, (8) phenylpropanoids (benzodioxoles, cucuminoids, hydroxyphenylpropenes) and (9) phenols (alkylphenols, methoxyphenols) (Murthy & Bapat, 2020). Flavonoids are the major category of phenolics found in nature, given their scientific importance in different domains, especially food and health.

2 Structure, Occurrence and Function of Flavonoids

Flavonoids, defined as ubiquitous plant pigments, are liable to yellow, orange, red, purple, and black colors of the flowers, fruits, grains, and other plant tissues and are involved in plant growth and defense. Flavonoids are synthesized through the *phenylpropanoid pathway* (also called the shikimic acid pathway) and subdivided into six subclasses: *flavonols*, found in cherries, onion, garlic, and broccoli; *flavones* found in artichokes, celery, parsley, and green pepper; *flavanones* found in oranges, lemons, grapefruits, and oregano; *flavanols* found in cocoa, tea, apples, nuts and blackberries; *isoflavones* found mainly in soybeans and soy products; and *anthocyanins* found in grapes, blueberries, black currants, red cabbage and chickpeas (Prior, 2003; Raffa et al., 2017). Their chemical structure is based on a C_6-C_3-C_6 carbon skeleton consisting of two benzene rings (A and B rings) linked by a heterocyclic pyran ring (C ring). Flavonoids are often hydroxylated in positions 3, 5, 7, 2′, 3′, 4′, and 5′, and could form methyl ethers or acetyl esters (Kumar & Pandey, 2013). Most flavonoids are O- or C-linked glycosides, and when glycosides are formed, the glycosidic linkage often occurs in positions 3 or 7 with D-glucose, L-rhamnose, rutinose, galactose, or arabinose (Kumar & Pandey, 2013). Other entities found in nature are also considered flavonoids, such as biflavonoids, prenyl-flavonoids, glycosidic ester flavonoids, flavonolignans, chalcones, and proanthocyanins (Brodowska, 2017).

Flavonoids were first discovered in 1930 by Albert Szent-Györgyi (Nobel Prize laureate in 1937) as new bioactive substances in *Citrus* extract classified as "vitamin P" which stabilized the biological activity of ascorbic acid and cured scurvy. Vitamin P was later identified as a mixture of hesperidin and eriodictyol-glycoside (flavanones) (Perez-Vizcaino & Fraga, 2018). Flavonoids and their natural derivatives are found in commercialized dietary supplements and numerous pharmaceutical preparations. They also possess multiple applications (1) in the food industry as flavoring, sweetening, and coloring agents, (2) in leather production, as tanning agents, (3) in pigment industry, as dyes or mordant-dyes, (4) in horticulture and flower industry, as efficient agents against plant viruses, microbial pathogens and wood-destroying fungi (Brodowska, 2017).

The increasing interest in flavonoids (shown by the growing number of scientific papers found in this field) could be explained by the fact that these compounds (1) are highly diversified, widely distributed in the plant kingdom, and easily separable with chromatographic methods, (2) are natural pigments with an exceptional safety score and a very wide range of hues (from yellow, orange, red, blue, violet to black) which make them good substituents of industrial synthetic food colorings, (3) possess interesting structural and functional properties that could serve as a base for the development of pharmaceutical drugs, (4) have a long history of medical use for the treatment of various ailments which make them a dominant class of natural therapeutic agents (Ayaz et al., 2019; Havsteen, 2002).

3 Biologically Active Flavonoids

Bioflavonoids identified as biologically active flavonoid compounds constitute an important part of the human diet. Their intake, as estimated in Australian, European, and US adult populations, varies considerably from 209 to 1017 mg/day (Peterson et al., 2015), a variability that can be related to differences in the food supply and cultural eating patterns between countries (Kent et al., 2018). Flavonoid intake in the Asian population could be higher (up to 2000 mg/day) because of the high consumption of legumes, soy products, and tea (Kozlowska & Szostak-Wegierek, 2014). Although from the total intake, only 5–10% (mostly monomeric and dimeric flavonoids) can be absorbed in the small intestine after deglycosylation reactions, the remaining molecules are metabolized in the colon by the gut microbiota: the microorganisms can break up the heterocyclic rings, and the formed hydroxylated phenyl carboxylic acids could be absorbed; flavonoids and the microbial-derived metabolites reach the plasma after 1 to 2 h after ingestion (Kamiloglu et al., 2020; Kopustinskiene et al., 2020).

Consumption of flavonoids could improve intestinal barrier function, modulate the intestinal microbiota and improve microbial dysbiosis in obesity by inhibiting the growth of pathogenic bacteria and increasing the probiotic genera such as

Bifidobacterium and *Lactobacillus* which could improve intestinal health by reducing endotoxin production, rising the conversion of primary into secondary bile acids, maintaining intestinal immune homeostasis and promoting nutrients absorption (Pei et al., 2020). Flavonoid-rich foods are associated with a lower incidence of cardiovascular diseases, cancer, and all-cause mortality: intake of 500 mg of flavonoids/day confers protection for the endothelial and platelet functions, maintains blood pressure and blood lipids, and prevents inflammation and thrombosis (Bondonno et al., 2019). Flavonoids consumption also enhances the cognitive capabilities and restrains the progression of neurodegenerative diseases like Alzheimer's: it seems that flavonoids inhibit the neuronal apoptosis induced by inflammation, reduce the oxidative stress and inhibit key enzymes involved in the fabrication of amyloid plaques, thus maintaining the number and quality of neurons (Ayaz et al., 2019; Bakoyiannis et al., 2019). Bioflavonoids are also thought to possess an important anticancer activity through the modulation of different enzymes and receptors involved in cancer development: it seems that flavonoids may inactivate carcinogens, reduce proliferation and metastasis, arrest the cell cycle, induce apoptosis, inhibit angiogenesis, and reverse multidrug resistance process associated with proteins (Ivey et al., 2017; Raffa et al., 2017; Wang et al., 2013). It has been suggested that flavonoids exert health benefits via antioxidant mechanisms, including direct scavenging of reactive oxygen species, inhibition of oxidases responsible for superoxide anion production, activation of antioxidant enzymes, chelation of metals, and mitigation of oxidative stress caused by nitric oxide accumulation (Raffa et al., 2017). Flavonoids exhibited intense antioxidant activity *in vitro*, but due to their low bioavailability, the *in vivo* antioxidant action of flavonoids is indirect and found to be mediated by cytosolic aryl hydrocarbon receptor (AhR)/nuclear factor erythroid 2-related factor 2 (Nrf2) pathway: flavan-3-ols, flavonols, flavones, and isoflavones bind to the AhR and induce the translocation of Nrf2 to the nucleus and the transcription of antioxidant enzymes (superoxide dismutase, glutathione peroxidase, catalase, peroxiredoxin, and heme oxygenase-1) (Pei et al., 2020).

The evidence for flavonoid health benefits accumulated in the last years has increased the scientific community's interest in identifying new molecules and searching for potential extraction sources to develop more effective therapies, new dietary products, natural food additives, herbal-based pharmaceuticals, and nutraceuticals. To date, interest was focused on the study of flavonoids of green and black teas (Imran et al., 2018; Raychaudhuri et al., 2019; Stangl et al., 2006), grapes, and wine (Fernandes et al., 2017; Jenkinson et al., 2012; Tanaka et al., 2019), citrus fruits (Gandhi et al., 2020; Hwang et al., 2012; Koolaji et al., 2020; Musumeci et al., 2020), berry fruits (Devore et al., 2012; Esposito et al., 2014; Kalt et al., 2020; Vauzour et al., 2010), vegetables (Kim et al., 2004; Pérez-Gregorio et al., 2014; Wu et al., 2019), culinary and medicinal herbs (Coppin et al., 2013; Gutiérrez-Grijalva et al., 2018; Slimestad et al., 2020; Yoo et al., 2009), cocoa and chocolate (Ahmed et al., 2020; Martins et al., 2020; Munguia et al., 2020; Socci et al., 2017), but for the extraction and identification of new molecules, exotic and underutilized fruits, and vegetables, the wastes and molasses of food and industrialized agricultural exploitations are considered as interesting alternatives to the food material.

Flavonoids content in food and plant materials varies widely among the species and between varieties within the same species. In addition, flavonoid content varies according to the plant organ, developmental stage, and growth conditions (Petrussa et al., 2013).

4 Flavonoids in *Ficus carica* L.

4.1 Total Flavonoids Content in Ficus carica *L.*

In the data available, flavonoids are found in different parts of *F. carica,* including the fruits (peels, pulps, and seeds), leaves, stem bark, and latex in variable amounts. In 100 g of fresh fruits, total flavonoid content varied considerably between 2.1 mg (Solomon et al., 2006) to 3694 mg (Mo et al., 2020), but most authors reported values within the range of 20–900 mg/100 g (Table 19.1). The variability observed in flavonoid content is related to the fig varieties, the geographical origin of fig samples, the agro-ecological factors, and the methods used in the extraction, determination, and expression of results (Aljane et al., 2020; Mahmoudi et al., 2018). Regarding the dried fig fruits, total flavonoid contents in 100 g of figs vary from 18.07 mg (Bachir Bey et al., 2016) to 275 mg (Soni et al., 2014) (Table 19.1). The drying process negatively affects the fruit‘s flavonoids, leading to considerable losses. Sun-drying of figs caused an important decrease in the flavonoid content of dark fig fruits reaching 86% (Bachir Bey et al., 2016). Sun-drying seems to affect more purple figs than yellow ones: a loss of 21% of the flavonoid content of purple figs was reported after drying, while an increase of about 75% was observed in yellow figs (Kamiloglu & Capanoglu, 2015). Dark figs are rich in anthocyanins, the most sensitive class of flavonoids to high temperatures. After drying of dark figs, anthocyanins disappeared completely, and only traces were detected, while 80% decrease was reported for flavonols (Bachir Bey et al., 2016).

Fig flavonoids are essentially concentrated in the external parts of the fruit (peels). Some studies reported the flavonoids content in pulp and peel of fig fruits separately. For example, 75 mg of flavonoids was found in 100 g of green peels from Portuguese figs, but only traces of flavonoids were reported in pulps (Palmeira et al., 2019). When comparing figs with different colors, black figs enclose higher amounts of flavonoids than green ones. Kamiloglu and Capanoglu (2015) found 234 and 44 mg of flavonoids in 100 g of peels and pulps of purple figs, respectively, against 63 and 7 mg/100 g of peels and pulps of yellow figs. Given the physiological role of flavonoids in plants (UV protection of plant cells and nucleic acids) and the physical protective role of peels for fruit grains, it is logical that flavonoids are concentrated in the fruit peels rather than the pulps.

F. carica leaves present higher flavonoid content than fruits (2.62 mg/g of leaves against 1.96 mg/g of fruits) (Trifunschi et al., 2015). In addition, the amount of rutin, the most abundant flavonoid compound in *F. carica*, was found to be two-fold

Table 19.1 Flavonoids extraction from different parts of *Ficus carica* L

Method	Sample	Optimal extraction procedure	Additional information	Target compounds	Content	References
Conventional methods						
SLE	Fruit	Solvent: MOH 80%, SSR: 1/25 (g/mL), Time: 2 h (agitation), Temperature: Ambient.	–	TP, TF, AC	TP = 69.5 mg GAE/100 g, TF = 7.73 mg QE/100 g, AC = 4.08 mg CGE/100 g.	Aljane et al. (2020)
SLE	Fruit	Solvent: ETOH 70%, SSR: 3/100 (g/mL), Time: 24 h (agitation 150 rpm), Temperature: 37 °C.	–	TP, TF	TP = 10.7 mg GAE/g, TF = 5.15 mg CAE/g.	Belguith-Hadriche et al. (2016)
SLE	Fruit	Solvent: ETOH 80%, SSR: 1/40 (g/mL), Time: 120 min (agitation, 200 rpm), Temperature: 80 °C.	Optimized parameters: – ETOH-H_2O ratio, – temperature.	TP, TF, PA	TP = 3.7 mg GAE/g, TF = 2.5 mg CE/g, PA = 0.87 mg CyE/g.	Bucic-Kojic et al. (2011)
SLE	Fruit	Solvent: MOH/H_2O/HCl (357/17/1.4, v/v/v), SSR: 10/25 (g/mL), Time: 60 min (agitation), Temperature: Ambient.	–	TP, AC	TP = 211.9 mg GAE/100 g, AC = 298.9 μg CRE/g.	Çalişkan and Polat (2011)
SLE	Fruit	Solvent: ACN 60%, SSR: 1/100 (g/mL), Time: 90 min (agitation), Temperature: 45 °C.	–	TP, TF	TP = 342 mg GAE/100 g, TF = 19.2 mg QE/100 g.	Meziant et al. (2015)
SLE	Fruit	Solvent: ACN 60%, SSR: 1/50 (g/mL), Time: 2 h (agitation), Temperature: 40 °C.	–	TP, TF, AC	TP = 493.4 mg GAE/100 g, TF = 48.31 mg QE/100 g, AC = 41.39 mg CGE/100 g.	Zidi et al. (2020)

SLE	Fruit (dried)	Solvent: ACN 60%, SSR: 1.34/100 (g/mL), Time: 2 h (agitation 400 rpm), Temperature: 40 °C.	Extraction repeated 2 times.	TP, TF, FO, AC, PA	TP = 644 mg GAE/100 g, TF = 126 mg QE/100 g, FO = 40.42 mg QGE/100 g, AC = 18.90 mg QGE/100 g, PA = 179.8 mg CyE/100 g.	Bachir bey and Louaileche (2015)
SLE	Fruit (dried)	Solvent: MOH 80%, SSR: 1/6 (g/mL), Time: 60 min (agitation, 200xg), Temperature: 4 °C.	–	TP, TF	TP = 201.7 mg GAE/100 g, TF = 112.2 mg QE/100 g.	Khadhraoui et al. (2019)
SLE	Fruit (dried)	Solvent: ACN/DCM/EAC/MOH (1/1/1/1, v/v/v/v), SSR: 5/14 (g/mL), Time: 48 h (agitation, 140 rpm), Temperature: 40 °C.	–	TP, TF	TP = 10.9 mg GAE/g, TF = 2.75 mg CE/g.	Soni et al. (2014)
SLE	Fruit peel	Solvent: ETOH, SSR: 1/20 (g/mL), Time: 13.74 min (agitation), Temperature: 35.64 °C.	Optimized parameters: – extraction time, – temperature, – ETOH-H_2O ratio.	AC	AC = 4.03 mg CRE/g.	Backes et al. (2018)
SLE	Fruit peel	Solvent: MOH 90% + citric acid 5% (10/90, v/v), SSR: 1/100 (g/mL), Time: 180 min (agitation), Temperature: Ambient.	Optimized parameters: – solvent, – SSR, – time, – acidification. Extraction repeated twice.	AC	AC = 346.6 mg CGE/100 g.	Meziant et al. (2018)

(continued)

Table 19.1 (continued)

Method	Sample	Optimal extraction procedure	Additional information	Target compounds	Content	References
SLE	Fruit peel	Solvent: MOH/H_2O/CA 5% (72/18/10, v/v/v), SSR: 1/100 (g/mL), Time: 150 min (agitation), Temperature: 25 °C.	Extraction repeated twice.	TP, TF, FO, AC	TP = 86.32 mg GAE/g, TF = 50.47 mg QE/g, FO = 37.45 mg RE/g, AC = 6.22 mg CRE/g.	Meziant et al. (2021)
SLE	Fruit seeds	Solvent: MOH 50%, SSR: 1/5 (g/mL), Time: 90 min (agitation), Temperature: 50 °C.	Optimized parameters: – solvent, – solvent/H_2O ratio.	TP, TF	TP = 71.4 mg GAE/100 g, TF = 31.2 mg CE/100 g.	Nakilcioglu-Tas and Ôtles (2021)
SLE	Leaf	Solvent: H_2O or ETOH 70%, SSR: 1/20 (g/mL), Time: 60 min (agitation), Temperature: 60 °C.	–	TF	Yield = 9% (ETOH).	Trifunschi and Ardelean (2013)
SLE	Leaf	Solvent: H_2O, SSR: 1/75 (g/mL), Time: 6 h, Temperature: 80 °C.	Steam explosion pretreatment: – sample rehydration 1/3 (g/mL) for 1 h, – steam explosion (0.2–0.4 MPa) for 3 min. Optimized parameters: – SSR, time, temperature. – Pretreatment	TF	TF = 32.73 mg/g (untreated sample), TF = 51.03 mg/g (steam explosion, 0.2 MPa).	Qin and Chen (2015)
SLE	Leaf + twig	Solvent: MOH 80%, SSR: 15/300 (g/mL), Time: 12 h (agitation 500 rpm), Temperature: 40 °C.	Re-extraction with 150 mL MOH 80% for 12 h at 40 °C.	TP, TF, AC	TP = 19.7 mg GAE/g (leaf), TP = 12.8 mg GAE/g (twig), TF = 9.72 mg QE/g (leaf), TF = 5.02 mg QE/g (twig), AC = 3.22 mg CGE/100 g (leaf), AC = 5.12 mg CGE/100 g (twig).	Boukhalfa et al. (2018)

SLE	Stem	Solvent: H_2O, SSR: 1/100 (g/mL), Time: 15–20 min (agitation), Temperature: Ambient.	–	TP, TF	TP = 133 mg GAE/g, TF = 43.2 mg QE/g.	Saoudi and El Feki (2012)
Maceration	Fruit	Solvent: ETOH, SSR: 1/50 (g/mL), Time: 24 h (agitation), Temperature: Ambient.	–	TP, TF, FO, PA	TP = 2190 mg GAE/100 g, TF = 858 mg QE/100 g, FO = 19.5 mg QE/100 g, PA = 496.6 mg CE/100 g.	Amessis-Ouchemoukh et al. (2017)
Maceration	Fruit	Solvent: ETOH 90%, SSR: 1/2 (g/mL), Time: 1 week, Temperature: Ambient.	Extraction repeated 4 times. Post-extraction fractionation: – extract suspended in H_2O (1/4, w/v), – partition with PTE (1/1, v/v) 5 times, – partition with EAC (1/1, v/v) 5 times.	Prenylated Isoflavone Derivatives	Compounds yield: (1): 0.38%, (2): 0.18%, (3): 0.27%, (4): 1.2%, (5): 1.85%, (6): 0.27%, (7): 0.13%, (8): 0.18%, (9): 2.1%, (10): 1.34%, (11): 0.26%, (12): 0.58%, (13): 0.13%, (14): 2.86%, (15): 0.69%, (16): 0.62%.	Liu et al. (2019)
Maceration	Fruit	Solvent: MOH, SSR: 1/20 (g/mL), Time: 12 h, Temperature: Ambient.	–	TP, TF	TP = 45.8 mg GAE/g, TF = 9.88 mg RE/g.	Shahbazi (2017)
Maceration	Fruit (dried)	Solvent: H_2O, ETOH 70%, MOH 80% or ACN 50%, SSR: 1/4 (g/mL), Time: 24 h (agitation), Temperature: Ambient.	–	TP, TF, TN	TP = 458 mg GAE/100 g (MOH), TF = 228.2 mg CE/100 g (ACN), TN = 254.1 mg CE/100 g (H_2O).	Benmaghnia et al. (2019)

(continued)

Table 19.1 (continued)

Method	Sample	Optimal extraction procedure	Additional information	Target compounds	Content	References
Maceration	Fruit (dried)	Solvent: H_2O or PTE, SSR: 1/10 (g/mL), Time: 24 h (agitation), Temperature: Ambient.	–	TP, TF, TN	TP = 756.6 mg GAE/100 g (H_2O), TP = 100 mg GAE/100 g (PTE), TF = 90 mg CE/100 g (H_2O), TF = 58 mg CE/100 g (PTE), TN = 40 mg GAE/100 g (ACN), TN < 10 mg GAE/100 g (MOH).	Debib et al. (2014)
Maceration	Fruit + Leaf	Solvent: ETOH 70%, SSR: 2/25 (g/mL), Time: 24 h, Temperature: Ambient.	–	TP, TF	TP = 25.2 mg CatE/g (leaves), TP = 18.6 mg CatE/g (fruits), TF = 2.62 mg RE/g (leaves), TF = 1.96 mg RE/g (fruits).	Trifunschi et al. (2015)
Maceration	Leaf	Solvent: MOH, SSR: 1/25 (g/mL), Time: 48 h (agitation), Temperature: Ambient.	–	TP, TF	TP = 96.4 mg GAE/g, TF = 33.5 mg QE/g.	Ayoub et al. (2019)
Maceration	Leaf	Solvent: H_2O or ETOH 70%, SSR: 1/20 (g/mL), Time: 3 days, Temperature: Ambient.	–	TF	Yield = 2.5% (H_2O), Yield = 8% (ETOH).	Trifunschi and Ardelean (2013)
Maceration	Latex	Solvent: MOH, SSR: 3/40 (g/mL), Temperature: Ambient.	Extraction repeated 3 times, Additional solid-phase extraction: Zorbax SPE C18 cartridge.	TP, TF	TP = 26.8 mg GAE/g, TF = 8.62 mg CE/g.	Pasayeva et al. (2020)

Soxhlet extraction	Leaf	Solvent: MOH, SSR: 1/10 (g/mL), Time: 8 h.	–	TP, TF	TP = 49.7 mg GAE/g, TF = 14.1 mg CE/g.	Mahmoudi et al. (2016)
Soxhlet extraction	Leaf	Solvent: ETOH or MOH or H_2O, SSR: 1/25 (g/mL), Time: 2.5 h.	Optimized parameters: – solvent-H_2O ratio. – fractionation with n-butanol (1/1, v/v).	TP, TF	TP = 150.7 mg GAE/g (ETOH-butanol fraction), TF = 350 mg RE/g (MOH-butanol fraction).	Radwan et al. (2020)
Innovative methods						
UAE	Fruit	Solvent: MOH 80%, SSR: 1/100 (g/mL), Time: 20 min, Ultrasound bath.	–	TP, TF	TP = 59 mg GAE/100 g, TF = 20.2 mg CE/100 g.	Marinova et al. (2005)
UAE	Fruit	Solvent: H_2O, SSR: 45/225 (g/mL), Time: 30 mins, Temperature: 25 °C, Ultrasound bath.	Extraction repeated 3 times. Pre-extraction using supercritical CO_2: – lipid extraction (4 h, 45 °C, 30 MPa, CO_2 flow rate 10 L/h), – separation (35 °C, 5 MPa).	TP, TF	TP = 12.5 mg GAE/g, TF = 1.84 mg CE/g.	Yang et al. (2009)
UAE	Fruit (fresh + Dried)	Solvent: MOH, SSR: 1/10 (g/mL), Time: 60 min (first extraction), 30 min (second extraction), 15 min (third extraction), Ultrasound bath (220 V, 40 kHz).	Extraction repeated 3 times.	TP, TF	TP = 181.6 mg/100 g (fresh), TP = 35.75 mg/100 g (dried), TF = 151.4 mg/100 g (fresh), TF = 18.07 mg/100 g (dried).	Bachir bey et al. (2016)

(continued)

Table 19.1 (continued)

Method	Sample	Optimal extraction procedure	Additional information	Target compounds	Content	References
UAE	Fruit (fresh + dried)	Solvent: MOH 75% + FA 0.1% (v/v), SSR: 2/5 (g/mL), Time: 15 min, Ultrasound bath.	Extraction repeated 3 times.	TP, TF, PA, AC	TP = 493 mg GAE/100 g (fresh), TP = 417 mg GAE/100 g (dried), TF = 66 mg CE/100 g (fresh), TF = 52 mg CE/100 g (dried), PA = 62 mg CyE/100 g (fresh), PA = 16 mg CyE/100 g (dried), AC = 83 mg CGE/100 g (fresh), AC = 14.5 mg CGE/100 g (dried).	Kamiloglu and Capanoglu (2015)
UAE	Fruit (dried)	Solvent: (MOH 80% or ACN 60%) + DTBMP 1%, SSR: 1/5 (g/mL), Ultrasound bath.	Extraction repeated 3 times.	TP, TF, TN	TP > 500 mg GAE/100 g (ACN), TP > 400 mg GAE/100 g (MOH), TF = 118 mg CE/100 g (ACN), TF = 137.4 mg CE/100 g (MOH), TN = 100 mg GAE/100 g (ACN), TN = 194 mg GAE/100 g (MOH).	Debib et al. (2014)

UAE	Fruit peel	Solvent: ETOH, SSR: 1/20 (g/mL), Time: 21.34 min, Sonicator (310.58 W).	Optimized parameters: – extraction time, – ultrasonic power, – ETOH-H_2O ratio.	AC	AC = 4.32 mg CRE/g.	Backes et al. (2018)
IL-UAE	Leaf + Fruit (peel + Pulp)	Solvent: BMIM (1 M), SSR: 1/50 (g/mL), Time: 30 min, Temperature: 30 °C, Ultrasound bath (50%).	Optimized parameters: – ionic liquid type, – ionic liquid concentration, – SSR, temperature, – ultrasonic extraction time.	Individual phenolic compounds, including Rutin	Rutin = 222.3 μg/0.1 g (leaves), Rutin = 133.7 μg/0.1 g (peels), Rutin = 107.9 μg/0.1 g (pulps).	Qin et al. (2015)
MAE	Fruit peel	Solvent: ETOH, SSR: 1/20 (g/mL), Time: 5 min, Temperature: 62.41 °C, Microwave power (400 W).	Optimized parameters: – extraction time, temperature, – ETOH-H_2O ratio.	AC	AC = 4.11 mg CRE/g.	Backes et al. (2018)
MAE	Leaf	Solvent: H_2O or ETOH 70%, SSR: 1/20 (g/mL), Three microwave cycles 3x(3 s power on, 60s power off), Temperature: 70 °C.	–	TF	Yield = 5.5% (H_2O), Yield = 10% (ETOH).	Trifunschi and Ardelean (2013)
Flash extraction	Fruit	Solvent: ETOH 50%, SSR: 1/50 (g/mL), Time: 80s, Voltage: 150 V.	Optimized parameters: – solvent concentration, – SSR, Extraction time, – Extraction voltage.	TF	TF = 36.94 mg/g	Mo et al. (2020).

(continued)

Table 19.1 (continued)

Method	Sample	Optimal extraction procedure	Additional information	Target compounds	Content	References
MCAE	Fruit	Solvent: MOH, SSR: 1/1 (g/mL), Ultra-Turrax homogenization.	–	TP, TF, TN, AC	TP = 74.1 mg GAE/g, TF = 12.7 mg CE/g, TN = 75.9 mg TAE/g, AC = 33.3 mg CGE/g.	Harzallah et al. (2016)

SLE Solid-liquid extraction, *UAE* Ultrasound-assisted extraction, *IL-UAE* Ionic liquid-based ultrasound-assisted extraction, *MAE* Microwave-assisted extraction, *MCAE* Mechanochemical-assisted extraction, *SSR* Solid-to-solvent ratio, *MOH* Methanol, *ETOH* Ethanol, *ACN* Acetone, *FA* Formic acid, *CA* Citric acid, *PTE* Petroleum ether, *DCM* Dichloromethane, *EAC* Ethyl acetate, *BMIM*1-Butyl-3-methylimidazolium hexafluorophosphate, *DTBMP* 26-di-tert-butyl-4-methylphenol, *TP* Total phenolics, *TF* Total flavonoids, *AC* Anthocyanins, *FO* Flavonols, *PA* Proanthocyanidins, *TN* Tannins, *GAE* Gallic acid equivalents, *QE* Quercetin equivalents, *CGE* Cyanidin-3-glucoside equivalents, *CAE* Caffeic acid equivalents, *CE* (+)-catechin equivalents, *CyE* Cyanidin equivalents, *CRE* Cyanidin-3-rutinoside equivalents, *QGE* Quercetin-3-glucoside equivalents, *RE* Rutin equivalents, *CatE* Catechol equivalents, *TAE* Tannic acid equivalents higher in leaves than both peels and pulps of fig fruits in the same extraction conditions (Qin et al., 2015). Total flavonoid contents of *F. carica* leaves reported in the literature range from 2.62 mg/g (Trifunschi et al., 2015) to 350 mg/g (Radwan et al., 2020) (Table 19.1). The examination of publications that treated *F. carica* leaves revealed many reports about leaf extracts' pharmacological and antioxidant properties, but only a few analyzed the flavonoid content. Variety, methods of extraction, experimental conditions, and expression of results influence considerably the estimation of flavonoid content in *F. carica* leaves (Konyalıoğlu et al., 2005; Mahmoudi et al., 2016; Radwan et al., 2020).

Regarding the other parts of *F. carica*, rare publications were conducted to determine flavonoid content in fig tree branches. Twig samples from *F. carica* cultivated in Algeria, extracted with aqueous methanol (80%), contained 5.02 mg of flavonoids per gram of dry matter (Boukhalfa et al., 2018). Aqueous extracts of stem samples from *F. carica* grown in Tunisia contain up to 43.2 mg of flavonoids per gram of dry extract (Saoudi & El Feki, 2012). Flavonoid content of *F. carica* latex gathered from the neck of fig fruits was also reported. Fig latex contains 43.2 mg of total flavonoids per milliliter of the crude sample (Aziz, 2012). In the same line, latex fractions enclosed between 2.227 mg (*n*-hexane fraction) and 8.615 mg (solid-phase extraction) of flavonoids per gram of extract (Paşayeva et al., 2020). To date, no report has been published on the presence of flavonoids in *F. carica* roots.

4.2 *Flavonoids Extraction from* Ficus carica *L.*

Preparation of bioactive flavonoids from any plant material used as food ingredients, pharmaceutical, and cosmetic products always begins with an extraction step. Extraction aims mainly to separate target soluble metabolites from a mixture of plant compounds, leaving behind the insoluble cellular residue (Azwanida, 2015). Different extraction methods may be applied, and various strategies depend on the origin of the biological material from which the target molecules are to be extracted (Stobiecki & Kachlicki, 2006). Extraction efficiency could be influenced by the extracting solvent, the solid-to-solvent ratio, time and temperature of extraction, pH, mode of stirring, number of extractions, and the particle size of plant material; these parameters influence the recovery of bioactive molecules significantly and should be considered in optimization procedures (Madureira et al., 2020).

Extraction methods evolved over the years to meet some requirements such as efficiency, versatility, ease of use, inexpensiveness, and preservation of the major fraction of bioactive compounds from structural modifications (Gullón et al., 2017). Conventional methods such as solid-liquid extraction (SLE), maceration, and Soxhlet extraction are commonly used in bioactive molecules research due to their simplicity, minimal processing, and the possibility of using several solvents of different polarities. However, they are time-consuming and could cause the degradation of thermo-sensitive molecules. Also, they are considered non-environmental respective methods due to large volumes of organic solvents (Azwanida, 2015). To overcome the limitations of the conventional methods, innovative extraction methods are developed, such as ultrasound-assisted extraction (UAE), microwave-assisted extraction (MAE), enzyme-assisted extraction (EAE), pulsed-electric field extraction (PEF), mechanochemical-assisted extraction (MCAE), infrared-assisted extraction (IRAE), pressurized-liquid extraction (PLE), ionic-liquids (ILs), deep eutectic solvents (DEEs) and supercritical-fluid extractions (SFE). In addition, the presence of carbohydrates and lipophilic substances in the obtained extracts may influence the qualitative and quantitative composition of flavonoids and their derivatives, so in many cases, additional cleaning based on solid-phase extraction (SPE) or liquid-liquid extraction (LLE) of the extracted compounds is required (Stobiecki & Kachlicki, 2006).

Several extraction procedures extracted flavonoids from *F. carica* plant material (Table 19.1). Generally, flavonoids are extracted together with phenolic compounds (crude phenolic extract), given the similarities in solubility and nature of these compounds. SLE seems to be widely used for the extraction of flavonoids from *F. carica* fresh fruits (Aljane et al., 2020; Belguith-Hadriche et al., 2016; Bucic-Kojic et al., 2011; Çalişkan & Polat, 2011; Meziant et al., 2015; Zidi et al., 2020), dried fruits (Bachir bey & Louaileche, 2015; Khadhraoui et al., 2019; Soni et al., 2014), fruit peels (Backes et al., 2018; Meziant et al., 2018; Meziant et al., 2021), fruit seeds (Nakilcioğlu-Taş & Ötleş, 2021), leaves (Qin & Chen, 2015; Trifunschi & Ardelean, 2013), twigs (Boukhalfa et al., 2018) and stem (Saoudi & El Feki, 2012), using distilled water, ethanol, methanol, and acetone at different concentrations as extracting solvents with or without heating. Maceration is also used in many extraction processes for the recovery of flavonoids from *F. carica* fresh fruits (Amessis-Ouchemoukh et al., 2017; Liu et al., 2019; Shahbazi, 2017), dried fruits (Benmaghnia et al., 2019; Debib et al., 2014), leaves (Ayoub et al., 2019; Trifunschi et al., 2015; Trifunschi & Ardelean, 2013) and latex (Paşayeva et al., 2020). Soxhlet extraction was only reported for the extraction of flavonoids from *F. carica* leaves (Mahmoudi et al., 2016; Radwan et al., 2020). Among the innovative methods, UAE is the most utilized method for the preparation of flavonoid extracts from *F. carica* fruits (Bachir Bey et al., 2016; Backes et al., 2018; Debib et al., 2014; Kamiloglu & Capanoglu, 2015; Marinova et al., 2005; Yang et al., 2009) and leaves (Qin et al., 2015). To a lesser extent, MAE was also reported to extract flavonoids from *F. carica* fruit peel (Backes et al., 2018) and leaves (Trifunschi & Ardelean, 2013). Flavonoids were also recovered from *F. carica* fruits by Ultra-Turrax-assisted extraction using methanol as solvent (Harzallah et al., 2016). Recently, a promising and ultra-rapid extraction method called flash extraction, already used at an industrial scale for the extraction of pigments from raisin skins for the fabrication of wine, was also assayed for the extraction of total flavonoids from *F. carica* fruits using aqueous ethanol (50%) as extracting solvent, and 80 seconds were sufficient to yield 36.9 mg of flavonoids from one gram of sample (Mo et al., 2020).

The efficiency of UAE over MAE and SLE for the extraction of cyanidin-3-rutinoside, the most abundant anthocyanin in dark *F. carica* fruit peels, was highlighted in a comparative study where results showed that UAE produced high purity extracts with slightly higher anthocyanin content (4.32 mg CRE/g of fig peel) compared to the MAE (4.11 mg CRE/g of peel) and SLE (4.03 mg CRE/g of peel) (Backes et al., 2018). MAE efficiency in extracting flavonoids from *F. carica* leaves using aqueous ethanol (70%) was compared to maceration and SLE methods, and the obtained results showed that MAE provides higher yields (10%, w/w) with the advantage of saving time (3 cycles of 3 seconds) compared to maceration (8%, w/w) conducted during 3 days and SLE (9%, w/w) during 60 minutes (Trifunschi & Ardelean, 2013). The UAE efficiency was compared with the maceration method for the recovery of phenolic compounds from *F. carica* leaves, and results showed that UAE produced higher extraction yields and higher phenolic contents than maceration (10.7–21.1 mg GAE/g for maceration and 12.9–23.0 mg GAE/g for UAE), but without statistically significant differences (Akhtar et al., 2019). Similarly, UAE was slightly more effective than maceration for extracting phenolic compounds from *F. carica* leaves (0.69 and 0.79 g GAE/100 g for maceration and UAE, respectively) and fruits (0.12 and 0.20 g GAE/100 g for maceration and UAE, respectively). However, both methods gave similar qualitative phenolic profiles (Ammar et al., 2015). The conventional extraction methods are still widely used in recent research despite their disadvantages, probably because they provided comparable extraction efficiency and similar flavonoid profiles as the innovative methods without requiring any specific expensive equipment.

4.3 *Flavonoid Profile of* Ficus carica *L.*

4.3.1 Methods of Characterization

After the extraction step necessary to isolate the flavonoid compounds from the plant material, the study of flavonoids is completed by identification and characterization steps of the extracted compounds and the elucidation of their structures (aglycones, glycosylated flavonoids, acylated flavonoids, and isomeric forms). Various analytical methods could be applied for the identification and quantification purpose, such as chromatographic systems (ex. thin-layer, liquid, and gas, micellar electro-kinetic, high-speed counter current, supercritical fluid chromatographies) equipped with different types of analyzers (electromagnetic, Fourier transform ion cyclotron resonance, time of flight, etc.), matrix-assisted laser desorption (MALDI), near-infrared reflectance spectroscopy (NIRS), capillary zone electrophoresis (CE), etc. (Gullón et al., 2017). Reverse-phase high performance (RP-HPLC) and ultra-high performance (UHPLC) liquid chromatographies coupled to UV-Visible spectrophotometry and mass spectrometry (MS) are established as the methods of choice for the separation and quantification of several individual flavonoids simultaneously by comparison of retention times as well as spectral properties to purified known standards. The characterization and identification of unknown compounds are performed with the aid of nuclear magnetic resonance (NMR) analyses (^{1}H and ^{13}C isotopes) which are required to provide structural information (Stobiecki & Kachlicki, 2006).

Research studies about the flavonoid profile of extracts of different *F. carica* parts and their main findings are assembled in Table 19.2. Most research works were

Table 19.2 Flavonoid compounds identified in *Ficus carica* L. extracts

Sample	Extract	Characterization	Identified compounds	Quantification	References
Fruits (fresh)	Acetone 25% + 1% trifluoroacetic acid	HPLC-UV and -DAD	Quercetin	0.655 mg/100 g	Gündesli et al. (2021)
			Kaempferol	2.396 mg/100 g	
			Myricetin	2.632 mg/100 g	
			Naringenin	0.054 mg/100 g	
			Catechin	0.590 mg/100 g	
			Epicatechin	7.809 mg/100 g	
Fruits (fresh)	Methanol	UHPLC-QTOF-MS	Quercetin	9.48–145.5 mg/kg	Soltana et al. (2018)
			Quercetin 3-O-rutinoside (rutin)	107.7–303.2 mg/kg	
			Quercetin 3-O-glucoside (isoquercetin)	5.20–69.1 mg/kg	
			Quercetin 3-O-(6″-malonyl) glucoside	0.16–1.90 mg/kg	
			Quercetin di-glycoside derivatives	2.74–20.7 mg/kg	
			Kaempferol	1.39–21.5 mg/kg	
			Kaempferol 3-O-glucoside	1.10–15.34 mg/kg	
			Isorhamnetin 3-O-glucoside	0.03–0.24 mg/kg	
			Syringetin	0.03–0.09 mg/kg	
			Tamarixetin	0.01–0.13 mg/kg	
			Luteolin 8-C-glucoside (orientin)	6.80–38.6 mg/kg	
			Luteolin 7-O-glucoside (cynaroside)	2.37–15.1 mg/kg	
			Luteolin C-glycoside derivatives	5.9–32.9 mg/kg	
			Apigenin 8-C-glucoside (vitexin)	14.53–71.1 mg/kg	
			Apigenin C-glycoside derivatives	6.04–20.6 mg/kg	
			Eriodictyol	7.07–25.5 mg/kg	
			Naringenin	7.13–44.4 mg/kg	

Fruits (fresh)	Ethanol 70%	RP-UHPLC-DAD -QTOF-MS/MS	Quercetin	2.0% *(relative area).*	Belguith-Hadriche et al. (2016)
			Quercetin 3-O-rutinoside (rutin)	53.9%.	
			Quercetin 3-O-glucoside (isoquercetin)	2.5%.	
			Quercetin 3-O-(6″-malonyl)-glucoside	0.4%.	
			Apigenin 8-C-glucoside (vitexin)	5.0%.	
			Apigenin C-glycoside derivatives	3.5%.	
			Luteolin	0.9%.	
			Luteolin 8-C-glucoside (orientin)	1.2%.	
			Luteolin 6-C-glucoside (isoorientin)	0.8%.	
			Naringenin	0.5%.	
			Eriodictyol	0.2%.	
			Taxifolin	1.4%.	
			Cyanidin 3,5-diglucoside	8.4%.	
			Cyanidin 3-rutinoside	17.6%.	
Fruits (fresh)	Methanol 30% + 1% ascorbic acid	UHPLC-PDA -QTOF/MS	Kaempferol 3-O-rutinoside	0.6–30.5 mg/100 g	Wojdylo et al. (2016)
			Quercetin 3-O-rutinoside	39.0–328 mg/100 g	
			Quercetin 3-O-glucoside	4.6–14.2 mg/100 g	
			Quercetin 3-O-(malonyl) glucoside	5.3–58.0 mg/100 g	
			Apigenin C-glycoside	0.5–24.3 mg/100 g	
			Cyanidin 3-O-rutinoside	0.4–116 mg/100 g	
			Cyanidin 3,5-O-diglucoside	0.0–1.9 mg/100 g	
			Pelargonidin 3-O-rutinoside	0.0–3.8 mg/100 g	
			(+)-Catechin	20.6–95.7 mg/100 g	
			(−)-Epicatechin	9.0–43.1 mg/100 g	
			Procyanidins	55.6–268 mg/100 g	

(continued)

Table 19.2 (continued)

Sample	Extract	Characterization	Identified compounds	Quantification	References
Fruits (fresh)	Methanol + 1% BHT	HPLC-DAD	Rutine	4.89–28.7 mg/100 g	Veberic et al. (2008)
			(+)-Catechin	1.07–4.03 mg/100 g	
			(−)-Epicatechin	0.34–0.97 mg/100 g	
Fruits (syrup)	Methanol + 1% BHT	HPLC-DAD-MS	Rutin	1.9 mg/g	Puoci et al. (2011)
			Catechin	0.34 mg/g	
			Epicatechin	0.007 mg/g	
Fruit seeds	Methanol 50%	HPLC-DAD	Rutin	50.7 mg/kg DM	Nakilcioglu-Tas & Ötles (2021)
			(−) Epicatechin	166.4 mg/kg DM	
Fruits (dried)	Methanol 80%	LC-ESI-MS	Quercetin	0.24–1.49 mg/100 g	Khadhraoui et al. (2019)
			Quercetin 3-O-rutinoside (rutin)	2.17–38.05 mg/100 g	
			Kaempferol	0.02–1.13 mg/100 g	
			Apigenin	0.00–4.85 mg/100 g	
			Apigenin 7-O-glucoside	0.00–0.91 mg/100 g	
			Luteolin	0.00–0.15 mg/100 g	
			Luteolin 7-O-glucoside	0.02–0.43 mg/100 g	
			Cirsiliol	0.95–2.72 mg/100 g	
			Naringenin	0.01–0.43 mg/100 g	
			Naringenin 7-rhamnoglucoside (naringin)	0.00–0.06 mg/100 g	
Fruits (dried)	Ethyl acetate fraction from ethanol 90% extract	HPLC and ^{1}H and ^{13}C NMR	Ficucaricone D	(*Not reported*).	Liu et al. (2019)
			Gancaonin N	(*Not reported*).	
			Isopiscerythrone	(*Not reported*).	
			Viridiflorin	(*Not reported*).	
			4’-Hydroxy-5,7-dimethoxy-6-(3-methyl-2-butenyl)-isoflavone	(*Not reported*).	

Fruits (dried)	Methanol 20%	RP-HPLC-DAD	Rutin	3.26 mg/100 g DM	Tawfik and Alhejy (2014)
			Quercetin	0.56 mg/100 g DM	
			Kaempferol	0.12 mg/100 g DM	
			Isorhamnetin	0.29 mg/100 g DM	
			Catechin	13.8 mg/100 g DM	
			Epicatechin	12.8 mg/100 g DM	
			Epigallocatechin	25.4 mg/100 g DM	
			Epicatechin gallate	0.85 mg/100 g DM	
			Epigallocatechin gallate	2.52 mg/100 g DM	
Fruits (dried)	Aqueous HCl (pH = 2)	HPLC-DAD	Quercetin 3-O-rutinoside	4.33–88.40 mg/kg	Faleh et al. (2012)
			Quercetin	0.10–8.76 mg/kg	
Fruits (fresh + dried)	Methanol	HPLC-DAD	Quercetin	6.76–16.5 mg/100 g (fresh), 0.64–3.43 mg/100 g (dried).	Bachir bey et al. (2016)
			Quercetin 3-rutinoside (rutin)	31.7–38.1 mg/100 g (fresh), 3.49–10.8 mg/100 g (dried).	
			Quercetin 3-glucoside (isoquercetin)	5.03–14.1 mg/100 g (fresh), 0.49–3.20 mg/100 g (dried).	
			Cyanidin 3-glucoside	1.34–6.86 mg/100 g (fresh).	
			Cyanidin 3-rutinoside	31.4–73.0 mg/100 g (fresh), 0.00–0.57 mg/100 g (dried).	
			Catechin	1.72–8.63 mg/100 g (fresh), 1.80–5.78 mg/100 g (dried).	
			Procyanidin B1	1.87–4.80 mg/100 g (fresh), 0.77–1.56 mg/100 g (dried).	
			Procyanidin B2	1.11–2.61 mg/100 g (fresh), 0.47–1.34 mg/100 g (dried).	
Fruits (fresh + dried)	Methanol + 1% BHT	HPLC-DAD	Rutin	15.1–68.2 mg/100 g (fresh), 5.75–12.35 mg/100 g (dried).	Nakilcioglu & Hisil (2013)
			(–) Epicatechin	42.2–76.9 mg/100 g (fresh), 14.7–23.4 mg/100 g (dried).	

(continued)

Table 19.2 (continued)

Sample	Extract	Characterization	Identified compounds	Quantification	References
Fruits (fresh + dried)	Methanol + 1% BHT + 3% formic acid	HPLC-DAD and -ESI-MS/MS	Rutin	0.61–1.86 mg/100 g (fresh), 1.38–12.06 mg/100 g (sun dried), 3.75–14.62 mg/100 g (oven dried).	Slatnar et al. (2011)
			Quercetin 3-O-glucoside	0.18–0.60 mg/100 g (fresh), 0.56–3.35 mg/100 g (sun dried), 1.10–2.98 mg/100 g (oven dried).	
			Kaempferol 3-O-glucoside	0.04–0.13 mg/100 g (fresh), 0.31–0.59 mg/100 g (sun dried), 0.56–1.43 mg/100 g (oven dried).	
			Luteolin 8-C-glucoside	0.00 mg/100 g (fresh), 0.13–0.16 mg/100 g (sun dried), 0.21–0.45 mg/100 g (oven dried).	
			Cyanidin 3-O-rutinoside	0.21–0.62 mg/100 g (fresh), 0.12–0.26 mg/100 g (sun dried), 0.12–0.31 mg/100 g (oven dried).	
			Catechin	1.36–2.88 mg/100 g (fresh), 5.88–11.4 mg/100 g (sun dried), 15.5–19.7 mg/100 g (oven dried).	
			Epicatechin	7.11–8.67 mg/100 g (fresh), 10.4–23.3 mg/100 g (sun dried), 26.6–36.6 mg/100 g (oven dried).	
Fruits (peel + pulp)	Ethanol 80%	LC-DAD-ESI-MS/MS	Quercetin 3-O-rutinoside (rutin)	0.513 mg/g ext. (peel).	Palmeira et al. (2019)
			Quercetin acetylglycoside-derivatives	0.038 mg/g ext. (peel).	
			Kaempferol glycoside derivatives	Traces (pulp).	
			Apigenin 2”-O-rhamnose-C-acetylglycoside	0.026 mg/g ext. (peel).	
			Apigenin C-glycosylated derivatives	0.072 mg/g ext. (peel).	
			Taxifolin glycoside derivative	0.097 mg/g ext. (peel).	

Fruits (peel + pulp)	Methanol 80%	HPLC-DAD	Quercetin 3-rutinoside (rutin)	14.2–15.3 mg/100 g (peel), 0.08–0.10 mg/100 g (pulp).	Viuda-Martos et al. (2015)
			Luteolin	1.07–2.82 mg/100 g (peel), 0.05–0.06 mg/100 g (pulp).	
			Luteolin 7-O-glucoside	2.18–3.80 mg/100 g (peel).	
			Apigenin	1.03–5.10 mg/100 g (peel), 0.10–0.11 mg/100 g (pulp).	
			Cyanidin 3-rutinoside	0.43–2.01 mg/100 g (peel).	
			Cyanidin 3-glucoside	0.12–1.33 mg/100 g (peel).	
			Catechin	6.10–12.48 mg/100 g (peel), 4.58–9.53 mg/100 g (pulp).	
			Epicatechin	2.25–5.74 mg/100 g (peel), 1.68–1.75 mg/100 g (pulp).	
Fruits (peel + pulp)	Methanol	HPLC-DAD	Quercetin 3-O-rutinoside (rutin)	527.9–1071 mg/kg (peel).	Del Caro and Piga (2008)
			Flavonol compound I	54.82–83.52 mg/kg (peel).	
			Flavonol compound II	114.6–296.1 mg/kg (peel).	
			Cyanidin 3-O-rutinoside	901.0 mg/kg (peel), 4.92–17.6 mg/kg (pulp).	
			Cyanidin 3-O-glucoside	28.24 mg/kg (peel).	

(continued)

Table 19.2 (continued)

Sample	Extract	Characterization	Identified compounds	Quantification	References
Fruits (peel + pulp)	Methanol +0.5% Trifluoroacetic acid	HPLC-DAD-MS	Cyanidin 3-rutinoside	1.542–7.832 mg/100 g (peel), 0.104–1.016 mg/100 g (pulp).	Duenas et al. (2008)
			Cyanidin 3-rutinoside dimer	0.044–0.093 mg/100 g (peel), 0.00–0.008 mg/100 g (pulp).	
			Cyanidin 3-glucoside	0.151–1.538 mg/100 g (peel), 0.019–0.215 mg/100 g (pulp).	
			Cyanidin 3-malonylglucoside	0.067–0.351 mg/100 g (peel), 0.002–0.009 mg/100 g (pulp).	
			Cyanidin 3, 5-diglucoside	0.199–0.523 mg/100 g (peel), 0.009–0.057 mg/100 g (pulp).	
			Cyanidin 3-malonylglycosyl-5-glucoside	0.053–0.139 mg/100 g (peel), 0.002–0.009 mg/100 g (pulp).	
			5-Carboxypyrano-Cyanidin 3-rutinoside	0.048–0.126 mg/100 g (peel), 0.004–0.094 mg/100 g (pulp).	
			Pelargonidin 3-glucoside	0.029–0.068 mg/100 g (peel), 0.001–0.009 mg/100 g (pulp).	
			Pelargonidin 3-rutinoside	0.047–0.352 mg/100 g (peel), 0.002–0.008 mg/100 g (pulp).	
			(Epi)catechin-(4 → 8)-Cyanidin 3-glucoside	0.00–0.048 mg/100 g (peel), 0.003–0.013 mg/100 g (pulp).	
			(Epi)catechin-(4 → 8)-Cyanidin 3-rutinoside I	0.059–0.123 mg/100 g (peel), 0.002–0.059 mg/100 g (pulp).	
			(Epi)catechin-(4 → 8)-Cyanidin 3-rutinoside II	0.044–0.093 mg/100 g (peel), 0.002–0.013 mg/100 g (pulp).	
			(Epi)catechin-(4 → 8)-Pelargonidin 3-rutinoside I	0.00–0.024 mg/100 g (peel).	
			(Epi)catechin-(4 → 8)-Pelargonidin 3-rutinoside II	0.00–0.020 mg/100 g (peel).	

Fruits (peel + pulp + dried)	Acetone/water /formic acid (80/19.5/0.5)	HPLC-DAD-ESI-MS and -MS/MS.	Quercetin rutinoside	2.9–15.8 mg/100 g (peel), 0.0–1.8 mg/100 g (pulp), 10.2–13.0 mg/100 g (dried).	Vallejo et al. (2012)
			Quercetin glucoside	0.2–3.2 mg/100 g (peel), 0.7–2.5 mg/100 g (dried).	
			Quercetin acetylglucoside	0.0–1.7 mg/100 g (peel), 2.1–2.8 mg/100 g (dried).	
			Kaempferol rutinoside	0.2–0.9 mg/100 g (peel), 1.0–2.0 mg/100 g (dried).	
			Apigenin 7-O-rutinoside	0.8–2.5 mg/100 g (peel).	
			Luteolin C-glycosides derivatives	0.1–1.9 mg/100 g (peel).	
			Cyanidin 3-rutinoside	0.0–108.9 mg/100 g (peel), 0.0–9.5 mg/100 g (pulp).	
			Cyanidin 3-glucoside	0.0–11.1 mg/100 g (peel).	
			Proanthocyanidins	0.1–1.0 mg/100 g (pulp).	
			Catechin/epicatechin	9.0–38.7 mg/100 g (pulp).	
Fruits (fresh) + Leaves	Ethanol 70%	HPLC-DAD	Quercetin	0.08% (fruits), 2.5% (leaves).	Trifunschi et al. (2015)
			Rutin	1.8% (fruits), 0.2% (leaves).	
			Kaempferol	Traces (fruits + leaves).	
			Luteolin	Traces (fruits), 0.07% (leaves).	
Fruits (dried) + leaves	Aqueous	HPLC-DAD	Quercetin	0.14 mg/100 g (leaves).	El-Shobaki et al. (2010)
			Galangin	24.76 mg/100 g (fruits).	
			Chrysin	7.64 mg/100 g (fruits), 7.14 mg/100 g (leaves).	
			Pinocembrin	46.94 mg/100 g (fruits), 52.85 mg/100 g (leaves).	
			Pinostrobin	19.47 mg/100 g (leaves).	

(continued)

Table 19.2 (continued)

Sample	Extract	Characterization	Identified compounds	Quantification	References
Fruits (peel + pulp) + Leaves	Methanol 80% Ethanol 70%	RP-UHPLC-DAD-QTOF -MS and -MS/MS	Quercetin	*(relative abundance)*: 0.69–1.06% (fruits), 0.59–1.34% (peel), 0.33–0.86% (pulp).	Ammar et al. (2015)
			Quercetin 3-O-rutinoside (rutin)	16.34–19.20% (fruits), 31.41–43.68% (peel), 0.71–0.75% (pulp), 14.46–14.96% (leaves).	
			Quercetin 3-O-glucoside (isoquercetin)	1.95–2.04% (fruits), 2.55–3.17% (peel), 0.86–1.05% (pulp), 4.28–4.67% (leaves).	
			Quercetin 3-O-(6''-malonyl) glucoside	0.48–1.60% (fruits), 1.45–3.00% (peel), 0.24–0.30% (pulp), 0.49–1.54% (leaves).	
			Quercetin O-glycoside derivatives	1.88–3.57% (fruits), 1.93–3.51% (peel), 0.50–0.55% (pulp), 0.72–1.27% (leaves).	
			Luteolin	0.75–0.76% (fruits), 0.50–1.07% (peel), 1.51–2.35% (pulp), 0.37–0.56% (leaves).	
			Luteolin 8-C-glucoside (orientin)	0.71–0.91% (fruits), 1.20–1.35% (peel), 0.49–0.55% (pulp), 0.92–1.59% (leaves).	

			Luteolin 6-C-glucoside (isoorientin)	0.40–0.87% (fruits), 0.97–1.01% (peel), 0.35–0.85% (pulp), 0.45–0.68% (leaves).	
			Luteolin 7-O-glucoside (cynaroside)	0.19–0.37% (fruits), 0.36% (peel), 0.62–0.72% (pulp), 0.17–0.24% (leaves).	
			Luteolin C-glycoside derivatives	0.50–0.53% (fruits), 0.52–0.63% (peel), 2.46–3.14% (leaves).	
			Apigenin	0.90–1.94% (fruits), 0.74–0.89% (peel), 1.32–1.48% (pulp), 0.14–0.18% (leaves).	
			Apigenin 8-C-glucoside (vitexin)	2.06–3.29% (fruits), 2.57–3.69% (peel), 1.26–1.91% (pulp), 1.53–4.21% (leaves).	
			Apigenin C-glycoside derivatives	2.66–8.71% (fruits), 4.53–9.68% (peel), 1.06–1.37% (pulp), 18.7–19.1% (leaves).	
			Eriodictyol glycoside derivatives	0.71–2.30% (fruits), 2.03–5.12% (peel), 0.58–0.87% (pulp).	
			Eriodictyol	0.73–1.46% (fruits), 0.81–3.41% (peel), 0.38–0.70% (pulp).	

(continued)

Table 19.2 (continued)

Sample	Extract	Characterization	Identified compounds	Quantification	References
			Naringenin	0.71–1.60% (fruits), 0.78–1.72% (peel), 0.61–0.79% (pulp), 0.06–0.07% (leaves).	
			Taxifolin (dihydroquercetin)	0.42–0.52% (fruits), 0.50–0.66% (peel), 0.79–0.95% (pulp).	
			Genistein	0.09–0.14% (leaves).	
			Genistein 4′-methyl ether (biochanin A)	0.09–0.74% (fruits), 0.08–0.14% (peel), 0.08–0.24% (pulp), 0.25–0.50% (leaves).	
			Hydroxygenistein methyl-ether-malonylglycoside	0.01–0.03% (leaves)	
			7-Methoxy 2′-hydroxy genistein (cajanin)	0.73–1.40% (fruits), 0.19–0.20% (peel), 1.37–4.97% (pulp), 0.49–1.09% (leaves).	
			Prenylgenistein I / II/ III	2.67–6.27% (fruits), 0.40–1.17% (peel), 7.56–17.89% (pulp), 4.57–6.92% (leaves).	
			Prenylhydroxygenistein I / II / III	3.79–12.1% (fruits), 0.65–0.71% (peel), 29.8–30.9% (pulp), 8.24–10.6% (leaves).	
			Cyanidin rutinoside I / II / III	1.85–52.7% (fruits), 33.2–82.5% (peel), 5.02–22.5% (pulp).	

			Cyanidin 3-rutinoside-glycoside	1.07% (fruits), 2.25% (peel).	
			Cyanidin 3,5-diglucoside	1.87–9.71% (fruits), 5.97–19.6% (peel), 5.19–24.7% (pulp).	
			Cyanidin 3-malonylglucosyl-5-glucoside	1.04% (fruits), 0.83% (peel), 1.34% (pulp).	
			Cyanidin 3-glucoside	1.70% (fruits), 1.70% (peel), 2.13% (pulp).	
			Petunidin + dimer cyanidin rutinoside	2.49% (fruits), 3.01% (peel).	
			Pelargonidin 3-rutinoside	2.00% (fruits), 2.83% (peel).	
			(+) Catechin	0.17–0.20% (fruits), 0.06–0.07% (peel), 0.27–0.72% (pulp), 0.03–0.04% (leaves).	
Fruits (peel + pulp) + Leaves	BMIM PF_6	HPLC-UV	Rutin	1.34 mg/g (peel), 1.08 mg/g (pulp), 2.22 mg/g (leaves).	Qin et al. (2014)
Fruits (peel + pulp) + Leaves	Aqueous	HPLC-DAD	Quercetin 3-rutinoside	629.6 mg/kg ext. (peel), 64.6 mg/kg ext. (pulp), 17,440 mg/kg ext. (leaves).	Oliveira et al. (2009)
			Quercetin 3-glucoside	30.8 mg/kg ext. (peel), 611 mg/kg ext. (leaves).	

(continued)

Table 19.2 (continued)

Sample	Extract	Characterization	Identified compounds	Quantification	References
Leaves	Ethyl acetate Fraction of Methanolic extract	GC-HRMS	Quercetin, Kaempferol, Chrysin	*(Not reported).*	Dureshahwar et al. (2019)
Leaves	Ethanol 70%	RP-UHPLC-MS and -MS/MS	Quercetin 3-O-rutinoside (rutin)	19.8–23.0% *(relative area).*	Belguith-Hadriche et al. (2017)
			Quercetin 3-O-glucoside (isoquercetin)	2.1–2.9%.	
			Apigenin 8-C-glucoside (vitexin)	1.0–1.2%.	
			Apigenin C-glycoside derivatives	8.1–12.7%.	
			Luteolin	0.8–2.5%.	
			Luteolin C-glycoside derivatives	1.8–2.4%.	
			7-Methoxy 2′-hydroxy genistein (cajanin)	1.2–1.4%.	
			Prenylgenistein	10.2–11.9%.	
			Prenylhydroxygenistein	8.7–10.2%.	
Leaves	Water/methanol/ Acetone (1/1/1)	HPLC-DAD-MS/ MS	Quercetin 3-O-rutinoside (rutin)	4.7–14.6 mg/g	Takahashi et al. (2014) and (2017)
			Quercetin 3-O-glucoside (isoquercetin)	*(Not reported).*	
			Quercetin 3-O-(6”-O-malonyl)-glucoside	*(Not reported).*	
			Kaempferol 3-O-glucoside (astragalin)	*(Not reported).*	
			Isoschaftoside	2.5–6.4 mg/g	
Leaves	Methanol 70%	HPLC-DAD and LC-ESI-MS/MS	Quercetin 3-O-rutinoside (rutin)	1233 mg/100 g	Teixeira et al. (2006, 2009)

Leaves	Ethanol 70%	HPLC-DAD and LC-ESI-MS	Quercetin	630 mg/kg extract.	Vaya and Mahmood (2006)
			Luteolin	680 mg/kg extract.	
			Biochanin A	17.4 mg/kg extract.	

BHT 2,6-di-tert-butyl 4-methyl phenol, *BMIM* PF_6 1-butyl-3-methylimidazolium hexafluorophosphate, *GC-HRMS*, gas chromatography coupled to high resolution mass spectrometry, *LC-ESI-MS* liquid chromatography coupled to electrospray ionization mass spectrometry, *RP-HPLC-DAD* reversed phase high performance liquid chromatography with diode array detection, *HPLC-UV* high performance liquid chromatography with ultraviolet detection, *HPLC-DAD-ESI-MS/MS* high performance liquid chromatography with diode array detection coupled to electrospray ionization and tandem mass spectrometry, *RP-UHPLC-DAD-QTOF-MS/MS* reversed phase ultra high performance liquid chromatography with diode array detection coupled to quadrupole time-of-flight tandem mass spectrometry, *UHPLC-PDA-QTOF-MS* ultrahigh performance liquid chromatography with photodiode array detection coupled to quadrupole time-of-flight mass spectrometry carried out on fruit extracts (fresh, dried, peels and pulps, juices and seeds) and leaves to a lesser extent, but no research was found on the identification of flavonoids in other plant parts (branches, wood, stem bark, roots, and latex). The analytical methods applied to *Ficus carica* extracts are mainly HPLC and UHPLC methods with diode array detectors (Bachir Bey et al., 2016; Del Caro & Piga, 2008; El-Shobaki et al., 2010; Faleh et al., 2012; Gündeşli et al., 2021; Nakilcioğlu & Hışıl, 2013; Nakilcioğlu-Taş & Ötleş, 2021; Oliveira et al., 2009; Tawfik & Alhejy, 2014; Trifunschi et al., 2015; Veberic et al., 2008; Viuda-Martos et al., 2015) or UV detector (Gündeşli et al., 2021; Qin et al., 2015). Higher sensitivity, more compound specificity, and larger flavonoid profiles are achieved by coupling the HPLC or UHPLC to mass spectrometry (Dueñas et al., 2008; Puoci et al., 2011), a tandem mass spectrometry (Belguith-Hadriche et al., 2017; Slatnar et al., 2011; Takahashi et al., 2014, 2017; Vallejo et al., 2012), an electrospray ionization mass spectrometry (Khadhraoui et al., 2019; Palmeira et al., 2019; Slatnar et al., 2011; Teixeira et al., 2006, 2009; Vallejo et al., 2012; Vaya & Mahmood, 2006) and a quadrupole time-of-flight mass spectrometry (Ammar et al., 2015; Belguith-Hadriche et al., 2016; Soltana et al., 2018; Wojdyło et al., 2016). Flavonoids from ethyl acetate extract of *Ficus carica* leaves were separated for the first time using gas chromatography equipped with high-resolution mass spectrometry (Dureshahwar et al., 2019). The major drawback of this method is that only identification of flavonoid aglycones is possible, and the intact flavonoid conjugates did not appear in the profile obtained since these compounds undergo derivatization before GC analysis.

4.3.2 General Flavonoid Profile

Six flavonoid subclasses were reported in *F. carica* extracts, namely flavonols, flavones, isoflavones, flavanones, flavan-3-ols, and anthocyanins. The qualitative and quantitative variations in flavonoid profiles reported in the literature are liable to the cultivars used, the seasonal differences, and agricultural land type and practices (Gündeşli et al., 2021). The flavonoid profile of *F. carica* also depends on the plant part considered in the study. Flavones are predominant in leaves with relative abundance between 26.1% (green varieties) and 28.35% (dark varieties), followed by flavonols (20.5–21.8%) and isoflavones (13.6–19.3%) (Ammar et al., 2015). Besides, flavonols are the major flavonoids in fruits of green varieties with a relative abundance of 21.3%, followed by flavones (17.3%), while anthocyanins predominate in fruits of dark varieties with a relative abundance of 70.8%, followed by flavonols (27.4%) and isoflavones (19.8%) (Ammar et al., 2015). Differences in flavonoid profiles are also reported between pulps and peels within the same variety. In effect, the repartition of flavonoids between these two parts is characterized by a high abundance of anthocyanins (52.8–99.1%), flavonols (38.6–53.9%), and flavones (17.3–18.3%) in peels, and a high abundance of isoflavones (42.6–50.2%) in pulps (Ammar et al., 2015). Recently, in singular research, flavonoids belonging to the subclasses of flavonols and flavan-3-ols were reported to occur also in fig seeds, with flavan-3-ols content being three-times higher than flavonols (Nakilcioğlu-Taş

& Ötleş, 2021). The drying process of figs was reported to affect different subclasses of flavonoids. After sun drying of figs, flavan-3-ols decreased from 9.87 mg/100 g (mean value) to 5.53 mg/100 g, and flavonols decreased from 53.7 mg/100 g (mean content of 3 dark varieties) to 11.2 mg/100 g. In comparison, anthocyanins showed a severe degradation from 58.75 mg/100 g (mean value) to less than 0.57 mg/100 g or disappeared completely after the sun-drying process, indicating the high thermo-sensitivity anthocyanins (Bachir Bey et al., 2016).

4.3.3 Flavonols in *Ficus carica* L.

Flavonols (3-hydroxy-2-phenylchromen-4-one) constitute the major group of flavonoid compounds distributed extensively in nature. They are characterized by a C_2-C_3 double bond and a hydroxyl group at C_3 with an Oxo group at C_4. The multiple hydroxyl, methoxyl, and glycoside groups on the basic structure give differences in biological and pharmacological properties of flavonols (Sharma et al., 2018). These compounds are found in their aglycone form and are also frequently glycosylated by attaching a monosaccharide (ex. glucose, rhamnose) or a disaccharide (ex. rutinose) to the hydroxyl group in position 3 of the C ring. The chemistry of flavonols provides them with high antioxidant activity; in addition to other astonishing properties such as the protective effect on the cardiovascular system, inhibition of key enzymes (adenosine deaminase, tyrosine kinase, aldose reductase, etc.), regulation of some hormones (estrogen, androgen, thyroid hormones, etc.), reduction of inflammation, decrease in blood cell aggregation, inhibition of microbial growth, and other activities (Brahmachari & Gorai, 2006).

Flavonols occur massively in fruits of *F. carica* as quercetin and kaempferol derivatives. Quercetin aglycone was identified in fresh fruits (Ammar et al., 2015; Bachir Bey et al., 2016; Belguith-Hadriche et al., 2016; Gündeşli et al., 2021; Soltana et al., 2018; Trifunschi et al., 2015), dried fruits (Bachir Bey et al., 2016; Faleh et al., 2012; Khadhraoui et al., 2019; Tawfik & Alhejy, 2014) and leaves (Ammar et al., 2015; Dureshahwar et al., 2019; El-Shobaki et al., 2010; Trifunschi et al., 2015; Vaya & Mahmood, 2006). Among quercetin derivatives, quercetin-3-*O*-rutinoside (rutin) is the predominant compound found in considerable amounts in *F. carica* fruit peels and leaves, with relative abundance up to 43.68% in peels and 14.96% in leaves (Ammar et al., 2015). Rutin is known to possess an intense antioxidant activity against free radicals and lipid peroxidation (Yang et al., 2008), important antimicrobial (Al-Majmaie et al., 2019), and anti-inflammatory and neuroprotective activities (Hao et al., 2016). Rutin was reported to have anticancer activity through different mechanisms, including cell growth and proliferation inhibition, apoptosis and autophagy induction, angiogenesis inhibition, carcinogen inactivation, and cytochrome P450 modulation, in addition to a high chemopreventive potential (Farha et al., 2020). Rutin is currently marketed as an active compound in many dietary supplements and prescribed to treat varicose veins, internal bleeding, hemorrhoids, etc. Oral doses of rutin (from 500 mg to 2000 mg/day) are taken for extended periods (up to 6 months) without any notable side effects (Gullón et al., 2017). Other quercetin derivatives were also identified in F. carica extracts (Table 19.2), namely quercetin-3-O-glucoside (isoquercetin) (Ammar et al., 2015; Bachir Bey et al., 2016; Belguith-Hadriche et al., 2016, 2017; Oliveira et al., 2009; Slatnar et al., 2011; Soltana et al., 2018; Takahashi et al., 2014;Vallejo et al., 2012; Wojdyło et al., 2016), quercetin-3-O-(6″-malonyl) glucoside (Ammar et al., 2015; Belguith-Hadriche et al., 2016; Soltana et al., 2018; Takahashi et al., 2014; Wojdyło et al., 2016), quercetin O-glycosides (Ammar et al., 2015), quercetin di-glycosides (Soltana et al., 2018), and quercetin acetyl-glycosides (Palmeira et al., 2019; Vallejo et al., 2012). Kaempferol aglycone was identified in *F. carica* fresh fruits (Gündeşli et al., 2021; Soltana et al., 2018; Trifunschi et al., 2015), dried fruits (Khadhraoui et al., 2019; Tawfik & Alhejy, 2014), and leaves (Dureshahwar et al., 2019; Trifunschi et al., 2015). Two kaempferol glycosides were identified in *F. carica* extracts, namely kaempferol 3-O-glucoside (astragalin) (Slatnar et al., 2011; Soltana et al., 2018; Takahashi et al., 2014) and kaempferol 3-*O*-rutinoside (Vallejo et al., 2012; Wojdyło et al., 2016). Myricetin was reported recently for the first time in fresh figs and revealed to be the most abundant flavonol aglycone (2.632 mg/100 g) (Gündeşli et al., 2021). Galangin (El-Shobaki et al., 2010) and isorhamnetin (Tawfik & Alhejy, 2014) were only identified in dried figs. In addition to the compounds mentioned above, minor flavonols, namely syringetin, tamarixetin and isohamnetin-3-O-glucoside were also reported in the methanolic extract of fresh fruits (Soltana et al., 2018).

4.3.4 Flavones in *Ficus carica* L.

Flavones (2-phenylchromen-4-one) are the principal pigments in white and cream-colored flowers and act as co-pigments with anthocyanins in blue flowers (Hostetler et al., 2017). Flavones and flavonols share the same structure and chemical properties except for the absence of the hydroxyl group at C_3. Flavones occur widely in nature as 7-O-glycosides and 6-C- and 8-C-glycosides, where the glycoside moiety is directly attached to the aromatic carbon atom, the thing that is not found in flavonols (Hollman & Arts, 2000). Natural flavones show several biological activities, including antioxidant, DNA protective, cardiovascular protective, anticancer, antiproliferative, anti-inflammatory, antimicrobial, anti-HIV, and anti-histaminic activities (Brahmachari & Gorai, 2006).

The main flavones found in *F. carica* are apigenin and luteolin in aglycone and glycosylated forms. Glycosylated compounds found in *F. carica* plant parts presented mostly C glycosidic links at positions 6, 7, and 8 with glucose, rhamnose, arabinose, or rutinose-O liked glycosides (exclusively with glucose) are also found at position 7. Apigenin derivatives are more abundant in fruits and leaves than luteolin derivatives (Table 19.2). Apigenin aglycone was found in fruits in abundance (0.90–1.94%) (Ammar et al., 2015); it is found in peels in higher amounts compared to pulps (Viuda-Martos et al., 2015). Apigenin glycosides were found in abundance in leaves: apigenin 8-C-glucoside (vitexin) (1.53–4.21%), apigenin C-diglycoside derivatives (18.7–19.1%) (Ammar et al., 2015) and apigenin-6-Carabinoside-8-C-glucoside (isochaftoside) (2.5–6.4 mg/g) (Takahashi et al., 2014). Apigenin-7-O-glucoside was reported in dried figs (Khadhraoui et al., 2019). Apigenin-7-O-rutinoside was identified in fig peels (Vallejo et al., 2012). Traces of an acetyl-glycosylated apigenin compound (apigenin-(2"-O-rhamnoside)-C-acetyl glycoside) was also reported in fig peels (Palmeira et al., 2019). Regarding luteolin, the aglycone was found to occur in *F. carica* fresh fruits (Ammar et al., 2015; Belguith-Hadriche et al., 2016; Viuda-Martos et al., 2015), dried fruits (Khadhraoui et al., 2019), and leaves (Ammar et al., 2015; Belguith-Hadriche et al., 2017; Trifunschi et al., 2015; Vaya & Mahmood, 2006). Luteolin glycosides identified in *F. carica* extracts are (1) luteolin-7-O-glucoside (cynaroside) found in fresh fruits (Ammar et al., 2015; Soltana et al., 2018; Viuda-Martos et al., 2015), in dried fruits (Khadhraoui et al., 2019), and leaves (Ammar et al., 2015), (2) luteolin-8-C-glucoside (orientin) found in fresh fruits (Ammar et al., 2015; Belguith-Hadriche et al., 2016; Soltana et al., 2018), in dried fruits (Slatnar et al., 2011), and leaves (Ammar et al., 2015), and (3) luteolin-6-C-glucoside (isoorientin) found in fresh fruits (Ammar et al., 2015; Belguith-Hadriche et al., 2016) and leaves (Ammar et al., 2015). Other C-glycosylated luteolin derivatives (structures are not fully elucidated yet) were reported in fresh figs (Ammar et al., 2015; Soltana et al., 2018; Vallejo et al., 2012) and leaves (Ammar et al., 2015; Belguith-Hadriche et al., 2017). Along with apigenin and luteolin, two other flavones were found in *F. carica*, chrysin was identified in dried fruits (El-Shobaki et al., 2010), and leaves (Dureshahwar et al., 2019; El-Shobaki et al., 2010), and cirsiliol was reported to occur in dried figs (Khadhraoui et al., 2019).

4.3.5 Isoflavones in *Ficus carica* L.

Isoflavones (3-phenylchromen-4-one) are a critical biologically active subclass of flavonoids that occur primarily in the *Fabaceae* family, particularly soybeans, considered the principal source of dietary isoflavones. These compounds showed potential antioxidant, estrogenic, LDL-cholesterol reducing, cardio-vascular protective, and anticancer activities (Vitale et al., 2013).

Isoflavones are reported in *F. carica* fruits and leaves, primarily as genistein and 2′-hydroxygenistein derivatives (aglycone and prenylated forms) (Table 19.2). Genistein aglycone and a glycosylated derivative of hydroxygenistein (hydroxygenistein methyl ether-malonylglycoside) were found exclusively in leaves (Ammar et al., 2015), while 4′-methyl ether genistein (biochanin A) and 7-methoxy-2′-hydroxygenistein (cajanin) were found in fig fruits (in both peels and pulps) (Ammar et al., 2015) and leaves (Ammar et al., 2015; Belguith-Hadriche et al., 2017; Vaya & Mahmood, 2006). Different isomeric forms of prenylgenistein and prenylhydroxygenistein were reported to occur in fig fruits (more abundant in pulps than in peels) (Ammar et al., 2015) and also in leaves (Belguith-Hadriche et al., 2017). Besides, other prenylated isoflavones were described for the first time in *F. carica* fruits, i.e., gancaonin N, isopiscerythrone, viridiflorin, 4′-hydroxy-5,7-dimethoxy-6-(3-meth yl-2-butenyl) isoflavone, in addition to a newly reported isoflavone named ficucaricone D, isolated from ethyl acetate extract fractionated on silica gel column chromatography and identified using the ^{1}H and ^{13}C NMR (Liu et al., 2019).

4.3.6 Anthocyanins in *Ficus carica* L.

Anthocyanins (2-phenyl-chromen-1-ylium) are the most colorful group of flavonoids. These pigments, contained in the flowers and the external part of fruit peels, confer the red, blue, purple, and black coloring depending on pH: in acidic conditions, anthocyanins appear as red pigments, while in alkaline conditions, anthocyanins show blue pigments. Anthocyanins are present in the reddish fruit pulps and leaves, grains, and tubers. Nevertheless, these pigments are highly sensitive to temperature, light, and pH (Khoo et al., 2017). Besides being strong antioxidants, anthocyanins and anthocyanidins (the aglycone forms of anthocyanins) are potential pharmaceutical ingredients, possessing various medicinal properties, including antidiabetic, anticancer, anti-inflammatory, anti-obesity, antimicrobial, and cardioprotective activities (Khoo et al., 2017).

Anthocyanins are found exclusively in *F. carica* fruits, distinctively abundant in peels of dark varieties. Most anthocyanins described in figs are cyanidin and pelargonidin derivatives (Table 19.2). Cyanidin-3-O-rutinoside is considered the main anthocyanin compound in figs, with content varying between 0.21 and 116.2 mg/100 g in fresh fruits (Ammar et al., 2015; Bachir Bey et al., 2016; Belguith-Hadriche et al., 2016; Slatnar et al., 2011; Solomon et al., 2006; Wojdyło et al., 2016), between 0.12 and 0.57 mg/100 g in dried fruits (Bachir Bey et al., 2016; Slatnar et al., 2011), between 0.43 and 108.9 mg/100 g in peels (Ammar et al., 2015; Del Caro & Piga, 2008; Dueñas et al., 2008; Vallejo et al., 2012; Viuda-Martos et al., 2015), and between 0.10 and 9.5 mg/100 g in pulps (Ammar et al., 2015; Del Caro & Piga, 2008; Dueñas et al., 2008; Vallejo et al., 2012). Dimers of cyanidin 3-rutinoside and other derivatives (cyanidin 3-rutinoside-glycoside and 5-carboxypyrano-cyanidin-3-rutinoside) were also reported in figs (Ammar et al., 2015; Dueñas et al., 2008). In addition, cyanidin-3,5-O-diglucoside and cyanidin-3-O-glucoside are found to be more abundant in fig pulps (5.19–24.74% and 2.13%, respectively) than in peels (5.97–19.6% and 1.70%, respectively) (Ammar et al., 2015). Other cyanidin derivatives were identified in fig peels and pulps, including cyanidin-3-malonylglucoside (Dueñas et al., 2008) and cyanidin-3-malonylglucosyl-5-glucoside (Ammar et al., 2015; Dueñas et al., 2008). Pelargonidin derivatives described in fig peels and pulps are pelargonidin-3-O-rutinoside (Dueñas et al., 2008) and pelargonidin-3-O-glucoside (Ammar et al., 2015; Dueñas et al., 2008; Wojdyło et al., 2016). The only anthocyanidin found in figs was petunidin, identified as a dimer of cyanidin rutinoside and reported for the first time in the peels of a dark fig variety (Ammar et al., 2015).

4.3.7 Flavanones in *Ficus carica* L.

Another subclass of flavonoids present in *F. carica* plant parts is flavanones. These compounds are the *Citrus* flavonoids, given their high prevalence in *Citrus* species (oranges, limes, lemon, mandarins, tangors, grapefruits, etc.) in their extracts and juices. Compounds belonging to the flavanones group share the basic structure (2-phenylchroman-4-one) that lacks the double bond in position C2-C3. This specificity in structure gives a non-planar molecular orientation to flavanones, and this feature could impact their biological properties and their interaction with biological receptors compared to flavonols and flavones. Different biological activities were reported for flavanone compounds, including antioxidant, anti-inflammatory, anticancer, anti-proliferative, anti-cholesterolemic, antimicrobial, and cardio-protective activities (Barreca et al., 2017; Khan et al., 2014).

The main flavanones described in *F. carica* plant parts are naringenin, taxifolin, and eriodictyol derivatives (Table 19.2). Naringenin aglycone was reported in fresh fruits (Ammar et al., 2015; Belguith-Hadriche et al., 2016; Gündeşli et al., 2021; Soltana et al., 2018), dried fruits (Khadhraoui et al., 2019), and leaves (Ammar et al., 2015). The only naringenin glycoside found in dried figs is naringin (naringenin-7-rhamnoglucoside) (Khadhraoui et al., 2019). Taxifolin (dihydroquercetin) aglycone was reported in fresh figs (in both peels and pulps) (Ammar et al., 2015; Belguith-Hadriche et al., 2016), while taxifolin glycoside derivatives were reported only in peels (Palmeira et al., 2019). Regarding eriodictyol, the aglycone was identified in fresh figs (Ammar et al., 2015; Belguith-Hadriche et al., 2016; Soltana et al., 2018), and glycoside derivatives were reported in fig peels and pulps (Ammar et al., 2015). In a single study, two additional flavanones were identified in *F. carica* leaves, namely pinostrobin (19.47 mg/100 g) and pinocembrin (52.8 mg/100 g); the latter also occurs in fruits (46.94 mg/100 g) (El-Shobaki et al., 2010).

4.3.8 Flavan-3-Ols in *Ficus carica* L.

The last flavonoid subgroup found in *F. carica* extracts is the group of flavan-3-ols (2-phenyl-3,4-dihydrochromen-3-ol) that constitute 83.5% of the total dietary flavonoids and are found in abundance in tea, cocoa, grapes, and apples (Lei et al., 2016). Flavan-3-ols form a diversified group of compounds with different molecular weights, from simple monomers such as catechin and epicatechin to complex polymers known as proanthocyanidins (Aron & Kennedy, 2008). Several biological properties were attributed to flavan-3-ols, such as antioxidant, anti-inflammatory, anti-cancer, antidiabetic, antibacterial, immuno-modulatory, and cardiovascular protective (Aron & Kennedy, 2008; Lei et al., 2016; Márquez Campos et al., 2020; Mena et al., 2014; Nawrot-Hadzik et al., 2021; Raman et al., 2018).

All flavan-3-ols identified in *F. carica* are catechin and epicatechin derivatives (Table 19.2). (+)-Catechin and (−)-epicatechin monomers were reported in fresh figs (Bachir Bey et al., 2016; Gündeşli et al., 2021; Nakilcioğlu & Hışıl, 2013; Puoci et al., 2011; Slatnar et al., 2011; Veberic et al., 2008; Viuda-Martos et al., 2015; Wojdyło et al., 2016) and dried figs (Bachir Bey et al., 2016; Nakilcioğlu & Hışıl, 2013; Slatnar et al., 2011; Tawfik & Alhejy, 2014; Vallejo et al., 2012). In fruit seeds, only (−)-epicatechin was reported in a higher amount (16.64 mg/100 g) than rutin (5.07 mg/100 g) (Nakilcioğlu-Taş & Ötleş, 2021). A small amount of (+)-catechin (0.04%) was also found in *F. carica* leaves (Ammar et al., 2015). Dimers of catechin and epicatechin formed with anthocyanins (i.e., cyanidin-3-glucoside, cyanidin-3-rutinoside, and pelargonidin-3-rutinoside) were also described in fig peels and pulps (Dueñas et al., 2008). Epicatechin gallate, epigallocatechin, and epigallocatechin gallate were reported in dried figs 7-times higher than flavonols (Tawfik & Alhejy, 2014). Regarding polymeric flavan-3-ols, up to 268 mg/100 g of procyanidins were reported in fresh figs (Bachir Bey et al., 2016; Vallejo et al., 2012; Wojdyło et al., 2016) and dried figs (Bachir Bey et al., 2016).

5 Bioaccessibility of Flavonoids in *Ficus carica* L.

The availability of flavonoids after digestion is crucial as the reduced bioavailability of certain antioxidants is considered a restraining aspect of their potential health benefits. The bioavailability of flavonoids depends on multiple factors. The first factor is the macro and micro-nutrients that surround flavonoids in the gastrointestinal tract: sugar increases considerably the bioavailability of flavan-3-ols in chocolate matrix, vitamin C increases the intestinal absorption of catechins in green tea, vitamin E enhances the bioavailability of anthocyanins in fruits and lipids increases both intestinal micellar absorption of flavonoids and plasma concentration of flavonoid metabolites; on the other hand, dietary fibers reduce the intestinal absorption of flavonoids and proteins interact with flavonoids and form complexes that decrease their bioavailability. The second factor is *the glycosylation of flavonoids*: flavonoid aglycones are lipophilic and could pass the intestinal epithelial cells *via* passive diffusion, while glycosylated flavonoids need epithelial transporters that limit their intestinal absorption. The third factor is *the metabolic transformations*: the formation of methyl, sulfate, and glucuronic derivatives of flavonoids after absorption in the liver and kidney could influence their biological activities. Finally, the last factor is the type of flavonoids: the highest bioavailability is recorded for quercetin glucosides, catechin, flavanones, and isoflavones, while anthocyanins and proanthocyanidins present the lowest bioavailability (Kamiloglu et al., 2020; Kopustinskiene et al., 2020).

Flavonoids'bioaccessibility differs from one species to another. For example, in the case of *Ficus carica*, recent research investigated the *in vitro* oro-gastrointestinal digestibility of 3 varieties of fresh and dried figs and their effect on total flavonoids and proanthocyanidin contents (Kehal et al., 2021). Results showed that the digestive process (oral, gastric, and intestinal phases) caused a gradual decrease in total flavonoid content of the fig varieties from 54–98% (before digestion) to 18–37% (after the intestinal phase) in fresh figs, and from 58–98% (before digestion) to

27–32% (after the intestinal phase) in dried figs, on the contrary, proanthocyanidins, in both fresh and dried figs, increased of 12–35% after the oral phase, then decreased gradually in the gastric and intestinal phases. Similarly, antioxidant activity was affected by the gastrointestinal digestion; fresh figs lost 69–82% of their initial reducing power and 39–54% of their radical scavenging activity, while losses estimated to 77–85% and 49–55% of the initial reducing power and radical scavenging activity, respectively were found in dried figs at the end of the intestinal phase of digestion. Whatever the nature of the figs, the decrease in flavonoid content during the digestive process was explained by an enzymatic degradation by PPO and peroxidases present in the samples and by the instability of these compounds in the simulated digestive conditions (mechanical action, enzymatic activity, and pH effect) that led to hydrolysis of the most sensitive compounds, in addition to the fact that gastrointestinal digestion release other compounds present in figs (ex. proteins and sugars) which interact with flavonoids and cause changes in structure and solubility leading to poor extractability.

Earlier, *in vitro* gastrointestinal digestion on antioxidant and some phenolic compounds of fresh and dried figs was determined (Kamiloglu & Capanoglu, 2013). Results showed that the initial antioxidant activity decreased considerably after gastrointestinal digestion by 74% and 83% in yellow and purple whole figs. Among the investigated compounds, rutin, cyanidin-3-glucoside (C3G), and cyanidin-3-rutinoside (C3R) were tested, and results revealed that only 12% and 16% of the initial rutin content was recovered after intestinal digestion of whole fresh yellow and purple figs, respectively. Although anthocyanin compounds (C3G and C3R) undergo a partial gastric degradation (39%) in yellow figs, an increase in anthocyanin contents was observed for dried purple figs, explained by the effect of pH and pepsin digestion that increase the flavylium cation in the solution. However, anthocyanins were barely detectable or not observed at all in the dialyzed intestinal fractions of purple figs (losses attained 95–100%), a situation explained by the transformation of the red flavylium cation formed in the gastric phase at $pH = 1.7$ to the colorless chalcone at $pH = 7$ in the intestinal phase, or by oxidation or degradation into other chemicals which cannot be detected.

The *in vitro* gastrointestinal digestibility is helpful as a rapid, safe, and ethically acceptable tool to study the bioaccessibility of food elements, but the generated results may not reflect the real *in vivo* digestion conditions, given the complexity of biological interactions, the physiological absorptions in the stomach and small intestine, and the transformations by the microbiota in the colon, conditions that are not taken into consideration in the *in vitro* simulated experiments. There is an apparent controversial situation between anthocyanins'real health beneficial effects and their low bioavailability (generally measured with *in vitro* digestibility assays). It was found that intact anthocyanins are the major form detected in blood; such a situation was explained by a specific mechanism of absorption or metabolism suggesting that anthocyanins may be absorbed directly from the stomach by transport through bilitranslocase at the gastric level because anthocyanins are good ligands for this carrier (Manach et al., 2004).

6 Impact of Post-harvest Processing on Flavonoids in *Ficus carica* L.

Fresh figs are highly perishable fruits because of their soft and delicate skin tissue (easily damaged), their climacteric behavior with moderate ethylene respiration (fruits continue to ripen and produce ethylene after harvest), and also because of their high humidity and nutrient contents (sugars, amino acids, minerals, etc.) which encourages microbial growth and spoilage, as a consequence, the shelf-life of fresh figs is limited. As a result, the commercial-quality is reduced (Zidi et al., 2020). In order to extend the shelf-life of fig fruits and avoid spoilage, the producers generally apply post-harvest drying of fruits as a practical and economical method of preservation: humidity of fresh fruits is reduced to below 26% by sun-drying, oven-drying, micro-wave drying, freeze-drying or osmotic dehydration (Bachir Bey et al., 2016; Naikwadi et al., 2010; Sharifian et al., 2012; Xanthopoulos et al., 2010). However, some of these methods lead to physic-chemical modifications of the fruits (non-enzymatic browning, color modification, etc.) and degradation of some thermal-sensitive bioactive compounds. Therefore, alternative methods of preserving fig fruits were proposed, such as edible coatings, modified atmosphere packaging, and product transformations (fig marmalade, juice, wine, etc.) to preserve the visual aspect of fresh figs and avoid chemical modifications and bioactive compounds degradation.

The effect of modified atmosphere packaging (MAP) on fresh fig bioactive compounds and antioxidant activity was investigated by Zidi et al. (2020). Flavonoid content decreased significantly in figs conditioned under MAP by about 38% after 10 days of storage; while a significant increase in anthocyanins content (23.65%) was reported for figs stored for 5 days under MAP as a result of continued maturation and respiration processes of the fruits at first, followed by a drop of 39% of anthocyanins content after 10 days of storage explained by the instability of these compounds under high CO_2 and low O_2 concentrations. As a consequence of flavonoid decrease, the free radical scavenging activity and chelating capacity of figs stored under MAP were decreased by about 76% and 18%, respectively. The authors concluded that besides the extended shelf-life of figs for up to 10 days and freshness maintained, the MAP technique significantly limited the ripening process by retarding the degradation of bioactive compounds and maintaining antioxidant activity during the storage time (10 days).

In an attempt to extend the shelf-life of fresh figs and maintain the nutritional quality and the antioxidant function, *Aloe vera* gel at different concentrations (25, 50, and 75%) was applied as a coating of fresh figs cold-stored for 15 days (Mirshekari et al., 2020). *Aloe vera* gel coating at 50% and 75% caused a significant increase in anthocyanins content (from 11.12 to 21.45 mg Cyanidin-3-*O*-glucoside Eq./100 g FW) of coated fresh figs after 15 days of storage; in addition to a higher total flavonoid content and antioxidant activity. These observations were explained by the persistence of flavonoids and anthocyanins biosynthesis in fig fruits after harvesting and that the Aloe vera gel coating activated enzymes involved in the biosynthesis process like phenylalanine ammonia-lyase (PAL). Results proved that the application of Aloe vera gel as a coating treatment to fresh figs could be a valuable alternative to the currently used preserving protocols in maintaining the freshness of figs for 15 days and preserving at the same time flavonoids antioxidant capacity of figs.

7 Concluding Remarks

The present chapter reviewed numerous scientific reports searching for the flavonoids found in *Ficus carica* plant parts. As a result, more than 70 compounds belonging to six different subclasses of flavonoids were identified in the fruits and leaves of *F. carica*, with appreciable contents that could form a solid base for industrial exploitation. Food industries that transform fig fruits into ready-to-use products (jam, syrup, confectioneries, wine & liquors, etc.) generate considerable quantities of co-products, including the peels (that contain high amounts of flavonols and anthocyanins), the whole non-conform and overripe fruits discarded during the processing that could be exploited as a valuable material for the flavonoids extraction. Moreover, a great variety of different figs currently exist, and each of them possesses its flavonoid profiles and, therefore, their adequate uses. Besides, exploiting the flavonoids of fig leaves could be more beneficial in pharmaceuticals. In contrast, fig fruits, which require further handing out (compound isolation, purification, and encapsulation to protect the compounds from degradation and improve their bioaccessibility), could be the base of high value-added nutraceutical production. Nonetheless, there is a need to fill the flagrant lack in flavonoids research observed in the literature published to date on the *F. carica* plant parts (latex, root, branches, stem, wood, etc.), and this for a complete vision of the real potential of an astonishing plant widely spread and used as food and medicine.

References

Ahmed, S., Ahmed, N., Rungatscher, A., Linardi, D., Kulsoom, B., Innamorati, G., Meo, S. A., Gebrie, M. A., Mani, R., & Merigo, F. (2020). Cocoa flavonoids reduce inflammation and oxidative stress in a myocardial ischemia-reperfusion experimental model. *Antioxidants, 9*(2), 167.

Akhtar, P., Yaakob, Z., Ahmed, Y., Shahinuzzaman, M., & Mar Contreras, M. (2019). Potential of leaves of eighteen cultivars of *Ficus carica* as antioxidants and profiling of phenolic compounds as an active molecules. *Iranian Journal of Pharmaceutical Sciences, 15*(2), 41–60.

Aljane, F., Neily, M. H., & Msaddak, A. (2020). Phytochemical characteristics and antioxidant activity of several fig (*Ficus carica* L.) ecotypes. *Italian Journal of Food Science, 32*(4), 755–768.

Al-Majmaie, S., Nahar, L., Sharples, G. P., Wadi, K., & Sarker, S. D. (2019). Isolation and antimicrobial activity of rutin and its derivatives from *Ruta chalepensis* (Rutaceae) growing in Iraq. *Records of Natural Products, 13*(1), 64–70.

Amessis-Ouchemoukh, N., Ouchemoukh, S., Meziant, N., Idiri, Y., Hernanz, D., Stinco, C. M., Rodríguez-Pulido, F. J., Heredia, F. J., Madani, K., & Luis, J. (2017). Bioactive metabolites involved in the antioxidant, anticancer and anticalpain activities of *Ficus carica* L., *Ceratonia siliqua* L. and *Quercus ilex* L. extracts. *Industrial Crops and Products, 95*, 6–17. https://doi.org/10.1016/j.indcrop.2016.10.007

Ammar, S., Contreras, M. D. M., Belguith-Hadrich, O., Segura-Carretero, A., & Bouaziz, M. (2015). Assessment of the distribution of phenolic compounds and contribution to the antioxidant activity in Tunisian fig leaves, fruits, skins and pulps using mass spectrometry-based analysis. *Food and Function, 6*(12), 3663–3677. https://doi.org/10.1039/c5fo00837a

Aron, P. M., & Kennedy, J. A. (2008). Flavan-3-ols: Nature, occurrence and biological activity. *Molecular Nutrition & Food Research, 52*(1), 79–104.

Ayaz, M., Sadiq, A., Junaid, M., Ullah, F., Ovais, M., Ullah, I., Ahmed, J., & Shahid, M. (2019). Flavonoids as prospective neuroprotectants and their therapeutic propensity in aging associated neurological disorders. *Frontiers in Aging Neuroscience, 11*, Article 155. https://doi.org/10.3389/fnagi.2019.00155

Ayoub, L., Hassan, F., Hamid, S., Abdelhamid, Z., & Souad, A. (2019). Phytochemical screening, antioxidant activity and inhibitory potential of *Ficus carica* and *Olea europaea* leaves. *Bioinformation, 15*(3), 226–232. https://doi.org/10.6026/97320630015226

Aziz, F. M. (2012). Protective effects of latex of Ficus carica L. against lead acetate-induced hepatotoxicity in rats. *Jordan Journal of Biological Sciences, 5*(3).

Azwanida, N. (2015). A review on the extraction methods use in medicinal plants, principle, strength and limitation. *Med Aromat Plants, 4*(196), 2167–0412.

Bachir bey, M., & Louaileche, H. (2015). A comparative study of phytochemical profile and *in vitro* antioxidant activities of dark and light dried fig (*Ficus carica* L.) varieties. *The Journal of Phytopharmacology, 4*(1), 41–48.

Bachir Bey, M., Richard, G., Meziant, L., Fauconnier, M.-L., & Louaileche, H. (2016). Effects of sun-drying on physicochemical characteristics, phenolic composition and *in vitro* antioxidant activity of dark fig varieties. *Journal of Food Processing and Preservation, 41*(5), e13164. https://doi.org/10.1111/jfpp.13164

Backes, E., Pereira, C., Barros, L., Prieto, M., Genena, A. K., Barreiro, M. F., & Ferreira, I. C. (2018). Recovery of bioactive anthocyanin pigments from *Ficus carica* L. peel by heat, microwave, and ultrasound based extraction techniques. *Food Research International, 113*, 197–209. https://doi.org/10.1016/j.foodres.2018.07.016

Bakoyiannis, I., Daskalopoulou, A., Pergialiotis, V., & Perrea, D. (2019). Phytochemicals and cognitive health: Are flavonoids doing the trick? *Biomedicine & Pharmacotherapy, 109*, 1488–1497.

Barreca, D., Gattuso, G., Bellocco, E., Calderaro, A., Trombetta, D., Smeriglio, A., Laganà, G., Daglia, M., Meneghini, S., & Nabavi, S. M. (2017). Flavanones: Citrus phytochemical with health-promoting properties. *BioFactors, 43*(4), 495–506.

Belguith-Hadriche, O., Ammar, S., del Mar Contreras, M., Turki, M., Segura-Carretero, A., El Feki, A., Makni-Ayedi, F., & Bouaziz, M. (2016). Antihyperlipidemic and antioxidant activities of edible Tunisian *Ficus carica* L. fruits in high fat diet-induced hyperlipidemic rats. *Plant Foods for Human Nutrition, 71*(2), 183–189.

Belguith-Hadriche, O., Ammar, S., del Mar Contreras, M., Fetoui, H., Segura-Carretero, A., El Feki, A., & Bouaziz, M. (2017). HPLC-DAD-QTOF-MS profiling of phenolics from leaf extracts of two Tunisian fig cultivars: Potential as a functional food. *Biomedicine & Pharmacotherapy, 89*, 185–193.

Benmaghnia, S., Meddah, B., Tir-Touil, A., & Gabaldón Hernández, J. A. (2019). Phytochemical analysis, antioxidant and antimicrobial activities of three samples of dried figs (*Ficus carica* L.) from the region of Mascara (West Algeria). *Journal of Microbiology, Biotechnology and Food Sciences, 9*(2), 208–215. https://doi.org/10.15414/jmbfs.2019.9.2.208-215

Bondonno, N. P., Dalgaard, F., Kyrø, C., Murray, K., Bondonno, C. P., Lewis, J. R., Croft, K. D., Gislason, G., Scalbert, A., & Cassidy, A. (2019). Flavonoid intake is associated with lower mortality in the Danish diet cancer and health cohort. *Nature Communications, 10*(1), 1–10.

Boukhalfa, F., Kadri, N., Bouchemel, S., Ait Cheikh, S., Chebout, I., Madani, K., & Chibane, M. (2018). Antioxidant activity and Hypolipidemic effect of *Ficus carica* leaf and twig extracts in triton WR-1339-induced hyperlipidemic mice. *Mediterranean Journal of Nutrition and Metabolism, 11*(1), 37–50. https://doi.org/10.3233/mnm-17180

Brahmachari, G., & Gorai, D. (2006). Progress in the research on naturally occurring flavones and flavonols: An overview. *Current Organic Chemistry, 10*(8), 873–898.

Brodowska, K. M. (2017). Natural flavonoids: Classification, potential role, and application of flavonoid analogues. *European Journal of Biological Research, 7*(2), 108–123.

Bucic-Kojic, A., Planinic, M., Tomas, S., Jokic, S., Mujic, I., Bilic, M., & Velic, D. (2011). Effect of extraction conditions on the extractability of phenolic compounds from lyophilised fig fruits (*Ficus carica* L.). *Polish Journal of Food and Nutrition Sciences, 61*(3).

Çalişkan, O., & Polat, A. A. (2011). Phytochemical and antioxidant properties of selected fig (*Ficus carica* L.) accessions from the eastern Mediterranean region of Turkey. *Scientia Horticulturae, 128*(4), 473–478.

Coppin, J. P., Xu, Y., Chen, H., Pan, M.-H., Ho, C.-T., Juliani, R., Simon, J. E., & Wu, Q. (2013). Determination of flavonoids by LC/MS and anti-inflammatory activity in Moringa oleifera. *Journal of Functional Foods, 5*(4), 1892–1899.

Debib, A., Tir-Touil, A., Mothana, R. A., Meddah, B., & Sonnet, P. (2014). Phenolic content, antioxidant and antimicrobial activities of two fruit varieties of Algerian *Ficus carica* L. *Journal of Food Biochemistry, 38*(2), 207–215.

Del Caro, A., & Piga, A. (2008). Polyphenol composition of peel and pulp of two Italian fresh fig fruits cultivars (*Ficus carica* L.). *European Food Research and Technology, 226*(4), 715–719. https://doi.org/10.1007/s00217-007-0581-4

Devore, E. E., Kang, J. H., Breteler, M. M., & Grodstein, F. (2012). Dietary intakes of berries and flavonoids in relation to cognitive decline. *Annals of Neurology, 72*(1), 135–143.

Dueñas, M., Pérez-Alonso, J. J., Santos-Buelga, C., & Escribano-Bailón, T. (2008). Anthocyanin composition in fig (*Ficus carica* L.). *Journal of Food Composition and Analysis, 21*(2), 107–115. https://doi.org/10.1016/j.jfca.2007.09.002

Dureshahwar, K., Mubashir, M., Upaganlwar, A., Sangshetti, J. N., Upasani, C. D., & Une, H. D. (2019). Quantitative assessment of tactile allodynia and protective effects of flavonoids of *Ficus carica* Lam. leaves in diabetic neuropathy. *Pharmacognosy Magazine, 15*(62), 128.

El-Shobaki, F., El-Bahay, A., Esmail, R., El-Megeid, A. A., & Esmail, N. (2010). Effect of figs fruit (*Ficus carica* L.) and its leaves on hyperglycemia in alloxan diabetic rats. *World Journal of Dairy & Food Sciences, 5*(1), 47–57.

Esposito, D., Chen, A., Grace, M. H., Komarnytsky, S., & Lila, M. A. (2014). Inhibitory effects of wild blueberry anthocyanins and other flavonoids on biomarkers of acute and chronic inflammation in vitro. *Journal of Agricultural and Food Chemistry, 62*(29), 7022–7028.

Faleh, E., Oliveira, A. P., Valentão, P., Ferchichi, A., Silva, B. M., & Andrade, P. B. (2012). Influence of Tunisian Ficus carica fruit variability in phenolic profiles and in vitro radical scavenging potential. *Revista Brasileira de Farmacognosia, 22*, 1282–1289.

Farha, A. K., Gan, R.-Y., Li, H.-B., Wu, D.-T., Atanasov, A. G., Gul, K., Zhang, J.-R., Yang, Q.-Q., & Corke, H. (2020). The anticancer potential of the dietary polyphenol rutin: Current status, challenges, and perspectives. *Critical Reviews in Food Science and Nutrition*, 1–28.

Fernandes, I., Pérez-Gregorio, R., Soares, S., Mateus, N., & De Freitas, V. (2017). Wine flavonoids in health and disease prevention. *Molecules, 22*(2), 292.

Gandhi, G. R., Vasconcelos, A. B. S., Wu, D.-T., Li, H.-B., Antony, P. J., Li, H., Geng, F., Gurgel, R. Q., Narain, N., & Gan, R.-Y. (2020). Citrus flavonoids as promising phytochemicals targeting diabetes and related complications: A systematic review of in vitro and in vivo studies. *Nutrients, 12*(10), 2907.

Gullón, B., Lú-Chau, T. A., Moreira, M. T., Lema, J. M., & Eibes, G. (2017). Rutin: A review on extraction, identification and purification methods, biological activities and approaches to enhance its bioavailability. *Trends in Food Science & Technology, 67*, 220–235.

Gündeşli, M. A., Kafkas, N. E., Güney, M., & Erzisli, S. (2021). Determination of phytochemicals from fresh fruits of fig (*Ficus carica* L.) at different maturity stages. *Acta Scientiarum Polonorum Hortorum Cultus, 20*(2), 73–81.

Gutiérrez-Grijalva, E. P., Picos-Salas, M. A., Leyva-López, N., Criollo-Mendoza, M. S., Vazquez-Olivo, G., & Heredia, J. B. (2018). Flavonoids and phenolic acids from oregano: Occurrence, biological activity and health benefits. *Plants, 7*(1), 2.

Hao, G., Dong, Y., Huo, R., Wen, K., Zhang, Y., & Liang, G. (2016). Rutin inhibits neuroinflammation and provides neuroprotection in an experimental rat model of subarachnoid hemorrhage, possibly through suppressing the RAGE-NF-κB inflammatory signaling pathway. *Neurochemical Research, 41*(6), 1496–1504.

Harzallah, A., Bhouri, A. M., Amri, Z., Soltana, H., & Hammami, M. (2016). Phytochemical content and antioxidant activity of different fruit parts juices of three figs (*Ficus carica* L.) varieties grown in Tunisia. *Industrial Crops and Products, 83*, 255–267. https://doi.org/10.1016/j.indcrop.2015.12.043

Havsteen, B. H. (2002). The biochemistry and medical significance of the flavonoids. *Pharmacology & Therapeutics, 96*(2–3), 67–202.

Hollman, P. C. H., & Arts, I. C. W. (2000). Flavonols, flavones and flavanols-nature, occurrence and dietary burden. *Journal of the Science of Food and Agriculture, 80*(7), 1081–1093.

Hostetler, G. L., Ralston, R. A., & Schwartz, S. J. (2017). Flavones: Food sources, bioavailability, metabolism, and bioactivity. *Advances in Nutrition, 8*(3), 423–435.

Hwang, S.-L., Shih, P.-H., & Yen, G.-C. (2012). Neuroprotective effects of citrus flavonoids. *Journal of Agricultural and Food Chemistry, 60*(4), 877–885.

Imran, A., Butt, M. S., Arshad, M. S., Arshad, M. U., Saeed, F., Sohaib, M., & Munir, R. (2018). Exploring the potential of black tea based flavonoids against hyperlipidemia related disorders. *Lipids in Health and Disease, 17*(1), 1–16.

Ivey, K. L., Jensen, M. K., Hodgson, J. M., Eliassen, A. H., Cassidy, A., & Rimm, E. B. (2017). Association of flavonoid-rich foods and flavonoids with risk of all-cause mortality. *British Journal of Nutrition, 117*(10), 1470–1477.

Jenkinson, C., Petroczi, A., & Naughton, D. P. (2012). Red wine and component flavonoids inhibit UGT2B17 in vitro. *Nutrition Journal, 11*(1), 1–7.

Kalt, W., Cassidy, A., Howard, L. R., Krikorian, R., Stull, A. J., Tremblay, F., & Zamora-Ros, R. (2020). Recent research on the health benefits of blueberries and their anthocyanins. *Advances in Nutrition, 11*(2), 224–236.

Kamiloglu, S., & Capanoglu, E. (2013). Investigating the in vitro bioaccessibility of polyphenols in fresh and sun-dried figs (*Ficus carica* L.). *International Journal of Food Science & Technology, 48*(12), 2621–2629.

Kamiloglu, S., & Capanoglu, E. (2015). Polyphenol content in figs (*Ficus carica* L.): Effect of sun-drying. *International Journal of Food Properties, 18*(3), 521–535. https://doi.org/10.1080/10942912.2013.833522

Kamiloglu, S., Tomas, M., Ozdal, T., & Capanoglu, E. (2020). Effect of food matrix on the content and bioavailability of flavonoids. *Trends in Food Science & Technology*.

Kehal, F., Chemache, L., Chaalal, M., Benbraham, M., Capanoglu, E., & Barkat, M. (2021). A comparative analysis of different varietal of fresh and dried figs by *in vitro* bioaccessibility of phenolic compounds and antioxidant activities. *Acta Universitatis Cinbinesis, Series E: Food Technology, 25*(1).

Kent, K., Charlton, K. E., Lee, S., Mond, J., Russell, J., Mitchell, P., & Flood, V. M. (2018). Dietary flavonoid intake in older adults: How many days of dietary assessment are required and what is the impact of seasonality? *Nutrition Journal, 17*(1), 7–7. https://doi.org/10.1186/s12937-017-0309-7

Khadhraoui, M., Bagues, M., Artés, F., & Ferchichi, A. (2019). Phytochemical content, antioxidant potential, and fatty acid composition of dried Tunisian fig (*Ficus carica* L.) cultivars. *Journal of Applied Botany and Food Quality, 92*, 143–150.

Khan, M. K., Zill, E. H., & Dangles, O. (2014). A comprehensive review on flavanones, the major citrus polyphenols. *Journal of Food Composition and Analysis, 33*(1), 85–104. https://doi.org/10.1016/j.jfca.2013.11.004

Khan, M. K., Paniwnyk, L., & Hassan, S. (2019). Polyphenols as natural antioxidants: Sources, extraction and applications in food, cosmetics and drugs. In *Plant based "green chemistry 2.0"* (pp. 197–235). Springer.

Khoo, H. E., Azlan, A., Tang, S. T., & Lim, S. M. (2017). Anthocyanidins and anthocyanins: Colored pigments as food, pharmaceutical ingredients, and the potential health benefits. *Food & Nutrition Research, 61*(1), 1361779.

Kim, D. O., Padilla-Zakour, O., & Griffiths, P. (2004). Flavonoids and antioxidant capacity of various cabbage genotypes at juvenile stage. *Journal of Food Science, 69*(9), C685–C689.

Konyalıoğlu, S., Sağlam, H., & Kıvçak, B. (2005). α-Tocopherol, flavonoid, and phenol contents and antioxidant activity of Ficus carica. Leaves. *Pharmaceutical Biology, 43*(8), 683–686.

Koolaji, N., Shammugasamy, B., Schindeler, A., Dong, Q., Dehghani, F., & Valtchev, P. (2020). Citrus peel flavonoids as potential cancer prevention agents. *Current Developments in Nutrition, 4*(5), nzaa025.

Kopustinskiene, D. M., Jakstas, V., Savickas, A., & Bernatoniene, J. (2020). Flavonoids as anticancer agents. *Nutrients, 12*(2), 457.

Kozlowska, A., & Szostak-Wegierek, D. (2014). Flavonoids-food sources and health benefits. *Roczniki Państwowego Zakładu Higieny, 65*(2).

Kumar, S., & Pandey, A. K. (2013). Chemistry and biological activities of flavonoids: An overview. *The Scientific World Journal, 2013*, Article 162750. https://doi.org/10.1155/2013/162750

Lei, L., Yang, Y., He, H., Chen, E., Du, L., Dong, J., & Yang, J. (2016). Flavan-3-ols consumption and cancer risk: A meta-analysis of epidemiologic studies. *Oncotarget, 7*(45), 73573–73592. https://doi.org/10.18632/oncotarget.12017

Liu, Y.-P., Guo, J.-M., Yan, G., Zhang, M.-M., Zhang, W.-H., Qiang, L., & Fu, Y.-H. (2019). Anti-inflammatory and antiproliferative prenylated isoflavone derivatives from the fruits of *Ficus carica. Journal of Agricultural and Food Chemistry, 67*(17), 4817–4823. https://doi.org/10.1021/acs.jafc.9b00865

Madureira, J., Barros, L., Cabo Verde, S., Margaça, F. M., Santos-Buelga, C., & Ferreira, I. C. (2020). Ionizing radiation technologies to increase the extraction of bioactive compounds from agro-industrial residues: A review. *Journal of Agricultural and Food Chemistry, 68*(40), 11054–11067.

Mahmoudi, S., Khali, M., Benkhaled, A., Benamirouche, K., & Baiti, I. (2016). Phenolic and flavonoid contents, antioxidant and antimicrobial activities of leaf extracts from ten Algerian *Ficus carica* L. varieties. *Asian Pacific Journal of Tropical Biomedicine, 6*(3), 239–245. https://doi.org/10.1016/j.apjtb.2015.12.010

Mahmoudi, S., Khali, M., Benkhaled, A., Boucetta, I., Dahmani, Y., Attallah, Z., & Belbraouet, S. (2018). Fresh figs (*Ficus carica* L.): Pomological characteristics, nutritional value, and phytochemical properties. *European Journal of Horticultural Science, 83*(2), 104–113. https://doi.org/10.17660/ejhs.2018/83.2.6

Manach, C., Scalbert, A., Morand, C., Rémésy, C., & Jiménez, L. (2004). Polyphenols: Food sources and bioavailability. *The American Journal of Clinical Nutrition, 79*(5), 727–747.

Marinova, D., Ribarova, F., & Atanassova, M. (2005). Total phenolics and total flavonoids in Bulgarian fruits and vegetables. *Journal of the University of Chemical Technology and Metallurgy, 40*(3), 255–260.

Márquez Campos, E., Jakobs, L., & Simon, M.-C. (2020). Antidiabetic effects of flavan-3-ols and their microbial metabolites. *Nutrients, 12*(6), 1592.

Martins, T. F., Palomino, O. M., Álvarez-Cilleros, D., Martín, M. A., Ramos, S., & Goya, L. (2020). Cocoa flavanols protect human endothelial cells from oxidative stress. *Plant Foods for Human Nutrition, 75*(2), 161–168.

Mena, P., Domínguez-Perles, R., Gironés-Vilaplana, A., Baenas, N., García-Viguera, C., & Villaño, D. (2014). Flavan-3-ols, anthocyanins, and inflammation. *IUBMB Life, 66*(11), 745–758.

Meziant, L., Saci, F., & Bachir bey, M., & Louaileche, H. (2015). Varietal influence on biological properties of Algerian light figs (*Ficus carica* L.). *International Journal of Bioinformatics & Biomedical Engineering, 1*(3), 237–243.

Meziant, L., Boutiche, M., Bachir bey, M., Saci, F., & Louaileche, H. (2018). Standardization of monomeric anthocyanins extraction from fig fruit peels (*Ficus carica* L.) using single factor methodology. *Journal of Food Measurement and Characterization, 12*(4), 2865–2873. https://doi.org/10.1007/s11694-018-9901-6

Meziant, L., Bachir-bey, M., Bensouici, C., Saci, F., Boutiche, M., & Louaileche, H. (2021). Assessment of inhibitory properties of flavonoid-rich fig (*Ficus carica* L.) peel extracts against tyrosinase, α-glucosidase, urease and cholinesterases enzymes, and relationship with antioxidant activity. *European Journal of Integrative Medicine, 43*, 101272. https://doi.org/10.1016/j.eujim.2020.101272

Mirshekari, A., Madani, B., Wall, M., & Biggs, A. R. (2020). Aloe vera coatings maintain antioxidants of fig (*Ficus carica* L.) fruit during storage. *Advances in Horticultural Science, 34*(2).

Mo, Y.-F., Yao, L.-Y., Feng, T., Song, S.-Q., & Sun, M. (2020). Optimization of flash extraction process of total flavonoids from fig (*Ficus carica* L.) and its antioxidant activities. *Science and Technology of Food Industry, 41*(12), 186–191. https://doi.org/10.13386/j.issn1002-0306.2020.12.030

Munguia, L., Ramirez-Sanchez, I., Meaney, E., Villarreal, F., Ceballos, G., & Najera, N. (2020). Flavonoids from dark chocolate and (−)-epicatechin ameliorate high-fat diet-induced decreases in mobility and muscle damage in aging mice. *Food Bioscience, 37*, 100710.

Murthy, H. N., & Bapat, V. A. (2020). Importance of underutilized fruits and nuts. In H. N. Murthy & V. A. Bapat (Eds.), *Bioactive compounds in underutilized fruits and nuts* (pp. 3–19). Switzerland Springer Nature.

Musumeci, L., Maugeri, A., Cirmi, S., Lombardo, G. E., Russo, C., Gangemi, S., Calapai, G., & Navarra, M. (2020). Citrus fruits and their flavonoids in inflammatory bowel disease: An overview. *Natural Product Research, 34*(1), 122–136.

Naikwadi, P., Chavan, U., Pawar, V., & Amarowicz, R. (2010). Studies on dehydration of figs using different sugar syrup treatments. *Journal of Food Science and Technology, 47*(4), 442–445.

Nakilcioğlu, E., & Hışıl, Y. (2013). Research on the phenolic compounds in Sarilop (*Ficus carica* L.) fig variety. *Gida, 38*(5), 267–274.

Nakilcioğlu-Taş, E., & Ötleş, S. (2021). Influence of extraction solvents on the polyphenol contents, compositions, and antioxidant capacities of fig (*Ficus carica* L.) seeds. *Anais da Academia Brasileira de Ciências, 93*.

Nawrot-Hadzik, I., Matkowski, A., Hadzik, J., Dobrowolska-Czopor, B., Olchowy, C., Dominiak, M., & Kubasiewicz-Ross, P. (2021). Proanthocyanidins and Flavan-3-Ols in the prevention and treatment of periodontitis-antibacterial effects. *Nutrients, 13*(1), 165.

Oliveira, A. P., Valentão, P., Pereira, J. A., Silva, B. M., Tavares, F., & Andrade, P. B. (2009). *Ficus carica* L.: Metabolic and biological screening. *Food and Chemical Toxicology, 47*(11), 2841–2846. https://doi.org/10.1016/j.fct.2009.09.004

Palmeira, L., Pereira, C., Dias, M. I., Abreu, R. M. V., Corrêa, R. C. G., Pires, T. C. S. P., Alves, M. J., Barros, L., & Ferreira, I. C. F. R. (2019). Nutritional, chemical and bioactive profiles of different parts of a Portuguese common fig (*Ficus carica* L.) variety. *Food Research International, 126*, 108572. https://doi.org/10.1016/j.foodres.2019.108572

Paşayeva, L., Özalp, B., & Fatullayev, H. (2020). Potential enzyme inhibitory properties of extracts and fractions from fruit latex of Ficus carica-based on inhibition of α-amylase and α-glucosidase. *Journal of Food Measurement and Characterization, 14*(5), 2819–2827.

Pei, R., Liu, X., & Bolling, B. (2020). Flavonoids and gut health. *Current Opinion in Biotechnology, 61*, 153–159.

Pérez-Gregorio, M., Regueiro, J., Simal-Gándara, J., Rodrigues, A., & Almeida, D. (2014). Increasing the added-value of onions as a source of antioxidant flavonoids: A critical review. *Critical Reviews in Food Science and Nutrition, 54*(8), 1050–1062.

Perez-Vizcaino, F., & Fraga, C. G. (2018). Research trends in flavonoids and health. *Archives of Biochemistry and Biophysics, 646*, 107–112.

Peterson, J. J., Dwyer, J. T., Jacques, P. F., & McCullough, M. L. (2015). Improving the estimation of flavonoid intake for study of health outcomes. *Nutrition Reviews, 73*(8), 553–576. https://doi.org/10.1093/nutrit/nuv008

Petrussa, E., Braidot, E., Zancani, M., Peresson, C., Bertolini, A., Patui, S., & Vianello, A. (2013). Plant flavonoids-biosynthesis, transport and involvement in stress responses. *International Journal of Molecular Sciences, 14*(7), 14950–14973.

Prior, R. L. (2003). Fruits and vegetables in the prevention of cellular oxidative damage. *The American Journal of Clinical Nutrition, 78*(3), 570S–578S. https://doi.org/10.1093/ajcn/78.3.570s

Puoci, F., Iemma, F., Spizzirri, U. G., Restuccia, D., Pezzi, V., Sirianni, R., Manganaro, L., Curcio, M., Parisi, O. I., Cirillo, G., & Picci, N. (2011). Antioxidant activity of a Mediterranean food product: "Fig syrup". *Nutrients, 3*(3), 317–329. https://doi.org/10.3390/nu3030317

Qin, L., & Chen, H. (2015). Enhancement of flavonoids extraction from fig leaf using steam explosion. *Industrial Crops and Products, 69*, 1–6.

Qin, H., Zhou, G., Peng, G., Li, J., & Chen, J. (2015). Application of ionic liquid-based ultrasound-assisted extraction of five phenolic compounds from fig (*Ficus carica* L.) for HPLC-UV. *Food Analytical Methods, 8*(7), 1673–1681.

Radwan, S., Handal, G., Rimawi, F., & Hanania, M. (2020). Seasonal variations in antioxidant activity, total flavonoids content, total phenolic content, antimicrobial activity and some bioactive components of Ficus carica L. *Palestine. International Journal of PharmTech Research, 13*(4), 329–340.

Raffa, D., Maggio, B., Raimondi, M. V., Plescia, F., & Daidone, G. (2017). Recent discoveries of anticancer flavonoids. *European Journal of Medicinal Chemistry, 142*, 213–228.

Raman, G., Shams-White, M., Avendano, E. E., Chen, F., Novotny, J. A., & Cassidy, A. (2018). Dietary intakes of flavan-3-ols and cardiovascular health: A field synopsis using evidence mapping of randomized trials and prospective cohort studies. *Systematic Reviews, 7*(1), 100. https://doi.org/10.1186/s13643-018-0764-z

Raychaudhuri, S., Ghosh, S., Roy, A., & Swarnakar, S. (2019). Protective role of black tea flavonoids against ethanol-induced gastropathy via matrix metalloproteinase pathway. *Indian Journal of Clinical Biochemistry, 34*(4), 379–394.

Saoudi, M., & El Feki, A. (2012). Protective role of Ficus carica stem extract against hepatic oxidative damage induced by methanol in male Wistar rats. *Evidence-based Complementary and Alternative Medicine, 2012*(150458), 1–8.

Shahbazi, Y. (2017). Antibacterial and antioxidant properties of methanolic extracts of apple (Malus pumila), grape (Vitis vinifera), pomegranate (*Punica granatum* L.) and common fig (Ficus carica L.) fruits. *Pharmaceutical Sciences, 23*(4), 308–315.

Sharifian, F., Motlagh, A. M., & Nikbakht, A. M. (2012). Pulsed microwave drying kinetics of fig fruit ('Ficus carica' L.). *Australian Journal of Crop Science, 6*(10), 1441–1447.

Sharma, A., Sharma, P., Tuli, H. S., & Sharma, A. K. (2018). Phytochemical and pharmacological properties of flavonols. *eLS, 2018*, 1–12.

Slatnar, A., Klancar, U., Stampar, F., & Veberic, R. (2011). Effect of drying of figs (*Ficus carica* L.) on the contents of sugars, organic acids, and phenolic compounds. *Journal of Agricultural and Food Chemistry, 59*(21), 11696–11702.

Slimestad, R., Fossen, T., & Brede, C. (2020). Flavonoids and other phenolics in herbs commonly used in Norwegian commercial kitchens. *Food Chemistry, 309*, 125678.

Socci, V., Tempesta, D., Desideri, G., De Gennaro, L., & Ferrara, M. (2017). Enhancing human cognition with cocoa flavonoids. *Frontiers in Nutrition, 4*, 19.

Solomon, A., Golubowicz, S., Yablowicz, Z., Grossman, S., Bergman, M., Gottlieb, H. E., Altman, A., Kerem, Z., & Flaishman, M. A. (2006). Antioxidant activities and anthocyanin content of fresh fruits of common fig (*Ficus carica* L.). *Journal of Agricultural and Food Chemistry, 54*(20), 7717–7723. https://doi.org/10.1021/jf060497h

Soltana, H., De Rosso, M., Lazreg, H., Vedova, A. D., Hammami, M., & Flamini, R. (2018). LC-QTOF characterization of non-anthocyanic flavonoids in four Tunisian fig varieties. *Journal of Mass Spectrometry, 53*(9), 817–823. https://doi.org/10.1002/jms.4209

Soni, N., Mehta, S., Satpathy, G., & Gupta, R. K. (2014). Estimation of nutritional, phytochemical, antioxidant and antibacterial activity of dried fig (Ficus carica). *Journal of Pharmacognosy and Phytochemistry, 3*(2).

Stangl, V., Lorenz, M., & Stangl, K. (2006). The role of tea and tea flavonoids in cardiovascular health. *Molecular Nutrition & Food Research, 50*(2), 218–228.

Stobiecki, M., & Kachlicki, P. (2006). Isolation and identification of flavonoids. In *The science of flavonoids* (pp. 47–69). Springer.

Takahashi, T., Okiura, A., Saito, K., & Kohno, M. (2014). Identification of phenylpropanoids in fig (Ficus carica L.) leaves. *Journal of Agricultural and Food Chemistry, 62*(41), 10076–10083.

Takahashi, T., Okiura, A., & Kohno, M. (2017). Phenylpropanoid composition in fig (Ficus carica L.) leaves. *Journal of Natural Medicines, 71*(4), 770–775.

Tanaka, T., Iuchi, A., Harada, H., & Hashimoto, S. (2019). Potential beneficial effects of wine flavonoids on allergic diseases. *Diseases, 7*(1), 8.

Tawfik, M. S., & Alhejy, M. (2014). Antioxidants in fig (*Ficus carica* L.) and their effects in the prevention of atherosclerosis in hamsters. *Journal of Food and Nutrition Sciences, 2*(4), 138. https://doi.org/10.11648/j.jfns.20140204.17

Teixeira, D. M., Patão, R. F., Coelho, A. V., & da Costa, C. T. (2006). Comparison between sample disruption methods and solid-liquid extraction (SLE) to extract phenolic compounds from Ficus carica leaves. *Journal of Chromatography A, 1103*(1), 22–28.

Teixeira, D. M., Canelas, V. C., do Canto, A. M., Teixeira, J., & Dias, C. B. (2009). HPLC-DAD quantification of phenolic compounds contributing to the antioxidant activity of Maclura pomifera, Ficus carica and Ficus elastica extracts. *Analytical Letters, 42*(18), 2986–3003.

Treutter, D. (2006). Significance of flavonoids in plant resistance: A review. *Environmental Chemistry Letters, 4*(3), 147–157. https://doi.org/10.1007/s10311-006-0068-8

Trifunschi, S. I., & Ardelean, D. G. (2013). Flavonoid extraction from Ficus carica leaves using different techniques and solvents. *Zbornik Matice srpske za prirodne nauke, 125*, 81–86.

Trifunschi, S. I., Munteanu, M. F. F., Ardelean, D. G., Orodan, M., Osser, G. M., & Gligor, R. I. (2015). Flavonoids and polyphenols content and antioxidant activity of *Ficus carica* L. extracts from Romania. *Zbornik Matice srpske za prirodne nauke, 128*, 57–65.

Vallejo, F., Marín, J. G., & Tomás-Barberán, F. A. (2012). Phenolic compound content of fresh and dried figs (*Ficus carica* L.). *Food Chemistry, 130*(3), 485–492. https://doi.org/10.1016/j.foodchem.2011.07.032

Vauzour, D., Vafeiadou, K., Rendeiro, C., Corona, G., & Spencer, J. P. (2010). The inhibitory effects of berry-derived flavonoids against neurodegenerative processes. *Journal of Berry Research, 1*(1), 45–52.

Vaya, J., & Mahmood, S. (2006). Flavonoid content in leaf extracts of the fig (*Ficus carica* L.), carob (*Ceratonia siliqua* L.) and pistachio (*Pistacia lentiscus* L.). *BioFactors, 28*(3–4), 169–175.

Veberic, R., Colaric, M., & Stampar, F. (2008). Phenolic acids and flavonoids of fig fruit (*Ficus carica* L.) in the northern Mediterranean region. *Food Chemistry, 106*(1), 153–157. https://doi.org/10.1016/j.foodchem.2007.05.061

Vitale, D. C., Piazza, C., Melilli, B., Drago, F., & Salomone, S. (2013). Isoflavones: Estrogenic activity, biological effect and bioavailability. *European Journal of Drug Metabolism and Pharmacokinetics, 38*(1), 15–25. https://doi.org/10.1007/s13318-012-0112-y

Viuda-Martos, M., Barber, X., Pérez-Álvarez, J. A., & Fernández-López, J. (2015). Assessment of chemical, physico-chemical, techno-functional and antioxidant properties of fig (*Ficus carica* L.) powder co-products. *Industrial Crops and Products, 69*, 472–479. https://doi.org/10.1016/j.indcrop.2015.03.005

Wang, S., Zhang, J., Chen, M., & Wang, Y. (2013). Delivering flavonoids into solid tumors using nanotechnologies. *Expert Opinion on Drug Delivery, 10*(10), 1411–1428.

Wojdyło, A., Nowicka, P., Carbonell-Barrachina, Á. A., & Hernández, F. (2016). Phenolic compounds, antioxidant and antidiabetic activity of different cultivars of *Ficus carica* L. fruits. *Journal of Functional Foods, 25*, 421–432. https://doi.org/10.1016/j.jff.2016.06.015

Wu, X., Zhao, Y., Haytowitz, D. B., Chen, P., & Pehrsson, P. R. (2019). Effects of domestic cooking on flavonoids in broccoli and calculation of retention factors. *Heliyon, 5*(3), e01310.

Xanthopoulos, G., Yanniotis, S., & Lambrinos, G. (2010). Study of the drying behaviour in peeled and unpeeled whole figs. *Journal of Food Engineering, 97*(3), 419–424.

Yang, J., Guo, J., & Yuan, J. (2008). In vitro antioxidant properties of rutin. *LWT-Food Science and Technology, 41*(6), 1060–1066.

Yang, X. M., Yu, W., Ou, Z. P., Liu, W. M., & Ji, X. L. (2009). Antioxidant and immunity activity of water extract and crude polysaccharide from *Ficus carica* L. fruit. *Plant Foods for Human Nutrition, 64*(2), 167–173.

Yoo, K. M., Hwang, I. K., & Moon, B. (2009). Comparative flavonoids contents of selected herbs and associations of their radical scavenging activity with antiproliferative actions in V79-4 cells. *Journal of Food Science, 74*(6), C419–C425.

Zidi, K., Kati, D. E., Benchikh, Y., Bey, M. B., Ouandjeli, D., & Yahiaoui, S. (2020). The use of modified atmosphere packaging as mean of bioactive compounds and antioxidant activities preservation of fresh figs (*Ficus carica* L.) from rare cultivars. *The Annals of the University Dunarea de Jos of Galati. Fascicle VI-Food Technology, 44*(1), 149–164.

Chapter 20
Fig Minerals

Sarfaraz Ahmed Mahesar, Hadia Shoaib, Abdul Rauf Khaskheli, Syed Tufail Hussain Sherazi, Abdul Hameed Kori, and Niaz Ali Malghani

1 Introduction

Ficus carica (*F. carica*) is a small deciduous tree native to Asia Minor, Persia, Syria, and the Mediterranean region known as Fig (Abu-Mustafa et al., 1963). *Ficus* belongs to the Moraceae family, which includes many hemiepiphytes, climbers, and creepers native to the tropical and subtropical regions. Figs are grown in several countries worldwide, including Italy, Turkey, Egypt, Morocco, Spain, Greece, the United States, and Brazil (Frodin, 2004; Rahmani & Aldebasi, 2017). Turkey is still the world's leading fig producer and exporter, with an average of 269,126 tonnes per year from 2007 to 2017 and exporting nearly 75% of its production (Turco et al., 2020).

Figs are commonly consumed both fresh and dried. Drying fruits for later use is common because they have a limited post-harvest shelf life of 7–8 days if refrigerated (Veberic et al., 2008). In terms of nutritional content, dried figs are the best of the dried fruits, providing a good supply of fiber, minerals, and vitamins (Badgujar et al., 2014). In addition, proteins, amino acids, polyphenols, carbohydrates, and lipids are all present in this food, low in fat and cholesterol. The fig cultivars and genotypes have a significant impact on the concentrations of these chemicals (Solomon et al., 2006; Veberic & Mikulic-Petkovsek, 2016). Fresh and dried figs and fig jam or chocolate coating are common ingredients in Mediterranean desserts

S. A. Mahesar (✉) · H. Shoaib · S. T. H. Sherazi · A. H. Kori · N. A. Malghani
National Centre of Excellence in Analytical Chemistry, University of Sindh, Jamshoro, Sindh, Pakistan
e-mail: sarfaraz.mahesar@usindh.edu.pk

A. R. Khaskheli
Institute of Pharmacy, Shaheed Mohtarma Benazir Bhutto Medical University, Larkana, Sindh, Pakistan

M. F. Ramadan (ed.), *Fig* (Ficus carica)*: Production, Processing, and Properties*, https://doi.org/10.1007/978-3-031-16493-4_20

such as Panettone and Colombia, often served around Christmas and Easter (Turco et al., 2020). It is an essential part of the Mediterranean diet because of the high concentrations of nutrients, vitamins, minerals, amino acids, and antioxidant substances like phenolic acids it contains (Amessis-Ouchemoukh et al., 2017; Bucic-Kojic et al., 2011; Oliveira et al., 2010).

Fig tree products have numerous therapeutic properties, such as the fruit's anti-angiogenic, anticancer, and anti-inflammation properties, which are well-known in traditional medicine. These include the treatment of cardiovascular, respiratory, and gastrointestinal disorders, including colitis, indigestion, nausea, and diarrhoea (Rubnov et al., 2001), antibacterial activity (Aref et al., 2010), cytotoxicity activity (Khodarahmi et al., 2011), antihelmintic activity (Amol et al., 2010), antimutagenic activity (Agabeĭli & Kasimova, 2005), antipyretic activity (Patil et al., 2010), antispasmodic activity, antiplatelet activity (Gilani et al., 2008), antiviral activity (Lazreg Aref et al., 2011), immunostimulant activity (Patil et al., 2010), anti-inflammatory and antispasmodic remedy (Amessis-Ouchemoukh et al., 2017; Idrus et al., 2018; Mawa et al., 2013), hypoglycemic activity (El-Shobaki et al., 2010), and hypolipidemic activity (Asadi et al., 2006).

Individuals who prefer a vegan diet or who have medical conditions that prevent them from consuming dairy or meat products, such as a milk allergy, may use fruits like pomegranates and olives as an alternative osteoporosis nutritional supplement (Chin & Ima-Nirwana, 2016; Nazrun Shuid & Naina Mohamed, 2013). It has been stated that the mineral composition of figs is similar to human milk, with iron as the essential mineral. The iron level is also believed to be 50% higher than beef liver (Lydia, 2009).

Several religious texts, including the Bible and the Quran, reference the herb's medicinal benefits. For example, in the Quranic sura "I swear by the fig and the olive," these plants' medical benefits are mentioned (Sura no. 95 verse no. 1). Furthermore, the Prophet Mohammad (Peace Be Upon Him) advised eating figs to treat piles and rheumatism, according to Hazrat Abu Darda R.H. (Farooqi, 1998).

2 Botanical Description

A big deciduous tree or large shrub (15–30 ft) with numerous branches from a single trunk is the most common form of *F. carica*. The tree's bark can be grey, smooth, dull-white, or hairy. Oval or nearly orbicular, the blades are 10–20 cm long, the petiole is 5–7.5 cm long, and the leaves are rough above and hairy below. Unisexual, tiny flowers cluster on the inner surface of a wide, hollow, outwardly pear-shaped receptacle. The receptacle is held at the base by several broad, smooth, spiky bracts, and the apex is perforated by an opening closed by numerous thin scales.

Male flowers are scarce and can be found near the bottom of the receptacles. Stamens: 1–5, usually three; sepals: four; filaments: longer than perianth. Female flower: Pedicellate, sepals-4, perianth very delicate and transparent, deeply cut into

Fig. 20.1 The ripe and unripe fig fruit. (Frodin, 2004; Rahmani & Aldebasi, 2017).

3–5 acute segments, ovary superior, hyaline, 1-celled with a single ovule, found on the upper part of receptacles, short-stalked.

To distinguish it from other ficus species, the fruit of *F. carica* has a pear-shaped interior cavity that produces an array of tiny fruits, which are similar in shape and size to the syconium of those other species (Salma et al., 2020). Figure 20.1 shows the ripe and unripe fruit of fig.

There are four basic pomological classifications of figs, depending on the nature of the flowers and the mode of pollination, Common fig, Capri fig, Smyrna fig, and San Pedro fig. The Smyrna fig is the most commercially important of the four varieties, and it is widely grown in the USA and Europe (Al-Snafi, 2017; Badgujar et al., 2014; Salma et al., 2020)

Detailed botanical information is discussed below (Joseph & Raj, 2011).

Kingdom: Plantae
Division: Magnoliophyta
Class: Magnoliopsida
Order: Rosales
Family: Moraceae
Genus: *Ficus*
Species: *Ficus carica*

3 Ingredients of *F. carica*

Fruits and leaves of *F. carica* contain a wide range of beneficial compounds. Phenolic compounds, organic acids, and volatile chemicals are found in most *F. carica* species (Gibernau et al., 1997; Oliveira et al., 2009). Insoluble carbohydrates, protein, vitamin A, vitamin C, calcium (Ca), and iron (Fe) (Teixeira, Patão, Coelho, & da Costa) can all be found in fruits, including cyanidin-3-*O*-glucoside, cyanidin-3-Orhamnoglucoside, saturated fat, cholesterol, and sodium (Na) (Oliveira et al., 2009; Teixeira et al., 2006; Solomon et al., 2006). Researchers found that dried fruit

Table 20.1 Chief ingredients of *F. carica*

Parts of *F. carica*	Constituent	References
Fruits	Cyanidin-3-O-glucoside, cyanidin-3-Orhamnoglucoside, saturated fat, cholesterol, Na, insoluble sugars, protein, vitamin A, vitamin C, Ca, Fe	Teixeira et al. (2006)
Ripe dried fruit	Alkaloids, flavonoids, coumarins, saponins, and terpenes	Abu-Mustafa et al. (1963) and Vaya and Mahmood (2006)
Dried seeds	Fixed oil, including fatty acids	Oliveira et al. (2009)
Leaves	Furanocoumarins including psoralen, bergapten, xanthotoxin, triterpenes such as calotropenyl acetate and lupeol acetate. It also includes 3-O- and 5-O-caffeoylquinic acids, ferulic acid, quercetin-3-O-glucoside, quercetin-3-O-rutinoside and psoralen	Ahmed et al. (1990), Çalişkan and Polat (2011), and Oliveira et al. (2009)

extracts contain alkaloids, flavonoids, coumarins, and saponins(Teixeira et al., 2006; Vaya & Mahmood, 2006). Furanocoumarins such as psoralen, bergapten, xanthotoxin, and triterpenes such as calotropenyl acetate and lupeol acetate have been found in the leaves (Abu-Mustafa et al., 1963; Ahmed et al., 1990). Phenolics, anthocyanins, fructose, glucose, and sucrose were also found in figs and a prior study found fruit to contain phytosterols (Çalişkan & Polat, 2011; Jeong & Lachance, 2001) (Table 20.1).

4 Mineral Composition of Fig

Minerals are inorganic substances that play an essential role in human and animals' health and disease states and are necessary for almost all metabolic functions in living cells (Ndukwe et al., 2013). Macro and micro minerals can be classified based on their natural distribution. For example, Na, Potassium (K), Ca, and Magnesium (Mg) are macro-minerals, while Chromium (Cr), Copper (Cu), Fe, Manganese (Mn), and Zinc (Zn) are micro-minerals. The minerals can be synthesized as conjugated proteins and/or attached to other micro molecules, i.e., phosphates, phytates, polyphenols, etc. (Sulaiman et al., 2011). Figure 20.2 shows the classification of macro- and micro-elements.

Fe is required to transport oxygen, oxidative metabolism, and cell development (Özcan & Akbulut, 2008). It is mostly absorbed in the duodenum, carried through the circulation and extracellular fluid bound to transferrin, and stored intracellularly in the form of ferritin in humans (Nnorom et al., 2007). Likewise, Cu and Cr are essential in various physiological processes (Karadaş & Kara, 2012; Khan et al., 2014). Zn is essential for cellular reproduction and the development of the immune system (Başgel & Erdemoğlu, 2006; Salgueiro et al., 2002).

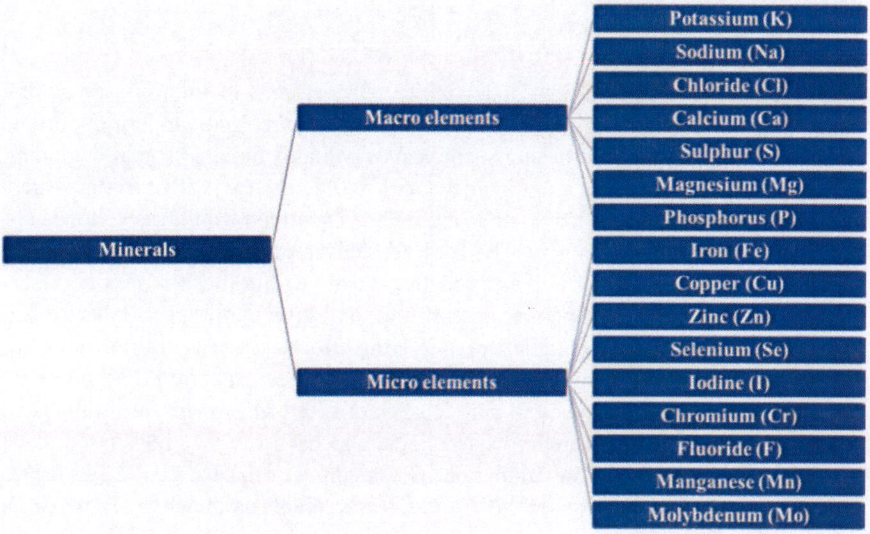

Fig. 20.2 Classification of minerals (macro- and micro-elements)

Many chronic, epidemic, endemic, and even malignant disorders are linked to specific elements, such as Cu, Fe, and Zn, which assist fight infection (Bogden & Oleske, 2007; Heghedus-Mindru et al., 2014). Anemia, for example, affects one-third of the world's population due to a deficit in Fe. Enzymes require trace elements to function properly. Therefore, a trace element deficit will result in a specific condition that reflects the nutrient's specific functions in the animal's metabolism (Mahmood et al., 2012; Duran et al., 2008).

These nutrients are generally received through diet, although not enough is obtained in most cases. Prelatent, latent, and manifested metal inadequacies are caused by poor dietetic habits and unhealthy food intake. There are many commercial supplements on the market today, but few of them contain the necessary ingredients in the appropriate natural form; thus, their absorption or bioavailability in the human body is not as intended (Ficsor et al., 2013).

Figs are the richest source of fiber, vitamin K, and minerals such as Cu, Mn, Mg, K, Ca, Fe, Na, Phosphorus (P), and Cr adequate for people's needs, as per USDA mission variety data. Because 5 dkg (dekagram) of dried figs provides more than 15% of a person's daily need for nutrients. The elements listed below are found in many sections of the fig, including the leaves, stems, and fruits, Mg, Mn, Fe, Ca, Cu, Na, K, P, Selenium (Se), Zn, Antimony (Sb) (Başgel & Erdemoğlu), Arsenic (As), Cr, Cobalt (Co), Lead (Pb), Lithium (Li), Molybdenum (Mo), Nickel (Ni), Titanium (Ti) (Turco et al.), Tin (Sn) (Al-Snafi), cadmium (Cd) and Aluminium (Al) (Ficsor et al., 2013; Soni et al., 2014).

4.1 The Plant

The fig plant can reach a height of 18–20 feet. The stem is medium in size, with a smattering of branches. Roots are not adventitious and are found underground. The milky white latex found in the stems and leaves contains ficin, a protein-degrading enzyme. The fig tree leaves are primarily green and petiolated, with curling borders and a hairy surface (Badgujar & Mahajan, 2009). The green fruit turns yellowish or red-black as it reaches maturity. The fruit and leaves of *F. carica* have long been used to treat throat ailments and as a sedative, laxative, emollientantitussive, resolvent, emmenagogue, constipation, hemorrhoid, and high cholesterol (Ghazi et al., 2012). The traditional use of fig leaves in treating diseases such as vitiligo, diabetes, coughs, asthma, constipation, and gingivitis has also been reported (Mohan et al., 2007). Furthermore, roots used to cure leukoderma and ringworms and fruits have antipyretic, purgative, and aphrodisiac effects and have been demonstrated to be beneficial in paralysis and inflammations (Rahmani & Aldebasi, 2017). Fig leaves contain the minerals are Mg, Mn, Fe, Ca, Cu, Na, K, and P, as shown in (Table 20.2).

4.2 Seeds

The fig is a fruit eaten with or without its peel and contains many seeds in its jelly-like flesh. The seeds are one of the most appealing portions of whole figs, and they impact both their nutritional value and their health consequences. The seeds are

Table 20.2 Mineral composition of fig leaves

Fig leaves	Elements (mg/100g)	Reported studies	
		(Ghazi et al., 2012)	(Essid et al., 2015)
	Mg	307.9–396.4	160.9–416.3
	Mn	21.9–23.1	0.84–6.09
	Fe	66.3–75.5	17.2–48.2
	Ca	1398.1–2551.3	169.2–779.1
	Na	–	45.76–391.6
	K	117.67–496.71	379.05–1412.8
	P	–	102.1–323.8

Table 20.3 Mineral composition of fig seeds

Fig seeds	Elements (mg/100g)	Reported studies	
		(Nakilcioğlu-Taş, 2019)	(Duman et al., 2018)
	Mg	4.64–5.37	–
	Mn	0.77–0.803	–
	Fe	1.83–2.26	0.19
	Ca	18.57–19.04	–
	Cu	0.525–0.534	0.619
	Na	6.945–7.082	–
	K	16.67–17.28	–
	P	17.83–18.62	–
	Zn	–	0.0219
	As	–	<0
	Cr	–	0.0785
	Co	–	0.0056
	Ni	–	0.158
	Al	–	0.0001

smaller, edible, and come in various sizes such as giant, medium, small, or minute, with 30–1600 seeds per fruit (Badgujar et al., 2014; Lansky et al., 2008). For this reason, the nutritional composition of figs is thought to be influenced mainly by the fig seeds (Nakilcioğlu-Taş, 2019). Table 20.3 shows the mineral composition of fig seeds.

4.3 The Ripe Fruits

The ripe fruits can be consumed fresh, dried, or preserved in other ways (Kamiloglu & Capanoglu, 2015). If refrigerated, fully ripened fresh figs have 2–3 days of shelf life. In the sun, they require 4–5 days to dry, and in a dehydrator, they take 10–12 h to dry. Dried figs are a good source of nutrients, particularly essential minerals and energy, and can be stored for six to eight months. Table 20.4 shows the mineral composition of fig fruit.

Table 20.4 Mineral composition of fig fruit

	Elements (mg/100g)	Reported studies				
		(Aljane & Ferchichi, 2009)	(M. N. Khan et al., 2011)	(Soni et al., 2014)	(Mahmoudi et al., 2018)	(Turco et al., 2020)
Fig Fruit	Mg	47.1–99.4	110.5–202.4	67.9	–	0.084–0.128
	Mn	0.21–0.43	–	0.475	–	0.353–1.4
	Fe	0.08–1.17	5.69–10.1	2.95	–	1.27–2.93
	Ca	167.6–403.2	78.7–132.8	154.6	–	0.146–0.449
	Cu	0.17–0.43	0.25–0.42	0.502	13.9–125.4	0.425–0.812
	Na	38.20–117.0	5.58–17.84	–	4.63–14.12	0.045–0.07
	K	688.8–1047	382.4–611.6	–	165.3–266.7	0.35–0.812
	P	–	31.91–76.96	36.575	7.62–16.5	–
	Zn	0.39–0.69	0.32–0.62	0.987	–	0.523–1.15
	Sb	–	–	0.0298	–	–
	As	–	–	0.1669	–	0.006–0.014
	Cr	–	–	0.147	–	0.003–0.008
	Co	–	–	0.0032	–	–
	Pb	–	–	0.0680	–	0–0.006
	Li	–	–	0.0025	–	0.003–0.008
	Mo	–	–	0.0026	–	–
	Ni	–	–	0.118	–	0.014–0.054
	Se	–	–	0.079	–	–
	Ti	–	–	0.037	–	–
	Sn	–	–	0.133	–	–
	Cd	–	–	0.0003	–	0–0.006

References

Abu-Mustafa, E., El-Tawil, B., & Fayez, M. (1963). Constituents of local plants-IV.: *Ficus carica* L., *F. sycomorus* L. and *F. salicifolia* L. leaves. *Phytochemistry, 3*(6), 701–703.

Agabeĭli, R., & Kasimova, T. (2005). Antimutagenic activity of *Armoracia rusticana, Zea mays* and *Ficus carica* plant extracts and their mixture. *Tsitologiia i Genetika, 39*(3), 75–79.

Ahmed, W., Ahmed, Z., & Malik, A. (1990). Triterpenes from the leaves of *Ficus carica. Fitoterapia, 61*(4).

Aljane, F., & Ferchichi, A. (2009). Postharvest chemical properties and mineral contents of some fig (*Ficus carica* L.) cultivars in Tunisia. *Journal of Food, Agriculture and Environment,* 7(209), e212.

Al-Snafi, A. (2017). Nutritional and pharmacological importance of *Ficus carica*-A review. *IOSR Journal of Pharmacy, 7*(3), 33–48.

Amessis-Ouchemoukh, N., Ouchemoukh, S., Meziant, N., Idiri, Y., Hernanz, D., Stinco, C. M., et al. (2017). Bioactive metabolites involved in the antioxidant, anticancer and anticalpain activities of *Ficus carica* L., *Ceratonia siliqua* L. and *Quercus ilex* L. extracts. *Industrial Crops and Products, 95*, 6–17.

Amol, P., Vikas, V., Vijay, R., & Rajesh, Y. (2010). Anthelmintic and preliminary phytochemical screening of leaves of *Ficus carica* Linn against intestinal helminthiasis. *International Journal of Research in Ayurveda and Pharmacy (IJRAP), 1*(2), 601–605.

Aref, H. L., Salah, K., Chaumont, J. P., Fekih, A., Aouni, M., & Said, K. (2010). In vitro antimicrobial activity of four *Ficus carica* latex fractions against resistant human pathogens (antimicrobial activity of *Ficus carica* latex). *Pakistan Journal of Pharmaceutical Sciences, 23*(1), 53–58.

Asadi, F., Pourkabir, M., Maclaren, R., & Shahriari, A. (2006). Alterations to lipid parameters in response to fig tree (*Ficus carica*) leaf extract in chicken liver slices. *Turkish Journal of Veterinary and Animal Sciences, 30*(3), 315–318.

Badgujar, S., & Mahajan, R. (2009). Evaluation of nematicidal properties of some laticiferous plants. *Green Farming, 2*(10), 680–684.

Badgujar, S. B., Patel, V. V., Bandivdekar, A. H., & Mahajan, R. T. (2014). Traditional uses, phytochemistry and pharmacology of *Ficus carica*: A review. *Pharmaceutical Biology, 52*(11), 1487–1503.

Başgel, S., & Erdemoğlu, S. (2006). Determination of mineral and trace elements in some medicinal herbs and their infusions consumed in Turkey. *Science of the Total Environment, 359*(1-3), 82–89.

Bogden, J. D., & Oleske, J. M. (2007). The essential trace minerals, immunity, and progression of HIV-1 infection. *Nutrition Research, 27*(2), 69–77.

Bucic-Kojic, A., Planinic, M., Tomas, S., Jokic, S., Mujic, I., Bilic, M., & Velic, D. (2011). Effect of extraction conditions on the extractability of phenolic compounds from lyophilised fig fruits (*Ficus carica* L.). *Polish Journal of Food and Nutrition Sciences, 61*(3).

Çalişkan, O., & Polat, A. A. (2011). Phytochemical and antioxidant properties of selected fig (*Ficus carica* L.) accessions from the eastern Mediterranean region of Turkey. *Scientia Horticulturae, 128*(4), 473–478.

Chin, K.-Y., & Ima-Nirwana, S. (2016). Olives and bone: A green osteoporosis prevention option. *International Journal of Environmental Research and Public Health, 13*(8), 755.

Duman, E., Şimşek, M., & Özcan, M. M. (2018). Monitoring of composition and antimicrobial activity of fig (*Ficus carica* L.) fruit and seed oil. *Journal of Agricultural Processes Technology, 24*(2), 75–80.

Duran, A., Tuzen, M., & Soylak, M. (2008). Trace element levels in some dried fruit samples from Turkey. *International Journal of Food Sciences and Nutrition, 59*(7-8), 581–589.

El-Shobaki, F., El-Bahay, A., Esmail, R., El-Megeid, A., & Esmail, N. (2010). Effect of figs fruit (*Ficus carica* L.) and its leaves on hyperglycemia in alloxan diabetic rats. *World Journal of Dairy & Food Sciences, 5*(1), 47–57.

Essid, A., Aljane, F., & Ferchichi, A. (2015). Mineral and phenolic content of leaves of Tunisian caprifig (*Ficus carica* L.) accessions. *Journal of New Sciences, 21*.

Farooqi, I. (1998). *Ahadith mein mazkoor nabatat, adwiya aur ghizain* (p. 38). Ilm-o-Irfan Publishers.

Ficsor, E., Szentmihályi, K., Lemberkovics, É., Blázovics, A., & Balázs, A. (2013). Analysis of *Ficus carica* L.-volatile components and mineral content.

Frodin, D. G. (2004). History and concepts of big plant genera. *Taxon, 53*(3), 753–776.

Ghazi, F., Rahmat, A., Yassin, Z., Ramli, N. S., & Buslima, N. A. (2012). Determination of total polyphenols and nutritional composition of two different types of *Ficus carica* leaves cultivated in Saudi Arabia. *Pakistan Journal of Nutrition, 11*(11), 1061.

Gibernau, M., Buser, H. R., Frey, J. E., & Hossaert-McKey, M. (1997). Volatile compounds from extracts of figs of Ficus carica. *Phytochemistry, 46*(2), 241–244.

Gilani, A. H., Mehmood, M. H., Janbaz, K. H., Khan, A.-u., & Saeed, S. A. (2008). Ethnopharmacological studies on antispasmodic and antiplatelet activities of *Ficus carica*. *Journal of Ethnopharmacology, 119*(1), 1–5.

Heghedus-Mindru, R., Heghedus-Mindru, G., Negrea, P., Sumalan, R., Negrea, A., & Stef, D. (2014). The monitoring of mineral elements content in fruit purchased in supermarkets and food markets in Timisoara, Romania. *Annals of Agricultural and Environmental Medicine, 21*(1).

Idrus, R., Sainik, N., Ansari, A. S., Zulfarina, M. S., Razali, R. A., Nordin, A., et al. (2018). *Ficus carica* and bone health: A systematic review. *Sains Malaysiana, 47*(11), 2741–2755.

Jeong, W. S., & Lachance, P. (2001). Phytosterols and fatty acids in fig (*Ficus carica*, var. Mission) fruit and tree components. *Journal of Food Science, 66*(2), 278–281.

Joseph, B., & Raj, S. J. (2011). Pharmacognostic and phytochemical properties of *Ficus carica* Linn-An overview. *International Journal of Pharmtech Research, 3*(1), 8–12.

Karadaş, C., & Kara, D. (2012). Chemometric approach to evaluate trace metal concentrations in some spices and herbs. *Food Chemistry, 130*(1), 196–202.

Khan, M. N., Sarwar, A., Adeel, M., & Wahab, M. (2011). Nutritional evaluation of *Ficus carica* indigenous to Pakistan. *African Journal of Food, Agriculture, Nutrition and Development, 11*(5), 5187–5202.

Khan, N., Choi, J. Y., Nho, E. Y., Habte, G., Jamila, N., Hong, J. H., et al. (2014). Determination of macronutrients in spices by inductively coupled plasma-optical emission spectrometry. *Analytical Letters, 47*(14), 2394–2405.

Khodarahmi, G. A., Ghasemi, N., Hassanzadeh, F., & Safaie, M. (2011). Cytotoxic effects of different extracts and latex of *Ficus carica* L. on HeLa cell line. *Iranian Journal of Pharmaceutical Research: IJPR, 10*(2), 273.

Lansky, E. P., Paavilainen, H. M., Pawlus, A. D., & Newman, R. A. (2008). Ficus spp. (fig): Ethnobotany and potential as anticancer and anti-inflammatory agents. *Journal of Ethnopharmacology, 119*(2), 195–213.

Lazreg Aref, H., Gaaliche, B., Fekih, A., Mars, M., Aouni, M., Pierre Chaumon, J., & Said, K. (2011). In vitro cytotoxic and antiviral activities of *Ficus carica* latex extracts. *Natural Product Research, 25*(3), 310–319.

Lydia, D. (2009). Wonders of figs. *Summaries and short reviews*.

Mahmood, T., Anwar, F., Iqbal, T., Bhatti, I. A., & Ashraf, M. (2012). Mineral composition of strawberry, mulberry and cherry fruits at different ripening stages as analyzed by inductively coupled plasma-optical emission spectroscopy. *Journal of Plant Nutrition, 35*(1), 111–122.

Mahmoudi, S., Khali, M., Benkhaled, A., Boucetta, I., Dahmani, Y., Attallah, Z., & Belbraouet, S. (2018). Fresh figs (*Ficus carica* L.): Pomological characteristics, nutritional value, and phytochemical properties. *European Journal of Horticultural Science, 83*(2), 104–113.

Mawa, S., Husain, K., & Jantan, I. (2013). *Ficus carica* L.(Moraceae): Phytochemistry, traditional uses and biological activities. *Evidence-Based Complementary and Alternative Medicine*.

Mohan, G. K., Pallavi, E., Kumar, R., Ramesh, M., & Venkatesh, S. (2007). Hepatoprotective activity of *Ficus carica* Linn leaf extract against carbon tetrachloride-induced hepatotoxicity in rats. *DARU Journal of Pharmaceutical Sciences, 15*(3), 162–166.

Nakilcioğlu-Taş, E. (2019). Biochemical characterization of fig (*Ficus carica* L.) seeds. *Journal of Agricultural Science, 25*(2), 232–237.

Nazrun Shuid, A., & Naina Mohamed, I. (2013). Pomegranate use to attenuate bone loss in major musculoskeletal diseases: An evidence-based review. *Current Drug Targets, 14*(13), 1565–1578.

Ndukwe, O., Awomukwu, D., & Ukpabi, C. (2013). Comparative evaluation of phytochemical and mineral constituents of the leaves of some medicinal plants in Abia State Nigeria. *International Journal of Academic Research in Progressive Education and Development, 2*(3), 244–252.

Nnorom, I., Osibanjo, O., & Ogugua, K. (2007). Trace heavy metal levels of some bouillon cubes, and food condiments readily consumed in Nigeria. *Pakistan Journal of Nutrition, 6*(2), 122–127.

Oliveira, A. P., Valentão, P., Pereira, J. A., Silva, B. M., Tavares, F., & Andrade, P. B. (2009). *Ficus carica* L.: Metabolic and biological screening. *Food and Chemical Toxicology, 47*(11), 2841–2846.

Oliveira, A. P., Silva, L. R., Andrade, P. B., Valentão, P., Silva, B. M., Pereira, J. A., & de Pinho, P. G. (2010). Determination of low molecular weight volatiles in *Ficus carica* using HS-SPME and GC/FID. *Food Chemistry, 121*(4), 1289–1295.

Özcan, M., & Akbulut, M. (2008). Estimation of minerals, nitrate and nitrite contents of medicinal and aromatic plants used as spices, condiments and herbal tea. *Food Chemistry, 106*(2), 852–858.

Patil, V. V., Bhangale, S. C., & Patil, V. R. (2010). Studies on immunomodulatory activity of Ficus carica. *International Journal of Pharmacy and Pharmaceutical Sciences, 2*(4), 97–99.

Rahmani, A. H., & Aldebasi, Y. H. (2017). Ficus carica and its constituents role in management of diseases. *Asian Journal of Pharmaceutical and Clinical Research, 10*(6), 49–53.

Rubnov, S., Kashman, Y., Rabinowitz, R., Schlesinger, M., & Mechoulam, R. (2001). Suppressors of cancer cell proliferation from fig (*Ficus carica*) resin: Isolation and structure elucidation. *Journal of Natural Products, 64*(7), 993–996.

Salgueiro, M. J., Zubillaga, M. B., Lysionek, A. E., Caro, R. A., Weill, R., & Boccio, J. R. (2002). The role of zinc in the growth and development of children. *Nutrition, 18*(6), 510–519.

Salma, S., Shamsi, Y., Ansari, S., & Nikhat, S. (2020). *Ficus carica* L.: A panacea of nutritional and medicinal benefits. *Cellmed, 10*(1), 1.1–1.6.

Solomon, A., Golubowicz, S., Yablowicz, Z., Grossman, S., Bergman, M., Gottlieb, H. E., et al. (2006). Antioxidant activities and anthocyanin content of fresh fruits of common fig (*Ficus carica* L.). *Journal of Agricultural and Food Chemistry, 54*(20), 7717–7723.

Soni, N., Mehta, S., Satpathy, G., & Gupta, R. K. (2014). Estimation of nutritional, phytochemical, antioxidant and antibacterial activity of dried fig (*Ficus carica*). *Journal of Pharmacognosy and Phytochemistry, 3*(2).

Sulaiman, S. F., Yusoff, N. A. M., Eldeen, I. M., Seow, E. M., Sajak, A. A. B., & Ooi, K. L. (2011). Correlation between total phenolic and mineral contents with antioxidant activity of eight *Malaysian bananas* (Musa sp.). *Journal of Food Composition and Analysis, 24*(1), 1–10.

Teixeira, D. M., Patão, R. F., Coelho, A. V., & da Costa, C. T. (2006). Comparison between sample disruption methods and solid–liquid extraction (SLE) to extract phenolic compounds from Ficus carica leaves. *Journal of Chromatography A, 1103*(1), 22–28.

Turco, V. L., Potortì, A. G., Tropea, A., Dugo, G., & Di Bella, G. (2020). Element analysis of dried figs (*Ficus carica* L.) from the Mediterranean areas. *Journal of Food Composition and Analysis, 90*, 103503.

Vaya, J., & Mahmood, S. (2006). Flavonoid content in leaf extracts of the fig (*Ficus carica* L.), carob (*Ceratonia siliqua* L.) and pistachio (*Pistacia lentiscus* L.). *Biofactors, 28*(3-4), 169–175.

Veberic, R., & Mikulic-Petkovsek, M. (2016). Phytochemical composition of common fig (*Ficus carica* L.) Cultivars. In *Nutritional composition of fruit cultivars* (pp. 235–255). Elsevier.

Veberic, R., Colaric, M., & Stampar, F. (2008). Phenolic acids and flavonoids of fig fruit (*Ficus carica* L.) in the northern Mediterranean region. *Food Chemistry, 106*(1), 153–157.

Chapter 21
Bioactive Compounds of Fig (*Ficus carica*)

Senem Kamiloglu and Banu Akgun

1 Introduction

The common fig (*Ficus carica* L.), belonging to the Moraceae family, is one of the most abundant fruits in the Mediterranean diet, promoting health by preventing-pathophysiological conditions related to chronic diseases (Kamiloglu & Capanoglu, 2013). Traditionally, *Ficus carica* has been used to treat disorders linked with the endocrine system (e.g., diabetes), respiratory system (e.g., asthma, cough, and liver diseases), gastrointestinal tract (e.g., vomiting and ulcer), and reproductive system (e.g., menstruation pain), as well as infectious diseases (e.g., scabies, gonorrhea, and skin disease) (Badgujar et al., 2014). These health-promoting effects of figs are associated with the presence of bioactive compounds.

The fig cultivation originated in the East Mediterranean region, which was later expanded into the West Mediterranean area (Teruel-Andreu et al., 2021). Today, with an annual production of 310 thousand tons, Türkiye is the world's leading fig producing country, accounting for approximately 25% of the world's total production. Other major fig producing countries include Egypt, Morocco, Iran, Algeria, and Spain. Therefore, the Mediterranean basin and the Near East remain crucial for

S. Kamiloglu (✉)
Department of Food Engineering, Faculty of Agriculture, Bursa Uludag University, Gorukle, Bursa, Türkiye

Science and Technology Application and Research Center (BITUAM), Bursa Uludag University, Gorukle, Bursa, Türkiye
e-mail: skamiloglu@uludag.edu.tr

B. Akgun
Food Authenticity and Allergen Department, Central Research Institute of Food and Feed Control, Bursa, Türkiye
e-mail: banu.akgun@tarimorman.gov.tr

M. F. Ramadan (ed.), *Fig* (Ficus carica)*: Production, Processing, and Properties*, https://doi.org/10.1007/978-3-031-16493-4_21

fig production. Although not listed among the top fig-producing countries, Australia and Netherlands are reported to be the largest exporters of this crop, with a combined value of over US$30 million in 2019 (FAOSTAT, 2021).

The peel color of fig fruits ranges from green to dark purple. Although figs are often consumed after removing the peels, they can also be consumed together with the peel (Solomon et al., 2006). Fresh figs have a short post-harvest life ranging between a week to 10 days. However, the combination of a cooled and carbon dioxide-enriched environment enhances the shelf life of the fruit by up to a month. Figs are also often consumed as dried fruits that can be stored for 6–8 months, ensuring proper preservation (Slatnar et al., 2011; Veberic et al., 2008). Some other processed fig products include jam, marmalade, or paste utilized in various bakery products (Saif et al., 2020).

Phytochemical studies on figs and their processed products revealed the presence of several bioactive compounds, including polyphenols, carotenoids, vitamins, organic acids, triterpenoids, phytosterols, and fatty acids. The concentration of these bioactive compounds is highly dependent on the fig variety, cultivation technique, environmental conditions, and processing parameters, among others. In this chapter, we provided an overview of the findings in the literature on the bioactive compounds of figs. The levels of polyphenols, carotenoids, vitamins, organic acids, triterpenoids, phytosterols, and fatty acids in figs were discussed in detail, along with the parameters that influence the levels of these compounds.

2 Polyphenols

Polyphenols are secondary plant metabolites that possess antioxidant, antimicrobial, antiviral, and anti-inflammatory properties. According to epidemiological studies and related meta-analyses, long-term intake of diets rich in plant polyphenols protects against the development of chronic diseases such as cardiovascular diseases, diabetes, certain types of cancers, and neurodegenerative diseases. Polyphenols are classified into different groups based on the number of phenol rings they include and the structural components (Ignat et al., 2011; Pandey & Rizvi, 2009). According to the findings in the literature, fig fruits are rich sources of polyphenols. However, the two most well-known sources of polyphenols, red wine and tea, contain lower levels of polyphenols than figs (Mawa et al., 2013; Vallejo et al., 2012). The major polyphenol groups present in fig fruits include flavonoids, phenolic acids, and coumarins (Fig. 21.1), which are discussed in detail in the following sections.

2.1 Flavonoids

Flavonoids are low molecular weight molecules with 15 carbon atoms organized in a C_6-C_3-C_6 pattern. In the flavonoid structure, two aromatic rings are connected by a 3-carbon bridge, which is typically in the form of a heterocyclic ring. Different

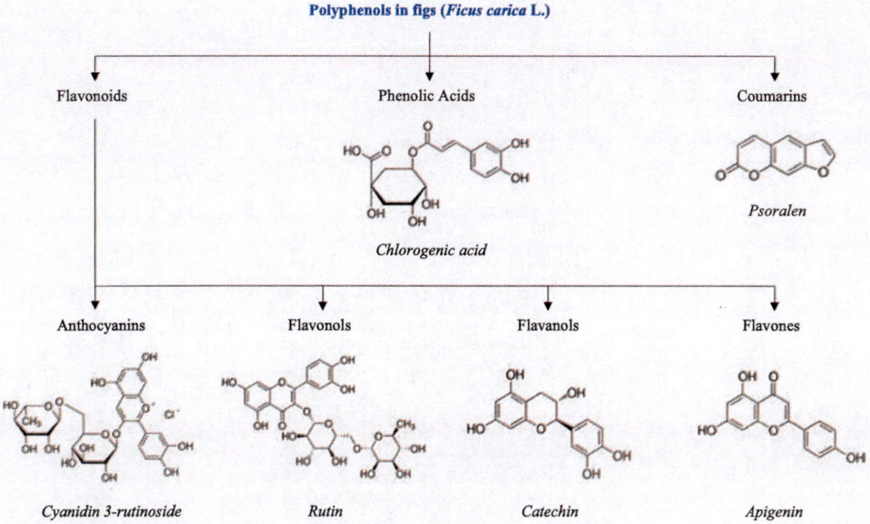

Fig. 21.1 Major classes of polyphenols in figs (*Ficus carica* L.)

subclasses arise from variations in the heterocyclic ring's substitution pattern (Balasundram et al., 2006). Among these subclasses, anthocyanins, flavonols, flavanols, and flavones are the ones that are reported to be present in figs.

2.1.1 Anthocyanins

Anthocyanins are water-soluble pigments responsible for the blue, purple, and red color of fruits (particularly berries), vegetables, legumes, roots, and cereals. They belong to the group of compounds known as flavonoids and are found as glycosides of their respective aglycones, called anthocyanidins. Cyanidin, delphinidin, malvidin, pelargonidin, peonidin, and petunidin are major anthocyanidins present in foods (Kamiloglu et al., 2015). As anthocyanins possess antioxidant, anti-inflammatory, antiatherogenic, and anti-apoptotic properties, their consumption is suggested to treat non-communicable diseases, including cardiovascular diseases, obesity, diabetes, and neurodegenerative diseases, to modulation bacteria and bone metabolism (Speer et al., 2020).

The anthocyanins present in fig fruits were quantified using spectrophotometric or chromatographic methods. The pH differential method is the most employed technique to determine the total anthocyanin content of figs. Cyanidin 3-rutinoside and cyanidin 3-glucoside are commonly used as a reference standard to express the results of total anthocyanin content of figs (Aljane et al., 2020; Çalişkan & Polat, 2011; Ercisli et al., 2012; Harzallah et al., 2016; Hssaini et al., 2019; Hssaini et al., 2020a; Martínez-García et al., 2013; Meziant et al., 2018, 2021; Petkova et al., 2019; Rodov et al., 2012; Saki et al., 2019; Solomon et al., 2006; Viuda-Martos et al., 2015) (Table 21.1). Although it is pretty difficult to compare the results

Table 21.1 Anthocyanin content of figs (*Ficus carica* L.)

Variety/Origin	Skin color	Form/Part	Compound(s)	Content	References
Mission/Israel	Black	Fresh	Total anthocyanins	10.9 mg C3G/100 g	Solomon et al. (2006)
			Cyanidin 3-rutinoside	n/a	
			Cyanidin 3-glucoside	n/a	
Chechick, Brown Türkiye, and Bursa/Israel	Purple, red	Fresh	Total anthocyanins	0.3–1.8 mg C3G/100 g	
Mattalona/Italy	Black	Peel	Cyanidin 3-rutinoside	901.03 mg/kg FW	del Caro and Piga (2008)
			Cyanidin 3-glucoside	28.24 mg/kg FW	
		Pulp	Cyanidin 3-rutinoside	4.92 mg/kg FW	
San Pietro/Italy	Green	Pulp	Cyanidin 3-rutinoside	17.60 mg/kg FW	
Colar, Granilla and Cuello de Dama/Salamanca, Spain; Bursa Siyahi/Türkiye	Dark purple	Skin	Cyanidin 3-rutinoside	15.42–78.32 μg/g FW	Dueñas et al. (2008)
			Cyanidin 3-glucoside	1.51–15.38 μg/g FW	
			Cyanidin 3,5-diglucoside	1.99–5.23 μg/g FW	
			Pelargonidin 3-rutinoside	0.47–3.52 μg/g FW	
			Cyanidin 3-malonylglucoside	0.67–3.51 μg/g FW	
			Cyanidin 3-malonylglycosyl-5-glucoside	0.53–1.39 μg/g FW	
			Carboxypyrano-cyanidin 3-rutinoside	0.48–1.26 μg/g FW	
			(Epi)catechin-(4→8)-cyanidin 3-rutinoside	0.53–1.23 μg/g FW	
			Cyanidin 3-rutinoside dimer	0.44–0.93 μg/g FW	
			Pelargonidin 3-glucoside	0.29–0.68 μg/g FW	
			(Epi)catechin-(4→8)-cyanidin 3-glucoside	0.38–0.48 μg/g FW	
			(Epi)catechin-(4→8)-pelargonidin 3-rutinoside	0.20–0.24 μg/g FW	
		Pulp	Cyanidin 3-rutinoside	3.58–10.16 μg/g FW	
			Cyanidin 3-glucoside	0.51–2.15 μg/g FW	
			Carboxypyrano-cyanidin 3-rutinoside	0.05–0.94 μg/g FW	
			(Epi)catechin-(4→8)-cyanidin 3-rutinoside	0.06–0.59 μg/g FW	
			Cyanidin 3,5-diglucoside	0.11–0.57 μg/g FW	
			(Epi)catechin-(4→8)-cyanidin 3-glucoside	0.05–0.13 μg/g FW	
			Cyanidin 3-malonylglycosyl-5-glucoside	0.04–0.09 μg/g FW	
			Cyanidin 3-malonylglucoside	0.05–0.09 μg/g FW	
			Pelargonidin 3-glucoside	0.01–0.09 μg/g FW	
			Pelargonidin 3-rutinoside	0.04–0.08 μg/g FW	
			Peonidin 3-rutinoside	0.02–0.06 μg/g FW	
			Cyanidin 3-rutinoside dimer	0.08 μg/g FW	

Cuello de Dama/Salamanca, Spain	Green	Pulp	Cyanidin 3-rutinoside Cyanidin 3-glucoside Cyanidin 3,5-diglucoside Carboxypyrano-cyanidin 3-rutinoside (Epi)catechin-(4→8)-cyanidin 3-glucoside (Epi)catechin-(4→8)-cyanidin 3-rutinoside Cyanidin 3-malonylglycosyl-5-glucoside Cyanidin 3-malonylglucoside Pelargonidin 3-rutinoside	1.04 μg/g FW 0.19 μg/g FW 0.09 μg/g FW 0.04 μg/g FW 0.03 μg/g FW 0.02 μg/g FW 0.02 μg/g FW 0.02 μg/g FW 0.02 μg/g FW	Dueñas et al. (2008)
76 local Mediterranean fig accessions/Hatay, Türkiye	Green	Fresh	Total anthocyanins	0.0–41.3 μg C3R/g FW	Çalişkan and Polat (2011)
	Yellow	Fresh	Total anthocyanins	0.6–29.7 μg C3R/g FW	
	Brown	Fresh	Total anthocyanins	2.0–49.6 μg C3R/g FW	
	Purple	Fresh	Total anthocyanins	2.5–46.5 μg C3R/g FW	
	Black	Fresh	Total anthocyanins	32.3–356.0 μg C3R/g FW	
Bela petrovka/Slovenia	n/a	Fresh	Cyanidin 3-rutinoside	0.21–0.62 mg/100 g	Slatnar et al. (2011)
		Oven-dried	Cyanidin 3-rutinoside	0.12–0.31 mg/100 g	
		Sun-dried	Cyanidin 3-rutinoside	0.12–0.26 mg/100 g	
24 local cultivars and Bursa Siyahi/Coruh valley, Türkiye	Black	Fresh	Total anthocyanins	31–42 mg C3R/100 g	Ercisli et al. (2012)
	Light purple	Fresh	Total anthocyanins	8–10 mg C3R/100 g	
	Purple	Fresh	Total anthocyanins	13–20 mg C3R/100 g	
	Dark purple	Fresh	Total anthocyanins	32 mg C3R/100 g	
Brown Türkiye/Bet Dagan, Israel	Light green-purple	Pulse lighted	Total anthocyanins	0.1–4 mg/100 g	Rodov et al. (2012)
18 cultivars/Alicante, Spain	n/a	Skin	Cyanidin 3-rutinoside Cyanidin 3-glucoside	16.9–108.9 mg/100 g FW 0.6–11.1 mg/100 g FW	Vallejo et al. (2012)
		Pulp	Cyanidin 3-rutinoside	1.0–9.5 mg/100 g FW	
Bursa siyahi, Morguz and Yesilguz/Türkiye	Green-purple	Fresh	Cyanidin 3-rutinoside Cyanidin 3-glucoside Cyanidin 3,5-diglucoside	21.27–91.23 mg/100 g FW 2.61–7.07 mg/100 g FW 2.91–6.24 mg/100 g FW	Yemiş et al. (2012)

(continued)

Table 21.1 (continued)

Variety/Origin	Skin color	Form/Part	Compound(s)	Content	References
Mission/Mexico	n/a	Fresh	Total anthocyanins	1.12–1.42 mg/100 g	Martínez-García et al. (2013)
Bouholi and Zidi/Tunisia	Dark purple	Fresh	Cyanidin 3-rutinoside Cyanidin 3-glucoside	7.6–13.9 mg/100 g FW 0.2–0.4 mg/100 g FW	Trad et al. (2014)
Thgagli, Bidhi, and Khendri/Tunisia	Yellow-green	Fresh	Cyanidin 3-rutinoside Cyanidin 3-glucoside	3.5–4.2 mg/100 g FW 0.1–0.7 mg/100 g FW	
Tounsi and Temri/Sfax, Tunisia	Purple	Dried	Cyanidin 3-rutinoside Cyanidin 3,5-diglucoside	n/a n/a	Ammar et al. (2015)
Sarilop/Aydin, Türkiye	Yellow	Skin, Pulp	Cyanidin 3-rutinoside Cyanidin 3-glucoside	0.1–2.2 mg/100 g DW 0.1 mg/100 g DW	Kamiloglu and Capanoglu (2015)
Bursa siyahi/Bursa, Türkiye	Purple	Skin, Pulp	Cyanidin 3-rutinoside Cyanidin 3-glucoside	18–180 mg/100 g DW 2.8–23 mg/100 g DW	
		Dried	Cyanidin 3-rutinoside Cyanidin 3-glucoside	1.5 mg/100 g DW 0.1 mg/100 g DW	
Colar and Cuello de dama/Spain	Purple	Peel	Total anthocyanins	6.21–6.63 mg C3G/g	Viuda-Martos et al. (2015)
Tounsi/Sfax, Tunisia	n/a	Dried	Cyanidin 3-rutinoside	n/a	Belguith-Hadriche et al. (2016)
Bidhi, Hamri and Kohli/Bekalta, Tunisia	Green, Purple, Black	Peel, Pulp	Total anthocyanins	144.62–3330.84 mg C3G/100 g FW	Harzallah et al. (2016)
10 different cultivars/Alicante, Spain	Black, Green, Greenish yellow, Brown-green, Intense purple	Fresh	Cyanidin 3-rutinoside Pelargonidin 3-rutinoside Cyanidin 3,5-diglucoside	0.4–116.2 mg/100 g DW 0.0–3.8 mg/100 g DW 0.0–1.9 mg/100 g DW	Wojdyło et al. (2016)

De Rey, Cuello Dama Negro, Colar Elche, San Antonio, and Brown Türkiye cultivars/Guadajira, Badajoz, Spain	Brown, Dark purple	Skin	Cyanidin 3-rutinoside Pelargonidin 3-rutinoside Cyanidin 3,5-diglucoside	12.19–97.4 mg/100 g 0.09–5.64 mg/100 g 0.12–6.39 mg/100 g	Pereira et al. (2017)
		Pulp	Cyanidin 3-rutinoside Cyanidin 3,5-diglucoside Pelargonidin 3-rutinoside	1.21–5.92 mg/100 g 0.01–0.66 mg/100 g 0.01–0.26 mg/100 g	
Azenjer/Bejaia, Algeria	Black	Peel	Total anthocyanins	175.76–326.82 mg/100 g DW	Meziant et al. (2018)
Sabz/Estahban, Iran	Yellow	Fresh	Cyanidin 3-malonyl-glucoside Cyanidin 3-rutinoside Pelargonidin 3-glucoside Cyanidin 3-glucoside	0.0–0.65 mg/g 0.25–0.43 mg/g 0.0–0.40 mg/g 0.0–0.28 mg/g	Sedaghat and Rahemi (2018)
135 genotypes/Ain-Taoujdate, Morocco	Green-yellow, purple	Fresh	Total anthocyanins	0.41–57.45 mg/100 g DW	Hssaini et al. (2019)
Croatian Yellow Giant/Plovdiv, Bulgaria	Yellow-green	Fresh Jam	Total anthocyanins Total anthocyanins	0.8 mg C3G/100 g 0.4 mg C3G/100 g	Petkova et al. (2019)
Unknown/Durango, Mexico	n/a	Uncoated	Cyanidin 3-rutinoside Cyanidin 3-glucoside	53.1–116.93 mg/100 g DW 4.80–13.00 mg/100 g DW	Reyes-Avalos et al. (2019)
		Coated	Cyanidin 3-rutinoside Cyanidin 3-glucoside	19.43–67.38 mg/100 g DW 2.79–6.73 mg/100 g DW	
Siah/Lorestan, Iran	Deep purple	Uncoated Coated	Total anthocyanins Total anthocyanins	~ 1.7–3.2 mg/100 g ~ 3.0–3.8 mg/100 g	Saki et al. (2019)
27 ecotypes/Medenine, Tunisia	Green yellowish, Green, Red greenish, Brown purplish, Purple greenish, Purple blackish	Fresh	Total anthocyanins	1.61–11.67 mg/100 g FW	Aljane et al. (2020)

(continued)

Table 21.1 (continued)

Variety/Origin	Skin color	Form/Part	Compound(s)	Content	References
Five local cultivars/ Morocco, six introduced cultivars from USA, Italy and Spain	Reddish purple	Fresh	Total anthocyanins	3.1–24.8 mg C3R/100 g DW	Hssaini et al. (2020a, b)
Brown Türkiye/Israel	Green-purple	Untreated	Cyanidin 3-glucoside Cyanidin 3-rutinoside	1.38–5.47 μg/g FW 0–2.45 μg/g FW	Lama et al. (2020)
		ABA-treated	Cyanidin 3-glucoside Cyanidin 3-rutinoside	14.13–1.66 μg/g FW 2.55–10.39 μg/g FW	
		ETP-treated	Cyanidin 3-glucoside Cyanidin 3-rutinoside	10.79–12.35 μg/g FW 2.85–4.84 μg/g FW	
Bpjihon, Masui Dauphine, Wild-type, Japanese purple fruit	Brown, purple, red, purple-brown	Peel	Cyanidin 3-glucoside Cyanidin 3,5-diglucoside Cyanidin 3-rutinoside	1.14–58.54 mg/100 g FW 0–30.73 mg/100 g FW 0–5.60 mg/100 g FW	Zhang et al. (2020)
		Pulp	Cyanidin 3-glucoside Cyanidin 3,5-diglucoside Cyanidin 3-rutinoside	7.72–42.11 mg/100 g FW 0–14.90 mg/100 g FW 0.40–3.01 mg/100 g FW	
Vasilika, Mission and MM/ Markopoulo, Greece	Green, purple	Peel	Cyanidin 3-rutinoside Cyanidin 3-glucoside	0.01–910.6 mg/100 g FW 0.01–38.2 mg/100 g FW	Karantzi et al. (2021)
		Flesh	Cyanidin 3-rutinoside Cyanidin 3-glucoside	14.2–212.2 mg/100 g FW 0.6–16.7 mg/100 g FW	
Bakor Noir, Bouankik, Azenjer, Tazegaght/Algeria	Greenish-purple, Reddish-purple, black, Reddish-violet	Fresh	Total anthocyanins	7.21–62.22 mg C3R/mL	Meziant et al. (2021)

ABA abscisic acid, *C3G* cyanidin 3-glucoside, *C3R* cyanidin 3-rutinoside, *DW* dry weight, *ETP* ethephon, *FW* fresh weight, *n/a* not available expressed using different standards, the outcomes obtained using the spectrophotometric pH differential method revealed that the total anthocyanin content in brown, purple and black fresh figs was higher than that of green and yellow figs. This observation is related to the anthocyanins in dark-colored figs primarily accumulated in peels. Food processing also affected the total anthocyanin content. According to the study carried out by Petkova et al. (2019), the total anthocyanin content of yellow-green figs from Bulgaria was reduced by 50% after jam processing. Moreover, freezing diminished the anthocyanins in figs. In another study (Saki et al., 2019), chitosan and thymol-coated purple figs from Iran exhibited higher total anthocyanin content than uncoated figs. Therefore, the authors proposed coating as an efficient method for quality maintenance and shelf life extension of fresh fig fruits.

High-performance liquid chromatography (HPLC) and ultra-performance liquid chromatography (UPLC) are the most common instruments used to separate, identify, and quantify anthocyanins. The major anthocyanins detected in figs were cyanidin-based with different sugar moieties (Ammar et al., 2015; Belguith-Hadriche et al., 2016; del Caro & Piga, 2008; Dueñas et al., 2008; Kamiloglu & Capanoglu, 2015; Karantzi et al., 2021; Lama et al., 2020; Pereira et al., 2017; Reyes-Avalos et al., 2019; Sedaghat & Rahemi, 2018; Slatnar et al., 2011; Solomon et al., 2006; Trad et al., 2014; Vallejo et al., 2012; Wojdyło et al., 2016; Yemiş et al., 2012; Zhang et al., 2020) (Table 21.1). In particular, cyanidin 3-rutinoside and cyanidin 3-glucoside were detected as the main anthocyanins in figs with different skin colors. Besides cyanidin-based anthocyanins, pelargonidin 3-rutinoside, pelargonidin 3-glucoside, and peonidin 3-rutinoside were also identified in different figs cultivars from Spain and Iran (Dueñas et al., 2008; Pereira et al., 2017; Sedaghat & Rahemi, 2018; Wojdyło et al., 2016). Similar to the results obtained using the spectrophotometric pH differential method, higher concentrations of anthocyanins were detected in dark-colored fruits than in light-colored figs. Moreover, peels of purple figs are found to contain higher levels of anthocyanins than pulps (del Caro & Piga, 2008; Dueñas et al., 2008; Kamiloglu & Capanoglu, 2015; Karantzi et al., 2021; Pereira et al., 2017; Vallejo et al., 2012; Zhang et al., 2020).

Anthocyanins are prone to chemical degradation by processing and storage conditions, such as temperature, light, pH, and oxygen (Oancea, 2021). In general, thermal processing results in reduced anthocyanin content and bioaccessibility. On the other hand, it has been indicated that thermal treatments damage cell walls, liberating the bound anthocyanins and thus enabling more accessible compounds for absorption (Eker et al., 2020). Accodingly, Slatnar et al. (2011) reported that both sun- and oven-drying caused degradation in cyanidin 3-rutinoside content of figs from Slovenia (26–61%). Similarly, significant decreases (96–98%) were observed in cyanidin 3-rutinoside and cyanidin 3-glucoside contents of Turkish figs after sun-drying (Kamiloglu & Capanoglu, 2015). Furthermore, no anthocyanins were detected in the dialyzed fraction after *in vitro* gastrointestinal digestion of sun-dried figs (Kamiloglu & Capanoglu, 2013). In another study (Reyes-Avalos et al., 2019), the effect of alginate-chitosan coating on the anthocyanin content of fresh figs was investigated. The results revealed lower contents of cyanidin 3-rutinoside and cyanidin 3-glucoside in coated figs than in uncoated ones. Reduced anthocyanin levels in the coated figs could be related to the internal atmosphere modification, as it has been reported that the biochemical reactions involved in the synthesis of anthocyanins can be regulated by a modified atmosphere (Reyes-Avalos et al., 2019). In addition, a recent study (Lama et al., 2020) showed that abscisic acid pre-treatment could initiate the biosynthesis of cyanidin 3-rutinoside and cyanidin 3-glucoside enhancing the nutritional value and the color of fig fruits. Overall, the studies conducted in the literature pointed out that anthocyanin composition and concentration in fig fruits depends on the variety, processing conditions, as well as pre- and post-harvest treatments.

2.1.2 Flavonols

Flavonols are the most common subgroup of the flavonoids. Quercetin, kaempferol, myricetin, and isorhamnetin are the major dietary flavonol aglycones, often occurring as *O*-glycosides in dietary sources. Flavonols possess several important biological activities, including antioxidant, anti-inflammatory, antiangiogenic, hypolipidemic, neuroprotective, and anticancer effects (Kamiloglu et al., 2020).

Many studies carried out in the literature identified rutin (quercetin 3-rutinoside) as the major flavonol present in figs fruits (Feng et al., 2015; Karantzi et al., 2021; Veberic et al., 2008; Viuda-Martos et al., 2015) and leaves (Ladhari et al., 2020; Li et al., 2021; Petruccelli et al., 2018; Qin et al., 2015; Takahashi et al., 2017; Teixeira et al., 2006). According to Wojdyło et al. (2016), the content of rutin corresponded to 55–80% of total flavonols identified in fig fruits. Besides rutin, quercetin, isoquercetin (quercetin 3-glucoside) (Gaaliche et al., 2019), quercetin 3-(6″-malonyl) glucoside, quercetin deoxyhexoside dihexoside, quercetin dihexoside, quercetin dideoxyhexoside hexoside (Ammar et al., 2015), quercetagetin 7-glucoside, kaempferol 7-glucoside (Francini et al., 2021), kaempferol 3-glucoside (Slatnar et al., 2011), kaempferol 3-rutinoside (Kamiloglu & Capanoglu, 2015; Wojdyło et al., 2016), kaempferol deoxyhexosyl-hexoside and taxifolin hexoside (Palmeira et al., 2019) were also detected in figs.

In general, higher concentrations of flavonols were found in the fig peels than in pulps (Palmeira et al., 2019). Accordingly, black figs contained higher levels of flavonols than green or yellow figs (del Caro & Piga, 2008; Kamiloglu & Capanoglu, 2015). Sedaghat and Rahemi (2018) reported that the concentration of flavonols, including rutin, isoquercetin, and kaempferol 3-glucoside, increases as figs mature. Potassium application also increased quercetin levels significantly (from 0.19 to 4.29 mg/g DW). However, potassium treatment also reduced the rutin content significantly (two times compared to the control) (Gaaliche et al., 2019). Drying is another parameter that influences the flavonol concentration in figs. According to Slatnar et al. (2011), statistically higher rutin contents, isoquercetin, and kaempferol 3-glucoside were determined in figs dried in the drying oven. On the other hand, Kamiloglu and Capanoglu (2015) showed that the sun-drying process caused decreases in all flavonols of yellow figs (24–50%). In contrast, rutin and kaempferol rutinoside contents increased in sun-dried purple figs (16–200%). Li et al. (2021)

indicated that flavonols from fig leaves possessed significant antioxidant activity. Furthermore, in a study investigating the bioactivity of a fig variety from the Shandong province of China, quercetin was identified as the major antiproliferative compound along with trace amounts of rutin (Mustafa et al., 2021).

2.1.3 Flavanols

Epicatechin and catechin are the major flavanols present abundantly in fruits, cocoa, red wine, and green tea (Hackman et al., 2008). The main physiological functions and regulating mechanisms of flavanols consist of antioxidant, anti-inflammatory, antidiabetic, anticancer, cardioprotective, and neuroprotective activities and enhanced skeletal muscle performance (Li et al., 2019).

From the group of flavanols, both catechin and epicatechin have been detected in fig fruits (Francini et al., 2021). Vallejo et al. (2012) also suggested the presence of additional polymeric procyanidins, up to tetramers, in fig fruits. The majority of the studies conducted in the literature identified catechin as the main flavanol (up to 37.8 mg/100 g FW) (Feng et al., 2015; Kamiloglu & Capanoglu, 2015; Pande & Akoh, 2010; Tsanova-Savova et al., 2005; Veberic et al., 2008; Viuda-Martos et al., 2015). On the other hand, Slatnar et al. (2011) reported epicatechin as the predominant compound (7.11–8.67 mg/100 g). In general, peels contained higher flavanols compared to pulps (Pereira et al., 2017; Viuda-Martos et al., 2015). However, Kamiloglu and Capanoglu (2015) indicated that although catechin was primarily located in the fruit skin for yellow figs, purple figs pulp contained higher compound levels. These differences may be cultivar-specific or related to the agroecological conditions of these studies (Wojdyło et al., 2016). Pereira et al. (2017) and Sedaghat and Rahemi (2018) found that the levels of flavanols increased during ripening. Effects of drying on the flavanol content of fig fruits varied. While Slatnar et al. (2011) observed an increase in epicatechin contents in Slovenian figs as a result of sun-drying (from 7.8 mg/100 g FW in fresh figs to 25.4 mg/100 g DW in dried figs), Kamiloglu and Capanoglu (2015) reported decreased levels for both catechin (35–45%) and epicatechin (45–68%) after sun-drying of Turkish figs.

2.1.4 Flavones

Apigenin and luteolin are the main flavone aglycones present in fruits, vegetables, herbs, and beverages. Flavones have demonstrated various potential health-beneficial activities in human trials and animal studies (Hostetler et al., 2017). For example, human trials reported improved effects on Alzheimer's disease, osteoarthritis, insomnia, depression, and anxiety. Similarly, animal models suggested therapeutic effects on diabetes, cancer, Alzheimer's disease, depression, and amnesia (Salehi et al., 2019).

Flavones primarily present in fig fruits include apigenin, luteolin, and their *C*- and *O*-glycosidic forms. In particular, Vallejo et al. (2012) reported the presence of luteolin 6C-hexose-8C-pentose and apigenin-rutinoside in figs from Spain, with the latter being the most abundant flavone in fruit peel (0.8–2.5 mg/100 g FW). Similarly, Ammar et al. (2015) and Soltana et al. (2018) detected apigenin, luteolin, and their *C*-glycosidic forms and luteolin 7-*O*-glucoside in different fig varieties from Tunisia. On the other hand, in another study investigating the flavones in purple figs from Spain, while apigenin and luteolin were detected both in peel and pulp, luteolin 7-*O*-glucoside was only present in the peel (Viuda-Martos et al., 2015). Palmeira et al. (2019) also determined *C*-glycosylated apigenin derivatives only in the peel of Portuguese figs. Apigenin *C*-hexoside-pentoside ranging from 0.53 to 24.4 mg/100 g DW was the only flavone detected according to the study of Wojdyło et al. (2016). Besides the fig fruits, fig leaves also contained apigenin, luteolin, and their glycosides (Ammar et al., 2015; Mustafa et al., 2021; Zhao et al., 2021). The drying process changed the flavone content of figs. For purple Turkish figs, the apigenin content was increased significantly (68%) due to drying, whereas for yellow figs, there was no statistically significant difference (Kamiloglu & Capanoglu, 2015).

2.2 Phenolic Acids

Phenolic acids can be categorized as hydroxycinnamic and hydroxybenzoic acids, having different molecular skeletons based on cinnamic and benzoic acids, respectively (Song et al., 2020). Hydroxycinnamic acids include caffeic, ferulic, *p*-coumaric, and sinapic acids, while gallic, protocatechuic, vanillic, and syringic acids are the most common hydroxybenzoic acids. In nature, hydroxycinnamic acids are more prevalent than hydroxybenzoic acids and commonly occur in conjugated forms. Various fruits, vegetables, spices, grains, and beverages contain phenolic acids (Rashmi & Negi, 2020). Chlorogenic acid is also one of the most common hydroxycinnamic acids with quinic acid that exerts various therapeutic functions such as antioxidant, antimicrobial, anti-inflammatory, cardioprotective, hepatoprotective, and central nervous system stimulator properties (Naveed et al., 2018).

The majority of the studies investigating the phenolic acid content of figs reported chlorogenic acid as the main phenolic acid in the fruits (del Caro & Piga, 2008; Feng et al., 2015; Slatnar et al., 2011; Vallejo et al., 2012; Wojdyło et al., 2016) and leaves (Qin et al., 2015; Teixeira et al., 2006). Besides chlorogenic acid, gallic acid (Qin et al., 2015; Reyes-Avalos et al., 2019), syringic acid (Veberic et al., 2008), ellagic acid (Pereira et al., 2017), *p*-coumaric acid (Kamiloglu & Capanoglu, 2015) and protocatechuic acid (Viuda-Martos et al., 2015) were also detected in figs. Ammar et al. (2015) performed a comprehensive identification of phenolics in Tunisian figs and reported the presence of 24 hydroxycinnamic acids, which were derivatives of coumaric, caffeic, ferulic, and sinapic acids. Chlorogenic acid and its other three isomers were also found.

Several factors are reported to influence the phenolic acid content in figs. For instance, skin color is one of the parameters affecting the levels of phenolic acids. Accordingly, purple figs were richer in phenolic acids than fellow figs (Kamiloglu

& Capanoglu, 2015). Similarly, the fruit part is another factor that affects the phenolic acid content. In general, higher phenolic acids were found in the fig peels than in pulps (Pereira et al., 2017; Vallejo et al., 2012). On the other hand, Kamiloglu and Capanoglu (2015) reported higher content of gallic acid in pulp compared to skin. Likewise, Viuda-Martos et al. (2015) also determined higher amounts of protocatechuic acid in pulp than in peel. According to Sedaghat and Rahemi (2018), the chlorogenic acid content of figs increased during ripening. Furthermore, Pourghayoumi et al. (2012) reported that levels of chlorogenic acid in figs were affected by the pollen sources, female cultivars, and their interaction. Another study showed that potassium treatment significantly enhanced the chlorogenic acid content (Gaaliche et al., 2019).

The drying process had different impacts on the phenolic acid content of fig fruits. According to Slatnar et al. (2011), statistically higher amounts of chlorogenic acid were measured in figs dried in the drying oven (1.3–4.9 mg/100 g and 3.4–32.4 mg/100 g in fresh and dried figs, respectively). On the other hand, Kamiloglu and Capanoglu (2015) indicated that drying reduced the chlorogenic acid and *p*-coumaric acid contents (34–87%), while an increase of 67% was observed in gallic acid content. A study investigating the effect of alginate-chitosan coating on the phenolic acid content of figs (Reyes-Avalos et al., 2019) revealed that coated figs exhibited higher gallic acid content (1.7 mg/100 g DW) compared to uncoated figs. (2.8 mg/100 g DW). During the storage period, chlorogenic acid levels were also increased for both coated and uncoating figs (from 1.21 to 3.14 mg/100 g DW and from 1.31 to 3.62 mg/100 g DW for uncoated and coated figs, respectively), whereas gallic acid content was stable for coated figs. These results imply that the modification of the internal atmosphere after applying alginate-chitosan coating affects chlorogenic acid more than gallic acid.

2.3 Coumarins

Coumarins are characterized by a furan ring attached to carbon 6 and 7 (i.e., linear type) or 7 and 8 (i.e., angular type) of a benzo-α-pyrone. Moraceae, Rutaceae, Leguminosae, and Apiaceae are the four primary families of higher plants that contain linear furanocoumarins, while the angular furanocoumarins are primarily found in the Apiaceae and Leguminosae families. Psoralen, bergapten, xanthotoxin, and isopimpinellin are the most abundant linear furanocoumarins existing in higher plants (Hung et al., 2017). Coumarins are widely utilized for their antimicrobial and anticancer activities (Hussain et al., 2019).

Psoralen and bergapten are the major coumarins detected in figs (*Ficus carica*). In particular, fig leaves are the primary source of these compounds (Britto et al., 2021; Chunyan et al., 2008; Kazemipoor et al., 2012; Teixeira et al., 2006; Wang et al., 2017; Yu et al., 2020). According to Oliveira et al. (2012), the level of psoralen and bergapten was relatively high in the fig leaves, corresponding to approximately 40–64% and 30–44% of total phenolic compounds, respectively. Besides fig leaves, fruits (Ammar et al., 2015; Qin et al., 2015), wood, and stem bark (Marrelli et al., 2014; Rouaiguia-Bouakkaz et al., 2013) were also reported to contain psoralen and bergapten. Psoralen content of fruit pulps was significantly higher than peels (Oliveira et al., 2009). Similarly, both psoralen and bergapten levels were higher in the wood than in the stem bark. Moreover, most dark-fruited fig trees produced these two coumarins more than the green ones (Rouaiguia-Bouakkaz et al., 2013). Besides psoralen and bergapten, Takahashi et al. (2014) identified a phenylpropanoid acid related to psoralen as psoralic acid glucoside. The authors suggested that this compound could be a precursor of psoralen.

Psoralen possesses limited antioxidant activity due to its structure, which does not contain any phenolic type of hydroxyl group that can act as a hydrogen donor (Teixeira et al., 2009). Although coumarins have been described as low effective antioxidants, these compounds were found to have an inhibitory effect on tumor cells (Li et al., 2011).

3 Carotenoids

Carotenoids are fat-soluble pigments that provide the characteristic yellow, orange, and red colors of many fruits, vegetables, and plant life. According to their chemical composition, they are classified into two main groups: (i) xanthophylls and (ii) carotenes. Lutein, α- and β-cryptoxanthin, zeaxanthin, canthaxanthin, and astaxanthin are essential members of the xanthophylls group, whereas α- and β-carotene and lycopene are the leading members of the carotene group (Kamiloglu et al., 2021). As humans do not synthesize carotenoids, they need to be received through the diet or *via* supplementation. The main potential benefits of carotenoids to human health include contribution to (i) pro-vitamin A function, (ii) eye health, (iii) cognitive performance, (iv) cardiovascular health, (v) bone health, (vi) sun protection, (vii) weight management, (viii) immune function, (ix) cancer prevention and (x) infant nutrition (Eggersdorfer & Wyss, 2018).

The carotenoids identified in figs include lutein, zeaxanthin, *β*-cryptoxanthin, *β*-carotene, *α*-carotene, and lycopene (Table 21.2) (Amessis-Ouchemoukh et al., 2017; Nawade et al., 2020; Petkova et al., 2019; Su et al., 2002; Yemiş et al., 2012). Studies investigating the carotenoid content of figs revealed that the abundance of these pigments depends on the variety. While the Australian figs contained lycopene as the major carotenoid (0.32 mg/100 g) (Su et al., 2002), yellow Turkish figs (var. Sarilop and Sarizeybek) (Yemiş et al., 2012) and green-purple figs from Israel (var. Brown Türkiye) (Nawade et al., 2020) are reported to contain lutein as the most abundant carotenoid (6.14–7.15 μg/g DW and 4.08–12.58 μg/g FW, respectively). On the other hand, *β*-carotene was the major carotenoid in white Algerian figs. (4.32 μg/100 g DW) (Amessis-Ouchemoukh et al., 2017). Overall, the total carotenoid content of fresh fig fruits (23.7–110 μg/g DW) was low compared to other fruits (Veberic & Mikulic-Petkovsek, 2016).

Table 21.2 Carotenoid content of figs (*Ficus carica* L.)

Variety/Origin	Skin color	Form	Compound(s)	Content	References
Unknown/Australia	Unknown	Fresh	Lycopene	0.32 mg/100 g	Su et al. (2002)
			Lutein	0.08 mg/100 g	
			β-carotene	0.04 mg/100 g	
			α-carotene	0.02 mg/100 g	
			Cryptoxanthin	0.01 mg/100 g	
Sarilop/Aydin, Türkiye	Yellow	Fresh	Lutein	6.14 μg/g DW	Yemiş et al. (2012)
			Zeaxanthin	3.32 μg/g DW	
			β-cryptoxanthin	2.04 μg/g DW	
			β-carotene	1.40 μg/g DW	
		Dried	Lutein	1.44–4.00 μg/g DW	
			Zeaxanthin	0.76–2.00 μg/g DW	
			β-cryptoxanthin	0.98–1.79 μg/g DW	
			β-carotene	0.45–0.97 μg/g DW	
Sarizeybek/Aydin, Türkiye	Yellow	Fresh	Lutein	7.15 μg/g DW	
			Zeaxanthin	2.20 μg/g DW	
			β-cryptoxanthin	1.85 μg/g DW	
			β-carotene	1.48 μg/g DW	
		Dried	Lutein	1.85–5.25 μg/g DW	
			Zeaxanthin	0.68–1.61 μg/g DW	
			β-cryptoxanthin	1.00–1.78 μg/g DW	
			β-carotene	0.47–1.22 μg/g DW	
Unknown/Beni Maouche, Algeria	White	Fresh	β-carotene	4.32 μg/100 g DW	Amessis-Ouchemoukh et al. (2017)
			β-cryptoxanthin	2.14 μg/100 g DW	
Croatian Yellow Giant/Plovdiv, Bulgaria	Yellow-green	Fresh	Total carotenoids	3.52 μg/g FW	Petkova et al. (2019)
		Frozen	Total carotenoids	1.73 μg/g FW	
		Jam	Total carotenoids	1.94 μg/g FW	
Brown Türkiye/Kibbutz, Machanaim, Israel	Green-purple	Fresh	Total carotenoids (calculated as the sum of lutein, β-carotene, zeaxanthin, and violaxanthin)	6.64–18.39 μg/g FW	Nawade et al. (2020)

DW dry weight, *FW* fresh weight

Besides the variety, several other factors affect the carotenoid content of fig fruits. For instance, a study by Nawade et al. (2020) indicated that the accumulation of carotenoid pigments decreases as the fruits mature. Accordingly, the total carotenoid contents of parthenocarpic and pollinated figs were reduced from 16.62–18.39 to 6.64–7.71 μg/g FW. Processing is another factor that affects the carotenoid content of fig fruits. Yemiş et al. (2012) monitored the changes in carotenoids during sun-drying of figs and reported that the drying process reduced the total carotenoid content of yellow Turkish figs by 78.5–81.5%. This reduction may be related to the susceptibility of carotenoids to oxidative changes during drying due to their high degree of unsaturation in chemical structure. In another study (Petkova et al., 2019), the carotenoid content of frozen figs and fig jam was compared with fresh figs. The results revealed that the carotenoid content in frozen fruits and jam was approximately 50% lower than that of fresh fruits. These results imply that both thermal and cold processing decreases the carotenoid content of fig fruits.

4 Vitamins

Vitamins are substances essential for the development and growth of the body. They are widely distributed in natural sources and have different chemical structures. Vitamins are classified into two groups; water- or fat-soluble. Vitamins of A, D, E, and K are fat-soluble. Vitamin C and the members of the vitamin B group, namely thiamine (vitamin B_1), riboflavin (vitamin B_2), niacin (vitamin B_3), pantothenic acid (vitamin B_5), vitamin B_6, folate (vitamin B_9), and vitamin B_{12} constitute the water-soluble vitamins. The human body cannot synthesize most vitamins (except for vitamin D and niacin), making them essential in the human diet. Vitamin D can be synthesized in the body with adequate exposure to sunlight, and niacin synthesis is related to a sufficient intake of tryptophan (Ball, 2005). Therefore, the optimum values of daily vitamin intake can promote health and eliminate deficiency diseases. Most foods can provide vitamins in variable quantities; nonetheless, there is no single food containing all vitamins. Fruits generally provide a wide variety of vitamins such as vitamin A, vitamin B, vitamin C, and vitamin E. Hence, for disease prevention, people should consume foods rich in vitamins (Uribe et al., 2017).

Nutritionally, fresh and dried figs are one of the richest plant sources in terms of vitamins. For example, Arvaniti et al. (2019) stated the vitamin content of 100 g of fresh figs as 142 IU of vitamin A, 2 mg of vitamin C, 0.060 mg of vitamin B_1, 0.050 mg of vitamin B_2, and vitamin content of 100 g dried figs as 10 IU vitamin A, 1.2 mg vitamin C, 0.085 mg vitamin B_1, 0.082 mg vitamin B_2.

According to Türkomp (2021), 100 grams of fresh Bursa black figs contain 2.1 mg vitamin C, 0.036 mg vitamin B_1, 0.038 mg vitamin B_2, 0.335 mg vitamin B_3, 0.174 mg vitamin B_6, 8 μg vitamin B_9 and 11 retinol equivalents (RE) vitamin A. In another database, it is shown that 100 grams of raw figs have 142 IU vitamin A, 2.0 mg vitamin C, 0.1 mg vitamin E, 4.7 mcg vitamin K, 0.1 mg vitamin B_1, 0.1 mg vitamin B_2, 0.4 mg vitamin B_3, 0.1 mg vitamin B_6, 6 mcg vitamin B_9 and 0.3 mg

Table 21.3 The vitamin content in 100 g raw figs (NutritionData, 2018a; Türkomp, 2021)

Vitamin	Amount (TürKomp database)	Amount (NutritionData)
Vitamin A	11 retinol equivalents (RE)	142 IU
Vitamin B_1	0.036 mg	0.1 mg
Vitamin B_2	0.038 mg	0.1 mg
Vitamin B_3	0.335 mg	0.4 mg
Vitamin B_5	–	0.3 mg
Vitamin B_6	0.174 mg	0.1 mg
Vitamin B_9	8 μg	6 mcg
Vitamin C	2.1 mg	2.0 mg
Vitamin E	–	0.1 mg

vitamin B_5 (NutritionData, 2018a) (Table 21.3). United States Department of Agriculture reported 2 mg vitamin C, 7 mcg vitamin A, 4.7 mcg vitamin K, 6 mcg vitamin B_9 in 100 g raw figs, and 1.2 mg vitamin C, 0 mcg vitamin A, 15.6 mcg vitamin K, 9 mcg vitamin B_9 in 100 g dried figs (Caporuscio, 2019). Popova (2019) determined vitamin C levels of commercially available fresh fruits as 1.1 mg/100 g fresh sample and 1.9 mg/100 g fresh sample using two different assays. Similarly, the vitamin C level of wild fig was stated as 3.3 mg/100 g of pulp by Kumari et al. (2017). According to Morton (1987), 100 g of fresh fig fruit contained 20–270 I.U. vitamin A, 0.034–0.06 mg vitamin B_1, 0.053–0.079 mg vitamin B_2, 0.32–0.412 mg vitamin B_3, 12.2–17.6 mg vitamin C. Guvenc et al. (2009) reported that fig fruit contained some lipophilic vitamins (γ-tocopherol, δ-tocopherol, vitamin D_2, vitamin D_3, α-tocopherol, α-tocopherol acetate, and vitamin K_1) and vitamin content varied in different sections. For example, vitamin K_1 was the highest in whole fig fruit, but it was relatively low in the skin. Also, the white part of the fig contained the highest level of γ-tocopherol.

In the world, consumers have started to prefer dehydrated fruits. Suitable drying techniques such as conventional hot air drying, sun drying, freeze-drying, and vacuum drying are commonly applied for obtaining dried figs. Dehydration removes active water from fresh figs, which improves the product's shelf life. However, high-temperature drying for a long time may negatively affect the nutritional value of figs (Abul-Fadl et al., 2015). This is because they were drying methods exposing the food to air cause vitamin A and vitamin C losses due to oxidation. On the other hand, Freeze-drying performed without oxygen does not result in vitamin C loss (Ball, 2005). Chimi et al. (2005) stated that sun-drying could destroy the vitamins of figs. Abul-Fadl et al. (2015) found that the L-ascorbic acid level of fresh figs decreased significantly when different drying methods were applied. In this study, L-ascorbic level was calculated as 11 mg/100 g dry weight basis of fresh fig, but air-convection drying, vacuum-drying, microwave-air convection, and microwave-vacuum drying reduced the amount of L-ascorbic acid by 90.2%, 76.6%, 63.1%, and 52.1%, respectively. According to NutritionData (2018b), 100 g dried figs contain 10.0 IU vitamin A, 1.2 mg vitamin C, 0.4 mg vitamin E, 15.6 mcg vitamin K, 0.1 mg vitamin B_1, 0.1 mg riboflavin, 0.6 mg niacin, 0.1 mg vitamin B_6, 9.0 mcg vitamin B_9 and 0.4 mg vitamin B_5, which proves that vitamin A and vitamin C level are affected by the dehydration process. Similarly, Caliskan (2015) reported that 100 g of dried figs contained 0.68 mg of vitamin C. Also, Morton (1987) reported 80 I.U. vitamin A, 0.1 mg vitamin B_1, 0.1 mg vitamin B_2, 0.7 mg vitamin B_3 and 0 mg vitamin C in 100 g dried figs.

By-products of figs (e.g., peel) can also be valuable in their vitamin contents. Mahmoudi et al. (2018) investigated the vitamin C content of fig peels and pulps for 9 different cultivars. It was found that the cultivar of Onk Elhamam had the highest vitamin C level as 10.67 ± 0.31 mg/100 fresh weight for peel and 9.33 mg/100 fresh weight for pulp. On the other hand, the peel of the Hamra cultivar had the lowest vitamin C level (1.33 mg/100 fresh weight).

In summary, vitamins are required nutrients for people, and fresh figs can be a good source of vitamin A, B, C, and E. Also, dried figs and fig by-products contain vitamins. However, dried figs have less vitamin A and C than fresh ones.

5 Organic Acids

Organic acids are acidic compounds significant for various industries such as food, beverage, animal feed, and human health. These compounds impact the taste and the microbiological stability of foods (Yang & Rainville, 2020). Organic acids may occur naturally present in plant-based foods, or they can be formed during fermentation. It is widespread to add these acids artificially to foods as acidulants, flavorants, preservatives, or spoilage microorganisms' growth inhibitors (Gurtler & Mai, 2014). Some organic acids commonly used in foods are acetic, benzoic, citric, formic, lactic, and propionic acid (Lianou et al., 2012). Organic acids can positively affect people's health. For instance, citric acid may prevent kidney stone formation. However, some organic acids (e.g., oxalic acid) may lead to kidney stone formation (Sukontaprapun et al., 2019). Malic and citric acid are the most widespread organic acids in most fruits. However, other types of organic acids, such as tartaric acid, can also be plentiful in some fruits. During the ripening of some fruits, the concentration of malate and/or citrate decreases (Famiani et al., 2015).

Like other fruit species, figs contain organic acids impacting their quality. The organic acid concentration of fruits is important since it affects the organoleptic properties of fruits (Lee, 1993). Russo et al. (2010) stated that the organic acids generally found in figs are tartaric, malic, citric, sorbic, and benzoic acids. Similarly, Hussain and Othman (2011) determined the concentration of organic acids (citric acid, fumaric acid, malic acid, and oxalic acid) in fig plants cultivated in orchard adjacent to serpentine soil. The plant fruit's highest total organic acid content was 178 µg/g dry weight. Also, the lower amount of total organic acids was measured in the branches with a 55.1 µg/g dry weight concentration. The contents of citric acid varied significantly between fruits (28.7 µg/g) and leaves (14.3 µg/g). Non-significant differences were detected between leaves (22.4 µg/g) and branches (20.0 µg/g) in terms of fumaric acid. The highest oxalic acid concentration was found in fruits (51 μg/g), and the branches (8.17 μg/g) contained the lowest levels of oxalic acid.

Kelebek et al. (2018) detected two organic acids (citric and malic acid) in different fig cultivars (Bursa Siyahi and Sarilop). In terms of organic acid profile, there was no significant difference between fig varieties. However, considering the distribution of organic acids, citric acid was dominant in fresh figs, while malic acid was the most abundant in dried figs. The highest amount of citric acid was detected in fresh Bursa siyahi figs. (4.99 g/kg), and dried Bursa siyahi figs contained the highest concentration of malic acid (4.09 g/kg).

Slatnar et al. (2011) investigated the effect of the sun- or oven-drying of figs on some organic acids (malic and citric acids) at different sampling dates. The level of malic acid varied between 0.52 g/kg and 0.76 g/kg in fresh figs. Malic acid content was found higher in dried figs, ranging between 3.11 g/kg and 9.07 g/kg. In fresh figs, the citric acid content changed between 1.36 g/kg and 1.83 g/kg. Similarly, the citric acid level increased in dried figs (changing between 3.33 g/kg and 10.5 g/kg).

Shiraishi et al. (1996) elucidated organic acid profiles in the juice of fig fruits from Japan (Brown Türkiye and Houraishi cultivars). In the juice of all figs, varieties studied, acetic, butyric, oxalic, malonic, succinic, fumaric, malic, and citric acids were detected. No propionic and lactic acids were found, but glyoxylic and tartaric acid was only determined in a few examined cultivars. In another study by Trad et al. (2014), organic acid contents of figs from Tunisia were determined. The concentrations of organic acids were relatively higher in the Zidi variety (0.43% citric acid and 0.17% malic acid). On the other hand, the lowest citric acid (0.23%) and malic acid (0.11%) were detected in the Bouhouli and Bidhi varieties.

Moreover, Melgarejo et al. (2001) investigated organic acids of first and second crop fig juices from 4 different cultivars (Colar, Florancha, Tio Antonio, and Gobernador). Acetic and fumaric acid were not generally detected in fig juices. On the other hand, malic (0.0093–0.0318%), oxalic (0.0036–0.0398%), citric (0.0056–0.0399%), and tartaric (0.0091–0.0346%) acids were reported at varying levels in analyzed fig juices.

Mohammed et al. (2020) identified organic acid levels of fig organs (leaf, stem, and fruit) using HPLC. Organic acids are distributed unequally in different parts of fig organs. Malic (61.4 μg/g) and oxalic (51.0 μg/g) acids were dominant organic acids in the studied fig fruits. Citric acid content showed a difference in different fig parts. The highest level of citric acid was found in fig fruit (28.7 μg/g), and the lowest level of citric acid was detected in the leaf (14.3 μg/g). The highest fumaric acid concentration was obtained in the fruit (36.9 μg/g). The total organic acid content was determined in fruits as 178 μg/g, leaves as 71.7 μg/g, and stems as 55.1 μg/g.

Shin et al. (2015) investigated changes in organic acids in 120 figs at six different maturation stages. In the final stage, figs mainly contained citric acid, malic acid, and tartaric acid. In another study, Zhang et al. (2020) determined organic acid components in figs at two ripening stages using HPLC system. The organic acids, namely tartaric acid, quinic acid, malic acid, citric acid, fumaric acid, and succinic acid, were detected in the peel and pulp of Sichuan fig from China. The malic acid, citric acid, and quinic acid ranges were 104.8–625.6 mg/100 g fresh weight,

9.78–555 mg/100 g fresh weight, and 104.3–311.1 mg/100 g fresh weight in the pulp, respectively. The range of fumaric (0.38–2.92 mg/100 g fresh weight) and tartaric acid (6.54–83.2 mg/100 g fresh weight) in the peel were considerably lower than other organic acids.

Overall, fig contains organic acids, which are significant for human health. Organic acids such as citric, malic, fumaric, tartaric, and oxalic acid are commonly found in different parts of fig organs. Different types of factors such as ripening and drying may affect the organic acid content of figs.

6 Terpenes

Terpenes are hydrocarbons with the formula $(C_5H_8)_n$ (Breitmaier, 2006), and according to their carbon units, they are classified as hemiterpenes (C5), monoterpenes (C10), sesquiterpenes (C15), diterpenes (C20), sesterterpenes (C25), triterpenes (C30) and carotenoids (C40) (Ullah & Khan, 2020). Terpenes (also known as terpenoids) are naturally occurring compounds that are the primary components of essential oils, and they play a role in the fragrance, taste, and pigments of plants (Cox-Georgian et al., 2019). Plant oil containing terpenes has traditionally been used to treat various diseases (Cho et al., 2017). Many terpenes are bioactive (Kupska et al., 2016), and figs are a significant source of various bioactive compounds, including terpenes (Solana & Romano, 2019). Studies on the aroma profile of figs showed that certain compounds such as aldehydes, alcohols, ketones, and terpenes constitute fig odor. Volatile compound quantities may change in different fig cultivars, fresh and dried figs (Palassarou et al., 2017).

Zidi et al. (2021) investigated the volatile organic compound (VOC) profiles of figs from 3 cultivars (Taamriwthe, Azegzaw, and Averkane) at 3 ripening stages (unripe, ripe, and fully ripe). Figs were harvested from Algeria in 2018. Six terpene compounds (limonene, linalool, epoxylinalol, α-Santalene, 1,8-cineol, and β-caryophyllene) were examined in volatile fig compounds. The most abundant terpene in the Azegzaw fig was linalool, and only 2 terpenes (linalool and 1,8-cineol) were detected in the odor of Averkene fig. Moreover, β-caryophyllene was only identified in unripe and ripe Taamriwthe fig, and α-Santalene was only found in ripe Taamriwthe fig.

Russo et al. (2017) characterized volatile compounds of white figs cv. "Dottato" from Southern Italy and commercial dried figs from Türkiye. In addition to aldehydes, alcohols, and ketones, terpenes were found as another class responsible for fig aroma. D-limonene, alpha-pinene, and 3-carene were identified in Dottato figs dried by the sun, Dottato figs oven-treated after the sun-drying, and Turkish samples. D-limonene was detected in all samples, and it was the most plentiful (3.64 mg/kg) in figs oven-treated after the sun-drying. Alpha-pinene (0.42 mg/kg) and 3-carene (0.17 mg/kg) were only detected in Turkish fig samples. It was noticed that thermal treatment led to a significant increase in terpene content.

Pereira et al. (2020) determined volatile compounds of fresh fig samples from 9 cultivars: early cultivars (San Antonio, Blanca Bétera, Brown Türkiye, Tres Voltas L'Any), mid cultivar (Banane), and late cultivars (Cuello Dama Blanco, Cuello Dama Negro, Colar Elche, and De Rey). They detected monoterpenes (unknown monoterpene (min-max: 0–0.39%), α-pinene (min-max: 0–0.65%), limonene (min-max: 0–1.18%) and linalool (min-max: 0–4.35%)). They found that terpenes were the best physicochemical property to assess the overall acceptability of figs, which was highly related to sensory attributes of flesh color and fruit flavor.

Gozlekci et al. (2011) investigated the aroma profile of peels and pulps of 4 fig cultivars ("Bursa Siyahi", "Karabakunya", "Sari Lop" and "Sultan Selim") from Türkiye by using Headspace Gas Chromatography-Mass Spectrophotometer (HS/GC-MS) technique. Among terpenes, β-caryophyllene and germacrene D were the main volatile compounds found in peel and pulp fig. The peel of "Karabakunya" contained the highest amount of β-caryophyllene (48.6%), and the peel of "Sari Lop" had the highest germacrene D composition (9.19%). Limonene content changed between not detected and 0.59% in a different part of figs. Total terpenes (60.57%) were the highest in the peel of figs from the Sultan Selim cultivar.

Mujić et al. (2012) analyzed the aroma profile of fresh figs (cultivar Petrovaca Crna) from Croatia, fresh figs frozen in liquid nitrogen, and dried figs (dark variety Petrovaca Crna) by HS-SPME followed by GC-MS. In tested fig samples, 80 volatile compounds were identified from different chemical groups (alcohols, aldehydes, esters, acids, terpenes, terpenic compounds, and other compounds). Linalool (3,7-dimethyl-1,6-octadien-3-ol) was the most abundant terpenic compound. They also found that the peak area of terpenes and terpenic compounds declined with dehydration.

The aroma profile of fresh fruits (pulp and peel) and leaves of Portuguese *Ficus carica* L. white ("Pingo de Mel" and "Branca Tradicional") and dark ("Borrasota Tradicional", "Verbera Preta" and "Preta Tradicional") varieties were determined by Oliveira et al. (2010a, b). 59 aroma compounds from different chemical classes (aldehydes, alcohols, ketones, esters, monoterpenes, sesquiterpenes, and norisoprenoids) were found by HS-SPME/GC-IT-MS in 5 Portuguese varieties of *F. carica* leaves and fruits. 2 monoterpenes named limonene and menthol and 17 sesquiterpenes such as copaene, germacrene D, (+)-Ledene and alloaromadendrene were reported in five Portuguese varieties of *F. carica* leaves. Also, 7 monoterpenes and 7 sesquiterpenes were found in five Portuguese varieties of *F. carica* fruits (peels and pulps). Furthermore, Trad et al. (2012) identified volatile compounds in pollinated and non-pollinated Tunisian figs from 3 cultivars (Bouhouli, Zidi, and Thgagli). γ-Sesquiterpene was only found in pollinated (1.27 μg/kg) and non-pollinated (2 μg/kg) figs from Bouhouli cultivar. Germacrene D (80 μg/kg) and linalool (8 μg/kg) found a higher amount in pollinated Bouhouli fig than in other samples.

Song et al. (2001) determined volatiles of figs (receptive, postpollinated, and postparasitized) and leaves of Ficus hispida. Linalool was the main component of steam-distilled oil of either male or female receptive figs. The volatile composition of figs changed from a receptive to a postpollinated state. It was thought that linalool, linalool oxide, alpha-terpeneol, and 2,6-dimethyl-1,7-octadiene-3,6-diol acted as the attractants of the wasps.

Yao et al. (2021) identified aroma compounds of Xinjiang dried figs by GC-MS and GC-O. Myrcene, styrene, α-copaene, β-caryophyllene, and δ-cadinene were reported under the terpenes and terpenic compounds studied in the figs. Moreover, Grison-Pigé et al. (2002) identified the blends of volatile compounds emitted by receptive figs of 20 *Ficus* species. Different terpenes (acyclic monoterpenes, cyclic monoterpenes, and sesquiterpenes) were detected. These studies showed that, figs may contain terpenes such as limonene, linalool, β-caryophyllene, and Germacrene D.

7 Phytosterols

Phytosterols comprise both plant sterols and plant stanols. Although the human body cannot synthesize them, they can be obtained from plant-based foods such as vegetables, fruits, nuts, and cereals (Martianto et al., 2021; Ras, 2014). In a typical western diet, the daily consumption of plant sterols is 200–400 mg. In the diet, the most widespread sterols are β-sitasterol, campesterol, and stigmasterol (Gupta et al., 2011). In addition, phytosterols offer different health benefits, such as lowering the serum cholesterol effect and preventing colon cancer (Han et al., 2008). Thus, phytosterols should be added to the diet.

In fruits, the total plant sterol contents generally varied from 1.6 to 32.6 mg/100 g (Han et al., 2008). Figs are good sources of phytosterols (Kumari et al., 2018), and different studies have been performed to identify the phytosterols in the figs and fig leaves. Gupta et al. (2011) stated total phytosterols in figs as 22 mg/100 g. Jeong and Lachance (2001) determined the sterol profiles of various parts of fig tree branches (bark, stem, and pith). It was found that all parts contained campesterol, stigmasterol, and sitosterol, which are the most common phytosterols in nature. β-Amyrin and psi-taraxasterol are also found in all parts. Moreover, Oliveira et al. (2010a) determined 7 phytosterols (betulol, lupeol, lanosterol, lupeol acetate, β-amyrin, β-sisterol, and α-amyrin) in a representative sample from *F. carica* latex by gas chromatography-ion trap mass spectrometry (GC-ITMS).

8 Fatty Acids

Fatty acids, an important component of lipids, can be categorized as saturated, monounsaturated, polyunsaturated, and *trans* fats. Foods contain over 20 types of fatty acids. Palmitic acid (saturated), stearic acid (saturated), oleic acid (monounsaturated), linoleic acid (polyunsaturated), linolenic acid (polyunsaturated), arachidonic acid (polyunsaturated), eicosapentaenoic acid (polyunsaturated) and docosahexaenoic acid (polyunsaturated) are most well-known fatty acids (White, 2009). Omega-3 (e.g. α-linolenic acid (18:3n-3), stearidonic acid (18:4n-3), eicosapentaenoic acid (EPA; 20:5n-3) and docosahexaenoic acid (DHA; 22:6n-3)) (Calder, 2013) and omega-6 (e.g. linoleic acid (LA; 18:2n-6), gamma-linolenic (18:3n-6) and arachidonic acid (ARA; 20:4n-6) are the essential fatty acids (Innes & Calder, 2018). They cannot be synthesized in the body; therefore, people must get these polyunsaturated fatty acids from foods. Monounsaturated fatty acids and polyunsaturated fatty acids serve many health benefits, such as decreasing coronary heart disease risk; thus, they are necessary for the human diet (White, 2009). There is an increase in saturated fat, omega-6 fatty acids, and *trans* fatty acids intake in industrialized societies and a decrease in omega-3, fruit and vegetable intake. This situation leads to an imbalance in the ratio of omega 6 to omega 3 (Simopoulos, 2009), which shows the necessity of changing dietary trends and food consumption habits.

For centuries, fig fruit has been an indispensable part of the health-promoting Mediterranean diet. The fruits, roots, and leaves of fig trees have a number of therapeutic properties and they are used to treat cardiovascular, respiratory, and gastrointestinal disorders (Nakilcioğlu-Taş, 2019). In addition, fruits are rich in nutrients such as vitamins, phenolic acids, and fatty acids (Icyer et al., 2017). Awareness of the importance of fatty acids in human health has increased recently (Kaur et al., 2014), and therefore several studies have been conducted to investigate the fatty acid content of figs.

Guvenc et al. (2009) reported that the fig fruits from Adiyaman district, Türkiye were rich in fatty acids. Gas chromatographic (GC) analysis of fatty acid methyl esters was used, and both non-essential (palmitic, stearic, and oleic acid) and essential (linoleic and linolenic acid) fatty acids were identified in fig fruit. Linoleic (18:2n-6), linolenic (18:3n-3), oleic (18:1n-9), and palmitic (16:0) acids were found at high levels in all fig parts (whole fig fruit, skin, white part, and red part). Linoleic acid levels changed between 19.35% and 24.38% in different fig parts. The highest linolenic acid content (39.3%) was detected in the red part of the fig fruit.

In another study, Palmeira et al. (2019) assessed the fatty acid content of parts of a Portuguese common fig variety by GC coupled to a flame ionization detector (FID). They detected 23 fatty acids in samples. There were differences in the fatty acid composition of peel and pulp extracts. The α-linolenic acid (18:3n3) was found to be the most abundant fatty acid in fig peel (28.0%) and pulp (51.8%), followed by linoleic (18:2n6), palmitic (16:0), and oleic (18:1n9) acids.

According to Solana and Romano (2019), polyunsaturated fatty acids constituted 84% and 69% of total fatty acids in dried and fresh fi. Oleic acid was the most plentiful in fig fruits concerning monounsaturated fatty acids. Moreover, Jeong and Lachance (2001) determined fatty acids in dried fig fruit as their methyl esters using the GC system. Myristic (14:0), palmitic (16:0), stearic (18:0), oleic (18:1), linoleic (18:2) and linolenic (18:3) acids formed its fatty acid composition. The most dominant fatty acid was linolenic acid (53.1%), followed by linoleic acid (21.1%), palmitic acid (13.8%), and oleic acid (9.8%). It was found that 84% of the total fatty acids were unsaturated fatty acids.

ANSES-CIQUAL food composition table showed that 100 g of dried figs contained 0.65% myristic acid, 12% palmitic, 3% stearic acid, 18% oleic acid, 22% linoleic acid, and 2% linolenic acid (Anonymous, 2021). Khadhraoui et al. (2019) found that fatty acid contents of dry fig cultivars were composed of saturated fatty acids, monounsaturated fatty acids, and polyunsaturated fatty acids. They identified palmitic acid (16:0), oleic acid (18:1), linoleic acid (18:2), linolenic acid (18:3), and arachidic acid (20:0) in 9 sun-dried fig cultivars by GC-MS. On average, 0.35 ± 0.02 g/100 g dry matter (DM) linoleic acid and 0.39 ± 0.04 g/100 g DM linolenic acid were found, which were the dominant fatty acids in all samples. Arachidic acid (C:20) was detected in all fig samples at a low level.

Depending on the maturity and size of a fruit, a fig may contain a large number of small seeds (Kolesnik et al., 1986). For example, Joseph and Raj (2011) stated that dried seeds have a substantial amount of oil containing linoleic (33.7%), linolenic (32.9%), oleic (18.9%), palmitic (5.23%), stearic (2.18%) and arachidic (1.05%) acid. Likewise, Icyer et al. (2017) found linolenic (41.1%), linoleic (30.9%), oleic (30.9%), and palmitic (7.16%), and stearic (2.9%) and no trans fatty acid in fig seed oil.

Güven et al. (2019) analyzed the fatty acid content of fig seed oil from Türkiye using GC. Nine major fatty acids (palmitic acid, stearic acid, oleic acid, linoleic acid, α-linolenic acid, γ-linolenic acid, arachidic acid, eichosenoic acid, and behenic acid) were identified in fig seed oil. Alpha-linolenic acid (18:3n-3) (32–50%) and linoleic acid (18:2n-6) (20–35%) were the dominant fatty acids. Moreover, Nakilcioğlu-Taş (2019) detected 14 fatty acids in fig seed oil. Fig seeds from a different location (Incirliova, Germencik, and Nazilli) could be a good source of γ-linolenic acid (18:3n-6), ranging from 37.8% to 41.8%. Also, the level of linoleic acid (18:2n-6) was the second-highest varying between 31.8% and 37.9%. On the other hand, the lowest amount (0.02–0.03%) was reported for arachidic acid.

Ergun and Bozkurt (2020) investigated the fatty acid composition of fig seed oil with GC and detected 19 different fatty acids in fig seed oil. The most abundant fatty acid was α-linolenic acid (26.3% of total fatty acids). Linoleic acid (18:2n-6) (24.2%) and oleic acid (18:1n-9) (19.6%) were the other dominant fatty acids in fig seed oil. Furthermore, Hssaini et al. (2020b) analyzed the fatty acid content of seeds oil of four fig cultivars. Eleven fatty acids were identified, and linolenic acid was the most dominant fatty acid varying between 38.4% and 43.5% in all samples. The average contents of linoleic acid were in the range of 28.9–34.5%. The third important fatty acid was oleic acid ranging from 13.4% to 15.6%.

Uysal et al. (2015) studied the fatty acid composition of fig tree leaves collected from the Mediterranean Region of Türkiye. It was reported that the predominant fatty acid in fig leaves was α-linolenic acid (18:3n-3) (25.6%), followed by palmitic acid (16:0) (25.2%), γ-linolenic acid (18:3n-6) (13.7%), linoleic acid (18:2n-6) (10.6%), stearic acid (18:0) (8.35%) and oleic acid (18:1n-9) (4.19%).

Studies demonstrated that different parts of fig fruits and fig by-products (e.g., seeds, leaves, and peel) could be important sources of some fatty acids. Generally, α-linolenic acid, γ-linolenic, linoleic acid from polyunsaturated fatty acids, oleic acid from monounsaturated fatty acids, and palmitic acid from saturated fatty acids are found in high quantity in figs and fig by-products. Thus, the consumption and use of these products could be good dietary essential fatty acid sources for humans.

9 Conclusions

Fig (*Ficus carica*) is one of the most abundant fruits in the Mediterranean diet that promote human health owing to its bioactive compounds such as polyphenols, carotenoids, vitamins, organic acids, triterpenoids, phytosterols, and fatty acids. Flavonoids, including anthocyanins, flavonols, flavanols and flavones, and phenolic acids and coumarins, are the main polyphenolic compounds present in figs. The levels of bioactive compounds in figs depend on the variety, cultivation, environmental conditions, and processing parameters. Dark-colored fruits contain higher bioactive compound concentrations than figs with light-colored skin. Therefore, the skins of fig fruit should not be discarded as they contain several health-promoting nutrients. Accumulation of polyphenols in fig fruits increases during fruit ripening, whereas the levels of carotenoids reduce as the fruits mature. Drying process results in the reduction of some bioactive compounds, especially anthocyanins and carotenoids, as well as vitamins. On the other hand, the content of specific bioactive compounds such as organic acids is reported to enhance in dried figs. Overall, both fresh and dried figs are good sources of bioactive compounds that possess health-promoting properties.

References

Abul-Fadl, M. M., Ghanem, T. H., El-Badry, N., & Nasr, A. (2015). Effect of some different drying methods on quality criteria of dried fig fruits. *Current Science International, 4*, 548–566.

Aljane, F., Neily, H., & Msaddak, A. (2020). Phytochemical characteristics and antioxidant activity of several fig (*Ficus carica* L.) ecotypes. *Italian Journal of Food Science, 32*, 755–768.

Amessis-Ouchemoukh, N., Ouchemoukh, S., Meziant, N., Idiri, Y., Hernanz, D., Stinco, C. M., Rodríguez-Pulido, F. J., Heredia, F. J., Madani, K., & Luis, J. (2017). Bioactive metabolites involved in the antioxidant, anticancer and anticalpain activities of *Ficus carica* L., *Ceratonia siliqua* L. and *Quercus ilex* L. extracts. *Industrial Crops and Products, 95*, 6–11. https://doi.org/10.1016/j.indcrop.2016.10.007

Ammar, S., del Contreras, M., Belguith-Hadrich, O., Bouaziz, M., & Segura-Carretero, A. (2015). New insights into the qualitative phenolic profile of *Ficus carica* L. fruits and leaves from Tunisia using ultra-high-performance liquid chromatography coupled to quadrupole-time-of-flight mass spectrometry and their antioxidant activity. *RSC Advances, 5*, 20035–20050. https://doi.org/10.1039/C4RA16746E

Anonymous. (2021). *Fig, dried.* https://ciqual.anses.fr/#/aliments/13013/fig-dried

Arvaniti, O. S., Samaras, Y., Gatidou, G., Thomaidis, N. S., & Stasinakis, A. S. (2019). Review on fresh and dried figs: Chemical analysis and occurrence of phytochemical compounds, antioxidant capacity and health effects. *Food Research International, 119*, 244–267. https://doi.org/10.1016/j.foodres.2019.01.055

Badgujar, S. B., Patel, V. V., Bandivdekar, A. H., & Mahajan, R. T. (2014). Traditional uses, phytochemistry and pharmacology of *Ficus carica* : A review. *Pharmaceutical Biology, 52*, 1487–1503. https://doi.org/10.3109/13880209.2014.892515

Balasundram, N., Sundram, K., & Samman, S. (2006). Phenolic compounds in plants and agri-industrial by-products: Antioxidant activity, occurrence, and potential uses. *Food Chemistry, 99*, 191–203. https://doi.org/10.1016/j.foodchem.2005.07.042

Ball, G. F. M. (2005). Nutritional aspects of vitamins. In *Vitamins in foods: Analysis, bioavailability, and stability* (pp. 9–10). CRC Press.

Belguith-Hadriche, O., Ammar, S., Contreras, M. D. M., Turki, M., Segura-Carretero, A., El Feki, A., Makni-Ayedi, F., & Bouaziz, M. (2016). Antihyperlipidemic and antioxidant activities of edible Tunisian *Ficus carica* L. fruits in high fat diet-induced Hyperlipidemic rats. *Plant Foods for Human Nutrition, 71*, 183–189. https://doi.org/10.1007/s11130-016-0541-x

Breitmaier, E. (2006). *Terpenes: Flavors, fragrances, pharmaca, pheromones*. Wiley.

Britto, I. O., Araújo, S. H. C., Toledo, P. F. S., Lima, G. D. A., Salustiano, I. V., Alves, J. R., Mantilla-Afanador, J. G., Kohlhoff, M., Oliveira, E. E., & Leite, J. P. V. (2021). Potential of *Ficus carica* extracts against *Euschistus heros*: Toxicity of major active compounds and selectivity against beneficial insects. *Pest Management Science, 77*, 4638–4647. https://doi.org/10.1002/ps.6504

Calder, P. C. (2013). Nutritional benefits of omega-3 fatty acids. In C. Jacobsen, N. S. Nielsen, A. F. Horn, & A.-D. M. Sørensen (Eds.), *Food enrichment with Omega-3 fatty acids*. Elsevier.

Caliskan, O. (2015). Mediterranean figs (*Ficus carica* L.) functional food properties. In *The Mediterranean diet* (pp. 629–637). Academic.

Çalişkan, O., & Polat, A. A. (2011). Phytochemical and antioxidant properties of selected fig (*Ficus carica* L.) accessions from the eastern Mediterranean region of Türkiye. *Scientia Horticulturae, 128*, 473–478. https://doi.org/10.1016/j.scienta.2011.02.023

Caporuscio, J. (2019). *Can figs be beneficial to our health?* Retrieved from https://www.medicalnewstoday.com/articles/327207

Chimi, H., Ouaouich, A., Semmar, M., & Tayebi, S. (2005). Industrial processing of figs by solar drying in Morocco. In *III International Symposium on Fig 798* (pp. 331–334).

Cho, K. S., Lim, Y., Lee, K., Lee, J., Lee, J. H., & Lee, I.-S. (2017). Terpenes from forests and human health. *Toxicological Research, 33*, 97–106. https://doi.org/10.5487/TR.2017.33.2.097

Chunyan, C., Bo, S., Ping, L., Jingmei, L., & Ito, Y. (2008). Isolation and purification of psoralen and Bergapten from *Ficus carica* L. leaves by high-speed countercurrent chromatography. *Journal of Liquid Chromatography & Related Technologies, 32*, 136–143. https://doi.org/10.1080/10826070802548747

Cox-Georgian, D., Ramadoss, N., Dona, C., & Basu, C. (2019). Therapeutic and medicinal uses of terpenes. In *Medicinal plants*. Springer.

del Caro, A., & Piga, A. (2008). Polyphenol composition of peel and pulp of two Italian fresh fig fruits cultivars (*Ficus carica* L.). *European Food Research and Technology, 226*, 715–719. https://doi.org/10.1007/s00217-007-0581-4

Dueñas, M., Pérez-Alonso, J. J., Santos-Buelga, C., & Escribano-Bailón, T. (2008). Anthocyanin composition in fig (*Ficus carica* L.). *Journal of Food Composition and Analysis, 21*, 107–115. https://doi.org/10.1016/j.jfca.2007.09.002

Eggersdorfer, M., & Wyss, A. (2018). Carotenoids in human nutrition and health. *Archives of Biochemistry and Biophysics, 652*. https://doi.org/10.1016/j.abb.2018.06.001

Eker, M. E., Aaby, K., Budic-Leto, I., Rimac Brnčić, S., El, S. N., Karakaya, S., Simsek, S., Manach, C., Wiczkowski, W., & de Pascual-Teresa, S. (2020). A review of factors affecting anthocyanin bioavailability: Possible implications for the inter-individual variability. *Foods, 9*, 2. https://doi.org/10.3390/foods9010002

Ercisli, S., Tosun, M., Karlidag, H., Dzubur, A., Hadziabulic, S., & Aliman, Y. (2012). Color and antioxidant characteristics of some fresh Fig (*Ficus carica* L.) genotypes from Northeastern Türkiye. *Plant Foods for Human Nutrition, 67*, 271–276. https://doi.org/10.1007/s11130-012-0292-2

Ergun, Z., & Bozkurt, T. (2020). Determination of fatty acid composition and antioxidant activity of fig seed oil. *International Journal of Agricultural and Natural Sciences, 13*, 101–107.

Famiani, F., Battistelli, A., Moscatello, S., Cruz-Castillo, J. G., & Walker, R. P. (2015). The organic acids that are accumulated in the flesh of fruits: Occurrence, metabolism and factors affecting their contents-a review. *Revista Chapingo Serie Horticultura, XXI*, 97–128. https://doi.org/10.5154/r.rchsh.2015.01.004

FAOSTAT. (2021). Food and agriculture organization corporate statistical database. *Crops*. Retrieved from: http://www.fao.org/faostat/en/#data/QC/visualize

Feng, Y.-C., Li, W.-L., He, F.-M., Kong, T.-T., Huang, X.-W., Gao, Z.-H., Lu, N.-H., & Li, H.-L. (2015). Aqueous two-phase system as an effective tool for purification of phenolic compounds from fig fruits (*Ficus carica* L.). *Separation Science and Technology, 50*, 1785–1793. https://doi.org/10.1080/01496395.2015.1014054

Francini, A., Sodini, M., Vicario, G., Raffaelli, A., Gucci, R., Caruso, G., & Sebastiani, L. (2021). Cations and phenolic compounds concentrations in fruits of fig plants exposed to moderate levels of salinity. *Antioxidants, 10*, 1865. https://doi.org/10.3390/antiox10121865

Gaaliche, B., Ladhari, A., Zarrelli, A., & Ben Mimoun, M. (2019). Impact of foliar potassium fertilization on biochemical composition and antioxidant activity of fig (*Ficus carica* L.). *Scientia Horticulturae, 253*, 111–119. https://doi.org/10.1016/j.scienta.2019.04.024

Gozlekci, S., Kafkas, E., & Ercisli, S. (2011). Volatile compounds determined by HS/GC-MS technique in Peel and pulp of fig (*Ficus carica* L.) cultivars grown in Mediterranean region of Türkiye. *Notulae Botanicae Horti Agrobotanici Cluj-Napoca, 39*, 105. https://doi.org/10.15835/nbha3926261

Grison-Pigé, L., Hossaert-McKey, M., Greeff, J. M., & Bessière, J.-M. (2002). Fig volatile compounds-a first comparative study. *Phytochemistry, 61*, 61–71. https://doi.org/10.1016/S0031-9422(02)00213-3

Gupta, A. K., Savopoulos, C. G., Ahuja, J., & Hatzitolios, A. I. (2011). Role of phytosterols in lipid-lowering: Current perspectives. *QJM, 104*, 301–308. https://doi.org/10.1093/qjmed/hcr007

Gurtler, J. B., & Mai, T. L. (2014). PRESERVATIVES | traditional preservatives-organic acids. In C. A. Batt & M. L. Tortorello (Eds.), *Encyclopedia of food microbiology* (2nd ed., pp. 119–130). Academic.

Güven, N., Gökyer, A., Koç, A., Temiz, N. N., Selvi, S., Koparal, B., Dedeoğlu, B. D., Büyükhelvacigil Öztürk, S., Büyükhelvacigil, H. F., Büyükhelvacıgil, R., & Erman, C. (2019). Physiochemical composition of fig seed oil from Türkiye. *Journal of Pharmacy and Pharmacology, 7*, 541–545.

Guvenc, M., Tuzcu, M., & Yilmaz, O. (2009). Analysis of fatty acid and some lipophilic vitamins found in the fruits of the Ficus carica variety picked from the Adiyaman District. *Research Journal of Biological Sciences, 4*, 320–323.

Hackman, R. M., Polagruto, J. A., Zhu, Q. Y., Sun, B., Fujii, H., & Keen, C. L. (2008). Flavanols: Digestion, absorption and bioactivity. *Phytochemistry Reviews, 7*, 195–208. https://doi.org/10.1007/s11101-007-9070-4

Han, J.-H., Yang, Y.-X., & Feng, M.-Y. (2008). Contents of Phytosterols in vegetables and fruits commonly consumed in China. *Biomedical and Environmental Sciences, 21*, 449–453. https://doi.org/10.1016/S0895-3988(09)60001-5

Harzallah, A., Bhouri, A. M., Amri, Z., Soltana, H., & Hammami, M. (2016). Phytochemical content and antioxidant activity of different fruit parts juices of three figs (*Ficus carica* L.) varieties grown in Tunisia. *Industrial Crops and Products, 83*, 255–257. https://doi.org/10.1016/j.indcrop.2015.12.043

Hostetler, G. L., Ralston, R. A., & Schwartz, S. J. (2017). Flavones: Food sources, bioavailability, metabolism, and bioactivity. *Advances in Nutrition: An International Review Journal, 8*, 423–435. https://doi.org/10.3945/an.116.012948

Hssaini, L., Charafi, J., Hanine, H., Ennahli, S., Mekaoui, A., Mamouni, A., & Razouk, R. (2019). Comparative analysis and physio-biochemical screening of an ex-situ fig (*Ficus carica* L.) collection. *Horticulture, Environment, and Biotechnology, 60*, 671–683. https://doi.org/10.1007/s13580-019-00170-4

Hssaini, L., Charafi, J., Razouk, R., Hernández, F., Fauconnier, M., Ennahli, S., & Hanine, H. (2020a). Assessment of morphological traits and fruit metabolites in eleven fig varieties (*Ficus Carica* L.). *International Journal of Fruit Science, 20*, 8–28. https://doi.org/10.1080/15538362.2019.1701615

Hssaini, L., Hanine, H., Charafi, J., Razouk, R., Elantari, A., Ennahli, S., Hernández, F., & Ouaabou, R. (2020b). First report on fatty acids composition, total phenolics and antioxidant activity in seeds oil of four fig cultivars (*Ficus carica* L.) grown in Morocco. *Oilseeds & Fats Crops and Lipids, 27*(8). https://doi.org/10.1051/ocl/2020003

Hung, W.-L., Suh, J. H., & Wang, Y. (2017). Chemistry and health effects of furanocoumarins in grapefruit. *Journal of Food and Drug Analysis, 25*, 71–83. https://doi.org/10.1016/j.jfda.2016.11.008

Hussain, M. M., & Othman, A. R. (2011). Determination of some heavy metals and organic acids in fig plant cultivated in orchard adjacent to serpentine soil. *The Egyptian Journal of Experimental Biology, 7*, 279–284.

Hussain, M. I., Syed, Q. A., Khattak, M. N. K., Hafez, B., Reigosa, M. J., & El-Keblawy, A. (2019). Natural product coumarins: Biological and pharmacological perspectives. *Biologia, 74*, 863–888. https://doi.org/10.2478/s11756-019-00242-x

Icyer, N. C., Toker, O. S., Karasu, S., Tornuk, F., Kahyaoglu, T., & Arici, M. (2017). Microencapsulation of fig seed oil rich in polyunsaturated fatty acids by spray drying. *Journal of Food Measurement and Characterization, 11*, 50–57. https://doi.org/10.1007/s11694-016-9370-8

Ignat, I., Volf, I., & Popa, V. I. (2011). A critical review of methods for characterisation of polyphenolic compounds in fruits and vegetables. *Food Chemistry, 126*, 1821–1835. https://doi.org/10.1016/j.foodchem.2010.12.026

Innes, J. K., & Calder, P. C. (2018). Omega-6 fatty acids and inflammation. *Prostaglandins, Leukotrienes and Essential Fatty Acids, 132*, 41–48. https://doi.org/10.1016/j.plefa.2018.03.004

Jeong, W.-S., & Lachance, P. A. (2001). Phytosterols and fatty acids in fig (*Ficus carica*, var. Mission) fruit and tree components. *Journal of Food Science, 66*, 278–281. https://doi.org/10.1111/j.1365-2621.2001.tb11332.x

Joseph, B., & Raj, S. J. (2011). Pharmacognostic and phytochemical properties of Ficus carica Linn-an overview. *International Journal of Pharmtech Research, 3*, 8–12.

Kamiloglu, S., & Capanoglu, E. (2013). Investigating the in vitro bioaccessibility of polyphenols in fresh and sun-dried figs (*Ficus carica* L.). *International Journal of Food Science and Technology, 48*, 2621–2629. https://doi.org/10.1111/ijfs.12258

Kamiloglu, S., & Capanoglu, E. (2015). Polyphenol content in figs (*Ficus carica* L.): Effect of Sun-drying. *International Journal of Food Properties, 18*, 521–535. https://doi.org/10.1080/10942912.2013.833522

Kamiloglu, S., Capanoglu, E., Grootaert, C., & Van Camp, J. (2015). Anthocyanin absorption and metabolism by human intestinal Caco-2 cells-a review. *International Journal of Molecular Sciences, 16*. https://doi.org/10.3390/ijms160921555

Kamiloglu, S., Tomas, M., & Capanoglu, E. (2020). Dietary Flavonols and O-glycosides. In J. Xiao, S. D. Sarker, & Y. Asakawa (Eds.), *Handbook of dietary phytochemicals*. Springer Singapore.

Kamiloglu, S., Tomas, M., Ozdal, T., Yolci-Omeroglu, P., & Capanoglu, E. (2021). Bioactive component analysis. In C. M. Galanakis (Ed.), *Innovative food analysis* (pp. 41–65). Elsevier.

Karantzi, A. D., Kafkaletou, M., Christopoulos, M. V., & Tsantili, E. (2021). Peel colour and flesh phenolic compounds at ripening stages in pollinated commercial varieties of fig (*Ficus carica* L.) fruit grown in southern Europe. *Journal of Food Measurement and Characterization, 15*, 2049–2063. https://doi.org/10.1007/s11694-020-00796-4

Kaur, N., Chugh, V., & Gupta, A. K. (2014). Essential fatty acids as functional components of foods- a review. *Journal of Food Science and Technology, 51*, 2289–2303. https://doi.org/10.1007/s13197-012-0677-0

Kazemipoor, M., Lorestani, M. A., & Ansari, M. (2012). Extraction and determination of biomarkers in Ficus carica L. leaves from various species and different cultivars by HPLC. *Journal of Liquid Chromatography & Related Technologies, 35*, 2831–2844. https://doi.org/10.1080/10826076.2011.639115

Kelebek, H., Dıblan, S., Kadiroğlu, P., Kola, O., & Selli, S. (2018). Effects of drying on phenolic compounds, sugars, organic acids and antioxidant properties of fig varieties (*Ficus carica* L.). *Çukurova Journal of Agricultural and Food Sciences, 3*, 127–136.

Khadhraoui, M., Bagues, M., Artés, F., & Ferchichi, A. (2019). Phytochemical content, antioxidant potential, and fatty acid composition of dried Tunisian fig (*Ficus carica* L.) cultivars. *Journal of Applied Botany and Food Quality, 92*, 143–150.

Kolesnik, A. A., Kakhniashvili, T. A., Zherebin, Y. L., Golubev, V. N., & Pilipenko, L. N. (1986). Lipids of the fruit of Ficus carica. *Chemistry of Natural Compounds, 22*, 394–397.

Kumari, K., Sharma, S., & Kaushik, R. (2017). Wild Himalayan fig: A nutraceutical under exploited fruit of Western Himalayan region- a review. *International Journal of Advanced Research, 5*, 833–839. https://doi.org/10.21474/IJAR01/5395

Kumari, K., Sharma, S., Joshi, V. K., & Sharma, S. (2018). Adding value to wild Himalayan fig (Ficus palmata): Composition, functional and sensory characteristics of jam. *The Journal of Phytopharmacology, 7*, 13–18.

Kupska, M., Wasilewski, T., Jędrkiewicz, R., Gromadzka, J., & Namieśnik, J. (2016). Determination of terpene profiles in potential Superfruits. *International Journal of Food Properties, 19*, 2726–2738. https://doi.org/10.1080/10942912.2016.1144066

Ladhari, A., Gaaliche, B., Zarrelli, A., Ghannem, M., & Ben Mimoun, M. (2020). Allelopathic potential and phenolic allelochemicals discrepancies in *Ficus carica* L. cultivars. *South African Journal of Botany*, 130. https://doi.org/10.1016/j.sajb.2019.11.026

Lama, K., Harlev, G., Shafran, H., Peer, R., & Flaishman, M. A. (2020). Anthocyanin accumulation is initiated by abscisic acid to enhance fruit color during fig (*Ficus carica* L.) ripening. *Journal of Plant Physiology, 251*, 153192. https://doi.org/10.1016/j.jplph.2020.153192

Lee, H. S. (1993). HPLC method for separation and determination of nonvolatile organic acids in orange juice. *Journal of Agricultural and Food Chemistry, 41*, 1991–1993. https://doi.org/10.1021/jf00035a033

Li, J., Tian, Y. Z., Sun, B. Y., Dan, Y. A. N. G., Chen, J. P., & Men, Q. M. (2011). Analysis on volatile constituents in leaves and fruits of Ficus carica by GC-MS. *Chinese Herbal Medicines, 4*, 63–69.

Li, P., Liu, A., Liu, C., Qu, Z., Xiao, W., Huang, J., Liu, Z., & Zhang, S. (2019). Role and mechanism of catechin in skeletal muscle cell differentiation. *The Journal of Nutritional Biochemistry, 74*, 108225. https://doi.org/10.1016/j.jnutbio.2019.108225

Li, C., Yu, M., Li, S., Yang, X., Qiao, B., Shi, S., Zhao, C., & Fu, Y. (2021). Valorization of fig (*Ficus carica* L.) waste leaves: HPLC-QTOF-MS/MS-DPPH system for online screening and identification of antioxidant compounds. *Plants, 10*, 2532. https://doi.org/10.3390/plants10112532

Lianou, A., Koutsoumanis, K. P., & Sofos, J. N. (2012). Organic acids and other chemical treatments for microbial decontamination of food. In A. Demirci & M. O. Ngadi (Eds.), *Microbial decontamination in the food industry*. Elsevier.

Mahmoudi, S., Khali, M., Benkhaled, A., Boucetta, I., Dahmani, Y., Attallah, Z., & Belbraouet, S. (2018). Fresh figs (*Ficus carica* L.): Pomological characteristics, nutritional value, and phytochemical properties. *European Journal of Horticultural Science, 83*, 104–113. https://doi.org/10.17660/eJHS.2018/83.2.6

Marrelli, M., Statti, G. A., Tundis, R., Menichini, F., & Conforti, F. (2014). Fatty acids, coumarins and polyphenolic compounds of *Ficus carica* L. cv. Dottato: Variation of bioactive compounds and biological activity of aerial parts. *Natural Product Research, 28*, 271–274. https://doi.org/10.1080/14786419.2013.841689

Martianto, D., Bararah, A., Andarwulan, N., & Średnicka-Tober, D. (2021). Cross-sectional study of plant sterols intake as a basis for designing appropriate plant sterol-enriched food in Indonesia. *Nutrients, 13*, 452. https://doi.org/10.3390/nu13020452

Martínez-García, J. J., Gallegos-Infante, J. A., Rocha-Guzmán, N. E., Ramírez-Baca, P., Candelas-Cadillo, M. G., & González-Laredo, R. F. (2013). Drying parameters of half-cut and ground figs (*Ficus carica* L.) var. Mission and the effect on their functional properties. *Journal of Engineering, 2013*, 1–8. https://doi.org/10.1155/2013/710830

Mawa, S., Husain, K., & Jantan, I. (2013). *Ficus carica* L. (Moraceae): Phytochemistry, traditional uses and biological activities. *Evidence-based Complementary and Alternative Medicine, 2013*, 1–8. https://doi.org/10.1155/2013/974256

Melgarejo, P., Hernández, F., Martínez, J. J., Sánchez, J., & Salazar, D. M. (2001). Organic acids and sugars from first and second crop fig juices. In *ISHS Acta Horticulturae 605: II International Symposium on Fig* (pp. 237–239). https://doi.org/10.17660/ActaHortic.2003.605.36

Meziant, L., Boutiche, M., Bachir Bey, M., Saci, F., & Louaileche, H. (2018). Standardization of monomeric anthocyanins extraction from fig fruit peels (*Ficus carica* L.) using single factor methodology. *Journal of Food Measurement and Characterization, 12*, 2865–2873. https://doi.org/10.1007/s11694-018-9901-6

Meziant, L., Bachir-bey, M., Bensouici, C., Saci, F., Boutiche, M., & Louaileche, H. (2021). Assessment of inhibitory properties of flavonoid-rich fig (*Ficus carica* L.) peel extracts against tyrosinase, α-glucosidase, urease and cholinesterases enzymes, and relationship with antioxidant activity. *European Journal of Integrative Medicine, 43*, 101272. https://doi.org/10.1016/j.eujim.2020.101272

Mohammed, O. A., Hussain, M. R., Shareef, O. H., Hama, A. A., Weli, S. M., Ali, F. M., & Salih, S. S. (2020). Analysis of some heavy metals and organic acids in *Ficus carica* growing adjacent in the serpentine soil in Sulaimani/Kurdistan, Iraq. *International Journal of Food Science, 2020*, 1–7. https://doi.org/10.1155/2020/8883517

Morton, J. F. (1987). Fig. In *Fruits of warm climates* (pp. 47–50). University of Miami.

Mujić, I., Bavcon Kralj, M., Jokić, S., Jarni, K., Jug, T., & Prgomet, Ž. (2012). Changes in aromatic profile of fresh and dried fig – the role of pre-treatments in drying process. *International Journal of Food Science & Technology, 47*, 2282–2288. https://doi.org/10.1111/j.1365-2621.2012.03099.x

Mustafa, K., Yu, S., Zhang, W., Mohamed, H., Naz, T., Xiao, H., Liu, Y., Nazir, Y., Fazili, A. B. A., Nosheen, S., Bai, X., & Song, Y. (2021). Screening, characterization, and in vitro-ROS dependent cytotoxic potential of extract from Ficus carica against hepatocellular (HepG2) carcinoma cells. *South African Journal of Botany, 138*, 217–226. https://doi.org/10.1016/j.sajb.2020.12.018

Nakilcioğlu-Taş, E. (2019). Biochemical characterization of fig (*Ficus carica* L.) seeds. *Journal of Agricultural Sciences*. https://doi.org/10.15832/ankutbd.398268

Naveed, M., Hejazi, V., Abbas, M., Kamboh, A. A., Khan, G. J., Shumzaid, M., Ahmad, F., Babazadeh, D., FangFang, X., Modarresi-Ghazani, F., WenHua, L., & XiaoHui, Z. (2018). Chlorogenic acid (CGA): A pharmacological review and call for further research. *Biomedicine & Pharmacotherapy, 97*, 67–74. https://doi.org/10.1016/j.biopha.2017.10.064

Nawade, B., Shaltiel-Harpaz, L., Yahyaa, M., Bosamia, T. C., Kabaha, A., Kedoshim, R., Zohar, M., Isaacson, T., & Ibdah, M. (2020). Analysis of apocarotenoid volatiles during the development of Ficus carica fruits and characterization of carotenoid cleavage dioxygenase genes. *Plant Science, 290*, 110292. https://doi.org/10.1016/j.plantsci.2019.110292

NutritionData. (2018a). *Figs, raw nutrition facts & calories*. Retrieved from https://nutritiondata.self.com/facts/fruits-and-fruit-juices/1884/2

NutritionData. (2018b). *Figs, dried, uncooked nutrition facts & calories*. Retrieved from https://nutritiondata.self.com/facts/fruits-and-fruit-juices/1889/2

Oancea, S. (2021). A review of the current knowledge of thermal stability of anthocyanins and approaches to their stabilization to heat. *Antioxidants, 10*, 1337. https://doi.org/10.3390/antiox10091337

Oliveira, A. P., Valentão, P., Pereira, J. A., Silva, B. M., Tavares, F., & Andrade, P. B. (2009). *Ficus carica* L.: Metabolic and biological screening. *Food and Chemical Toxicology, 47*, 2841–2846. https://doi.org/10.1016/j.fct.2009.09.004

Oliveira, A. P., Silva, L. R., Ferreres, F., Guedes de Pinho, P., Valentão, P., Silva, B. M., Pereira, J. A., & Andrade, P. B. (2010a). Chemical assessment and *in vitro* antioxidant capacity of *Ficus carica* latex. *Journal of Agricultural and Food Chemistry, 58*, 3393–3398. https://doi.org/10.1021/jf9039759

Oliveira, A. P., Silva, L. R., Guedes de Pinho, P., Gil-Izquierdo, A., Valentão, P., Silva, B. M., Pereira, J. A., & Andrade, P. B. (2010b). Volatile profiling of Ficus carica varieties by HS-SPME and GC-IT-MS. *Food Chemistry, 123*, 548–557. https://doi.org/10.1016/j.foodchem.2010.04.064

Oliveira, A. P., Baptista, P., Andrade, P. B., Martins, F., Pereira, J. A., Silva, B. M., & Valentão, P. (2012). Characterization of *Ficus carica* L. cultivars by DNA and secondary metabolite analysis: Is genetic diversity reflected in the chemical composition? *Food Research International, 49*, 710–719. https://doi.org/10.1016/j.foodres.2012.09.019

Palassarou, M., Melliou, E., Liouni, M., Michaelakis, A., Balayiannis, G., & Magiatis, P. (2017). Volatile profile of Greek dried white figs (*Ficus carica* L.) and investigation of the role of β-damascenone in aroma formation in fig liquors. *Journal of the Science of Food and Agriculture, 97*, 5254–5270. https://doi.org/10.1002/jsfa.8410

Palmeira, L., Pereira, C., Dias, M. I., Abreu, R. M. V., Corrêa, R. C. G., Pires, T. C. S. P., Alves, M. J., Barros, L., & Ferreira, I. C. F. R. (2019). Nutritional, chemical and bioactive profiles of different parts of a Portuguese common fig (*Ficus carica* L.) variety. *Food Research International, 126*. https://doi.org/10.1016/j.foodres.2019.108572

Pande, G., & Akoh, C. C. (2010). Organic acids, antioxidant capacity, phenolic content and lipid characterisation of Georgia-grown underutilized fruit crops. *Food Chemistry, 120*, 1067–1075. https://doi.org/10.1016/j.foodchem.2009.11.054

Pandey, K. B., & Rizvi, S. I. (2009). Plant polyphenols as dietary antioxidants in human health and disease. *Oxidative Medicine and Cellular Longevity, 2*, 270–278. https://doi.org/10.4161/oxim.2.5.9498

Pereira, C., López-Corrales, M., Serradilla, M. J., del Villalobos, M. C., Ruiz-Moyano, S., & Martín, A. (2017). Influence of ripening stage on bioactive compounds and antioxidant activity in nine fig (*Ficus carica* L.) varieties grown in Extremadura, Spain. *Journal of Food Composition and Analysis, 64*, 203–212. https://doi.org/10.1016/j.jfca.2017.09.006

Pereira, C., Martín, A., López-Corrales, M., de Córdoba, M. G., Galván, A. I., & Serradilla, M. J. (2020). Evaluation of the physicochemical and sensory characteristics of different fig cultivars for the fresh fruit market. *Foods, 9*, 619. https://doi.org/10.3390/foods9050619

Petkova, N., Ivanov, I., & Denev, P. (2019). Changes in phytochemical compounds and antioxidant potential of fresh, frozen, and processed figs (*Ficus carica* L.). *International Food Research Journal, 26*, 1881–1888.

Petruccelli, R., Ieri, F., Ciaccheri, L., & Bonetti, A. (2018). Polyphenolic profiling and chemometric analysis of leaves from Italian *Ficus carica* L. varieties. Polyphenol compounds in common fig. *European Journal of Horticultural Science, 83*, 94–103. https://doi.org/10.17660/eJHS.2018/83.2.5

Popova, A. (2019). Comparison of vitamin C content of commercially available fresh fruits. *Asian Food Science Journal*, 1–6. https://doi.org/10.9734/afsj/2019/v13i230100

Pourghayoumi, M., Bakhshi, D., Rahemi, M., & Jafari, M. (2012). Effect of pollen source on quantitative and qualitative characteristics of dried figs (*Ficus carica* L.) cvs 'Payves' and 'Sabz' in Kazerun-Iran. *Scientia Horticulturae, 147*, 98–104. https://doi.org/10.1016/j.scienta.2012.08.026

Qin, H., Zhou, G., Peng, G., Li, J., & Chen, J. (2015). Application of ionic liquid-based ultrasound-assisted extraction of five phenolic compounds from fig (*Ficus carica* L.) for HPLC-UV. *Food Analytical Methods, 8*, 1673–1681. https://doi.org/10.1007/s12161-014-0047-9

Ras, R. T. (2014). *Phytosterols and blood lipid risk factors for cardiovascular disease*. Wageningen University.

Rashmi, H. B., & Negi, P. S. (2020). Phenolic acids from vegetables: A review on processing stability and health benefits. *Food Research International, 136*, 109298. https://doi.org/10.1016/j.foodres.2020.109298

Reyes-Avalos, M. C., Minjares-Fuentes, R., Femenia, A., Contreras-Esquivel, J. C., Quintero-Ramos, A., Esparza-Rivera, J. R., & Meza-Velázquez, J. A. (2019). Application of an alginate-chitosan edible film on figs (*Ficus carica*): Effect on bioactive compounds and antioxidant capacity. *Food and Bioprocess Technology, 12*, 499–511. https://doi.org/10.1007/s11947-018-2226-y

Rodov, V., Vinokur, Y., & Horev, B. (2012). Brief postharvest exposure to pulsed light stimulates coloration and anthocyanin accumulation in fig fruit (*Ficus carica* L.). *Postharvest Biology and Technology, 68*, 43–46. https://doi.org/10.1016/j.postharvbio.2012.02.001

Rouaiguia-Bouakkaz, S., Amira-Guebailia, H., Rivière, C., Delaunay, J.-C., Waffo-Téguo, P., & Mérillon, J.-M. (2013). Identification and quantification of furanocoumarins in stem bark and wood of eight Algerian varieties of *Ficus carica* by RP-HPLC-DAD and RP-HPLC-DAD-MS. *Natural Product Communications, 8*, 1934578X1300800. https://doi.org/10.1177/1934578X1300800420

Russo, M. T., di Sanzo, R., Cefaly, V., Fuda, S., Postorino, S., & Suraci, F. (2010). Quality and nutritional composition of dried figs: A traditional Calabria region (Italy) product. In *ISHS Acta Horticulturae 939: XXVIII international horticultural congress on science and horticulture for people (IHC2010): International symposium on emerging health topics in fruits and vegetables* (pp. 189–196). https://doi.org/10.17660/ActaHortic.2012.939.24

Russo, F., Caporaso, N., Paduano, A., & Sacchi, R. (2017). Characterisation of volatile compounds in Cilento (Italy) figs (*Ficus carica* L.) cv. Dottato as affected by the drying process. *International Journal of Food Properties*, 1–11. https://doi.org/10.1080/10942912.2017.1344991

Saif, S., Hanif, M. A., Rehman, R., Hanif, M., Khan, O., & Khan, S. (2020). Figs. In *Medicinal plants of South Asia* (pp. 273–286). Elsevier.

Saki, M., ValizadehKaji, B., Abbasifar, A., & Shahrjerdi, I. (2019). Effect of chitosan coating combined with thymol essential oil on physicochemical and qualitative properties of fresh fig (*Ficus carica* L.) fruit during cold storage. *Journal of Food Measurement and Characterization, 13*, 1147–1158. https://doi.org/10.1007/s11694-019-00030-w

Salehi, B., Venditti, A., Sharifi-Rad, M., Kręgiel, D., Sharifi-Rad, J., Durazzo, A., Lucarini, M., Santini, A., Souto, E., Novellino, E., Antolak, H., Azzini, E., Setzer, W., & Martins, N. (2019). The therapeutic potential of Apigenin. *International Journal of Molecular Sciences, 20*, 1305. https://doi.org/10.3390/ijms20061305

Sedaghat, S., & Rahemi, M. (2018). Effects of physio-chemical changes during fruit development on nutritional quality of fig (*Ficus carica* L. var. 'Sabz') under rain-fed condition. *Scientia Horticulturae, 237*, 44–50. https://doi.org/10.1016/j.scienta.2018.04.003

Shin, T.-S., Park, J.-A., & Jung, B.-M. (2015). Changes in organic acids, free sugars, and volatile flavor compounds in fig (*Ficus carica* L.) by maturation stage. *Journal of the Korean Society of Food Science and Nutrition, 44*, 1016–1027. https://doi.org/10.3746/jkfn.2015.44.7.1016

Shiraishi, S. I., Kawakami, K., Widodo, S. E., Shiraishi, M., & Kitazaki, M. (1996). Organic acid profiles in the juice of fig fruits. *Journal-Faculty of Agriculture Kyushu University, 41*, 29–34.

Simopoulos, A. P. (2009). Omega-6/omega-3 essential fatty acids: Biological effects. *World Review of Nutrition and Dietetics, 99*, 1–16.

Slatnar, A., Klancar, U., Stampar, F., & Veberic, R. (2011). Effect of drying of figs (*Ficus carica* L.) on the contents of sugars, organic acids, and phenolic compounds. *Journal of Agricultural and Food Chemistry, 59*, 11696–11702. https://doi.org/10.1021/jf202707y

Solana, R. R., & Romano, A. (2019). Chemical and biological characteristics of *Ficus carica* L. fruits, leaves, and derivatives (Wine, Spirit, and Liqueur). In *Modern fruit industry*. IntechOpen.

Solomon, A., Golubowicz, S., Yablowicz, Z., Grossman, S., Bergman, M., Gottlieb, H. E., Altman, A., Kerem, Z., & Flaishman, M. A. (2006). Antioxidant activities and anthocyanin content of fresh fruits of common fig (*Ficus carica* L.). *Journal of Agricultural and Food Chemistry, 54*, 7717–7723. https://doi.org/10.1021/jf060497h

Soltana, H., de Rosso, M., Lazreg, H., Vedova, A. D., Hammami, M., & Flamini, R. (2018). LC-QTOF characterization of non-anthocyanic flavonoids in four Tunisian fig varieties. *Journal of Mass Spectrometry, 53*, 817–823. https://doi.org/10.1002/jms.4209

Song, Q., Yang, D., Zhang, G., & Yang, C. (2001). Volatiles from Ficus hispida and their attractiveness to fig wasps. *Journal of Chemical Ecology, 27*, 1929–1942. https://doi.org/10.1023/A:1012226400586

Song, H., Zhang, L., Wu, L., Huang, W., Wang, M., Zhang, L., Shao, Y., Wang, M., Zhang, F., Zhao, Z., Mei, X., Li, T., Wang, D., Liang, Y., Li, J., Xu, T., Zhao, Y., Zhong, Y., Chen, Q., & Lu, B. (2020). Phenolic acid profiles of common food and estimated natural intake with different structures and forms in five regions of China. *Food Chemistry, 321*, 126675. https://doi.org/10.1016/j.foodchem.2020.126675

Speer, H., D'Cunha, N. M., Alexopoulos, N. I., McKune, A. J., & Naumovski, N. (2020). Anthocyanins and human health-a focus on oxidative stress, inflammation and disease. *Antioxidants, 9*. https://doi.org/10.3390/antiox9050366

Su, Q., Rowley, K., Itsiopoulos, C., & O'Dea, K. (2002). Identification and quantitation of major carotenoids in selected components of the Mediterranean diet: Green leafy vegetables, figs and olive oil. *European Journal of Clinical Nutrition, 56*, 1149–1154. https://doi.org/10.1038/sj.ejcn.1601472

Sukontaprapun, B., Charoenkiatkul, S., Thiyajai, P., Sukprasansap, M., Saetang, P., & Judprasong, K. (2019). Key organic acids in indigenous plants in Thailand. *American Journal of Plant Sciences, 10*, 1855–1870. https://doi.org/10.4236/ajps.2019.1010131

Takahashi, T., Okiura, A., Saito, K., & Kohno, M. (2014). Identification of Phenylpropanoids in fig (*Ficus carica* L.) leaves. *Journal of Agricultural and Food Chemistry, 62*, 10076–10083. https://doi.org/10.1021/jf5025938

Takahashi, T., Okiura, A., & Kohno, M. (2017). Phenylpropanoid composition in fig (*Ficus carica* L.) leaves. *Journal of Natural Medicines, 71*, 770–775. https://doi.org/10.1007/s11418-017-1093-6

Teixeira, D. M., Patão, R. F., Coelho, A. V., & da Costa, C. T. (2006). Comparison between sample disruption methods and solid-liquid extraction (SLE) to extract phenolic compounds from Ficus carica leaves. *Journal of Chromatography A, 1103*, 22–28. https://doi.org/10.1016/j.chroma.2005.11.047

Teixeira, D. M., Canelas, V. C., do Canto, A. M., Teixeira, J. M. G., & Dias, C. B. (2009). HPLC-DAD quantification of phenolic compounds contributing to the antioxidant activity of *Maclura pomifera, Ficus carica* and *Ficus elastica* extracts. *Analytical Letters, 42*, 2986–3003. https://doi.org/10.1080/00032710903276646

Teruel-Andreu, C., Andreu-Coll, L., López-Lluch, D., Sendra, E., Hernández, F., & Cano-Lamadrid, M. (2021). *Ficus carica* fruits, by-products and based products as potential sources of bioactive compounds: A review. *Agronomy, 11*, 1834. https://doi.org/10.3390/agronomy11091834

Trad, M., Ginies, C., Gaaliche, B., Renard, C. M. G. C., & Mars, M. (2012). Does pollination affect aroma development in ripened fig [*Ficus carica* L.] fruit? *Scientia Horticulturae, 134*, 93–99. https://doi.org/10.1016/j.scienta.2011.11.004

Trad, M., le Bourvellec, C., Gaaliche, B., Renard, C. M. G. C., & Mars, M. (2014). Nutritional compounds in figs from the southern Mediterranean region. *International Journal of Food Properties, 17*, 491–499. https://doi.org/10.1080/10942912.2011.642447

Tsanova-Savova, S., Ribarova, F., & Gerova, M. (2005). (+)-Catechin and (−)-epicatechin in Bulgarian fruits. *Journal of Food Composition and Analysis, 18*, 691–698. https://doi.org/10.1016/j.jfca.2004.06.008

Türkomp. (2021). *Fig, Bursa Siyahi*. http://www.turkomp.gov.tr/food-incir-bursa-siyahi-367

Ullah, H., & Khan, H. (2020). Epigenetic drug development for autoimmune and inflammatory diseases. In *Histone modifications in therapy* (pp. 395–413). Elsevier.

Uribe, N. G., García-Galbis, M. R., & Espinosa, R. M. M. (2017). New advances about the effect of vitamins on human health: Vitamins supplements and nutritional aspects. In *Functional food-improve health through adequate food* (pp. 57–75).

Uysal, S., Zengin, G., Aktumsek, A., & Karatas, S. (2015). Fatty acid composition, Total sugar content and anti-diabetic activity of methanol and water extracts of nine different fruit tree leaves collected from Mediterranean region of Türkiye. *International Journal of Food Properties, 18*, 2268–2276. https://doi.org/10.1080/10942912.2014.973964

Vallejo, F., Marín, J. G., & Tomás-Barberán, F. A. (2012). Phenolic compound content of fresh and dried figs (*Ficus carica* L.). *Food Chemistry, 130*, 485–492. https://doi.org/10.1016/j.foodchem.2011.07.032

Veberic, R., & Mikulic-Petkovsek, M. (2016). Phytochemical composition of common fig (*Ficus carica* L.) cultivars. In M. S. J. Simmonds & V. R. Preedy (Eds.), *Nutritional composition of fruit cultivars* (pp. 235–255). Elsevier.

Veberic, R., Colaric, M., & Stampar, F. (2008). Phenolic acids and flavonoids of fig fruit (Ficus carica L.) in the northern Mediterranean region. *Food Chemistry, 106*, 153–157. https://doi.org/10.1016/j.foodchem.2007.05.061

Viuda-Martos, M., Barber, X., Pérez-Álvarez, J. A., & Fernández-López, J. (2015). Assessment of chemical, physico-chemical, techno-functional and antioxidant properties of fig (*Ficus carica* L.) powder co-products. *Industrial Crops and Products, 69*, 472–479. https://doi.org/10.1016/j.indcrop.2015.03.005

Wang, T., Jiao, J., Gai, Q.-Y., Wang, P., Guo, N., Niu, L.-L., & Fu, Y.-J. (2017). Enhanced and green extraction polyphenols and furanocoumarins from Fig (*Ficus carica* L.) leaves using deep eutectic solvents. *Journal of Pharmaceutical and Biomedical Analysis, 145*, 339–345. https://doi.org/10.1016/j.jpba.2017.07.002

White, B. (2009). Dietary fatty acids. *American Family Physician, 80*, 345–350.

Wojdyło, A., Nowicka, P., Carbonell-Barrachina, Á. A., & Hernández, F. (2016). Phenolic compounds, antioxidant and antidiabetic activity of different cultivars of *Ficus carica* L. fruits. *Journal of Functional Foods, 25*, 421–432. https://doi.org/10.1016/j.jff.2016.06.015

Yang, J., & Rainville, P. D. (2020). *Analysis of organic acids using a mixed- mode LC column and an ACQUITY QDa mass detector*. Application Note. https://www.waters.com/webassets/cms/library/docs/720006938en.pdf

Yao, L., Mo, Y., Chen, D., Feng, T., Song, S., Wang, H., & Sun, M. (2021). Characterization of key aroma compounds in Xinjiang dried figs (*Ficus carica* L.) by GC-MS, GC-olfactometry, odor activity values, and sensory analyses. *LWT, 150*, 111982. https://doi.org/10.1016/j.lwt.2021.111982

Yemiş, O., Bakkalbaşı, E., & Artık, N. (2012). Changes in pigment profile and surface colour of fig (*Ficus carica* L.) during drying. *International Journal of Food Science & Technology, 47*, 1710–1719. https://doi.org/10.1111/j.1365-2621.2012.03025.x

Yu, L., Meng, Y., Wang, Z.-L., Cao, L., Liu, C., Gao, M.-Z., Zhao, C.-J., & Fu, Y.-J. (2020). Sustainable and efficient surfactant-based microwave-assisted extraction of target polyphenols and furanocoumarins from fig (*Ficus carica* L.) leaves. *Journal of Molecular Liquids, 318*, 114196. https://doi.org/10.1016/j.molliq.2020.114196

Zhang, X., Wang, X., Liu, L., Wang, W., Liu, Y., Deng, Q., Zhang, H., Wang, X., & Xia, H. (2020). Evaluation of the comparative morphological and quality in fig (Sichuan, China) with different colors under different ripening stages. *Scientia Horticulturae, 265*, 109256. https://doi.org/10.1016/j.scienta.2020.109256

Zhao, C., Li, S., Li, C., Wang, T., Tian, Y., & Li, X. (2021). Flavonoids from fig (Ficus carica Linn.) leaves: The development of a new extraction method and identification by UPLC-QTOF-MS/MS. *Applied Sciences, 11*, 7718. https://doi.org/10.3390/app11167718

Zidi, K., Kati, D. E., Bachir-bey, M., Genva, M., & Fauconnier, M.-L. (2021). Comparative study of fig volatile compounds using headspace solid-phase microextraction-gas chromatography/mass spectrometry: Effects of cultivars and ripening stages. *Frontiers in Plant Science, 12*, 667809. https://doi.org/10.3389/fpls.2021.667809

Chapter 22
Fig Volatiles

Mustafa Kiralan, Sündüz Sezer Kiralan, and Onur Ketenoglu

1 Introduction

Fig (*Ficus carica* L.) is a member of the botanical family Moraceae and is grown in many parts of the world with moderate climates. Fig has worldwide importance for dry and fresh consumption (Dueñas et al., 2008), and the consumers prefer the fresh form. According to the reported data by FAO in 2019 (FAO, 2021), the worldwide fig production is 1.315.588 tonnes. Turkey is a major producer of edible figs with 310,000 tonnes, followed by Egypt and Morocco (the production quantity of these countries are 225,295 tonnes and 153,472 tonnes, respectively).

The different parts of figs have many positive effects on human health, such as anti-inflammatory, antipyretic, purgative, aphrodisiac, anti-cancer, antidiabetic, hepatoprotective, and anti-angiogenic properties (Ghanbari et al., 2019; Patil & Patil, 2011; Rahmani & Aldebasi, 2017). Figs are an excellent source of nutrition due to their rich composition in minerals, vitamins, and dietary fiber (Badgujar et al., 2014; Crisosto et al., 2011). Carbohydrates constitute a large part (73.5%) of the dried figs, while protein content is reported to be 4.67%, and the dietary fiber content of dried figs is 3.68% (Soni et al., 2014). Fructose and glucose are the most common sugars in fig fruit (Slatnar et al., 2011). Çalişkan and Aytekin Polat (2011) emphasized that major sugars were fructose (~56%) and glucose (~43%) in seventy-six Turkish fig accessions. The highest values in fructose and glucose were 11.9 g/100 g and 9.7 g/100 g, respectively.

M. Kiralan (✉) · S. S. Kiralan
Department of Food Engineering, Balikesir University, Faculty of Engineering, Balikesir, Turkey

O. Ketenoglu
Faculty of Agriculture, Department of Food Engineering, Eskisehir Osmangazi University, Eskisehir, Turkey

M. F. Ramadan (ed.), *Fig* (Ficus carica)*: Production, Processing, and Properties*, https://doi.org/10.1007/978-3-031-16493-4_22

Besides these nutritional compounds, fig; contains different bioactive compounds such as phenolic compounds, phytosterols, volatiles, organic acids, hydrocarbons, and aliphatic alcohols. The presence and quantity of these phytochemicals depend on the different parts of the fig. Most of these compounds are present in latex, followed by leaves, fruit, and root (Badgujar et al., 2014; Mawa et al., 2013). The content of total phenolics, total anthocyanins, and total flavonoids in Tunisian fig ecotypes varied from 51.5 to 100 mg gallic acid equivalents (GAE) on a fresh weight (FW) basis, 2.57 to 11.6 mg of cyanidin-3-glucoside equivalents (CGE) per g FW, 0.33 to 17.59 mg of quercetin equivalent in mg per g FW, respectively (Aljane et al., 2020).

Among phenolic acids, chlorogenic acid was the major phenolic acid in three fig cultivars grown in Slovenia, followed by gallic acid (Veberic et al., 2008). As for organic acids, malic acid was the main compound in fig samples, representing 24.7–58.7% of the total organic acids content (Slatnar et al., 2011). Cyanidin-3-glucoside, cyanidin-3,5-diglucoside, cyanidin-3-rutinoside (major anthocyanin), and pelargonidin-3-glucoside were identified in three fresh common fig cultivars grown in Turkey, additionally; lutein, zeaxanthin, β-cryptoxanthin, and β-carotene were the identified carotenoids in yellow fig cultivars grown in Turkey (Yemiş et al., 2012).

Latex is a part of fresh fig, consumed in small amounts with fresh fig. Phytosterols, especially β-sitosterol and lupeol, are the major sterols present in the latex of fresh figs. Cysteine and tyrosine are present as the major amino acids among 13 amino acids identified in *F. carica* latex. Besides, *F. carica* latex was reported to contain saturated fatty acids such as palmitic, arachidic, and behenic acids (Oliveira, et al., 2010a).

Alongside nutrients and functional compounds, volatile compounds are important constituents that contribute to the unique aroma of fig. This section will discuss the chemical composition of volatiles, extraction methods, and usage of fig volatiles in different industries.

2 Chemical Composition of Volatiles

Volatile compounds contribute to the characteristic aroma and flavor of fruits. Numerous chemical compounds such as esters, alcohols, aldehydes, ketones, lactones, terpenoids, and apocarotenoids constitute the aroma profile of fruits (El Hadi et al., 2013). Volatile compounds have been used to determine fruit quality. Especially, individual volatile compounds could be monitored and evaluated in fruits (Plutowska & Wardencki, 2007). Volatile compounds are also an important part of fig composition. The chemical composition of fig volatiles changes with climatological conditions, cultivar, maturity, processing, and storage conditions (Grison-Pigé et al., 2002; Mujić et al., 2012, 2014).

Gozlekci et al. (2011) reported that aldehydes and terpenes were the major chemical groups responsible for the aroma of peels and pulps in fig fruits from four

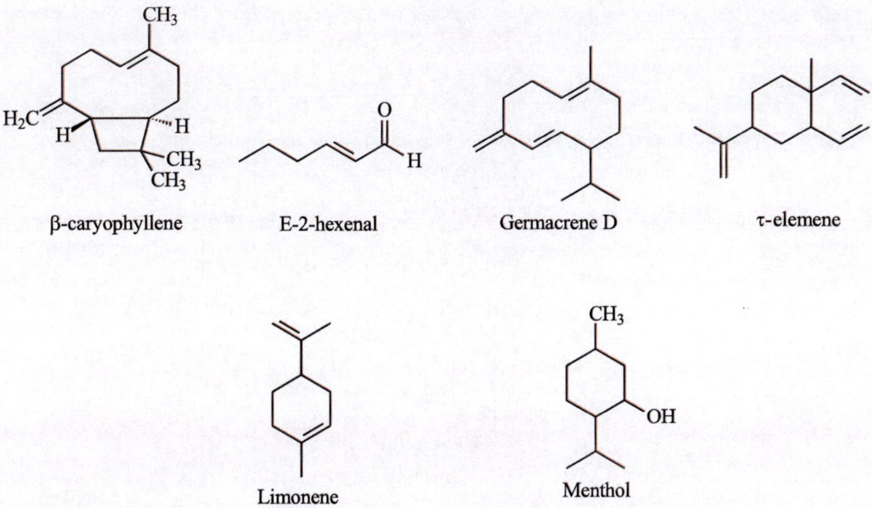

Fig. 22.1 Chemical structures of important volatiles in different parts of figs

commercial fig cultivars (Bursa Siyahi, Karabakunya, Sari Lop, and Sultan Selim) growing in Turkey. According to the researchers, total terpenes were higher in the peel than in the pulp. β-Caryophyllene was the major terpene in all samples, as also found in high amounts in the peels of figs (up to 48.65% of total volatiles). E-2-Hexenal was found as the major aldehyde, and its level increased up to 20.0%.

Oliveira et al. (2010c) identified 59 volatile and semi-volatile compounds in the leaves and fruits (peels and pulps) from Portuguese fig cultivars. The number of identified compounds was found to be higher in leaves (40 different compounds) than in pulps (30) and peels (27). Aldehydes and monoterpenes were abundantly present in pulp and peels, while these compounds were present at lower leaf concentrations. Sesquiterpenes were the major class of compounds in the leaves. Germacrene D, β-caryophyllene, and τ-elemene (Fig. 22.1) were the predominant sesquiterpenes in the leaves of all cultivars. On the other hand, monoterpenes were found at higher concentrations in pulps and peels of figs. The major monoterpenes present in pulps and peels were limonene and menthol.

3 Extraction Techniques

Different extraction techniques influence volatile compounds in figs. However, most conventional methods require large volumes of organic solvents and result in significant waste, requiring longer extraction time and trace levels of various compounds. From these aspects, solid-phase microextraction (SPME) is a solvent-free, cost-effective, rapid, sensitive technique (Jalili et al., 2020; Kumar et al., 2008).

Table 22.1 Composition of fig volatiles depends on different parts of the fruit and different extraction techniques

Part of figs	Extraction method	Major volatiles	Reference
Leaves	Hydrodistillation	(Z)-3-Hexenyl benzoate (19.8%), n-Tetracosane (11.6%), n-Hexadecanoic acid (9.2%), (E)-2-Hexenal (7.2%)	Ayoub et al. (2010)
Leaves	Hydrodistillation	Ficusin (32.8%), Caryophyllene oxide (9.0%), Bergapten (6.1%)	Nafis et al. (2019)
Fruits Leaves	Hydrodistillation	**Fruits** Furfural (10.55%), 5-methyl-2-furaldehyde (10.1%), and benzeneacetaldehyde (6.59%) **Leaves** Psoralen (10.12%), β-damascenone (10.17%), benzyl alcohol (4.56%), behenic acid (4.79%), and bergapten (1.99%)	Li et al. (2012)
Fruits Leaves	Hydrodistillation	**Fruits** Cedrol (43.8%), α-terpinyl acetate (22.5%), manoyl oxide (12.9%), α-pinene (9.3%), abietadiene (8.1%),trans-calamenene (3.9%) and nheneicosane (3.5%) **Leaves** Cedrol (38.9%), manoyl oxide (24.8%), α-terpineol acetate (21.7%), abietatriene (11.8%), γ-muurolene (7.4%), α-pinene (6.1%), pentadecanal (5.2%) and nonadecanal (2.3%).	Soltana et al. (2017)
Latex	Headspace Solid-Phase Microextraction (HS-SPME)	**(1) sesquiterpenes** (*ca.* 91% of identified compounds) germacrene D cadinene τ-muurolene **(2) alcohols** (ca. 4%) 1-hexanol 1-heptanol **(3) Ketones** represented the minor components (<0.1%) 6-Methyl-5-hepten-2-one	Oliveira, et al. (2010b)
Pulp	Headspace Solid-Phase Microextraction (HS-SPME)	β-Caryophyllene (15.9–37.4%), (E)-2-Hexanal (11.0–20.0%), 2-Furancarboxaldehyde, 5(hydroxymethyl), (5.96–13.5%), benzaldehyde (5.05–8.45%), furfural (0.30–3.19%)	Gozlekci et al. (2011)
Peel	Headspace Solid-Phase Microextraction (HS-SPME)	β-Caryophyllene (21.1–48.6%), 2-Furancarboxaldehyde, 5(hydroxymethyl) (not detected-4.56%), furfural (2.56–3.65%), (E)-2-Hexanal (not detected-1.77%), benzaldehyde (not detected-0.24%)	Gozlekci et al. (2011)

(continued)

Table 22.1 (continued)

Part of figs	Extraction method	Major volatiles		Reference
Whole	Headspace Solid-Phase Microextraction (HS-SPME)	**(1) Ester** ethyl acetate (0.67–65.26%) **(2) Furans** 5-Hydroxymethylfurfural (0–33.38%) **(3) Pyranones** 3-Hydroxy-2,3-dihydromaltol (0.36–25.06%) **(4) Aldehydes** Benzaldehyde (1.45–32.44%)		Pereira et al. (2020)
Flesh	Headspace Solid-Phase Microextraction (HS-SPME)	<u>Cuello Dama Negro</u> Hydrocarbons (2.10%) Hexane (0.90%) Aldehydes (12.15%) 3-Methyl butanal (0.06%) Hexanal (7.64%) Esters (1.14%) Ethyl acetate (0.59%) **Pyranones** (70.88%) 3-Hydroxy-2,3-dihydromaltol (54.24%)-Maltol (0.52%)	<u>San Antonio</u> Hydrocarbons (17.92%) Hexane (12.16%) Aldehydes (17.88%) 3-Methyl butanal (11.18%) Hexanal (1.36%) **Esters (43.96%)** Ethyl acetate (43.15%) Pyranones (2.08%) 3-Hydroxy-2,3-dihydromaltol (0.02%) Maltol (2.01%)	Villalobos et al. (2018)

Both traditional and modern extraction techniques have been used to evaluate the chemical composition of figs. Table 22.1 summarizes the volatile compounds depending on the part of the fruit and extraction technique. For example, Li et al. (2012) emphasized that furfural and 5-methyl-2-furaldehyde were the major volatiles in the fruit part, while psoralen and β-damascenone were predominant volatiles in the leaves of figs. In another study in which hydrodistillation was used, the fruits were rich in cedrol (43.8%), α-terpinyl acetate (22.5%), and manoyl oxide (12.9%); likewise, cedrol (38.9%), manoyl oxide (24.8%), α-terpineol acetate (21.7%) were found as major compounds in the leaves of figs (Soltana et al., 2017).

Headspace solid-phase microextraction (HS-SPME) significantly improves the determination sensitivity of volatile compounds in figs. Gozlekci et al. (2011) stated that a bicyclic sesquiterpene, namely β-caryophyllene, was present in high percentages of pulp and peel parts of figs. (15.9–37.4%, 21.1–48.6%, respectively). β-Caryophyllene is reported to contribute a unique aroma to the essential oils of most plants (Scandiffio et al., 2020). The wooden odor is a characteristic of β-caryophyllene, making this volatile preferable in cosmetic and food additives. Besides, this substance is approved as a flavoring agent by official authorities such as the Food and Drug Administration (FDA) and the European Food Safety Authority (EFSA). β-caryophyllene is also reported to have such biological activities as anti-inflammatory, anticarcinogenic, antimicrobial, antioxidant, antihyperglycemic effects, and analgesic activities (Fidyt et al., 2016; Hashiesh et al., 2020; Moo et al. 2020; Scandiffio et al., 2020).

Villalobos et al. (2018) revealed that esters were the most abundant chemical group in the flesh of figs from the San Antonio cultivar, while the major chemical group was pyranones in the Cuello Dama Negro cultivar using the HS-SPME method. The most abundant compound was 3-hydroxy-2,3-dihydromaltol, at 54.24% of the total area in the Cuello Dama Negro cultivar, while ethyl acetate (43.15%) was the predominant volatile compound present in the San Antonio cultivar.

Latex is widely distributed and produced from the fig tree and consists of secondary metabolites such as terpenoids, phenolics, and essential fatty acids (Lazreg Aref et al., 2011). Oliveira et al. (2010b) used the HS-SPME method to identify volatiles in fig latex. The major chemical group was sesquiterpenes which comprised 91% of the identified compounds. According to these authors, germacrene D, cadinene, and τ-muurolene were the major sesquiterpenes.

4 Factors Affecting Fig Volatiles

Different factors such as processing conditions influence the volatiles of figs. Mujić et al. (2014) examined the effect of pre-treatments such as sulfur dioxide, citric acid, and ascorbic acid on the volatiles of figs before drying. Benzaldehyde was found as the major volatile in all treated and untreated dried figs (56–82.7 mg/kg), followed by hexanal (7.20–55.7 mg/kg). Although the pre-treatments provided some advantages, such as decreasing the drying time, they caused a decrease in the aldehydes part of the aroma components. On the other hand, ascorbic acid application preserved esters which are responsible for the desired aroma in figs from Bružetka Bijela variety, whereas, in the case of Zimnica, sulfur dioxide showed a protective effect on esters compared to ascorbic acid (Mujić et al., 2014).

Russo et al. (2017) evaluated the effect of drying methods on the volatiles of figs. Aldehydes were found as the major chemical group in sun-dried (approximately 70% of the total volatiles) and oven-treated samples after sun drying (approximately 80% of the total volatiles), especially hexanal and benzaldehyde were the most abundant volatiles. A significant increase was observed in the samples treated with the oven after sun-drying, while a decrease was observed in alcohols and esters content. Regarding individual major compounds, a significant decrease in hexanal content and a contemporary increase in benzaldehyde were recorded in the samples treated with oven drying after sun drying.

Storage is also an important effect on the volatiles of figs. Villalobos et al. (2018) studied the effect of modified atmosphere packaging (MAP) on the volatiles of two fig cultivars 'Cuello Dama Negro' (CDN) and 'San Antonio' (SA) during post-harvest storage. CDN exhibited slight changes in the volatiles of the control and modified atmosphere-packaged samples during cold storage at 0 °C. However, the volatiles of cultivar SA was affected by cold storage and packaging. An increase was observed in esters, pyranones, and aldehydes during the cold storage period of control samples. In addition, the ethyl acetate, 3-methyl butanal, and benzaldehyde were largely influenced by the cold storage. In their study, the microperforated M50 film for the cultivar SA preserved volatile compounds of figs without negatively affecting the flavor of the figs.

5 Biological Activities of Fig Volatiles

Biological compounds in foods possess different functions, such as antioxidant and antimicrobial effects. Because essential oils are rich in such compounds, they play an important role in protecting human health.

Nafis et al. (2019) evaluated essential oil's antioxidant and antimicrobial activities from fig leaves. The results from three antioxidant tests (DPPH· radical scavenging, β-carotene acid bleaching test, and reducing power assay) showed that essential oil had moderate antioxidant activity (IC_{50} values from DPPH·, reducing power, and β-carotene/linoleic acid assay were 1.45 mg/mL, 1.92 mg/mL and 1.56 mg/mL, respectively). Like antioxidant activity, the essential oil of figs also exhibited moderate antimicrobial activity. Yeast strains and Gram-positive bacteria were generally more sensitive to the essential oil when compared to Gram-negative bacteria.

6 Conclusion

Fig (*Ficus carica* L.) is an important fruit grown in many countries with moderate climates. Such countries as Turkey, Morocco, and Egypt are the major producers of fig fruit. Since the fruit contains many bioactive components having different activities such as antimicrobial, antioxidant, anti-inflammatory, anticarcinogenic, and antidiabetic, the consumption of the fresh fig plays an important role in preserving human health. In addition to the fresh fruit, the essential oil obtained from figs is also a rich source of many bioactive components. Several studies revealed that fig essential oil is a mixture of different chemical components, making the essential oil a complex extract with numerous health-promoting impacts. The composition of the fig essential oil depends on the part of the fruit and the extraction method used in essential fig oil production. The ongoing studies will enlighten different bioactive effects of figs, and consecutively the worldwide production and the consumption of this beneficial fruit will increase.

References

Aljane, F., Neily, M. H., & Msaddak, A. (2020). Phytochemical characteristics and antioxidant activity of several fig (*Ficus carica* L.) ecotypes. *Italian Journal of Food Science, 32*(4), 755–768. https://doi.org/10.14674/IJFS.1884

Ayoub, N., Singab, A. N., Mostafa, N., & Schultze, W. (2010). Volatile constituents of leaves of *Ficus carica* Linn. Grown in Egypt. *Journal of Essential Oil Bearing Plants, 13*(3), 316–321. https://doi.org/10.1080/0972060X.2010.10643827

Badgujar, S. B., Patel, V. V., Bandivdekar, A. H., & Mahajan, R. T. (2014). Traditional uses, phytochemistry and pharmacology of *Ficus carica*: A review. *Pharmaceutical Biology, 52*(11), 1487–1503. https://doi.org/10.3109/13880209.2014.892515

Çalişkan, O., & Aytekin Polat, A. (2011). Phytochemical and antioxidant properties of selected fig (*Ficus carica* L.) accessions from the Eastern Mediterranean Region of Turkey. *Scientia Horticulturae, 128*(4), 473–478. https://doi.org/10.1016/j.scienta.2011.02.023

Crisosto, H., Ferguson, L., Bremer, V., Stover, E., & Colelli, G. (2011). 7 – Fig (Ficus carica L.). In E. M. Yahia (Ed.), *Postharvest biology and technology of tropical and subtropical fruits* (pp. 134–160e). Woodhead Publishing.

Dueñas, M., Pérez-Alonso, J. J., Santos-Buelga, C., & Escribano-Bailón, T. (2008). Anthocyanin composition in fig (*Ficus carica* L.). *Journal of Food Composition and Analysis, 21*(2), 107–115. https://doi.org/10.1016/j.jfca.2007.09.002

El Hadi, M. A., Zhang, F. J., Wu, F. F., Zhou, C. H., & Tao, J. (2013). Advances in fruit aroma volatile research. *Molecules, 18*(7), 8200–8229. https://doi.org/10.3390/molecules18078200

FAO. (2021). *Food and Agriculture Organization of the United Nations*. FAOSTAT Data: Crops and livestock products. https://www.fao.org/faostat/en/#data/. Accessed on 23 Sept 2021.

Fidyt, K., Fiedorowicz, A., Strządała, L., & Szumny, A. (2016). β-Caryophyllene and beta-caryophyllene oxide-natural compounds of anticancer and analgesic properties. *Cancer Medicine, 5*(10), 3007–3017. https://doi.org/10.1002/cam4.816

Ghanbari, A., Le Gresley, A., Naughton, D., Kuhnert, N., Sirbu, D., & Ashrafi, G. H. (2019). Biological activities of *Ficus carica* latex for potential therapeutics in Human Papillomavirus (HPV) related cervical cancers. *Scientific Reports, 9*(1), 1013. https://doi.org/10.1038/s41598-018-37665-6

Gozlekci, S., Kafkas, E., & Ercisli, S. (2011). Volatile compounds determined by HS/GC-MS technique in Peel and pulp of fig (*Ficus carica* L.) cultivars grown in Mediterranean region of Turkey. *Notulae Botanicae Horti Agrobotanici Cluj-Napoca, 39*(2). https://doi.org/10.15835/nbha3926261

Grison-Pigé, L., Hossaert-McKey, M., Greeff, J. M., & Bessière, J. M. (2002). Fig volatile compounds-a first comparative study. *Phytochemistry, 61*(1), 61–71. https://doi.org/10.1016/s0031-9422(02)00213-3

Hashiesh, H. M., Meeran, M. F. N., Sharma, C., Sadek, B., Kaabi, J. A., & Ojha, S. K. (2020). Therapeutic potential of β-Caryophyllene: A dietary cannabinoid in diabetes and associated complications. *Nutrients, 12*(10), 2963. https://doi.org/10.3390/nu12102963

Jalili, V., Barkhordari, A., & Ghiasvand, A. (2020). A comprehensive look at solid-phase microextraction technique: A review of reviews. *Microchemical Journal, 152*, 104319. https://doi.org/10.1016/j.microc.2019.104319

Kumar, A., Gaurav, N., Malik, A. K., Tewary, D. K., & Singh, B. (2008). A review on development of solid phase microextraction fibers by sol-gel methods and their applications. *Analytica Chimica Acta, 610*(1), 1–14. https://doi.org/10.1016/j.aca.2008.01.028

Lazreg Aref, H., Gaaliche, B., Fekih, A., Mars, M., Aouni, M., Pierre Chaumon, J., & Said, K. (2011). In vitro cytotoxic and antiviral activities of *Ficus carica* latex extracts. *Natural Product Research, 25*(3), 310–319. https://doi.org/10.1080/14786419.2010.528758

Li, J., Tian, Y.-z., Sun, B.-y., Yang, D., Chen, J.-p., & Men, Q.-m. (2012). Analysis on volatile constituents in leaves and fruits of *Ficus carica* by GC-MS. *Chinese Herbal Medicines, 4*(1), 63–69. https://doi.org/10.3969/j.issn.1674-6384.2012.01.010

Mawa, S., Husain, K., & Jantan, I. (2013). *Ficus carica* L. (Moraceae): Phytochemistry, traditional uses and biological activities. *Evidence-based Complementary and Alternative Medicine*. https://doi.org/10.1155/2013/974256

Moo, C. L., Yang, S. K., Osman, M. A., Yuswan, M. H., Loh, J. Y., Lim, W. M., et al. (2020). Antibacterial activity and mode of action of beta-caryophyllene on *Bacillus cereus*. *Polish Journal of Microbiology, 69*(1), 49–54. https://doi.org/10.33073/pjm-2020-007

Mujić, I., Bavcon Kralj, M., Jokić, S., Jarni, K., Jug, T., & Prgomet, Ž. (2012). Changes in aromatic profile of fresh and dried fig-the role of pre-treatments in drying process. *International Journal of Food Science & Technology, 47*(11), 2282–2288. https://doi.org/10.1111/j.1365-2621.2012.03099.x

Mujić, I., Bavcon Kralj, M., Jokić, S., Jug, T., Šubarić, D., Vidović, S., & Jarni, K. (2014). Characterisation of volatiles in dried white varieties figs (*Ficus carica* L.). *Journal of Food Science and Technology, 51*(9), 1837–1846. https://doi.org/10.1007/s13197-012-0740-x

Nafis, A., Kasrati, A., Jamali, C. A., Samri, S. E., Mezrioui, N., Abbad, A., & Hassani, L. (2019). Antioxidative effect and first evidence of synergistic antimicrobial effects of *Ficus carica* (L.) leaf essential oil with conventional antibiotics. *Journal of Essential Oil Bearing Plants, 22*(5), 1289–1298. https://doi.org/10.1080/0972060X.2019.1684386

Oliveira, A. P., Silva, L. R., Andrade, P. B., Valentão, P., Silva, B. M., Gonçalves, R. F., & Guedes de Pinho, P. (2010a). Further insight into the latex metabolite profile of *Ficus carica*. *Journal of Agricultural and Food Chemistry, 58*(20), 10855–10863. https://doi.org/10.1021/jf1031185

Oliveira, A. P., Silva, L. R., Ferreres, F., Guedes de Pinho, P., Valentão, P., Silva, B. M., et al. (2010b). Chemical assessment and *in vitro* antioxidant capacity of *Ficus carica* latex. *Journal of Agricultural and Food Chemistry, 58*(6), 3393–3398. https://doi.org/10.1021/jf9039759

Oliveira, A. P., Silva, L. R., Guedes de Pinho, P., Gil-Izquierdo, A., Valentão, P., Silva, B. M., et al. (2010c). Volatile profiling of *Ficus carica* varieties by HS-SPME and GC–IT-MS. *Food Chemistry, 123*(2), 548–557. https://doi.org/10.1016/j.foodchem.2010.04.064

Patil, V. V., & Patil, V. R. (2011). Ficus carica Linn. – An overview. *Research Journal of Medicinal Plants, 5*(3), 246–253. https://doi.org/10.3923/rjmp.2011.246.253

Pereira, C., Martín, A., López-Corrales, M., Córdoba, M. D. G., Galván, A. I., & Serradilla, M. J. (2020). Evaluation of the physicochemical and sensory characteristics of different fig cultivars for the fresh fruit market. *Food, 9*(5), 619. https://doi.org/10.3390/foods9050619

Plutowska, B., & Wardencki, W. (2007). Aromagrams-aromatic profiles in the appreciation of food quality. *Food Chemistry, 101*(2), 845–872. https://doi.org/10.1016/j.foodchem.2005.12.028

Rahmani, A. H., & Aldebasi, Y. H. (2017). Ficus carica and its constituents role in management of diseases. *Asian Journal of Pharmaceutical and Clinical Research, 10*(6), 49–53. https://doi.org/10.22159/ajpcr.2017.v10i6.17832

Russo, F., Caporaso, N., Paduano, A., & Sacchi, R. (2017). Characterisation of volatile compounds in Cilento (Italy) figs (*Ficus carica* L.) cv. Dottato as affected by the drying process. *International Journal of Food Properties, 20*(sup2), 1366–1376. https://doi.org/10.1080/10942912.2017.1344991

Scandiffio, R., Geddo, F., Cottone, E., Querio, G., Antoniotti, S., Gallo, M. P., et al. (2020). Protective effects of (E)-β-Caryophyllene (BCP) in chronic inflammation. *Nutrients, 12*(11), 3273. https://doi.org/10.3390/nu12113273

Slatnar, A., Klancar, U., Stampar, F., & Veberic, R. (2011). Effect of drying of figs (*Ficus carica* L.) on the contents of sugars, organic acids, and phenolic compounds. *Journal of Agricultural and Food Chemistry, 59*(21), 11696–11702. https://doi.org/10.1021/jf202707y

Soltana, H., Flamini, G., & Hammami, M. (2017). Volatile compounds from six varieties of Ficus carica from Tunisia. *Records of Natural Products, 2*, 157–165.

Soni, N., Mehta, S., Satpathy, G., & Gupta, R. (2014). Estimation of nutritional, phytochemical, antioxidant and antibacterial activity of dried fig (*Ficus carica*). *Journal of Pharmacognosy and Phytochemistry, 3*, 158–165.

Veberic, R., Colaric, M., & Stampar, F. (2008). Phenolic acids and flavonoids of fig fruit (*Ficus carica* L.) in the northern Mediterranean region. *Food Chemistry, 106*(1), 153–157. https://doi.org/10.1016/j.foodchem.2007.05.061

Villalobos, M. C., Serradilla, M. J., Martín, A., Aranda, E., López-Corrales, M., & Córdoba, M. G. (2018). Influence of modified atmosphere packaging (MAP) on aroma quality of figs (*Ficus carica* L.). *Postharvest Biology and Technology, 136*, 145–151. https://doi.org/10.1016/j.postharvbio.2017.11.001

Yemiş, O., Bakkalbaşı, E., & Artık, N. (2012). Changes in pigment profile and surface colour of fig (*Ficus carica* L.) during drying. *International Journal of Food Science & Technology, 47*(8), 1710–1719. https://doi.org/10.1111/j.1365-2621.2012.03025.x

Chapter 23
Fig Enzymes: Characterization, Biological Roles, and Applications

Hesham A. El Enshasy, Bassam Abomoelak, Roshanida A. Rahman, Ong Mei Leng, Dalia Sukmawati, and Zaitul Iffa Rasid

1 Introduction

Fig (*Ficus carica*) is an important source of many functional nutrients supporting human health. Therefore, it was considered a superfood/nutrient-dense and energy-dense food (Trad et al., 2014; Caliskan, 2015; Khadhraoui et al., 2019). This is based on its high contents of sugar, fiber, organic acids, micronutrients, and other biofunctional molecules such as polysaccharides, phenolics, flavonoids, volatile compounds, and many other molecules which play a significant role in fig taste, fragrance, and biological activities (Gibernau et al., 1997; Oliveira et al., 2009; Chawia et al., 2012; Yang et al., 2015; Wang et al., 2017; Arvaniti et al., 2019; Zhao et al., 2021). Different studies have reported on the bioactivities of fig fruits and leave with potential applications as antioxidant, hypoglycemic, antidiabetic, anti-inflammatory, anthelmintic, antibacterial, antifungal, nematocidal, and antiviral (Solomon et al., 2006; Liu et al., 2011; Hosainzadegan et al., 2012; Debin et al., 2018; Yousef et al., 2019; Sieniawska et al., 2022). In addition, fruit extract, leave

H. A. El Enshasy (✉) · R. A. Rahman · Z. I. Rasid
Institute of Bioproduct Development, Universiti Teknologi Malaysia, Skudai, Malaysia
e-mail: henshasy@ibd.utm.my; r-anida@utm.my; zaituliffa@ibd.utm.my

B. Abomoelak
Arnold Palmer Hospital Pediatric Specialty Diagnostic Laboratory, Orlando, FL, USA
e-mail: Bassam.Abomoelak@orlandohealth.com

O. M. Leng
Harita Go Green Sdn Bhd, Ulu Tiram, Malaysia

D. Sukmawati
Biology Department, Faculty of Mathematics and Natural Sciences, Universitas Negeri Jakarta, Jakarta, Indonesia
e-mail: Dalia-Sukmawati@unj.ac.id

M. F. Ramadan (ed.), *Fig* (Ficus carica)*: Production, Processing, and Properties*, https://doi.org/10.1007/978-3-031-16493-4_23

extracts, and latex exhibited inhibitory effects on the proliferation of cancer cells in *in vitro* study using HeLa, stomach, and esophagus cancer cell lines (Hashemi et al., 2011; Khodarahami et al., 2011; Amessis-Ouchemoukh et al., 2017; Rahmani and Aldebasi, 2017; Mopuri et al., 2018). Besides the potential medical applications, fig leave extract has also been used in food industries as a preservative besides the potential medical applications. For example, recent research reported on the potential use of fig leaves extract alone or in combination with olive leaves extract as a natural preservative of pasteurized buffalo milk which can extend the shelf life without altering the milk property (Abdel-Aziz et al., 2020).

In addition to a large number of research deals with the organic and inorganic contents of the fruits, other studies reported on the presence of a wide range of enzymes that play a regulatory and functional role during the development and ripening of many fruits. The overall physiochemical and metabolic profiles of fruits are highly affected by many environmental factors which exhibit a strong influence on the performance of the enzyme system and thus affecting the fruit ripening process (Jain et al., 2001; Abu-Goukh et al., 2010; Bae et al., 2014; Gill et al., 2017; López-Huertas & Palma, 2020). In fig, different enzyme systems have been reported, including amylases, proteases, invertases, and lipase (Lazreg-Aref et al., 2018). The nutritional content and bioactive compound profile of fig fruit are governed by the activities of the enzyme system in each stage of development, which also determines the carbohydrate, protein, and organic acid profile (Sedaghat & Rahemi, 2018; Cui et al., 2019). Of different enzymes studied, ficin is considered the most attractive biocatalyst and one of the well-studied enzymes in fig. This is based on its unique characteristics and high potential applications in industrial, food, and pharmaceutical industries.

2 Fig Enzymes

Proteolytic enzymes were the first group reported in the latex of fig tree (*Ficus carica* L). The early studies were focused on the seasonal variation effect on the enzyme content of latex and initial kinetic characterization for the optimal pH of the enzyme (Robbins, 1935a, b). Since then, much research has been focused on the enzyme content of fig latex in different plant parts as an attractive natural resource for enzymes of diverse industrial applications.

Figs include different types of enzymes that play a regulatory role during fruit development, growth, and ripening. However, amylase (both of α- and β-types), protease, peroxidase, invertase, and lipases are the main reported enzymes (Lazreg-Aref et al., 2018; Sedaghat and Rahemi, 2018). In general, the chemical and enzymatic profile of the fig latex are varied and influenced mainly by the type of cultivar, harvesting time, and the maturity stage (Lazreg-Aref et al., 2017, 2018). The fig fruit ripening process is carried out as a function of enzyme-mediated cell wall hydrolysis, especially polysaccharides. Both α- and β-amylase activities increased significantly up to the ripening phase and decreased gradually once the fruit entered the senescence stage. This plays a regulatory role in the starch conversion to sugar during the ripening stage (Sedaghat and Rahemi, 2018).

On the other hand, it can also be assumed that the invertase enzyme could play a regulatory role in determining the developed fruit's metabolic profile and glucose/fructose ratio. Of different enzymes present in the fig latex, proteases exhibited the highest activity compared to amylase and lipase in almost all types of studied fig cultivars. It has also been reported that, for most of the fig cultivars studied, the highest latex content and maximal protease activity have been reported in fig fruits collected from the mid-maturity to the early ripe stage. However, some fig cultivars exhibited the highest protease activity at the fully ripe stage (Lazreg-Aref et al., 2017). On the other hand, lipase activity is almost absent at the unripe fruit stage in most cultivars, and the increase of lipase activity was parallel to the increase of protease activity during the receptivity period. The increase in lipase activity catalyzed the pathway of volatile compound production during the fruit maturity stage, which significantly attracted more pollinators to improve the quality of produced fig fruit (Gaaliche et al., 2011). Based on lipase activity, many volatile compounds considered fingerprints of fig ripening are produced from lipids and fatty acids. These include ketonic compounds and specific esters like isoamyl acetate and butyl acetate (Lazreg-Aref et al., 2018).

It is also interesting to note that recent research has been reported on the presence of enzyme inhibitors fractions in the fruit latex of *Ficus carica* L. subsp. *carica.* The composition of the isolated fraction was determined by LC-MS/MS and showed that it includes xanthone, organic acid, chromone derivative, fatty acid, and phenolic acid. Furthermore, this semi-purified fraction exhibited an inhibitory effect on α-amylase and α-glucosidase enzymes (Paşayeva et al., 2020). Thus, it could be concluded that these inhibitors could play a regulatory role in the enzymatic activities during the fruit ripening process.

In addition to the previous enzymes, more recently, three *Ficus carica* peroxidases isoenzymes were isolated from fruit latex and denoted as (FP1, FP2, and FP3). These enzymes have been isolated using CM-sepharose, DEAE-sepaharose, and Sephacryl S-200. The isoenzyme FP1 is monomeric with a molecular weight of 30,000 Dalton. Both FP1 and FP3 exhibited almost the same optimal pH and temperature of 5.5 and 40 °C, respectively. On the other hand, the optimal catalytic activities for PF2 were observed at pH 7.0 and a temperature of 30 °C.

Furthermore, the biocatalytic activities of FP1 and FP2 were found to be increased by 11% and 53% in the presence of a high calcium concentration of 10 mM, respectively. At the same time, the presence of calcium ions at this concentration decreases the activity of FP3 by 67%. On the other hand, applying copper ion at a concentration of 10 mM inhibits the activity of FP1 completely and increases the catalytic activities of FP2 and FP3 by 9% and 269%, respectively (Elsayed et al., 2018). Based on the biocatalytic properties of these enzymes in terms of temperature and pH range, they could have high potential industrial applications. However, the fig latex and its associated enzymes play a regulatory role in fruit maturity and ripening and help defend against different plant pathogens, including microbes and insects.

3 Ficin: The Most Important Enzyme of Fig Latex

Ficin (EC 3.4.22.3; CAS # 9001-33-6), also known as ficain, Debricin, or higueroxyl delabarre, is a nonspecific sulfhydryl protease that could be extracted from the latex sap of different parts of the plant such as unripe fruit, leaves, and stem. This enzyme is characterized by the presence of a single reactive cysteine (Cys) at its active site. The amino acid sequence in the enzyme active site is similar to papain. Ficin biocatalytic reaction involves three steps. First, the rapid development of loose substrate-enzyme complex; Second, acylation of the -SH group of the enzyme active site by the carbonyl group of the substrate; Third, the decomposition of the acylated enzyme complex to produce the enzyme and product. However, besides the well-known proteolytic activities of ficin, a recent study also showed this enzyme's intrinsic peroxidase-like activity. The enzyme showed the capacity to act as protease and peroxidase based on the presence of two active sites in each biocatalytic reaction and thus could catalyze more than one kind of substrate (Yang et al., 2017).

The first report of ficin isolation in the pure crystal structure is back in the late 1930s (Walti, 1938). This enzyme was initially reported as an enzymatic mixture made of several types of cysteine proteases. It can be found in the fig tree latex in different isoforms. However, the most commonly studied ficins are endophytopeptidase which exist in the latex of *Ficus carica, Ficus glabrata*, and many other species. In early research, ficin from *Ficus carica* was purified, characterized, and reported as a homogenous enzyme after isolation using SDS-PAGE electrophoresis and gel filtration chromatography. It is composed of a single polypeptide chain of molecular weight of 23,100 ± 300 Dalton when determined using the MALDI-TOF technique, with optimal pH between 6.5 and 8.5 (Devaraj et al., 2008). Another study also reported that the ficin obtained from *Ficus racemose* has a molecular weight of 55,500 Dalton with optimal pH rage 4.5–6.5 at 60 °C (Arshad et al., 2016). Another study reported three ficin isoforms (A, B, and C) in *Ficus carica* cv. Sabz latex (Zare et al., 2013). All ficin isoforms undergo an autolysis process when stored at high temperatures. However, ficins B and C isoforms were less stable when stored at cold temperatures than ficin A. It was also reported on the contribution of a cysteine residue of the active site in the biocatalytic activity of ficin. This has been proven by losing enzyme activity by adding a cysteine inhibitor. Based on the published data in the RCSB Protein Database bank, ficins are present in 5 isoforms (Fig. 23.1), namely (Ficin A, BI, BII, CI, CII, and D2). The PDB code for the enzymes given in the database is: for ficin A (4YYQ), for ficin BI (4YYR), for ficin BII (4YYS), for ficin CI (4YYU), for ficin CII (4YYV), and ficin D2 (4YYW). However, other research has also reported the presence of a glycosylated proteinase from *Ficus carica* var. Höraishi and denoted as ficin S. This enzyme isoform was electrophoretically homogeneous with optimal pH of 8.0 and optimal temperature of 60 °C (Sugiura and Sasaki, 1974). The main feature of this enzyme is the wide range of pH stability between 2 and 8 when stored for 20 h at 4 °C. The activity of this enzyme increases in the presence of cysteine and mercaptoethanol. However,

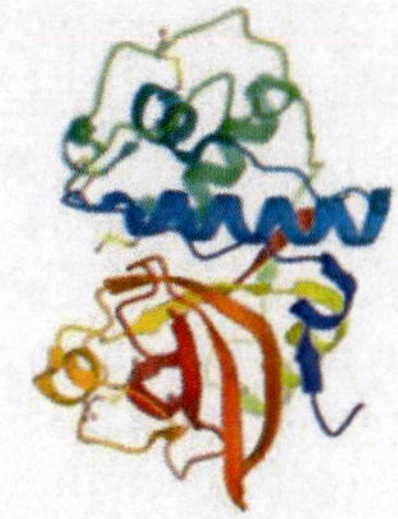 Ficin A (4YYQ) characterized by X-Ray diffraction at a resolution of 1.594 Å	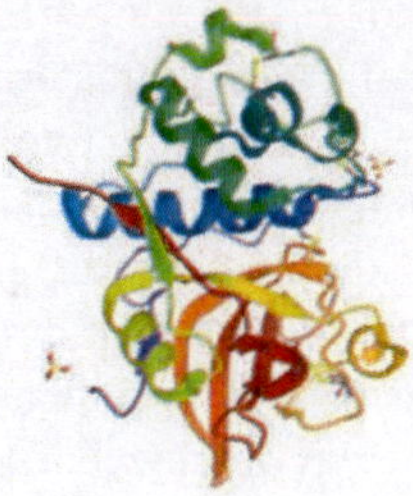 Ficin B (4YYR), crystal form I, characterized by X-Ray diffraction at a resolution of 1.349 Å
 Ficin B (4YYS), crystal form II, characterized by X-Ray diffraction at a resolution of 1.35 Å	 Ficin C (4YYU), crystal form I, characterized by X-Ray diffraction at a resolution of 1.18 Å
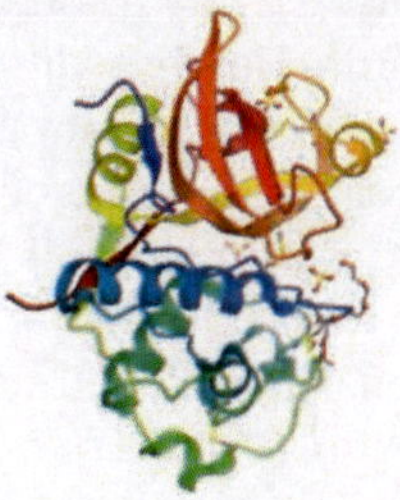 Ficin C (4YYV), crystal form II, Characterized by X-Ray diffraction at a resolution of 1.902 Å	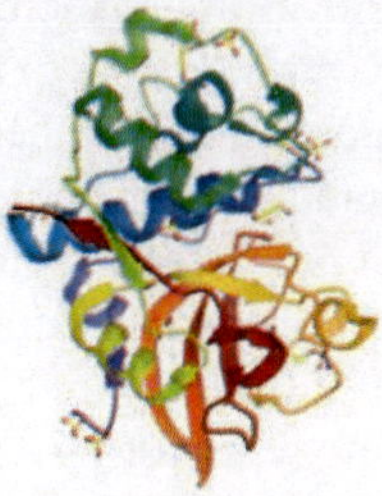 Ficin D2 (4YYW) characterized by X-Ray diffraction at a resolution of 1.446 Å

Fig. 23.1 Crystallographic structure of different ficin isoforms isolated from *Ficus carica,* N-terminus blue, C-terminus red. (Obtained from RCSB Protein Database bank)

the main differences between ficin S and other isoforms are only in the isoelectric point and the sugar residue.

However, another study also reported the presence of new isoforms Ficin C III, and Ficin D is further divided into isoforms D2-I and D2-II (Haesaerts et al., 2015).

Another research group also claimed the discovery of a new isoform and gave the name of Ficin E, a homogeneous protease of molecular weight of 24,294 Dalton. This ficin isoform is characterized by its maximal activity at pH 6.0 and temperate of 50 °C, making it attractive with high potential applications in many industries (Baeyens-Volant et al., 2015). However, based on the increased market demand for ficin, a new attempt has been made to produce this important endoprotease in microbial biofactory. The ficin production gene has been successfully cloned and expressed using a lactose induction system in *Escherichia coli* BL21. The production medium was optimized, and the bioactive enzyme was produced in a submerged cultivation system.

4 Industrial Applications of Ficin

Based on its strong proteolytic properties, availability at a reasonable cost, and attractive biocatalytic properties, ficin has been used as an attractive enzyme candidate in many industries. In general, the use of plant enzymes, especially in the food and pharmaceutical industries, is highly interested based on their safety and ease of approval as Generally Regarded as Safe (GRAS) biocatalysts according to Food and Drug Administration (FDA). In addition, the production of ficin from figs can be carried out using non-edible plant parts such as leaves and stems, which do not interfere with the food bioresources. These plant parts are also considered wastes in fig plantations. Based on all these advantages ficins are widely used as biocatalysts for bioactive compounds production, food, and pharmaceutical industries, with many new novel applications in medical and cosmetic fields (Morellon-Sterling et al., 2020). In general, ficin is widely used in free form. However, different immobilization approaches have been applied to increase enzyme efficiency and stability. Enzyme immobilization improves the ficin kinetics behavior in terms of activity and stability and decreases the biocatalysts cost in the conversion process based on its potential reusability in operation when applied in repeated batch or continuous modes of operation. Figure 23.2 summarizes the main current applications of ficins in biotechnology industries.

4.1 Bioactive Compounds Synthesis

Ficin is a cysteine endopeptidase enzyme with a highly efficient capacity to break protein. It was able to cleave protein at the glycosylic side of Gly, Ser, Thri, Met, Lys, Arg, Tyr, Ala, Asn, and Val. Thus, it can be widely used in peptide hydrolysis to produce a low molecular weight-specific polypeptide chain. Therefore, milk protein hydrolysate of shorter peptide chain production for infant feeding is one of the

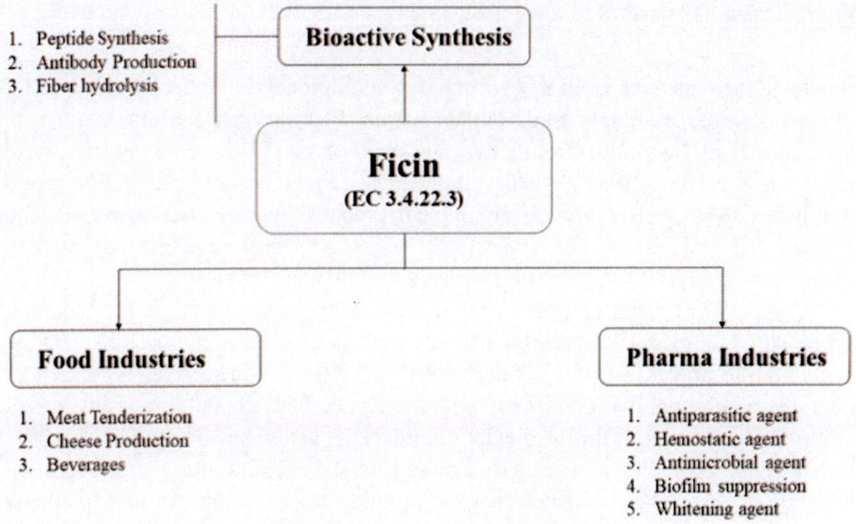

Fig. 23.2 Different applications of ficin in biotechnology industries

main fields of the current ficin applications. It is well known that cow milk is unsuitable for infant feeding because of the presence of large casein micelles that infants are not digestible. In addition, the presence of casein and calcium in high concentrations can inhibit the absorption of dietary non-heme iron, affecting iron nutrition (Ziegler, 2007). Therefore, for the development of infant milk, enzymatic hydrolysis is necessary to produce high bioavailable nutrients of high nutritional value and low allergic effects (Xie et al., 2013). However, milk protein hydrolysis can produce hydrolysate of various molecular weight peptides with low/zero free amino acid content or a high amount of free amino acids with short-chain peptides (depending on the enzyme used). When ficin is applied in this process, it produces a highly hydrolyzed protein with a significant reduction of allergenicity based on the reduction of epitopes concentration which IgG and IgE can recognize in the infant immune system (Verhoeckx et al., 2015). In another research, ficin was able to hydrolyze bovine casein and produced eight peptides of antioxidant properties (De Pierro et al., 2014). The main advantage of ficin over pepsin in this process is the ability of ficin to hydrolyze protein under neutral and alkaline conditions (Aider, 2021).

Ficin has also been reported as the most suitable protease to produce a high yield of active bivalent $F(ab)_2$ fragment from the murine monoclonal IgG1 with high quality suitable for clinical trials. Therefore, it has high potential in the antibody production industry as a safe and specific protease (Mariant et al., 1991).

4.2 Food Industries

Plant enzymes have been widely accepted in food processing in many cultures. Of different plant proteases used, Papain (from Papaya fruit), Bromelain (from Pineapple fruit), actinidin (from Kiwi), and ficin (from Fig) have a long successful history in food industries (Troncoso et al., 2022). Ficin, in particular, is often used to improve the properties and texture of meat, prepare cheese, clarify beverages, and prevent chill haze. In meat industries, ficin has been used as a tenderizer alone or in a combination of other plant proteases such as papain or bromelain.

In beverage industries, fig enzyme drink is manufactured using a fermentation process by adding easily fermented sugars such as glucose or sucrose to speed the fermentation process. As fig is microbial-rich fruit, the added sugar is fermented ethanol and carbon dioxide directly without any exogenous addition of any starter culture. The product of this fermentation process is acidic and composed of a mixture of fig syrup and fig vinegar with a more sourish taste if left for a longer fermentation time. It has been claimed that the fig enzyme-rich drink can provide many functional roles in the human body by promoting digestion, helping to regulate blood pressure and sugar level, reducing cholesterol, and reducing the cancer risk due to the presence of many antioxidants. In particular, ficin is also used to hydrolyze protein in beer to decrease the turbidity developed during cold storage.

In meat industries, ficin is considered among the best plant enzymes in the tenderization process when used alone or in combination with other enzymes (Madhusankha & Thilakarathna, 2021). This is based on its ability to hydrolyze the muscle protein and collagen to soften the meat tissue. Ficin is among the best suitable for collagen and elastin degradation at an acceptable range for fresh meat processing (pH 5–8) and (temperature 45–45 °C) (Tántamacharik et al., 2018). Therefore, the solubility of meat proteins can be increased after ficin marination. In addition, ficin is able to degrade elastin at low temperatures with less reactivity with myofibrillar proteins and collagen (Singh et al., 2019). It has also been used in the sausage industry as it has been used for the efficient cleaning of animal intestines. Another study also reported that adding ficin to soy proteins helps to improve the final sausage quality and customer acceptability (Ramezani et al., 2003). In seafood industries, ficin was added alone or in mixtures of other enzymes such as amylase and cellulase shrimp processing industries to ease the separation of the shell from the meat (Venugopal et al., 2000; Fernandes, 2016).

In the Cheese industry, ficin help in the milk coagulation process, which is important for milk clotting in dairy industries. It has been reported that plant proteases such as ficin and papain had a more significant effect on crud milk proteolysis than calf rennin. Other studies also reported that ficin could improve ewes milk crud by producing less viscous with a better compact structure (Shabani et al., 2018). It has also been reported that adding ficin to ovine milk can produce a better-coagulated product with a higher yield than traditionally used chymosin (Bornaz et al., 2010).

4.3 Pharmaceutical, Medical and Cosmetic Industries

In general, plant proteases have a wide range of applications in pharmaceutical, medical, and cosmetic industries based on their unique enzyme behaviors in terms of unique substrate specificity, high stability, the capacity to function at a wide range of pH, and the ability to work at relatively higher temperature (Balakireva et al., 2019). Of different plant proteases, ficin can act as an antiparasitic agent based on its high proteolytic activity with the capacity to digest parasite eggs and mature worms (Behnke et al., 2008). Ficin has high potential when used alone or in combination with papaya latex to act as antiparasitic against the adult worms of *Heligmosomoides bakeri.* Clear damage was observed to the worm cuticle when exposed for 30 min to the saline solution of only 30 μm ficin or 200 μm crude papaya latex or a combination thereof (Stepek et al., 2007). Ficin was also successfully used as a hemostatic agent. It can reduce the prothrombin time of normal plasma or coagulation factor deficient plasma, activate partial thromboplastin, and act as a factor X activator. This effect is mediated through the successive hydrolysis in the heavy chain between Leu178 and Asp 179, Arg 187 and Gly 188, Arg 194 and Ile 195, and the release of a carboxy-terminal peptide (Richter et al., 2002; Morellon-Sterling et al., 2020). On the other hand, recent research reported the presence of fibrinolytic serine protease in fig latex which can control the ficin coagulation stimulatory effect (Hamed et al., 2020).

Ficin has also been applied in skin treatment based on its activity as an inhibitor of skin inflammation. This is mediated through the inhibition of mRNA and protein expression level of inducible NO synthase and inhibition of the phosphorylation of inflammatory signal proteins (Cho et al., 2019). The same study also showed the potential use of ficin in cosmetics as a whitening agent as it can act as a melanin pigment inhibitor using a mouse melanoma cell model. Another exciting study reported the high potential application of free and chitosan-immobilized ficin as an antibacterial agent with wound healing properties. Adding ficin in the free or immobilized form in the rat *in vivo* model reduced the skin wound area caused by *Staphylococcus aureus* twofold after 4 days of treatment instead of 6 days. In addition, the application of immobilized enzymes reduced the cell count of *S. aureus* by 3-log on the wound surface after only 6 days compared to 10 days in untreated skin (Baidamshina et al., 2020). This clearly shows the high potential use of chitosan immobilized ficin as part of wound-dressing materials to accelerate wound healing and prevent contamination during the healing process.

Other interesting research also reported the potential use of ficin in cell banking of camel semen samples (Niasari-Naslaji et al., 2016). In this research, ficin has decreased camel semen viscosity, limiting the processing and preservation process. However, adding ficin in a concentration of 0.05 mg/ml followed by stirring for 5 min can decrease the semen viscosity without adversely affecting sperm viability.

In immunohistochemical research, ficin was also applied as a potential trypsin substitution and ease antigen site exposure during the pre-treatment phase before the aldehyde fixation process (Taschini et al., 1987). In clinical diagnosis industries, ficin has been applied to increase the sensitivity of in vitro diagnostic tests (IVD). An early study showed that using ficin-treated cells in a serological test of blood grouping can improve the blood group's determination by enabling antibody detection (Giles, 1960). In another study, in testing the plasma using ficin-treated reagents, the enzyme proteolytic activities increase the sensitivity of the assay, assisting in better identification of the antibody in ABO, Rh, Kidd, Lewis P, and I antigen in the blood group system. On the other hand, ficin inhibit Duffy and MNS system antigens. Thus, adding ficin to the diagnosis process allowed to identify 25 new alloantibodies that were masked and could not be detected without enzyme addition (Hill et al., 2016).

Recent research also reported on the high potential application of ficin as an efficient protease enzyme for biofilm removal. It is widely known that microbial biofilm formation creates problems in treating many microbial infections. The biofilm creates an extra protection layer of a complex structure that prevent the direct exposure of microbes to antibiotics and thus creates a specific sort of microbial resistance. Ficin has been effectively used to disrupt skin biofilm matrix made of *Staphylococcus aureus* and *Staphylococcus epidermidis*. The presence of ficin in the treatment protocol leads to a twofold decrease in the concentration of the anti-infectives (ciprofloxacin and benzalkonium chloride) used in skin treatment without any cytotoxic activity (Baidamshina et al., 2017). Recent research also reports the high potential biofilm suppressant application when used in concentrations as low as 4 mg/mL. At this concentration, it exhibited a significant inhibitory effect on *Streptococcus* mutans biofilm formation and virulence mechanism involving lactic acid production and exopolysaccharide synthesis. Therefore, ficin could be used as an anticaries agent in dental treatment (Sun et al., 2021).

5 Ficin Immobilization to Improve Industrial Applications

Most research on ficin applications in the industry has been carried out using a free enzyme system. However, in recent years, many studies focused on the immobilization of this enzyme to increase stability and reusability and decrease the enzyme cost in many industrial processes. One of the early attempts was carried out for ficin immobilization using diatomaceous earth (Celite) as an adsorbent. The immobilized enzyme was used in traditional Turkish cheese (Teleme) production. As a result, the optimal temperature was increased from 60 °C, both free and immobilized enzymes exhibited high milk clotting activity, and the produced cheese by immobilized enzyme gave better chemical and sensory properties compared to a free enzyme (Fadýloğlu, 2001). In another research effort, ficin was covalently immobilized on poly (γ-methyl-D-glutamate) fibers and produced a highly active and stable enzyme suitable which is more suitable for continuous operation (Jokei et al., 1999). Another study also reported the successful immobilization of ficin using polyvinyl alcohol (PVA), with high retention of enzyme activity up to 92%, and was used successfully for 9 cycles without losses in biocatalytic activities (Rojas-Mercado et al., 2018). In other research, glyoxyl-agarose was successfully used as an immobilization matrix. The immobilized enzyme exhibited higher stability, a wider range of pH and temperature activity curves, and catalyzed casein hydrolysis five times without losing its initial activity (Siar et al., 2017). Another recent study also reported the successful use of immobilized ficin using the same matrix to improve the milk coagulation process by reducing the clotting time, producing a higher coagulation yield than the free enzyme (Siar et al., 2020).

6 Conclusion and Future Perspectives

In addition, with the increased awareness of the importance of natural enzymes in human health, plant enzymes have become more widely used in the food and pharmaceutical industries. So far, ficin has exhibited unique characteristics as a dual active enzyme of both protease and peroxidase activities. Therefore, we expect an increase in its application range, especially in the medical field as a high-end market for plant enzymes. This is because edible plant parts produce it, considered one of the safest natural resources according to the FDA. Every year, ficin applications increase in many industries. It is expected that improving enzyme kinetics using protein engineering techniques and immobilization methods will increase the acceptability and allow the penetration of ficin to many industries, replacing in large part the currently used chemical catalysts. To cope with the increased demand, new suspension culture cultivation technology in large-scale bioreactors produces a high-quality enzyme in a shorter time. In addition, this process is not affected by environmental variables or climate changes. In addition, with the increased knowledge about molecular biology and biosynthesis of ficin enzymes, the ficin production genes were recently expressed in microbial biofactory. Therefore, we expect to see a large-scale production process of ficin using the microbial system shortly. Further research in all these directions will reduce the production cost of ficin and make it available to replace other endoproteases in the market.

References

Abdel-Aziz, M., Darwish, M. S., Mohamed, A. H., El-Khateeb, A. Y., & Hamed, S. E. (2020). Potential activity of aqueous fig leaves extract, olive leaves extract and their mixture as natural preservatives to extend the shelf life of pasteurized buffalo milk. *Food, 9*, 615.

Abu-Goukh, A. A., Shattir, A. T., & El Fatih, M. M. (2010). Physio-chemical changes during growth and development of papaya fruit. II: Chemical changes. *Agriculture and Biology Journal of North America, 5*, 871–887.

Aider, M. (2021). Potential applications of ficin in the production of traditional cheeses and protein hydrolysates. *JDS Communications, 2*, 233–237.

Amessis-Ouchemoukh, N., Ouchemoukh, S., Meziant, N., Idiri, Y., Hernanz, D., Stinco, C. M., Rodríguez-Pulido, F. J., Heredia, F., Madani, K., & Luis, J. (2017). Bioactive metabolites involved in the antioxidant, anticancer and anticalpain activities of *Ficus carica* L., *Caratonia siliqua* L. and *Quercus ilex* L. extracts. *Industrial Crops and Products, 95*, 6–17.

Arshad, M. S., Kwon, J.-H., Imran, M., Sohaib, M., Aslam, A., Nawa, I., & Javed, M. (2016). Plant and bacterial proteases: A key towards improving meat tenderization, a mini review. *Cogent Food Agriculture, 2*, 1261780.

Arvaniti, O. S., Smaras, Y., Gatidou, G., Thomaidis, N. S., & Stasinakis, A. S. (2019). Review on fresh and dried figs: Chemical analysis and occurance of phytochemical compounds, antioxidant capacity and health effects. *Food Research International, 119*, 244–267.

Bae, H., Yun, S., Hun, J. H., Yoon, I. K., Nam, E., & Kwon, J. (2014). Assessment of organic acid and sugar composition in apricot, plumcot, and peach during fruit development. *Journal of Applied Botany and Food Quality, 87*, 24–29.

Baeyens-Volant, D., Matagne, A., El Mahyaoui, R., Wattiez, R., & Azarkan, M. (2015). A novel form of ficin from *Ficus carica* latex: Purification and characterization. *Phytochemistry, 117*, 154–167.

Baidamshina, D. R., Trizna, E. Y., Holyavka, M. G., Bogachev, M. I., Artyukhov, V. G., Akhatova, F. S., Rozhina, E. V., Fakhrullin, R. F., & Rayumov, A. R. (2017). Targeting microbial biofilms using ficin, a nonspecific plant protease. *Scientific Reports, 7*, 46068.

Baidamshina, D. R., Koroleva, V. A., Trizna, E. Y., Pankova, S. M., Agafonova, M. N., Chirkova, M. N., Vasileva, O. S., Akhmetov, N., Shubina, V. V., Porfiryev, A. G., Semenova, E. V., Sachenokov, O. A., Bogachev, M. I., Artyukhov, V. G., Baltina, T. V., Holyavka, M. G., & Kayumov, A. R. (2020). Anti-biofilm and wound-healing activity of chitosan-immobilized ficin. *International Journal of Biological Macromolecules, 164*, 4205–4217.

Balakireva, A. V., Kuznetsova, N. V., Petushkova, A. I., Savvateeva, L. V., & Zamyatnin, A. A., Jr. (2019). Trends and prospects of plant proteases in therapeutics. *Current Medicinal Chemistry, 26*, 465–486.

Behnke, J. M., Buttle, D. J., Stepek, G., Lowe, A., & Duce, I. R. (2008). Developing novel anthelmintics from plant cysteine proteinases. *Parasities & Vectors, 1*, 29.

Bornaz, S., Guizani, N., Feilah, N., Sahli, A., Ben Slama, M., & Attia, H. (2010). Effect of plant originated coagulants and chymosin on ovine milk coagulation. *International Journal of Food Properties, 13*, 10–22.

Caliskan, O. (2015). Mediterranean figs (*Ficus carica* L.) functional food properties. In V. R. Preedy & R. S. Watson (Eds.), *The Mediterranean diet. An evidence-based approach* (pp. 629–637). Elsevier Press.

Chawia, A., Kaur, R., & Sharma, A. K. (2012). *Ficus carica* Linn.: A review on its pharmacognostic, phytochemical and pharmacological aspects. *International Journal of Pharmaceutical and Phytopharmacological Research, 1*(4), 215–232.

Cho, U. M., Choi, D. H., Yoo, D. S., Park, S. J., & Hwang, H. S. (2019). Inhibitory effect of ficin derived from fig latex on inflammation and melanin production in skin cells. *Biotechnology and Bioprocess Engineering, 24*, 288–297.

Cui, Y., Wang, Z., Chen, S., Vainstein, A., & Ma, H. (2019). Proteome and transcriptome analysis reveal key molecular differences between quality parameters of commercial-ripe and tree-ripe fig (*Ficus carica* L.). *BMC Plant Biology, 19*, 146.

De Pierro, G., O'Keeffe, M. B., Poyarkov, A., Lomolino, R. J., Fitzgerald, R. J., Di Pierro, R., O'Keeffe, G., Poyarkov, M. B., Lomolino, A., & Fitzgerald, G. (2014). Antioxidant activity of bovine casein hydrolysates produced by *Ficus carica* L. -derived proteinase. *Food Chemistry, 156*, 305–311.

Debin, A., Tir-Touil, M., Meddah, B., Hamaidi-Chergui, F., Menadi, S., & Alsayadi, M. (2018). Evaluation of antimicrobial and antioxidant activities of oily macerates of Algerian dried figs (*Ficus carica* L.). *International Food Science Research Journal, 25*, 351–356.

Devaraj, K. B. D., Kumar, P. R., & Prakash, V. (2008). Purification, characterization, and solvent-induced thermal stabilization of ficin from *Ficus carica*. *Journal of Agriculture Food Chemistry., 56*(23), 11417–11423.

Elsayed, A. M., Hegazy, U. M., Hegazy, M. G. A., Abdel-Ghany, S. S., Salama, W. H., Salem, A. M. H., & Fahmy, A. S. (2018). Purification and biochemical characterization of peroxidase isoenzyme from *Ficus carica* latex. *Biocatalysis and Agricultural Biotechnology, 16*, 1–9.

Fadýloğlu, S. (2001). *Immobilization and characterization of ficin. Nahrung/Food, 45*, 143–146.

Fernandes, P. (2016). Enzymes in fish and seafood processing. *Frontiers in Bioengineering and Biotechnology, 4*, 59.

Gaaliche, B., Trad, M., & Mars, M. (2011). Effect of pollination intensity frequency and pollen source on fig. (*Ficus carica* L.) productivity and fruit quality. *Scientia Horticulturae, 130*, 737–742.

Gibernau, M., Buser, H. R., Frey, J. E., & Hossaert-McKey, M. (1997). Volatile compounds from extracts of figs of *Ficus carica*. *Phytochemistry, 46*, 241–244.

Giles, C. M. (1960). Survey of uses for ficin in blood group serology. *Vax Sanguinis, 5*, 467–471.

Gill, P. P. S., Jawandha, S. K., Kaur, N., & Singh, N. (2017). Physio-chemical changes during progressive ripening of mango (*Mangifera indica* L.) *cv.* Dashehari under different temperature regimes. *Journal of Food Science Technology, 54*(7), 1964–1970.

Haesaerts, S., Buitrago, J. A. R., Loris, R., Baeyens-Volant, D., & Azarkan, M. (2015). Crystallization and preliminary X-ray analysis of four cysteine proteases from *Ficus carica* latex. *Acta Cryst, F71*, 459–465.

Hamed, M. B., El-Badry, M. O., Kandil, E. I., Borai, I. H., & Sahmy, A. S. (2020). A contradictory action of procoagulant ficin by a fibrinolytic serine protease from Egyptian *Ficus carica* latex. *Biotechnology Reports, 27*, e00492.

Hashemi, S. A., Abediankenari, S., Ghasemi, M., Azadbakht, M., Yousefzadeh, Y., & Dehpour, A. A. (2011). The effect of fig tree latex (*Ficus carica*) on stomach cancer line. *Iranian Red Crescent Medical Journal, 13*(4), 272–275.

Hill, B. C., Hanna, C. A., Adamski, J., Pham, H. P., Marques, M. B., & Williams, L. A. (2016). Ficin-Treated red cells help identify clinically significant alloantibodies masked as reactions of undetermined specificity in gel microtubes. *Laboratory Medicine, 48*, 24–28.

Hosainzadegan, H., Alizadeh, M., Karimi, F., & Pakzad, P. (2012). Study of antibacterial effects of ripped and raw fig alone and in combination. *Journal of Medicinal Plant Research, 6*, 2864–2867.

Jain, N., Dhawan, K., Malhotra, S. P., Siddiqui, S., & Singh, R. (2001). Compositional and enzymatic changes in guava (*Psidium guajava* L.) fruits during ripening. *Acta Physiologiae Plantarum, 23*, 357–362.

Jokei, K., Furuta, M., Oka, M., & Hayashi, T. (1999). Protease immobilization onto poly(GMMA-methyl-D-glutamate) fibers. *Sen'i Gakkaishi, 55*, 166–171.

Khadhraoui, M., Bagues, M., Artés, F., & Ferchichi, A. (2019). Phytochemical content, antioxidant potential, and fatty acid composition of dried Tunisian fig (*Ficus carica* L.) cultivars. *Journal of Applied Botany and Food Quality, 92*, 143–150.

Khodarhami, G. A., Ghasemi, N., Hassanzadeh, F., & Safaie, M. (2011). Cytotoxic effects of different extracts and latex of *Ficus carica* L. on HeLa cell line. *Iranian Journal of Pharmaceutical Research, 10*(2), 273–277.

Lazreg-Aref, H., Zaouay, F., Hammami, M., & Mars, M. (2017). Fig fruit latex yield and protease activity as influenced by cultivar, maturity stage and harvesting time. *Acta Horticulturae, 1173*, 323–328.

Lazreg-Aref, H., Gaaliche, B., Ladhari, A., Hammami, M., & Hammami, S. O. (2018). Co-evaluation of enzyme activities and latex in fig (*Ficus carica* L.) during fruit maturity process. *South African Journal of Botany, 115*, 143–152.

Liu, F., Yang, Z., Zheng, X., Luo, S., Zhang, K., & Li, G. (2011). Nematicidal coumarin from *Ficus caris* L. *Journal of Asia-Pacific Entomology, 14*, 79–81.

López-Huertas, E., & Palma, J. M. (2020). Changes in glutathione, ascorbate, and antioxidant enzymes during olive fruit ripening. *Journal of Agricultural and Food Chemistry, 68*(44), 12221–12228.

Madhusankha, G. D. M. P., & Thilakarathna, R. C. N. (2021). Meat tenderization mechanism and the impact of plant exogenous proteases: A review. *Arabian Journal of Chemistry, 14*, 102967.

Mariant, M., Camagna, M., Tarditi, L., & Seccamani, E. (1991). A new enzymatic method to obtain high-yield F(ab)$_2$ suitable for clinical use from mouse IgG1. *Molecular Immunology, 28*, 69–77.

Mopuri, R., Ganjai, M., Meriga, B., Moorbanally, N. A., & Islam, M. S. (2018). The effect of *Ficus carica* on the activity of enzymes related to metabolic syndrome. *Journal of Food and Drug Analysis, 26*, 201–210.

Morellon-Sterling, R., El-Siar, H., Tavano, O. L., Berenguer-Murcia, A., & Fernández-Lafuente, R. (2020). Ficin: A protease extract with relevance in biotechnology and biocatalysis. *International Journal of Biological Macromolecules, 162*, 394–404.

Oliveira, A. A. P., Valentão, P., Pereira, J. A., Silva, B. M., Tavares, F., & Andrade, P. B. (2009). *Ficus carica* L.: Metabolic and biological screening. *Food and Chemical Toxicology, 47*, 2841–2846.

Paşayeva, L., Özalp, B., & Fatullayev, H. (2020). Potential enzyme inhibitory properties of extracts and fractions from fruit latex of *Ficus carica*-based on inhibition of α-amylase and α-glucosidase. *Journal of Food Measurement and Characterization, 14*, 2819–2827.

Rahmani, A. H., & Aldebasi, Y. H. (2017). *Ficus carica* and its constituent's role in management of diseases. *Asian Journal of Pharmaceutical and Clinical Research, 10*(6), 49–53.

Ramezani, R., Aminlari, M., & Fallahi, H. (2003). Effect of chemically modified soy proteins and ficin-tenderized meat on the quality attributes of sausage. *Food Chemistry and Toxicology, 68*, 85–88.

RCSB. *Protein Database bank*. https://www.rcsb.org/. Accessed on 20 June 2022.

Richter, G., Schwarz, H. P., Dorner, F., & Turecek, P. L. (2002). Activation and inactivation of human factor X by proteases derived from *Ficus carica*. *British Journal of Haematology, 119*, 1042–1051.

Robbins, B. H. (1935a). Proteolytic enzyme content of latex from the fig tree (*Ficus carica* L.). seasonal variation. *Proceedings of the Society for Experimental Biology and Medicine, 32*, 892–893.

Robbins, B. H. (1935b). Proteolytic enzyme in the latex from the fig tree (*Ficus glabrata*). The pH of optimal activity. *Proceeding of the Society of Experimental Biology and Medicine, 32*, 894–896.

Rojas-Mercado, S. A., Moreno-Cortez, I. E., Lucio-Porto, R., & Pavón, L. L. (2018). Encapsulation and immobilization of ficin extract in electrospun polymeric nanofibers. *International Journal of Biological Macromolecules, 118*, 2287–2295.

Sedaghat, S., & Rahemi, M. (2018). Enzyme activity regarding sugar and organic acid changes during development stages in reinfed fig (*Ficus carica* L.cv Sabz). *International Journal of Fruit Science, 18*(1), 14–28.

Shabani, R., Shahidi, S. A., & Rafe, A. (2018). Rheological and structural properties of enzyme induced gelation of milk proteins by ficin and *Polyporus badius*. *Food Science & Nutrition, 13*, 287–294.

Siar, E., Zaak, H., Kornecki, J. F., Zidoune, M. N., Barbosa, O., & Fernandez-Lafuente, R. (2017). Stabilization of ficin extract by immobilization on glyoxyl agarose. Preliminary characterization of the biocatalyst performance in hydrolysis of proteins. *Process Biochemistry, 58*, 98–104.

Siar, E., Morellon-Sterling, R., Zidoune, M. N., & Fernendez-Lafuente, R. (2020). Use of glyoxyl-agarose immobilized ficin extract in milk coagulation: Unexpected importance of the ficin loading on the biocatalysts. *International Journal of Biological Macromolecules, 144*, 419–426.

Sieniawska, E., Świątek, L., Sinan, K. I., Zengin, G., Boguszewska, A., Polz-Dacewicz, M., Sadeer, N. B., Etienne, O. K., & Mahomoodally, M. F. (2022). Phytochemical insights into *Ficus sur* extracts and their biological activity. *Molecules, 27*, 1863.

Singh, P. K., Shrivastava, N., & Ojha, B. K. (2019). Enzymes in the meat industry. In M. Kuddos (Ed.), *Enzymes food biotechnology: Production, applications, and future prospects* (pp. 111–128). Elsevier Press.

Solomon, A., Golubowicz, S., Yablowicz, Z., Grossman, S., Bergman, M., Gottlieb, H. E., Altman, A., Kerem, Z., & Flaishman, M. A. (2006). Antioxidant activities and anthocyanin content of fresh fruits of common fig (*Ficus carica* L.). *Journal of Agricultural and Food Chemistry, 54*, 7717–7723.

Stepek, G., Lowe, A. E., Buttle, D. J., Duce, I. R., & Dehnke, J. M. (2007). *In vitro* anthelmintic effects of cysteine proteinases from plants against intestinal helminths of rodents. *Journal of Helminthology, 81*, 353–360.

Sugiura, M., & Sasaki, M. (1974). Studies on proteinases from *Ficus carica* var. Höraishi. V, purification and properties of a sugar-containing proteinase (Ficin S). *Biochemical Biophysica Acta. Enzymology., 350*, 38–47.

Sun, Y., Jiang, W., Zhang, M., Zhang, L., Shen, Y., Huang, S., Li, M., Qiu, W., Pan, Y., Zhou, L., & Zhang, K. (2021). The inhibitory effects of ficin on *Streptococcus mutans* biofilm formation. *BioMed Research International, 2021*, 6692328.

Tántamacharik, T., Carne, A., Agyeri, D., Birch, J., & Bekhit, A. E. A. (2018). Use of plant proteolytic enzyme for meat processing. In M. G. Guevara & G. R. Daleo (Eds.), *Biotechnological applications of plant proteolytic enzymes* (pp. 43–67). Springer Nature.

Taschini, P. A., Doris, M., & Mac Donald, H. T. (1987). Protease digestion step in immunohistochemical procedures: Ficin as a substitute for trypsin. *Laboratory Medicine, 18*, 532–536.

Trad, M., Le Bourvellec, C., Gaaliche, B., Ronard, C. M. G. C., & Mars, M. (2014). Nutritional compounds in figs from southern Mediterranean region. *International Journal of Food Properties, 17*, 491–499.

Troncoso, F. D., Sánchez, D. A., & Ferreira, M. L. (2022). Production of plant protreases and new biotechnological applications: An updated review. *Chemistry Open, 11*, e202200017.

Venugopal, V., Lakshmanan, R., Doke, S. N., & Bongirwar, D. R. (2000). Enzymes in fish processing, biosensors and quality control. *Food Biotechnology, 14*, 21–27.

Verhoeckx, K. C. M., Vissers, Y. M., Baumert, J. L., Faludi, R., Feys, M., Flanagan, S., Herouet-Guicheney, C., Holzhauser, T., Shimojo, R., van der Bolt, N., Wichers, H., & Kimber, I. (2015). Food processing and allergenicity. *Food Chemistry and Toxicology, 80*, 223–240.

Walti, A. (1938). Crystalline ficin. *Journal of the American Chemical Society, 60*, 493.

Wang, Z., Cui, Y., Vainstein, A., Chen, S., & Ma, H. (2017). Regulation of fig (*Ficus carica* L.) fruit color: Metabolomic and transcriptomic analyses of the flavonoid biosynthetic pathway. *Frontiers. Plant Science, 8*, 1990.

Xie, N., Wang, C., Ao, J., & Li, B. (2013). Non-gastrointestinal-hydrolysis enhances bioavailability and antioxidant efficacy of casein as compared with its in vitro gastrointestinal digest. *Food Research International, 51*, 114–122.

Yang, X., Guo, J. L., Ye, J. Y., Zhang, Y. X., & Wang, W. (2015). The effects of *Ficus carica* polysaccharide on immune response and expression of some immune-related genes in grass carp, *Ctenopharyngodon Idella. Fish & Shellfish Immunology, 42*, 132–137.

Yang, Y., Shen, D., Long, Y., Xie, Z., & Zhang, H. (2017). Intrinsic peroxidase-like activity of ficin. *Scientific Reports, 7*, 43141.

Yousef, N. S., El-Ghandour, A., & El-Shershaby, S. S. A. (2019). Antimicrobial activity of fig and olive extracts. *Journal of Food and Dairy Science, 10*, 503–508.

Zare, H., Moosavi-Movahedi, A. A., Salami, M., Mirzaei, M., Saboury, A. A., & Sheibani, N. (2013). Purification and autolysis of the ficin isoforms from fig (*Ficus carica cv.* Sabz) latex. *Phytochemistry, 87*, 16–22.

Zhao, C., Li, S., Li, C., Wang, T., Tian, Y., & Li, X. (2021). Flavonoid from fig (*Ficus carica* Linn.) leaves: The development of a new extraction method and identification by UPLC-QTOF-MS/MS. *Applied Sciences, 11*, 7718.

Chapter 24
Preventive Roles of Phytochemicals from *Ficus carica* in Diabetes and Its Secondary Complications

Additiya Paramanya, Nimisha Patel, Dinesh Kumar, Fatima Zahra Kamal, Belkıs Muca Yiğit, Priya Sundarrajan, Prairna Balyan, Johra Khan, and Ahmad Ali

1 Introduction

Ficus carica L. (Moraceae), or the common fig, is the most important member of the 23 species of the genus *Ficus*. It is one of the first plants cultivated by humans and is native to the Mediterranean and a few parts of Asia (Mawa et al., 2013). Previously, only the fruit of the plant, i.e., the fig, was consumed only for nutritional purposes due to its high vitamin and mineral content, while the bark, latex, and roots were used for topical applications in traditional medicine (Misbah et al., 2013).

A. Paramanya · D. Kumar · P. Balyan · A. Ali (✉)
Department of Life Sciences, University of Mumbai, Mumbai, India
e-mail: ahmadali@mu.ac.in

N. Patel
Department of Life Sciences, J C Bose University of Science and Technology, YMCA, Faridabad, Haryana, India

F. Z. Kamal
Faculty of Sciences and Techniques, Laboratory of Physical Chemistry of Processes and Materials, Hassan First University, Settat, Morocco

B. M. Yiğit
Vocational School of Technical Sciences, Department of Forestry, Igdir University, Igdir, Turkey

P. Sundarrajan
Department of Life Science and Biochemistry, St. Xavier's College (Autonomous), Mumbai, India

J. Khan
Department of Medical Laboratory Sciences, College of Applied Medical Sciences, Majmaah University, Al Majmaah, Saudi Arabia

M. F. Ramadan (ed.), *Fig* (Ficus carica)*: Production, Processing, and Properties*, https://doi.org/10.1007/978-3-031-16493-4_24

Figs are an excellent source of proanthocyanidins, a powerful antioxidant with potential antidiabetic effects. The fruit's anthocyanin content depends on its skin color, ranging from yellow, green, violet, purple, and black. These pigments are found to be highest in black-colored figs and are proven efficient in reducing oxidative stress, inflammation, and carcinogenic effects (Tall et al., 2004; Ercisli et al., 2012).

Oxidative stress is one of the major contributors to diabetes, antioxidants in fig might prove beneficial in combating it. Organic extracts of fruits and leaves have demonstrated induction of insulin secretion and inhibition of enzyme levels such as pancreatic lipase, α-glucosidase, and α-amylase (Irudayaraj et al., 2016; Mopuri et al., 2018). In addition, the plants have been proven to show varying degrees of hypoglycemic activity, suggesting that their use can greatly reduce the effects and complications of diabetes.

2 Management of Diabetes

Diabetes, commonly known as diabetes mellitus, is a chronic metabolic condition marked by a shortage of insulin, a high blood glucose level, and insulin resistance. Diabetes develops when the body's insulin production is very low or non-existent or when the body's response to the released insulin is inadequate (Shoback & Gardner, 2011). Type 1, type 2, and gestational diabetes are the three most common kinds of diabetes. Type 1 diabetes, also known as insulin-dependent diabetes mellitus (IDDM) or juvenile diabetes, occurs when the pancreas fails to produce enough insulin due to beta cell loss (Norman & Henry, 2015). Type 2 is a condition wherein the body fails to give insulin response appropriately; it is commonly known as non-insulin-dependent diabetes mellitus (NIDDM) or adult-onset diabetes. Type 2 diabetes, also known as non-insulin-dependent diabetes mellitus (NIDDM) or adult-onset diabetes, is a condition in which the body fails to respond to insulin adequately. Lack of exercise and weight increase is typical causes (Moller, 2001). The third type of diabetes is gestational diabetes, which occurs when a pregnant woman who has never had diabetes gets excessive blood sugar levels, which usually resolve itself once the baby is born (Coustan, 1993).

Diabetes mellitus has such modest symptoms that around eight million people are unaware they have it. Frequent urination, hazy vision, irritation, increased hunger and thirst, and weight loss are common symptoms (Cooke & Plotnick, 2008). The glycated hemoglobin (A1C) test, which indicates the sugar content in the blood bound to hemoglobin, is one of the diagnostic tests for type 1 diabetes. The goal of diabetes management is to maintain a normal blood sugar level (Fig. 24.1). This can be accomplished using appropriate drugs, exercise, dietary changes, and other methods (Nathan et al., 2005). A sedentary lifestyle is one of the most prominent risk factors for type 2 diabetes. To keep blood sugar levels under control, it is critical to improving your lifestyle by exercising and making dietary changes. In a research study, patients who exercised moderately for 50 min three times a week saw their

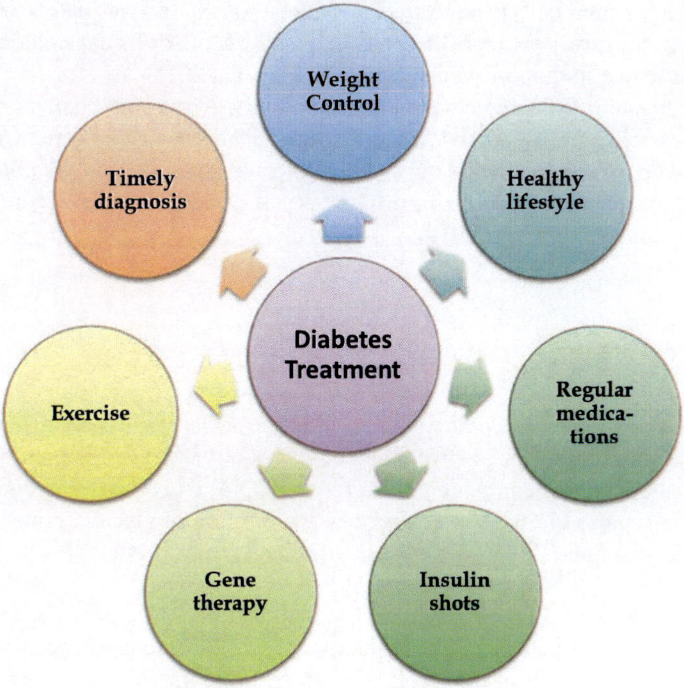

Fig. 24.1 Strategies to treat diabetes

HbA1c levels drop by 0.75% after 8 weeks (Boulé et al., 2001). Diabetes raises the risk of cardiovascular disease and stroke. Weight loss combined with a healthy diet, such as a low-carbohydrate diet or a diet rich in fruits, vegetables, whole grains, low-fat dairy products, and artificial sweeteners, has been shown to help manage hyperglycemia (Evert et al., 2019).

Diabetes can be treated with a variety of medications. Insulin shots, preferably long-acting insulin to match the basal rate, are given subcutaneously in the case of type 1 diabetes. In the case of type 2 diabetes, medications such as metformin are commonly taken orally Metformin is the first-line treatment for T2DM and gestational diabetes that works by lowering glucose production in the liver. It reduces calorie intake by lowering glucose synthesis in the liver, increasing insulin sensitivity, and exerting anorexiant effects (Krentz & Bailey, 2005). Other anti-diabetic medications include thiazolidinediones (TZDs), alpha-glucosidase inhibitors, SGLT2 inhibitors, etc. TZDs lower blood glucose levels by improving insulin sensitivity in cells. Glucosidase inhibitors work by blocking the enzyme alpha-glucosidase, which delays carbohydrate absorption in the intestines. SGLT2 inhibitors prevent glucose from reabsorbing in the kidneys, resulting in increased secretion and lower blood sugar levels (Wang & Perri, 2018). In recent years, nanotechnology has become increasingly essential in treating diabetes, with glucose sensors delivering insulin to the desired location. A microcapsule containing auxiliary islets of Langerhans cells is implanted beneath the skin of patients in this procedure, which aids in the restoration of the body's glucose regulator feedback loop without the use of severe immunosuppressants (DiSanto et al., 2015).

Gene therapies that use non-viral and viral transduction could help people with diabetes by reducing islet cell destruction or replacing the insulin gene. Gene therapy could be done in epithelial cells, hepatocytes, or stem cells *in vivo* or *in vitro*. Islet allotransplantation could also restore normal blood glucose levels and prevent graft rejection (Wong et al., 2010).

3 Glycation

The word glycation has become a well-known term nowadays in biological research. The glycation process has involved the non-enzymatic interaction of amine groups of nitrogen-containing compounds viz. proteins, DNA, etc., with the carbonyl group of reactive sugars. Maillard first described the reaction of glycation interaction in 1912. This reaction is multistep and requires days, weeks, or months to complete (Yeh et al., 2017).

Heating and storage have enhanced the Maillard reaction during the processing of food. It leads to unstable Schiff's base and Amadori products, mainly fructosamine, in the early stage of the glycation process. In the intermediate stage of glycation, these unstable products get converted either into stable adducts or reversed. The redox reactions, precipitation, aggregation and various other physiological processes play a major role in the further generation of advanced glycation end-products (AGEs), which are considered late-stage glycation products (Kikuchi et al., 2003; Kumar et al., 2021). AGEs are known as irreversible and deleterious products. The reactivity of sugars plays an important role in increasing glycation. The monocarbonyl sugars, such as glucose, fructose, etc., were less reactive than the dicarbonyl species, viz. methylglyoxal. The generated AGEs in glycation observed their interference in diabetes secondary complications, neurodegenerative disorders, and cardiovascular disorders (Jha et al., 2018).

The generated AGEs are classified into fluorescent crosslinking AGEs, nonfluorescent crosslinking AGEs, and nonfluorescent non-crosslinking AGEs. The crossline, pentosidine imidazolium dilysine crosslinks, N-ε-carboxyethyllysine (CEL), and N-ε-carboxymethyl-lysine (CML) are the few well-characterized AGEs generated in the process of glycation. Most AGEs are irreversible in their production except for pyrraline and pentosidine (Yeh et al., 2017). In the last few decades, scientists have been dedicated to developing therapeutic drugs/compounds that could hinder or stop the production of AGEs. For example, aminoguanidine, a synthetic drug, is widely researched for inhibiting the glycation process. In addition, the natural compounds and secondary metabolites are also explored for the treatment of glycation-mediated complications and inhibition of the generation of glycation-mediated products in either stage of glycation (Abbas et al., 2015).

4 Oxidative Stress

Oxidative stress corresponds to the systemic inefficiency in balancing the level of reactive oxygen species produced in cells due to normal metabolic activities (Zahra et al., 2021). This imbalance affects cells' redox state, compromising their defense mechanisms to detoxify the reactive species (Nita & Grzybowski, 2016). In addition, prolonged oxidative stress causes damage to all components of the cell, including proteins, lipids, and DNA (Kamal et al., 2022).

Because of their high content of unsaturated fatty acids, lipid structures, particularly cell membranes, are vulnerable to (ROS/RNS) assault (Zahra et al., 2021). The oxidative lipid damage leads to what is known as "lipid peroxidation". It has been found that lipid peroxidation causes oxidative changes in the structure, function, and permeability of the biological membranes (Niki, 2008; Zahra et al., 2021). It also produces potentially toxic byproducts like malondialdehyde (MDA), 4-hydroxy-2-nonenal (HNE), and reactive carbonyl species (RCS), which are chemically reactive and covalently modify vital macromolecules such as DNA bases, proteins and low-density lipoproteins (LDL) leading to cytotoxic events (Niki, 2008; Zahra et al., 2021).

DNA is also influenced by oxidative stress. Oxidative stress causes additional DNA damage, including mismatch, inter-strand crosslink, single-strand breaks, intra-strand crosslink, oxidation of specific bases, and double-strand breaks (Fig. 24.2). Inefficient DNA repair mechanisms cause genome instability, which

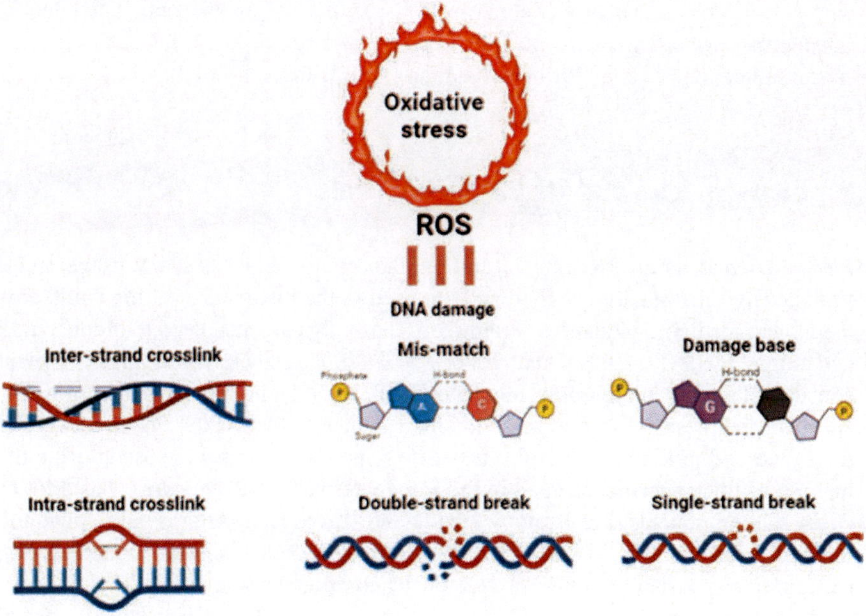

Fig. 24.2 DNA oxidative damage

leads to altered cell behavior or cell death (Alhmoud et al., 2020). The base alteration of 8-oxoguanine is one of the most important byproducts of DNA oxidation (8-oxo-G) (Zahra et al., 2021). It is mildly mutagenic and causes a G: C to T: A translocation (Zahra et al., 2021). It has been linked to cell transformation and cancer initiation (Zahra et al., 2021). Proteins have not escaped oxidative damage. In fact, because of their high concentrations, they are directly targeted by ROS/RNS (Niki, 2008). Deamination, decarboxylation, conformation modifications, changes in electrical charges, cross-linking, and side-chain modification may all occur as a result of protein oxidation, resulting in to halt of enzyme activity and accumulation within cells and the extracellular environment (Niki, 2008; Zahra et al., 2021). For example, histidine can be modified to oxo-histidine, "a damaged form of histidine" after exposure to oxidative stress (Ezraty et al., 2017). As a result of its toxic insult to biomolecules, oxidative stress can lead to grave diseases.

The accumulation of oxidative stress is the primary cause of aging on a cellular level (Romano et al., 2010). It is also associated with metabolic diseases like diabetes, neurodegenerative diseases like Parkinson's, Alzheimer's, depression, and cardiac diseases like myocardial infarction and atherosclerosis (Roberts & Sindhu, 2009). Besides, several diseases like cancer, vitiligo, and increased severity of microbial infections have been traced to oxidative stress (Chakraborti et al., 2019). Although the generation of free radicals, as a metabolic process, is the primary cause of oxidative stress, they do not always induce the same. In their production, they are instantly neutralized with the help of antioxidants (Kamal et al., 2022). However, stress, chemical exposure, and genetic/environmental factors more extensively trigger oxidative stress mechanisms. Therefore, the effects do not show pathological symptoms instantly (Zahra et al., 2021). Instead, they accumulate silently, causing extensive damage to macromolecules resulting in dreadful outcomes like cancer, aging, and death (Burton & Jauniaux, 2011; Preiser, 2012).

5 Ficus Species and Its Phytochemicals

Ficus carica L. is an economical and valuable plant. Fig has many usage areas. Fruits of figs are valuable as food and are used in the medicinal, getting paint, cosmetic, and medicine industries. Apart from these, figs are also used to produce raki and animal feed. (Kavak & Önal, 2006; Kıyak, 2009; Özbek, 1978). Figs is a plant considered sacred in religion and dates back to ancient times. In addition, the Hittites used these plants and recorded their names in their texts. Fig is among the trees recorded in Hittite cuneiform texts. It is known that figs were used during the making of thick sacrificial bread in these texts (Ünar, 2019). *F. carica* has a lot of chemical components. For example, fruits and leaves have Anthocyanin, flavonol, flavones, phenolic acid, volatile, aldehydes, alcohols, ketones, esters, monoterpens, sesquiterpens, norisoprenoids, miscellaneous compounds (Fig. 24.3).

At that time, leaves and roots have coumarin and sterol. *F. carica* latex is special latex. In addition, it has aldehydes (pentanal, hexanal, heptanal, and benzaldehyde),

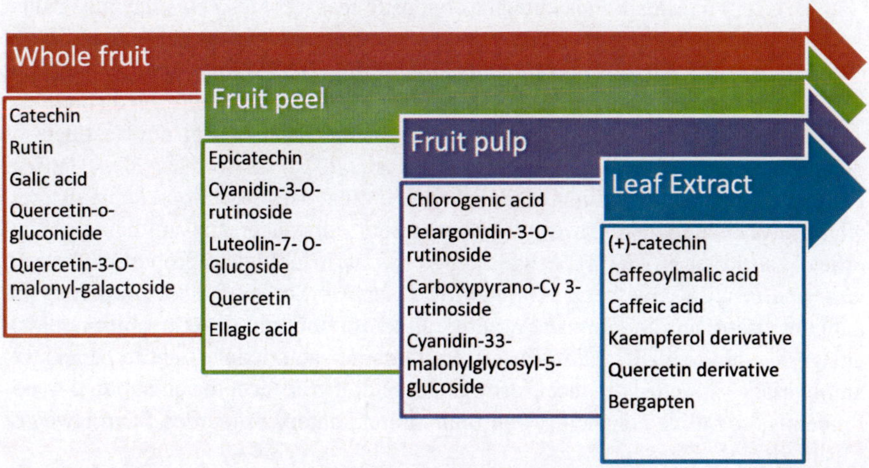

Fig. 24.3 Phytochemicals from different parts of *F. carica* plant

alcohols (1-butanol-3-methyl, 1-butanol-2-methyl, 1-pentanol, 1-hexanol, 1-heptanol, phenylethyl alcohol, phenylpropyl alcohol), Ketones (6-methyl-5-hepten-2-one), monoterpenes (α-thujene, α-pinene, β-pinene, limonene, eucalyptol, terpinolene, *cis*-linalool oxide, linalool, epoxylinalol), sesquiterpenes (α-guaiene, α-bourbonene, β-caryophyllene, *trans*-α-bergamotene, α-caryophyllene, s-muurolene, germacrene D, cadinene, a-calacorene), miscellaneous (methyl salicylate, quinoline, psoralene), organic acids (oxalic, citric, malic, quinic, shikimic, fumaric), fatty acids (myristic, pentadecanoic, palmitic, heptadecanoic, *cis*-10-heptadecenoic, stearic, oleic, elaidic, linoleic, arachidic, heneicosanoic, behenic, tricosanoic, lignoceric), steroids and amino acids (Barolo et al., 2014; Duman & Yazıcı, 2018). When the studies have been examined, it is common to find studies on the fruit and latex properties of figs. However, it is possible to come across studies on the wood of the fig plant in recent years. Moreover, the use of fig plants for landscaping has been emphasized in the research (Serin & Penezoğlu, 2020).

6 Natural Products from Ficus Species as Antidiabetic Agents

Various species of *Ficus* provide various valuable products such as edible fruits, fodder, and fuel wood, traditionally used as religious, ornamental, and medicinal plants by humans. It has been globally documented that the *Ficus* plants are significantly spiritual and provide material resources for humans (Shi et al., 2014). In the genus of *Ficus*, there are as many as 18 species, of which the species having importance in traditional medicines are *Ficus benghalensis,* popularly known as the Indian

Banyan tree, *Ficus hispida* is known as opposite leaf fig, *Ficus elastica,* also called rubber fig, *Ficus religiosa* which is also popularly known as peepal tree and *Ficus carica* also referred to as Figs or locally as Anjir in India. Among the different plant species, one of the prominent and frequently used plants in folk medicines like Siddha, Unani, homoeopathy, and Ayurveda, as well as used for the treatment of diabetes, is *Ficus benghalensis* (Kumari et al., 2016) as well as *Ficus carica.* Pharmacological studies support the ethnomedicinal uses of various *Ficus* species, which have been found to have anti-inflammatory, anticancer, and antidiabetic properties (Lansky et al., 2008). A wide range of pharmacological properties, both *in vitro* and *in vivo,* have been exhibited by *F. carica.* Some of these properties are anticonvulsant, anti-inflammatory, antiviral, antitumor, antidiabetic, antimicrobial, analgesic, antioxidant, antiulcer, antiasthmatic, antianxiety, antihypertensive, anthelmintic, apoptosis inducer, estrogenic, endothelin receptor antagonist, hypolipidemic, cognitive enhancing, and immunomodulatory properties (Kurniawan & Yusuf, 2021).

The indigenous people worldwide have traditionally been using the different parts of various *Ficus* species to cure various diseases, including diabetes. Several studies using *in vivo* and *in vitro* models have reported that the extracts are made from different parts of *Ficus* species (particularly *F. carica, F. benghalensis, F. deltoidei, F. glumosa, F. religiosa, and F. racemosa*), have potent anti-diabetic properties. Traditionally the fruits, seeds, leaves, bark, tender shoots, and latex *of F. carica* leaves have been used as an anti-inflammatory agent to treat diarrhea, jaundice, and nutritional anemia (Canal et al., 2000).

In African medicine (Olaokun et al., 2014), Ayurveda (Rangika et al., 2015), and Siddha medicine, diabetes is treated using the edible fruits and leaves of *Ficus carica* (Irudayaraj et al., 2017). The antidiabetic effects of organic extracts of different parts like fruits and leaves of *Ficus carica* were demonstrated in alloxan- and streptozotocin-induced diabetic animal models. Inducing insulin secretion, inhibiting absorption and increasing glucose uptake, and inhibiting levels of α-glucosidase, α-amylase, and pancreatic lipase (Irudayaraj et al., 2016; Mopuri et al., 2018), the extracts helped lessen any complications arising from diabetes or related conditions. Antifungal, antioxidant, anthelmintic, antimutagenic, antipyretic, antispasmodic, hepatoprotective, and hypoglycemic properties have been observed (Vikas & Vijay, 2011).

To elucidate the treatment of diabetes, the decoction prepared from the bark of *Ficus carica* was used (Perez et al., 1996). The bioactive agents believed to be present in *Ficus* are sterols, flavonoids, saponins, polyphenolic compounds, and tannins, to name a few. Hypoglycaemic activity in rat models was found to be elicited by ethanolic extracts of leaves of *Ficus carica.* In addition, the various important bioactive compounds imparting antidiabetic effect in *Ficus carica* flavonoid compounds, β-sitosterol, 6-*O*-acyl-b-D-glucosyl β-sitosterol, and polyphenols are reported to have antidiabetic effects (Mopuri et al., 2018; Kurniawan & Yusuf, 2021).

Antihyperglycemic activity has been observed in the extracts made from leaves of *Ficus carica* (Kurniawan & Yusuf, 2021). In animal models, when ethanol and ethyl acetate extract of *Ficus carica* was administered orally, it reduced blood glucose levels and increased insulin levels significantly. Fig extract was observed to correct any decrease in skeletal muscle (a site where glucose uptake is stimulated by insulin) and hepatic glycogen in diabetes (Irudayaraj et al., 2017). It has been reported that *F. carica* fruit extracts lower oxidative stress since oxidative stress is a major cause of the complications and pathogenesis of type 2 diabetes. The extracts also modulate the enzymes of antioxidant defense systems, especially α-glucosidase and α-amylase. Hence, from these studies, the extensive Ayurvedic use of plants such as *Ficus carica* for the treatment of diabetes has been corroborated by scientific evidence in recent years.

7 Prevention of Advanced Glycation End Products (AGEs) by Phytochemicals from *Ficus carica*

AGEs, a heterogeneous, chemically diverse, and complicated set of chemicals generated either exogenously or endogenously, have piqued the scientific community's interest in recent decades due to mounting evidence of their involvement in various disorders like diabetes mellitus (Davis et al., 2016). Inhibition of the generation of ROS compounds during the glycation process; inhibition of the Schiff base pathway and the production of Amadori products, inhibiting the formation of dicarbonyl compounds; detoxifying AGE precursors, such as methylglyoxal and influencing AGEs to bind to their receptors (RAGE) are the fundamental processes that prevent the formation of AGEs (Gkogkolou & Böhm, 2012).

Clinical trials have been conducted on aminoguanidine, the first chemical to suppress AGE production, but it is unlikely to be used for medicinal purposes due to safety concerns and a lack of efficacy (Dömötör et al., 2022). Therefore, AGE inhibitors derived from natural compounds with a low toxicity profile are more attractive options for the development of functional additives and even medications to cure and manage diabetes complications and other AGE-related disorders. In addition, several natural substances with antioxidant characteristics currently show good inhibitory action against AGE development with minimum toxicity (Song et al., 2021).

Phytochemical investigations of the plant's leaves and fruits revealed that they are high in phenolics, flavonoids, polysaccharides, vitamin C, alkaloids, saponins, coumarins, tannins, and organic acids, and volatile chemicals, making them a potent antioxidant (Caliskan & Polat, 2011). Another study demonstrated that the 96% ethanol extract of *F. carica* leaf from Indonesia had an IC_{50} value of less than 1000 μg/ml in the DPPH· assay, indicating it to be effective in suppressing the development of AGEs compounds (Abdillah And & Prakoso, 2018).

Polyphenol compounds suppress the development of AGEs and provide resistance to the pathogenic effects of AGEs, according to *in vitro* glycation studies. Several investigations have found a strong link between antiglycation action and phenolic extracts (Riyaphan et al., 2021). *F. carica* is an excellent source of phenolic compounds, quercetin rutinoside was the major individual phenolic, and luteolin (187 mg/kg extract) was the most abundant non-glycosylated flavonoid in *F. carica* (Vaya & Mahmood, 2006). An isolated rat heart examined the antioxidant activity and effects of *F. carica* leaf extract on ischemia/reperfusion injury. The treatment groups received enriched solutions containing the extract (0.04, 0.2, and 1 mg/ml, respectively) during stabilization and reperfusion (after 30 min of global ischemia). The dry sample had a total phenolic content of 12.2 mg GAE/100 g and flavonoid content of 40.7 mg/g (Allahyari et al., 2014). According to Mujić et al. (2010), in *F. carica*, phenolics ranged from 7.24 to 11.1 mg GAE/g dry extract, and the highest TPC was found in methanolic extracts. The most abundant phenolic acids and flavonoids in fig; include proanthocyanidins, gallic acid, chlorogenic acid, ferulic acid, rutin, quercetin-3-*O*-rutinoside, and epicatechin.

Most of the polysaccharides present in plants have low toxicity; therefore, they are a valuable source for developing glycation inhibitors, but their mechanism of action is still unknown. Therefore, the antioxidative activities of water extract and crude hot-water soluble polysaccharide (PS) from the fruit were investigated, and it was discovered that both water extract and crude hot-water soluble PS had significant scavenging activities at effective concentrations of 0.72 and 0.61 mg/mL respectively (Yang et al., 2009). Furthermore, it was discovered by Zou and his colleagues (2020) that the *F. carica* PS contains antioxidant and immunity-enhancing properties. PS, one of the most influential and significant components in *F. carica* L., has long been thought to be a favorable immunostimulant that might be exploited in animal and human immunotherapy (Wang et al., 2022).

F. carica leaves are rich in vitamin C. Vitamin C is a water-soluble vitamin that preserves the biomolecules while reducing oxidative stress. It makes ionic connections with proteins, allowing them to bind to them competitively and protect their physiological structure. *In vitro*, ascorbic acid can compete with glucose for protein binding, lowering AGE formation. In diabetic individuals, vitamin C inhibits the sorbitol pathway and reduces the production of fructose, a highly active sugar in glycation (Song et al., 2021).

Terpenoids are natural compounds with a wide range of structural and functional properties. Some terpenoids have been shown to reduce AGE production, for example, pentacyclic triterpenoids. They have various biological effects, including antioxidant, anti-atherosclerotic, antihyperglycemic, and anti-inflammatory effects (Cox-Georgian et al., 2019). Bauerenol, methyl maslinate, lupeol acetate, and oleanolic acid are the four triterpenoids that have been identified from the leaves of *F. carica* (Mawa et al., 2013). Some terpenoids have excellent antiglycation activities, but only a few investigations of their antiglycation effects have been conducted, and no clinical applications have been documented to date.

NMR data confirmed that cyanidin-3-*O*-rutinoside (C3R) was the main anthocyanin in all fruits of *F. carica* (Solomon et al., 2006). C3R is a naturally occurring anthocyanin with antioxidant, anti-hyperglycaemic, anti-glycation, and cardioprotective activities. In addition, C3R was found to be a mixed-type competitive inhibitor of pancreatic lipase with an IC_{50} of 59.4 1.41 M, according to Thilavech and Adisakwattana (2019).

Coumarins are heterocyclic chemicals with antiglycative activity that has the potential to cure insulin-dependent diabetic mellitus. Coumarin was extracted from the methanol extract of the leaves of *F. carica* L. using bioassay-guided isolation (Mawa et al., 2013). Coumarin improves bone turnover in diabetic osteoblasts and osteoclasts by reducing the production of advanced glycation end products (Pan et al., 2022).

8 Antioxidants Derived from Ficus in Diabetes Management

The substances that can break free radicals and decrease the oxidative stress found in living organisms are Antioxidants (Ayinde et al., 2007). These antioxidants can donate electrons which can inhibit free radical-related oxidation reactions. Antioxidants are either synthesized endogenously or can be taken from different naturally occurring sources from plants (Rout et al., 2009; Egbuna et al., 2021). Different studies are carried out to study the antioxidant property of different parts of the *Ficus* species. Researchers reported antioxidant activity against Trolox, ferric, lipid, and lipid peroxidation, reducing power in Ficus plant parts. Some researchers also found the free radical scavenging capacity of these antioxidants of the *Ficus* plant against DPPH· and 2,2-aninobis (3- ethylbenzothiazoline-6-sulfonic acid) in dose control studies as the *Ficus* plant extract contains a high amount of phenolic components (Mazumder et al., 2009). *Ficus* plant extract was found to possess anticancer, antidiabetic, Hepatoprotective, antitumor, analgesic, antimicrobial, hypoglycemic, gastro protective, anti-inflammatory, antifungal, antimalarial, and antiparasitic activity (Table 24.1). Most species of *Ficus* showed antimicrobial activity (Singh & Goel, 2009).

Different species of *Ficus* are used as traditional or alternate medicine for diseases including stomachic, hypotension, anti-dysentery drugs, and diabetes. Some species of *Ficus* like *F.glomerata*, *F.glumosa*, *F. carica*, *F. racemosa*, and *F. benghalensis* are effective in treating diabetes (Li et al., 2021). The aqueous extract of *F. benghalensis* is found to regulate insulin secretion and enzymatic activities, increase insulin secretion, carbohydrate absorption, hepatic glycogen synthesis, and antioxidant activity in the diabetic patient body (Ren et al., 2022). *In vivo* study on male rate found that aqueous extract of *F. glumosa*, *F. carica*, *F. racemosa*, and *F. benghalensis* reduces oxidative stress and increases the weight of these rats. Another study with methanol extract of *F. carica* leaves helps to control the increase in lipid peroxide by functioning as a hepatoprotective agent (Kerian, 2021; Sahardi et al., 2021).

In vitro ethanol extract of fruits of *Ficus racemosa* Linn. on blood glucose in experimental animals promote glucose uptake and metabolism or inhibit hepatic gluconeogenesis since alloxan treatment causes permanent destruction of β-cells. Furthermore, the DPPH· radical scavenging activity of this extract prevents the ill effects of diabetes and oxidative stress in vivo studies. Another study by Siddique and Saleem (2011) on ethanol extract of *F. racemosa* showed the presence of lupeol

Table 24.1 Antioxidant and biological activities of different parts of plants

Species of *Ficus*	Part of plant	Solvent type	Antioxidant activity	Biological activity	Reference
Ficus carica	Leaves	Hexane, water	TEAC: 14.0%, 23.5 acetate/g DW FRAC: 7.9–16.1 mmol/kg FW DPPH-RSC: 11.4 mmol/100 g DW ABT-RSC: 6.48 mmol/100 g DW	Hepatoprotective activity with a decrease in lipid peroxides with cytochrome p450 complex inhibition	Mondal et al. (2022)
	Fruit	Dichloromethane	ILP: 0.02 mg/mL		
		N hexane	ILP: 1.64 mg/mL		
Ficus racemosa	Stem	Methanol	DPPH-RSC: 16.2%, ABT-RSC: 8615.3 mmol/g DM	Hepatoprotective activity by reducing the activities of ALT, AST, and ALP	Alzahrani et al. (2021) and Tewari et al. (2021)
	Bark	Ethanol	DPPH-RSC: 79%	Anticancer activity by reduction of lipid peroxidation, γ-glutamyl transpeptidase, and xanthine oxidase and by generation of hydrogen peroxide Hypoglycemic activity by decreasing blood glucose level	
	Roots	Water	FRAC: 0.5–0.26 mg/mL		
Ficus benghalensis	Aerial roots	Methanol	DPPH-RSC: 71%, ABT-RSC: 10884.6 mmol/g DM	Antioxidant and hypolipidemic activity by reduction in lipid peroxidation, cholesterol level and triacylglycerol	Joshi et al. (2021)
		Acetone, Water	FRAC: 0.5–0.26 mg/mL	Anticancer and antibacterial activity but no antifungal activity	Kapile et al. (2022)

Ficus sur	Bark	Water	TEAC: 14.0%, 23.5 acetate/g DW, FRAC:7.9–16.1 mmol/kg FW, 6.48 mmol/100 g DW, DPPH-RSC: 11.4 mmol/100 g DW, ABT-RSC: 10884 mmol/g DM	Antifungal and antibacterial activities	Sieniawska et al. (2022)
	Unripe fruit		TEAC: 62.3 GAE/g DW, FRAC:19.6 μmol FSE/mg DW, DPPH-RSC:7.3 QE/mg DE		
Ficus religiosa	Fruit	Water	DPPH-RSC: 16.2%, FRAC:0.5–0.26 mg/mL, ABT-RSC:8615 mmol/g DM	Antitumor activity due to blockage of calcium uptake in pituitary cells	Ghadigaonkar et al. (2021)
	Whole	Water, Methanol		Antioxidant and antidiabetic activity with lowering the superoxide dismutase exaggerated activity	Randima et al. (2022)
	Bark	Water	DPPH-RSC: 79%, ABT-RSC:10884.6 μmol/g DM	Antimicrobial activity with inhibition zone against *B. subtilis*	Murugesu et al. (2021)
Ficus polita Vahl	Whole	Water	TEAC:2.45%, 5.40 mg GAE/g DW	Antiviral activity due to inhibition of reverse transcriptase activity of HIV-1	Sieniawska et al. (2022)
	Leaves	Water		Antimalarial action against Plasmodium falciparum	Ben Nasr et al. (2021)

as a super antioxidant (Siddique & Saleem, 2011). Lupeol decreases peroxidation and is effective against diabetes, cardiac, and reno-protective activities (Joshi et al., 2016).

9 Natural Products from *Ficus carica* for Prevention of Secondary Complications of Diabetes

- **Cardiovascular diseases** include diseases/disorders of the heart and blood clots. Alamgeer et al. (2017) deduced that the presence of phenols, flavonoids, and potassium contributed to the antihypertensive and cardioinhibitory effects. However, the beta-adrenergic and calcium channel blocking activity were not detected, implying the antihypertensive effect was due to some other mechanism. In another study, 70% methanolic extracts of *F. carica* leaves were used to prevent ischemia/reperfusion I/R-induced myocardial infarction in isolated rat hearts (Allahyari et al., 2014).
- **Neurodegenerative disorders** imply progressive damage to the structure and function of neurons. It includes diseases like prion disease, Parkinson's disease, Alzheimer's disease, and Huntington's disease. Since oxidative stress contributes to the diseases, antioxidants from *F. carica,* namely hydroxycinnamic acids, flavonoids (e.g., rutin), etc. (Szwajgier et al., 2017). According to a study, the n-hexane and acetone extract of the leaves could inhibit activities of the enzyme acetylcholinesterase (AChE) by 62.9% and 50.8%, and that of butyrylcholinesterase (BChE) by 76.9% and 45.6%, respectively (Orhan et al., 2011).
- **Kidney damage (nephropathy)**
  Ghaffar et al. (2015) induced nephrotoxicity in adult albino mice using gentamicin, thereby increasing their serum urea and creatinine levels. *Ficus carica* L. leaf extract was fed orally to a group of mice at a concentration of 400 mg/kg/day along with gentamicin 200 mg/kg/day intraperitoneally. Their histological structure was compared with healthy mice and mice with induced nephropathy, and a marked improvement in the renal structure was observed. In another study, *F. carica* supplementation improved the serum ALT (alanine aminotransferase) and AST (aspartate aminotransferase) levels and ameliorated the metabolic alterations induced by gamma irradiation in male albino rats (Fouad et al., 2019).

  F. carica leaves extract was coated with Au nanoparticles to increase the sensitivity and circumvent the potential drug resistance. Au-NPs and extract at the ratio of 3:2 could significantly decrease blood urea, MDA, homocysteine levels, serum creatinine, and hydroxyproline content in kidney tissue homogenate of Cisplatin-induced acute kidney injury (El-Sayed et al., 2019)
- **Eye damage (retinopathy)**
  Güven et al., 2019 extracted fig seed oil and found that the oil was high in α-linolenic (n-3) acid, one of the *Omega* 3 fatty acids recommended for ocular

health. Vitamin E and α-linolenic acids from *F. carica* are used in supplementary therapy to treat yellow spot damage or macular degeneration (Richer et al., 2016)

- **Foot damage**
  Most diabetic patients experience peripheral artery disease (PAD), leading to decreased foot blood flow. If these patients suffer from neuropathy, they lose sensations in their feet altogether. This subjects the feet to a plethora of problems ranging from foot ulcers and calluses to increased skin infections, eventually leading to amputation of the feet (Holt et al., 2021). Phytochemicals from *F. carica* could prevent neuropathy and prove an essential remedy for diabetes-induced foot damage.
- **Skin conditions**
  According to American Diabetes Association, skin problems are usually the first warning signs of diabetes. Bacterial and fungal infections are usually triggered in diabetic patients. However, several skin conditions are observed chiefly or only in people with diabetes (Fig. 24.4).

Traditionally, leaf paste and latex of *F. carica* have been used for treating a wide array of skin infections. According to Park et al. (2013), ethanolic extract from fig branches could considerably reduce the production of TNF-α, i.e., a pleiotropic inflammatory cytokine. In another study, the flavonoids from fig's skin extract could inhibit tyrosinase enzyme at IC_{50} values from 0.09 to 0.45 mg/m: (Meziant et al., 2021). This would imply it can be used for skin depigmentation observed in skin conditions resulting from diabetes.

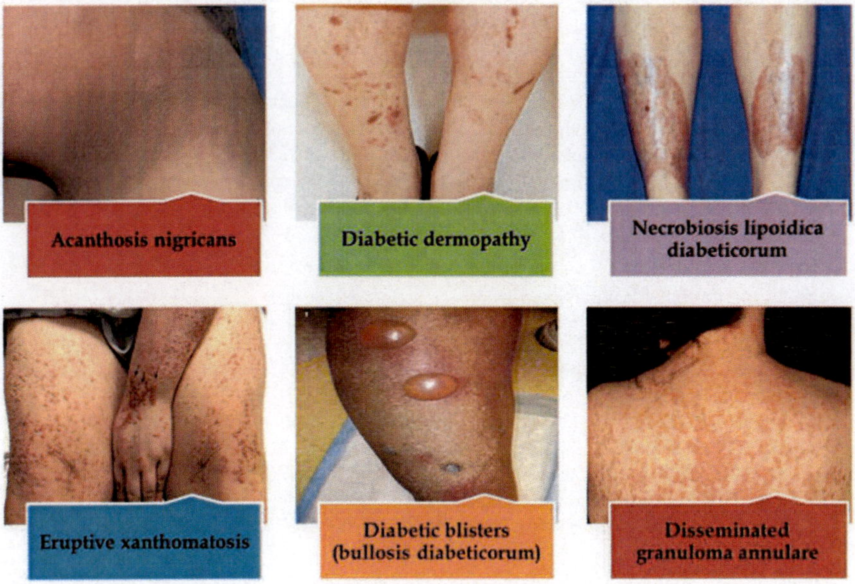

Fig. 24.4 Skin conditions observed in diabetic patients

References

Abbas, G., Al-Harrasi, A. S., Hussain, H., Hussain, J., Rashid, R., & Choudhary, M. I. (2015). Antiglycation therapy: Discovery of promising antiglycation agents for the management of diabetic complications. *Pharma Bio, 54*(2), 198–206.

Alamgeer, I. S., Asif, H., & Saleem, M. (2017). Evaluation of antihypertensive potential of *Ficus carica* fruit. *Pharmaceutical Biology, 55*(1), 1047–1053. https://doi.org/10.1080/13880209.2017.1278611

Alhmoud, J. F., Woolley, J. F., Al Moustafa, A. E., & Malki, M. I. (2020). DNA damage/repair management in cancers. *Cancers, 12*(4), 1050.

Allahyari, S., Delazar, A., & Najafi, M. (2014). Evaluation of general toxicity, antioxidant activity and effects of *Ficus carica* leaves extract on ischemia/reperfusion injuries in isolated heart of rat. *Advanced Pharmaceutical Bulletin, 4*(2), 577–582.

Alzahrani, F. O., Ahmed, A. A., Al-Robai, S. A., Mohamed, H. A., Abdallah, M. E., & Abdalla, A. N. (2021). Total phenol, flavonoid, and individual phenolic acid contents and antioxidant and cytotoxic activities of methanol extracts from *Ficus cordata* ssp. salicifolia. *Canadian Journal of Plant Science, 101*(4), 517–524.

And, A. S., & Prakoso, A. W. (2018). Antioxidant and anti-ages (advanced glycation product) activities of ethanolic extract of *Ficus carica* Linn from Indonesia. *Journal of Pharmacognosy and Phytochemistry, 7*(5), 598–601.

Ayinde, B. A., Omogbai, E., & Amaechina, F. C. (2007). Pharmacognosy and hypotensive evaluation of Ficus exasperata Vahl (Moraceae) leaf. *Acta Poloniae Pharmaceutica, 64*(6), 543–546.

Barolo, M. I., Mostacero, N. R., & López, S. N. (2014). *Ficus carica* L. (Moraceae): An ancient source of food and health. *Food Chemistry, 164*, 119–127.

Ben Nasr, R., Baudelaire, E. D., Dicko, A., & Ouarda, H. E. F. (2021). Phytochemicals, antioxidant attributes and larvicidal activity of *Mercurialis annua* L. (Euphorbiaceae) leaf extracts against *Tribolium confusum* (Du Val) larvae (Coleoptera; Tenebrionidae). *Biology, 10*(4), 344.

Boulé, N. G., Haddad, E., & Kenny, G. P. (2001). Effects of exercise on glycemic control and body mass in type 2 diabetes mellitus: A meta-analysis of controlled clinical trials. *JAMA, 286*, 1218–1227.

Burton, G. J., & Jauniaux, E. (2011). Oxidative stress. *Best Practice & Research: Clinical Obstetrics & Gynaecology, 25*(3), 287–299.

Caliskan, O., & Polat, A. A. (2011). Phytochemical and antioxidant properties of selected fig (*Ficus carica* L.) accessions from the eastern Mediterranean region of Turkey. *Scientia Horticulture, 128*(4), 473–478. https://doi.org/10.1016/j.scienta.2011.02.023

Canal, J. R., Torres, D. M., Romero, A., & Perez, C. (2000). A chloroform extract obtained from a decoction *of Ficus carica* L. leaves improve the cholesterolaemic status of rats with streptozotocin induced diabetes. *Acta Veterinaria Hungarica, 87*, 71–76.

Chakraborti, S., Chakraborti, T., Chattopadhyay, D., & Shaha, C. (Eds.). (2019). *Oxidative stress in microbial diseases*. Springer.

Cooke, D. W., & Plotnick, L. (2008). Type 1 diabetes mellitus in pediatrics. *Pediatrics in Review, 29*(11), 374–384.

Coustan, D. R. (1993). Gestational diabetes. *Diabetes Care, 16*(3), 8–15.

Cox-Georgian, D., Ramadoss, N., Dona, C., & Basu, C. (2019). Therapeutic and medicinal uses of terpenes. In *Medicinal plants: From farm to pharmacy* (pp. 333–359). Springer. https://doi.org/10.1007/978-3-030-31269-5_15

Davis, K. E., Prasad, C., Vijayagopal, P., Juma, S., & Imrhan, V. (2016). Advanced glycation end products, inflammation, and chronic metabolic diseases: Links in a chain? *Critical Reviews in Food Science and Nutrition, 56*(6), 989–998. https://doi.org/10.1080/10408398.2012.744738. PMID: 25259686.

DiSanto, R. M., Subramanian, V., & Gu, Z. (2015). Recent advances in nanotechnology for diabetes treatment. *Wiley Interdisciplinary Reviews: Nanomedicine and Nanobiotechnology, 7*(4), 548–564.

Dömötör, O., May, N. V., Gál, G. T., Spengler, G., Dobrova, A., Arion, V. B., & Enyedy, É. A. (2022). Solution equilibrium studies on salicylidene amino guanidine Schiff Base metal complexes: Impact of the hybridization with L-proline on stability, redox activity and cytotoxicity. *Molecules, 2044*, 27. https://doi.org/10.3390/molecules27072044

Duman, E., & Yazıcı, A. S. (2018). Yaş İncir (Mor Güz-Sarı Lop) Çekirdek ve Çekirdek Yağlarının Fiziko-Kimyasal Özellikleri. *Anadolu Ege Tarımsal Araştırma Enstitüsü Dergisi, 28*(1), 69–76.

Egbuna, C., Awuchi, C. G., Kushwaha, G., Rudrapal, M., Patrick-Iwuanyanwu, K. C., Singh, O., Odoh, U. E., Khan, J., Jeevanandam, J., & Kumarasamy, S. (2021). Bioactive compounds effective against type 2 diabetes mellitus: A systematic review. *Current Topics in Medicinal Chemistry, 21*(12), 1067–1095.

El-Sayed, S. M., El-Naggar, M. E., Hussein, J., Medhat, D., & El-Banna, M. (2019). Effect of *Ficus carica* L. leaves extract loaded gold nanoparticles against cisplatin-induced acute kidney injury. *Colloids and Surfaces B: Biointerfaces, 184*, 110465.

Ercisli, S., Tosun, M., Karlidag, H., et al. (2012). Color and antioxidant characteristics of some fresh fig (*Ficus carica* L.) genotypes from northeastern Turkey. *Plant Foods for Human Nutrition, 67*, 271–276. https://doi.org/10.1007/s11130-012-0292-2

Evert, A. B., Dennison, M., Gardner, C. D., Garvey, W. T., Lau, K. H., & MacLeod, J. (2019). Nutrition therapy for adults with diabetes or prediabetes: A consensus report. *Diabetes Care, 42*(5), 731–754.

Ezraty, B., Gennaris, A., Barras, F., & Collet, J.-F. (2017). Oxidative stress, protein damage and repair in bacteria. *Nature Reviews Microbiology, 15*(7), 385–396.

Fouad, D., Alhatem, H., Abdel-Gaber, R., & Ataya, F. (2019). Hepatotoxicity and renal toxicity induced by gamma-radiation and the modulatory protective effect of *Ficus carica* in male albino rats. *Research in Veterinary Science, 125*, 24–35.

Ghadigaonkar, S., Reddy, A., Kalakumar, B., Lakshman, M., & Rajkumar, U. (2021). Quantification of total phenolic content, total flavonoid content and evaluation of *in vitro* free radical scavenging activities in *Ficus religiosa* Linn. *Pharma Innovation Journal, 10*(3), 84–88.

Ghaffar, A., Tahir, M., Lone, K. P., Faisal, B., & Latif, W. (2015). The effect of *Ficus carica* L. (anjir) leaf extract on gentamicin induced nephrotoxicity in adult male albino mice. *The Journal of Ayub Medical College, Abbottabad: JAMC, 27*(2), 398–401.

Gkogkolou, P., & Böhm, M. (2012). Advanced glycation end products: Key players in skin ageing? *Dermato-Endocrinology, 4*(3), 259–270. https://doi.org/10.4161/derm.22028

Güven, N., Gökyer, A., Koç, A., Temiz, N. N., Selvi, S., Koparal, B., et al. (2019). Physiochemical composition of fig seed oil from Turkey. *The Journal of Pharmacy and Pharmacology, 7*, 541–545.

Holt, R. I., DeVries, J. H., Hess-Fischl, A., Hirsch, I. B., Kirkman, M. S., Klupa, T., et al. (2021). The management of type 1 diabetes in adults. A consensus report by the American Diabetes Association (ADA) and the European Association for the Study of diabetes (EASD). *Diabetes Care, 44*(11), 2589–2625.

Irudayaraj, S. S., Stalin, A., Sunil, C., Duraipandiyan, V., Al-Dhabi, N. A., & Ignacimuthu, S. (2016). Antioxidant, antilipidemic and antidiabetic effects of ficusin with their effects on GLUT4 translocation and PPARγ expression in type 2 diabetic rats. *Chemico-Biological Interactions, 256*, 85–93.

Irudayaraj, S. S., Christuda, S., Antony, S., Duraipandiyan, V., Al-Dhabi, N. A., & Ignacimuthu, S. (2017). Protective effects of *Ficus carica* leaves on glucose and lipids levels, carbohydrate metabolism enzymes and b-cells in type 2 diabetic rats. *Pharmaceutical Biology, 55*(1), 1074–1081.

Jha, P., Momin, A. R., Kumar, D., & Ali, A. (2018). Reversal of glycoxidative damage of DNA and protein by antioxidants. *Annals of Phytomedicine, 7*(1), 101–105.

Joshi, H., Vaishnav, D., Sanghvi, G., Rabadia, S., Airao, V., Sharma, T., Parmar, S., & Sheth, N. (2016). *Ficus recemosa* bark extract attenuates diabetic complications and oxidative stress in STZ-induced diabetic rats. *Pharmaceutical Biology, 54*(9), 1586–1595.

Joshi, D. G., Jat, R. K., & Patil, S. B. (2021). In vitro protein denaturation and membrane stabilising anti-arthritic activity of aqueous extracts of bark of *Ficus benghalensis* L. against methotrexate. *The Pharma Innovation Journal, 10*(4), 689–692.

Kamal, F. Z., Stanciu, G. D., Lefter, R., Cotea, V. V., Niculaua, M., Ababei, D. C., et al. (2022). Chemical composition and antioxidant activity of *Ammi visnaga* L. essential oil. *Antioxidants, 11*(2), 347.

Kapile, C., Kulkarni, A., Pardeshi, P., Sayed, A., & Nehe, A. (2022). *Ficus religiosa*: A beneficial medicinal plant. *Journal of Drug Delivery and Therapeutics, 12*(2-S), 210–218.

Kavak, F., & Önal, A. (2006). *Yün İncir (Ficus carica) Yaprağından Boyar Madde Ekstraksiyonu İle, Tüylü Deri Ve Pamuklu Kumaş Boyamadaki Kullanılabilirliği*. XX. Ulusal Kimya Kongresi, Erciyes Üniversitesi, 4–8 Eylül 2006.

Kerian, K. (2021). Polysaccharides as wound healing agent: A mini review. *Food Research, 5*(2), 31–37.

Kikuchi, S., Shinpo, K., Takeuchi, M., Yamagishi, S., Makita, Z., Sasaki, N., & Tashiro, K. (2003). Glycation: A sweet tempter for neuronal death. *Brain Research Brain Research Reviews, 41*, 306–323.

Kıyak, B. (2009). *Güney Ege Bölgesinde Meyve Üretimi, İlçelerin Performansları ve Çiftçinin Eline Geçen Fiyatlar*. Güney Ege Kalkınma Ajansı.

Krentz, A. J., & Bailey, C. J. (2005). Oral antidiabetic agents: Current role in type 2 diabetes mellitus. *Drugs, 65*(3), 385–411.

Kumar, D., Desa, A., Chougle, S., Bhatkalkar, S. G., Sachar, S., Selvaa, K. C., & Ali, A. (2021). Evaluation of the antiglycating potential of thymoquinone and its interaction with BSA. *Journal of Biomolecular Structure and Dynamics, 40*(18), 8455–8463. https://doi.org/10.1080/07391102.2021.1912642

Kumari, S. J., Sangeetha, M., & Pavithra, R. (2016). A retrospective review on Indian traditional herbs and its biocompounds in diabetes. *The International Journal of Pharmacology Research, 9*, 444–460.

Kurniawan, M. F., & Yusuf, F. A. (2021). The possible antidiabetic effect of Ficus carica L. tablet on lloxan-induced diabetes model in rats. *Open Access Macedonian Journal of Medical Sciences, 9*(A), 727–734.

Lansky, E. P., Paavilainen, H. M., Pawlus, A. D., & Newman, R. A. (2008). Ficus spp. (fig): Ethnobotany and potential as anticancer and anti-inflammatory agents. *Journal of Ethnopharmacology, 119*, 195–213.

Li, Y.-M., Zhong, R.-f., Chen, J., & Luo, Z.-G. (2021). Structural characterization, anticancer, hypoglycemia and immune activities of polysaccharides from *Russula virescens*. *International Journal of Biological Macromolecules, 184*, 380–392.

Mawa, S., Husain, K., & Jantan, I. (2013). *Ficus carica* L. (Moraceae): Phytochemistry, traditional uses and biological activities. *Evidence-Based Complementary and Alternative Medicine: Ecam, 2013*, 974256. https://doi.org/10.1155/2013/974256

Mazumder, P. M., Farswan, M., & Parcha, V. (2009). Hypoglycaemic effect of -*Ficus arnottiana* Miq. bark extracts on streptozotocin induced diabetes in rats. *Natural Product Radiance, 8*(5), 478–482.

Meziant, L., Bachir-bey, M., Bensouici, C., Saci, F., Boutiche, M., & Louaileche, H. (2021). Assessment of inhibitory properties of flavonoid-rich fig (*Ficus carica* L.) peel extracts against tyrosinase, α-glucosidase, urease and cholinesterases enzymes, and relationship with antioxidant activity. *The European Journal of Integrative Medicine, 43*, 101272.

Misbah, H., Aziz, A. A., & Aminudin, N. (2013). Antidiabetic and antioxidant properties of *Ficus deltoidea* fruit extracts and fractions. *BMC Complementary and Alternative Medicine, 13*, 118. https://doi.org/10.1186/1472-6882-13-118

Moller, D. E. (2001). New drug targets for type 2 diabetes and metabolic syndrome. *Nature, 414*(6865), 821–827.

Mondal, M., Sarkar, C., Saha, S., Hossain, M. N., Norouzi, R., Mubarak, M. S., Siyadatpanah, A., Wilairatana, P., Hossain, R., & Islam, M. T. (2022). Hepatoprotective activity of andrographolide possibly through antioxidative defense mechanism in Sprague-Dawley rats. *Toxicology Reports, 9*, 1013.

Mopuri, R., Ganjayi, M., Meriga, B., Koorbanally, N. A., & Islam, M. S. (2018). The effects of Ficus carica on the activity of enzymes related to metabolic syndrome. *Journal of Food and Drug Analysis, 26*(1), 201–210.

Mujić, I., Dudas, S., Skutin, H. M., Perusic, D., Zeković, Z., Lepojević, Z., Radojković, M., Vidović, S., Milošević, S., & Mesic, E. O. (2010). *Determination of antioxidant properties of fig fruit extract (Ficus carica L.)*. ISHS Acta Horticulturae 940: XXVIII international horticultural congress on science and horticulture for people (IHC2010): International symposium on the challenge for a sustainable production, protection and consumption of Mediterranean fruits and nuts.

Murugesu, S., Selamat, J., & Perumal, V. (2021). Phytochemistry, pharmacological properties, and recent applications of *Ficus benghalensis* and *Ficus religiosa*. *Plants, 10*(12), 2749.

Nathan, D. M., Cleary, P. A., Backlund, J. Y., Genuth, S. M., Lachin, J. M., Orchard, T. J., Raskin, P., Zinman, B., & Diabetes Control and Complications Trial/Epidemiology of Diabetes Interventions and Complications (DCCT/EDIC) Study Research Group (2005). Intensive diabetes treatment and cardiovascular disease in patients with type 1 diabetes. *The New England Journal of Medicine, 353*(25), 2643–2653. https://doi.org/10.1056/NEJMoa052187

Niki, E. (2008). Lipid peroxidation products as oxidative stress biomarkers. *BioFactors, 34*(2), 171–180.

Nita, M., & Grzybowski, A. (2016). The role of the reactive oxygen species and oxidative stress in the pathomechanism of the age-related ocular diseases and other pathologies of the anterior and posterior eye segments in adults. *Oxidative Medicine and Cellular Longevity, 2016*, 3164734.

Norman, A., & Henry, H. (2015). *Hormones* (pp. 136–137). Academic Press. ISBN 9780123694447.

Olaokun, O. O., McGaw, L. J., Awouafack, M. D., Eloff, J. N., & Naidoo, V. (2014). The potential role of GLUT4 transporters and insulin receptors in the hypoglycaemic activity of *Ficus lutea* acetone leaf extract. *BMC Complementary and Alternative Medicine, 14*, 269.

Orhan, I. E., Üstün, O., & Şener, B. (2011). Estimation of cholinesterase inhibitory and antioxidant effects of the leaf extracts of Anatolian *Ficus carica* var. domestica and their total phenol and flavonoid contents. *Natural Product Communications, 6*(3), 1934578X1100600315.

Özbek, S. (1978). *Özel Meyvecilik*. Çukurova Üniversitesi, Ziraat Fakültesi Yayınları, No: 128. Ders Kitabı, Adana.

Pan, Y., Liu, T., Wang, X., & Sun, J. (2022). Research progress of coumarins and their derivatives in the treatment of diabetes. *Journal of Enzyme Inhibition and Medicinal Chemistry, 37*(1), 616–628. https://doi.org/10.1080/14756366.2021.2024526

Park, S., Han, J., Im, K., Whang, W. K., & Min, H. (2013). Antioxidative and anti-inflammatory activities of an ethanol extract from fig (*Ficus carica*) branches. *Food Science and Biotechnology, 22*(4), 1071–1075.

Perez, C., Dominguez, E., Ramiro, J. M., Romero, A., Campillo, J. E., & Torres, M. D. (1996). A study on the glycaemic balance in streptozotocin diabetic rats treated with aqueous extract of *Ficus carica* (Fig tree) leaves. *Phytotherapy Research, 10*, 82–83.

Preiser, J. C. (2012). Oxidative stress. *Journal of Parenteral and Enteral Nutrition, 36*(2), 147–154.

Randima, T., Gajaman, P., Wimalasiri, Y., Perera, P., Navaratne, V., & Sanjeewani, N. (2022). In vitro evaluation and comparison of antibacterial activity of the novel herbal aqueous creams containing aqueous extracts of the leaves of *Ficus religiosa* L. *Sri Lanka Journal of Indigenous Medicine (SLJIM), 6*(02), 501–513. Retrieved from http://192.248.19.137/journal/index.php/sljim/article/view/29

Rangika, B. S., Dayananda, P. D., & Peiris, D. C. (2015). Hypoglycemic and hypolipidemic activities of aqueous extract of flowers from *Nycantus arbor-tristis* L. in male mice. *BMC Complementary and Alternative Medicine, 15*, 289.

Ren, R., Zeng, H., Mei, Q., Xu, Z., Mazhar, M., & Qin, L. (2022). Effects of *Monascus purpureus*-fermented tartary buckwheat extract on the blood lipid profile, glucose tolerance and antioxidant enzyme activities in KM mice. *Journal of Cereal Science, 105*, 103465.

Richer, S., Ulanski, L., & Popenko, N. A. (2016). Age-related macular degeneration beyond the age-related eye disease study II. *Advances in Ophthalmology and Optometry, 1*, 335–369.

Riyaphan, J., Pham, D. C., Leong, M. K., & Weng, C. F. (2021). In silico approaches to identify polyphenol compounds as α-glucosidase and α-amylase inhibitors against type-II diabetes. *Biomolecules, 11*(12), 1877. https://doi.org/10.3390/biom11121877

Roberts, C. K., & Sindhu, K. K. (2009). Oxidative stress and metabolic syndrome. *Life Sciences, 84*(21–22), 705–712.

Romano, A. D., Serviddio, G., De Matthaeis, A., Bellanti, F., & Vendemiale, G. (2010). Oxidative stress and aging. *Journal of Nephrology, 23*, S29–S36.

Rout, S., Panda, T., & Mishra, N. (2009). Ethno-medicinal plants used to cure different diseases by tribals of Mayurbhanj district of North Orissa. *Studies on Ethno-Medicine, 3*(1), 27–32.

Sahardi, N. F. N. M., Jaafar, F., Zakaria, S. N. A., TAN, J., Nordin, M. F. M., & Makpol, S. (2021). Comparison of the antioxidant activity of Malaysian ginger (*Zingiber officinale* roscoe) extracts with that of selected natural products and its effect on the viability of myoblast cells in culture. *Sains Malaysiana, 50*(5), 1445–1456.

Serin, Z. O. S., & Penezoğlu, M. K. (2020). İncir (*Ficus carica*) odunun kimyasal, fiziksel ve morfolojik özellikleri. *Avrupa Bilim ve Teknoloji Dergisi, 19*, 843–849.

Shi, Y., Hu, H., Xu, Y., & Liu, A. (2014). An ethnobotanical study of the less known wild edible figs (genus Ficus) native to Xishuangbanana, Southwest China. *Journal of Ethnobiology and Ethnomedicine, 10*, 68.

Shoback, D. G., & Gardner, D. (2011). *Greenspan's basic & clinical endocrinology* (9th ed.). McGraw-Hill Medical. ISBN 978-0-07-162243-1.

Siddique, H. R., & Saleem, M. (2011). Beneficial health effects of lupeol triterpene: A review of preclinical studies. *Life Sciences, 88*(7–8), 285–293.

Sieniawska, E., Świątek, Ł., Sinan, K. I., Zengin, G., Boguszewska, A., Polz-Dacewicz, M., Bibi Sadeer, N., Etienne, O. K., & Mahomoodally, M. F. (2022). Phytochemical insights into Ficus sur extracts and their biological activity. *Molecules, 27*(6), 1863.

Singh, D., & Goel, R. K. (2009). Anticonvulsant effect of *Ficus religiosa*: Role of serotonergic pathways. *Journal of Ethnopharmacology, 123*(2), 330–334.

Solomon, A., Golubowicz, S., Yablowicz, Z., Grossman, S., Bergman, M., Gottlieb, H. E., Altman, A., Kerem, Z., & Flaishman, M. A. (2006, October). Antioxidant activities and anthocyanin content of fresh fruits of common fig (*Ficus carica* L.). *Journal of Agricultural and Food Chemistry, 54*(20), 7717–7723. https://doi.org/10.1021/jf060497h

Song, Q., Liu, J., Dong, L., Wang, X., & Zhang, X. (2021). Novel advances in inhibiting advanced glycation end-product formation using natural compounds. *Biomed Pharmacotherapy, 140*, 111750. https://doi.org/10.1016/j.biopha.2021.111750

Szwajgier, D., Borowiec, K., & Pustelniak, K. (2017). The neuroprotective effects of phenolic acids: Molecular mechanism of action. *Nutrients, 9*(5), 477.

Tall, J. M., Seeram, N. P., Zhao, C., Nair, M. G., Meyer, R. A., & Raja, S. N. (2004). Tart cherry anthocyanins suppress inflammation-induced pain behaviour in rat. *Behavioural Brain Research, 153*, 181–188.

Tewari, D., Zengin, G., Ak, G., Sinan, K. I., Cziáky, Z., Mishra, S. T., & Jekő, J. (2021). Phenolic profiling, antioxidants, multivariate, and enzyme inhibitory properties of wild Himalayan fig (*Ficus palmata* Forssk.): A potential candidate for designing innovative nutraceuticals and related products. *Analytical Letters, 54*(9), 1439–1456.

Thilavech, T., & Adisakwattana, S. (2019, September 5). Cyanidin-3-rutinoside acts as a natural inhibitor of intestinal lipid digestion and absorption. *BMC Complementary and Alternative Medicine, 19*(1), 242. PMID: 31488210; PMCID: PMC6727418. https://doi.org/10.1186/s12906-019-2664-8

Ünar, Ş. (2019). Fruıt Trees In Anatolıa Of Hıttıte Perıod (Hitit Dönemi Anadolu'sunda Meyve Ağaçları). *Anasay, 9*, 11–30.

Vaya, J., & Mahmood, S. (2006). Flavonoid content in leaf extracts of the fig (*Ficus carica* L.), carob (*Ceratonia siliqua* L.) and pistachio (*Pistacia lentiscus* L.). *BioFactors, 28*(3–4), 169–175.

Vikas, V. P., & Vijay, R. P. (2011). *Ficus carica Linn-* an overview. *Research Journal of Medicinal Plant, 5*, 246–253.

Wang, Y., & Perri, M. (2018). A systematic review of patient-reported satisfaction with oral medication therapy in patients with type 2 diabetes. *Value in Health, 21*(11), 1346–1353.

Wang, E., Chen, X., Liu, T., & Wang, K. (2022). Effect of dietary *Ficus carica* polysaccharides on the growth performance, innate immune response, and survival of crucian carp against *Aeromonas hydrophila* infection. *Fish & Shellfish Immunology, 120*, 434–440., ISSN 1050-4648. https://doi.org/10.1016/j.fsi.2021.12.018

Wong, M. S., Hawthorne, W. J., & Manolios, N. (2010). Gene therapy in diabetes. *Self Nonself, 1*(3), 165–175.

Yang, X. M., Yu, W., Ou, Z. P., Ma, H. L., Liu, W. M., & Ji, X. L. (2009). Antioxidant and immunity activity of water extract and crude polysaccharide from *Ficus carica* L. fruit. *Plant Foods for Human Nutrition, 64*(2), 167–173.

Yeh, W. J., Hsia, S. M., Lee, W. H., & Wu, C. H. (2017). Polyphenols with antiglycation activity and mechanisms of action: A review of recent findings. *Journal of Food and Drug Analysis, 25*(1), 84–92.

Zahra, K. F., Lefter, R., Ali, A., Abdellah, E. C., Trus, C., Ciobica, A., & Timofte, D. (2021). The involvement of the oxidative stress status in cancer pathology: A double view on the role of the antioxidants. *Oxidative Medicine and Cellular Longevity, 2021*, 9965916. https://doi.org/10.1155/2021/9965916

Zou, Q., Zhang, X., Liu, X., Li, Y., Tan, Q., Dan, Q., Yuan, T., Liu, X., Liu, R. H., & Liu, Z. (2020, July 1). *Ficus carica* polysaccharide attenuates DSS-induced ulcerative colitis in C57BL/6 mice. *Food & Function, 11*(7), 6666–6679. https://doi.org/10.1039/d0fo01162b

Chapter 25
Composition and Health-Promoting Effects of Fig (*Ficus carica*) Extracts

Toyosi Timilehin George, Ayodeji B. Oyenihi, Omolola R. Oyenihi, and Anthony O. Obilana

1 Introduction

The leading causes of death or morbidity in humans have been projected to be the development or progression of chronic, non-communicable diseases due to various consumer lifestyle practices ranging from food consumption habits and environmental and genetic factors. These disorders and life-threatening conditions are envisaged to be more pronounced among the poor and the most vulnerable societies and developed countries and societies with unhealthy lifestyles and eating habits. Much of the curative measures for these disorders widely available are synthetic drugs, although plant compounds have been reported to be viable sources of important healthy compounds against many human diseases (Ajeigbe et al., 2021; Oyenihi & Smith, 2019). While advances have been made over the years in medicine, pharmacology, and pharmacy culminating in discovering therapeutic drugs, the side effects associated with some of these (adverse drug reactions) have been a cause for

T. T. George (✉)
Department of Food Science and Technology, Texas A&M University, College Station, TX, USA
e-mail: ttgeorge93@tamu.edu

A. B. Oyenihi
Functional Foods Research Unit, Faculty of Applied Sciences, Cape Peninsula University of Technology, Bellville, South Africa

O. R. Oyenihi
Oxidative Stress Research Centre, Department of Biomedical Sciences, Faculty of Health and Wellness Sciences, Cape Peninsula University of Technology, Bellville, South Africa

A. O. Obilana
Department of Food Science and Technology, Faculty of Applied Sciences, Cape Peninsula University of Technology, Bellville, South Africa

M. F. Ramadan (ed.), *Fig* (Ficus carica)*: Production, Processing, and Properties*, https://doi.org/10.1007/978-3-031-16493-4_25

concern (Oyenihi & Smith, 2019). This has led to the continuous awareness of the need for increased consumption of plant-based foods, generally believed to be less toxic or have well-tolerated safety profiles from centuries of anecdotal usage.

Furthermore, empirical pieces of scientific evidence continue to indicate positive results in many *in vitro* and *in vivo* experimental trials involving these medicinal plants (Lazreg Aref et al., 2011; Shahinuzzaman et al., 2020; Wojdyło et al., 2016). These have been ascribed to the plethora of bioactive phyto-compounds with health benefits against chronic diseases such as obesity, cancer, infertility, and cardiovascular and neurodegenerative disorders. Fig (*Ficus carica*) is one of such plants that has been reported to contain rich and healthy compounds that can benefit consumers as functional food components.

Fig (*Ficus carica*) tree is one of the earliest cultivated crops in history, belonging to the *Moraceae* family (Mulberry family), and originated from the Middle Eastern region of the world. Because of its health, nutritional and economic importance, the fruit is now being cultivated in many world areas (Abdel-Rahman et al., 2021; Wojdyło et al., 2016) and consumed in fresh, dried, and processed forms (Dueñas et al., 2008). Fig is mainly cultivated by countries such as Turkey, Egypt, Tunisia, Iran, Morocco, and many other countries on the west and the east Mediterranean. Countries such as Brazil, the United States, China, and Portugal have increased their fig production capacity (Veberic & Mikulic-Petkovsek, 2015). The color of the fig fruit (*Ficus carica*) varies depending on planting location and sub-species in which they belong, with purple and green being the most common colors. In addition, the fig tree has immense cultural and religious significance in areas of the world where they originated, as they are used as objects of religious and cultural worship (Badgujar et al., 2014). Beyond their use in various cultural settings, they have good economic potential due to their desirable sensory and organoleptic properties and their health-promoting ability. Empirical evidence from scientific research has detailed the importance of the fig fruits and leaves, especially in human diets and health, with various parts of the tree reported possessing health-promoting compounds and attributes, potent against many known lifestyle diseases (Abdel-Rahman et al., 2021; Amessis-Ouchemoukh et al., 2017). To further reinforce this, the medicinal importance of this tree has previously been demonstrated in the traditional Ayurveda and Siddha system of medicine common in India and other middle eastern countries (Badgujar et al., 2014).

Different parts of the fig tree have been studied previously and reported for their health-promoting importance. For example, the fig seed has been reported to contain bioactive compounds that have antioxidant activities and can mitigate oxygen imbalances in the body (Nakilcioğlu-Taş & Ötleş, 2021). This may be attributed to phytosterol and its components, such as β-sitosterol and stigmasterol, which possess cholesterol-lowering properties (Barolo et al., 2014; Jeong & Lachance, 2001). In addition to the seed, the fig fruit and leaf have been reported for various therapeutic importance against chronic and lifestyle diseases such as cancer, diabetes, etc. (Boyacıoğlu et al., 2021; Wojdyło et al., 2016). For instance, a previous report on the anticancer property of fig leaf extract shows about 67–80% inhibition of Hep2 and HepG2 cell lines (Abdel-Rahman et al., 2021). In another study, the antiproliferative potential of latex from *Ficus carica* on various HT-29 cancer has been reported and found promising (Boyacıoğlu et al., 2021).

Generally, the health-promoting function of fig has been attributed to the presence of bioactive compounds (majorly polyphenols) in different tree parts and may vary from one part to the other (Kamiloglu & Capanoglu, 2015). Polyphenols are secondary metabolite compounds present in plants for defense against predators but possess therapeutic, nutritional, and nutraceutical functions (Veberic & Mikulic-Petkovsek, 2015). They come in various structural forms and are the largest class of bioactive compounds in plants hence presenting complexity in their characterization (Oyenihi & Smith, 2019). Examples of these compounds include; anthocyanins, flavonoids, phenolic acids, and others that have previously been related to the antioxidant activity of the plant. Polyphenols possess strong chemoprevention and anti-cancer attributes, and their chemoprevention pathways and mechanisms have been comprehensively discussed in a study by Oyenihi and Smith (2019). Beyond their anticancer property, bioactive compounds in figs also possess antioxidants (Boyacıoğlu et al., 2021; Polat, 2011), antidiabetic (Deepa et al., 2018), and anti-inflammatory actions.

In this chapter, we focused on the composition and health-promoting attributes of *Ficus carica*. We discussed their importance in human health and the economic benefits of these natural healthy compounds going forward. We briefly mentioned how the reach bioactive compounds in fig could be preserved and maintained and recommended the possible inclusion of the health-promoting compounds in fig in commonly consumed staples.

2 Nutritional and Phytochemical Composition of Fig

*2.1 Nutritional Composition of Figs (*Ficus carica*)*

Figs (*Ficus carica*) are nutritious, containing significant macro- and micro-nutrients (Al-Snafi, 2017; Chauhan & Tanwar, 2015; Khan et al., 2011; Pal, 2020; Sadia et al., 2014; Verma et al., 2015; Vora et al., 2017). The quantities and concentration of these micro and macronutrients depend on differences in varieties, state (dry or fresh fruit), geographical location, type of soil, and general climate of the fruit's environment. The nutritional composition of fresh and dried figs is summarised in Table 25.1. The dried fruit has been reported to contain a high quantity of carbohydrates ranging from 65.2% to 73.5%, while fresh fig fruits contain about 8–20% (Table 25.1). Most of the carbohydrates in figs are glucose, fructose, and sucrose, with the balance being dietary fiber, insoluble cellulose in the skin, and soluble pectin (Pal, 2020). Other proximate compositions of the fruit reported for different accessions and varieties in different locations are presented in Table 25.1. The nutritional content of the fig is distributed between the peel and the pulp, which are both edible and have been reported to provide adequate nutritional and biochemical

Table 25.1 Nutritional content of fig fruit (g/100 g)

Fig fruit	Fresh fruit	Dried fruit	References
Moisture	79.9–88.1	12.9–25.9	Al-Snafi (2017), Chauhan & Tanwar (2015), Khan et al. (2011), Pal (2020), Sadia et al. (2014) and Vora et al. (2017)
Ash	0.60–4.00	0.89–4.65	Al-Snafi (2017), Chauhan & Tanwar (2015), Khan et al. (2011), Pal (2020), Sadia et al. (2014) and Vora et al. (2017)
Protein	0.53–1.30	1.80–8.60	Al-Snafi (2017), Chauhan & Tanwar (2015), Khan et al. (2011), Pal (2020), Sadia et al. (2014), Verma et al. (2015) and Vora et al. (2017)
Lipid	0.20–0.53	0.14–2.70	Al-Snafi (2017), Chauhan & Tanwar (2015), Khan et al. (2011), Pal (2020), Sadia et al. (2014), Verma et al. (2015) and Vora et al. (2017)
Crude fibre	2.10–2.20	3.68–69.4	Al-Snafi (2017), Khan et al. (2011), Pal (2020), Sadia et al. (2014) and Vora et al. (2017)
Carbohydrates	7.60–20.0	65.2–73.5	Al-Snafi (2017), Chauhan & Tanwar (2015), Khan et al. (2011), Pal (2020), Sadia et al. (2014), Verma et al. (2015) and Vora et al. (2017)
Energy (Kcal/100 g)	37.0	308.5–332.7	Al-Snafi (2017), Pal (2020), Sadia et al. (2014) and Verma et al. (2015)

composition required for consumers (Mahmoudi et al., 2018). These authors reported that fruit pulps were rich in carbohydrates, and the peel of the 'Onk Elhamam', the pulp of the 'Boughandjo', and the peel of the 'Bakkor Khal' cultivars had the highest vitamin C and phosphorus content, as well as rich potassium and calcium composition.

Figs are rich in various minerals, including Ca, Fe, P, K, and Na, necessary for the physiological functioning of the human body (Al-Snafi, 2017; Chauhan & Tanwar, 2015; Mahmoudi et al., 2018; Pal, 2020). As a result of the nutrient content, Khan et al. (2011) recommended that the fig fruit, which is available all year round, be used as a natural food supplement to provide instant energy for healthy and unhealthy persons. This was further buttressed by Vora et al. (2017), who reported that the concentration of nutrients in dried figs makes them a rich energy source with a high calorific value and may be consumed as a snack.

2.2 *Phytochemical Composition of Figs (*Ficus carica*)*

There are over a hundred bioactive compounds that have been isolated from both dry and fresh figs (Gibernau et al., 1997; Oliveira et al., 2009; Pal, 2020); some of these include phenolic acids, flavonoid compounds, anthocyanins, carotenoids, fatty acids, phytosterols, etc. (Fig. 25.1). The phytochemical composition of figs majorly depends on the extraction solvent, time, medium, and conditions (Mahmoudi et al.,

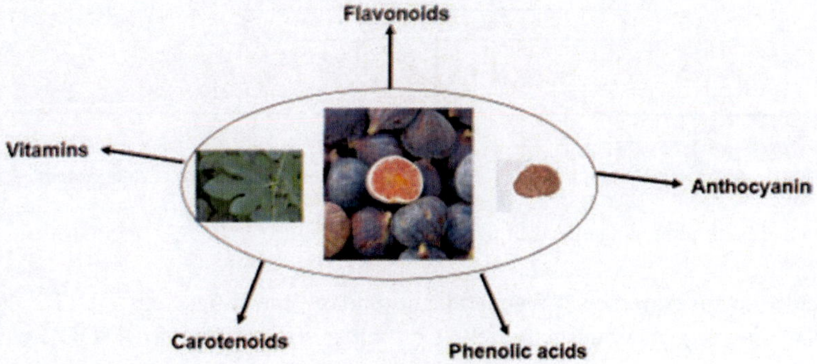

Fig. 25.1 Main phytochemical component of figs (*Ficus carica*)

2018; Verma et al., 2015). Additionally, different planting conditions, locations, and varieties may result in variations in phytochemical composition (Çalişkan & Aytekin, 2011; Veberic et al., 2008). Phytochemical compounds extracted from fig have been reported to have significant antioxidant activity capable of scavenging free radicals, thus inhibiting the oxidative mechanisms that lead to degenerative illnesses and life-threatening conditions (Oliveira et al., 2009). Phytochemicals in figs are distributed in various parts such as peels, pulps, seeds, etc. Mahmoudi et al. (2018) observed that variable levels of bioactive compounds, including polyphenols, flavonoids, anthocyanins, and tannins, were distributed in the peels and pulps of some cultivars and concluded that generally, the peels, especially those with a dark color, contained the high concentrations of phytochemicals and exhibited the highest antioxidant activity compared to fig fruit pulps. They also recommended the consumption of the whole fig fruit based on the phytochemical content of the tested fig cultivars. Shamsi et al. (2020) concluded that figs are a good source of essential nutrients and bioactive compounds that play a vital role in mitigating the occurrence of degenerative and non-degenerative diseases and illnesses. Therefore, the consumption and usage of figs as inclusion in daily diets is essential for the proper functioning of human health.

3 Health-Promoting Effects of Fig

3.1 Antioxidant and Anti-inflammatory

The interconnections between the antioxidant and anti-inflammatory qualities of natural products, including herbs used as foods or medicines to help prevent or slow down the development of chronic diseases, have been well described (Serafini & Peluso, 2016). The biochemical and molecular mechanisms of actions of medicinal plants commonly utilize the antioxidant and anti-inflammatory pathways to achieve these therapeutic effects. Assessment of antioxidant activities is usually evaluated

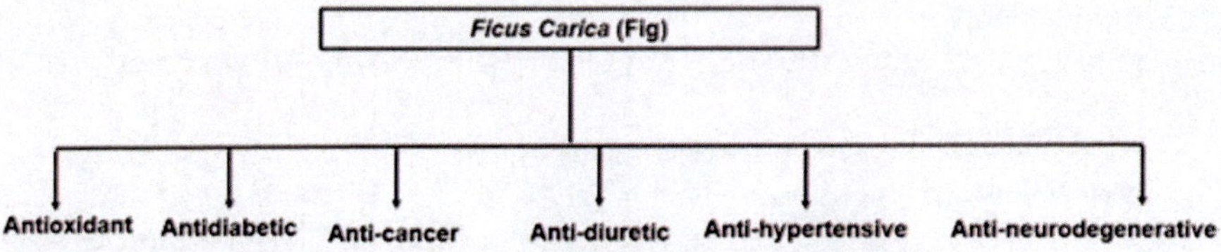

Fig. 25.2 Health-promoting effects of fig (*Ficus carica*)

through the measurement of general antioxidant capacities (e.g., ABTS, FRAP, DPPH·, ORAC), enzyme activities (e.g., superoxide dismutase (SOD), catalase (CAT), glutathione peroxidase, etc.) and amount of antioxidant compounds (e.g., reduced glutathione (GSH), vitamins C & E, α-lipoic acid, etc.). Additionally, the expression patterns of antioxidant genes or proteins (notably, Nr-f2 and antioxidant response elements) are also commonly determined. The ability of fig (*Ficus carica*) to influence these antioxidant indices has been well-defined in the literature (Fig. 25.2). For example, scavenging of DPPH·, ABTS, and hydroxyl superoxide radicals and inhibition of hydrogen peroxide or xanthine oxidase by fig ethanol extracts were reported (Amessis-Ouchemoukh et al., 2017; Saoudi & El Feki, 2012). Seasonal variations possibly play a role in the antioxidant potency of fig leaves, as those collected in July exhibited stronger activity than those from June or September (Marrelli et al., 2014). The ABTS, FRAP, and DPPH· values also correspond with the polyphenol and carotenoid constituents and indicate a preponderance of antioxidant capacity in the fig peel/skin than pulp or whole fruit (Arvaniti et al., 2019), although the leaf was initially shown to be more potent than fruit, wood or bark (Belguith-Hadriche et al., 2017). The antioxidant quality of dark-colored fig varieties may also be superior to their light-colored counterparts (Arvaniti et al., 2019; Pereira et al., 2017). This was also demonstrated previously where the peel and total fruit juices from dark-colored variety had the most DPPH· and reduced power activities (Harzallah et al., 2016). Furthermore, water extract of fig leaves was demonstrated to be 3× and 4× higher in ABTS radical scavenging compared to the ethanol and ultrasound-assisted extracts, respectively (Alcántara et al., 2020). The flavonoid-rich fig peel extracts were recently shown to exhibit potent free radical scavenging and metal-chelating activities associated with their ability to inhibit main enzymes implicated in many chronic diseases such as tyrosinase, α-glucosidase, urease, and cholinesterase (Meziant et al., 2021). The pretreatment with fig stem extract ameliorated the decreases in hepatic SOD, CAT, and glutathione peroxidase activities in methanol intoxicated rats (Saoudi & El Feki, 2012). A similar result was obtained for the antioxidant enzymes in carbon tetrachloride-induced hepatotoxicity in mice (Hira et al., 2021). These studies exemplify the antioxidant attributes of fig.

Inflammatory processes may be initiated by the recruitment of B cells and T cells; infiltration of macrophages, leukocytes, and monocytes into cells, followed by the expression of pro-inflammatory chemokines, cytokines, and cell adhesion molecules (Tasneem et al., 2019). The elevation of pro-inflammatory cytokines and chemokines (e.g., IL-1, IL-6, IL-18, C-reactive protein, TNF-α, MCP-1) is accompanied by the decrease in those of the anti-inflammatory mediators (e.g., IL-4, IL-10 IL-13, TGF-β) have been the major hallmarks of inflammation-induced cellular injury (Tasneem et al., 2019). Thus, these markers represent major targets of many therapeutics and medicinal plants, such as fig. For example, the anti-inflammatory effects of fig leaf extract have been demonstrated by inhibiting leukocyte accumulation and production of TNF-α, PGE2, or VEGF in a rat air pouch model (Eteraf-Oskouei et al., 2015). Furthermore, the fig leaf extract was also reported to downregulate VEGF, TNF-α, and IL-1α gene expressions in human keratinocytes (Turkoglu et al., 2017). Besides the treatment of immuno-compromised mice with fermented fig extracts enhanced serum levels of IL-4, IL-6, TNF-α, and IFN-γ, suggesting their immune system-modulatory capacity (Zhao et al., 2020).

Oxidative stress has often correlated with many inflammatory or immune-modulatory assessments as both phenomena mainly act in concert to propagate chronic diseases (Ahmed et al., 2017). Many pro-inflammatory processes can be triggered by excess free radicals or vice versa, thus illustrating the medicinal utility of fig extracts which could act in both antioxidant and anti-inflammatory mechanisms to achieve efficacy. For instance, the administration of fig seed oil reduced tissue levels of TNF-α and IL-1β and elevated those of GSH, SOD, and CAT, resulting in the amelioration of ischemia and reperfusion injury in rats (Orak et al., 2021). The silver nanoparticles from fig extracts were also shown to reduce lipid peroxidation and nitric oxide levels and down-regulate the gene expression of TNF-α, and IL-1β, in a leishmaniasis-induced inflammation mouse model (Almayouf et al., 2020). In a recent study, the protective effect of fig extract against tebuconazole-mediated toxicity in bats was attributed to its antioxidant and anti-inflammatory activities.

3.2 *Antidiabetic*

Diabetes, a chronic disease with heterogeneous pathogenesis and pathophysiology, continues to ravage the global population, with adult prevalence expected to reach a whopping 784 million in 2045 (International Diabetes Federation, 2021). The disease can be distinguished by the loss of glucose and/or insulin homeostasis regulation in the body. While the administration of insulin has been the mainstay in managing type 1 diabetes, that of type 2 diabetes requires a combination of diet and lifestyle modifications, exercise, and the use of glucose-lowering drugs. However the difficulty in adhering to the daily dosage regimen and issues associated with the clinical side effects of these drugs have hampered their effectiveness, prompting the increased global acceptance of medicinal plants with antidiabetic properties (Oyenihi et al., 2020). In this regard, the traditional use of fig to manage diabetes and its complications has been reported in folklores (Chawla et al., 2012). This traditional antidiabetic efficacy has been confirmed in many *in vitro* and *in vivo* studies with various postulated mechanisms of action as comprehensively reviewed elsewhere (Deepa et al., 2018). Fig extracts were demonstrated to inhibit the activities of key enzymes implicated in the progression of diabetes, such as α-amylase and α-glucosidase (Mopuri et al., 2018; Wojdyło et al., 2016). In a high-fat diet + streptozotocin-induced type 2 diabetic rat model, fig extracts ameliorated the altered activities of metabolizing enzymes, including hepatic glucose-6-phosphatase, fructose-1,6-bisphosphatase, and hexokinase (Irudayaraj et al., 2017) and restored the lipid parameters to near-normal values (Arafa et al., 2020). The blood glucose-lowering properties of fig extracts were also exhibited via the suppression of hepatic gluconeogenesis in an AMPK-dependent manner (Zhang et al., 2019). Abscisic acid and ficusin compounds isolated from the fruits and leaves of figs, respectively, have also shown impressive antidiabetic potentials (Atkinson et al., 2019; Irudayaraj et al., 2016), with the administration of the former reported to lower the glycemic and insulinemic indices in healthy adults (Atkinson et al., 2019).

3.3 *Anti-cancer Effect*

The incidence and prevalence of cancer continue to increase worldwide, posing immense health, social, and economic burdens to humans, requiring novel and concerted efforts to reverse this disturbing trend (Mattiuzzi & Lippi, 2019). Therefore, medicinal plant extracts and other natural products, either stand-alone or supplementary to the current anticancer management strategies, with varying efficacy issues and toxicity concerns, have been encouraged (Oyenihi & Smith, 2019). The traditional anticancer effects of fig and derived compounds have been well-reviewed previously, especially those relating to their cancer preventative and therapeutic, antioxidant, and anti-inflammatory qualities (Lansky et al., 2008). Fig extracts displayed cytotoxicity against a variety of cancer types, including breast (Zhang et al., 2018b), prostate, colon (Boyacıoğlu et al., 2021), hepatocellular carcinoma (Abdel-Rahman et al., 2021; Mustafa et al., 2021), and human laryngeal carcinoma (Abdel-Rahman et al., 2021). In these studies, the anticancer activities were thought to occur through the elevation of oxidative stress, apoptosis, cell cycle arrest; decrease in cell migration; and regulation of molecular signaling. Bergapten and psoralen from the leaves (Zhang et al., 2018a) and ficin from the latex (Boyacıoğlu et al., 2021) are some of the important anticancer components in figs. Besides, the silver nanoparticles from dried fig fruits were reported to retain the cytotoxicity against the MCF-7 breast cell line (Jacob et al., 2017). The benefits of using fig extracts as an adjuvant to chemo-photodynamic therapy were also recently demonstrated on rhabdomyosarcoma cells (Aziz et al., 2021).

3.4 Anti-diuretic Effects

Diuretic drugs modulate the volume and composition of body fluids, hence used for treating different clinical conditions involving oedema or volume overloads like hypertension, heart failure, ascites, nephrotic syndrome, and cirrhosis. Herbal diuretics are considered safe and cost-effective alternatives to synthetic drugs. Sruthi et al. (2012) investigated the diuretic activity of ethanolic extract of *Ficus carica* whole plant at doses of 200 and 400 mg/kg of body weight in rats based on parameters such as total urine volume and urine concentration of sodium (Na+), potassium (K+), and chloride (Cl−). Results showed a significant diuretic effect of the ethanolic extract treatment based on increased urine volume and urinary excretion of electrolytes; Na+, K+, and Cl-.

Munteanu et al. (2016) also reported the diuretic effects of aqueous and ethanolic leaf extracts of *Ficus carica*. The extracts showed a dose-dependent increase in urine excretion. The aqueous extract (1.4 mg/kg) demonstrated a uricosuric effect which could benefit patients with elevated blood uric acid levels. The chemical composition of extracts revealed the presence of flavonoids which may be responsible for inhibiting tubular reabsorption of water and anion and its diuretic activity. However, it is still not clear if the diuretic activity is attributed solely to the flavonoid constituents or the synergistic effects of the phytoconstituents. Therefore, there is a need to investigate the active principles responsible for the diuretic effects. Also, these findings agree with the traditional claim of *Ficus carica* fruits as a diuretic agent in Unani, Ayurvedic, and Chinese-traditional systems of medicines (Badgujar et al., 2014; Chawla et al., 2012; Salma et al., 2020). The scientific investigations regarding the diuretic effects of *Ficus carica* still require novel in-depth analysis that can ascertain the needed dosage level for safety.

3.5 Anti-neurodegenerative Effects

Alzheimer's disease (AD) and other neurodegenerative disorders (NDDs) are characterized by a progressive decline in neuronal function and structure, which eventually results in neuronal death in the nervous system (Karnati et al., 2015; Koric et al., 2016; Pugazhenthi et al., 2017, Zhang et al., 2019). The pathological cellular and molecular events in NDDs include; oxidative stress, neuroinflammation, aberrant RNA processing, DNA damage, protein oligomerization and aggregation, axonal transport deficiency, calcium deregulation, impairment in the mitochondrial function and structure, and abnormal neuron-glia interactions (Dugger & Dickson, 2017). One of the characteristic features of AD is a progressive loss of memory. Also, the neuropathology of AD is characterized initially by the deposition of senile plaques composed of amyloid beta-protein (Aβ) and neurofibrillary tangles containing hyperphosphorylated tau protein brain and later by the loss of neurons and their processes. Treatment with *Ficus carica* fruit improved learning and memory in experimental rat models (Sumanth & Mamatha, 2014). Subash et al. (2014) reported the neuroprotective effect of *Ficus carica* fruits supplemented diet in a transgenic (Tg) mouse AD model. *Ficus carica* treatment improved memory-related behavioral deficits and reduced oxidative damage and plasma Aβ (1–40 and 1–42) levels in Tg mice.

Cholinergic mechanisms are involved in memory formation, and cholinergic cells are susceptible to death in AD patients (Maurer & Williams, 2017). *Ficus carica* fruit and leaf extracts showed potent inhibitory activities against acetylcholinesterase and butyrylcholinesterase enzymes (Subash et al., 2014; Sumanth & Mamatha, 2014). The inhibitory action of *Ficus carica* on these enzymes prolongs the activity of acetyl- and butyryl-choline neurotransmitters in the synapse (Breithaupt & Weismann, 2004; Marshall, 2004; Orhan et al., 2011). Also, Aβ peptides could induce Ca^{2+} influx that leads to increased activity of these enzymes and Ca^{2+} mediated oxidative stress (Barbosa et al., 2002). The mitigating effect of figs against oxidative stress was evident by the inhibition of malondialdehyde accumulation and a reduction in protein carbonyl levels in the cortex and hippocampus of T_g mice. In addition, fig supplemented-diet prevented a decrease in the antioxidant enzymes- superoxide dismutase and catalase and attenuated the elevated levels of glutathione peroxidase and glutathione reductase in brain regions of AD-Tg mice (Subash et al., 2014). Also, the APPSw2576 transgenic mouse model of AD was exposed to a *Ficus carica* fruit-supplemented diet. Essa et al. (2015) reported a decrease in plasma inflammatory cytokines levels, suppression of Aβ peptide levels in both cortex and hippocampus, and a significant recovery in brain ATP levels. Altogether, data from these studies showed the promising therapeutic benefits of *Ficus carica* fruits against neurodegenerative diseases such as AD.

3.6 Anti-lipidemic/Cholesterol Limiting Effect

Dyslipidaemia (characterized by abnormal levels of blood lipids) is a high-risk factor for cardiovascular derangements such as coronary heart diseases, atherosclerosis, and myocardial infarction (Hedayatnia et al., 2020). The antihyperlipidemic effects of *Ficus carica* leaf extracts are well-reported in rats and humans (Perez et al., 1999; Serraclara et al., 1998). *Ficus carica* significantly improved the lipid profile and decreased adipogenic risk factors in high-fat diet (HFD) rats (Joerin et al., 2014). HFD-induced obese male rats fed with *Ficus carica* leaf extract had significantly decreased triglycerides and elevated high-density lipoprotein cholesterol (HDL-C). Another study reported that hydro-methanolic leaf extracts of *Ficus carica* and its fractions significantly lowered cholesterol levels in the serum and liver of experimentally-induced nutritional hyperlipidemic rats (Rassouli et al., 2010).

Studies also suggest that figs treatment reduced cholesterol levels in an obesity-linked type 2 diabetes mellitus induced with a high-fat diet (HFD) plus streptozocin (Arafa et al., 2020). Hydroethanolic extract of the fruit of *Ficus carica* administered at 250 mg/kg/day and 500 mg/kg/day lowered cholesterol, triglycerides, low-density lipoprotein (LDL), and very-low-density lipoprotein (VLDL) while increasing the protective effects of high-density lipoprotein (HDL). Furthermore, ethanolic fruit extract of *Ficus carica* inhibited pancreatic lipase- an enzyme that digests dietary fats and enables absorption into the intestinal lumen (Birari & Bhutari, 2007). Altogether, these studies indicate that *Ficus carica* protects against cholesterol and triglycerides accumulation and has the potential to treat or manage atherosclerosis and coronary heart disorders. These reports also support the traditional application and benefits of *Ficus carica* in treating metabolic syndrome-related conditions (Mawa et al., 2013). The lipid-lowering effects of *Ficus carica* may be due to the flavonoids and tannins that play a significant role in the mobilization and metabolism of lipids (Rassouli et al., 2010). Future studies evaluating the mechanisms by which *Ficus carica* mediates its antihyperlipidemic activities would provide additional relevant information.

3.7 Hepatoprotective Effect

Figs have exerted organ-level protection, in the liver in particular, from various injuries caused by chemicals such as ethanol (Turan & Celik, 2016) and carbon tetrachloride (Aghel et al., 2010; Singab et al., 2010), and rifampicin (Gond & Khadabadi, 2008). The hepatoprotective effects were suggested to occur via the amelioration of hepatic marker enzymes (mostly alanine and aspartate transaminases), antioxidant, and anti-inflammatory actions. In a gamma-radiation toxicity rat model, the daily administration of fig extract for 21 days elicited decreases in the extent of hepatic and renal damage, probably through its antioxidant and anti-apoptotic actions (Fouad et al., 2019).

3.8 Anti-hypertensive Effect

In many parts of the world, most medications available for hypertension and cardiovascular diseases are synthetic drugs; however, some have exhibited side effects. For example, the use of some anti-hypertensive drugs has been linked to cases of erectile dysfunction (Ajeigbe et al., 2021). The use of medicinal plants and fruits to treat this life-threatening disease is currently being explored, and the fig is one of such fruits. This has been linked to the rich presence of antioxidant compounds present in its fruits, leaves, and seeds (Badgujar et al., 2014; Olomon et al., 2006). A recent study by Alamgeer et al. (2017) has shown the effectiveness of methanol extract of fig fruit in decreasing hypertension in normotensive and glucose-induced hypertensive rats. This study showed that this effect is dose-dependent, and the effective dosage level was reported to be 1000 mg/kg body weight. The study attributed this effectiveness of the fig extract to the rich presence of phenolic acids, flavonoids, and potassium in the fruit. This indicates the promising potential of fig in treating one of the world's leading causes of death. Since different parts of the fig tree have been established to contain rich antioxidant compounds, the leaves, seeds, and barks should be further explored in treating hypertension. A comparison of the efficacy of the extracts obtained from these different plant parts will also assist in the isolation and elucidation of the specific compound(s) most responsible for the anti-hypertensive activity. The paucity of clinical trials on the effectiveness of the fig extracts in treating hypertension and related health disorders is observed, limiting their clinical usage.

4 Storage and Preservation of Bioactive Contents of *Ficus carica*

The shelf life of most fruits and leaves of importance is low primarily due to continuous respiration during storage and the presence of ethylene (Kaur et al., 2021). Some of these fruits are rich in bioactive compounds that are essential for promoting consumers' health. Hence, they must be adequately stored to maximize their importance. Figs are commonly preserved by freezing, storage in modified atmosphere packages, dehydration, processing into puree for complete freezing, and, more recently, encapsulating these bioactive contents into nano- and micro-capsules. These methods have been stated to be a viable preservatives for bioactive plants (Kaur et al., 2021). Having a system in place that caters, maintains, and preserves the integrity of healthy compounds in figs will improve its economic value and the use of the crop in various parts of the world, especially in areas where they are not native.

5 Conclusion

Although figs have been reported to contain a high amount of essential phytochemicals with health-promoting functions, the medical benefits of the crop are still largely untapped, especially in areas of the world where they are not native. With continuous growth in many life-threatening diseases, it is expedient that more medical and pharmacological research studies are conducted using this nutrient-rich and health-benefiting plant either *via* direct supplementation or inclusion in different human foods. This inclusion can be achieved by fortifying and enriching commonly consumed staple foods with figs. Hence, research should be placed on determining the optimal dose required for an effective or efficient fortification protocol. Additionally, the medicinal benefits of figs should be confirmed in comprehensive, well-designed clinical studies that include clinical end-points and therapeutic dose-response and toxicity profile assessments. Also, since most phytochemical-rich fruits are prone to faster degradation due to increased respiration and ethelyne production, it is important to find sustainable, cost-friendly, and safe storage and preservation of figs beyond the current drying technique is frequently employed. This is important in making them available all year round with their bioactive and phytochemical contents kept intact.

References

Abdel-Rahman, R., Ghoneimy, E., Abdel-Wahab, A., Eldeeb, N., Salem, M., Salama, E., & Ahmed, T. (2021). The therapeutic effects of *Ficus carica* extract as antioxidant and anticancer agent. *South African Journal of Botany, 141*, 273–277.

Aghel, N., Kalantari, H., & Hezazadeh, S. (2010). Hepatoprotective effect of *Ficus carica* leaf extract on mice intoxicated with carbon tetrachloride. *Iranian Journal of Pharmaceutical Research, 10*, 63–67.

Ahmed, S. M., Luo, L., Namani, A., Wang, X. J., & Tang, X. (2017). Nrf2 signaling pathway: Pivotal roles in inflammation. *Biochimica et Biophysica Acta - Molecular Basis of Disease, 1863*, 585–597.

Ajeigbe, O. F., Oboh, G., & Ademosun, A. O. (2021). Ficus plants in the co-management of hypertension and erectile dysfunction. *Phytomedicine Plus, 1*, 100096.

Alamgeer, I. S., Asif, H., & Saleem, M. (2017). Evaluation of antihypertensive potential of *Ficus carica* fruit. *Pharmaceutical Biology, 55*, 1047–1053.

Alcántara, C., žugčić, T., Abdelkebir, R., García-pérez, J. V., Jambrak, A. R., Lorenzo, J. M., Collado, M. C., Granato, D., & Barba, F. J. (2020). Effects of ultrasound-assisted extraction and solvent on the phenolic profile, bacterial growth, and anti-inflammatory/antioxidant activities of Mediterranean olive and fig leaves extracts. *Molecules (Basel, Switzerland), 25*, 1718.

Almayouf, M. A., El-khadragy, M. F., Awad, M. A., & Al-olayan, E. M. (2020). The effects of silver nanoparticles biosynthesized using fig and olive extracts on cutaneous leishmaniasis induced inflammation in female balb/c mice. *Bioscience Reports, 40*.

Al-Snafi, A. E. (2017). Nutritional and pharmacological importance of *Ficus carica* – A review. *IOSR Journal of Pharmacy (IOSRPHR), 07*, 33–48.

Amessis-Ouchemoukh, N., Ouchemoukh, S., Meziant, N., Idiri, Y., Hernanz, D., Stinco, C. M., Rodríguez-Pulido, F. J., Heredia, F. J., Madani, K., & Luis, J. (2017). Bioactive metabolites involved in the antioxidant, anticancer and anticalpain activities of *Ficus carica* L., Ceratonia siliqua L. and Quercus ilex L. extracts. *Industrial Crops and Products, 95*, 6–17.

Arafa, E. S. A., Hassan, W., Murtaza, G., & Buabeid, M. A. (2020). *Ficus carica* and *Sizigium cumini* regulate glucose and lipid parameters in high-fat diet and streptozocin-induced rats. *Journal of Diabetes Research, 2020*, 6745873.

Arvaniti, O. S., Samaras, Y., Gatidou, G., Thomaidis, N. S., & Stasinakis, A. S. (2019). Review on fresh and dried figs: Chemical analysis and occurrence of phytochemical compounds, antioxidant capacity and health effects. *Food Research International, 119*, 244–267.

Atkinson, F. S., Villar, A., Mulà, A., Zangara, A., Risco, E., Smidt, C. R., Hontecillas, R., Leber, A., & Bassaganya-riera, J. (2019). Abscisic acid standardized fig (*Ficus carica*) extracts ameliorate postprandial glycemic and insulinemic responses in healthy adults. *Nutrients, 11*.

Aziz, B., Khurshid, A., Mahmood, R., Khan, J. A., Javaid, S., Alam, M., Mujtaba Ul Hassan, S., & Ikram, M. (2021). Study of synergistic effects of *Ficus carica* leaves extract mediated chemo-photodynamic therapy on rhabdomyosarcoma cells. *Photodiagnosis and Photodynamic Therapy, 36*, 102565.

Badgujar, S. B., Patel, V. V., Bandivdekar, A. H., & Mahajan, R. T. (2014). Traditional uses, phytochemistry and pharmacology of *Ficus carica* : A review. *Pharmaceutical Biology, 52*, 1–17.

Barbosa, J., Jr., Ferreira, L. T., Martins-Silva, C., Santos, M. S., Torres, G. E., Caron, M. G., et al. (2002). Trafficking of the vesicular acetylcholine transporter in SN56 cells: A dynamin-sensitive step and interaction with the AP-2 adaptor complex. *Journal of Neurochemistry, 82*(5), 1221–1228.

Barolo, M. I., Ruiz Mostacero, N., & López, S. N. (2014). *Ficus carica* L. (Moraceae): An ancient source of food and health. *Food Chemistry*.

Belguith-hadriche, O., Ammar, S., Contreras, M. D. M., Fetoui, H., Segura-Carretero, A., El Feki, A., & Bouaziz, M. (2017). HPLC-DAD-QTOF-MS profiling of phenolics from leaf extracts of two Tunisian fig cultivars: potential as a functional food. *Biomedicine & Pharmacotherapy, 89*, 185–193.

Birari, R. B., & Bhutani, K. K. (2007). Pancreatic lipase inhibitors from natural sources: Unexplored potential. *Drug Discovery Today, 12*(19–20), 879–889.

Boyacıoğlu, O., Kara, B., Tecimen, S., Kılıç, M., Delibaş, M., Erdoğan, R., Özdemir, E., Bahadır, A., & Örenay-Boyacıoğlu, S. (2021). Antiproliferative effect of *Ficus carica* latex on cancer cell lines is not solely linked to peroxidase-like activity of ficin. *European Journal of Integrative Medicine, 45*, 101348.

Breithaupt, H., & Weigmann, K. (2004). Manipulating your mind. What will science discover about our brains, and how are we going to deal with it? *EMBO Reports, 5*(3), 230–232.

Çalişkan, O., & Aytekin Polat, A. (2011). Phytochemical and antioxidant properties of selected fig (Ficus carica L.) accessions from the eastern Mediterranean region of Turkey. *Scientia Horticulturae, 128*, 473–478.

Chauhan, A., & Tanwar, B. (2015). Influence of processing on physicochemical, nutritional and phytochemical composition of *Ficus carica* (fig) fruit. *Research Journal of Pharmaceutical, Biological and Chemical Sciences Influence, 6*, 1474–1489.

Chawla, A., Kaur, R., & Sharma, A. K. (2012). *Ficus carica* Linn. A review on its pharmacognostic, phytochemical and pharmacological aspects. *International Journal of Pharmaceutical and Phytopharmacological Research, 1*, 215–232.

Deepa, P., Sowndhararajan, K., Kim, S., & Park, S. J. (2018). A role of Ficus species in the management of diabetes mellitus: A review. *Journal of Ethnopharmacology, 215*, 210–232.

Dueñas, M., Pérez-Alonso, J. J., Santos-Buelga, C., & Escribano-Bailón, T. (2008). Anthocyanin composition in fig (*Ficus carica L.*). *Journal of Food Composition and Analysis, 21*, 107–115.

Dugger, B. N., & Dickson, D. W. (2017). Pathology of neurodegenerative diseases. *Cold Spring Harbor Perspectives in Biology, 9*(7), a028035.

Essa, M. M., Subash, S., Akbar, M., Al-Adawi, S., & Guillemin, G. J. (2015). Long-term dietary supplementation of pomegranates, figs and dates alleviate neuroinflammation in a transgenic mouse model of Alzheimer's disease. *PLoS One, 10*(3), e0120964.

Eteraf-oskouei, T., Allahyari, S., Akbarzadeh-atashkhosrow, A., Delazar, A., Pashaii, M., Gan, S. H., & Najafi, M. (2015). Methanolic extract of *Ficus carica* linn. Leaves exerts antiangiogenesis effects based on the rat air pouch model of inflammation. *Evidence-based Complementary and Alternative Medicine, 2015*, 760405.

Fouad, D., Alhatem, H., Abdel-gaber, R., & Ataya, F. (2019). Hepatotoxicity and renal toxicity induced by gamma-radiation and the modulatory protective effect of *Ficus carica* in male albino rats. *Research in Veterinary Science, 125*, 24–35.

Gibernau, M., Buser, H. R., Frey, J. E., & Hossaert-McKey, M. (1997). Volatile compounds from extracts of figs of *Ficus carica*. *Phytochemistry, 46*, 241–244.

Gond, N. Y., & Khadabadi, S. S. (2008). Hepatoprotective activity of *Ficus carica* leaf extract on rifampicin-induced hepatic damage in rats. *Indian Journal of Pharmaceutical Sciences, 70*, 364–366.

Harzallah, A., Bhouri, A. M., Amri, Z., Soltana, H., & Hammami, M. (2016). Phytochemical content and antioxidant activity of different fruit parts juices of three figs (*Ficus carica* L.) varieties grown in Tunisia. *Industrial Crops and Products, 83*, 255–267.

Hedayatnia, M., Asadi, Z., Zare-Feyzabadi, R., Yaghooti-Khorasani, M., Ghazizadeh, H., Ghaffarian-Zirak, R., et al. (2020). Dyslipidemia and cardiovascular disease risk among the MASHAD study population. *Lipids in Health and Disease, 19*(1), 1–11.

Hira, S., Mulfraz, M., Naqvi, S. S., Qureshi, R. U., & Gul, H. (2021). Protective effect of leaf extract of ficus carica l. against carbon tetrachloride-induced hepatic toxicity in mice and hepg2 cell line. *Tropical Journal of Pharmaceutical Research, 20*, 113–119.

International Diabetes Federation. (2021). *Idf diabetes atlas 10th edition* .

Irudayaraj, S. S., Stalin, A., Sunil, C., Duraipandiyan, V., Al-dhabi, N. A., & Ignacimuthu, S. (2016). Antioxidant, antilipidemic and antidiabetic effects of ficusin with their effects on glut4 translocation and pparγ expression in type 2 diabetic rats. *Chemico-Biological Interactions, 256*, 85–93.

Irudayaraj, S. S., Christudas, S., Antony, S., Duraipandiyan, V., Abdullah, A. D. N., & Ignacimuthu, S. (2017). Protective effects of *Ficus carica* leaves on glucose and lipids levels, carbohydrate metabolism enzymes and β-cells in type 2 diabetic rats. *Pharmaceutical Biology, 55*, 1074–1081.

Jacob, S. J. P., Prasad, V. L. S., Sivasankar, S., & Muralidharan, P. (2017). Biosynthesis of silver nanoparticles using dried fruit extract of ficus carica - screening for its anticancer activity and toxicity in animal models. *Food and Chemical Toxicology, 109*, 951–956.

Jeong, W. S., & Lachance, P. A. (2001). Phytosterols and fatty acids in fig (*Ficus carica*, var. Mission) fruit and tree components. *Journal of Food Science, 66*, 278–281.

Joerin, L., Kauschka, M., Bonnländer, B., Pischel, I., Benedek, B., & Butterweck, V. (2014). Ficus carica leaf extract modulates the lipid profile of rats fed with a high-fat diet through an increase of HDL-C. *Phytotherapy Research, 28*(2), 261–267.

Kamiloglu, S., & Capanoglu, E. (2015). Polyphenol content in figs (*Ficus carica* L.): Effect of sun-drying. *International Journal of Food Properties, 18*, 521–535.

Karnati, H. K., Panigrahi, M. K., Gutti, R. K., Greig, N. H., & Tamargo, I. A. (2015). miRNAs: Key players in neurodegenerative disorders and epilepsy. *Journal of Alzheimer's Disease, 48*(3), 563–580.

Kaur, G., Kaur, P., & Kaur, A. (2021). Preserving bioactive quality and color of novel frozen lemongrass puree tablets. *Journal of Food Processing and Preservation.*

Khan, M., Sarwar, A., Adeel, M., & Wahab, M. (2011). Nutritional evaluation of *ficus carica* indigenous to Pakistan. *African Journal of Food, Agriculture, Nutrition and Development, 11.*

Koric, L., Guedj, E., Habert, M. O., Semah, F., Branger, P., Payoux, P., et al. (2016). Molecular imaging in the diagnosis of Alzheimer's disease and related disorders. *Revue Neurologique, 172*(12), 725–734.

Lansky, E. P., Paavilainen, H. M., Pawlus, A. D., & Newman, R. A. (2008). *Ficus spp.* (fig): Ethnobotany and potential as anticancer and anti-inflammatory agents. *Journal of Ethnopharmacology, 119*, 195–213.

Lazreg Aref, H., Gaaliche, B., Fekih, A., Mars, M., Aouni, M., Chaumon, J. P., & Said, K. (2011). In vitro cytotoxic and antiviral activities of Ficus carica latex extracts. *Natural Product Research, 25*, 310–319.

Mahmoudi, S., Khali, M., Benkhaled, A., Boucetta, I., Dahmani, Y., Attallah, Z., & Belbraouet, S. (2018). Fresh figs (*Ficus carica* L.): Pomological characteristics, nutritional value, and phytochemical properties. *European Journal of Horticultural Science, 83*, 104–113.

Marrelli, M., Statti, G. A., Tundis, R., Menichini, F., & Conforti, F. (2014). Fatty acids, coumarins and polyphenolic compounds of *Ficus carica* l. Cv. Dottato: Variation of bioactive compounds and biological activity of aerial parts. *Natural Product Research, 28*, 271–274.

Marshall, E. (2004). A star-studded search for memory-enhancing drugs. *Science, 304*, 36–38.

Mattiuzzi, C., & Lippi, G. (2019). Current cancer epidemiology. *The Journal of Epidemiology and Global Health, 9*, 217–222.

Maurer, S. V., & Williams, C. L. (2017). The cholinergic system modulates memory and hippocampal plasticity via its interactions with non-neuronal cells. *Frontiers in Immunology, 8*, 1489.

Mawa, S., Husain, K., & Jantan, I. (2013). *Ficus carica* L.(Moraceae): Phytochemistry, traditional uses, and biological activities. *Evidence-Based Complementary and Alternative Medicine, 2013*.

Meziant, L., Bachir-bey, M., Bensouici, C., Saci, F., Boutiche, M., & Louaileche, H. (2021). Assessment of inhibitory properties of flavonoid-rich fig (*Ficus carica* l.) Peel extracts against tyrosinase, α-glucosidase, urease and cholinesterases enzymes, and relationship with antioxidant activity. *European Journal of Integrative Medicine, 43*, 101272.

Mopuri, R., Ganjayi, M., Meriga, B., Koorbanally, N. A., & Islam, M. S. (2018). The effects of *Ficus carica* on the activity of enzymes related to metabolic syndrome. *Journal of Food and Drug Analysis, 26*, 201–210.

Munteanu, M. F., Gligor, R., Pogurschi, E. N., & Trifunschi, S. (2016). Diuretic activity of *Ficus carica* L. extracts. *Analele Universităţii din Oradea, Fascicula: Ecotoxicologie, Zootehnie şi Tehnologii de Industrie Alimentară, 15*(A), 75–82.

Mustafa, K., Yu, S., Zhang, W., Mohamed, H., Naz, T., Xiao, H., Liu, Y., Nazir, Y., Fazili, A. B. A., Nosheen, S., Bai, X., & Song, Y. (2021). Screening, characterization, and in vitro-ros dependent cytotoxic potential of extract from *Ficus carica* against hepatocellular (hepg2) carcinoma cells. *South African Journal of Botany, 138*, 217–226.

Nakilcioğlu-Taş, E., & Ötleş, S. (2021). Influence of extraction solvents on the polyphenol contents, compositions, and antioxidant capacities of fig (*Ficus carica* L.) seeds. *Anais da Academia Brasileira de Ciências, 93*, 1–11.

Oliveira, A. P., Valentão, P., Pereira, J. A., Silva, B. M., Tavares, F., & Andrade, P. B. (2009). *Ficus carica* L.: Metabolic and biological screening. *Food and Chemical Toxicology, 47*, 2841–2846.

Olomon, A.N.A.T.S., Olubowicz, S.A.R.A.G., Ablowicz, Z.E.E.V.Y., Rossman, S.H.G., Ergman, M.A.B., Ottlieb, H.U.G.O.E.G., Ltman, A.R.I.E.A., Erem, Z.O.K., & Laishman, M.O.A.F. (2006). Antioxidant activities and anthocyanin content of fresh fruits of common fig (*Ficus carica* L.).

Orak, C., Sirinyıldız, F., Gökmen Yılmaz, E., Cesur, G., & Ek, R. O. (2021). Protective effects of *Ficus carica* seed oil on ischemia and reperfusion injury in a rat model of acute mesenteric ischemia. *Ulusal Travma ve Acil Cerrahi Dergisi, 27*, 402–409.

Orhan, I. E., Üstün, O., & Şener, B. (2011). Estimation of cholinesterase inhibitory and antioxidant effects of the leaf extracts of Anatolian *Ficus carica* var. domestica and their total phenol and flavonoid contents. *Natural Product Communications, 6*(3), 1934578X1100600315.

Oyenihi, A. B., & Smith, C. (2019). Are polyphenol antioxidants at the root of medicinal plant anti-cancer success? *Journal of Ethnopharmacology, 229*, 54–72.

Oyenihi, A. B., Ahiante, B. O., Oyenihi, B. R., & Masola, B. (2020). Chapter 21 – *Centella asiatica*: Its potential for the treatment of diabetes. In V. R. Preedy (Ed.), *Diabetes* (2nd ed.). Academic Press.

Pal, I. (2020). A comprehensive review on *Ficus carica* L. – An unexplored medicinal plant. *International Journal of Advanced Research, 8*, 876–881.

Pereira, C., López-corrales, M., Serradilla, M. J., Villalobos, M. D. C., Ruiz-moyano, S., & Martín, A. (2017). Influence of ripening stage on bioactive compounds and antioxidant activity in nine fig (*Ficus carica* L.) varieties grown in extremadura, Spain. *Journal of Food Composition and Analysis, 64*, 203–212.

Perez, C., Canal, J. R., Campillo, J. E., Romero, A., & Torres, M. D. (1999). Hypotriglyceridaemic activity of *Ficus carica* leaves in experimental hypertriglyceridaemic rats. *Phytotherapy Research, 13*, 188–191.

Polat, A. A. (2011). Scientia Horticulturae phytochemical and antioxidant properties of selected fig (*Ficus carica* L.) accessions from the eastern Mediterranean region of Turkey. *Scientia Horticulturae, 128*, 473–478.

Pugazhenthi, S., Qin, L., & Reddy, P. H. (2017). Common neurodegenerative pathways in obesity, diabetes, and Alzheimer's disease. *Biochimica et Biophysica Acta (BBA) - Molecular Basis of Disease, 1863*(5), 1037–1045.

Rasouli, A., Fatemi, A. A., Asadi, F., & Salehi, M. H. (2010). Effects of Fig tree (*Ficus carica*) leaf extracts on serum and liver cholesterol levels in hyperlipidemic rats. *International Journal of Veterinary Science, 2*, 77.

Sadia, H., Ahmad, M., Sultana, S., Abdullah, A. Z., Teong, L. K., Zafar, M., & Bano, A. (2014). Nutrient and mineral assessment of edible wild fig and mulberry fruits. *Fruits, 69*, 159–166.

Salma, S., Shamsi, Y., Ansari, S., & Nikhat, S. (2020). *Ficus carica* L.: A panacea of nutritional and medicinal benefits. *Cell, 10*(1), 1–1.

Saoudi, M., & El Feki, A. (2012). Protective role of *Ficus carica* stem extract against hepatic oxidative damage induced by methanol in male wistar rats. *Evidence-based Complementary and Alternative Medicine, 2012*, 150458.

Serafini, M., & Peluso, I. (2016). Functional foods for health: The interrelated antioxidant and anti-inflammatory role of fruits, vegetables, herbs, spices and cocoa in humans. *Current Pharmaceutical Design, 22*, 6701–6715.

Serraclara, A., Hawkins, F., Perez, C., Dominguez, E., Campillo, J. E., & Torres, M. D. (1998). Hypoglycemic action of an oral fig-leaf decoction in type-I diabetic patients. *Diabetes Research and Clinical Practice, 39*, 19–22.

Shahinuzzaman, M., Yaakob, Z., Anuar, F. H., Akhtar, P., Kadir, N. H. A., Hasan, A. K. M., Sobayel, K., Nour, M., Sindi, H., Amin, N., Sopian, K., & Akhtaruzzaman, M. (2020). In vitro antioxidant activity of Ficus carica L. latex from 18 different cultivars. *Scientific Reports, 10*, 1–14.

Shamsi, Y., Ansari, S., Nikhat, S., Hamdard, J., & Hamdard, J. (2020). *Ficus carica* L.: A panacea of nutritional and medicinal benefits, *10*, 2–7.

Singab, A. N. B., Ayoub, N. A., Ali, E. N., & Mostafa, N. M. (2010). Antioxidant and hepatoprotective activities of Egyptian moraceous plants against carbon tetrachloride-induced oxidative stress and liver damage in rats. *Pharmaceutical Biology, 48*, 1255–1264.

Sruthi, B., Sunny, G., Hajra, N., & Sakthivel, S. (2012). Diuretic activity of ethanolic extracts of *Ficus carica* L. fruits. *International Journal of Research Pharmacol Pharmacother, 1*, 25–28.

Subash, S., Essa, M. M., Al-Asmi, A., Al-Adawi, S., & Vaishnav, R. (2014). Chronic dietary supplementation of 4% figs on the modification of oxidative stress in Alzheimer's disease transgenic mouse model. *BioMed Research International, 2014*.

Sumanth, M., & Mamatha, K. (2014). Learning and memory-enhancing activity of *Ficus carica* (Fig): An experimental study in rats. *Drug Design, Development and Therapy, 5*(123), 2394–2002.

Tasneem, S., Liu, B., Li, B., Choudhary, M. I., & Wang, W. (2019). Molecular pharmacology of inflammation: Medicinal plants as anti-inflammatory agents. *Pharmacological Research, 139*, 126–140.

Turan, A., & Celik, L. (2016). Antioxidant and hepatoprotective properties of dried fig against oxidative stress and hepatotoxicity in rats. *International Journal of Biological Macromolecules, 91*, 554–559.

Turkoglu, M., Pekmezci, E., Kilic, S., Dundar, C., & Sevinc, H. (2017). Effect of *Ficus carica* leaf extract on the gene expression of selected factors in hacat cells. *Journal of Cosmetic Dermatology, 16*, e54–e58.

Veberic, R., & Mikulic-Petkovsek, M. (2015). Phytochemical composition of common fig (*Ficus carica* L.) cultivars. In *Nutritional composition of fruit cultivars* (pp. 235–255). Elsevier Inc.

Veberic, R., Colaric, M., & Stampar, F. (2008). Phenolic acids and flavonoids of fig fruit (*Ficus carica* L.) in the northern Mediterranean region. *Food Chemistry, 106*, 153–157.

Verma, I., Gupta, R. K., Rajinder, C., & Gupta, K. (2015). Estimation of phytochemical, nutritional, antioxidant and antibacterial activity of dried fruit of sacred figs (*Ficus religiosa*) and formulation of value-added product (Hard Candy). *Journal of Pharmacognosy and Phytochemistry JPP, 4*, 257–267.

Vora, J. D., Vora, D., Pednekar, S. R., Patwardhan, A. U., & Shaikh, S. (2017). Biochemical, organoleptic assessment of fig (*Ficus carica*). *IOSR Journal of Biotechnology and Biochemistry, 03*, 95–104.

Wojdyło, A., Nowicka, P., Carbonell-Barrachina, Á. A., & Hernández, F. (2016). Phenolic compounds, antioxidant and antidiabetic activity of different cultivars of *Ficus carica* L. fruits. *Journal of Functional Foods, 25*, 421–432.

Zhang, Y., Wan, Y., Huo, B., Li, B., Jin, Y., & Hu, X. (2018a). Extracts and components of *Ficus carica* leaves suppress survival, cell cycle, and migration of triple-negative breast cancer mda-mb-231 cells. *Oncotargets and Therapy, 11*, 4377–4386.

Zhang, X., Fu, Z., Meng, L., He, M., & Zhang, Z. (2018b). The early events that initiate β-amyloid aggregation in Alzheimer's disease. *Frontiers in Aging Neuroscience, 10*, 359.

Zhang, Y., Chen, J., Zeng, Y., Huang, D., & Xu, Q. (2019). Involvement of ampk activation in the inhibition of hepatic gluconeogenesis by *Ficus carica* leaf extract in diabetic mice and hepg2 cells. *Biomedicine & Pharmacotherapy, 109*, 188–194.

Zhao, J., Gong, L., Wu, L., She, S., Liao, Y., Zheng, H., Zhao, Z., Liu, G., & Yan, S. (2020). Immunomodulatory effects of fermented fig (*Ficus carica* L.) fruit extracts on cyclophosphamide-treated mice. *Journal of Functional Foods, 75*, 104219.

Chapter 26
Genotoxic and Antimutagenic Activity of *Ficus carica* Extracts

Nusrath Yasmeen, Gondrala Usha kiranmai, and Aga Syed Sameer

Abbreviations

NER	Nucleotide excision repair
BER	base excision repair
CA	chromosomal Aberration
SCE	sister chromatid exchange
UDS	unscheduled DNA synthesis
MN	Micronucleus test
HPRT	Hypoxanthine-guanine phosphoribosyl transferase
XPRT	xanthine guanine phosphoribosyl transferase
MI	Mitotic Index
ERA	E. coli WP2 reversion assay
SMART	Somatic Mutations and Recombination Test
AC	Allium cepa Assay
RXL	Ring-X-Loss test
SHA	Sperm Head Anomaly
AI	Apoptosis-Inducing

N. Yasmeen
Department of Biotechnology, Amity University, Jaipur, Rajasthan, India

G. Usha kiranmai
Faculty of Pharmaceutics, University College of Pharmaceutical Sciences, Kakatiya University, Warangal, Telangana, India

A. S. Sameer (✉)
Department of Basic Medical Sciences, Quality Assurance Unit, College of Medicine, King Saud Bin Abdul Aziz University for Health Sciences (KSAU-HS), King Abdullah International Medical Research Centre (KAIMRC), Jeddah, Saudi Arabia

National Guard Health Affairs (NGHA), King Abdulaziz Medical City, King Abdullah International Medical Research Centre (KAIMRC), Jeddah, Saudi Arabia
e-mail: agas@ksau-hs.edu.sa; agasy@ngha.med.sa

M. F. Ramadan (ed.), *Fig* (Ficus carica)*: Production, Processing, and Properties*, https://doi.org/10.1007/978-3-031-16493-4_26

RI	Cell Replication Indice
PDB	Plasmid DNA Breakage
FM	Forward Mutagenesis
ICPM	Induction of Cytoplasmic Petite Mutations
CYP450	cytochrome P450

1 Introduction

Plant phytochemicals and their secondary metabolites have a tremendous potential to treat numerous diseases. They are considered safer and economical hence relied upon by many people globally (Gorlenko et al., 2020). According to the World Health Organization (WHO), people using traditional medicine to cure their diseases vary between 70% and 80% of the global population. Numerous drugs we are currently using are derived from natural products, i.e., folk remedies, semi-synthetic or synthetic (Oyebode et al., 2016). Despite rapid technological advancements and novel methodologies involved in the production of pharmaceuticals, plant phytochemicals remain an essential source of lead molecular entities for the development of new drugs. This could be attributed to the fact that phytochemicals are complex and challenging to synthesize by chemical processes (Forni et al., 2019; Thomford et al., 2018). Since time immemorial, drugs from plants and animals have been used based on trial and error methods that exhibited specific therapeutic properties signified by the cause-and effect-relationship (Atanasov et al., 2015). Traditional medicines used were complex concoctions of naturally occurring substances of varied origins (plants, animals, mineral sources), etc., whereas modern pharmaceutical drugs are substances synthetic in origin with a single active entity (Mushtaq et al., 2018). Existing literature supports that the drugs from crude natural products are less toxic than single entity synthetic drugs; this may be attributed to several synergistic/antagonistic components in the concoctions.

The biological significance of herbal medicine has led to the resurgence of scientific interest among ethno pharmacologists striving to develop new drugs using active phytochemical ingredients as a lead molecule. Furthermore, the lack of a strong regulatory system involved in evaluating, testing, monitoring, and regularizing natural products as medicines led to the popularity of natural products. The people's acceptance of traditional or folklore medicine is based on the belief that natural products are much safer than their alternative allopathic forms. However, current scientific data contrasts this general perception and suggests that traditional herbal medicines can be exceptionally harmful to humans; they can potentially interact with food and other ingredients and lead to adverse events (Fokunang et al., 2020; Ekor, 2014). Natural product safety profiling is further jeopardized by inappropriate manufacturing, labeling, deceptive claims, and misleading patients. Therefore, safety evaluation and regulatory policies must be standardized, and stringent regulatory policies regarding herbal medicine usage must be implemented globally (Alostad et al., 2020).

The tremendous therapeutic potential of herbal medicine and the belief that "Natural means safe" clubbed with their long-term usage has led to a lack of significant scientific evidence related to herbal drug safety and toxicity parameters. However, current data and reports about the toxicity effects of herbal medicines show a significant upraise in acute and chronic intoxications. Additionally, the availability of antidotes to treat such intoxications and the toxicological impact of herbal medicine among pediatrics and insensitive groups is unavailable. Subsequently, it becomes necessary to evaluate herbal medicine (Ghorani-Azam et al., 2018). Therefore, traditional or folklore medicine must undergo rigorous screening procedures via several in vitro and in vivo experiments regarding their safety, efficiency, and toxicity, similar to allopathic drugs.

Furthermore, the preclinical evaluation of herbal medicine must be performed to verify its mutagenic, genotoxic, and cytotoxic effects (Moreira et al., 2014). Assessment of the genotoxicity effects of natural medicine is crucial as genotoxins, if present in phytochemicals, may cause alteration in DNA via mutations and increase the risk of disease progression of pathologies such as cancer (Qari et al., 2021). Additionally, toxicological risk assessment can prove beneficial in identifying plant products that possess antimutagenic, anticarcinogenic effects, which can be considered crucial in the research for developing novel therapeutic agents. Furthermore, as suggested by Graz (2013), reverse pharmacology, also known as the "bedside-to-bench" approach, might enhance the knowledge of the safety of folklore medicines. This chapter summarizes the toxicity risk assessment regarding genotoxicity and antimutagenic effects of *Ficus* species, focusing on *Ficus carica*.

2 Genotoxicity and Its Assessment Techniques

Genetic toxicology (genotoxicity) is a kind of toxicity that describes the ability of chemical substances to induce changes in DNA or chromosome, enzymes, or cellular organelles of somatic or germ cells, either through direct or indirect interactions, that cause damage/harm to genetic information, but does not necessarily cause a mutation (Hasselgren et al., 2019). Genotoxicity is involved in specific correction procedures like the direct repair mutations utilizing nucleotide excision repair (NER) and base excision repair (BER), or they may exist in somatic cells or remain as an altered gene in germ cells and cause genetic diseases (Qari et al., 2021). At the same time, mutagenicity refers to the induction of variations in the genetic material of organisms that are permanent and transmissible by chemicals called mutagens. Mutagenic chemicals interact with DNA resulting in alteration in DNA base pair sequence via addition, deletion, or replacement of bases, thus altering their sequence in the DNA and conformity of the genetic information, consequently enhancing the mutation frequency, loss of heterozygosity, chromosomal aberrations (Honma, 2020). These events have a crucial role in various malignancies, including cancer. Therefore, identifying genotoxic, mutagenic, and carcinogenic effects of traditional medicine, dietary substances, pharmaceuticals, actually every new compound is essential for safety concerns. Plants that possess mutagenic properties are regarded as potentially unsafe. Hence, toxicity assessment must be performed as per several regulatory authorities' terms, and the compounds must be deemed safe for human use only if these toxicity parameters are ruled out (Wiesner, 2014). To assess genotoxicity and/or mutagenicity, one single endpoint does not exist, it involves an array of endpoints. The three crucial endpoints which must be evaluated for genotoxicity assessment are Gene mutation (mutagenicity) frequency, chromosomal aberration in structure (Clastogenicity), and chromosomal aberration in number (Aneugenicity). A battery of the most common *in vivo* and *in vitro* testing techniques for genotoxicity utilized are the bacterial Salmonella mutagenicity test (Ames test), the VITOTOX Assay (VIT), alkaline comet assay, chromosomal Aberration (CA) assay, sister chromatid exchange (SCE) assay, unscheduled DNA synthesis (UDS) test, Micronucleus test (MN), Hypoxanthine-guanine phosphoribosyl transferase (HPRT) test, a transgene of xanthine guanine phosphoribosyl transferase (XPRT) test, Assessment of Mitotic Index (MI), *E. coli* WP2 reversion assay (ERA), Somatic Mutations and Recombination Test (SMART), *Allium cepa* Assay (AC), Umu-C Assay(UC), SOS Induc test(SOS), Assessment of Somatic Segregation in *Aspergillus nidulans* (AN), Gene Mutation in *Saccharomyces cerevisiae* (SC), Assessment of Recessive Mutation Lethal X-linked (SLRL) in *Drosophilla melanogaster*, Ring-X-Loss test (RXL), Sperm Head Anomaly (SHA), Apoptosis-Inducing (AI), Cell Replication Indice (RI), Plasmid DNA Breakage (PDB), Forward Mutagenesis (FM), and Induction of Cytoplasmic Petite Mutations (ICPM) (Sponchiado et al., 2016; Hasselgren et al., 2019; Turkez et al., 2017). Most commonly used genotoxicity assays are shown in Fig. 26.1.

Conversely, plants with antimutagenic potential are therapeutically significant due to their chemo-preventive properties (Ntuli et al., 2018). Chemical agents that reduce the mutagenicity of chemical or physical substances, either by inhibiting the

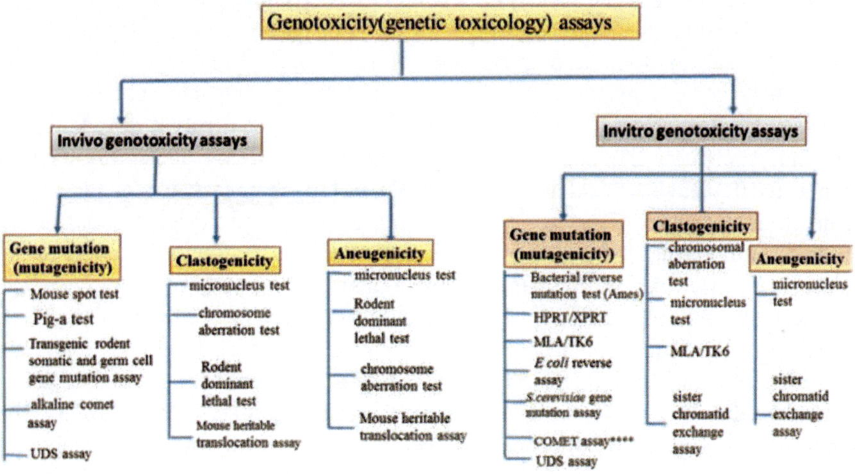

Fig. 26.1 Commonly used genetic toxicity tests (*in vitro* and *in vivo*)

action between DNA and mutagens or by neutralizing the mutagen itself, are known as Antimutagens. The term antimutagens were coined by Novick and Szilard (1952). The antimutagenic agents were classified as desmutagens and bioantimutagens. Desmutagens are agents that suppress mutation by blocking the chemical interactions, whereas, Bioantimutagens act intracellularly after the DNA damage occurs, processing its "repair and replication", subsequently decreasing the mutation frequency (Choudhuri et al., 2021; Dashwood, 2002). Natural products exhibit their antimutagenic properties via any one of the following mechanisms (1) Inactivation/sequestration of mutagens directly, (2) enhanced deactivation or inhibition of metabolic activation of mutagen, (3) protection of DNA against toxic effects of mutagens, (4) Rectification of DNA damage via repair and replication to make it less mutagenic, (5) cell cycle modulation, (6) interference with detoxification system of the liver (Akram et al., 2020; Zeiger, 2011). The investigation of antigenotoxic natural products must be done to develop a novel anti-cancer drug, as severe side effects and drug resistance accompany the current therapeutic strategies for cancer. Hence, replacement with a less hazardous and more beneficial product is anticipated.

Interestingly, certain compounds possess both mutagenic and antimutagenic effects. These compounds are known as "Janus mutagens" (Słoczyńska et al., 2014). For instance, the lignin dibenzyl butyrolactone called hinokinin (HK) obtained from the dry seeds of Piper cubeba act as a "janus" compound, i.e., at low doses, behaves as an antioxidant/antimutagen (scavenge ROS). In contrast, at high doses, it is a prooxidant/mutagen (which produces free radicals) (Resende et al., 2012). Unfortunately, these biphasic properties of phytochemical compounds are not tested appropriately for their mutagenicity/genotoxicity/carcinogenicity, yet they are recommended as potent anticarcinogens for human use. This warrants a thorough screening of natural products before being used as medicines, and the screening reports must be interpreted with extreme care and caution (Errol Zeiger, 2003).

3 Overview of *Ficus carica*

Ficus is considered the largest genus of angiosperms, with approximately 850 species identified and distributed worldwide, covering tropical and sub-tropical zones (Yang et al., 2015). Plants belonging to this genus are deciduous shrubs or small evergreen trees, herbs, creepers or climbers, epiphytes or hemi epiphytes, lithophytes, and vines rheophytes (Suttisansanee et al., 2021). Plants under the genus *Ficus* belong to the order Urticales and the family Moraceae (Mulberry). They have nurtured crucial niches in economic and therapeutic areas. Numerous *Ficus* spp., with their parts like stem, roots, bark, leaves, fruits, latex, resins, etc., are deemed medicinally important in Ayurveda, Chinese, and Arabic folklore medicine as several pharmacological activities have been reported (Irudayaraj et al., 2017). Among the several *Ficus* species, *Ficus carica* Linn. (FC) has gained significant popularity due to its edible fruit, either in dry or fresh form, either peeled or unpeeled, and is commonly known as "fig". It is the oldest known cultivated fruit native to the

Mediterranean Basin, the Middle East, and southwest Asia. *F. carica* grows up to a height of 6.9–10 m, has a small grey bark and numerous branches which secrete milky white latex, enriched with enzyme Ficin (Hamed et al., 2020), and fragrant leaves that are oval, orbicular in shape, the pubescent undersurface and above the rough surface with 3–5 lobes (Patil & Patil, 2011). The inflorescence of *Ficus carica* is complex and consists of Synconium with unisexual flowers (Peniwidiyanti et al., 2021). However, fruits are mature synconia, fleshy, hollow, and involuted with a closed receptacle. They are bulbus with ostiole (opening) at the distal end for pollination by female wasp *Blastophaga psenes* (L.) Grav. Their peel color varies as it ripens from green to purple, attributed to the presence of anthocyanins (Teixeira et al., 2018; Wang et al., 2017). In recent times figs have gained enormous attention as they constitute an integral part of the Mediterranean diet. Fig is a highly nutritious fruit, low in sodium, abundant in vitamins (A, B1, B2, B3, and C), amino acids (aspartic acid and glutamine), minerals (potassium, calcium, magnesium, zinc, iron), natural sugars; they have no cholesterol or fat, polypeptides and phenolic compounds (Mohammed et al., 2020; Pereira et al., 2017).

Ficus carica biosynthesizes several primary and secondary metabolites that can be isolated from its plant parts such as leaves, bark, dried and fresh fruit, i.e., fig, latex. Approximately 126 chemical constituents have been identified in *F. carica* to date, many more are yet to be revealed. The major type of phytochemicals reported in *Ficus carica* is phenolic acids, flavonoids (mostly in fruits and leaves), carotenoids, anthocyanins (fruit pulp and peel), coumarins (leaves and roots), sterols(leaves), and terpenoids, alkaloids, tannins, sterols and proteins (latex), arabinose, β-carotenes, β-amyrins, glycosides, xanthotoxol, and β-sitosterol. Arvaniti et al., in their study dealing with fresh and dried figs, reported the presence of gallic acid, rutin, chlorogenic acid, epicatechin, and quercetin-3-*O*-rutinoside (Arvaniti et al., 2019). Anthocyanins of type, pelargonidin-3-*O*-rutinoside, cyanidin-3,5-diglucoside, and cyanidin-3-O-rutinoside were predominantly present in fig peel, whereas epicatechin and cyanidin-3-*O*-rutinoside were present in pulp extracts (Hssaini et al., 2021). Various phytochemicals of therapeutic significance identified in plant parts of *Ficus carica* like root, fruit, leaves, latex, etc., are enlisted in Table 26.1, while their chemical structure is depicted in Fig. 26.2.

4 Therapeutic Potential of *Ficus carica*

Various parts of *Ficus carica* like leaves, fruit, latex, and root are deemed medicinally valuable in folklore medicine and traditionally used to treat ailments. *Ficus carica* treats diarrhea, expectorant, diuretic, colic, intestinal spasm, and gastrointestinal tract indigestion. Similarly, respiratory disorders such as sore throats, bronchial problems and inflammation, and cardiovascular and metabolic disorders are treated (Baygeldi et al., 2021). The utilization of "fig" fruit as food and medicine is as old as the human civilization. In the ancient Indian medicine system called Ayurveda, sore throat was relieved by a gargle of fig decoction, also used to treat

Table 26.1 Phytochemical constituents of therapeutic significance identified in various parts (roots, leaves, fig, latex) of *Ficus carica*

Part used	Class of phytochemical constituent	Example
Leaf	Coumarin	4′,5′-Dihydropsoralen
		Umbelliferone
		Marmesin
		Bergapten
	Flavonoid	Rutin
	Sterol	Baurenol
		24-methylenecycloartanol
		Lupeol
		ψ-taraxasterol ester
	Triterpenoid	Ficusogenin
Leaf, root	Coumarin	Psoralen
	Sterol	β-Sitosterol
Fruit	Coumarin	Umbelliferone
		Scopoletin
	Alkaloid	Quinines
	Anthocyanin	Cyanidin-3-O-glucoside
		Cyanidin-3- O-rhamnoglucoside
	Hydrocarbon	Stilbenes
Latex	Triterpenoid	6-O-Linoleyl-b-D-glucosyl-b-sitosterol
		6-O-Oleyl-b-D-glucosyl-b-sitosterol
		6-O-palmitoyl-b-D-glucosyl-b-sitosterol
	Sterols	Betulol
		Lupeol
		Lanosterol
		Lupeol acetate
		β-amyrin
		β-sitosterol
		α-amyrin
	Acyl steryl glucosides	24-Methylenecycloartanol
		ψ-taraxasterol ester

dysentery, menorrhagia, and gonorrhea, and used as an aphrodisiac. Not only this, Ayurveda recommends ingestion of entire fig for treatment of diarrhea. Evidence from traditional Chinese medicine recommends a decoction of fig for local application and cooked fig for ingestion to cure hemorrhoids. Traditionally, the decoction of the *Ficus* leaves was used to treat diabetes (Konyalıoğlu et al., 2005). Italian people used a decoction of dried leaves along with apple peel as a cough suppressant; this decoction can have walnut hulls, leaves of Malva *sylvestris,* and heads of *Matricaria chamomilla* (Mautone et al., 2019).

F. carica latex was applied to treat warts and viruses. This traditional use was substantiated by the evidence provided by Bohlooli et al., who reported no adverse effects and safety in the use of folklore medicine (Bohlooli et al., 2007). Additionally,

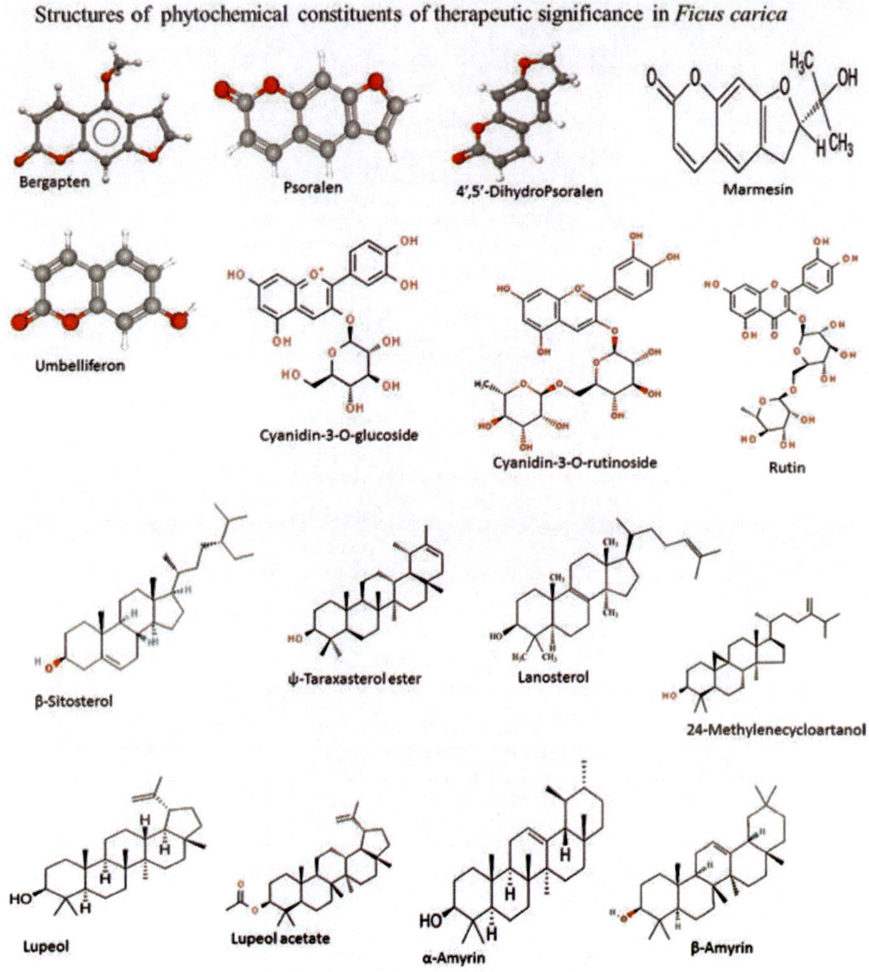

Fig. 26.2 Chemical structures of phytochemical constituents of therapeutic significance in *Ficus carica*. (Structures were obtained from https://pubchem.ncbi.nlm.nih.gov/compound/; http://www.chemspider.com/Chemical-Structure.XX.html)

dried figs soaked in olive oil were a traditional cure for cough and bronchial problems. Manjula et al. reported that jaundice could be treated when 20 mL of fig leaf juice is administered with a cup of goat milk once a day for 3 days (Manjula et al., 2011). The green unripe fig fruits, when cooked, can serve as galactogogue and tonic; upon roasting, the fruit acts as an emollient and a poultice that can be used to treat gumboils, dental abscesses (Ali et al., 2009). Fig is a folkloric emmenagogue used to treat amenorrhea (Salehi et al., 2021). Although traditional phytotherapy is based on experience and is transferred to generations by word of mouth, very meager evidence exists in the form of documents. Undoubtedly, fig has many more uses which need to be investigated. Based on the available folkloric use, scientists have conducted experiments and tried to prove the uses of *Ficus carica*. Consequently, it was discovered that *Ficus carica* L. (fig) possesses several pharmacological properties such as antispasmodic, anti-cancer, anti-inflammatory, antioxidant, antiviral, hemostatic, etc. (Mawa et al., 2013). The most prominent therapeutic properties of fig are summarized in Table 26.2. Although the exact mechanisms of action of *Ficus carica* used for the treatment of various disorders are not fully elucidated, it is apparent that the ubiquity of polyphenols, high content of total phenols, and flavonoids are responsible for the actions. The therapeutically significant phytoconstituents of fig might exert their action through any of the following mechanisms: radical scavenging antioxidant effects, antiproliferative effects, apoptotic induction, inducing cytochrome P450 (CYP450) monooxygenase enzyme system, which ultimately enhances the body's innate detoxification system. About its varied uses, *Ficus carica* has occupied ethnopharmacological niches and has garnered a place in occidental Pharmacopeias, in PDR for Herbal Medicines, and is a compendium of herbology.

As one of the oldest cultivated crops, *Ficus carica* fruits are considered safe. However, specific toxicity issues have been reported related to FC leaves, as to moderate toxicity and causation of phototoxicity. Furthermore, exposure to latex may initiate inflammatory reactions like asthma, dermatitis, asthma, and/or anaphylaxis. Furthermore, this oral administration of latex may initiate hallucinations (Lansky et al., 2008). Hence evidence must be gathered based on scientific investigations, and the entire toxicity profile must be updated; meanwhile, FC leaves must be consumed with precaution. This is possible only by a series of toxicological investigations.

Table 26.2 The most prominent therapeutic properties of *Ficus carica*

Therapeutic property	Plant part used	Mechanism	Reference
Anticonvulsant activity	Stem bark ethanolic extract	Potentiation of the inhibitory glycine receptor	Raafat and Wurglics (2019)
Hepatoprotective activity	Petroleum ether extract of leaves; ethanolic leaf extract; methanolic leaf extract; 80%aqueous ethanolic leaf extract	Presence of flavonoids; detailed mechanism yet to be known	Gond and Khadabadi (2008), Mujeeb et al. (2011), Mohan et al. (2007), and Aghel et al. (2011)
Antioxidant activity	Methanolic leaves extract free-radical scavenging effect, hydroalcoholic extract of leaves	Free radical scavenging potential	Ayoub et al. (2019) and Ali et al. (2012)
Anti-inflammatory activity	Hydro-alcoholic extract of *F. carica* leaves; ethanolic extract and fractionation of fig branches	Scavenge the free radicals like superoxide generated by photoreduction of riboflavin; inhibit NO and TNF-α in RAW264.7 cells	Ali et al. (2012) and Park et al. (2013)

(continued)

Table 26.2 (continued)

Therapeutic property	Plant part used	Mechanism	Reference
Alzheimer's disease	Hexane leaf extract; a mixture of fig and olive oil; dietary supplementation phenolic compounds and flavonoids	AChE inhibition; phenolic and flavonoid-yet unclear	Saxena et al. (2013), Alharthy and Bawazir (2019), and Subash et al. (2016)
Anti-diabetic activity	Methanolic extract of stem bark; ethanolic fruit extract; freeze-dried leaf extract	The insulinotropic effect of the extract; inhibition of α-amylase and α-glucosidase enzymes; inhibits apoptosis in pancreatic β-cell via inhibition of AMPK/JNK/caspase-3 pathway	Bhat et al. (2013), Mopuri and Islam (2016), and Zhang et al. (2020)
Hypoglycaemic activity	Ethyl acetate extract of leaves; aqueous leaf extract	Insulin sensitization activity; prevention of B-cell degeneration; enhanced insulin secretion	Stephen Irudayaraj et al. (2017) and Perez et al. (2000)
Anticancer activity	Methanolic leave extract; methanolic leaf and fruit extract on Huh7it liver cancer cells; Peel, pulp, leaves, whole fruit, and latex extracts on HT-29 and HCT-116 colorectal cancer cells; fig tree latex; *F. carica* fruit latex; synthesized silver nanoparticles of dried fig; latex;	Preventing cellular proliferation, enhancing apoptosis in melanoma cancer cells; luteolin and quercetin induce cell cycle arrest and enhance apoptosis by activating p53 and p21; apoptosis inducers; antiproliferative; downregulation of p16 and HPV oncoproteins; cytotoxic effect on MCF7 breast cancer Cell lines; downregulate ERK2, CREB, and AKT2 pathways in MDA-Mb-231 breast cancer cells	Sadeghizade et al. (2021), Purnamasari et al. (2019), Soltana et al. (2019), Hashemi et al. (2011), Jacob et al. (2017), Ghanbari et al. (2019), and AlGhalban et al. (2021)
Antibacterial and antifungal activity	Methanolic extracts of fruit, leaf, root, bark	Not clear	Shafique et al. (2021)
Nematicidal activity	Methanolic leaf extract	Presence of psoralen	Liu et al. (2011)
Antispasmodic and antiplatelet activity	80% ethanolic dried fruit extract	K+ ATP channel activation	Gilani et al. (2008)

(continued)

Table 26.2 (continued)

Therapeutic property	Plant part used	Mechanism	Reference
Hemostatic effect	latex	Ficin shows a haemostatic effect through activation of Factor X	Richter et al. (2002)
Anti-nociceptive effect delayed effect	Petroleum ether extract of fruits	Yet to be elucidated	Mirghazanfari et al. (2019)
Antihypertensive and cardioinhibitory activity	Aqueous-methanol extract of fruits	Polyphenols leading to vasorelaxation might be involved in antihypertensive effects yet to be elucidated	Alamgeer et al. (2017)

5 Genotoxicity and Antimutagenic Evidence for *Ficus carica*

The global trend is to use natural phytochemicals in food and medicine. *Ficus carica* (Fig) products can be considered the best examples of natural products used in food and medicine. *Ficus carica* consists of several phytoconstituents of therapeutic value. Despite being deemed safe since its usage from ancient times, it can also exhibit toxicity effects. Phytoconstituents can be genoprotective if they offer protection against DNA damage, cause inactivation of carcinogens, or enhance DNA repair processes. Genoprotective/antimtagenic compounds can be anticarcinogenic/chemopreventive. Hence, it is becoming imperative to investigate compounds with the above-mentioned activities, so they can be used to develop new drugs.

The anti-mutagenic potential of figs is not thoroughly studied, and the research towards this toxicity assessment is still naïve. The first antimutagenic action of *Ficus carica* extract was on environment xenobiotics reported. Agabeili and kasimova observed a decrease in mutation frequency induced by environmental mutagens such as N-metil-N′-nitro-N-nitrozoguanidin (MNNG) when treated with extracts of *Armoracia rusticana, Ficus carica, Zea mays*, and their mixture decreased the level of mutations in *Vicia faba* cells, chlorophyll mutations in Arabidopsis thaliana, and NaF-induced mutations in rat marrow cells (Agabeili & Kasimova, 2005). Furthermore, Radwan et al. (2018) investigated the genotoxic effects of *Ficus carica* and *Schinus molle* in varied climatic conditions (i.e., in healthy and polluted) in the Asir region of Saudi Arabia. They carried out Genotoxicity tests in mice using Comet assay, micronucleus test, and chromosomal analysis. They concluded from a comet assay that *F. carica* and *S. molle* collected from polluted areas significantly damaged the DNA in bone marrow cells and increased the number of micro-nucleated cells in the mice. Briefly, plants from polluted areas are genotoxic to bone marrow cells in mice. They also deduced that metals like Ni and Co are essential for certain cellular processes. However, their bioaccumulation leads to chromosomal aberrations, breaks in single or double DNA strands, and a cross-link between chromatid exchange DNA-protein or DNA damage (Radwan et al., 2018).

Cisplatin is a platinum-based antineoplastic agent successfully employed to treat several cancer types like breast cancer, ovarian cancer, neuroblastoma, brain tumors, etc. The main disadvantage of current cancer chemotherapies is resistance and toxic side effects. Cisplatin also subjects the patients using it to nephrotoxicity, genotoxicity, Hepatotoxicity, and reproductive toxicity (Makovec, 2019). Fahmy et al. evaluated the genoprotective/antimutagenic effect of *Ficus carica* (fig) utilizing methanolic leaf extracts against cisplatin-induced genotoxicity and testicular damage. Genetic toxicology endpoints worked on here are bone marrow micronuclei in polychromatic erythrocytes (MPEs), chromosomal aberrations (CAs) in the bone marrow, and primary spermatocytes with the expression of the genes TNF-α, iNOS and NF-kB. Cisplatin treatment caused a significant increase in the percentage of chromosomal abnormalities, increased gene NF-kB levels, increased micronuclei in polychromatic erythrocytes, and enhanced toxicity to bone marrow cells. However, treatment with *Ficus carica* extract reversed all the abnormalities in a dose-dependent manner proving it to be a promiscuous chemopreventive/genoprotective agent (Fahmy et al., 2020).

Lightbourn and Thomas further investigated the protective effect of ficus leaf extract against DES-induced DNA damage in MCF10A cells which are benign human breast epithelial cells also they investigated the nuclear insult induced by environmental estrogens and their metabolites on MCF10A cells using Single-cell gel electrophoresis (SCGE)/COMET assay. Diethylstilbestrol (DES) is a carcinogenic agent. It is a synthetic estrogenic hormone that is acted upon by CYP450 enzymes converting it to an intermediary metabolite DES-4,4′-quinone (DESQ) which forms DNA adducts and strand breaks.

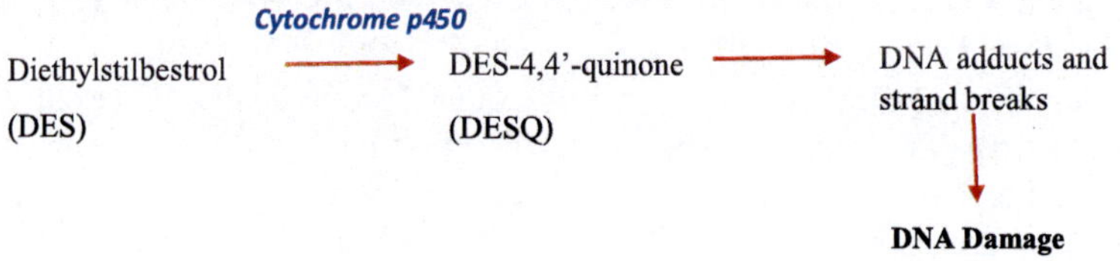

In this study, they concluded that *Ficus carica* L. leaf extract inhibits DNA strand breaks by interfering with the nuclear complex in MCF10A cells either in the presence/absence of diethylstilbestrol (DES) and 4′,4″-diethylstilbestrol quinine (DESQ) metabolite (Lightbourn & Thomas, 2019).

Hadžic et al. studied the inhibitory effects of delphinidin and luteolin phytoconstituents on genotoxicity induced by halogenated boroxine [$K_2(B_3O_3F_4OH)$] in human lymphocytes *in vitro.* Luteolin (3′,4′,5,7-tetrahydroxyflavone) is a yellow pigmented flavone predominantly seen in the leaves of *Ficus carica* and to a lesser proportion in the fruit figs chemical structure is depicted in Fig. 26.3. Chromosomal aberrations were analyzed by the cytokinesis-block micronucleus cytome assay in human peripheral blood lymphocytes. Luteolin inhibits genotoxic effects by decreasing cellular proliferation and also induces cell cycle arrest in G1/S and G2/M checkpoints (Hadžić et al., 2015).

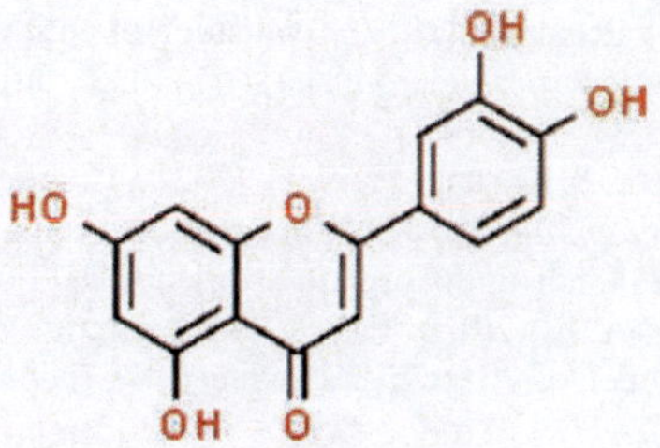

Fig. 26.3 Structure of luteolin found in *Ficus carica* leaves and fruit (National centre for biotechnology information (2021) Pub Chem ID:5280445)

6 Conclusion

Plant phytochemicals are the crucial source for identifying and isolating lead compounds that can be formulated into a novel therapeutic agent that can be used to cure all kinds of diseases. Evidence from ethnobotanical and pharmacological studies reveals that the administration of natural products as medicine or food/diet supplements is inversely related to disease progression. They represent a promising therapeutic strategy for cancer prevention. However, as they are intended to be used as novel drugs, they must be tested for their toxicological parameters and deemed safe. The genotoxicity, reproductive toxicology, mutagenicity testing must be performed using a battery of *in vivo* and *in vitro* experiments authorized by a regulatory agency. Anitmutagenicity tests commonly use Chromosome aberration assay, micronuclei test, and comet assay. Of these, comet assay is most preferred due to ease of availability and reliability. The presence of flavonoids and polyphenols might be regarded as the prominent phytochemicals crucial for varied therapeutic uses. With the advent of modern technologies, several strategies can be applied for the toxicological assessment of folklore medicine, the most powerful being the genomic approach. It serves as a powerful tool acknowledging the pharmaco-toxicological activities of folklore medicinal products. The involvement of "omics" technologies provides ample opportunity and techniques to elucidate the underlying mechanisms of various phytochemicals and create a comprehensive basis for better hazard identification and more relevant genotoxicity assessment.

References

Agabeili, R. A., & Kasimova, T. E. (2005). Antimutagenic activity of *Armoracia rusticana, Zea mays* and *Ficus carica* plant extracts and their mixture. *Tsitologiia i Genetika, 39*, 75–79.

Aghel, N., Kalantari, H., & Rezazadeh, S. (2011). Hepatoprotective effect of *Ficus carica* leaf extract on mice intoxicated with carbon tetrachloride. *Iranian Journal of Pharmaceutical Research: IJPR, 10*(1), 63–68.

Akram, M., et al. (2020). Medicinal plants with anti-mutagenic potential. *Biotechnology & Biotechnological Equipment, 34*(1), 309–318. https://doi.org/10.1080/13102818.2020.1749527

Alamgeer, I. S., Asif, H., & Saleem, M. (2017). Evaluation of antihypertensive potential of *Ficus carica* fruit. *Pharmaceutical Biology, 55*(1), 1047–1053. https://doi.org/10.1080/13880209.2017.1278611

AlGhalban, F. M., Khan, A. A., & Khattak, M. N. K. (2021). Comparative anticancer activities of *Ficus carica* and *Ficus salicifolia* latex in MDA-MB-231 cells. *Saudi Journal of Biological Sciences, 28*(6), 3225–3234. https://doi.org/10.1016/j.sjbs.2021.02.061

Alharthy, N. A., & Bawazir, A. E. (2019). Effects of the mixture dried figs (*Ficus carica*) and olive oil on amnesia model of Alzheimer's induced by scopolamine in male albino rats. *Pharmacophore, 10*(4), 62–71.

Ali, H., Monga, J., Gupta, L., Singh, A., Narayan, S., & Siddiqui, Z. A. (2009). Anti-inflammatory effects of hydro-methanolic extract of *Ficus carica*. *Biomedical & Pharmacology Journal, 2*(1), 129–132.

Ali, B., Mujeeb, M., Aeri, V., Mir, S. R., Faiyazuddin, M., & Shakeel, F. (2012). Anti-inflammatory and antioxidant activity of *Ficus carica* Linn. leaves. *Natural Product Research, 26*(5), 460–465. https://doi.org/10.1080/14786419.2010.488236

Alostad, A. H., Steinke, D. T., & Schafheutle, E. I. (2020). Herbal medicine classification: Policy recommendations. *Frontiers in Medicine, 7*, 31.

Arvaniti, O. S., Samaras, Y., Gatidou, G., Thomaidis, N. S., & Stasinakis, A. S. (2019). Review on fresh and dried figs: Chemical analysis and occurrence of phytochemical compounds, antioxidant capacity and health effects. *Food Research International (Ottawa, Ont.), 119*, 244–267. https://doi.org/10.1016/j.foodres.2019.01.055

Atanasov, A. G., et al. (2015). Discovery and resupply of pharmacologically active plant-derived natural products: A review. *Biotechnology Advances, 33*(8), 1582–1614. https://doi.org/10.1016/j.biotechadv.2015.08.001

Ayoub, L., Hassan, F., Hamid, S., Abdelhamid, Z., & Souad, A. (2019). Phytochemical screening, antioxidant activity and inhibitory potential of *Ficus carica* and *Olea europaea* leaves. *Bioinformation, 15*(3), 226–232. https://doi.org/10.6026/97320630015226

Baygeldi, N., Küçükerdönmez, Ö., Akder, R. N., & Çağındı, Ö. (2021). Medicinal and nutritional analysis of fig (*Ficus carica*) seed oil; a new gamma tocopherol and omega-3 source. *Progress in Nutrition, 23*(2), 1–6. https://doi.org/10.23751/pn.v23i2.9980

Bhat, M. Z. A., Ali, D. M., & Mir, S. R. (2013). Anti-diabetic activity of *Ficus carica* L. stem barks and isolation of two new flavonol esters from the plant by using spectroscopical techniques. *Asian Journal of Biomedical and Pharmaceutical Sciences, 3*(18), 22–28.

Bohlooli, S., Mohebipoor, A., Mohammadi, S., Kouhnavard, M., & Pashapoor, S. (2007). Comparative study of fig tree efficacy in the treatment of common warts (*Verruca vulgaris*) vs. cryotherapy. *International Journal of Dermatology, 46*(5), 524–526.

Choudhuri, S., Kaur, T., Jain, S., Sharma, C., & Asthana, S. (2021). A review on genotoxicity in connection to infertility and cancer. *Chemico-Biological Interactions, 109531*. https://doi.org/10.1016/j.cbi.2021.109531

Dashwood, R. H. (2002). Modulation of heterocyclic amine-induced mutagenicity and carcinogenicity: An 'A-to-Z' guide to chemopreventive agents, promoters, and transgenic models. *Mutation Research-Reviews in Mutation, 511*(2), 89–112. https://doi.org/10.1016/s1383-5742(02)00005-4

Ekor, M. (2014). The growing use of herbal medicines: Issues relating to adverse reactions and challenges in monitoring safety. *Frontiers in Pharmacology, 4*, 177. https://doi.org/10.3389/fphar.2013.00177

Fahmy, M. A., Hassan, E. E., Ibrahim, N. E., Hassan, E. M., Hassan, Z. M., & Omara, E. A. (2020). Protective role of Ficus carica extract against hepato-testicular side effects and genotoxicity induced by cisplatin. *Pharmacognosy Journal, 12*(3). https://doi.org/10.5530/pj.2020.12.96

Fokunang, et al. (2020). Pharmacovigilance of natural herbal medicines research for efficacy, safety and quality assurance of phytomedicine products. *Journal of Complementary and Alternative Medical Research, 12*, 21–37.

Forni, C., et al. (2019). Beneficial role of phytochemicals on oxidative stress and age-related diseases. *BioMed Research International, 2019*, 8748253. https://doi.org/10.1155/2019/8748253

Ghanbari, A., Le Gresley, A., Naughton, D., Kuhnert, N., Sirbu, D., & Ashrafi, G. H. (2019). Biological activities of *Ficus carica* latex for potential therapeutics in Human Papillomavirus (HPV) related cervical cancers. *Scientific Reports, 9*(1), 1–11. https://doi.org/10.1038/s41598-018-37665-6

Ghorani-Azam, A., et al. (2018). Plant toxins and acute medicinal plant poisoning in children: A systematic literature review. *Journal of Research in Medical Sciences: The Official Journal of Isfahan University of Medical Sciences, 23*, 26. https://doi.org/10.4103/jrms.JRMS_629_17

Gilani, A. H., Mehmood, M. H., Janbaz, K. H., Khan, A. U., & Saeed, S. A. (2008). Ethnopharmacological studies on antispasmodic and antiplatelet activities of *Ficus carica*. *Journal of Ethnopharmacology, 119*(1), 1–5. https://doi.org/10.1016/j.jep.2008.05.040

Gond, N. Y., & Khadabadi, S. S. (2008). Hepatoprotective activity of *Ficus carica* leaf extract on rifampicin-induced hepatic damage in rats. *Indian Journal of Pharmaceutical Sciences, 70*(3), 364–366. https://doi.org/10.4103/0250-474X.43003

Gorlenko, C. L., Kiselev, H. Y., Budanova, E. V., Zamyatnin, A. A., Jr., & Ikryannikova, L. N. (2020). Plant secondary metabolites in the battle of drugs and drug-resistant bacteria: New heroes or worse clones of antibiotics? *Antibiotics (Basel, Switzerland), 9*(4), 170. https://doi.org/10.3390/antibiotics9040170

Graz, B. (2013). What is "clinical data"? Why and how can they be collected during field surveys on medicinal plants? *Journal of Ethnopharmacology, 150*(2), 775–779. https://doi.org/10.1016/j.jep.2013.08.036

Hadžić, M., Haverić, S., Haverić, A., & Galić, B. (2015). Inhibitory effects of delphinidin and luteolin on genotoxicity induced by $K_2(B_3O_3F_4OH)$ in human lymphocytes in vitro. *Biologia, 70*(4), 553–558. https://doi.org/10.1515/biolog-2015-0066

Hamed, M. B., El-Badry, M. O., Kandil, E. I., Borai, I. H., & Fahmy, A. S. (2020). A contradictory action of procoagulant ficin by a fibrinolytic serine protease from Egyptian Ficus carica latex. *Biotechnology Reports, 27*, e00492. https://doi.org/10.1016/j.btre.2020.e00492

Hashemi, S. A., Abediankenari, S., Ghasemi, M., Azadbakht, M., Yousefzadeh, Y., & Dehpour, A. A. (2011). The effect of fig tree latex (*Ficus carica*) on stomach cancer line. *Iranian Red Crescent Medical Journal, 13*(4), 272–275.

Hasselgren, C., et al. (2019). Genetic toxicology in silico protocol. *Regulatory Toxicology and Pharmacology: RTP, 107*, 104403. https://doi.org/10.1016/j.yrtph.2019.104403

Honma, M. (2020). An assessment of mutagenicity of chemical substances by (quantitative) structure-activity relationship. *Genes and Environment, 42*(1), 1–13. https://doi.org/10.1186/s41021-020-00163-1

Hssaini, L., et al. (2021). Survey of phenolic acids, flavonoids and in vitro antioxidant potency between fig peels and pulps: Chemical and chemometric approach. *Molecules (Basel, Switzerland), 26*(9), 2574. https://doi.org/10.3390/molecules26092574

Jacob, S. J. P., Prasad, V. S., Sivasankar, S., & Muralidharan, P. (2017). Biosynthesis of silver nanoparticles using dried fruit extract of *Ficus carica* -screening for its anticancer activity and toxicity in animal models. *Food and Chemical Toxicology, 109*, 951–956. https://doi.org/10.1016/j.fct.2017.03.066

Konyalıoğlu, S., Sağlam, H., & Kıvçak, B. (2005). α-Tocopherol, flavonoid, and phenol contents and antioxidant activity of *Ficus carica* leaves. *Pharmaceutical Biology, 43*(8), 683–686. https://doi.org/10.1080/13880200500383538

Lansky, E. P., Paavilainen, H. M., Pawlus, A. D., & Newman, R. A. (2008). *Ficus* spp. (fig): Ethnobotany and potential as anticancer and anti-inflammatory agents. *Journal of Ethnopharmacology, 119*(2), 195–213. https://doi.org/10.1016/j.jep.2008.06.025

Lightbourn, A. V., & Thomas, R. D. (2019). Crude edible fig (*Ficus carica*) leaf extract prevents diethylstilbestrol (DES)-induced DNA strand breaks in single-cell gel electrophoresis (SCGE)/comet assay: Literature review and pilot study. *Journal of Bioequivalence & Bioavailability, 11*(2), 19. https://doi.org/10.35248/0975-0851.19.11.389

Liu, F., Yang, Z., Zheng, X., Luo, S., Zhang, K., & Li, G. (2011). Nematicidal coumarin from *Ficus carica* L. *Journal of Asia-Pacific Entomology, 14*(1), 79–81. https://doi.org/10.1016/j.aspen.2010.10.006

Makovec, T. (2019). Cisplatin and beyond: Molecular mechanisms of action and drug resistance development in cancer chemotherapy. *Radiology and Oncology, 53*(2), 148–158. https://doi.org/10.2478/raon-2019-0018

Manjula, R. R., Rao, J. K., & Reddi, T. S. (2011). Ethnomedicinal plants used to cure jaundice in Kammam district of Andhra Pradesh, India. *Journal of Phytology, 3*(10), 33–35.

Mautone, M., De Martino, L., & De Feo, V. (2019). Ethnobotanical research in Cava de' Tirreni area, Southern Italy. *Journal of Ethnobiology and Ethnomedicine, 15*(1), 50. https://doi.org/10.1186/s13002-019-0330-3

Mawa, S., Husain, K., & Jantan, I. (2013). *Ficus carica* L. (Moraceae): Phytochemistry, traditional uses, and biological activities. *Evidence-Based Complementary and Alternative Medicine, 2013*. https://doi.org/10.1155/2013/974256

Mirghazanfari, S. M., Alameh, H. A., Parviz, M., & Ahmadi Basir, Z. (2019). Evaluation of *Ficus carica* Linn fruit extracts on formalin induced pain in the male rat paw. *Annals of Military and Health Sciences Research, 17*(3), 1–6. https://doi.org/10.5812/amh.97383

Mohammed, O. A., Hussain, M. R., Shareef, O. H., Hama, A. A., Weli, S. M., Ali, F. M., & Salih, S. S. (2020). Analysis of some heavy metals and organic acids in *Ficus carica* growing adjacent in the serpentine soil in Sulaimani/Kurdistan, Iraq. *International Journal of Food Science, 2020*. https://doi.org/10.1155/2020/8883517

Mohan, G. K., Pallavi, E., Kumar, R., Ramesh, M., & Venkatesh, S. (2007). Hepatoprotective activity of *Ficus carica* Linn leaf extract against carbon tetrachloride-induced hepatotoxicity in rats. *DARU Journal of Pharmaceutical Sciences, 15*(3), 162–166.

Mujeeb, M., Khan, S. A., Aeri, V., & Ali, B. (2011). Hepatoprotective Activity of the Ethanolic Extract of Ficus caricaLinn. LeavesinCarbon Tetrachloride-Induced Hepatotoxicityin Rats. *Iranian journal of pharmaceutical research: IJPR, 10*(2), 301.

Mopuri, R., & Islam, M. S. (2016). Antidiabetic and anti-obesity activity of Ficus carica: In vitro experimental studies. *Diabetes & Metabolism, 42*(4), 300. https://doi.org/10.1016/j.diabet.2016.07.020

Moreira, D. D. L., Teixeira, S. S., Monteiro, M. H. D., De-Oliveira, A. C. A., & Paumgartten, F. J. (2014). Traditional use and safety of herbal medicines. *Revista Brasileira de Farmacognosia, 24*, 248–257.

Mushtaq, S., Abbasi, B. H., Uzair, B., & Abbasi, R. (2018). Natural products as reservoirs of novel therapeutic agents. *EXCLI Journal, 17*, 420–451. https://doi.org/10.17179/excli2018-1174

National Center for Biotechnology Information. (2021). *PubChem compound summary for CID 5280445, Luteolin*. Retrieved December 30, 2021, from https://pubchem.ncbi.nlm.nih.gov/compound/Luteolin

Novick, A., & Szilard, L. (1952). Anti-mutagens. *Nature, 170*(4335), 926–927. https://doi.org/10.1038/170926a0

Ntuli, S., Gelderblom, W., & Katerere, D. R. (2018). The mutagenic and antimutagenic activity of *Sutherlandia frutescens* extracts and marker compounds. *BMC Complementary and Alternative Medicine, 18*(1), 93. https://doi.org/10.1186/s12906-018-2159-z

Oyebode, O., Kandala, N. B., Chilton, P. J., & Lilford, R. J. (2016). Use of traditional medicine in middle-income countries: A WHO-SAGE study. *Health Policy and Planning, 31*(8), 984–991. https://doi.org/10.1093/heapol/czw022

Park, S., et al. (2013). Antioxidative and anti-inflammatory activities of an ethanol extract from fig (*Ficus carica*) branches. *Food Science and Biotechnology, 22*, 1071–1075. https://doi.org/10.1007/s10068-013-0185-7

Patil, V. V., & Patil, V. R. (2011). *Evaluation of anti-inflammatory activity of Ficus carica Linn. leaves*.

Peniwidiyanti, P., Irsyam, A. S. D., Dewi, A. P., Hariri, M. R., Irwanto, R. R., & Al Anshori, Z. (2021). Newly recorded alien species of *Ficus* L. (Moraceae) in Java, Indonesia. *Journal of Tropical Biodiversity and Biotechnology, 6*(2), 65313.

Pereira, C., López Corrales, M., Martín, A., Villalobos, M. D. C., Córdoba, M. D. G., & Serradilla, M. J. (2017). Physicochemical and nutritional characterization of brebas for fresh consumption from nine fig varieties (*Ficus carica* L.) grown in Extremadura (Spain). *Journal of Food Quality, 2017*, 1–12. https://doi.org/10.1155/2017/6302109

Perez, C., Dominguez, E., Canal, J. R., Campillo, J. E., & Torres, M. D. (2000). Hypoglycaemic activity of an aqueous extract from *Ficus carica*. https://doi.org/10.1076/1388-0209(200007)3831-SFT181

Purnamasari, R., Winarni, D., Permanasari, A. A., Agustina, E., Hayaza, S., & Darmanto, W. (2019). Anticancer activity of methanol extract of *Ficus carica* leaves and fruits against proliferation, apoptosis, and necrosis in Huh7it cells. *Cancer Informatics, 18*. https://doi.org/10.1177/1176935119842576

Qari, S. H., Alrefaei, A. F., Ashour, A. B., & Soliman, M. H. (2021). Genotoxicity and carcinogenicity of medicinal herbs and their nanoparticles. *Nutraceuticals, 1*(1), 31–41. https://doi.org/10.3390/nutraceuticals1010005

Raafat, K., & Wurglics, M. (2019). Phytochemical analysis of *Ficus carica* L. active compounds possessing anticonvulsant activity. *Journal of Traditional and Complementary Medicine, 9*(4), 263–270. https://doi.org/10.1016/j.jtcme.2018.01.007

Radwan, A. M., Reyad, N. F., & Ganaie, M. A. (2018). Comparative studies on the effect of environmental pollution on secondary metabolite contents and genotoxicity of two plants in Asir area, Saudi Arabia. *Tropical Journal of Pharmaceutical Research, 17*(8), 1599–1605. https://doi.org/10.4314/tjpr.v17i8.19

Resende, F. A., et al. (2012). Mutagenicity and antimutagenicity of (-)-hinokinin a trypanosomicidal compound measured by salmonella microsome and comet assays. *BMC Complementary and Alternative Medicine, 12*, 203. https://doi.org/10.1186/1472-6882-12-203

Richter, G., Schwarz, H. P., Dorner, F., & Peter, L. (2002). Activation and inactivation of human factor X by proteases derived from *Ficus carica*. *British Journal of Haematology, 119*, 1042–1051. https://doi.org/10.1046/j.1365-2141.2002.03954.x

Sadeghizade, M., Baharara, J., Salek, F., & Amini, E. (2021). Assessment the effect of *Ficus carica* leaf extract on B16F10 melanoma cancer cells: The roles of p53, Caspase-3 & Caspase-9 on induction of intrinsic apoptosis cascade. *Jundishapur Journal of Natural Pharmaceutical Products*. https://doi.org/10.5812/jjnpp.117429. (in press).

Salehi, B., Prakash Mishra, A., Nigam, M., Karazhan, N., Shukla, I., Kiełtyka-Dadasiewicz, A., . & Sharifi-Rad, J. (2021). Ficus plants: State of the art from a phytochemical, pharmacological, and toxicological perspective. Phytotherapy Research, 35(3), 1187–1217. https://doi.org/10.1002/ptr.6884.

Saxena, V., Ahmad, H., & Gupta, R. (2013). Memory enhancing effects of *Ficus carica* leaves in hexane extract on interoceptive behavioral models. *Asian Journal of Pharmaceutical and Clinical Research, 6*(Suppl 3), 109–113.

Shafique, F., et al. (2021). Antibacterial and antifungal activity of *Ficus carica* plant extract. *Journal of Pharmaceutical Research International, 33*, 1–9. https://doi.org/10.9734/jpri/2021/v33i1831311

Słoczyńska, K., Powroźnik, B., Pękala, E., & Waszkielewicz, A. M. (2014). Antimutagenic compounds and their possible mechanisms of action. *Journal of Applied Genetics, 55*(2), 273–285. https://doi.org/10.1007/s13353-014-0198-9

Soltana, H., et al. (2019). Antitumoral activity of *Ficus carica* L. on colorectal cancer cell lines. *Cellular and Molecular Biology, 65*(6), 6–11. https://doi.org/10.14715/cmb/2019.65.6.2

Sponchiado, G., et al. (2016). Quantitative genotoxicity assays for analysis of medicinal plants: A systematic review. *Journal of Ethnopharmacology, 178*, 289–296. https://doi.org/10.1016/j.jep.2015.10.026

Stephen Irudayaraj, S., et al. (2017). Protective effects of *Ficus carica* leaves on glucose and lipids levels, carbohydrate metabolism enzymes and β-cells in type 2 diabetic rats. *Pharmaceutical Biology, 55*(1), 1074–1081.

Subash, S., et al. (2016). Consumption of fig fruits grown in Oman can improve memory, anxiety, and learning skills in a transgenic mice model of Alzheimer's disease. *Nutritional Neuroscience, 19*(10), 475–483. https://doi.org/10.1016/j.jep.2008.05.040

Suttisansanee, U., et al. (2021). Health-promoting bioactivity and in vivo genotoxicity evaluation of a hemiepiphyte fig, *Ficus dubia*. *Food Science & Nutrition, 9*(4), 2269–2279. https://doi.org/10.1002/fsn3.2205

Teixeira, S. P., Costa, M. F., Basso-Alves, J. P., Kjellberg, F., & Pereira, R. A. (2018). Morphological diversity and function of the stigma in *Ficus* species (Moraceae). *Acta Oecologica, 90*, 117–131.

Thomford, N. E., Senthebane, D. A., Rowe, A., Munro, D., Seele, P., Maroyi, A., & Dzobo, K. (2018). Natural products for drug discovery in the 21st century: Innovations for novel drug discovery. *International Journal of Molecular Sciences, 19*(6), 1578. https://doi.org/10.3390/ijms19061578

Turkez, H., Arslan, M. E., & Ozdemir, O. (2017). Genotoxicity testing: Progress and prospects for the next decade. *Expert Opinion on Drug Metabolism & Toxicology, 13*(10), 1089–1098. https://doi.org/10.1080/17425255.2017.1375097

Wang, Z., Cui, Y., Vainstein, A., Chen, S., & Ma, H. (2017). Regulation of fig (*Ficus carica* L.) fruit color: Metabolomic and transcriptomic analyses of the flavonoid biosynthetic pathway. *Frontiers in Plant Science, 8*, 1990. https://doi.org/10.3389/fpls.2017.01990

Wiesner, J. (2014). Challenges of safety evaluation. *Journal of Ethnopharmacology, 158*, 467–470. https://doi.org/10.1016/j.jep.2014.08.013

Yang, X., Guo, J. L., Ye, J. Y., Zhang, Y. X., & Wang, W. (2015). The effects of *Ficus carica* polysaccharide on immune response and expression of some immune-related genes in grass carp, *Ctenopharyngodon idella*. *Fish & Shellfish Immunology, 42*(1), 132–137.

Zeiger, E. (2003). Illusions of safety: Antimutagens can be mutagens, and anticarcinogens can be carcinogens. *Mutation Research, 543*(3), 191–194. https://doi.org/10.1016/s1383-5742(02)00111-4

Zeiger, E. (2011). Article commentary: Cell death is antimutagenic, and other considerations when testing for antimutagenicity. *International Journal of Toxicology, 30*(3), 267–271. https://doi.org/10.1177/1091581811401351

Zhang, Y., Li, Y., Ma, P., Chen, J., & Xie, W. (2020). Ficus carica leaves extract inhibited pancreatic β-cell apoptosis by inhibiting AMPK/JNK/caspase-3 signaling pathway and antioxidation. *Biomedicine & Pharmacotherapy, 122*, 109689. https://doi.org/10.1016/j.biopha.2019.109689

Chapter 27
Composition and Biological Activities of *Ficus carica* Latex

Mostafa M. Hegazy, Reham Hassan Mekky, Wael M. Afifi, Ahmad E. Mostafa, and Hatem S. Abbass

Abbreviations

^{13}C NMR	Carbon-13 Nuclear Magnetic Resonance
^{1}H NMR	Proton Nuclear Magnetic Resonance
ACE	Angiotensin I-Converting Enzyme
AGS	6-O-acyl-β-D-glucosyl-β-sitosterol
CAE	Chlorogenic Acid Equivalents
CE	Catechin Equivalents.
Cv.	Cultivar
EI-MS	Electron Ionization Mass Spectrometry
FLP-1	Ficus Latex Peptide-1
FLP-2	Ficus Latex Peptide-2
FLP-3	Ficus Latex Peptide-3
FP1	Ficus Peroxidase 1
FP2	Ficus Peroxidase 2
FP3	Ficus Peroxidase 3
FPI	Ficus Cysteine Protease I
FPII	Ficus Cysteine Protease II
FPIII	Ficus Serine Protease III

M. M. Hegazy (✉)
Department of Pharmacognosy and Medicinal Plants, Faculty of Pharmacy, Al-Azhar University, Cairo, Egypt
e-mail: MostafaHegazy@azhar.edu.eg

R. H. Mekky
Department of Pharmacognosy, Faculty of Pharmacy, Egyptian Russian University, Badr City, Cairo, Egypt
e-mail: reham-mekky@eru.edu.eg

W. M. Afifi · A. E. Mostafa · H. S. Abbass
Department of Pharmacognosy and Medicinal Plants, Faculty of Pharmacy, Al-Azhar University, Cairo, Egypt
e-mail: wael.afify@su.edu.eg; Aemostafa@azhar.edu.eg; hsam8406@azhar.edu.eg

M. F. Ramadan (ed.), *Fig* (Ficus carica)*: Production, Processing, and Properties*, https://doi.org/10.1007/978-3-031-16493-4_27

GAE	Gallic Acid Equivalents
GC-IT-MS	Gas Chromatography Ion Trap Mass Spectrometry
GC-MS	Gas Chromatography-Mass Spectrometry
HPLC	High-Performance Liquid Chromatography
HPLC-DAD	High-Performance Liquid Chromatography- Diode-Array Detector
LC-MS/MS	Liquid Chromatography-Tandem Mass Spectrometry
MeOH	Methanol
pI	Isoelectric Point
PLCP	Papain-Like Cysteine Proteases
PR-4/type II	Pathogenesis-Related Protein 4 type II
SARS-CoV-2	Severe Acute Respiratory Syndrome Coronavirus 2
TFC	Total Flavonoid Content
TPC	Total Phenolic Content

1 Introduction

Latex is a natural sticky polymer from different plant parts and contains various phytochemicals and antioxidants. It shows diverse biological activities such as antifungal, antibacterial, antiviral, anti-inflammatory, and anticancer (Santos & Van Ree, 2011; Upadhyay, 2011). All *Ficus* spp. possess latices within their vasculatures, affording protection and self-healing from physical assaults (Lansky et al., 2008). Latex had been reported as a treatment for warts (Mabberley, 1997; Martínez, 2008). Although latex is not ingested *per se*, some precautions regarding its direct use are needed because it is known to have keratolytic and corrosive properties (Hemmatzadeh et al., 2003). Latex is used as expectorant, diuretic, anthelmintic and anemia (Kirtikar & Basu, 1935). Interestingly, latex is naturally occurring in Laticifers, defined as cells or series of connected cells containing a fluid called latex and forming systems that permeate various tissues of the plant body (Evert, 2006). Latex consists of a cytoplasmatic fluid of laticiferous tissues containing the usual organelles of plant cells, such as nuclei, mitochondria, vacuoles, and ribosomes (Kim et al., 2008). It is an aqueous suspension of a complex mixture of molecules found in specialized secretory cells of plants, known as laticifers, which synthesize and store a diversity of primary metabolites such as enzymatic and non-enzymatic proteins and secondary metabolites of different classes such as flavonoids, phenolic acids, coumarins, xanthones, terpenoids, and sterols in appreciable amounts (Agrawal & Konno, 2009). A laticifer is a plant cell unique in shape, differentiation, and physiological function, and its cytoplasm is a sticky fluid called latex (Kitajima et al., 2018). In addition to the industrial importance of plant latex as a rubber source, latex is a component of plant defense against microbes and herbivores. Laticifers form long tubular or branched structures running throughout the plant's body. Owing to this structure, much latex is exuded from the cut site when the plant body is cut, and toxic proteins and metabolites attack pests. Even though latices share a common biological role in pest defense, their protein, and chemical constituents are highly variable among plant species (Hagel et al., 2008; Konno, 2011). Similarly, protein constituents of latices are variable even among organs in a single species (Kitajima et al., 2012, 2013). Latex of *F. carica* (Fig. 27.1) contains volatile components; aldehydes such as Pentanal, hexanal, heptanal, and benzaldehyde; alcohols such as 1-butanol-3-methyl, 1-butanol-2-methyl, and 1-heptanol; ketones such as 6-methyl-5-hepten-2-one; monoterpenes such as α-pinene, β-pinene, limonene, and eucalyptol; sesquiterpenes such as α-guaiene, α-bourbonene, β-caryophyllene, and germacrene D; and miscellaneous such as methyl salicylate, quinoline, and psoralene (Oliveira et al., 2010b). It also contains non-volatile components; organic acids such as oxalic, citric, quinic, and shikimic (Oliveira et al., 2010b); fatty acids such as myristic, palmitic, stearic, oleic, and linoleic (Oliveira et al., 2010a); triterpenoids such as lupeol, lupeol acetate, and β-amyrin; steroids such as β-sitosterol, 6-*O*-palmitoyl-β-D-glucosyl-β-sitosterol, and 6-*O*-linoleyl-β-D-glucosyl-β-sitosterol (Oliveira et al., 2010a); and amino acids such as leucine, tryptophan, serine, ornithine, tyrosine, and cysteine (Oliveira et al., 2010a). *Ficus carica* latex contains large amounts of isoforms of ficin, a cysteine protease, which is toxic to the caterpillars of *Lepidoptera* (Konno et al., 2004), and fungi (Karnchanatat et al., 2011; López-García et al., 2012), as well as isoforms of trypsin inhibitor, which is also known to be toxic to insects (Hilder et al., 1987), and fungi (Huynh et al., 1992; Terras et al., 1993). Ficins to trypsin inhibitor ratio differed between latices from immature fruit and young petioles, suggesting that, despite these organs being young and unlignified, their latices have adopted different defense strategies. By comparing the proteomes, metabolomes, and transcriptomes in the latex of various *F. carica* organs to find out the diversity of defense strategies, three different organs containing latex have been selected; immature fruit, which are economically important as food; young and unlignified petioles, whose laticifers are expected to be connected to those in leaf veins; and > 1-year-old trunks, which are lignified and thus may have different pests from unlignified organs, Trypsin inhibitors were the most abundant proteins in petiole latex, while cysteine proteases (ficins) were the most abundant in immature fruit and trunk latices (Kitajima et al., 2018). The expression levels of pathogenesis-related proteins were highest in the

Fig. 27.1 *Ficus carica* latex

latex of trunk, suggesting that this latex had adapted a defensive role against microbe attacks (Kitajima et al., 2018). Although young petioles and immature fruit are both unlignified soft organs, and potential food for herbivorous insects, unigenes for the sesquiterpenoid pathway, which likely produces defense-associated volatiles, and the phenylpropanoid pathway, which produces toxic furanocoumarins, were expressed less in immature fruit latex (Kitajima et al., 2018).

2 Bioactive Metabolites Described in Fig (*Ficus carica* L.) Latex

A wide range of bioactive metabolites was observed in *F. carica* L. latex, indicating a new insight into the importance of fig latex as an undervalued source of bioactive metabolites. In this sense, the utilization of state-of-art techniques such as gas chromatography (GC) and/or high-performance liquid chromatography (HPLC) coupled with high-resolution mass spectrometry (MS) and tandem mass spectrometry (MS/MS) allowed the annotation of many phenolic compounds, terpenoids, sterols, xanthones, alkanes, alkenes, alkynes, alcohols, aldehydes, ketones, ethers, fatty acids, aromatic compounds, amino acids, and nitrogenous compounds. Generally, latex was collected manually from either the fruits or the leaves and stored in sterile containers till analysis (Oliveira et al., 2010a, b; Menichini et al., 2012; Abdel-Aty et al., 2019). It bears noting that (Menichini et al., 2012) conducted a study on the latex collected from both the fruits and the leaves of the Italian cultivar Dottato where the latex was collected from the leaves in June, July, and September and from the fruits in June and July (Table 27.1 and 27.2). The process of metabolites extraction from fig latex was performed by applying multiple solvents extraction and fractionation (Lazreg-Aref et al., 2012a; Menichini et al., 2012; Abdel-Aty et al., 2019; Paşayeva et al., 2020), solid-phase extraction (Paşayeva et al., 2020) and/or, headspace solid-phase microextraction (Oliveira et al., 2010b), Table 27.2. Totally, 130 metabolites were reported in fig latex.

2.1 *Phenolic Compounds*

The detected phenolic compounds in fig latex were classified into hydroxybenzoic acids, hydroxycinnamic acids, flavonoids, and coumarins, considering each candidate's name, molecular formula, occurrence, geographical origins, extraction, and analytical methods. Table 27.2 and Fig. 27.2 illustrate the observed phenolic compounds in fig latex. Besides identifying compounds, total phenolic content (TPC) and total flavonoid content (TFC) of latex were estimated in different organs, seasons, and geographical origins, as shown in Table 27.1.

Table 27.1 Total phenolic and total flavonoid contents of different latices

Sample	Geographical origin	Extraction procedure after latex collection	TPC	TFC	Reference
Leaves of Cv. Dottato (June)	Calabria, Italy	(Acetone: methanol: water: acetic acid, 40:40:20:0.1, v/v) followed by heating and sonication	34 ± 0.7[a] mg CAE/ g extract	–	Menichini et al. (2012)
Leaves of Cv. Dottato (July)	Calabria, Italy	(Acetone: methanol: water: acetic acid, 40:40:20:0.1, v/v) followed by heating and sonication	26 ± 0.6[a] mg CAE/g extract	–	Menichini et al. (2012)
Leaves of Cv. Dottato (September)	Calabria, Italy	(Acetone: methanol: water: acetic acid, 40:40:20:0.1, v/v) followed by heating and sonication	36 ± 0.8[a] mg CAE/g extract	–	Menichini et al. (2012)
Fruits of Cv. Dottato (June)	Calabria, Italy	(Acetone: methanol: water: acetic acid, 40:40:20:0.1, v/v) followed by heating and sonication	29 ± 0.6[a] mg CAE/g extract	–	Menichini et al. (2012)
Fruits	Al Sharqia government, Egypt	Hydroalcoholic extract (80% MeOH) of defatted latex	50.20 ± 2.50[b] mg GAE/g extract	12.50 ± 1.10[c] mg CE/ g extract	Abdel-Aty et al. (2019)
Unripe fruits of *F. carica* subsp. *Carica*	Adana, Turkey	Methanol extract	1.879 ± 0.004[b] mg GAE/ g extract	2.965 ± 0.006[c] mg CE/ g extract	Paşayeva et al. (2020)
Unripe fruits *F. carica* subsp. *Carica*	Adana, Turkey	*n*-hexane fraction of the methanol extract	2.910 ± 0.005[b] mg GAE/ g extract	2.227 ± 0.001[c] mg CE/ g extract	Paşayeva et al. (2020)
Unripe fruits *F. carica* subsp. *Carica*	Adana, Turkey	*n*-hexane-ethyl acetate fraction of the methanol extract	1.016 ± 0.003[b] mg GAE/g extract	2.914 ± 0.020[c] mg CE/ g extract	Paşayeva et al. (2020)

(continued)

Table 27.1 (continued)

Sample	Geographical origin	Extraction procedure after latex collection	TPC	TFC	Reference
Unripe fruits *F. carica* subsp. *Carica*	Adana, Turkey	Methanol fraction of the methanol extract	11.297 ± 0.001[b] mg GAE/g extract	3.063 ± 0.011[c] mg CE/ g extract	Paşayeva et al. (2020)
Unripe fruits *F. carica* subsp. *Carica*	Adana, Turkey	Methanol followed by solid-phase extraction	26.816 ± 0.007[b] mg GAE/g extract	8.615 ± 0.009[c] mg CE/ g extract	Paşayeva et al. (2020)
Leaves and fruits	Jalgaon, India	Latex was diluted in water (1:5, v/v)	98.75[b] μg GAE/ ml	–	Borase et al. (2013)

[a] CAE values are chlorogenic acid equivalents
[b] GAE values are gallic acid equivalents
[c] (CE) values are catechin equivalents.

2.1.1 Phenolic Acids

The reported phenolic acids in fig latex were grouped into hydroxybenzoic acids (compounds; 1–5, Table 27.2 and Fig. 27.2) and hydroxycinnamic acids (compounds; 6–11, Table 27.2 and Fig. 27.2). Hydroxybenzoic acids were classified into monohydroxybenzoic acids, dihydroxybenzoic acids, including *O*-methylated derivatives, and trihydroxybenzoic acids conjugation with quinic acid (Oliveira et al., 2010b; Lazreg Aref et al., 2011a; Tezcan et al., 2015; Abdel-Aty et al., 2019; Paşayeva et al., 2020). Regarding hydroxycinnamic acids, they were either free (*p*-coumaric acid, caffeic acid, ferulic acid, and sinapic acid) or conjugated with quinic acid (chlorogenic acid) or 3,4-dihydroxyphenyllactic acid (Table 27.2 and Fig. 27.2).

2.1.2 Flavonoids and Coumarins

Five flavonoid derivatives were observed in fig latex (compounds 12–16, Table 27.2 and Fig. 27.2). They were analyzed either by HPLC or LC-MS/MS. They represented different flavonoids subclasses as flavones (luteolin), flavonols (quercetin and rutin), flavan-3-ols (catechin), and anthocyanins (5-carboxypryano-pelargonidin 3-*O*-β-glucopyranoside). As for coumarins (compounds 17–18), psoralen and bergapten were observed.

2.2 *Terpenoids and Sterols*

Qualitatively, terpenoids were the primary class of the reported metabolites in fig latex (36 compounds), being classified into monoterpenes (10), sesquiterpenes (11), and triterpenes (15), Table 27.2 and Fig. 27.2. Terpenoids possess several biological

Table 27.2 Phenolic compounds, terpenoids, and sterols reported in fig latex

#	Compound	Molecular formula	Occurrence of the latex	Geographical Origin	Extraction procedure after latex collection	Method of analysis	Reference
Phenolic compounds: Hydroxybenzoic acids							
1	Methyl salicylate	$C_8H_8O_3$	Fruits	Al Sharqia government, Egypt	Hydroalcoholic extract (80% MeOH) of defatted latex	GC-MS	Abdel-Aty et al. (2019)
			Immature green fruits from *F. carica* Cv. Pingo de Mel	Mirandela region, Portugal	Headspace Solid-Phase Microextraction (HS-SPME).	GC-IT-MS	Oliveira et al. (2010b)
2	Protocatechuic acid	$C_7H_6O_4$	Unripe fruits	Sousse, Tunisia	n-hexane and n-hexane -ethyl acetate fractions of the methanolic extract	LC-MS/MS	Lazreg Aref et al. (2011a)
			Fruits	Al Sharqia government, Egypt	Hydroalcoholic extract (80% MeOH) of defatted latex	HPLC	Abdel-Aty et al. (2019)
			Leaves	Bursa, Turkey	Water fraction after defatting and multiple solvents fractionation	HPLC	Tezcan et al. (2015)
3	Vanillic acid	$C_8H_8O_4$	Fruits	Al Sharqia government, Egypt	Hydroalcoholic extract (80% MeOH) of defatted latex	HPLC	Abdel-Aty et al. (2019)
4	Methyl vanillate	$C_9H_{10}O_4$					
5	Trigalloylquinic acid	$C_{28}H_{24}O_{18}$	Unripe fruits *F. carica* subsp. *carica*	Adana, Turkey	Methanol followed by solid-phase extraction	LC-MS/MS	Paşayeva et al. (2020)
Phenolic compounds: Hydroxycinnamic acids							

(continued)

Table 27.2 (continued)

#	Compound	Molecular formula	Occurrence of the latex	Geographical Origin	Extraction procedure after latex collection	Method of analysis	Reference
6	*p*-coumaric acid	$C_9H_8O_3$	Unripe fruits	Sousse, Tunisia	n-hexane fraction of the methanolic extract	LC-MS/MS	Lazreg Aref et al. (2011a)
			Leaves	Bursa, Turkey	Water fraction after defatting and multiple solvents fractionation	HPLC	Tezcan et al. (2015)
7	Caffeic acid	$C_9H_8O_4$	Unripe fruits	Sousse, Tunisia	n-hexane and n-hexane -ethyl acetate fractions of the methanolic extract	LC-MS/MS	Lazreg Aref et al. (2011a)
			Fruits	Al Sharqia government, Egypt.	Hydroalcoholic extract (80% MeOH) of defatted latex	HPLC	Abdel-Aty et al. (2019)
			Leaves	Bursa, Turkey	Water fraction after defatting and multiple solvents fractionation	HPLC	Tezcan et al. (2015)
8	Chlorogenic acid	$C_{16}H_{18}O_9$	Fruits	Al Sharqia government, Egypt	Hydroalcoholic extract (80% MeOH) of defatted latex	HPLC	Abdel-Aty et al. (2019)
9	Ferulic acid	$C_{10}H_{10}O_4$	Unripe fruits	Sousse, Tunisia	n-hexane and n-hexane-ethyl acetate fractions of the methanolic extract	LC-MS/MS	Lazreg Aref et al. (2011a)
			Fruits	Al Sharqia government, Egypt	hydroalcoholic extract (80% MeOH) of defatted latex	HPLC	Abdel-Aty et al. (2019)
			Leaves	Bursa, Turkey	Water fraction after defatting and multiple solvents fractionation	HPLC	Tezcan et al. (2015)

10	Sinapic acid	$C_{11}H_{12}O_5$	Fruits	Al Sharqia government, Egypt	Hydroalcoholic extract (80% MeOH) of defatted latex	HPLC	Abdel-Aty et al. (2019)
11	Rosmarinic acid	$C_{18}H_{16}O_8$	Unripe fruits *F. carica* subsp. *carica*	Adana, Turkey	Methanol followed by solid-phase extraction	LC-MS/MS	Paşayeva et al. (2020)
Phenolic compounds: Flavonoids							
12	Luteolin	$C_{15}H_{10}O_6$	Unripe fruits	Sousse, Tunisia	n-hexane and n-hexane -ethyl acetate fractions of the methanolic extract	LC-MS/MS	Lazreg Aref et al. (2011a)
13	Quercetin	$C_{15}H_{10}O_7$	Leaves	Bursa, Turkey	Water fraction after defatting and multiple solvents fractionation	HPLC	Tezcan et al. (2015)
14	Quercetin 3-*O*-rutinoside (rutin)	$C_{27}H_{30}O_{16}$	Fruits	Al Sharqia government, Egypt	Hydroalcoholic extract (80% MeOH) of defatted latex	HPLC	Abdel-Aty et al. (2019)
15	Catechin	$C_{15}H_{14}O_6$	Fruits	Al Sharqia government, Egypt	Hydroalcoholic extract (80% MeOH) of defatted latex	HPLC	Abdel-Aty et al. (2019)
16	5-carboxypryano-pelargonidin 3-*O*-β-glucopyranoside	$C_{24}H_{21}O_{12}^+$	Unripe fruits *F. carica* subsp. *carica*	Adana, Turkey	Methanol followed by solid-phase extraction	LC-MS/MS	Paşayeva et al. (2020)
Phenolic compounds: Coumarins							

(continued)

Table 27.2 (continued)

#	Compound	Molecular formula	Occurrence of the latex	Geographical Origin	Extraction procedure after latex collection	Method of analysis	Reference
17	Psoralen	$C_{11}H_6O_3$	Fruits of Cv. Dottato (June, July)	Calabria, Italy	Methanol followed by n-hexane	GC-MS	Menichini et al. (2012)
			Leaves of Cv. Dottato (June, July, September)				
			Fruits	Al Sharqia government, Egypt	Hydroalcoholic extract (80% MeOH) of defatted latex	GC-MS	Abdel-Aty et al. (2019)
			Immature green fruits from *F. carica* Cv. Pingo deMel	Mirandela region, Portugal	Headspace Solid-Phase Microextraction (HS-SPME)	GC-IT-MS	Oliveira et al. (2010b)
18	Bergapten	$C_{12}H_8O_4$	Fruits of Cv. Dottato (June, July)	Calabria, Italy	Methanol followed by n-hexane	GC-MS	Menichini et al. (2012)
			Leaves of Cv. Dottato (June, July, September)				
Terpenoids: Monoterpenes							
19	α-thujene	$C_{10}H_{16}$	Immature green fruits from *F. carica* Cv. Pingo de Mel	Mirandela region, Portugal	Headspace Solid-Phase Microextraction (HS-SPME)	GC-IT-MS	Oliveira et al. (2010b)
20	α-pinene	$C_{10}H_{16}$					
21	β-pinene	$C_{10}H_{16}$					
22	Limonene	$C_{10}H_{16}$					
23	Terpinolene	$C_{10}H_{16}$					
24	3-Bornanone	$C_{10}H_{16}O$	Caprifig Jrani inedible unripe fruit	Mesjed Aissa, Tunisia	n-hexane	GC-MS	Lazreg-Aref et al. (2012a)

25	Eucalyptol	$C_{10}H_{18}O$	Immature green fruits from *F. carica* Cv. Pingo de Mel	Mirandela region, Portugal	Headspace Solid-Phase Microextraction (HS-SPME)	GC-IT-MS	Oliveira et al. (2010b)
26	Linalool	$C_{10}H_{18}O$					
27	cis-linalool oxide	$C_{10}H_{18}O_2$					
28	Epoxylinalol	$C_{10}H_{18}O_2$					
Terpenoids: Sesquiterpenes							
29	Epi-bicyclosesquiphellandrene	$C_{15}H_{24}$	Fruits of Cv. Dottato (July)	Calabria, Italy	Methanol followed by n-hexane	GC-MS	Menichini et al. (2012)
			Leaves of Cv. Dottato (June, July, September)				
30	Germacrene-D	$C_{15}H_{24}$	Caprifig Jrani inedible unripe fruit	Mesjed Aissa, Tunisia	n-hexane	GC-MS	Lazreg-Aref et al. (2012a)
			Immature green fruits from *F. carica* Cv. Pingo de Mel	Mirandela region, Portugal	Headspace Solid-Phase Microextraction (HS-SPME).	GC-IT-MS	Oliveira et al. (2010b)
31	Delta-Cadinene	$C_{15}H_{24}$	Caprifig Jrani inedible unripe fruit	Mesjed Aissa, Tunisia	n-hexane	GC-MS	Lazreg-Aref et al. (2012a)
			Immature green fruits from *F. carica* Cv. Pingo de Mel	Mirandela region, Portugal	Headspace Solid-Phase Microextraction (HS-SPME)	GC-IT-MS	Oliveira et al. (2010b)
32	α-guaiene	$C_{15}H_{24}$	Immature green fruits from *F. carica* Cv. Pingo de Mel	Mirandela region, Portugal	Headspace Solid-Phase Microextraction (HS-SPME)	GC-IT-MS	Oliveira et al. (2010b)
33	α-bourbonene	$C_{15}H_{24}$					
34	α -caryophyllene	$C_{15}H_{24}$					
35	β-caryophyllene	$C_{15}H_{24}$					
36	τ-muurolene	$C_{15}H_{24}$					
37	α-calacorene	$C_{15}H_{24}$					
38	Trans-α-bergamotene	$C_{15}H_{24}$					

(continued)

Table 27.2 (continued)

#	Compound	Molecular formula	Occurrence of the latex	Geographical Origin	Extraction procedure after latex collection	Method of analysis	Reference
39	Aristolone	$C_{15}H_{22}O$	Caprifig Jrani inedible unripe fruit	Mesjed Aissa, Tunisia	n-hexane	GC-MS	Lazreg-Aref et al. (2012a)
Terpenoids: Triterpenes							
40	Squalene	$C_{30}H_{50}$	Caprifig Jrani inedible unripe fruit	Mesjed Aissa, Tunisia	n-hexane	GC-MS	Lazreg-Aref et al. (2012a)
41	1, 6, 10, 14, 18, 22-Tetracosahexaen-3-ol	$C_{30}H_{50}O$					
42	β-Amyrin	$C_{30}H_{50}O$	Fruits of Cv. Dottato (June, July)	Calabria, Italy	Methanol followed by n-hexane	GC-MS	Menichini et al. (2012)
			Leaves of Cv. Dottato (June, July, September)				
			Caprifig Jrani inedible unripe fruit	Mesjed Aissa, Tunisia	n-hexane	GC-MS	Lazreg-Aref et al. (2012a)
			Green fruits	Mirandela region, Portugal	Different extracts of n-hexane, dichloromethane, alkaline hydrolysis	GC-IT-MS/HPLC-DAD	Oliveira et al. (2010a)
43	α-Amyrin	$C_{30}H_{50}O$	Green fruits	Mirandela region, Portugal	Different extracts of n-hexane, dichloromethane, alkaline hydrolysis	GC-IT-MS/HPLC-DAD	Oliveira et al. (2010a)
44	α-Amyrenyl acetate	$C_{32}H_{52}O_2$	Caprifig Jrani inedible unripe fruit	Mesjed Aissa, Tunisia	n-hexane	GC-MS	Lazreg-Aref et al. (2012a)
45	Maragenin I acetate	$C_{31}H_{48}O_3$					

46	Lupeol (Lup-20(29)-en-3-ol (3β)-)	$C_{30}H_{50}O$	Fruits of Cv. Dottato (June, July)	Calabria, Italy	Methanol followed by n-hexane	GC-MS	Menichini et al. (2012)
			Leaves of Cv. Dottato (June, July, September)				
			Caprifig Jrani inedible unripe fruit	Mesjed Aissa, Tunisia	n-hexane	GC-MS	Lazreg-Aref et al. (2012a)
			Green fruits	Mirandela region, Portugal	Different extracts of n-hexane, dichloromethane, alkaline hydrolysis	GC-IT-MS/ HPLC-DAD	Oliveira et al. (2010a)
47	Lupeol acetate	$C_{32}H_{52}O_2$	Green fruits (July)	Sari, Mazandaran Province, Iran	n-hexane followed by chloroform	^1H, ^{13}C NMR, EI-MS	Jeivad et al. (2020)
			Green fruits	Mirandela region, Portugal	Different extracts of n-hexane, dichloromethane, alkaline hydrolysis	GC-IT-MS/ HPLC-DAD	Oliveira et al. (2010a)
48	Lupeol palmitate	$C_{46}H_{80}O_2$	Green fruits in July 2015	Sari, Mazandaran Province, Iran	n-hexane followed by chloroform	^1H, ^{13}C NMR, EI-MS	Jeivad et al. (2020)
49	D:C-Friedours-7-en-3-ol, (3β)-	$C_{30}H_{52}O$	Fruits of Cv. Dottato (June, July)	Calabria, Italy	Methanol followed by n-hexane	GC-MS	Menichini et al. (2012)
			Leaves of Cv. Dottato (June, July, September)				
50	Olean-12-en-3-ol, acetate, (3β)-	$C_{32}H_{52}O_2$	Fruits of Cv. Dottato (June, July)	Calabria, Italy	Methanol followed by n-hexane	GC-MS	Menichini et al. (2012)
			Leaves of Cv. Dottato (June, July, September)				
			Caprifig Jrani inedible unripe fruit	Mesjed Aissa, Tunisia	n-hexane	GC-MS	Lazreg-Aref et al. (2012a)

(continued)

Table 27.2 (continued)

#	Compound	Molecular formula	Occurrence of the latex	Geographical Origin	Extraction procedure after latex collection	Method of analysis	Reference
51	Urs-12-en-24-oic acid, 3-oxo-, methyl ester, (+)-	$C_{31}H_{48}O_3$	Fruits of Cv. Dottato (June, July)	Calabria, Italy	Methanol followed by n-hexane	GC-MS	Menichini et al. (2012)
			Leaves of Cv. Dottato (June, July, September)				
			Caprifig Jrani inedible unripe fruit	Mesjed Aissa, Tunisia	n-hexane	GC-MS	Lazreg-Aref et al. (2012a)
52	A-Neooleana-3(5), 12-diene	$C_{30}H_{48}$	Fruits of Cv. Dottato (June, July)	Calabria, Italy	Methanol followed by n-hexane	GC-MS	Menichini et al. (2012)
			Leaves of Cv. Dottato (June, July, September)				
53	A′-Neogammacer-22(29)-en-3-ol, acetate, (3β, 21β)-	$C_{32}H_{52}O_2$	Fruits of Cv. Dottato (June, July)				
			Leaves of Cv. Dottato (June, July, September)				
54	Betulol	$C_{30}H_{50}O_2$	Green fruits	Mirandela region, Portugal	Different extracts of n-hexane, dichloromethane, alkaline hydrolysis	GC-IT-MS/ HPLC-DAD	Oliveira et al. (2010a)
Sterols							
55	Lanosterol	$C_{30}H_{50}O$	Green fruits	Mirandela region, Portugal	Different extracts of n-hexane, dichloromethane, alkaline hydrolysis	GC-IT-MS/ HPLC-DAD	Oliveira et al. (2010a)

56	Lanosta-8,24-dien-3-ol,acetate(3β)-	$C_{32}H_{52}O_2$	Fruits of Cv. Dottato (June, July)	Calabria, Italy	Methanol followed by n-hexane	GC-MS	Menichini et al. (2012)
			Leaves of Cv. Dottato (June, July, September)				
			Caprifig Jrani inedible unripe fruit	Mesjed Aissa, Tunisia	n-hexane	GC-MS	Lazreg-Aref et al. (2012a)
57	5α-Lanosta-7,24-dien-3β-ol acetate	$C_{32}H_{52}O_2$	Fruits of Cv. Dottato (June, July)	Calabria, Italy	Methanol followed by n-hexane	GC-MS	Menichini et al. (2012)
			Leaves of Cv. Dottato (June, July, September)				
58	5α-Lanosta-9(11), 25-dien-3β-ol acetate	$C_{34}H_{56}O_2$	Fruits of Cv. Dottato (June, July)				
			Leaves of Cv. Dottato (June, July, September)				
59	β-Sitosterol	$C_{29}H_{50}O$	Fruits of Cv. Dottato (June, July)				
			Leaves of Cv. Dottato (June, July, September)				
			Green fruits	Mirandela region, Portugal	Different extracts of n-hexane, dichloromethane, alkaline hydrolysis	GC-IT-MS/ HPLC-DAD	Oliveira et al. (2010a)
60	6-aza-B-Homo-5αcholestane	$C_{27}H_{49}N$	Caprifig Jrani inedible unripe fruit	Mesjed Aissa, Tunisia	n-hexane	GC-MS	Lazreg-Aref et al. (2012a)
61	9, 19-Cyclolanost-24-en-3-ol	$C_{30}H_{50}O$					

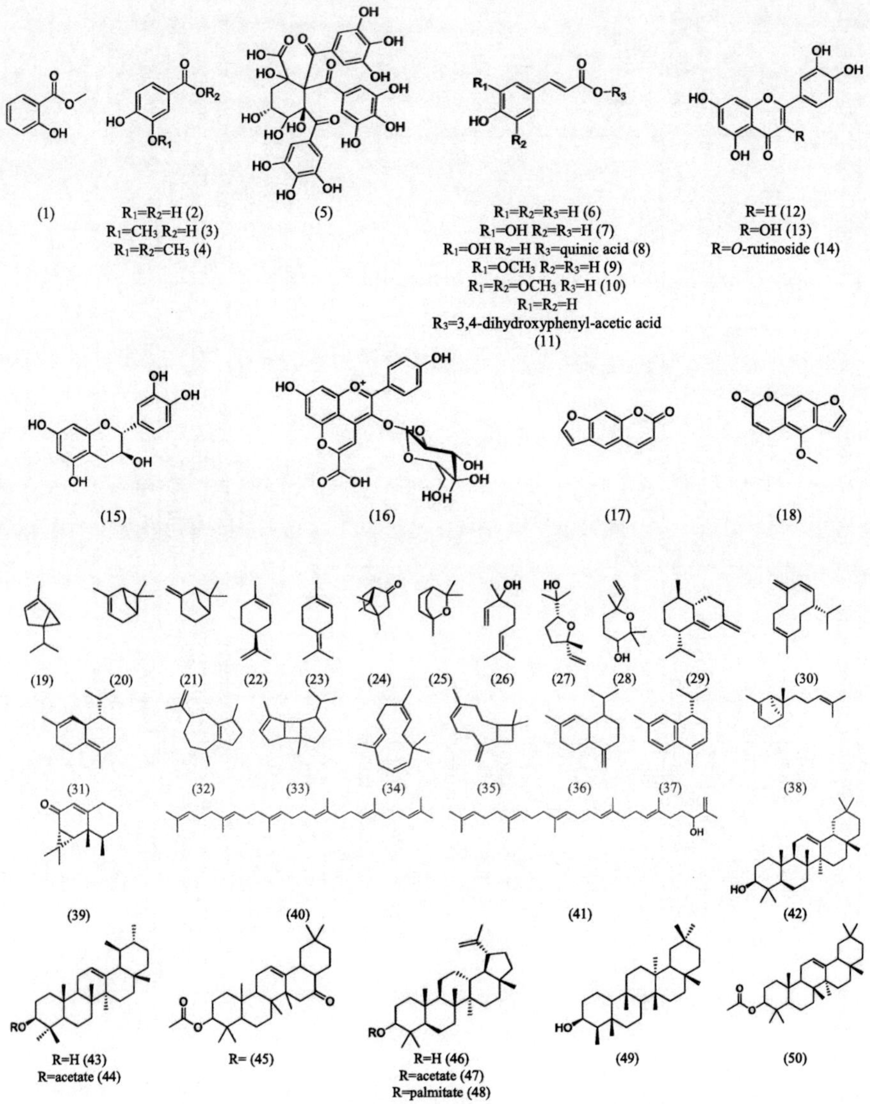

Fig. 27.2 Chemical structures of phenolic compounds, terpenoids, and sterols r eported in the latex of figs

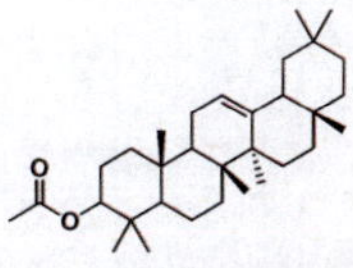

(51)

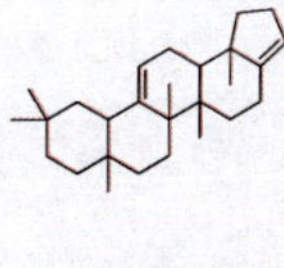

(52)

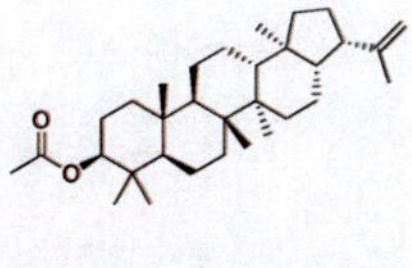

(53)

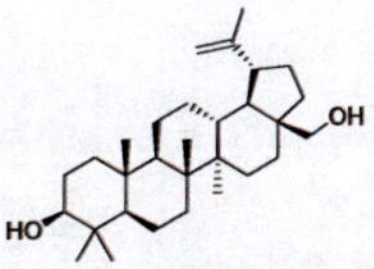

(54)

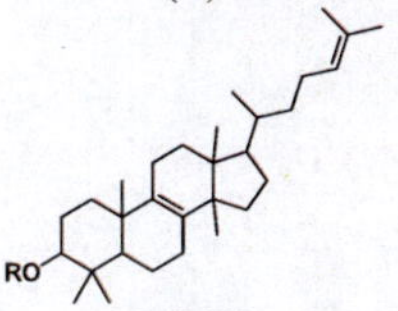

R=H (55)
R= acetate (56)

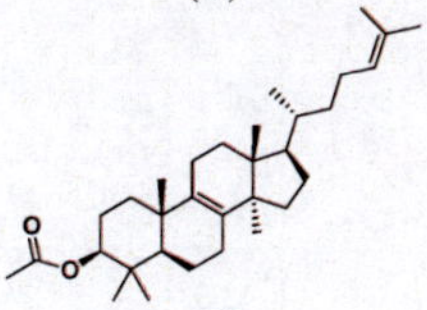

(57)

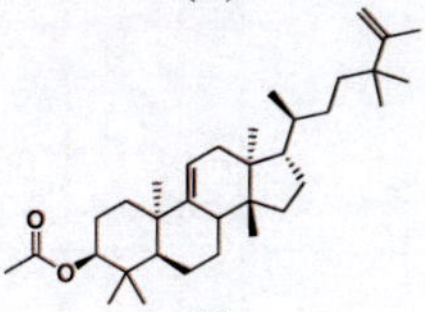

(58)

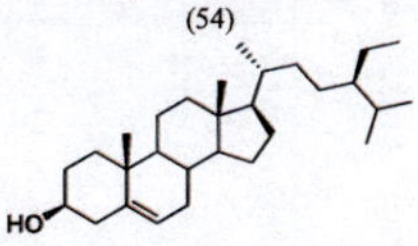

(59)

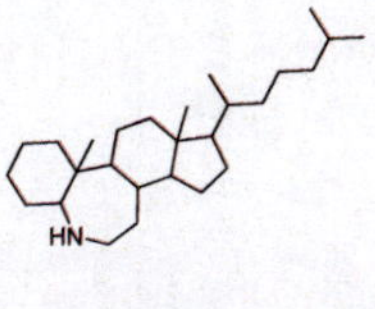

(60)

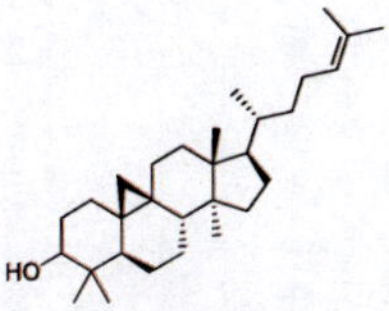

(61)

Fig. 27.2 (continued)

activities, and hence, their presence in the latex of figs contributes significantly to enhancing its biological potential (Wang et al., 2005). Besides, sterols occurrence was observed with seven derivatives (Table 27.2 and Fig. 27.2).

2.3 *Xanthones, Hydrocarbons, Fatty Acids, Aromatic Compounds, Amino Acids, and Other Nitrogenous Compounds*

Xanthones were represented by two derivatives, namely xanthotoxin and mangiferin which could be an addition to the biological potential of fig latex. Xanthones have a wide range of biological activities such as anti-inflammatory, anti-cancer, anti-Alzheimer, and cardioprotective (El-Seedi et al., 2010). Also, several hydrocarbons were reported. They are classified into alkanes (8), alkenes (2), alkynes (1), alcohols (7), aldehydes (5), ketones (3), and ethers (1). Aromatic compounds were present with eight derivatives. Besides, fatty acids were widely present with 16 derivatives, including unsaturated fatty acids incorporating the antihyperlipidemic and cardioprotective activity of fig latex (Oliveira et al., 2010a). In line with nitrogenous compounds, 16 derivatives were reported in fig latex with essential amino acids, providing a new nutritive prospect for such an undervalued plant part (Oliveira et al., 2010a), Table 27.3 and Fig. 27.3.

Table 27.3 Xanthones, alkanes, alkenes, alkynes, alcohols, aldehydes, ketones, ethers, fatty acids, aromatic compounds, amino acids, and nitrogenous compounds reported in fig latex

#	Compound	Molecular formula	Occurrence of the latex	Geographical origin	Extraction procedure after latex collection	Method of analysis	Reference
Xanthones							
62	Xanthotoxin	$C_{16}H_8O_4$	Fruits	Al Sharqia government, Egypt	Hydroalcoholic extract (80% MeOH) of defatted latex	GC-MS	Abdel-Aty et al. (2019)
63	Mangiferin	$C_9H_{18}O_{11}$	Unripe fruits *F. carica* subsp. *carica*	Adana, Turkey	*n*-hexane-ethyl acetate fraction of the methanol extract	LC-MS/MS	Paşayeva et al. (2020)
Alkanes							
64	Pentadecane	$C_{15}H_{32}$	Caprifig Jrani inedible unripe fruit	Mesjed Aissa, Tunisia	*n*-hexane	GC-MS	Lazreg-Aref et al. (2012a)
65	Eicosane	$C_{20}H_{42}$	Leaves of Cv. Dottato (June)	Calabria, Italy	Methanol followed by *n*-hexane	GC-MS	Menichini et al. (2012)
66	Tricosane	$C_{23}H_{48}$	Caprifig Jrani inedible unripe fruit	Mesjed Aissa, Tunisia	*n*-hexane	GC-MS	Lazreg-Aref et al. (2012a)
67	Hexacosane	$C_{26}H_{54}$					
68	Heptacosane	$C_{27}H_{56}$	Leaves of Cv. Dottato (June)	Calabria, Italy	Methanol followed by *n*-hexane	GC-MS	Menichini et al. (2012)
69	Octacosane	$C_{28}H_{58}$	Caprifig Jrani inedible unripe fruit	Mesjed Aissa, Tunisia	*n*-hexane	GC-MS	Lazreg-Aref et al. (2012a)
70	Nanocosane	$C_{29}H_{60}$					
71	Triacontane	$C_{30}H_{62}$					
Alkenes							
72	2, 6, 10-Dodecatrien-1-ol	$C_{12}H_{20}O$	Caprifig Jrani inedible unripe fruit	Mesjed Aissa, Tunisia	*n*-hexane	GC-MS	Lazreg-Aref et al. (2012a)
73	1, E-8, Z-10-Hexadecatriene	$C_{16}H_{28}$					
Alkynes							
74	7-Pentadecyne	$C_{15}H_{28}$	Caprifig Jrani inedible unripe fruit	Mesjed Aissa, Tunisia	*n*-hexane	GC-MS	Lazreg-Aref et al. (2012a)

Alcohols							
75	1-butanol-2-methyl	$C_5H_{12}O$	Immature green fruits from *F. carica* cultivar Pingo de Mel	Mirandela region, Portugal	Headspace Solid-Phase Microextraction (HS-SPME)	GC-IT-MS	Oliveira et al. (2010b)
76	1-butanol-3-methyl	$C_5H_{12}O$					
77	1-pentanol	$C_5H_{12}O$					
78	1-hexanol	$C_6H_{14}O$					
79	1-heptanol	$C_7H_{16}O$					
80	Phenylethyl alcohol	$C_8H_{10}O$					
81	Phenylpropyl alcohol	$C_9H_{12}O$					
Aldehydes							
82	Pentanal	$C_5H_{10}O$	Immature green fruits from *F. carica* cultivar Pingo de Mel	Mirandela region, Portugal	Headspace Solid-Phase Microextraction (HS-SPME)	GC-IT-MS	Oliveira et al. (2010b)
83	Hexanal	$C_6H_{12}O$					
84	Heptanal	$C_7H_{14}O$					
85	Octanal	$C_8H_{16}O$					
86	Benzaldehyde	C_7H_6O					
Ketones							
87	6-methyl-5-hepten-2-one	$C_8H_{14}O$	Immature green fruits from *F. carica* cultivar Pingo de Mel	Mirandela region, Portugal	Headspace Solid-Phase Microextraction (HS-SPME)	GC-IT-MS	Oliveira et al. (2010b)
88	Dodecan-2-on	$C_{11}H_{18}O_2$	Caprifig Jrani inedible unripe fruit	Mesjed Aissa, Tunisia	*n*-hexane	GC-MS	Lazreg-Aref et al. (2012a)
89	2,5,8-Trimethyl-1-tetralone	$C_{13}H_{16}O$	Fruits	Al Sharqia government, Egypt	Hydroalcoholic extract (80% MeOH) of defatted latex	GC-MS	Abdel-Aty et al. (2019)
Ethers							
90	Oxirane	C_2H_4O	Caprifig Jrani inedible unripe fruit	Mesjed Aissa, Tunisia	*n*-hexane	GC-MS	Lazreg-Aref et al. (2012a)
Fatty acids							

(continued)

Table 27.3 (continued)

#	Compound	Molecular formula	Occurrence of the latex	Geographical origin	Extraction procedure after latex collection	Method of analysis	Reference
91	Myristic acid	$C_{14}H_{28}O_2$	Green fruits	Mirandela region, Portugal	Methanol followed by alkaline hydrolysis and derivatization	GC-IT-MS	Oliveira et al. (2010a)
92	Pentadecanoic acid	$C_{15}H_{30}O_2$					
93	Palmitic acid (Hexadecanoic acid)	$C_{16}H_{32}O_2$	Caprifig Jrani inedible unripe fruit	Mesjed Aissa, Tunisia	*n*-hexane	GC-MS	Lazreg-Aref et al. (2012a)
94	Heptadecanoic acid	$C_{17}H_{34}O_2$	Green fruits	Mirandela region, Portugal	Methanol followed by alkaline hydrolysis and derivatization	GC-IT-MS	Oliveira et al. (2010a)
95	cis-10-heptadecenoic acid	$C_{17}H_{32}O_2$					
96	Stearic acid	$C_{18}H_{36}O_2$					
97	Oleic acid	$C_{18}H_{34}O_2$					
98	Elaidic acid	$C_{18}H_{34}O_2$					
99	Linoelaidic acid	$C_{18}H_{32}O_2$	Unripe fruits *F. carica* subsp. *carica*	Adana, Turkey	Methanol followed by solid-phase extraction	LC-MS/MS	Paşayeva et al. (2020)
100	Linoleic acid	$C_{18}H_{32}O_2$	Green fruits	Mirandela region, Portugal	Methanol followed by alkaline hydrolysis and derivatization	GC-IT-MS	Oliveira et al. (2010a)
101	*Cis*-Linoleic acid methyl ester	$C_{19}H_{34}O_2$	Caprifig Jrani inedible unripe fruit	Mesjed Aissa, Tunisia	*n*-hexane	GC-MS	Lazreg-Aref et al. (2012a)
102	Arachidic acid	$C_{20}H_{40}O_2$	Green fruits	Mirandela region, Portugal	Methanol followed by alkaline hydrolysis and derivatization	GC-IT-MS	Oliveira et al. (2010a)
103	Heneicosanoic acid	$C_{21}H_{42}O_2$					
104	Behenic acid	$C_{22}H_{44}O_2$					
105	Tricosanoic acid	$C_{23}H_{46}O_2$					
106	Lignoceric acid	$C_{24}H_{48}O_2$					
Aromatic compounds							
107	1,8-Dimethoxynaphthalene	$C_{12}H_{12}O_2$	Caprifig Jrani inedible unripe fruit	Mesjed Aissa, Tunisia	*n*-hexane	GC-MS	Lazreg-Aref et al. (2012a)

108	6-methyl 4-chromanone	$C_{10}H_{10}O_2$	Unripe fruits *F. carica* subsp. *carica*	Adana, Turkey	Methanol followed by solid-phase extraction	LC-MS/MS	Paşayeva et al. (2020)
					n-hexane-ethyl acetate fraction of the methanol extract		
109	*p*-OH-Phenylacetic acid	$C_8H_8O_2$	Unripe fruits	Sousse, Tunisia	*n*-hexane fraction of the methanolic extract	LC-MS/MS	Lazreg Aref et al. (2011a)
110	Oxalic acid, 2-methylphenylpentyl ester	$C_{14}H_{18}O_4$	Unripe fruits *F. carica* subsp. *carica*	Adana, Turkey	*n*-hexane-ethyl acetate fraction of the methanol extract	LC-MS/MS	Paşayeva et al. (2020)
111	Phthalic acid	$C_8H_6O_4$	Fruits	Al Sharqia government, Egypt.	Hydroalcoholic extract (80% MeOH) of defatted latex	GC-MS	Abdel-Aty et al. (2019)
112	Dibutyl phthalate	$C_{16}H_{22}O_4$					
113	bis (2-Ethylhexyl) phthalate	$C_{24}H_{38}O_4$	Caprifig Jrani inedible unripe fruit	Mesjed Aissa, Tunisia	*n*-hexane	GC-MS	Lazreg-Aref et al. (2012a)
114	[3.2] Metacyclophane-10-ene	$C_{16}H_{14}$					
Amino acids							
115	Argenine	$C_6H_{14}N_4O_2$	Unripe fruits	Sousse, Tunisia	*n*-hexane and *n*-hexane-ethyl acetate fractions of the methanolic extract	LC-MS/MS	Lazreg Aref et al. (2011a)

(continued)

Table 27.3 (continued)

#	Compound	Molecular formula	Occurrence of the latex	Geographical origin	Extraction procedure after latex collection	Method of analysis	Reference
116	Glycine	$C_2H_5NO_2$	Green fruits	Mirandela region, Portugal	0.1M HCl	HPLC-DAD	Oliveira et al. (2010a)
117	Alanine	$C_3H_7NO_2$					
118	Serine	$C_3H_7NO_2$					
119	Asparagine	$C_4H_8N_2O_3$					
120	Ornithine	$C_5H_{12}N_2O_2$					
121	Glutamine	$C_5H_{10}N_2O_3$					
122	Leucine	$C_6H_{13}NO_2$					
123	Tryptophan	$C_{11}H_{12}N_2O_2$					
124	Phenylalanine	$C_9H_{11}NO_2$					
125	Tyrosine	$C_6H_{14}N_2O_2$					
126	Histidine	$C_6H_9N_3O_2$					
127	Lysine	$C_9H_{11}NO_3$					
128	Cysteine	$C_3H_7NO_2S$					
Nitrogenous compounds							
129	4-(4-Methoxyphenyl)-2,6-diphenylpyridine	$C_{24}H_{19}NO$	Caprifig Jrani inedible unripe fruit	Mesjed Aissa, Tunisia	*n*-hexane	GC-MS	Lazreg-Aref et al. (2012a)
130	Quinoline	C_9H_7N	Immature green fruits from F. carica cultivar Pingo deMel	Mirandela region, Portugal	Headspace Solid-Phase Microextraction (HS-SPME).	GC-IT-MS	Oliveira et al. (2010b)

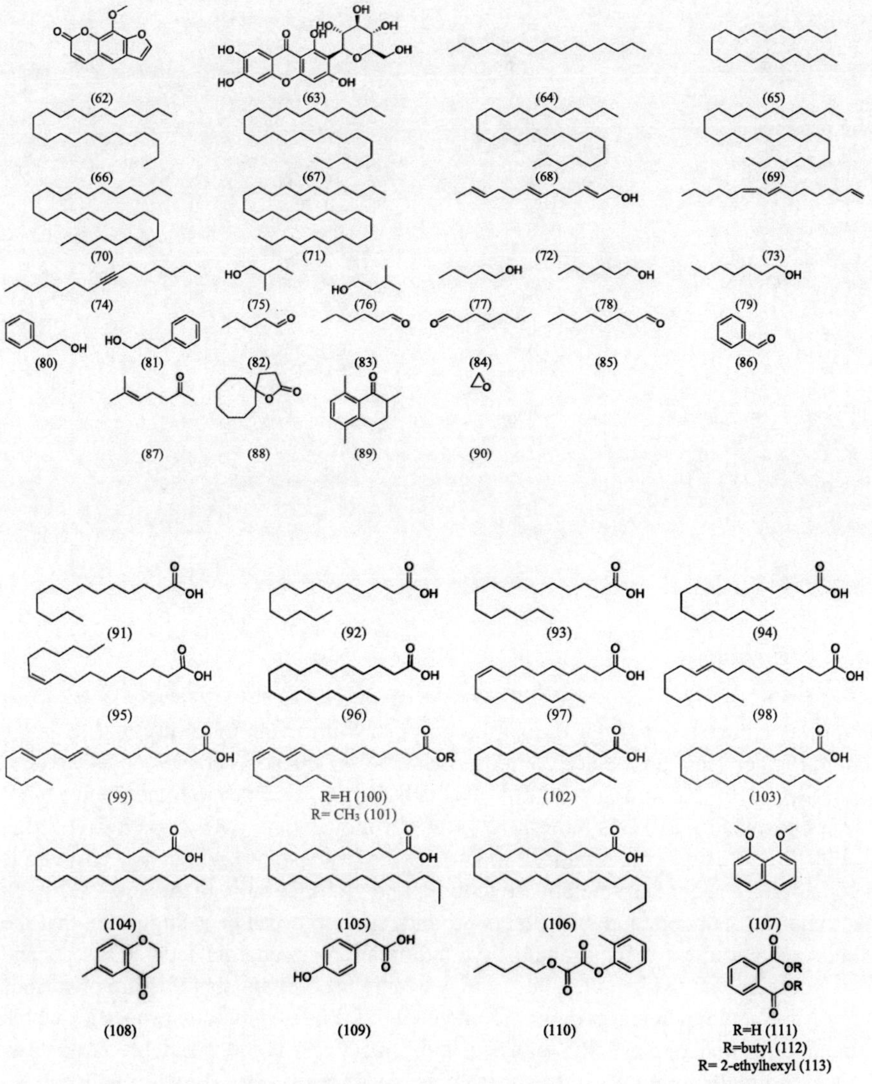

Fig. 27.3 Chemical structures of xanthones, hydrocarbons, fatty acids, aromatic compounds, amino acids, and other nitrogenous compounds

(continued)

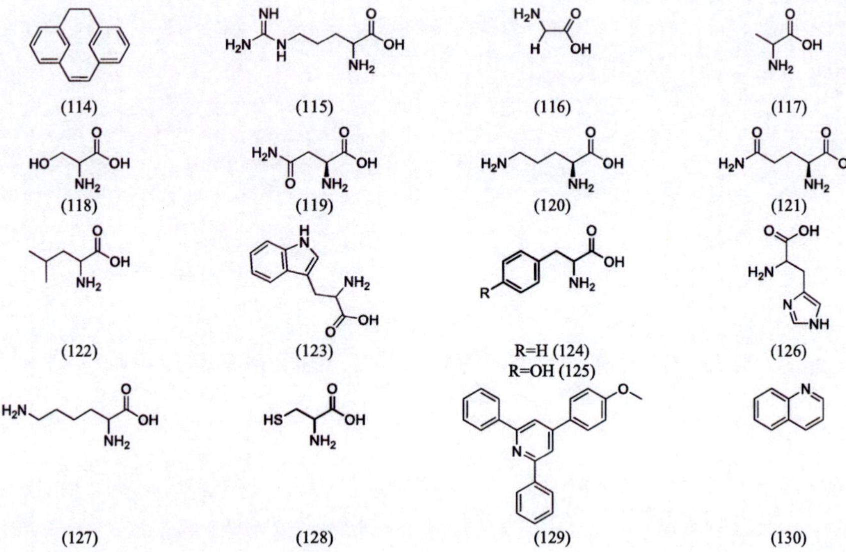

Fig. 27.3 (continued)

2.4 *Rubber*

Natural rubber is a type of crucial biopolymer with meaningful use and strategic importance in human society, composed of 320–35,000 isoprene molecules (*cis*-1,4-polyisoprene) (Kang et al., 2000); of which the constituent monomer isopentenyl pyrophosphate (IPP) is synthesized from the mevalonate (MVA) pathways (Men et al., 2018). Rubber is produced in latex at the expense of high-energy cost and is considered a secondary metabolite with no known function in plant cells. Although it is not fully understood why plants produce rubber, it has been suggested that the latex secretion is a defense against wounding and/or predators such as insects and microorganisms (Kim et al., 2003) and is widely used in the manufacture of various industrial and medicinal products (Kang et al., 2000), have unique properties which include resilience, elasticity, abrasion and impact resistance, efficient heat dispersion, and malleability at cold temperatures. As a result, natural rubber is irreplaceable in many applications (heavy-duty tires, medical devices, surgical gloves, etc.) (Men et al., 2018). Therefore, to understand the molecular mechanism of rubber biosynthesis, it is helpful to investigate other rubber-producing plants and characterize and compare the enzymes related to rubber biosynthesis in the different plant species. Recently, we have identified the fig tree (*Ficus carica*) as an alternative rubber-producing temperate plant and reported that the high level of divalent metal ion in the latex serum of the fig tree exerts different rubber biosynthetic activities in *F. carica*. The amounts of latex in plants and latex rubber content vary depending on the physiological conditions of plants influenced by environments. The molecular size of rubber is determined by the action of enzymes and/or factors such as rubber transferase and rubber elongation factor (Kang et al., 2000).

2.5 Proteins

Latex of *F. carica* is rich in proteins of different classes, biological functions, and industrial applications (Mohammad & Alzweiri, 2021). Fig tree latex contains between 10% and 17% enzyme and non-enzyme proteins (Lazreg-Aref et al., 2018). Zhai et al. (2021) reported the presence of 38.9% papain-like cysteine proteases in the total protein content as the most abundant protein of ripe fig fruit latex (Zhai et al., 2021). Furthermore, many enzyme activities of the latex were characterized, such as protease, peroxidase, chitinase, amylase, and lipase, which play vital roles for the fig plant. For instance, chitinases and peroxidases serve as defensive proteins protecting the fig plant against insects and fungi (Mohammad & Alzweiri, 2021). In addition, many non-enzymatic peptides and proteins are involved as a part nonspecific system of protection against microbes, such as a low molecular weight chitin-binding protein of antifungal activity isolated from latex collected from young branches of fig tree cultivated in Uzbekistan at the beginning of fruiting stage. The antifungal protein was classified as a pathogenesis-related protein 4 type II (PR-4/type II) according to its partial amino acid sequence and chitin binding ability (Table 27.4) (Mavlonov et al., 2008). Moreover, three peptides of angiotensin I-converting enzyme (ACE) inhibitory activity were isolated from stalks latex of a fig tree grown in Japan. Its amino acid sequences were identified as Ala-Val-Asn-Pro-Ile-Arg (FLP-1), Leu-Tyr-Pro-Val-Lys (FLP-2), and Leu-Val-Arg (FLP-3) (Table 27.4) (Maruyama et al., 1989).

2.5.1 Protease Activity

Proteases are the part of latex responsible for breaking down proteins into smaller polypeptides or amino acids. They are classified according to the catalytic amino acid of the active site into four classes: cysteine proteases, serine proteases, aspartic proteases, and metalloproteases (Mohammad & Alzweiri, 2021). Generally, plant proteases participate in many vital functions, such as nutrient mobilization, fruit maturation, protection of fruit during ripening against plant pathogens, and wound healing by latex coagulation injuries caused by a biotic or abiotic agent (Mohammad & Alzweiri, 2021; Azarkan et al., 2011). The latex of *F. carica* has several proteases of two classes: cysteine and serine proteases. Most latex proteases are members of the cysteine protease class, commonly named ficins. While collagenolytic, fibrinolytic, and subtilase activities are serine protease classes (Mohammad & Alzweiri, 2021). Changes in the proteolytic enzyme content all over the year reported by Robbins, 1935 for the latex of the fig tree growing in Alabama, USA. The thin and watery latex collected in June contained the lowest concentration of proteolytic enzymes, then the concentration dramatically increased ten folds after 2 months in August.

On the contrary, the thick and creamy latex collected from October to April contained the highest levels of proteolytic enzymes (Robbins, 1935). The proteolytic activity of the fig latex is a summation of cysteine and serine proteases activities. Vatić et al. (2020) showed that the fig latex has higher efficiency than papain and trypsin in releasing *Listeria monocytogenes* after dissociating the fermented meat products for enumeration of foodborne pathogenic bacteria. Furthermore, latex activity was not affected by Triton X-100 as a strong non-ionic detergent with more cost-efficiency than other alternatives since it could be used efficiently in concentration100 times than previously reported (Vatić et al., 2020).

Cysteine Protease Activity

Ficin or ficain (EC 3.4.22.3) is the cysteine protease of *F. carica* latex and is widely present in *Ficus* species in many isoforms (Zhai et al., 2021). The isolated ficin isoforms showed stability over a wide pH range with maximum catalytic activity at pH = 7. Ficin has more than one enzyme active; it showed a protease activity and an intrinsic peroxidase activity performed on two separate active sites (Mohammad & Alzweiri, 2021). Ficin is a member of the papain-like cysteine proteases (PLCP) family, characterized by strong proteolytic activity and broad substrate specificity (Zhai et al., 2021). Ficins as a major PLCP of ripe fig fruit, present in receptacles more than in inflorescences (Zhai et al., 2021). PLCP, including ficin, is of high commercial value and is widely used in cheese making, meat tenderization, biscuit baking, and producing bioactive peptides as versatile biocatalysts and digestive drugs (Di Pierro et al., 2014; Zhai et al., 2021). Zhai et al. (2021) identified seventy-four proteins in the latex of the fig fruit grown in China with molecular wights ranging from 8.9 to 206.1 kDa and isoelectric point (pI) values of 4.84–10.7. The latex contained trypsin-like protease inhibitor, pathogenesis-related (PR) protein isoforms, chitinase, and endochitinase, collectively stress-response proteins that play an essential role in fruit protection against different plant pathogens (Zhai et al., 2021). The unprotected cysteine proteases undergo degradation through irreversible oxidation and/or autolysis, making chromatographic profiles complex and challenging to resolve. Azarkan et al. (2011) used a thiol-pegylation strategy by specific and reversible monomethoxypolyethylene glycol (mPEG) reagent to purify the fully active ficins more easily (Azarkan et al., 2011). Protection of ficins thiol functions is a prerequisite before any fractionation to prevent autolysis and/or irreversible oxidation of these proteases and improve chromatographic separation. Catalytic cysteine residues of ficins are protected with Selective and reversible reagent such as methylmethanethiol sulfonate (MMTS) (Table 27.4) (Azarkan et al., 2011). Unlike papain, ficin is more acidic with higher substrate-binding efficiency. It was necessary to find other ways to produce this industrially valuable protease in a way that preserves fig tree from branch cutting. So the large-scale production of ficin could be possible using cloned *Escherichia coli* BL21, which can express 0.67 mg/mL of recombinant ficin under optimized conditions (Sattari et al., 2020).

Serine Protease Activity

Collagenase is a serine protease that catalyzes collagen and gelatin breakdown (Mohammad & Alzweiri, 2021). Collagenolytic activity in the latex of *F. carica* var. Brown Turkey grown in Montenegro was studied by Raskovic and Polovic in 2016 in an attempt to explain its uses for skin in the Western Balkans. Collagenolytic serine protease of the fig latex can cleave native collagen *triple helix*. Therefore, *the treatment with fig latex collagenase could promote* connective tissue remodelings such as debridement of wounds and ulcers. Also, collagenase improved the penetration of low molecular weight molecules across the gelatin hydrogel.

Consequently, the collagenase of fig latex could be used to enhance the delivery of the anti-proliferative drugs in cancer treatment. Moreover, latex collected in spring has two-fold the collagenolytic activity of the summer. In addition, the collagenase of the fig latex was stable for up to 1 h in the simulated gastric conditions (Raskovic et al., 2016). A novel collagenolytic activity was purified and characterized by Raskovic et al. (2014) in the latex collected from shoots of *F. carica* var. brown Turkey grown in Montenegro (Table 27.4). The purified collagenase was acidic and showed optimum catalyzing activity at pH 8.0–8.5 and 60 °C with a pI value of about 5. Based on the specificity and high stability of the purified collagenase, the authors suggest that the enzyme could be useful for applications in many fields such as biotechnology and medicine (Raskovic et al., 2014). Hamed et al. (2020) purify and characterize a fibrinolytic serine protease (FPIII) from fruit latex of the Egyptian *F. carica* (Table 27.4). Authors suggest that FPIII is responsible for or participates in the fibrinogenolytic and anticoagulant activities that control ficin by counteracting its procoagulant activity. It showed thermal stability up to 60 °C and good catalytic efficiency on different protein substrates with optimum pH of 8.5 (Hamed et al., 2020).

2.5.2 Peroxidase Activity

Peroxidases are heme-containing glycoproteins that catalyze oxidation reactions. Therefore, they are involved in plant protection against microorganisms (Mohammad & Alzweiri, 2021). Latex collected from *F. carica* branches was used to purify three peroxidase isoenzymes named FP1, FP2, and FP3. The isoenzyme FP1 was completely purified with optimum activity at pH 5.5 and 40 °C (Table 27.4). On the other hand, FP2 and FP3 isoenzymes were partially purified because of their low concentration and level of activity. The high concentration of calcium ions increases the activity of FP1 and FP2 isoenzymes. The three isoenzymes showed a broad specificity to some phenolic substrates and *O*-Phenylenediamine. These isoenzymes' properties satisfy the prerequisites needed for industrial and environmental applications (Elsayed et al., 2018).

Table 27.4 Enzyme activities from fig latex

Name and/or class	Isolation	MW	References
Chitin binding protein (PR-4/type II)	Latex was processed and subjected to successive chromatographic purification over chitin as an affinity matrix, SP Sephadex C-50 as cation exchanger, and reversed-phase high-performance liquid chromatography (RP-HPLC)	6481 kDa	Mavlonov et al. (2008)
FLP-1, FLP-2, and FLP-3 (ACE inhibitory peptides)	The boiled latex was filtered, ultra-filtered, and fractionated over SP-Sephadex C-25 as a cation exchanger, and then the active fractions were chromatographed over Sephadex LH-20 followed by RP-HPLC	668, 618, and 386 Da, respectively	Maruyama et al. (1989)
Ficin A, B, C, D1, and D2 (PLCP)	Fresh latex was collected in the presence of MMTS followed by fast protein liquid chromatography over an SP-Sepharose as a cation exchanger, and then the eluted ficins were pegylated by mPEG reagent	About 24 kDa for all ficin forms	Azarkan et al. (2011)
Collagenase (serine protease)	Latex was centrifuged to remove insoluble gum. Then, the clear water fraction was chromatographed over a Sephadex G-50 column as an ion exchanger matrix, followed by further purification over a thiol-Sepharose column for the fractions with caseinolytic and gelatinolytic and the unbound fractions with collagenolytic and gelatinolytic activities were collected	41 kDa	Raskovic et al. (2014)
FPIII (fibrinolytic serine protease)	Latex fractionated over Diethylaminoethyl cellulose as ion exchange resin, and the fractions with highest proteolytic activity purified over *carboxymethyl-cellulose column followed further purification over* Sephacryl S-200 column for the concentrated fractions with highest proteolytic activity	48 kDa	Hamed et al. (2020)
FP1 isoenzyme (peroxidase)	Insoluble materials of the processed latex were removed, and aqueous latex with peroxidase activity was fractionated over Carboxymethyl (CM)-Sepharose column as a cation exchanger; then, the fractions with peroxidase activity were subjected to diethylaminoethyl (DEAE)-Sepharose column as an anion-exchanger and Sephacryl S-200 column as a size-exclusion matrix	30 kDa	Elsayed et al. (2018)
Lipase	Latex was ultra-centrifuged and then fractionated over silica gel column chromatography eluting with acetone. Then the polarity was increased with water	29 kDa	Lazreg-Aref et al. (2012b)

2.5.3 Chitinase Activity

Chitinase is the enzyme that catalyzes the breakdown of chitin, which is a structural component of insects' exoskeleton and fungal cell walls. Chitinase has the ability of retarding insect development and destruction of the fungal cell wall. Therefore, chitinases serve as one of the plant defensive strategies, which also could be applied as biopesticides and antifungal agents. However, chitinases of *F. carica* latex were isolated mainly from tree trunks (Mohammad & Alzweiri, 2021). The chitinolytic activity of the latex collected from *F. carica* var. brown Turkey from the beginning of the flowering to the fruit ripening stages was examined, which showed 6.5 times higher activity at the beginning of flowering than ripened fruit. Accordingly, the optimal fig protection against plant pathogens at the beginning of the flowering stage depends mainly on chitinolytic activity, which is a critical period for the fig plant and then gradually replaced over time with other strategies (Raskovic et al., 2016).

2.5.4 Amylase Activity

Amylases are among industrially essential enzymes involved in the hydrolysis of starch (Gupta et al., 2003). Latices collected from unripe fruit stalks of *F. carica* var. Kahli and Bidhi grown in Tunisia were investigated for their amylolytic activities. Bidhi and Kahli amylases showed optimum activity in pH of 6.5 and 7 at 45 °C and a half-life of 85 and 60 min at 80 °C, respectively. They were highly stable in a wide pH range (4–12). amylase activity of Bidhi variety increased 260% by addition of ferrous ion or cupric ion and was strongly decreased by magnesium ion and EDTA. In the same way, the presence of calcium and magnesium ions caused a dramatic increase of Kahli amylase activity up to 220% and 260%, respectively. Amylases of both varieties were insensitive to inhibition by glucose with high thermostability in the presence of starch. Therefore, the properties of these amylases could be very efficient for glucose production from starch (Lazreg Aref et al., 2011c).

2.5.5 Lipase Activity

Lipases are enzymes that hydrolyze oils and fats into di- and monoacylglycerol, fatty acids, and glycerol. Lipase enzyme (ZL) was purified and characterized in Latex collected from unripe fruit stalks of *F. carica* var. Zidi is grown in Tunisia (Table 27.4). Lipase activity was 12-fold higher for long-chain triacylglycerols than the short-chain. The purified lipase was stable in pH range 4–8 and displayed a half-life of 90 min at 60 °C. The purified lipase activity was strongly reduced by ferrous, magnesium, and zinc ions but significantly increased by calcium and cupric ions. Features of purified lipase activity such as stability over a wide pH range and at high temperature, surfactant resistance, and storage stability make it a suitable potential additive for many fields such as pharmaceutical, leather, and detergent industries (Lazreg-Aref et al., 2012b).

3 Biological Activities of Fig Latex

The biological activities of the latex of *F. carica* belong to its diversity of secondary metabolites of different chemical classes, as mentioned earlier as phenolics, flavonoids, terpenoids, etc. These different metabolites showed many biological activities, such as antibacterial, antifungal, antivirals, anticancer, anti-angiogenic, antioxidant, anti-inflammatory, anticoagulant, antihypertensive, and hepatoprotective antiparasitic activities.

3.1 Antimicrobial Activity

Microorganisms are found everywhere around us in soil, water, air, skin, and most man-made environments throughout the world. The symptoms of such infections are generalized inflammation and sepsis, and when they occur in critical body organs, such as the lungs, the urinary tract, and kidneys, the results can be fatal (Al-Sokari & El Sheikha, 2015). Latex of *F. carica* exhibited strong activities against tested microorganisms, including gram-positive bacteria, gram-negative bacteria, and yeast. These activities are represented in Tables 27.5 and 27.6.

These results revealed that caprifig latex had inhibition of the growth of all bacterial and yeast species represented in this chapter. The pharmacological properties are probably due to the high content of enzymes, flavonoids, and furanocoumarins from fig latex (Rashid et al., 2014). Antimicrobial activities of the hexane extract are generally due to the high content of coumarins. Free 6-OH in the coumarin nucleus is essential for antifungal activity, and the free hydroxyl group at position 7 is essential for antibacterial activity (Lazreg-Aref et al., 2012a). Antimicrobial activities are also assigned to low molecular weight short polypeptides in fig latex, which are shown to possess antibacterial and antifungal activities. It seems that ficin, chitinases, and chitinase-like proteins abundantly accumulated in fig's latex are the major defense mechanism against insects and fungi rely on its proteolytic and chitinolytic, explaining the potential antifungal activities. The antifungal activity of the isolated protein can be explained by its chitin-binding ability as a protein binding to the chitin component of the cell wall that suppresses fungal hyphae growth. (Mavlonov et al., 2008; Raskovic et al., 2016). Our data showed no uniform response between bacterial and fungal strains in terms of susceptibility to antimicrobial compounds in different extracts; thus, different solvents have been found to have the varied capacity to extract different phytoconstituents depending on their solubility or polarity in the solvent. The ethyl acetate and chloroform fractions have a medium polarity, so the substances that have the most important biological activities were the medium

Table 27.5 Antibacterial activities of *F. carica* latex different fractions

Microorganism and type	Activities of different extracts	Methodology	Reference
Aeromonas hydrophila (Gram-negative bacteria)	Weak (L.)	Well diffusion method (Inhibition zone)	Ahmed (2016)
Citobacter freundei (Gram-positive bacteria)	Weak (LH.) Strong (L., LC., LE.)	Disk diffusion assay (Inhibition zone) Paper diffusion assay (Inhibition zone) Microwell dilution assay (MIC)	Lazreg Aref et al. (2010, 2011b)
Enterococcus fecalis (Gram-positive bacteria)	Weak (LC.) Moderate (LE., LH.)	Paper diffusion assay (Inhibition zone) Microwell dilution assay (MIC)	Lazreg Aref et al. (2010, 2012a)
Enterobacter cloacae (Gram-negative bacteria)	Moderate (L.)	Disc diffusion method (Inhibition zone)	Lazreg-Aref et al. (2012a)
Escherichia coli (Gram-negative bacteria)	Moderate (LC., LE., LH.) Strong (L.)	Paper diffusion assay (Inhibition zone) Well diffusion method (Inhibition zone) Microwell dilution assay (MIC)	Lazreg Aref et al. (2010, 2012a) and Rashid et al. (2014)
Klebsiella pneumonae (Gram-negative bacteria)	Strong (L.)	Well diffusion method (Inhibition zone)	Rashid et al. (2014) and Ahmed (2016)
Klebsiella spp. (Gram-negative bacteria)	Strong (L.)	Well diffusion method (Inhibition zone)	Ahmed (2016)
Proteus mirabilis (Gram-negative bacteria)	Strong (L., LC., LE., LH.)	Paper diffusion assay (Inhibition zone) Disc diffusion method (Inhibition zone) Microwell dilution assay (MIC)	Lazreg Aref et al. (2010, 2011b)
Pseudomonas aeruginosa (Gram-negative bacteria)	Weak (LC., LE.) Moderate (LH.) Strong (L.)	Paper diffusion assay (Inhibition zone) Disc diffusion method (Inhibition zone) Well diffusion method (Inhibition zone) Microwell dilution assay (MIC)	Lazreg Aref et al. (2010, 2012a) and Rashid et al. (2014)

(continued)

Table 27.5 (continued)

Microorganism and type	Activities of different extracts	Methodology	Reference
Salmonella spp (Gram-negative bacteria)	Strong (L.)	Well diffusion method (Inhibition zone)	Ahmed (2016)
Salmonella typhi (Gram-negative bacteria)	Strong (L.)	Well diffusion method (Inhibition zone)	Rashid et al. (2014)
Seratia spp (Gram-negative bacteria)	Moderate (L.)	Well diffusion method (Inhibition zone)	Ahmed (2016)
Staphylococcus aureus (Gram-positive bacteria)	Strong (L., LH.) Moderate (LC.)	Paper diffusion assay (Inhibition zone) Disc diffusion assay (Inhibition zone) Well diffusion method (Inhibition zone) Microwell dilution assay (MIC)	Lazreg Aref et al. (2010, 2012a) and Rashid et al. (2014)
Staphylococcus epidermidis (Gram-positive bacteria)	Moderate (L.) Strong (LH.)	Paper diffusion method (Inhibition zone) Disc diffusion method (Inhibition zone)	Lazreg-Aref et al. (2012a) and Al-Sokari and El Sheikha (2015)
Staphylococcus saprophyticus (Gram-positive bacteria)	Strong (LH.)	Disc diffusion method (Inhibition zone)	Lazreg-Aref et al. (2012a)
Streptococcus pyogenes (Gram-positive bacteria)	Strong (L.)	Well diffusion method (Inhibition zone)	Rashid et al. (2014)

L Latex, *LH* hexanoic fraction, *LE* Ethyl acetate fraction, *LC* chloroformic fraction

polar ones, and so both extracts showed the highest activities (Lazreg Aref et al., 2010). These kinds of differences in susceptibility among microorganisms against antimicrobial substances in *F. carica* latex different extracts may be explained by the differences in cell well composition and or inheritance genes on plasmids that can be easily transferred among bacterial strains or even due to the existence of different plant varieties (Lazreg Aref et al., 2010, 2011b). Finally, a great interest in acquiring more knowledge regarding the antimicrobial effects ascribed to caprifig latex exists among the scientific community since there is increasing evidence from prospective observational studies that caprifig latex lowers the risk of infectious diseases along with being effective for the treatment of superficial fungal and bacterial infections, particularly in cutaneous mycosis. Moreover, another interesting potential application of *F. carica* against fungus-induced diseases of crops results in substituting phytotoxic and environmentally harmful substances with non-toxic caprifig latex. Latex could also be used as a natural food preservative in the food

Table 27.6 Antifungal activities of *F. carica* latex different fractions

Microorganism and type	Activities of different extracts	Methodology	Reference
Aspergellus fumigatus (Hyphamycets yeast)	Weak (L., LH.) Moderate (LC.) Strong (LE.)	Agar incorporation (Percentage of inhibition)	Lazreg Aref et al. (2010)
Aspergillus nigar (Fungi)	Strong (L.)	Well diffusion method (Inhibition zone)	Rashid et al. (2014)
Candida albicans (Opportunists pathogenic yeast)	Weak (LH.) Strong (L., LC., LE.)	Agar incorporation (Percentage of inhibition) Well diffusion method (Inhibition zone)	Lazreg Aref et al. (2010) and Rashid et al. (2014)
Candida cruzi (Yeast)	Moderate (L.)	Well diffusion method (Inhibition zone)	Ahmed (2016)
Candida glabrata (Yeast)	Weak (LE.) Moderate (L.) Strong (LC.)	Agar incorporation (Percentage of inhibition) Wells punched method (Inhibition zone)	Lazreg Aref et al. (2011b)
Candida kefyr (Yeast)	Moderate (L.)	Well diffusion method (Inhibition zone)	Ahmed (2016)
Candida sojae (Yeast)	Strong (L.)	Well diffusion method (Inhibition zone)	Ahmed (2016)
Candida tropicalis (Yeast)	Strong (L.)	Well diffusion method (Inhibition zone)	Ahmed (2016)
Cryptococcus neoformans (Opportunists pathogenic yeast)	Weak (LH.) Moderate (L.) Strong (LC., LE.)	Well diffusion method (Inhibition zone) Agar incorporation (Percentage of inhibition)	Lazreg Aref et al. (2010) and Ahmed (2016)
Fusarium oxysporum (Fungi)	Strong (L.)	Well diffusion method (Inhibition zone)	Rashid et al. (2014)
Microsporum canis (Dermatophytes yeast)	Moderate (LC.) Strong (L., LE., LH.)	Agar incorporation (Percentage of inhibition) Paper diffusion method (Inhibition zone)	Lazreg Aref et al. (2010) and Al-Sokari and El Sheikha (2015)
Trichophyton rubrum (Dermatophytes yeast)	Weak (L., LC., LH.) Moderate (LE.)	Agar incorporation (Percentage of inhibition)	Lazreg Aref et al. (2010)
Trichophyton soudanense (Dermatophytes yeast)	Weak (L., LC., LE., LH.)	Agar incorporation (Percentage of inhibition)	Lazreg Aref et al. (2010)

L Latex, *LH* hexanoic fraction, *LE* Ethyl acetate fraction, *LC* chloroformic fraction processing industry, especially when we consider that fig latex has a high content of minerals, vitamins, antioxidants, fiber, and proteins, which makes it valuable functional food (Raskovic et al., 2016).

3.2 Antiviral Activity

The fig tree latex could be helpful in the treatment of bacterial and viral infections caused by strains resistant to conventional drugs (Lazreg Aref et al., 2011a). Extracts of *F. carica* latex have been found to inhibit the replication of Herpes simplex type 1 (HSV-1), Echovirus type 11 (ECV-11), and adenovirus (ADV). Moreover, they have been found to act against several drug-resistant pathogens; among compounds isolated from *F. carica*, it is found that lupeol (46), α-amyrin (43), and luteolin (12) could be used as promising inhibitors against SARS-CoV-2 main protease which plays a dynamic role in mediating viral replication and transcription. The molecular dynamics simulation study revealed that among all the phytochemicals, α-amyrin (43) is the most stable phytochemical, which could inhibit SARS-CoV-2 main protease strongly. The binding free energy analysis shows that α-amyrin (43) and lupeol (46) have higher binding free energy in contrast to the known SARS-CoV-2 Mpro inhibitor α-ketoamide (Ali et al., 2020). The antiviral activity of *F. carica* latex is statistically significantly against Herpes simplex type 2 (HSV-2) lower than that of acyclovir. In particular, the synergy produced by the antiviral activity of *F. carica* and acyclovir combined had a more promising effect against HSV-2 than acyclovir alone. It has been determined that the antiviral activity of *F. carica* is increased due to the polysaccharides. The activity of *F. carica* latex against HSV-2 was confirmed by a significant decrease in the number of viral copies (Ay & Duran, 2018). The fig tree latex application on pox lesions showed continuous regression and shrinking of the nodules from the fourth day of treatment (Abid & Ali, 2014). The hexanic and hexane-ethyl acetate extracts inhibited the multiplication of HSV-1, ECV-11, and ADV at 78 mg/ml concentrations. These two extracts are the most potent candidates for herbal medicines to treat viral infectious diseases such as herpes virus, echovirus, and adenovirus (Badgujar et al., 2014). Fig tree latex had a short duration therapy to treat warts (*Verruca vulgaris*), flanking, and shrinking of responsive warts after initiation of fig tree latex therapy (Bohlooli et al., 2007).

3.3 Anticancer Activity

Earlier studies have reported the antiproliferative effects of *F. carica* latex against cells derived from esophageal and stomach cancers separately. The overall knowledge suggests that fig latex could decrease tumor growth without adversely affecting hematological and histological factors (Ghandehari & Fatemi, 2018; AlGhalban et al., 2021). Latex of *F. carica* uses the different molecular mechanisms of action, including antiproliferative and anti-metastatic effects, along with significant effects on cell shape and fighting against side effects of oxidative stress of free radicals generated by the immune system cells (Ghandehari & Fatemi, 2018). Activities were expected due to phytochemicals like 6-*O*-acyl-β-D-glucosyl-β-sitosterol and its palmitoyl derivatives and antioxidant compounds, including flavonoids and polyphenolic compounds. Proteolytic enzymes abundantly accumulated in fig's latex are the major antiproliferative agents explaining the anticancer potential (Mavlonov et al., 2008; Hashemi et al., 2011; Raskovic et al., 2016). The phenolic compounds in fig latex are the main inhibitors of tumor growth, like protocatechuic acid derived from latex which is partially responsible for the induction of cell death and inhibition of invasion in these cell lines. The fig latex possesses a group of proteases such as ficin, caseinolytic, and gelatinolytic enzymes that induce apoptosis in cancer cells without causing any side effects on normal cells (Ghandehari & Fatemi, 2018; AlGhalban et al., 2021). Interestingly, the fig latex could inhibit the proliferation of cancer cell lines without any cytotoxic effect on normal human cells with several mechanisms (Hashemi et al., 2011). The antiproliferative actions were described regarding fig latex; mainly its potential for apoptosis induction via interaction with cell cycle causing rises of G0/G1phase cell population and decreases in S and G2/M phase cell population (Ghandehari & Fatemi, 2018). Another supposed mechanism for such effects might be associated with the inhibition of DNA synthesis, apoptosis induction, cell cycle arrest of cancer cells, and extrinsic death receptor and intrinsic mitochondrial-dependent apoptotic signaling pathway (Wang et al., 2008; Shin et al., 2017). Given the fact that angiogenesis and inflammations are playmakers in cancer progression, therefore the inhibitory effects of both factors by fig latex are a major cytotoxicity mechanism. As known, platelets protected cancer cells from the immune system, and they exacerbated metastasis by producing fibrin that caused connection of cells to the endothelium of vessels. Therefore, fig latex preventable effect of platelets increases in the number and/or activity further confirm its anti-angiogenesis and anti-inflammatory potentials. Although mean platelet volume (MPV) and platelet distribution width (PDW) are early indices of platelet activation, while no significant difference was observed in the MPV, PDW and/or platelet (PLT) upon administration of fig latex (Ghandehari & Fatemi, 2018). The anti-angiogenesis and anti-inflammatory effects also were confirmed through a significant decrease in the production of tumor necrosis factor-alpha, prostaglandin E_2 and vascular endothelial growth factor (Ghandehari & Fatemi, 2018). Fig latex also prevents invasion through induction of lethal-7d (let-7d) expression (Tezcan et al., 2015). Moreover, fig latex inhibited the proliferation of endothelial cells of central vein cultured in a three-dimensional medium with collagen matrix and suppressed the production of capillary tubules from these cells without damaging them (Ghandehari & Fatemi, 2018). Interestingly, fig latex showed preventative potential against cancer induced anemia which is the most common side effects caused by neoplastic diseases via inhibition of erythropoietin secretion. The anti-inflammatory effect of the fig latex prevented the destruction of erythropoietin-secreting tissues and, accordingly, prevented a decrease in RBC and blood cell indices (Ghandehari & Fatemi, 2018). Furthermore, consistent changes in cell shape after treatment with the latex were observed by microscopical analysis. The cells showed morphological changes such as round shape and shrinkage, which indicated a stoppage of proliferation and turning on apoptosis. Although a round shape was the dominant morphology, other cell shapes like cell blebbing, cytoplasmic vacuolation, and uneven cell structure and distribution confirmed apoptosis (AlGhalban et al., 2021). Reduced expression of phosphorylated map kinase (ERK2), cAMP-response element-binding protein (CREB), and glycogen synthase kinase-3a/b (GSK-3a/b) with fig latex treatment which is actively involved in cellular proliferation processes (AlGhalban et al., 2021). To evaluate the cytotoxic activities of fig latex, different methodologies were applied. Examples include DNA damage assay, MTT assay, wound healing assay, tumors mean volumes measurement, WST-1 assay, chick chorioallantoic membrane assay, and morphological studies (Ghandehari & Fatemi, 2018; AlGhalban et al., 2021). Antiproliferative activities were reported in human colon cancer cell line (HT-29) (Soltana et al., 2019), human glioma (HCC) (Wang et al., 2008), human oral cancer cells (Shin et al., 2017), glioblastoma multiforme (GBM) (Tezcan et al., 2015), stomach cancer (Hashemi et al., 2011), Melanoma cells (A375) (Menichini et al., 2012), human hepatocellular carcinoma (HepG-2), human breast cancer cells (MCF-7), and human colon cancer cell line (HCT-116) (Yahiaoui et al., 2022). In conclusion, the administration of fig latex prevented the oversized growth of cancerous tumors without causing side effects on hematologic parameters and induction of the inflammation in tissues that can be attributed to antioxidant, anti-inflammatory, and pro-apoptotic effects this natural product.

3.4 Anti-Angiogenic Activity

Angiogenesis, a complex multi-step process resulting in new blood vessels from preexisting ones, represents a critical process for oxygen and nutrient supply to proliferating cells. Therefore, promoting tumor growth and metastasis. Inhibition of tumor angiogenesis is therefore considered a promising anticancer target. It was reported that latex extracts of *F. carica* possess strong antiangiogenic and antiproliferative activities (Mostafaei et al., 2011). Latex from immature fruits of several different *F. carica* varieties were tested for anti-angiogenic activity in a unique system involving the use of segments of placental veins and human tumor biopsies. Three out of six of these varieties demonstrated significant anti-angiogenic activity yet were non-toxic. This non-toxic, anti-angiogenic activity was proved using a human umbilical vein endothelial cell (HUVEC) tubule formation assay and MTT antiproliferation assay (Pawlus et al., 2008).

Bioactive compounds isolated from *F. carica* latex-like 6-O-acyl-β-D-glucosyl-β-sitosterol derivatives, i.e., AGS (acyl moiety: palmitoyl, linoleoyl, stearyl, and oleyl). Palmitoyl derivative of AGS acts as the most active inhibitor for various cancer cell lines compared with other derivatives. For example, AGS is reported for in vitro inhibition of Burkitt lymphoma cell line (DG-75) and prostatic cancer cell lines (DU-145). Thus, AGS is the most influential anticancer agent. Therefore, latex extract could be an ideal candidate as a prospective agent for preventing angiogenesis in cancer and other chronic disorders (Badgujar et al., 2014).

Furthermore, fig latex can prevent the diffusion of cancer cells into new areas and the formation of metastatic tumors. Treatment with fig latex has allowed the cell cycle exit by inducing the cell cycle arrest in the Sub G1 phase in cervical cancer cell lines. This finding demonstrates the ability of fig latex to prevent cell progression and proliferation. Cell cycle deregulation is known to be a significant element in cervical cancer carcinogenesis hence regulation of cancer cells division through induction of apoptosis is an efficient anti-cancer strategy (Ghanbari et al., 2019). Latex of *F. carica* inhibited the proliferation of cancer cell lines but did not indicate any cytotoxic activity against normal cells in vitro. Latex of *F. carica* had a promising effect in inhibiting stomach cancer cell line growth but without any clear effect on peripheral blood mononuclear cells. Ficin, a cysteine proteinase isolated from the latex of *F. carica* tree. Cysteine proteinases are a group of enzymes leading to the apoptosis of cancer cells. Furthermore, the anticancer effects may be associated with antioxidant properties due to its phenolic components. Additionally, We concluded that *F. carica* has an anticancer cell activity while cancer cells were more sensitive to this latex than normal cells (Hashemi et al., 2011).

3.5 Antioxidant Activity

Antioxidant compounds, such as phenolics, and organic acids, protect against oxidation or cellular damage caused by reactive species, preventing the initiation of several diseases, like many types of cancer, heart disease, diabetes, and neurodegenerative illnesses (García-Alonso et al., 2004). The phenolic content of latex extracted from the fruits of *F. carica,* collected from Ministry of Agriculture, Sharkia government, Egypt, was effective in scavenging free radicals of DPPH (2,2-diphenyl-1-picrylhydrazyl), and ABTS (2,2′-azino-bis(3-ethylbenzothiazoline-6-sulfonic acid)), with IC_{50} values of 13.6, and 4.5, μg GAE/ml, respectively (Abdel-Aty et al., 2019). Its total antioxidant activity, using phosphomolybdenum assay, had been reported with EC_{50} values of 39 μg GAE/mL (Abdel-Aty et al., 2019). The antioxidant ability of latex extracted from immature green fruits of *F. carica*, cultivar pingo de Mel trees growing in the Mirandela region (Northeast Portugal), was screened by DPPḤ assay, which allowed observation of a concentration-dependent potential (IC_{25} = 1049 μg/mL) (Oliveira et al., 2010b). The analyzed sample presented a protective effect against superoxide radical, in a concentration-dependent way, with an IC_{25} at 291 μg/mL (Oliveira et al., 2010b). It displayed nitric oxide scavenging capacity, which was concentration-dependent (IC_{25} = 1768 μg/mL) (Oliveira et al., 2010b). Polyphenols and flavonoids extracted from the latex collected from the neck of *F. carica* L. fruit before ripening from Erbil city, north of Iraq, showed a significant reduction in the rate of superoxide dismutase and glutathione reductase enzymes in liver cells of mature male Wistar albino rats (*Rattus norvegicus*) (Aziz, 2012). Ficin had a DPPḤ radical scavenging ability of about 78.7% compared to

L-ascorbic acid at the highest concentration of 1000 μg/mL. In addition, ficin had an H_2O_2 scavenging ability of about 50.5% compared to L-ascorbic acid at the highest concentration of 1000 μg/mL. These results suggest that Polyphenols, flavonoids, and ficin in the latex of *F. carica* are responsible for its anti-oxidant activity (Aziz, 2012; Cho et al., 2019).

3.6 Anti-Inflammatory Activity

Ficus carica has been exhibited as an anti-inflammatory remedy in traditional medicine (Werbach, 1993; Duke, 2002; Mawa et al., 2013). Experimentally, it was found that NO (nitric oxide) production was significantly decreased by ficin in a concentration-dependent manner: at a concentration of 10, 25, 50, and 100 μg/mL, NO production was significantly reduced to 90.5%, 77.3, 67.9, and 59.6%, respectively, compared to that in the lipopolysaccharide (LPS) treated group without ficin treatment (Cho et al., 2019). As well as ficin could decrease the amount of phosphorylated IκB and Nuclear factor κB (NF-κB) protein induced by LPS in a concentration-dependent manner (Cho et al., 2019).

3.7 Anticoagulant Activity

Latices of several *Ficus* species contain ficin. Ficin (a single polypeptide chain active at neutral pH while inactive at pH below 3.0) is a heterogeneous group of proteases. The procoagulant activity of *F. carica* crude extract could be explained by cysteine proteases FPI, FPII, and serine protease FPIII, which are types of ficin. This procoagulants characteristics of cysteine proteases might be due to their ability to cleave the Aα and Bβ chains, but not the γ chain of fibrinogen, releasing fibrinopeptide A and B, respectively, forming fibrin clot that is soft and friable or their ability to activate human factor X by cleave it to create FXa. However, the anticoagulant activity of *F. carica* crude extract might be due to the complete digestion of fibrinogen or the presence of serine protease inhibitor that inhibited coagulation factors or antiplatelet effect so that *F. carica* latex has procoagulant and anticoagulant properties at the same time. Latex of *F. carica* can control its prominent coagulation stimulant nature caused by ficin (more than 5 isoenzymes) by fibrinolytic serine proteases. The anticoagulant property of the latex at a higher protein concentration suggested the presence of other anti-proteases and small peptides that may inhibit ficin (Hamed et al., 2020). Ficin shortened the activated partial thromboplastin time and the prothrombin time of normal plasmas and plasmas deficient in coagulation factors; this showed that the hemostatic potency of *Ficus* proteases was based on the activation of human coagulation factor X (FX), which could explain the use of these latices as a local hemostatic agent in natural medicine (Richter et al., 2002).

3.8 Antihypertensive Activity (Angiotensin I-Converting Enzyme Inhibitors Activity)

Angiotensin I-converting enzyme catalyzes the production of the vasoconstrictor angiotensin II and the inactivation of the vasodilator bradykinin. Three peptides were isolated from the latex of *Ficus carica*; sequences of these peptides were: Ala-Val-Asn-Pro-Ile-Arg (FLP-1), Leu-Tyr-Pro-Val-Lys (FLP-2), and Leu-Val-Arg (FLP-3) are compared with ACE isolated from rabbit lung Hippuryl-L-histidyl-L-leucine (Hip-His-Leu) with IC_{50} 13, 4.5 and 14 μM, respectively. They revealed the Angiotensin I-converting enzyme inhibitor effect by catalyzing His-Leu release from the carboxyl-terminal angiotensin. Therefore, ACE inhibitors derived from *F. carica* latex may be categorized along with natural peptide inhibitors (Maruyama et al., 1989).

3.9 Hepatoprotective Activity

The extract of latex, collected from the neck of *F. carica* L. fruit before ripening, from Erbil city, north of Iraq, showed attenuated lead hepatotoxicity in rats; through one or more of these suggested mechanisms of action; even by lowering the oxidative stress, second: increasing the oxidant enzymes level, and third: acting as a chelating agent for lead ions. In addition, it was evaluated by induction of oxidative stress using lead acetate in rats. The *F. carica* latex was highly successful in attenuating lead hepatotoxicity due to its high total phenol and flavonoid contents (Aziz, 2012).

3.10 Antiparasitic Activity

Latex of *F. carica* collected from the garden of Kufa University, Iraq, in a comparative study with sodium stibogluconate, showed promising antileishmanial activity in a concentration of 500 μg/mL. Complete healing in the ninth week was (85.9%) and without side effects on the *Leishmania tropica* parasite (Shlash et al., 2021). On the other hand, the fig latex had anthelmintic activity (de Amorin et al., 1999), especially against Ascaris and Tricharus. This effect has been attributed to the presence of ficin in fig latex (an enzyme that has been recognized for its anthelmintic activity) (Joseph & Raj, 2011).

3.11 Adverse Reactions

Although the fruit of *F. carica* is edible, it is rarely described as a cause of food allergy. The latex contains ficin, a proteolytic enzyme that causes pruritus (Arthur & Shelley, 1955) and acts as an irritant on inflamed skin (Budavari et al., 1996). Latex found in the leaves and stems seems to contain irritant and phototoxic chemicals, although one case report suggests that the fruit also contains photosensitizers (Ippen, 1982). Skin contact with latex may produce allergic reactions like dermatitis, asthma, and anaphylaxis (Bohlooli et al., 2007).

4 Conclusion

Several studies have investigated fig latex bioactive components and pharmacological and industrial activities. In conclusion, the administration of fig latex provides multitask curative features without causing side effects attributed to its phytochemical constituents. On the other hand, oxidative stress, inflammation, viral infection, microbial infection, hepatotoxicity, hypertension, parasitic invasion, blood coagulation, angiogenesis, and oversized growth of cancerous tumors inhibition were reported for fig latex. Furthermore, the phenolic content is responsible for the most pharmacological activities, specially mangiferin potentials as antidiabetic, antimicrobial, anticancer antioxidant phytoconstituent. Moreover, fig latex is considered a rich source of proteases such as ficin, caseinolytic, and gelatinolytic enzymes involved in several activities, including anticancer without side effects on normal cells and as a promising alternative to animal enzymes in milk clotting. Finally, fig latex needs further standardization procedures as a pharmaceutical product.

References

Abdel-Aty, A. M., Hamed, M. B., Salama, W. H., Ali, M. M., Fahmy, A. S., & Mohamed, S. A. (2019). *Ficus carica, Ficus sycomorus* and *Euphorbia tirucalli* latex extracts: Phytochemical screening, antioxidant and cytotoxic properties. *Biocatalysis and Agricultural Biotechnology, 20*, 101199.

Abid, T., & Ali, K. A. (2014). Proteolytic versus surgical removal: The therapeutic effect of fig tree latex (*Ficus carica* L) on cutaneous and diphtheric forms of avian pox in pigeons (*Columba domestica*). *Iraqi Journal of Veterinary Sciences, 28*(1), 49–53.

Agrawal, A. A., & Konno, K. (2009). Latex: A model for understanding mechanisms, ecology, and evolution of plant defense against herbivory. *Annual Review of Ecology, Evolution, and Systematics, 40*, 311–331.

Ahmed, N. M. (2016). *Antifungal and antibacterial activity of fig fruit latexes from two Ficus species plants*. 3rd scientific conference-College of Veterinary Medicine-University of Tikrit.

AlGhalban, F. M., Khan, A. A., & Khattak, M. N. K. (2021). Comparative anticancer activities of Ficus carica and *Ficus salicifolia* latex in MDA-MB-231 cells. *Saudi Journal of Biological Sciences, 28*(6), 3225–3234.

Ali, M. C., Nur, A. J., Khatun, M. S., Dash, R., Rahman, M. M., & Karim, M. M. (2020). Identification of potential SARS-CoV-2 main protease inhibitors from *Ficus carica* latex: An in-silico approach. *Journal of Advanced Biotechnology and Experimental Therapeutics, 3*, 57–67.

Al-Sokari, S. S., & El Sheikha, A. F. (2015). In vitro antimicrobial activity of crude extracts of some medicinal plants from Al-Baha region in Saudi Arabia. *Journal of Food and Nutrition Sciences, 3*(1–2), 74–78.

Arthur, R. P., & Shelley, W. B. (1955). The role of proteolytic enzymes in the production of pruritus in man. *The Journal of Investigative Dermatology, 25*(5), 341–346.

Ay, E., & Duran, N. (2018). *Investigation of the antiviral activity of Ficus carica L. latex against HSV-2*. International Conference on Advanced Materials and Systems (ICAMS), The National Research & Development Institute for Textiles and Leather-INCDTP.

Azarkan, M., Matagne, A., Wattiez, R., Bolle, L., Vandenameele, J., & Baeyens-Volant, D. (2011). Selective and reversible thiol-pegylation, an effective approach for purification and characterization of five fully active ficin (iso) forms from *Ficus carica* latex. *Phytochemistry, 72*(14–15), 1718–1731.

Aziz, F. M. (2012). Protective effects of latex of *Ficus carica* L. against lead acetate-induced hepatotoxicity in rats. *Jordan Journal of Biological Sciences, 5*(3), 175–182.

Badgujar, S. B., Patel, V. V., Bandivdekar, A. H., & Mahajan, R. T. (2014). Traditional uses, phytochemistry and pharmacology of *Ficus carica*: A review. *Pharmaceutical Biology, 52*(11), 1487–1503.

Bohlooli, S., Mohebipoor, A., Mohammadi, S., Kouhnavard, M., & Pashapoor, S. (2007). Comparative study of fig tree efficacy in the treatment of common warts (*Verruca vulgaris*) vs. cryotherapy. *International Journal of Dermatology, 46*(5), 524–526.

Borase, H. P., Patil, C. D., Suryawanshi, R. K., & Patil, S. V. (2013). *Ficus carica* latex-mediated synthesis of silver nanoparticles and its application as a chemophotoprotective agent. *Applied Biochemistry and Biotechnology, 171*(3), 676–688.

Budavari, S., O'Neil, M. J., Smith, A., Heckelman, P. E., & Kinneary, J. F. (1996). *"The Merck Index." An encyclopedia of chemicals, drugs, and biologicals* (12th ed.). Merck and Co.

Cho, U. M., Choi, D. H., Yoo, D. S., Park, S. J., & Hwang, H. S. (2019). Inhibitory effect of ficin derived from fig latex on inflammation and melanin production in skin cells. *Biotechnology and Bioprocess Engineering, 24*(2), 288–297.

de Amorin, A., Borba, H. R., Carauta, J. P. P., Lopes, D., & Kaplan, M. A. C. (1999). Anthelmintic activity of the latex of Ficus species. *Journal of Ethnopharmacology, 64*(3), 255–258.

Di Pierro, G., O'Keeffe, M. B., Poyarkov, A., Lomolino, G., & FitzGerald, R. J. (2014). Antioxidant activity of bovine casein hydrolysates produced by *Ficus carica* L.-derived proteinase. *Food Chemistry, 156*, 305–311.

Duke, J. A. (2002). *Handbook of medicinal herbs*. CRC Press.

Elsayed, A. M., Hegazy, U. M., Hegazy, M. G., Abdel-Ghany, S. S., Salama, W. H., Salem, A. M., & Fahmy, A. S. (2018). Purification and biochemical characterization of peroxidase isoenzymes from Ficus carica latex. *Biocatalysis and Agricultural Biotechnology, 16*, 1–9.

El-Seedi, H. R., El-Barbary, M., El-Ghorab, D., Bohlin, L., Borg-Karlson, A.-K., Goransson, U., & Verpoorte, R. (2010). Recent insights into the biosynthesis and biological activities of natural xanthones. *Current Medicinal Chemistry, 17*(9), 854–901.

Evert, R. F. (2006). *Esau's plant anatomy: Meristems, cells, and tissues of the plant body: Their structure, function, and development*. Wiley.

García-Alonso, M. a., de Pascual-Teresa, S., Santos-Buelga, C., & Rivas-Gonzalo, J. C. (2004). Evaluation of the antioxidant properties of fruits. *Food Chemistry, 84*(1), 13–18.

Ghanbari, A., Le Gresley, A., Naughton, D., Kuhnert, N., Sirbu, D., & Ashrafi, G. H. (2019). Biological activities of *Ficus carica* latex for potential therapeutics in Human Papillomavirus (HPV) related cervical cancers. *Scientific Reports, 9*(1), 1–11.

Ghandehari, F., & Fatemi, M. (2018). The effect of Ficus carica latex on 7, 12-dimethylbenz (a) anthracene-induced breast cancer in rats. *Avicenna Journal of Phytomedicine, 8*(4), 286.

Gupta, R., Gigras, P., Mohapatra, H., Goswami, V. K., & Chauhan, B. (2003). Microbial α-amylases: A biotechnological perspective. *Process Biochemistry, 38*(11), 1599–1616.

Hagel, J. M., Yeung, E. C., & Facchini, P. J. (2008). Got milk? The secret life of laticifers. *Trends in Plant Science, 13*(12), 631–639.

Hamed, M. B., El-Badry, M. O., Kandil, E. I., Borai, I. H., & Fahmy, A. S. (2020). A contradictory action of procoagulant ficin by a fibrinolytic serine protease from Egyptian *Ficus carica* latex. *Biotechnology Reports, 27*, e00492.

Hashemi, S., Abediankenari, S., Ghasemi, M., Azadbakht, M., Yousefzadeh, Y., & Dehpour, A. (2011). The effect of fig tree latex (*Ficus carica*) on stomach cancer line. *Iranian Red Crescent Medical Journal, 13*(4), 272.

Hemmatzadeh, F., Fatemi, A., & Amini, F. (2003). Therapeutic effects of fig tree latex on bovine papillomatosis. *Journal of Veterinary Medicine, Series B, 50*(10), 473–476.

Hilder, V. A., Gatehouse, A. M. R., Sheerman, S. E., Barker, R. F., & Boulter, D. (1987). A novel mechanism of insect resistance engineered into tobacco. *Nature, 330*(6144), 160–163.

Huynh, Q. K., Borgmeyer, J. R., & Zobel, J. F. (1992). Isolation and characterization of a 22 kDa protein with antifungal properties from maize seeds. *Biochemical and Biophysical Research Communications, 182*(1), 1–5.

Ippen, H. (1982). Phototoxic reaction to figs. *Der Hautarzt; Zeitschrift fur Dermatologie, Venerologie, und Verwandte Gebiete, 33*(6), 337–339.

Jeivad, F., Yassa, N., Ostad, S. N., Hassannejad, Z., Hassanzadeh Gheshlaghi, G., & Sabzevari, O. (2020). *Ficus Carica* L. latex: Possible chemo-preventive, apoptotic activity and safety assessment. *Iranian Journal of Pharmaceutical Research, 19*(3), 231–240.

Joseph, B., & Raj, S. J. (2011). Pharmacognostic and phytochemical properties of *Ficus carica* Linn-an overview. *International Journal of Pharmtech Research, 3*(1), 8–12.

Kang, H., Kim, Y. S., & Chung, G. C. (2000). Characterization of natural rubber biosynthesisin Ficus benghalensis. *Plant Physiology and Biochemistry, 38*(12), 979–987.

Karnchanatat, A., Tiengburanatam, N., Boonmee, A., Puthong, S., & Sangvanich, P. (2011). Zingipain, a cysteine protease from *Zingiber ottensii* valeton rhizomes with antiproliferative activities against fungi and human malignant cell lines. *Preparative Biochemistry & Biotechnology, 41*(2), 138–153.

Kim, J. S., Kim, Y. O., Ryu, H. J., Kwak, Y. S., Lee, J. Y., & Kang, H. (2003). Isolation of stress-related genes of rubber particles and latex in fig tree (*Ficus carica*) and their expressions by abiotic stress or plant hormone treatments. *Plant and Cell Physiology, 44*(4), 412–414.

Kim, Y. S., Park, S. J., Lee, E. J., Cerbo, R. M., Lee, S. M., Ryu, C. H., Kim, G. S., Kim, J. O., & Ha, Y. L. (2008). Antibacterial compounds from rose Bengal-sensitized photooxidation of β-caryophyllene. *Journal of Food Science, 73*(7), C540–C545.

Kirtikar, K., & Basu, B. D. (1935). *Indian medicinal plants*. Lalit Mohan Publication.

Kitajima, S., Taira, T., Oda, K., Yamato, K. T., Inukai, Y., & Hori, Y. (2012). Comparative study of gene expression and major proteins' function of laticifers in lignified and unlignified organs of mulberry. *Planta, 235*(3), 589–601.

Kitajima, S., Yamamoto, Y., Hirooka, K., Taki, C., & Hibino, S. (2013). Laticifers in mulberry exclusively accumulate defense proteins related to biotic stresses. *Plant Biotechnology, 30*(4), 399–402.

Kitajima, S., Aoki, W., Shibata, D., Nakajima, D., Sakurai, N., Yazaki, K., Munakata, R., Taira, T., Kobayashi, M., & Aburaya, S. (2018). Comparative multi-omics analysis reveals diverse latex-based defense strategies against pests among latex-producing organs of the fig tree (*Ficus carica*). *Planta, 247*(6), 1423–1438.

Konno, K. (2011). Plant latex and other exudates as plant defense systems: Roles of various defense chemicals and proteins contained therein. *Phytochemistry, 72*(13), 1510–1530.

Konno, K., Hirayama, C., Nakamura, M., Tateishi, K., Tamura, Y., Hattori, M., & Kohno, K. (2004). Papain protects papaya trees from herbivorous insects: Role of cysteine proteases in latex. *The Plant Journal, 37*(3), 370–378.

Lansky, E. P., Paavilainen, H. M., Pawlus, A. D., & Newman, R. A. (2008). Ficus spp. (fig): Ethnobotany and potential as anticancer and anti-inflammatory agents. *Journal of Ethnopharmacology, 119*(2), 195–213.

Lazreg Aref, H., Salah, K., Chaumont, J. P., Fekih, A., Aouni, M., & Said, K. (2010). In vitro antimicrobial activity of four *Ficus carica* latex fractions against resistant human pathogens (antimicrobial activity of *Ficus carica* latex). *Pakistan Journal of Pharmaceutical Sciences, 23*(1), 53–58.

Lazreg Aref, H., Gaaliche, B., Fekih, A., Mars, M., Aouni, M., Pierre Chaumon, J., & Said, K. (2011a). In vitro cytotoxic and antiviral activities of *Ficus carica* latex extracts. *Natural Product Research, 25*(3), 310–319.

Lazreg Aref, H., Karima, B. H. S., Fekih, A., Chemli, R., Mars, M., Aouni, M., Chaumon, J. P., & Said, K. (2011b). Variability in antimicrobial activity of latex from two varieties of *Ficus carica. African Journal of Microbiology Research, 5*(12), 1361–1367.

Lazreg Aref, H., Mosbah, H., Louati, H., Said, K., & Selmi, B. (2011c). Partial characterization of a novel amylase activity isolated from Tunisian *Ficus carica* latex. *Pharmaceutical Biology, 49*(11), 1158–1166.

Lazreg-Aref, H., Mars, M., Fekih, A., Aouni, M., & Said, K. (2012a). Chemical composition and antibacterial activity of a hexane extract of Tunisian caprifig latex from the unripe fruit of *Ficus carica. Pharmaceutical Biology, 50*(4), 407–412.

Lazreg-Aref, H., Mosbah, H., Fekih, A., Mars, M., & Said, K. (2012b). Purification and biochemical characterization of lipase from Ficus carica latex of Tunisian East Coast Zidi variety. *Journal of the American Oil Chemists' Society, 89*(10), 1847–1855.

Lazreg-Aref, H., Gaaliche, B., Ladhari, A., Hammami, M., & Hammami, S. (2018). Co-evolution of enzyme activities and latex in fig (*Ficus carica* L.) during fruit maturity process. *South African Journal of Botany, 115*, 143–152.

López-García, B., Hernández, M., & Segundo, B. S. (2012). Bromelain, a cysteine protease from pineapple (*Ananas comosus*) stem, is an inhibitor of fungal plant pathogens. *Letters in Applied Microbiology, 55*(1), 62–67.

Mabberley, D. J. (1997). *The plant-book: A portable dictionary of the vascular plants*. Cambridge University Press.

Martínez, G. J. (2008). Farmacopea natural y tratamiento de afecciones de la piel en la medicina tradicional de los campesinos de las sierras de Córdoba (República Argentina). *Dominguezia, 24*(1), 27–46.

Maruyama, S., Miyoshi, S., & Tanaka, H. (1989). Angiotensin I-converting enzyme inhibitors derived from *Ficus carica. Agricultural and Biological Chemistry, 53*(10), 2763–2767.

Mavlonov, G., Ubaidullaeva, K. A., Rakhmanov, M., Abdurakhmonov, I. Y., & Abdukarimov, A. (2008). Chitin-binding antifungal protein from *Ficus carica* latex. *Chemistry of Natural Compounds, 44*(2), 216–219.

Mawa, S., Husain, K., & Jantan, I. (2013). Ficus carica L.(Moraceae): phytochemistry, traditional uses and biological activities. *Evidence-Based Complementary and Alternative Medicine, 2013*, 974256.

Men, X., Wang, F., Chen, G.-Q., Zhang, H.-B., & Xian, M. (2018). Biosynthesis of natural rubber: Current state and perspectives. *International Journal of Molecular Sciences, 20*(1), 50.

Menichini, G., Alfano, C., Provenzano, E., Marrelli, M., Statti, G. A., Somma, F., Menichini, F., & Conforti, F. (2012). Fig latex (*Ficus carica* L. cultivar Dottato) in combination with UV irradiation decreases the viability of A375 melanoma cells in vitro. *Anti-Cancer Agents in Medicinal Chemistry- Anti-Cancer Agents, 12*(8), 959–965.

Mohammad, H., & Alzweiri, M. (2021). Phytochemistry and pharmacological activities of *Ficus carica* latex: A systematic review. *Journal of Chinese Pharmaceutical Sciences, 31*(2), 81–96.

Mostafaei, A., Mansouri, K., Norouznezhad, A., & Mohammadi, M. H. (2011). *Anti-angiogenic activity of Ficus carica latex extract on human umbilical vein endothelial cells*.

Oliveira, A. P., Silva, L. R., Andrade, P. B., Valentão, P., Silva, B. M., Gonçalves, R. F., Pereira, J. A., & Guedes de Pinho, P. (2010a). Further insight into the latex metabolite profile of *Ficus carica*. *Journal of Agricultural and Food Chemistry, 58*(20), 10855–10863.

Oliveira, A. P., Silva, L. R., Ferreres, F., Guedes de Pinho, P., Valentão, P., Silva, B. M., Pereira, J. A., & Andrade, P. B. (2010b). Chemical assessment and in vitro antioxidant capacity of *Ficus carica* latex. *Journal of Agricultural and Food Chemistry, 58*(6), 3393–3398.

Paşayeva, L., Özalp, B., & Fatullayev, H. (2020). Potential enzyme inhibitory properties of extracts and fractions from fruit latex of *Ficus carica*-based on inhibition of α-amylase and α-glucosidase. *Journal of Food Measurement and Characterization, 14*(5), 2819–2827.

Pawlus, A., Cartwright, C., Vijjeswarapu, M., Liu, Z., Woltering, E., & Newman, R. (2008). Anti-angiogenic activity from the fruit latex of *Ficus carica* (Fig). *Planta Medica, 74*(09), PA97.

Rashid, K. I., Mahdi, N. M., Alwan, M. A., & Khalid, L. B. (2014). Antimicrobial activity of fig (*Ficus carica* Linn.) leaf extract as compared with latex extract against selected bacteria and fungi. *Journal of Babylon University/Pure and Applied Sciences, 5*(22), 1620–1626.

Raskovic, B., Bozovic, O., Prodanovic, R., Niketic, V., & Polovic, N. (2014). Identification, purification and characterization of a novel collagenolytic serine protease from fig (*Ficus carica* var. Brown Turkey) latex. *Journal of Bioscience and Bioengineering, 118*(6), 622–627.

Raskovic, B., Lazic, J., & Polovic, N. (2016). Characterisation of general proteolytic, milk clotting and antifungal activity of *Ficus carica* latex during fruit ripening. *Journal of the Science of Food and Agriculture, 96*(2), 576–582.

Richter, G., Schwarz, H. P., Dorner, F., & Turecek, P. L. (2002). Activation and inactivation of human factor X by proteases derived from *Ficus carica*. *British Journal of Haematology, 119*(4), 1042–1051.

Robbins, B. H. (1935). Proteolytic enzyme content of latex from the fig tree (Ficus Carica L'). Seasonal variation. *Proceedings of the Society for Experimental Biology and Medicine, 32*(6), 892–893.

Santos, A., & Van Ree, R. (2011). Profilins: Mimickers of allergy or relevant allergens? *International Archives of Allergy and Immunology, 155*(3), 191–204.

Sattari, F., Rigi, G., & Ghaedmohammadi, S. (2020). The first report on molecular cloning, functional expression, purification, and statistical optimization of *Escherichia coli*-derived recombinant Ficin from Iranian fig tree (*Ficus carica* cv. Sabz). *International Journal of Biological Macromolecules, 165*, 2126–2135.

Shin, B. S., Lee, S. A., Moon, S. M., Han, S. H., Hwang, E. J., Kim, S.-G., Kim, D. K., Kim, J.-S., Park, B.-R., & Kim, C. S. (2017). Latex of *Ficus carica* L. induces apoptosis through caspase and Bcl-2 family in FaDu human hypopharynx squamous carcinoma cells. *International Journal of Oral Biology, 42*(4), 183–190.

Shlash, D. S. A., Neama, M. S. N. A., & Alfatlawy, D. E. T. N. (2021). Efficiency of Ficus carica latex in therapy of Leishmania tropica in patients. *International Journal of Pharmaceutical Research, 13*(1), 348–352.

Soltana, H., Pinon, A., Limami, Y., Zaid, Y., Khalki, L., Zaid, N., Salah, D., Sabitaliyevich, U. Y., Simon, A., & Liagre, B. (2019). Antitumoral activity of *Ficus carica* L. on colorectal cancer cell lines. *Cellular and Molecular Biology, 65*(6), 6–11.

Terras, F. R. G., Schoofs, H. M. E., Thevissen, K., Osborn, R. W., Vanderleyden, J., Cammue, B. P. A., & Broekaert, W. F. (1993). Synergistic enhancement of the antifungal activity of wheat and barley thionins by radish and oilseed rape 2S albumins and by barley trypsin inhibitors. *Plant Physiology, 103*(4), 1311–1319.

Tezcan, G., Tunca, B., Bekar, A., Yalcin, M., Sahin, S., Budak, F., Cecener, G., Egeli, U., Demir, C., Guvenc, G., Yilmaz, G., Erkan, L. G., Malyer, H., Taskapilioglu, M. O., Evrensel, T., & Bilir, A. (2015). *Ficus carica* latex prevents invasion through induction of let-7d expression in GBM cell lines. *Cellular and Molecular Neurobiology, 35*(2), 175–187.

Upadhyay, R. K. (2011). Plant latex: A natural source of pharmaceuticals and pesticides. *International Journal of Green Pharmacy (IJGP), 5*(3), 169.

Vatić, S., Mirković, N., Milošević, J. R., Jovčić, B., & Polović, N. Đ. (2020). Broad range of substrate specificities in papain and fig latex enzymes preparations improve enumeration of Listeria monocytogenes. *International Journal of Food Microbiology, 334*, 108851.

Wang, G., Tang, W., & Bidigare, R. R. (2005). *Terpenoids as therapeutic drugs and pharmaceutical agents. Natural products* (pp. 197–227). Springer.

Wang, J., Wang, X., Jiang, S., Lin, P., Zhang, J., Lu, Y., Wang, Q., Xiong, Z., Wu, Y., & Ren, J. (2008). Cytotoxicity of fig fruit latex against human cancer cells. *Food and Chemical Toxicology, 46*(3), 1025–1033.

Werbach, M. (1993). *Healing with food: 281 nutritional healing plans for 50 common ailments*. HarperPerennial, A Division of Harper Collins Publishers.

Yahiaoui, S., Kati, D.-E., Ali, L. M., El Cheikh, K., Morére, A., Menut, C., Bachir-bey, M., & Bettache, N. (2022). Assessment of antioxidant, antiproliferative, anti-inflammatory, and enzyme inhibition activities and UPLC-MS phenolic determination of *Ficus carica* latex. *Industrial Crops and Products, 178*, 114629.

Zhai, Y., Cui, Y., Song, M., Vainstein, A., Chen, S., & Ma, H. (2021). Papain-like cysteine protease gene family in fig (*Ficus carica* L.): Genome-wide analysis and expression patterns. *Frontiers in Plant Science, 12*, 994.

Chapter 28
Extraction and Analysis of Polyphenolic Compounds in *Ficus carica* L.

Babra Moyo and Nikita T. Tavengwa

1 Introduction

Ficus carica L, a plant that belongs to the Moraceae family, is one of the most widely farmed plants in the world because of its richness in phytochemicals that can have potential applications in pharmaceutical, cosmetic, and food industries (Amessis-Ouchemoukh et al., 2017; Khadhraoui et al., 2019; Maghsoudlou et al., 2017). This species is distinguished by a variety of chemical components which have been beneficial to human health for millennia (Ammar et al., 2015; Bachir Bey et al., 2017; Mopuri et al., 2018). Figs are mainly recognized for their medicinal characteristics, such as their usage in traditional medicine to treat cardiovascular (Alamgeer et al., 2017) and anti-inflammatory problems (Allahyari et al., 2014; Amessis-Ouchemoukh et al., 2017).

Phenolics, including phenolic acids and flavonoids, are responsible for the health benefits of these compounds found in various fruits and vegetables, including figs (Arvaniti et al., 2019; Nakilcioğlu-Taş & Ötleş, 2021). Interest in extracting phenolic compounds from plant materials such as fruits and vegetables has grown in the past years as demand for natural antioxidants increases. Therefore, there is a need to develop efficient extraction methods that can be used to isolate these compounds from plant materials.

B. Moyo
Department of Biochemistry and Microbiology, Faculty of Science, Engineering and Agriculture, University of Venda, Thohoyandou, South Africa

Department of Food Science and Technology, Faculty of Science, Engineering and Agriculture, University of Venda, Thohoyandou, South Africa
e-mail: barbara.moyo@univen.ac.za

N. T. Tavengwa (✉)
Department of Chemistry, Faculty of Science, Engineering and Agriculture, University of Venda, Thohoyandou, South Africa
e-mail: nikita.tavengwa@univen.ac.za

M. F. Ramadan (ed.), *Fig (Ficus carica): Production, Processing, and Properties*, https://doi.org/10.1007/978-3-031-16493-4_28

The structural diversity of secondary metabolites coupled with the high complexity of the plant metabolome poses a challenge in selecting a suitable extraction technique for the pre-concentration of these compounds. This implies that an inefficient extraction or a single-step method might influence the isolation of these compounds from the plant samples. As a result, it is critical to select a suitable extraction technique to isolate structurally diverse phytocompounds (Alara et al., 2021). Solvent extraction is the commonly used technique for extracting fig crude extracts (Ersoy et al., 2017; Konak et al., 2017; Meziant et al., 2018; Pereira et al., 2017). Selecting a suitable solvent for this extraction technique is of uttermost importance since the amount and the nature of phytochemicals extracted will depend on the extraction solvent. Therefore, the physicochemical properties of the analytes of interest and the analysis instrument should be considered (Choi et al., 2021; Kang et al., 2019). Other methods that have been applied in the past years for the extraction of various polyphenols from various plant samples include pressurized liquid extraction (Barrales et al., 2018; Benito-Román et al., 2015), ultrasound-assisted extraction (Bakirtzi et al., 2016; Bosiljkov et al., 2017; Um et al., 2018) and microwave-assisted extraction (Alara et al., 2018; Sarfarazi et al., 2020). However, the application of these extraction techniques in figs is still limited.

Although few reviews have been published on the extraction of polyphenols from figs, to the best of our knowledge, none have reviewed extraction methods in detail. Arvaniti et al. (2019) reviewed the extraction of phytocompounds from figs using different techniques such as solid-liquid extraction (SLE), solid-phase extraction, and subcritical water extraction, to mention a few. However, the principles of these methods were not discussed in depth. In a review authored by Solana and Romano (2019), the analysis of polyphenols in beverages produced from figs using colorimetric assays. However, in this review, extraction methods were not discussed in detail, given that sample preparation is a paramount step in analyzing phytochemicals. In this context, the application of conventional SLE in extracting polyphenols from fig fruits was discussed in this chapter. In addition, non-conventional extraction based on SLE, such as subcritical water extraction and microwave-assisted extraction, were discussed briefly, given that their application in this particular matrix is still limited.

1.1 Nutritional and Health Benefits of Ficus carica *L.*

The fig (*Ficus carica* L.) is one of the earliest fruit trees cultivated globally (Ersoy et al., 2017). It is a popular crop worldwide and is consumed widely in Mediterranean diets (Alexandre et al., 2017; Barolo et al., 2014; Harzallah et al., 2016). The different parts of the fig plant are used for food and therapeutic purposes (Ammar et al., 2015; Debib et al., 2016). Fig fruits have long been used in traditional medicine for their laxative, pulmonary, antispasmodic, and anti-inflammatory properties (Barolo et al., 2014; Nadeem & Zeb, 2018; Rodríguez-Solana et al., 2018; Viuda-Martos et al., 2015). These health-promoting benefits are primarily because figs contain significant levels of polyphenols, which have antioxidant properties and can help prevent diseases caused by oxidative stress (Amessis-Ouchemoukh et al., 2017;

Bahrin et al., 2018; Harzallah et al., 2016; Nakilcioğlu-Taş & Ötleş, 2021). These compounds help reduce cellular damage caused by reactive species such as hydrogen peroxide while acting as scavengers for free radicals produced by oxygen or nitrogen (Amessis-Ouchemoukh et al., 2017; Solana & Romano, 2019). Moreover, figs are high in crude fiber, minerals, and vitamins, and they are lipid, sodium, and cholesterol free (Buenrostro-Figueroa et al., 2017; Nakilcioğlu-Taş & Ötleş, 2021).

The pharmaceutical industry is increasingly becoming interested in natural products to produce drugs as alternatives to synthetic drugs (Barolo et al., 2014). For this reason, the therapeutic qualities of the genus *Ficus* have been widely explored during the last few decades through biological activity tests. Antibacterial (Benmaghnia et al., 2021; Palmeira et al., 2019), antidiabetic (Mopuri et al., 2018; Wojdyło et al., 2016), anticonvulsant (Raafat & Wurglics, 2019), and hepatoprotection (Debib et al., 2016) properties are some of the biological activities that have been reported for *Ficus carica* L. The by-products resulting from fig processing, such as the fig skin, have been reported to possess high quantities of polyphenol compounds (Cheynier et al., 2013; Harzallah et al., 2016; Kamiloglu & Capanoglu, 2015). Dark-colored figs have been reported to have high polyphenols due to their higher anthocyanin content (Kamiloglu & Capanoglu, 2015; Maghsoudlou et al., 2017; Meziant et al., 2018). Therefore, it is advisable to consume the whole fruit as the skin has more phenolic compounds. Fig seeds are also thought to play a significant role in the nutrient content and health benefits of the fruit (Nakilcioğlu-Taş & Ötleş, 2021).

1.2 Plant Secondary Metabolites

Secondary metabolites are a diverse group of naturally occurring compounds synthesized by plants for their reproduction and defense against biotic (viruses, bacteria, fungi, and vertebrates) and abiotic stress (UV light, drought, contamination with heavy metals, and temperature changes) (Khare et al., 2020; Tungmunnithum et al., 2018). These compounds have a high potential for use as drugs by public healthcare systems (Boy et al., 2018; Choi et al., 2021). Secondary metabolites are classified according to their biosynthetic pathway, chemical structure, composition, or solubility (Soukup & Soukup, 2015; Thirumurugan et al., 2018). Alkaloids, terpenes, and phenolics are the major compounds produced by plants (Kabera et al., 2014; Katz & Baltz, 2016; Alamgeer et al., 2017). However, only polyphenols will be discussed in this paper as they are widely studied in fig fruits.

1.2.1 Phenolics

Secondary metabolites known as phenolics are structurally diverse compounds that are found in most plant species, including *Ficus carica* L. (Ammar et al., 2015; Nakilcioğlu-Taş & Ötleş, 2021). They can exist as aglycones, glycosides, polymerized or monomer structures, and matrix or free-bound molecules (Alara et al., 2021). Phenolics are the most abundant group of secondary metabolites found in figs, and they are responsible for the color, taste, and flavor of these fruits (Ammar et al., 2015; Garagounis et al., 2021; Vaishnav & Demain, 2011). The presence of one or more phenol groups distinguishes them, and they range in complexity from basic molecules with one aromatic ring to highly complex polymeric compounds (Pandey & Rizvi, 2009). Some phenolics found in figs, like quercetin and naringenin, are pharmacologically essential compounds because of their anti-inflammatory properties (Cheng et al., 2019; Karuppagounder et al., 2016). Genistein, also found in figs, possesses phytoestrogenic activity (Farmer et al., 2016; Morales-Otal et al., 2016). In addition, various phenolic compounds, like flavonoids, are potent antioxidants because of their free radical scavenging effect (Panche et al., 2016; Rodríguez-García et al., 2019). Phenolics are classified according to their biosynthetic pathway or structure into simple phenolic acids, flavonoids, tannins, coumarins, lignans, and stilbenes, depending on the number of phenol rings that they contain and the moieties that bind these rings (Abegaz & Kinfe, 2019; Hussein & El-Anssary, 2019). The two types of phenolic acids are hydroxycinnamic and hydroxybenzoic acids, and these have been reported in figs (Ammar et al., 2015; Nakilcioğlu-Taş & Ötleş, 2021). An example of simple benzoyl phenolic acid is gallic acid, a parent compound of gallotannins, whereas *p*-coumaric, caffeic, and ferulic acids are examples of hydroxycinnamic acids (Saibabu et al., 2015). Ammar et al. (2015) reported the presence of simple phenolic acids in figs, including gallic, vanillic, caffeic, and ferulic acids.

Chalcones, flavans, flavanols, flavanone, flavone, isoflavone, anthocyanidins, and xanthones are the different types of flavonoids. Hundreds of compounds in each group vary in the number and positioning of hydroxyl groups on the flavan scaffold (Abegaz & Kinfe, 2019). Anthocyanins are responsible for the coloration of figs, and these compounds can be used as food colorants (Ammar et al., 2015; Backes et al., 2018). Quercetin is a flavonol that is one of nature's most effective antioxidants and a key component in figs (Maghsoudlou et al., 2017; Wojdyło et al., 2016). Tannins exist as hydrolyzable tannins and condensed tannins. Hydrolyzable tannins consist of several molecules of phenolic acids, such as gallic and hexahydroxydiphenic acids, attached to a central glucose molecule through an ester bond (Durazzo et al., 2019).

On the other hand, condensed tannins, or proanthocyanidins, are compounds with structures based on oligomeric flavonoid precursors that differ in the type of linkage between flavonoid units (Fraga-Corral et al., 2020). However, few studies have investigated the presence of tannins in figs (Alexandre et al., 2017). Coumarins are derivatives of benzo-α-pyrone, which is an *o*-hydroxycinnamic acid lactone (Camara et al., 2020). Examples of coumarins detected in fig leaves and fruits include umbellifone, psoralen, and methoxypsolaren (Ammar et al., 2015; Qin et al., 2015). Figure 28.1 shows the general chemical structures of the common polyphenols found in figs.

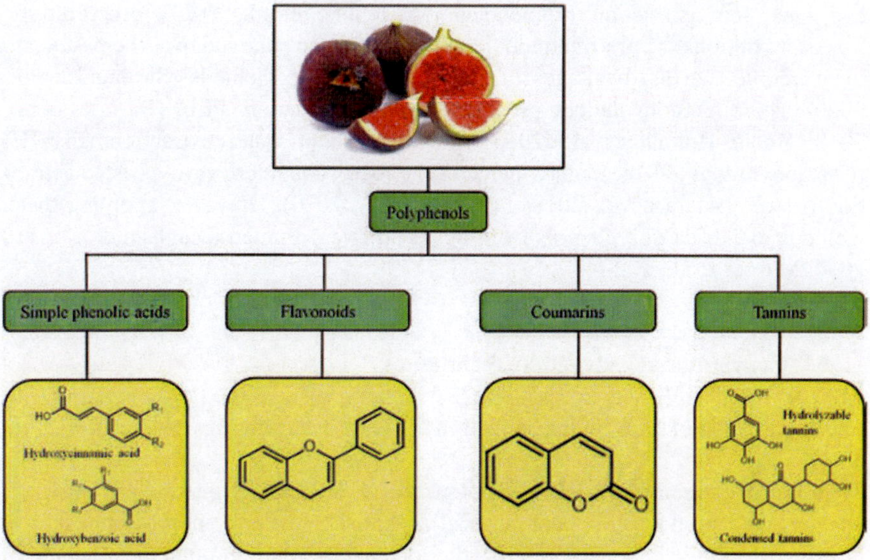

Fig. 28.1 Chemical structures of the different classes of polyphenols

1.3 Extraction Methods of Polyphenols from Figs

Extraction of polyphenols from figs is the first step in removing interferences and isolating these compounds from this matrix (Hikmawanti et al., 2021). The downstream analysis only identifies metabolites previously extracted; therefore, developing a proper sample preparation technique is critical for obtaining accurate results (Ibañez et al., 2012). Developing efficient sample preparation techniques in plant metabolomics is particularly challenging due to the complexity of plant biological systems, making it challenging to determine target analytes reliably (Mastellone et al., 2021). Moreover, given their occurrence in low concentrations in plant matrices, it is critical to adopt extraction methods that enable the efficient recovery of these compounds from bulk matrices while also removing matrix interferences (Câmara et al., 2020). In the last few decades, detection technologies have grown at an unparalleled rate (Aron et al., 2020; Nothias et al., 2020; Stefanucci et al., 2020). Phenomenal developments have been made in high-resolution analytical techniques, such as chromatography and spectroscopy, making phytochemical analysis much easier than before. Nevertheless, sample preparation is still a bottleneck in analyzing these compounds (Moyo & Tavengwa, 2021a). Moreover, there is no widely established method for recovering all phenolics or those belonging to a specific class from plant materials, including figs. As a result, an optimized procedure for recovering phenolic compounds from plant sources is required (Alara et al., 2021). The most widely used traditional extraction methods, including liquid-liquid extraction (LLE), solid-liquid extraction (SLE), Soxhlet extraction, and hydro-distillation, are time-consuming, labor-intensive, consume large amounts of organic solvents and have low extraction efficiencies (Mastellone et al., 2021). Accordingly, "greener" options for pre-treatment plant samples have emerged over the past years to overcome the disadvantages of classical extraction methods (Sarfarazi et al., 2020). These methods include pressurized liquid extraction (PLE) (Barrales et al., 2018; Benito-Román et al., 2015), pressurized hot water extraction (PHWE) (Cvetanović et al., 2018; Lachos-Perez et al., 2020), and microwave-assisted extraction (MAE) (Alara et al., 2018; Sarfarazi et al., 2020). However, applying these methods to extract polyphenols from figs is still very limited (Maghsoudlou et al., 2017; Masota et al., 2020).

1.3.1 Conventional Extraction Techniques

The Application of SLE in the Extraction of Phenolics from Figs

Solid-liquid extraction (SLE), a technique where analytes of interest are extracted using a liquid from a solid (plant material), is based on the principles of osmosis and diffusion (Naviglio et al., 2019). First, the solvent penetrates through the plant material matrix, and analytes of interest (polyphenols) are released into the solvent through osmosis (Zhang et al., 2018). Subsequently, polyphenols disperse into the solvent via diffusion (Milescu et al., 2020). Figure 28.2 shows the general SLE procedure, including the most reported sample pre-treatment steps from studies reviewed in this chapter. Although SLE has been known for a long time, it is still widely employed to extract phytochemicals from various plant materials, including figs. Notwithstanding the simplicity of the SLE technique, it has drawbacks such as the consumption of large volumes of organic solvents, inadequate extraction efficiency, chemical transformation of extracts, and lengthy extraction times (Alexandre et al., 2017; Amessis-Ouchemoukh et al., 2017; Mopuri et al., 2018; Vernès et al., 2020). Furthermore, the high concentration of free sugars and other matrices can challenge analyzing phenolic compounds from figs using this technique (Arvaniti et al., 2019).

Despite the mentioned challenges associated with SLE, it is notable that it is still the most widely utilized technique for extracting polyphenols from the whole fruit as well as other portions of figs, such as the peel and pulp, as given in Table 28.1 (Alamgeer et al., 2017; Amessis-Ouchemoukh et al., 2017; Ammar et al., 2015; Bachir Bey et al., 2017; Bachir Bey & Louaileche, 2015; Benmaghnia et al., 2021; Debib et al., 2016; Ersoy et al., 2017; Harzallah et al., 2016; Hoxha et al., 2015; Hoxha & Kongoli, 2016; Kamiloglu & Capanoglu, 2015; Khadhraoui et al., 2019; Konak et al., 2017; Mahmoudi et al., 2018; Meziant et al., 2018; Mopuri et al., 2018; Nakilcioğlu-Taş & Ötleş, 2021; Pereira et al., 2017; Viuda-Martos et al., 2015; Wojdyło et al., 2016). Prior to the SLE procedure, sample pre-treatment steps that have been reported in the studies reviewed in this chapter involve drying the figs using different methods of drying, such as shade drying (Maghsoudlou et al., 2017), freeze-drying (Bachir Bey et al., 2017; Wojdyło et al., 2016) and sun drying (Kamiloglu & Capanoglu, 2015; Khadhraoui et al., 2019; Konak et al., 2017). Dried fig samples were then ground to increase the contact area between the sample and the solvent (Njila et al., 2017), facilitating the analytes' mass transfer. In some

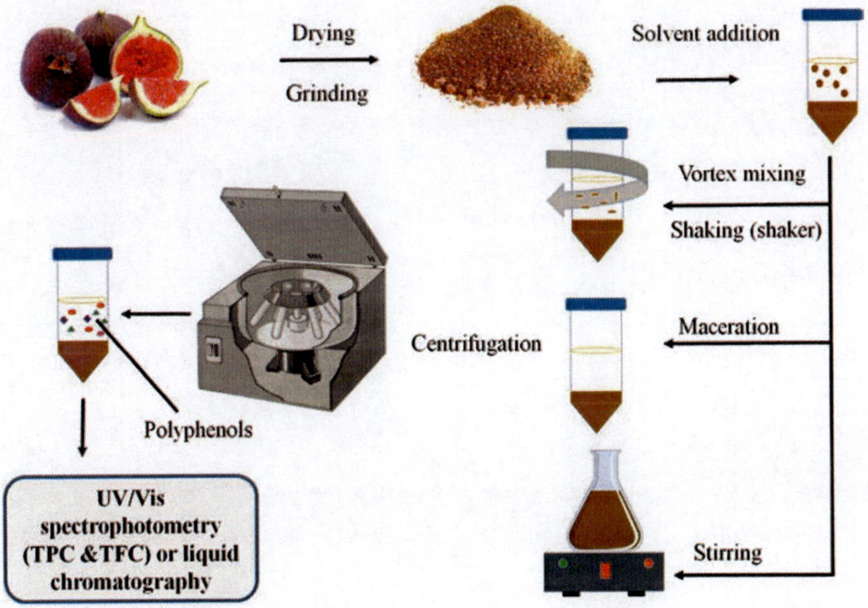

Fig. 28.2 General SLE procedure

studies, fresh fig samples were homogenized with the solvent (Ersoy et al., 2017). After extraction, samples were centrifuged to remove suspended solids (Alexandre et al., 2017; Amessis-Ouchemoukh et al., 2017; Nakilcioğlu-Taş & Ötleş, 2021). Table 28.1 summarizes the application of different extraction methods to isolate polyphenols from figs. Even though there is no universally accepted optimal solvent for polyphenol extraction, it is often assumed that solvents with higher polarity are suitable due to the increased solubility of polyphenols in such solvents (Alara et al., 2021). The varying physicochemical properties of polyphenols significantly impact the solubility of these compounds in solvents. Methanol (Bachir Bey et al., 2017; Harzallah et al., 2016; Mahmoudi et al., 2018), ethanol (Amessis-Ouchemoukh et al., 2017; Konak et al., 2017; Mopuri et al., 2018), and acetone (Pereira et al., 2017) are the most commonly used solvents for extracting polyphenols from figs. Different compositions of ethanol (Alexandre et al., 2017) and methanol, particularly 80% methanol (Ammar et al., 2015; Benmaghnia et al., 2021; Hoxha et al., 2015; Khadhraoui et al., 2019; Viuda-Martos et al., 2015), were also widely used as extraction solvents. The selection of the solvent is crucial for solvent extraction methods. Nonetheless, few studies were reviewed in this chapter that investigated the effect of different solvents in extracting polyphenols (Backes et al., 2018; Meziant et al., 2018; Mopuri et al., 2018). Extraction methods can be improved by optimizing parameters that may affect the extraction efficiency of the analytes of interest. Acidified organic solvents have also been recommended to prevent the degradation of polyphenols (Meziant et al., 2018). Some of the acids that have been reported include formic acid (Kamiloglu & Capanoglu, 2015; Pereira et al., 2017), hydrochloric acid (Pereira et al., 2017), ascorbic acid (Wojdyło et al., 2016),

Table 28.1 Application of different extraction methods in the analysis of polyphenols in figs

Matrix	Polyphenols analyzed	Extraction method	Extraction solvent	Extraction procedure after solvent addition	Method of detection	Concentration	References
Dried figs	TPC and TFC	SLE	Ethanol	Maceration and samples were stirred for 24 h at room temperature	UV/Vis spectrophotometry	2190 mg GAE 100 g^{-1} (TPC) 858 mg QE 100 g^{-1} (TFC)	Amessis-Ouchemoukh et al. (2017)
Fermented fig by-products	TPC, TFC and anthocyanidins	High pressure assisted extraction	9% ethanol (TPC), 47% ethanol (TFC)	High pressure assisted extraction	UV/Vis spectrophotometry (TPC and TFC) and UHPLC-PDA (phenolic acids)	–	Alexandre et al. (2017)
Fig seeds	TPC, TFC and 5 individual polyphenols	SLE	50% methanol (TPC, TFC and polyphenols)	The sample mixture was shaken at 50 °C for 90 min in a shaking water bath	UV/Vis spectrophotometry (TPC and TFC) HPLC-DAD (individual polyphenols)	714 mg GAE kg^{-1} (TPC) 312 mg CE kg^{-1} (TFC)	Nakilcioğlu-Taş and Ötleş (2021)
Fresh figs	TPC	SLE	50% acetone and 80% methanol	Homogenization of fresh samples with the solvent in a mixer	UV/Vis spectrophotometry	25.4–37.2 mg GAE g^{-1} (TPC)	Ersoy et al. (2017)
Fig juice	TPC and TFC	SLE	Methanol	Homogenization of the samples with the solvent in a homogenizer	UV/Vis spectrophotometry	50.5–74.1 mg GAE g^{-1}(TPC) 11.3– 12.7 mg CE kg^{-1} (TFC)	Harzallah et al. (2016)
Dried figs	14 phenolic acids and 8 flavonoids	SLE	Methanol	Ultrasonication for 1 h 45 min	HPLC-FLD	0.13–73 mg 100 g^{-1}	Bachir Bey et al. (2017)
Dried figs	TPC and TFC	SLE	80% methanol	Extraction on an orbital shaker for 1 h	UV/Vis spectrophotometry and LC-ESI-MS (polyphenols)	73.7–201 mg GAE g^{-1} (TPC) 57.9– 112 mg CE kg^{-1} (TFC)	Khadhraoui et al. (2019)

Dried figs	TPC	SLE	Ethanol	The sample was homogenized for 4 min	UV/Vis spectrophotometry	55.5–574 mg GAE g^{-1}	Konak et al. (2017)
Fig pulp and skin	TPC and TFC	SWE	Water	Extraction at 160 °C and at a pressure of 6.89 MPa for 30 min	UV/Vis spectrophotometry	–	Maghsoudlou et al. (2017)
Fig pulp, skin and leaves	91 phenolic compounds (profiling) and TPC	SLE	80% methanol	Ultrasonication	UHPLC-QTOF-MS (profiling) and UV/Vis spectrophotometry (TPC)	0.26–0.32 mg GAE 100 g^{-1} (TPC)	Ammar et al. (2015)
Dried figs	TPC	SLE	80% methanol containing 1% trifluoracetic acid	Maceration in the solvent at 4 °C for 16 h	UV/Vis spectrophotometry	700 mg GAE 100 kg^{-1}	Debib et al. (2015)
Dried figs	TPC and TFC	SLE	80% methanol containing 0.1% formic acid	Ultrasonication at low temperatures for 15 min	UV/Vis spectrophotometry	167–930 mg GAE 100 g^{-1} (TPC) 7–234 mg CE 100 g^{-1} (TFC)	Kamiloglu and Capanoglu (2015)
Dried fig peels	Monomeric anthocyanins	SLE	80% methanol containing 0.1% citric acid	Stirring at room temperature on a magnetic stirrer for 180 min	UV/Vis spectrophotometry	346 mg 100 g^{-1}	Meziant et al. (2018)
Dried figs	TPC and TFC	SLE	Ethanol	Extraction in the solvent for 48–72 h	UV/Vis spectrophotometry	14.6 mg GAE g^{-1} (TPC) 81.6 mg QE g^{-1} (TFC)	Mopuri et al. (2018)
Fig pulp and peel	TPC and TFC	SLE	80% ethanol containing 1% hydrochloric acid	Stirring at room temperature on a magnetic stirrer for 20 min	UV/Vis spectrophotometry	26.7–169 mg GAE 100 g^{-1} (TPC) 81.6 mg QE 100 g^{-1} (TFC)	Pereira et al. (2017)

(continued)

Table 28.1 (continued)

Fig pulp and peal	TPC and TFC	SLE	80% methanol	The sample was shaken for 2 min, followed by ultrasonication for 2 h at room temperature	UV/Vis spectrophotometry	1.92–5.76 mg GAE g^{-1} (TPC) 9.24–19.1 mg RE g^{-1} (TFC)	Viuda-Martos et al. (2015)
Dried figs	10 polyphenols	SLE	30% methanol acidified with 1% ascorbic acid	Ultrasonication for 15 min	UHPLC-QTOF-MS	0.5–328 mg 100 g^{-1}	Wojdyło et al. (2016)
Dried figs	TPC	SLE	80% methanol	Vortex mixing for 15 mins	UV/Vis spectrophotometry	45.2–160 mg GAE 100 g^{-1}	Hoxha et al. (2015)
Fig pulp and peel	TPC and TFC	SLE	80% ethanol containing 2% hydrochloric acid	Vortex mixing for 15 min	UV/Vis spectrophotometry	56.3–285 mg GAE 100 g^{-1} (TPC) and 42.47–269.54 mg CE g^{-1} (TFC)	Hoxha and Kongoli (2016)
Dried figs	TPC and 4 individual polyphenols	Aqueous two phase extraction	Ethanol and dipotassium hydrogen phosphate solution	Stirring on a magnetic stirrer	UV/Vis spectrophotometry (TPC) and micellar electrokinetic chromatography	22.5–288 μg g^{-1} (polyphenols)	Feng et al. (2015)
Dried figs	TPC and TFC	SLE	70% methanol	Maceration	UV/Vis spectrophotometry (TPC and TFC) and HPLC-UV (6 polyphenols)	31.8 mg GAE 100 g^{-1} (TPC) and 538 mg QE 100 g^{-1} (TFC)	Alamgeer et al. (2017)
Dried figs	TPC and TFC	SLE	60% acetone	The sample mixture was shaken at 40 °C for 2 h in a shaking water bath	UV/Vis spectrophotometry	482–644 mg GAE 100 g^{-1} 87.2–126 mg QE 100 g^{-1}	Bachir Bey and Louaileche (2015)
Fig peels	Cyanidin 3-rutinoside	Ultrasound assisted extraction	Citric acid acidified 100% ethanol	The sample was ultrasonicated at 30-35 °C for approximately 22 min	HPLC-DAD-ESI/MS	–	Backes et al. (2018)

Dried figs	TPC and TFC	SLE	80% methanol (TPC) and 50% acetone (TFC)	Samples were macerated under agitation for 24 h	UV/Vis spectrophotometry	78–458 mg GAE 100 g^{-1} 38.8–228 mg CE 100 g^{-1}	Benmaghnia et al. (2021)
Fig pulp and peels	TPC and TFC	SLE	100% methanol	Maceration at room temperature for 24 h	UV/Vis spectrophotometry	0.59–3.82 mg GAE g^{-1} 149–254 mg QE g^{-1}	Mahmoudi et al. (2018)
Fig pulp and peels	5 phenolic compounds	Ionic liquid-based ultrasonic-assisted extraction	1-butyl-3-methylimidazolium hexafluorophosphate	Ultrasonication for 30 min at 30 °C	HPLC-DAD	1.99–133 μg 0.1 g^{-1}	Qin et al. (2015)

– Not given

trifluoroacetic acid (Debib et al., 2016) and citric acid (Backes et al., 2018; Meziant et al., 2018).

In order to facilitate the mass transfer of the polyphenols to the solvent when using SLE, some studies have employed maceration (Alamgeer et al., 2017; Amessis-Ouchemoukh et al., 2017; Benmaghnia et al., 2021; Mahmoudi et al., 2018), vortex mixing (Hoxha et al., 2015; Hoxha & Kongoli, 2016), magnetic stirring (Meziant et al., 2018; Pereira et al., 2017) and shaking on a shaker (Bachir Bey & Louaileche, 2015; Khadhraoui et al., 2019; Nakilcioğlu-Taş & Ötleş, 2021) (Fig. 28.2). It is noteworthy that the extraction times of these mechanical means of facilitating extraction varied. Although maceration is a simple method, long extraction times and low extraction efficiencies are its main disadvantages (Amessis-Ouchemoukh et al., 2017; Debib et al., 2016). Moreover, the sample needs to be shaken occasionally to facilitate the mass transfer in this procedure (Njila et al., 2017). Ultrasound-assisted SLE was the commonly used method to facilitate extraction in most studies (Ammar et al., 2015; Bachir Bey et al., 2017; Kamiloglu & Capanoglu, 2015; Viuda-Martos et al., 2015; Wojdyło et al., 2016). Ultrasound radiation can improve the mass transfer of analytes from the matrix by breaking the plant material's cellular walls, subsequently allowing the solvent to penetrate the matrix (Chen et al., 2007). Furthermore, ultrasonication reduces the matrix particle size, allowing improved interaction between the solvent and the sample. Consequently, improving mass transfer results in higher extraction yields in shorter times (Um et al., 2018). This procedure can also reduce the degradation of thermolabile phytochemicals during long extraction processes (Backes et al., 2018).

It is noteworthy that most of the studies reviewed in this paper analyzed total phenolic content (TPC) and total flavonoid content (TFC). However, these assays do not give information on the individual polyphenols in the samples. Therefore, to fully understand the health benefits of figs, the first step is to uncover the phytochemicals that they contain using more sensitive analytical instruments such as liquid chromatography. Moreover, the quantitative analysis of polyphenols from figs by liquid chromatography is still restricted to a few selected compounds (Alexandre et al., 2017; Nakilcioğlu-Taş & Ötleş, 2021; Wojdyło et al., 2016).

1.3.2 Modern SLE Based Extraction Methods

Alternative methods to conventional methods, such as high pressure-assisted extraction, have emerged in the past years to overcome the challenges associated with traditional techniques for extracting polyphenols from figs (Alexandre et al., 2017). The main advantage of this technique is that extraction is done at room temperature hence preventing the thermal degradation of thermo-labile phytochemical compounds. Subcritical assisted extraction (SWE) has also been employed to extract TPC and TFC. In a study done by Maghsoudlou et al. (2017), SWE was employed to extract TPC and TFC from fig pulp and skin. As a polar solvent that is inexpensive, available in abundance, and environmentally friendly, water was used in this study. SWE offers significant advantages over other extraction techniques, including greater quality extracts, short extraction times, a lower volume of solvents needed, and a low operation cost (Ko et al., 2020). In a study done by Feng et al. (2015), an aqueous two-phase system was used to recover and partially purify phenolic compounds from fig fruits in a single step. The superiority of this method lies in the minimization of separation stages and high purification of natural products compared to conventional extraction methods (de Oliveira et al., 2018). Pressurized MAE was used for extracting bioactive molecules from *Ficus carica* L. bark (Masota et al., 2020). However, this method has never been reported for fig fruits. MAE reduces the thermal gradient created by traditional heating, which augments the degradation risk of heat-sensitive bioactive compounds (Chaves et al., 2020). In this technique, pressure builds up in the sample matrix, increasing the porosity of the matrix and allowing for improved extraction through solvent penetration (Ameer et al., 2017).

Due to compliance with the principles of green analytical chemistry, green solvents, including ionic liquids (Qin et al., 2015; Xu et al., 2016) and deep eutectic solvents (Bubalo et al., 2016; Peng et al., 2016; Zannou & Koca, 2020) are increasingly gaining interests in the extraction of phytochemicals. In a study conducted by Qin et al. (2015), five phenolic acids were extracted using ionic liquid-based ultrasound-assisted extraction. A small sample mass and volume of the solvent were used in this study compared to all the studies reviewed in this paper. Extraction was suspected to be facilitated by multi-interactions between phenolic compounds and imidazolium cation, including π-π, ionic/charge-charge, and hydrogen bonding. Remarkably, the application of these non-conventional extraction techniques is still limited in analyzing polyphenols from fig fruits.

1.4 Methods of Detection of Polyphenols in Fig Fruits

Sample collection, sample extraction, analytical measurement, data processing and analysis, and biochemical interpretation are common aspects of phytochemical analysis studies (Aron et al., 2020; Nothias et al., 2020). The analytical workhorses of phytochemical analysis studies are mass spectrometry (MS) and nuclear magnetic resonance (NMR) spectroscopy (Stefanucci et al., 2020; Amaar et al., 2015; Nhoek et al., 2021). The sophisticated analytical technique of liquid chromatography-mass spectrometry (LC-MS) identifies and quantifies a broad spectrum of metabolites. Its benefits include sensitivity and a quick and easy procedure for large-scale sample analysis, making it the instrument of choice for untargeted metabolomics of plant-derived natural compounds (Salem et al., 2020). Despite these developments in analytical instrumentation, spectrophotometric methods are still widely used to date for the determination of polyphenols from figs (Ammar et al., 2015; Bachir Bey & Louaileche, 2015; Debib et al., 2016; Feng et al., 2015; Hoxha et al., 2015; Kamiglou et al., 2015; Viuda-Martos et al., 2015; Harzallah et al., 2016; Hoxha & Kongoli, 2016; Alamgeer et al., 2017; Amessis-Ouchemoukh et al., 2017; Alexandre et al., 2017; Ersoy et al., 2017; Konak et al., 2017; Maghsoudlou et al., 2017; Pereira et al., 2017; Mahmoudi et al., 2018; Meziant et al., 2018; Mopuri et al., 2018; Khadhraoui et al., 2019; Benmaghnia et al., 2021; Nakilcioğlu-Taş & Ötleş, 2021). The Folin-Ciocalteu method, a colometric assay, is the commonly used method for

TPC, whereas the content of flavonoids is usually determined as the total flavonoid content, spectrophotometrically also (Kamiloglu & Capanoglu, 2015; Khadhraoui et al., 2019; Maghsoudlou et al., 2017). It is noteworthy that, despite its widespread usage, the Folin-Ciocalteu method is susceptible to interferences contained in plant samples, including sugars, ascorbic acid, proteins, and non-phenolic organic compounds that react with the Folin-Ciocalteu reagent (Arvaniti et al., 2019). A limited number of studies have used liquid chromatography to survey the polyphenols present in figs (Ammar et al., 2015). In a study done by Amaar et al. (2015), 91 phenolic compounds were identified using the metabolic profiling approach. It is noteworthy that this is the only study where the metabolite profile of polyphenols was investigated in-depth of all the reviewed studies in this chapter.

2 Challenges and Future Trends

Organic solvents are still extensively used to extract various polyphenols in the SLE procedure, and these solvents are associated with various challenges, including toxic residues, chemical alteration of extracts, and generation of toxic waste (Amessis-Ouchemoukh et al., 2017; Mopuri et al., 2018). Moreover, traditional extraction methods, including SLE, have low extraction efficiencies. As a result, innovative techniques that prevent harmful solvents while enhancing efficiency and sustainability are required (Vernès et al., 2020). Significant efforts have been made in this direction, and to date, a significant number of innovative extraction techniques using green and environmentally friendly solvents, such as deep eutectic solvents (Bubalo et al., 2016; Peng et al., 2016; Zannou & Koca, 2020) and supramolecular solvents (Cai et al., 2022; Torres-Valenzuela et al., 2020) have been made. Moreover, modern extraction techniques that use water as an extraction solvent, such as PHWE, have been introduced in the past years (Cvetanović et al., 2018; Lachos-Perez et al., 2020). Furthermore, miniaturized extraction techniques such as dispersive liquid-liquid micro-extraction (DLLME) (Diuzheva et al., 2018; Fusari et al., 2019; Ramirez et al., 2021), hollow fiber liquid micro-extraction (Mlunguza et al., 2020; Moyo & Tavengwa, 2021b) were developed in the past decade, and they consume less volumes of organic solvents. Additionally, DLLME has short extraction times. Therefore, future research on the extraction of phytochemicals from figs should consider employing these modern extraction techniques and green solvents.

The high concentration of free sugars and other matrices can challenge analyzing phenolic compounds from figs using the SLE technique (Arvaniti et al., 2019). Therefore, compounds other than the analytes of interest and other interfering substances may need to be removed via an additional clean-up technique such as solid-phase extraction. Moreover, selective sorbents such as molecularly imprinted polymers may reduce the matrix interferences (Selvamuthukumaran & Shi, 2017). Currently, there is scanty information on the specific phytochemicals found in fig fruits (Alamgeer et al., 2017; Ammar et al., 2015; Feng et al., 2015; Wojdyło et al., 2016). As a result, more profiling studies should be conducted, and this may uncover new metabolites that may be potent besides the widely studied polyphenolic classes.

3 Conclusion

The most studied phytochemical compounds present in fresh and dried figs are phenolic acids and flavonoids. SLE is still the widely employed extraction technique for polyphenols from figs, despite its disadvantages. Future research should consider modern extraction techniques and green solvents as they offer various advantages over conventional methods. Colorimetric assays are still widely used to determine total polyphenolic compounds in these samples. Despite being well-studied, the chemical make-up of the fig tree needs to be further investigated; mainly, the metabolic profiling approach can be used to decipher diverse phytochemicals present in these fruits. Supplementing traditional assays with more specific and selective analytical instrumentation might help understand the health benefits of these fruits and potentially potent drugs that might be discovered via this route. Concentrations of phenolic compounds vary, and this might be due to the methods of extraction and detection used, season, level of maturity, and fig variety.

Conflict of Interest The authors declare no conflict of interest.

References

Abegaz, B. M., & Kinfe, H. H. (2019). Secondary metabolites, their structural diversity, bioactivity, and ecological functions: An overview. *Physical Sciences Reviews, 4*.

Alamgeer, Iman, S., Asif, H., & Saleem, M. (2017). Evaluation of antihypertensive potential of Ficus carica fruit. *Pharmaceutical Biology, 55*, 1047–1053.

Alara, O. R., Abdurahman, N. H., Mudalip, S. A., & Olalere, O. A. (2018). Microwave-assisted extraction of *Vernonia amygdalina* leaf for optimal recovery of total phenolic content. *Journal of Applied Research on Medicinal and Aromatic Plants, 10*, 16–24.

Alara, O. R., Abdurahman, N. H., & Ukaegbu, C. I. (2021). Extraction of phenolic compounds: A review. Current *Research in Food Science, 4*, 200–214.

Alexandre, E., Araújo, P., Duarte, M. F., de Freitas, V., Pintado, M., & Saraiva, J. A. (2017). High-pressure assisted extraction of bioactive compounds from industrial fermented fig by-product. *Journal of Food Science and Technology, 54*, 2519–2531.

Allahyari, S., Delazar, A., & Najafi, M. (2014). Evaluation of general toxicity, anti-oxidant activity and effects of *Ficus carica* leaves extract on ischemia/reperfusion injuries in isolated heart of rat. *Advanced Pharmaceutical Bulletin, 4*(Suppl 2), 577.

Ameer, K., Shahbaz, H. M., & Kwon, J. H. (2017). Green extraction methods for polyphenols from plant matrices and their byproducts: A review. *Comprehensive Reviews in Food Science and Food Safety, 16*(2), 295–315.

Amessis-Ouchemoukh, N., Ouchemoukh, S., Meziant, N., Idiri, Y., Hernanz, D., Stinco, C. M., et al. (2017). Bioactive metabolites involved in the antioxidant, anticancer and anticalpain activities of *Ficus carica* L., *Ceratonia siliqua* L. and *Quercus ilex* L. extracts. *Industrial Crops and Products, 95*, 6–17.

Ammar, S., del Mar Contreras, M., Belguith-Hadrich, O., Segura-Carretero, A., & Bouaziz, M. (2015). Assessment of the distribution of phenolic compounds and contribution to the antioxidant activity in Tunisian fig leaves, fruits, skins and pulps using mass spectrometry-based analysis. *Food & Function, 6*, 3663–3677.

Aron, A. T., Gentry, E. C., McPhail, K. L., Nothias, L. F., Nothias-Esposito, M., Bouslimani, A., et al. (2020). Reproducible molecular networking of untargeted mass spectrometry data using GNPS. *Nature Protocols, 15*(6), 1954–1991.

Arvaniti, O. S., Samaras, Y., Gatidou, G., Thomaidis, N. S., & Stasinakis, A. S. (2019). Review on fresh and dried figs: Chemical analysis and occurrence of phytochemical compounds, antioxidant capacity and health effects. *Food Research International, 119*, 244–267.

Bachir Bey, M. B., & Louaileche, H. (2015). A comparative study of phytochemical profile and in vitro antioxidant activities of dark and light dried fig (*Ficus carica* L.) varieties. *The Journal of Phytological, 4*, 41–48.

Bachir Bey, M., Richard, G., Meziant, L., Fauconnier, M. L., & Louaileche, H. (2017). Effects of sun-drying on physicochemical characteristics, phenolic composition and in vitro antioxidant activity of dark fig varieties. *Journal of Food Processing and Preservation, 41*(5), e13164.

Backes, E., Pereira, C., Barros, L., Prieto, M. A., Genena, A. K., Barreiro, M. F., & Ferreira, I. C. (2018). Recovery of bioactive anthocyanin pigments from *Ficus carica* L. peel by heat, microwave, and ultrasound based extraction techniques. *Food Research International, 113*, 197–209.

Bahrin, N., Muhammad, N., Abdullah, N., Talip, B. H. A., Jusoh, S., & Theng, S. W. (2018). Effect of processing temperature on antioxidant activity of *Ficus carica* leaves extract. *Journal of Science and Technology, 10*, 99–103.

Bakirtzi, C., Triantafyllidou, K., & Makris, D. P. (2016). Novel lactic acid-based natural deep eutectic solvents: Efficiency in the ultrasound-assisted extraction of antioxidant polyphenols from common native Greek medicinal plants. *Journal of Applied Research on Medicinal and Aromatic Plants, 3*, 120–127.

Barolo, M. I., Mostacero, N. R., & López, S. N. (2014). *Ficus carica* L.(Moraceae): An ancient source of food and health. *Food Chemistry, 164*, 119–127.

Barrales, F. M., Silveira, P., Barbosa, P. D. P. M., Ruviaro, A. R., Paulino, B. N., Pastore, G. M., et al. (2018). Recovery of phenolic compounds from citrus by-products using pressurized liquids-An application to orange peel. *Food and Bioproducts Processing, 112*, 9–21.

Benito-Román, Ó., Alvarez, V. H., Alonso, E., Cocero, M. J., & Saldaña, M. D. (2015). Pressurized aqueous ethanol extraction of β-glucans and phenolic compounds from waxy barley. *Food Research International, 75*, 252–259.

Benmaghnia, S., Meddah, B., Tir-touil, A., & Hernández, J. A. G. (2021). Phytochemical analysis, antioxidant and antimicrobial activities of three samples of dried figs (*Ficus carica* L.) from the region of mascara. *Journal of Microbiology, Biotechnology and Food Sciences, 2021*, 208–215.

Bosiljkov, T., Dujmić, F., Bubalo, M. C., Hribar, J., Vidrih, R., Brnčić, M., et al. (2017). Natural deep eutectic solvents and ultrasound-assisted extraction: Green approaches for extraction of wine lees anthocyanins. *Food and Bioproducts Processing, 102*, 195–203.

Boy, H. I. A., Rutilla, A. J. H., Santos, K. A., Ty, A. M. T., Alicia, I. Y., Mahboob, T., et al. (2018). Recommended medicinal plants as source of natural products: A review. *Digital Chinese Medicine, 1*, 131–142.

Bubalo, M. C., Ćurko, N., Tomašević, M., Ganić, K. K., & Redovniković, I. R. (2016). Green extraction of grape skin phenolics by using deep eutectic solvents. *Food Chemistry, 200*, 159–166.

Buenrostro-Figueroa, J. J., Velázquez, M., Flores-Ortega, O., Ascacio-Valdés, J. A., Huerta-Ochoa, S., Aguilar, C. N., & Prado-Barragán, L. A. (2017). Solid state fermentation of fig (*Ficus carica* L.) by-products using fungi to obtain phenolic compounds with antioxidant activity and qualitative evaluation of phenolics obtained. *Process Biochemistry, 62*, 16–23.

Cai, Z. H., Wang, J. D., Wang, L. T., Zhang, S., Yan, X. Y., Wang, Y. Q., et al. (2022). Green efficient octanoic acid based supramolecular solvents for extracting active ingredients from *Zanthoxylum bungeanum* Maxim. peels. *Journal of Cleaner Production, 331*, 129731.

Câmara, J. S., Albuquerque, B. R., Aguiar, J., Corrêa, R. C., Gonçalves, J. L., Granato, D., et al. (2020). Food bioactive compounds and emerging techniques for their extraction: Polyphenols as a case study. *Food, 10*, 37.

Chaves, J. O., De Souza, M. C., Da Silva, L. C., Lachos-Perez, D., Torres-Mayanga, P. C., Machado, A. P. D. F., et al. (2020). Extraction of flavonoids from natural sources using modern techniques. *Frontiers in Chemistry, 8*, 864.

Chen, X. J., Guo, B. L., Li, S. P., Zhang, Q. W., Tu, P. F., & Wang, Y. T. (2007). Simultaneous determination of 15 flavonoids in Epimedium using pressurized liquid extraction and high-performance liquid chromatography. *Journal of Chromatography A, 1163*, 96–104.

Cheng, S. C., Wu, Y. H., Huang, W. C., Pang, J. H. S., Huang, T. H., & Cheng, C. Y. (2019). Anti-inflammatory property of quercetin through downregulation of ICAM-1 and MMP-9 in TNF-α-activated retinal pigment epithelial cells. *Cytokine, 116*, 48–60.

Cheynier, V., Comte, G., Davies, K. M., Lattanzio, V., & Martens, S. (2013). Plant phenolics: Recent advances on their biosynthesis, genetics, and ecophysiology. *Plant Physiology and Biochemistry, 72*, 1–20.

Choi, S. R., Lee, M. Y., Kim, S. A., Oh, J., Hyun, D. W., Lee, S., et al. (2021). Nontargeted metabolomics as a screening tool for estimating bioactive metabolites in the extracts of 50 indigenous Korean plants. *Metabolites, 11*, 585.

Cvetanović, A., Švarc-Gajić, J., Zeković, Z., Gašić, U., Tešić, Ž., Zengin, G., et al. (2018). Subcritical water extraction as a cutting edge technology for the extraction of bioactive compounds from chamomile: Influence of pressure on chemical composition and bioactivity of extracts. *Food Chemistry, 266*, 389–396.

de Oliveira, A. C., Sosa, F. H. B., da Costa, M. C., de Souza Monteiro Filho, E., & Ceriani, R. (2018). Study of liquid-liquid equilibria in aqueous two-phase systems formed by poly (ethylene glycol) (PEG) and sodium thiosulfate pentahydrate ($Na_2S_2O_3$. $5H_2O$) at different temperatures. *Fluid Phase Equilibria, 476*, 118–125.

Debib, A., Dueñas, M., Boumediene, M., Mothana, R. A., Latifa, A., & Tir-Touil, M. A. (2016). Synergetic Hepatoprotective effect of phenolic fractions obtained from *Ficus carica* dried fruit and extra virgin olive oil on CCl_4-induced oxidative stress and hepatotoxicity in rats. *Journal of Food Biochemistry, 40*, 507–516.

Diuzheva, A., Carradori, S., Andruch, V., Locatelli, M., De Luca, E., Tiecco, M., et al. (2018). Use of innovative (micro) extraction techniques to characterise Harpagophytum procumbens root and its commercial food supplements. *Phytochemical Analysis, 29*, 233–241.

Durazzo, A., Lucarini, M., Souto, E. B., Cicala, C., Caiazzo, E., Izzo, A. A., et al. (2019). Polyphenols: A concise overview on the chemistry, occurrence, and human health. *Phytotherapy Research, 33*, 2221–2243.

Ersoy, N., Gozlekci, S., Gok, V., & Yilmaz, S. (2017). Fig (*Ficus carica* L.) fruit: Some physical and chemical properties. *Acta Horticulture*.

Farmer, C., Robertson, P., Xiao, C. W., Rehfeldt, C., & Kalbe, C. (2016). Exogenous genistein in late gestation: Effects on fetal development and sow and piglet performance. *Animal, 10*, 1423–1430.

Feng, Y. C., Li, W. L., He, F. M., Kong, T. T., Huang, X. W., Gao, Z. H., et al. (2015). Aqueous two-phase system as an effective tool for purification of phenolic compounds from fig fruits (*Ficus carica* L.). *Separation Science and Technology, 50*, 1785–1793.

Fraga-Corral, M., García-Oliveira, P., Pereira, A. G., Lourenço-Lopes, C., Jimenez-Lopez, C., Prieto, M. A., & Simal-Gandara, J. (2020). Technological application of tannin-based extracts. *Molecules, 25*, 614.

Fusari, C. M., Ramirez, D. A., & Camargo, A. B. (2019). Simplified analytical methodology for glucosinolate hydrolysis products: A miniaturized extraction technique and multivariate optimization. *Analytical Methods, 11*, 309–316.

Garagounis, C., Delkis, N., & Papadopoulou, K. K. (2021). Unraveling the roles of plant specialized metabolites: Using synthetic biology to design molecular biosensors. *New Phytologist, 231*, 1338–1352.

Harzallah, A., Bhouri, A. M., Amri, Z., Soltana, H., & Hammami, M. (2016). Phytochemical content and antioxidant activity of different fruit parts juices of three figs (*Ficus carica* L.) varieties grown in Tunisia. *Industrial Crops and Products, 83*, 255–267.

Hikmawanti, N. P. E., Ramadon, D., Jantan, I., & Mun'im, A. (2021). Natural deep eutectic solvents (NADES): Phytochemical extraction performance enhancer for pharmaceutical and nutraceutical product development. *Plants, 10*, 2091.

Hoxha, L., & Kongoli, R. (2016). Evaluation of antioxidant potential of Albanian fig varieties "Kraps Zi" and "Kraps Bardhe" cultivated in the region of Tirana. *Journal of Hygienic Engineering and Design, 16*, 70–74.

Hoxha, L., Kongoli, R., & Hoxha, M. (2015). Antioxidant activity of some dried autochthonous Albanian fig (*Ficus carica*) cultivars. *International Journal of Crop Science and Technology, 1*, 20–26.

Hussein, R. A., & El-Anssary, A. A. (2019). Plants secondary metabolites: The key drivers of the pharmacological actions of medicinal plants. *Herbal Medicine, 1*, 13.

Ibañez, E., Herrero, M., Mendiola, J. A., & Castro-Puyana, M. (2012). Extraction and characterization of bioactive compounds with health benefits from marine resources: Macro and micro algae, cyanobacteria, and invertebrates. In *Marine bioactive compounds* (pp. 55–98). Springer.

Kabera, J. N., Semana, E., Mussa, A. R., & He, X. (2014). Plant secondary metabolites: Biosynthesis, classification, function and pharmacological properties. *The Journal of Pharmacy and Pharmacology, 2*, 377–392.

Kamiloglu, S., & Capanoglu, E. (2015). Polyphenol content in figs (*Ficus carica* L.): Effect of sun-drying. *International Journal of Food Properties, 18*, 521–535.

Kang, K. B., Ernst, M., van der Hooft, J. J., da Silva, R. R., Park, J., Medema, M. H., et al. (2019). Comprehensive mass spectrometry-guided phenotyping of plant specialized metabolites reveals metabolic diversity in the cosmopolitan plant family Rhamnaceae. *The Plant Journal, 98*, 1134–1144.

Karuppagounder, V., Arumugam, S., Thandavarayan, R. A., Sreedhar, R., Giridharan, V. V., & Watanabe, K. (2016). Molecular targets of quercetin with anti-inflammatory properties in atopic dermatitis. *Drug Discovery Today, 21*, 632–639.

Katz, L., & Baltz, R. H. (2016). Natural product discovery: Past, present, and future. *Journal of Industrial Microbiology and Biotechnology, 43*, 155–176.

Khadhraoui, M., Bagues, M., Artés, F., & Ferchichi, A. (2019). Phytochemical content, antioxidant potential, and fatty acid composition of dried Tunisian fig (*Ficus carica* L.) cultivars. *Journal of Applied Botany and Food Quality, 92*, 143–150.

Khare, S., Singh, N. B., Singh, A., Hussain, I., Niharika, K., Yadav, V., et al. (2020). Plant secondary metabolites synthesis and their regulations under biotic and abiotic constraints. *Journal of Plant Biology, 63*, 203–216.

Ko, M. J., Nam, H. H., & Chung, M. S. (2020). Subcritical water extraction of bioactive compounds from *Orostachys japonicus* A. Berger (Crassulaceae). *Scientific Reports, 10*, 1–10.

Konak, R., Kösoğlu, İ., & Yemenıcıoğlu, A. (2017). Effects of different drying methods on phenolic content, antioxidant capacity and general characteristics of selected dark colored Turkish fig cultivars. *Acta Horticulturae, 1173*, 335–340.

Lachos-Perez, D., Baseggio, A. M., Torres-Mayanga, P. C., Avila, P. F., Tompsett, G. A., Marostica, M., et al. (2020). Sequential subcritical water process applied to orange peel for the recovery flavanones and sugars. *The Journal of Supercritical Fluids, 160*, 104789.

Maghsoudlou, E., Esmaeilzadeh Kenari, R., & Raftani Amiri, Z. (2017). Evaluation of antioxidant activity of fig (*Ficus carica*) pulp and skin extract and its application in enhancing oxidative stability of canola oil. *Journal of Food Processing and Preservation, 41*, e13077.

Mahmoudi, S., Khali, M., Benkhaled, A., Boucetta, I., Dahmani, Y., Attallah, Z., & Belbraouet, S. (2018). Fresh figs (*Ficus carica* L.): Pomological characteristics, nutritional value, and phytochemical properties. *European Journal of Horticultural Science, 83*, 104–113.

Masota, N. E., Vogg, G., Heller, E., & Holzgrabe, U. (2020). Comparison of extraction efficiency and selectivity between low-temperature pressurized microwave-assisted extraction and prolonged maceration. *Archiv der Pharmazie, 353*, 2000147.

Mastellone, G., Marengo, A., Sgorbini, B., Rubiolo, P., & Cagliero, C. (2021). New phases for analytical scale extraction from plants: Current and future trends. *TrAC Trends in Analytical Chemistry, 141*, 116288.

Meziant, L., Boutiche, M., Bachir Bey, M., Saci, F., & Louaileche, H. (2018). Standardization of monomeric anthocyanins extraction from fig fruit peels (*Ficus carica* L.) using single factor methodology. *Journal of Food Measurement and Characterization, 12*, 2865–2873.

Milescu, R. A., Segatto, M. L., Stahl, A., McElroy, C. R., Farmer, T. J., Clark, J. H., & Zuin, V. G. (2020). Sustainable single-stage solid-liquid extraction of hesperidin and rutin from agro-products using cyrene. *ACS Sustainable Chemistry & Engineering, 8*, 18245–18257.

Mlunguza, N. Y., Ncube, S., Mahlambi, P. N., Chimuka, L., & Madikizela, L. M. (2020). Determination of selected antiretroviral drugs in wastewater, surface water and aquatic plants using hollow fibre liquid phase microextraction and liquid chromatography-tandem mass spectrometry. *Journal of Hazardous Materials, 382*, 121067.

Mopuri, R., Ganjayi, M., Meriga, B., Koorbanally, N. A., & Islam, M. S. (2018). The effects of Ficus carica on the activity of enzymes related to metabolic syndrome. *Journal of Food and Drug Analysis, 26*, 201–210.

Morales-Otal, A., Ferreira-Nuño, A., Olayo-Lortia, J., Barrios-González, J., & Tarragó-Castellanos, R. (2016). Effects of neonatal treatment with two phytoestrogens on male rat sexual behavior and partner preference. *Behavioural Pharmacology, 27*, 570–578.

Moyo, B., & Tavengwa, N. T. (2021a). Critical review of solid phase extraction for multiresidue clean-up and pre-concentration of antibiotics from livestock and poultry manure. *Food Additives & Contaminants: Part A, 39*, 229–241.

Moyo, B., & Tavengwa, N. T. (2021b). Enrichment of tetracycline residues from honey samples using carrier-mediated hollow fibre liquid-phase micro-extraction and quantification by LC-Q-TOF/MS. *Journal of the Science of Food and Agriculture., 102*, 3204–3212.

Nadeem, M., & Zeb, A. (2018). Impact of maturity on phenolic composition and antioxidant activity of medicinally important leaves of *Ficus carica* L. *Physiology and Molecular Biology of Plants, 24*, 881–887.

Nakilcioğlu-Taş, E., & Ötleş, S. (2021). Influence of extraction solvents on the polyphenol contents, compositions, and antioxidant capacities of fig (*Ficus carica* L.) seeds. *Anais da Academia Brasileira de Ciências, 93*.

Naviglio, D., Scarano, P., Ciaravolo, M., & Gallo, M. (2019). Rapid Solid-Liquid Dynamic Extraction (RSLDE): A powerful and greener alternative to the latest solid-liquid extraction techniques. *Food, 8*, 245.

Nhoek, P., Ahn, S., Park, I. G., Pel, P., Huh, J., Kim, H. W., et al. (2021). Salicinoyl Quinic acids and their prostaglandin E2 production inhibitory activities from the fruits of *Casearia grewiifolia*. *Journal of Natural Products, 84*, 2437–2446.

Njila, M. N., Mahdi, E., Lembe, D., Nde, Z., & Nyonseu, D. (2017). Review on extraction and isolation of plant secondary metabolites. In *Proceedings of the 7th international conference on agricultural, chemical, biological and environmental sciences (ACBES-2017), Kuala Lumpur, Malaysia* (pp. 22–24).

Nothias, L. F., Petras, D., Schmid, R., Dührkop, K., Rainer, J., Sarvepalli, A., et al. (2020). Feature-based molecular networking in the GNPS analysis environment. *Nature Methods, 17*, 905–908.

Palmeira, L., Pereira, C., Dias, M. I., Abreu, R. M., Corrêa, R. C., Pires, T. C., et al. (2019). Nutritional, chemical and bioactive profiles of different parts of a Portuguese common fig (*Ficus carica* L.) variety. *Food Research International, 126*, 108572.

Panche, A. N., Diwan, A. D., & Chandra, S. R. (2016). Flavonoids: An overview. *Journal of Nutritional Science, 5*, e47.

Pandey, K. B., & Rizvi, S. I. (2009). Plant polyphenols as dietary antioxidants in human health and disease. *Oxidative Medicine and Cellular Longevity, 2*, 270–278.

Peng, X., Duan, M. H., Yao, X. H., Zhang, Y. H., Zhao, C. J., Zu, Y. G., & Fu, Y. J. (2016). Green extraction of five target phenolic acids from *Lonicerae japonicae* Flos with deep eutectic solvent. *Separation and Purification Technology, 157*, 249–257.

Pereira, C., López-Corrales, M., Serradilla, M. J., del Carmen Villalobos, M., Ruiz-Moyano, S., & Martín, A. (2017). Influence of ripening stage on bioactive compounds and antioxidant activity in nine fig (*Ficus carica* L.) varieties grown in Extremadura, Spain. *Journal of Food Composition and Analysis, 64*, 203–212.

Qin, H., Zhou, G., Peng, G., Li, J., & Chen, J. (2015). Application of ionic liquid-based ultrasound-assisted extraction of five phenolic compounds from fig (Ficus carica L.) for HPLC-UV. *Food Analytical Methods, 8*, 1673–1681.

Raafat, K., & Wurglics, M. (2019). Phytochemical analysis of *Ficus carica* L. active compounds possessing anticonvulsant activity. *Journal of Traditional and Complementary Medicine, 9*, 263–270.

Ramirez, D. A., Altamirano, J. C., & Camargo, A. B. (2021). Multi-phytochemical determination of polar and non-polar garlic bioactive compounds in different food and nutraceutical preparations. *Food Chemistry, 337*, 127648.

Rodríguez-García, C., Sánchez-Quesada, C., & Gaforio, J. J. (2019). Dietary flavonoids as cancer chemopreventive agents: An updated review of human studies. *Antioxidants, 8*, 137.

Rodríguez-Solana, R., Galego, L. R., Pérez-Santín, E., & Romano, A. (2018). Production method and varietal source influence the volatile profiles of spirits prepared from fig fruits (*Ficus carica* L.). *European Food Research and Technology, 244*, 2213–2229.

Saibabu, V., Fatima, Z., Khan, L. A., & Hameed, S. (2015). Therapeutic potential of dietary phenolic acids. *Advances in Pharmacological Sciences, 2015*, 823539.

Salem, M. A., Perez de Souza, L., Serag, A., Fernie, A. R., Farag, M. A., Ezzat, S. M., & Alseekh, S. (2020). Metabolomics in the context of plant natural products research: From sample preparation to metabolite analysis. *Metabolites, 10*, 37.

Sarfarazi, M., Jafari, S. M., Rajabzadeh, G., & Galanakis, C. M. (2020). Evaluation of microwave-assisted extraction technology for separation of bioactive components of saffron (*Crocus sativus* L.). *Industrial Crops and Products, 145*, 111978.

Selvamuthukumaran, M., & Shi, J. (2017). Recent advances in extraction of antioxidants from plant by-products processing industries. *Food Quality and Safety, 1*, 61–81.

Solana, R. R., & Romano, A. (2019). Chemical and biological characteristics of *Ficus carica* L. Fruits, Leaves, and derivatives (Wine, Spirit, and Liqueur). In *Modern fruit industry*. IntechOpen.

Soukup, R. W., & Soukup, K. (2015). The series "Progress in the chemistry of organic natural products": 75 years of service in the development of natural product chemistry. In *Progress in the chemistry of organic natural products 100* (pp. 453–588). Springer.

Stefanucci, A., Zengin, G., Llorent-Martinez, E. J., Dimmito, M. P., Della Valle, A., Pieretti, S., et al. (2020). *Viscum album* L. homogenizer-assisted and ultrasound-assisted extracts as potential sources of bioactive compounds. *Journal of Food Biochemistry, 44*, e13377.

Thirumurugan, D., Cholarajan, A., Raja, S. S., & Vijayakumar, R. (2018). An introductory chapter: Secondary metabolites. In *Secondary metabolites – Sources and Applications* (pp. 1–21). InTechOpen.

Torres-Valenzuela, L. S., Ballesteros-Gomez, A., Serna, J., Arango, A., & Rubio, S. (2020). Supramolecular solvents for the valorization of coffee wastewater. *Environmental Science: Water Research & Technology, 6*, 757–766.

Tungmunnithum, D., Pinthong, D., & Hano, C. (2018). Flavonoids from *Nelumbo nucifera* Gaertn., a medicinal plant: Uses in traditional medicine, phytochemistry and pharmacological activities. *Medicines, 5*, 127.

Um, M., Han, T. H., & Lee, J. W. (2018). Ultrasound-assisted extraction and antioxidant activity of phenolic and flavonoid compounds and ascorbic acid from rugosa rose (*Rosa rugosa* Thunb.) fruit. *Food Science and Biotechnology, 27*, 375–382.

Vaishnav, P., & Demain, A. L. (2011). Unexpected applications of secondary metabolites. *Biotechnology Advances, 29*, 223–229.

Vernès, L., Vian, M., & Chemat, F. (2020). Ultrasound and microwave as green tools for solid-liquid extraction. In *Liquid-phase extraction* (pp. 355–374). Elsevier.

Viuda-Martos, M., Barber, X., Pérez-Álvarez, J. A., & Fernández-López, J. (2015). Assessment of chemical, physico-chemical, techno-functional and antioxidant properties of fig (Ficus carica L.) powder co-products. *Industrial Crops and Products, 69*, 472–479.

Wojdyło, A., Nowicka, P., Carbonell-Barrachina, Á. A., & Hernández, F. (2016). Phenolic compounds, antioxidant and antidiabetic activity of different cultivars of Ficus carica L. fruits. *Journal of Functional Foods, 25*, 421–432.

Xu, J. J., Li, Q., Cao, J., Warner, E., An, M., Tan, Z., et al. (2016). Extraction and enrichment of natural pigments from solid samples using ionic liquids and chitosan nanoparticles. *Journal of Chromatography A, 1463*, 32–41.

Zannou, O., & Koca, I. (2020). Optimization and stabilization of the antioxidant properties from alkanet (*Alkanna tinctoria*) with natural deep eutectic solvents. *Arabian Journal of Chemistry, 13*, 6437–6450.

Zhang, Q. W., Lin, L. G., & Ye, W. C. (2018). Techniques for extraction and isolation of natural products: A comprehensive review. *Chinese Medicine, 13*, 1–26.

Part III
Fig (*Ficus carica*): Technology, Processing, and Applications

Chapter 29
Fig (*Ficus carica*) Drying Technologies

Olfa Rebai, Oumayma Ghaffari, and Sami Fattouch

Abbreviations

1-MCP	1-Methylcyclopropene
AD	Air-convection drying
MAP	Modified atmosphere packaging
MW	Microwave
MWAD	Microwave-air convection
MWVD	Microwave vacuum drying
T.S.S	Total soluble solid
TA	Titratable acidity
VD	Vacuum-drying

1 Introduction

Fig (*Ficus carica* L.) belongs to the order of *Urticales* and the family of *Moraceae* with over 1400 species. One of the first worldwide cultivated trees for its edible fruit. It is supposed to originate from Western Asia, spread to the Mediterranean by humans, and grow in many parts of the world with moderate climates (Baraket et al., 2009). Besides its primary metabolites content, several phytochemical studies on *F. carica* revealed the presence of a panoply of bioactive phenolics, phytosterols, organic acids, anthocyanins, triterpenoids, coumarins, and volatile compounds such

O. Rebai (✉) · O. Ghaffari · S. Fattouch
EcoChem Laboratory, Department of Biological and Chemical Engineering,
National Institute of Applied Sciences and Technology (INSAT),
University of Carthage (UCAR), Tunis, Tunisia
e-mail: oumayma.ghaffari@insat.u-carthage.tn; sami.fattouch@insat.ucar.tn

M. F. Ramadan (ed.), *Fig* (Ficus carica)*: Production, Processing, and Properties*, https://doi.org/10.1007/978-3-031-16493-4_29

as aliphatic alcohols, and few other classes of secondary metabolites (Oliveira et al., 2009). Figs contain essential amino acids and are cholesterol-free. They are an excellent source of minerals, vitamins, carbohydrates, and dietary fiber.

Thanks to their wide variety of chemical constituents, *Ficus carica* L. fruits are used in traditional medicine to remedy many health problems and biological activities. *F. carica* has been traditionally used for its medicinal benefits as a metabolic, cardiovascular, respiratory, antispasmodic, and anti-inflammatory remedy (Duke et al., 2002). Leaves and roots of *F. carica* are used in different disorders such as gastrointestinal (colic, indigestion, loss of appetite, and diarrhea), respiratory (sore throats, cough, and bronchial problems), inflammatory, and cardiovascular disorders. Figs have a laxative effect due to their content of many antioxidants. Moreover, they are a good source of flavonoids and other polyphenols. The phenolic antioxidant qualitative and quantitative composition of dried fig fruits was determined and analyzed by several studies and investigators. The *F. carica* juice mixed with honey is commonly used for hemorrhage. In Indian medicine, fig fruits are a mild laxative, expectorant, and diuretic (Solomon et al., 2006).

F. carica fruits could be used fresh, dried, canned, or as jam (Tous & Fergueson, 1996). Recurrent consumer anxiety is rising about using preservatives and chemical additives for food storage. Thus, many efforts have been made to find safe preservation methods during the last decades. Fruits and vegetables contain a large quantity of water, due to which they rapidly deteriorate. Drying is among the most used processes to remove the food's moisture, so bacteria, yeast, and mold cannot grow and spoil the food. The nutritional value of food is only minimally affected by drying (Ahmed et al., 2013). During water evaporation, the fruit decreases in weight and shrink; however, its beneficial substances do not disappear and are "preserved". Generally, dried foods are tasty, nutritious, lightweight, and easy to prepare, store and use. A fresh fig contains total sugar (mainly glucose and fructose), but it rises to about 50% when dried. It is commercialized in the market as sweet due to its high level of sugars and provides an excellent dietary substitute (Veberic et al., 2008). Paradoxically, fig preserved forms, especially dried fruits, are used as supplement food for diabetics in some eastern countries. Moreover, dried flesh has been known as a source of medicine or a part of medicine in many Mediterranean countries, i.e., in Turkey, Egypt, Tunisia, Algeria, Morocco, Spain, Greece, and Italy, since the old time.

2 Criteria to Consider for Dried Fruits

Food drying is one of the oldest methods of preserving food for later use. It could either be an alternative to canning, freezing, or complimenting these methods. Drying removes excess (free) water, compromising the sensory and nutritive values. By definition, food dehydration removes free water from food, generally by circulating hot air through it, which prohibits the growth of spoiling microbials, particularly bacteria. Drying also slows down the action of enzymes (naturally occurring substances that cause food to ripen) but does not inactivate them. When water is removed from a material, its structure changes and becomes harder and less likely to spoil. When dried, the food becomes smaller and lighter in weight. When needed, in some cases, the water is added back to have food ready for use, allowing the food to return to its original shape and texture.

Drying empirically uses heat to remove water from a raw or processed material. This technological method can be done in several ways, including solar drying or using an oven, a microwave, or a sophisticated dehydrator. The sun, the wind, and a smoky fire removed water from fruits, meats, grains, and herbs. The energy input is less than needed to freeze, and the storage space is minimal compared with that needed for canning jars and frozen containers. Foods can be dried using warm temperatures, low humidity, and air current. In drying, warm temperatures cause the moisture to evaporate. The higher the temperature gradient between the food and the ambient air is, the increased speed in the evaporation of food water. However, when considering hysteresis phenomena, an excess temperature gradient could harm the food texture. Generally, a food-specific temperature regime should be set and optimized. Commonly, heating at medium (50–55 °C) or relatively high (60 °C) temperatures is used. In addition, lower air humidity allows moisture to move quickly from the food to the air. The heat and mass (water) transfer depend on the velocity of the circulating air. The air current speeds up drying by moving the surrounding moist air away from the food.

3 Fig Drying Technologies

Fig fruit (*Ficus carica* L.) could be dried either by traditional methods (sun or solar drying) or in conventional hot-air dryers (Babalis et al., 2006). Drying might be classified into two groups: natural drying and artificial drying. Mechanical dehydration or artificial dehydration can be further classified into atmospheric and subatmospheric types based on the conditions employed in the drying process. Artificial drying is done with the help of mechanical or electrical equipment. Based on the model of the drying process, drying at atmospheric pressure conditions could be further divided into batch and continuous types. Mechanical drying includes the methods of drying by (1) heated air, (2) direct contact with a heated surface, e.g., drum drying, and (3) application of energy from a radiating microwave or dielectric source. According to Ahrens et al. (2000), artificial methods could efficiently remove a large amount of moisture. In addition, controlling various factors such as temperature, drying air flux, and drying time is also possible when artificial methods are used (Maisnam et al., 2017).

3.1 Natural Sun-Drying

Drying is one of the oldest preservation processes available to humankind that we could track since prehistoric times. People have sun-dried fruit and vegetables for thousands of years to preserve for leaner times. The relatively high sugar and acid content of fruits was enough to make them safe to dry in the sun for a few days. New technologies brought changed techniques, but the increasing demand for healthy, low-cost natural foods and the need for sustainable income, are bringing solar drying to the fore as a valuable alternative for surplus products. Sun-drying works best in hot, dry climates, but it can also be successful in more humid zones. In today's food market, dried food plays an essential role in the food supply chain. As for fruits and vegetables, it can be estimated that they constitute about 1% of the total drying in the food industry, by large being the grains the most important (Ahmed et al., 2013).

Sun driers are classed according to the systems of heating the product. They are not designed based on heat and mass transfer calculations or principles since they remain traditional and artisanal.

(1) **Natural drier** is the traditional common means which uses the direct sun action on the materiel itself and the ambient air to which the product is exposed. In these driers, the raw products are spread out in an enclosed space where they are protected against dust contamination and subjected to limited airflow control (just using some physical obstacles to break strong air current, if needed).
(2) **Direct driers:** The material to be dried in these units is placed in an enclosure with a transparent cover. Heat is generated by sun radiation of the product itself and the internal surfaces of the enclosure. The heat causes the moisture to evaporate from the product. In addition, it expands the air in the enclosure, creating air circulation, which facilitates mass and energy (heat) transfers.
(3) **Indirect driers:** The sun radiation heats the air exclusively in these driers. The hot air is then led by thermocirculation or by forced circulation into a chamber or cabinet holding the products to be dried. Different instruments of this type are available in the market; some are associated with a fan to intensify and control air circulation.
(4) **Mixed driers:** are indirect driers where the hot air from the exterior solar heater is combined with direct solar heating of the products through a transparent roof of the drying cabinet (FAO, 1985).

The drying process of food products is to leave them outside in the sun and the windy position and allow them to dry within 7–10 days, depending on the air current, temperature, and humidity, for good dehydration. As far as drying temperature is concerned, the most important thing is to dry them in sunny and dry weather, particularly in a hot summer. Drying in the sun is very economical. Operators only have to spread, as much as possible, the produce on a suitable surface and let it dry in the sun. The nutritional benefits of sun drying, or solar drying, are that it takes just enough moisture out of fruits to prevent them from spoiling while being stored. Sun-drying retains large amounts of vitamins, fiber, and minerals. For people with diabetes, dried fruit prepared without adding sugar (water activity, *aw*, depressor) is a healthy choice instead of desserts (Dhaouadi et al., 2014).

Sun-dried fig (Fig. 29.1) can be eaten from starter to dessert. It can be eaten plain or stuffed with almonds or walnuts. It also goes perfectly with the flavors of Mediterranean cuisine, especially in North African "couscous" and "tajines". Cut into small pieces; it is a wonderful addition to salads. In addition, it improves the bargaining position of farmers. Sometimes farmers sell figs at very low prices

Fig. 29.1 From left to right, whole *F. carica* fruit, the fruit cut open, and sun-dried fruits (Sukowati et al., 2019)

during the harvest season because they cannot store or preserve their surplus products, thus encouraging people to establish their sun-dried figs.

However, disadvantages associated with these traditional sun dryers are well known and should be highlighted here, that it is an uncontrolled method. Mainly, fig fruits did not dry uniformly, thus leading to heterogeneously stabilized/unstabilized products during storage. The drying is slow and sometimes incomplete under unfavorable climatological conditions. When the sun is not available, this method is not possible. Moreover, this process requires many unskilled laborers, takes a long time, and requires a large area. Constantly, sun-dried figs are subjected to noxious effects of dust, dirt, and insect infestation. Regular and incessant supervision of the dryer is necessary as the product must be gathered and moved undercover in the event of rain, and predators must be chased away. Indeed, possible losses of crops by birds, hens, or other animals are inevitable.

3.2 *Artificial Drying*

The method in which excess water of figs is removed artificially offers better control than natural drying, thus, resulting in greater product uniformity and quality. However, the initial investment in equipment and expenditures on energy inputs are high and may not always be justified. Therefore, different types of artificial drying are used: (1) mechanical drying, (2) infrared or (3) dielectric drying, and (4) chemical drying.

3.2.1 Solar Drying

Using a solar drying system for fruits is becoming an attractive alternative food preservation technique over conventional ones. Solar drying also uses the sun as the heat source. Therefore, solar drying remains a viable drying and food preservation technology alternative with the potential to reduce post-harvest losses, which can be as high as 25–30% in food grains and 30–50% in fruits and vegetables, as reported

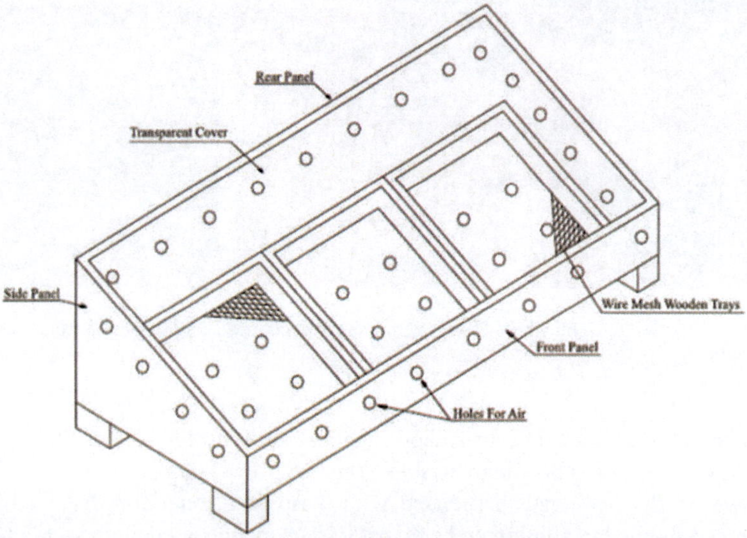

Fig. 29.2 Experimental setup of a cabinet type drier (Phadke et al., 2015)

by Patel et al. (2013). This is due to many advantages of the solar drying system compared to others: low cost, high quality of dried products, and environmentally friendly (Heredia et al., 2007). Here, instruments are designed based on heat and mass transfer calculations and optimization (Fig. 29.2). Generally, three types of artificial solar driers are used, (1) the absorption or hot-box type driers in which the product is directly heated by the sun, (2) the indirect or convection driers in which the product is exposed to warm air which is heated using a solar absorber or heat exchanger and (3) drier, which is a combination of the first and the second types. In solar drying, the energy source is the sun. Therefore, it is a very cheap drying method, but on the other hand, it has many drawbacks because the food is exposed to contamination sources (insects, birds, and other animals) and is also strongly susceptible to weather conditions. Indirect solar drying, the product is exposed to solar radiation, whereas, in indirect solar drying, the sun's energy is harnessed by collectors used to heat the air that will contact the fig fruits to dry (Patel et al., 2009).

3.2.2 Heat Pump Dryer

Heat pump fruit drying is used to dry all kinds of food products in hygienic ambient at economic conditions with low energy consumption. Due to the increasing prices of fossils and electricity and the emission of CO_2 in conventional drying methods, green energy-saving and other heat recovery methods for processing and drying food have become very important. Heat pump technology has been successfully used for drying agricultural products and other domestic dehumidification/heating applications. It has excellent prospects and revolutionary ability. However, heat pump drying (HPD) of fruits and vegetables has been largely unexploited in South

Africa and, by extension, in the sub-Saharan African region (Fayose & Huan, 2016). Fig drying using a heat pump reduces energy cost with relatively low air temperature and relative humidity. This process could be achieved by condensing moisture from the drying air and then reheating, and product quality can be improved in terms of color and smell (Soponronnarit et al., 2007).

HPD technique is more effective in drying material with more free moisture in some fruits. The application of heat pump drying contributes positively to dried fruit and vegetables' quality attributes, including improved microbial safety, better color, vitamin C retention, enhanced volatile compounds, aroma and flavor substances, rehydration, and texture. In addition, compared to sun drying, product quality is improved in preserving the product's natural sensorial properties (Soponronnarit et al., 2007).

3.2.3 Pilot Cabinet Dryer

Cabinet dryers are the most popular devices for fruit drying. However, one of the drawbacks of this dryer can be nonuniformity in the desired end product moisture content (Darabi et al., 2015). The cabinet dryer tested is of the type pioneered by the Brace Research Institute in Canada (Anon, 1965). It is essentially a rectangular container insulated at its base and covered with a glass or transparent plastic roof. Holes drilled through the base permit fresh air to enter the cabinet. Outlet holes are located on the upper parts of the cabinet side and rear panels. In this case, the cabinet was constructed of thick plywood panels.

Mujić et al. (2012) used a pilot plant cabinet dryer to compare the aroma profiles of fresh figs and figs frozen in liquid nitrogen and non-treated and pre-treated dried figs of the Croatian cultivar. This study offers an insight into volatiles preserved during drying, and results have shown that pretreated dried figs with sulfur dioxide were capable of preserving many aromatic compounds that are important for consumers.

3.2.4 Microwave Drying

Called also drying by radiation. Generally, long drying time and high temperature are the factors responsible for the loss of heat-sensitive components of the food. However, microwave drying has been considered excellent in retaining nutritional and sensory values besides saving energy and time in recent years. Hence, microwave drying is used by several modern food industries worldwide. Microwaves (MWs) are electromagnetic (EM) waves synchronized with perpendicular oscillations of electric and magnetic fields in a frequency between 300 MHz and 300 GHz, ranging from 1 m to 1 mm. Microwave heating mechanisms are based on the oscillation of ions and polar molecules mainly water when a material is exposed to the EM waves which cause the internal friction and conversion of kinetic energy into heat (Laguerre et al. 2018).

During microwave drying, the quality preservation of the product is of utmost importance (Khraisheh et al., 2004). Product quality includes three principal areas: nutritional value, acceptability, and safety. The reasons for uneven heating with MWs are due to several factors: (i) large product size; (ii) resonance phenomena; (iii) heterogeneous material composition, and (iv) shape of the product (Radojčin et al., 2021). Exposure of food to infra-red led to charge build-up in the electronic state and the vibrational and rotational state at the atomic and molecular levels. This causes the food to heat up without any changes in the air temperature surrounding the food.

Sharifian et al. (2013) investigated the qualitative indices for the pulsed microwave dried figs (*Ficus carica* L.) through image processing techniques. Kinetic parameters for the color change were determined using the total color change parameter, chroma, hue angle, and browning index. Also, in the same study, Sharifian et al. (2013) demonstrated that the MW power intensity and pulsing ratio significantly influenced the color change of fig fruit during drying. Increasing MW power intensity can provide a desirable contribution to the light yellowish color of dried figs. Nevertheless, from the growing total color changes, the authors concluded that 2.5 W/g is the optimum MW power intensity level concerning fig color criteria (Sharifian et al., 2013). However, depending on the type of material, MWs, in some cases, cannot fully complete drying and are usually combined with hot air drying or vacuuming.

Sharifian et al. (2012) investigated the effects of microwave power intensity and pulsing ratio on the drying behavior of fig fruits in a laboratory-scale microwave dryer and has shown that the drying time of products increased by approximately 200% under the pulsing ratio of 1.5–4. In contrast, drying time decreased by approximately 500% under microwave power intensity of 0.5–2.5 W/g; microwave power intensity resulted in the raised sample temperature leading to better moisture removal. Sample temperature decreased with the pulsing ratio while increased with MW power intensity. The warming-up period was short, and subsequent drying only occurred during the continuous temperature.

3.2.5 Microwave Hot Air Drying

Microwave heating is used in many processes such as baking, pre-cooking, and drying industry. This method associates the MW process and hot air circulation as a mass and heat transfer vector. The microwave energy is absorbed by food materials and converted into heat due to the ionic interaction and the dipolar rotation. Salt and water are common molecules in most foods. Therefore, salt molecules are vibrated due to ionic interaction. The application of microwave heating in conjunction with hot air drying has led to a significant reduction in the drying time of fig fruits (Abul-Fadl et al., 2015).

3.2.6 Microwave Vacuum Drying

Giri and Prasad (2007) reported that food products' quality could be better maintained by vacuum drying. However, vacuum drying requires high investment and operating costs and thus is only suitable for high-value products such as medicinal herbs. In addition, vacuum drying has a poor mass transfer rate and leads to an extended drying time (Wang et al., 2018). Correspondingly, Xu et al. (2006) reported that vacuum drying has high operating costs due to the need to maintain a vacuum over long drying periods.

3.2.7 Heat-Drying Ovens

An oven is a direct-fired chamber or cabinet holding the products to be dried that distributes heat effectively. Ovens are widely used in the food industry to preserve perishable commodities, especially foodstuff and heating and drying. The oven consists of an insulator box that serves as the housing for which the food is placed for baking, a heating element, and a thermostat that regulates the heat generation rate and regulates the fuel or current flow to the heating element. Modern ovens are fueled by gas, electricity, or both and are commonly used for cooking, frying, baking, and roasting. However, the daily need, high market demands, and problems often associated with the high cost of procuring the gas or kerosene type of oven for cooking, drying and baking prompt us to develop an electric-powered oven for baking bread cake etc.

3.2.8 Electric Drying

Many studies have shown that nonthermal food processing techniques can negatively affect food quality. The first commercial electric drier was developed in 2008, which has been improved to exceed hundreds of kg of product to dry in one batch (Martínez-García et al., 2013a, b). No fuel is used, no CO_2 emission, thus making this process eco-friend. The electric drier is a heating chamber meant for evaporating food's water by converting electrical energy to heat or thermal energy. These novel pretreatment and treatment techniques have been developed further to improve product quality (Deng et al., 2017). With no waste heat loss, the circulating air volume can be reduced by 20 to 40% compared to other conventional driers. When miniaturized, an electric dryer can be portable, relatively cheap, readily available, and easy to install, operate and maintain (Adegbola et al., 2012). Nowadays, these instruments are more and more electronically controlled.

Electric dryers work on the principle of convection or infrared radiation. The hot air convection makes the food moisture gradually released. Even though this dryer has an affordable cost, it may induce fruit deformation, change its taste, and lose some vitamins. Infrared-based electric dryers use radiations that penetrate deeply into fruits, so the drying occurs at sufficiently low temperatures, not exceeding 60 °C. The fruit's practically preserved taste, color, shape, and vitamins (preserved up to 90%) make this electric drying process ideal.

3.2.9 Tray Dryer

A tray dryer is convectional drying equipment with enclosed insulated chambers and trays in a trolley placed on top of each other. The application of tray dryer is widely used in agricultural drying because of its simple design and capability to dry products at high volume. However, the greatest drawback of the tray dryer is uneven drying because of poor airflow distribution in the drying chamber (Misha et al., 2013). Gwala and Padmavati (2016) compared the drying kinetics of an electric tray drying, a solar cabinet drying, and an open sun drying. The tray dryer was superior in heat transfer but weaker in mass transfer, explaining the little difference observed in drying time compared to solar and sun drying. The evidence was suggested by drying models, drying constants, and the effective moisture diffusivity model (Gwala & Padmavati, 2016).

3.2.10 Spray Drying Technology

Spray drying equipment and techniques developed over several decades, from the 1870s through the early 1900s. Spray drying comes of age during World War II, with the sudden need to reduce the transport weight of foods and other materials (Sonone et al., 2016). This method transforms feed from a fluid state into dried particulate form by spraying the feed into a hot drying medium (Andreou et al., 2021). It is a continuous particle processing drying operation. The feed can be a solution, suspension, dispersion, or emulsion. The dried product could be in the form of powders, granules, or agglomerates, depending upon the physical and chemical properties of the feed, the dryer design, and the final powder properties desired (Killeen, 2000).

The spray drying process has advantages that could be designed to virtually any capacity required. Feed rates range from a few pounds per hour to over 100 tons per hour. Operation is continuous and adaptable to fully automatic control (Gharsallaoui et al., 2007). It can be used with both heat-resistant and heat-sensitive products (Chegini & Ghobadian, 2007). Nearly spherical particles can be produced. Some limitations include limited versatility in producing particles or structures with complex morphologies and rapid drug release rates often exhibiting a burst effect.

Due to existing challenges in the field of syrup or powder production and the presence of strong demand for turning waste into high-value products. Kalantari et al. (2018) examined the production of fig extract from waste figs under mild extraction conditions and attempted to optimize fig syrup's extraction and drying conditions. Fig extract was used to feed the spray dryer to prepare a powder with suitable physicochemical properties to produce seedless fig paste or drink. Furthermore, the effects of drying aids and spray drying operating conditions were investigated on the quality of the obtained powder. Results show that the optimal extraction conditions for producing fig extract are suggested to be 5 °C at 72 h. Fig extract powder was produced from the natural fig extract using a pilot-scale two-fluid nozzle spray dryer.

Furthermore, increasing the inlet air temperature led to increased porosity, dispersibility, and particle size distribution. In addition, an increase in the airflow rate increased the moisture content, dispersibility, and porosity of the powders. The best powder was produced at a temperature of 170 °C using an airflow rate of 400 m^3/h, conditions that can be a good strategy for converting waste figs to high-value products capable of being used to formulate healthy food products Kalantari et al., 2018.

The quality characteristics of spray-dried powders from unripe fig extract were investigated by Chae and Hong (2016). The spray-dried powder showed the highest protease activity with (0.84 unit/g), moisture content, and L value of the spray-dried powder than those the freeze-dried powder. Furthermore, in the same study, in vitro digestion experience, spray-dried powders of the unripe fig showed a protease survival until approximately 25% (Chae & Hong, 2016).

4 Advantages and Disadvantages of Fig Drying

Drying is one of the oldest unit operations and has recently become widespread. It has been used in different industries to gain different utilities. The methods of drying are diversified according to the purpose of the process. There are more than 200 types of dryers. For every dryer, the process conditions, such as drying chamber temperature, pressure, air velocity (if the carrier gas is air), relative humidity, and product retention time, have to be determined according to feed, product, purpose, different drying methods used to dry and retain the quality of the dried fig fruit. Therefore, there is a greater need to select an appropriate drying method for good quality figs considering prevailing conditions. The objective is the different drying methods to dry the fig fruit to increase its shelf life and select appropriate drying methods for figs of good quality.

4.1 Impact of Fig Drying on Its Technological Characteristics

Drying food is a process by which water is removed from the food by vaporization or sublimation, thus reducing the water available for chemical, enzymatic or microbial degradation reactions. The acceptability (visual appeal, taste, aroma, flavor, and texture), structural property, and nutritional value of fruits and vegetables are also highly affected (Zhang et al., 2005; Zhang & Xu, 2003). Thus, the main objective of any drying process is to produce a dried product of desired quality at minimum cost and maximum throughput and to optimize these factors consistently (Chua et al., 2001). Technological treatment changes the structure of the raw fruit by modifying enzymatic reactions occurring in the tissue and affecting the heat and mass exchange conditions in the plant material (Janowicz et al., 2008). Each drying technique depends on various factors, such as the required type of product, size, level of ripeness, structure, color, aroma, chemical composition, nutritional composition, expected final quality, availability of a dryer, and costs. These changes and influencing factors are physical quality, chemical quality, nutrition, and sensory quality. The main objectives of drying include preserving food and increasing its shelf life by reducing the water content and water activity; avoiding the need for the use of refrigeration systems for transport and storage (expensive); reducing space requirements for storage and transport; diversifying the supply of foods with different flavors and textures, thus offering the consumers a great choice when buying foods (Guiné, 2018).

The drying conditions used, mainly heat and humidity can significantly influence fig fruit's technological and functional characteristics. Effect of drying on physicochemical, functional, and morphological characteristics of the fig. Generally, the process of drying in spouted beds has been presented in the literature as an attractive alternative for drying pastes and suspensions, producing a high-quality powder at and low cost. It is widely used in the dehydration of heat-sensitive materials to preserve bioactive compounds and other structures of interest, offering advantages over other methods of drying, especially short drying times (Bezerra et al., 2013).

The intensity of changes in quality parameters during drying depends on the nature of the material, pretreatment, used process, and their parameters. These changes include color, shape, volume, density, texture, flavor, nutrients, water activity, microbiology, rehydration ability, etc. Ong & Law, 2010. Convective drying kinetics is affected by many factors, such as fig size and composition and thermodynamics properties of drying air such as temperature, humidity, and velocity. With drying, fig mass and volume are reduced, and packaging and handling costs. It also reduces the microbiological activity and chemical changes during the storage of the final dehydrated product (Martínez-García et al., 2013a, b).

Nagaraja et al. (2016) evaluated different drying methods for pre-treated fig fruits. Fig fruit was dried under different drying methods (sun drying, solar cabinet dryer, microwave oven, and cabinet tray dryer) to elicit their effect on drying rate, drying time, biochemical properties, and sensory attributes. Microwave oven drying showed the highest drying rate and took less time to reach the required final moisture content. A higher drying rate was observed in the pre-treated samples than in the untreated samples under all drying methods. The drying time for cut-fig fruits was nearly half of that for whole fig fruits. The sulphited samples showed higher biochemical properties irrespective of drying methods. The samples dried in a microwave oven, solar cabinet, and sun-drying had higher values of biochemical properties than the cabinet tray dried samples. Acceptable sensory quality was obtained in products dried with a microwave oven, solar cabinet, and sun-drying (Nagaraja et al., 2016). Babalis studied the influence of drying air temperature and air velocity, and Babalis and Belessiotis (2004) demonstrated a strong influence of air temperature and velocity at the early drying stages. The relative insensitivity of the drying process at the later stages. Higher air temperatures affect significantly the drying rate at the first stages of drying, but their relative effect diminishes after a period of about 10–15 h. Then, a range of 1–2 m/s for the drying air velocity has to be preferably established inside an industrial dryer for agricultural products, giving the best cost/efficiency performance for the drying of the figs. These investigations revealed that the drying kinetics is most significantly affected by temperature, with the air flow velocity having a limited influence on the drying process. Correlations expressing the drying constants and effective moisture diffusivity dependence on the drying-air parameters are also reported (Babalis & Belessiotis, 2004).

Another trend in drying technology is Osmotic dehydration; OD is the process of water removal by immersion of water-containing cellular solid in a concentrated aqueous solution (Ponting, 1973). The fundamental purpose of food dehydration is to lower the water content to minimize rates of chemical reactions and facilitate distribution and storage. In osmotic dehydration, foods are immersed or soaked in a saline or sugar solution. This results in three counter mass transfer phenomena (Ponting, 1973).

Martínez-García et al. (2013a, b) investigated drying parameters on figs and their functional properties. They demonstrated that the drying process increased the total phenolic content, degraded the anthocyanin content of figs, and enhanced the dried fig's antioxidant activity. Furthermore, dehydrating ground figs was faster and maintained their functional properties better than half-cut figs.

Physical properties of food products are responsible for perceived quality; the color of food is measured directly and non-destructively using apparatus like colorimeters. They use several color spaces for the expression of color values. The usually used color space is CIELAB (CIE L*a*b*, color system), and this method helped monitor color changes during the drying of fig. Texture properties include structural and mechanical characteristics that significantly influence dried products' quality. Abul-Fadl et al. (2015) investigated the effect of different drying methods on the quality criteria of dried Fig fruits. In this study, the relation between moisture content and elapsed time to reach the required moisture content (drying curves) of dried tested fig fruits for air convection (AD) were compared with combined microwave-air convection (MWAD) drying methods. A combined microwave-hot air drying (MWHA) technique may be a better alternative. This method provides a higher drying rate and a better product quality than microwave drying or hot air drying (Chandrasekaran et al., 2013).

The final drying time of fruits fig using air convection drying AD and vacuum drying VD methods compared to the combined microwave-air convection MWAD and microwave- vacuum drying MWVD methods indicated that the much longer drying time was obtained by using the vacuum drying method for fig. These results may be due to a slight mass transfer rate in the vacuum drying method leading to a higher drying time in tested samples than in other drying methods used (Abul-Fadl et al., 2015).

The chemical composition of fresh and dried fig was also reviewed by Arvanitia et al. (2019), presents the main phytochemical compounds found in fresh and dried figs of different varieties, describes the analytical methods used for their determination, and discuss the antioxidant capacity and the potential effects of figs in human health.

Abul-Fadl et al. (2015) have investigated results concerning the effect of different drying methods on the chemical composition of dried figs produced. No significant differences were observed between fig samples dried by AD and VD compared with the samples dried by the combined MWAD and MWVD in moisture, ash, and crude fiber contents. On the other side, significant differences were found between the dried fig by using the AD method and the samples dried by MWAD, and also with the tested samples dried by using VD and MWVD of protein, lipid, and total sugars significantly reducing sugars. At the same time, the tested samples dried by using AD have the lowest content of the previous components of protein, lipid, total sugars, and reducing sugars.

4.2 Effects on the Chemical Composition and Nutraceutical Content

It is already known that drying fruit and vegetables using high temperature and for a long drying time by conventional heating damages the quality of the final dried products (Viswanathan et al., 2003). Almost all dried fruits provide essential nutrients and an array of health protective bioactive ingredients that help to reduce the risk of illness by preventing chronic diseases. Natural products have the potential to be used as therapeutic drugs for humans and livestock species. Along with their analogs, such compounds can also act as intermediates to produce valuable drugs (Makkar et al., 2009).

Fig (*Ficus carica* L.) is one of the most important agricultural products of the tropic and subtropics areas. In the Middle East and the Mediterranean region, the fig has been included in the diet since the ancient years, and it is considered the symbol of longevity. In the northern Mediterranean region, fig trees produce one or two crops per year, depending on the cultivar.

In general, most of the monitoring studies have demonstrated that the phytochemical composition is often affected by the fig variety, but it is also strongly dependent on other factors such as the color, the part of the fruit, the level of maturity, and also the drying process that has been used (Harzallah et al., 2016).

The fig is a healthy fruit with numerous medicinal properties that may reduce the risk of cancer and cardiovascular diseases. Fig fruit is consumed fresh, dried, preserved, canned, and candied. It is used for alcohol and wine production in the Mediterranean region and Europe for fig-coffee preparation. Fresh and dried figs are especially rich in fiber, trace minerals, antioxidant polyphenols, proteins, sugars, organic acids, and volatile compounds that provide a pleasant characteristic aroma. Dried figs can be stored for 6–8 months (Oliveira et al., 2009). There are many physicochemical quality characteristics of dried fig produced, such as color (O.D. at 340 nm), T.S.S. (%), titratable acidity (T.A. % as citric acid), pH value, rehydration ratio, and shrinkage (%), which are played an essential role in assessing their quality and palatability as well as the consumer acceptability of this product.

Various studies have estimated the phytochemical composition in whole fruit of fresh and dried fig varieties and their skin and pulp fractions. Fresh figs were subjected to two different drying processes: sun drying and oven drying. To assess their effect on the nutritional and health-related properties of figs, sugars, organic acids, single phenolics, total phenolics, and antioxidant activity were determined before and after processing Slatna et al., 2011.

The nutritional profile of dried fig fruit has shown its potential health benefits (Table 29.1). It has been revealed that the dried fruit of fig has carbohydrates as a major component (73.5%) that corresponds to its high energy value (317.7 kcal). Dried fig fruit has a very low amount of fat (0.56%), so it can be helpful for weight loss. A moderate amount of protein (4.67%) was found in the dried fruit, while dietary fiber content (3.68%) was good. Figs contain both soluble and insoluble dietary fiber with many health benefits. Dried figs contained moisture (16.63%) and high ash content (4.65%). Moisture content affects the texture, taste, appearance, and stability of foods related to the storage attributes of the dried fruit. The ash content measures the total amount of minerals present within a food. Mineral content was also analyzed using ICP-OES.

Dried fig was an excellent source of minerals like Sr, Ca, Mg, P, and Fe (Table 29.2). A relatively high amount of Strontium was found in fig. Strontium has been found to contribute to good bone health. A patented form of strontium called strontium ranelate is used to treat postmenopausal osteoporosis that reduces the risk of vertebral and hip fractures. It is the first antiosteoporotic agent that appears to simultaneously increase bone formation and decrease bone resorption, thus resulting in the creation of new bone (Reginster et al., 2007). Calcium is crucial for bones,

Table 29.1 The nutritional profile of dried fig fruit (Soni et al., 2014)

Sample	Energy (Kcal/100 g)	Total Carbohydrate	Fat	Protein	Dietary fiber	Moisture	Ash
Dried fig fruit	317.7	73.5	0.56	4.67	3.68	16.63	4.65

Table 29.2 Mineral content in dried fig fruit (Soni et al., 2014)

Analyte	Concentration (ppm)
Sr	Saturated
Ca	1545.46
Mg	679.04
P	365.75
Fe	29.49
Zn	9.87
Cu	5.02
Mn	4.75
Sb	0.298
As	1.669
Be	N.D.
Cd	0.0034
Cr	1.47
Co	0.032
Pb	0.680
Li	0.245
Mo	0.026
Ni	1.178
Se	0.790
Tl	1.5686
Ti	0.3727
Sn	1.329

N.D not determined

maintaining overall health, and essential for strong bones and teeth. Magnesium is needed for enzyme action, strong bones and teeth, balanced hormones, and a healthy nervous and cardiovascular system. Phosphorous is responsible for the growth and repair of body cells and tissues. Iron is needed for the production of red blood cells and enzymes (Soni et al., 2014).

The major types of phytochemical compounds in fresh and dried figs include phenolic acids, flavonoids, and carotenoids. Phenolic acids and flavonoids are the major types of phytochemical compounds that have been found in both fresh and dried figs. Their levels are strongly influenced by factors such as the color, the part of the fruit, maturity, and the drying process. Fresh fig was known to be contained an adequate amount of antioxidants, and thus, it is necessary to research known a drying method used led to the excessive retention of antioxidant compounds that attain the high nutritive and healthy effect of fig products. All tested antioxidants

(L-ascorbic acid, total phenolic, and flavonoid compounds) were significantly reduced by all drying methods used to determine their original fresh-fig fruit level. The results indicate that adequately dried figs can be used as a good source of phenolic compounds. Gallic acid, chlorogenic acid, rutin, quercetin-3-*O*-rutinoside, and epicatechin are the most predominant phenolic acids and flavonoids in dried and fresh fig varieties. On the other hand, the literature has reported contradictory results regarding air- and sun-drying effects on the total content of phytochemical compounds and the concentrations of individual phenolic compounds and carotenoids in figs (Arvaniti et al., 2019). Both fresh and dried figs contain various carotenoid compounds; the structures of the most commonly found compounds are lutein, zeaxanthin, β-cryptoxanthin, α-carotene, β-carotene, and lycopene.

4.3 Nutritional Traits Modification

Many studies have demonstrated that daily intake of fruit and vegetables reduces chronic-degenerative diseases. In other investigations, it has been observed that fruit- and vegetable-rich diets protect against different diseases, including cancer and cardiovascular diseases. Etiology for these diseases points to the free radicals as promoters of protein, nucleic acids, and cellular lipids oxidations that damage biological systems; fruits and vegetables contain a significant number of components with antioxidant activity, such as flavonoids, carotenoids, and vitamins C and E (Lam et al., 2005).

Furthermore, it has been demonstrated that figs are an essential source of minerals and vitamins such as iron, calcium, potassium, thiamin, and riboflavin. Furthermore, figs are sodium, fat, and cholesterol free; they contain at least 17 amino acids, with aspartic and glutamic acids at the highest concentration. In addition, fig contains a relatively high fiber content (5.8% w/w), and more than 28% of it is soluble, so it helps control blood sugar and cholesterol and reduce weight (Sadhu, 1990).

Fig varieties with dark skin contain high levels of polyphenols, anthocyanin, and flavonoids, together with higher antioxidant activity than fig varieties of lighter skin (Solomon et al., 2006). Various plant parts like fruit, root, and figs leaves have numerous therapeutic benefits. They are used in traditional medicines to treat gastrointestinal disorders (colic, indigestion, loss of appetite, and diarrhea), respiratory disorders (sore throats, coughs, and bronchial problems), and anti-inflammatory cardiovascular disorders as an anti-inflammatory, antispasmodic remedy. *Ficus carica* has also been found to have antidiabetic, hypolipidemic, hepatoprotective, antispasmodic antipyretic, antibacterial, antifungal, scavenging activity, and immune response (Duke et al., 2002).

Today, the fig is a remarkable world crop due to its consumption as fresh or dry fruit.

The nutritional profiling of the dried fig fruit indicates that it is a good source of carbohydrates and minerals like strontium, calcium, magnesium, phosphorus, and iron. It has average protein and dietary fiber content with very low fat. Phytochemistry of the fruit revealed the presence of total phenolics, flavonoids, alkaloids, saponins, and other secondary metabolites that contribute to its high antioxidant activity (Soni et al., 2014). Dried figs are high in fiber and carbohydrates and low in fat, making them healthy food choices. Vitamin A is retained during drying; however, because vitamin A is light sensitive, food containing it should be stored in dark places. Figs are one of the highest sources of calcium and fiber in a plant. Furthermore, dried figs are the richest in fiber and contain essential minerals and vitamins.

The antioxidant activity reported that the antioxidant capacity is frequently due to phenolic acids, namely gallic acid, caffeic acid, chlorogenic acid, and flavonoids such as quercetin, catechin rutin, and kaempferol. So far, there are controversial results in the literature on the role of the drying process on the antioxidant capacity of figs. In some studies, it has been reported that the high antioxidant capacity is due to dried figs were higher than those of fresh figs (Slatna et al., 2011; Chang et al., 2016), while in others, it has been reported that the sun-drying procedure reduced the antioxidant capacity of figs (Bachir Bey et al., 2016; Kamiloglu et al., 2015).

5 Shelf-Life Extension Using Modified Atmosphere Packaging

The development of satisfactory methods for the shelf-life extension that ensure quality maintenance of products with minimum loss has drawn the attention of food technologists. Modified Atmosphere Packaging MPA or Controlled Atmosphere Packaging is making a big impact in produce packaging. This advanced process uses micro-perforation technology to alter the package atmosphere. The ability to effectively manage respiration gasses helps to improve product shelf life considerably. Over the last years, modified atmosphere packaging (MAP) has received increasing attention as a method of food preservation. MAP is defined as the enclosure of food products in gas-barrier materials in which the gaseous environment has been changed (Sivertsvik et al., 2002). Packaging technologies are important to protect products against deteriorative effects, including microbial, biochemical, and physical activities from environmental influences. This involves retardation of spoilage, the extension of shelf-life, and quality maintenance in packed food. Other packaging functions include containment, convenience, marketing, and communication (Restuccia et al., 2010). The rate of deterioration during the storage of a product depends on the biochemical compositions of substrates and metabolites in the tissue, the microbial contamination, and the condition of storage (Masniyom, 2011).

Modified Atmosphere Packaging is an advanced packaging technology that creates an optimal environment to extend produce shelf life. Herein, the question is How Modified Atmosphere Packaging Extends the Shelf Life of the product? The modified atmosphere created by MAP packaging limits oxygen flow and moisture evaporation. The controlled gas exchange reduces respiration rates by limiting the oxygen available in the package. Too much oxygen also increases bacterial decay and rapid spoilage. Excess oxygen also encourages sprouting, discoloration, and blemishing. A low oxygen environment produces lower respiration rates. Lower respiration rates extend shelf life by preventing aging and ensuring moisture retention in picked produce.

Generally, MAP is used for extending the shelf-life period of fresh or minimally processed foods. Apart from other perishable products, attempts have been carried out to prolong the shelf-life of figs by using MAP. It is already known that the fig is a nutritious fruit, richer in fiber, potassium, calcium, and iron, and is free of sodium, fat, and cholesterol. Figs also are an essential source of vitamins, amino acids, and antioxidants. However, the fig is very sensitive to microbial growth even under cold storage; thus, it is essential to consider alternative processes to extend its shelf life. Most commercial production is in dried or otherwise processed forms since the ripe fruit does not transport well and, once picked, does not keep well. The ripening stages at harvest and post-harvest preservation conditions greatly influence the fig fruit's quality (Ayhan & Kara cay, 2011). Modified atmosphere packaging could provide an alternative method of fruit preservation, longer shelf life, and added value. Ayhan and Kara cay (2011) showed that considering the sensory, physical, and chemical qualities, the commercial shelf life of fresh figs is 15 days in the air and low oxygen applications at 4 °C.

Alturki (2013) examined and monitored the quality parameters to determine the shelf life of fresh figs under refrigerated Modified Atmosphere Packaging (MAP) conditions. The fruit quality attributes such as decay, fruit size, external color, firmness, weight loss, water activity, figs fruits' water content, and total soluble solids were measured during storage at 4 °C for more than 42 days. Modified atmosphere packaging with 20% CO_2 successfully reduced the rate of ripening process while the control treatment figs showed signs of ripening development such as significant weight loss and change in figs color. However, the decay incidence was reduced and delayed by all the MAP treatments. At the end of storage, the weight loss was less than 1% for all the MAP treatments. The fruit firmness decreased with an increase in weight loss in each treatment after 28 days of storage. There was no significant difference in total soluble solids among the MAP treatments. The shelf life of fresh figs packed in the unsealed cardboard box was less than 14 days.

After storage, all MAP treatments had lost less than 1% of their weight. After 28 days of storage, the fruit firmness reduced as weight loss increased in each condition. Total soluble solids did not differ significantly across the MAP treatments. Fresh figs placed in an unsealed cardboard box had less than 14 days of shelf life. Finally, because MAP improved the external look of the fresh figs, it should be used commercially in the fresh fig sector for fruit preservation (Alturki, 2013).

Fruits of *Ficus carica L.* have an exceptional flavor, color, and scent. It does, however, have a short shelf life due to its morphological peculiarities. Martinez-Damian et al. (2020) attempt to determine the effect of environment packing on the nutritional quality and look of figs held at 1 °C. Fruits with decreased weight loss and firmness were seen using packaging and cold, although citric acid and Vitamin C levels remained the same. There was no significant fluctuation in the content of Total Anthocyanins, total phenols, or antioxidant capacity except during the last evaluation period. On the other hand, a considerable impact on the reduction of respiration and ethylene generation was seen, which was substantiated by the preservation of sensory features of the fruit (texture, color, appearance, marketing level, scent, and flavor) from excellent too good. The use of cold (1 °C) and the creation of passive atmospheres (packing) are effective aids in keeping the nutraceutical quality and look of fig fruits, both of which are highly valued by consumers and marketers of this fruit (Martínez-Damián et al., 2020).

In another case, 1-methylcyclopropene (1-MCP) has been frequently used to improve the shelf life and quality of fruits and vegetables. Researchers discovered that using paper containing 1-MCP at room temperature can significantly suppress cell wall degrading enzymes, reduce cell wall polysaccharide disintegration, and delay softening in Younai plums.

Furthermore, to increase the storage duration of fruits, 1-MCP is frequently combined with other preservation methods such as UV-C irradiation (Pristijono et al., 2018), calcium chloride (Li et al., 2014), and methyl jasmonate (Ku et al., 2013). Thus, compared to either therapy alone, a combination of 1-Methylcyclopropene (1-MCP) and modified atmosphere (MA) treatment has been proven to be more effective in preserving particular fruits.

The impact of MA and MA+1-MCP treatments on fig fruit postharvest quality after 30 days at 10.5 °C. Compared to the control and MA groups, the MA+1-MCP treatment significantly enhanced fruit texture, reduced weight loss and MDA accumulation, and suppressed ethylene production and respiration rate (Fig. 29.3). In conclusion, the MA+1-MCP treatment will be an effective way to preserve the fruit quality of figs during postharvest storage.

The color of the peel and flesh is the most evident and dependable indicator for customers to determine the quality of fruits and vegetables. Figure 29.3 shows that after 30 days of storage, the fig fruit had varying degrees of degradation. The flesh

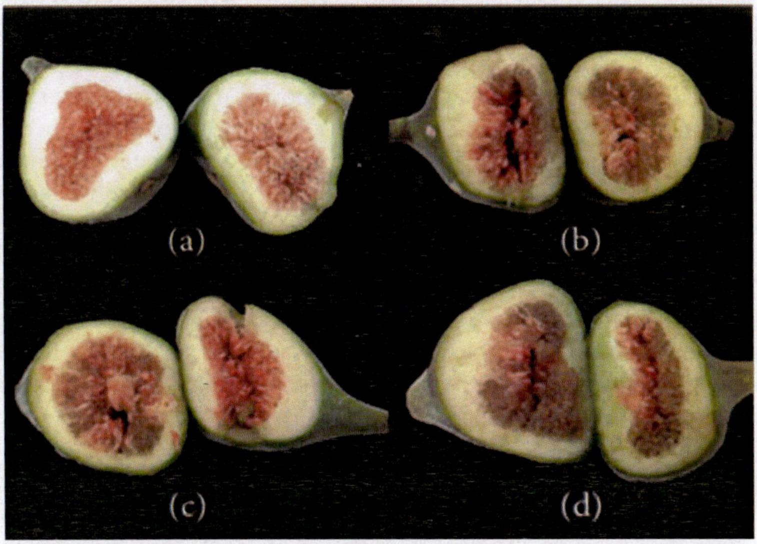

Fig. 29.3 Effects of MA and MA + 1-MCP treatments on deterioration development in fig fruit after storage for 30 days at −1 °C: (**a**) fresh fruits; (**b**) MA treatment and stored for 30 days; (**c**) MA + 1-MCP treatment and stored for 30 days; (**d**) the control fruit stored for 30 days (Song et al., 2019)

of MA+1-MCP-treated fruits is superior to that of MA-treated and control fruits. The 2e flesh of MA+1-MCP-treated fruits is the best, whereas control fruits are nearly inedible.

In conclusion, MAP resulted in a better external appearance, and therefore the commercial application of MAP in the fresh fig industry should be introduced for fruit preservation.

6 Conclusions and Future Prospects of Fig Drying

There is a large amount of research, and available knowledge, carried out by academia, but the industry is not taking full advantage even in the more developed countries. A lack of interaction between researchers and industry is observed worldwide; consequently, industrial advancement is slow because of a mismatch between research and industrial needs. In less developed countries where the industry is not very important, there is a general feeling that drying is an easy operation and not too much input is needed, and anybody can do it. A consequence of this attitude has been the failure of many drying projects. Drying foods in general and figs, in particular, is a complex business, and a mere translation from other fields is not often advisable. Energy efficiency linked to environmentally friendly processes and products also appear as a growing trend, mainly in some developed countries. Therefore, there is a need to check existing data on drying kinetics to establish their applicability regarding the prevailing resistance. Indeed, there is a need to study products (isotherms and drying kinetics) not common in more developed countries where the leading research in the field has been carried out. This will lead to discovering new niches of activity. This is also true for countries where drying research is an active field where the use of off-quality fresh market products is often disregarded. One can say that there is a general concern about improving the final consumer products quality. Consequently, pretreatments (including enrichment), optimization and control of operating conditions, and a combination of fig drying strategies according to the drying stage should be studied better to migrate water, storage, and nutritional profile consumption.

Though good feeds have been achieved in drying technology, there is a need to develop cost-effective and energy-efficient drying methods further. The desirable aspects of fruits and vegetables need to be preserved during the drying process. Combinations of drying techniques produce products that are more appreciated by the consumers and reduce the heat-induced deterioration in fruits and vegetables. On the other hand, the consumption of high energy by instruments used for drying adds to the cost of the finished end product. The application of the natural drying method reduces the cost and is much easier to use. The major drawback of the natural drying method is that it is time-consuming, and the drying conditions cannot be manipulated compared to the artificial drying method. Future work could focus on developing natural drying solutions with greater efficiency. Drying kinetics has helped achieve good and effective results with minimal resource utilization. However, drying models that consider all the variables are still non-existing, making it a potential research area. Advances in dehydration techniques and the development of novel drying methods have enabled the preparation of a wide range of dehydrated products and convenience foods from fruits and vegetables, meeting the quality, stability, and functional requirements coupled with the economy. This has been made possible by sustained experimental studies to understand the theoretical and fundamental aspects of the process and optimization of the techniques to achieve a favorable combination of cost and quality. This paper reviews and highlights the developments of the most popular drying methods of fig and their products during the last decade covering theoretical aspects and practical applications with major emphasis on techniques that have received the maximum attention (Ahmed et al., 2013).

References

Abul-Fadl, M. M., Ghanem, T. H., & N. EL-Badry & Nasr, A. (2015). Effect of some different drying methods on quality criteria of dried fig fruits. *Current Science International, 4*, 548–566.

Adegbola, A. A., Adogbeji, O. V., Abiodun, O. I., & Olaoluwa, S. (2012). Design, construction and performance evaluation of low-cost electric baking oven. *Innovative Systems Design and Engineering, 3*(11), 38–49.

Ahmed, N., Singh, J., Chauhan, H., Gupta, P., Anjum, A., & Kour, H. (2013). Different drying methods: their applications and recent advances. *International Journal of Food Nutrition and Safety, 4*(1), 34–42.

Ahrens, D. C., Villela, F. A., & Doni Filho, L. (2000). Physiological and industrial of quality white oat seeds (*Avena sativa*) in intermittent drying. *Revista Brasileria de Sementes, 22*, 12–20.

Alturki, S. (2013). Utilization of modified atmosphere packaging to extend the shelf-life of fresh figs. *Biotechnology, 12*, 81–86.

Andreou, V., Thanou, I., Giannoglou, M., Giannakourou, M. C., & Katsaros, G. (2021). Dried figs quality improvement and process energy savings by combinatory application of osmotic pretreatment and conventional air drying. *Foods, 10*, 1846.

Anon, R. (1965). *How to make a solar cabinet dryer for agricultural produce do it yourself leaflet L6* (p. 9). Brace Research Institute.

Arvanitia, O. S., Samarasa, Y., Gatidoub, G., Thomaidis, N. S., & Stasinakisb, A. S. (2019). Review on fresh and dried figs: chemical analysis and occurrence of phytochemical compounds, antioxidant capacity and health effects. *Food Research International, 119*, 244–267.

Ayhan, Z., & Kara cay, E. (2011). Preservation of the Bursa siyahı fresh fig under modified atmosphere packaging (MAP) and cold storage. *International Journal of Agriculture Sciences, 1*, 1–9.

Babalis, S. J., & Belessiotis, V. G. (2004). Influence of the drying conditions on the drying constants and moisture diffusivity during the thin-layer drying of figs. *Journal of Food Engineering, 65*, 449–458.

Babalis, S. J., Papanicolaou, E., Kyriakis, N., & Belessiotis, V. G. (2006). Evaluation of thin-layer drying models for describing drying kinetics of figs (*Ficus Carica*). *Journal of Food Engineering, 75*, 205–214.

Bachir Bey, M., Richard, G., Meziant, L., Fauconnier, M. L., & Louaileche, H. (2016). Effects of sun-drying on physicochemical characteristics, phenolic composition and in vitro antioxidant activity of dark fig varieties. *Journal of Food Processing and Preservation, 41*, 1–8.

Baraket, G., Saddoud, O., Chatti, K., Mars, M., Marrakchi, M., Trifi, M., & Salhi-Hannachi, A. (2009). (ITSs) region of the nuclear ribosomal DNA (nrDNA) in fig cultivars (*Ficus carica* L.). *Scientia Horticulturae, 120*, 34–40.

Bezerra, C. V., Amante, E. R., de Oliveira, D. C., Rodrigues, A. M., & da Silva, L. H. M. (2013). Green banana (*Musa cavendishii*) flour obtained in spouted bed-effect of drying on physicochemical, functional and morphological characteristics of the starch. *Industrial crops and products, 41*, 241–249.

Chae, H. Y., & Hong, J. H. (2016). Quality characteristics of spray dried powder from unripe fig extract. *Korean Journal of Food Preservation, 23*(3), 355–360.

Chandrasekaran, S., Ramanathan, S., & Basak, T. (2013). Microwave food processing a review. G time of dehydrated fig fruits by using different drying methods. *Food Research International, 52*(1), 243–261.

Chang, S. K., Alasalvar C, & Shahidi, F. (2016). Review of dried fruits: Phytochemicals, antioxidant efcacies, and health benefts. *Journal of Functionnal Foods, 21*,113–132.

Chegini, G. R., & Ghobadian, B. (2007). Spray dryer parameters for fruit juice drying. *World Journal of Agricultural Sciences, 3*(2), 230–236.

Chua, K. J., Mujumdar, A. S., Hawlader, M. N. A., Chou, S. K., & Ho, J. C. (2001). Batch drying of Bannana pieces-effect of stepwise change in drying air temperature on drying kinetics and product colour. *Food Research International, 34*, 721–731.

Darabi, H., Zomorodian, A., Akbari, M. H., & Lorestani, A. N. (2015). Design a cabinet dryer with two geometric configurations using CFD. *Journal of Food Science and Technology, 52*(1), 359–366.

Deng, L. Z., Mujumdar, A. S., Zhang, L. Z., Yang, X. H., Wang, J., Zheng, Z. A., Gao, Z. J., & Xiao, H. W. (2017). Chemical and physical pretreatments of fruits and vegetables: Effects on drying characteristics and quality attributes-a comprehensive review. *Critical Reviews in Food Science and Nutrition, 59*(9), 1408–1432.

Dhaouadi, K., Belkhir, M., Akinocho, I., Raboudi, F., Pamies, D., Barrajón, E., Estevan, C., & Fattouch, S. (2014). Sucrose supplementation during traditional carob syrup processing affected its chemical characteristics and biological activities. *LWT-Food Science and Technology, 57*, 1–8.

Duke, J. A., Bugenschutz-godwin, M. J., & Du collier, J. & Duke, P. K. (2002). *Hand book of medicinal herbs* (2nd ed.). CRC Press.

FAO. (1985). Expert Consultation on Planning the Development of Sun-drying Techniques in Africa

Fayose, F., & Huan, Z. (2016). Heat pump drying of fruits and vegetables: principles and potentials for sub-Saharan Africa. *International Journal of Food Science, 16*, 1–8.

Gharsallaoui, A., Rouaut, G., Chambin, O., Voilley, A., & Saurel, R. (2007). Applications of spray-drying in microencapsulation of food ingredients: An overview. *Food Research International, 40*(9), 1107–1121.

Giri, S. K., & Prasad, S. (2007). Drying kinetics and rehydration characteristics of microwave vacuum and convective hot-air dried mushrooms. *Journal of Food Engineering, 78*, 512–521.

Guiné, R. (2018). The drying of foods and its effect on the physical-chemical, sensorial and nutritional properties. *International Journal of Food Engineering, 2*(4), 93–100.

Gwala, W., & Padmavati, R. (2016). *Comparative study of indirect solar drying, electric tray drying and open sun drying of pineapple slices using drying kinetics and drying models.* International Journal of Latest Technology in Engineering, Management.

Harzallah, A., Bhouri, A. M., Amri, Z., Soltana, H., & Hammami, M. (2016). Phytochemical content and antioxidant activity of different fruit parts juices of three figs varieties grown in Tunisia. *Industrial Crops and Products, 83*, 255–267.

Heredia, A., Barrera, C., & Andrés, A. (2007). Drying of cherry tomato by a combination of different dehydration techniques. Comparison of kinetics and other related properties. *Journal of Food Engineering, 80*(1), 111–118.

Janowicz, M., Domian, E., Lenart, A., & Pomarańska-Łazuka, W. (2008). Profile of convective drying process of osmotically dehydrated apples in sucrose solution. *Zywnosc Nauka Technologia Jakosc, 4*(59), 190–198.

Kalantari, M., Niakousari, M., Haghighi-Manesh, S., & Rasouli, M. (2018). Fig extract drying: The relationship between the main operating parameters of a pilot-scale spray dryer and product specifications. *Food Science Nutrition*, 325–333.

Kamiloglu, S., Toydemir, G., Boyacioglu, D., Beekwilder, J., Robert, D., Hall, R. D., & Capanoglu, E. (2015). A review on the effect of drying on antioxidant potential of fruits and vegetables. *Critical Review in Food Science and Nutrition, 56*, 110–129.

Khraisheh, M. A. M., McMinn, W. A. M., & Magee, T. R. A. (2004). Quality and structural changes in starchy foods during microwave and convective drying. *Food Research International, 37*, 497–503.

Killeen, M. J. (2000). Spray drying and spray congealing of pharmaceuticals. In J. Swarbrick & J. C. Boylan (Eds.), *Encyclopedia of pharmaceutical technology*. PharmaceuTech, Inc.

Ku, K. M., Choi, J. H., Kim, H. S., Kushad, M. M., Jeffery, E. H., & Juvik, J. A. (2013). Methyl jasmonate and 1-methylcyclopropene treatment effects on quinone reductase inducing activity and post-harvest quality of broccoli. *PLoS One, 8*(10), e77127.

Laguerre, J. C., & Hamoud Agha, M. M. (2018). *Microwave heating for food preservation*. Food Preservation -In book: Food Preservation - From Basics to Advanced Technologies. https://doi.org/10.5772/intechopen.82543.

Lam, H. S., Proctor, A., Howard, L., & Cho, M. J. (2005). Rapid fruit extracts antioxidant capacity determination by Fourier transforms infrared spectroscopy. *Journal of Food Science, 70*(9), 545–549.

Li, L., Zhaojun, B., Xihong, L & Ting Xue. (2014). Effect of 1-methylcyclopropene and calcium chloride treatments on quality maintenance of 'Lingwu Long' Jujube fruit. *Journal of Food Sciences Technology, 51*(4), 700–707.

Maisnam, D., Prasa, R., Anirban, D., Sawinder, K., & Chayanika, S. (2017). Recent advances in conventional drying of foods. *Journal of Food Technology and Preservation, 1*, 25–34.

Makkar, H. P. S., Norvsambuu, T., Lkhavatsere, S., & Becker, K. (2009). Plant secondary metabolites in some medicinal plants of Mongolia used for enhancing animal health and production. *Tropicultura, 3*, 159–167.

Martinez-Damian, M. T., Cruz-Arvizu, O., & Cruz-Alvare, O. (2020). Effect of modified atmosphere packaging on nutraceutical quality and overall appearance of figs stored at 1 °C. *Notulae Botanicae Horti Agrobotanici Cluj-Napoca, 48*(4), 2292–2305.

Martínez-García, J. J., Gallegos-Infante, J. A., Rocha-Guzmán, N. E., Ramírez-Baca, P., Candelas-Cadillo, M. G., & González-Laredo, R. F. (2013a). Drying parameters of half-cut and ground figs (*Ficus carica* L.) var. mission and the effect on their functional properties. *Journal of Engineering*, 1–8.

Martínez-García, J. J., Gallegos-Infante, J. A., Rocha-Guzmán, N. E., Ramírez-Baca, P., Candelas-Cadillo, M. G., & González-Laredo, R. F. (2013b). Drying parameters of half-cut and ground figs (*Ficus carica* L.) var. mission and the effect on their functional properties. *Journal of Engineering*, 710–830.

Masniyom, P. (2011). Deterioration and shelf-life extension of fish and fishery products by modified atmosphere packaging. *Songklanakarin Journal of Science & Technology, 33*(2).

Misha, S., Mat, S., Ruslan, M. H., Sopian, K., & Salleh, E. (2013). Review on the application of a tray dryer system for agricultural products. *World Applied Sciences Journal, 22*(3), 424–433.

Mujić, I., Bavcon Kralj, M., Jokić, S., Jarni, K., Jug, T., & Prgomet, Ž. (2012). Changes in aromatic profile of fresh and dried fig-the role of pre-treatments in drying process. *International Journal of Food Science & Technology, 47*(11), 2282–2288.

Nagaraja, K., Sunil, C. K., Chidanand, D. V., & Ramachandra, M. (2016). Drying kinetics of fig (*Ficus carica*. L) under various drying methods. *Journal of Agricultural Engineering, 53*(4), 42–50.

Oliveira, A. P., Valentao, P., Pereira, J. A., Silva, B. M., Tavares, F., & Andrade, P. B. (2009). *Ficus carica L.*: Metabolic and biological screening. *Food and Chemical Toxicololgy, 47*, 2841–2846.

Ong, S. P., & Law, C. L. (2010). Hygrothermal properties of various foods, vegetables and fruits. *Drying of Foods, Vegetables and Fruits, 1*, 31–58.

Patel, R. P., Patel, M. P., & Suthar, A. M. (2009). Spray drying technology: An overview. *Indian journal of Science and Technology, 2*(10), 44–47.

Patel, A. H., Shah, S. A., & Hitesh, B. (2013). Review on solar dryers for grain vegetables and fruits. *International Journal of Engineering Research Technology, 2*(1).

Phadke, P., Walke, P., & Kriplani, V. M. (2015). Direct type natural convection solar dryer: A review. *International Journal of Advanced Research in Science and Engineering, 4*, 256–262.

Ponting, J. D. (1973). Osmotic dehydration of fruits – recent modifications and application. *Process Biochemistry, 8*, 18–20.

Pristijono, P., Bowyer, M.C., Scarlett, C. J., Vuong, Q. V. & John B. Golding, J.B(2018). Combined postharvest UV-C and 1-methylcyclopropene (1-MCP) treatment, followed by storage continuously in low level of ethylene atmosphere improves the quality of Tahitian limes. Journal of Food Science and Technology-Mysore, 55, 2467–2475

Radojčin, M., Pavkov, I., Bursać Kovačević, D., Putnik, P., Wiktor, A., Stamenković, Z., & Gere, A. (2021). Effect of selected drying methods and emerging drying intensification technologies on the quality of dried fruit: A review. *Processes, 9*(1), 132.

Reginster, J. Y., Malaise, O., Neuprez, A., & Bruyere, O. (2007). Strontium ranelate in the prevention of osteoporotic fractures. *International Journal of Clinical Practice, 61*, 324–328.

Restuccia, D., Spizzirri, G. U., Parisi, I. O., Cirillo, G., Curcio, M., Iemma, F., Puoci, F., Vinci, G., & Picci, N. (2010). New EU regulation aspects and global market of active and intelligent packaging for food industry applications. *Food Control, 21*, 1425–1435.

Sadhu, M. K. Fig. (1990). *In: Fruits: Tropical and Subtropical*, Edited by: Kose, T. K. and Mitra, S. K. 650–663. Calcutta, India: Naya Prokash.

Sharifian, F., Motlagh, A. M., & Nikbakht, A. M. (2012). Pulsed microwave drying kinetics of fig fruit (*Ficus carica L.*). *Australian Journal of Crop Science, 6*(10), 1441–1447.

Sharifian, F., Modarres-Motlagh, A., Komarizade, M. H., & Nikbakht, A. M. (2013). Colour change analysis of fig fruit during microwave drying. *International Journal of Food Engineering, 9*(1), 107–114.

Sivertsvik, M., Jeksrud, W. K., & Rosnes, J. T. (2002). A review of modified atmosphere packaging of fish and fishery products-significance of microbial growth, activities and safety. *International Journal of Food Science & Technology, 37*, 107–127.

Slatna, A., Klancar, U., Stampar, F., & Veberic, R. (2011). Effect of drying of figs (*Ficus carica* L.) on the contents of sugars, organic acids, and phenolic compounds. *Journal of Food Agriculture and Food Chemistry, 59*(21), 11696–11702.

Solomon, A., Golubowicz, S., Yablowicz, Z., Grossman, S., Bergman, M., Gottlieb, H. E., & Flaishman, M. A. (2006). Antioxidant activities and anthocyanin content of fresh fruits of common fig (*Ficus carica* L.). *Journal of Agricultural and Food Chemistry, 54*(20), 7717–7723.

Song, C., Li, A., Chai, Y., Li, Q., Lin, Q., & Duan, Y. (2019). Effects of 1-Methylcyclopropene combined with modified atmosphere on quality of fig (*Ficus carica* L.) during postharvest storage. *Journal of Food Quality*.

Soni, N., Mehta, S., Satpathy, S., & Gupta, R. K. (2014). Estimation of nutritional, phytochemical, antioxidant and antibacterial activity of dried fig (*Ficus carica*). *Journal of Pharmacognosy and Phytochemistry, 3*(2), 158–165.

Sonone, E. V. S., Unde, P. A., & Kad, V. P. (2016). Effect of spray dryer parameters on different properties of fruit juice powder. *International Journal of Advanced Engineering, Management and Science, 2*(8), 1301–1312.

Soponronnarit, S., Nathakaranakule, A., Wetchacama, S., Swasdisevi, T., & Rukprang, P. (2007). Fruit drying using heat pump. *International Energy Journal, 20*(1).

Sukowati, Y. K., Johan, A., & Murwani, R. (2019). Ethanol extracts of *Ficus carica* fruit and leaf normalize high serum lipid profile, TNF-α, and MDA due to high fat diet in Sprague Dawley rat. *Current Research in Nutrition and Food Science, 07*(3), 772–782.

Tous, J., & Fergueson, L. (1996). Mediterranean fruits. In J. Janick (Ed.), *Progress in New crops*. Atlas Press.

Veberic, R., Jakopic, J., & Stampar, F. (2008). Internal fruit quality of figs (*Ficus carica* L.) in the Northern Mediterranean Region. *Italian Journal of Food Science, 20*(2), 255–262.

Viswanathan, R., Jayas, D. S., & Hulasare, R. B. (2003). Sorption isotherms of tomato slices and onion shreds. *Biosystems Engineering, 86*(4), 465–472.

Wang, J., Law, C. L., Nema, P. K., Zhao, J. H., Liu, Z. L., Deng, L. Z., Gao, Z. J., & Xiao, H. W. (2018). Pulsed vacuum drying enhances drying kinetics and quality of lemon slices. *Journal of Food Engineering, 224*, 129–138.

Xu, Y. Y., Zhang, M., Mujumdar, A. S., Duan, X., & Jin-Cai, S. (2006). A two-stage vacuum freeze and convective air drying method for strawberries. *Drying Technology, 24*, 1019–1023.

Zhang, M., & Xu, Y. Y. (2003). Research developments of combination drying technology for fruits and vegetables at home and abroad. *Journal of Wuxi University of Light Industry, 22*(6), 103–106.

Zhang, M., Li, C. L., & Ding, X. L. (2005). Effects of heating conditions on the thermal denaturation of white mushroom suitable for dehydration. *Drying Technology, 23*(5), 1119–1125.

Chapter 30
Chemistry and Functionality of Processed Figs

Asad Nawaz, Noman Walayat, Ali Hassan, Maryam Chaudhary, and Ibrahim Khalifa

1 Introduction

Figs are widespread dry fruits usually grown in warm and dry climates. The best condition for intensive figs cultivation is a semi-arid climate and irrigation. The figs production worldwide is more than one million tons, and it is most common in the Mediterranean region. In this region, figs have been cultivated for centuries. The trees of figs produce one or two crops yearly, depending on the cultivar type. The primary crop is grownup from flowers introduced in the previous year, and the fruit gets ripened at the start of summer. The second main crop is produced from flowers that appear in the present season, and the fruit gets ripened in late summer. Hence, the growth of both crops is evident in different climatic conditions. Fruits cultivated from the two crops may also vary in shape and size. Figs are commonly consumed fresh and also peeled in dried form. Naturally, fresh fruits have a short post-harvest life span of 7–10 days, but a combination of favourable conditions and carbon dioxide in the atmosphere enhances the fruit shelf life and can be stored for up to

A. Nawaz
Shenzhen Key Laboratory of Marine Microbiome Engineering, Institute for Advanced Study, Shenzhen University, Shenzhen, China

N. Walayat
College of Food Science and Technology, Zhejiang University of Technology, Hangzhou, China

A. Hassan · M. Chaudhary
National Institute of Food Science and Technology, University of Agriculture, Faisalabad, Pakistan

I. Khalifa (✉)
Food Technology Department, Faculty of Agriculture, Benha University, Moshtohor, Egypt
e-mail: ibrahiem.khalifa@fagr.bu.edu.eg

M. F. Ramadan (ed.), *Fig* (Ficus carica)*: Production, Processing, and Properties*, https://doi.org/10.1007/978-3-031-16493-4_30

2–4 weeks. Figs are also commonly consumed as dried fruit, as the drying method extends their storability and maintains quality characteristics (Veberic et al., 2008).

Figs are used in primary and processed forms to produce traditional and industrial products like jams, infusions, wines, liqueurs, spirits, etc. Fig fruit is a perishable food item, primarily consumed in dried form. Fresh or dry figs are considered very important for their high nutritive value, high mineral content, fats, sugars, calcium, and presence of other non-nutritive elements such as fiber, water, and antioxidants like phenolic compounds that pose numerous health benefits. In addition, the decoctions, infusions, and processing of fig leaves have traditionally been used to treat diseases because of their therapeutic effect and chemical characteristics (FR, 2013).

The processing of figs in dried form is nutritious with high potential benefits. In addition, dried fig is a rich source of bioactive compounds with health-promoting benefits. However, dried figs' chemistry and quality characteristics depend on different factors like harvesting practices, stages of ripening, geographical region, and processing techniques. Before harvesting figs, the fruits are left on the tree until they get fully ripe, dried partially, and fall to the ground. Then, semi-dried figs are placed on trays made of wood for further drying and kept for 2–3 days to achieve 22–24% dry matter. Finally, these dried figs are transferred to industries for further processing to increase and maintain the final product quality. The processing steps involved in drying figs are sorting, fumigation, cleaning, washing, proper shaping, and final packaging.

It is essential to process or wash figs under controlled conditions as different treatments during washing of figs may affect the textural, sensorial, chemical, and color properties and may lead to loss of texture, flavor, and essential nutrients of the final product. Furthermore, the washing of figs is helpful in the removal of pesticides or microorganisms, therefore minimizing the adverse effects on human health. The most used technique for washing figs in food industries is with chlorinated water (using sodium chloride) because it is cost-effective and efficiently decontaminates the microorganisms. Although chlorine is used as a popular sanitizing agent, improper usage may cause deterioration of quality by reacting with other organic compounds present in food items. Therefore, it is imperative to watch or monitor the free level of chlorine, as if it is not monitored properly, it may lead to the generation of carcinogenic by-products. Recently, different techniques such as ultrasound and pulse electric field have been used as an alternative treatment to chemicals used during washing. The most promising technology is the US technique which is non-thermal technology as it led towards microbial safety and shelf-life extension. It also retains the nutritional value, sensory, chemical, and functionality of food items.

Additionally, this technique has less use of the chemical sanitizing agent, minimal human contact, and consistent and uniform cleaning. However, the US technique alone is not sufficient for proper surface decontamination. Therefore, different studies suggested the combined use of the US technique and sanitizing agents in washing during the processing of dried figs to retain their chemical and physical properties (Gençdağ et al., 2021).

2 Types of Processed Figs

One of the Mediterranean basin's oldest fruit species, the common fig may produce two harvests every year. However, the quality of the crops may vary regarding season, variety, and harvesting stage (Pereira et al., 2017). Figs may be consumed fresh, dried, or processed into various goods, including jams, jelly, squashes, smoothies, wines, and bakeries (Barolo et al., 2014). Generally, processed figs are consumed compared to fresh ones due to the high moisture content in fresh fruits (Teruel et al., 2021). Fruit drying is the simplest method of preserving fig fruit that is usually processed utilizing some unit operations, such as drying and freezing (Yao et al., 2021). The conventional sun-drying procedure produces a wide range of quality food owing to its ability to retain phenolic and flavonoids compound with minimum loss of nutrition (Manoj et al., 2018).

Meanwhile, automated air dehydration offers various benefits, including producing fruit with high sugar content and total phenolics. More recently, the freeze-drying method has widely been used for the drying of figs in order to preserve their original phenolic compounds and bioactive compounds (Martínez et al., 2013). Besides this, the processing of figs through fermentation has also been reported to retain a high amount of phenolic and bioactive compounds (Buenrostro-Figueroa et al., 2017). Interestingly, fig fruits have been frozen at −18 °C to preserve their nutritional content and make the best use of figs in off seasons (Petkova et al., 2019). Another study (Palmeira et al., 2019) also reported the freeze-drying of fig fruit parts (peel and pulp) at −80 °C and found that antioxidant and phenolic compounds were higher in the peel as compared to the pulp.

Similarly, another study (Bachir bey et al., 2014) also reported the freeze-drying of a dark-purple skin and red pulp to preserve the phenolic compounds. Thus, the freeze-drying method is widely used to preserve the phenolic and bioactive compounds. Furthermore, because fresh or dried fig fruits have a high concentration of beneficial chemicals, their use should be promoted as a possible healthy substitute for sweets. Thus, the right choice of processing is preferred to maintain its native nutrients.

Besides processing, fig fruits have been utilized in numerous processing foods, such as biscuits, jam, jelly, confectionery, soups, etc. (Teruel et al., 2021). It has been established that the use of fig fruits in different products resulted in a higher nutritional value and the accumulation of various bioactive compounds. For instance, the study reported that the addition of fig seed powder in bakery products, such as buns and muffins, increased fiber content and total phenolic and antioxidant activity (Chauhan, 2016). Another study reported the addition of fig seed powder replacing wheat flour in biscuits, which resulted in increased protein and fiber content and an upsurge in total phenolic compounds (Bölek, 2021). In addition, the use of fruits in desserts and smoothies has been encouraged or reported as an alternative to sugar, along with the increased phytochemicals (Jahromi & Niakousari, 2018). However, in some studies, the processed figs, especially the dried ones, resulted in lower total phenolic compounds in wines when dried figs were added compared to the fresh figs (Lu et al., 2021). In conclusion, the use of fig fruits to develop functional food has potential and practical significance.

3 Composition of Processed Figs

Figs have remained the main part of the diet and symbol of health in Middle East Countries. However, a complete understanding of the composition of processed figs is inevitable. Fresh and dried figs are high in fiber, amino acids (aspartic acid and glutamine), vitamins (thiamine, riboflavin), carotenoids (lutein, cryptoxanthin, lycopene, -carotene), minerals (iron, calcium, potassium), antioxidant polyphenols, sugars, and organic acids (Chauhan, 2016). Furthermore, figs are low in sodium, fat, and cholesterol. These Ficus species have been shown to have positive health-related effects due to their phytochemical composition, primarily flavonoids, which appears promising in lowering blood lipid levels and preventing conditions such as obesity, diabetes, cardiovascular disease, neurodegenerative disorders, and even certain types of cancer (Caliskan, 2015; Wojdyło et al., 2016). It is understood that the processing of figs results in compositional changes that may be beneficial, or sometimes they losses nutritional values. Usually, figs are processed to increase their shelf life and nutritional values. Because of their high sugar concentration and low organic acid level, they have a sweet flavor. Their phenolic content is moderate, with red varieties containing larger quantities of anthocyanins. Compared to other fruits, the fig has low total carotenoids, concentrated mainly in the fruit's skin (Pereira et al., 2017). Table 30.1 shows the antioxidant capacity, total phenols, and total flavonoids of various types of processed figs. Several studies have reported that fig fruit has multiple health benefits, including antibacterial, antidiabetic, anti-inflammatory, and anticancer (Caliskan, 2015; Li et al., 2021). However, the processing of figs, because of heat treatment, may cause a reduction in these characteristics, such as loss of phenolic and flavonoid compounds.

Arvaniti et al. (2019) reported the composition of fresh and dried figs, and results showed that protein fat and total sugars were found to be highest in dried figs (3.30%, 0.93%, and 47.9%, respectively) as compared to fresh one (0.75, 0.30 and 16.2, respectively), whereas water was highest (79.3%) in fresh one compared to the dry one. Moreover, the study reported that dried fig fruits had ample amounts of minerals (Na, K, P, Zn, Mg, Fe, and Ca) and vitamins (A, C, B1, and B2). Petkova et al. (2019) studied the composition of figs fruits in fresh and frozen storage (−18 °C for 4 months) and reported that moisture and pH were increased in frozen storage, whereas ash content was decreased during frozen storage. Slatnar et al. (2011) studied the composition of processed figs dried by sun drying and oven drying. The results showed fructose (~52%) was highest among the sugars, followed by glucose (~46%) and sucrose (~2%). Interestingly, it was found that the oven drying method resulted in a high number of total sugars compared to sun-drying and fresh. Another study found higher total sugar content in fresh figs. (23.5 g/100 fw) as compared to frozen figs. (19.2 g/100 g fw) (Petkova et al., 2019). Similarly,

Table 30.1 Total phenols, flavonoids, and antioxidant capacities of processed figs

Types/ variety of figs	Processing method	Total phenolics	Total flavonoids	Antioxidant capacity	References
Bela petrovka	Drying oven	530 mg.GAE kg^{-1} dw	N.D	N.D	Slatnar et al. (2011)
Dry figs from Serbia	N.A	195.33 mg/100 g dw	N.D	0.388 mmol/100 g dw	Miletić et al. (2014)
Dried figs from Turkey	Oven drying	19.2 mg/100 g fw	N.D	N.D	Vallejo et al. (2012)
Dried Saoudi Douiret	Sun-drying	201.7 mg (GAE/100 dw)	112.28 mg (QE/100 g dw)	131.55 mg (TEAC/100 g dw)	Khadhraoui et al. (2019)
Mission dried figs	Freeze drying	3.08 mg CE/g	N.D	2.0 µM eq Trolox/g	Martínez et al. (2013)
Mission dried figs	Drying at 55 °C	3.23 mg CE/g	N.D	3.7 µM eq Trolox/g	Martínez et al. (2013)
Mission dried figs	Drying at 65 °C	3.72 mg CE/g	N.D	3.83.7 µM eq Trolox/g	Arvaniti et al. (2019)
Fermented figs	Fungal strains *Rhizopus oryzae* at 30 °C for 5 days	4.77 mg GAE/g. dw	N.D	0.53 mg of GAE/g of dm	Buenrostro-Figueroa et al. (2017)
Cuello dama Powder figs (peel)	60 °C during 24 h	2.67 mg GAE/g	17.61 mg RE/g	DPPH 9.50%/g	Viuda-Martos et al. (2015)

N.D means not detected, dw means dry weight

regarding organic acids, it was found that malic acid (25–59% of total organic acids) was highest compared to citric acid, while it was highest in processed figs compared to fresh fruits. Pande and Akoh (2010) also reported the highest malic acid content in figs, especially in peel and pulp, compared to whole fruits. Regarding phenolic compounds, it was found that figs contain many phenolic compounds, including hydroxycinnamic acids, flavan-3-ols, flavanols, and anthocyanins. However, it has been reported that crop season also differs the phenolic contents. In addition to this, it has been found that processing methods significantly affect the composition of phenolic compounds (Slatnar et al., 2011), lower in fresh and higher in processed figs, especially oven drying compared to sun drying. It is believed that oven drying results in the loss of phenolic compounds, but recently, it has been revealed that they are better preserved by the drying process than hydroxycinnamic acids or monomeric catechin (Devic et al., 2010). This is because monomeric catechins are responsible for enzymatic browning, but they are also easily diffused due to their lower molecular weight. On the other hand, Arvaniti et al. (2019); Slatnar et al.

(2011) reported the degradation of phenolic compounds in dried figs. Arvaniti et al. (2019) also found the loss of 15% of total phenolic compounds due to drying compared to a fresh one in figs "Cuello Dama".

Besides this, safety is the major concern during the processing of figs. Myotoxicity and mould growth are the major toxins that may produce during processing. Iqbal et al. (2018) reported the presence of aflatoxin and ochratoxin in dried figs. Similarly, another study also reported the presence of aflatoxin in dried figs (Petrić et al., 2018). However, pre-treatments can help reduce these toxins, an important safety concern. Several types of food additives, such as citric acid, emulsifiers, amino acids (e.g., L-cysteine), and antioxidants, may help reduce toxins.

4 Bioactive Components of Processed Fig

Fig fruits have excellent sources of bioactive compounds reported to cure various diseases, such as atopic dermatitis, skin warts, and cervical cancer. Teruel et al. (2021) comprehensively studied a detailed meta-analysis of by-products of fig fruits and found 19, 21, and 22 types of phenolic compounds in whole fruits, peel, and pulp, respectively. Moreover, these compounds were from various families, such as flavonoids, phytosterols, coumarins, hydroxycinnamic acids, anthocyanins, and phenolic acids. Slavin (2006) reported that figs are the richest source of benzaldehyde used to cure skin diseases. Weibin et al. (2001) reported that dried figs, especially by sun-drying, contained bioactive compounds, such as flavone, routine, and quercetin used for the formulation of medicines used for curing cardiovascular disease. It has been found that figs are excellent sources of Laxatives bioactive compounds that help increase stool motility and reduce constipation (Tabrizi et al., 2020).

Regarding the processing of individual parts, it was found that peel contained more phenolic compounds as compared to the pulp of freeze-dried fig fruits. The study found 18 phenolic compounds in fig peel of green fig, among which 8 were flavonoids, and the remaining were phenolic acids and their derivatives (Palmeira et al., 2019). The same study reported that caffeic and vanillic acid derivatives were abundant among the phenolic acids, whereas quercetin and C-glycosylated apigenin derivatives were the most abundant flavonoid compounds. Similar results were reported by another study when peel and pulp were evaluated for phenolic compounds. Agreeing with these findings, Harzallah et al. (2016) also reported the more phenolic compounds in the peel of three varieties (green, purple and black figs); however, the black variety showed more antioxidant capacities. Therefore, it is inferred that bioactive compounds differ greatly in varieties having darker colors will have more phenolic and flavonoid compounds. These compounds also depend on the ripening and/or harvesting stages: more ripened fruits have more phenolic compounds than the unripen ones.

5 Stability of Bioactive Components

5.1 *Bioactive Components*

"Bioactive components play an important role in the body and help to promote good health. This term refers to non-essential biomolecules in foods that exhibit the potential to modulate metabolic processes and different mechanisms in the human body to optimize human health".

Plants secondary metabolites or phytochemicals are non-nutritive metabolites of plants that are important for plant growth, survival, and reproduction. Most of these components are biologically active compounds and provide health benefits. *F. carica* plants exhibit the highest number of compounds. Phytochemical studies showed the presence of bioactive compounds and other secondary metabolites in figs plant (Barolo et al., 2014). Fig fruits have been consumed primarily in dried and fresh form; they have also traditionally been processed and preserved. Nowadays, consumers demand products based on fig fruit because of their high nutritional profile and different bioactive components in figs (Slavin, 2006).

Figs comprise many bioactive compounds in the flesh, peel, leaves, and whole fruits. The bioactive components present in figs are chlorogenic acid, cyaniding, luteolin, rutin, and catechin. These bioactive compounds posed potential health benefits such as hepatoprotective, antibacterial, antidiabetic, antioxidant, anti-inflammatory, and anticancer properties. Therefore, the demand for figs and fig-based products has increased by consumers.

The bioactive components present in whole figs are catechin, epicatechin, polymeric procyanidins, rutin, chlorogenic acid, cyanidin-3,5-*O*-diglycoside, cyanidin-3-*O*-rutinoside, pelargonidin-3-*O*-rutinoside, kaempferol-3-rutinoside, kaempferol-3-glucoside, quercetin-3-glucoside, quercetin-3-*O*-rutioside, quercetin-3-galactoside, quercetin-3-*O*-malonyl-galactoside, apigenin-C-hexoside-pentoside, gallic acid, syringic acid, and ellagic acid. In whole figs, Quercetin-3-*O*-rutioside is considered the major individual phenolic compound followed by quercentin-3-glucoside, chlorogenic acid, and polymeric procyanidins cyaniding-3-*O*-rutinoside. The most abundant bioactive compounds considered in the peel are cyanidin-3-*O*-rutinoside followed by cyanidin-3-*O*-diglucoside, cyanidin-3,5-diglucoside, catechin, epicatechin, and quercetin-rutinoside. Cyanidin-3-rutinoside and epicatechin are the main bioactive compounds in pulp of fig, whereas caftaric acid in the form of kaempferol-3-*O*-glucoside is considered to be the main component in leaves of figs (Teruel et al., 2021).

Contents of the bioactive compound and antioxidant activity in figs depend on the type of cultivator in both processed and fresh figs. Fresh and dried figs are a good source of carbohydrates, amino acids, fiber, sugars, vitamins, minerals (manganese, copper, magnesium, calcium, and potassium), phenolic compounds, and organic acids. Phytochemicals present in figs are arabinose, β-carotenes, β-amyrins, β-sitosterols, glycosides, xanthotoxol, flavonoids, alkaloids, saponins, coumarins, and terpenes.

There is an impact of different processing conditions on the bioactive properties of figs. The research was conducted by some researchers who evaluated phytochemical components in frozen, fresh, and processed figs. Proximate analysis and total chlorophyll, total phenols, total carotenoids, total anthocyanins, and carbohydrates contents were evaluated. In addition, they determined the antioxidant content of frozen, fresh, and processed figs by FRAP and DPPH· assays. The study's findings depicted a high range of antioxidant activity and bioactive components in fresh fruit compared to frozen fruits in which the levels of bioactive components decreased significantly. To preserve bioactive components, mainly phenolic and carotenoids, processing figs in the form of jams is the best option. However, after a few months of storage at a temperature of −18 °C, significant losses in bioactive compounds were observed. Regardless, figs processing in the form of the jam was considered to retain a better amount of bioactive compounds than storage in freezing (Petkova et al., 2019).

The antioxidant activity and bioactive compound content depend on the type of cultivator in both processed and fresh figs. One of the research projects was conducted in which main phenolic compounds were studied in nine cultivars of sun-dried figs with different skin colors. For all parameters, a considerable difference was observed between the cultivars. Dark-colored cultivars contained a high level of phenolic and flavonoids and high oxidant potential than light-colored cultivars (Khadhraoui et al., 2019).

One of the studies was conducted in which the impact of figs processing on nutritional, physicochemical, and phytochemical properties was investigated. They washed the figs and performed the other operations to process them into jam or nectar. The physicochemical parameters, iron, phosphorus, and calcium content were evaluated. Proximate analysis, vitamin content, beta-carotene, and total carbohydrates were also analyzed for nutritional analysis after processing. Different phytochemicals determined were total flavonoids, total phenolic, anthocyanins, and tannins. The findings of this study depicted that the processing of fig pulp into nectar and jam caused a significant increase in physicochemical properties, but a significant decrease in iron, pH, phosphorus, and calcium was observed. Processing the pulp of fig into jam and nectar caused a decrease in nutritional properties and phytochemical composition compared to fig pulp. Proximate composition, vitamin C, β-carotene, total flavonoids, total phenolic, tannins, and total anthocyanins decreased significantly due to processing, whereas a significant increase in carbohydrate was observed that caused an increase in energy value (Tanwar et al., 2014).

6 Functionality of Processed Fig

6.1 Functional Properties

Food's functional characteristics are the important physicochemical properties that reflect the interactions between the molecular compositions, conformation, structures, and physicochemical properties of different components present in the food according to the environmental conditions in which they are associated and measured. Functional properties are required to predict and evaluate how fats, proteins, carbohydrates, and fiber act on specific food items. Functional characteristics also define the action of ingredients during processing, preparation, and cooking and their effect on the final product after processing, depending on its taste, appearance, and texture. Food functionality is characterized by food products' quality, structure, nutritional composition, appearance, and acceptability (Awuchi et al., 2019) (Table 30.2).

The proper processing and storage of figs are required to retain their nutritional and functional properties, as figs are very sensitive to microbial spoilage or growth under cold storage conditions. Therefore, it is essential to use another alternative process to increase their shelf life. The processing of figs, including the drying method, causes many changes, such as the size of figs, compositional properties, and thermodynamics parameters of drying air such as humidity, temperature, and velocity. The volume and mass of figs get reduced by the drying method, which also

Table 30.2 Nutritional composition of figs

Dietary element	Fresh fig (per 100 g)	Dried fig (per 100 g)
Total calories (kcal)	74.0	249.0
Water (grams)	79.11	30.05
Protein (grams)	0.75	3.30
Total fat (grams)	0.30	0.93
Saturated fat (grams)	0.06	0.93
Sugars (grams)	16.26	47.92
Fiber (grams)	2.9	9.8
Calcium (milligrams)	35.0	162
Cholesterol (milligrams)	0	0
Phosphorus (milligrams)	14.0	67.0
Magnesium (milligrams)	17.0	68.0
Iron (milligrams)	0.37	2.03
Potassium (milligrams)	232	680
Zinc (milligrams)	0.15	0.55
Vitamin C (milligrams)	2.0	1.2
Vitamin A (IU)	142.0	10
Riboflavin (milligrams)	0.050	0.085
Thiamin (milligrams)	0.060	0.085

Source: USDA National Nutrient Data Base for Standard References Haytowitz et al. (2018)

affects the handling and packaging cost. The drying process also causes a reduction in microbial activity and retains the chemical properties of the final product (Martínez et al., 2013).

6.2 Fig as a Functional Food

Functional foods are "foods containing physiologically active compounds that provide benefits along with basic nutrition and prevent us from diseases or promote health".

Figs are harvested worldwide and consumed in fresh and dry forms. Its edible portion is the fruit, hollow, fleshy, and receptacle. Figs are an essential source of minerals, vitamins, carbohydrates, organic acids, sugars, and phenolic compounds. Dried, fresh, and processed figs contain high amounts of polyphenols and fiber. Figs have a high concentration of phenolic compounds such as proanthocyanidins. The phenolic compounds present in figs are higher in content than tea and red wine, which are good sources of phenolic compounds. Fig roots, fruit, and leaves are used in traditional medicines to treat different diseases such as gastrointestinal, cardiovascular, and respiratory disorders and are used as antispasmodic and anti-inflammatory remedies (Mawa et al., 2013).

Fig is considered a remarkable crop globally because of its consumption on a dry and fresh basis. Fig fruit is an essential source of vitamins and minerals such as calcium, potassium, iron, riboflavin, and thiamine. Figs are rich in fat, sodium, and cholesterol-free and provide the body with all the vital nutrients needed. Figs contain nearly 17 essential amino acids, of which glutamic and aspartic acid are present in the highest concentration. The high fiber content is present in figs, and most of the fiber present is soluble. Therefore, it helps regulate blood sugar levels, improve cholesterol level, and help maintain a healthy weight. The concentration of polyphenols is high in dry figs compared to other beverages and fruits. Because of these important health benefits, figs are considered an important functional food (Solomon et al., 2006).

Antioxidant compounds such as vitamin E, phenolic, organic acids, and carotenoids scavenge free radicals and inhibit the oxidative mechanism that leads to degenerative disorders. Phenolic compounds are common secondary metabolites of plants that provide physiological and health benefits to human health because of their antioxidant properties. Phenolic compounds help to regulate body mechanisms by reducing or donating hydrogen atoms, acting as free radical scavengers. These phenolic compounds are essential components of the flavor, color, and aroma of fresh vegetables, fruits, and other food products. In addition, phenolic compounds also play anti-inflammatory, anti-mutagenic, anti-carcinogenic, and exhibit antimicrobial properties.

F. carica comprises a high level of antioxidants and phenolic compounds that play an important role in promoting human health. These phenolic compounds act as antioxidants in different ways, as hydrogen donators, reducing agents, free radicals scavenging, and singlet oxygen quenchers. High levels of flavonoids, polyphenols, antioxidant potential, and anthocyanins' profile in figs help treat many diseases. The antioxidant potential in figs may help protect lipoproteins in plasma from oxidation and enhance the antioxidant capacity of plasma. Figs are high in essential minerals required to optimize body mechanisms. Figs also consist of sugars, mainly glucose and fructose. Anthocyanin content in different parts of figs helps to regulate body metabolism (Çalişkan & Polat, 2011). A combination of compounds 6-O-acyl-β-d-glucosyl-β-sitosterols in figs was found to have an inhibitory effect on the proliferation of cancer (Rubnov et al., 2001).

In many countries, figs have been used as a therapeutic medicine. Figs plant exhibits antioxidant, antifungal, anti-carcinogenic and anti-helminthic activities and inhibits acetylcholinesterase. Figs are also considered to be effective against oesophageal cancer cell lines. It also exhibits antifungal and anti-bacterial properties. Leaf extracts of figs increased the protection against CCl_4 hepatic damage and are considered a hepatoprotective. It also exhibits antipyretic potential. Figs are also considered anti-hypertensive and reduce coronary heart disease in hypertensive patients. Bioactive compounds in figs increase insulin secretion and reduce blood glucose and have been anti-diabetic (Hashemi & Abediankenari, 2013).

Flavonoids, alkaloids, coumarins, terpenes, and saponins are found in an aqueous extract of dried ripe fruit of *F. carica*. Phenolic compounds isolated from the leaves of figs are reported to exhibit pharmacological properties. These compounds have diverse biological mechanisms and provide health benefits. The extract of methanol from *F. carica* is effective against the activity of oral bacteria and acts as an anti-bacterial agent. The combination of methanol with gentamicin and ampicillin was found to be synergistic against oral bacteria, which shows that figs act as natural antibacterial agents (Jeong et al., 2009).

7 The General Impact of Figs on Health

Figs are considered very important for the proper functionality of the human body. Some important benefits of figs are summarized below.

- Figs satisfy the cravings for sweets and are considered a healthy alternative compared to other sweet food items and can be taken as a snack and used in dairy products.
- Figs are rich in potassium and help overcome potassium imbalance in the body. High sodium levels in figs help maintain sodium balance in the body. Therefore, figs also help maintain blood pressure by maintaining a sodium potassium imbalance in the body.
- Figs also help improve digestive issues, prevent diarrhea, constipation, and aid in digestion.
- Figs are considered a good source of prebiotics that improves overall gut health.

- Figs help increase bone density as they are an excellent source of potassium and calcium. These essential minerals help to improve bone health and prevent osteoporosis.
- Figs are rich in bioactive compounds and polyphenols that act as antioxidants in the body and prevent cancer-producing cells.

8 Conclusion

In this chapter, the processing, chemistry, and functionality of fig fruits have been discussed. Various processing methods have been discussed, by which freeze-drying method is the best choice to retain its bioactive compounds. On the other hand, the oven drying method has been proposed as an alternative to sun-drying due to safety concerns, e.g., aflatoxins. Regarding composition, it has been established that peel contains more phenolic and bioactive compounds than seeds and/or whole fruits. In addition, varieties, such as dark-colored fruits, have more phenolic and bioactive compounds than green whereas ripen fruits have more sugars and flavonoids than unripe and green ones. The bioactive compounds in processed figs have the potential to cure numerous disorders, such as skin problems, digestion disorders, cancers, and heart diseases. Furthermore, processed figs have the potential to be used in numerous food products, which can be served as functional foods.

References

Arvaniti, O. S., Samaras, Y., Gatidou, G., Thomaidis, N. S., & Stasinakis, A. S. (2019). Review on fresh and dried figs: Chemical analysis and occurrence of phytochemical compounds, antioxidant capacity and health effects. *Food Research International, 119*, 244–267. https://doi.org/10.1016/j.foodres.2019.01.055

Awuchi, C. G., Igwe, V. S., & Echeta, C. K. (2019). The functional properties of foods and flours. *International Journal of Advanced Academic Research, 5*(11), 139–160.

Bachir bey, M., Meziant, L., Benchikh, Y., & Louaileche, H. (2014). Deployment of response surface methodology to optimize recovery of dark fresh fig (*Ficus carica* L., var. Azenjar) total phenolic compounds and antioxidant activity. *Food Chemistry, 162*, 277–282. https://doi.org/10.1016/j.foodchem.2014.04.054

Barolo, M. I., Mostacero, N. R., & López, S. N. (2014). *Ficus carica* L.(Moraceae): An ancient source of food and health. *Food Chemistry, 164*, 119–127. https://doi.org/10.1016/j.foodchem.2014.04.112

Bölek, S. (2021). Effects of waste fig seed powder on quality as an innovative ingredient in biscuit formulation. *Journal of Food Science, 86*(1), 55–60. https://doi.org/10.1111/1750-3841.15548

Buenrostro-Figueroa, J. J., Velázquez, M., Flores-Ortega, O., Ascacio-Valdés, J. A., Huerta-Ochoa, S., Aguilar, C. N., & Prado-Barragán, L. A. (2017). Solid state fermentation of fig (*Ficus carica* L.) by-products using fungi to obtain phenolic compounds with antioxidant activity and qualitative evaluation of phenolics obtained. *Process Biochemistry, 62*, 16–23. https://doi.org/10.1016/j.procbio.2017.07.016

Caliskan, O. (2015). Chapter 56 – Mediterranean figs (*Ficus carica* L.) functional food properties. In V. R. Preedy & R. R. Watson (Eds.), *The Mediterranean diet* (pp. 629–637). Academic.

Çalişkan, O., & Polat, A. A. (2011). Phytochemical and antioxidant properties of selected fig (*Ficus carica* L.) accessions from the eastern Mediterranean region of Turkey. *Scientia Horticulturae, 128*(4), 473–478.

Chauhan, A. (2016). Development of value-added products (bun, muffin, noodles and nuggets) by substitution with carissa spinarum and *Ficus carica* powder. *Asian Journal of Pharmaceutical Clinical Research, 9*, 95–101.

Devic, E., Guyot, S., Daudin, J.-D., Bonazzi, C. J. F., & Technology, B. (2010). Kinetics of polyphenol losses during soaking and drying of cider apples. *Food and Bioprocess Technology: An International Journal, 3*(6), 867–877.

FR, A. S. B. (2013). Khaliq FH, Irshad S, Madni A. A review on the prosperous phytochemical and pharmacological effects of *Ficus carisa*. *International Journal of Bioassay, 2*(5), 843–849.

Gençdağ, E., Görgüç, A., Okuroğlu, F., & Yılmaz, F. M. (2021). The effects of power-ultrasound, peroxyacetic acid and sodium chloride washing treatments on the physical and chemical quality characteristics of dried figs. *Journal of Food Processing and Preservation, 45*(1), e15009.

Harzallah, A., Bhouri, A. M., Amri, Z., Soltana, H., & Hammami, M. (2016). Phytochemical content and antioxidant activity of different fruit parts juices of three figs (*Ficus carica* L.) varieties grown in Tunisia. *Industrial Crops and Products, 83*, 255–267. https://doi.org/10.1016/j.indcrop.2015.12.043

Hashemi, S. A., & Abediankenari, S. (2013). Suppressive effect of fig (*Ficus carica*) latex on esophageal cancer cell proliferation. *Acta Facultatis Medicae Naissensis, 30*(2), 93.

Haytowitz, D., Ahuja, J., Wu, X., Khan, M., Somanchi, M., Nickle, M., et al. (2018). USDA National Nutrient Database for standard reference, legacy. *USDA National Nutrient Database for Standard Reference*.

Iqbal, S. Z., Mehmood, Z., Asi, M. R., Shahid, M., Sehar, M., & Malik, N. (2018). Co-occurrence of aflatoxins and ochratoxin A in nuts, dry fruits, and nuty products. *Journal of Food Safety, 38*(4), e12462.

Jahromi, M., & Niakousari, M. (2018). Development and characterisation of a sugar-free milk-based dessert formulation with fig (*Ficus carica* L.) and carboxymethylcellulose. *International Journal of Dairy Technology, 71*(3), 801–809. https://doi.org/10.1111/1471-0307.12484

Jeong, M.-R., Kim, H.-Y., & Cha, J.-D. (2009). Antimicrobial activity of methanol extract from *Ficus carica* leaves against oral bacteria. *Journal of Bacteriology and Virology, 39*(2), 97–102.

Khadhraoui, M., Bagues, M., Artés, F., & Ferchichi, A. (2019). Phytochemical content, antioxidant potential, and fatty acid composition of dried Tunisian fig (*Ficus carica* L.) cultivars. *Journal of Applied Botany and Food Quality, 92*, 143–150.

Li, Z., Yang, Y., Liu, M., Zhang, C., Shao, J., Hou, X., et al. (2021). A comprehensive review on phytochemistry, bioactivities, toxicity studies, and clinical studies on *Ficus carica* Linn. leaves. *Biomedicine & Pharmacotherapy, 137*, 111393. https://doi.org/10.1016/j.biopha.2021.111393

Lu, Y., Xu, K., Song, Y., Zhao, Z., & Guo, M. (2021). Effect of thermal processing on the phenolic profiles, antioxidant capacity, and volatile compounds of fig wines. *Journal of Food Processing and Preservation, 45*(4), e15321.

Manoj, K., Bornare, D., & Kalyan, B. (2018). Effect of drying on physicochemical and nutritional quality of *Ficus carica* (fig). *International Journal of Engineering, 7*(6), 117–121.

Martínez-García, J. J., Gallegos-Infante, J. A., Rocha-Guzmán, N. E., Ramírez-Baca, P., Candelas-Cadillo, M. G., & González-Laredo, R. F. (2013). Drying parameters of half-cut and ground figs (*Ficus carica* L.) var. mission and the effect on their functional properties. *Journal of Engineering, 2013*, 710830. https://doi.org/10.1155/2013/710830

Mawa, S., Husain, K., & Jantan, I. (2013). *Ficus carica* L.(Moraceae): Phytochemistry, traditional uses and biological activities. *Evidence-Based Complementary and Alternative Medicine, 2013*, 1–8.

Miletić, N., Popović, B., Mitrović, O., Kandić, M., & Leposavić, A. (2014). Phenolic compounds and antioxidant capacity of dried and candied fruits commonly consumed in Serbia. *Czech Journal of Food Sciences, 32*(4), 360–398.

Palmeira, L., Pereira, C., Dias, M. I., Abreu, R. M., Corrêa, R. C., Pires, T. C., et al. (2019). Nutritional, chemical and bioactive profiles of different parts of a Portuguese common fig

(*Ficus carica* L.) variety. *Food Research International, 126*, 108572. https://doi.org/10.1016/j.foodres.2019.108572

Pande, G., & Akoh, C. C. (2010). Organic acids, antioxidant capacity, phenolic content and lipid characterisation of Georgia-grown underutilized fruit crops. *Food Chemistry, 120*(4), 1067–1075. https://doi.org/10.1016/j.foodchem.2009.11.054

Pereira, C., López-Corrales, M., Serradilla, M. J., Villalobos, M. d. C., Ruiz-Moyano, S., & Martín, A. (2017). Influence of ripening stage on bioactive compounds and antioxidant activity in nine fig (*Ficus carica* L.) varieties grown in Extremadura, Spain. *Journal of Food Composition and Analysis, 64*, 203–212. https://doi.org/10.1016/j.jfca.2017.09.006

Petkova, N., Ivanov, I., & Denev, P. (2019). Changes in phytochemical compounds and antioxidant potential of fresh, frozen, and processed figs (*Ficus carica* L.). *International Food Research Journal, 26*(6), 1881–1888.

Petrić, J., Šarkanj, B., Mujić, I., Mujić, A., Sulyok, M., Krska, R., et al. (2018). Effect of pretreatments on mycotoxin profiles and levels in dried figs. *Arhiv za higijenu rada i toksikologiju, 69*(4), 328.

Rubnov, S., Kashman, Y., Rabinowitz, R., Schlesinger, M., & Mechoulam, R. (2001). Suppressors of cancer cell proliferation from fig (*Ficus carica*) resin: Isolation and structure elucidation. *Journal of Natural Products, 64*(7), 993–996.

Slatnar, A., Klancar, U., Stampar, F., & Veberic, R. (2011). Effect of drying of figs (*Ficus carica* L.) on the contents of sugars, organic acids, and phenolic compounds. *Journal of Agricultural and Food Chemistry, 59*(21), 11696–11702.

Slavin, J. L. (2006). Figs: Past, present, and future. *Nutrition Today, 41*(4), 180–184.

Solomon, A., Golubowicz, S., Yablowicz, Z., Grossman, S., Bergman, M., Gottlieb, H. E., et al. (2006). Antioxidant activities and anthocyanin content of fresh fruits of common fig (*Ficus carica* L.). *Journal of Agricultural and Food Chemistry, 54*(20), 7717–7723.

Tabrizi, A., Dargahi, R., Tehrani Ghadim, S., Javadi, M., Rasouli Pirouzian, H., Azizi, A., & Homayouni Rad, A. (2020). Chapter 11- functional laxative foods: Concepts, trends and health benefits. In R. Atta ur (Ed.), *Studies in natural products chemistry* (Vol. 66, pp. 305–330). Elsevier.

Tanwar, B., Andallu, B., & Modgil, R. (2014). Influence of processing on physicochemical, nutritional and phytochemical composition of *Ficus carica* L.(fig) products. *Asian Journal of Dairying & Foods Research, 33*(1), 37–43.

Teruel-Andreu, C., Andreu-Coll, L., López-Lluch, D., Sendra, E., Hernández, F., & Cano-Lamadrid, M. (2021). *Ficus carica* fruits, by-products and based products as potential sources of bioactive compounds: A review. *Agronomy, 11*(9), 1834. https://doi.org/10.3390/agronomy11091834

Vallejo, F., Marín, J., & Tomás-Barberán, F. A. (2012). Phenolic compound content of fresh and dried figs (*Ficus carica* L.). *Food Chemistry, 130*(3), 485–492.

Veberic, R., Colaric, M., & Stampar, F. (2008). Phenolic acids and flavonoids of fig fruit (*Ficus carica* L.) in the northern Mediterranean region. *Food Chemistry, 106*(1), 153–157.

Viuda-Martos, M., Barber, X., Pérez-Álvarez, J. A., & Fernández-López, J. (2015). Assessment of chemical, physico-chemical, techno-functional and antioxidant properties of fig (*Ficus carica* L.) powder co-products. *Industrial Crops and Products, 69*, 472–479. https://doi.org/10.1016/j.indcrop.2015.03.005

Weibin, J., Kai, M., Zhifeng, L., Yelin, W., & Lianju, W. (2001). *The production and research of fig (Ficus carica L.) in China.* Paper presented at the II international symposium on fig 605.

Wojdyło, A., Nowicka, P., Carbonell-Barrachina, Á. A., & Hernández, F. (2016). Phenolic compounds, antioxidant and antidiabetic activity of different cultivars of *Ficus carica* L. fruits. *Journal of Functional Foods, 25*, 421–432. https://doi.org/10.1016/j.jff.2016.06.015

Yao, L., Mo, Y., Chen, D., Feng, T., Song, S., Wang, H., & Sun, M. (2021). Characterization of key aroma compounds in Xinjiang dried figs (*Ficus carica* L.) by GC-MS, GC-olfactometry, odor activity values, and sensory analyses. *LWT, 150*, 111982. https://doi.org/10.1016/j.lwt.2021.111982

Chapter 31
Fig (*Ficus carica*) Syrup as a Natural Sugar Substitute

Akram Sharifi, Elham Taghavi, and Sara Khoshnoudi-Nia

1 Fig and Fig Compounds

Fig (*Ficus carica* L.) is an important crop worldwide and one of the most abundant fruits in the Mediterranean diet for dry and fresh consumption. Fig industrialization produces many co-products. These co-products come from fresh figs discarded because of inadequate ripening or over-ripening, spoiled, size, texture, or low quality as table fruit, but they are safe for human consumption. The non-use of fig co-product constitutes an economic loss since it is rich in nutrients and bioactive compounds, which can be extracted and used as value-added materials. In addition, figs and figs co-products are an excellent source of minerals, vitamins, and dietary fiber; they are fat and cholesterol-free and contain many amino acids (Viuda-Martos et al., 2015).

Fresh fig is well known for its attractive taste and excellent nutritional composition. In addition, it is a highly nutritious fruit and is rich in calories and carbohydrates, lipids, phenolics, and enzymes. Figs are known to contain numerous phenolic compounds with potential antioxidative activity (Veberic et al., 2008). Sugars,

A. Sharifi (✉)
Department of Food Science and Technology, Faculty of Industrial and Mechanical Engineering, Qazvin Branch, Islamic Azad University, Qazvin, Iran
e-mail: asharifi@qiau.ac.ir

E. Taghavi
Faculty of Fisheries and Food Science, Universiti Malaysia Terengganu, Kuala Nerus, Terengganu, Malaysia
e-mail: elham.taghavi@umt.edu.my

S. Khoshnoudi-Nia
Seafood Processing Research Group, School of Agriculture, Shiraz University, Shiraz, Iran
e-mail: s-khoshnoudi@shirazu.ac.ir

M. F. Ramadan (ed.), *Fig* (Ficus carica)*: Production, Processing, and Properties*, https://doi.org/10.1007/978-3-031-16493-4_31

acids, and phenolic compoundscontribute to the taste and color of figs and the flavor characteristic, which mainly depends on the proper balance of the volatile chemical constituents (Pereira et al., 2020). For example, 100 g of fresh Fig (*Ficus carica* L.) contains 86% water, 310 kJ energy, 19.2 g carbohydrate, 16.3 g sugars, 0.3 g total fat, 0.7 g protein, and T.S.S.(°Bx) = 18. (Hussain et al., 2021; Veberic & Mikulic-Petkovsek, 2016; Pawase et al., 2017). Vitamins and minerals of fresh fig (*Ficus carica* L.) fruit are shown in Table 31.1.

1.1 Carbohydrate

Fig is a high carbohydrate fruit and a good source of dietary fiber. The carbohydrate content of fresh fig is 19.2 g/100 g of fresh fruit weight. About 92% of the carbohydrates in figs are in the form of sugars. Rest consists of dietary fiber, insoluble cellulose in the skin, and soluble protein in fruit. The total sugar content of fig is 16.3 g/100 g, while dietary fiber constitutes about 5 g/100 g. Some fig varieties contain β-D glucans with substantial antimicrobial activity (Hussain et al., 2021).

Glucose, fructose, and sucrose are the main sugars present in fig fruits. Sorbitol is present at low concentrations, and, therefore, the fig is considered a sorbitol-poor species (Mascellani et al., 2021). The content levels of glucose, fructose, and sucrose are 1216–11,050 mg/100 g, 1916–11,900 mg/100 g, and 98–320 mg/100 g, respectively. Only traces of sucrose have been measured in analyzed fig fruit and juice extracted from figs. Çalişkan and Polat (2011) showed sucrose content in 76 local Mediterranean fig accessions. In some accessions, sucrose was not determined; the highest value was measured in 'Mor 3' (320 mg/kg). Other authors also report that this sugar is not present in fig juice or not distinguishable from other sugars (Çalişkan & Polat, 2011). Viuda-Martos et al. (2015) showed sucrose, fructose, and glucose in fig powder 5.91, 67.3, and 68.2 (mg/g sample).

In contrast, sucrose is the prevalent sugar in fig leaves. Sucrose is the main translocation carbohydrate from leaves to other plant parts, including fruit. The low sucrose concentration in fig fruit could result from anabolic processes and

Table 31.1 Vitamins and minerals of fresh fig (*Ficus carica* L.) (Hussain et al., 2021)

Component	Nutritional value/100 g	Component	Nutritional value/100 g
Vitamin A Equiv.	142 IU	Vitamin E	0.1 mg
Thiamine (B1)	0.1 mg	Phosphorus	14 mg
Riboflavin (B2)	0.1 mg	Calcium	35 mg
Niacin (B3)	0.4 mg	Iron	0.4 mg
Pantothenic acid (B5)	0.3 mg	Magnesium	17 mg
Vitamin B6	0.1 mg	Manganese	0.1 mg
Folate (B9)	6 mcg	Potassium	232 mg
Vitamin C	2 mg	Sodium	1 mg
Vitamin K	4.7 mcg	Zinc	0.2 mg

respiration during fruit development. The low sucrose content might also be caused by high invertase activity during fruit ripening. In contrast, several fruit species are characterized by the accumulation of both sucrose and fructose in ripe fruit. The latter is formed from sorbitol, translocated from leaves to fruit as sucrose (Vemmos et al., 2013).

The sugar composition can influence fruit sweetness. For example, fructose has a higher relative sweetness than glucose. Therefore, the perception of sweetness of different fig accessions can be linked to their fructose content. The values in fig are comparable to those reported in some other fruits, such as apples and strawberries. Total sugar contentvaries among different fig cultivars (Veberic & Mikulic-Petkovsek, 2016).

1.2 *Nutrition and Health Considerations of Sugar Substitutes*

1.2.1 Glycemic Index (GI)

Potassium is known to regulate blood sugar in the body. Fig is high in potassium and keeps a check on blood sugar levels. Thus, the fig is said to help in maintaining diabetes. Low potassium levels cause less insulin production, which leads to a spike in blood sugar. Potassium affects the pancreatic β-cells, which are involved in insulin secretions. Researchers have also found the lowering effect of chlorogenic acid present in fig on blood sugar level. Chlorogenic acid reduces the blood glucose concentration by inhibiting the G-6 phase, the two main metabolic pathways responsible for the release of glucose from the liver (Ahrens & Thompson, 2013). Relative glycaemic response to bulk sweeteners and alternatives show in Table 31.2.

Table 31.2 Glycaemic and insulinaemic responses to bulk sweeteners and alternatives (O'Donnell & Kearsley, 2012)

Sugar or alternative	Relative glycaemic response	Categorisation
Maltodextrin	91	High
Maltose	105	High
Trehalose	72	High
Sucrose (present in fig fruits)	68	Intermediate
Lactose	46	Low
Isomaltulose	32	Very low
Glucose (main sugar of fig)	100	High
Fructose (main sugar of fig)	19	Very low
Tagatose	3	Very low
Xylitol	12	Very low
Sorbitol (present in fig fruits)	9	Very low
Maltitol	45	Low
Isomalt	9	Very low
Lactitol	5	Very low

While non-communicable diseases are set to overburden private and governmental health budgets, preventive methods must maintain health. Dietary change provides one such method and explicitly reducing the glycaemic response to the diet via food selection, food modification and ingredient choices and development are valuable objectives. There is now growing evidence from clinical data confirms the potential for reduced severity of disease, and from epidemiological data for reducing the risk of developing a variety of non-communicable diseases. Incorporation of the low-glycemic approach as one component of a 'better diet' is likely to remain important. Strategies that combine reduced GI or GL reduced saturated fat and reduced energy are likely to be most effective, and these attributes are found in alternative sugars and sweeteners (O'Donnell & Kearsley, 2012). The glycemic index of figs is 51, which falls under the low GI category.

1.2.2 Dental Health

Fig syrup as a natural sugar substitute and low-glycemic sweeteners should have little or no effect on a person's blood sugar levels. However, applying natural sweeteners like fig syrup as a natural sugar substitute in food products can improve oral health.

Low-calorie sweeteners are non-cariogenic; however, their low bulk in food, drink, and medical products are often compensated by the inclusion of potentially cariogenic bulking agents. Manufacturers, regulators, and consumers must recognize any potential dental health risk. Therefore, investigation of the dental properties of sweeteners and sugar alternatives needs to continue and should be encouraged for all potentially favorable new sweetening and bulking agents, especially those likely to be used in food, drink, and medicine products likely to be consumed between meals (O'Donnell & Kearsley, 2012).

1.2.3 Calorie Control and Weight Management

Some people choose to limit their food energy intake by replacing high energy sugar with other sweeteners having little or no food energy (sugar substitutes). This allows them to eat the same foods they usually would while losing weight and avoiding other problems associated with excessive calorie intake (Tandel, 2011).

Non-nutritive, intense sweeteners have a valuable role in the diet providing sweetness without calories. Using intense sweeteners to replace sugar can reduce energy in foods and beverages, potentially energy intake. However, this depends on the nutritional composition of particular foods and drinks and the physiological, psychological, and behavioral effects of consuming foods with sweeteners in contrast to those with sugar. This chapter will look at the scientific evidence behind these complex issues (O'Donnell & Kearsley, 2012).

Overall, evidence suggests that foods formulated with non-nutritive intense sweeteners, reduced-calorie bulk sweeteners and bulking agents can play an interesting and valuable role in helping consumers to improve the energy balance of their diets and as part of weight reducing diets. Contrary to some reports, the main body of published data shows that these ingredients, when incorporated into foods with lower caloric density and/or reduced glycaemic impact, can help consumers to eat fewer calories overall. A balanced approach to weight loss and maintenance is essential for long-term success (O'Donnell & Kearsley, 2012).

1.2.4 Diabetes Mellitus

People with diabetes have difficulty regulating their blood sugar levels. However, limiting their sugar intake by substituting sugar with artificial sweeteners allows them to enjoy a varied diet; some sugar substitutes release energy but are metabolized more slowly, allowing blood sugar levels to remain more stable (Tandel, 2011).

1.2.5 Hypoglycemic Effect

Hyperglycemia is any disease characterized by excessive urine secretion. The most a common form of hyperglycemia is diabetes mellitus, the world's largest endocrine disease (Selvin et al., 2011). Fig fruit has an amazing property of regulating blood glucose levels. Fig consumption results in a reduction in total cholesterol and a decline in total cholesterol to HDL cholesterol ratio, which ultimately leads to a decrease in hyperglycemia (Hussain et al., 2021).

1.3 Dried Fig Fruits

Dried fig is one of the most produced fruit products in the world. It is used in many forms: as whole fruit for consumers; or in industrial forms, such as paste, concentrate, powder, and chopped fruit. Dried fig consumption manifests a positive effect on human health. This beneficial effect has been ascribed to many micronutrients reserved in dry fruit. Dried figs are rich in fiber, trace minerals, polyphenols, proteins, sugars, and volatile compounds that provide a pleasant characteristic aroma and are thus a good choice for a healthy snack (Bekatorou et al., 2002).

In addition to the energy provided by the high sugar content, dried figs have a high nutrient content compared with other dried fruits. It is considered an important source of minerals (potassium and calcium) and vitamins (B3 and nicotinic acid). Dried fig fruits have a statistically higher individual and total sugar content than fresh fruit (Table 31.3) (Veberic & Mikulic-Petkovsek, 2016).

Gebhardt et al. (1982) reported 52.90% total sugar and 51.3% reducing sugar content in the dried fig fruit (Gebhardt et al., 1982). Thonte and Patil (1988) reported that sugar-treated dried fig fruits contained 41.50–59.00% sugars. They reported that ethereal-treated and dried figs contained 32.50% reduced sugar, whereas

Table 31.3 Levels of individual sugars in dried figs (Veberic & Mikulic-Petkovsek, 2016)

Sugar	Whole dried fruit (mg/100 g DW)
Sucrose	249–740
Fructose	8253–19,557
Glucose	9690–21,588
Total sugars	18,518-41,885

control fruits had 33.71% (Thonta & Patil, 1988). Pawar et al. (1992) reported that the non-reducing sugar content in dried Poona fig was 6.64–9.82% (Pawar et al., 1992).

Dried figs have one of the highest concentrations of polyphenols, a highly effective antioxidant. Fig syrup is just as good as the fresh and dried fruits if the right preparation process is followed. When you opt for fresh fruit, you may have to wait until it is in season and contend with the short storage period. The dried fruits may stay for longer even though they can occupy a large storage space if you want to keep a regular supply.

1.4 Sugar Replacement in Food Product

Sugars and sweeteners have an essential role in the human diet, and choosing the right ones in the right amounts can influence health. Knowledge will enable good choices, and further research and understanding of the literature will confirm or deny how good our choices are and where improvements are possible. The choice is not simply the healthier or healthiest since the technological properties and economics of sugars and sweeteners impact which of them can be used suitably in a particular food. A wide range of potential influences on health is offered by sugars and sweeteners when selected appropriately (O'Donnell & Kearsley, 2012).

1.4.1 Synthetic Sugar Replacement in a Food Product (High-Potency Sweeteners)

Six high-intensity sweeteners are FDA-approved as food additives in the United States: saccharin (Hampton, 2008), aspartame (Tandel, 2011), acesulfame potassium (Ace-K) (Tandel, 2011), sucralose (Tandel, 2011), neotame (Food & Administration, 2018), and advantame (Tandel, 2011) (Table 31.4). High-intensity sweeteners are commonly used as sugar substitutes or sugar alternatives because they are often sweeter than sugar but contribute only a few to no calories when added to foods. High-intensity sweeteners, like all other ingredients added to food in the United States, must be safe for consumption (Food & Administration, 2018).

Table 31.4 Synthetic sugar replacement in food products

Sweetener	Structure	Regulatory status	MSICS	ADI	NTSPE to ADI[a]	CV	EBNCS	Reference
Acesulfame Potassium (Ace-K)	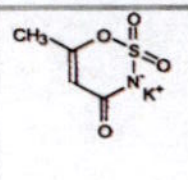	Approved as a sweetener and flavor enhancer in foods generally (except in meat and poultry) 21 CFR 172.800	200×	15	23	0	Sweet One®Sunett®	1, 8, 3
Advantame	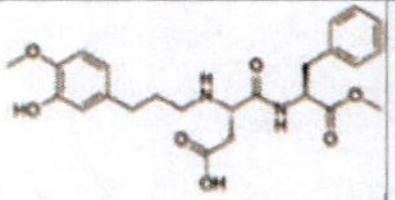	Approved as a sweetener and flavor enhancer in foods generally (except in meat and poultry) 21 CFR 172.803	20,000×	32.8	4920	0		1, 5, 7, 8
Aspartame	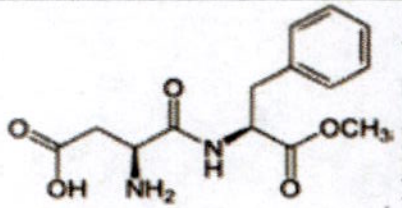	Approved as a sweetener and flavor enhancer in foods generally 21 CFR 172.804	200×	50	75	4	Nutrasweet®Equal®Sugar Twin®	1, 2, 3, 8
Neotame	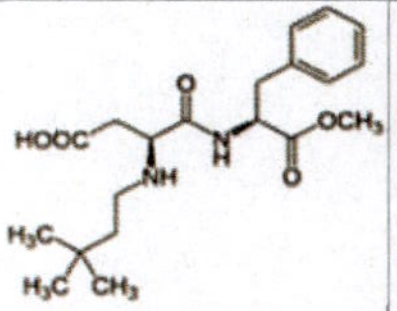	Approved as a sweetener and flavor enhancer in foods generally (except in meat and poultry) 21 CFR 172.829	7000–13,000×	0.3	23(sweetness intensity at 10,000 × sucrose)	0	Newtame®	1, 6, 8

(continued)

Table 31.4 (continued)

Sweetener	Structure	Regulatory status	MSICS	ADI	NTSPE to ADI[a]	CV	EBNCS	Reference
Saccharin	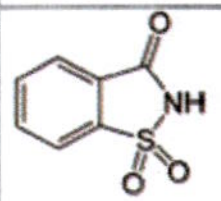	Approved as a sweetener only in certain special dietary foods and as an additive used for specific technological purposes 21 CFR 180.37	200–700×	15	45(sweetness intensity at 400× sucrose)	0	Sweet and Low® Sweet Twin® Sweet'N Low® Necta Sweet®	1, 5, 2, 8
Sucralose	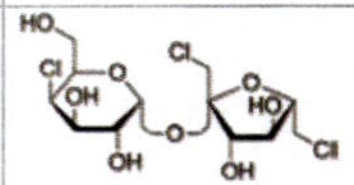	Approved as a sweetener in foods generally 21 CFR 172.831	600×	5	23	0	Splenda®	1, 4, 5, 8
Cyclamates			30–60×	11		0		1, 8

MSICS Multiplier of Sweetness Intensity Compared to Sucrose, *ADI* Acceptable Daily Intake: milligrams per kilogram body weight per day (mg/kg bw/d), *NTSPE to ADI* Number of Tabletop Sweetener Packets Equivalent to ADI[a], *CV* Caloric value (Cal/g), *EBNCS* Examples of Brand Names Containing Sweetener

References:

1-Carocho et al. (2017), 2- Meena et al. (2012), 3-George et al. (2010), 4-Furlán and Campderrós (2017), 5-Reis et al. (2011), 6-Morais et al. (2014), 7-Kobayashi et al. (2015). 8-Food and Administration (2018)

[a]Number of Tabletop Sweetener Packets a 60 kg (132 pound) person would need to consume to reach the ADI. Calculations assume a packet of high-intensity sweetener is as sweet as two teaspoons of sugar

® Registered trademark

The safety of High-Potency Sweeteners is demonstrated by a substantial body of evidence and continued review by independent regulatory agencies/committees, including JECFA/Codex, FDA, and EFSA. These organizations have considered the decades of positive and negative human and animal studies to draw their conclusions (Ashwell et al., 2020). Table 31.4 summarizes the most important synthetic sugar replacement in food products.

1.4.2 Natural High-Potency Sweeteners

Many high-potency sweeteners of diverse chemical structures are known to occur naturally. Natural sweeteners show many similarities to their synthetic counterparts' overall taste characteristics. They range in potencies from approximately 10 times that of sucrose to 1000s of times the sweetening power of sucrose. Overall sweet taste quality, measured by traditional methods that assess temporal characteristics and others that quantify the intensity of side- and aftertastes also range across the spectrum in much the same way as has been found with synthetic sweeteners. As a consequence, many of the challenges inherent in formulating sugar-reduced foods and beverages so that they will deliver taste experiences fully acceptable to consumers are similar, irrespective of the origin of the potent sweetener (O'Donnell & Kearsley, 2012).

Natural high-potency sweeteners contain Brazzein and Curculin, which are low-calorie sweeteners and are stable to heat and pH variation (DuBois & Prakash, 2012; Kant, 2005). Erythritol is a bulk sweetener for cooking, confectionery, and beverages. It is stable to light and temperature, has a clean taste similar to sugar, and has a cooling effect when tasted in solids (DuBois & Prakash, 2012; Kurihara & Nirasawa, 1994; Priya et al., 2011). Glycyrrhizic acid is a natural flavor and flavor enhancer. Glycyrrhizic acid has technological properties: soluble in hot water, anti-inflammatory and anti-viral; bitter and licorice-like off-tastes; delayed onset of sweetness and sweetness linger (Grenby, 1991; Priya et al., 2011). Other natural high-potency sweeteners include the following: Mabinlin, Miraculin, Monatin, Monellin, Pentadin, Steviol glycosides, stevioside and rebaudioside A, and Thaumatin (DuBois & Prakash, 2012; Gibbs et al., 1996; Grenby, 1991; Kant, 2005; Khattab et al., 2017; Kurihara & Nirasawa, 1994; Kroger et al., 2006; Priya et al., 2011; Rodrigues et al., 2016). Table 31.5 shows the Potency of natural sweeteners.

Polyols (sugar alcohols) are nutritive sweeteners obtained by the catalytic hydrogenation of the oxo-group of natural sugars, i.e., by substituting an aldehyde or keto group with hydroxyl. The sweetness of sugar alcohols (polyols) is shown in Table 31.6. The sweetness of polyols is lower than sucrose. Therefore, polyols might be used as a bulk sweetener. Food products' desired sweetness and flavor are achieved by combining polyols and non-nutritive, usually artificial, sweeteners. Polyols are responsible for texture, preservation, filling, moisture capture, and cooling effect in the mouth. Polyol sweetness, such as maltitol, is up to 90% of the sucrose sweetness (Petković, 2019).

1.4.3 Plants and Fruits as a Source of Natural Sugar Replacement in Food Products

The production of sugar-free foods, in which a sweet taste is obtained by using non-caloric intensive sweetening substances, is becoming increasingly popular. Natural high-intensity sweeteners are presented according to the source of origin. The plant parts from They were obtained, and their sweetness potency was also considered (Table 31.7). The sweetening properties, advantages, and disadvantages resulting from the possibility of natural high-intensity sweeteners used in the production of food is described, including sweeteners already in use or whose commercialization is in progress (Świąder et al., 2019).

1.5 Fig Syrup

Fig industrialization produces many co-products. These co-products come from fresh figs discarded because of inadequate ripening or over-ripening, spoiled, size, texture, or low quality as table fruit, but they are safe for human consumption. The non-use of fig co-product constitutes a real economic loss since it is rich in nutrients and bioactive compounds, which can be extracted and used as value-added materials. In addition, figs and figs co-products are an excellent source of minerals, vitamins, and dietary fiber; they are fat and cholesterol-free and contain many amino acids (Viuda-Martos et al., 2015).

In formulating healthy food products, the fig fruit is an excellent source. The fig is rich in vitamins, minerals, and dietary fiber and contains many natural antioxidant compounds, according to the USDA. Furthermore, figs' high fructose and glucose content make them suitable for formulating foods with natural sweeteners (Chang et al., 2016). However, because of economic reasons, damaged and deformed figs are more applicable in the industry as they can be converted to fig extract and fig powder. So it is vital to select those extraction conditions that may not cause heat damage to nutrients and inhibit temperature-dependent adverse reactions such as the Maillard reaction (Razavi et al., 2010).

The main challenge of fig syrup extraction is to provide an appropriate gentle process. Based on documented assays, there are few reports concerning the production process of fig extract (Farahnaky et al., 2010). Fig extract, however, does not have a long shelf life. Therefore, it needs to be concentrated (Slatnar et al., 2011).

Iran produced 98,990 tons of fig fruit in 2020. The Sabz cultivar belonging to the Smyrna type figs is the most common exportable Iranian local fig, among other varieties, and Fars province, notably Estahban district estimated to be the most important producer of sabz fig in the country (Kalantari et al., 2018).

Table 31.5 The potency of natural sweeteners (Potency is expressed as times sweeter than sucrose) (Carniel Beltrami et al., 2018)

Sweetener	Potency*	Sweetener	Potency*
Brazzein	500–2000	Mogrol glycosides, Luo Han Guo sweetener, and mogroside V	~250 (1, 8, 10)
Curculin	550–9000	Monatin	2700 (1)
Erythritol	0.6–0.7	Monellin	2500–3000 (8, 10)
Glycyrrhizic acid	30–110	Pentadin	~500 (4, 6, 8)
Mabinlin	100–400	Steviol glycosides, stevioside and rebaudioside A	200–400 (1, 8)
Miraculin	400,000	Thaumatin	1600–9800 (1, 7)

*Potency is expressed as times sweeter than sucrose

Table 31.6 The basic important physical and chemical parameters of different polyols (sugar alcohols)

Polyol	Sweetness (2–5)	Food application (1)
Xylitol	1.0	Jellies, chewing gums, coatings for gum, mint-flavor candies
Maltitol	0.9	Chocolate, spread, hard candies, chewing gums, coating for gums
Sorbitol	0.6	Chewing gums, tablets, candies, humectants, plasticizers, hard candies, baked goods
Erythritol	0.6	Hard/soft candies, chocolate, beverages, bakery products, chewing gums
Mannitol	0.6	Dusting power, chewing gums, effervescent products
Isomalt	0.5	Chewing gum, dusting powder
Lactitol	0.4	Candies, frozen desserts, jams and jellies, chocolate, dusting powder, bulking agent, baked products
Sucrose	1.0	

References: 1: Petković (2019). 2: O'Brien-Nabors (2016). 3: Wheeler and Pi-Sunyer (2008). 4: Fitch and Keim (2012). 5: Grabitske and Slavin (2008)

1.5.1 Fig Syrup Production

Several steps are involved in the production of fig syrup. First, figs are soaked in water before a special mixer chops them up. Then, the mixture is lightly steamed while careful tending helps check its consistency and keep it from overcooking. This process is repeated several times to ensure that it gets to the right thickness. Finally, the mixture will be sieved off and packed in storage containers ready for distribution. Typical syrup can last for about a month when stored in a cool, dry place without refrigeration. Therefore, the syrup is mainly safe for most people and can be added to different foods. Fig syrup, owing to the particular sweet and sour taste, can be adopted either in cake production (as filler and topping) or as a sort of 'barbecue sauce'; therefore, it is important to guarantee the rheological characteristics suitable for different applications (Gabriele et al., 2010).

Table 31.7 Characterization of plants and high-intensity sweeteners obtained from them (Świąder et al., 2019)

Plant species	Plant parts	Sweeteners	Sweetness potency*	Compound	FDA/EFSA sweetener status
Capparis masaikai lev.	Fruit Seeds	Mabinlin I-1 Mabinlin II Mabinlin III Mabinlin IV	10–400	Protein	Not approved
Curculigo latifolia Dryand (Lemba)	Fruit	Curculin Curculin	500** 430–2070	Protein	Not approved
Dioscoreophyllum cumminsii (Stapf) Diels (SerendipityBerry)	Fruit	Monellin	3000	Protein	Not approved
Pentadiplandra brazzeana Baill.	Fruit	Pentadin brazzein	500–2000	Protein	Not approved
Synsepalum dulcificum Daniell (miracle fruit)	Fruit	Miraculin	400,000**	Protein	Not approved
Thaumatococcus daniellii (Benn.) Benth (Katemfe)	Fruit	Thaumatin I thaumatin II thaumatin a thaumatin b thaumatin c	1600–3000	Protein	GRAS flavor enhancer/E957 sweetener, flavor enhancer
Glycyrrhiza glabra Linn (Licorice)	Root	Glycyrrhizin	50	Terpenoid (triterpenoid glycoside)	GRAS flavor Enhancer/not approved EU
Siraitia grosvenorii Swingle (Luo Han Guo, monk fruit)	Fruit	Siamenoside I 11-oxomogroside V Mogroside V mogroside IVa mogroside IVe	500	Terpenoid (triterpenoid glycoside)	GRAS/not approved EU
Stevia rebaudiana Bertoni (stevia)	Leaf	DulcosideA RebaudiosideA RebaudiosideB RebaudiosideC RebaudiosideD RebaudiosideE RebaudiosideF RebaudiosideM Rubusoside Stevioside Steviolbioside	30 200 150 30 221 174 200 250 114 210 90	Terpenoid (diterpenoid glycoside) steviol glycosides sweetener	GRAS steviol glycosides/E 960
Sclerochiton ilicifolius A. Meeuse (Arruva)	Root	Monatin2R,4R Monatin 2S,4S	3000 1200	Indole	Not approved

*relative to sucrose
**modifying sour taste into sweet taste

Kalantari et al. (2018) optimize extraction conditions and fig syrup production at three different temperatures (5, 25, and 40 °C), Fig to water ratios of (1:1.0, 1:1.5, and 1:2.0 w/w) and storage times of 12, 24, 48, and 72 h. The optimum extraction condition of fig extract was provided by a temperature of 5 °C, a fig to water ratio of 1:1.5 w/w, and an extraction time of 72 h, which produced fig extract with 14% ± 0.5°Brix. The extract was then passed through a sieve (with a mesh size of 400 μ) so that large suspended solids were removed from the liquid (Kalantari et al., 2018). Therefore, fig extract with 14% °Brix needs to be concentrated to increase its shelf life. Table 31.8 shows the physicochemical properties of fig extract.

During the production of fig syrup, fig fruits were ground and then mixed with boiling water for the time requested for the solution to reach a standard concentration (14% °Brix). After filtration, the extracted solution is concentrated by evaporation, reducing the water content to the desired level. Under usual industrial conditions, during concentration, the atmospheric pressure is maintained at 1 atm, and the temperature of the boiling solution increases above 100 °C because of boiling point increase with syrup concentration. These process conditions are used for simplicity and low cost, even though either flexibility or control of final syrup properties are penalized (Gabriele et al., 2010).

Process innovation is oriented toward correcting weak points by modifying the process conditions to either a more selective extraction of components from fruits or better process control, improving flexibility. The efficiency of solid-liquid extraction can be improved by reducing raw material dimensions: owing to the increase in surface/volume ratio, the contact between solid and solvent is more efficient, and some components can be more easily extracted. High-temperature conditions are commonly used because they can increase mass transfer rate and thermodynamic solubility limits of some components, yielding to faster processes (Gabriele et al., 2010).

Table 31.8 Physico-chemical properties of fig (*Ficus carica* L.*cv.* Sabz) extract (Kalantari et al., 2018)

Properties	Fig extract
Total soluble solids (1 g/100 g of the fig extract)	14.0 ± 0.3
Total solids (1 g/100 g of the fig extract)	15.34 ± 0.04
Protein	0.28 ± 0.04
Fat	0.09 ± 0.03
Total sugar (1 g/100 g of the fig extract)	6.81 ± 0.11
Glucose	3.7 ± 0.2
Fructose	2.8 ± 0.1
Acidity (based on the concentration of citric acid)	0.30 ± 0.02
pH	4.88 ± 0.12
Ash (1 g/100 g of the fig extract)	0.67 ± 0.09
L* a* b*	50.66 ± 2.06 −1.33 ± 0.76 20.33 ± 0.11

L^*(Lightness)
a^* (Red/Green Value)
b^* (Blue/Yellow Value)

However, high temperatures cause thermal degradation of some fruit components, mainly polysaccharides, leading to highly reactive intermediates that can form colored polymers (Coca et al., 2004). As a consequence, nutritional properties, rheological properties and appearance are affected by drastic thermal conditions, and milder processes should be used to preserve some product characteristics even though a longer operating time could be necessary. The main components extracted from fruit are simple sugars (mainly fructose and sucrose) and complex polysaccharides. In addition, it is known that long-chain branched molecules build up a 3-D network into the material as a result of the gelation process and chain-chain interactions; therefore, consistency is mainly related to the amount of extracted polysaccharides like starch, a mixture of amylose and amylopectin, and pectin, a complex polymer containing D-galacturonic acid units (Gabriele et al., 2010).

When considering the latter, owing to the chemical behavior of its carboxylic groups, it is possible to improve the efficiency of extraction by changing the pH value and adopting a slightly alkaline solvent. Consequently, a more or less structured syrup could be obtained by varying the extraction pH. Therefore, the traditional process was modified from these general considerations, and the following changes were introduced (Fig. 31.1).

1. Addition of sodium bicarbonate and ammonium bicarbonate as alkaline agents.
2. Addition of a preliminary fruit grinding to increase the extraction efficiency.
3. Adoption of low temperature to reduce thermal degradation reactions.
4. Under-vacuum concentration to reduce operating temperature and oxygen availability for oxidation processes. Adopting the proposed changes, a bench-scale process was set up as a scale-down of traditional industrial systems, and different syrup samples were prepared to evaluate the effects of each adopted modification (Gabriele et al., 2010).

Gabriele et al. (2010) investigated the effects of selected operating parameters on the fig syrup rheology. Process conditions have been changed, either adding alkaline salts, as pH modifiers to improve pectin extraction efficiency, or decreasing operative temperature, reducing thermal damage of the material. It has been found that alkaline salt addition yields structured products because of the improved efficiency

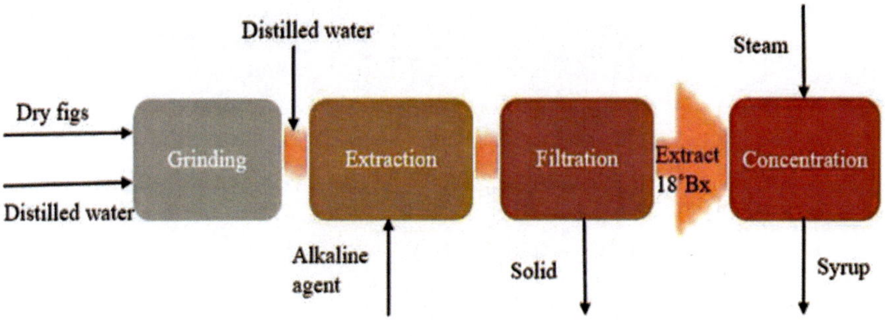

Fig. 31.1 Process flow sheet for figs syrup production. (Gabriele et al., 2010)

of pectin extraction and the enhancement of the gelation phenomena. In addition, mild conditions during either extraction or concentration lead to a more structured syrup, whereas more drastic operations bring liquid-like products (Gabriele et al., 2010).

In addition, it is interesting to notice that it is possible to adopt the process conditions to produce an industrial-standard syrup always having the same rheological properties, even when raw materials properties can enormously change because of external factors such as weather conditions, fruit cultivar, and harvesting period. This 'rheology-aided' process analysis allows for directly choosing the proper conditions, reducing significant time- and money-consuming 'trial and error' procedures focused on standardization of high-quality product production (Gabriele et al., 2010).

1.5.2 Powder Production of Fig or Fig Syrup

The ready-to-eat foods in the food market are mostly powdered; therefore, an appropriate method to dry figs will increase their usage in various products. As a result, powder technology is gaining importance in the food industry and among food producers (Koç & Ertekin, 2016). Spray drying, tray drying, freeze-drying, foam mat drying, and microwave drying are the most used drying methods to produce powdered products. However, after drying (except for spray drying), the obtained product is scraped and milled to obtain powder (Varhan et al., 2019).

Fig extract does not have a long shelf life. Therefore, it must be concentrated or converted to powder with longer-lasting storage characteristics. Drying is the major food processing operation that can enhance the shelf life (Slatnar et al., 2011).

Spray drying has a short-timed process and operational conditions that can be maintained. It is broadly used to produce high-quality powders regarding water activity, color, flavor, and nutritional values from various foods and fruit and vegetable extracts. For example, spray drying of fig extract can be used to produce healthy foods because it is considered some encapsulation (Kalantari et al., 2018). The spray drying process is very effective in preparing any easy-to-use and long-lasting powder that, after reconstitution, is more or less similar to the original juice, commonly known to be directly related to drying conditions (Encina et al., 2016). Sharifi et al. (2015), Zareifard et al. (2012), Jamdar et al. (2021), and Shahidi et al. (2020) have published some valuable reports regarding the operating conditions and qualitative characteristics of spray-dried juice/extract powders (Sharifi et al., 2015; Zareifard et al., 2012; Jamdar et al., 2021; Shahidi et al., 2020).

The foam mat drying method consists of converting liquid or semiliquid foods into stable foams via intense agitation and air and using foaming agents and stabilizers. As liquid foams are metastable, stabilizing agents are needed, usually surfactant molecules. Stabilizers delay the several mechanisms of foam aging, such as drainage, coalescence, and coarsening. Some foam stabilizers include xanthan gum, propylene glycol alginate, methylcellulose, Arabic gum, and maltodextrin. In foam-mat drying, drying usually refers to dehydration of the thin foam layer by hot air. Besides foam mat hot-air drying, other drying methods, including freeze-drying, vacuum-drying, infrared-drying, and microwave drying, have also been reported (Hajiaghaei & Sharifi, 2022).

Varhan et al. (2019) optimized foam mat-dried fig powder formulation. The fig foam was dried through hot air and microwave separately. Their results showed that the microwave drying method was more efficient than the hot air in time. Besides, the higher foam thickness provided more hydroxymethylfurfural content in the fig powder. Therefore, the drying method and conditions were important to produce a high-quality foam mat-dried fig powder (Varhan et al., 2019).

Khapre et al. (2015) produced fig (*Ficus carica* L.) fruits powder, and the fig powder was utilized in a value-added product like burfi (Indian cookie). Fig powder incorporated burfi was nutritionally rich in terms of fiber (3.7%), potassium (0.46%), and protein (13.1%). The prepared product was low-cost compared to similar market products (Khapre et al., 2015). Khapre et al. (2011) also produced toffee with fig fruits powder (Khapre et al., 2011). Yeganehzad et al. (2020) investigated the possibility of developing a fruit snack formulation based on dried fig powder and chocolate-coated (Yeganehzad et al., 2020).

2 Conclusions

Consumer interest in natural high-potency sweeteners has grown in recentyears, followed by concerns about the use of artificial additives in foods. Knowledge of potent sweeteners in plants is not new; there have been literature references to examples dating back to the mid-nineteenth century. However, most references to potent natural sweeteners also note overall taste characteristics incompatible with successful commercialization. Fig syrup is a rare example of a super-food that also tastes great. It has been appreciated for centuries in the Mediterranean and the Middle East and is now appreciated worldwide. The health benefits of figs in general and of fig syrup, in particular, have been known for thousands of years. Figs are far more nutritious than some commonly used natural sweeteners like dates. Fig syrup can be used as the potential substitute for sugar in various food products such as fruit juices, biscuits, cakes, muffins, desserts, and confectionery.

References

Ahrens, M. J., & Thompson, D. L. (2013). Effect of emulin on blood glucose in type 2 diabetics. *Journal of Medicinal Food, 16*(3), 211–215.

Ashwell, M., et al. (2020). Expert consensus on low-calorie sweeteners: Facts, research gaps and suggested actions. *Nutrition Research Reviews, 33*(1), 145–154.

Bekatorou, A., et al. (2002). Low-temperature brewing using yeast immobilized on dried figs. *Journal of Agricultural and Food Chemistry, 50*(25), 7249–7257.

Çalişkan, O., & Polat, A. A. (2011). Phytochemical and antioxidant properties of selected fig (*Ficus carica* L.) accessions from the eastern Mediterranean region of Turkey. *Scientia Horticulturae, 128*(4), 473–478.

Carniel Beltrami, M., et al. (2018). Sweeteners and sweet taste enhancers in the food industry. *Food Science and Technology, 38*, 181–187.

Carocho, M., et al. (2017). Sweeteners as food additives in the XXI century: A review of what is known, and what is to come. *Food and Chemical Toxicology, 107*, 302–317.

Chang, S. K., et al. (2016). Review of dried fruits: Phytochemicals, antioxidant efficacies, and health benefits. *Journal of Functional Foods, 21*, 113–132.

Coca, M., et al. (2004). Study of coloured components formed in sugar beet processing. *Food Chemistry, 86*(3), 421–433.

DuBois, G. E., & Prakash, I. (2012). Non-caloric sweeteners, sweetness modulators, and sweetener enhancers. *Annual Review of Food Science and Technology, 3*, 353–380.

Encina, C., et al. (2016). Conventional spray-drying and future trends for the microencapsulation of fish oil. *Trends in Food Science & Technology, 56*, 46–60.

Farahnaky, A., et al. (2010). The effects of glucose syrup and glycerol on some physical properties of dried figs. *Journal of Texture Studies, 41*(5), 633–650.

Fitch, C., & Keim, K. S. (2012). Position of the academy of nutrition and dietetics: Use of nutritive and nonnutritive sweeteners. *Journal of the Academy of Nutrition and Dietetics, 112*(5), 739–758.

Food, U., & Administration, D. (2018). *Additional information about high-intensity sweeteners permitted for use in food in the United States*. US Food and Drug Administration.

Furlán, L. T. R., & Campderrós, M. E. (2017). The combined effects of stevia and sucralose as sugar substitute and inulin as fat mimetic on the physicochemical properties of sugar-free reduced-fat dairy dessert. *International Journal of Gastronomy and Food Science, 10*, 16–23.

Gabriele, D., et al. (2010). Innovation in fig syrup production process: A rheological approach. *International Journal of Food Science & Technology, 45*(9), 1947–1955.

Gebhardt, S. E., et al. (1982). *Composition of foods: Fruits and fruit juices: Raw, processed, prepared.* US Department of Agriculture, Human Nutrition Information Service.

George, V., et al. (2010). Optimisation of sweetener blends for the preparation of lassi. *International Journal of Dairy Technology, 63*(2), 256–261.

Gibbs, B. F., et al. (1996). Sweet and taste-modifying proteins: A review. *Nutrition Research, 16*(9), 1619–1630.

Grabitske, H. A., & Slavin, J. L. (2008). Low-digestible carbohydrates in practice. *Journal of the American Dietetic Association, 108*(10), 1677–1681.

Grenby, T. (1991). Intense sweeteners for the food industry: An overview. *Trends in Food Science & Technology, 2*, 2–6.

Hajiaghaei, M., & Sharifi, A. (2022). Physicochemical properties of red beetroot and quince fruit extracts instant beverage powder: Effect of drying method and Maltodextrin concentration. *Journal of Food Quality, 2022*, 1–8.

Hampton, T. (2008). Sugar substitutes linked to weight gain. *JAMA, 299*(18), 2137–2138.

Hussain, S. Z., et al. (2021). *Fig (Ficus carica)-morphology, taxonomy, composition and health benefits. Fruits grown in Highland regions of the Himalayas: Nutritional and health benefits* (pp. 77–90). Springer.

Jamdar, F., et al. (2021). Physicochemical properties and enzymatic activity of wheat germ extract microencapsulated with spray and freeze drying. *Food Science & Nutrition, 9*(2), 1192–1201.

Kalantari, M., et al. (2018). Fig extract drying: The relationship between the main operating parameters of a pilot-scale spray dryer and product specifications. *Food Science & Nutrition, 6*(2), 325–333.

Kant, R. (2005). Sweet proteins-potential replacement for artificial low calorie sweeteners. *Nutrition Journal, 4*(1), 1–6.

Khapre, A., et al. (2011). Development of technology for preparation of fig (*Ficus carica* L.) fruit powder and its utilization in toffee. *Journal of Dairying Foods & Home Sciences, 30*(4), 267–270.

Khapre, A., et al. (2015). Studies on processing technology and cost estimation of fig (*Ficus carica* L.) fruit powder enriched Burfi (Indian cookie). *Journal of Applied and Natural Science, 7*(2), 621–624.

Khattab, S. N., et al. (2017). Physico-chemical and sensory characteristics of steviolbioside synthesized from stevioside and its application in fruit drinks and food. *Journal of Food Science and Technology, 54*(1), 185–195.

Kobayashi, M., et al. (2015). "Determination method of ultra-high-intensity sweetener, advantame, in processed foods by HPLC and LC-MS/MS." Shokuhin eiseigaku zasshi. *Journal of the Food Hygienic Society of Japan, 56*(1), 14–18.

Koç, M., & Ertekin, F. K. (2016). Şeker içeriği yüksek gıdaların püskürtülerek kurutulması: ürün kazanımı ve toz ürün özelliklerinin geliştirilmesi. *Türk Tarım-Gıda Bilim ve Teknoloji dergisi, 4*(5), 336–344.

Kroger, M., et al. (2006). Low-calorie sweeteners and other sugar substitutes: A review of the safety issues. *Comprehensive Reviews in Food Science and Food Safety, 5*(2), 35–47.

Kurihara, Y., & Nirasawa, S. (1994). Sweet, antisweet and sweetness-inducing substances. *Trends in Food Science & Technology, 5*(2), 37–42.

Mascellani, A., et al. (2021). Moderate salinity stress affects expression of main sugar metabolism and transport genes and soluble carbohydrate content in ripe fig fruits (*Ficus carica* L. cv. Dottato). *Plants, 10*(9), 1861.

Meena, M. K., et al. (2012). Formulation optimisation of a whey lemon beverage using a blend of the sweeteners aspartame and saccharin. *International Journal of Dairy Technology, 65*(1), 146–151.

Morais, E., et al. (2014). Development of chocolate dairy dessert with addition of prebiotics and replacement of sucrose with different high-intensity sweeteners. *Journal of Dairy Science, 97*(5), 2600–2609.

O'Brien-Nabors, L. (2016). *Alternative sweeteners*. CRC Press.

O'Donnell, K., & Kearsley, M. (2012). *Sweeteners and sugar alternatives in food technology*. Wiley.

Pawar, S., et al. (1992). Effect of pretreatments on chemical composition and drying rates of solar dried figs. *Indian Food Packer, 46*, 39–39.

Pawase, P., et al. (2017). Effect of addition of pectin on textural and sensorial quality characteristics of fig mango bar. *Journal of Pharmacognosy and Phytochemistry, 6*(5), 469–471.

Pereira, C., et al. (2020). Evaluation of the physicochemical and sensory characteristics of different fig cultivars for the fresh fruit market. *Food, 9*(5), 619.

Petković, M. (2019). Alternatives for sugar replacement in food technology: Formulating and processing key aspects. In *Food engineering*. IntechOpen.

Priya, K., et al. (2011). Natural sweeteners: A complete review. *Journal of Pharmacy Research, 4*(7), 2034–2039.

Razavi, S. M., et al. (2010). The physical properties of fig (*Ficus carica* L.) as a function of moisture content and variety. *Philippine Agricultural Scientist, 93*(2), 170–181.

Reis, R. C., et al. (2011). Sweetness equivalence of different sweeteners in strawberry-flavored yogurt. *Journal of Food Quality, 34*(3), 163–170.

Rodrigues, J. F., et al. (2016). Miracle fruit: An alternative sugar substitute in sour beverages. *Appetite, 107*, 645–653.

Selvin, E., et al. (2011). Performance of A1C for the classification and prediction of diabetes. *Diabetes Care, 34*(1), 84–89.

Shahidi, B., et al. (2020). Phenolic content and antioxidant activity of flixweed (*Descurainia sophia*) seeds extracts: Ranking extraction systems based on fuzzy logic method. *Sustainable Chemistry and Pharmacy, 16*, 100245.

Sharifi, A., et al. (2015). Effect of spray drying conditions on the physicochemical properties of barberry (*Berberis vulgaris*) extract powder. *International Food Research Journal, 22*(6), 2364.

Slatnar, A., et al. (2011). Effect of drying of figs (*Ficus carica* L.) on the contents of sugars, organic acids, and phenolic compounds. *Journal of Agricultural and Food Chemistry, 59*(21), 11696–11702.

Świąder, K., et al. (2019). Plants as a source of natural high-intensity sweeteners: A review. *Journal of Applied Botany and Food Quality, 92*, 160–171.

Tandel, K. R. (2011). Sugar substitutes: Health controversy over perceived benefits. *Journal of Pharmacology & Pharmacotherapeutics, 2*(4), 236.

Thonte, G., & Patil, V. (1988). Studies on drying of fig fruits. *Indian Food Packer, 42*(4), 94–99.

Varhan, E., et al. (2019). Foam mat drying of fig fruit: Optimization of foam composition and physicochemical properties of fig powder. *Journal of Food Process Engineering, 42*(4), e13022.

Veberic, R., & Mikulic-Petkovsek, M. (2016). *Phytochemical composition of common fig (Ficus carica L.) cultivars. Nutritional composition of fruit cultivars* (pp. 235–255). Elsevier.

Veberic, R., et al. (2008). Phenolic acids and flavonoids of fig fruit (*Ficus carica* L.) in the northern Mediterranean region. *Food Chemistry, 106*(1), 153–157.

Vemmos, S. N., et al. (2013). Seasonal changes in photosynthetic activity and carbohydrate content in leaves and fruit of three fig cultivars (*Ficus carica* L.). *Scientia Horticulturae, 160*, 198–207.

Viuda-Martos, M., et al. (2015). Assessment of chemical, physicochemical, techno-functional and antioxidant properties of fig (*Ficus carica* L.) powder co-products. *Industrial Crops and Products, 69*, 472–479.

Wheeler, M. L., & Pi-Sunyer, F. X. (2008). Carbohydrate issues: Type and amount. *Journal of the American Dietetic Association, 108*(4), S34–S39.

Yeganehzad, S., et al. (2020). Formulation, development and characterization of a novel functional fruit snack based on fig (*Ficus carica* L.) coated with sugar-free chocolate. *Heliyon, 6*(7), e04350.

Zareifard, M. R., et al. (2012). A feasibility study on the drying of lime juice: The relationship between the key operating parameters of a small laboratory spray dryer and product quality. *Food and Bioprocess Technology, 5*(5), 1896–1906.

Chapter 32
Fig (*Ficus carica*) Shelf Life

Elham Taghavi, Akram Sharifi, Navideh Anarjan, and Mohd Nizam Lani

Abbreviations

1-MCP	1-Methylcyclopropene
APE	Aqueous soy polyphenolic antimicrobial extract
APX	Ascorbate peroxidase
CA	Controlled atmosphere
CDB	Cuello Dama Blanco
CDN	Cuello Dama Negro
CP	Cold plasma
DNA	Deoxyribonucleic acid
EW	Electrolyzed water
FDA	Food and Drug Administration
GRAS	Generally Recognized As Safe
Gy	Gray
MAP	Modified atmosphere packaging
OA	Ozonated air
OTR	Oxygen transmission rate
RH	Relative humidity
ROS	Reactive oxygen species
TSS	Total soluble solids

E. Taghavi (✉) · M. N. Lani
Faculty of Fisheries and Food Science, Universiti Malaysia Terengganu,
Kuala Nerus, Malaysia
e-mail: elham.taghavi@umt.edu.my

A. Sharifi
Department of Food Science and Technology, Faculty of Industrial and Mechanical Engineering, Qazvin Branch, Islamic Azad University, Qazvin, Iran

N. Anarjan
Department of Engineering, Tabriz Branch, Islamic Azad University, Tabriz, Iran

M. F. Ramadan (ed.), *Fig* (Ficus carica)*: Production, Processing, and Properties*, https://doi.org/10.1007/978-3-031-16493-4_32

WVTR Water vapor transmission rate

1 Shelf Life

Shelf life is the time for which a stored item remains usable (International Food Information Service, 2007), or in other words, shelf life is the length of time food could be kept before it deteriorates (Lombarda, 2006). Fruits are unique among food products in terms of shelf life since they remain as living tissues until consumed or processed. All living tissues respire, and the consequences are quite profound for these products' shelf-life and storage stability. Slowing respiration can slow senescence and thus prolong shelf life; however, some respiration must continue, or the products will rapidly senesce and die. Furthermore, each fruit is unique in genetic makeup (species, cultivar, clone, etc.), its stage of development (maturation, stage of ripening, etc.), and the pre-and post-harvest conditions it has experienced. These facts make a shelf-life prediction of fresh produce particularly difficult compared to products with more uniform composition and stability (Aked, 2001).

2 Fig Shelf Life

Fresh figs should be plump and relatively soft but free of bruises (Murdock, 2002). Fig fruit is highly delicate at harvest and quite perishable after harvest and during handling and cold storage (Byeon & Lee, 2020). It should be refrigerated for 7 days (Murdock, 2002). After harvest, the quality of fig fruit is relatively unstable, and the fruit's fresh weight loss and respiration rate are sharply increased during cold storage, regardless of the cultivar (Byeon & Lee, 2020). This fruit is generally categorized as climacteric fruit as they show a sharp increase in respiration rate and ethylene production at the end of phase II of the double sigmoid growth curve (Rosianski et al., 2016). When fig fruit mature, their dimensions (diameter and length) abruptly increase, contributing to the rupturing of the ostiole-end region and peel tissues (Kong et al., 2013). These fig fruit physiological disorders are referred to as ostiole-end splitting and peel side cracking, respectively (Villalobos et al., 2016), and they deteriorate fruit quality after harvest and during cold storage by favoring conditions that induce fruit decay (Byeon & Lee, 2020; Villalobos et al., 2016).

3 Factors Influencing the Shelf Life of Fig

Various factors influence the shelf life of the fig. These factors include fig characteristics, microbiological spoilage, chemical deterioration, physical deterioration, transportation and distribution, temperature, storage atmosphere, insects and packages, and moisture. Figure 32.1 shows the factors that influence the shelf life of fig.

3.1 Fig Characteristics

Appearance is the key factor for consumers in making purchases of fresh produce. Vital components of visual quality include color and color uniformity, glossiness, absence of defects in shape or skin finish, and freedom from disease. Textural properties and flavor and aroma of fig are other criteria that define fig as fruit (Aked, 2001). Fig shape is usually obovoid, turbinate, or pear-shaped. Its size may range from 2.5 to 10 cm in length. Fig color varies depending on the variety; some of the typical colors are yellow, green-yellow, copper color, red and purple. The skin is tender and thin, and the syconium wall is fleshy and either pale yellow, amber, light pink, red, or purple (Crisosto et al., 2011). Generally, the fig is a pleasantly sweet

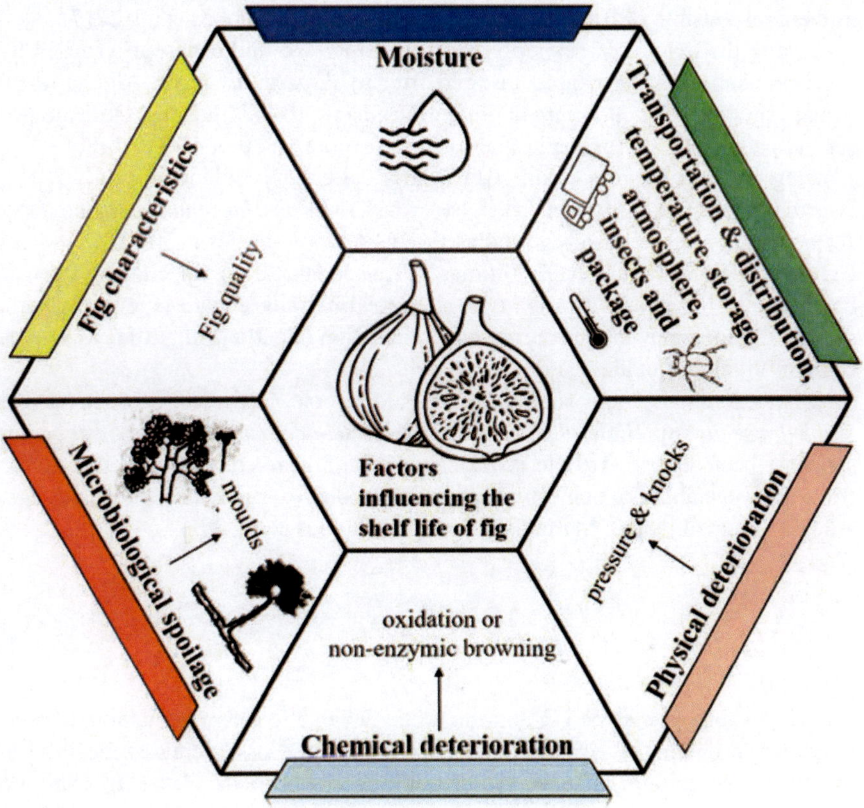

Fig. 32.1 Factors influencing the shelf life of fig

fruit that consists of soft flesh pursed around a large number of tiny edible seeds (Murdock, 2002). Fig texture is a very important parameter for consumer acceptance. a texture that is too soft may be rejected, and it can increase the degree of mechanical injury (Bahar & Lichter, 2018). In addition, the quality of the fig could affect its shelf life of the fig.

3.2 Microbiological Spoilage

Fresh figs are very sensitive to microbial spoilage, even in cold storage conditions. Thus, fresh figs are high perishable products during postharvest with microbiological decay that induces an unpleasant taste and smell due to rot, and suitable conservation methods must be applied (Paolucci et al., 2020). The growth and proliferation of microorganisms cause the main postharvest losses in fig fruits. The predominant microflora of fresh fig is mainly composed of moulds such as *Botrytis cinerea*, *Monilinia laxa*, *Alternaria alternata*, *Fusarium moniliforme*, *Rhizopus stolonifer*, *Aspergillus niger*, *Cladosporium herbarum*, and *Phytophora palmivora*, being most of them responsible of fruit decay (María del Carmen Villalobos et al., 2017).

Among the main diseases reported in fig fruits, we find endosepsis caused by *Fusarium* spp. or smut caused in dried fruit *by Aspergillus niger*, Alternaria rot (caused by *Alternaria alternata* or other *Alternaria* spp). Additionally, these moulds can be pathogenic or toxigenic due to their mycotoxins production ability while growing on fruits (Cantín et al., 2011; María del Carmen Villalobos et al., 2017), even during its cold storage. Likewise, yeast such as *Hanseniaspora* spp., and *Torulopsia* spp. have been reported as the main yeast species in fruits (María del Carmen Villalobos et al., 2017; Tournas & Katsoudas, 2005). Specifically, species belonging to the genus *Hanseniaspora*, *Saccharomyces*, *Pichia* or even bacteria such as *Bacillus* may cause souring or fermentation (Cantín et al., 2011; María del Carmen Villalobos et al., 2017).

Other microbial groups such as mesophilic aerobic, lactic acid bacteria (LAB), *Staphylococcus* spp., *Enterobacteriaceae* spp., *Pseudomonas* spp., and *Acetobacter* spp. have been described to a lesser extent as part of the microbial population in ripe fruits and vegetables, which can also be responsible for fruit decay and health hazard (Badosa et al., 2008; María del Carmen Villalobos et al., 2017).

3.3 Chemical Deterioration

Chemical spoilage may be *via* non-microbial enzymatic action, oxidation, or non-enzymatic browning. Physical spoilage includes water loss, increased moisture in dry food, freezer burn, and recrystallization of frozen foods. However, there are occasions when the cause and/or manifestation of spoilage is a combination of these different types of hazards, e.g., physical changes to the product or its container caused by chemical reactions or microbial growth and metabolism (Blackburn, 2006).

3.4 Physical Deterioration

Fresh figs are very perishable (Crisosto et al., 2011), sensitive to physical damage, and highly susceptible to postharvest decay infections (Crisosto et al., 2011; Venditti et al., 2005). Fresh fig quality (taste and aroma) increases with ripening; however, its sensitivity to damage also increases, reducing its market life (Crisosto et al., 2011). Figs have concise storage and shelf life. This is mainly due to their fast ripening and high susceptibility to decay (Crisosto et al., 2011; Venditti et al., 2005), which results from their easily damaged epidermis and high sugar content. The fruit epidermis is very sensitive to pressure and knocks, which can cause fig bruising, splitting and injuries, favoring pathogen infections. Sensitivity to physical damage is affected by the fig cultivar and the maturity stage at harvest (Crisosto et al., 2011).

3.5 Transportation, Distribution, Temperature, Storage Atmosphere, Insects and Package

- *Transportation & distribution*

The effective distribution of perishable commodities like fresh figs is of utmost importance when transporting. Currently, mismanagement of premium fruit and vegetables leads to a considerable amount of capital loss, such that logistics service providers apply refrigerated trucks to deliver them to minimize the food spoilage rate during transportation (Tsang et al., 2018). Moreover, fruits and vegetables retain their physiological functions, such as respiration and transpiration, after harvesting and remain fresh only as long as their normal metabolism continues (Prasad et al., 2018). So, attention to these issues is quite a concern during transportation and distribution.

- *Temperature*

Low cold storage is one of the strategies used to maintain quality and extend the shelf life of fruit (Mditshwa et al., 2020). Postharvest temperatures are critical for fresh figs (Crisosto et al., 2011). The number of studies dealing with postharvest technologies for fresh figs is somewhat limited, but it is generally accepted that storage temperature should be low (0–2 °C) and relative humidity (RH) should be high (Bahar & Lichter, 2018; Gözlekçi et al., 2005). For instance, in California, 'Black Mission' and 'Calimyrna' figs, two different types of figs stored in optimum conditions (−1 to 0 °C), have an expected postharvest life of 1–2 weeks in air and 3–4 weeks in a controlled atmosphere (Crisosto & Kader, 2007).

- *Storage atmosphere*

Storage in a controlled atmosphere (CA) with 5–10% oxygen and 15–20% carbon dioxide is beneficial for reducing decay, maintaining firmness, and reducing respiration and ethylene production rates. Reducing respiration and ethylene production rates will lengthen fresh fruit life (Crisosto & Kader, 2007; Crisosto et al., 2011). However, extended storage in CA could lose characteristic flavor (Crisosto et al., 2011). Bahar and Lichter (2018) studied CA as a potential tool to extend the shelf life of Ottomanit figs. They found that storing Ottomanit figs under CA conditions of 5 kPa O_2 and 5 kPa CO_2 maintains fruit quality better than the ambient atmosphere or a CA with higher CO_2 levels. This study improved the quality concerning firmness, integrity, and decay control.

- *Insects*

Souring or fermentation spoilage of the figs occurs when the fruit is still on the tree. The agents that cause souring are yeasts and bacteria, which insects transport. Therefore, souring symptoms are only noticeable when fruits are ripe and have the ostiole wide open. The ostiole aperture is necessary for the presence of this disease since the ostiole is the way of entrance for the vector insects. Therefore to control this disease, it is necessary to eradicate the vector insects (Crisosto et al., 2011).

- *Package*

The package protects its contents from outside environmental effects, such as water, water vapor, gases, odors, microorganisms, dust, shocks, vibrations, compressive forces, etc., and protects the product's environment. For many food products, the protection afforded by the package is an essential part of the preservation process. In general, once the integrity of the package is breached, the product is no longer preserved (Robertson, 2016).

3.6 Moisture

Water is the most abundant constituent of food. It is a key determinant of food storage and has many roles in processing (Hssaini et al., 2020; Kong & Singh, 2016). In macroscale chemistry, the water impact on food reactions and quality is more important than any chemical constituent (Rahman, 2007). The importance of water function in biological products goes far beyond its quantitative aspect. Water contributes to the texture and appearance of vegetables and plays an essential role in the occurrence of biochemical reactions and microbial growth (Berk, 2018). Controlling the food moisture content during processing is an ancient preservation technique that is still widely practiced today (Desa et al., 2019). This is achieved by either moisture removal or binding to ensure long food stability to microbial and chemical deterioration (Kong & Singh, 2016). During post-harvest, fresh fruits and vegetables continue to undergo changes leading to their senescence and deterioration, such as biochemical reactions and microbial growth. These two factors are the most common limitation on food shelf life. Hence, several preservation processes aim to achieve the stability of foods by maintaining the moisture content under the levels required for microbial development and biochemical reactions. This stability mainly depends on the relationship between the food equilibrium moisture content and its corresponding water activity at a constant temperature (Hssaini et al., 2020; Peng et al., 2007). Figs are an important source of essential dietary nutrients such as fibers, carbohydrates, minerals, and vitamins (Hssaini et al., 2019; Hssaini et al., 2020). Since the moisture content of fresh figs is generally greater than 80% (dry basis), they are classified as highly perishable commodities (Desa et al., 2019; Hssaini et al., 2020). Fig water loss can be reduced with early cold storage. However, longer periods of delayed cooling tended to produce a lower percentage of quality fruit during the storage life and shelf life of figs (Crisosto et al., 2011).

4 Techniques on Shelf Life Extension of the Fig

Various practical technology can reduce the fig losses and shelf-life extension of figs. These include common or traditional techniques (drying, using chemical preservatives such as 1-methylcy-clopropene, calcium chloride, chlorine dioxide, and packaging) and novel techniques (edible coating, irradiation, ozone, cold plasma, electrolyzed water, and antimicrobial). Figure 32.2 illustrates the Techniques for shelf life extension of fig.

4.1 Common Techniques on Shelf Life Extension of Fig

4.1.1 Drying

As mentioned earlier, the moisture content of fresh figs is generally greater than 80% (dry basis); they are classified as highly perishable commodities (Desa et al., 2019; Hssaini et al., 2020). Moreover, fresh and dried figs can adsorb or desorbs water during storage depending on the environmental conditions, which, if not controlled, this water loss and gain can negatively affect their quality. Therefore, their valorization depends on maintaining the moisture content under the levels required for its spoilage or deterioration. Therefore, nowadays, an important interest is given to drying and monitoring the moisture content changes during and after this process as one of the most suitable fruit preservation methods (Hssaini et al., 2020). Moreover, drying can extend the fig shelf life remarkably. More information can be found on fig drying technologies in this book.

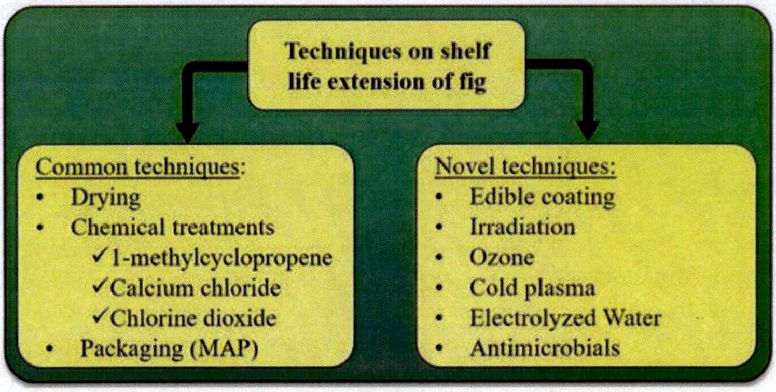

Fig. 32.2 Techniques on shelf life extension of fig

4.1.2 Chemical Treatments

Chemical treatments have been shown to affect weight loss during the storage and shelf life (Mditshwa et al., 2020). Therefore, some studies investigate the effects of chemical substances applied on figs to improve their postharvest conservation, such as 1-methylcy-clopropene (1-MCP) and calcium chloride.

- *1-Methylcyclopropene* (C_4H_6)

1-MCP is an unsaturated cyclic olefin that acts as a competitive ethylene antagonist, i.e., it blocks ethylene receptors and can be used to control ethylene production, respiration rate, and softening and extends the shelf-life of a wide range of fruits (Brasil Dias Tofanelli et al., 2018). However, 1-MCP applications on fruits may not always have similar results on the postharvest quality since its effect has been shown to vary according to the climacteric fruit species, cultivar, maturation and ripening stages of the fruit, and 1-MCP application forms (Brasil Dias Tofanelli et al., 2018; Junhua Zhang et al., 2017). 1-MCP was patented in 1996, followed by rapid registration and commercialization because of its non-toxic mode of action, effectiveness at low concentrations, and easy application as a gas. It has also been used extensively as a research tool to study the effects of ethylene on a range of climacteric and nonclimacteric fruits, vegetables, and flowers (Li et al., 2016; Jing Zhang et al., 2020). The impact of 1-MCP on the physiological and biochemical processes related to fruit ripening and consequent effects on quality during storage and shelf-life periods could vary greatly among different types of fruits, storage durations, and storage conditions. The majority of commercial use of 1-MCP is on apples, with less use on other products, in part because it is harder to delay rather than totally inhibit ripening of fruits such as avocado, banana, pear, and tomato (Jing Zhang et al., 2020).

There are few studies on the postharvest behavior of figs treated with 1-MCP. However, Gözlekçi et al. (2005) revealed that 10 μg/L of 1-MCP slowed down the softening of Bardakci figs and retained fruit firmness during a period of

15 days of storage in styrofoam trays wrapped with commercial cling film at 0 °C and 90–92% RH. Also, D'Aquino et al. (2003) treated Bianca figs with 400 μg/L of 1-MCP at 20 °C for 24 h and concluded that 1-MCP treatment of the figs and storage at 15 °C for 7 days might improve their storability.

In the other study, Brasil Dias Tofanelli et al. (2018) investigate the effects of different concentrations of 1-MCP on postharvest quality of Roxo-de-Valinhos fresh ripe figs. Their research was conducted by applying 4 various concentrations of 1-MCP (0 as control, 5, 10, and 20 μg/L) on figs, after which the fruits were evaluated at five different storage times (0, 4, 8, 12, and 16 days). The result showed that the1-MCP treatment improved the quality (firmness and acidity) of ripe Roxo-de-Valinhos figs during storage. In this regard, 20 μg/L of 1-MCP provided the best fruit firmness at day 12 after harvest. Higher acidity was observed in the figs without a 1-MCP treatment at 8 days of storage and with 20 μg/L of 1-MCP after 12 days of storage. Moreover, 1-MCP did not affect fruit weight loss. Overall, the figs treated with 1-MCP showed promising and rising values of firmness, acidity, and total soluble solids (TSS) at the highest dose. The TSS concentration tended to increase at the higher doses of 1-MCP.

- *Calcium chloride ($CaCl_2$)*

The postharvest calcium treatment effectively reduced physiological diseases and delayed senescence of fruits (Gao et al., 2020; Zhang et al., 2019). In addition, calcium treatment effectively delays ripening and maintains the quality of many fruits (Gao et al., 2020). The postharvest calcium treatment is a safe and non-toxic alternative to chemical treatment for fruit storage and has become an important focus for research in fruit preservation (Gao et al., 2020; Mishra et al., 2018). It maintains cell shape, cell membrane integrity, tissue hardness, and lipid metabolism and prolongs fruits'storage period. However, the possible mechanism of calcium regulating fruit ripening is still unclear (Gao et al., 2020).

Irfan et al. (2013) studied the response evaluation of fig fruit to dip treatments with calcium and sodium salts to improve the fruit texture and retain initial fruit quality attributes during storage and extend storage life. They found that $CaCl_2$ extends the keeping quality of fig fruit during storage and shelf-life. $CaCl_2$ in 4% concentration was notable in retention of fruit color, texture, and increased accumulation of ascorbic acid, compared to untreated control figs. Besides, pretreatment with $CaCl_2$ in 4% concentration was most effective in growing both mesophilic aerobic bacteria and yeast and molds at low temperature (1 ± 0.5 °C, 95–98% RH) storage; it caused further delays in ripening and senescence of figs. Therefore 4% $CaCl_2$ was beneficial in prolonging the postharvest life twofold.

- *Chlorine dioxide (ClO_2)*

Chlorine dioxide is a potent oxidizing agent and has a broad antimicrobial activity. It has about 2.5 times the oxidation capacity of chlorine. In addition, it does not react with nitrogen-containing compounds or ammonia to form dangerous chloramine compounds. Therefore, the US Food and Drug Administration (FDA) (1998) has allowed the use of aqueous chlorine dioxide in washing fruit and vegetables (Karabulut et al., 2009).

Karabulut et al. (2009) examined the possibility of using chlorine dioxide by fogging to control postharvest diseases fig. In their research, the fruit was fogged with various concentrations of chlorine dioxide in a cold storage unit for 60 min at room temperature. Treated fruit was stored either in the air or in modified atmosphere bags for 7 d at 1 °C followed by 2 days shelf-life at 20 °C. Fogging at 300–1000 μL/L significantly reduced the natural incidence of decay, most of which was gray mold. Modified atmosphere packaging did not improve the efficacy of fogging in reducing decay incidence. The epiphytic population on the fruit surface was similarly reduced by chlorine dioxide fogging. All treatments significantly reduced total microorganisms, fungal and bacterial populations in fig fruit. In addition, microorganisms in the storage atmosphere were significantly reduced. None of the treatments affected the visual quality and taste of fruit.

4.1.3 Packaging

- *Modified atmosphere packaging (MAP)*

MAP is a technique used for prolonging the shelf life of fresh or minimally processed food by changing the composition of the air surrounding the food in the package (Mangaraj & Goswami, 2009). In other words, MAP is the imposition of a gas atmosphere, typically containing an inert gas, such as nitrogen, combined with an antimicrobial active gas, such as carbon dioxide, upon a packaged food product to extend its shelf life (Spencer, 2005). In MAP, the modification of the atmosphere inside the package is achieved by the natural interplay between the product's respiration and the permeation of gases through the packaging. Also, the gas composition is modified initially and changes dynamically depending on the respiration rate of the food product and the permeability of the film surrounding the food product (Mangaraj & Goswami, 2009).

Furthermore, modifying the atmosphere within the package can be established either passively by commodity or intentionally via active packaging or a combination of the both (Mahajan et al., 2007; Mangaraj & Goswami, 2009). In the case of active MAP, an atmosphere is established by pulling a slight vacuum and replacing the atmosphere of the package with the desired gas mixture of O_2, CO_2, and N_2. On the other hand, in the case of passive MAP, the modification of the atmosphere inside the package is achieved by natural interplay between two processes, the respiration of products and the transfer of the gases through the packaging that leads to an atmosphere richer in CO_2 and poorer in O_2 and it depends on the characteristics of the commodity and packaging film (Mangaraj & Goswami, 2009).

Fresh figs have high respiratory intensity (64 mg CO_2 /kg.h) at ambient temperature (25 °C) (Liao et al., 2016), which causes their daily water loss rate to exceed 5% (Ye, 2014). Hence, modified atmosphere packaging (MAP) is one of the most appropriate ways to maintain a fig's quality. Studies by Ma et al. (2019) have designed two types of MAP: DMAP and TMAP. DMAP consists of O (O film is a modified PE film by styrene butadiene styrene block polymer) and W film (W film is a modified PE film by nano vermiculite), whereas TMAP consists of B, M, and W film. B and M film refer to modified O film (B film) whose OTR is 4.44×10^5 mL/m^2 and 1-MCP releasing the film, respectively. O and B film has a high oxygen transmission rate (OTR) and low water vapor transmission rate (WVTR). On the other hand, W film has high WVTR and low OTR. Therefore, low oxygen, high carbon dioxide, and high humidity atmosphere were formed in DMAP and TMAP. Their studies showed that DMAP and TMAP could prolong shelf life of figs from 2 days to 4 days and up to 6 days at 25 °C. In addition, figs with DMAP and TMAP maintained a lower loss of water, firmness, decay rate, TA content, vitamin C content, and high sensory quality during storage.

Also, the use of packaging under a modified atmosphere has been reported to reduce the respiration rate, with the benefit of delaying senescence by reducing metabolic activity and microbial proliferation (Valero et al., 2008). These effects of MAP have been previously reported by Maria del Carmen Villalobos et al. (2014) for fresh breba fruits and figs. (2016), who reported that the use of microperforated films to create passive modified atmospheres was effective in reducing the incidence of decay caused by fungal growth and in retaining quality characteristics. Thus, atmospheres with low levels of O_2 and high levels of CO_2 inhibit the growth of most aerobic microorganisms due to the well-known antimicrobial activity of CO_2 at high concentrations (María del Carmen Villalobos et al., 2017; Serradilla et al., 2013).

Martinez-Damian et al. (2020) report on MAP's effect on nutraceutical quality and overall appearance of figs stored at 1 °C. The combined use of packaging and cold in their investigation allowed them to observe fig with less weight loss and firmness, maintaining the citric acid and Vitamin C values without significant changes. Except for the last evaluation period, no statistical variation was found in the content of total anthocyanins, total phenols, and antioxidant capacity. Moreover, it was also possible to observe a significant impact on the reduction of respiration and ethylene production, which could be corroborated by the conservation of sensory aspects of the fig (texture, color, appearance, marketing level, aroma, and flavor). The generation of passive atmospheres (packaging) and cold (1 °C) are valuable tools in maintaining the nutraceutical quality and appearance of fig fruits, aspects highly appreciated by consumers and marketers of this fruit.

4.2 *Novel Techniques on Shelf Life Extension of Fig*

4.2.1 Edible Coating

Edible coatings are mainly used to increase food products appearance and extend the shelf life of the fruits (Salehi, 2020). By definition, edible coatings are natural food-safe and ecologically friendly substitutes applied to reduce water transfer, gaseous exchange, and oxidation of fresh produce (Tesfay et al., 2017). A sustainable approach to extending the shelf-life of figs can be constituted by applying an edible coating able to maintain the quality of the fruit during storage (Paolucci et al., 2020). Also, edible coatings have a high potential to carry active ingredients such as anti-browning agents, colorants, flavors, nutrients, and antimicrobial compounds that can extend product shelf life and reduce the risk of pathogen growth on the food surface (Galus et al., 2020; Vieira et al., 2021).

Paolucci et al. (2020) studied the active edible polysaccharide-based coating to preserve fresh figs. They compared fresh figs in a commercial preservation system with the figs preserved in an edible coating (Alginic acid/Agar, 70/30) and an active edible coating to preserve their quality characteristics. The coating efficacy was enhanced with the addition of pomegranate peel extract at two different concentrations (0.25% and 0.5%). Including a component with high antioxidant activity in an edible coating proved to be an excellent method for preserving the quality of this highly perishable fruit, such as fig. In the other research, Vieira et al. (2021) investigated the enhancement of postharvest quality and shelf life of fresh figs *via* composite coatings of chitosan and alginate emulsions with olive oil during storage at 4 °C. Their results revealed that coatings effectively delayed postharvest ripening indicators (i.e., respiration rate, mass loss, softening, total soluble solids/titratable acidity ratio) and fungal decay. The results foresee a fruit's shelf life between 14 and 19 days under refrigeration at 4 °C that may be followed up to 2 days at ambient temperature, higher than that estimated for uncoated fruits (less than 14 days at 4 °C plus to 2 days at ambient temperature).

Adiletta et al. (2019) studied the chitosan coating to preserve the physicochemical (weight loss, soluble solid content, and titratable acidity) and nutraceutical traits (total polyphenol, anthocyanina, flavonoid, ascorbic acid content, and antioxidant capacity) in fresh fig (Troiana). They compared fruits treated with 1% chitosan and 1% ascorbic acid coating and subjected to refrigerated storage (4 °C) for 9 days. Their results have shown that alginate-chitosan bilayer coating had a significant effect on weight loss of fresh fig since weight loss for coated only 5–7% is much lower than uncoated fig, which 18%. Their finding also demonstrated that chitosan coating delayed the decrease in ascorbic acid content in fresh fig fruits during cold storage. Moreover, chitosan-coated figs also maintained a higher antioxidant capacity than uncoated ones during 9 days of cold storage. The catalase activity had decreased significantly in uncoated fig while showing a stable trend toward storage in coated fig. Ascorbate peroxidase (APX) activity had increased significantly in chitosan-coated fruits compared to the uncoated fig during cold storage. Their study highlighted the beneficial effects of the antioxidant system that allowed to delay oxidative damage during postharvest treatment.

4.2.2 Irradiation

The application of irradiation can improve safety and extend the shelf of fruit like fig. Irradiation can use for various purposes such as: eliminating foodborne illness effectively, destroying or inactivating organisms that cause spoilage and consequently extending the shelf life of foods, controlling the insects by decreasing the need for other pest-control practices that may harm the fruit, and delaying in the ripening of fruit to increase longevity. There are three sources of radiation approved for use in foods by the FDA: Gamma rays, X-rays, and electron beams (or e-beam) (Administration, 2016).

In a study related to fig irradiation, Silva et al. (2009) investigated the physicochemical characteristics changes caused by γ-irradiation in pre-ready green figs to increase the useful shelf life of fig products. In their study, the samples were processed (washing, making hygienic, cooking for 15 minutes, and cooling), wrapped in a plastic sack, and stored at 8 °C in an OBD camera for 7 days. Later samples were irradiated with doses of 0 (control), 1.0, and 2.0 kGy, under a dose of 0.601 kGy/h, in a Gammacell-220 irradiator and stored for 24 h to 8 °C in OBD. These results confirmed that the irradiation did not significantly alter the physicochemical properties (i.e., pH, soluble solids content, color peel, color pulp, texture, chlorophyll A, chlorophyll B, and total carotenoids) of pre-ready green fig, with the exception for texture in all doses. In this regard, more research on irradiation needs to be undertaken for fig shelf life.

4.2.3 Ozone

Ozone (O_3) is a powerful sanitizer that is Generally Recognized as Safe (GRAS). In 2001, ozone, in gas and aqueous phases, was approved by the US Food and Drug Administration (FDA) as an antimicrobial additive for direct contact with foods. In the handling process, ozone can be applied in two forms. First, gaseous ozone is added continuously or intermittently to the storage atmosphere of the harvested product. Aqueous ozone is added immediately after the harvest or during the washing treatment. Extrinsic and intrinsic factors influence the efficacy of ozone. Extrinsic factors include pH, organic matter, pressure and temperature (in water quality), air relative humidity (in air quality), concentration, and treatment time application method (in ozone treatment); while intrinsic factors include the type of fruit and vegetable, weight, characteristics of the product surface, and surface area (in a food product), the activity of water (a_w), characteristics of microbial strains, physical state of bacterial strains, natural microflora, artificially inoculated microorganisms, and population size (in microbial load) (Sarron et al., 2021).

Some researchers believe that fresh figs cannot be washed as their skin can be easily damaged during the washing process, and water may induce the germination of fungal spores on the surface of the fruit (Afsah-Hejri et al., 2021; Karabulut et al., 2009). In this context, Afsah-Hejri et al. (2021) investigated the effect of ozonated air (OA) on weight loss, volume reduction (shriveling), and skin firmness on fresh figs and monitored the changes in the epidermis of OA-treated fruits within 14 days of storage at 4 °C and 65% RH. For this purpose, Kadota and Black Mission figs were treated with OA (15 ppm for 3 h), and their aging parameters were evaluated. Their study showed that OA treatment of fresh figs could be used to minimize the volume and weight loss and improve the overall quality of fresh figs.

In the other study, Zorlugenç et al. (2008) studied the effectiveness of gaseous ozone and ozonated water on microbial flora and aflatoxin B_1 content of dried figs. Dried figs were exposed to13.8 mg/L of ozone gas and 1.7 mg/L ozonated water for 7.5, 15, and 30 min. During the ozonated water process, *E. coli*, coliform, yeast, and molds were completely inactivated, whereas approximately an 88% reduction was observed in aerobic mesophilic bacteria count at the end of the 30 min. The *E. coli* and molds population was destroyed in gaseous ozone treatment. However, a substantial reduction in aerobic mesophilic bacteria, coliform, and yeast counts was determined. In addition, *Aspergillus flavus* and *Aspergillus parasiticus,* which cause aflatoxin formation, were isolated from non-ozonated dried figs. Due to the inactivation of all molds in dried figs, aflatoxin formation potential was decreased after the ozonation process. Overall, the experimental results of their research indicated that the application of gaseous ozone and ozonated water had a significant effect on the degradation of aflatoxin B_1 in contaminated dried figs.

4.2.4 Cold Plasma (CP)

Cold plasma (CP), also known as non-thermal discharge plasma, is an emerging non-thermal processing technique that ensures food quality and safety (Waghmare, 2021). CP is an emerging green process technology in today's industrialized era since it is eco-friendly and chemical-free (Chen et al., 2019; Waghmare, 2021). Plasma is generally characterized as "the fourth state of matter". Plasma consists of complete or partly ionized gases with a range of charged and neutral reactive species that drive its chemistry (Varilla et al., 2020). CP is defined as a source of high-energy electrons at ambient temperature and pressure that interact in an open or controlled environment (Varilla et al., 2020; Whitehead, 2016). As a novel non-thermal technology, CP can be applied to keep the freshness of the fruit. Therefore, it was practical to extend the shelf life of fruit by inhibiting its softening. Also, the low temperature and active particles during CP processing play a significant role in reducing food nutrient loss and biomacromolecules modification. Therefore many studies about applications of CP for food processing have been reported for decontamination, enzyme inactivation, starch modification, toxin degradation, food allergy, and food packaging (Pan et al., 2021).

The action mood of CP for food decontamination is back to the principal active components identified for the antimicrobial activity of CP systems are reactive oxygen species (ROS) (Varilla et al., 2020; Whitehead, 2016). CP also produces hydrogen peroxide and nitrogen species with antimicrobial properties (Maisch et al., 2012; Varilla et al., 2020). Furthermore, the effect of plasma treatment on microbial cells is mainly due to the plasma ions and cell interactions. The reactive species in plasma have been widely associated with the direct oxidative effects on the outer surface of microbial cells. The effect of plasma is highly dependent on the presence of water. The highest effect was observed in moist organisms compared to the lowest in the dry organism. The potential application of plasma in inactivation is based on the fact that reactive plasma species damage the deoxyribonucleic acid (DNA) in the chromosomes (Thirumdas et al., 2015).

Burana (2021) reported the effect of plasma treatments and low temperature on some fresh fig fruit's physical and physiology. In this investigation, various times of plasma treatment, including 0 (control, untreated), 2, 5, and 10 hours during storage at 4 °C, were considered. It was found that plasma treatments reduced the metabolism of fresh figs, including weight loss, respiration, and ethylene production during storage. Also, the decrease of firmness was reduced by plasma treatment, especially during 2 h plasma treatment. Moreover, plasma treatments considerably inhibited fungal incidents that prolong the storage life of fresh figs under storage at 4 °C. Unfortunately, though, these treatments did not affect the color change in this fruit. The findings recommend that plasma treatment combined with storage at low temperature has commercial potential and helps reduce postharvest decay, maintaining the quality and prolonging the storage life of fresh fig fruit.

In the other research, Abbaszadeh et al. (2018) studied the application of CP technology in the quality preservation of fresh fig fruit. In this study, dielectric barrier discharge plasma was applied to fruits in two steps. Durations of the first treatments were 1 and 5 min. Based on these results, main experiments were conducted for 30, 90, and 180 s, and CP was applied to packed and unpacked fig samples. Subjective measurements showed significant shelf-life improvement for the treated figs compared to the control samples, while objective quality attributes were not altered except for pH and a* index which were not undesired. Therefore, direct application of plasma for 90 s and in-package treatment for 30 s were suggested for further investigation in their research. Furthermore, pulsed plasma processing of packed figs for 30 s is recommended considering the practical aspects. Overall they conclude that atmospheric CP has the potential to be applied to fresh fig fruits to prolong their shelf life.

4.2.5 Electrolyzed Water (EW)

EW refers to the acidic or alkaline solution produced by the electrolysis of various neutral salts in an electrolytic cell (Zhang et al., 2021). In this technology, by using tap water and salt, the system generates two-stream simultaneously, one is alkaline, and the other is acidic, with the characteristics of cleaning and sanitizing solutions, respectively. Researchers evaluated the effectiveness of these solutions as sanitizing and cleaning agents (Yamaner et al., 2016). The EW is a sustainable and green concept and has several advantages over traditional cleaning systems, including cost-effectiveness, ease of application, effective disinfection, on-the-spot production, and safety for human beings and the environment (Rahman et al., 2016). The main applications of EW in improving the quality of postharvest fruits and vegetables include controlling the microorganisms in postharvest, suppressing browning, delaying senescence, maintaining bioactive compounds, alleviating chilling injury, and removing pesticide residues from fresh fruits and vegetables. Therefore, EW treatment seems to be a promising method for preserving postharvest fruits and vegetables (Zhang et al., 2021).

There is no study on the application of EW in postharvest figs so far. However, Yamaner et al. (2016) studied the efficacy of neutralized EW and mild heat against foodborne pathogens isolated from fig. In their study, spores of *Aspergillus flavus* and *Penicillium expansum* were isolated from the surface of fig fruits. In addition, *Escherichia coli* and *Bacillus cereus,* known on figs' surfaces, were also evaluated. Spores and vegetative cells of fungi and bacterium were exposed to five different concentrations of neutralized EW (100, 75, 25, 5, and 1%) at three different temperatures (22, 50, and 70 °C) for 1, 3, and 5 min. According to the results, at 22 °C, 1% neutralized EW exposure for 1 min effectively decreased the number of E. coli and *B. cereus* vegetative cells by approximately 8.5 log cfu/mL and 6.3 log cfu/mL, respectively. At 50 °C, 5% neutralized EW exposure for 1 min significantly reduced the *P. expansum* and *A. flavus* spore numbers by 7 log cfu/mL and 5.54 log cfu/mL, respectively. With mild temperature (22–50 °C), low chlorine neutralized EW (9.22 mg/L - 33.85 mg/L available chlorine concentrations) showed a strong antibacterial and antifungal activity against foodborne pathogens. They conclude that neutralized EW can be used widely as a sanitizer in fig enterprises instead of high-cost chlorine-based disinfectants.

4.2.6 Antimicrobials

Only a few investigations and limited studies have been focused on using natural antimicrobial agents from plants and microorganisms to extend the shelf life of fig. Fig crops are distinguished by their perishability and searching for antimicrobial agents that can extend the fruits'shelf life are very challenging. Besides mold, mesophilic aerobic, lactic acid bacteria (LAB), *Staphylococcus* spp., *Enterobacteriaceae*, *Pseudomonas* spp., and *Acetobacter* spp. are also responsible for fruit decay of fig; however, they have been described in a little attention (Badosa et al., 2008).

The use of aqueous extracts obtained from the roots, leaves, or seeds of plants has been proven to extend the shelf life of fruits. Maria del Carmen Villalobos et al. (2016) have studied the soybean meal extract was applied by dipping two fig cultivars; Cuello Dama Blanco (CDB) and Cuello Dama Negro (CDN), in combination with microperforated M50 film (1/50 mm; ø = 100 μm) as well as with macroperforated film in order to evaluate the gas composition, weight loss, percentage of fruits with disorders, microbial counts, physicochemical parameters and sensory quality during cold storage for 21 days. Their studies showed the synergistic effects when combining soybean meal extract with MAP can reduce mold and yeast levels, reduce disorders by fungal growth and maintain the quality parameters of fig cultivars. For the CDN cultivar, cold storage could be extended to 21 days regarding physicochemical quality and control of postharvest decay, while the use of M50 film without natural compounds application allowed the shelf life up to 14–17 days for the CDN cultivar and 7 days for CDB cultivar.

Other technologies such as biocontrol agents and natural compounds such as essential oils or aqueous phenolic extracts have been used in food preservation due to their antimicrobial properties (María del Carmen Villalobos et al., 2017; Isman, 2000; Janisiewicz et al., 2001). In addition, these compounds inhibit the mycelial growth of molds and spore germination, affecting the cellular metabolism of the pathogens (María del Carmen Villalobos et al., 2017; Regnier et al., 2010; Maria Serrano et al., 2005; Tzortzakis, 2007a, b). Moreover, the combination of both technologies renders them more effective in prolonging the postharvest life and the safety of the horticultural produce (María Serrano et al., 2008; Villalobos et al., 2015).

5 Future Trend

Various practical technology can reduce the fig losses and shelf-life extension of figs. These include common or traditional techniques (drying, using chemical preservatives such as 1-methylcy-clopropene, calcium chloride, chlorine dioxide, and packaging) and novel techniques (edible coating, irradiation, ozone, cold plasma, electrolyzed water, and antimicrobial agents). Future work could evaluate the possibility of increasing the figs shelf life using traditional and novel techniques.

References

Abbaszadeh, R., Alimohammad, K., & Zarrabi Ekbatani, R. (2018). Application of cold plasma Technology in Quality Preservation of fresh fig fruit (Siyah): A feasibility study. *International Journal of Horticultural Science Technology, 5*(2), 165–173.

Adiletta, G., Zampella, L., Coletta, C., & Petriccione, M. (2019). Chitosan coating to preserve the qualitative traits and improve antioxidant system in fresh figs (*Ficus carica* L.). *Agriculture, 9*(4), 84.

Administration, U. S. F. a. D. (2016). *Food irradiation: What you need to know*. Retrieved from https://www.fda.gov/food/buy-store-serve-safe-food/food-irradiation-what-you-need-know

Afsah-Hejri, L., Toudeshki, A., Homayouni, T., Mehrazi, S., Pareh, A. G., Gordon, P., & Ehsani, R. (2021). Potential of ozonated-air (OA) application to reduce the weight and volume loss in fresh figs (*Ficus carica* L.). *Postharvest Biology Technology, 180*, 111631.

Aked, J. (2001). Fruits and vegetables. In D. Kilcast & P. Subramaniam (Eds.), *Food shelf life stability* (pp. 249–278). CRC Press LLC.

Badosa, E., Trias, R., Parés, D., Pla, M., & Montesinos, E. (2008). Microbiological quality of fresh fruit and vegetable products in Catalonia (Spain) using normalised plate-counting methods and real time polymerase chain reaction (QPCR). *Journal of the Science of Food and Agriculture, 88*(4), 605–611.

Bahar, A., & Lichter, A. (2018). Effect of controlled atmosphere on the storage potential of Ottomanit fig fruit. *Scientia Horticulturae, 227*, 196–201.

Berk, Z. (2018). *Food process engineering and technology*. Academic.

Blackburn, C. D. W. (2006). Managing microbial food spoilage: An overview. In *Food spoilage microorganisms* (pp. 147–170). Woodhead Publishing

Brasil Dias Tofanelli, M., Cuquel, F. L., Sousa Medeiros, J. G., & Welinski De Oliveira D'angelo, J. (2018). Effects of 1-methylcyclopropene on postharvest quality of Roxo-de-Valinhos fresh ripe figs. *Revista Colombiana de Ciencias Hortícolas, 12*(1), 75–81.

Burana, C. (2021). Effect of plasma treatments and low temperature to some physical and some physiology of fresh fig fruit (*Ficus carica* L.) cv. Black Jack. *International Scientific Journal of Engineering Technology, 5*(1), 6–10.

Byeon, S.-E., & Lee, J. (2020). Fruit quality and major primary metabolites differ across production systems in cold-stored figs (*Ficus carica* L.). *Scientia Horticulturae, 274*, 109669.

Cantín, C. M., Palou, L., Bremer, V., Michailides, T. J., & Crisosto, C. H. (2011). Evaluation of the use of sulfur dioxide to reduce postharvest losses on dark and green figs. *Postharvest Biology and Technology, 59*(2), 150–158.

Chen, G., Dong, S., Zhao, S., Li, S., & Chen, Y. (2019). Improving functional properties of zein film via compositing with chitosan and cold plasma treatment. *Industrial Crops Products, 129*, 318–326.

Crisosto, C., & Kader, A. (2007). *Figs Postharvest quality maintenance guidelines*. Retrieved from https://ucanr.edu/sites/kac/files/123822.pdf

Crisosto, H., Ferguson, L., Bremer, V., Stover, E., & Colelli, G. (2011). Fig (*Ficus carica* L.). In *Postharvest biology and technology of tropical and subtropical fruits* (pp. 134-160e). Woodhead Publishing

D'Aquino, S., Palma, A., Dore, A., & Agabbio, M. (2003). *Non conventional treatments to reduce figs decay.* Paper presented at the International conference on quality in chains. An Integrated View on Fruit and Vegetable Quality 604.

del Carmen Villalobos, M., Serradilla, M. J., Martín, A., Hernández-León, A., Ruíz-Moyano, S., & de Guía Córdoba, M. (2017). Characterization of microbial population of breba and main crops (*Ficus carica*) during cold storage: Influence of passive modified atmospheres (MAP) and antimicrobial extract application. *Food Microbiology, 63*, 35–46.

del Carmen Villalobos, M., Serradilla, M. J., Martín, A., Ruiz-Moyano, S., Pereira, C., & de Guia Cordoba, M. (2016). Synergism of defatted soybean meal extract and modified atmosphere packaging to preserve the quality of figs (*Ficus carica* L.). *Postharvest Biology Technology, 111*, 264–273.

del Carmen Villalobos, M., Serradilla, M. J., Martín, A., Ruiz-Moyano, S., Pereira, C., & de Guía Córdoba, M. J. (2014). Use of equilibrium modified atmosphere packaging for preservation of 'San Antonio' and 'Banane' breba crops (*Ficus carica* L.). *Postharvest Biology and Technology, 98*, 14–22.

Desa, W. N. M., Mohammad, M., & Fudholi, A. (2019). Review of drying technology of fig. *Trends in Food Science & Technology, 88*, 93–103.

Galus, S., Arik Kibar, E. A., Gniewosz, M., & Kraśniewska, K. (2020). Novel materials in the preparation of edible films and coatings-a review. *Coatings, 10*(7), 674.

Gao, Q., Tan, Q., Song, Z., Chen, W., Li, X., & Zhu, X. (2020). Calcium chloride postharvest treatment delays the ripening and softening of papaya fruit. *Journal of Food Processing Preservation, 44*(8), e14604.

Gözlekçi, S., Erkan, M., Karaşahin, I., & Şahin, G. (2005). *Effect of 1-methylcyclopropene (1-MCP) on fig (Ficus carica cv. Bardakci) storage*. Paper presented at the III International symposium on Fig 798.

Hssaini, L., Charafi, J., Hanine, H., Ennahli, S., Mekaoui, A., Mamouni, A., & Razouk, R. (2019). Comparative analysis and physio-biochemical screening of an ex-situ fig (*Ficus carica* L.) collection. *Horticulture, Environment, Biotechnology, 60*(5), 671–683.

Hssaini, L., Ouaabou, R., Charafi, J., Idlimam, A., Lamharrar, A., Razouk, R., & Hanine, H. (2020). Hygroscopic proprieties of fig (*Ficus carica* L.): Mathematical modelling of moisture sorption isotherms and isosteric heat kinetics. *South African Journal of Botany, 145*, 265.

Irfan, P., Vanjakshi, V., Prakash, M. K., Ravi, R., & Kudachikar, V. (2013). Calcium chloride extends the keeping quality of fig fruit (*Ficus carica* L.) during storage and shelf-life. *Postharvest Biology Technology, 82*, 70–75.

Isman, M. B. (2000). Plant essential oils for pest and disease management. *Crop Protection, 19*(8–10), 603–608.

Janisiewicz, W., Tworkoski, T., & Kurtzman, C. (2001). Biocontrol potential of *Metchnikowia pulcherrima* strains against blue mold of apple. *Phytopathology, 91*(11), 1098–1108.

Karabulut, O. A., Ilhan, K., Arslan, U., & Vardar, C. (2009). Evaluation of the use of chlorine dioxide by fogging for decreasing postharvest decay of fig. *Postharvest Biology Technology, 52*(3), 313–315.

Kong, F., & Singh, R. (2016). Chemical deterioration and physical instability of foods and beverages. In *The stability and shelf life of food* (pp. 43–76). Elsevier

Kong, M., Lampinen, B., Shackel, K., & Crisosto, C. H. (2013). Fruit skin side cracking and ostiole-end splitting shorten postharvest life in fresh figs (*Ficus carica* L.) but are reduced by deficit irrigation. *Postharvest Biology and Technology, 85*, 154–161.

Li, L., Lichter, A., Chalupowicz, D., Gamrasni, D., Goldberg, T., Nerya, O., et al. (2016). Effects of the ethylene-action inhibitor 1-methylcyclopropene on postharvest quality of non-climacteric fruit crops. *Postharvest Biology Technology, 111*, 322–329.

Liao, L., Li, J., Ma, H., Wei, F., & Li, X. (2016). Effects of storage temperature and maturation on postharvest physiology of Xinjiang Zaohuang fig fruits. *Journal of Nuclear Agricultural Sciences, 30*(2), 282.

Lombarda, R. (2006). *Dictionary of food science and nutrition*. A & C Black Publishers Ltd..

Ma, J., Li, D., Yang, D., Xu, W., Fu, Y., Liao, R., et al. (2019). Effects of packaging designs with multiple pieces of function films on the quality of figs stored at ambient temperature. *Scientia Horticulturae, 251*, 32–38.

Mahajan, P., Oliveira, F., Montanez, J., & Frias, J. (2007). Development of user-friendly software for design of modified atmosphere packaging for fresh and fresh-cut produce. *Innovative Food Science Emerging Technologies, 8*(1), 84–92.

Maisch, T., Shimizu, T., Isbary, G., Heinlin, J., Karrer, S., Klämpfl, T. G., et al. (2012). Contact-free inactivation of *Candida albicans* biofilms by cold atmospheric air plasma. *Applied Environmental Microbiology, 78*(12), 4242–4247.

Mangaraj, S., & Goswami, T. K. (2009). Modified atmosphere packaging of fruits and vegetables for extending shelf-life-a review. *Fresh Produce, 3*(1), 1–31.

Martinez-Damian, M. T., Omegar, C.-A., & Oscar, C.-A. (2020). Effect of modified atmosphere packaging on nutraceutical quality and overall appearance of figs stored at 1°C. *Notulae Botanicae Horti Agrobotanici Cluj-Napoca, 48*(4), 2292–2305.

Mditshwa, A., Magwaza, L. S., Tesfay, S. Z., & Mbili, N. C. (2020). Use of Chemicals for Shelf-Life Enhancement of fruits. In B. N. Dar & S. A. Mir (Eds.), *Emerging Technologies for Shelf-Life Enhancement of Fruits* (1st ed., pp. 265–288). Apple academic Press Inc.

Mishra, V., Abrol, G. S., & Dubey, N. (2018). Sodium and calcium hypochlorite as postharvest disinfectants for fruits and vegetables. In M. W. Siddiqui (Ed.), *Postharvest disinfection of fruits and vegetables* (pp. 253–272). Elsevier.

Murdock, D. (2002). *Encyclopedia of foods: A guide to healthy nutrition "prepared by medical nutrition experts from Mayo Clinic, Univ. of California, LA, dole food co"*. Academic. An Imprint of Elsevier.

Pan, Y.-W., Cheng, J.-H., & Sun, D.-W. (2021). Inhibition of fruit softening by cold plasma treatments: Affecting factors and applications. *Critical Reviews in Food Science Nutrition, 61*(12), 1935–1946.

Paolucci, M., Di Stasio, M., Sorrentino, A., La Cara, F., & Volpe, M. G. (2020). Active edible polysaccharide-based coating for preservation of fresh figs (*Ficus carica* L.). *Foods, 9*(12), 1793.

Peng, G., Chen, X., Wu, W., & Jiang, X. (2007). Modeling of water sorption isotherm for corn starch. *Journal of Food Engineering, 80*(2), 562–567.

Prasad, K., Jacob, S., & Siddiqui, M. W. (2018). Fruit maturity, harvesting, and quality standards. In *Preharvest modulation of postharvest fruit and vegetable quality* (pp. 41–69). Elsevier

Rahman, M. S. (2007). *Handbook of food preservation*. CRC press.

Rahman, S., Khan, I., & Oh, D. H. (2016). Electrolyzed water as a novel sanitizer in the food industry: Current trends and future perspectives. *Comprehensive Reviews in Food Science Food Safety, 15*(3), 471–490.

Regnier, T., Combrinck, S., Du Plooy, W., & Botha, B. (2010). Evaluation of *Lippia scaberrima* essential oil and some pure terpenoid constituents as postharvest mycobiocides for avocado fruit. *Postharvest Biology and Technology, 57*(3), 176–182.

Robertson, G. L. (2016). Packaging and food and beverage shelf life. In *The stability and shelf life of food* (pp. 77–106). Woodhead Publishing

Rosianski, Y., Freiman, Z. E., Cochavi, S. M., Yablovitz, Z., Kerem, Z., & Flaishman, M. A. (2016). Advanced analysis of developmental and ripening characteristics of pollinated common-type fig (*Ficus carica* L.). *Scientia Horticulturae, 198*, 98–106.

Salehi, F. (2020). Edible coating of fruits and vegetables using natural gums: A review. *International Journal of Fruit Science, 20*(sup2), S570–S589. https://doi.org/10.1080/15538362.2020.1746730

Sarron, E., Gadonna-Widehem, P., & Aussenac, T. (2021). Ozone treatments for preserving fresh vegetables quality: A critical review. *Foods, 10*(3), 605.

Serradilla, M. J., del Carmen Villalobos, M., Hernández, A., Martín, A., Lozano, M., & de Guía Córdoba, M. (2013). Study of microbiological quality of controlled atmosphere packaged 'Ambrunés' sweet cherries and subsequent shelf-life. *International Journal of Food Microbiology, 166*(1), 85–92.

Serrano, M., Martinez-Romero, D., Castillo, S., Guillén, F., & Valero, D. (2005). The use of natural antifungal compounds improves the beneficial effect of MAP in sweet cherry storage. *Innovative Food Science & Emerging Technologies, 6*(1), 115–123.

Serrano, M., Martínez-Romero, D., Guillén, F., Valverde, J. M., Zapata, P. J., Castillo, S., & Valero, D. (2008). The addition of essential oils to MAP as a tool to maintain the overall quality of fruits. *Trends in Food Science & Technology, 19*(9), 464–471.

Service, I. F. I. (2007). *IFIS dictionary of food science and technology*. Wiley.

Silva, L. C., Harder, M. N., Arthur, P. B., Lima, R. B., Modlo, D. M., & Arthur, V. (2009). *Physical-chemical characteristics of figs (Ficus carica) preready to submitted to ionizing radiation.* Paper presented at the 2009 International nuclear atlantic conference, Rio de Janeiro, Brazil.

Spencer, K. C. (2005). Modified atmosphere packaging of ready-to-eat foods. In *Innovations in food packaging* (pp. 185–203). Elsevier

Tesfay, S. Z., Magwaza, L. S., Mbili, N., & Mditshwa, A. (2017). Carboxyl methylcellulose (CMC) containing moringa plant extracts as new postharvest organic edible coating for Avocado (*Persea americana* Mill.) fruit. *Scientia Horticulturae, 226*, 201–207.

Thirumdas, R., Sarangapani, C., & Annapure, U. S. (2015). Cold plasma: A novel non-thermal technology for food processing. *Food Biophysics, 10*(1), 1–11.

Tournas, V., & Katsoudas, E. (2005). Mould and yeast flora in fresh berries, grapes and citrus fruits. *International Journal of Food Microbiology, 105*(1), 11–17.

Tsang, Y., Choy, K., Wu, C., Ho, G., Lam, H., & Tang, V. (2018). An intelligent model for assuring food quality in managing a multi-temperature food distribution centre. *Food Control, 90*, 81–97.

Tzortzakis, N. G. (2007a). Maintaining postharvest quality of fresh produce with volatile compounds. *Innovative Food Science & Emerging Technologies, 8*(1), 111–116.

Tzortzakis, N. G. (2007b). Methyl jasmonate-induced suppression of anthracnose rot in tomato fruit. *Crop Protection, 26*(10), 1507–1513.

Valero, A., Begum, M., Hocking, A., Marín, S., Ramos, A., & Sanchis, V. (2008). Mycelial growth and ochratoxin a production by *Aspergillus* section Nigri on simulated grape medium in modified atmospheres. *Journal of Applied Microbiology, 105*(2), 372–379.

Varilla, C., Marcone, M., & Annor, G. A. (2020). Potential of cold plasma technology in ensuring the safety of foods and agricultural produce: A review. *Foods, 9*(10), 1435.

Venditti, T., Molinu, M., Dore, A., D'hallewin, G., Fiori, P., Tedde, M., & Agabbio, M. (2005). Treatments with gras compounds to keep fig fruit (*Ficus carica* L.) quality during cold storage. *Communications in Agricultural Applied Biological Sciences, 70*(3), 339–343.

Vieira, T. M., Moldão-Martins, M., & Alves, V. D. (2021). Composite coatings of chitosan and alginate emulsions with olive oil to enhance postharvest quality and shelf life of fresh figs (*Ficus carica* L. cv. 'Pingo De Mel'). *Foods, 10*(4), 718.

Villalobos, M. D. C., Serradilla, M. J., Martín, A., Lopez Corrales, M., Pereira, C., & Córdoba, M. D. G. (2016). Preservation of different fig cultivars (*Ficus carica* L.) under modified atmosphere packaging during cold storage. *Journal of the Science of Food and Agriculture, 96*(6), 2103–2115.

Villalobos, M. D. C., Serradilla, M. J., Martín, A., Ordiales, E., Ruiz-Moyano, S., & Córdoba, M. D. G. (2015). Antioxidant and antimicrobial activity of natural phenolic extract from defatted soybean flour by-product for stone fruit postharvest application. *Journal of the Science of Food and Agriculture, 96*(6), 2116–2124.

Waghmare, R. (2021). Cold plasma technology for fruit based beverages: A review. *Trends in Food Science Technology, 114*, 60–69.

Whitehead, J. C. (2016). The chemistry of cold plasma. In *Cold plasma in food and agriculture* (pp. 53–81): Elsevier.

Yamaner, C., Ayvaz, M., Konak, R., Tan, N., Kosoglu, I., & Dimoglo, A. (2016). Efficacy of neutralised electrolysed water and mild heat against foodborne pathogens isolated from *Ficus carica*. *Italian Journal of Food Science, 28*(2), 208.

Ye, W. B. (2014). Effects of edible film coating of fenugreek gum with Chinese herbal medicine Codonopisis treatment on storage quality of *Ficus carica* L. *Xibei Nongye Xuebao, 23*, 160–166.

Zhang, J., Cheng, D., Wang, B., Khan, I., & Ni, Y. (2017). Ethylene control technologies in extending postharvest shelf life of climacteric fruit. *Journal of Agricultural Food Chemistry, 65*(34), 7308–7319.

Zhang, J., Ma, Y., Dong, C., Terry, L. A., Watkins, C. B., Yu, Z., & Cheng, Z.-M. M. (2020). Meta-analysis of the effects of 1-methylcyclopropene (1-MCP) treatment on climacteric fruit ripening. *Horticulture Research, 7*(1), 1–16.

Zhang, L., Wang, J.-W., Chen, J.-Y., Song, T., Jiang, Y.-G., Zhang, Y.-F., et al. (2019). Preharvest spraying calcium ameliorated aroma weakening and kept higher aroma-related genes expression level in postharvest 'Nanguo'pears after long-term refrigerated storage. *Scientia Horticulturae, 247*, 287–295.

Zhang, W., Cao, J., & Jiang, W. (2021). Application of electrolyzed water in postharvest fruits and vegetables storage: A review. *Trends in Food Science Technology, 114*, 599–607.

Zorlugenç, B., Zorlugenç, F. K., Öztekin, S., & Evliya, I. B. (2008). The influence of gaseous ozone and ozonated water on microbial flora and degradation of aflatoxin B1 in dried figs. *Food and Chemical Toxicology, 46*(12), 3593–3597.

Chapter 33
Use of Proteolytic Activity of *Ficus carica* in Milk Coagulation

Hasitha Priyashantha, C. S. Ranadheera, Tharindu R. L. Senadheera, H. T. M. Hettiarachchi, Shishanthi Jayarathna, and Janak K. Vidanarachchi

Abbreviations

Cys	Cysteine
MMTS	Methylmethanethiol sulfonate
mPEG	Monomethoxypolyethylene glycol
ORAC	Oxygen radical absorbance capacity
PLCPs	Papain like cysteine proteases

1 Introduction

Proteases are hydrolytic enzymes that could hydrolysis the peptide bonds in proteins into smaller polypeptides or amino acids, and at present, approximately 60% of the commercialized enzymes are proteases (Mazorra-Manzano et al., 2018). This

H. Priyashantha (✉) · S. Jayarathna
Department of Molecular Sciences, Swedish University of Agricultural Sciences, Uppsala, Sweden
e-mail: hasi.tvp@slu.se; shishanthi.jayarathna@slu.se

C. S. Ranadheera
School of Agriculture & Food, Faculty of Veterinary & Agricultural Sciences, The University of Melbourne, Parkville, VIC, Australia
e-mail: Senaka.ranadheera@unimelb.edu.au

T. R. L. Senadheera
Department of Biochemistry, Memorial University of Newfoundland, St. John's, NL, Canada
e-mail: trlsenadheer@mun.ca

H. T. M. Hettiarachchi · J. K. Vidanarachchi
Department of Animal Science, Faculty of Agriculture, University Peradeniya, Peradeniya, Sri Lanka
e-mail: asf15046@agri.pdn.ac.lk; janakvid@agri.pdn.ac.lk

M. F. Ramadan (ed.), *Fig* (Ficus carica)*: Production, Processing, and Properties*, https://doi.org/10.1007/978-3-031-16493-4_33

group of enzymes plays a vital role in many industries, including food, pharmaceutical, detergent, leather, etc., and therefore, has received significant commercial interest. Even though the various synthetic, animal, plant, and microbial-derived proteases are abundant, much focus has been recently directed towards plant-derived proteases as they possess broad substrate specificities and activities over a wide range of temperature and pH (Kumari et al., 2010; González-Rábade et al., 2011).

Plant proteases play an essential role in plant growth, reproduction, digestion, seed germination, defense mechanism, apoptosis, etc. Hence, contribute to every aspect of the plant life-cycle, from mobilizing storage proteins during seed germination to cell death (Schaller, 2004). The quantity, quality, and type of plant proteases vary from one species to another and/ or within the same plant itself (Ben Amira et al., 2017). This greater diversity among thousands of plants has resulted in various plant proteases having different functionalities. Nearly all parts of plants, *e.g.*, fruits, flowers, latex, leaves, stems, seeds, and roots, contain proteases and have been extracted using different methods (Shah et al., 2014). MEROPS, the peptide database (http://merops.sanger.ac.uk), has classified proteases into seven families according to their active sites. Among them, plant proteases belong to five families: (I) Serine, (II) Cysteine (Cys), (III) Aspartic, (IV) Metallo, and (V) Threonine (Rawlings et al., 2006).

Papain, bromelain, and ficin derived from *Carica papaya*, *Ananas comosus,* and *Ficus* spp., respectively, are the most widely used commercialized plant endoproteases that belong to the Cys-family (Devaraj et al., 2008a, b). In the food industry, they are mainly used in meat tenderization, dairy processing, brewing, beverage production, flavored-protein hydrolysates, emulsifiers, etc. (Devaraj et al., 2008a, b; González-Rábade et al., 2011). Among those applications, plant proteases such as milk coagulants have become a great interest in the dairy industry.

Milk coagulation is the most crucial primary step in the cheese-making process, with much economic interest among cheese producers. Generally, milk coagulation achieves using calf rennet, a mixture of chymosin and pepsin (Anusha et al., 2014). However, the increasing world population, economic growth, and changes in eating behaviours have considerably raised the need to expand cheese production capacities and efficiencies. Nevertheless, the supply of rennet is not sufficient to meet the current demand due to the low availability of ruminant abomasa (Roseiro et al., 2003; Jacob et al., 2011; Ben Amira et al., 2017). Furthermore, the emerging vegan market, due to the ban of recombinant products in some countries (e.g., The Netherlands, France, Germany, etc.), animal welfare concerns, the trend of vegetarianism, religious taboos, and health concerns coupled with the limited supply of rennet have encouraged the use of alternative plant proteases in milk coagulation (Roseiro et al., 2003; Aider, 2021).

Moreover, easy access, availability, low cost, and simple extraction and purification procedures have attracted more attention to discovering novel plant proteases as cheap alternatives for the calf rennet. As a result, numerous plant proteases have been studied worldwide for their milk clotting properties. Among those various plant proteases, ficins (also known as ficain) derived from *Ficus carica* L. have gained much attention as milk coagulants due to their well-characterized structures, functions, and properties diversified and precise usage in milk coagulation at the industrial level. Therefore, this chapter examines the use of ficins derived from *F. carica* L. in milk coagulation by paying detailed attention to its properties, extraction, mechanisms, and characteristics.

2 An Overview of Use of Plant Proteases in Milk Coagulation

Milk coagulation is essential in the cheese-making process and has long been studied due to its commercial and research interests. Proteolytic enzymes (*e.g.*, chymosin) destabilize the casein micelles sterically stabilized in their native form in milk due to the "hairy" κ-casein surface layer. Chymosin (EC 3.4.23.4) cleaves at the specific peptide bond of Phe_{105}-Met_{106} of protruding κ-casein protein by releasing caseinomacropeptide (κ-casein peptide 106–169) to the aqueous phase of the milk (serum) and thereby weakening the steric and electrostatic stabilization of micelles (Fox et al., 2017). Milk coagulation properties directly influence the properties of the resulting cheese.

Rennet is a mixture of chymosin and pepsin (EC 3.4.23.1) discovered long before from the abomasum of suckling calves (Roseiro et al., 2003). High specificity towards κ-casein and low calf rennet proteolytic activity made it the most suitable enzyme for cheese production (Anusha et al., 2014). However, the high price and the limited supply of calf rennet and ethical, religious, and dietary protests led to discovering possible alternatives to coagulate milk. An alternative milk coagulant should have similar biochemical properties, such as greater specificity towards κ-casein, high milk clotting index, proteolytic activity at the cheese making pH, and temperature to calf rennet (Liburdi et al., 2019). Moreover, it should have sufficient thermostability to ensure proper whey drainage without residual coagulant activity. Even though microbial and recombinant chymosins are available, plant proteases as milk coagulants have become a growing interest in the cheese industry due to their great diversity and clean-label environment in usage.

Plant proteases have been used as milk coagulants in artisanal cheese production over a long period in the Mediterranean, West African, and Southern European countries (Liburdi et al., 2019). Almost all parts of plants contain proteases that can clot the milk under appropriate conditions (Roseiro et al., 2003; Shah et al., 2014). Many plant proteases which show milk clotting properties belong to aspartic proteases (Table 33.1). However, some cysteine and serine proteases have also shown milk clotting properties. These proteases have been extracted from different parts of plants (Table 33.1), such as latex, leaves, flowers, stems, fruits, and seeds (Roseiro et al., 2003; Shah et al., 2014).

Although numerous proteases are in the findings, only a few are on the industrial applications, such as papain, bromelain, cardosins, and ficins, as their structures, functionalities, and required conditions for various food processing operations have been thoroughly evaluated. Among them, ficins from *F. carica* have gained much

Table 33.1 Types and sources of plant proteases having milk clotting properties

Type of protease	Source	Tissue	References
Aspartic	*Balanites aegyptiaca*	Fruits	Beka et al. (2014)
	Bromelia hieronymi	Fruits	Bruno et al. (2010)
	Centaurea calcitrapa	Germinated seeds	Salvador et al. (2006)
	Cirisum vulgare	Flowers	Lufrano et al. (2012)
	Citrus aurantium	Flowers	Tripathi et al. (2011) and Mazorra-manzano et al. (2018)
	Cynara humilis	Flowers	Esteves et al. (2003)
	Cynara scolymus	Flowers	Chazarra et al. (2007)
	Cynara cardunculus	Flowers	Roseiro et al. (2003), Silva et al. (2003), Gomes et al. (2018)
	Moringa oleifera	Flowers	Pontual et al. (2012)
	Onopordum acanthium	Flowers	Brutti et al. (2012)
	Oryza sativa	Seeds	Asakura et al. (1997)
	Solanum elaeagnifolium	Fruits	Guiama (2010) and Chávez-garay et al. (2016)
	Solanum tuberosum	Leaves, tubers	Tito et al. (2020)
	Streblus asper	Stems	Salehi et al. (2017)
	Withania coagulans	Fruits	Cavalli et al. (2008)
Cysteine	*Actinidia chinensis*	Fruits	Puglisi et al. (2014)
	Albizia lebbeck	Seeds	Egito et al. (2007)
	Asclepias curassavica	Latex	Liggieri et al. (2009)
	Bromelia pinguin	Fruits	
	Calotropis gigantea	Latex, stems	Anusha et al. (2014)
	Calotropis procera	Latex, leaves	Aworh and Muller (1987) and Abebe and Emire (2020)
	Euphorbia nivulia	Latex	Badgujar and Mahajan (2014)
	F. carica sylvestris	Latex	Faccia et al. (2012)
	Ficus racemosa	Latex	Devaraj et al. (2008a, b)
	Helianthus annus	Seeds	Egito et al. (2007) and Jain et al. (2020)
	Sideroxylon obtusifolium	Latex	
	Zingiber officinale	Rhizome	Huang et al. (2011)
Serine	*Crinum asiaticum*	Latex	Singh et al. (2010)
	Cucumis melo	Fruits	Shah et al. (2014)
	Cucurbita moschata	Seeds	Kumar and Sasmal (2020)
	Ficus religiosa	Latex	Kumari et al. (2010)
	Lactuca sativa	Leaves	Lo Piero et al. (2002)
	Solanum dubium	Seeds	Ahmed et al. (2009)

attention since their milk clotting properties have been characterized and assessed thoroughly.

3 Chemistry, Composition, and Properties of Proteases in *F. carica*

3.1 Ficins (EC 3.4.22.3); the Major Proteolytic Enzyme in F. carica

F. carica L., generally known as "Common Fig" is a tree with deeply lobed (three or five lobes) leaves (Fig. 33.1a), which belong to the family Moraceae. Fig is a latex-producing fruit crop and consists of numerous varieties with considerable genetic diversity (Badgujar et al., 2014; Barolo et al., 2014). Although figs are native to Southwest Asia and the Mediterranean region, fig cultivations are widespread globally (Badgujar et al., 2014). Figs are among the earliest cultivated fruits, are highly nutritious, and possess cultural, medicinal, and industrial values; thus have a high demand (Devaraj et al., 2008a, b; Badgujar et al., 2014). Apart from the fruit (Fig. 33.1b, c), fig latex is the essential component in *Ficus* spp., possessing a wide range of industrial usage due to its inherent proteolytic activity.

Generally, plant latices are complex emulsions rich in proteases, alkaloids, starches, sugars, oils, tannins, resins, gums, etc. (Kumari et al., 2010). For example, fig latex consisted of 70% water, 20% gum, 9% ficins, and 1% other dry materials (Zare et al., 2013). The reason for latex having such a high number of proteases and their exact roles is still unclear. However, findings suggested that they are involved

Fig. 33.1 *F. carica*, also known as Common Fig plant's healthy leaves (**a**), immature fruits (**b**), and matured/ripened fruits (**c**) (Photo: Hasitha Priyashantha)

in defense mechanisms by protecting fruits against plant pathogens (Kumari et al., 2010; Azarkan et al., 2011; Haesaerts et al., 2015). In addition to ficin (EC:3.4.22.3), which is the main protease in latex of fig (Barolo et al., 2014), various other secondary metabolites, i.e., *cis*-1,4-polyisoprene, alkaloids, terpenoids, tannins, sterols are also found in fig latex (Milošević et al., 2020). Most of these compounds are toxic to pathogens, including insets and fungi, and therefore, act as defensive chemicals in fig fruit (Zare et al., 2013).

The majority of proteases found within plant latices belong to the cysteine and serine proteases, and they encompass several proteases such as papain, bromelain, ficin, calotropins, etc., in industrial applications. Ten proteolytic components named A, B, C,…, J from the lattices of different *F. carica* varieties have been identified (Sgarbieri et al., 1964). Among them, the Cys endo-proteases, commonly termed ficins (as illustrated in Fig. 33.2), which are closely related to the papain family (family C1, clan CA), are the most exploited industrial proteolytic component (Haesaerts et al., 2015). The proteases in latex of *F. carica* are available in multiple forms of proteolytic enzymes, including ficin, lysozyme, class-II chitinases, and a thaumatin-like protein (Baeyens-Volant et al., 2015). Proteases or proteolytic enzymes play a fundamental role in hydrolyzing or cleaving peptide bonds of protein, which is used as a beneficial enzymatic activity in food processing, e.g., milk coagulation, meat tenderization, dough conditioning agent, bioactive peptide production, etc. (Aider, 2021).

Fig. 33.2 Latex of fig contains 9% of ficin (EC:3.4.22.3), the major proteolytic enzyme. Its chemical structure ($C_9H_{14}N_4O_3$) (**a**) and 3D structure (**b**) are visualized using 4YYU in the protein databank; https://www.rcsb.org/structure/4YYU. (3D image is made with PyMOL v1.8.6.0 (Schrödinger LLC). Photo: Hasitha Priyashantha)

Purified ficin is a unique single polypeptide chain with a molecular mass of 23.1 kDa (Devaraj et al., 2009). Moreover, most papain-like cysteine proteases (PLCPs) are polypeptides with small molecular masses ranging from 20 to 30 kDa (Rawlings et al., 2016). Nevertheless, it shares some common properties like substrate specificity, transpeptidase reactions, activation by reducing agent, and esterase activity similar to papain's (Hamed et al., 2020). Therefore, ficin is also known as one of the main enzymes in the group of PLCPs that includes bromelain, calpain, cathepsin B, and chymopapain (Aider, 2021). Ficin consists of 201 amino acid residues (Aider, 2021). The amino acid sequence present in the active site residues of ficin resembles the corresponding sequence of the papain (Milošević et al., 2019). Ficin is a nonspecific sulfhydryl protease classified as a cysteine endopeptidase enzyme (Zhai et al., 2021). Compositional analysis of purified ficin further indicates the presence of three disulfide bonds and a single free cysteine residue at the active site (Devaraj et al., 2008a, b; Rawlings et al., 2016). The active site of the ficin specifically consists of cysteine (Cys-25) and histidine (His-159) (Aider, 2021).

Several ficin isoforms from *F. galabrata* were isolated and purified in the early 1940s, and their structures have been thoroughly evaluated. Multiple ficin isoforms from *F. carica* have also been isolated; however, some of these isoforms' functions, properties, and structures are yet to be discovered (Zare et al., 2016). Moreover, the number and the relative amounts of ficin types available within different *F. carica* cultivars and varieties also differed (Sgarbieri et al., 1964; Zare et al., 2016). To the current knowledge, *F. carica* contains 6 ficin isoforms, namely Ficin A, B, C, D1, D2 (Azarkan et al., 2011), and E (Baeyens-Volant et al., 2015). They all are non-glycosylated molecules. Among them, Ficin A is the most abundantly available ficin isoform. The recently discovered ficin E co-eluted with ficin A has a different N-terminal sequence (LPETMDWRSK) and electrophoretic behavior compared to other isoforms (Baeyens-Volant et al., 2015). All the other five isoforms have similar molecular mass and N-terminal ends; however, their crystallization structures differ considerably. Ficin isoforms B and C share a common crystal formation, indicating a close sequence and a structural similarity, while ficin A and D showed some distant formations (Haesaerts et al., 2015). The protein data bank codes for ficin A, B, C, and D were 4YYQ, 4YYR, 4YYV, and 4YYW, respectively (Milošević et al., 2020). Recent investigations on ficin S revealed that it possesses a different isoelectric point and a separate sugar content compared to other isoforms (Morellon-Sterling et al., 2020). However, limited studies have been conducted to date to characterize the physicochemical, kinetic, three-dimensional structures, and other functional properties of different ficin isoforms (Aider, 2021). Ficin is mainly free form, and ficin cocktail applications are becoming popular as an industrial biocatalyst (Morellon-Sterling et al., 2020).

Generally, ficins from *F. carica* have a wide specificity range toward essential and neutral amino acids such as Gly, Val, Leu, Ala, Ser, Asn, Arg, His, and they hydrolyze Tyr, Phe, and Val peptide bonds in proteins (Yang et al., 2017; Aider, 2021). Ficin has been recognized as sulfhydryl protease and contains two sulfhydryl groups. Only one group directly reacts with the catalytic site, while the other group occasionally shows its activity (Devaraj et al., 2008a, b; Zare et al., 2013). Moreover, it is necessary to have -SH, $-NH_3^+$, and $-COO^-$ reactive groups to have catalytic action of ficin (Zare et al., 2013). The active site of ficin consists of two amino acids: Cys-25 and His-159, and up-to-date, only three fragments of ficin have been studied, (I) a fragment around the Cys catalytic site, (II) a His catalytic site, and (III) the N-terminal fragment (Aider, 2021). Further, ficin has broad substrate specificity and intense proteolytic activity toward the basic and neutral amino acids (Zhai et al., 2021).

Ficin functions around a wide range of pH from 3 to 10 and the optimum activity have been reported around neutral pH, generally around 6.5–8.5. It loses its activity completely below pH 3, resulting in a partially folded structure (Devaraj et al., 2008a, b; Milošević et al., 2019). The optimum temperature range for the ficins is around 45–55 °C; however, its general activity could be observed from 30–90 °C (Aider, 2021). Above 70 °C, ficins are prone to get denatured, and the apparent denaturation temperature of ficin has been reported as 73 °C (Devaraj et al., 2008a, b). According to Devaraj et al. (2008a, b), ficin loses its activity by 37% when incubated for 10 minutes at 70 °C and loses 50% of its activity for 20 min at the same temperature. The activity of the ficin enzyme has been reported to be enhanced in the presence of cysteine and mercaptoethanol, whereas p-chloromercuribenzoate and $HgCl_2$ were identified as ficin enzymes inhibitors (Morellon-Sterling et al., 2020).

The major problem associated with the ficins is the autolysis, which affects their storage time and stability. According to Zare et al. (2016), the amount of peptide production due to autolysis of ficin A, B, C, and the unretained fraction extracted from *F. carica* cv. Sabz was increased when the storage time increased from 1 to 10 days. However, their autolysis rate differed significantly, and the ficin A and unretained fraction showed much lower autolytic activities.

3.2 *Other Proteolytic Components in* F. carica

Besides Cys proteases, several other proteolytic components have also been identified from the *F. carica* latex. Hamed et al. (2020) discovered a serine protease named FPIII, which has a specific activity of 2430 units/mg of protein. It showed optimum activity around 8.5 pH and 60 °C. According to Hamed et al. (2020), the relative activity of FPIII is highest on gelatin, following fibrin, albumin, hemoglobin, casein, and collagen in the decreasing order. In addition to ficin, subtilase, peroxidase, chitinase, and mandelonitrile lyase are also present in *F. carica* latex (Kitajima et al., 2018). A recent investigation on collagenolytic serine proteases and chitinolytic enzymes present in fig latex has gained much attention for their extensive proteolytic properties (Raskovic et al., 2016).

Further studies, however, are needed to configure the characteristics and properties of these fig latex-derived proteases. Raskovic et al. (2014) identified and purified another collagenolytic serine protease from the latex of *F. carica* variety Brown

Turkey using gel and covalent chromatography techniques. It is a highly stable monomeric protease having a molecular mass of 41 kDa, showing activity around a wide range of temperatures and pH. Further, this protease showed its specific activity toward gelatin and collagen. It possesses shallow non-specific proteolytic activity; hence, it is an excellent source to replace microbial collagenolytic enzymes in industrial applications (Raskovic et al., 2014). Moreover a chitin-binding protease with antimicrobial properties has also been identified and purified from the latex of *F. carica*, which could be used in pharmaceutical and biotechnological industries (Mavlonov et al., 2008).

4 Enzymatic Extractions and Sample Preparations

Pioneering studies on isolating different forms of ficin from fig latex were reported in the early 1960s (Kramer & Whitaker, 1964). Ficin isolated from different cultivars of fig consists of closely related multiple isoforms (Milošević et al., 2019). However, these ficin isoforms may have different conformational characteristics due to the variation in the amino acid sequence in each isoform (Zare et al., 2013). In addition, these ficin isoforms possess different isoelectric points, which facilitates their separation by chromatographic methods (Milošević et al., 2019). As described earlier in this chapter, fig latex mainly contains an aqueous solution and a gum portion (i.e., 70% water and 20% gum) (Zare et al., 2013). Therefore, the first step of the isolation process is to remove the gum using a centrifugation process, followed by dialysis to concentrate the supernatant (Kitajima et al., 2018). Ion-exchange chromatography or cationic gel mediuma has been commonly used as a separation technique to distinguish the different active fractions of purified ficin (Azarkan et al., 2011; Zare et al., 2013). The main concern during the isolation process of ficin is its stability concerning the structure and biological activity (Baeyens-Volant et al., 2015). Zare et al. (2013) studied the homogeneity of the peaks of the fractions eluted from the ion exchange chromatography using gel filtration using High-Performance Liquid Chromatography, electrophoresis using sodium dodecyl sulfate-polyacrylamide gel electrophoresis, and recycling-chromatography of the peaks using sulphopropyl-sepharose resins. Most of these fractions have exhibited heterogenous peaks suggesting that ficin may have autolyzed. Ficin A showed the lowest heterogenous peaks compared to other ficin isoforms, suggesting that ficin A is a suitable isoform for industrial applications. Therefore, the use of autolysis inhibitors, i.e., iodoacetamide and potassium tetrathionate, could be a viable approach to maintain the stability of ficin during production, isolation, purification, storage, and other industrial applications (Zare et al. 2013; Aider, 2021).

Early studies on fig lattices have used carboxymethyl (CM)-cellulose chromatography to identify the proteolytic components of ficin based on its activity profiles (Sgarbieri et al., 1964). However, such an approach may not provide detailed and clear evidence for molecular mass information that may directly influence autolysis

(Azarkan et al., 2011). Disulfide bonds in proteases are resistant to autolysis and key to protease stability. Therefore, it is recommended to use protective and reversible measures to preserve free thiol groups present in ficin to avoid the adverse effects of autolysis. The use of reversible inhibitors of the thiol group, including S-methyl methanethiosulfonate or 2,20-dithiodipyridine, has been identified as an effective method to produce simplified chromatograms. The thiol-pegylation strategy is another practical approach to suppressing the autolysis and producing highly purified ficin forms (Azarkan et al., 2011). This includes rapid fractionation of ficin isoforms followed by chemical modification using a monomethoxypolyethylene glycol (mPEG) reagent (Azarkan et al., 2011). A recent study on ficin isoforms investigated the potential of performing *in situ* chemical modification of the catalytic cysteine to prevent the autolysis and other possible protein degradation of latex proteases (Baeyens-Volant et al., 2015). This method includes collecting latex from fig using methylmethanethiol sulfonate (MMTS) to prevent the possible denaturation of ficin isoforms during processing (Baeyens-Volant et al., 2015). Mass spectroscopy is generally performed to characterize the *de novo* sequencing of purified peptides to confirm the ficin isoforms (Azarkan et al., 2011). Moreover, Milošević et al. (2020) reported that performing chromatography at an acidic pH (~5 pH) could retard the autolysis process of ficin since cysteine proteases show lower activity at acidic pH values.

Furthermore, studying the comparative stability of cysteine proteases revealed that the presence of multiple isoforms of ficin greatly influenced the stability of papain. These findings suggested that ficin isoforms are more suitable for industrial usages, i.e., therapeutics, diagnostics, bioreactors, biosensors, and fine chemical applications, than other papain-like proteases (Milošević et al., 2019). Furthermore, it was evident that ficin isoforms in the mixture could be more functionally stable than isolated ficin isoforms (Milošević et al., 2020). In contrast to these findings, separating ficin from other enzymes in the latex improved its effectiveness in proteolytic activity. Finally, the complexity of ficin identification and characterization may be associated with the genomic/sequential differences among varieties and cultivars (Milošević et al., 2020). Thus, predominantly, ficin isoforms have been identified using approaches based on proteome and transcriptome analyses (Kitajima et al., 2018; Aider, 2021).

5 Milk Clotting Properties of Ficins from *F. carica*

The capability of proteases to hydrolyze the κ-casein is referred to as the milk clotting activity (Jacob et al., 2011). It is an essential parameter in determining the suitability of protease in cheese-making. The milk clotting index is the ratio between milk clotting activity and proteolytic activity of proteases. Higher the milk clotting index, higher the suitability as a rennet substitute since it possesses high milk clotting activity with low general proteolytic activity. High proteolytic activity is the major drawback of plant proteases since it leads to excessive hydrolysis of milk caseins resulting in bitter flavors, textural defects, and yield loss of cheese (Roseiro et al., 2003; Jacob et al., 2011; Anusha et al., 2014; Shah et al., 2014). Although numerous plant proteases have been proposed as rennet substitutes, they do not possess a high enough milk clotting index to use at the industrial level (Anusha et al., 2014).

Among several proteolytic components in the fig latex, only ficins have been reported to have milk clotting activity (Raskovic et al., 2016) and visualized in Fig. 33.3, using crude extract of fig. Therefore, despite their high proteolytic activity, numerous attempts have been made to produce various cheese types from different types of milk, using ficins as milk coagulants. Bornaz et al. (2010) studied the proteolytic activity of ficin, cardosin, and chymosin on fresh cheese production using ewe's milk. According to their findings, ficin resulted in a great amount of curd loss in the whey, which is a negative perspective regarding cheese-making, and the total solids, total nitrogen, and the casein amount in the whey also appeared to be highest when ficin is used. However, despite the significant loss of caseins, ficin yielded the highest cheese yield due to its high water retention ability and low syneresis capacity compared to cheese obtained from cardosin and chymosin. Nevertheless, no significant differences in sensory properties were observed among the three types of curds (Bornaz et al., 2010). Daffri et al. (2019) compared the milk clotting activity of ficin from two *F. carica* varieties: Marseillaise and Dauphine, and found that both types possess similar milk clotting activity (1100 UAC) even though the Marseillaise variety had higher proteolytic activity than Dauphine.

Raskovic et al. (2016) reported the changes in the protease profile of fig latex over the fruit ripening process. According to their findings, the ficins concentration

Fig. 33.3 Milk clotting ability of *F. carica*. Raw milk (**a**) is coagulated (within 30 min) into a semi-solid coagulum (**b**) using crude extract of *F. carica* whole fruit. (Photo: Hasitha Priyashantha)

in the latex altered over time. At the beginning of fruit ripening, fig latex contains high concentrations of ficin forms, which possess high caseinolytic activity, leading to a low milk clotting index. The authors predicted that, with the fruit ripening, ficin isoforms with broad specificity for caseins are gradually replaced with isoforms having high specificity towards κ-casein. Therefore, the authors suggested that the ficins derived from ripened fruits or harvested in summer are the best source for milk coagulation as they possess high milk clotting activity and milk clotting index (Raskovic et al., 2016). An early study by Kamer & Whitaker (1964) also reported the milk clotting properties of ten isoforms of ficins derived from *F. carica* var. Kadota found that different isoforms have different specificity toward caseins. According to their findings, all ten isoforms showed caseinolytic activity, and among them, the isoforms termed B and D showed more or less similar and the highest caseinolytic activity. All the isoforms except isoform A showed milk clotting activity; however, only the isoforms C, D, E, and G resulted in considerably higher milk clotting activity than other isoforms. Moreover, the authors reported that the isoforms C, D, E, F, and G are more heat and pH-stable than isoforms. Among them, isoform G appeared to be the best form for milk coagulation, as it possessed comparatively lower caseinolytic activity and higher milk clotting activity, resulting in the highest milk clotting index of all other isoforms (Kamer & Whitaker, 1964).

Akar and Fadiloglu (1999) studied the properties of a "Teleme" cheese and found that the Teleme cheese made with purified ficin had better chemical and sensory properties than that made with crude fig latex. This might be due to removing non-casein-specific proteases with the purification process. According to their study, the purified ficin using ion-exchanged chromatography gave about two-fold higher milk clotting index than the crude latex. Moreover, further purification of ficin using gel filtration resulted in about four times higher milk clotting index than the crude latex. However, purified ficins took much longer to coagulate milk than crude latex. A study of using ultra-filtered bovine skim milk in cheese production revealed that ficin performed better in more concentrated milk (4 times ultra-filtrated) than regular milk, giving less free amino acids and less clotting time and firmer curd. However, ficin resulted in lower cheese yield than other proteases (cardosin and chymosin), making it unsuitable for bovine milk, especially when using regular milk (Low et al., 2006). Siar et al. (2020) reported that ficin immobilized using glyoxyl-agarose led to dense coagulum and gave a higher yield than the free-ficin when coagulating skim milk. For that, milk coagulation should separate into two steps, (I) enzymatic hydrolysis of casein at 4 °C to prevent hydrolysate aggregation, which causes separation of the aggregate and biocatalysts, and (II) force hydrolysate aggregation by heating up to 40 °C. According to the authors, this method increases clotting yield by up to 27% (Siar et al., 2020). Ficin activity on camel milk has been studied by Fguiri et al. (2021), and according to their findings, ficin gave more acidic cheese rich in fat, dry matter, ash, and proteins than the chymosin. Moreover, ficin resulted in higher microbiological quality in cheese made with camel and cow milk than chymosin (Fguiri et al., 2021).

Overall, ficin appeared to be more suitable in coagulating ovine and caprine milk than bovine milk, as ficin resulted in poor milk clotting properties in bovine milk. However, ultrafiltration of milk could enhance the activity of ficins; as the concentration of milk increases, the mean distance between casein micelles decreases, resulting in increment of aggregation velocity and reduced clotting time, thus lesser milk protein hydrolysis. Moreover, purification of ficin often leads to better coagulation properties. Current knowledge of the activity of ficins against coagulation of different milk types facilitates a clear understanding of the potential and limitations of using ficins as a commercial calf rennet replacer. Nevertheless, further studies will enhance its precise usage in milk coagulation as a cheap yet, efficient milk coagulant.

6 Other Beneficial Roles of Using *F. carica* in Milk Coagulation

Ficin has great potential for use in commercial applications and is considered an emerging plant-based enzyme due to its strong catalytic activity. In particular, ficin has been successfully used in the dairy industry as a coagulant for the cheese manufacturing process as an alternative to animal-originated coagulants (Aider, 2021). In addition to the coagulant activity, ficin has exhibited unique proteolytic activity in the presence of milk proteins (Baeyens-Volant et al., 2015). This characteristic feature of ficin has the potential to produce milk protein hydrolysates as a functional food ingredient (Milošević et al., 2020). The milk protein hydrolysis process results in an array of bioactive peptides that can be used as functional food ingredients for promoting health and reducing the risk of certain diseases (Rasika et al., 2015; Abdel-Hamid et al., 2017). Bioactive peptides are generally encrypted within the precursor protein structure as inactive peptide fragments that can be released *in vivo* or *in vitro* enzyme hydrolysis (Rasika et al., 2015; Abdel-Hamid et al., 2017). Ficin exhibits consistent and rapid proteolytic hydrolysis maps of proteins compared to other plant-derived enzymes (Siar et al., 2020). This specific characteristic of ficin has been extensively studied in milk casein protein to determine the bioactive potentials of resulting peptides. Antioxidant activity was one of the most investigated bioactive properties in food-derived peptides. Antioxidant peptides can interact with the free radicals and reactive oxidative species to prevent or retard the oxidation process in biological materials and food systems (Shahidi & Zhong, 2008). Potent antioxidative peptide sequences have been identified from bovine casein hydrolysates produced using *F. carica* L. derived proteinase (Di Pierro et al., 2014). The radical scavenging activity of produced peptides was evaluated using an oxygen radical absorbance capacity (ORAC) assay that quantified the percentage and degree of inhibiting free radicals. *F. carica* L. latex proteinase has the potential to hydrolyze casein substrate and produce low molecular weight peptides that may be responsible for its radical scavenging ability (Di Pierro et al., 2014).

In addition to antioxidant activity, antimicrobial activities of milk protein hydrolysates produced from ficin have been reported in recent literature. Goat milk casein was hydrolyzed using ficin and trypsin and resultant peptides were analyzed for their inhibitory activity against representative Gram-positive and Gram-negative bacteria (Esmaeilpour et al., 2016). Among the enzyme treatments, extensive proteolysis of goat milk substrate was observed with ficin treatment. Moreover, the highest antimicrobial activity against *E.coli* was reported in ficin-assisted hydrolyzed samples. The study's findings further demonstrated that peptides composed of hydrophobic and charged (polar) residues exert excellent antimicrobial activities. This indicates that ficin could be a promising plant-derived enzyme to produce potent antimicrobial peptides (Esmaeilpour et al., 2016). Furthermore, Raskovic et al. (2016) investigated the antifungal activity of fig latex and its proteolytic and milk clotting properties. They reported that different isoforms of ficin with various casein specificities were present in the fig latex during different fruit ripening phases, and the highest antifungal activity of ficin was exhibited in the early phases of flowering (Raskovic et al., 2016). The authors also indicated that *F. carica* latex has the potential to be utilized as a natural food preservative and a functional food ingredient due to its unique compositional characteristics, including antioxidant compounds, vitamins, minerals, fibre, and proteins (Raskovic et al., 2016).

Shabani et al. (2018) studied the rheological and microstructural characteristics of ewes' milk curd obtained by ficin and *Polyporus badius* enzymes. They reported that increasing the concentration of ficin has a pronounced effect on gel elasticity and other beneficial microstructural properties (Shabani et al., 2018). Furthermore, the use of latex peptidases in reducing the allergenicity of milk proteins has also been investigated to develop novel hypoallergenic formulas for children with a milk allergy (Oliveira et al., 2019). In this regard, ficin could be a promising candidate for producing extensively hydrolyzed milk proteins with reduced antigenicity and allergenicity (Abd El-Salam & El-Shibiny, 2017). This characteristic feature of ficin is closely linked with its proteolytic activity, which can adversely affect IgG or IgE antibody-mediated responses (Aider, 2021). However, further clinical studies are recommended to illustrate the mechanism of action related to the immunomodulatory activity of milk protein hydrolysates derived from ficin. Enzymatic hydrolysis of milk proteins is also considered an effective approach to enhancing bioactive peptides' bioavailability and beneficial amino acids. The produced milk hydrolysates have gained much attention as a key ingredient in food supplements, infant formula, sports nutrition, therapeutic diets, and geriatric nutrition (Abd El-Salam & El-Shibiny, 2017; Aider, 2021). However, up-to-date, limited studies have been conducted to determine the sensorial properties of milk protein hydrolysates produced from plant-derived latex proteases. Thus, extensive investigations of the distinct proteolytic activity of ficin on milk proteins will pave the pathways to identify its diverse applications in functional food, nutraceutical, and pharmaceutical industries.

References

Abd El-Salam, M. H., & El-Shibiny, S. (2017). Preparation, properties, and uses of enzymatic milk protein hydrolysates. *Critical Reviews in Food Science and Nutrition, 57*(6), 1119–1132.

Abdel-Hamid, M., Otte, J., De Gobba, C., Osman, A., & Hamad, E. (2017). Angiotensin I-converting enzyme inhibitory activity and antioxidant capacity of bioactive peptides derived from enzymatic hydrolysis of buffalo milk proteins. *International Dairy Journal, 66*, 91–98.

Abebe, B., & Emire, S. (2020). Manufacture of fresh cheese using east African *Calotropis procera* leaves extract crude enzyme as milk coagulant. *Food Science & Nutrition, 8*(9), 4831–4842.

Ahmed, I. A. M., Morishima, I., Babiker, E. E., & Mori, N. (2009). Characterization of partially purified milk-clotting enzyme from *Solanum dubium* Fresen seeds. *Food Chemistry, 116*(2), 395–400.

Aider, M., 2021. Potential applications of ficin in the production of traditional cheeses and protein hydrolysates. JDS Communications.

Akar, B., & Fadiloglu, S. (1999). Teleme production by purified ficin. *Journal of Food Quality, 22*(6), 671–680.

Anusha, R., Singh, M. K., & Bindhu, O. S. (2014). Characterization of potential milk coagulants from *Calotropis gigantea* plant parts and their hydrolytic pattern of bovine casein. *European Food Research and Technology, 238*(6), 997–1006.

Asakura, T., Watanabe, H., Abe, K., & Arai, S. (1997). Oryzasin as an aspartic proteinase occurring in rice seeds: Purification, characterization, and application to milk clotting. *Journal of Agricultural and Food Chemistry, 45*(4), 1070–1075.

Aworh, O. C., & Muller, H. G. (1987). Cheese-making properties of vegetable rennet from Sodom apple (*Calotropis procera*). *Food Chemistry, 26*(1), 71–79.

Azarkan, M., Matagne, A., Wattiez, R., Bolle, L., Vandenameele, J., & Baeyens-Volant, D. (2011). Selective and reversible thiol-pegylation, an effective approach for purification and characterization of five fully active ficin (iso)forms from *Ficus carica* latex. *Phytochemistry, 72*(14–15), 1718–1731.

Badgujar, S. B., & Mahajan, R. T. (2014). International journal of biological macromolecules Nivulian-II a new milk clotting cysteine protease of *euphorbia nivulia* latex. *International Journal of Biological Macromolecules, 70*, 391–398.

Badgujar, S. B., Patel, V. V., Bandivdekar, A. H., & Mahajan, R. T. (2014). Traditional uses, phytochemistry and pharmacology of *Ficus carica*: A review. *Pharmaceutical Biology, 52*(11), 1487–1503.

Baeyens-Volant, D., Matagne, A., El Mahyaoui, R., Wattiez, R., & Azarkan, M. (2015). A novel form of ficin from *Ficus carica* latex: Purification and characterization. *Phytochemistry, 117*, 154–167.

Barolo, M. I., Ruiz Mostacero, N., & López, S. N. (2014). *Ficus carica* L. (Moraceae): An ancient source of food and health. *Food Chemistry, 164*, 119–127.

Beka, R. G., Krier, F., Botquin, M., Guiama, V. D., Donn, P., Libouga, D. G., Mbofung, C. M., Dimitrov, K., Slomianny, M., Guillochon, D., & Vercaigne-marko, D. (2014). Characterization of a milk-clotting extract from *Balanites aegyptiaca* fruit pulp. *International Dairy Journal, 34*(1), 25–31.

Ben Amira, A., Besbes, S., Attia, H., & Blecker, C. (2017). Milk-clotting properties of plant rennets and their enzymatic, rheological, and sensory role in cheese making: A review. *International Journal of Food Properties, 20*(1), S76–S93.

Bornaz, S., Guizani, N., Fellah, N., Sahli, A., Slama, M. B., & Attia, H. (2010). Effect of plant originated coagulants and chymosin on ovine milk coagulation. *International Journal of Food Properties, 13*(1), 10–22.

Bruno, M. A., Lazza, C. M., Errasti, M. E., López, L. M., Caffini, N. O., & Pardo, M. F. (2010). Milk clotting and proteolytic activity of an enzyme preparation from *Bromelia hieronymi* fruits. *LWT- Food Science and Technology, 43*(4), 695–701.

Brutti, C. B., Pardo, M. F., Caf, N. O., & Natalucci, C. L. (2012). *Onopordum acanthium* L. (Asteraceae) flowers as coagulating agent for cheese-making. *LWT- Food Science and Technology, 45*, 172–179.

Cavalli, S. V., Silva, S. V., Cimino, C., Malcata, F. X., & Priolo, N. (2008). Hydrolysis of caprine and ovine milk proteins, brought about by aspartic peptidases from *Silybum marianum* flowers. *Food Chemistry, 106*(3), 997–1003.

Chávez-Garay, D. R., Gutiérrez-Méndez, N., Valenzuela-Soto, M. E., & García-Triana, A. (2016). Partial characterization of a plant coagulant obtained from the berries of *Solanum elaeagnifolium. CyTA Journal of Food, 14*(2), 200–205.

Chazarra, S., Sidrach, L., Lopez-Molina, D., & Rodríguez-López, J. N. (2007). Characterization of the milk-clotting properties of extracts from artichoke (*Cynara scolymus*, L.) flowers. *International Dairy Journal, 17*(12), 1393–1400.

Daffri, T., & A. A. and B. (2019). International journal of agriculture and biosciences opportunities in Ethiopia. *International Journal of Agriculture and Biosciences, 8*(2), 89–98.

Devaraj, K. B., Gowda, L. R., & Prakash, V. (2008a). An unusual thermostable aspartic protease from the latex of *Ficus racemosa* (L.). *Phytochemistry, 69*(3), 647–655.

Devaraj, K. B., Kumar, P. R., & Prakash, V. (2008b). Purification, characterization, and solvent-induced thermal stabilization of ficin from *Ficus carica. Journal of Agricultural and Food Chemistry, 56*(23), 11417–11423.

Devaraj, K. B., Kumar, P. R., & Prakash, V. (2009). Characterization of acid-induced molten globule like state of ficin. *International Journal of Biological Macromolecules, 45*(3), 248–254.

Di Pierro, G., O'Keeffe, M. B., Poyarkov, A., Lomolino, G., & Fitzgerald, R. J. (2014). Antioxidant activity of bovine casein hydrolysates produced by *Ficus carica* L. derived proteinase. *Food Chemistry, 156*, 305–311.

Egito, A. S., Girardet, J. M., Laguna, L. E., Poirson, C., Mollé, D., Miclo, L., Humbert, G., & Gaillard, J. L. (2007). Milk-clotting activity of enzyme extracts from sunflower and albizia seeds and specific hydrolysis of bovine κ-casein. *International Dairy Journal, 17*(7), 816–825.

Esmaeilpour, M., Ehsani, M. R., Aminlari, M., Shekarforoush, S., & Hoseini, E. (2016). Antimicrobial activity of peptides derived from enzymatic hydrolysis of goat milk caseins. *Comparative Clinical Pathology, 25*(3), 599–605.

Esteves, C. L. C., Lucey, J. A., Wang, T., & Pires, E. M. V. (2003). Effect of pH on the gelation properties of skim milk gels made from plant coagulants and chymosin. *Journal of Dairy Science, 86*(8), 2558–2567.

Faccia, M., Picariello, G., Trani, A., Loizzo, P., Gambacorta, G., Lamacchia, C., & Di Luccia, A. (2012). Proteolysis of Cacioricotta cheese made from goat milk coagulated with caprifig (*Ficus carica* sylvestris) or calf rennet. *European Food Research and Technology, 234*(3), 527–533.

Fguiri, I., Sboui, A., Naziha, A., Ziadi, M., Guemri, M., Arroum, S., Dbara, M., & Khorchani, T. (2021). Camel milk-clotting properties of latex protease from *Ficus carica. Journal of Animal and Veterinary Advances, 19*, 99.

Fox, P. F., Guinee, T. P., Cogan, T. M., & McSweeney, P. L. (2017). Enzymatic coagulation of milk. In *Fundamentals of cheese science* (pp. 185–229). Springer.

Gomes, S., Belo, A. T., Alvarenga, N., Dias, J., Lage, P., Cruz, C., Brás, T., Duarte, M. F., & Martins, A. P. L. (2018). Characterization of *Cynara cardunculus* L. flower from Alentejo as coagulant agent for cheese making. *International Dairy Journal, 91*, 178–184.

González-Rábade, N., Badillo-Corona, J. A., Aranda-Barradas, J. S., & del Carmen Oliver-Salvador, M. (2011). Production of plant proteases *in vivo* and *in vitro* – A review. *Biotechnology Advances, 29*(6), 983–996.

Guiama, V. D., Libouga, D. G., Ngah, E., Beka, R. G., Ndi, K. C., Maloga, B., Bindzi, J. M., Donn, P., & Mbofung, C. M. (2010). Milk-clotting potential of fruit extracts from Solanum esculentum, *Solanum macrocarpon* L. and *Solanum melongena. African Journal of Biotechnology, 9*(12).

Haesaerts, S., Rodriguez Buitrago, J. A., Loris, R., Baeyens-Volant, D., & Azarkan, M. (2015). Crystallization and preliminary X-ray analysis of four cysteine proteases from *Ficus carica* latex. *Acta Crystallographica Section F: Structural Biology Communications, 71*(4), 459–465.

Hamed, M. B., El-Badry, M. O., Kandil, E. I., Borai, I. H., & Fahmy, A. S. (2020). A contradictory action of procoagulant ficin by a fibrinolytic serine protease from Egyptian *Ficus carica* latex. *Biotechnology Reports, 27*, e00492.

Huang, X. W., Chen, L. J., Luo, Y. B., Guo, H. Y., & Ren, F. Z. (2011). Purification, characterization, and milk coagulating properties of ginger proteases. *Journal of Dairy Science, 94*(5), 2259–2269.

Jacob, M., Jaros, D., & Rohm, H. (2011). Recent advances in milk clotting enzymes. *International Journal of Dairy Technology, 64*(1), 14–33.

Jain, S., Gupta, S. K., Kumar, A., Bangar, Y. C., & Ahlawat, S. S. (2020). Extraction and partial purification of a novel enzyme from oil seed cakes of sunflower and assessing its use for milk clotting activity. *The Pharma Innovation Journal, 9*, 224–229.

Kitajima, S., Aoki, W., Shibata, D., Nakajima, D., Sakurai, N., Yazaki, K., et al. (2018). Comparative multi-omics analysis reveals diverse latex-based defense strategies against pests among latex-producing organs of the fig tree (*Ficus carica*). *Planta, 247*(6), 1423–1438.

Kramer, D. E., & Whitaker, J. R. (1964). Ficus enzymes. Ii. Properties of the Proteolytic Enzymes from the Latex. *The Journal of Biological Chemistry, 239*(7), 2178–2183.

Kumar, A., & Sasmal, S. (2020). Food hydrocolloids rheological and physico-chemical properties of milk gel using isolate of pumpkin (*Cucurbita moschata*) seeds: A new source of milk clotting peptidase. *Food Hydrocolloids, 106*(March), 105866.

Kumari, M., Sharma, A., & Jagannadham, M. V. (2010). Decolorization of crude latex by activated charcoal, purification and physico-chemical characterization of religiosin, a milk-clotting serine protease from the latex of *Ficus religiosa. Journal of Agricultural and Food Chemistry, 58*(13), 8027–8034.

Liburdi, K., Boselli, C., Giangolini, G., Amatiste, S., & Esti, M. (2019). An evaluation of the clotting properties of three plant rennets in the milks of different animal species. *Food, 8*(12), 600.

Liggieri, C., Obregón, W., Trejo, S., & Priolo, N. (2009). Biochemical analysis of a papain-like protease isolated from the latex of *Asclepias curassavica* L. *Acta Biochimica et Biophysica Sinica, 41*(2), 154–162.

Lo Piero, A. R., Puglisi, I., & Petrone, G. (2002). Characterization of "Lettucine", a serine-like protease from *Lactuca sativa* leaves, as a novel enzyme for milk clotting. *Journal of Agricultural and Food Chemistry, 50*(8), 2439–2443.

Low, Y. H., Agboola, S., Zhao, J., & Lim, M. Y. (2006). Clotting and proteolytic properties of plant coagulants in regular and ultrafiltered bovine skim milk. *International Dairy Journal, 16*(4), 335–343.

Lufrano, D., Faro, R., Castanheira, P., Parisi, G., Veríssimo, P., Vairo-cavalli, S., Simões, I., & Faro, C. (2012). Phytochemistry molecular cloning and characterization of procirsin, an active aspartic protease precursor from *Cirsium vulgare* (Asteraceae). *Phytochemistry, 81*, 7–18.

Mavlonov, G. T., Ubaidullaeva, K. A., Rakhmanov, M. I., Abdurakhmonov, I. Y., & Abdukarimov, A. (2008). Chitin-binding antifungal protein from *Ficus carica* latex. *Chemistry of Natural Compounds, 44*(2), 216–219.

Mazorra-Manzano, M. A., Moreno-Hernández, J. M., & Ramírez-Suarez, J. C. (2018). Milk-clotting plant proteases for cheesemaking. In *Biotechnological applications of plant proteolytic enzymes* (pp. 21–41). Springer.

McSweeney, P. L., Ottogalli, G., & Fox, P. F. (2017). *Diversity and classification of cheese varieties: An overview* (pp. 781–808). Cheese.

Milošević, J., Janković, B., Prodanović, R., & Polović, N. (2019). Comparative stability of ficin and papain in acidic conditions and the presence of ethanol. *Amino Acids, 51*(5), 829–838.

Milošević, J., Vrhovac, L., Đurković, F., Janković, B., Malkov, S., Lah, J., & Polović, N. Đ. (2020). Isolation, identification, and stability of Ficin 1c isoform from fig latex. *New Journal of Chemistry, 44*(36), 15716–15723.

Morellon-Sterling, R., El-Siar, H., Tavano, O. L., Berenguer-Murcia, Á., & Fernández-Lafuente, R. (2020). Ficin: A protease extract with relevance in biotechnology and biocatalysis. *International Journal of Biological Macromolecules, 162*, 394–404.

Moreno-Hernández, J. M., Pérez, M. D. J. B., Osuna-Ruiz, I., Salazar-Leyva, J. A., Ramirez-Suarez, J. C., & Mazorra-Manzano, M. Á. (2021). Exploring the milk-clotting properties of extracts from *Bromelia pinguin* fruit. *Journal of Microbiology, Biotechnology and Food Sciences, 2021*, 62–66.

Oliveira, J. P. B., Candreva, A. M., Rizzo, G., Ramos, M. V., Oliveira, J. S., Oliveira, H. D., & Freitas, C. D. T. (2019). Allergenicity reduction of cow's milk proteins using latex peptidases. *Food Chemistry, 284*(September 2018), 245–253.

Payrol, J. A., Obregón, W. D., Trejo, S. A., & Caffini, N. O. (2008). Purification and characterization of four new cysteine endopeptidases from fruits of *Bromelia pinguin* L. grown in Cuba. *The Protein Journal, 27*(2), 88–96.

Pontual, E. V., Carvalho, B. E. A., Bezerra, R. S., Coelho, L. C. B. B., Napoleão, T. H., & Paiva, P. M. G. (2012). Caseinolytic and milk-clotting activities from *Moringa oleifera* flowers. *Food Chemistry, 135*(3), 1848–1854.

Puglisi, I., Petrone, G., & Lo Piero, A. R. (2014). A kiwi juice aqueous solution as coagulant of bovine milk and its potential in Mozzarella cheese manufacture. *Food and Bioproducts Processing, 92*(1), 67–72.

Rasika, D. M. D., Toshihisa, U., Jayakody, L. N., Suriyagoda, L. D. B., Silva, K. F. S. T., Ando, S., & Vidanarachchi, J. K. (2015). ACE-inhibitory activity of milk fermented with *Saccharomyces cerevisiae* K7 and *Lactococcus lactis* subsp. *lactis* NBRC 12007. *Journal of the National Science Foundation of Sri Lanka, 43*(2), 141–151.

Raskovic, B., Bozovic, O., Prodanovic, R., Niketic, V., & Polovic, N. (2014). Identification, purification and characterization of a novel collagenolytic serine protease from fig *(Ficus carica* var. Brown Turkey) latex. *Journal of Bioscience and Bioengineering, 118*(6), 622–627.

Raskovic, B., Lazic, J., & Polovic, N. (2016). Characterisation of general proteolytic, milk clotting and antifungal activity of *Ficus carica* latex during fruit ripening. *Journal of the Science of Food and Agriculture, 96*(2), 576–582.

Rawlings, N. D., Morton, F. R., & Barrett, A. J. (2006). MEROPS: The peptidase database. *Nucleic Acids Research, 34*, D270–D272.

Rawlings, N. D., Barrett, A. J., & Finn, R. (2016). Twenty years of the MEROPS database of proteolytic enzymes, their substrates and inhibitors. *Nucleic Acids Research, 44*(D1), D343–D350.

Roseiro, L. B., Barbosa, M., Ames, J. M., & Wilbey, R. A. (2003). Cheese-making with vegetable coagulants – The use of *Cynara* L. for the production of ovine milk cheeses. *International Journal of Dairy Technology, 56*(2), 76–85.

Salehi, M., Aghamaali, M. R., Sajedi, R. H., Asghari, S. M., & Jorjani, E. (2017). Purification and characterization of a milk-clotting aspartic protease from *Withania coagulans* fruit. *International Journal of Biological Macromolecules, 98*, 847–854.

Salvador, S. M., Novo, C., & Domingos, A. (2006). Evaluation of the presence of aspartic proteases from *Centaurea calcitrapa* during seed germination. *Enzyme and Microbial Technology, 38*(7), 893–898.

Schaller, A. (2004). A cut above the rest: The regulatory function of plant proteases. *Planta, 220*(2), 183–197.

Sgarbieri, V. C., Gupte, S. M., Kramer, D. E., & Whitaker, J. R. (1964). Ficus enzymes. I. Separation of the Proteolytic Enzymes of *Ficus Carica*. *The Journal of Biological Chemistry, 239*(7), 2170–2177.

Shabani, R., Shahidi, S. A., & Rafe, A. (2018). Rheological and structural properties of enzyme-induced gelation of milk proteins by ficin and *Polyporus badius*. *Food Science & Nutrition, 6*(2), 287–294.

Shah, M. A., Mir, S. A., & Paray, M. A. (2014). Plant proteases as milk-clotting enzymes in cheese-making: A review. *Dairy Science and Technology, 94*(1), 5–16.

Shahidi, F., & Zhong, Y. (2008). Bioactive peptides. *Journal of AOAC International, 91*(4), 914–931.

Siar, E. H., Morellon-Sterling, R., Zidoune, M. N., & Fernandez-Lafuente, R. (2020). Use of glyoxyl-agarose immobilized ficin extract in milk coagulation: Unexpected importance of the ficin loading on the biocatalysts. *International Journal of Biological Macromolecules, 144*, 419–426.

Silva, S. V., Allmere, T., Malcata, F. X., & Andrén, A. (2003). Comparative studies on the gelling properties of cardosins extracted from *Cynara cardunculus* and chymosin on cow's skim milk. *International Dairy Journal, 13*(7), 559–564.

Silva, A. C. D., Nascimento, T. C. E. D. S., Silva, S. A. D., Herculano, P. N., & Moreira, K. A. (2013). Potential of Quixaba (*Sideroxylon obtusifolium*) latex as a milk-clotting agent. *Food Science and Technology, 33*, 494–499.

Singh, K. A., Kumar, R., Rao, G. R. K., & Jagannadham, M. V. (2010). Crinumin, a chymotrypsin-like but glycosylated serine protease from *Crinum asiaticum*: Purification and physicochemical characterization. *Food Chemistry, 119*(4), 1352–1358.

Tito, F. R., Pepe, A., Tonon, C. V., Daleo, G. R., & Guevara, M. G. (2020). Determination and characterisation of milk-clotting activity of two *Solanum tuberosum* aspartic proteases (StAPs). *International Dairy Journal, 104*, 104645.

Tripathi, P., Tomar, R., & Jagannadham, M. V. (2011). Purification and biochemical characterization of a novel protease streblin. *Food Chemistry, 125*(3), 1005–1012.

Yang, Y., Shen, D., Long, Y., Xie, Z., & Zheng, H. (2017). Intrinsic peroxidase-like activity of ficin. *Scientific Reports, 7*(1), 1–8.

Zare, H., Moosavi-Movahedi, A. A., Salami, M., Mirzaei, M., Saboury, A. A., & Sheibani, N. (2013). Purification and autolysis of the ficin isoforms from fig (*Ficus carica* cv. Sabz) latex. *Phytochemistry, 87*, 16–22.

Zare, H., Moosavi-Movahedi, A. A., Salami, M., Sheibani, N., Khajeh, K., & Habibi-Rezaei, M. (2016). Autolysis control and structural changes of purified ficin from Iranian fig latex with synthetic inhibitors. *International Journal of Biological Macromolecules, 84*, 464–471.

Zhai, Y., Cui, Y., Song, M., Vainstein, A., Chen, S., & Ma, H. (2021). Papain-like cysteine protease gene family in fig (*Ficus carica* L.): Genome-wide analysis and expression patterns. *Frontiers in Plant Science, 12*(May), 1–14.

Chapter 34
The Potential of Fig (*Ficus carica*) for New Products

Sara Khoshnoudi-Nia, Akram Sharifi, and Elham Taghavi

1 Introduction

The fig (*Ficus carica* L.) is a tree or shrub with a pear-shaped and fleshy fruit from the *Moraceae* family. Fig is native to southwestern Asia and the eastern part of the Mediterranean due to suitable climate conditions (Hossain et al., 2010; Mawa et al., 2013). Fig is a very delicious fruit and rich in fiber, proteins, amino acids, carbohydrates, sugars, minerals (copper, manganese, magnesium, potassium, and calcium), organic acids, vitamins, and bioactive compounds (e.g., antioxidant polyphenols, flavonoids, anthocyanin pigment) which are the reasons could explain the growing demand for this fruit in the world (Petkova et al., 2019; Shamin-Shazwan et al., 2019). Furthermore, due to the high concentration of calcium (133 mg/100 g; a higher than 10% of the RDI or recommended dietary intake) of fig fruits, figs could be considered "rich in calcium" (Barolo et al., 2014).

Moreover, the phytochemicals have been found in figs as follows: arabinose, β-amyrins, β-carotenes, glycosides, β-sitosterols, xanthotoxol, alkaloids, flavonoids (e.g., rutin, catechin, and epicatechin), coumarins, phenolic acid (chlorogenic acid,

S. Khoshnoudi-Nia (✉)
Seafood Processing Research Group, School of Agriculture, Shiraz University, Shiraz, Iran
e-mail: s-khoshnoudi@shirazu.ac.ir

A. Sharifi
Department of Food Science and Technology, Faculty of Industrial and Mechanical Engineering, Qazvin Branch, Islamic Azad University, Qazvin, Iran
e-mail: asharifi@qiau.ac.ir

E. Taghavi
Faculty of Fisheries and Food Science, Universiti Malaysia Terengganu, Terengganu, Malaysia
e-mail: elham.taghavi@umt.edu.my

M. F. Ramadan (ed.), *Fig* (Ficus carica)*: Production, Processing, and Properties*, https://doi.org/10.1007/978-3-031-16493-4_34

hydroxycinnamic acids (3-*O*- and 5-*O*-caffeoylquinic acids, ferulic acid, gallic acid and syringic acid), flavonoid glycosides (quercetin-3-*O*-glucoside and quercetin-3-*O*-rutinoside), furanocoumarins (psoralen and bergapten), saponins, and terpenes (Çalişkan & Polat, 2011; Petkova et al., 2019; Pourghayoumi et al., 2016; Tanwar et al., 2014; Veberic & Mikulic-Petkovsek, 2016). Because of these natural bio-compounds, several health benefits have been reported for fig fruit, and it has an important place in the Mediterranean diet. For example, fresh and dried figs are recommended for enhancing eyesight, liver, and spleen diseases. In addition, it could be used as a good laxative, expectorant, and diuretic agent (Badgujar et al., 2014; Shahrajabian et al., 2021).

Fig is consumed either in fresh or dried fruit. Fresh Fig has a short post-harvest life (around one week) and is highly sensitive to microbial activity due to high moisture content. A combination of modified atmosphere packaging and cold storage conditions may extend this time up to 2–4 weeks (del Carmen Villalobos et al., 2016; Villalobos et al., 2016). However, the drying procedure can reduce the microbial load of fruits by decreasing the water activity and consequently minimizes physical and biochemical changes during storage. Therefore, figs are commercially available in dried form. The annual production of dried figs is about 1.332813 million tons in Turkey (310,000 t), Egypt (225,295 t), Morocco (153,472 t), Iran (130,328 t), Algeria (114,092 t), and Spain (51,600) are the main producer of this fruit (FAOSTAT, 2020). In most fig-producing countries, fig fruit is sun-dried to save energy and cost. Moreover, the sun-drying method is an environment-friendly process. The suitable fresh fig fruit for drying is at full maturity and after wrinkling and dehydrating (over-ripe stage) on the tree (Aksoy, 2015).

Fresh or dried fig fruit can be processed to produce varied food products such as jam, jelly, fig powder, marmalade, wine, liquor, juice, etc. Moreover, several by-products (such as low-quality figs and wastes of juice) are generated during fig processing, which is a valuable source of bioactive compounds like phenolic compounds, colorants, sterols, etc. Therefore, the current chapter aims to discuss several fig products and by-products, evaluate the influence of the food processing methods and storage conditions on antioxidants, and review the nutritional composition of fig products and by-products. Furthermore, recent findings on the potential of figs to develop new functionality and nutraceutical products will be highlighted.

2 Fig Products

2.1 Dried Fig

One of the main forms of consumption of figs is dried form. Fig fruits are usually dried by a natural sun drying method. However, artificial drying systems can also be used. The energy cost is the most significant factor in selecting an artificial technique. In addition, the fig quality and maturity, drying time and temperature regime, airflow rate, and relative humidity affect the final dried fig quality (Freiman et al., 2015). Fig fruits with thin and fleshy skin containing high sugar and low acidity level are the most suitable fresh fig fruit for drying, and these dried figs are softer and sweeter. Fresh fig fruit can be peeled and sun-dried to produce dried fruits with a softer texture and lighter color. Also, fresh fruit may be sliced or cubed before drying (Aksoy, 2015). Cut or cubed dried figs mixed with nuts and cereals for producing snacks, muesli, and granola. The cubed figs may be coated with chocolate (Isa et al., 2020).

The bioactive compound content of fresh or processed fig fruit strongly depends on the cultivar type (De Pilli et al., 2019). Khradhraoui et al. (2019) investigated the phytochemical content, antioxidant potential, and fatty acid composition of nine sun-dried fig cultivars originating from South-Eastern and Middle Eastern Tunisia. Dark fruits exhibited a higher total polyphenol content (Saoudi Douiret cultivar: 201 mg GAE 100 g^{-1} dm) than green ones (Bayoudhi Douiret cultivar: 73.74 mg GAE 100 g^{-1} dm). Fatty acid methyl esters contained (C16: 0), (C18: 1), ((C18: 2) 9, 12), ((C18: 3) 9, 12, 15) and (C20: 0). Strong correlations were found between the amounts of various phenolic compounds and fatty acids with antioxidant capacity (Khadhraoui et al., 2019).

The method of the drying process may affect the nutritional and health-related properties. In this regard, Slatnar et al. (2011) showed that the highest sugar/organic acid ratio was obtained for the sun-dried sample. Furthermore, the higher content of all individual phenolic compounds (chlorogenic acid, catechin, epicatechin, kaempferol-3-*O*-glucoside, luteolin-8-C-glucoside, rutin, quercetin-3-*O*-glucoside, and cyanidin-3-*O*-rutinoside) was determined after the oven-drying process. Similarly, higher total phenolic content and antioxidant activity were detected after the oven drying process. The results indicated that dried figs are a good source of phenolic compounds and organic acids. However, these authors expressed the nutritional properties based on moisture weight, not dried ones (Slatnar et al., 2011). Moreover, Vallejo et al. showed that after fig drying, the losses of polyphenols did not exceed 15% (Vallejo et al., 2012).

Not only is the quality important, but safety is an essential factor in fig dried. Many studies reported the risk of the presence of mycotoxins in dry fruits, especially in dry figs (Sulyok et al., 2020; Trucksess & Scott, 2008; Yujiao Wang et al., 2018). Aflatoxins are considered the most toxigenic metabolites of the mycotoxin classes produced mainly by *Aspergillus flavus* and *Aspergillus parasiticus* (Ait Mimoune et al., 2018). Treating the fresh fig with Sulphur before drying caused a lighter color and prevented pest damage. The maximum residue limit for SO_2 in dried figs is 1000 ppm (Codex Alimentarius, 1989). Moreover, the labeling of food containing sulfites (>10 ppm) is obliged by US Food and Drug Administration (FDA). Methyl bromide was the main fumigant for controlling pests before 2015. However, it has been banned in developed and developing countries since 2005 and 2015. To obtain sun-dried figs, figs are exposed to sunlight to complete the dehydration process, increasing fungal contamination and aflatoxins development (Ait Mimoune et al., 2018). A controlled drying process helps to reach a safety level to enhance the protection of public health. Petrić et al. (2018) treated the figs with various solutions as followed: 0.5% citric acid; 0.5% ascorbic acid; 0.3% L-cysteine; 0.2% chestnut extract; 0.15% Echinacea extract before drying in a cabinet dryer. They found nine metabolites (i.e., aflatoxin B (AFB1), ochratoxin A, ochratoxin alpha, kojic acid, emodin, altenuene, alternariol methyl ether, brevianamide F, and tryptophol in fig samples. However, all pretreatment solutions reduced AFB1, being a major fig contaminant. The most efficient pretreatment was L-cysteine (15% reduction), while ascorbic acid increased mycotoxin (158% increase). However, all pretreatment solutions reduced AFB1 as a major fig (Petrić et al., 2018). Ozone is the green and effective alternative for Methyl bromide (Akbas & Ozdemir, 2008; Sadeghi et al., 2017).

2.2 Fig Paste and Puree

Dried fig fruit that is not fit for direct human consumption was classified as Class 2 or substandard or surplus. These figs are processed to develop new products, decrease waste and make value-added. These figs may be ground into a paste. In fig paste production, the seeds can be crushed or not, depending on the demand. The fig paste or puree can be used to fill cookies and cakes and as a raw matter in making ice cream, jam, marmalades, or juice. Fig bars and fig balls also may be prepared with fig paste. Small fig balls can also be coated with chocolate or mixed with various nuts and then coated with roasted sesame seeds, chia seed, coconut, or cocoa powder (Aksoy, 2015).

2.3 Fig Jam

Jam production is one of the most popular ways to enhance shelf-life and preserve fig fruit. The steps of jam processing were as follows: the fresh figs were sorted, washed, and cut (the fruit can also be used without cutting). For the inactivation of the enzymes, the fruits were heated at 80 °C. Then, sugar dissolved in water was added and boiled to reach a final soluble solid concentration (600–700 g/kg). Then, pectin was added, and the mixture was heated for 10 min to complete the hydration of the pectin. Finally, hot-packing (85 °C) and sterilization were conducted. Although jams are popular fruit products, they contain high levels of sugar. Instead, preserves are produced without additional sugar being better alternatives for conventional jams (Levaj et al., 2010).

Jam formulation affects the physicochemical, nutritional, and sensory properties of fig jams. Kumari et al. (2018) evaluate the effect of various concentrations of pectin (0.3–0.7%) and citric acid (0.3–0.7%) on physicochemical and sensory properties of wild Himalayan fig jam. The fig jams contained 0.7% pectin, and 0.3% citric acid was the best formulation. This jam showed higher titratable acidity (12.0%: based on citric acid), ascorbic acid content (1.24 mg/100 g), total sugar

(66.4%; reducing sugar: 58.8%), total phenol (39.8 mg/100 g), anthocyanin (17.0 mg/100 g) and sensory scores. The FTIR curve of fig jam and its pulp showed no alteration in physicochemical parameters during jam processing on fig pulp (Kumari et al., 2018).

The ripeness of fig fruit can also affect the properties of the jam. Levaj et al., 2010 evaluated the gel strength and sensory properties of jams and preserves made from figs with different ripeness. During ripening, the total solids, sugars, and pH values were increased in jams and preserves. Investigation of the amount of three pectin fractions (soluble in water, soluble in oxalate, and alkali) showed that the most abundant pectin fraction of figs was soluble in oxalate one. It was not affected by the stage of ripeness. The amount of soluble in water and soluble in alkali fractions (pectin fractions responsible for the strength of gels) decreased with increasing fruit ripeness. Gels made soluble in oxalate pectin were significantly softer than other gels. While the ripeness stage had not affected the strength of jams, with increased fruit ripeness, the preserve's strength decreased. The sensory properties (on taste, firmness, and color) of jams or preserves were not affected by the stage of ripeness. However, the sensory scores of jam were higher than preserves (Levaj et al., 2010).

Furthermore, the literature review showed that the bioactivity of fig fruit is affected by processing conditions. Tanwar et al. (2014) also evaluated the effect of jam and nectar processing on fig fruit products' physicochemical and nutritional properties. The results revealed that conversion of fig fruit pulp into jam and nectar caused a significant increase in total soluble solids and titratable acidity, carbohydrate, and energy value but a decrease in pH, iron, calcium and phosphorus, moisture, ash, crude fiber, crude protein, crude fat, vitamin C, ß carotene, total phenolics, total flavonoids, total anthocyanins and tannins (Tanwar et al., 2014). Moreover, these authors (2015) assessed the effect of similar processes on the antioxidant activity of fig fruit. Fig pulp showed higher antioxidant activity than fig jam and fig nectar. Total phenolics decreased by 10 and 55%, flavonoids by 98 and 45%, anthocyanins by 79 and 33%, and tannins by 83 and 77% in fig jam and fig nectar as compared to fig pulp, respectively (Tanwar et al., 2015).

The effect of cold storage and jam processing on physicochemical properties and phytochemical compounds of yellow-colored figs (*Ficus carica* L.) has been evaluated by Petkova et al. (2019). The highest and lowest bioactive compounds content and antioxidant activity were found in fresh and frozen fruits, respectively (after 4 months of storage at −18 °C) The jam processing was a better technique to preserve bioactive compounds as compared freezing method (especially for carotenoids and phenolic compounds). After the jam and freezing process, the losses of total phenolic content were around 10% and 45%, respectively (Petkova et al., 2019). Rababah et al. (2011) found that total phenolics and anthocyanins of fig jam decreased by 68.6% and 60.2% after 5 months of storage. Despite reducing these compounds in jam processing, it is a good method to enhance the shelf-life of fig fruit.

Abd-El-Hak et al. (2016) blended the sycamore and fig fruit to formulate a nutritional and value-added jam. Sycamore and fig fruit were blended in the ratios of

100:0, 0:100, 20:80, 80:20, and 50:50, respectively. Various jam samples' chemical and sensory properties were evaluated during 12 months of storage at room temperature. The results showed that sycamore was rich in ash and crude fiber. The sensory score of jam samples was improved by increasing the fig content of jam up to 50%. During storage, the acidity of samples decreased, and their physiochemical properties improved. Therefore, the blend of sycamore and fig in jam production was recommended to increase nutritional value.

2.4 Fig Juice

As a healthy drink, Fig juice is an excellent source of vitamin C and natural polyphenols (e.g., flavonoids and proanthocyanidin), helping boost the immune system and balance blood pressure (Joseph & Raj, 2011). preparation of fig juice is done directly by applying pressure to fresh fig or boiling the dried fig in water and extracting the procedure. Soft and juicy fresh figs are more suitable for obtaining fig juice directly. Furthermore, fig juice and concentrate can be consumed directly or used in the pastry industry as glazing and bio-sweetener (Soni et al., 2014).

Although fig juice contains a high level of phenolic compounds, most of these compounds are difficult to be absorbed by the human body. To improve the absorption, hydrolysis by enzymes or microbes in the digestive tract is a suitable solution (Filannino et al., 2016). Fermentation can enhance the nutritional value and antioxidant activity of food products. Various biochemical changes (especially conversion of glycoside into the aglycone and increasing the flavonoid release) occurred during the fermentation causing the nutrition component and anti-nutrition ratio bioactivity and digestibility changes (Zhang et al., 2012). On the other hand, beverages are an appropriate delivery system for probiotics. There are several health benefits related to the probiotics, such as antimicrobial, antimutagenic, anticarcinogenic, antidiarrheal, and antihypertensive properties, enhancing lactose digestion, reducing cholesterol and mineral malabsorption Therefore, probiotic fig juice can be considered a healthy and nutraceutical beverage for vegetarians and consumers with lactose-allergy (Khezri et al., 2016). Khezri et al. (2016) compared the survivability of three species of lactic acid bacteria (*Lactobacillus casei*, *L. plantarum*, and *L. delbrueckii*) in fig juice. Heat-treated fig juices were treated by three species (6 log CFU mL^{-1}) separately and incubated at 30 °C for 72 h. Among various species, *L. delbrueckii* showed the best survivability in high acidity and reached nearly 9 log CFU mL^{-1} after 48 h of fermentation. After 4 weeks of cold storage (~4 °C), the viable cell counts of *L. delbrueckii* and *L. plantarum* were 6 and 5 log CFU mL^{-1}, respectively. However, *L. casei* was just survived for 2 weeks. Therefore *L. delbrueckii* was introduced the most suitable strain among other species at the consumption time. Therefore, probiotic fig juice can serve as a healthy beverage for vegetarians and consumers with lactose allergies (Khezri et al., 2016).

Wijayanti et al. (2017) used *Lactobacillus acidophilus, Lactobacillus bulgaricus, Lactobacillus casei,* and *Lactobacillus plantarum* as a starter (at 37 °C for 24 h) for producing fermented fig juice. The fermented fig juice demonstrated higher total phenolic content. *L. bulgaricus* showed the highest increase in total phenolic content (0.45%) and antioxidant activity (IC_{50} = 76.55 ppm) as compared to the control ones (Total phenolic compound = 0.09%; IC_{50} = 76.7 ppm) (Wijayanti et al., 2017). To enhance the viability of probiotics in an acidic medium, Khezri et al. (2018) formulated synbiotic fig juice. They used *L. delbrueckii* as a probiotic culture and inulin as a prebiotic ingredient and evaluated the effect of various formulations on microbial survivability, physicochemical, total phenolic content, antioxidant capacity, and sensory parameters of fig juices. After 48 h of fermentation, the cell contents of lactic acid bacteria reached the maximum level in both probiotic (8.41 Log CFU mL^{-1}) and symbiotic (8.83 Log CFU mL^{-1}) samples. The total phenolic content and antioxidant capacity of fermented fig juices were significantly higher than the control samples. The viability of *Lactobacillus delbrueckii* significantly decreased during storage time. However, the rate of decline was slower in the synbiotic juice (7.49 Log CFU mL^{-1} after 4 weeks of cold storage) than in the probiotic one (6.59 Log CFU mL^{-1}). The highest overall sensory score was obtained for control juice (7.8), followed by synbiotic juice (6.1) and probiotic juice (5.7). Synbiotic fig juice was better in terms of probiotic and antioxidant properties. At the end of the storage period (4 weeks), the viable count of *Lactobacillus delbrueckii* in synbiotic samples remained about 1 Log CFU mL^{-1} higher than probiotic ones being above the minimum requirement (6 Log CFU mL^{-1}) for probiotic food products (Khezri et al., 2018). Overall, fig juice could be a suitable carrier for probiotics, and prebiotics and fermentation could enhance the bioavailability of polyphenols.

2.5 *Fig Jelly*

To obtain the fig jelly, fig pulp, sugar, around 0.5% citric acid, and 1% pectin were mixed. The fruit mixture was heated until boiling and reaching ~65 Brix. The finished product was pasteurized in jars with lids (Bof et al., 2012; Curi et al., 2019). Fig variety can be affected the nutritional and physicochemical characteristics of fig jelly. Therefore, Curi et al. (2019) assessed the effect of fig cultivars on the properties of fig jelly. The results showed that the various fig cultivars obtained jellies with different physical, chemical, and rheological properties. However, the fig cultivar had no significant effect on the acceptability of fig jelly (Curi et al., 2019). The effects of freezing and jelly processing on the antioxidant capacity of various fruit (e.g., grape, apple, strawberry, pear, guava, and fig (*Ficus carica* L.) for 90 days was also investigated by Bof et al. (2012). Freezing (−15 °C) did not significantly differ in the antioxidant capacity of grape and fig pulp. However, fig pulp showed the lowest antioxidant activity. This lower antioxidant activity may refer to the variety and harvest dates. However, the antioxidative compounds of strawberry and fig jelly were heat-resistant and antioxidant activity remained stable in fig jelly (Bof et al., 2012).

2.6 Wine and Vinegar

Wine is an alcoholic beverage (alcohol content: 5 to 13%) made from fermented fruits. The high sugar content of fig fruit makes it a good candidate for producing wine and vinegar. Yeast converts the juice sugar to ethanol and carbon dioxide (Sheng et al., 2019). Fruit wine contains most of the fruit nutrients, alcohol, vitamin, and other compounds produced by the yeast activation that moderate consumption of it may protect people against certain cancers and improve mental and heart health (Artero et al., 2015).

The properties of fruit (fresh or dried fruit, pH values, sugar, nitrogen contents, etc.), type of yeast, and fermentation conditions can affect the wine quality (Chilaka et al., 2010). The juice was mixed with 0.05% Diammonium hydrogen phosphate (DAHP) and pasteurized at 85 °C for 30 min. Pasteurized juice was inoculated with *Saccharomyces cerevisiae* (at 24 °C temperature for 20 days). The wine properties are as follows: Brix: 8.5°; TSS: 0.4 to 4.2%; total sugar: 0.2 to 4.0%; alcohol: 9.9 to 11.8% alcohol. During the fermentation period, total soluble solids, titratable acidity, reduced sugar, total sugar decreased, and the alcohol percentage of wine increased. They introduced the wine prepared at 10% inoculums and 24 °C temperature for 20 fermentation days as the best quality sample (PotdarVrushali et al., 2020).

The fig processing conditions can be affected the quality of the wine. For example, Lu et al. (2021) reported that dried fig wines had lower anthocyanin contents and antioxidant activity than wines prepared with fresh samples due to the thermal degradation of anthocyanins during the drying process. However, the drying process had no significant effect on the total polyphenol and flavonoid contents (Lu et al., 2021).

For wine production processes, one of the most important steps is the choice of yeast. Since *Saccharomyces* showed a relatively high fermentation rate and yield (alcohol production), it is still the most frequent yeast in wine production (Lu et al., 2021). However, it was shown that *non-Saccharomyces might* improve the flavor complexity of wines by producing various desirable volatile compounds (Binati et al., 2020; Hranilovic et al., 2018). Liu et al. (2021) evaluated the quality of fig wines co-fermented by three *non-Saccharomyces (Wickeramomyces anomala; Hanseniaspora uvarum,* and *Pichia fermentans).* The results showed that fig wines fermented by *non-Saccharomyces* exhibited higher contents of organic acids and mono-phenols. Furthermore, co-fermentation by *non-Saccharomyces* could obtain up to 38 aroma compounds, while fermentation by *Saccharomyces* obtained around 30 substances. Therefore, the aroma of the different samples showed significant differences. The result revealed that pleasant fig wine could be obtained by co-fermentation of *non-Saccharomyces* without a *Saccharomyces* starter culture (Liu et al., 2021).

3 Value-Added Fig Products

By increasing the world's population and the global food demand, the food industry and agriculture sector have increased their production. As a result of the increased production and consumption, a considerable amount of agro-food waste and by-products are produced, which are generally left unused and unattended (Sharma et al., 2020). Therefore, it is crucial to establish ways to recycle these valuable wastes. They can be used as sustainable and renewable resource materials for synthesizing cost-effective, value-added products for various applications. In addition, the products are in line with global health, food security, and sustainable agriculture strategies (Dey et al., 2021).

Fig fruit by-products (peel, seeds, and no-optimal fruits) are generated during fig processing to produce fig jam, fig juice, wine, etc. These by-products contain various bioactive compounds and functional ingredients. Recycling these by-products to produce individual compounds and/or extracts and additives for the food industry can improve food products' functional, nutritional, and techno-functional characteristics (Teruel-Andreu et al., 2021). In this sub-section, some of the value-added products obtained from the peel, seed, and fig pulp were reviewed.

3.1 *Extraction of Phenolic Compounds*

Phenolic compounds are important constituents of fruits that contribute to the taste, color, and human health effects fruits (Khoshnoudi-Nia et al., 2020; Veberic et al., 2008). Phenolic compounds and carotenoids are considered the most important natural antioxidants (Tsao & Deng, 2004). Fig by-products are a rich source of phenolic compounds. Phenolic acids (*e.g.,* chlorogenic acid, syringic acid, caffeic acid, ferulic acid, coumaric acid, quinol, and gallic acid) and flavonoids (*e.g.* (+)-catechin, (−)epicatechin, anthocyanin, kaempferol, quercitin, and myricetin) are the two major classes of fig fruits phenols (Bucic-Kojic et al., 2011; Dueñas et al., 2008). In this regard, based on the conventional and modern isolation and characterization methods, 126 chemical constituents in the eight major groups (hydroxybenzoic acids, hydroxycinnamic acids, flavonoids, coumarins, furanocoumarins, volatile constituents, triterpenoids, and miscellaneous) were found for *Ficus carica* (Badgujar et al., 2014). Several phenolic compounds were found in fig fruit. The type and level of phenolic compounds can be varied depending on the variety, part of the fruit, time of harvest, maturity level, and extraction conditions (Vallejo et al., 2012). Vallejo et al. (2012) collected 18 fig cultivars commonly grown in south-eastern Spain in spring and summer and evaluated their polyphenolic profile. Depending on the season crop, the total phenolic compounds ranged between 19.1 to 140 mg/100 g. Fruit harvested in spring generally showed higher phenolic values than samples obtained in summer. Flavonoids are considered "environmental compounds". Because phenolic compounds were directly produced in response to environmental conditions (such as ultraviolet light and CO_2 levels) (Caldwell et al., 2005). Moreover, Vallejo et al. (2012) reported that a high concentration of phenolic compounds is present either in the skin (mainly anthocyanins, especially cyanidin-3- rutinoside) or pulp (mainly proanthocyanidins, especially (epi)catechin) of fig (Vallejo et al., 2012). Analyses of the potential health-promoting compositions of 10 Spain cultivars of fig fruits showed that the phytochemical profiling of fig varied based on cultivar and the varietal type. In fig fruit, eleven polyphenolic compounds (in flavan-3-oils, phenolic acids, flavonols, flavons, and anthocyanins classes) were reported. The total polyphenol content of the fig fruits was ranged from 715 ('Verdal' brevas) to 186 mg 100 g^{-1} dry matter ('Campera'). Betulinic acid (content ranging from 57 to 97%) and oleanolic acid were the main triterpenoids of fig fruits. The cultivar 'Verdal' showed the highest antioxidant activity. The 'Tiberio', 'Campera', 'Calabacita', and 'Cuello Dama Blanca' cultivars had significant antidiabetic effect (Wojdyło et al., 2016).

Extraction conditions (the solvent type and the temperature) and the sample type affect the quality and quantity of bioactive compounds extracted from fruits. Sun et al. (2015) isolated phenolic compounds from fresh peeled figs with acidified ethanol at room temperature. The combination of alcohol-water showed a better performance than mono-component solvents to isolate phenolic compounds. Ethanol and water are safe and suitable solvents for the food industry than other organic solvents (Sun et al., 2015). In this regard, Bucic-Kojic et al. (2011) evaluated the effect of extraction conditions (50–80%, v/v aqueous-ethanol and temperature of 25–80 °C) and fig variety on phenolic compounds of lyophilized fig fruits. The content of phenolic compounds (2.4–3.7 mg GAE g^{-1} dw), total flavonoids (0.44–2.5 mg CE g^{-1} dw), and total proanthocyanidins (0.68–0.87 mg g^{-1} dw) increased by increasing the ethanol volume in the solvent from 50 to 80%. The phenolic compound content increased by increasing temperature (25 to 80 °C). The total phenolic compounds content in fig fruit was 2.5 to 3.7 mg GAE g^{-1} dw, while the total flavonoids and proanthocyanidins content were 0.68–2.5 mg CE g^{-1} dw and 0.67–0.87 mg g^{-1} dw, respectively. Therefore, the best extraction conditions were introduced 80% v/v aqueous ethanol at 80 °C for 120 min. The highest level of phenolic compounds was obtained in a variety of Crnica (4.7 mg GAE g^{-1} dw), followed by Brežutka bijela (3.9 mg GAE g^{-1} dw), Šaraguja (3.7 mg GAE g^{-1}dw), Termenjača (3.1 mg GAE g^{-1} dw), and Bjelica (2.6 mg GAE g^{-1} dw) varieties (Bucic-Kojic et al., 2011). Meziant et al. (2014) optimize the extraction conditions (acetone concentration (40–80%), temperature (25–65 °C), and time (60–120 min) of the total phenolic compounds of fresh dark fig (*Ficus carica* L.) by response surface methodology (RSM: The Box-Behnken design). The optimal extraction parameters were 63.4% acetone, 115.1 min, and 48.66 °C. In this condition, the total phenolic was 536.4 mg GAE 100 g^{-1}, and antioxidant activity was obtained at 71.8 mg GAE 100 g^{-1} dw (Meziant et al., 2014). However, organic solvents (e.g., acetone, methanol, ethyl acetate, and hexane) are related to concerns about environmental pollution, toxicology, and safety. Developing methods to improve the extraction yield and efficiency of phenolic compounds based on eco-friendly solvents is important. Some papers have reported that the use of fermentation and enzymatic processes can enhance the performance of bioactive compounds extraction from agro-industrial by-products (Azabou et al., 2016; Larios-Cruz et al., 2019; Torres-León et al., 2021). Fermentation can enhance the level of bioactive compounds in fruits and vegetables (Martins et al., 2013). Buenrostro-Figueroa et al. (2017) optimized the solid-state fermentation (SSF) of fig (*Ficus carica* L.) by-products by four fungal strains to enhance the release of total phenolic compounds. The optimized SSF conditions were as follows: *A. niger* HT4 for 36 h at 40 °C; moisture content: 60%; pH: 5.0; mineral composition (g L^{-1}): $NaNO_3$: 0.57; KH_2PO_4: 3.04; $MgSO_4$.7H2O: 1.52; and KCl: 5.37. Under the optimal conditions, the total phenolic compound was 10.1 mg of GAE g^{-1} of dm, and the TPC release and antioxidant activity values increased up to 5.48 and 98.5 times, respectively (Buenrostro-Figueroa et al., 2017). Overall, quercetin-3-*O*-rutioside, polymeric procyanidins, quercetin-3-glucoside, chlorogenic acid, and cyanidin-3-*O*-rutinoside were introduced as the major individual phenolic compound in whole figs, while picatechin and cyanidin-3-rutinoside were the main phenolic compounds of fig pulp (Teruel-Andreu et al., 2021).

3.2 Fig Seed Oil

Along with the progress in the vegetable oils industry, there is a great interest in using new and cost-effective oil sources. The fig seed oil is healthy. Because it contains ω-3 and rich in unsaturated fatty acids is (Nakilcioğlu-Taş, 2019). The α-linolenic acid of the fig seed oil is a β-sitosterol and an important omega source (Güven et al., 2019). Therefore, in recent years, several studies have focused on the extraction of fig seed oil as a cost-effective source of vegetable oil and evaluated its quality properties. In this regard, Chougui et al. (2013) evaluated the fig seed composition of four varieties of *Opuntia ficus-indica*. Fig seeds were ground into a fine powder, and their oil was extracted. The oil content of various varieties varied and ranged from 7.3% (from the red variety) to 9.3% (from the yellow one). In all the samples, linoleic acid was the dominating fatty acid (58.7 (red variety) to 63.1% (orange variety), followed by oleic (15.2% (orange) to 24.3% (red)), palmitic acid (12.7% (red) to 13.4% (orange)) and stearic (3.3% (orange) to 4.2% (red)). Vaccenic acid was also only found in the green and orange ones. Gas chromatography analysis of the defatted fig seed powder showed more than 20 phenolic compounds detected in the defatted fig seed extract (Chougui et al., 2013).

Duman et al. (2018) reported that the oil content of fig seed was 29.1% (acidity: 4.6 mg KOH g^{-1}; iodine value: 176 I_2 g100g^{-1}; peroxide level: 4.5 meq O_2 kg^{-1}; refractive index: 1.480 (nD); density: 0.925 mg mL^{-1}; viscosity: 52.1 mPa; melting point: −15.1 °C; total phenol: 21.7 mg GAE g^{-1} and oxidative stability: 110 °C h^{-1}). Linolenic (41.8%), linoleic (29.0%), oleic (18.6%), and Palmitic (6.76%) acids were reported as major fatty acids of fig seeds oil. Moreover, their results showed that the fig seed oil had a suitable antimicrobial effect against *Staphylococcus aureus*, *Pseudomonas aeruginosa*, *Escherichia coli*, *Enterococcus faecalis*, and *klebsiella pneumoniae* bacterial strains besides *Candida albicans* and *Aspergillus flavus* as fungal strains (Duman et al., 2018). Hssaini et al. (2020) evaluated the effect of fig cultivars on fig seed oil. They extracted fig seed oil from four fig cultivars. They reported that fig seeds contained 21.5 to 28.5% yellow-colored oil. Gas-liquid chromatography analysis of the seed oils showed high percentages of linolenic acid (38.4–43.5) and linoleic acid (28.9–34.5%). The frequent saturated fatty acids in all fig seed oils were Palmitic acid (8.54 to 9.05%) and stearic acid (2.59 to 3.3%) (Hssaini et al., 2020). Also, in another work, the total oil yield was 14.08%, and the antioxidant capacity was 140.1 mg Trolox, equivalent to 100 g^{-1}. The highest fatty acid was α-linolenic acid (26.3%), followed by linoleic acid (24.2%) and oleic acid (19.6%) (Ergun & Bozkurt, 2020).

Since fig oil is rich in polyunsaturated fatty acids, it is sensitive to oil oxidation. Therefore, for more protection, Icyer et al. (2017) encapsulated fig seed oil in gum arabic (GA) and maltodextrin (MD) as wall material and optimized the encapsulation condition. According to the results, the optimum encapsulation ratio of MD: GA: oil and spray-drying temperatures were 8:1:3 and 150 °C, respectively (Icyer et al., 2017).

3.3 *Fig Powder*

Low-quality fig fruit can be grounded to produce fig powder. The processing technology used for manufacturing fig powder was simple, and the powder can be used to develop various functional and value-added food products. Khapre et al. (2015) prepared the fig (*Ficus carica* L.) fruits powder (*Deanna* variety) and used it in the formulation of burfi (Indian cookie). Fig powder enhanced the nutritional value of burfi in terms of fiber (3.7%), potassium (0.46%), and protein (13.1%). In addition, the fig burfi had good taste and low cost compared to similar products (Khapre et al., 2015). Viuda-Martos et al.(2015) determined the chemical, physicochemical and techno-functional properties of the fig powder co-products obtained from the pulp and peel of two cultivars (colar and cuello de dama). The results showed that the main sugars were fructose and glucose. The sugar, organic acid, and phenolic acid content of fig powder obtained from the pulp were higher than from fig peel. However, the flavonoids were predominant in fig powder obtained from fig peel, and this powder showed higher antioxidant activity than that obtained from the pulp (Viuda-Martos et al., 2015).

There are many edible seeds in a fig which are a rich source of protein, fat, carbohydrates, dietary fiber, antioxidants, phenolic compounds, *omega*-3 (*α*-linolenic and cis- 5, 8, 11, 14, 17-eicosapentaenoic acid), and *omega*-6 (linoleic acid) fatty acids and minerals (e.g., Mg, Zn, and Cu) (Baygeldi et al. 2021; Hssaini et al., 2020). The fig seeds are valuable health-promoting food ingredients. However, during the production of various fig products, these seeds are discarded. These wastes have a good potential to produce value-added products. In this regard, Bölek (2021) enriched the biscuits with fig seed powder and evaluated its effect on quality parameters (proximate composition, color values, total phenolic contents, antioxidant activity, texture, and sensory properties) of biscuits. He replaces wheat flour with fig seed powder at four levels (0, 5, 10, and 15%). Fig seed powder contains suitable content of fiber (56.6%), protein (14.3%), fat (20.3%), and total phenolic (TPC: 665.1 mg GAE 100 g-1). The addition of this powder to the biscuit formulation improved the fiber content of the sample and increased the total phenolic compounds and antioxidant activity of biscuits significantly. The addition of fig seed powder to the biscuit formulation caused an increase in the water absorption capacity, dough stability, hardness, and cohesiveness of the biscuit dough due to the high content of dietary fiber. The increased dough stability can be referred to their H-bonding capability obtained by amino acids (especially glutamine and proline) of fig seeds.

Moreover, the interaction between fiber, fat, and gluten, besides the lower gluten and higher water holding capacity, leads to a decrease in the gas retention capacity and spread ratio of biscuits (Kohajdová et al., 2011; Seevaratnam et al., 2012). The addition of fig seed powder to the formulation of up to 10% enhanced the biscuits' sensory scores (odors, flavor, and overall score). However, a sensory score of appearance and texture decreases with the increasing substation ratio of fig seed powder due to the dilution of gluten caused by wheat flour replacement with fig seed powder (Bölek, 2021).

3.4 Food Colorant

Food colorants or pigments are natural or synthetic substances used to create, recover, and standardize food color. Food colorants make food more attractive to consumers. Anthocyanins are one of the most important water-soluble pigments with antioxidant activity. They are responsible for the attractive colors of various fruits and flowers (Yuntao Wang et al., 2017). Therefore, anthocyanins are ideal alternatives for synthetic dyes with antioxidant activity, which can easily incorporate into aqueous-based food (Sharif et al., 2020). In addition, the antioxidant activity of anthocyanin induces various health benefits in preventing cancer (Santos et al., 2013), neuronal and cardiovascular illnesses (He & Giusti, 2010), inflammation (Lee et al., 2011), and other diseases.

Peel of fig has a significant potential to isolate phenolic compounds, especially anthocyanins (Buenrostro-Figueroa et al., 2017). In this regard, a high concentration of phenolic compounds and anthocyanins also were reported in the peel of mature figs (Barolo et al., 2014). Backes et al. (2020) used natural purple colorants extracted from fig peel in doughnuts (icing) and a typical Brazilian pastry called "beijinho" formulation. The extract of fig peel had no significant changes in nutrition value and rheological features. Even the dough firmness and consistency were improved, and "beijinhos" became softer and chewier. Moreover, the extract showed antioxidant and antimicrobial activities in the products (Backes et al., 2020).

3.5 Fig Pectin

On an industrial scale, Pectin, an abundant structural acidic polysaccharide, is commonly produced from citrus, beet, and apple wastes. Due to gelling and emulsifying properties of pectin, it is an excellent candidate for stabilizing jams, jellies, and acidic dairy drinks (Brejnholt, 2009). Moreover, it can also be used to formulate micro-and nano-capsules and edible films/coating (Kurek et al., 2021). In addition, various health benefits were introduced for pectin, such as decreasing cholesterol and serum glucose levels, preventing cancer, and improving the immune system (Wikiera et al., 2014). Therefore, there is a growing interest in isolating this valuable and multifunctional polysaccharide from new and non-expensive sources, especially fruit peels and pomace. In this regard, Gharibzahedi et al. (2019) used four extraction methods (i.e., hot-water, ultrasound-assisted, microwave-assisted, and ultrasound-microwave assisted) to extract pectin fig (*Ficus carica* L.) peel. The most yield was obtained for ultrasound-assisted extraction (11.7%). The fig peel pectin, especially the sample extracted by the ultrasound-assisted, exhibits a relatively good antioxidant activity-dependent. Moreover, the pectin extracted by the ultrasound-assisted method presented the maximum galacturonic acid content (76.8%), molecular weight (6.91 × 103 kDa), the highest emulsifying activity (61.2–61.3%), and emulsion stability (94.3–95.2%) (Gharibzahedi et al., 2019).

4 Conclusion

Nowadays, there is an increased interest in developing novel nutrient-rich food products. There is scientific research about the potential use of fresh, dried, and underutilized fig fruit and figs by-products to produce new and value-added food products. The use of low-quality figs and their waste to produce food ingredients and bioactive compounds could benefit the food industry, reduce environmental issues, and benefit consumers' health. Although fig-based products were reported, the effect of novel technologies for extraction of fig bioactive compounds and using these valuable extracted compounds in food formulation to improve the quality, shelf-life, and nutritional value of food products needs further study. In addition, the use of fig waste to produce novel products such as fig coffee and fermented fig milk should be investigated.

References

Abd-El-Hak, N. A., Abd-El-Rahman, H., & Rizk, A. (2016). Production and evaluation of sycamore and fig blends jam. *Egyptian Journal of Agricultural Research, 94*(3), 647–658.

Ait Mimoune, N., Arroyo-Manzanares, N., Gámiz-Gracia, L., García-Campaña, A. M., Bouti, K., Sabaou, N., & Riba, A. (2018). Aspergillus section flavi and aflatoxins in dried figs and nuts in Algeria. *Food Additives & Contaminants: Part B, 11*(2), 119–125.

Akbas, M. Y., & Ozdemir, M. (2008). Application of gaseous ozone to control populations of *Escherichia coli, Bacillus cereus* and *Bacillus cereus* spores in dried figs. *Food Microbiology, 25*(2), 386–391.

Aksoy, U. (2015). *The dried fig management and the potential for new products*. Paper presented at the V International Symposium on Fig 1173.

Artero, A., Artero, A., Tarín, J. J., & Cano, A. (2015). The impact of moderate wine consumption on health. *Maturitas, 80*(1), 3–13.

Azabou, S., Abid, Y., Sebii, H., Felfoul, I., Gargouri, A., & Attia, H. (2016). Potential of the solid-state fermentation of tomato by products by *Fusarium solani* pisi for enzymatic extraction of lycopene. *LWT-Food Science and Technology, 68*, 280–287.

Backes, E., Leichtweis, M. G., Pereira, C., Carocho, M., Barreira, J. C., Genena, A. K., Baraldi, I. J., Barreiro, M. F., Barros, L., & Ferreira, I. C. (2020). *Ficus carica* L. and *Prunus spinosa* L. extracts as new anthocyanin-based food colorants: A thorough study in confectionery products. *Food Chemistry, 333*, 127457.

Badgujar, S. B., Patel, V. V., Bandivdekar, A. H., & Mahajan, R. T. (2014). Traditional uses, phytochemistry and pharmacology of *Ficus carica*: A review. *Pharmaceutical Biology, 52*(11), 1487–1503.

Barolo, M. I., Mostacero, N. R., & López, S. N. (2014). *Ficus carica* L. (Moraceae): An ancient source of food and health. *Food Chemistry, 164*, 119–127.

Baygeldi, N., Küçükerdönmez, Ö., Akder, R. N., & Çağındı, Ö. (2021). Medicinal and nutritional analysis of fig (*Ficus carica*) seed oil; a new gamma tocopherol and omega-3 source. *Progress in Nutrition, 23*(2), 1–6.

Binati, R. L., Junior, W. J. L., Luzzini, G., Slaghenaufi, D., Ugliano, M., & Torriani, S. (2020). Contribution of non-Saccharomyces yeasts to wine volatile and sensory diversity: A study on *Lachancea thermotolerans*, *Metschnikowia* spp. and *Starmerella bacillaris* strains isolated in Italy. *International Journal of Food Microbiology, 318*, 108470.

Bof, C. M. J., Fontana, R. C., Piemolini-Barreto, L. T., & Sandri, I. G. (2012). Effect of freezing and processing technologies on the antioxidant capacity of fruit pulp and jelly. *Brazilian Archives of Biology and Technology, 55*, 107–114.

Bölek, S. (2021). Effects of waste fig seed powder on quality as an innovative ingredient in biscuit formulation. *Journal of Food Science, 86*(1), 55–60.

Brejnholt, S. M. (2009). Pectin. *Food stabilisers, thickeners and gelling agents*, 237–265.

Bucic-Kojic, A., Planinic, M., Tomas, S., Jokic, S., Mujic, I., Bilic, M., & Velic, D. (2011). Effect of extraction conditions on the extractability of phenolic compounds from lyophilised fig fruits (*Ficus carica* L.). *Polish Journal of Food and Nutrition Sciences, 61*(3).

Buenrostro-Figueroa, J., Velázquez, M., Flores-Ortega, O., Ascacio-Valdés, J., Huerta-Ochoa, S., Aguilar, C., & Prado-Barragán, L. (2017). Solid state fermentation of fig (*Ficus carica* L.) by-products using fungi to obtain phenolic compounds with antioxidant activity and qualitative evaluation of phenolics obtained. *Process Biochemistry, 62*, 16–23.

Caldwell, C. R., Britz, S. J., & Mirecki, R. M. (2005). Effect of temperature, elevated carbon dioxide, and drought during seed development on the isoflavone content of dwarf soybean [*Glycine max* (L.) Merrill] grown in controlled environments. *Journal of Agricultural and Food Chemistry, 53*(4), 1125–1129.

Çalişkan, O., & Polat, A. A. (2011). Phytochemical and antioxidant properties of selected fig (*Ficus carica* L.) accessions from the eastern Mediterranean region of Turkey. *Scientia Horticulturae, 128*(4), 473–478.

Carmen Villalobos, M., Serradilla, M. J., Martín, A., Ruiz-Moyano, S., Pereira, C., & de Guia Cordoba, M. (2016). Synergism of defatted soybean meal extract and modified atmosphere packaging to preserve the quality of figs (*Ficus carica* L.). *Postharvest Biology and Technology, 111*, 264–273.

Chilaka, C., Uchechukwu, N., Obidiegwu, J., & Akpor, O. (2010). Evaluation of the efficiency of yeast isolates from palm wine in diverse fruit wine production. *African Journal of Food Science, 4*(12), 764–774.

Chougui, N., Tamendjari, A., Hamidj, W., Hallal, S., Barras, A., Richard, T., & Larbat, R. (2013). Oil composition and characterisation of phenolic compounds of *Opuntia ficus-indica* seeds. *Food Chemistry, 139*(1–4), 796–803.

Codex Alimentarius. (1989). *Food Additives, Joint FAO/WHO Food Standards Program.* Rome, Italy.

Curi, P. N., Albergaria, F. C., Pio, R., Schiassi, M. C. E. V., de Sousa Tavares, B., & De Souza, V. R. (2019). Characterisation and jelly processing potential of different fig cultivars. *British Food Journal.*

De Pilli, T., Lopriore, G., Montemitro, M., & Alessandrino, O. (2019). Effects of two sweet cherry cultivars (*Prunus avium* L., cvv.'Ferrovia'and 'Lapins') on the shelf life of an innovative bakery product. *Journal of Food Science and Technology, 56*(1), 310–320.

Dey, T., Bhattacharjee, T., Nag, P., Ghati, A., & Kuila, A. (2021). Valorization of agro-waste into value added products for sustainable development. *Bioresource Technology Reports, 16*, 100834.

Dueñas, M., Pérez-Alonso, J. J., Santos-Buelga, C., & Escribano-Bailón, T. (2008). Anthocyanin composition in fig (*Ficus carica* L.). *Journal of Food Composition and Analysis, 21*(2), 107–115.

Duman, E., Şimşek, M., & Özcan, M. M. (2018). Monitoring of composition and antimicrobial activity of fig (*Ficus carica* L.) fruit and seed oil. *Journal of Agricultural Science and Technology, 24*(2), 75–80.

Ergun, Z., & Bozkurt, T. (2020). Determination of fatty acid composition and antioxidant activity of fig seed oil. *International Journal of Agricultural and Natural Sciences, 13*(2), 101–107.

FAOSTAT. (2020). *Crops and livestock products*. Retrieved from https://www.fao.org/faostat/en/#data/QCL

Filannino, P., Di Cagno, R., Addante, R., Pontonio, E., & Gobbetti, M. (2016). Metabolism of fructophilic lactic acid bacteria isolated from the Apis mellifera L. bee gut: Phenolic acids as external electron acceptors. *Applied and Environmental Microbiology, 82*(23), 6899–6911.

Freiman, Z. E., Rosianskey, Y., Dasmohapatra, R., Kamara, I., & Flaishman, M. A. (2015). The ambiguous ripening nature of the fig (*Ficus carica* L.) fruit: A gene-expression study of potential ripening regulators and ethylene-related genes. *Journal of Experimental Botany, 66*(11), 3309–3324.

Gharibzahedi, S. M. T., Smith, B., & Guo, Y. (2019). Pectin extraction from common fig skin by different methods: The physicochemical, rheological, functional, and structural evaluations. *International Journal of Biological Macromolecules, 136*, 275–283.

Güven, N., Gökyer, A., Koç, A., Temiz, N. N., Selvi, S., Koparal, B., Deniz, B., Dedeoğlu, S. B. Ö., Büyükhelvacigil, H. F., & Büyükhelvacıgil, R. (2019). Physiochemical composition of fig seed oil from Turkey. *Journal of Pharmacy and Pharmacology, 7*, 541–545.

He, J., & Giusti, M. M. (2010). Anthocyanins: Natural colorants with health-promoting properties. *Annual Review of Food Science and Technology, 1*, 163–187.

Hossain, A., Salleh, A., Mekhled, M., & Al-Saif, A. (2010). Chlorophyll fluorescence intensity, photosynthetic yield and flowering in fig fruit trees as affected by phloem stress. *Bulgarian Journal of Agricultural Science, 16*(5), 547–552.

Hranilovic, A., Li, S., Boss, P. K., Bindon, K., Ristic, R., Grbin, P. R., Van der Westhuizen, T., & Jiranek, V. (2018). Chemical and sensory profiling of shiraz wines co-fermented with commercial non-saccharomyces inocula. *Australian Journal of Grape and Wine Research, 24*(2), 166–180.

Hssaini, L., Hanine, H., Charafi, J., Razouk, R., Elantari, A., Ennahli, S., Hernández, F., & Ouaabou, R. (2020). First report on fatty acids composition, total phenolics and antioxidant activity in seeds oil of four fig cultivars (*Ficus carica* L.) grown in Morocco. *OCL, 27*, 8.

Icyer, N. C., Toker, O. S., Karasu, S., Tornuk, F., Kahyaoglu, T., & Arici, M. (2017). Microencapsulation of fig seed oil rich in polyunsaturated fatty acids by spray drying. *Journal of Food Measurement and Characterization, 11*(1), 50–57.

Isa, M., Jaafar, M., Kasim, K., & Mutalib, M. (2020). *Cultivation of fig (Ficus carica L.) as an alternative high value crop in Malaysia: A brief review.* Paper presented at the IOP Conference Series: Materials Science and Engineering.

Joseph, B., & Raj, S. J. (2011). Pharmacognostic and phytochemical properties of *Ficus carica* Linn – An overview. *International journal of pharmtech research, 3*(1), 8–12.

Khadhraoui, M., Bagues, M., Artés, F., & Ferchichi, A. (2019). Phytochemical content, antioxidant potential, and fatty acid composition of dried Tunisian fig (*Ficus carica* L.) cultivars. *Journal of Applied Botany and Food Quality, 92*, 143–150.

Khapre, A., Satwadhar, P., & Syed, H. (2015). Studies on processing technology and cost estimation of fig (*Ficus carica* L.) fruit powder enriched Burfi (Indian cookie). *Journal of Applied and Natural Science, 7*(2), 621–624.

Khezri, S., Dehghan, P., Mahmoudi, R., & Jafarlou, M. (2016). Fig juice fermented with lactic acid bacteria as a nutraceutical product. *Pharmaceutical Sciences, 22*(4), 260–266.

Khezri, S., Mahmoudi, R., & Dehghan, P. (2018). Fig juice fortified with inulin and *Lactobacillus delbrueckii*: A promising functional food. *Applied Food Biotechnology, 5*(2), 97–106.

Khoshnoudi-Nia, S., Sharif, N., & Jafari, S. M. (2020). Loading of phenolic compounds into electrospun nanofibers and electrosprayed nanoparticles. *Trends in Food Science & Technology, 95*, 59–74.

Kohajdová, Z., Karovicova, J., Jurasová, M., & Kukurová, K. (2011). Effect of the addition of commercial apple fibre powder on the baking and sensory properties of cookies. *Acta Chimica Slovaca, 4*(2), 88–97.

Kumari, K., Sharma, S., Joshi, V., & Sharma, S. (2018). Adding value to wild Himalayan fig (*Ficus palmata*): Composition, functional and sensory characteristics of jam. *The Journal of Phytopharmacology, 7*(1), 13–18.

Kurek, M., Benbettaieb, N., Ščetar, M., Chaudy, E., Elez-Garofulić, I., Repajić, M., Klepac, D., Valić, S., Debeaufort, F., & Galić, K. (2021). Novel functional chitosan and pectin bio-based packaging films with encapsulated *Opuntia-ficus indica* waste. *Food Bioscience, 41*, 100980.

Larios-Cruz, R., Londoño-Hernández, L., Gómez-García, R., García-Galindo, I., Sepulveda, L., Rodríguez-Herrera, R., & Aguilar, C. N. (2019). Extraction of bioactive molecules through fermentation and enzymatic assisted technologies. *High Value Fermentation Products: Human Health, 1*, 27–59.

Lee, J. H., Lee, H.-J., & Choung, M.-G. (2011). Anthocyanin compositions and biological activities from the red petals of Korean edible rose (*Rosa hybrida* cv. Noblered). *Food Chemistry, 129*(2), 272–278.

Levaj, B., Bunić, N., Dragović-Uzelac, V., & Kovačević, D. B. (2010). Gel strength and sensory attributes of fig (*Ficus carica*) jams and preserves as influenced by ripeness. *Journal of Food Science, 75*(2), S120–S124.

Liu, X., Xu, S., Wang, M., Wang, L., & Liu, J. (2021). Effect of mixed fermentation with Pichia fermentans, Hanseniaspora uvarum, and Wickeramomyces anomala on the quality of fig (*Ficus carica* L.) wines. *Journal of Food Processing and Preservation, 45*(2), e15169.

Lu, Y., Xu, K., Song, Y., Zhao, Z., & Guo, M. (2021). Effect of thermal processing on the phenolic profiles, antioxidant capacity, and volatile compounds of fig wines. *Journal of Food Processing and Preservation, 45*(4), e15321.

Martins, S., Teixeira, J. A., & Mussatto, S. I. (2013). Solid-state fermentation as a strategy to improve the bioactive compounds recovery from *Larrea tridentata* leaves. *Applied Biochemistry and Biotechnology, 171*(5), 1227–1239.

Mawa, S., Husain, K., & Jantan, I. (2013). *Ficus carica* L. (Moraceae): Phytochemistry, traditional uses and biological activities. *Evidence-Based Complementary and Alternative Medicine, 2013*.

Meziant, L., Benchikh, Y., & Louaileche, H. (2014). Deployment of response surface methodology to optimize recovery of dark fresh fig (*Ficus carica* L., var. Azenjar) total phenolic compounds and antioxidant activity. *Food Chemistry, 162*, 277–282.

Nakilcioğlu-Taş, E. (2019). Biochemical characterization of fig (*Ficus carica* L.) seeds. *Journal of Agricultural Sciences, 25*(2), 232–237.

Petkova, N., Ivanov, I., & Denev, P. (2019). Changes in phytochemical compounds and antioxidant potential of fresh, frozen, and processed figs (*Ficus carica* L.). *International Food Research Journal, 26*(6), 1881–1888.

Petrić, J., Šarkanj, B., Mujić, I., Mujić, A., Sulyok, M., Krska, R., Šubarić, D., & Jokić, S. (2018). Effect of pretreatments on mycotoxin profiles and levels in dried figs. *Arhiv za Higijenu Rada i Toksikologiju, 69*(4), 328.

PotdarVrushali, B., Dandagavhal Gauri, R., & More Anjali, J. (2020). Optimum parameters for wine production from fig fruit (*Ficus carica*) juice. *IOSR Journal of Environmental Science, Toxicology and Food Technology, 14*(8), 28–34.

Pourghayoumi, M., Bakhshi, D., Rahemi, M., Noroozisharaf, A., Jafari, M., Salehi, M., Chamane, R., & Hernandez, F. (2016). Phytochemical attributes of some dried fig (*Ficus carica* L.) fruit cultivars grown in Iran. *Agriculturae Conspectus Scientificus, 81*(3), 161–166.

Rababah, T. M., Al-Mahasneh, M. A., Kilani, I., Yang, W., Alhamad, M. N., Ereifej, K., & Al-u'datt, M. (2011). Effect of jam processing and storage on total phenolics, antioxidant activity, and anthocyanins of different fruits. *Journal of the Science of Food and Agriculture, 91*(6), 1096–1102.

Sadeghi, R., Mirabi Moghaddam, R., & Taghizadeh, M. (2017). Application of ozone to control dried fig pests-*Oryzaephilus surinamensis* (Coleoptera: Silvanidae) and Ephestia kuehniella (Lepidoptera: Pyralidae)-and its organoleptic properties. *Journal of Economic Entomology, 110*(5), 2052–2055.

Santos, I. S., Ponte, B. M., Boonme, P., Silva, A. M., & Souto, E. B. (2013). Nanoencapsulation of polyphenols for protective effect against colon-rectal cancer. *Biotechnology Advances, 31*(5), 514–523.

Seevaratnam, V., Banumathi, P., Premalatha, M., Sundaram, S., & Arumugam, T. (2012). Functional properties of *Centella asiatica* (L.): A review. *International Journal of Pharmacy & Pharmaceutical Sciences, 4*(5), 8–14.

Shahrajabian, M. H., Sun, W., & Cheng, Q. (2021). A review of chemical constituents, traditional and modern pharmacology of fig (*Ficus carica* L.), a super fruit with medical astonishing characteristics. *Polish Journal of Agronomy, 44*, 22–29.

Shamin-Shazwan, K., Shahari, R., Amri, C. N. A. C., & Tajuddin, N. S. M. (2019). Figs (*Ficus carica* L.): Cultivation method and production based in Malaysia. *Engineering Heritage Journal (GWK), 3*(2), 06–08.

Sharif, N., Khoshnoudi-Nia, S., & Jafari, S. M. (2020). Nano/microencapsulation of anthocyanins; a systematic review and meta-analysis. *Food Research International, 132*, 109077.

Sharma, G., Kaur, M., Punj, S., & Singh, K. (2020). Biomass as a sustainable resource for value-added modern materials: A review. *Biofuels, Bioproducts and Biorefining, 14*(3), 673–695.

Sheng, H., Tang, L., Xin, S., Wang, Z., Chen, S., & Jiang, H. (2019). Effects of dried fruit fermentation on fig wine quality. *Food and Fermentation Industries, 45*(7), 165–172.

Slatnar, A., Klancar, U., Stampar, F., & Veberic, R. (2011). Effect of drying of figs (*Ficus carica* L.) on the contents of sugars, organic acids, and phenolic compounds. *Journal of Agricultural and Food Chemistry, 59*(21), 11696–11702.

Soni, N., Mehta, S., Satpathy, G., & Gupta, R. K. (2014). Estimation of nutritional, phytochemical, antioxidant and antibacterial activity of dried fig (*Ficus carica*). *Journal of Pharmacognosy and Phytochemistry, 3*(2).

Sulyok, M., Krska, R., & Senyuva, H. (2020). Profiles of fungal metabolites including regulated mycotoxins in individual dried Turkish figs by LC-MS/MS. *Mycotoxin Research, 36*(4), 381–387.

Sun, C., Wu, Z., Wang, Z., & Zhang, H. (2015). Effect of ethanol/water solvents on phenolic profiles and antioxidant properties of Beijing propolis extracts. *Evidence-Based Complementary and Alternative Medicine, 2015*.

Tanwar, B., Andallu, B., & Modgil, R. (2014). Influence of processing on physicochemical, nutritional and phytochemical composition of *Ficus carica* L.(fig) products. *Asian Journal of Dairying & Foods Research, 33*(1), 37–43.

Tanwar, B., Andallu, B., & Modgil, R. (2015). Influence of processing on phytochemical characteristics and in vitro antioxidant actvity of *Ficus carica* L. (fIg) products. *International Journal of Food and Fermentation Technology, 5*(2), 223.

Teruel-Andreu, C., Andreu-Coll, L., López-Lluch, D., Sendra, E., Hernández, F., & Cano-Lamadrid, M. (2021). *Ficus carica* fruits, by-products and based products as potential sources of bioactive compounds: A review. *Agronomy, 11*(9), 1834.

Tsao, R., & Deng, Z. (2004). Separation procedures for naturally occurring antioxidant phytochemicals. *Journal of Chromatography B, 812*(1-2), 85–99.

Torres-León, C., Chávez-González, M. L., Hernández-Almanza, A., Martínez-Medina, G. A., Ramírez-Guzmán, N., Londoño-Hernández, L., & Aguilar, C. N. (2021). Recent advances on the microbiological and enzymatic processing for conversion of food wastes to valuable bioproducts. *Current Opinion in Food Science, 38*, 40–45.

Trucksess, M., & Scott, P. (2008). Mycotoxins in botanicals and dried fruits: A review. *Food Additives and Contaminants, 25*(2), 181–192.

Vallejo, F., Marín, J., & Tomás-Barberán, F. A. (2012). Phenolic compound content of fresh and dried figs (*Ficus carica* L.). *Food Chemistry, 130*(3), 485–492.

Veberic, R., & Mikulic-Petkovsek, M. (2016). Phytochemical composition of common fig (*Ficus carica* L.) cultivars. In *Nutritional composition of fruit cultivars* (pp. 235–255). Elsevier.

Veberic, R., Colaric, M., & Stampar, F. (2008). Phenolic acids and flavonoids of fig fruit (*Ficus carica* L.) in the northern Mediterranean region. *Food Chemistry, 106*(1), 153–157.

Villalobos, M. D. C., Serradilla, M. J., Martín, A., Lopez Corrales, M., Pereira, C., & Córdoba, M. D. G. (2016). Preservation of different fig cultivars (*Ficus carica* L.) under modified atmosphere packaging during cold storage. *Journal of the Science of Food and Agriculture, 96*(6), 2103–2115.

Viuda-Martos, M., Barber, X., Pérez-Álvarez, J. A., & Fernández-López, J. (2015). Assessment of chemical, physico-chemical, techno-functional and antioxidant properties of fig (*Ficus carica* L.) powder co-products. *Industrial Crops and Products, 69*, 472–479.

Wang, Y., Li, J., & Li, B. (2017). Chitin microspheres: A fascinating material with high loading capacity of anthocyanins for colon specific delivery. *Food Hydrocolloids, 63*, 293–300.

Wang, Y., Nie, J., Yan, Z., Li, Z., Cheng, Y., & Chang, W. (2018). Occurrence and co-occurrence of mycotoxins in nuts and dried fruits from China. *Food Control, 88*, 181–189.

Wijayanti, E. D., Setiawan, N. C. E., & Cristi, J. P. (2017). Effect of lactic acid fermentation on total phenolic content and antioxidant activity of fig fruit juice (*Ficus carica*). *Advances in Health Sciences Research, 2*, 282–289.

Wikiera, A., Irla, M., & Mika, M. (2014). Health-promoting properties of pectin. *Postepy higieny i medycyny doswiadczalnej (Online), 68*, 590–596.

Wojdyło, A., Nowicka, P., Carbonell-Barrachina, Á. A., & Hernández, F. (2016). Phenolic compounds, antioxidant and antidiabetic activity of different cultivars of *Ficus carica* L. fruits. *Journal of Functional Foods, 25*, 421–432.

Zhang, Z., Lv, G., Pan, H., Fan, L., Soccol, C. R., & Pandey, A. (2012). Production of powerful antioxidant supplements via solid-state fermentation of wheat (*Triticum aestivum* Linn.) by Cordyceps militaris. *Food Technology and Biotechnology, 50*(1), 32–39.

Chapter 35
Fig Production and Processing: A Pakistan Perspective

Aijaz Hussain Soomro and Tahseen Fatima Miano

Abbreviations

AJK	Azad Jammu and Kashmir
ha	Hectare
KP	Khyber Pakhtunkhwa
USA	United States of America

1 Introduction

Nature has bestowed Pakistan with a wide range of agro-ecological regions such as tropical, subtropical, and temperate regions and thus a diversification in growing many horticultural produces (Ahmad et al., 2021). Among different horticultural crops, the fruits grown in Pakistan have considered highly economical and important crops for potential exportation, local consumption in raw or processed form, and food insecurity (Shahzad et al., 2019). Moreover, fruits are the oldest form of natural staple food for humanity (Khadivi et al., 2018; Alam et al., 2021).

Fig is one such fruit that is supposed to be the oldest fruit domesticated in Mediterranean regions (Kharadi et al., 2005) and is native to western Asia and the eastern parts of Mediterranean countries (Hmimsa et al., 2012). It is believed that the fig fruit was domesticated 5000 years earlier than wheat and millet (Khan et al., 2022). The consumption of fig fruit dates back to the beginning of civilization (Ferraz et al., 2021). This fruit is valuable agricultural produce grown in tropical and subtropical regions worldwide. Since ancient times, the fig has been a part of the diet and is considered the symbol of heaven and longevity. It is a popular fruit

A. H. Soomro (✉) · T. F. Miano
Institute of Food Sciences and Technology, Sindh Agriculture University, Tandojam, Sindh, Pakistan
e-mail: ahsoomro@sau.edu.pk

M. F. Ramadan (ed.), *Fig* (Ficus carica)*: Production, Processing, and Properties*, https://doi.org/10.1007/978-3-031-16493-4_35

worldwide and is known by different names in different languages. The fruit is named Teen in Arabic, Anjeer in Urdu, Gujrati, Hindi, Kashmiri, Figu in French, and Anjir in Turkish (Nuri & Shahabuddin, 2021). There are 470 varieties of edible fig. Commercially grown varieties are Celeste, Brown Turkey, Brunswick, Marseilles, Adriatic, Genoa, Purple Genca, and Black Ischia. These varieties are grown in different regions of the world. Among them, Black Mission and Brown Turkey are highly produced varieties. In Pakistan, the fig cultivation was 162 ha, with around 487 tons. The average fig production is about 3781 Kg per ha. This indicates that the yield of fig fruit in Pakistan is very low compared to other countries. Despite having suitable agro-climatic conditions in the prospect of fig fruit production, Pakistan is far enough due to the lack of interest of growers and other stakeholders. However, countries like Turkey, Egypt, and Iran are the leading producer of fig fruit globally. The fig fruit is also cultivated in Algeria, Morocco, Syria, the USA, Brazil, Spain, India, Albania, and Tunisia (Arpaci et al., 2018).

2 Fig Tree and Its Cultivation

Fig plant is a member of the Moraceae family, mainly cultivated for its valuable fruit (Kim et al., 2007). Fig tree bears various horticultural properties, and it offers fresh fig fruit twice a year during the spring and summer seasons (Aksoy, 2017). This tree may ideally grow in rural areas; however, it is badly neglected and underutilized, especially around Mediterranean regions from where it originated. The fig trees are mainly cultivated in Pakistan's KP, Punjab, and Sindh Provinces. In Azad Jammu and Kashmir (AJK), the fig fruits are grown in the natural forests, and marginal arable land with lower economic returns since the fruit is mainly consumed locally (Qureshi et al., 2007). However, the climatic condition in Pakistan provides an ultimate opportunity for the growth and development of fig trees. Different fig varieties, namely English fig, Wild fig, Dark brown, and Dark black figs, are among some common fig varieties in Haramosh Valley, Gilgit, Pakistan (Abbas et al., 2016). The common fig varieties grown in different countries are presented in Table 35.1.

The fig tree is a deciduous shrub or a small tree that is mainly adaptive to different soil types and climatic conditions and is widely grown in many parts of the world. The tree grows fast with a spreading habit, the flowers of the fig tree are tiny enough, and the fruit is hollow (Stover et al., 2007). The seeds of fig fruit are non-viable; therefore, fig trees are propagated via cutting or grafting. A fig tree needs sunlight for around eight hours a day for proper fruit ripening; however, fruits must be allowed on the tree for ripening. The nutrient uptake of the fig tree increases during the developmental stage of fig fruit. Fig trees occur in bisexual forms, i.e., a functional male caprifig and other unisexual females which produces edible fig fruit (Kim et al., 2007).

Table 35.1 Fig varieties and their regions of cultivation

Fig varieties	Region of cultivation	Reference (s)
Wild Himalayan fig	AJK, Pakistan	Khan et al. (2022)
Green, Purple, and Black figs	Tunisia Mediterranean coast	Harzallah et al. (2016)
Mission, Vasilika, and Mavra Markopoulou figs	Southern Europe	Karantzi et al. (2021)
Sarilop, Bursa Siyahi, Goklop, Yediveren, Yesilguz, Morguz, Sari Zeybek, and Ufak Yesil figs	Dortyol, Hatay, Turkey	Çalişkan and Polat (2008)
Khafrak, Rowno, Sigoto, Shahanjir. Seyah, Matti figs	Fars province, Iran	Pourghayoumi et al. (2016)
Conadria, Deanna, Excel, Poona, Dinkar figs	India	Gawade and Waskar (2005)
Roxo de Valinhos, White Genova, PI-189, Troyano figs	Brazil	Ferraz et al. (2021)
Mission, Chechick, Bursa, Brown-Turkey, Brunswick, Kadota figs	Bet-Dagan, Israel	Solomon et al. (2006)
Aberkane, Azandjar, Bouankik figs	Algeria	Bachir Bey et al. (2017)

3 The Scenario of Fig Production in Pakistan

In Pakistan, the fig trees are mainly cultivated in KP, Punjab, and Sindh Provinces. In Azad Jammu and Kashmir (AJK), the fig fruits are grown in the natural forests and marginal arable land, with lower economic returns since the fruit is mainly consumed locally. The total area under fig cultivation in Pakistan is about 125 hectares which is very low compared to the rest of the world (Qureshi et al., 2007). However, it is factual that the climatic condition in Pakistan provides an ultimate opportunity for the proper growth and development of fig trees. Different fig varieties, namely *English* fig, *Wild* fig, *Dark brown,* and *Dark black* figs, are among some common fig varieties in Haramosh Valley, Gilgit, Pakistan (Abbas et al., 2016), while varieties like Black Turkey and Black Mission are also two prominent varieties of fig grown in Pakistan (Amjal et al., 2016).

4 Fig Fruit

Fig fruit is an edible part of the fig tree which is pear-shaped, fleshy, receptive, hollow fruit (Dueñas et al., 2008), sweet with gelatinous pulp (Stover et al., 2007) is termed as syconium. This fruit may be consumed in fresh or dried form while both forms are nutritionally superior and deliciously palatable. The developmental stages of growing fig fruits are different from the other fruits (Heperkan et al., 2012). Usually, the fruit may vary in size and color depending on the variety (Farhangi et al., 2014). The fruit size may range from 3 to 5 cm, while a mature fig fruit has a tough brown skin that often cracks when the fruit ripens/softens and exposes the fruit pulp in beneath. The color of figs varies from dark purple to green. The fig fruit varieties with dark skin color are rich in polyphenols, i.e., flavonoids and anthocyanins (Solomon et al., 2006). This fleshy fruit bears many subtle and crunchy seeds and is embedded in the fruit pulp so that the seed mass is bound with the jelly-like thick material within the inner skin of fig fruit. The pulp of fig fruit is mainly water, minerals, volatile compounds, and vitamins, while the seeds have a nutty taste. The skin of fig fruit is loaded with many polyphenols and exerts antioxidant activity in darker fig varieties than lighter ones (Solomon et al., 2006). In addition, the fruit contains volatile components providing a pleasant and characteristic aroma (Bachir Bey et al., 2017).

5 Nutritional Composition of Fig Fruit

Fruits from the fig tree are a true gift from nature. The fruit contains many essential components associated with maintaining health and wellbeing (Soni et al., 2014). The fruit bears significant varietal differences. Thereby composition of fruit varies significantly amongst cultivars. The fig fruit bears medicinal properties, has an excellent nutritional profile, and is low in calories, i.e., 74 calories/100 g of fresh fruit (Farhangi et al., 2014). Both fresh and dried fig fruits are rich in fiber, minerals, organic acids, and antioxidant vitamins (A, E, and K) while are free from cholesterol (Stover et al., 2007). Fruit is also the richest source of amino acids (Bharathkumar et al., 2018), particularly aspartic acid and glutamine. The fruit is loaded with minerals and is considered a healthy fruit since it is rich in many essential minerals, i.e., iron, potassium, calcium, magnesium, sodium, and zinc (Sadia et al., 2014). Fresh fig fruit has an average dietary fiber and protein content while very low-fat content (Soni et al., 2014). According to Vinson et al. (2005), fresh and dried figs contain more fiber and polyphenols. Solomon et al. (2006) also reported that the higher the polyphenols contents, especially anthocyanins in fig fruit, the higher was their antioxidant activity. Figs also comprised sugars and organic acids that influence their quality and exhibit one of the highest concentrations of polyphenols among the commonly consumed fruits and beverages that contribute positively to human health (Veberic et al., 2008). Mahmoudi et al. (2018) found that the peels and pulps of figs have good nutritional value. The fruit pulps are rich in carbohydrates, and the peel of the 'Onk Elhamam' cultivar had the highest vitamin C content. The pulp of the 'Boughandjo' cultivar contained the highest concentration of phosphorus, and the peel of the 'Bakkor Khal' cultivar was richest in potassium and calcium.

Moreover, Ersoy et al. (2007) suggested that fig fruit contains sugars such as maltose, xylose ribose, arabinose,e, etc. Fig fruit is a good source of health-promoting bioactive components (Harzallah et al., 2016). However, dried fig fruit contains a relatively higher concentration of crude fibers (approximately 5.8%), sugars, and polyphenols (Vinson et al., 2005). It also has a reasonable proportion of

B-complex vitamins such as niacin, pyridoxine, pantothenic acid, folates (Farhangi et al., 2014), and minerals, i.e., potassium, and calcium, magnesium, and sodium (Turco et al., 2020). In Azad Jammu and Kashmir, edible figs are generally grown in natural forests and marginal lands, with very low economic returns as most of the fruit is consumed Since figs are perishable, they are dried naturally under sunlight in Kashmir. Figs produced in central Kashmir have a very high fiber content. They are also reported to be an excellent source of calcium and potassium. Dried Kashmiri figs contain omega-3 and omega-6 fatty acids, which are considered helpful in preventing heart disease and can cure sore throats due to their high mucilage content (Mir et al., 2018).

6 Medicinal Properties of Fig Fruit

The consumption of figs has positive health effects due to the numerous nutraceutical compounds that may help prevent cardiovascular diseases and the growth of carcinoma cells (Allegra et al., 2017). Some fig varieties contain many therapeutic compounds, thereby helping traditional ethnomedical remedies. Almost all parts of its tree (leaf, fruit, latex, root, etc.) have medicinal effects against innumerable diseases. Many species of *Ficus carica* are potent sources of antioxidants with strong anti-microbial, anti-diabetic, anti-inflammatory, anti-spasmodic, and anti-obesogenic activities (Shahrajabian et al., 2021). The fig tree's fruit plays an essential role against numerous health disorders related to oxidative stress (Ahmad et al., 2013). Thereby fig fruit immensely contributes to good health and wellbeing. The phytochemical components in the fruit scavenge toxic free radicals from the body and exhibit protection against cancer, diabetes, and infectious and degenerative diseases (Farhangi et al., 2014). According to Shahrajabian et al. (2021), dark-skinned fig varieties are rich sources of polyphenols, flavonoids, anthocyanins, and pro-anthocyanidins in exerting antioxidant activity than light-skinned fruit. Steam distillation of fig fruit contains benzaldehyde which exerts considerable antitumor action.

Interestingly, Solomon et al. (2006) suggested that fig fruit is a useful mild laxative, diuretic, and expectorant in Indian Ayurveda. At the same time, its paste is often used against swellings and for relieving pain. Fresh fig is valued for its laxative effects, treating skin infections, and maintaining acid-base balance in the body. Apart from the fig fruit, the other parts of the fig tree (roots, leaves, etc.) are reportedly used as traditional medicines and are immensely effective against specific cardiovascular, gastrointestinal, and respiratory ailments (Khan et al., 2022). Moreover, fig leaves' decoction (a method of extracting a plant's medicinal constituents by boiling the plant material) is effective against ameliorating post-prandial hyperglycemia. A reduction in hypercholesterolemia and hyperglycemia has been further utilized to treat the aqueous decoction of fig tree leaves.

On the other hand, the fig leave extract may induce a significant hypoglycemic effect after oral or intraperitoneal administration. Patients with Diabetic Mellitus may benefit from fig tree leaves; therefore, researchers should explore its influential role in Diabetic Mellitus (Ajmal et al., 2016). Many pathogenic bacteria pose health problems because of the multidrug resistance that may have developed due to the antibiotics used against infections. The fig possesses antimicrobial properties against some pathogenic microorganisms. For example, the dried fig extracts inhibited the growth of *Bacillus subtilis* and *Proteus mirabilis* (Soni et al., 2014).

7 Perishability of Fig Fruit

The fig fruit has tremendous commercial importance (Irget et al., 2008), but unluckily the fruit is highly perishable and has a scanty shelf-life. The major reasons for its perishable nature are fungal attack, weight loss, mechanical damage, etc. (Colelli & Amodio, 2020). It has a short shelf-life compared to other fruit due to its high moisture content and sugar concentration. Fresh fig fruit's moisture content may range from 70.5% to 82.3% (Bachir Bey et al., 2017). In its new form, the fruit is highly vulnerable to microbial attack even under the cold-store facility or refrigeration conditions (Farahnaky et al., 2009). Due to its immense perishable nature, the fruit is consumed in its fresh raw form or dried. A fully ripened fresh fig fruit has a shelf-life of 2–3 days under refrigeration conditions. Fruit takes about 4–5 days for sun drying while 10–12 h under dehydrator. The United Nations Economic Commission for Europe-Standard (2004) suggests that moisture content in commercial-grade dried fig fruit should not exceed 26%. The dried figs have prolonged storability of about 6–8 months.

8 Fungal Invasion and Mycotoxin Production in Fig Fruit

Fig is an economically important fruit crop and is susceptible to fungal infection. Fig fruits are at high risk of toxigenic fungi. The fungal invasion may occur in the fig tree after fruit shriveling or falls and even during their drying. The fungal growth is often associated with the subsequent production of mycotoxin (Heperkan et al., 2012). The water activity of fig fruits during their production or processing affects the growth of toxigenic fungi and the formation of mycotoxins. The climatic conditions like high humidity %, mild temperatures, and rain promote fungal growth. The dominant fungal species that occur in dried figs may vary widely, and their types depend on different factors such as geographical regions, processing methods, etc. Some common types of reported toxigenic fungi are *Aspergillus spp.*, *Fusarium* spp., and *Penicillium spp.* Aflatoxins and ochratoxins occur more widely in fig fruit (Heperkan, 2006).

9 Fig Fruit Processing and Value Addition in Pakistan

The perishable nature of fig fruit tends to fruit growers and processors for its suitable processing and effective value-addition. A main proportion of the fruit is subjected to drying or dehydration to develop dried figs. The processing technologies are employed on fig fruit to attain more imperative and valuable commercial fruit-based commodities. However, a significant number of scientific studies have been conducted by many researchers for developing fig fruit-based food products (Table 35.2). Pakistan lacks suitable fig processing industries since the fruit is produced or cultivated in marginal quantities. The fruit's industrial processing and technological aspects may attain a prime place in the country's economy if serious measures for a sustainable fig production system are taken. It is worthful that increased demand for fig fruit (fresh or processed) worldwide is gaining efforts to expand its production; therefore, Pakistan is also paying keen attention to increased cultivation of figs in different arable areas of the country. However, it is also imperative for the food processing industries in Pakistan to mitigate the post-harvest losses of figs by applying technological practices during processing and preservation with improved drying techniques and preparation of fig-based value-added products.

10 Fig-Based Value-Added Food Products

Efforts are needed to increase the production of fig fruits to meet market demand in Pakistan since the production of the commodity is associated with its processing. However, the comprehensive research study for exploiting the diversity in wild figs

Table 35.2 Utilization of fig fruit for the development of value-added food products

Fig fruit form	Use in value-addition	References
Fresh fig fruit	Fig jam and preserve	Levaj et al. (2010)
	Fig jam and nectar	Tanwar et al. (2014)
	Dried fig	Slatnar et al. (2011)
	Novel functional fig snack	Yeganehzad et al. (2020)
	Canned figs	Curi et al. (2019)
	Fig wine	Jeong et al. (2005)
	Fig fruit-based fermented milk	Abd-Eltawab and Ebid (2019)
Dried fig	Fig powder	Khapre et al. (2011)
	Fig acnes/seeds oil	Soltana et al. (2016)
	Hard candy	Verma and Gupta (2015)
	Fig wine	Kadam et al. (2011)
Fig powder	Toffee	Khapre et al. (2011)
	Cookies	Khapre et al. (2015a)
	Burfi (Indian cookie)	Khapre et al. (2015b)
Fig paste	Fat replacer in baked foods and sauces	Benjamin (2020)

is not conducted for their possible use in sustainable fruit production. Subsequently, propagation and production of the species in the region may contribute to upgrading the local economy and may significantly influence socio-economic and ecological stability. Therefore, the local farmers have good opportunities for producing wild figs to increase income generation and provide a sustainable natural food source to the consumers (Khan et al., 2022). However, the perishable nature of the fig fruit also causes significant loss to the country's economy.

Moreover, it is also direly essential to reduce its waste (or utilize surpluses) by preventing or controlling the post-harvest losses of the fruit. The fruit thereby needs to receive proper preservation/ processing techniques that should be simple and easily adaptable. Benjamin (2020) suggested that fig-based ingredients of different grades have been established to have attributes like natural coloring and flavoring agents, humectants, and anti-staling agents. Fig fruits may be used to develop many value-added food products. Sher et al. (2016) suggest that presently, the fruit is being utilized for producing various food commodities (jams and jellies, etc.). In addition, the fig fruit may be utilized as an optional ingredient in baked items (cakes, bread, biscuits, and pies) and for garnishing.

10.1 Dried Fig

Drying is the most popular and effective way of processing/preserving figs known from prehistoric times (Kislev et al., 2006). Drying and dehydration have proven to be reliable preservation techniques for fig fruit in terms of nutritional quality and technical feasibility (Desa et al., 2019). Drying has many advantages for food quality decreasing water activity, reducing microbiological activity, and minimizing physical and chemical changes (Mujic et al., 2014). During fig drying, the temperature-time relationship plays a crucial role. The global market of dried fig fruit is expanding; however market always demands safe, hygienic, and high-quality dried figs. Dried figs comprise about 90% of the world's production. Dried figs are distinguished by high nutritional value and possess functional food properties (Vinson, 1999). Galván et al. (2021) suggest that the processing of dried fig fruit has increased by 44% in the last decade. The high energy demanding artificial drying methods are unreliable and expensive, while in contrast, slower methods like sun drying or solar drying are suitable alternatives. Therefore, natural sun drying is a widely practiced drying method in tropical and subtropical regions. During their drying, storage, and retailing, dried fig processing should be considered since fruits may be susceptible to fungal attacks or mycotoxin production (Devreese et al., 2013). The suggested ambient temperatures, i.e., 16 and 25 °C, may be considered reliable against fungal growth for both processors and consumers for storing dried figs (Galván et al., 2021). Dried figs are a good source of carbohydrates, sugars, minerals, vitamins, organic acids, and phenolic compounds (Veberic et al., 2008). Both fresh and dried figs have high fiber and polyphenols (Vinson, 1999; Vinson et al., 2005).

10.2 Fig Powder

The technology for producing fruit powder is gaining significant importance. Fig fruit is generally subjected to drying and then milled to obtain fig powder. The fig powder is considered ready-to-eat food (Varhan et al., 2019), whereas it may also be used as a food ingredient. Varhan et al. (2019) suggested that drying technologies preferred to produce food powder are spray-drying, freeze-drying, tray-drying, foam-mat-drying, microwave drying, etc. Fig powder is nutritionally rich and contains reasonable quantities of proteins, fats, minerals, and sugar (Khapre et al., 2011). Nowadays, there is an increase in nutrient-rich value-added products' development by partially replacing its ingredients with others, such as underutilized fruits and added value by-products (pectins, colorants, emulsifiers, and antioxidants) from leaves and peels. As for fig by-products, the use of fig powder as a colorant in the production of buns and muffins (Chauhan & Tanwar, 2016). The addition of fig seed powder to the formulation of a cookie also improved its fiber content and increased the total phenolic content and antioxidant activity (Bolek, 2020). Fig fruit cake is a widely used snack by children and adults (Dhankar, 2013).

Additionally, fig by-products' sweet extracts have been used for making traditional desserts without adding sugar, for example, "Shir Anjir", an Iranian dessert (Jahromi & Niakousari, 2018). The fig powder supplement, used as a natural sweetener and flavoring agent in goat's milk yogurt, has improved the taste, texture, and aroma and covered the unpleasant flavor of goat's milk (Mahmoudi et al., 2021). Fig powder may be utilized for making various food products (Table 35.2).

10.3 Fig Jam

Fig jam making is a popular approach for preserving fig fruit. The fig jam prepared from the freshly harvested and fully ripened fig fruit is a rich source of nutrients. The fig jam is a palatable and nutritious food item and has tremendous market potential. The figs are usually available commercially after drying because fresh fruits are only available during the season. The fresh fruits have a short shelf life, usually 7–10 days. Therefore, figs could be preserved through jam processing. In Bulgaria, fig jam with whole green and ripened fruits is a very popular type of preservation of these seasonal fruits. (Rababah et al., 2011; Tanwar et al., 2014).

10.4 Fig Paste

Dried figs may be utilized to produce versatile food products, for instance, fig concentrate, fig butter, and fig paste. Fig paste is a slurry that may be incorporated in many food formulations. The quality of fig paste may vary depending upon the variety of fig, its particle size, and the final moisture content of the paste slurry (Benjamin, 2020). The fig paste is a versatile ingredient in terms of technical functionalities. Fig paste form colloidal bonds with components of the food matrix (fats, protein, water, etc.). Fig pastes and water slurry ratio of 50:50 is considered excellent emulsification with suitable sensory appeal and physicochemical properties. Therefore, fig paste may be used as an ingredient in food products. Traditionally, fig paste is prepared by selecting mature and fresh fig fruits. After washing, figs are cooked with the addition of sugar on a slow flame by occasionally stirring for a few hours till a homogenous slurry of paste is achieved (Lee et al., 2012).

10.5 Fig Seed Oil

Fig seeds may be separated from the fruit, dried, and processed to obtain fig oil. Dry seeds contain 30% oil (Joseph & Raj, 2011). The fatty acid compositions of the fig seed contain a significant amount of α-linolenic, linoleic, and oleic acids. In addition, fig seed also contains a small quantity of palmitic acid. Linolenic and linoleic acids are essential fatty acids and constitute about 70% of the fig seed fatty acids (Baygeldi et al., 2021). In addition, omega-3 fatty acid reduces the risk factors for cardiovascular and age-related macular degeneration due to their anti-inflammatory effect (Richer et al., 2016).

10.6 Fig Pickle and Fig Chutney

Fig fruit pickle can be prepared by selecting good quality fig fruits using different food ingredients such as oil, vinegar, sugar, and spices, while fig fruit chutney is prepared by adding whole fig fruit paste, brown sugar, apple cider vinegar, chopped onion, and spices.

10.7 Canned Figs

Many forms of perishable fruits are usually canned to prolong their shelf life. Canning is an old method of preserving fragile food commodities. Fruits with shorter shelf life may be canned to increase their shelf life. In general, figs are characterized by their high moisture content. Usually, 82% are considered a great source of carbohydrates of which the principal sugars are soluble, responsible for their sweet taste, and present high perishability, making their fast transportation to the centers of consumption necessary. The leading causes of losses in fig quality are inadequate harvest and packing and the lack of standardization in the classification of the product (Caetano et al., 2017). The canned fruits prepared from the different fig cultivars may have different textures. The amount of sugar present in each cultivar, pH, and acidity can influence gelling and the final product's texture (Souza et al., 2014). In addition, aspects such as moisture content and the chemical composition of the fruit may also change texture, as it may affect cooking time, yield, and, therefore, the moisture content of the processed product (Curi et al., 2017). Curi et al. (2019) prepared eight different formulations of canned fig fruits and evaluated them for their quality attributes. Conclusively, fig fruit supplementation in the diet has been beneficial to coping with various metabolic disorders due to their high nutritional potential and antioxidant activity.

References

Abbas, T., Khatoon, S., Alam, R., Hussain, B., Hussain, Z., Gonzalez, M. Y., Abbas, Y., Ali, N., & Hussain, N. (2016). A physico-chemical study of different Fig (*Ficus carica* L.) varieties in Haramosh valley, Gilgit-Pakistan. *International Journal of Environment, Agriculture and Biotechnology, 1*(3), 516–525.

Abd-Eltawab, S., & Ebid, W. (2019). Production and evaluation of stirred synbiotic fermented milk fortified with fig fruit (*Ficus carica* L.). *Egyptian Journal of Food Science, 47*(2), 201–212.

Ahmad, J., Khan, I., Khan, S., & Iqbal, D. (2013). Evaluation of antioxidant and antimicrobial activity of *Ficus carica* leaves: An in vitro approach. *Journal of Plant Pathology and Microbiology, 4*(1), 1–4.

Ahmad, K., Afridi, M., Khan, N. A., & Sarwar, A. (2021). Quality deterioration of postharvest fruits and vegetables in developing country Pakistan: A mini overview. *Asian Journal of Agriculture and Food Sciences, 9*(2), 83–90.

Ajmal, M., Arshad, M. U., Saeed, F., Ahmed, T., Khan, A. U., Bader-ul-Ain, H., & Suleria, H. A. R. (2016). Exploring the nutritional characteristics of different parts of fig in relation to hypoglycemic potential. *Pakistan Journal of Life & Social Sciences, 14*(2), 115–122.

Aksoy, U. (2017). The dried fig management and the potential for new products. *Acta Horticulture, 1173*, 377–382.

Alam, M. W., Malik, A., Rehman, A., Sarwar, M., Muhammad, S., Hameed, A., Alsamadany, H., Alzahrani, Y., & Ahmed, Z. (2021). First report of *Alternaria alternata* causing fruit rot on Fig (*Ficus carica*) in Pakistan. *Plant Disease, 105*(7), 2015.

Allegra, A., Alfeo, V., Gallotta, A., & Todaro, A. (2017). Nutraceutical content in 'Melanzana' and 'Dottato' fig fruit (*Ficus carica* L.). *Acta Horticulturae, 1173*, 319–322.

Arpacı, S., Konak, R., & Çiçek, E. (2018). *A national value: Turkish figs*. In XXX. International Horticultural Congress August 12–16, 2018, İstanbul, Turkey, Book of Proceedings, pp. 16–22.

Bachir Bey, M., Richard, G., Meziant, L., Fauconnier, M. L., & Louaileche, H. (2017). Effects of sun-drying on physicochemical characteristics, phenolic composition and in vitro antioxidant activity of dark fig varieties. *Journal of Food Processing and Preservation, 41*(5), e13164.

Baygeldi, N., Küçükerdönmez, Ö., Akder, R. N., & Çağındı, Ö. (2021). Medicinal and nutritional analysis of fig (*Ficus carica*) seed oil; A new gamma tocopherol and omega-3 source. *Progress in Nutrition, 23*(2), 1–6.

Benjamin, T. (2020). *Use of fig paste as a fat replacement in baked goods and sauces*. https://digitalcommons.fiu.edu/fiu-undergraduate-research-conference

Bharathkumar, A., Jagadeesh, S. L., Netravati, V. H., Bhuvaneshwari, G., & Bindu, H. (2018). A study on fruit preparation on quality of fig fruits (cv. Bellary) osmotic-dehydrated under solar tunnel dryer. *Journal of Pharmacognosy and Phytochemistry, 7*(3), 3177–3180.

Bölek, S. (2020). Effects of waste fig seed powder on quality as an innovative ingredient in biscuit formulation. *Journal of Food Science, 86*, 55–60.

Caetano, P. K., Vieites, R. L., Daiuto, E. R., Moura, S. C. S. R., & de. (2017). Processamento e qualidade de compotas de figo diet e convencional. *Brazilian Journal of Food Technology, 20*, e2016026.

Çalişkan, O., & Polat, A. A. (2008). Fruit characteristics of fig cultivars and genotypes grown in Turkey. *Scientia Horticulturae, 115*(4), 360–367.

Chauhan, A., & Tanwar, B. (2016). Development of value added products (bun, muffin, noodles, and nuggets) by substitution with Carissa spinarum and *Ficus carica* powder. *Asian Journal of Pharmaceutical and Clinical Research, 9*, 95–101.

Colelli, G., & Amodio, M. L. (2020). Subtropical fruits: Figs. In: *Controlled and modified atmospheres for fresh and fresh-cut produce* (pp. 427–434). Academic.

Curi, P. N., Tavares, B. S., Almeida, A. B., Pio, R., Pasqual, M., Peche, P. M., & Souza, V. R. (2017). Characterization and influence of subtropical persimmon cultivars on juice and jelly characteristics. *Anais da Academia Brasileira de Ciências, 89*, 1205–1220.

Curi, P. N., Locatelli, G., Albergaria, F. C., Pio, R., Pádua, L. A. D., & Souza, V. R. D. (2019). Potential of figs from cultivars grown in subtropical regions for canning purposes. *Pesquisa Agropecuária Brasileira, 54*, e00154.

Desa, W. N. M., Mohammad, M., & Fudholi, A. (2019). Review of drying technology of fig. *Trends in Food Science & Technology, 88*, 93–103.

Devreese, M., De Baere, S., De Backer, P., & Croubels, S. (2013). Quantitative determination of the Fusarium mycotoxins beauvericin, enniatin A, A1, B and B1 in pig plasma using high performance liquid chromatography-tandem mass spectrometry. *Talanta, 106*, 212–219.

Dhankar, P. (2013). A Study on development of coconut based gluten free cookies. *International Journal of Engineering Science Invention, 2*, 10–19.

Dueñas, M., Pérez-Alonso, J. J., Santos-Buelga, C., & Escribano-Bailón, T. (2008). Anthocyanin composition in fig (*Ficus carica* L.). *Journal of Food Composition and Analysis, 21*(2), 107–115.

Ersoy, N., Gözlekçi, Ş., & Kaynak, L. (2007). Changes in sugar contents of fig fruit (*Ficus carica* l. Cv. Bursa Siyahı) during development. *Süleyman Demirel Üniversitesi Ziraat Fakültesi Dergisi, 2*(2), 22–26.

Farahnaky, A., Ansari, S., & Majzoobi, M. (2009). Effect of glycerol on the moisture sorption isotherms of figs. *Journal of Food Engineering, 93*(4), 468–473.

Farhangi, H., Ajilian, M., Saeidi, M., & Khodaei, G. H. (2014). Medicinal fruits in holy Quran. *International Journal Of Pediatrics, 2*, 89–102.

Ferraz, R. A., Leonel, S., Souza, J. M. A., Modesto, J. H., Ferreira, R. B., & de Souza Silva, M. (2021). Agronomical and quality differences of four fig cultivars grown in Brazil. *Semina: Ciências Agrárias, 42*(2), 619–634.

Galván, A. I., Rodríguez, A., Martín, A., Serradilla, M. J., Martínez-Dorado, A., & Córdoba, M. D. G. (2021). Effect of temperature during drying and storage of dried figs on growth, gene expression and aflatoxin production. *Toxins, 13*(2), 134.

Gawade, M. H., & Waskar, D. P. (2005). Effect of different varieties and pretreatments on yield and qualify of dried fig fruits. *Indian Journal of Agricultural Research, 39*(2), 138–141.

Harzallah, A., Bhouri, A. M., Amri, Z., Soltana, H., & Hammami, M. (2016). Phytochemical content and antioxidant activity of different fruit parts juices of three figs (*Ficus carica* L.) varieties grown in Tunisia. *Industrial Crops and Products, 83*, 255–267.

Heperkan, D. (2006). The importance of mycotoxins and a brief history of mycotoxin studies in Turkey. *ARI Bulletin of Istanbul Technical University, 54*, 18–27.

Heperkan, D., Moretti, A., Dikmen, C. D., & Logrieco, A. F. (2012). Toxigenic fungi and mycotoxin associated with figs in the Mediterranean area. *Phytopathologia Mediterranea, 51*(1), 119–130.

Hmimsa, Y., Aumeeruddy-Thomas, Y., & Ater, M. (2012). Vernacular taxonomy, classification and varietal diversity of fig (*Ficus carica* L.) among Jbala cultivators in northern Morocco. *Human Ecology, 40*(2), 301–313.

Irget, M. E., Aksoy, U., Okur, B., Ongun, A. R., & Tepecik, M. (2008). Effect of calcium based fertilization on dried fig (*Ficus carica* L. Cv. Sarılop) yield and quality. *Scientia Horticulturae, 118*(4), 308–313.

Jahromi, M., & Niakousari, M. (2018). Development and characterisation of a sugar-free milk-based dessert formulation with fig (*Ficus carica* L.) and carboxymethylcellulose. *International Journal of Dairy Technology, 71*, 801–809.

Jeong, M. R., Cha, J. D., Yun, S. I., Han, J. H., & Lee, Y. E. (2005). Manufacturing of wine with Korean figs (*Ficus carica* L.) and quality improvement by adding fig leaves. *Journal of the East Asian Society of Dietary Life, 15*(1), 112–118.

Joseph, B., & Raj, S. J. (2011). Pharmacognostic and phytochemical properties of *Ficus carica* Linn -An overview. *International Journal of Pharm Tech Research, 3*(1), 8–12.

Kadam, N. U., Upadhye, A. A., & Ghosh, J. S. (2011). Fermentation and characterization of wine from dried *Ficus carica* (L) using *Saccharomyces cerevisiae* NCIM 3282. *International Food Research Journal, 18*(4), 1569–1571.

Karantzi, A. D., Kafkaletou, M., Christopoulos, M. V., & Tsantili, E. (2021). Peel colour and flesh phenolic compounds at ripening stages in pollinated commercial varieties of fig (*Ficus carica* L.) fruit grown in Southern Europe. *Journal of Food Measurement and Characterization, 15*(2), 2049–2063.

Khadari, B., Grout, C., Santoni, S., & Kjellberg, F. (2005). Contrasted genetic diversity and differentiation among Mediterranean populations of *Ficus carica* L.: A study using mtDNA RFLP. *Genetic Resources and Crop Evolution, 52*(1), 97–109.

Khadivi, A., Anjam, R., & Anjam, K. (2018). Morphological and pomological characterization of edible fig (*Ficus carica* L.) to select the superior trees. *Scientia Horticulturae, 238*, 66–74.

Khan, M. R., Khan, M. A., Habib, U., Maqbool, M., Rana, R. M., Awan, S. I., & Duralija, B. (2022). Evaluation of the characteristics of native wild Himalayan fig (*Ficus palmata* Forsk.) from Pakistan as a potential species for sustainable fruit production. *Sustainability, 14*(1), 468.

Khapre, A. P., Satwadhar, P. N., & Deshpande, H. W. (2011). Development of technology for preparation of fig (*Ficus carica* L.) fruit powder and its utilization in toffee. *Journal of Dairying Foods & Home Sciences, 30*(4), 267–270.

Khapre, A. P., Satwadhar, P. N., & Deshpande, H. W. (2015a). Studies on standardization of fig fruit (*Ficus carica* L.) powder enriched cookies and its composition. *Asian Journal of Dairy and Food Research, 34*(1), 71–74.

Khapre, A. P., Satwadhar, P. N., & Syed, H. M. (2015b). Studies on processing technology and cost estimation of fig (*Ficus carica* L.) fruit powder enriched Burfi (Indian cookie). *Journal of Applied and Natural Science, 7*(2), 621–624.

Kim, K. M., Kim, M. Y., Yun, P. Y., Chandrasekhar, T., Lee, H. Y., & Song, P. S. (2007). Production of multiple shoots and plant regeneration from leaf segments of fig tree (*Ficus carica* L.). *Journal of. Plant Biology, 50*(4), 440–446.

Kislev, M. E., Hartmann, A., & Bar-Yosef, O. (2006). Early domesticated fig in the Jordan Valley. *Science, 312*, 1372–1374.

Lee, H. Y., Kim, J. H., Jeung, H. W., Lee, C. U., Kim, D. S., Li, B., & Chae, S. W. (2012). Effects of *Ficus carica* paste on loperamide-induced constipation in rats. *Food and Chemical Toxicology, 50*(3-4), 895–902.

Levaj, B., Bunić, N., Dragović-Uzelac, V., & Kovačević, D. B. (2010). Gel strength and sensory attributes of fig (*Ficus carica*) jams and preserves as influenced by ripeness. *Journal of Food Science, 75*(2), S120–S124.

Mahmoudi, S., Khali, M., Benkhaled, A., Boucetta, I., Dahmani, Y., Attallah, Z., & Belbraouet, S. (2018). Fresh figs (*Ficus carica* L.): Pomological characteristics, nutritional value, and phytochemical properties. *European Journal of Horticultural Science, 83*(2), 104–113.

Mahmoudi, S., Dias, C. B., Manhita, A., Boutoumi, H., & Charif, R. (2021). Formulation of goat's milk yogurt with fig powder: Aromatic profile, physicochemical and microbiological characteristics. *Food Science and Technology International, 27*(8), 712–725.

Mir, M. M., Amit, K., Umar, I., Mir, S. A., Rehman, M. U., Banday, S. A., Rather, G. H., & Sibat, F. (2018). Characterization of fig (*Ficus carica* L.) germplasm in central Kashmir of North Western Himalayan region. *Indian Journal of Plant Genetic Resources, 31*, 57–63.

Mujic, I., Kralj, M. B., Jokic, S., Jug, T., Subaric, D., Vidovic, S., Zivkovic, J., & Jarni, K. (2014). Characterization of volatiles in dried white varieties figs (*Ficus carica* L.). *Journal of Food Science and Technology, 51*(9), 1837–1846.

Nuri, Z. N., & Shahabuddin, M. (2021). A review on nutritional values and pharmacological importance of *Ficus carica*. *Journal of Current Research in Food Science, 2*(1), 07–11.

Pourghayoumi, M., Bakhshi, D., Rahemi, M., Noroozisharaf, A., Jafari, M., Salehi, M., & Hernandez, F. (2016). Phytochemical attributes of some dried fig (*Ficus carica* L.) fruit cultivars grown in Iran. *Agriculturae Conspectus Scientificus, 81*(3), 161–166.

Qureshi, R. A., Ghufran, M. A., Gilani, S. A., Sultana, K., & Ashraf, M. (2007). Ethnobotanical studies of selected medicinal plants of Sudhan Gali and Ganga Chotti hills, district Bagh, Azad Kashmir. *Pakistan Journal of Botany, 39*(7), 2275–2283.

Rababah, T. M., Al-Mahasneh, M. A., Kilani, I., Yang, W., Alhamad, M. N., Ereifej, K., & Al-u'datt, M. (2011). Effect of jam processing and storage on total phenolics, antioxidant activity, and anthocyanins of different fruits. *Journal of the Science of Food and Agriculture, 91*(6), 1096–1102.

Richer, S., Ulanski, L., & Popenko, N. A. (2016). Age-related macular degeneration beyond the age-related eye disease study II. *Advances in Ophthalmology and Optometry, 1*, 335–369.

Sadia, H., Ahmad, M., Sultana, S., Abdullah, A. Z., Teong, L., Zafar, M., & Bano, A. (2014). Nutrient and mineral assessment of edible wild fig and mulberry fruits. *Fruits, 69*(2), 159–166.

Shahrajabian, M. H., Sun, W., & Cheng, Q. (2021). A review of chemical constituents, traditional and modern pharmacology of fig (*Ficus carica* L.), a super fruit with medical astonishing characteristics. *Polish Journal of Agronomy, 44*, 22–29.

Shahzad, M. A., Razzaq, A., Aslam, M., Gulzar, M. F., & Nisar, N. (2019). Opportunities for agricultural trade in the context of the China-Pakistan Economic Corridor. *Business and Economic Research, 9*(1), 263–282.

Sher, H., Bussmann, R. W., Hart, R., & de Boer, H. J. (2016). Traditional use of medicinal plants among Kalasha, Ismaeli and Sunni groups in Chitral District, Khyber Pakhtunkhwa province, Pakistan. *Journal of Ethnopharmacology, 188*, 57–69.

Slatnar, A., Klancar, U., Stampar, F., & Veberic, R. (2011). Effect of drying of figs (*Ficus carica* L.) on the contents of sugars, organic acids, and phenolic compounds. *Journal of Agricultural and Food Chemistry, 59*(21), 11696–11702.

Solomon, A., Golubowicz, S., Yablowicz, Z., Grossman, S., Bergman, M., Gottlieb, H. E., Altman, A., Kerem, Z., & Flaishman, M. A. (2006). Antioxidant activities and anthocyanin content of fresh fruits of common fig (*Ficus carica* L.). *Journal of Agricultural and Food Chemistry, 54*(20), 7717–7723.

Soltana, H., Tekaya, M., Amri, Z., El-Gharbi, S., Nakbi, A., Harzallah, A., & Hammami, M. (2016). Characterization of fig achenes' oil of *Ficus carica* grown in Tunisia. *Food Chemistry, 196*, 1125–1130.

Soni, N., Mehta, S., Satpathy, G., & Gupta, R. K. (2014). Estimation of nutritional, phytochemical, antioxidant and antibacterial activity of dried fig (*Ficus carica*). *Journal of Pharmacognosy and Phytochemistry, 3*(2), 158–165.

Souza, V. R. D., Pereira, P. A. P., Pinheiro, A. C. M., Lima, L. C. D. O., Pio, R., & Queiroz, F. (2014). Analysis of the subtropical blackberry cultivar potential in jelly processing. *Journal of Food Science, 79*, S1776–S1781.

Stover, E., Aradhya, M., Ferguson, L., & Crisosto, C. H. (2007). The fig: Overview of an ancient fruit. *Horticultural Science, 42*(5), 1083–1087.

Tanwar, B., Andallu, B., & Modgil, R. (2014). Influence of processing on physicochemical, nutritional and phytochemical composition of *Ficus carica* L.(fig) products. *Asian Journal of Dairying & Foods Research, 33*(1), 37–43.

Turco, V. L., Potortì, A. G., Tropea, A., Dugo, G., & Di Bella, G. (2020). Element analysis of dried figs (*Ficus carica* L.) from the Mediterranean areas. *Journal of Food Composition and Analysis, 90*, 103503.

UNECE. (2004). *United Nations Economic Commission for Europe (UNECE)-Standard, DDP-14, concerning the marketing and commercial quality control of Dried Figs*. United Nations Edition.

Varhan, E., Elmas, F., & Koç, M. (2019). Foam mat drying of fig fruit: Optimization of foam composition and physicochemical properties of fig powder. *Journal of Food Process Engineering, 42*(4), e13022.

Veberic, R., Colaric, M., & Stampar, F. (2008). Phenolic acids and flavonoids of fig fruit (*Ficus carica* L.) in the northern Mediterranean region. *Food Chemistry, 106*, 153–157.

Verma, I., & Gupta, R. K. (2015). Estimation of phytochemical, nutritional, antioxidant and antibacterial activity of dried fruit of sacred figs (*Ficus religiosa*) and formulation of value added product (Hard Candy). *Journal of Pharmacognosy and Phytochemistry, 4*(3), 257.

Vinson, J. A. (1999). The functional food properties of fig. *Cereal Foods World, 44*, 82–87.

Vinson, J. A., Zubik, L., Bose, P., Samman, N., & Proch, J. (2005). Dried fruits: Excellent *in vitro* and *in vivo* antioxidants. *Journal of the American College of Nutrition, 24*(1), 44–50.

Yeganehzad, S., Kiumarsi, M., Nadali, N., & Ashkezary, M. R. (2020). Formulation, development and characterization of a novel functional fruit snack based on fig (*Ficus carica* L.) coated with sugar-free chocolate. *Heliyon, 6*(7), e04350.

Chapter 36
Wound Healing and *Ficus carica* (Fig)

Nahla A. Tayyib

Abbreviations

FC	*Ficus carica*
HE	Hematoxylin and Eosin
MT	Masson's Trichome
SWA	Scratch wound assay
BHK 21	Baby Hamster Kidney
MDCK	Maddison-Darby Canine Kidney

1 Healing Powers of Fig

Figs are rich in calcium, iron, and vitamin A in the form of carotene, while niacin and riboflavin are found in fewer amounts. Three fresh figs contain 140 IU of vitamin A, 0.5 mg of iron, 15 mg of calcium, 0.2 mg of niacin, and 0.03 mg of riboflavin. Two dried figs contain 20 IU of vitamin A, 1.4 mg of iron, 35 mg of calcium, 0.5 mg of niacin, and 0.04 mg of riboflavin (Ben-Noun et al., 2003).

2 *Ficus carica* and Its Enzymes

Figs are among the sweetest widely available fruits, a member of the mulberry family (Arvaniti et al., 2019). Figs were employed as a sweetener in ancient times, well before the discovery and production of refined sugar. In the first verses of Sura

N. A. Tayyib (✉)
Nursing Practice Department, College of Nursing, Umm Al-Qura University, Makkah, Saudi Arabia
e-mail: natayyib@uqu.edu.sa

M. F. Ramadan (ed.), *Fig* (Ficus carica)*: Production, Processing, and Properties*, https://doi.org/10.1007/978-3-031-16493-4_36

Al-Teen, the Quran discusses the medicinal benefit of the figs: "I swear by the Fig and the Olive". In addition to pomegranates, olives, grapes, bananas, and dates, the Quran mentions the fig plant as a fruit (Asma Arshad, 2014; Mehdi Trad et al. 2014; Yadav et al., 2015). The fig is the fruit of *Ficus carica* (FC), a tiny tree in the Moraceae family of flowering plants (Barolo et al., 2014). In the Mediterranean and western Asia, it has been growing since antiquity and is widely grown for its fruit and as a decorative plant worldwide (Mehdi Trad et al. 2014). In addition, the fig tree is extensively mentioned in the Bible and has existed for thousands of years in the Middle East and North Africa. Figs are not only delicious but also incredibly healthy fruit (Dafni & Böck, 2019).

An investigation by Gagaoua et al. (2014) found that figs are one of the few fruits with an enzyme capable of protein digestion. Moreover, several researchers (Kirtikar & Basu, 1975; Warrier, 1996; Chopra et al., 1992) argued that Ayurveda had employed various components of this plant and latex, either raw or in different formulations, to cure wounds. Findings revealed the properties of fig leaf extract, an enzyme of fig that degrades proteins in enhancing the efficacy of medicines against infections such as *Staphylococcus*. Furthermore, ficin, found in the latex of fig, is one of the most promising proteases developed from plants (*Ficus carica*), capable of metabolizing protein into amino acids.

3 Wound Healing Properties of Fig in Eliminating Biofilms

The wound healing properties of figs have generated a broader interest in investigating the role of the specific parts of the fruits (Harish et al., 2008). Studies have shown that treatment with ficin reduces the wound's half-closure time by two days and accelerates its microbial decontamination by four days (Baidamshina et al., 2020). Model wounds of laboratory animals treated with immobilized ficin were cleansed of microorganisms and healed more quickly. Moreover, the structure of the newly created tissue treated with protease possesses properties most similar to those of the original, intact tissue, Kayumov et al. (2015) emphasized. In another study, Baidamshina et al. (2017) stated that *Staphylococcus*-formed biofilms frequently impede the treatment of wounds. A study from the University of Kazan discovered that ficin, an enzyme from fig latex, is effective for eliminating these biofilms, which could be a significant barrier to wound care. Within the biofilm, antibiotics cannot reach the bacterium. Interestingly, increasing the antibiotic concentration by as much as a thousandfold in wound treatment is one option. Bacteriophages, an antibacterial treatment regaining favor, can function within biofilms.

In an *in vitro* study, chitosan-immobilized ficin destroyed staphylococcal (S) biofilms, thereby enhancing the antibacterial activity against biofilm-embedded bacteria. *In vivo*, in the presence of ficin (either soluble or immobilized), the skin wound areas of rats infected with *S. aureus* decreased by twofold after four days as opposed to six. Moreover, topical treatment of the immobilized enzyme resulted in a 3-log reduction of *S. aureus* cell count on wound surfaces in 6 days. In contrast, the control group required more than ten days to achieve the same outcome. Therefore, additional benefits include smoother re-epithelization and the generation of new tissue with collagen structural properties that closely resemble those observed in the native tissue. Both soluble and immobilized ficin appear effective for the treatment of biofilm-associated infections, as well as for accelerating wound healing and microbial decontamination (Baidamshina et al., 2020).

Bacterial cells are particularly resistant to antibiotics and immune system responses in biofilm, resulting in chronic reinfections. In addition, numerous opportunistic bacteria (*Staphylococcus, Micrococcus, Klebsiella, Pseudomonas*, etc.) develop biofilms on chronic and acute cutaneous wounds, preventing their healing and causing reinfection and sepsis (Worthington et al., 2013). Ficin is a safe and effective medication for treating external wounds to inhibit biofilm formation and reduce the risk of reinfection (Baidamshina et al., 2017). Biofilms are a thin but resilient collection of surface-associated microbial cells encased in an extracellular matrix, and bacteria are protected from antibiotics within these biofilms. Moreover, one option in wound care is to increase the antibiotic concentration by as much as a thousand-fold to push bacteriophages off the ledge in an antimicrobial therapy that could function within biofilms (Harper et al. 2014).

Numerous studies have been conducted on plants used to hasten wound healing. According to patents and articles, many herbal compositions expedite wound healing and could be used to treat wounds. The fig leaf is one of the therapeutic herbs used (*Ficus carica* Linn). Fig leaves contain flavonoids, terpenoids, tannins, alkaloids, and saponins. It is recognized that these chemicals in tin leaves have biological action as antioxidants, anticancer agents, anti-inflammatory agents, antiviral agents, and antibacterial agents. There are hundreds of different types of figs. In Tib al Nabawi, this fruit is utilized as a medicine in the form of large tablets resembling crushed fruit (Mutiara Figs, 2010).

4 Ficin and Its Medicinal Profile from Different Parts of Fig

Different parts, particularly the latex, of *Ficus racemosa* L. have been employed as a wound-healing medication in Ayurveda and traditional medical systems worldwide. As a result, there are large populations of *Ficus racemosa* L. (Family Moraceae) worldwide, including Asia, Africa, America, and Australia. They are known as Cluster Figs in English (Yadav et al., 2015). Ayurveda also employs a variety of raw and prepared forms of this plant's components and latex in healing wounds. Medicinal herbs for wound healing and skin illnesses are the key to fighting pathogen resistance to medicines and allopathic treatment. Consistent with Baidamshina et al. (2017), researchers in Russia have explored the influence of ficin, an enzyme that degrades proteins, on the efficacy of medicines against *Staphylococcus* infections. In Russia, researchers at Kazan Federal University and Voronezh University have focused on an understudied protease isolated from fig latex. As reported in scientific reports, the enzyme successfully removed biofilms from Staphylococcus bacteria in preclinical trials.

Another report (Labiotech, 2017) documented that *Staphylococcus* bacteria, which could cause severe infections and are widespread in skin lesions, find it easier to degrade biofilms due to their comparatively high protein content. Proteases such as trypsin and papain extracted from papaya were previously utilized in wound care. However, ficin appears to be more effective than the regularly used trypsin. Plant proteins, such as papain and ficin, have a low allergenic profile and are typically less cytotoxic than animal or bacterial enzymes. Murti et al. (2010) highlighted that figs play a significant role in the treatment or maintenance of wounds. They also mentioned that herbal remedies, such as plant-based figs, could significantly enhance wound disinfection, debridement, and the provision of a moist environment, which helps the creation of an adequate natural healing climate. The traditional Indian medical system considers the entire plant's parts (leaves, fruits, bark, latex, and root sap) medicinally significant (Murti et al., 2010).

Bopage et al. (2018) evaluated the properties of the trunk bark that is abundant in sterols, including sitosterol, lupenol, and stigmasterol. In addition, they stated that the fruits have gluanol acetate, which is the principal constituent. The remaining components include hydrocarbons, glucanol, tiglic acid, taraxasterol, lupeol acetate, friedelin, and lupeol acetate. The many forms of steroids included in latex include euphol, isoeuphorbol, sitosterol, 4-deoxyphorbol, cycloartenol, and cycloeuphordenol. It was revealed that the hexanes and dichloromethane extract of the stem bark of *F. racemosa* have a statistically significant effect on wound healing, and the percentage of wound closure has increased over both tested cell lines 24 h after treatment. In terms of antioxidant activity, in a nanosecond pulse radiolysis and stopped-flow spectrophotometric examination, the ethanol and water extracts of the stem bark of *F. racemosa* exhibited significant free radical scavenging activity in the wound healing process. In addition, the ethanol extract demonstrated considerably more antioxidant potential at steady-state than the water extract.

In addition, Veerapur et al. (2009) conducted an antioxidant assay using conventionally accepted techniques. Using a micronucleus assay on V79 cells, they also examined the radioprotective ability of the same extract and discovered that the optimal concentration for radioprotection was 20 g/mL. In the same experiment, the cytokines did not inhibit the proliferative index. Based on the preceding observation, the ethanol extract of *F. racemosa* exhibited substantial antioxidant and radioprotective traits.

5 Fig Latex and Mechanism of Tissue Repair

Investigation of latex applied externally to wounds reduces inflammation, discomfort, and edema and aids healing (Joseph & Raj, 2010). It was noted that several enzymes, such as trypsin, chymotrypsin, and collagenase, are currently utilized to heal wounds; these enzymes remove necrotic masses and fibrin clots. It was stated that healing is accelerated when enzymes or proteases are used to treat wounds (Madhumathi, 2017).

Since antiquity, proteolytic enzymes have been utilized to aid tissue repair. Since the 1960s, the oral proteolytic enzyme preparation trypsin: chymotrypsin has been used in clinical settings. It provides superior resolution of inflammatory symptoms and promotes faster healing of acute tissue injury compared to many currently available enzyme formulations (Chandanwale et al., 2017). In Ayurveda and Sri Lanka's indigenous system of medicine, many portions of *Ficus racemosa* L., particularly its latex, have been employed as a wound-healing remedy. Using animal models, this plant's wound-healing capacity has been examined (Bopage et al., 2018). Three sequential but temporally overlapping steps comprise the wound repair process: inflammation, cell proliferation, and tissue regeneration. The method of wound healing is fundamentally a connective tissue response. This process begins with an acute inflammatory phase, followed by the manufacture of collagen and other extracellular macromolecules, which are subsequently reshaped to form scars.

Wound healing active constituents were studied using both *in vivo* and human trials; however, now, wound healing active plant secondary metabolites are studied using both *in vitro* and *in vivo* methodologies based on cell cultures of fibroblasts, keratinocytes, epithelial cells, and endothelial cells (Phan et al., 2000). The breaking strength of the incision wounds was significantly raised in the drug-treated groups, i.e. (257.502.81) in control was increased up to (497.203.33) with *F. racemosa* aqueous extract, and (394.706.61) with *F. racemosa* ethanol extract. The results were equivalent to that of the traditional medicine, povidone-iodine, with a breaking strength. The difference between the drug-treated and control rats was statistically significant. The aqueous extract ointment treated groups demonstrated wound healing comparable to that of the reference treatment, povidone-iodine lotion, beginning on the fourth day. However, with the aqueous extract ointment-treated group, the wound closure time was shorter, and the percentage of wound contraction was much higher [12 days for 100% contraction, which was practically superior to that of the povidone iodine-treated group (17 days)]. Animals treated with an ointment containing ethanol extract exhibited considerable wound contraction beginning on the fourth day and wound closure in 14.17 days.

It was reported that the potential wound healing active constituents are the extracts of stem bark of *F. racemose*. Lupeol acetate acts as a prodrug for wound healing through bioactivity-directed fractionation using scratch wound assay (SWA) over Baby Hamster Kidney (BHK 21) and Maddison-Darby Canine Kidney (MDCK) epithelial cells. It is recognized that keratinocytes, which create keratin, are one of the primary cell types that initiate cell proliferation and migration, resulting in wound closure. In addition, MDCK cells are capable of generating keratin. Consequently, even though the MDCK cell line is a broad model of epithelial cells, it has been utilized as an *in vitro* model in wound healing assays.

6 Incision and Excision Medel: Wound Healing Activity

The incision and excision models were utilized to investigate the wound healing activity. In the incision model, the measured parameter was the injured skin's breaking strength. In the excision model, both extracts' percentage of wound contraction and duration of epithelialization were determined. The reference standard medicine for comparison with other groups was povidone iodine ointment. In this model, a 6 cm incision was made through the entire thickness of the skin on either side of the spinal column of rats. The wound was subsequently closed with 1 cm apart interrupted sutures. The animals were subsequently divided into seven groups at random. Rats in Group I (control) received no treatment, rats in Group II (ointment control) received a topical application of 50 mg of simple ointment base (BP), rats in Group III were treated topically with standard drug (povidone-iodine), rats in Group IV were given AQEFR orally (200 mg/kg body weight), rats in Group V were treated with topical AQEFR ointment (5%), rats in Group VI were given (5%). The skin's breaking strength was tested after 10 days (Kokane et al., 2009; Reddy & Elanchezhian 2008).

In terms of breaking strength in the incision model and percentage of wound contraction and duration of epithelialization in the excision model, it was determined that the aqueous extract of *F. racemosa* had more vigorous wound healing activity than other groups in both models. In terms of breaking strength in the incision model and percentage of wound contraction and duration of epithelialization in the excision model, it was determined that the aqueous extract of *F. racemosa* had more vital wound healing activity than other groups in both models (Murti & Kumar, 2012). In a model of excision wound healing, the aqueous and ethanol extracts of the root of *F. racemosa* increase the percentage of wound closure via promoting epithelialization. This improved epithelialization could result from *F. racemosa* extracts' effect on collagen production. Higher tensile strength indicates faster wound healing. Therefore, it promotes the wound-healing properties of *F. racemosa*. The increased tensile strength of treated incision wounds may be related to increased collagen concentration and fiber stability. Collagen, a component of developing cells, is synthesized by a healing tissue.

The platelet-aggregating activity of the aqueous extract of *F. racemosa* was investigated in healthy human volunteers. Compared to the controls, the platelet-aggregating activity of both extracts ranged between 3% and 51%. The dose-dependent platelet-aggregating activity of both extracts was the same and dose-dependent. The cold water extract of the bark had bergenin as its primary chemical ingredient, while the hot water extract also contained ferulic acid, kaempferol, and coumarin (Ahmed et al., 2012). Numerous extracellular matrix (collagen) components are synthesized by fibroblast cells, the most frequent cells found in connective tissue. This study included laboratory mice, separated into control groups that were not administered in leaf extract and treatment groups that were administered in the leaf extract. The experimental animals were incised and not administered tin leaf extracts; the control group was examined on days 3 and 7. As a treatment group, the other group was incised with tin leaf extract and monitored on days 3 and 7. The extraction of the tin leaf was performed once every day. The animal was treated with ketamine HCl and diazepam before incision. All incision wounds on animals used in experiments were treated with wound plaster (hypafix) to prevent infection and contamination.

7 Evidence Clinical Study of Wound Healing with Fig

The tissue is removed and placed in a sterile tube containing a 10% formalin buffer solution. The tissue is then placed in liquid paraffin for two to twenty-four hours. Using a rotary microtome, 4 m-thick paraffin blocks were sliced to a thickness of 4 m. The tissue is stained with hematoxylin and eosin (HE) to determine the number of fibroblasts and macrophages. In addition, Masson's Trichome (MT) staining was carried out to assess the thickness of collagen. When applied to incision wounds, tin leaf extract increases the quantity of macrophage and fibroblast cells. It denotes a quickening of macrophage activation and infiltration into the wounded tissue and a quickening of the inflammatory phase. In addition to macrophage cells, macrophages' production and decreased pro-inflammatory cytokines increased fibroblast cell proliferation, accelerating wound healing. The extract contains several anti-inflammatory chemicals, including flavonoids, terpenoids, and tannins.

This plant has been examined by Bopage et al. (2018) to investigate its wound-healing ability using animal models. This study aimed to acquire insight into the wound healing process and discover the putative wound healing active substances present in *F. racemosa* L. bark utilizing scratch wound assay (SWA) as the *in vitro* test. Stem bark extracts of *F. racemosa* were tested for cell migration boosting ability and antibacterial activity utilizing the scratch wound assay (SWA) on Baby Hamster Kidney (BHK 21) and Madin-Darby Canine Kidney (MDCK) cell lines and the Kirby Bauer disc diffusion assay on common bacteria and fungi. Dichloromethane and hexane extract that enhanced cell movement on SWA was fractionated using column chromatography and preparative thin-layer chromatography (TLC) to identify the active components. Dichloromethane and hexane extract enhanced cell migration in both cell lines, while EtOAc and MeOH extracts had antibacterial and antifungal action against *Staphylococcus* and *Bacillus*. It was concluded that chemical components of *Ficus racemosa* L stem bark stimulate cell motility and have antibacterial activity. This dual effect supports *F. racemosa*'s traditional usage in wound healing. Studies have also shown that the chlorogenic acid in figs aids in lowering blood sugar and controlling blood glucose levels in type II diabetes. In diabetic rats, chlorogenic acid (CGA) exhibited antioxidant and pro-oxidant properties while speeding wound healing (Bagdas et al., 2015).

8 Conclusion

Fig as a medicinal plant that could be scientifically proven for treating wounds was highlighted in this work. Additional attention has been placed on wound healing in its natural state, pharmacological activities, the role of fig fruits in wound care, and metrics used to gauge the effectiveness of those efforts.

References

Ahmed, F., Kumar, M. S., Urooj, A., & Kemparaju, K. (2012). Platelet aggregation inducing activity of *Ficus racemosa* stem bark extracts. *Journal of Pharmacology & Pharmacotherapeutics, 3*(4), 329–330. https://doi.org/10.4103/0976-500X.103692

Arvaniti, O. S., Samaras, Y., Gatidou, G., Thomaidis, N. S., & Stasinakis, A. S. (2019). Review on fresh and dried figs: Chemical analysis and occurrence of phytochemical compounds, antioxidant capacity and health effects. *Food Research International, 119*, 244–267. https://doi.org/10.1016/j.foodres.2019.01.055

Asma Arshad. (2014). *Food for thoughts, Islamic voice*. Accessed online https://islamicvoice.com/health/food-for-thought/

Bagdas, D., Etoz, B. C., Gul, Z., Ziyanok, S., Inan, S., Turacozen, O., Gul, N. Y., Topal, A., Cinkilic, N., Tas, S., Ozyigit, M. O., & Gurun, M. S. (2015). In vivo systemic chlorogenic acid therapy under diabetic conditions: Wound healing effects and cytotoxicity/genotoxicity profile. *Food and Chemical Toxicology, 81*, 54–61. https://doi.org/10.1016/j.fct.2015.04.001

Baidamshina D.R., Trizna E.Y., Holyavka M.G, et al. (2017). Targeting microbial biofilms using Ficin, a nonspecific plant protease. Sci Rep. 7:46068. Published 2017 Apr 7. https://doi.org/10.1038/srep46068

Baidamshina, D. R., Koroleva, V. A., Trizna, E. Y., Pankova, S. M., Agafonova, M. N., Chirkova, M. N., Vasileva, O. S., Akhmetov, N., Shubina, V. V., Porfiryev, A. G., Semenova, E. V., Sachenkov, O. A., Bogachev, M. I., Artyukhov, V. G., Baltina, T. V., Holyavka, M. G., & Kayumov, A. R. (2020). Anti-biofilm and wound-healing activity of chitosan-immobilized Ficin. *International Journal of Biological Macromolecules, 164*, 4205–4217. https://doi.org/10.1016/j.ijbiomac.2020.09.030

Barolo, M. I., Ruiz Mostacero, N., & López, S. N. (2014). *Ficus carica* L. (Moraceae): An ancient source of food and health. *Food Chemistry, 164*, 119–127. https://doi.org/10.1016/j.foodchem.2014.04.112

Ben-Noun, L. L. (2003). Figs-the earliest known ancient drug for cutaneous anthrax. *The Annals of Pharmacotherapy, 37*(2), 297–300. https://doi.org/10.1177/106002800303700224

Bopage, N. S., Kamal Bandara Gunaherath, G. M., Jayawardena, K. H., Wijeyaratne, S. C., Abeysekera, A. M., & Somaratne, S. (2018). Dual function of active constituents from bark of *Ficus racemosa* L in wound healing. *BMC Complementary and Alternative Medicine, 18*(1), 29. https://doi.org/10.1186/s12906-018-2089-9

Chandanwale, A., Langade, D., Sonawane, D., & Gavai, P. (2017). A randomized, clinical trial to evaluate efficacy and tolerability of trypsin: Chymotrypsin as compared to Serratiopeptidase and trypsin: Bromelain: Rutoside in wound management. *Advances in Therapy, 34*(1), 180–198. https://doi.org/10.1007/s12325-016-0444-0

Chopra, R. N., Chopra, I. C., Varma, B. S. (1992). *Supplement to glossary of Indian medicinal plants* (reprinted ed, p. 29). CSIR.

Dafni, A., & Böck, B. (2019). Medicinal plants of the bible-revisited. *Journal of Ethnobiology Ethnomedicine, 15*, 57. https://doi.org/10.1186/s13002-019-0338-8

Gagaoua, N., Boucherba, A., Bouanane-Darenfed, F., Ziane, S., Nait-Rabah, K., Hafid, H.-R., & Boudechicha, N. (2014). Three-phase partitioning as an efficient method for the purification and recovery of ficin from Mediterranean fig (*Ficus carica* L.) latex. *Separation and Purification Technology, 132*, 461–467.

Harish B, Krishna V, Kumar HS, Ahamed BK, Sharath R, Swamy HK et al. (2008). Wound healing activity and docking of glycogen-synthase-kinase-3-β-protein with isolated triterpenoid lupeol in rats. *Phytomedicine, 15*(9), 763–767.

Harper, D. R., Parracho, H., Walker, J., Sharp, R., Hughes, G., Werthén, M., Lehman, S., & Morales, S. (2014). *Bacteriophages and Biofilms. Antibiotics, 3*(3), 270–284. https://doi.org/10.3390/antibiotics3030270.

Joseph, B., & Raj, S. J. (2010). Phytopharmacological and phytochemical properties of three ficus species-an overview. *International Journal of Pharma and Bio Sciences, 1*, 246–253.

Kayumov, A. R., Nureeva, A. A., Trizna, E. Y., Gazizova, G. R., Bogachev, M. I., Shtyrlin, N. V., Pugachev, M. V., Sapozhnikov, S. V., & Shtyrlin, Y. G. (2015). New derivatives of pyridoxine exhibit high antibacterial activity against biofilm-embedded staphylococcus cells. *BioMed Research International, 2015*, 890968. https://doi.org/10.1155/2015/890968

Kirtikar, K. R., & Basu, B. D. (1975). *Indian medicinal plants* (Dehra Dun and M/S Periodical experts, 2327-8, reprint ed). M/S Bishen Singh Mahendra Pal Singh.

Kokane D.D., More R.Y., Kale M.B., Nehete M.N., Mehendale P.C., Gadgoli C.H. (2009) Evaluation of wound healing activity of root of Mimosa pudica. *Journal of Ethnopharmacology, 124*, 311–315.

Labiotech. (2017). https://www.labiotech.eu/more-news/russia-ficin-staphylococcus-infections-wounds/. Accessed 13 June 2022.

Madhumathi, P. (2017). *Wound healing properties of Ficin discovered.* https://www.medindia.net/news/wound-healing-properties-of-ficin-discovered-171255-1.htm, https://healthyeating.sfgate.com/enzymes-figs-1097.html. Accessed 13 June 2022.

Mehdi Trad, C. L. B., Gaaliche, B., Renard, C. M. G. C., & Mars, M. (2014). Nutritional compounds in Figs from the southern Mediterranean region. *International Journal of Food Properties, 17*(3), 491–499. https://doi.org/10.1080/10942912.2011.642447

Murti, K., Kumar, U., Lambole, V., PB, S., & AP, M. (2010). Pharmacological properties of Ficus racemosa-a review. *Pharmacologyonline, 2*, 802–807.

Murti, K., & Kumar, U. (2012). Enhancement of wound healing with roots of *Ficus racemosa* L. in albino rats. *Asian Pacific Journal of Tropical Biomedicine, 2*(4), 276–280. https://doi.org/10.1016/S2221-1691(12)60022-7

Mutiara Figs. (2010). https://www.theguardian.com/lifeandstyle/2018/apr/01/fig-trees-for-urban-gardens-james-wong. Accessed 13 June 2022.

Phan, T. T., Lee, S. T., Chan, S. Y., Hughes, M. A., & Cherry, G. W. (2000). Investigating plant-based medicines for wound healing with the use of cell culture technologies and *in vitro* models: A review. *Annals of the Academy of Medicine, Singapore, 29*(1), 27–36.

Reddy D.V., Elanchezhian, N. (2008). Evaluation of tropical tree leaves as ruminant feedstuff based on cell contents, cell wall fractions, and polyphenolic compounds. *Livestock Research For Rural Development, 20*(5), 77.

Veerapur, V. P., Prabhakar, K. R., Parihar, V. K., Kandadi, M. R., Ramakrishana, S., Mishra, B., Satish Rao, B. S., Srinivasan, K. K., Priyadarsini, K. I., & Unnikrishnan, M. K. (2009). *Ficus racemosa* stem bark extract: A potent antioxidant and a probable natural Radioprotector. *Evidence-based Complementary and Alternative Medicine: Ecam, 6*(3), 317–324. https://doi.org/10.1093/ecam/nem119

Warrier, P. K. (1996). *Indian medicinal plants – A compendium of 500 species* (Vol. 13, pp. 34–35). Orient Lengman Ltd.

Worthington, R. J., Blackledge, M. S., & Melander, C. (2013). Small-molecule inhibition of bacterial two-component systems to combat antibiotic resistance and virulence. *Future Medicinal Chemistry, 5*, 1265–1284. https://doi.org/10.4155/fmc.13.58

Yadav, R. K., Nandy, B. C., Maity, S., Sarkar, S., & Saha, S. (2015). Phytochemistry, pharmacology, toxicology, and clinical trial of *Ficus racemosa. Pharmacognosy Reviews, 9*(17), 73–80. https://doi.org/10.4103/0973-7847.156356

Index

M. F. Ramadan (ed.), *Fig* (Ficus carica)*: Production, Processing, and Properties*, https://doi.org/10.1007/978-3-031-16493-4

Q

R

S